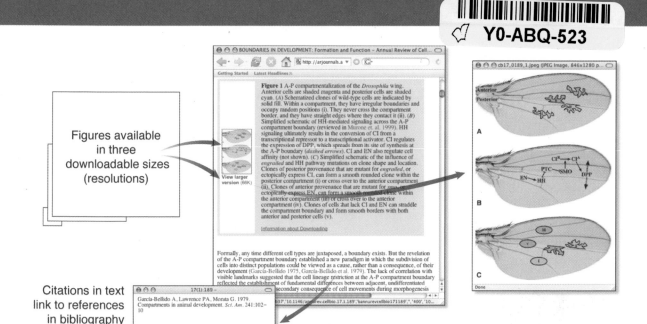

Figures available in three downloadable sizes (resolutions)

Citations in text link to references in bibliography

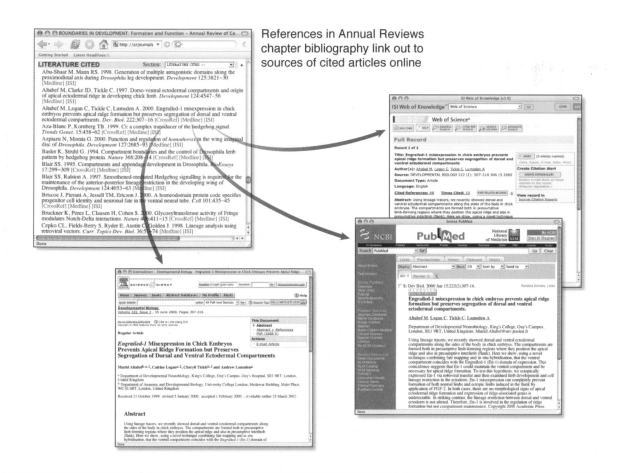

References in Annual Reviews chapter bibliography link out to sources of cited articles online

Annual Review of Cell and
Developmental Biology

Editorial Committee (2005)

Peter Cresswell, Yale University
Anirvan Ghosh, University of California, San Diego
Larry Goldstein, University of California, San Diego
Natasha Raikhel, University of California, Riverside
Jasper Rine, University of California, Berkeley
Janet Rossant, University of Toronto
Randy Schekman, University of California, Berkeley
Jean Schwarzbauer, Princeton University
S. Lawrence Zipursky, University of California, Los Angeles

**Responsible for the Organization of Volume 21
(Editorial Committee, 2003)**

Peter Cresswell
Larry Goldstein
Susan McConnell
R. Scott Poethig
Janet Rossant
Randy Schekman
Jean Schwarzbauer
Bruce Spiegelman
S. Lawrence Zipursky
Jasper Rine (Guest)

Production Editor: Sandra H. Cooperman, Shirley S. Park
Bibliographic Quality Control: Mary A. Glass
Electronic Content Coordinator: Suzanne K. Moses
Illustration Coordinator: Doug Beckner
Subject Indexer: Kyra Kitts

Annual Review of Cell and Developmental Biology

Volume 21, 2005

Randy Schekman, *Editor*
University of California, Berkeley

Larry Goldstein, *Associate Editor*
University of California, San Diego

Janet Rossant, *Associate Editor*
University of Toronto

www.annualreviews.org • science@annualreviews.org • 650-493-4400

Annual Reviews
4139 El Camino Way • P.O. Box 10139 • Palo Alto, California 94303-0139

Annual Reviews
Palo Alto, California, USA

COPYRIGHT © 2005 BY ANNUAL REVIEWS, PALO ALTO, CALIFORNIA, USA. ALL RIGHTS RESERVED. The appearance of the code at the bottom of the first page of an article in this serial indicates the copyright owner's consent that copies of the article may be made for personal or internal use, or for the personal or internal use of specific clients. This consent is given on the condition that the copier pay the stated per-copy fee of $20.00 per article through the Copyright Clearance Center, Inc. (222 Rosewood Drive, Danvers, MA 01923) for copying beyond that permitted by Section 107 or 108 of the U.S. Copyright Law. The per-copy fee of $20.00 per article also applies to the copying, under the stated conditions, of articles published in any *Annual Review* serial before January 1, 1978. Individual readers, and nonprofit libraries acting for them, are permitted to make a single copy of an article without charge for use in research or teaching. This consent does not extend to other kinds of copying, such as copying for general distribution, for advertising or promotional purposes, for creating new collective works, or for resale. For such uses, written permission is required. Write to Permissions Dept., Annual Reviews, 4139 El Camino Way, P.O. Box 10139, Palo Alto, CA 94303-0139 USA.

International Standard Serial Number: 1081-0706
International Standard Book Number: 0-8243-3121-4

All Annual Reviews and publication titles are registered trademarks of Annual Reviews.

⊗ The paper used in this publication meets the minimum requirements of American National Standards for Information Sciences—Permanence of Paper for Printed Library Materials, ANSI Z39.48-1992.

Annual Reviews and the Editors of its publications assume no responsibility for the statements expressed by the contributors to this *Annual Review*.

TYPESET BY TECHBOOKS, FAIRFAX, VA
PRINTED AND BOUND BY QUEBECOR WORLD, KINGSPORT, TN

Preface

This year we celebrate milestones in the lives of two people who were instrumental to the growth of this series. Sandra Cooperman, our Production Editor since 1988, has decided to move on with her busy life to new challenges in retirement. Sandra was enormously important in maintaining high standards, imposing some restraint on the enthusiasm of our contributors, and keeping the members of the Editorial Committee on schedule. I learned a lot about the editorial process, and even a bit of grammar, during the more than 10 years we worked together. Fortunately, we have enjoyed a smooth transition with the appointment of Shirley Park, our new Production Editor.

Bruce Alberts, a founding editor of this series, has just stepped down from 12 years as President of the National Academy of Sciences. Bruce will resume his faculty appointment at UCSF and no doubt continue his quest to reinvigorate primary and secondary science education in this country. Nothing could be more important. We are witnessing a renewed assault on the application of science and the scientific method in matters of public policy and teaching. Eighty years after the Scopes trial in Tennessee, we find ourselves no more advanced in public understanding and acceptance of evolutionary principles. As biologists, we in the United States have failed in our responsibility to provide sufficient support to the beleaguered high school teachers who represent us at the most important interface between science and the public. Things would be worse were it not for the indefatigable and wise council of leaders like Bruce Alberts and organizations such as the National Center for Science Education (http://www.ncseweb.org).

During the past five years, our Editorial Committee has been enriched by the spirited contributions of Susan McConnell and Scott Poethig. To ensure renewal in all the areas represented in our discipline, we have appointed two new members covering plant cell and developmental biology and neurobiology. Natasha Raikhel of UC Riverside and Anirvan Ghosh of UC San Diego have joined our Committee for a five-year term.

In a new feature for this series, we have solicited a Prefatory Chapter from one of the founders of modern cell biology: David Sabatini of New York University. Sabatini is a visionary leader who was instrumental in the development of glutaraldehyde as a fixative to preserve biological materials for transmission electron microscopy. His research at Rockefeller University and New York University led to an appreciation of the vectorial translocation of secretory proteins across the membrane of the endoplasmic reticulum. I take deep personal pleasure in having David's chapter lead off what we intend to be an annual feature of this series.

Randy Schekman
Berkeley, California

**Annual Review of
Cell and
Developmental
Biology**

Volume 21, 2005

Contents

INDEXES

ERRATA

An online log of corrections to *Annual Review of Cell and Developmental Biology*
chapters may be found at http://cellbio.annualreviews.org/errata.shtml

Related Articles

Annual Reviews is a nonprofit scientific publisher established to promote the advancement of the sciences. Beginning in 1932 with the *Annual Review of Biochemistry*, the Company has pursued as its principal function the publication of high-quality, reasonably priced *Annual Review* volumes. The volumes are organized by Editors and Editorial Committees who invite qualified authors to contribute critical articles reviewing significant developments within each major discipline. The Editor-in-Chief invites those interested in serving as future Editorial Committee members to communicate directly with him. Annual Reviews is administered by a Board of Directors, whose members serve without compensation.

2005 Board of Directors, Annual Reviews

Richard N. Zare, *Chairman of Annual Reviews, Marguerite Blake Wilbur Professor of Chemistry, Stanford University*
John I. Brauman, *J.G. Jackson–C.J. Wood Professor of Chemistry, Stanford University*
Peter F. Carpenter, *Founder, Mission and Values Institute, Atherton, California*
Sandra M. Faber, *Professor of Astronomy and Astronomer at Lick Observatory, University of California at Santa Cruz*
Susan T. Fiske, *Professor of Psychology, Princeton University*
Eugene Garfield, *Publisher, The Scientist*
Samuel Gubins, *President and Editor-in-Chief, Annual Reviews*
Steven E. Hyman, *Provost, Harvard University*
Daniel E. Koshland Jr., *Professor of Biochemistry, University of California at Berkeley*
Joshua Lederberg, *University Professor, The Rockefeller University*
Sharon R. Long, *Professor of Biological Sciences, Stanford University*
J. Boyce Nute, *Palo Alto, California*
Michael E. Peskin, *Professor of Theoretical Physics, Stanford Linear Accelerator Center*
Harriet A. Zuckerman, *Vice President, The Andrew W. Mellon Foundation*

Management of Annual Reviews

Samuel Gubins, President and Editor-in-Chief
Richard L. Burke, Director for Production
Paul J. Calvi Jr., Director of Information Technology
Steven J. Castro, Chief Financial Officer and Director of Marketing & Sales

Annual Reviews of

Anthropology
Astronomy and Astrophysics
Biochemistry
Biomedical Engineering
Biophysics and Biomolecular
   Structure
Cell and Developmental Biology
Clinical Psychology
Earth and Planetary Sciences
Ecology, Evolution, and
   Systematics
Entomology
Environment and Resources

Fluid Mechanics
Genetics
Genomics and Human Genetics
Immunology
Law and Social Science
Materials Research
Medicine
Microbiology
Neuroscience
Nuclear and Particle Science
Nutrition
Pathology: Mechanisms of
   Disease

Pharmacology and Toxicology
Physical Chemistry
Physiology
Phytopathology
Plant Biology
Political Science
Psychology
Public Health
Sociology

SPECIAL PUBLICATIONS
Excitement and Fascination of
   Science, Vols. 1, 2, 3, and 4

In Awe of Subcellular Complexity: 50 Years of Trespassing Boundaries Within the Cell

David D. Sabatini

New York University School of Medicine, New York, NY 10016-6497;
email: david.sabatini@med.nyu.edu

Annu. Rev. Cell Dev. Biol.
2005. 21:1–33

First published online as a
Review in Advance on
July 29, 2005

The *Annual Review of
Cell and Developmental
Biology* is online at
http://cellbio.annualreviews.org

doi: 10.1146/
annurev.cellbio.21.020904.151711

Copyright © 2005 by
Annual Reviews. All rights
reserved

1081-0706/05/1110-
0001$20.00

Key Words

glutaraldehyde fixation, membrane-bound ribosomes, signal
hypothesis, MDCK cells, epithelial cell polarity

Abstract

In this review I describe the several stages of my research career,
all of which were driven by a desire to understand the basic mechanisms responsible for the complex and beautiful organization of the
eukaryotic cell. I was originally trained as an electron microscopist in
Argentina, and my first major contribution was the introduction of
glutaraldehyde as a fixative that preserved the fine structure of cells,
which opened the way for cytochemical studies at the EM level. My
subsequent work on membrane-bound ribosomes illuminated the
process of cotranslational translocation of polypeptides across the
ER membrane and led to the formulation, with Gunter Blobel, of
the signal hypothesis. My later studies with many talented colleagues
contributed to an understanding of ER structure and function and
aspects of the mechanisms that generate and maintain the polarity
of epithelial cells. For this work my laboratory introduced the now
widely adopted Madin-Darby canine kidney (MDCK) cell line, and
demonstrated the polarized budding of envelope viruses from those
cells, providing a powerful new system that further advanced the
field of protein traffic.

Contents

MY ARGENTINE BEGINNINGS AND THE ENIGMA OF ORGANELLES

The educational system that prevailed in Argentina in the late 1940s, when I finished high school, required that I then choose a specific university track in which to continue my studies. I found this an extremely difficult decision to make, as I had a wide range of interests both within the sciences and the humanities. I decided to enter medical school, not moved by a passionate yearning to be a physician, but rather because I believed that the study and practice of medicine offered almost limitless opportunities for both scientific exploration and a full appreciation of the human condition.

I obtained my education as a physician in Rosario, then the second most populous city in Argentina, where I graduated with an M.D. from the University of Litoral in 1954, a few months before the forced end of Peron's populist regime, brought about by a military coup that occurred in September of 1955. The beginning of my scientific career was made possible by that event because (following years of obscurantism and intolerance) after the fall of Peron, Argentine universities went into a sort of academic renaissance, and teaching positions were again, for a time, opened to all those qualified, without regard for political affiliation.

In medical school I had read several excellent classical text books, including the histology ones by Giuseppe Levy, Alexander Maximow, and William Bloom, and those on pathology by Herwig Hamperl, H. Ribbert, and G. Roussy, and became interested in the structure and function of cells. At such an early stage of my scientific life I also became acquainted with E.B. Wilson, through his marvelous introduction to the 1925 edition of his book *The Cell in Development and Heredity*, which I found at the medical school library. I was soon fascinated by questions such as whether the mysterious different cellular organelles illustrated in those books, although apparently cooperating with each other to sustain the activity of the cell, had a certain life of their own. Or, to phrase it in a more scientific manner, by what mechanisms did they perpetuate themselves as distinct structures within the cell and from one cellular generation to another, while retaining their individuality. So little was known about these questions that I used to think of them as somewhat akin to those concerning the structure of matter itself. I made some failed attempts to begin a research career in Rosario, and in 1956 I decided to move to Buenos Aires.

The late 1950s were splendid and exciting times at the University of Buenos Aires, and I was very fortunate to become an instructor there in 1956, at the Institute of General Anatomy and Embryology of the School of Medicine. Under the direction of Eduardo De Robertis—a pioneer electron microscopist and the discoverer of synaptic vesicles, as well as the author of the first textbook entitled *Cell Biology*, which was published in English in 1948[1]—this institute was later to become a celebrated center of cell biology research and training in South America. I had little experience in research and obtained my position through an open competition, involving written and oral examinations, that was announced in the newspapers and was intended to recruit aspiring scientists to the basic science departments to which the professors who had been forced out of the universities in the previous decade were now returning. I look back with amazement at the fact that at that time—following an academically regressive political regime that lasted for 12 years—it was possible by those methods to assemble at the University of Buenos Aires a group of very capable young scientists, many of them physicians with an interest in basic research, that included several individuals who would later become highly successful researchers. Most of them, regrettably, because of subsequent political events, are now scattered throughout the world.

Some of the outstanding physiologists and biochemists who left their university positions during the Peronist era, such as Bernardo Houssay and Luis Leloir (who received Nobel prizes in 1947 and 1970, respectively) or Eduardo Braun Menendez (the discoverer of hypertensin, now called angiotensin), had continued their research in Argentina because they were able to establish private laboratories, outside the official academic system, using their own financial resources, as well as those from private donors.[2]

Other investigators, who had been cast out of the universities, for example, Eduardo De Robertis, had taken the route of exile, which, unfortunately, was also to be well traveled by the next generation of Argentine scientists during the military regimes that, with sporadic interruptions, dominated the political landscape until the end of the 1980s. De Robertis—who had worked in the United States at the University of Chicago (1939–1941) with R.R. Bentley, a pioneer in cell fractionation, and subsequently developed a brilliant scientific career in Argentina—left Buenos Aires in 1947. He joined the laboratory of F.O. Schmitt, the founder of the Biology Department at the Massachusetts Institute of Technology, who had created the first electron microscopy (EM) facility devoted to the study of cellular ultrastructure. After gaining international recognition for his contributions to cellular neurobiology, De Robertis settled in Montevideo in 1949, at the Institute directed by Clemente Estable (a disciple of the Spanish School of Ramon y Cajal, and also a patrician figure in Latin American Science) where he found the generous hospitality and tolerance for which the natives of Uruguay are well known to Argentines.

As junior faculty members, we waited anxiously in Buenos Aires for almost a year until De Robertis was able, with the help of the Rockefeller Foundation, to establish at the Institute the first EM facility in the country, to which he could transfer a functional laboratory from Montevideo. I owe a great debt to De Robertis, without whom, most likely, I would not be a scientist today. His keen powers of observation and technical skills set for us standards of world-class caliber for EM and laboratory experimentation. He was

---

[1] The first Spanish edition, entitled *Citologia General*, was published in 1946.

[2] Some of those laboratories, including the laboratory established by the Fundacion Campomar for Leloir's group, have prospered and survived to this day. This was an unwanted and beneficial effect of the previous regime.

also enterprising, imaginative, and capable of remarkable insights. Most importantly, he rejected the notion that the quality or significance of research carried out in Latin America could be curtailed by environmental limitations. He was stern and highly valued discipline and hard work, but he also cared deeply for those whose commitment to science was demonstrated in the laboratory. He instilled in his students lofty ambitions, and a good measure of his expectations was apparent from the high caliber of the laboratories to which he chose to send them abroad for further research experience. In sum, I consider my training with "el maestro," as we all called him affectionately among ourselves, comparable in its rigor and excitement to what I could

have had in the best of laboratories abroad. In De Robertis's laboratory I carried out extensive EM studies on the adrenal gland and began to study secretion in both the adrenal medula and cortex (De Robertis & Sabatini 1958, 1960; Sabatini & De Robertis 1961; Sabatini et al. 1962). In fact, as far back as 1960 we proposed a role for the Golgi apparatus in the formation of the cathecolamine-containing granules of the adrenal medulla and noted that the granules release their content at the cell surface by an exocytotic membrane fusion event (De Robertis & Sabatini 1960) (**Figure 1**). These notions, although a component of the dogma of cell biology today, were not widely accepted at the time.

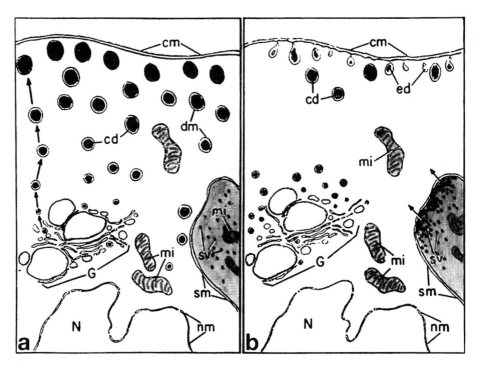

Figure 1

A diagramatic interpretation of the secretory process in a chromaffin cell of the adrenal medulla. (*a*) A chromaffin cell in the resting state. Cathecolamine-containing granules (cd) bounded by a membrane (dm) form (*arrows*) from the Golgi apparatus (G). The cell makes a synapse (sm) with a nerve ending containing synaptic vesicles (sv) and mitochondria (mi). (*b*) A chromaffin cell after electric stimulation of the splanchnic nerve. An exocytic discharge of cathecolamines takes place at the intercellular cleft resulting from fusion of the granule membranes with the plasma membrane (cm). Drawing from E.D.F. De Robertis, D.D. Sabatini. 1960. With permission from *Fed. Proc.* 19:70–78.

THE LURE OF PROTEINS, THEIR ORIGIN, AND TRAVAILS

By the end of the 1950s, when my first educational stage as a physician and a budding scientist was being completed in Buenos Aires, the grounds had already been set in the United States and Europe for the explosive developments that, during the remainder of the century, were to transform the biological sciences. As my turn was coming in De Robertis's group to seek training abroad—to which we all looked forward as necessary for the ripening of our scientific careers—I could scan an ample horizon of emerging fields in my search for a laboratory in which to invest the coming years in a foreign land.

A decade was closing that had witnessed a revolution in genetics and the explosive growth of molecular biology that followed the discovery of the double helix in 1953. Proteins were, of course, known to be the final products of gene expression, and their diversity as building blocks and as molecular machines had been recognized as responsible for the myriad of metabolic and physiological processes that sustain life. The first primary sequence of a protein, insulin, had already been determined by F. Sanger (who had visited our laboratory in Buenos Aires) and the first glimpses at the three-dimensional structure of some proteins by L. Pauling, M. Perutz, and J. Kendrew were making it clear that much would soon be learned about the particular architectural features that endowed proteins with specific functional activities. I could see that chemistry and crystallography held the key to unlock these secrets, and was attracted to these fields.

On the other hand, electron microscopy studies, initially concerned with macromolecular complexes, had by the mid-1950s already revealed a previously unimaginable complexity in the organization of the cytoplasm of the eukaryotic cell. The notion that the distinct compartments or organelles, whose existence had long intrigued me, defined chemical compositions and carried out specific cellular functions was also being established by a combination of histochemistry, electron microscopy, cell fractionation, and biochemical and enzymatic analysis. This added considerably to the fascination the eukaryotic cell exerted over those, like myself, who had begun to peek at it through the powerful eyes of the electron microscope.

Cell biology pioneers such as T. Caspersson and J. Brachet had already demonstrated a relationship between the protein biosynthetic activity of the cell and its capacity for RNA synthesis, manifested in prominent nucleoli and the intense basophilic staining of regions of the cytoplasm. The process by which protein themselves were synthesized, however, remained a mystery that was also alluring to me, as it lay at the boundary of biochemistry and cell biology. By 1960 it had been recognized that protein synthesis took place in ribosomes, but the puzzle of how these particles carried out the decoding of genetic information had not been solved.

Of course at that time few were concerned with the even greater mystery of the mechanisms and pathways that ensure that proteins, once synthesized, become the building blocks of organelles or become functional within them. Although our knowledge of the properties of membranes was meager, it was indeed difficult to imagine how proteins, once released from ribosomes, could freely traverse the ubiquitous intracellular membrane barriers to find their correct destination within an organellar lumen or in the membrane that limited a specific compartment. For anyone whose interests straddled the boundaries of cell biology and biochemistry this should have seemed an insurmountable problem to tackle.

In 1959 I had to make the difficult decision of whether I would get training in protein crystallography—which promised to give the ultimate understanding of biological processes at the molecular and even the atomic level, and therefore go to Great Britain, the undisputed mecca of crystallography at that time—or go to the United States to work with Palade and Siekevitz, who were beginning to unravel the pathway responsible for protein secretion, one of the most complex cellular processes.

Cytological studies extending back to the nineteenth century had made it clear that secretion requires the concerted and probably sequential action of several organelles, and efforts to increase our understanding of the secretory process were moving toward the vanguard of biological research. During the 1950s Palade and Siekevitz had pioneered in developing an approach to study the secretory pathway that combined the techniques of cell fractionation with biochemical and morphological analyses, which today is still the underpinning for most of the work being carried out in the field of molecular cell biology. I believe that it is fair to say that it is from those studies that the field of organelle and membrane biogenesis and the broad current concern with protein traffic sprung to life.

Moreover, the work of Palade and Siekevitz had focused foremost on the protein biosynthetic role of the endoplasmic reticulum (ER), an organelle unknown before EM. The discovery that the rough, ribosome-studded portions of the ER (Palade 1955) corresponded to the highly basophilic ribonucleoprotein-rich areas of the cytoplasm, characteristic of cells with intense protein biosynthetic activity, had placed the ER at center stage in the emerging field of cell biology.

Palade had discovered the ribosome in 1953 (Palade 1953, 1955) as a small particulate component of the cytoplasm visible in his electron micrographs, and Siekevitz had a long-standing interest in protein synthesis since his times in Zamecnik's laboratory in Boston, where he had pioneered in demonstrating that a microsomal fraction was capable of carrying out the process in vitro (Siekevitz 1952). The identification of the ribosome as a molecular machine that carries out protein synthesis had be-

gun with the finding that polypeptide chains, labeled in vivo during short incubations of *Escherichia coli* with radioactive amino acid precursors, were recovered in association with ribosomes when these particles were purified from the remains of disrupted bacteria (McQuillen 1959). Yet, until 1961 (the same year I began to work at Rockefeller), when Jacob & Monod (1961) proposed the mRNA hypothesis—derived primarily from observations made on ribosome reprogramming in *E. coli* after mating and after bacteriophage-infection—the prevalent notion was that individual ribosomes were "congenitally specialized" to synthesize specific proteins.

I decided to apply for a postdoctoral position with Palade, expecting that in Siekevitz's laboratory, where I had asked Palade to place me, I would be able to investigate the function of the membrane-bound ribosomes that characterize the rough ER. Meanwhile, with the support of De Robertis and Houssay, I obtained a Rockefeller Foundation Fellowship. I am still impressed by the fact that I was interviewed twice by officers of the Foundation who came to Buenos Aires and devoted a substantial amount of their time to assess my potential as a scientist. Palade agreed to take me into his laboratory, but he wrote that it would be best if I could postpone my arrival to New York for six months, which was planned for January 1961, because Keith Porter and his group were leaving the Rockefeller Institute for Harvard and more space would then become available. Since my Rockefeller Foundation Fellowship had been arranged to begin in January, Palade suggested that I first spend a few months at Yale with Russell Barrnett, a well-known histochemist with whom he had previously collaborated. I knew of Barrnett's work, since he had developed with Seligman a histochemical technique for the detection of sulfhydryl groups at the light microscope level, and in my first paper I had used a related method to localize cysteine-rich neuropeptides in the toad hypothalamus (Lasansky & Sabatini 1957).

A CYTOCHEMICAL INTERLUDE: THE POWER OF DIALDEHYDES

On January 26th of 1961 my wife and I arrived in New Haven during a severe snow storm to find that Barrnett was out of town for a week. We were snowbound, confined to the now extinct Taft Hotel, and in danger of running out of funds, when we were rescued by Mrs. E. DeVane, a daughter of the then Yale University President, who somehow, learning of the plight of the stranded South American couple, kindly brought us to her home and helped us find an apartment.

My stay in Barrnett's lab, short as it was by the usual standards, was immensely productive. I was the most experienced electron microscopist in his group and was coming from a laboratory using cutting edge technology not yet widely adopted in the United States. Thus in Argentina we had already replaced methacrylate with epoxy resin embedding and had added to our old RCA EMU2 EM a new Siemens Elmiskope. Barrnett had inherited an RCA EMU3 from Sanford Palay, a former Yale professor who had trained with Porter and Palade and had recently moved to Harvard, and I was put in charge of that instrument.

The focus of Barrnett's laboratory was on the application of histochemical techniques to electron microscopy, something that was being eagerly attempted in several labs, but with little success. Since the 1940s histochemistry had been a thriving field of research and a powerful tool for histophysiological studies at the light microscope level, but its application to EM faced considerable challenges. The fine structure of unfixed tissues and cells sadly decayed to almost an unrecognizable state during the harsh conditions of incubation needed for some histochemical reactions. Conversely, prefixation in OsO_4—the universal fixative for EM, mainly used in a veronal-acetate buffered form introduced by Palade (1952)—rapidly and irreversibly inactivated the enzymatic activities that histochemists wanted to relate to

subcellular structures. This was also the case with other metal-containing reagents, such as $KMnO_4$ and $K_2Cr_2O_7$, that were used to increase contrast in the EM.

At Yale, with my Fieser & Fieser *Organic Chemistry* textbook (Fieser & Fieser 1956) open on my desk, I decided to examine the usefulness as fixatives of a series of dialdehydes, agents that in a single molecule incorporate a double dose of the group responsible for the efficacy of formaldehyde, the preferred fixative for light microscopy. I hoped that some dialdehydes would serve as good cross-linking reagents that, at appropriate concentrations, could prevent the disintegration of subcellular structures, while perhaps maintaining some enzymatic or cytochemical activities. By a great coincidence, I found a source of glutaraldehyde nearby. Klaus Bensch—a young resident in pathology who was a frequent visitor to Barrnett's lab and was working with Don King, then a pathology professor at Yale—had used it following an industrial recipe to produce gelatin microcapsules containing *E. coli* DNA, which was administered to cultured mammalian cells in attempts to transform them into amino acid prototrophs. My very first experiments showed that glutaraldehyde gave an excellent preservation of cellular structures (**Figure 2**), while allowing the demonstration in situ of several specific enzymatic activities. I soon extended my search for new fixatives to other dialdehydes and related reactive organic compounds, but none was better than glutaraldehyde. It was a simple matter to choose concentrations and conditions that gave a compromise between acceptable tissue preservation and retention of enzymatic activities. The first paper we published (Sabatini et al. 1963) had 38 figures and illustrated the cytochemical detection at the EM level of the products of more than a dozen enzymes. Glutaraldehyde had distinct advantages over OsO_4 as a fixative in that it penetrated rapidly and deeply into tissues and was not nearly as noxious and difficult an agent to handle as OsO_4. However,

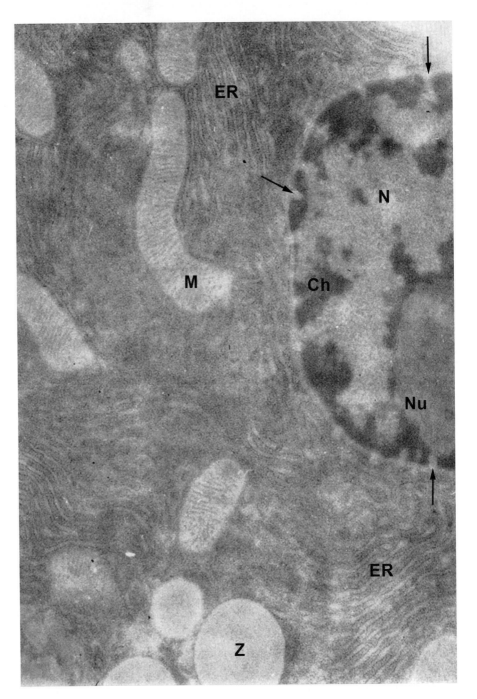

Figure 2

Partial view of a pancreatic acinar cell fixed in glutaraldehyde alone. The structure of the nucleus and of cytoplasmic elements, such as mitochondria and rough endoplasmic reticulum and zymogen granules, is well preserved. Membranes, however, are not visible unless the sample is post-fixed in OsO₄. From D.D. Sabatini, K. Bensch, R. Barnett. Reproduced from *J. Cell Biol.* 1963, 17:19–59 by copyright permission of the Rockefeller Univ. Press.

glutaraldehyde by itself provided little electron contrast (**Figure 2**), a drawback easily remedied by the introduction of a post-fixation step with OsO₄ that could be carried out after the histochemical reaction had been completed and be followed by staining with the usual contrast-enhancing uranyl or lead-containing reagents. Post-fixation in OsO₄ not only revealed membranes and stained other lipid-containing structures that were seen as negative images after glutaraldehyde fixation alone, it also reacted with some of the organic deposits generated during the histochemical reactions, making their products more visible under the electron beam. The advantages of the double-fixation procedure I developed to preserve fine structural details were well illustrated in a second paper (Sabatini et al. 1964) in which I included primarily the work I did at Rockefeller in collaboration with Fritz Miller, then a visiting scientist in Palade's laboratory who, on his return to Europe, assumed the Chair of Histology in Munich.

I was later particularly gratified that one of the main bonuses of glutaraldehyde was its preservation of cytoplasmic microtubules, which, as components of the cytoskeleton, Keith Porter first described in detail in doubly fixed specimens (Ledbetter & Porter 1963, Porter 1965). I believe that the warm and amiable life-long relationship I had with Keith owed much to his appreciation of the very indirect contribution I made to one of his many important discoveries.

In 1961, just before I moved to Rockefeller, I presented our results with glutaraldehyde fixation at a meeting of the New York Society for Electron Microscopy, which was held at an NYU town house in Washington Square. The other presenter that night was Alex Novikoff, a leading cytochemist at the Albert Einstein College of Medicine, and a fiery speaker and equally challenging critic. I had great respect for Novikoff, who was already an impressive figure in the emerging field of cell biology. While at the University of Vermont (from which he was dismissed during the McCarthy period and from which he received an honorary degree not too long before his death), he had been a pioneer in cell fractionation, one of the first to use some of the same marker enzymes we now employ to define subcellular compartments. While visiting Christian De Duve's laboratory in Belgium, he had obtained the first electron micrographs of subcellular fractions containing lysosomes (Novikoff 1956). These or-

ganelles were at first pure mental constructs of De Duve's brilliant analytical mind, which had deduced their existence from biochemical experiments demonstrating the compartmentalization and latency of acid hydrolases. De Duve and his collaborators later further purified the subcellular particles that contained the enzymes. Novikoff, who examined them at the EM, concluded that they corresponded to a type of electron-dense body that he and others had recognized in many cell types and, in hepatocytes, were abundant near bile canaliculi (Beaufay et al. 1956, Novikoff 1956). After my presentation at Washington Square, Novikoff engaged me and Barrnett in a heated discussion on whether the small blocks of tissue that we had used for some of the histochemical reactions were more prone to generate artifacts, owing to diffusion of the reaction product within the block, than the thin frozen sections that he preferred. Amusingly, after returning to New Haven, Barrnett wrote to Novikoff to admonish him for mercilessly challenging in public a novice like me who had an imperfect command of English. But Novikoff responded by praising my vigorous defense and insisted that I should not be pitied. This first encounter inured me to Alex's sometimes intimidating attacks and marked the beginning of a life-long friendly relationship with him.

Not long after I arrived in New Haven, I met Palade for the first time, when he came to present a lecture on his work on capillary permeability, and I was fortunate to be invited to the dinner honoring him that night, where I also met Jon Singer, then a Yale faculty member. Thus, in a single day, as a beginning postdoctoral fellow, I had the chance to meet and talk with two pioneers in cell biology whose discoveries were to shape our field, and I was deeply impressed by both. Singer was then laying the basis for his development of immunoelectron microscopy and was introducing a procedure to link ferritin molecules to antibodies, which, thus marked, could be used to detect specific proteins within the cell (Singer 1959). He was later to develop the fluid mosaic model for membrane structure (Singer & Nicolson 1972) that has almost completely passed the test of time and inspired much of our later thoughts on membrane biogenesis. In collaboration with Tokuyasu (Painter et al. 1973, Tokuyasu & Singer 1976), Singer also advanced the technique of cryo-ultramicrotomy, which made immunoelectron microscopy the key tool for molecular cell biology studies that it is today.[3]

GETTING STARTED AT ROCKEFELLER

At the Rockefeller I was assigned a desk and a bench in Siekevitz's laboratory in the fifth floor of the South Lab, since renamed the Bronk Lab. This was a wonderful place to do science. Siekevitz's enthusiasm was infectious, and he set a marvelous example with his own intense involvement in experiments, readiness to discuss novel ideas, and willingness to give his junior associates considerable freedom in any direction they took. In the same room, Yutaka Tashiro, a Rockefeller Foundation fellow from Japan, who has since made many contributions to cell biology and became president of the Kansai University Medical College in Japan, was already working. David Luck, an M.D., who was to make seminal discoveries on mitochondrial biogenesis and to become a professor at Rockefeller, was finishing a Ph.D. thesis that he had begun with Keith Porter on the association of glycogen particles, so abundant in hepatocytes, with glycogen synthetase, the enzyme that carries out their synthesis. Len Sauer, another M.D.

working toward his Ph.D. with Siekevitz, was studying the regulation of electron transport and oxidative phosphorylation in isolated mitochondria. In other laboratories on the same floor Lucien Caro was perfecting and applying autoradiography to tissues of animals injected with radioactive amino acids, attempting to trace the intracellular pathway followed by newly synthesized secretory proteins after their segregation in the ER. James Jamieson, a student from Canada with an M.D. degree, was attempting to isolate the secretory granules of heart muscle that we now know contain the atrial natriuretic factor. He was later to develop with Palade the system of pancreatic tissue slices, which in pulse-chase labeling experiments, employing autoradiography at the EM level and cell fractionation, allowed the elucidation of the steps involved in the transfer of newly synthesized proteins from the ER to zymogen granules (see Palade 1975). Faculty members in the cell biology group directed by Palade included Walter Stoeckenius, Sam Dales, and Marilyn Farquhar, who held a visiting appointment at that time. Stoeckenius was elucidating the arrangement of lipid and protein molecules within the bilayer of natural and synthetic membranes at the highest level of resolution attainable with the electron microscope. Sam Dales was using EM to study virus-cell interactions, as well as viral replication and assembly in infected tissue culture cells, which represented an avant garde approach at that time. Marilyn Farquhar, who as a postdoctoral fellow with Palade had studied the glomerular capillaries of the nephron, was embarking on the landmark work (Farquhar & Palade 1963) that revealed the organization and structural details of the components of the junctional complexes that hold together epithelial cells and allow intercellular communication between them. Marilyn and I became good friends and, because of our shared interest in fine structure and histochemistry, we met frequently for chats during lunch, and I was privileged to be one of the first to see some of her new and exciting findings.

---

[3] Upon my arrival at the Rockefeller Institute in the summer of 1961, I used Singer's procedure to successfully attach ferritin molecules to the outer surface of isolated liver mitochondria. After disrupting the surface-labeled mitochondria by sonication, I was able to separate the denser ferritin-bearing outer membrane fragments from those derived from the inner membrane. Unfortunately, I failed to preserve and detect biochemically any of the activities that we now know characterize the outer mitochondrial membrane, and dropped the project.

Although I came to the laboratory as a postdoctoral fellow, after one year I was seduced by the opportunity to join the Ph.D. program at Rockefeller, following the example of other young physicians, who had already done or were doing so, and later developed stellar careers, such as H. Rasmussen, Gerald Edelman, Chuck Stevens, Ed Reich, David Luck, James Jamieson, Scott Grundy, and others. Palade supported my application and, after an interview with Detlev Bronk, the Rockefeller University President, I was admitted into the program with Palade as my mentor. This change of status gave me the opportunity of repairing some of my educational deficiencies by arranging for some excellent tutorials in math (with E. Kogbetliantz), chemistry (with T.P. King and W. Agosta), and physical chemistry (with D. Yphantis). It also provided for much coveted free xeroxing privileges but did not change the direction of my research. What it did, however, was to greatly enrich my intellectual and personal life, as I entered into daily contact with the exceptional group of brilliant aspiring young scientists who were my classmates, including David Baltimore, Robert Barlow, Tony Cerami, Bert Hille, David Hirsch, Bob Klug, Harvey Lodish, Dan Rifkin, and others.

It is impossible for me to transmit in a few pages the extraordinarily stimulating environment that prevailed at the Rockefeller when I arrived, or to dwell on the scientific achievements that in 1974 brought George Palade the Nobel Prize, together with his former mentor, Albert Claude—a pioneer in biological EM and in the development of cell fractionation procedures—and his Belgian colleague at Rockefeller, Christian De Duve—the discoverer of lysosomes and peroxisomes. I also regret not being able to dwell, as I did in a published profile (Sabatini 1999), on Palade's admirable personal attributes that make him one of the most admired and beloved figures of today's scientific scene. Palade had an awesome capacity to assimilate and elaborate new ideas, and every discussion I had with him increased my enthusiasm for the work we were doing, helping to reinforce my conviction that nowhere else could I find such a fertile environment to make innovative discoveries about membrane and organelle biogenesis. My later career as a research mentor also benefited greatly from the example Palade set by always being available and eager to communicate with his younger associates and being able to maintain a calm, patient, and reassuring attitude, even when one's optimistic expectations were being dashed by the dismal reality of experimental results.

ENTERING THE ROUGH ENDOPLASMIC RETICULUM

In Siekevitz's laboratory I became a close friend and admirer of Tashiro. He was a very experienced biochemist from whom I had much to learn. He and I shared a profound interest in ribosomes, which he had begun to study in Japan using the analytical ultracentrifuge. Tashiro was interested in the molecular architecture of the ribosome and was using the analytical centrifuge in the laboratory of David Yphantis, a major contributor to technical improvements of that instrument, to characterize rat liver ribosomes. He introduced me to the intricacies of the technique of analytical centrifugation, as well as to the use of the sucrose gradient centrifugation methods. These had been recently applied to the study of ribosomes by Richard Roberts at the Carnegie Institution, and with them we could separate and obtain preparative amounts of ribosome monomers and subunits. We frequently visited the laboratory of Mary Peterman at the Sloan Kettering Institute, who was also studying the physical and chemical properties of eukaryotic ribosomes, and we followed closely the work of J.D. Watson and his colleagues at Harvard, who had shown earlier that *E. coli* ribosomes were composed of two unequal subunits (Tissieres et al. 1959) and that the larger one contained the nascent polypeptide chain (Gilbert 1963).

The workers at Harvard had taken advantage of the fact that *E. coli* ribosomes spontaneously dissociate into functional subunits when Mg^{2+} ions are removed by dialysis. Tashiro, on the other hand, had to use the Mg^{2+} chelating agent ethylenediaminetetraacetic acid (EDTA) to induce the dissociation of rat liver monomeric ribosomes into the subunits (Tashiro & Siekevitz 1965, Tashiro & Yphantis 1965). Unfortunately, this treatment also caused the partial unfolding and irreversible inactivation of the subunits. Nevertheless, Tashiro was able to demonstrate that in rat liver ribosomes nascent polypeptide chains are also associated with the large subunits (Tashiro & Siekevitz 1965).

I was impressed by the asymmetry and obvious specialization of the two sub-ribosomal domains and asked myself whether the association of the ribosomes with the ER membrane that characterizes the rough portions of the ER involves one or both ribosomal subunits, and how this might relate to the function of the ribosomes. I wondered if EDTA treatment of rough microsomes would also dissociate the bound ribosomes into subunits and whether, in that case, one or the other subunit would remain bound to the membrane.

The first experiments I did worked like a charm (Sabatini et al. 1966). At low concentrations, EDTA preferentially released small ribosomal subunits, whereas most large subunits remained associated with the membrane. When microsomes containing radioactive nascent chains labeled after very short pulses in vivo were treated with EDTA, I found that the ribosomal subunits recovered after detergent dissolution of the membranes contained a higher level of radioactive nascent chains than those subunits removed by EDTA. I had also begun an EM study of free ribosomes and of rough microsomes using the technique of negative staining. It was clear that ribosomal particles contained a cleft that divided them into two unequal regions and that, in general, in membrane-bound ribosomes the cleft lay parallel to the microsomal membrane to which the larger region was adjacent (**Figure 3**). Of course, these observations immediately suggested that the nascent chain in the large subunit might play a role in anchoring the ribosome to the membrane. At that time, however, we cautiously stated, "It seems plausible to consider the possibility that strong attachment of 'active ribosomes' is related to passage of the protein through the membrane. However, our work does not indicate which of the following possibilities holds. (*a*) The ribosomes which are strongly attached are, because of this situation, more active in protein synthesis and, therefore, become labeled in vivo. (*b*) The presence of the product of protein synthesis on the ribosomes is what makes them stick to the membrane" (Sabatini et al. 1966).

The initial evidence for a role of membrane-bound ribosomes in the synthesis of secretory proteins had been provided by Siekevitz & Palade in 1960. They reported that very soon after the injection of a radioactive amino acid into an animal, the purified pancreatic secretory enzyme chymotrypsinogen, having the highest specific radioactivity, was found in association with the ribosomes, which could be isolated by sedimentation from rough microsomes treated with detergent to dissolve the membranes. At later times after the injection, however, chymotrypsinogen was found in the detergent-soluble microsomal subfraction, which included the content of the microsomal lumen. Curiously, the route by which the chymotrypsinogen molecules released from the membrane-bound ribosomes found their way to the lumen of the cisterna was not considered, perhaps because it was regarded as a problem to be dealt with later. Indeed, a scheme drawn in 1961 (Palade et al. 1961) showed the site of passage of the protein through the membrane at some distance from the ribosomal membrane junction, as if it could be effected by an independent transport mechanism that operated on completed polypeptides released from the ribosome.

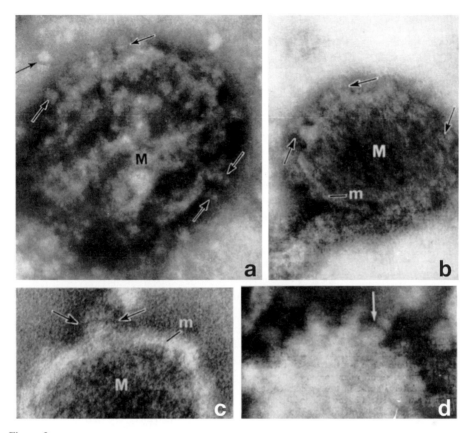

Figure 3

Ribosomes in rough microsomes are bound to the membrane through the large ribosomal subunit. Electron micrographs of rat liver rough microsomes examined after negative staining. Many ribosomes show a groove, penetrated by the stain (*arrows*), parallel to the membrane surface. The large ribosomal subunit, which contains the nascent chain, lies close to the membrane. M: microsome; m: microsomal membrane. Reprinted from *J. Mol. Biol.* 19, D.D. Sabatini, Y. Tashiro, G.E. Palade. On the attachment of ribosomes to microsomal membranes, pp. 503–24. Copyright 1966 with permission from Elsevier.

LAYING THE BASIS FOR THE SIGNAL HYPOTHESIS

It was left to Colvin Redman, who joined the Siekevitz and Palade laboratory in 1964, after obtaining his Ph.D. in Canada with Lowell Hokins—the scientist who first discovered the involvement of phosphoinositides in controlling secretion—to demonstrate that completed polypeptides, in this case amylase, which was synthesized in vitro by pigeon pancreas microsomes, were not released into the incubation medium but remained associated with the microsomes, most likely sequestered in their luminal cavities (Redman et al. 1966). These experiments, however, could not exclude that polypeptides discharged from the ribosomes into the medium were rapidly taken up into the microsomes by an uptake and transporting mechanism. Therefore, I, in collaboration with Redman, decided to examine the fate of in vitro-labeled incomplete polypeptides that were synthesized in a system of liver microsomes and were released from the ribosomes by the action of puromycin. This aminoglycoside antibiotic substitutes for aminoacyl tRNA and its incorporation at the C-terminal end of a nascent polypeptide causes premature termination of

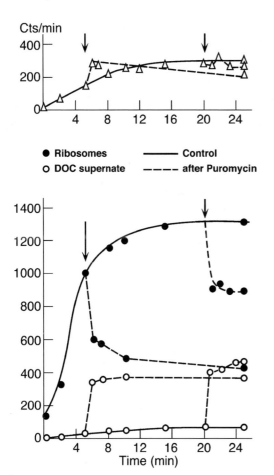

Figure 4

Nascent polypeptides released from membrane-bound ribosomes after incubation with puromycin in vitro are vectorially discharged into the microsomal lumen. Microsomes were incubated for protein synthesis with ^{14}C-leucine. Puromycin was added to the system at the times indicated by the arrows. At various times thereafter the microsomes were separated from the incubation medium by sedimentation and resuspended. The microsome suspension was treated with the detergent sodium deoxycholate (DOC) to solubilize the membranes and release the content of the microsomal lumen. A subsequent centrifugation separated the ribosomes from the final supernatant. The top panel shows that a small fraction of the radioactivity is recovered in the incubation medium obtained after sedimenting the microsomes and that that fraction does not increase after puromycin treatment. In the bottom panel, the closed circles represent the radioactivity associated with the ribosomes, which decreases rapidly (*dashed lines*) when the nascent chains are released by puromycin. The open circles correspond to the radioactivity present in the detergent-soluble fraction (DOC supernatant after ribosome removal). This fraction, which increases rapidly after puromycin treatment, represents nascent chains translocated into the lumen or inserted into the membrane. Drawing from C.M. Redman, D.D. Sabatini, 1966. Vectorial discharge of peptides released by puromycin from attached ribosomes. *Proc. Natl. Acad. Sci. USA* 56:608–615.

polypeptide synthesis, purging the ribosomes of nascent chains, which are released in the form of peptidyl-puromycin molecules. We found that the peptidyl-puromycin molecules lost from the ribosomes were not released into the surrounding medium, but remained associated with the microsomes, from which they could be recovered in soluble form after dissolution of the membranes by detergent and sedimentation of the ribosomes (Redman & Sabatini 1966) (**Figure 4**).

These observations, together with the finding that a significant fraction of the peptidyl puromycin molecules could also be released from the microsomes by a mild sonication procedure, suggested that such molecules had become sequestered within the microsomal lumen. This was confirmed by the finding that they could also be released using a low concentration of detergent that permeabilized the membranes without causing their extensive disassembly. This method to dissect the microsomes, which worked much better with liver microsomes than with the pigeon pancreas microsomes used by Redman, had been introduced by Lars Ernster (Ernster et al. 1962), a distinguished Swedish biochemist who had spent a sabbatical year in Siekevitz's laboratory.

We suggested that nascent chains synthesized in bound ribosomes become engaged with a translocation machinery in the membrane before their synthesis is completed and, therefore, the site of translocation must be in close proximity to the ribosome, most likely at the ribosome-membrane junction (Redman & Sabatini 1966). Hence, transport across the membrane appeared to be effected by a non-discriminating mechanism that does not distinguish between complete and artificially terminated incomplete chains, when either one is released from the ribosome. I had become convinced that translocation was a cotranslational phenomenon and that it resulted from the fact that in a membrane-bound ribosome growth of the nascent chain occurred in an environment that precludes its release into the surrounding medium, which in the cell is represented by the cytosol. We described a model, illustrated in a later paper (Sabatini & Blobel 1970) that explained the vectorial, unidirectional discharge of the polypeptide into the lumen of the ER as a result of the polypeptide growing in a tunnel or central cavity within the large ribosomal subunit (**Figure 5**). We suggested that because the ribosome sits on the membrane, this tunnel is made continuous with the ER cisternal space through a permanent or intermittent discontinuity in the microsomal membrane. We wrote, "As visualized at present the transfer mechanism relies primarily on release from the large subunit and on structural restrictions at the ribosome membrane junction and, hence, is non discriminatory and possibly passive" (Redman & Sabatini 1966). In a note to this paper we indicated, "Unpublished electron microscopic observations by D. Sabatini, Y. Tashiro, and G.E. Palade are the basis of this model. The existence of a channel is suggested by electron microscopy of negatively stained large subunits. Discontinuities in the membrane under the large subunits can be detected in some instances in sectioned specimens." Electron micrographs of isolated large ribosome subunits in which a central depression or cavity was apparent because it accumulated negative stain (as expected from the entrance to a tunnel in the subunit) were published in 1971 (Sabatini et al. 1971). The passive character of the transport of the peptidyl puromycin molecule into the microsomal lumen was supported by a paper in which Redman showed that ATP was not required for this process (Redman 1967).

The two papers representing my work on membrane-bound ribosomes became the basis for my doctoral thesis at Rockefeller, where I received a Ph.D. in 1966. I felt that my work had opened an important avenue of cell biology research, which I very much wanted to follow, hoping to provide a molecular description of the relationship between the ribosome and the translocation apparatus that I presumed existed in the membrane. Of course, very little was known about the basic organization of cellular membranes

Figure 5

A model depicting the basic structural features of the ribosome-membrane junction. The relationship of nascent polypeptides and ribosomes, with the membranes of the endoplasmic reticulum, account for the protection of the polypeptides from the attack of added proteases. The structural arrangement proposed is also compatible with the known features of the process of transfer of secretory polypeptides into the cisternal cavity. Drawing from D.D. Sabatini, G. Blobel. Reproduced from *J. Cell Biol.* 1970, 45:146–57, by copyright permission of the Rockefeller Univ. Press.

at that time, but it seemed most likely that the passageway or pore in the membrane through which the nascent polypeptide was transported was of proteinaceous nature or was constructed of protein subunits, although given our meager knowledge of membrane structure, I was hesitant to present such a speculation. Obviously, this was a topic of great interest to Palade, and I was, therefore, delighted when he offered me an Assistant Professorship and a laboratory being vacated by Sam Dales, who was leaving Rockefeller to take a position at the Public Health Institute of New York.

In my new laboratory I began to examine more directly the notion that nascent polypeptides indeed grow within a ribosomal tunnel and, in the case of those synthesized in the ER, upon emerging from the ribosome enter in a relationship with the underlying membrane that precludes their release into the medium or its cellular equivalent, the cytosol, as proposed in the model I had generated. I was successful with my first grant application and had started to determine to what extent nascent polypeptides in free ribosomes and in microsomes, which I could label in vivo or in vitro, were protected from the attack of exogenous proteolytic enzymes, when a new postdoctoral fellow, Günter Blobel, an M.D. from Germany who had recently completed his Ph.D. in Wisconsin with Van Potter (Siekevitz's old mentor), joined the Palade and Siekevitz laboratory.

GÜNTER AND I AND THE FORMULATION OF THE HYPOTHESIS

During his Wisconsin sojourn, Günter had worked on rough microsomes and was, therefore, already interested in the subject of my research. He was a frequent visitor to my laboratory and we soon began to hold many animated discussions on this somewhat esoteric subject. I was pleased to find that he had read my two papers on membrane-bound ribosomes carefully and that he was anxious to collaborate with me in experiments where I had begun to examine the accessibility to exogenous proteases of nascent polypeptide chains in free and membrane ribosomes. Together (Blobel & Sabatini 1970) we found that nascent polypeptides labeled in vitro in free ribosomes were easily and completely digested when the ribosomes were incubated with a mixture of trypsin and chymotrypsin at 37°C. But when the incubation was carried out at 0°C, most of the label incorporated in a brief pulse was protected from proteolysis and appeared in protease-resistant fragments of approximately 39 amino acids in length that remained associated with the partially proteolyzed ribosomes. Pulse-chase and continuous labeling experiments indicated that the protected segment contained the growing end of the nascent chain, which Dintzis had demonstrated several years before was at its C terminus (Dintzis 1961). Everyone recognized that this end of the nascent chain should be intimately associated with the ribosome because it serves as the substrate to which the peptidyltransferase, an integral component of the ribosome, adds amino acids to the growing chain. Similar results on a ribosome-protected fragment of the growing polypeptide had been obtained in Alex Rich's laboratory at M.I.T.

(Malkin & Rich 1967) with polyribosomes from rabbit reticulocytes, which produce almost exclusively globin chains that could be labeled in vivo. All these findings, together with the hydrodynamic properties of large subunits examined by analytical centrifugation (Petermann & Pavlovec 1969), provided strong support for the notion that a cavity or tunnel within the large ribosomal subunit contains the nascent polypeptide, as I had proposed for membrane-bound ribosomes to explain the vectorial discharge of the incomplete polypeptides released by puromycin (Redman & Sabatini 1966).

We suggested that the nascent polypeptide chain in the membrane-bound ribosomes would remain protease inaccessible as it grew beyond the 39-amino acid length sequestered within the ribosome provided that the exogenous protease did not break the membrane barrier (Redman & Sabatini 1970, Blobel & Sabatini 1970).

Omura, Sato, and their collaborators (Ito & Sato 1969, Omura et al. 1967) had already studied the effect of proteolytic enzymes on microsomes, showing that they were capable of digesting or dissecting out from the cytoplasmic face of the membranes several sets of proteins—later shown to be anchored in the membrane by C- or N-terminal segments. Their work had also shown that microsomal vesicles incubated with proteolytic enzymes at 0°C remain largely intact and apparently impermeable to the proteases.

Günter and I found that during proteolytic digestion of microsomes at 0°C, the bound ribosomes were released from the membranes (Sabatini & Blobel 1970). Concomitantly, the individual nascent polypeptides that they contained underwent cleavage, generating two sets of segments that remained largely protected from proteolysis. One set of segments remained associated with the detached ribosomes, which, although bearing partially proteolyzed ribosomal proteins, still sedimented as intact particles. These segments were of the same length as those protected in free ribosomes and corresponded to the C-terminal portions of the polypeptides. The other set of segments generated by proteolysis corresponded to the N-terminal portions of the growing polypeptides, and these were of variable length, mostly larger than the ribosome-protected ones. The N-terminal segments remained associated with the membranes and were also inaccessible to the proteases as long as the membranes remained intact.

This work was originally submitted to the *Journal of Cell Biology* as a single paper, but the editor found it too long and asked that we split it into two, which we did. One paper dealt specifically with free ribosomes (Blobel & Sabatini 1970); the second (Sabatini & Blobel 1970) presented evidence for the protection of the N-terminal portions of the nascent polypeptides by the microsomal membrane. In this paper, the model proposing the existence of a tunnel within the ribosome that is continuous with a passageway for the nascent polypeptide through the membrane was graphically illustrated (**Figure 5**).

Günter and I soon learned that by using puromycin in a medium of relatively high ionic strength we could dissociate ribosomes within polysomes into subunits that, contrary to those obtained by chelation of Mg^{2+} with EDTA, remained properly folded and were capable of protein synthesis when reprogrammed with mRNA (Blobel & Sabatini 1971). This allowed Yoshiaki Nonomura, who came to my laboratory as a postdoctoral fellow from Ebashi's group in Tokyo, to undertake an EM study by negative staining of the structure of individual active ribosomal subunits, of the monomers formed by their association, and of the relationship of the subunits to mRNA within polysomes (Nonomura et al. 1971). This work was followed by a study by Takashi Morimoto (a former student of Tashiro who had joined my laboratory at Rockefeller and whom I later recruited to NYU where he became a valued member of the cell biology department) (Morimoto et al. 1972a,b) of the mechanism by which ribosomes assemble into the tetramers that Breck Byers, then a student of Keith Porter at Harvard, had shown form

crystalline arrays in chicken embryos upon slow cooling (Byers 1967).

Soon after, we discovered the utility of puromycin to generate functionally capable ribosomal subunits. Mark Adelman—who had completed his Ph.D. in Chicago working with Ed Taylor in one of the first demonstrations of the presence of actin and myosin in non muscle cells—joined my laboratory as a postdoctoral fellow to study the role of the nascent chain in the association of the ribosome with the membrane. He developed a new procedure to obtain large amounts of highly purified rough microsomes from rat liver (Adelman et al. 1973a), which he used to study the role of the nascent chain in maintaining the association of ribosomes with the membrane. He showed that, even after the microsomes were treated with puromycin to release the nascent chains, the ribosomes remained associated with binding sites on the microsomal membrane. However, they now could be effectively detached from the membranes simply by raising the ionic strength, which was not possible without previous puromycin treatment (Adelman et al. 1973b, Sabatini et al. 1971). We concluded, therefore, that at least two molecular interactions are responsible for maintaining the ribosome-membrane junction: a direct one between the large subunit and a putative receptor in the membrane, which is disrupted by high salt treatment, and a second one that is provided by the nascent chain linking the polypeptide exit site in the large ribosomal subunit to the passageway in the underlying microsomal membrane, which leads to the microsomal lumen (**Figure 6**).

By 1971 it seemed clear to us that free and membrane-bound ribosomes were structurally identical and functionally interchangeable, something for which Nica Borgese, then a student in my laboratory, was accumulating evidence in subunit exchange experiments (Borgese et al. 1973). The only difference between free and membrane-bound ribosomes seemed to be that they were translating different classes of mRNAs, with the bound polysomes from secretory glands synthesizing

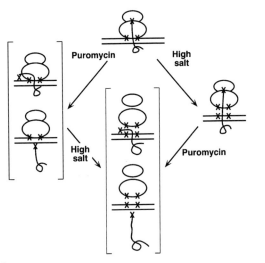

Figure 6

Interactions that maintain the association of active ribosomes with the ER membrane. Membrane-bound ribosomes are held on the membrane by two types of bonds: (*a*) ionic interactions between sites in the large ribosomal subunit and complementary sites in the membrane and (*b*) a link provided by the nascent polypeptide chain. The latter can be disrupted by puromycin, which releases the nascent chain from the ribosome. The ionic interactions are broken in media of high salt concentration. A combination of both treatments is necessary to release all ribosomes from the membrane. Drawing from D.D. Sabatini, N. Borgese, M. Adelman, G. Kreibich, G. Blobel. 1972. Studies on the membrane-associated protein synthesis apparatus of eukaryotic cells. RNA viruses/ribosomes. North Holland, Amsterdam. *FEBS Symp.* 27:147–71.

In the figure, labels read: Puromycin, High salt, High salt, Puromycin.

mainly secretory proteins. We had obtained a plethora of evidence implicating the N-terminal portion of the nascent polypeptide in establishing and maintaining the association of a bound ribosome with the membrane. Most salient were (*a*) the vectorial discharge resulting from puromycin treatment (Redman & Sabatini 1966), (*b*) the role of the membrane in protecting the N-terminal portion of the polypeptide from the proteolytic attack by exogenous enzymes (Sabatini & Blobel 1970), and (*c*) the fact that ribosomes containing nascent chains could not be detached by simply raising the ionic strength, which effectively dissociates from the membrane those ribosomes lacking nascent polypeptides (Adelman et al. 1973a, Sabatini et al. 1971, Sabatini 1972).

The remaining burning question concerned the features in the N-terminal re-

gion of a nascent secretory polypeptide that were responsible for the association of the ribosome-nascent chain complex with the membrane. In 1971, it was already established that, as was known for prokaryotes, methionine was the initiating amino acid for eukaryotic proteins. It then seemed reasonable to assume that because no known secretory protein retained an N-terminal methionine, the membrane-binding information that we postulated existed in the N-terminal sequence was removed during or after insertion of the polypeptide into the membrane. We considered the possibility that N-terminal amino acids were modified by acylation with fatty acids or by the addition of cholesterol or other hydrophobic moieties. There are now examples of such modifications in proteins, but none was known at that time other than the formylation of the initiator methionine, which occurs only in prokaryotes. In fact, we suspected that the feature in the nascent polypeptide that served to trigger the association of the ribosome with the membrane could be a stretch of hydrophobic amino acids, because we found that polyphenylalamine synthesized in vitro by microsomes programmed with poly U, upon termination with puromycin, remained membrane associated (Sabatini et al. 1971).

After many hours of enthralling argumentation in front of my office blackboard, Günter and I arrived at the first formulation of what was later to be known as the signal hypothesis (Blobel & Sabatini 1971) (**Figure 7**). We proposed that "all mRNAs to be translated on bound ribosomes have a common feature, such as several codons near their 5' end, not present in mRNAs which are to be translated on free ribosomes. The resulting common sequence of amino acids near the N-terminal of the nascent chains or a modification of it (indicated by X) would then be recognized by a factor mediating the binding to the membrane. This binding factor could be a soluble protein, which recognizes both a site on the large ribosomal subunit and a site on the membrane. After release of the chain from

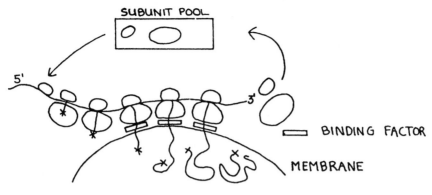

Figure 7

The first formulation of what came to be called the signal hypothesis. The crosses represent signal peptides at the N termini of the nascent chains, which were proposed to be recognized by a factor that initiates binding of the ribosome-nascent chain complex to the membrane and, therefore, has properties of SRP. In *Biomembranes*, Vol. 2, 1971, pp. 193–95. Ribosome-membrane interactions in eukaryotic cells. G. Blobel, D.D. Sabatini, with kind permission of Springer Science and Business Media.

the ribosome into the intracisternal space, the ribosomal subunits and perhaps the binding factor dissociate from the membrane and enter their respective soluble pools. The pool of ribosomal subunits would also be fed from subunits generated from free polysomes after chain completion and release and a new cycle would be started." This was a rather prophetic statement that predicted not only the presence of a signal peptide sequence at the N terminus of the nascent chain, but also the existence of a signal-binding factor with the behavior later demonstrated by Blobel and his associates and former disciples (Walter & Blobel 1981, Walter et al. 1981) for the signal recognition particle [SRP and its cognate receptor in the membrane (Gilmore et al. 1982a,b; Meyer et al. 1982)].

THE FULL FLOWERING OF THE INITIAL MODEL: MOLECULES AND MECHANISMS

Before leaving Rockefeller in 1972, my laboratory had nearly completed other studies that established the presence in ER membranes of a finite number of specific ribosome receptors to which ribosomes, even after being purged of their nascent chains, are capable of binding (Borgese et al. 1974). We deduced that such

binding sites contained an essential protein component because the ribosome-binding capacity of ER membranes was sensitive to heat treatment and was abolished by mild proteolytic digestion, and proposed that those sites were spatially close to those through which cotranslational translocation occurs.

With the development of cell-free systems for the translation of natural mRNAs, several laboratories observed that the primary translation products of mRNAs for secretory proteins produced in systems lacking microsomal membranes were somewhat larger than those produced in microsomes (Kemper et al. 1974, Mach et al. 1973, Milstein et al. 1972, Tonegawa & Baldi 1973). It was also shown that the extra segments were at the N termini of the primary translation products. But, of course, it was Günter's laboratory, employing a reconstituted in vitro protein synthesizing system containing microsomal membranes and ribosomal subunits programmed with immunoglobulin mRNA, that provided definitive proof for the signal hypothesis (Blobel & Dobberstein 1975a,b). They demonstrated the role of a cleavable signal sequence in initiating insertion of nascent immunoglobulin light chains into the membrane, characterized the signal sequence, and revealed the action of a membrane-associated signal peptidase in its cotranslational removal. In these experiments, the protease protection assay that we had developed was used to demonstrate the sequestration in the microsomal lumen of completed immunoglobulin light chains whose signal sequences were removed cotranslationally.

Over the next decade, in now classical experiments, Günter's laboratory went on to illuminate nearly all aspects of the complex molecular interactions that lead to the insertion of specific polypeptides into ER membranes. Years later, his laboratory provided electrophysiological proof for the existence of a protein-conducting channel in the microsomal membrane that appears to be opened by the signal peptide and to be kept in that state by the bound ribosome (Simon & Blobel 1991). Soon thereafter, the existence of an aqueous channel was also detected using microsomes with in vitro-synthesized, fluorescently labeled nascent chains whose accessibility to selective quenching procedures was assessed under various conditions (Crowley et al. 1994).

Major steps toward the molecular identification of the components of the translocation complex were made in the late 1980s in the laboratories of Randy Schekman and Tom Rapoport, beginning with the discovery that a yeast mutant (*sec61*) was defective in an early stage in the translocation of secretory proteins into the ER (Deshaies & Schekman 1987). Subsequently, cross-linking experiments showed that the Sec61p polypeptide of yeast, as well as its mammalian orthologue, was in close association with translocating nascent chains (Gorlich et al. 1992, Musch et al. 1992, Sanders et al. 1992). Soon thereafter, the Sec61 complex, which was found to consist of three polypeptides, was shown to be essential to confer translocation activity to reconstituted proteoliposomes (Oliver et al. 1995).

In 1977 we had carried out freeze-fracture studies of rough microsomes (Ojakian et al. 1977) that gave us a first glimpse at the size and distribution of the components of the protein translocation channel and associated ribosome-binding sites. These revealed the presence of intramembranous particles (IMPs), approximately 10 nm in diameter, that were associated with the bound ribosomes because they were displaced in the plane of the membrane with the ribosomes when the latter were forced to undergo aggregation on the membrane surface.

Later cryoelectron microscopy studies employing image processing techniques, in which Günter collaborated with Joachim Frank (Beckmann et al. 1997), showed that the large ribosomal subunit, indeed, contains a tunnel that, when the ribosome is bound to Sec61p, is continuous with the transmembrane channel provided by this protein, as we had predicted decades earlier for bound ribosomes (Redman & Sabatini 1966, Sabatini & Blobel 1970, Sabatini et al. 1971).

A full elucidation of the structure of a eukaryotic translocation channel is yet to come. However, the structure of SecY, the archeabacteria equivalent of the main subunit (α subunit) of the eukaryotic Sec61 complex, has been solved through crystallography at a resolution of 3.2 Å (van den Berg et al. 2004). The deduced structure makes several tantalizing suggestions, including sites for signal sequence recognition, for the lateral exit into the bilayer of transmembrane segments of nascent membrane polypeptides, and for binding of ribosomes.

It is more than 30 years since Günter and I delineated the basic features of the signal hypothesis. We have now gained enough perspective to view that work—which began with a rather focused concern with the early stages of the secretory process—within a larger frame of reference, one that encompasses more general questions related to membrane and organelle biogenesis. From this we can derive considerable satisfaction as some of the concepts and ideas we put forward have had a wide impact in the field of protein traffic.

The years that followed the presentation and verification of the signal hypothesis revealed that the cotranslational mechanism first envisaged for the passage of secretory polypeptides across the membrane also applies to resident luminal and integral membrane proteins of the entire endomembrane system, including the ER itself, the Golgi apparatus, the plasma membrane, endosomes, and lysosomes. It is also now clear that a wide variety of signal sequences exist in polypeptides that are inserted into ER membranes, which, however, are all decoded by the same translocation apparatus. Some such signal sequences are transient features of the polypeptide that serve only to initiate its insertion into the membrane, whereas others also serve as permanent transmembrane anchors responsible for specific transmembrane dispositions of the mature polypeptide. Moreover, studies, first on simple transmembrane proteins, and later on proteins that traverse the membrane multiple times, brought to light the signaling function of other largely hydrophobic, interior segments of translocating polypeptides that serve to arrest the translocation process and to determine the proper disposition of the polypeptide relative to the phospholipid bilayer in the membrane. Both Günter and I envisaged the complex topology of certain membrane proteins as resulting from the sequential action of insertion and halt transfer signals operating cotranslationally, before substantial evidence for this mechanism was obtained (Blobel 1980, Sabatini et al. 1982). Several subsequent papers from our group clarified important aspects of the nature of these signals (Finidori et al. 1987, Monier et al. 1988, Rizzolo et al. 1985).

The notion we introduced in 1971 of a signal peptide containing information decoded by soluble and membrane receptors that informs the cell of the polypeptide destination was a harbinger of the discovery of signals that determine the post-translational importation of polypeptides into other membranes and organelles, including chloroplasts, mitochondria, and peroxisomes. Moreover, the role of the signal peptide as the first sorting signal that initiates the journey of a variety of proteins toward diverse destinations,

as distant from the site of their insertion into the ER as the surface of the cell or lysosomes, opened the way to the realization that other sorting signals and cognate receptors must exist that mediate the successive sorting events needed to steer each polypeptide along the one pathway that leads to its site of function.

In the light of current knowledge, our original model for cotranslational translocation, also illustrates, once more, how natural phenomena invariably are more complex than one initially imagines. As the translocation machinery whose existence we recognized was dissected into its molecular components, the model, rather stylish at the beginning, grew much more elaborate and ornate, with the inclusion of a host of additional proteins that either assist in the translocation or modify the nascent polypeptide during this process. The Sec61p protein that constitutes the channel consists of at least three polypeptide subunits (Gorlich & Rapoport 1993). Two other ER transmembrane polypeptides, ribophorins I and II, that we, along with Gert Kreibich, identified as putative ribosome receptors (Kreibich et al. 1978a,b) are closely associated with the translocation site (Yu et al. 1990). They are now known to constitute, with two other subunits, the oligosaccharyl transferase that transfers a glycan moiety to asparagine residues in nascent glycoproteins (Kelleher et al. 1992). Similarly, SRP—the complex factor that targets the ribosomes to the ER membrane—consists of an RNA molecule and six polypeptides (Walter & Blobel 1983), one of which is a GTP-binding protein (Bernstein et al. 1989) that is recognized by the SRP receptor in the membrane. The receptor itself consists of two subunits, both of which are also GTP-binding proteins (Miller et al. 1995).

Amazingly, a retrotranslocation machinery is now known to also function in the ER in cooperation with a cytosolic ATPase and to transfer misfolded polypeptides, already segregated in the lumen of the ER or incorporated into the membrane, back to the cytosol for degradation by the proteasome (Lilley & Ploegh 2004, Ye et al. 2004). This machinery is an essential component of a quality control mechanism in the ER that employs luminal chaperones and an elegant transmembrane signaling system that controls levels of protein synthesis and the expression of genes whose products make this organelle a nearly perfect manufacturing plant. Remarkably, certain viruses have learned to evade the immune system by encoding proteins that promote the retrotranslocation and, hence, degradation of MHC class I molecules that otherwise would have presented viral antigens at the cell surface and thus called into action cytotoxic lymphocytes to eliminate the infected cell (Lilley & Ploegh 2004, Ye et al. 2004).

It is gratifying for me to think that our initial model provided a framework that facilitated further progress and that, although it has undergone considerable elaboration and fine details have been elucidated, its fundamental aspects are still in place.

A POST-TRANSLATIONAL TRANSFER TO NYU

In September 1972, a few months before Palade, Farquhar, Jamieson, and their groups left Rockefeller for Yale, my research group moved to the NYU School of Medicine, where I assumed the chairmanship of the Cell Biology Department—succeeding Howard Green, who had moved to M.I.T. I not only transferred my laboratory, but was able to appoint several new faculty members interested in protein traffic, including Milton Adesnik, who had finished postdoctoral training with Jim Darnell at Columbia, as well as Gert Kreibich and Takashi Morimoto, who had been postdoctoral fellows with me at Rockefeller. Jim Lake, who was working on ribosomal structure at Rockefeller, also decided to join us. With these young and generous people, I developed productive collaborations. With Gert Kreibich we continued for some time to characterize biochemically the membranes and content of

microsomal vesicles, demonstrating that specific sets of proteins sequestered within the microsomal lumen, including serum and lysosomal proteins, are transient components of the ER (Kreibich et al. 1973, Kreibich & Sabatini 1974). We also showed that the numerous integral components of the ER membranes have characteristic transmembrane dispositions (Kreibich et al. 1974), with their mannose-rich oligosaccharide chains luminally disposed (Rodriguez Boulan et al. 1978). We also continued our research on ribosome structure (Ivanov & Sabatini 1981) and on the ribosome-membrane association (Kreibich et al. 1978a, 1982; Kruppa & Sabatini 1977; Lande et al. 1975; Lewis & Sabatini 1977; Marcantonio et al. 1982; Ojakian et al. 1977) as well as on the role of bound ribosomes in the synthesis of proteins other than secretory proteins, such as proteins of the ER itself (Bar-Nun et al. 1980, Chyn et al. 1979, Harnik-Ort et al. 1987, Monier et al. 1988, Okada et al. 1982, Rosenfeld et al. 1984), the plasma membrane (Colman et al. 1982, Finidori et al. 1987, Mentaberry et al. 1986, Sabban et al. 1981, Sherman & Sabatini 1983), or lysosomes (Croze et al. 1989, Nishimura et al. 1986, Rosenfeld et al. 1982). In that period we also saw the virtue of utilizing cultured cells infected with enveloped viruses, such as Sindbis, influenza or vesicular stomatitis virus (VSV), to investigate the post-ER sorting of plasma membrane proteins, for which other laboratories (Katz et al. 1977, Wirth et al. 1977) had shown viral glycoproteins serve as facile models.

MDCK CELLS, THE POLARIZED BUDDING OF VIRUSES, AND THE SORTING OF ORGANELLAR PROTEINS

Several years after I came to NYU, I benefited once more from the windfall of emigres that usually follows the repressive policies of military regimes. The distinguished physiologist Marcelino Cereijido, who was Dean of the School of Biochemistry in Buenos

Aires, and is now in Mexico, had to leave Argentina, and I invited him to become a visiting professor in my department. He was interested in epithelial physiology and, together, we sought to develop a cell culture system that could be used to study, with a combination of biophysical and cell biological techniques, the generation and maintenance of the transport properties of polarized epithelial monolayers. After some searching, we settled for a dog kidney-derived cell line (Madin-Darby canine kidney; MDCK), some of whose epithelial features had been recognized a few years earlier by J. Leighton, then Professor of Pathology at Pittsburgh (Leighton et al. 1970). Confluent monolayers of these cells exhibit some of the properties of distal convoluted tubules of the kidney, including the capacity of vectorially transporting water and electrolytes across the cell layer, which reflects the functional polarization of individual cells. We grew the cells on collagen-coated disks of a nylon mesh, which were then placed between two fluid compartments, and found that as the monolayers reached confluency, a transepithelial electrical resistance developed. We also showed that the establishment of the resistance occurs concomitantly with the development of an extensive system of tight junctions that restricts the passage of molecules and ions through the intercellular spaces and defines morphologically, biochemically, and functionally distinct apical and basolateral domains on the surface of the cells. Calcium ions were found to be necessary for maintaining the integrity of the junctions, which could be reversibly opened and closed by the removal of calcium by the chelating agent EDTA and calcium readdition. Our findings (Cereijido et al. 1978a,b) and the almost contemporaneous work from Pitelka's laboratory (Misfeldt et al. 1976) established that MDCK cells are a suitable system to study the sorting processes, which in polarized epithelial cells effect the segregation of specific polypeptides into the two plasma membrane domains, as well as to investigate other phenomena characteristic of epithelial cells, such as the formation of tight junctions, transcytosis, and the polarized organization of the cytoskeleton (see Griepp et al. 1983). Following our extensive characterization of the properties of MDCK cells, this cell line has become a widely adopted paradigm for studies of epithelial cell physiology and intracellular protein traffic in general.

Soon after we developed the MDCK cell system, Enrique Rodriguez-Boulan, a young Argentine physician-scientist who had begun his training in Argentina with Cereijido, joined my laboratory, where he initially studied the asymmetric disposition of integral membrane proteins in the ER (Rodriguez-Boulan et al. 1978). With him we later discovered a striking manifestation of the polarized nature of MDCK cells (Rodriguez-Boulan & Sabatini 1978). We found that when these cells were infected with enveloped viruses, the virions assembled selectively on either one or the other plasma membrane domain, from where the virus subsequently buds. We saw that influenza (**Figure 8**), Sendai, and Simian virus 5 budded exclusively from the apical surface of the cells, whereas other virions, such as vesicular stomatitis virus (VSV), assembled only on the basolateral regions of the plasma membrane (**Figure 9**). Enrique observed (Boulan & Pendergast 1980) that, before budding takes place, the viral glycoproteins accumulate in the respective cell surfaces, as expected if their site of accumulation determines the site of budding. With Michael Rindler and Ivan Ivanov we were able to show that in MDCK cells the glycoproteins are segregated in the *trans* region of the Golgi apparatus into different membrane-bound carriers that are directly delivered to the specific plasma membrane domain, which highlighted the role of the Golgi in the sorting of cargo proteins (Rindler et al. 1984, 1985). We also demonstrated that in single cells, attachment to a substrate is sufficient to trigger the expression of plasma membrane polarity, which is manifested in the asymmetric budding of viruses (Rodriguez-Boulan et al. 1983).

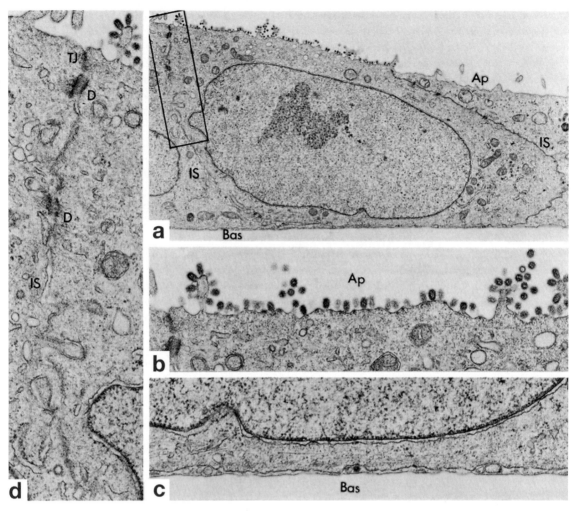

Figure 8

Exclusive budding of influenza virions from the apical surface of MDCK cells in a confluent monolayer.
(*a*) View of one cell with abundant virion budding from the apical surface, shown at higher magnification
in (*b*). Virus budding does not take place from the basal (*c*) and lateral surfaces (*d*) of the same cell. From
E. Rodriguez-Boulan, D.D. Sabatini. 1978. Asymmetric budding of viruses in epithelial monolayers:
A model system for study of epithelial polarity. *Proc. Natl. Acad. Sci. USA* 75:5071–75.

Soon it became possible to use transfected
cells to express the individual viral glycoproteins in the absence of infection, which allowed us to demonstrate that the viral glycoproteins are effectively segregated to the
appropriate cell surface domain even in the
absence of other viral components (Gottlieb
et al. 1986). This definitively proved that the
sorting information necessary for their asymmetric distribution is contained in the gly-

coproteins themselves. Because we had also
observed that truncated forms of the viral
glycoproteins consisting only of their luminal domains were secreted in a nonpolarized form from MDCK cells, we concluded
that their sorting information is contained
within their membrane or cytoplasmic tails
(Gonzalez et al. 1987).

The simple experimental system of virus-
infected polarized MDCK cells that emerged

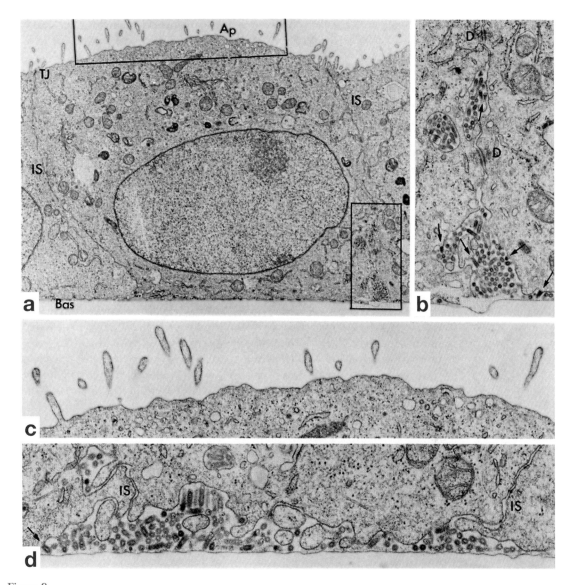

Figure 9

Exclusive budding of vesicular stomatitis virions from the basal and lateral surfaces of an MDCK cell in a confluent monolayer. Note that the apical surface of a cell (*a* and *c*), with abundant microvilli, shows no virus particles. On the other hand, virions are abundant in the intercellular space (*b*) between the lateral membranes of two adjacent cells and at the basal surface (*d*). The arrows show budding virions. From E. Rodriguez-Boulan, D.D. Sabatini. 1978. Asymmetric budding of viruses in epithelial monolayers: A model system for study of epithelial polarity. *Proc. Natl. Acad. Sci. USA* 75:5071–75.

from this work was soon adopted by many investigators to study the mechanisms of intracellular protein sorting and plasma membrane biogenesis and became the basis for the sprouting of a whole new area of cell bio-

logical investigation. In our own laboratory we showed that the integrity of the microtubule apparatus in MDCK cells was essential for the proper sorting of the influenza glycoprotein to the apical surface and the subsequent polarized budding of the virus (Rindler et al. 1987). The polarized organization of MDCK cells also brought to the fore the existence of specialized endocytic machineries operating at the two surfaces. This was strikingly demonstrated when we found that the actin cytoskeleton plays a critical role in the process of apical endocytosis (which, for example, is required for infection with influenza virions), but plays no role in endocytosis from the basolateral surface (Gottlieb et al. 1993). We also discovered the curious fact that whereas the VSV-G glycoprotein, after reaching the cell surface, undergoes endocytosis, the influenza HA glycoprotein is not internalized to any significant extent (Gottlieb et al. 1986). Interestingly, we had obtained evidence from the behavior of chimeric proteins that the features determining the endocytic behavior of VSV-G are contained in the cytoplasmic domain of the protein (Rizzolo et al. 1985). The role of signals in the cytoplasmic tails of transmembrane proteins in determining the capacity of the proteins to be endocytosed was soon definitely established by the now classical studies (Davis et al. 1987) from Brown and Goldstein's laboratories on the LDL receptor.

Thanks to a partnership with my colleagues, Gert Kreibich, Takashi Morimoto, and in particular Milton Adesnik, with whom I have shared a common interest in protein traffic for over two decades and who helped me keep abreast of the new techniques of molecular cell biology, the work in my laboratory moved with the times. Now, using organelles isolated or prepared by variants of the classical techniques of cell fractionation pioneered at Rockefeller, we are engaged in attempts to reproduce in in vitro systems complex sorting and transport processes (Gravotta et al. 1990; Mayer et al. 1996; Simon et al. 1996a,b, 1998). We analyze the results with refined methods of immune-electron microscopy and biochemistry that I had the privilege of seeing born and evolving. We employ recombinant DNA-derived cargo molecules and similarly generated elements of the transport machinery and do not cease to be amazed by the detailed knowledge that is being attained about processes that seemed so impenetrable just a few decades ago.

LITERATURE CITED

Adelman MR, Blobel G, Sabatini DD. 1973a. An improved cell fractionation procedure for the preparation of rat liver membrane-bound ribosomes. *J. Cell Biol.* 56:191–205

Adelman MR, Sabatini DD, Blobel G. 1973b. Ribosome-membrane interaction. Nondestructive disassembly of rat liver rough microsomes into ribosomal and membranous components. *J. Cell Biol.* 56:206–29

Bar-Nun S, Kreibich G, Adesnik M, Alterman L, Negishi M, Sabatini DD. 1980. Synthesis and insertion of cytochrome P-450 into endoplasmic reticulum membranes. *Proc. Natl. Acad. Sci. USA* 77:965–69

Beaufay H, De Duve C, Novikoff AB. 1956. Electron microscopy of lysosome rich fractions from rat liver. *J. Biophys. Biochem. Cytol.* 2:179–84

Beckmann R, Bubeck D, Grassucci R, Penczek P, Verschoor A, et al. 1997. Alignment of conduits for the nascent polypeptide chain in the ribosome-Sec61 complex. *Science* 278:2123–26

Bernstein HD, Poritz MA, Strub K, Hoben PJ, Brenner S, Walter P. 1989. Model for signal sequence recognition from amino-acid sequence of 54 K subunit of signal recognition particle. *Nature* 340:482–86

Blobel G. 1980. Intracellular protein topogenesis. *Proc. Natl. Acad. Sci. USA* 77:1496–500

Blobel G, Dobberstein B. 1975a. Transfer of proteins across membranes. I. Presence of proteolytically processed and unprocessed nascent immunoglobulin light chains on membrane-
bound ribosomes of murine myeloma. *J. Cell Biol.* 67:835–51

Blobel G, Dobberstein B. 1975b. Transfer of proteins across membranes. II. Reconstitution of
functional rough microsomes from heterologous components. *J. Cell Biol.* 67:852–62

Blobel G, Sabatini D. 1971. Dissociation of mammalian polyribosomes into subunits by
puromycin. *Proc. Natl. Acad. Sci. USA* 68:390–94

Blobel G, Sabatini DD. 1970. Controlled proteolysis of nascent polypeptides in rat liver cell
fractions. I. Location of the polypeptides within ribosomes. *J. Cell Biol.* 45:130–45

Borgese D, Blobel G, Sabatini DD. 1973. In vitro exchange of ribosomal subunits between
free and membrane-bound ribosomes. *J. Mol. Biol.* 74:415–38

Borgese N, Mok W, Kreibich G, Sabatini DD. 1974. Ribosomal-membrane interaction: in
vitro binding of ribosomes to microsomal membranes. *J. Mol. Biol.* 88:559–80

Boulan ER, Pendergast M. 1980. Polarized distribution of viral envelope proteins in the plasma
membrane of infected epithelial cells. *Cell* 20:45–54

Byers B. 1967. Structure and formation of ribosome crystals in hypothermic chick embryo
cells. *J. Mol. Biol.* 26:155–67

Cereijido M, Robbins ES, Dolan WJ, Rotunno CA, Sabatini DD. 1978a. Polarized monolayers
formed by epithelial cells on a permeable and translucent support. *J. Cell Biol.* 77:853–80

Cereijido M, Rotunno CA, Robbins ES, Sabatini DD. 1978b. Polarized epithelial membranes
produced in vitro. In *Membrane Transport Processes*, pp. 433–61. New York: Raven

Chyn TL, Martonosi AN, Morimoto T, Sabatini DD. 1979. In vitro synthesis of the Ca^{2+}
transport ATPase by ribosomes bound to sarcoplasmic reticulum membranes. *Proc. Natl.
Acad. Sci. USA* 76:1241–45

Colman DR, Kreibich G, Frey AB, Sabatini DD. 1982. Synthesis and incorporation of myelin
polypeptides into CNS myelin. *J. Cell Biol.* 95:598–608

Crowley KS, Liao S, Worrell VE, Reinhart GD, Johnson AE. 1994. Secretory proteins move
through the endoplasmic reticulum membrane via an aqueous, gated pore. *Cell* 78:461–71

Croze E, Ivanov IE, Kreibich G, Adesnik M, Sabatini DD, Rosenfeld MG. 1989. Endolyn-78, a
membrane glycoprotein present in morphologically diverse components of the endosomal
and lysosomal compartments: implications for lysosome biogenesis. *J. Cell Biol.* 108:1597–
613

Davis CG, van Driel IR, Russell DW, Brown MS, Goldstein JL. 1987. The low density lipoprotein receptor. Identification of amino acids in cytoplasmic domain required for rapid endocytosis. *J. Biol. Chem.* 262:4075–82

De Robertis E, Sabatini D. 1958. Mitochondrial changes in the adrenocortex of normal ham-
sters. *J. Biophys. Biochem. Cytol.* 4:667–68

De Robertis ED, Sabatini DD. 1960. Submicroscopic analysis of the secretory process in the
adrenal medulla. *Fed. Proc.* 19(Suppl. 5):70–78

Deshaies RJ, Schekman R. 1987. A yeast mutant defective at an early stage in import of secretory
protein precursors into the endoplasmic reticulum. *J. Cell Biol.* 105:633–45

Dintzis HM. 1961. Assembly of the peptide chains of hemoglobin. *Proc. Natl. Acad. Sci. USA*
47:247–61

Ernster L, Siekevitz, P, Palade GE. 1962. Enzyme-structure relationships in the endoplasmic
reticulum of rat liver. A morphological and biochemical study. *J. Cell Biol.* 15:541–62

Farquhar MG, Palade GE. 1963. Junctional complexes in various epithelia. *J. Cell Biol.* 17:375–
412

Fieser LF, Fieser M. 1956. *Organic Chemistry*. Boston: Reinhold

Finidori J, Rizzolo L, Gonzalez A, Kreibich G, Adesnik M, Sabatini DD. 1987. The influenza hemagglutinin insertion signal is not cleaved and does not halt translocation when presented to the endoplasmic reticulum membrane as part of a translocating polypeptide. *J. Cell Biol.* 104:1705–14

Gilbert W. 1963. Polypeptide synthesis in *Escherichia coli*. II. The polypeptide chain and S-RNA. *J. Mol. Biol.* 6:389–403

Gilmore R, Blobel G, Walter P. 1982a. Protein translocation across the endoplasmic reticulum. I. Detection in the microsomal membrane of a receptor for the signal recognition particle. *J. Cell Biol.* 95:463–69

Gilmore R, Walter P, Blobel G. 1982b. Protein translocation across the endoplasmic reticulum. II. Isolation and characterization of the signal recognition particle receptor. *J. Cell Biol.* 95:470–77

Gonzalez A, Rizzolo L, Rindler M, Adesnik M, Sabatini DD, Gottlieb T. 1987. Nonpolarized secretion of truncated forms of the influenza hemagglutinin and the vesicular stomatitus virus G protein from MDCK cells. *Proc. Natl. Acad. Sci. USA* 84:3738–42

Gorlich D, Prehn S, Hartmann E, Kalies KU, Rapoport TA. 1992. A mammalian homolog of SEC61p and SECYp is associated with ribosomes and nascent polypeptides during translocation. *Cell* 71:489–503

Gorlich D, Rapoport TA. 1993. Protein translocation into proteoliposomes reconstituted from purified components of the endoplasmic reticulum membrane. *Cell* 75:615–30

Gottlieb TA, Gonzalez A, Rizzolo L, Rindler MJ, Adesnik M, Sabatini DD. 1986. Sorting and endocytosis of viral glycoproteins in transfected polarized epithelial cells. *J. Cell Biol.* 102:1242–55

Gottlieb TA, Ivanov IE, Adesnik M, Sabatini DD. 1993. Actin microfilaments play a critical role in endocytosis at the apical but not the basolateral surface of polarized epithelial cells. *J. Cell Biol.* 120:695–710

Gravotta D, Adesnik M, Sabatini DD. 1990. Transport of influenza HA from the *trans*-Golgi network to the apical surface of MDCK cells permeabilized in their basolateral plasma membranes: energy dependence and involvement of GTP-binding proteins. *J. Cell Biol.* 111:2893–908

Griepp EB, Dolan WJ, Robbins ES, Sabatini DD. 1983. Participation of plasma membrane proteins in the formation of tight junctions by cultured epithelial cells. *J. Cell Biol.* 96:693–702

Harnik-Ort V, Prakash K, Marcantonio E, Colman DR, Rosenfeld MG, et al. 1987. Isolation and characterization of cDNA clones for rat ribophorin I: complete coding sequence and in vitro synthesis and insertion of the encoded product into endoplasmic reticulum membranes. *J. Cell Biol.* 104:855–63

Ito A, Sato R. 1969. Proteolytic microdissection of smooth-surfaced vesicles of liver microsomes. *J. Cell Biol.* 40:179–89

Ivanov IE, Sabatini DD. 1981. Surface features and handedness of a model for the eukaryotic small ribosomal subunit. *J. Ultrastruct. Res.* 76:263–76

Jacob F, Monod J. 1961. Genetic regulatory mechanisms in the synthesis of proteins. *J. Mol. Biol.* 3:318–56

Katz FN, Rothman JE, Knipe DM, Lodish HF. 1977. Membrane assembly: synthesis and intracellular processing of the vesicular stomatitis viral glycoprotein. *J. Supramol. Struct.* 7:353–70

Kelleher DJ, Kreibich G, Gilmore R. 1992. Oligosaccharyltransferase activity is associated with a protein complex composed of ribophorins I and II and a 48 kd protein. *Cell* 69:55–65

Kemper B, Habener JF, Mulligan RC, Potts JT Jr, Rich A. 1974. Pre-preparathyroid hormone: a direct translation product of parathyroid messenger RNA. *Proc. Natl. Acad. Sci. USA* 71:3731–35

Kreibich G, Debey P, Sabatini DD. 1973. Selective release of content from microsomal vesicles without membrane disassembly. I. Permeability changes induced by low detergent concentrations. *J. Cell Biol.* 58:436–62

Kreibich G, Freienstein CM, Pereyra BN, Ulrich BL, Sabatini DD. 1978a. Proteins of rough microsomal membranes related to ribosome binding. II. Cross-linking of bound ribosomes to specific membrane proteins exposed at the binding sites. *J. Cell Biol.* 77:488–506

Kreibich G, Hubbard AL, Sabatini DD. 1974. On the spatial arrangememt of proteins in microsomal membranes from rat liver. *J. Cell Biol.* 60:616–27

Kreibich G, Ojakian G, Rodriguez-Boulan E, Sabatini DD. 1982. Recovery of ribophorins and ribosomes in "inverted rough" vesicles derived from rat liver rough microsomes. *J. Cell Biol.* 93:111–21

Kreibich G, Sabatini DD. 1974. Selective release of content from microsomal vesicles without membrane disassembly. II. Electrophoretic and immunological characterization of microsomal subfractions. *J. Cell Biol.* 61:789–807

Kreibich G, Ulrich BL, Sabatini DD. 1978b. Proteins of rough microsomal membranes related to ribosome binding. I. Identification of ribophorins I and II, membrane proteins characteristic of rough microsomes. *J. Cell Biol.* 77:464–87

Kruppa J, Sabatini DD. 1977. Release of poly A$^+$ messenger RNA from rat liver rough microsomes upon disassembly of bound polysomes. *J. Cell Biol.* 74:414–27

Lande MA, Adesnik M, Sumida M, Tashiro Y, Sabatini DD. 1975. Direct association of messenger RNA with microsomal membranes in human diploid fibroblasts. *J. Cell Biol.* 65:513–28

Lasansky A, Sabatini DD. 1957. Distribution of sulfhydryl and disulfide groups in the neurohypophysis and hypothalamus of the toad. *Rev. Soc. Argent. Biol.* 33:177–82

Ledbetter MC, Porter KR. 1963. A "microtubule" in plant cell fine structure. *J. Cell Biol.* 19:239–50

Leighton J, Estes LW, Mansukhani S, Brada Z. 1970. A cell line derived from normal dog kidney (MDCK) exhibiting qualities of papillary adenocarcinoma and of renal tubular epithelium. *Cancer* 26:1022–28

Lewis JA, Sabatini DD. 1977. Accessibility of proteins in rat liver-free and membrane-bound ribosomes to lactoperoxidase-catalyzed iodination. *J. Biol. Chem.* 252:5547–55

Lilley BN, Ploegh HL. 2004. A membrane protein required for dislocation of misfolded proteins from the ER. *Nature* 429:834–40

Mach B, Faust C, Vassalli P. 1973. Purification of 14S messenger RNA of immunoglobulin light chain that codes for a possible light-chain precursor. *Proc. Natl. Acad. Sci. USA* 70:451–55

Malkin LI, Rich A. 1967. Partial resistance of nascent polypeptide chains to proteolytic digestion due to ribosomal shielding. *J. Mol. Biol.* 26:329–46

Marcantonio EE, Grebenau RC, Sabatini DD, Kreibich G. 1982. Identification of ribophorins in rough microsomal membranes from different organs of several species. *Eur. J. Biochem.* 124:217–22

Mayer A, Ivanov IE, Gravotta D, Adesnik M, Sabatini DD. 1996. Cell-free reconstitution of the transport of viral glycoproteins from the TGN to the basolateral plasma membrane of MDCK cells. *J. Cell Sci.* 109(Pt 7):1667–76

McQuillen K, Roberts RB, Britten RJ. 1959. Synthesis of nascent protein by ribosomes in *Escherichia coli*. *Proc. Natl. Acad. Sci. USA* 45:1437–47

Mentaberry A, Adesnik M, Atchison M, Norgard EM, Alvarez F, et al. 1986. Small basic proteins of myelin from central and peripheral nervous systems are encoded by the same gene. *Proc. Natl. Acad. Sci. USA* 83:1111–14

Meyer DI, Krause E, Dobberstein B. 1982. Secretory protein translocation across membranes—the role of the "docking protein." *Nature* 297:647–50

Miller JD, Tajima S, Lauffer L, Walter P. 1995. The beta subunit of the signal recognition particle receptor is a transmembrane GTPase that anchors the alpha subunit, a peripheral membrane GTPase, to the endoplasmic reticulum membrane. *J. Cell Biol.* 128:273–82

Milstein C, Brownlee GG, Harrison TM, Mathews MB. 1972. A possible precursor of immunoglobulin light chains. *Nat. New Biol.* 239:117–20

Misfeldt DS, Hamamoto ST, Pitelka DR. 1976. Transepithelial transport in cell culture. *Proc. Natl. Acad. Sci. USA* 73:1212–16

Monier S, Van Luc P, Kreibich G, Sabatini DD, Adesnik M. 1988. Signals for the incorporation and orientation of cytochrome P450 in the endoplasmic reticulum membrane. *J. Cell Biol.* 107:457–70

Morimoto T, Blobel G, Sabatini DD. 1972a. Ribosome crystallization in chicken embryos. I. Isolation, characterization, and in vitro activity of ribosome tetramers. *J. Cell Biol.* 52:338–54

Morimoto T, Blobel G, Sabatini DD. 1972b. Ribosome crystallization in chicken embryos. II. Conditions for the formation of ribosome tetramers in vitro. *J. Cell Biol.* 52:355–66

Musch A, Wiedmann M, Rapoport TA. 1992. Yeast Sec proteins interact with polypeptides traversing the endoplasmic reticulum membrane. *Cell* 69:343–52

Nishimura Y, Rosenfeld MG, Kreibich G, Gubler U, Sabatini DD, et al. 1986. Nucleotide sequence of rat preputial gland beta-glucuronidase cDNA and in vitro insertion of its encoded polypeptide into microsomal membranes. *Proc. Natl. Acad. Sci. USA* 83:7292–96

Nonomura Y, Blobel G, Sabatini D. 1971. Structure of liver ribosomes studied by negative staining. *J. Mol. Biol.* 60:303–23

Novikoff AB. 1956. Electron microscopy: cytology of cell fractions. *Science* 124:969–72

Ojakian GK, Kreibich G, Sabatini DD. 1977. Mobility of ribosomes bound to microsomal membranes. A freeze-etch and thin-section electron microscope study of the structure and fluidity of the rough endoplasmic reticulum. *J. Cell Biol.* 72:530–51

Okada Y, Frey AB, Guenthner TM, Oesch F, Sabatini DD, Kreibich G. 1982. Studies on the biosynthesis of microsomal membrane proteins. Site of synthesis and mode of insertion of cytochrome b5, cytochrome b5 reductase, cytochrome P-450 reductase and epoxide hydrolase. *Eur. J. Biochem.* 122:393–402

Oliver J, Jungnickel B, Gorlich D, Rapoport T, High S. 1995. The Sec61 complex is essential for the insertion of proteins into the membrane of the endoplasmic reticulum. *FEBS Lett.* 362:126–30

Omura T, Siekevitz P, Palade GE. 1967. Turnover of constituents of the endoplasmic reticulum membranes of rat hepatocytes. *J. Biol. Chem.* 242:2389–96

Painter RG, Tokuyasu KT, Singer SJ. 1973. Immunoferritin localization of intracellular antigens: the use of ultracryotomy to obtain ultrathin sections suitable for direct immunoferritin staining. *Proc. Natl. Acad. Sci. USA* 70:1649–53

Palade G. 1975. Intracellular aspects of the process of protein synthesis. *Science* 189:347–58

Palade GE. 1952. A study of fixation for electron microscopy. *J. Exp. Med.* 95:285–98

Palade GE. 1953. A small particulate component of the cytoplasm. *J. Appl. Phys.* 24:1419 (Abstr.)

Palade GE. 1955. A small particulate component of the cytoplasm. *J. Biophys. Biochem. Cytol.* 1:59–68

Palade GE, Siekevitz P, Caro LG. 1961. Structure, chemistry and function of the pancreatic exocrine cell. In *Ciba Foundation Symposium on the The Exocrine Pancreas*, pp. 23–55. Boston/New York: Little Brown

Petermann ML, Pavlovec A. 1969. Effects of magnesium and formaldehyde on the sedimentation behavior of rat liver ribosomes. *Biopolymers* 7:73–81

Porter KR. 1965. Cytoplasmic microtubules and their functions. In *Ciba Foundation Symposium, Principles of Biomolecular Organization*, ed. GEW Wolstenholme, M O'Connor, pp. 308–45. Boston: Little

Redman CM. 1967. Studies on the transfer of incomplete polypeptide chains across rat liver microsomal membranes in vitro. *J. Biol. Chem.* 242:761–68

Redman CM, Sabatini DD. 1966. Vectorial discharge of peptides released by puromycin from attached ribosomes. *Proc. Natl. Acad. Sci. USA* 56:608–15

Redman CM, Siekevitz P, Palade GE. 1966. Synthesis and transfer of amylase in pigeon pancreatic micromosomes. *J. Biol. Chem.* 241:1150–58

Rindler MJ, Ivanov IE, Plesken H, Rodriguez-Boulan E, Sabatini DD. 1984. Viral glycoproteins destined for apical or basolateral plasma membrane domains traverse the same Golgi apparatus during their intracellular transport in doubly infected Madin-Darby canine kidney cells. *J. Cell Biol.* 98:1304–19

Rindler MJ, Ivanov IE, Plesken H, Sabatini DD. 1985. Polarized delivery of viral glycoproteins to the apical and basolateral plasma membranes of Madin-Darby canine kidney cells infected with temperature-sensitive viruses. *J. Cell Biol.* 100:136–51

Rindler MJ, Ivanov IE, Sabatini DD. 1987. Microtubule-acting drugs lead to the nonpolarized delivery of the influenza hemagglutinin to the cell surface of polarized Madin-Darby canine kidney cells. *J. Cell Biol.* 104:231–41

Rizzolo LJ, Finidori J, Gonzalez A, Arpin M, Ivanov IE, et al. 1985. Biosynthesis and intracellular sorting of growth hormone-viral envelope glycoprotein hybrids. *J. Cell Biol.* 101:1351–62

Rodriguez Boulan E , Sabatini DD, Pereyra BN, Kreibich G. 1978. Spatial orientation of glycoproteins in membranes of rat liver rough microsomes. II. Transmembrane disposition and characterization of glycoproteins. *J. Cell Biol.* 78:894–909

Rodriguez-Boulan E, Paskiet KT, Sabatini DD. 1983. Assembly of enveloped viruses in Madin-Darby canine kidney cells: polarized budding from single attached cells and from clusters of cells in suspension. *J. Cell Biol.* 96:866–74

Rodriguez-Boulan ER, Sabatini DD. 1978. Asymmetric budding of viruses in epithelial monlayers: a model system for study of epithelial polarity. *Proc. Natl. Acad. Sci. USA* 75:5071–75

Rosenfeld MG, Kreibich G, Popov D, Kato K, Sabatini DD. 1982. Biosynthesis of lysosomal hydrolases: their synthesis in bound polysomes and the role of co- and post-translational processing in determining their subcellular distribution. *J. Cell Biol.* 93:135–43

Rosenfeld MG, Marcantonio EE, Hakimi J, Ort VM, Atkinson PH, et al. 1984. Biosynthesis and processing of ribophorins in the endoplasmic reticulum. *J. Cell Biol.* 99:1076–82

Sabatini DD. 1999. George E. Palade: charting the secretory pathway. *Trends Cell Biol.* 9:413–17

Sabatini DD, Bensch K, Barrnett RJ. 1963. Cytochemistry and electron microscopy. The preservation of cellular ultrastructure and enzymatic activity by aldehyde fixation. *J. Cell Biol.* 17:19–58

Sabatini DD, Blobel G. 1970. Controlled proteolysis of nascent polypeptides in rat liver cell fractions. II. Location of the polypeptides in rough microsomes. *J. Cell Biol.* 45:146–57

Sabatini DD, Blobel G, Nonomura Y, Adelman MR. 1971. Ribosome-membrane interaction: structural aspects and functional implications. *Adv. Cytopharmacol.* 1:119–29

Sabatini DD, Borgese N, Adelman MR, Kreibich G, Blobel G. 1972. Studies on the membrane associated protein synthesis apparatus of eukaryotic cells. RNA viruses/ribosomes. *FEBS Symp.* 27:147–71

Sabatini DD, De Robertis ED. 1961. Ultrastructural zonation of adrenocortex in the rat. *J. Biophys. Biochem. Cytol.* 9:105–19

Sabatini DD, De Robertis ED, Bleichmar HB. 1962. Submicroscopic study of the pituitary action on the adrenocortex of the rat. *Endocrinology* 70:390–406

Sabatini DD, Kreibich G, Morimoto T, Adesnik M. 1982. Mechanisms for the incorporation of proteins in membranes and organelles. *J. Cell Biol.* 92:1–22

Sabatini DD, Miller F, Barrnett RJ. 1964. Aldehyde fixation for morphological and enzyme histochemical studies with the electron microscope. *J. Histochem. Cytochem.* 12:57–71

Sabatini DD, Tashiro Y, Palade GE. 1966. On the attachment of ribosomes to microsomal membranes. *J. Mol. Biol.* 19:503–24

Sabban E, Marchesi V, Adesnik M, Sabatini DD. 1981. Erythrocyte membrane protein band 3: its biosynthesis and incorporation into membranes. *J. Cell Biol.* 91:637–46

Sanders SL, Whitfield KM, Vogel JP, Rose MD, Schekman RW. 1992. Sec61p and BiP directly facilitate polypeptide translocation into the E.R. *Cell* 69:353–65

Sherman J. MT, Sabatini D 1983. Biosynthesis of the Na, K-ATPase in MDCK cells. In *Current Topics in Membrane and Transport*, ed. F Bonner, A Kleinzeller, 19:753–64. New York: Academic

Siekevitz P. 1952. Uptake of radioactive alanine in vitro into the proteins of rat liver fractions. *J. Biol. Chem.* 195:549–65

Siekevitz P, Palade GE. 1960. A cytochemical study on the pancreas of the guinea pig. 5. In vivo incorporation of leucine-1-C14 into the chymotrypsinogen of various cell fractions. *J. Biophys. Biochem. Cytol.* 7:619–30

Simon JP, Ivanov IE, Adesnik M, Sabatini DD. 1996a. The production of post-Golgi vesicles requires a protein kinase C-like molecule, but not its phosphorylating activity. *J. Cell Biol.* 135:355–70

Simon JP, Ivanov IE, Shopsin B, Hersh D, Adesnik M, Sabatini DD. 1996b. The in vitro generation of post-Golgi vesicles carrying viral envelope glycoproteins requires an ARF-like GTP-binding protein and a protein kinase C associated with the Golgi apparatus. *J. Biol. Chem.* 271:16952–61

Simon JP, Morimoto T, Bankaitis VA, Gottlieb TA, Ivanov IE, et al. 1998. An essential role for the phosphatidylinositol transfer protein in the scission of coatomer-coated vesicles from the *trans*-Golgi network. *Proc. Natl. Acad. Sci. USA* 95:11181–86

Simon SM, Blobel G. 1991. A protein-conducting channel in the endoplasmic reticulum. *Cell* 65:371–80

Singer SJ. 1959. Preparation of an electron-dense antibody conjugate. *Nature* 183:1523–24

Singer SJ, Nicolson GL. 1972. The fluid mosaic model of the structure of cell membranes. *Science* 175:720–31

Tashiro Y, Siekevitz P. 1965. Ultracentrifugal studies on the dissociation of hepatic ribosomes. *J. Mol. Biol.* 11:149–65

Tashiro Y, Yphantis DA. 1965. Molecular weights of hepatic ribosomes and their subunits. *J. Mol. Biol.* 11:174–86

Tissieres A, Watson JD, Schlessinger D, Hollingworth BR. 1959. Ribonucleoprotein particles from *Escherichia coli*. *J. Mol. Biol.* 1:221–33

Tokuyasu KT, Singer SJ. 1976. Improved procedures for immunoferritin labeling of ultrathin frozen sections. *J. Cell Biol.* 71:894–906

Tonegawa S, Baldi I. 1973. Electrophoretically homogeneous myeloma light chain mRNA and its translation in vitro. *Biochem. Biophys. Res. Commun.* 51:81–87

van den Berg B, Clemons WM Jr, Collinson I, Modis Y, Hartmann E, et al. 2004. X-ray structure of a protein-conducting channel. *Nature* 427:36–44

Walter P, Blobel G. 1981. Translocation of proteins across the endoplasmic reticulum. II. Signal recognition protein (SRP) mediates the selective binding to microsomal membranes of in-vitro-assembled polysomes synthesizing secretory protein. *J. Cell Biol.* 91:551–56

Walter P, Blobel G. 1983. Disassembly and reconstitution of signal recognition particle. *Cell* 34:525–33

Walter P, Ibrahimi I, Blobel G. 1981. Translocation of proteins across the endoplasmic reticulum. I. Signal recognition protein (SRP) binds to in-vitro-assembled polysomes synthesizing secretory protein. *J. Cell Biol.* 91:545–50

Wirth DF, Katz F, Small B, Lodish HF. 1977. How a single Sindbis virus mRNA directs the synthesis of one soluble protein and two integral membrane glycoproteins. *Cell* 10:253–63

Ye Y, Shibata Y, Yun C, Ron D, Rapoport TA. 2004. A membrane protein complex mediates retro-translocation from the ER lumen into the cytosol. *Nature* 429:841–47

Yu YH, Sabatini DD, Kreibich G. 1990. Antiribophorin antibodies inhibit the targeting to the ER membrane of ribosomes containing nascent secretory polypeptides. *J. Cell Biol.* 111:1335–42

Mechanisms of Apoptosis Through Structural Biology

Nieng Yan and Yigong Shi

Department of Molecular Biology, Lewis Thomas Laboratory, Princeton University, Princeton, New Jersey 08544; email: nyan@molbio.princeton.edu; yshi@molbio.princeton.edu

Annu. Rev. Cell Dev. Biol. 2005. 21:35–56

First published online as a Review in Advance on May 5, 2005

The *Annual Review of Cell and Developmental Biology* is online at http://cellbio.annualreviews.org

doi: 10.1146/ annurev.cellbio.21.012704.131040

Copyright © 2005 by Annual Reviews. All rights reserved

1081-0706/05/1110-0035$20.00

Key Words

caspase, IAP, Bcl-2 family of proteins, death receptor, signaling

Abstract

Apoptosis plays a central role in the development and homeostasis of metazoans. Research in the past two decades has led to the identification of hundreds of genes that govern the initiation, execution, and regulation of apoptosis. An earlier focus on the genetic and cell biological characterization has now been complemented by systematic biochemical and structural investigation, giving rise to an unprecedented level of clarity in many aspects of apoptosis. In this review, we focus on the molecular mechanisms of apoptosis by synthesizing available biochemical and structural information. We discuss the mechanisms of ligand binding to death receptors, actions of the Bcl-2 family of proteins, and caspase activation, inhibition, and removal of inhibition. Although an emphasis is given to the mammalian pathways, a comparative analysis is applied to related mechanistic information in *Drosophila* and *Caenorhabditis elegans*.

Contents

INTRODUCTION

Apoptosis, an ancient word coined in the fifth century BC (Andre 2003) and commissioned in 1972 (Kerr et al. 1972), refers to a specific form of programmed cell death. Apoptosis is central to the development and homeostasis of metazoans (Danial & Korsmeyer 2004, Hay et al. 2004, Horvitz 2003, Rathmell & Thompson 2002, Wang 2001), and its dysregulation causes a number of human pathologies, including cancer, autoimmune diseases, and neurodegenerative disorders (Green & Evan 2002, Hanahan & Weinberg 2000, Thompson 1995, Vaux & Flavell 2000, Yuan

& Yankner 2000). Research in the past two decades has led to the identification of hundreds of genes that govern the initiation, execution, and regulation of apoptosis in several species (Danial & Korsmeyer 2004). The pathways of apoptosis are evolutionarily conserved, culminating in the activation of death proteases, the caspases (named after cysteine protease with Asp substrate specificity) (Riedl & Shi 2004, Thornberry & Lazebnik 1998).

In mammalian cells, apoptosis comes in two forms, intrinsic and extrinsic, which are triggered by cell death stimuli from intra- and extracellular environments, respectively (**Figure 1**). The intracellular stimuli, such as DNA damage, generally result in the activation of the BH3-only Bcl-2 family of proteins, which causes the release of proapoptotic factors from the intermembrane space of mitochondria into the cytoplasm (Cory & Adams 2002, Wang 2001). The release process is mediated by Bax and Bak and antagonized by the antiapoptotic proteins Bcl-2 and Bcl-xL. One of the released protein factors, cytochrome *c*, directly activates Apaf-1 and, in the presence of dATP or ATP, induces the formation of a multimeric complex dubbed the "apoptosome." The apoptosome mediates the activation of the initiator caspase, caspase-9, which subsequently activates the effector caspases, caspase-3 and caspase-7, which are responsible for the dismantling of an apoptotic cell. The active caspases are subject to inhibition by the inhibitor of apoptosis (IAP) family of proteins (Salvesen & Duckett 2002). Smac/DIABLO, another protein released from mitochondria during apoptosis, interacts with multiple IAPs and counters IAP-mediated caspase inhibition.

The extracellular death stimuli, such as the Fas ligand, directly activate the death receptors through ligand-induced assembly of a death-inducing signaling complex (DISC) at the plasma membrane (Nagata 1999, Peter & Krammer 2003) (**Figure 1**). An adapter protein, Fas-associated death domain (FADD), appears to be the obligate factor, which recruits the initiator caspase,

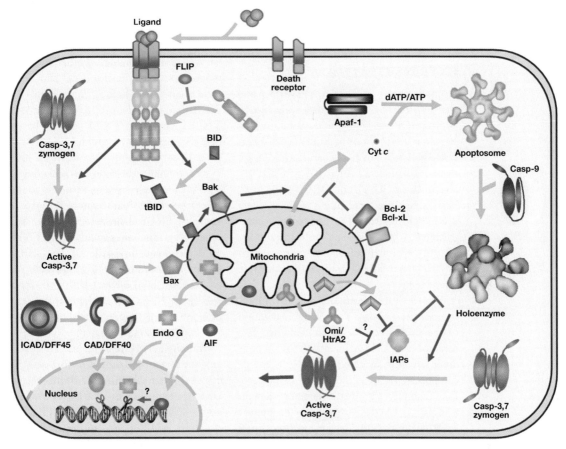

Figure 1

Schematic diagram of apoptotic pathways in mammalian cells. The cyan arrows indicate signal flow. Proapoptotic and antiapoptotic activities are colored magenta and green, respectively. See text for details.

procaspase-8 or -10, to DISC for activation. The activated caspase-8 subsequently cleaves and activates caspase-3 and -7. Thus extrinsic and intrinsic cell deaths converge at the point of caspase-3 or -7 activation. An important physiological target of the activated caspase-8 is Bid, a BH3-only member of the Bcl-2 family of proteins. After cleavage, the C-terminal fragment of Bid (truncated Bid or tBid) translocates to the outer membrane of mitochondria and induces the release of proapoptotic factors (Li et al. 1998, Luo et al. 1998). Thus Bid mediates the crosstalk from the extrinsic to intrinsic form of cell death.

One of the most important caspase targets is DFF45/ICAD, which forms a tight inhibitory complex with DFF40/CAD, a potent DNase when freed. The cleavage of ICAD/DFF45 by caspase-3 or -7 unleashes the free DFF40/CAD, which is responsible for the degradation of chromosomes into nucleosomal fragments during apoptosis. Another nuclease, endonuclease G, is also released from mitochondria and participates in DNA degradation during apoptosis.

One decade of systematic investigation by structural biology has revealed significant insights into the cell death pathways, giving rise to an unprecedented level of clarity in many aspects of apoptosis. In this review, we focus exclusively on the molecular mechanisms of apoptosis by synthesizing known biochemical and structural information.

EXTRINSIC PATHWAY

Death receptors, located on cell surface and characterized by extracellular cysteine-rich domains (CRDs), belong to the tumor necrosis factor (TNF) family of proteins (Ashkenazi & Dixit 1998). The best-characterized death receptors include TNFR1 (p55 or CD120a), Fas (Apo1 or CD95), and DR3 (also Apo3 or Wsl1). Each death receptor contains a single death domain (DD) in the intracellular compartment, which is thought to be responsible for the recruitment of adapter proteins, such as FADD and TRADD, through homotypic interactions.

The activated death ligands are homotrimeric and thus induce oligomerization of the death receptors upon binding. The receptor-associated adapter molecules recruit other effector proteins, forming a multicomponent DISC at the plasma membrane. For example, the death effector domain (DED) of FADD is thought to interact with the prodomain (which has two copies of DED) of procaspase-8, thus bringing three molecules of procaspase-8 into close proximity of one another and facilitating their auto-activation through an unknown mechanism.

Efforts to understand death receptor signaling by structural biology have been hampered by the technical difficulty in working with membrane-associated protein complexes. Further confounding this effort is the uncertainty on whether we have a complete list of the protein components for DISC. Nonetheless, structural characterization on the components of DISC has begun to reveal mechanistic insights.

Mechanism of Ligand Binding to Death Receptors

Structure of the death ligand is exemplified by those of TNFα and TNFβ (Eck & Sprang 1989, Eck et al. 1992, Jones et al. 1989), which were determined in their native trimeric forms. Each monomer forms a β-sandwich with a jellyroll topology (**Figure 2a**); three monomers stack against one another to form a bell-shaped homo-trimer (**Figure 2a**). The structure of the extracellular domain of TNFR1 bound to TNFβ reveals the specific recognition of a death ligand by the activated death receptor (Banner et al. 1993). In this complex, the four CRDs within one TNFR1 molecule stack up vertically to form an elongated rod-like structure (**Figure 2b**). Each TNFR1 rod interacts with two adjacent TNFβ monomers and the three TNFR1 molecules do not directly bind to one another. Protruding loops from CRD2 (50s loop) and CRD3 (90s loop) interact with two distinct regions of TNFβ. On the basis of this structure, the cytoplasmic regions of the receptors are proposed to cluster together.

Although TNFR1 contains four CRDs, only the second and third directly bind to the ligand (**Figure 2b**). Another death receptor, DR5, contains only two CRDs, corresponding to CRD2 and CRD3 of TNFR1. The structure of DR5 bound to the death ligand TRAIL reveals a distinct mode of recognition (Hymowitz et al. 1999, Mongkolsapaya et al. 1999). Although CRD1 of DR5 interacts with TRAIL in a fashion similar to CRD2 of TNFR1, CRD2 of DR5 binds to TRAIL in a different conformation compared with that of CRD3 of TNFR1. These differences underlie distinct signaling specificity by different death ligands and may have important ramifications for the design of specific therapeutic agents.

In addition to TNFβ-bound activated receptors, the structure of free TNFR1 is also available (Naismith et al. 1995, 1996), which reveals two distinct types of crystallographic dimers. In one scenario, the two receptors are arranged in an antiparallel fashion that would result in the separation of their cytoplasmic domains by a distance of over 100 Å. This arrangement would prevent the intracellular adapter proteins from forming a productive signaling complex. Thus the structure of this

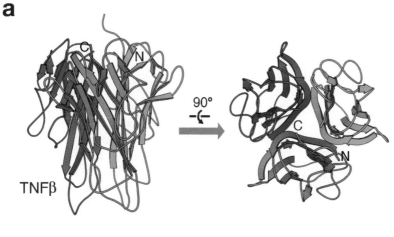

a

C N

TNFβ

90°

C

N

b

50s loop

CRD1

CRD2

CRD3

90s

CRD4

N

90°

90s loop

TNFR1

50s loop

TNFβ

Plasma membrane

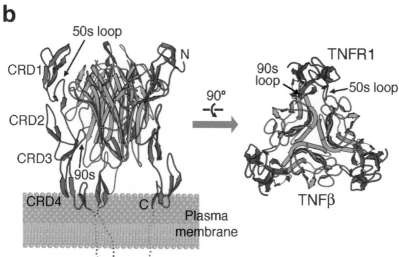

Figure 2

Structure of death ligand and its complex with receptor. (*a*) Structure of the trimeric TNFβ. Two perpendicular views are shown. (*b*) Structure of a complex between TNFβ and the extracellular domain of TNFR1. The four CRDs as well as the two important loops are labeled. This and other figures were prepared using MOLSCRIPT (Kraulis 1991) and PYMOL (DeLano 2002).

dimeric TNFR1 might represent the inactive form. In the other scenario, the two receptors are placed parallel to each other, with their TNF-binding surfaces fully exposed. Each of the two receptors is capable of forming a trimeric assembly upon ligand binding; thus this arrangement would result in clustering of TNF/TNFR1 trimers, a scenario that may enhance signaling efficiency.

Three Classes of Signaling Motifs

Apoptotic signaling in the extrinsic pathway is thought to rely on homotypic interactions among DDs and DEDs (Fesik 2000). A third class of conserved signaling motif, caspase recruitment domain (CARD), is present in a number of mammalian proteins including Apaf-1, caspase-2 and caspase-9, and the IAP proteins c-IAP1 and c-IAP2.

Structures of representative members of the three classes of signaling motifs reveal a conserved structural arrangement (Chou et al. 1998, Eberstadt et al. 1998, Huang et al. 1996). In each case, the structure consists of six antiparallel helices, with minor variation in interhelical packing (**Figure 3***a–c*). Mutational analyses on these domains revealed critical residues important for homotypic interactions, of which examples of CARD-CARD

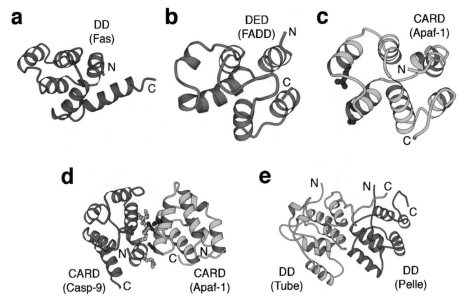

Figure 3

Structure of signaling motifs in apoptosis. (*a*) Structure of the DD from Fas. (*b*) Structure of the DED from FADD. (*c*) Structure of Apaf-1 CARD. The acidic residues important for caspase-9 binding are colored blue. (*d*) Structure of a heterodimer between the CARD domains of Apaf-1 (*cyan*) and caspase-9 (*magenta*). Critical interface residues are shown. (*e*) Structure of a heterodimer between the DDs of Pelle (*blue*) and Tube (*green*) in *Drosophila*.

and DD-DD interactions have been documented in mechanistic detail.

The recognition of procaspase-9 by Apaf-1, primarily through a CARD-CARD interaction, is essential to the formation of the apoptosome holoenzyme and subsequent activation of caspase-9. The positively charged helices H1 and H4 of procaspase-9 CARD are recognized by the negatively charged helices H2 and H3 of Apaf-1 CARD (Qin et al. 1999) (**Figure 3d**). In *Drosophila*, recruitment of the Ser/Thr kinase Pelle to the plasma membrane by the adapter protein Tube is important for embryogenesis. Compared with the CARD-CARD recognition (**Figure 3d**), the structure of a death domain complex between Pelle and Tube reveals a quite different recognition mechanism, in which the C-terminal tail of the Tube DD makes significant and indispensable contacts with Pelle (Xiao et al. 1999) (**Figure 3e**). It remains to be seen whether the structural arrangements in the Apaf-1-caspase-9 and Pelle-

Tube complexes are generally applicable to other pairs of CARD-CARD and DD-DD interactions.

Similar to other signaling motifs such as SH2, the sole purpose of the three classes of signaling motifs in apoptosis is to recruit downstream effector proteins. In contrast to other signaling motifs that usually lead to phosphorylation of downstream proteins, death signaling results in the auto-cleavage and activation of initiator caspases. The large and specific protein-protein interface in death signaling may play an important role in safeguarding the exquisite specificity required for initiating an apoptotic response.

INTRINSIC PATHWAY

Mechanisms of Action by the Bcl-2 Family of Proteins

Bcl-2 family of proteins. The Bcl-2 family of proteins plays a central role in the regulation of apoptosis. They are the gatekeepers in the mitochondria-initiated intrinsic apoptosis. In type II cells, they are also essential for death receptor-initiated extrinsic cell death. More than two dozen Bcl-2 family members have been identified (Cory & Adams 2002, Danial & Korsmeyer 2004). On the basis of function and sequence similarity, Bcl-2 proteins are grouped into three subfamilies. The antiapoptosis subfamily, represented by Bcl-2 and Bcl-xL in mammals, and CED-9 in nematodes, inhibits cell death through distinct mechanisms. Bcl-2 and Bcl-xL prevent the release of mitochondrial proteins, whereas CED-9 prevents CED-4 from activating CED-3 (the only apoptotic caspase in nematodes). The proapoptosis proteins constitute two subfamilies, the so-called multidomain subfamily represented by Bax and Bak and the BH3-only subfamily. Bax and Bak exist as monomers in the absence of apoptotic signaling but can form homo-oligomers upon activation. These large homo-oligomers are thought to form pores that mediate the release of mitochondrial proteins such as cytochrome c, although conclusive evidence is lacking. Members in the Bcl-2/Bcl-xL subfamily contain all four conserved Bcl-2 homology (BH) domains, BH4, BH3, BH1, and BH2, whereas the Bax/Bak subfamily lacks the BH4 domain. Most members of the Bcl-2 family contain a single, membrane-spanning region at their C-termini. Members of the opposing subfamilies, as well as the two proapoptotic subfamilies, can dimerize in aqueous solution, mediated by the amphipathic BH3 helix.

Structure of Bcl-2 family of proteins. Structural information is available on all three subfamilies of Bcl-2 proteins. The first structure of the Bcl-2 family, determined on Bcl-xL (Muchmore et al. 1996), reveals two centrally located hydrophobic α helices (α5 and α6), surrounded by five amphipathic helices (**Figure 4a**). The Bcl-xL structure resembles the pore-forming domains of bacterial toxins such as diphtheria toxin, suggesting

that the Bcl-2 family of proteins may form pores at the mitochondrial outer membrane to regulate ion exchange. This conjecture was subsequently shown to be true using an in vitro reconstituted assay for a few members of the Bcl-2 family. However, it is unclear whether the pH-dependent, ion-conducting property of Bcl-2 proteins occurs in vivo and, if so, how it contributes to the regulation of apoptosis.

Despite very weak sequence homology with Bcl-xL, the BH3-only protein Bid exhibits a conserved structure very similar to that of Bcl-xL (Chou et al. 1999, McDonnell et al. 1999). The minor differences include the length and relative orientation of several helices as well as an extra α helix between helices α1 and α2 (**Figure 4b**). This structure suggests a model, in which caspase-8-mediated cleavage of Bid leads to the exposure of the BH3 domain, which then mediates binding to the other two subfamilies of Bcl-2 proteins. Direct binding to tBid is thought to result in the activation and oligomerization of Bax.

Similar to Bid, the structure of the full-length Bax in aqueous solution closely resembles that of Bcl-xL (Suzuki et al. 2000). Intriguingly, the C-terminal membrane-spanning region (α9) folds back to bind a hydrophobic groove that is believed to normally accommodate the BH3 domain of a Bcl-2 family member (**Figure 4c**). Thus this structure likely represents the inactive form of Bax, providing a plausible explanation to the observation that Bax remains largely as a cytosolic protein in the absence of apoptotic signaling. This structure also suggests how binding to BH3-only protein may facilitate the translocation of Bax to mitochondria, as the BH3 domain is likely to displace the C-terminal membrane-spanning region (α9).

Because Bcl-2 family proteins function in the context of lipid membrane, it is important to examine their structure in the presence of membrane. No such structure is yet available due to technical difficulty. Toward

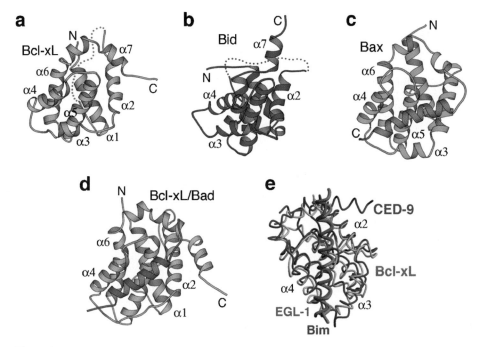

Figure 4

Structure of the Bcl-2 family of proteins. (*a*) Structure of Bcl-xL. The flexible loop linking helices α1 and α2 are represented by a dotted line. (*b*) Structure of the uncleaved form of Bid. Cleavage after Asp59 results in the activation of Bid, presumably through exposure to the BH3 helix. (*c*) Structure of the full-length Bax. Note that the C-terminal amphipathic helix (*orange*) folds back to bind a hydrophobic surface groove, resembling the Bcl-xL-bound Bad peptide. (*d*) Structure of Bcl-xL bound to a BH3 peptide from Bad (*purple*). This Bad peptide exists as an amphipathic helix, with the hydrophobic side binding to Bcl-xL. (*e*) Superposition of the structures of CED-9/EGL-1 and Bcl-xL/Bim complexes. The BH3 regions of EGL-1 and Bim are shown in magenta and gray, respectively.

this goal, Bcl-xL (Losonczi et al. 2000) and Bid (Oh et al. 2005) have been characterized under membranous conditions and, compared with the aqueous solution, exhibited significant structural differences.

Recognition of the BH3 domain. The antiapoptotic proteins, such as Bcl-xL in mammals and CED-9 in nematodes, interact with proapoptotic members by recognizing their BH3 domains. The specific recognition was revealed by the structure of Bcl-xL bound to a BH3 peptide derived from Bak (Sattler et al. 1997), Bad (Petros et al. 2000), or Bim (Liu et al. 2003) (**Figure 4*d***). The BH3 peptide forms an amphipathic α helix and in-

teracts with a deep hydrophobic groove on the surface of Bcl-xL. The binding of the BH3 domain is accompanied by a significant conformational change in Bcl-xL. These observations were further confirmed by the structure of CED-9 bound to an extended BH3 domain from EGL-1 (Yan et al. 2004a). Despite limited sequence similarity between Bcl-xL and CED-9, their structures exhibit identical features. The molecular recognition between CED-9 and EGL-1 closely parallels those between Bcl-xL and Bim (**Figure 4*e***). Comparative analysis of these protein complexes will likely reveal some general code that governs the interaction among the Bcl-2 family of proteins.

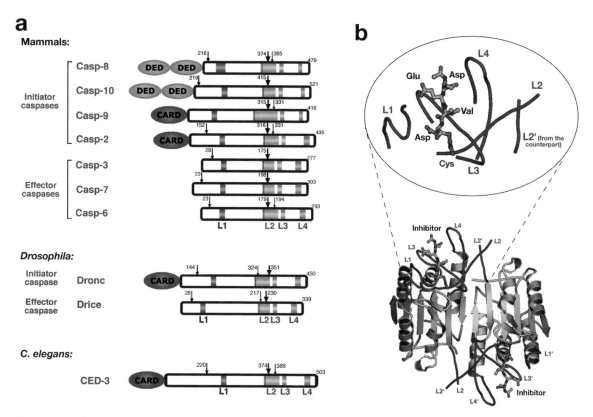

Figure 5

(*a*) Representative caspases in mammals, *Drosophila*, and *C. elegans*. The position of the first activation cleavage is highlighted by a large arrow; additional sites of cleavage are represented by medium and small arrows. The four surface loops (L1–L4) that shape the catalytic groove are indicated. The catalytic residue Cys is shown as a red line at the beginning of loop L2. (*b*) Structural features of an inhibitor-bound caspase-3. The bound peptide inhibitor is colored gold. The five surface loops that constitute the substrate-binding groove of each caspase unit are labeled. The apostrophe denotes the other caspase unit. Note that L2′ stabilizes the active site of the adjacent caspase unit. The substrate-binding groove is shown in detail on the right. The catalytic residue Cys is covalently linked to the peptide inhibitor.

Caspases—the Executioners of Apoptosis

Initiator and effector caspases. The critical involvement of a caspase in apoptosis was first documented in 1993 (Yuan et al. 1993), in which CED-3 was found to be indispensable for the programmed cell death in the nematode *Caenorhabditis elegans*. At least 14 distinct mammalian caspases have been identified (Shi 2002). Caspases involved in apoptosis make up two groups: the initiators, which include caspases-2, -8, -9, and -10, and the effectors, which include caspases-3, -6, and -7

(**Figure 5*a***). An initiator caspase invariably contains an extended N-terminal prodomain (>90 amino acids) important for its function, whereas an effector caspase contains only 20–30 residues in its prosequence. All caspases are synthesized in cells as catalytically inactive zymogens and must undergo proteolytic activation during apoptosis. The activation of an effector caspase, such as caspase-3 or -7, is performed by an initiator caspase, such as caspase-9, through internal cleavages to separate the large and small subunits. The initiator caspases, however, are auto-activated under apoptotic conditions.

Structural features of caspases. The first caspase structure was determined on caspase-1 (or ICE, interleukin 1β-converting enzyme) bound to a covalent peptide inhibitor (Walker et al. 1994, Wilson et al. 1994). Structural information is now available on caspase-2 (Schweizer et al. 2003), caspase-3 (Mittl et al. 1997, Rotonda et al. 1996), caspase-7 (Wei et al. 2000), caspase-8 (Blanchard et al. 1999, Watt et al. 1999), and caspase-9 (Renatus et al. 2001). In most cases, the functional caspase unit is a homodimer, with each monomer having a large (~20 kDa) and a small (~10 kDa) subunit (**Figure 5b**). The active site, highly conserved among all caspases, is formed by five protruding loops, loops L1, L2, L3, and L4 from one monomer and loop L2′ from the adjacent monomer. The L1 and L4 loops constitute two sides of the substrate-binding groove (**Figure 5b**). Loop L3 is located at the base of the groove. Loop L2, which harbors the catalytic Cys, is positioned at one end of the groove with Cys poised for catalysis. These four loops determine the sequence specificity of the substrates. Importantly, the active site conformation is stabilized by the L2′ loop from the adjacent monomer, which interacts with loops L2 and L4 to form a so-called loop-bundle (Chai et al. 2001b).

Mechanism of Effector Caspase Activation

Why are the zymogens of effector procaspases catalytically inactive? A plausible answer was provided by the crystal structure of procaspase-7 (Chai et al. 2001b, Riedl et al. 2001a), which reveals significant conformational changes in the five active site loops compared with that of the inhibitor-bound structure (**Figure 6**). With the exception of L1, all other loops move away from their productive positions, unraveling the substrate-binding groove. Loop L2 is twisted, shifting the catalytic cysteine away from its active conformation. The loop-bundle seen in the inhibitor-bound caspases is missing in the procaspase-7 zymogen as the L2′ loop is flipped by 180°, existing in a closed conformation. The closed conformation is necessitated by the unprocessed nature of the procaspase-7 zymogen. This conformational arrangement in procaspase-7 zymogen disallows formation of a substrate-binding groove, thereby explaining why the procaspase-7 zymogen does not possess detectable catalytic activity. The mechanistic understanding on caspase-7 activation provides a plausible explanation to why effector caspases exist as a

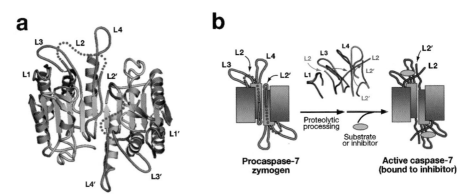

Figure 6

Molecular mechanism of procaspase-7 activation. (*a*) Structure of a procaspase-7 zymogen. Compared with that of the inhibitor-bound caspase-7, the conformation of the active site loops (*salmon*) does not support substrate-binding or catalysis. The L2′ loop, locked in a closed conformation by covalent linkage, is occluded from adopting its productive and open conformation. (*b*) A cartoon showing the conformational changes of the active site loops upon activation. Compared with the procaspase-7 zymogen, the L2′ loop is flipped 180° in the inhibitor-bound caspase-7 to stabilize loops L2 and L4.

homo-dimer—because the provision of the L2′ loop is needed for the catalytic activity.

Inhibitors of Apoptosis

The inhibitor of apoptosis (IAP) family of proteins suppresses apoptosis by negatively regulating caspases (Deveraux & Reed 1999, Salvesen & Duckett 2002). There are eight distinct mammalian IAPs: XIAP, c-IAP1, c-IAP2, ML-IAP/Livin, ILP-2, NAIP, Bruce/Apollon, and survivin; and three *Drosophila* IAPs: DIAP1, DIAP2, and dBruce (**Figure 7a**). In mammals, caspases-3, -7, and -9 are subject to inhibition by IAPs (Shi 2002). Although caspase-9 binds to several IAPs, it is primarily inhibited by XIAP. By contrast, caspase-3 and -7 are inhibited by XIAP, and to a lesser extent, by c-IAP1, c-IAP2, or NAIP (Maier et al. 2002, Salvesen & Duckett 2002). The hallmark of an IAP is the presence of at least one conserved zinc-binding BIR

(baculoviral IAP repeat) domain. The structures of various BIR domains (Hinds et al. 1999; Sun et al. 1999, 2000) reveal a conserved topology, with a central three-stranded antiparallel β sheet surrounded by four α helices (**Figure 7b**). Four invariant residues, three cysteines and one histidine, coordinate a zinc atom.

Most IAPs contain more than one BIR domain, with the different BIR domains exhibiting distinct functions. In XIAP, the third BIR domain (BIR3) potently inhibits caspase-9, whereas the linker region between BIR1 and BIR2 specifically targets caspase-3 and -7. Survivin, which contains only one BIR domain and forms a homo-dimer, does not inhibit caspase activity in vitro. ML-IAP/Livin was reported to inhibit both caspase-3 and -9 (Ashhab et al. 2001, Kasof & Gomes 2001, Vucic et al. 2000), although this conclusion is not supported by sequence analysis.

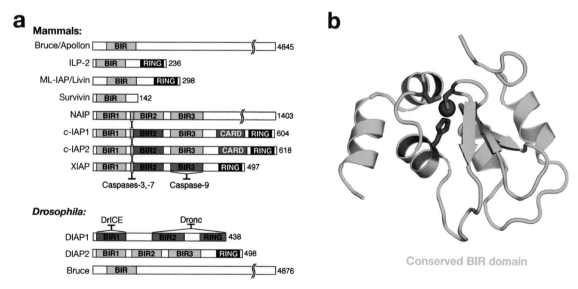

Figure 7

(*a*) IAPs in mammals and *Drosophila*. A conserved peptide sequence preceding the BIR2 domain of XIAP, c-IAP1, c-IAP2, or NAIP (*green*) is responsible for inhibiting caspase-3 and -7 in mammals. Only the BIR3 domain of XIAP has been confirmed to inhibit caspase-9. In *Drosophila*, the BIR1 and BIR2 domains of DIAP1, but not DIAP2, are responsible for inhibiting the caspase-3 and -9 homologues Drice and Dronc, respectively. (*b*) Structure of the BIR domain. The bound zinc atom as well as the four conserved Cys/His residues are labeled.

Mechanisms of Caspase Inhibition

Inhibition of effector caspases. The mechanism of IAP-mediated inhibition of effector caspases was revealed by the structure of caspase-3 (Riedl et al. 2001b) or caspase-7 (Chai et al. 2001a, Huang et al. 2001) bound to an inhibitory peptide fragment preceding the BIR2 domain of XIAP. This peptide fragment fills the active site groove of caspase-3 or -7, thus preventing substrate entry and subsequent catalysis (**Figure 8a**). Compared with the covalent peptide inhibitors, the XIAP fragment occupies the active site groove in a reverse orientation. The interactions of XIAP with caspase-3 or -7 resemble that of the peptide inhibitor Asp-Glu-Val-Asp at the P4-P3-P2-P1 positions, respectively. Asp148 of XIAP, previously shown to be essential for caspase-3 inhibition (Sun et al. 1999), binds to the groove similarly as the P4 residue (Asp). The hydrophobic contact Val146 makes to surrounding residues closely resembles the P2 residue (Val). The inhibition mechanism

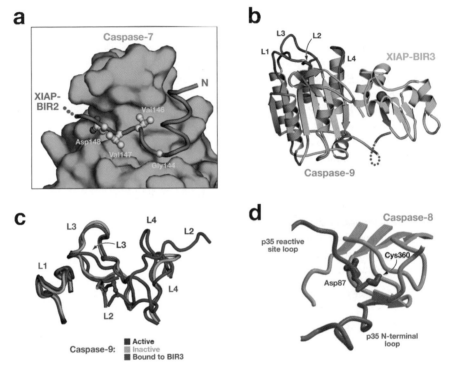

Figure 8

Mechanisms of caspase inhibition. (*a*) Mechanism of inhibition of effector caspases by XIAP. A close-up view of the active site of caspase-7 (in a surface representation) bound to an XIAP fragment (*gold*). Important residues of XIAP, Gly144, Val146, Val147, and Asp148, are highlighted in yellow. (*b*) Structure of caspase-9 bound to the BIR3 domain of XIAP. XIAP-BIR3 binds to a large caspase-9 surface that is normally required for its homodimerization. The active-site loops are shown in blue. The catalytic residue, Cys287 on loop L2, and the zinc atom in XIAP-BIR3 are colored red. (*c*) Superposition of four active-site loops from the BIR3-bound caspase-9 (*blue*) and the active (*yellow*) and inactive (*magenta*) monomers of the caspase-9 homo-dimer. Note that the active-site conformation of the BIR3-bound caspase-9 closely resembles that of the inactive caspase-9 monomer. (*d*) Mechanism of p35-mediated pan-caspase inhibition. A close-up view of the covalent inhibition of caspase-8 by p35. The thioester intermediate is shown between Asp87 of p35 and Cys360 (active site residue). The N terminus of p35 restricts solvent access to this intermediate.

is likely applicable to c-IAP1, c-IAP2, and NAIP.

Although the peptide fragment that precedes XIAP-BIR2 plays a dominant role in inhibiting caspase-3 or -7, this fragment in isolation is insufficient (Chai et al. 2001a, Sun et al. 1999). Compelling evidence suggests that this peptide fragment needs to be presented in a competent conformation by a surrounding BIR domain. Consistent with this notion, XIAP-BIR2 also makes direct interactions to caspase-3 (Riedl et al. 2001b).

Inhibition of caspase-9. Only processed caspase-9 is subject to inhibition by the BIR3 domain of XIAP because the exposed N terminus of the small subunit of caspase-9 is required for binding to BIR3 (Srinivasula et al. 2001). A mechanistic explanation for the inhibition of caspase-9 by XIAP was revealed by the structure of caspase-9 bound to the BIR3 domain of XIAP (Shiozaki et al. 2003) (**Figure 8b**). In the uninhibited state, the processed caspase-9 exists exclusively as a monomer. The interaction between a conserved surface groove on XIAP-BIR3 and the N-terminal tetrapeptide of the small subunit of caspase-9 anchors their mutual recognition, as the isolated tetrapeptide is sufficient for binding to caspase-9. The BIR3 domain uses another surface patch to heterodimerize with caspase-9 through an interface that is usually required for the homodimerization of caspases. Together, these interactions result in the formation of a catalytically incompetent conformation at the active site of caspase-9 (**Figure 8c**). This structural observation provides an explanation to a body of published biochemical data.

An important conclusion from the structural and biochemical analyses is that binding to caspase-9 is necessary but not sufficient for its inhibition. The specific recognition of the caspase-9 homodimerization interface requires four amino acids in the BIR3 domain of XIAP, which are not conserved in c-IAP1 or c-IAP2 (Shiozaki et al.

2003). This observation explains why c-IAP1 and c-IAP2 can bind to but do not inhibit caspase-9.

Inhibition of caspases by viral proteins. In contrast to XIAP, which specifically targets caspases-3, -7, and -9, the baculoviral protein p35 is a broad-spectrum caspase inhibitor. The mechanism of p35-mediated caspase inhibition is revealed by the structure of caspase-8 bound to p35 (Xu et al. 2001), which surprisingly shows a covalent thioester linkage between Asp87 of p35 and the catalytic residue Cys360 of caspase-8 (**Figure 8d**). Although a thioester bond is generally susceptible to hydrolysis, this bond is protected by the neighboring N terminus of p35, which occludes water access (**Figure 8d**). Another baculoviral protein, p49, functions similarly to p35 and may exhibit the same mechanism of inhibition. In addition, the cowpox virus-derived serpin CrmA has also been shown to inhibit several caspases, probably through the disruption of the active site of the enzyme, as shown for the inhibition of serine proteases by serpins (Huntington et al. 2000). These distinct mechanisms add to the complexity of caspase inhibition by natural proteins.

Smac/DIABLO and an IAP-Binding Motif

Smac/DIABLO. Smac/DIABLO is synthesized in the cytoplasm and targeted to the intermembrane space of mitochondria. Upon apoptotic stimulation, Smac is released into the cytosol, together with cytochrome *c*. Whereas cytochrome *c* directly activates Apaf-1 and hence caspase-9, Smac interacts with multiple IAPs and counters their inhibitory effect on both initiator and effector caspases. Mature Smac is an elongated homodimer (Chai et al. 2000). Owing to cleavage of its mitochondria-targeting sequence, the mature Smac protein contains a tetrapeptide at its N terminus, Ala-Val-Pro-Ile, which binds to the conserved surface groove on the BIR3 domain of XIAP (Liu et al. 2000, Wu et al. 2000). A missense mutation of the N-terminal residue Ala to Met in Smac led to a complete loss of interactions with XIAP and the concomitant loss of Smac function (Chai et al. 2000).

A tetrapeptide IAP-binding motif. Smac is a member of a growing family of proteins that share an IAP-binding tetrapeptide motif. Another mitochondrial protein, HtrA2/Omi, also contains an IAP-binding tetrapeptide motif at its N terminus (**Figure 9a**) and can counter XIAP-mediated inhibition of caspase-9 at high concentrations. However, the bovine homologue of HtrA2/Omi lacks this motif (Li et al. 2002), calling into question whether HtrA2/Omi can antagonize XIAP inhibition under physiological conditions. In *Drosophila*, there are at least four functional homologues of Smac/DIABLO: Reaper, Hid, Grim, and Sickle, which are collectively referred to as the RHG proteins. The RHG proteins bind to and antagonize DIAP1-mediated suppression of Dronc and Drice (**Figure 9a**). Another *Drosophila* protein, Jafrac2, contains a somewhat divergent tetrapeptide

motif but has nonetheless been shown to bind to DIAP1 and to counter DIAP1-mediated Dronc suppression (Tenev et al. 2002).

BIR Recognition by conserved IAP-binding motifs. The molecular explanation for the indispensable role of the Smac N-terminal sequences is provided by structure of XIAP-BIR3 bound to either a monomeric Smac protein (Wu et al. 2000) or a Smac peptide (Liu et al. 2000). The Smac N-terminal tetrapeptide recognizes a conserved surface groove on BIR3, with the first residue, Ala, binding to a hydrophobic pocket and making hydrogen bonds to neighboring XIAP residues (**Figure 9b**). To accommodate these interactions, the N terminus of Smac must be free, thus explaining why only mature Smac can bind to IAPs. This analysis also explains why mutation of Ala to Met abrogated interactions with the BIR domains. The binding groove for the Smac tetrapeptide is conserved on the BIR2 and BIR3 domains of XIAP, c-IAP1, and c-IAP2, or on the BIR1 and BIR2 domains of DIAP1.

a

Mammals:

Smac/DIABLO	A	V	P	I	A	Q	K	S
hCasp-9	A	T	P	F	Q	E	G	L
mCasp-9	A	V	P	Y	Q	E	G	P
xCasp-9	A	T	P	V	F	S	G	E
hHtrA2/Omi	A	V	P	S	P	P	P	A
mHtrA2/Omi	A	V	P	A	P	P	P	T

Drosophila:

Reaper	A	V	A	F	Y	I	P	D
Grim	A	I	A	Y	F	L	P	D
Hid	A	V	P	F	Y	L	P	E
Sickle	A	I	P	F	F	E	E	E
Jafrac2	A	K	P	E	D	N	E	S

b

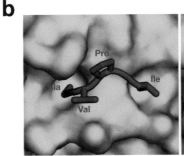
AVPI (Smac) bound to BIR3 of XIAP

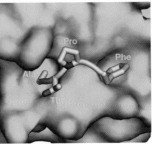
ATPF (Caspase-9) bound to BIR3 of XIAP

Figure 9

Molecular mechanism of Smac-mediated removal of caspase-9 inhibition by XIAP. (*a*) A conserved IAP-binding tetrapeptide motif across species. The tetrapeptide motif has the consensus sequence A-(V/T/I)-(P/A)-(F/Y/I/V/S). Preferred and allowed amino acids are indicated by magenta and yellow, respectively. (*b*) Mechanism of Smac-mediated removal of caspase-9 inhibition by XIAP. The IAP-binding tetrapeptide motifs from both Smac and from caspase-9 bind to the same conserved surface groove on the BIR3 domain of XIAP. It is this mutual exclusion that allows Smac to remove XIAP-mediated inhibition of caspase-9.

Mechanism of Smac-Mediated Removal of Caspase Inhibition

Removal of caspase-9 inhibition. The N terminus (Ala-Thr-Pro-Phe) of the small subunit of caspase-9 conforms to the IAP-binding tetrapeptide motif (Srinivasula et al. 2001) (**Figure 9a**) and anchors its binding to the BIR3 domain of XIAP (**Figure 8b**). During apoptosis, Smac is released from mitochondria into the cytoplasm, where it uses its own IAP-binding tetrapeptide to bind to the BIR3 domain of XIAP, hence competitively displacing caspase-9. So, a conserved IAP-binding motif in caspase-9 and Smac/DIABLO serves opposing purposes on caspase activity. The dimeric scaffold of Smac allows it to bind more tightly to XIAP than the monomeric caspase-9 protein, hence giving Smac a competitive edge (Huang et al. 2003).

Removal of effector caspase inhibition. Although the Smac tetrapeptide in isolation can remove XIAP-mediated inhibition of caspase-9 (Chai et al. 2000), it is incapable of removing IAP-mediated inhibition of effector caspases. The reason is that the XIAP moiety that is responsible for inhibiting caspase-3 or -7 is a peptide fragment preceding the BIR2 domain of XIAP, away from the conserved binding groove for the tetrapeptide. Modeling studies suggest that steric clashes preclude XIAP from simultaneously binding to caspase-3 and Smac/DIABLO (Chai et al. 2001a, Huang et al. 2003). Binding to both BIR2 and BIR3 domains is thought to be required for Smac-mediated removal of caspase-7 inhibition (Huang et al. 2003).

Activation of Caspase-9

The activation of procaspase-9 relies on the apoptosome (**Figure 1**). Importantly, the processed caspase-9 is marginally active in the absence of the apoptosome, prompting the concept of a holo-enzyme (Rodriguez & Lazebnik 1999). Through association with the apoptosome, the catalytic activity of caspase-9 is enhanced by two to three orders of magnitude compared with that of isolated caspase-9.

We do not yet understand how caspase-9 is activated by the apoptosome. Two models have been proposed. The induced proximity model states that the initiator caspases auto-process themselves when brought into close proximity of each other (Salvesen & Dixit 1999). However, this model merely describes a process but does not address the mechanism of initiator caspase activation. The proximity-induced dimerization model (Boatright & Salvesen 2003) represents a qualitative refinement of the induced proximity model and proposes that the initiator caspases, caspase-9 and -8, are activated upon dimerization, which is facilitated by the oligomeric complexes, apoptosome and DISC, respectively (Boatright et al. 2003, Donepudi et al. 2003, Renatus et al. 2001). At present, the accuracy of this model awaits further testing, as the supporting evidence has not been definitive.

Mechanism of DFF40/CAD Activation

DFF40/CAD plays an essential role in the degradation of chromosomal DNA during apoptosis. Prior to apoptosis, DFF40/CAD is sequestered by its inhibitor DFF45/ICAD as a heterodimer. DFF45 and DFF40 are thought to contain three domains each (I1–I3 for DFF45 and C1–C3 for DFF40) (**Figure 11**). The isolated C1 and I1 domains interact with each other (Otomo et al. 2000), although additional interactions involving the other domains greatly enhance the overall association for the DFF40/45 heterodimer (Woo et al. 2004). During apoptosis, caspase-3 or -7 cleaves DFF45 into three fragments, resulting in the release of DFF40 from the heterodimer and its subsequent homodimerization. The dimeric DFF40 is potent DNase that cleaves chromosomes into nucleosomal fragments. The structure of the activated DFF40 suggests the DFF40 homodimer may bind to and cleave internucleosomal DNA like a pair of scissors (Woo et al. 2004).

LESSONS FROM FLIES AND WORMS

A Conserved Pathway of Caspase Activation

The apoptotic pathway shares considerable similarity among worms, flies, and mammals (**Figure 10**). Genetic studies have identified four genes, *ced-3*, *ced-4*, *ced-9*, and *egl-1*, that function sequentially to control the onset of apoptosis in *C. elegans* (Horvitz 2003). CED-3 is the only apoptotic caspase in nematodes and functions as both an initiator and an effector caspase. The activation of CED-3 is mediated by CED-4, which is thought to involve its oligomerization (Yang et al. 1998). Prior to apoptosis, CED-4 is sequestered by the antiapoptotic protein CED-9 through direct interactions. During apoptosis, the negative regulation of CED-4 by CED-9 is countered by EGL-1, which is transcriptionally activated by cell death stimuli. CED-9 is a functional and structural homologue of the mammalian protein Bcl-2/Bcl-xL, whereas EGL-1 is a BH3-only member of the Bcl-2

family of proteins. CED-4 and CED-3 are the functional orthologs of the mammalian proteins Apaf-1 and caspase-9, respectively. In *Drosophila*, Dark (also known as Dapaf-1 or HAC-1) shares significant sequence similarity with Apaf-1 and is important for the activation of the initiator caspase Dronc. Dronc, in turn, cleaves and activates the effector caspase Drice. Dronc and Drice share considerable sequence similarity with, and are the functional homologues of, caspase-9 and caspase-3, respectively.

Mechanisms of Caspase Regulation in *Drosophila*

Mechanism of dronc regulation. Given the extraordinary conservation of apoptotic pathway between mammals and fruit flies (**Figure 10**), it was anticipated that the underlying mechanisms of caspase regulation should be conserved. Surprisingly, however, biochemical and structural studies reveal very different mechanisms for the regulation of caspases in *Drosophila* (Chai et al. 2003, Yan et al. 2004b).

DIAP1 directly binds to and suppresses Dronc, whereas the RHG proteins counter DIAP1-mediated suppression of Dronc. The N-terminal IAP-binding motifs of the RHG

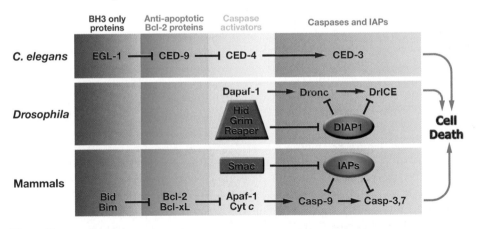

Figure 10

A conserved apoptotic pathway in *C. elegans*, *Drosophila*, and mammals. Caspase-9 in mammals and Dronc in *Drosophila* are initiator caspases, whereas caspase-3 and -7 in mammals and Drice in *Drosophila* belong to effector caspases. CED-3 in *C. elegans* functions both as the initiator and the effector caspase.

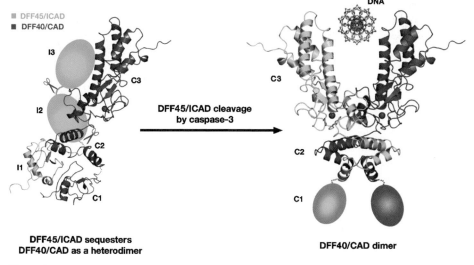

DFF45/ICAD sequesters
DFF40/CAD as a heterodimer

DFF40/CAD dimer

DFF45/ICAD cleavage
by caspase-3

Figure 11

A structure-based model of DFF40/CAD activation. DFF40/CAD is usually sequestered as a monomer by DFF45/ICAD. During apoptosis, caspase-3 or -7 cleaves DFF45/ICAD, resulting in the release of DFF40/CAD and its subsequent dimerization. The dimeric DFF40/CAD is a potent DNase.

proteins bind to a conserved surface groove on the BIR2 domain of DIAP1 (Wu et al. 2001), which suggests that Dronc might bind to DIAP1 in a way that is similar to the caspase-9-XIAP interactions. However, Dronc does not contain any sequences that resemble the IAP-binding tetrapeptide motif. Biochemical analyses revealed that the BIR2 domain of DIAP1 recognizes a 5-amino acid sequence in the linker region between the prodomain and the caspase unit of Dronc (Chai et al. 2003). This recognition is essential for DIAP1-mediated negative regulation of Dronc. Remarkably, structural analysis revealed that the Dronc-binding surface on DIAP1-BIR2 coincides with that required for binding to the N-terminal sequences of the RHG proteins (**Figure 8c**), thus explaining how the RHG proteins competitively eliminate DIAP1-mediated negative regulation of Dronc.

In mammals, XIAP potently inhibits the catalytic activity of caspase-9. In *Drosophila*, however, DIAP1 exhibits no effect on the catalytic activity of Dronc (Yan et al. 2004b). Rather, DIAP1 acts as an E3 ubiquitin lig-

ase to recognize and to ubiquitinate Dronc (Chai et al. 2003, Wilson et al. 2002), hence targeting Dronc for proteasome-mediated degradation.

Mechanisms of Drice regulation. In mammals, the inhibition of caspase-3 or -7 entails a linker segment preceding the BIR2 domain of XIAP, c-IAP1, or c-IAP2. However, this linker sequence is not conserved in DIAP1. Biochemical analyses showed that DIAP1 directly inhibits the catalytic activity of Drice through its BIR1 domain and that this inhibition can be countered effectively by the RHG proteins (Yan et al. 2004b). DIAP1 binds to and inhibits Drice only after the cleavage of its N-terminal 20 amino acids (Yan et al. 2004b), which could serve as an auto-inhibitory sequence for DIAP1 function. DIAP1-mediated inhibition of Drice involves a highly conserved surface groove on BIR1 (Yan et al. 2004b). The same surface groove is the binding site for the N-terminal IAP-binding motifs of the RHG proteins, hence explaining how the RHG proteins counter DIAP1-mediated inhibition of Drice. Despite these biochemical advances, we do not yet understand the exact mechanism of how DIAP1 inhibits Drice. A recent report suggests that the N-terminal sequences of the large subunit of Drice are important for the interactions between Drice and BIR1.

Cleavage after residue Asp20 of Drice has also been proposed to accelerate the degradation of the full-length DIAP1 protein through the N-end rule (Ditzel et al. 2003). Because Dronc activation precedes Drice activation and DIAP1 is readily cleaved by Dronc (Yan et al. 2004b), this proposed mechanism may also lead to the removal of the N-terminal domain of DIAP1. Alternatively, because Drice forms a stable complex with DIAP1-BIR1 (Yan et al. 2004b), the proposed N-end rule could facilitate the degradation of Drice as well.

PROSPECTS

A systematic characterization of apoptosis by structural biology has significantly enhanced our understanding of the underlying molecular mechanisms. However, many questions remain. For example, despite vigorous effort, we still know very little about the activation mechanisms of the initiator caspases or how the Bcl-2 family of proteins controls the permeability of mitochondria. Our mechanistic insights in the *Drosophila* and *C. elegans* apoptotic pathways are also very limited. The apoptotic mechanisms will undoubtedly become more complex with the discovery and characterization of additional players and pathways, which will present structural biologists with exciting new challenges for many years to come.

LITERATURE CITED

Andre N. 2003. Hippocrates of Cos and apoptosis. *Lancet* 361:1306

Ashhab Y, Alian A, Polliack A, Panet A, Yehuda DB. 2001. Two splicing variants of a new inhibitor of apoptosis gene with different biological properties and tissue distribution pattern. *FEBS Lett.* 495:56–60

Ashkenazi A, Dixit VM. 1998. Death receptors: signaling and modulation. *Science* 281:1305–8

Banner DW, D'Acry A, Janes W, Gentz R, Schoenfeld H-J, et al. 1993. Crystal structure of the soluble human 55 kd TNF receptor-human TNFb complex: implications for TNF receptor activation. *Cell* 73:431–45

Blanchard H, Kodandapani L, Mittl PRE, Di Marco S, Krebs JK, et al. 1999. The three dimensional structure of caspase-8: an initiator enzyme in apoptosis. *Structure* 7:1125–33

Boatright KM, Renatus M, Scott FL, Sperandio S, Shin H, et al. 2003. A unified model for apical caspase activation. *Mol. Cell* 11:529–41

Boatright KM, Salvesen GS. 2003. Mechanisms of caspase activation. *Curr. Opin. Cell Biol.* 15:725–31

Chai JJ, Du CY, Wu J-W, Kyin S, Wang XD, Shi Y. 2000. Structural and biochemical basis of apoptotic activation by Smac/DIABLO. *Nature* 406:855–62

Chai JJ, Shiozaki E, Srinivasula SM, Wu Q, Datta P, et al. 2001a. Structural basis of caspase-7 inhibiton by XIAP. *Cell* 104:769–80

Chai JJ, Wu Q, Shiozaki E, Srinivasula SM, Alnemri ES, Shi Y. 2001b. Crystal structure of a procaspase-7 zymogen: mechanisms of activation and substrate binding. *Cell* 107:399–407

Chai JJ, Yan N, Huh JR, Wu J-W, Li WY, et al. 2003. Molecular mechanism of Reaper/Grim/Hid-mediated suppression of DIAP1-dependent Dronc ubiquitination. *Nat. Struct. Biol.* 10:892–98

Chou JJ, Li H, Salvesen GS, Yuan J, Wagner G. 1999. Solution structure of BID, an intracellular amplifier of apoptotic signaling. *Cell* 96:615–24

Chou JJ, Matsuo H, Duan H, Wagner G. 1998. Solution structure of the RAIDD CARD and model for CARD/CARD interaction in Caspase-2 and Caspase-9 recruitment. *Cell* 94:171–80

Cory S, Adams JM. 2002. The Bcl2 family: regulators of the cellular life-or-death switch. *Nat. Rev. Cancer* 2:647–56

Danial NN, Korsmeyer SJ. 2004. Cell death: critical control points. *Cell* 116:205–19

DeLano WL. 2002. *The PyMOL molecular graphics system.* **http://www.pymol.org**

Deveraux QL, Reed JC. 1999. IAP family proteins—suppressors of apoptosis. *Genes Dev.* 13:239–52

Ditzel M, Wilson R, Tenev T, Zachariou A, Paul A, et al. 2003. Degradation of DIAP1 by the N-end rule pathway is essential for regulating apoptosis. *Nat. Cell Biol.* 5:467–73

Donepudi M, Mac Sweeney A, Briand C, Grutter MG. 2003. Insights into the regulatory mechanism for caspase-8 activation. *Mol. Cell* 11:543–49

Eberstadt M, Huang B, Chen Z, Meadows RP, Ng S-C, et al. 1998. NMR structure and mutagenesis of the FADD (Mort1) death-effector domain. *Nature* 392:941–45

Eck MJ, Sprang SR. 1989. The structure of tumor necrosis factor alpha at 2.6 Å resolution: implications for receptor binding. *J. Biol. Chem.* 264:17595–605

Eck MJ, Ultsch M, Rinderknecht E, de Vos AM, Sprang SR. 1992. The structure of human lymphotoxin (tumor necrosis factor beta) at 1.9 Å resolution. *J. Biol. Chem.* 267:2119–22

Fesik SW. 2000. Insights into programmed cell death through structural biology. *Cell* 103:273–82

Green DR, Evan GI. 2002. A matter of life and death. *Cancer Cell* 1:19–30

Hanahan D, Weinberg RA. 2000. The hallmarks of cancer. *Cell* 100:57–70

Hay BA, Huh JR, Guo M. 2004. The genetics of cell death: approaches, insights and opportunities in *Drosophila*. *Nat. Rev. Genet.* 5:911–22

Hinds MG, Norton RS, Vaux DL, Day CL. 1999. Solution structure of a baculoviral inhibitor of apoptosis (IAP) repeat. *Nat. Struct. Biol.* 6:648–51

Horvitz HR. 2003. Worms, life, and death (Nobel lecture). *Chembiochem.* 4:697–711

Huang B, Eberstadt M, Olejniczak ET, Meadows RP, Fesik SW. 1996. NMR structure and mutagenesis of the Fas (APO-1/CD95) death domain. *Nature* 384:638–41

Huang YH, Park YC, Rich RL, Segal D, Myszka DG, Wu H. 2001. Structural basis of caspase inhibition by XIAP: differential roles of the linker versus the BIR domain. *Cell* 104:781–90

Huang YH, Rich RL, Myszka DG, Wu H. 2003. Requirement of both the second and third BIR domains for the relief of X-linked inhibitor of apoptosis protein (XIAP)-mediated caspase inhibition by Smac. *J. Biol. Chem.* 278:49517–22

Huntington JA, Read RJ, Carrell RW. 2000. Structure of a serpin-protease complex shows inhibition by deformation. *Nature* 407:923–26

Hymowitz SG, Christinger HW, Fuh G, Ultsch M, O'Connell M, et al. 1999. Triggering cell death: the crystal structure of Apo2L/TRAIL in a complex with death receptor 5. *Mol. Cell* 4:563–71

Jones EY, Stuart DI, Walker NP. 1989. Structure of tumour necrosis factor. *Nature* 338:225–28

Kasof GM, Gomes BC. 2001. Livin, a novel inhibitor of apoptosis protein family member. *J. Biol. Chem.* 276:3238–46

Kerr JFF, Wylie AH, Currie AR. 1972. Apoptosis: a basic biological phenomenon with wide-ranging implications in tissue kinetics. *Br. J. Cancer* 26:239–57

Kraulis PJ. 1991. Molscript: a program to produce both detailed and schematic plots of protein structures. *J. Appl. Crystallogr.* 24:946–50

Li H, Zhu H, Xu CY, Yuan J. 1998. Cleavage of BID by caspase-8 mediates the mitochondrial damage in the Fas pathway of apoptosis. *Cell* 94:491–501

Li WY, Srinivasula SM, Chai JJ, Li PW, Wu J-W, et al. 2002. Structural insights into the pro-apoptotic function of mitochondrial serine protease HtrA2/Omi. *Nat. Struct. Biol.* 9:436–41

Liu XQ, Dai SD, Zhu YN, Marrack P, Kappler JW. 2003. The structure of a Bcl-xL/Bim fragment complex: implications for Bim function. *Immunity* 19:341–52

Liu ZH, Sun C, Olejniczak ET, Meadows RP, Betz SF, et al. 2000. Structural basis for binding of Smac/DIABLO to the XIAP BIR3 domain. *Nature* 408:1004–8

Losonczi JA, Olejniczak ET, Betz SF, Harlan JE, Mack J, Fesik SW. 2000. NMR studies of the anti-apoptotic protein Bcl-xL in micelles. *Biochemistry* 39:11024–33

Luo X, Budihardjo I, Zou H, Slaughter C, Wang X. 1998. Bid, a Bcl2 interacting protein, mediates cytochrome *c* release from mitochondria in response to activation of cell surface death receptors. *Cell* 94:481–90

Maier JK, Lahoua Z, Gendron NH, Fetni R, Johnston A, et al. 2002. The neuronal apoptosis inhibitory protein is a direct inhibitor of caspases 3 and 7. *J. Neurosci.* 22:2035–43

McDonnell JM, Fushman D, Milliman CL, Korsmeyer SJ, Cowburn D. 1999. Solution structure of the proapoptotic molecule BID: a structural basis for apoptotic agonists and antagonists. *Cell* 96:625–34

Mittl PR, Di Marco S, Krebs JF, Bai X, Karanewsky DS, et al. 1997. Structure of recombinant human CPP32 in complex with the tetrapeptide acetyl-Asp-Val-Ala-Asp fluoromethyl ketone. *J. Biol. Chem.* 272:6539–47

Mongkolsapaya J, Grimes JM, Chen N, Xu X-N, Stuart DI, et al. 1999. Structure of the TRAIL-DR5 complex reveals mechanisms conferring specificity in apoptotic initiation. *Nat. Struct. Biol.* 6:1048–53

Muchmore SW, Sattler M, Liang H, Meadows RP, Harlan JE, et al. 1996. X-ray and NMR structure of human Bcl-xL, an inhibitor of programmed cell death. *Nature* 381:335–41

Nagata S. 1999. Fas ligand-induced apoptosis. *Annu. Rev. Genet.* 33:29–55

Naismith JH, Devine TQ, Brandhuber BJ, Sprang SR. 1995. Crystallographic evidence for dimerization of unliganded tumor necrosis factor receptor. *J. Biol. Chem.* 270:13303–7

Naismith JH, Devine TQ, Kohno T, Sprang SR. 1996. Structures of the extracellular domain of the type I tumor necrosis factor receptor. *Structure* 4:1251–62

Oh KJ, Barbuto S, Meyer N, Kim RS, Collier RJ, Korsmeyer SJ. 2005. Conformational changes in BID, a pro-apoptotic BCL-2 family member, upon membrane binding: a site-directed spin labeling study. *J. Biol. Chem.* 280:753–67

Otomo T, Sakahira H, Uegaki K, Nagata S, Yamazaki T. 2000. Structure of the heterodimeric complex between CAD domains of CAD and ICAD. *Nat. Struct. Biol.* 7:658–62

Peter ME, Krammer PH. 2003. The CD95(APO-1/Fas) DISC and beyond. *Cell Death Differ.* 10:26–35

Petros AM, Nettesheim DG, Wang Y, Olejniczak ET, Meadows RP, et al. 2000. Rationale for Bcl-xL/Bad peptide complex formation from structure, mutagenesis, and biophysical studies. *Protein Sci.* 9:2528–34

Qin H, Srinivasula SM, Wu G, Fernandes-Alnemri T, Alnemri ES, Shi Y. 1999. Structural basis of procaspase-9 recruitment by the apoptotic protease-activating factor 1. *Nature* 399:547–55

Rathmell JC, Thompson CB. 2002. Pathways of apoptosis in lymphocyte development, homeostasis, and disease. *Cell* 109:S97–107

Renatus M, Stennicke HR, Scott FL, Liddington RC, Salvesen GS. 2001. Dimer formation drives the activation of the cell death protease caspase 9. *Proc. Natl. Acad. Sci. USA* 98:14250–55

Riedl SJ, Fuentes-Prior P, Renatus M, Kairies N, Krapp S, et al. 2001a. Structural basis for the activation of human procaspase-7. *Proc. Natl. Acad. Sci. USA* 98:14790–95

Riedl SJ, Renatus M, Schwarzenbacher R, Zhou Q, Sun C, et al. 2001b. Structural basis for the inhibition of caspase-3 by XIAP. *Cell* 104:791–800

Riedl SJ, Shi Y. 2004. Molecular mechanisms of caspase regulation during apoptosis. *Nat. Rev. Mol. Cell. Biol.* 5:897–907

Rodriguez J, Lazebnik Y. 1999. Caspase-9 and Apaf-1 form an active holoenzyme. *Genes Dev.* 13:3179–84

Rotonda J, Nicholson DW, Fazil KM, Gallant M, Gareau Y, et al. 1996. The three-dimensional structure of apopain/CPP32, a key mediator of apoptosis. *Nat. Struct. Biol.* 3:619–25

Salvesen GS, Dixit VM. 1999. Caspase activation: The induced-proximity model. *Proc. Natl. Acad. Sci. USA* 96:10964–67

Salvesen GS, Duckett CS. 2002. IAP proteins: blocking the road to death's door. *Nat. Rev. Mol. Cell Biol.* 3:401–10

Sattler M, Yoon HS, Nettesheim D, Meadows RP, Harlan JE, et al. 1997. Structure of Bcl-xL-Bak peptide complex: recognition between regulators of apoptosis. *Science* 275:983–86

Schweizer A, Briand C, Grutter MG. 2003. Crystal structure of caspase-2, apical initiator of the intrinsic apoptotic pathway. *J. Biol. Chem.* 278:42441–47

Shi Y. 2002. Mechanisms of caspase inhibition and activation during apoptosis. *Mol. Cell* 9:459–70

Shiozaki EN, Chai JJ, Rigotti DJ, Riedl SJ, Li PW, et al. 2003. Mechanism of XIAP-mediated inhibition of caspase-9. *Mol. Cell* 11:519–27

Srinivasula SM, Saleh A, Hedge R, Datta P, Shiozaki E, et al. 2001. A conserved XIAP-interaction motif in caspase-9 and Smac/DIABLO mediates opposing effects on caspase activity and apoptosis. *Nature* 409:112–16

Sun C, Cai M, Gunasekera AH, Meadows RP, Wang H, et al. 1999. NMR structure and mutagenesis of the inhibitor-of-apoptosis protein XIAP. *Nature* 401:818–22

Sun C, Cai M, Meadows RP, Xu N, Gunasekera AH, et al. 2000. NMR structure and mutagenesis of the third BIR domain of the inhibitor of apoptosis protein XIAP. *J. Biol. Chem.* 275:33777–81

Suzuki M, Youle RJ, Tjandra N. 2000. Structure of Bax: coregulation of dimer formation and intracellular localization. *Cell* 103:645–54

Tenev T, Zachariou A, Wilson R, Paul A, Meier P. 2002. Jafrac2 is an IAP antagonist that promotes cell death by liberating Dronc from DIAP1. *EMBO J.* 21:5118–29

Thompson CB. 1995. Apoptosis in the pathogenesis and treatment of disease. *Science* 267:1456–62

Thornberry NA, Lazebnik Y. 1998. Caspases: enemies within. *Science* 281:1312–16

Vaux DL, Flavell RA. 2000. Apoptosis genes and autoimmunity. *Curr. Opin. Immunol.* 12:719–24

Vucic D, Stennicke HR, Pisabarro MT, Salvesen GS, Dixit VM. 2000. ML-IAP, a novel inhibitor of apoptosis that is preferentially expressed in human melanomas. *Curr. Biol.* 10:1359–66

Walker NP, Talanian RV, Brady KD, Dang LC, Bump NJ, et al. 1994. Crystal structure of the cysteine protease interleukin-1b-converting enzyme: a (p20/p10)2 homodimer. *Cell* 78:343–52

Wang X. 2001. The expanding role of mitochondria in apoptosis. *Genes Dev.* 15:2922–33

Watt W, Koeplinger KA, Mildner AM, Heinrikson RL, Tomasselli AG, Watenpaugh KD. 1999. The atomic-resolution structure of human caspase-8, a key activator of apoptosis. *Structure* 7:1135–43

Wei Y, Fox T, Chambers SP, Sintchak J-A, Coll JT, et al. 2000. The structures of caspases-1, -3, -7 and -8 reveal the basis for substrate and inhibitor selectivity. *Chem. Biol.* 7:423–32

Wilson KP, Black J-A, Thomson JA, Kim EE, Griffith JP, et al. 1994. Structure and mechanism of interleukin-1b converting enzyme. *Nature* 370:270–75

Wilson R, Goyal L, Ditzel M, Zachariou A, Baker DA, et al. 2002. The DIAP1 RING finger mediates ubiquitination of Dronc and is indispensable for regulating apoptosis. *Nat. Cell Biol.* 4:445–50

Woo EJ, Kim YG, Kim MS, Han WD, Shin S, et al. 2004. Structural mechanism for inactivation and activation of CAD/DFF40 in the apoptotic pathway. *Mol. Cell* 14:531–39

Wu G, Chai JJ, Suber TL, Wu J-W, Du CY, et al. 2000. Structural basis of IAP recognition by Smac/DIABLO. *Nature* 408:1008–12

Wu J-W, Cocina AE, Chai JJ, Hay BA, Shi Y. 2001. Structural analysis of a functional DIAP1 fragment bound to grim and hid peptides. *Mol. Cell* 8:95–104

Xiao T, Towb P, Wasserman SA, Sprang SR. 1999. Three-dimensional structure of a complex between the death domains of Pelle and Tube. *Cell* 99:545–55

Xu G, Cirilli M, Huang Y, Rich RL, Myszka DG, Wu H. 2001. Covalent inhibition revealed by the crystal structure of the caspase-8/p35 complex. *Nature* 410:494–97

Yan N, Gu LC, Kokel D, Chai JJ, Li WY, et al. 2004a. Structural, biochemical, and functional analyses of CED-9 recognition by the proapoptotic proteins EGL-1 and CED-4. *Mol. Cell* 15:999–1006

Yan N, Wu J-W, Chai JJ, Li W, Shi Y. 2004b. Molecular mechanisms of DrICE inhibition by DIAP1 and removal of inhibition by Reaper, Hid, and Grim. *Nat. Struct. Mol. Biol.* 11: 420–28

Yang X, Chang HY, Baltimore D. 1998. Essential role of CED-4 oligomerization in CED-3 activation and apoptosis. *Science* 281:1355–57

Yuan J, Shaham S, Ledoux S, Ellis HM, Horvitz HR. 1993. The *C. elegans* cell death gene Ced-3 encodes a protein similar to mammalian interleukin-1 beta-converting enzyme. *Cell* 75:641–52

Yuan J, Yankner BA. 2000. Apoptosis in the nervous system. *Nature* 407:802–9

Regulation of Protein Activities by Phosphoinositide Phosphates

Verena Niggli

Department of Pathology, University of Bern, CH-3010 Bern, Switzerland;
email: verena.niggli@pathology.unibe.ch

Annu. Rev. Cell Dev. Biol.
2005. 21:57–79

First published online as a
Review in Advance on
May 6, 2005

The *Annual Review of
Cell and Developmental
Biology* is online at
http://cellbio.annualreviews.org

doi: 10.1146/
annurev.cellbio.21.021704.102317

Copyright © 2005 by
Annual Reviews. All rights
reserved

1081-0706/05/1110-
0057$20.00

Key Words

cytoskeleton, phosphatidylinositol 4,5-bisphosphate,
phosphatidylinositol 3,4,5-trisphosphate, plasma membrane,
actin-membrane linkage

Abstract

Phosphoinositide phosphates (PIPs) correspond to phosphorylated
derivatives of phosphatidylinositol (PI). Despite their relatively low
abundance in the plasma membrane, PIPs play a crucial role as precursors of second messengers and are themselves important signaling
and targeting molecules. Indeed, modulation of levels of PIPs affects,
for example, cortical actin organization, membrane dynamics, and
cell migration. The focus of this review is on selected interesting
targets of PIPs. Those proteins that bind PIPs and are involved in
regulation of actin assembly, actin membrane linkage, and actin contractility are discussed, as well as those that are involved in signaling,
such as small GTPases, protein kinases, and phosphatases, or in regulation of membrane dynamics.

Contents

INTRODUCTION

Phosphoinositide phosphates (PIPs) correspond to phosphorylated derivatives of phosphatidylinositol (PI). Major PIPs are phosphatidylinositol 4-phosphate and phosphatidylinositol 4,5-bisphosphate (PI-4,5-P_2), although they constitute only 0.5% of the total lipids in eukaryotic membranes. Even less abundant are the products of the enzyme phosphatidylinositol 3-kinase (PI 3-kinase), PI-3,4,5-P_3 and PI-3,4-P_2, whose synthesis is strictly controlled by signaling (Doughman et al. 2003, Takenawa & Itoh 2001). Thus proteins that interact in vitro with equal affinity with PI-4,5-P_2, PI-3,4,-P_2 and PI-3,4,5-P_3 are thought to interact in situ mainly with PI-4,5-P_2, due to its higher abundance (Cozier et al. 2003).

PI-4,5-P_2, PI-3,4-P_2 and PI-3,4,5-P_3 (the focus of this review) are mainly found in the plasma membrane and may thus act as markers of the cell boundary. Despite their relatively low abundance, these PIPs play a crucial role as precursors of second messengers and are themselves important signaling and targeting molecules. Indeed, modulation of levels of PI-4,5-P_2 affect cortical actin organization, membrane ruffling, endocytosis, and synaptic vesicle recycling, whereas products of PI 3-kinase are thought to be involved in cell migration and exocytosis (Czech 2003, DiPaolo et al. 2004, Huang et al. 2004, Hurley & Meyer 2001, Insall & Weiner 2001, Kanzaki et al. 2004, Suchy & Nussbaum 2002, Tolias et al. 2000, Wenk & De Camilli 2004, Yamamoto et al. 2001). Insall & Weiner (2001) propose that PI-4,5-P_2 restricts actin polymerization to the plasma membrane playing a permissive role in this process, whereas PI-3,4,5-P_3 is an instructive molecule specifying spatial and temporal dynamics of actin polymerization. However, this may be an oversimplification as specific isoforms of enzymes generating PI-4,5-P_2 (phosphatidylinositol 4-phosphate 5-kinase, PIP 5-kinase) are localized, for example, in focal contacts and are also controlled by signaling proteins such as small GTPases (Doughman et al. 2003, Wang et al. 2004). PIPs themselves may also be locally concentrated in the plasma membrane by enrichment in lipid rafts, by formation of small clusters of PIPs stabilized by hydrogen bonds, or by proteins (for a lucid discussion, see Janmey & Lindberg 2004). However, the size and abundance of such clusters in living cells is

Protein	Motif(s)		Specificity
α-Actinin	158 TAPY**K**NVNIQNFHISW**K** 174		$PI\text{-}4,5\text{-}P_2 \approx PI\text{-}3,4,5\text{-}P_3$
Ezrin	58 WL**K**LD**KK**VSAQEV**RK** 72		$PI\text{-}4,5\text{-}P_2 \approx PI\text{-}3,4,5\text{-}P_3$
	253 **KK**FVI**K**PID**KK** 263		
Profilin	83 FSMDL**R**T**K**ST 92		$PI\text{-}3,4,5\text{-}P_3 > PI\text{-}4,5\text{-}P_2$
	126 **K**CYEMASHL**RR** 136		
N-WASP	183 SHT**KEKKK**G**K**A**KKK**RLT**K** 200		$PI\text{-}4,5\text{-}P_2 \approx PI\text{-}3,4,5\text{-}P_3$
WAVE2	170 DIM**KEKRKH RKEKK**DNPN 187		$PI\text{-}3,4,5\text{-}P_3 > PI\text{-}4,5\text{-}P_2$

Figure 1

Examples of amino acid sequence motifs rich in basic residues in cytoskeletal proteins implicated in binding to PIPs (basic residues highlighted). α-Actinin: (Fukami et al. 1996, Fraley et al. 2003); ezrin: (Barret et al. 2000, Hamada et al. 2000); profilin: (Yu et al. 1992, Sohn et al. 1995, Lambrechts et al. 2002); N-WASP: (Rohatgi et al. 2000, Oikawa et al. 2004); WAVE2: (Oikawa et al. 2004).

currently under debate. Thus we actually do not know the real local concentration of PIPs in the plasma membrane of a given cell at a given moment.

Many of the proteins discussed here bind PIPs via the so-called pleckstrin homology (PH) domains, but cytoskeletal proteins in particular show a remarkable diversity in the structural organization of regions mediating interaction with PIPs (Itoh & Takenawa 2002, Lemmon et al. 2002, Niggli 2001). For example, binding motifs other than PH domains are FERM domains, clusters of basic residues (**Figure 1**), and amphipathic helices (see below).

REGULATION OF ACTIN ASSEMBLY BY PIPs

Gelsolin

Gelsolin is a widely distributed 84-kDa actin-binding, severing, and capping protein whose function is regulated by calcium and PIPs. Gelsolin caps the fast growing barbed ends of actin filaments. The in vitro interaction of gelsolin with PIPs and its functional consequences have been studied in detail. Gelsolin binds in vitro $PI\text{-}4,5\text{-}P_2$ and $PI\text{-}3,4,5\text{-}P_3$ equally well (Feng et al. 2001). Gelsolin residues 135–149 and 150–169, containing clusters of basic amino acids, mediate interaction with PIPs. Interestingly, they overlap with the G-actin (residues 1–149) and the F-actin (residues 150–406) binding domains (Xian & Janmey 2002). Residues 620–634 may also contribute to binding to PIPs (Feng et al. 2001). Both electrostatic interactions and hydrophobic interactions involving fatty acid tails of the lipids appear to be involved in these interactions, which results in pulling PIPs from the bilayer (Feng et al. 2001, Liepina et al. 2003). The latter findings suggest that gelsolin, in addition to being regulated by PIPs, can also reduce the PIP content of the plasma membrane. Binding of actin-bound gelsolin to bilayers containing PIPs results in conformational changes in the protein. This conformational change (rather than competition of actin and PIPs for

FERM domain: 4,1-ezrin-radixin-moesin domain that is implicated in both protein-protein and protein lipid interactions

PH domain: pleckstrin homology domain that can mediate binding of proteins to specific types of PIPs

PI 3-kinase: phosphatidylinositol 3-kinase, a family of enzymes that phosphorylate PIPs at position 3 of the inositol ring

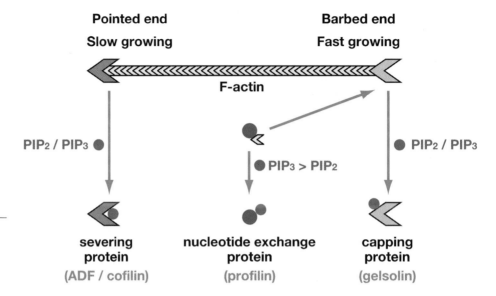

PIPS INHIBIT

Pointed end
Slow growing

Barbed end
Fast growing

F-actin

PIP₂ / PIP₃ ●

● PIP₃ > PIP₂

● PIP₂ / PIP₃

**severing
protein**
(ADF / cofilin)

**nucleotide exchange
protein**
(profilin)

**capping
protein**
(gelsolin)

Figure 2

Regulation of
actin-associated
proteins involved
in controlling actin
assembly by PIPs.

PIP 5-kinase:
phosphatidylinositol
4-phosphate
5-kinase; a family of
enzymes that
generate
phosphatidylinositol
4,5-bisphosphate
from
phosphatidylinositol
4-monophosphate

PIPs:
phosphoinositide
phosphates;
phosphorylated
derivatives of the
phospholipid
phosphatidylinositol

the same binding site) induces dissociation of gelsolin from actin and allows actin filament assembly at the barbed end (**Figure 2**; Liepina et al. 2003).

Despite detailed information on molecular mechanisms of in vitro interactions, no conclusive data are yet available on the physiological role of binding of gelsolin to PIPs at the plasma membrane. A few studies concern the association of gelsolin with PIPs in cells and the role of PIPs in modulating the location of gelsolin. In osteoclasts, for example, gelsolin immunoprecipitated from cellular lysates contains PIPs such as PI-4,5-P₂ and PI-3,4,5-P₃. Anti-PI-4,5-P₂ antibodies react with immunoprecipitated gelsolin (Chellaiah & Hruska 1996). Interestingly, stimulation of osteoclasts with osteopontin increases the amount of PIPs associated with gelsolin, correlating with increased actin polymerization (Chellaiah & Hruska 1996). Treatment of Sertoli cells with a synthetic peptide consisting of the PI-4,5-P₂-binding region (amino acids 160–169) of gelsolin displaces gelsolin and actin from these PI-4,5-P₂–rich adhesion complexes (Guttman

et al. 2002). Such peptides also disrupt the interaction of gelsolin with PI 3-kinase, as detected by coimmunoprecipitation from cellular lysates (Biswas et al. 2004). Because these peptides cannot discriminate between PI-4,5-P₂ and PI-3,4,5-P₃ (Feng et al. 2001), it is not clear which of these lipids regulates the cellular localization of gelsolin and its interaction with PI 3-kinase. In summary, there is some evidence that gelsolin interacts with PIPs in intact cells and that PIPs regulate its location, but evidence for a specific functional regulation of gelsolin by PIPs or for a role of gelsolin in modulating the availability of PIPs at the plasma membrane in intact cells is lacking.

Cofilin

The 18-19-kDa protein cofilin promotes disassembly of actin filaments at the slow-growing pointed ends, severs actin filaments, thus creating new barbed ends for polymerization, and also binds to actin monomers (preferably ADP-actin). Structurally, it shows similarity to gelsolin. Cofilin is involved in cytokinesis and cell migration (DesMarais et al. 2005). In vitro studies showed that, comparable to gelsolin, yeast cofilin binds PI-4,5-P_2 and PI-3,4,5-P_3 equally well, resulting in inhibition of its activity (**Figure 2**; Ojala et al. 2001). Yeast cofilin is a polar molecule, with a positively charged surface at one side of the molecule, that is implicated in interaction with both F-actin and PIPs. These binding sites overlap, but Ojala et al. (2001) have succeeded in the identification by site-directed mutagenesis of basic residues 109 and 110, which are required for interaction with PIPs but are not involved in interaction with F-actin and which play a minor role in G-actin binding. Replacement of wild-type cofilin by cofilin with these two residues mutated to alanines in yeast resulted in abnormal actin organization, but the cells were viable, indicating that this lipid interaction is not crucial for cell survival, at least in yeast.

Profilin

Profilin is a small (14 to 17 kDa), widely distributed protein whose disruption is embryonic-lethal in insects and mice (Verheyen & Cooley 1994, Witke et al. 2001). Profilin binds to actin monomers and promotes exchange of ADP for ATP. These actin-profilin complexes bind to the free barbed ends of actin filaments, which results in dissociation of profilin and incorporation of actin into the filament. In vitro binding of PIPs to profilin precludes interaction with actin, keeping the protein in an inactive state (**Figure 2**). Products of PI 3-kinase (PI-3,4-P_2, PI-3,4,5-P_3) bind in vitro with higher affinity to profilin than does PI-4,5-P_2 (Lu et al. 1996), but the physiological profilin ligand has not yet been identified. The protein also contains binding sites for poly (L-proline) (PLP), stretches present in signaling proteins, such as the enabled/vasodilator-stimulated phosphoprotein Ena/VASP. At least two domains (residues 83–92, 126–136) that contain clusters of basic residues have been implicated

in contributing to interaction with PIPs in profilin (**Figure 1**), thus creating a binding pocket involving N- and C-terminal helices. Double mutations of arginines 88 and 136 to alanines or glutamic acid and aspartic acid, respectively, strongly reduce binding to PI-4,5-P_2 micelles. The first site (residues 83–92) overlaps with the actin-binding site and the second site (residues 126–136) with the PLP-binding site. As expected, binding of profilin to PIPs thus interferes with both actin and PLP binding (Lambrechts et al. 2002). Co-valently linked profilin-actin complexes still interact with PIPs, confirming the presence of a second PI-4,5-P_2 binding site clearly separated from the actin-binding site (Skare & Karlsson 2002). Because both lipid binding sites overlap with sites for other proteins, site-directed mutagenesis of the residues involved in binding to PIPs may not exclusively affect lipid interaction. Therefore, it is difficult to delineate in situ the functional role of binding of profilin to PIPs in the plasma membrane. The hypothesis, based on in vitro data, that profilin controls PI-4,5-P_2 turnover (Goldschmidt-Clermont et al. 1990, 1991), as proposed similarly for gelsolin, has also not been verified in situ.

A recent study by Wittenmayer et al. (2004) investigated the capacity of human profilin I wild-type and mutants to induce a nontumorigenic phenotype in a human breast cancer cell line expressing low levels of endogenous profilin. Profilin wild-type and mutants Y59A (significantly lower affinity for actin; normal binding of PIPs and PLP); H133S (loss of PLP binding; normal binding to actin and PIPs); and R88L (less than 30% of wild-type binding to PIPs; normal PLP binding, somewhat reduced actin interaction) were expressed in this cell line. The results indicated that the actin-binding site, but not the capacity of profilin to interact with PIPs and PLP, is required for tumor suppression. These data do not exclude a functional role of interaction with PIPs, as the residual PI-4,5-P_2 binding capacity of the R88L mutant may be sufficient for normal function.

N-Wasp, Wasp, Wave

Proteins of the Wiskott-Aldrich syndrome protein (WASP) family are multidomain proteins acting as molecular switches to activate actin assembly via Arp2/3 (actin-related protein) in response to intracellular signals. They bind G-actin and Arp2/3, recruiting Arp2/3 to actin filaments and inducing branching. Interaction with Arp2/3 occurs via the VCA (verprolin-homology, cofilin-homology, and acidic region) domain. WASP isoforms are thought to be involved in movement of endocytic vesicles and formation of filopodia. The related protein WAVE (WASP family verprolin homologous protein) also activates Arp2/3 via a related VCA domain and is required for lamellipodia formation in the leading edge of motile cells (Takenawa & Miki 2001). WASP proteins are activated by Cdc42, and WAVE proteins are involved in Rac-induced actin dynamics. Interestingly, both proteins contain related, highly basic domains thought to be involved in binding of PIPs, which modulates their function (**Figures 1, 3a**). Despite these similarities, mechanism of activation of N-WASP (neuronal Wiskott-Aldrich syndrome protein)/WASP and WAVE isoforms appears to be quite different, as summarized by Bompard & Caron (2004).

N-WASP and WASP, in the absence of stimuli, are in an autoinhibited conformation owing to intramolecular interactions between C and N termini. These interactions are relieved by cooperative binding of upstream activators such as PI-4,5-P_2, Cdc42, and SH3-domain-containing proteins. In vitro assays showed that PI-4,5-P_2 and PI-3,4,5-P_3 (Oikawa et al. 2004) bind to N-WASP, thereby reducing the affinity between C and N termini. Elimination of residues 186–200 (**Figure 1**) in N-WASP or mutation of basic residues 186, 189, 192, and 195 to glutamic acid results in abolishment of binding to PI-4,5-P_2 (Rohatgi et al. 2000). Despite these clearcut in vitro data, the role of PIPs in in situ activation of WASP proteins is not well established. For example, the

actin comet-dependent movement of endosomes in fibroblasts is stimulated by overexpression of phosphatidylinositol-4-phosphate 5-kinase (PIP5K) and requires N-WASP. The basic N-WASP domain is recruited to endosomes, but N-WASP constructs that lack the basic domain restore formation of actin comet tails as efficiently as do wild-type proteins, arguing against a crucial role of PIPs in regulating WASP function in situ (Benesch et al. 2002). However, N-WASP lacking the basic domain may be partially activated (Rohatgi et al. 2000).

In contrast to N-WASP, WAVE proteins are constitutively active in vitro and occur in multidomain complexes. WAVE activation requires its proper localization, as well as the presence of F-actin and/or the release of additional proteins. Rac induces translocation of the WAVE complex to lamellipodia (Bompard & Caron 2004). A recent interesting study by Oikawa et al. (2004), involving both in vitro and in situ experiments, indicates an important role for PIPs in this process. Oikawa et al. (2004) showed that WAVE2 interacts in vitro preferentially with PI-3,4,5-P_3 (K_d 185 nM) and with lower affinity with PI-4,5-P_2 (K_d 1.2 μM). A truncated WAVE2 peptide consisting of residues 1–183 still binds to PIPs, whereas peptide 1–170 has lost this capacity, implicating residues 171–183 in this interaction (**Figure 1**). Mutation of five basic residues in this domain to alanines results in loss of binding of the full-length protein to PIPs, which clearly confirms the truncation experiments.

Inhibition of PI 3-kinase with wortmannin inhibits growth factor-induced translocation of WAVE2 to lamellipodia. Moreover, WAVE2 mutants with mutated basic residues are no longer recruited to the plasma membrane in stimulated cells, and overexpression of these mutants reduces Rac-dependent formation of lamellipodia (Oikawa et al. 2004). Thus PI-3,4,5-P_3, together with Rac, may localize activated WAVE2 at the membrane. Information on how these mutations in the PI-3,4,5-P_3-binding domain of WAVE2 affect the integrity or the activity of the WAVE-containing complex is lacking.

REGULATION OF ACTIN-MEMBRANE LINKAGE AND ACTIN-CROSSLINKING BY PIPs

Ezrin/Radixin/Moesin

The ezrin/radixin/moesin (ERM) protein family contains a conserved N-terminally located membrane-binding domain of approximately 300 amino acids, called the FERM (4.1-ezrin-radixin-moesin) domain, whose crystal structure was recently determined (Hamada et al. 2000, Smith et al. 2003). This domain mediates both protein-protein interactions and protein-lipid interactions at the plasma membrane. ERM proteins also contain F-actin-binding sites in their C terminus and are thus capable of linking actin filaments to the membrane, directly to PIPs, indirectly to cytoplasmic tails of transmembrane proteins, or to scaffolding proteins linked to transmembrane proteins (Bretscher et al. 2002).

Studies based on site-directed mutagenesis of ezrin and on analysis of crystals of the radixin FERM domain complexed to $Ins(1,4,5)P_3$ implicate clusters of basic residues exposed on the surface of the FERM domain (**Figure 1**; lysines 60, 63, 253, 254, 262, 263) to be involved in interaction with PIPs (Barret et al. 2000, Hamada et al. 2000, Niggli 2001). We investigated lipid specificity of in vitro binding of ezrin to PIPs incorporated into large liposomes. Our preliminary data show that ezrin interacts preferentially with $PI\text{-}4,5\text{-}P_2$ and $PI\text{-}3,4,5\text{-}P_3$ when compared with $PI\text{-}3,4\text{-}P_2$ and does not discriminate between the first two lipids (V. Niggli, A. Bahloul & A. Houdusse, unpublished data). Thus, its main physiological interaction partner very likely is $PI\text{-}4,5\text{-}P_2$.

Functions of ERM proteins are tightly regulated by signaling. In the absence of appropriate signals, ERM proteins are maintained in an inactive conformation through an intramolecular interaction between the N-terminal ERM association domain (N-ERMAD) and the C-terminal ERM association domain (C-ERMAD), which are linked by a central α-helix-rich domain (Hoeflich et al. 2003). In the inactive state, membrane and F-actin binding sites are masked by C- and N-ERMAD association (Bretscher et al. 2002). Phosphorylation of a conserved threonine (T567 in ezrin) and binding of the protein to PIPs are thought to induce conformational changes resulting in dissociation of C- and N-ERMAD, thus unmasking the binding sites. Mutation of lysines 253, 254, 262, and 263 to asparagine results in almost complete abolishment of in vitro interactions of ezrin with PIPs and precludes its recruitment to the plasma membrane in living cells (Barret et al. 2000). Importantly, no other tested interactions were significantly affected by these mutations. In a recent elegant study, Fievet et al. (2004) have explored the sequence of events resulting in signal-dependent ezrin activation in intact cells. They expressed wild-type or ezrin defective in $PI\text{-}4,5\text{-}P_2$ binding with lysines 253, 254, 262, and 263 mutated to asparagine in epithelial cells. In contrast to wild-type ezrin, mutated ezrin failed to localize to apical microvilli and was no longer retained in the detergent-insoluble cytoskeleton. However, PIPs do not appear to be required for localization of ezrin at the plasma membrane because the N-terminal domain of mutated ezrin localizes to the membrane in a manner similar to that of wild-type protein. Rather it appears that interaction with PIPs is a prerequisite for phosphorylation of ezrin, as only wild type but not mutant ezrin was phosphorylated in cells on threonine 567. Phosphomimetic mutation at threonine 567 of a mutant ezrin defective in binding to PIPs resulted in membrane association. In conclusion, PIPs are required for unmasking a phosphorylation site in ezrin, allowing stable activation, but they do not seem to be involved in ezrin's precise localization at the membrane (Fievet et al. 2004). Interestingly, the cytosplasmic tail of the transmembrane protein ICAM-2, which binds to subdomain C of ezrin FERM, also interacts with PIPs. PIPs thus may recruit and locally activate masked ezrin in the vicinity of its interaction partner ICAM-2 (Hamada et al. 2003).

Vinculin

Vinculin, a protein capable of interacting with F-actin and several other proteins and with lipid bilayers, has a head of 80 Å, corresponding to a 95-kDa N-terminal domain, and a tail domain corresponding to the C-terminal residues 879–1066. The tail domain contains binding sites for acidic phospholipids, actin, and paxillin, whereas the head domain interacts with talin and α-actinin. Similar to ezrin, vinculin head and tail domains can participate in an intramolecular interaction that masks other binding sites. This intramolecular interaction is thought to be disrupted by interaction with PIPs, which results in the exposure of ligand-binding sites. Vinculin may couple cell adhesion and membrane protrusion by inducing actin assembly in focal adhesions (DeMali 2004). Certainly its function is vital; homozygous inactivation of the vinculin gene in mice causes embryonic lethality (Zemljic-Harpf et al. 2004).

Recently, the crystal structure of intact vinculin has been determined at 3.1 Å resolution (Bakolitsa et al. 2004). The protein is characterized by a bundle of helical bundles structure with overall dimensions of $100 \times 100 \times 50$ Å. The vinculin head holds the tail in an autoinhibited structure (Bakolitsa et al. 2004). The isolated tail domain consists of a helical bundle, 60 Å long and 20–30 Å in diameter. Five amphipathic helices are connected by short loops, adopting an antiparallel topology with hydrophobic residues buried inside the bundle. The last helix is followed by a C-terminal arm, which ends in a five-residue hydrophobic hairpin (residues 1062–1066, HPWYQ) (Bakolitsa et al. 1999). Interaction of vinculin with lipid bilayers is complex. A collar of basic residues (including amino acids 910, 911,

1049, 1060, and 1061), a basic ladder centered on helix 3 (residues 945–973), and a hydrophobic hairpin (residues 1062–1066) are thought to be involved. Data based on site-directed mutagenesis support the model, that the first contact of the vinculin tail with the lipid bilayer is mediated by the basic collar, the basic ladder, and the hydrophobic hairpin at the C-terminal end. This C-terminal hairpin (residues 1062–1066) can then insert into the bilayer, triggering unfurling of the helical bundle and resulting in closer association of helices H2 and H3, in particular, with the bilayer (Johnson et al. 1998; Bakolitsa et al. 1999, 2004).

According to the available data, the vinculin tail interacts with both acidic phospholipids such as phosphatidylserine and PIPs. Further studies are needed to determine whether specific separate binding sites exist for these different lipids. Until very recently it was not clear whether binding to PIPs is indeed sufficient to fully activate autoinhibited vinculin (Niggli 2001). The beautiful structural study by Bakolitsa et al. (2004) now advances our knowledge on this point. In intact vinculin, the basic ladder is mostly exposed to solvent, but the basic collar is partially occluded, and conformational changes induced by PIPs in the tail are inhibited. The authors propose a kinetic pathway to activation, in which binding to PIPs transiently releases the tail from the head, allowing interaction with other ligands, which then stabilize the active conformation. Vinculin would thus be activated by a spatial colocalization of lipid- and protein-binding partners rather than by one partner alone. Indeed, we observed that lipid bilayer insertion of intact vinculin is enhanced in the presence of its binding partner α-actinin (Niggli & Gimona 1993). Also in support of this model, the direct interaction of vinculin with Arp2/3 in cell lysates is synergistically enhanced by activated Arp2/3 and PtdIns(4,5)P$_2$ (DeMali et al. 2002).

Despite the detailed insight into the molecular mechanisms of interaction of vinculin with PIPs, little is known about its physiological relevance. According to a preliminary report, expression of a vinculin mutant lacking residues 1052–1066 in vinculin null fibroblasts was effective in suppressing cell migration, comparable to the effect of expressing wild-type protein, but was unable to restore normal rates of cell spreading. The mutant protein also failed to localize to peripheral focal contacts (Saunders et al. 2003). In contradiction to this report, sequestration of PI-4,5-P_2 does not seem to disturb vinculin localization in focal adhesions (Martel et al. 2001). It is possible that functions other than binding to PIPs are affected in the mutated protein or that sequestration of PIPs in the latter study was not fully effective.

Talin

Talin, another member of the protein 4.1 family, plays a crucial role in integrin activation (Cram & Schwarzbauer 2004). It is an interesting molecule because it both binds to and is regulated by PIPs and itself controls synthesis of PI-4,5-P_2. The 47-kDa head of globular head of talin, which has about 24% sequence homology with the ERM family FERM domain, contains binding sites for PIPs, focal adhesion kinase (FAK), cytosolic tails of $\beta1$ and $\beta3$ integrins, F-actin, and a type I neuronal isoform of phosphatidylinositol 4-P-5-kinase γ (PIPKIγ). The 190 kDa rod domain interacts with vinculin and F-actin (Lee et al. 2004, Nayal et al. 2004). Talin is thought to disrupt the interaction of the α- and β-subunits of integrins, thereby activating these adhesion receptors. Interestingly, in vitro PI-4,5-P_2 induces a conformational change in talin and enhances its interaction with integrins (Martel et al. 2001). However, the precise location and molecular architecture of the binding site for this lipid in the FERM domain of talin are unclear. Alignment of the talin and ERM FERM domains shows little sequence homologies concerning positions of basic amino acid residues crucial for lipid binding of ezrin and radixin (Hamada et al. 2000, Barret et al.

2000). An amphipathic talin peptide (residues 385–406) has been shown to insert into acidic lipid monolayers and bilayers, forming an α-helix in the membrane (Seelig et al. 2000), possibly similar to the amphipathic α-helices in the tail of vinculin. More direct evidence is required to confirm the role of this sequence in interaction with PIPs. Recent studies of Garcia-Alvarez et al. (2003) involving crystallization of domains F2 and F3 of the talin FERM domain and site-directed mutagenesis implicate residues R358, W359, and I396 in interaction with integrin cytoplasmic β tails. Binding sites for PIPs and integrins in talin may thus overlap, but on the basis of the above data, binding to those two interaction partners may not be mutually exclusive. As found with ezrin and vinculin, PIPs may relieve intramolecular inhibitory interactions in talin (Garcia-Alvarez et al. 2003). In support of the physiological relevance of binding of talin to PIPs, sequestration of PI-4,5-P_2 in intact cells delocalizes talin from focal adhesions (Martel et al. 2001).

Intriguingly, the talin FERM domain also interacts with and activates type I PIPKγ. This interaction is required for recruitment of both proteins to focal adhesions, where PIPKIγ is further activated by focal adhesion kinase, which results in a localized increased production of PI-4,5-P_2. A talin molecule can simultaneously bind F-actin and PIPKIγ, but interactions with integrins and PIPKIγ are mutually exclusive. However, the antiparallel talin homodimer could simultaneously bind integrins and PIPKIγ (Di Paolo et al. 2002, Ling et al. 2002, Morgan et al. 2004). The following scenario emerges from these studies: First, complexes of talin and PIPKIγ are recruited to focal adhesions. Second, PIPKIγ is phosphorylated and activated by FAK, thereby resulting in increased synthesis of PI-4,5-P_2. This lipid then activates talin and enhances its interaction with cytosolic tails of integrins, which results in stabilization of the high-affinity state of integrins and formation of stable focal adhesions. Of course, PI-4,5-P_2 enriched in focal adhesions could also modify functions of other cytoskeletal proteins such as vinculin, α-actinin, or Arp2/3 (see above). The interaction of talin with PIPKIγ may indeed be of physiological relevance because a peptide that disrupts interaction between these two proteins interferes with accumulation of F-actin at synapses and impairs clathrin-mediated synaptic vesicle endocytosis (Morgan et al. 2004).

Annexin

Annexins are proteins characterized by calcium-dependent binding to negatively charged phospholipids. Calcium induces membrane association of annexins, but the functional role of this interaction is not well understood. Annexin 2 also interacts with F-actin and thus may function as a reversible actin-membrane linker. Annexin 2 localizes to early endosomes, pinosomes, and phagosomes and has been implicated in endosome trafficking and modulation of lipid raft function (Babiychuk & Draeger 2000, Gerke & Moss 2002). Recently, this protein was shown to interact in vitro with some selectivity with PI-4,5-P_2, when compared to PI-3,4,5-P_3 or PI-3,4-P_2. This interaction also occurs in the absence of calcium but is markedly enhanced by this ion. Half-maximal interactions were observed at 5 μM of PI-4,5-P_2, comparable to the affinity of the PH domain of phospholipase δ. Interestingly coexpression of annexin 2 and active Arf6, a GTPase that regulates endosomal trafficking and activates PI 4-P-5-kinase, results in relocalization of annexin 2 to intracellular membranes (Hayes et al. 2004, Rescher et al. 2004). Annexin 2 could thus be involved in dynamic membrane remodeling involved in formation of endocytic vesicles depending on PIPs.

α-Actinin

α-Actinin is a rod-like, antiparallel homodimer with actin-binding domains, spectrin repeats, and C-terminal EF-hands. Its main functions concern linkage of actin stress fibers to integrins and bundling of actin filaments (Djinovic-Carugo et al. 2002). Its crystal structure has not yet been solved. α-Actinin contains a binding site for PIPs in the calponin-homology domain 2 (CH2), which, together with CH1, is involved in actin binding. This binding site has been mapped to residues 168–184 in chicken α-actinin and to residues 158–174 in human α-actinin as determined by in vitro lipid binding of truncated α-actinin or of different peptides rich in basic residues derived from α-actinin and on site-directed mutagenesis (**Figure 1**; Fraley et al. 2003, Fukami et al. 1996). PI-4,5-P_2 and PI-3,4,5-P_3 bind with equal affinity to this domain. Interactions with PI-4,5-P_2 or PI-3,4,5-P_3 differently affect the structure of the protein; the former lipid appears to stabilize α-actinin, whereas the latter lipid increases protein flexibility (Corgan et al. 2004). Young & Gautel (2000) propose that interaction with PI-4,5-P_2 relieves an intramolecular interaction of the C-terminal calmodulin-like domain with the actin-binding domain of α-actinin, thereby activating the molecule similar to the effects of PIPs on vinculin and ezrin. The effects of PIPs on in vitro F-actin bundling activities of α-actinin are controversial. According to Fukami et al. (1992), this function is stimulated by PIPs, whereas according to Fraley et al. (2003) and Corgan et al. (2004), F-actin bundling is attenuated by PIPs. In situ evidence using expression of GFP-tagged α-actinin defective in binding to PIPs suggests that PIPs do indeed negatively affect the actin-bundling activity of α-actinin (Fraley et al. 2003).

PI-3,4,5-P_3 selectively disrupts in vitro the interaction of α-actinin with the cytoplasmic tail of β1 integrin (**Figure 3b**; Greenwood et al. 2000). The novel concept that PI-4,5-P_2 interacts with α-actinin in resting cells and controls the amount of free protein available, whereas PI-3,4,5-P_3 interacts transiently and locally with α-actinin in focal adhesions in stimulated cells, resulting in loss of α-actinin-containing focal adhesions (Fraley

a

PIPS ACTIVATE

Proteins initiating actin
assembly at the PM

Proteins linking actin
filaments to the PM

b

PIPS INHIBIT

Actin filament
crosslinking by α-actinin

Association of
α-actinin with integrin

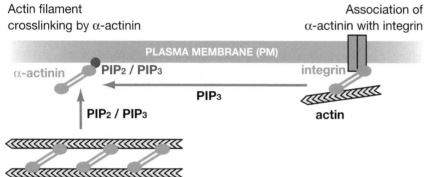

Figure 3

Regulation of
actin-associated
proteins involved in
initiating actin
assembly at the
plasma membrane
(*a*), in mediating
actin-membrane
linkage (*a*, *b*) and in
controlling actin
crosslinking (*b*) by
PIPs.

et al. 2003, Greenwood et al. 2000), will have to be confirmed.

REGULATION OF ACTIN CONTRACTILITY BY PIPs

Myosin Isoforms

Several unconventional myosin isoforms interact with PIPs. Such interactions may serve to target these motors to sites of actin assembly, or they could be the basis of transport of PIP-containing vesicles by myosin. With one exception (myosin 10), the binding sites for PIPs have not been well characterized.

Myosin 1c has been reported to contain two sites interacting with acidic phospholipids and PIPs: one in the tail and one in the calmodulin-binding IQ domains. Whether this interaction is indeed relevant in situ

for localization and regulation of the protein awaits experimental proof (Hirono et al. 2004).

NINAC, the eye-enriched myosin 3, interacts in vitro with PI-4,5-P_2 and PI-3,4,5-P_3. PIPs, moreover, induce indirect association of NINAC and arrestin, a rhodopsin regulatory protein. Arrestin itself also binds PIPs, and this association is required for its light-dependent trafficking in photoreceptor cells. An attractive hypothesis envisages that photoreceptor illumination induces formation of PI-3,4,5-P_3-rich vesicles, which bind arrestin and are transported by NINAC into microvilli where arrestin terminates rhodopsin signaling (Lee & Montell 2004, Strissel & Arshavsky 2004).

Interestingly, the isoform myosin 10 contains in its tail three PH domains with selectivity for PIPs. This structural feature might serve to target this isoform to regions of actin turnover, as it is enriched in cellular membrane ruffles, in lamellipodia (Berg et al. 2000), and at phagocytic cups (Cox et al. 2003). It is the only vertebrate myosin with this structural feature. Using expression of GFP-tagged myosin 10 constructs in living cells, it could be shown that only the PH domains exhibited membrane localization and that all three PH domains appear to be required for tight membrane association (Mashanov et al. 2004, Yonezawa et al. 2003). In vitro the PH domains of myosin 10 interact preferentially with PI-3,5-P_2 and PI-3,4,5-P_3 and with lower affinity with PI-4,5-P_2 (Mashanov et al. 2004). Inhibition of PI 3-kinase by wortmannin inhibits recruitment of myosin 10 to phagocytic cups in macrophages, suggesting a physiologically relevant regulation of localization of this motor protein by PI-3,4,5-P_3. In support of this notion, a point mutation in the PH domain, predicted to abolish PI-3,4,5-P_3 binding, suppresses the inhibitory effect of a truncation fragment of myosin 10 (which contains its tail) on phagocytosis (Cox et al. 2003). Myosin 10 could thus serve to couple movement of actin filaments with outward movement of the plasma membrane.

REGULATION OF SIGNALING PROTEINS BY PIPs

Phospholipase C Isoforms

The phospholipase C (PI-PLC) family plays an important role in receptor-linked signal transduction by hydrolyzing PI-4,5-P_2, thereby generating diacylglycerol, which in turn activates protein kinase C isoforms, and inositol-1,4,5-trisphosphate (IP$_3$), which induces an increase in cytosolic calcium. Interestingly these enzymes are themselves regulated by PIPs. The three major families (PI-PLCβ, δ, and γ) all contain N-terminal PH domains that interact with PIPs with different specificity. PI-PLCβ shows in vitro specificity for PI-3-P; PI-PLCδ for PI-4,5-P_2;

and PI-PLCγ for PI-3,4,5-P_3 (Katan 1998, Rhee 2001). It is now well accepted that the PH domain of PI-PLCδ1 is a sensor of its substrate PI-4,5-P_2 and allows the enzyme to be tethered to the membrane so that it can catalyze hydrolysis of many PI-4,5-P_2 molecules without being released from the membrane. The product IP$_3$ competes with PI-4,5-P_2 for the same binding site on PI-PLCδ1 and displaces it from the membrane (Katan 1998, Rhee 2001).

Concerning PI-PLCγ, PI-3,4,5-P_3 binds to both its N-terminal PH domain and the C-terminal SH2 domain, contributing to the membrane translocation of this enzyme and to more efficient hydrolysis of PI-4,5-P_2. Indeed wortmannin treatment of PDGF-stimulated NIH 3T3 cells results in a 40% decrease in hydrolysis of PI-4,5-P_2 (Rhee 2001). Interestingly, the SH3 domain of PI-PLCγ has recently been shown to act as a guanine nucleotide exchange factor (GEF) for the nuclear GTPase PIKE-S (PI 3-kinase enhancer), suggesting a positive feedback loop (Ye & Snyder 2004).

Proteins Regulating Activity of Small Cytosolic GTP-Binding Proteins

Guanine nucleotide exchange factors (GEFs), which stimulate activity of small GTP-binding proteins, and GTPase-activating proteins (GAPs), which promote inactivation of the GTP-binding proteins, are modulated by PIPs. In all cases, binding to PIPs is mediated by PH domains (Burridge & Wennerberg 2004). Selected examples are discussed below.

SWAP-70 is a GEF that specifically activates Rac, but not Rho or Cdc42, and is involved in cell migration and adhesion (Sivalenka & Jessberger 2004). Elegant work by Shinohara et al. (2002) provides evidence for its specific activation by PI-3,4,5-P_3 in vitro and in situ. These authors could show that SWAP-70, which is expressed mainly in B lymphocytes and mast cells, interacts in vitro specifically with beads coated with

PI-3,4,5-P_3. This interaction requires the PH domain of SWAP-70 and can be prevented by mutation of the basic residues R230 and K291 located in the PH domain. Indeed, the marked enhancement of the in vitro GEF activity of SWAP-70 on Rac by PI-3,4,5-P_3 was also abolished by mutation of these residues. SWAP-70 expressed in NIH 3T3 cells translocated transiently to membrane ruffles upon stimulation of cells with PDGF. Membrane ruffling induced by PDGF was abolished by expression of mutants lacking the capacity to interact with PI-3,4,5-P_3. When transfected in COS7 cells, only wild-type, but not the mutants lacking the capacity to bind PI-3,4,5-P_3, induced Rac1 activation. SWAP-70 thus is an interesting PI-3,4, 5-P_3-dependent GEF for Rac activated downstream of tyrosine kinase receptors. Interestingly, both GTP loading and GTP hydrolysis of the small GTP-binding protein Arf (ADP-ribosylation factor) are controlled by PI 3-kinase products via the GEF ARNO (Arf nucleotide-binding site opener) and the GAP ARAP3 (Arf GAP and Rho GAP with ankyrin repeats and PH domains). PI 3-kinase products thus impact both initiation and termination of the Arf-mediated responses. ARAP3 interacts in vitro specifically with PI-3,4,5-P_3-coated beads and contains five predicted PH domains of which the most N terminal appears to be crucial for interaction with this lipid (Krugmann et al. 2002, 2004). The mutations R307/8A in this domain completely abolish in vitro interactions with PI-3,4,5-P_3. The in vitro GAP activity of ARAP3 on Arf requires PI-3,4,5-P_3. Mutants that are unable to interact with this lipid do not induce redistribution of Arf from the cell periphery to an intracellular compartment (Krugman et al. 2002). It is not clear whether interaction with PI-3,4,5-P_3 is required for membrane targeting or for enzyme activation.

GAPs for Ras may preferentially interact with PI-3,4,5-P_3, e.g., GAP1^m, or bind with similar affinity to both PI-3,4,5-P_3 and PI-4,5-P_2, e.g., GAP1^{IP4BP} (Cozier et al. 2000, 2003). Cozier et al. (2003), using molecular modeling, identified residues in the PH domain of the latter protein that are involved in interaction with PI-4,5-P_2. Introduction of the mutation K591T in GAP1^{IP4BP} generated a protein that retained its high-affinity interaction with PI-3,4,5-P_3 (K_d approximately 14 μM), whereas its affinity for PI-4,5-P_2 was reduced by sixfold (K_d 190 μM). Interestingly, cellular location was clearly different for wild-type and mutated proteins. The wild-type protein was enriched at the plasma membrane of resting cells, indicating interaction with the more abundant PI-4,5-P_2, whereas the mutant protein was located predominantly in the cytosol and was recruited to the plasma membrane only in growth factor-stimulated cells, depending on the activity of PI 3-kinase. This work now enables correlation of in vitro affinities of protein interactions with different PIPs with the in situ impact of manipulation of synthesis of specific PIPs on protein location.

Serine/Threonine Protein Kinases

Well-described examples of protein kinases requiring binding to PIPs via PH domains for their proper localization and activation are 3-phosphoinositide-dependent protein kinase 1 (PDK1) and protein kinase B (PKB/Akt). PDK1 is a central signaling molecule that phosphorylates and activates at least 24 different protein kinases, among them PKC isoforms and also PKB. The PH domain of PDK1, which has been crystallized recently, displays an unusually spacious ligand-binding site that may account for its special ligand-binding properties. PDK1 binds in vitro with high affinity (K_d: 4 nM) to the PI 3-kinase products PI-3,4,5-P_3 and PI-3,4-P_2, and, with approximately 15-fold lower affinity, to PI-4,5-P_2. Interestingly, it also interacts with nanomolar affinity with InsP$_6$ and Ins(1,3,4,5,6)P$_5$ (Komander et al. 2004). These inositol phosphates are present at high micromolar levels in cells. Neither PIPs nor inositol phosphates induce a conformational change in the protein, which suggests that these molecules affect its location rather than its catalytic activity. Findings by Komander et al. (2004) suggest that interaction of PDK1 with these inositol phosphates may anchor part of PDK1 in the cytosol where it could activate substrates such as ribosomal S6 kinase (RSK) independently of PIPs. Elegant work by McManus et al. (2004), using a knockin strategy, convincingly demonstrates that interaction of the PH domain of PDK1 with PIPs plays a crucial role in enabling PDK1 to activate PKB in vivo, whereas activation of RSK does not require this interaction. In this work, PDK1 was mutated in the PH domain (RRR472-474LLL), which resulted in a protein with full catalytic activity that was unable to interact with PIPs. Replacement of the wild-type PDK1 gene by the mutated gene in embryonic stem cells resulted in abolishment of stimulus-dependent activation of PKB in these cells, whereas RSK was activated normally. PIPs thus function here by bringing into close proximity a selected substrate and the enzyme acting on it at the plasma membrane.

Interaction of the protein kinase PKB with membranes via PIPs is a prerequisite for its subsequent activation by multisite phosphorylation (Scheid & Woodgett 2003). The PH domain of PKB interacts in vitro selectively with PI-3,4,5-P$_3$ and PI-3,4-P$_2$ but not with PI-4,5-P$_2$. This interaction does not affect its enzyme activity directly but has recently been shown to induce a marked conformational change in the protein, as shown by comparing crystals of the PH domain of PKB alone with those complexed to Ins(1,3,4,5)P$_4$. Such changes have not been observed previously for other PH domains. These conformational changes may be transmitted to the catalytic region of PKB facilitating subsequent phosphorylation of the activation loop by PDK1. This finding could explain why PDK1 cannot phosphorylate wild-type PKB in absence of PIPs, whereas PDK1 efficiently phosphorylates mutant forms of PKB that lack the PH domain (Milburn et al. 2003).

Tyrosine Kinases and Phosphatases

Btk is a member of the Tec family of protein tyrosine kinases and contains an N-terminal PH domain interacting specifically with PI-3,4,5-P$_3$. This interaction is required for targeting the enzyme to the membrane but is not necessary for enzyme activity. Btk then phosphorylates critical tyrosine residues on PLCγ2, which itself also binds PI-3,4,5-P$_3$ (see above), resulting in its activation and increased hydrolysis of PI-4,5-P$_2$. Btk also functions as a shuttle to bring the enzyme phosphatidylinositol 4-phosphate 5-kinase to the membrane. Thus Btk regulates the production of the substrate required by the downstream target of Btk (Carpenter 2004).

The protein tyrosine phosphatase-like protein 1 (PTPL1) is characterized by the presence of a FERM domain. This domain is required for enrichment of the enzyme in dorsal microvilli when expressed in HeLa cells (Bompard et al. 2003). Interestingly, this FERM domain contains clusters of basic residues similar to those of the FERM domain of ezrin. Specifically, basic residues 63, 253, 254, and 262 in ezrin (Barret et al. 2000) are also conserved in PTPL1 (Bompard et al. 2003). However, whether PTPL1 interacts specifically with PIPs in vitro is controversial. According to Kimber et al. (2003), another protein, the tandem-PH-domain-containing protein-1 (TAPP1), which interacts specifically with PI 3,4-P$_2$, binds PTPL1 and targets it to the membrane where this enzyme then switches off signaling pathways by dephosphorylation of substrates.

REGULATION OF MEMBRANE TRAFFICKING BY PIPs

Maintenance of Golgi Apparatus (Spectrin)

Non-erythrocyte spectrin, a rod-like protein related to α-actinin and dystrophin, consists of an α- and a β-subunit. It contains a PH domain in the C-terminal region of its

β-subunit. PI-4,5-P$_2$ and PI-3,4,5-P$_3$ bind to the spectrin PH domain with similar affinities (Hyvönen et al. 1995, Kavran et al. 1998).

Spectrin has been suggested to stabilize the Golgi apparatus and to affect endocytosis negatively. The PH domain of human spectrin targets this protein to the plasma membrane in intact cells and may also be involved in recruiting spectrin to PI-4,5-P$_2$-rich domains in Golgi membranes (Lorra & Huttner 1999). Interestingly, fragmentation of the Golgi in the absence of ongoing PI-4,5-P$_2$ synthesis correlates with phosphorylation and redistribution of spectrin to the cytoplasm (Siddhanta et al. 2003). Experiments are required to determine whether mutated spectrin lacking the ability to bind PIPs has an altered cellular location and/or altered functions.

Exocytosis (Synaptotagmin)

PI-4,5-P$_2$ is clearly required for exocytosis in neuroendocrine cells. For example, the GFP-tagged PH domain of PLCδ1, which interacts specifically with this lipid (see above), when expressed in adrenal chromaffin cells locates exclusively to the plasma membrane and prevents exocytosis (Holz et al. 2000). The secretory vesicle protein synaptotagmin has been identified as a major interaction partner of PI-4,5-P$_2$ involved in this process (Bai et al. 2004, Schiavo et al. 1996). Synaptotagmin is an integral membrane protein with a single, membrane-spanning domain near the N terminus and two cytosolic domains, C2A and C2B, consisting of distinct eight-stranded β-sandwich structures. C2A binds calcium and acidic phospholipids, whereas C2B interacts with PIPs. The in vitro specificity of this interaction is modified by calcium. According to Schiavo et al. (1996), the C2B domain binds best to liposomes containing PI-3,4,5-P$_3$ below 1 nM free calcium, whereas at 100 μM free calcium, PI-4,5-P$_2$ and to a lesser extent also PI-3,4-P$_2$ are the preferred interaction partners (Schiavo et al. 1996). Some-

what in contradiction to this work, Bai et al. (2004) postulate an important role of calcium-independent interaction of C2B with PI-4,5-P$_2$. This interaction is mediated by a polybasic region in a groove on the side of C2B. Mutation of lysines 326 and 327 (located in this groove) to alanine decreases interaction of C2B with PI-4,5-P$_2$. On the basis of their in vitro data, Bai et al. (2004) now present an attractive model whereby synaptotagmin directs secretory vesicles specifically to the plasma membrane, which is marked by containing PI-4,5-P$_2$, thereby inducing close apposition of membranes and facilitating their fusion. The latter process is mediated by the calcium-independent prebinding of C2B domain to bilayers containing PI-4,5-P$_2$. An increase in cytosolic calcium then induces reorientation and membrane insertion of C2B, followed by passive insertion of C2A. This process reinforces the close apposition of the membranes. Introduction of synaptotagmin carrying the mutations mentioned above into flies that otherwise lack the wild-type protein results in a 30% decrease of evoked transmitter release. However, as such mutations also affect interaction of the protein with other ligands, these findings do not conclusively confirm the in situ relevance of the interaction of synaptotagmin with PIPs for exocytosis (Mackler & Reist 2001).

Endocytosis (Dynamin)

The large molecular mass GTPase dynamin is a force-generating molecule responsible for membrane fission during endocytosis. Dynamin assembles around the neck of invaginated pits, cleaving the vesicle from the parent membrane. PIPs play a key role in endocytosis and may recruit dynamin and other proteins to the bilayer as a prerequisite for their function in promoting endocytosis. Dynamin, comparable to spectrin, binds in vitro preferentially to PIPs, when compared with other acidic phospholipids, but it cannot discriminate between PI-4,5-P$_2$, PI-3,4-P$_2$, and PI-3,4,5-P$_3$. PIPs stimulate its GTPase activity.

The affinity of the in vitro interaction of the isolated monomeric PH domain of dynamin with PIPs is weak, but oligomerization of dynamin seems to increase the apparent affinity of interaction about sevenfold (Klein et al. 1998). Dynamin2 also promotes the association of actin filaments, nucleated by Arp2/3 and cortactin, to lipid vesicles containing PI-4,5-P_2 (Schafer et al. 2002).

The following observations indicate a physiological relevance of these interactions. A reduction in cellular levels of PI-4,5-P_2 results in a loss of dynamin from cell membranes (Hill et al. 2001). Interestingly, transfection of cells with a construct encoding the PH domain of dynamin with a point mutation that abolishes in vitro interaction with PIPs blocks endocytosis (Achiriloaie et al. 1999).

SUMMARY POINTS

1. PIPs activate functions of proteins (N-WASP, WAVE, ezrin, vinculin, talin and myosin 10) that promote localized actin polymerization, actin-membrane linkage, and contractility via membrane targeting or promotion of conformational changes (**Figure 3a**).

2. Proteins that promote actin filament depolymerization, sever or cap actin filaments (cofilin, gelsolin) and thus induce remodeling of actin filaments, in contrast, are inhibited by PIPs (**Figure 2**).

3. Most of the cytoskeletal proteins, with exception of WAVE, myosin 10, and possibly profilin, bind with similar affinity to PI-4,5-P_2 and PI-3,4,5-P_3 and thus are very likely regulated in situ by PI-4,5-P_2, correlating with in situ data indicating that increased synthesis of PI-4,5-P_2 results in actin assembly.

4. Proteins involved in signaling (phospholipases, hydrolyzing PIPs, proteins regulating functions of small GTP binding proteins, protein kinases) are, with very few exceptions, regulated specifically by binding of products of PI 3-kinase (PI-3,4,5-P_3, PI-3,4-P_2) to their PH domains.

5. The cytoskeleton may be regulated indirectly by products of PI 3-kinase via signaling proteins with specificity for these products.

6. Proteins that activate GTP-binding proteins and those that promote formation of the inactive forms are stimulated by products of PI 3-kinase, presumably ensuring a localized and transient activation of small GTPases.

7. PIPs affect functions of proteins involved in membrane trafficking (spectrin, dynamin, synaptotagmin, myosin 10, annexin 2), correlating with data showing that membrane insertion and retrieval can be regulated by plasma membrane concentrations of PIPs (Czech 2003, Wenk & De Camilli 2004).

FUTURE ISSUES TO BE RESOLVED

1. The role of lipid rafts in organizing PIPs will have to be explored.

2. Imaging techniques that allow direct visualization of protein interactions with specific PIPs in living cells during migration, exocytosis, etc, will have to be developed.

GLOSSARY

FERM domain: 4.1-ezrin-radixin-moesin domain that is implicated in both protein-protein and protein lipid interactions

PH domain: pleckstrin homology domain that can mediate binding of proteins to specific types of PIPs

PI 3-kinase: phosphatidylinositol 3-kinase, a family of enzymes that phosphorylate PIPs at position 3 of the inositol ring

PIP 5-kinase: phosphatidylinositol 4-phosphate 5-kinase; a family of enzymes that generate phosphatidylinositol 4,5-bisphosphate from phosphatidylinositol 4-monophosphate

PIPs: phosphoinositide phosphates; phosphorylated derivatives of the phospholipid phosphatidylinositol

ACKNOWLEDGMENTS

I thank E. Sigel for preparing the figures and for careful reading of the manuscript. My work is supported by the Swiss National Science Foundation, by the Novartis Foundation (Basel, Switzerland), and by the Bernese Cancer League.

LITERATURE CITED

Achiriloaie M, Barylko B, Albanesi JP. 1999. Essential role of the dynamin pleckstrin homology domain in receptor-mediated endocytosis. *Mol. Cell Biol.* 19:1410–15

Babiychuk EB, Draeger A. 2000. Annexins in cell membrane dynamics. Ca^{2+}-regulated association of lipid microdomains. *J. Cell Biol.* 150:1113–24

Bai J, Tucker WC, Chapman ER. 2004. PIP_2 increases the speed of response of synaptotagmin and steers its membrane-penetration activity toward the plasma membrane. *Nat. Struct. Mol. Biol.* 11:36–44

Bakolitsa C, Cohen DM, Bankston LA, Bobkov AA, Cadwell GW, et al. 2004. Structural basis for vinculin activation at sites of cell adhesion. *Nature* 430:583–86

Bakolitsa C, de Pereda JM, Bagshaw CR, Critchley DR, Liddington RC. 1999. Crystal structure of the vinculin tail suggests a pathway for activation. *Cell* 99:603–13

Barret C, Roy C, Montcourrier P, Mangeat P, Niggli V. 2000. Mutagenesis of the phosphatidylinositol 4,5-bisphosphate (PIP_2) binding site in the NH_2-terminal domain of ezrin correlates with its altered cellular distribution. *J. Cell Biol.* 151:1067–80

Benesch S, Lommel S, Steffen A, Stradal TE, Scaplehorn N, et al. 2002. Phosphatidylinositol 4,5-biphosphate (PIP_2)-induced vesicle movement depends on N-WASP and involves Nck, WIP, and Grb2. *J. Biol. Chem.* 277:37771–76

Berg JS, Derfler BH, Pennisi CM, Corey DP, Cheney RE. 2000. Myosin-X, a novel myosin with pleckstrin homology domains, associates with regions of dynamic actin. *J. Cell Sci.* 113:3439–51

Biswas RS, Baker D, Hruska KA, Chellaiah MA. 2004. Polyphosphoinositides-dependent regulation of the osteoclast actin cytoskeleton and bone resorption. *BMC Cell Biol.* 5:19–39

Bompard G, Caron E. 2004. Regulation of WASP/WAVE proteins: making a long story short. *J. Cell Biol.* 166:957–62

In this study the crystal structure of full-length vinculin is presented. The data provide important novel information on the mechanism of activation of vinculin.

Bompard G, Martin M, Roy C, Vignon F, Freiss G. 2003. Membrane targeting of protein tyrosine phosphatase PTPL1 through its FERM domain via binding to phosphatidylinositol 4,5-biphosphate. *J. Cell Sci.* 116:2519–30

Bretscher A, Edwards K, Fehon RG. 2002. ERM proteins and merlin: integrators at the cell cortex. *Nat. Rev. Mol. Cell Biol.* 3:586–99

Burridge K, Wennerberg K. 2004. Rho and Rac take center stage. *Cell* 116:167–79

Carpenter CL. 2004. Btk-dependent regulation of phosphoinositide synthesis. *Biochem. Soc. Trans.* 32:326–29

Chellaiah M, Hruska K. 1996. Osteopontin stimulates gelsolin-associated phosphoinositide levels and phosphatidylinositol triphosphate-hydroxyl kinase. *Mol. Biol. Cell* 7:743–53

Corgan AM, Singleton C, Santoso CB, Greenwood JA. 2004. Phosphoinositides differentially regulate α-actinin flexibility and function. *Biochem. J.* 378:1067–72

Cox D, Berg JS, Cammer M, Chinegwundoh JO, Dale BM, et al. 2003. Myosin X is a downstream effector of PI(3)K during phagocytosis. *Nat. Cell Biol.* 4:469–77

Cozier GE, Bouyoucef D, Cullen PJ. 2003. Engineering the phosphoinositide-binding profile of a class I pleckstrin homology domain. *J. Biol. Chem.* 278:39489–96

Cozier GE, Lockyer PJ, Reynolds JS, Kupzig S, Bottomley JR, et al. 2000. GAP1[IP4BP] contains a novel group I pleckstrin homology domain that directs constitutive plasma membrane association. *J. Biol. Chem.* 275:28261–68

Cram EJ, Schwarzbauer JE. 2004. The talin wags the dog: new insights into integrin activation. *Trends Cell Biol.* 14:55–57

Czech MP. 2003. Dynamics of phosphoinositides in membrane retrieval and insertion. *Annu. Rev. Physiol.* 65:791–815

DeMali KA. 2004. Vinculin-a dynamic regulator of cell adhesion. *Trends Biochem. Sci.* 29:565–67

DeMali KA, Barlow CA, Burridge K. 2002. Recruitment of the Arp2/3 complex to vinculin: coupling membrane protrusion to matrix adhesion. *J. Cell Biol.* 159:881–91

DesMarais V, Ghosh M, Eddy R, Condeelis J. 2005. Cofilin takes the lead. *J. Cell. Sci.* 118:19–26

Di Paolo G, Moskowitz HS, Gipson K, Wenk MR, Voronov S, et al. 2004. Impaired PtdIns(4,5)P$_2$ synthesis in nerve terminals produces defects in synaptic vesicle trafficking. *Nature* 431:415–22

Di Paolo G, Pellegrini L, Letinic K, Cestra G, Zoncu R, et al. 2002. Recruitment and regulation of phosphatidylinositol phosphate kinase type 1γ by the FERM domain of talin. *Nature* 420:85–89

Djinovic-Carugo K, Gautel M, Ylanne J, Young P. 2002. The spectrin repeat: a structural platform for cytoskeletal protein assemblies. *FEBS Lett.* 513:119–23

Doughman RL, Firestone AJ, Anderson RA. 2003. Phosphatidylinositol phosphate kinases put PI4,5P$_2$ in its place. *J. Membr. Biol.* 194:77–89

Feng L, Mejillano M, Yin HL, Chen J, Prestwich GD. 2001. Full-contact domain labeling: identification of a novel phosphoinositide binding site on gelsolin that requires the complete protein. *Biochemistry* 40:904–13

Fievet BT, Gautreau A, Roy C, Del Maestro L, Mangeat P, et al. 2004. Phosphoinositide binding and phosphorylation act sequentially in the activation mechanism of ezrin. *J. Cell Biol.* 164:653–59

Fraley TS, Tran TC, Corgan AM, Nash CA, Hao J, et al. 2003. Phosphoinositide binding inhibits α-actinin bundling activity. *J. Biol. Chem.* 278:24039–45

Fukami K, Furuhashi K, Inagaki M, Endo T, Hatano S, et al. 1992. Requirement of phosphatidylinositol 4,5-bisphosphate for alpha-actinin function. *Nature* 359:150–52

In this work, the phosphoinositide binding selectivity of a PH domain of a Ras GTPase activating protein is modified by mutation of one residue, resulting in an altered cellular location.

This study demonstrates that in vivo interaction of ezrin with PIPs is required for phosphorylation and subsequent conformational activation rather than for plasma membrane targeting of this protein.

Fukami K, Sawada N, Endo T, Takenawa T. 1996. Identification of a phosphatidylinositol 4,5-bisphosphate-binding site in chicken skeletal muscle alpha-actinin. *J. Biol. Chem.* 271:2646–50

Garcia-Alvarez B, de Pereda JM, Calderwood DA, Ulmer TS, Critchley D, et al. 2003. Structural determinants of integrin recognition by talin. *Mol. Cell* 11:49–58

Gerke V, Moss SE. 2002. Annexins: from structure to function. *Physiol. Rev.* 82:331–71

Goldschmidt-Clermont PJ, Kim JW, Machesky LM, Rhee SG, Pollard TD. 1991. Regulation of phospholipase C-γ1 by profilin and tyrosine phosphorylation. *Science* 251:1231–33

Goldschmidt-Clermont PJ, Machesky LM, Baldassare JJ, Pollard TD. 1990. The actin-binding protein profilin binds to PIP$_2$ and inhibits its hydrolysis by phospholipase C. *Science* 247:1575–78

Greenwood JA, Theibert AB, Prestwich GD, Murphy-Ullrich JE. 2000. Restructuring of focal adhesion plaques by PI 3-kinase. Regulation by PtdIns (3,4,5)-P$_3$ binding to α-actinin. *J. Cell Biol.* 150:627–42

Guttman JA, Janmey P, Vogl AW. 2002. Gelsolin—evidence for a role in turnover of junction-related actin filaments in Sertoli cells. *J. Cell Sci.* 115:499–505

Hamada K, Shimizu T, Matsui T, Tsukita S, Hakoshima T. 2000. Structural basis of the membrane-targeting and unmasking mechanisms of the radixin FERM domain. *EMBO J.* 19:4449–62

Hamada K, Shimizu T, Yonemura S, Tsukita S, Hakoshima T. 2003. Structural basis of adhesion-molecule recognition by ERM proteins revealed by the crystal structure of the radixin-ICAM-2 complex. *EMBO J.* 22:502–14

Hayes MJ, Merrifield CJ, Shao D, Ayala-Sanmartin J, Schorey CD, et al. 2004. Annexin 2 binding to phosphatidylinositol 4,5-bisphosphate on endocytic vesicles is regulated by the stress response pathway. *J. Biol. Chem.* 279:14157–64

Hill E, van Der Kaay J, Downes CP, Smythe E. 2001. The role of dynamin and its binding partners in coated pit invagination and scission. *J. Cell Biol.* 152:309–23

Hirono M, Denis CS, Richardson GP, Gillespie PG. 2004. Hair cells require phosphatidylinositol 4,5-bisphosphate for mechanical transduction and adaptation. *Neuron* 44:309–20

Hoeflich KP, Tsukita S, Hicks L, Kay CM, Ikura M. 2003. Insights into a single rod-like helix in activated radixin required for membrane-cytoskeletal cross-linking. *Biochemistry* 42:11634–41

Holz RW, Hlubek MD, Sorensen SD, Fisher SK, Balla T, et al. 2000. A pleckstrin homology domain specific for phosphatidylinositol 4, 5-bisphosphate (PtdIns-4,5-P$_2$) and fused to green fluorescent protein identifies plasma membrane PtdIns-4,5-P$_2$ as being important in exocytosis. *J. Biol. Chem.* 275:17878–85

Huang S, Lifshitz L, Patki-Kamath V, Tuft R, Fogarty K, Czech MP. 2004. Phosphatidylinositol-4,5-bisphosphate-rich plasma membrane patches organize active zones of endocytosis and ruffling in cultured adipocytes. *Mol. Cell Biol.* 24:9102–23

Hurley JH, Meyer T. 2001. Subcellular targeting by membrane lipids. *Curr. Opin. Cell Biol.* 13:146–52

Hyvönen M, Macias MJ, Nilges M, Oschkinat H, Saraste M, et al. 1995. Structure of the binding site for inositol phosphates in a PH domain. *EMBO J.* 14:4676–85

Insall RH, Weiner OD. 2001. PIP$_3$, PIP$_2$, and cell movement-similar messages, different meanings? *Dev. Cell* 1:743–47

Itoh T, Takenawa T. 2002. Phosphoinositide-binding domains: functional units for temporal and spatial regulation of intracellular signalling. *Cell. Signal.* 14:733–43

Janmey PA, Lindberg U. 2004. Cytoskeletal regulation: rich in lipids. *Nat. Rev. Mol. Cell Biol.* 5:658–66

Johnson RP, Niggli V, Durrer P, Craig SW. 1998. A conserved motif in the tail domain of vinculin mediates association with and insertion into acidic phospholipid bilayers. *Biochemistry* 37:10211–22

Kanzaki M, Furukawa M, Raab W, Pessin JE. 2004. Phosphatidylinositol 4,5-bisphosphate regulates adipocyte actin dynamics and GLUT4 vesicle recycling. *J. Biol. Chem.* 279:30622–33

Katan M. 1998. Families of phosphoinositide-specific phospholipase C: structure and function. *Biochim. Biophys. Acta* 1436:5–17

Kavran JM, Klein DE, Lee A, Falasca M, Isakoff SJ, et al. 1998. Specificity and promiscuity in phosphoinositide binding by pleckstrin homology domains. *J. Biol. Chem.* 273:30497–508

Kimber WA, Deak M, Prescott AR, Alessi DR. 2003. Interaction of the protein tyrosine phosphatase PTPL1 with the PtdIns(3,4)P_2-binding adaptor protein TAPP1. *Biochem. J.* 376:525–35

Klein DE, Lee A, Frank DW, Marks MS, Lemmon MA. 1998. The pleckstrin homology domains of dynamin isoforms require oligomerization for high affinity phosphoinositide binding. *J. Biol. Chem.* 273:27725–33

Komander D, Fairservice A, Deak M, Kular GS, Prescott AR, et al. 2004. Structural insights into the regulation of PDK1 by phosphoinositides and inositol phosphates. *EMBO J.* 23:3918–28

Krugmann S, Anderson KE, Ridley SH, Risso N, McGregor A, et al. 2002. Identification of ARAP3, a novel PI3K effector regulating both Arf and Rho GTPases, by selective capture on phosphoinositide affinity matrices. *Mol. Cell* 9:95–108

Krugmann S, Williams R, Stephens L, Hawkins PT. 2004. ARAP3 is a PI3K- and Rap-regulated GAP for RhoA. *Curr. Biol.* 14:1380–84

Lambrechts A, Jonckheere V, Dewitte D, Vandekerckhove J, Ampe C. 2002. Mutational analysis of human profilin I reveals a second PI(4,5)-P_2 binding site neighbouring the poly(L-proline) binding site. *BMC Biochem.* 3:12–23

Lee HS, Bellin RM, Walker DL, Patel B, Powers P, et al. 2004. Characterization of an actin-binding site within the talin FERM domain. *J. Mol. Biol.* 343:771–84

Lee SJ, Montell C. 2004. Light-dependent translocation of visual arrestin regulated by the NINAC myosin III. *Neuron* 43:95–103

Lemmon MA, Ferguson KM, Abrams CS. 2002. Pleckstrin homology domains and the cytoskeleton. *FEBS Lett.* 513:71–76

Liepina I, Czaplewski C, Janmey P, Liwo A. 2003. Molecular dynamics study of a gelsolin-derived peptide binding to a lipid bilayer containing phosphatidylinositol 4,5-bisphosphate. *Biopolymers* 71:49–70

Ling K, Doughman RL, Firestone AJ, Bunce MW, Anderson RA. 2002. Type Iγ phosphatidylinositol phosphate kinase targets and regulates focal adhesions. *Nature* 420:89–93

Lorra C, Huttner WB. 1999. The mesh hypothesis of Golgi dynamics. *Nat. Cell. Biol.* 1:E113-15

Lu PJ, Shieh WR, Rhee SG, Yin HL, Chen CS. 1996. Lipid products of phosphoinositide 3-kinase bind human profilin with high affinity. *Biochemistry* 35:14027–34

Mackler JM, Reist NE. 2001. Mutations in the second C2 domain of synaptotagmin disrupt synaptic transmission at *Drosophila* neuromuscular junctions. *J. Comp. Neurol.* 436:4–16

This review puts together recent data on localized production of PIPs and their role in actin assembly.

Martel V, Racaud-Sultan C, Dupe S, Marie C, Paulhe F, et al. 2001. Conformation, localization, and integrin binding of talin depend on its interaction with phosphoinositides. *J. Biol. Chem.* 276:21217–27

Mashanov GI, Tacon D, Peckham M, Molloy JE. 2004. The spatial and temporal dynamics of pleckstrin homology domain binding at the plasma membrane measured by imaging single molecules in live mouse myoblasts. *J. Biol. Chem.* 279:15274–80

McManus EJ, Collins BJ, Ashby PR, Prescott AR, Murray-Tait V, et al. 2004. The in vivo role of PtdIns(3,4,5)P$_3$ binding to PDK1 PH domain defined by knockin mutation. *EMBO J.* 23:2071–82

Milburn CC, Deak M, Kelly SM, Price NC, Alessi DR, Van Aalten DM. 2003. Binding of phosphatidylinositol 3,4,5-trisphosphate to the pleckstrin homology domain of protein kinase B induces a conformational change. *Biochem. J.* 375:531–38

Morgan JR, Di Paolo G, Werner H, Shchedrina VA, Pypaert M, et al. 2004. A role for talin in presynaptic function. *J. Cell Biol.* 167:43–50

Nayal A, Webb DJ, Horwitz AF. 2004. Talin: an emerging focal point of adhesion dynamics. *Curr. Opin. Cell Biol.* 16:94–98

Niggli V. 2001. Structural properties of lipid-binding sites in cytoskeletal proteins. *Trends Biochem. Sci.* 26:604–11

Niggli V, Gimona M. 1993. Evidence for a ternary interaction between alpha-actinin, (meta)vinculin and acidic-phospholipid bilayers. *Eur. J. Biochem.* 213:1009–15

Oikawa T, Yamaguchi H, Itoh T, Kato M, Ijuin T, et al. 2004. PtdIns(3,4,5)P$_3$ binding is necessary for WAVE2-induced formation of lamellipodia. *Nat. Cell Biol.* 6:420–26

Ojala PJ, Paavilainen V, Lappalainen P. 2001. Identification of yeast cofilin residues specific for actin monomer and PIP$_2$ binding. *Biochemistry* 40:15562–69

Rescher U, Ruhe D, Ludwig C, Zobiack N, Gerke V. 2004. Annexin 2 is a phosphatidylinositol (4,5)-bisphosphate binding protein recruited to actin assembly sites at cellular membranes. *J. Cell Sci.* 117:3473–80

Rhee SG. 2001. Regulation of phosphoinositide-specific phospholipase C. *Annu. Rev. Biochem.* 70:281–312

Rohatgi R, Ho HY, Kirschner MW. 2000. Mechanism of N-WASP activation by CDC42 and phosphatidylinositol 4, 5-bisphosphate. *J. Cell Biol.* 150:1299–310

Saunders R, Jennings L, Sutton DH, Barsukov I, Holt MR, et al. 2003. Vinculin mutations that decrease PIP$_2$ binding lead to protein mislocalisation and failure to rescue cell spreading defects in vinculin null fibroblasts. *Mol. Biol. Cell* 14:63a (Abstr.)

Schafer DA, Weed SA, Binns D, Karginov AV, Parsons JT. 2002. Dynamin2 and cortactin regulate actin assembly and filament organization. *Curr. Biol.* 12:1852–57

Scheid MP, Woodgett JR. 2003. Unravelling the activation mechanisms of protein kinase B/Akt. *FEBS Lett.* 546:108–12

Schiavo G, Gu QM, Prestwich GD, Sollner TH, Rothman JE. 1996. Calcium-dependent switching of the specificity of phosphoinositide binding to synaptotagmin. *Proc. Natl. Acad. Sci. USA* 93:13327–32

Seelig A, Blatter XL, Frentzel A, Isenberg G. 2000. Phospholipid binding of synthetic talin peptides provides evidence for an intrinsic membrane anchor of talin. *J. Biol. Chem.* 275:17954–61

Shinohara M, Terada Y, Iwamatsu A, Shinohara A, Mochizuki N, et al. 2002. SWAP-70 is a guanine-nucleotide-exchange factor that mediates signalling of membrane ruffling. *Nature* 416:759–63

This work provides evidence for the in vivo relevance of the specific interaction of the protein kinase PDK1 with PI-3,4,5-P$_3$ using knockin technology.

This study provides in situ evidence for the role of the specific interaction of WAVE2 with PI-3,4,5-P$_3$ in membrane recruitment and Rac-induced lamellipodia formation.

Evidence is provided for the specific activation of the GEF SWAP-70 by PI-3,4,5-P$_3$ in vitro and in situ and its role in Rac-dependent membrane ruffling.

Siddhanta A, Radulescu A, Stankewich MC, Morrow JS, Shields D. 2003. Fragmentation of the Golgi apparatus. A role for βIII spectrin and synthesis of phosphatidylinositol 4,5-bisphosphate. *J. Biol. Chem.* 278:1957–65

Sivalenka RR, Jessberger R. 2004. SWAP-70 regulates c-kit-induced mast cell activation, cell-cell adhesion, and migration. *Mol. Cell Biol.* 24:10277–88

Skare P, Karlsson R. 2002. Evidence for two interaction regions for phosphatidylinositol(4,5)-bisphosphate on mammalian profilin I. *FEBS Lett.* 522:119–24

Smith WJ, Nassar N, Bretscher A, Cerione RA, Karplus PA. 2003. Structure of the active N-terminal domain of Ezrin. Conformational and mobility changes identify keystone interactions. *J. Biol. Chem.* 278:4949–56

Sohn RH, Chen J, Koblan KS, Bray PF, Goldschmidt-Clermont PJ. 1995. Localization of a binding site for phosphatidylinositol 4,5-bisphosphate on human profilin. *J. Biol. Chem.* 270:21114–20

Strissel KJ, Arshavsky VY. 2004. Myosin III illuminates the mechanism of arrestin translocation. *Neuron* 43:2–4

Suchy SF, Nussbaum RL. 2002. The deficiency of PIP$_2$ 5-phosphatase in Lowe syndrome affects actin polymerization. *Am. J. Hum. Genet.* 71:1420–27

Takenawa T, Itoh T. 2001. Phosphoinositides, key molecules for regulation of actin cytoskeletal organization and membrane traffic from the plasma membrane. *Biochim. Biophys. Acta* 1533:190–206

Takenawa T, Miki H. 2001. WASP and WAVE family proteins: key molecules for rapid rearrangement of cortical actin filaments and cell movement. *J. Cell Sci.* 114:1801–9

Tolias KF, Hartwig JH, Ishihara H, Shibasaki Y, Cantley LC. 2000. Type Iα phosphatidylinositol-4-phosphate 5-kinase mediates Rac-dependent actin assembly. *Curr. Biol.* 10:153–56

Verheyen EM, Cooley L. 1994. Profilin mutations disrupt multiple actin-dependent processes during *Drosophila* development. *Development* 120:717–28

Wang YJ, Li WH, Wang J, Xu K, Dong P, et al. 2004. Critical role of PIP5KIγ87 in InsP$_3$-mediated Ca^{2+} signaling, *J. Cell Biol.* 167:1005–10

Wenk MR, De Camilli P. 2004. Protein-lipid interactions and phosphoinositide metabolism in membrane traffic: insights from vesicle recycling in nerve terminals. *Proc. Natl. Acad. Sci. USA* 101:8262–69

Witke W, Sutherland JD, Sharpe A, Arai M, Kwiatkowski DJ. 2001. Profilin I is essential for cell survival and cell division in early mouse development. *Proc. Natl. Acad. Sci. USA* 98:3832–36

Wittenmayer N, Jandrig B, Rothkegel M, Schluter K, Arnold W, et al. 2004. Tumor suppressor activity of profilin requires a functional actin binding site. *Mol. Biol. Cell* 15:1600–8

Xian W, Janmey PA. 2002. Dissecting the gelsolin-polyphosphoinositide interaction and engineering of a polyphosphoinositide-sensitive gelsolin C-terminal half protein. *J. Mol. Biol.* 322:755–71

Yamamoto M, Hilgemann DH, Feng S, Bito H, Ishihara H, et al. 2001. Phosphatidylinositol 4,5-bisphosphate induces actin stress-fiber formation and inhibits membrane ruffling in CV1 cells. *J. Cell Biol.* 152:867–76

Ye K, Snyder SH. 2004. PIKE GTPase: a novel mediator of phosphoinositide signaling. *J. Cell Sci.* 117:155–61

Yonezawa S, Yoshizaki N, Sano M, Hanai A, Masaki S, et al. 2003. Possible involvement of myosin-X in intercellular adhesion: importance of serial pleckstrin homology regions for intracellular localization. *Dev. Growth Differ.* 45:175–85

Young P, Gautel M. 2000. The interaction of titin and α-actinin is controlled by a phospholipid-regulated intramolecular pseudoligand mechanism. *EMBO J.* 19:6331–40

Yu F-X, Sun H-Q, Janmey PA, Yin HL. 1992. Identification of a polyphosphoinositide-binding sequence in an actin-monomer-binding domain of gelsolin. *J. Biol. Chem.* 267:14616–21

Zemljic-Harpf AE, Ponrartana S, Avalos RT, Jordan MC, Roos KP, et al. 2004. Heterozygous inactivation of the vinculin gene predisposes to stress-induced cardiomyopathy. *Am. J. Pathol.* 165:1033–44

Principles of Lysosomal Membrane Digestion: Stimulation of Sphingolipid Degradation by Sphingolipid Activator Proteins and Anionic Lysosomal Lipids

Thomas Kolter and Konrad Sandhoff

Kekulé-Institut für Organische Chemie und Biochemie der Universität, 53121 Bonn, Germany; email: tkolter@uni-bonn.de; sandhoff@uni-bonn.de

Annu. Rev. Cell Dev. Biol.
2005. 21:81–103

First published online as a
Review in Advance on
May 16, 2005

The *Annual Review of
Cell and Developmental
Biology* is online at
http://cellbio.annualreviews.org

doi: 10.1146/
annurev.cellbio.21.122303.120013

Copyright © 2005 by
Annual Reviews. All rights
reserved

1081-0706/05/1110-
0081$20.00

Key Words

glycosphingolipids, GM2-activator, lysosomes, saposins, sphingolipids

Abstract

Sphingolipids and glycosphingolipids are membrane components of eukaryotic cell surfaces. Their constitutive degradation takes place on the surface of intra-endosomal and intra-lysosomal membrane structures. During endocytosis, these intra-lysosomal membranes are formed and prepared for digestion by a lipid-sorting process during which their cholesterol content decreases and the concentration of the negatively charged bis(monoacylglycero)phosphate (BMP)—erroneously also called lysobisphosphatidic acid (LBPA)—increases. Glycosphingolipid degradation requires the presence of water-soluble acid exohydrolases, sphingolipid activator proteins, and anionic phospholipids like BMP. The lysosomal degradation of sphingolipids with short hydrophilic head groups requires the presence of sphingolipid activator proteins (SAPs). These are the saposins (Saps) and the GM2 activator protein. Sphingolipid activator proteins are membrane-perturbing and lipid-binding proteins with different specificities for the bound lipid and the activated enzyme-catalyzed reaction. Their inherited deficiency leads to sphingolipid- and membrane-storage diseases. Sphingolipid activator proteins not only facilitate glycolipid digestion but also act as glycolipid transfer proteins facilitating the association of lipid antigens with immunoreceptors of the CD1 family.

Contents

PRINCIPLES OF LYSOSOMAL MEMBRANE DIGESTION

Lysosomal Degradation of Complex Biomolecules

The constitutive degradation of macromolecules and of smaller substances that are composed of cleavable building blocks occurs in the acidic subcellular compartments, the endosomes and the lysosomes. Cellular and foreign components reach these organelles via different routes, by endocytosis, phagocytosis, autophagy, or direct transport. Inside the lysosomes, hydrolytic enzymes with acidic pH-optima cleave macromolecules such as proteins, polysaccharides, and nucleic acids, but also substances with complex structures such as glycoconjugates and phospholipids. The building blocks formed during degradation are able to leave the lysosomes either via diffusion or with the aid of specialized transport systems. Outside the lysosomal compartment, for instance in the cytosol, the building blocks can be utilized for the resynthesis of complex molecules or can be further degraded to provide metabolic energy. Inheritable disorders affecting proteins acting in these degradation pathways lead to lysosomal storage diseases and are characterized by the accumulation of nondegradable enzyme substrates. They can be classified according to the stored substances, e.g., as sphingolipidoses, mucopolysaccharidoses, mucolipidoses, glycoprotein and glycogen storage diseases (Suzuki 1994).

Lysosomal Degradation of Membranes

Membranes are essentially composed of amphiphilic lipids and proteins. Both components have to be degraded in the course of membrane digestion inside lysosomes. Proteins and glycoproteins, glycerolipids, sphingolipids, and glycosphingolipids are cleaved into their building blocks: amino acids, monosaccharides, sialic acids, glycerol, fatty acids, and a sphingoid base, e.g., sphingosine. Also, cholesterol is liberated during this process. In contrast to the lysosomal degradation of soluble macromolecules such as proteins and oligosaccharides, in the case of membrane digestion the question arises: Why and in which way are some membranes within the lysosomes degraded, whereas other membranes, the limiting membranes of endosomes and lysosomes, remain unaffected and survive. An examination of the molecular mechanisms of membrane digestion reveals a complex machinery composed of lipids of inner lysosomal membranes and lysosomal proteins that ensures selective degradation of inner membranes containing foreign material and components of the former plasma membrane. Sorting of membrane lipids during endocytosis and the maturation of endosomes, together with the selectivity of degrading enzymes and activator proteins, turns out to be crucial for this process. Basic insight into this highly complex process came from the investigation of inherited diseases that result from defects of glycosphingolipid catabolism; essentials of this metabolic pathway are summarized below.

Sphingolipids and Glycosphingolipids

Major insight into the process of membrane digestion came from the investigation of glycosphingolipid catabolism. Glycosphingolipids (**Figure 1**) are ubiquitously expressed on eukaryotic cell surfaces. They are composed of a hydrophobic ceramide moiety and an extracytoplasmic oligosaccharide chain (Kolter & Sandhoff 1999). Combination of different carbohydrate residues, anomeric linkages, and additional modifications of the carbohydrate and lipid moiety lead to a variety of naturally occurring glycosphingolipids that are biosynthetically formed in a combinatorial manner (Kolter et al. 2002). Glycosphingolipid structures depend on species and cell type. They can be classified into series that are characteristic for a group of evolutionarily related organisms. Neuronal cells, especially of the central nervous system, are rich in glycosphingolipids of the ganglio-series, the sialic acid-containing gangliosides. Their lysosomal degradation is particularly well understood on the molecular level and is discussed in more detail below.

It is believed that glycosphingolipids of the plasma membrane, together with cholesterol, the phosphosphingolipid sphingomyelin, and glycosylphosphatidyl-anchored proteins, segregate into functional microdomains, often called rafts (Simons & Ikonen 1997, Brown & London 2000, Munro 2003). The raft

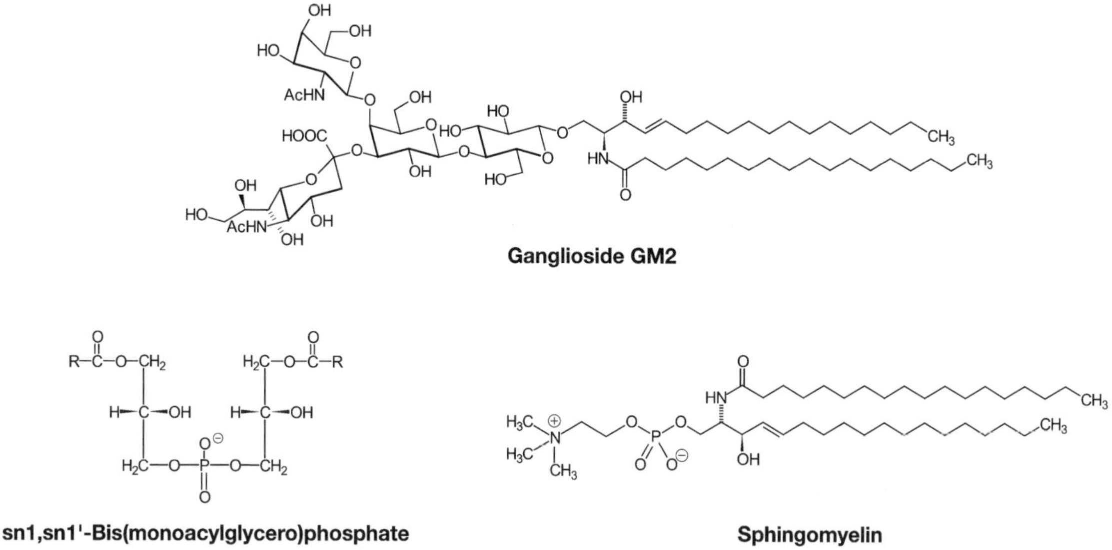

Ganglioside GM2

sn1,sn1'-Bis(monoacylglycero)phosphate

Sphingomyelin

Figure 1

Structures of ganglioside GM2, bis(monoacylglycero)phosphate (BMP), and sphingomyelin.

concept originated from the differential solubilities of membrane components in detergent-containing solutions, but there is still no convincing proof for their existence under physiological conditions (Heerklotz 2002, Heerklotz et al. 2003).

Glycosphingolipid biosynthesis starts with the formation of ceramide at the membranes of the endoplasmic reticulum (ER) (Merrill 2002) and continues at the membranes of the Golgi apparatus with the stepwise addition of single carbohydrate residues (Kolter et al. 2002). Addition of a glucose residue in β-glycosidic linkage to ceramide occurs at the cytoplasmic face of the Golgi apparatus; then, glucosylceramide undergoes a transversal membrane translocation, and the carbohydrate chain is elongated by membrane-resident glycosyltransferases with their active sites in the lumen of the Golgi apparatus. As a consequence, the oligosaccharide moieties of most complex glycosphingolipids face the extracytoplasmic space on the plasma membrane and the lumen of cellular organelles.

Glycosphingolipid biosynthesis is coupled to the intracellular movement of its biosynthetic intermediates and final products to the plasma membrane (van Meer & Lisman 2002). The combinatorial variety of naturally occurring glycolipids can be largely attributed to the combination of glycosyltransferase activities found in different species and cell types.

Similar to phospholipids (Opekarová & Tanner 2003), glycosphingolipids can influence the activity of integral membrane proteins (Yamashita et al. 2003). The activities of peripheral membrane proteins also can be influenced by the lipid composition of the membrane. Because such membrane-binding proteins are crucial for lysosomal sphingolipid degradation, the influence of the lipid composition of the internal membranes of the lysosomal compartment on their functions has to be discussed.

The investigation of lysosomal sphingolipid degradation led to the discovery of principles governing membrane digestion, and key topics of this pathway are discussed in more detail below.

LYSOSOMAL SPHINGOLIPID DEGRADATION

The lysosomal degradation of glycosphingolipids is a sequential pathway that starts with the stepwise release of monosaccharide units from the nonreducing end of the oligosaccharide chain (**Figure 2**). These reactions are catalyzed by exohydrolases with acidic pH-optima. Several of these enzymes need the assistance of small glycoprotein cofactors, the SAPs (Sandhoff et al. 2001). In addition to enzymes and activator proteins, the membrane must have the right lipid composition in order to be degraded. This means a low cholesterol content and the presence of the negatively charged lysosomal lipid BMP.

The stepwise cleavage of the hydrophilic head groups from glycosphingolipids ultimately generates sphingosine, fatty acids, monosaccharides, sialic acids, and sulfate. These final degradation products are able to leave the lysosome. Members of other glycosphingolipid series enter the degradation pathway of gangliosides at the lactosylceramide stage. Glycosphingolipids of the gala-series, but also of sphingomyelin, are degraded to ceramide.

Most hydrolases are water-soluble polycations at a lysosomal pH of less than 5.0. They bind to negatively charged membranes, but hardly work on their membrane-bound substrates. A notable exception is acid sphingomyelinase, which slowly hydrolyzes membrane-bound sphingomyelin even in the absence of an activator protein, presumably due to its N-terminal saposin-homology domain (Linke et al. 2001a).

Non-glycosylated sphingolipids, such as ceramide and sphingomyelin (Goni & Alonso 2002), have non-lysosomal degradation steps that apparently do not need the assistance of an activator protein. A cytoplasmic glucosylceramide-cleaving enzyme, which is not deficient in Gaucher's disease,

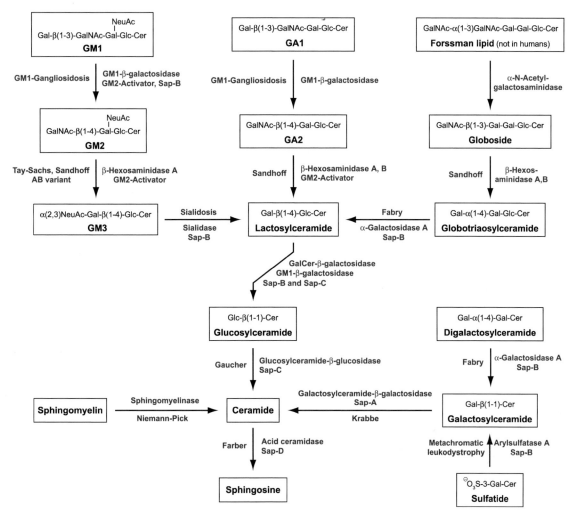

Figure 2

Degradation of selected sphingolipids in the lysosomes of the cells (modified from Kolter & Sandhoff 1998). The eponyms of individual inherited diseases (shown in red) are given. Activator proteins required for the respective degradation step in vivo are indicated. Variant AB: AB variant of GM2 gangliosidosis (deficiency of GM2-activator protein).

contributes to the degradation of the cytoplasmic glucosylceramide pool (van Weely et al. 1993).

Topology

For lysosomal membrane digestion, plasma membrane components are transported to the lysosomes within the process of endocytotic membrane flow. Vesicles of the endosomal/lysosomal compartment are formed via different routes, starting with the formation of clathrin-coated pits, non-clathrin-coated pits, caveolae, and others (Maxfield & McGraw 2004). The observation that the integrity of lysosomal- and endosomal-limiting membranes is preserved during the process of lysosomal degradation led necessarily to the assumption that two distinct pools of membranes must be present in the endosomal/lysosomal compartment (Fürst & Sandhoff 1992). Early reports about the ultrastructural examination of cells derived from patients with defects of glycosphingolipid catabolism indicated storage of intralysosomal lipid aggregates: nondegradable lipids accumulate as multivesicular storage bodies (MVB) in diseases such as the GM1-gangliosidosis (Suzuki & Chen 1968) or combined Sap deficiency (Harzer et al. 1989). These reports gave the first hint of a topo-logical differentiation of the two membrane pools (Fürst & Sandhoff 1992).

The following model for the topology of endocytosis and membrane digestion was initially proposed in 1992 (Fürst & Sandhoff 1992) and has been further supported by a series of observations (Sandhoff & Kolter 1996). According to this view, plasma membrane components, lipids and proteins, reach the lysosomal compartment either as intra-endosomal membranes or as part of the limiting membrane (**Figure 3**). Both membranes differ in their lipid and protein

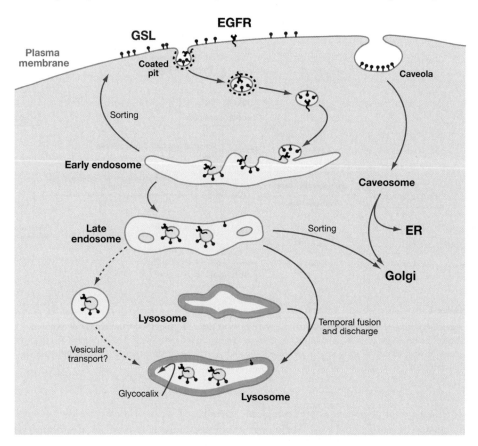

Figure 3

Model of endocytosis and lysosomal digestion of membranes (modified from Sandhoff & Kolter 1996). Glycosphingolipids (GSL) are highlighted on the plasma membrane and on internal membranes. Parts of the plasma membrane, including GSLs, are incorporated into the membranes of intra-endosomal vesicles and membrane structures during endocytosis. The vesicles reach the lysosomal compartment when late endosomes are transiently fused with primary lysosomes and are degraded there. The lysosomal perimeter membrane is protected from degradation by a thick glycocalix. EGFR: epidermal growth factor receptor.

composition. The lysosomal leaflet of the limiting membrane is covered with a thick glycocalix that protects the membrane from attack by the membrane-degrading enzymes present in the lysosol. This glycocalix is formed by lysosomal integral and peripheral membrane proteins, which are highly N-glycosylated with polylactosamine units and therefore highly resistant toward lysosomal digestion (Eskelinen et al. 2003). Apparently, the degrading enzymes present within the lumen of the lysosomes cannot easily access their substrates through this glycocalix. Indeed, more than 30 years ago it was demonstrated that after its incorporation into the limiting lysosomal membrane, plasma membrane-derived ganglioside GM3 is protected from degradation (Henning & Stoffel 1973), whereas it is easily degraded during the constitutive process of membrane turnover (Tettamanti 2004). In addition to this membrane pool resistant to degradation, components originating from the plasma membrane transiently occur as components of intra-endosomal/intra-lysosomal membranes. The major part of membrane digestion has to proceed on the surface of these internal membrane structures. This second pool originates from the plasma membrane and reaches the lumen of the endosomes as intra-endosomal vesicles or other lipid aggregates and not as part of the perimeter membrane. Vesicles of this type accumulate in patients with sphingolipid-storage diseases (see above) and have been described by other groups in normal cells (e.g., van der Goot & Gruenberg 2002).

The occurrence of intra-endosomal membranes is not necessarily restricted to cells with impaired membrane digestion; they are regularly visible in the microscope as MVBs. MVB formation starts with inward budding of the limiting endosomal membrane (Hopkins et al. 1990). Lipids, as well as proteins, are sorted to either the internal or the limiting membrane. One sorting signal sufficient for protein targeting to internal vesicles of MVBs is the ubiquitinylation

of cargo proteins (Katzmann et al. 2001). This signal appears to be conserved between yeast and higher eukaryotes, but ubiquitin-independent factors have also been reported (Umebayashi 2003). This membrane segregation is accompanied by lipid sorting that prepares the internal membranes for the attack by the lysosomal degradation system (see below).

SPHINGOLIPID ACTIVATOR PROTEINS STIMULATE SPHINGOLIPID DEGRADATION

In vivo, the lysosomal degradation of membranes is dependent on certain proteins and lipids. It has been known for more than 30 years that glycosphingolipid degradation requires the presence of SAPs (for review, see Sandhoff et al. 2001). Whereas other membrane components, for example phospholipids, can apparently be degraded without cofactors that make them accessible to the degrading enzymes, degradation of glycosphingolipids with short carbohydrate chains of four or fewer sugars is critically dependent not only on lysosomal glycosidases but also on activator proteins and negatively charged lysosomal lipids. According to the topology of endocytosis discussed above, lysosomal enzymes cleave sphingolipid substrates that are part of intra-endosomal and intra-lysosomal membrane structures. In the absence of detergents that are able to solubilize the lipids, glycosphingolipids with short carbohydrate chains are not sufficiently accessible to the water-soluble enzymes present in the lysosol in the absence of membrane-perturbing activator proteins (Wilkening et al. 1998, 2000). In vitro, synthethic water-soluble GSLs with either short-chain fatty acids or no fatty acids (lysoGSLs) can already be hydrolyzed by the water-soluble enzymes in the absence of SAPs. Two genes are known to encode the SAPs: one encodes the GM2-activator protein, the other encodes the Sap-precursor protein, also called prosaposin (Sandhoff et al. 2001). This protein is post-translationally processed to four homologous mature proteins, Saps A-D, or saposins A-D. These activator proteins act on the intra-endosomal/intra-lysosomal membrane pool and lead to the selective degradation of membrane lipids without impairment of lysosomal integrity. Inherited deficiency of either lysosomal enzymes or SAPs leads to the accumulation of nondegradable membranes within the lysosomal compartment and to the development of sphingolipid-storage diseases (Kolter & Sandhoff 1998, Suzuki & Vanier 1999, Platt & Walkley 2004, Winchester 2004).

The GM2-Activator

The GM2-activator is a glycoprotein with a molecular mass of 17.6 kDa in its deglycosylated form (Sandhoff et al. 2001). It acts as a cofactor essential for the in vivo degradation of ganglioside GM2 (**Figures 1, 2, 4**) by β-hexosaminidase A (Conzelmann & Sandhoff 1979). β-hexosaminidases are dimeric isoenzymes formed by combination of two subunits, α and β, which differ in their substrate specificity (Hepbildikler et al. 2002). β-hexosaminidase A can cleave glycolipid substrates on membrane surfaces only if they extend far enough into the aqueous phase. Therefore, in the absence of detergents, the degradation of ganglioside GM2 occurs only in the presence of the GM2 activator protein. The crystal structure of β-hexosaminidase B (Maier et al. 2003, Mark et al. 2003) has been solved, so now the question of how the members of the three-component system might interact can be addressed on the molecular level.

The inherited deficiency of the GM2-activator protein leads to the AB variant of GM2-gangliosidoses, in which lipid accumulation in neuronal cells leads to the early death of the patients (Conzelmann & Sandhoff 1978). An X-ray crystallographic structure of the non-glycosylated protein expressed in *Escherichia coli* is available (Wright & Rastinejad 2000, Wright et al. 2003). According to this, the GM2-activator contains a hydrophobic cavity that harbors the ceramide moiety of ganglioside GM2. A detailed model of this mechanism, based on earlier considerations (Fürst & Sandhoff 1992), structural information (Wright et al. 2003), and photoaffinity labeling (Wendeler et al. 2004), is shown in **Figure 4**. To present ganglioside GM2 or related glycosphingolipids (e.g., GM1; Wilkening et al. 2000) to the active site of the corresponding degrading enzyme, the GM2-activator has to insert into the bilayer of intra-lysosomal lipid vesicles and lift the glycolipid out of the membrane. Therefore, it can be regarded as a weak detergent with high selectivity, which as a "liftase" (Fürst & Sandhoff 1992) forms stoichiometric, water-soluble glycolipid-protein complexes that are the physiological Michaelis-Menten substrates of β-hexosaminidase A (Conzelmann & Sandhoff 1979). Similar to other SAPs, the GM2-activator protein acts as a lipid transfer protein in vitro (**Figure 5a, Table 1**) that can carry lipids from donor to acceptor liposomes (Conzelmann et al. 1982). The transfer properties of activator proteins are crucial for the loading of lipid antigens to the immunoreceptors of the CD1 family (see below, **Table 1**). As is discussed below, the lipid composition of the GM2-containing membrane, as well as their lateral pressure, is important for degradation and for ensuring that only ganglioside GM2 in the internal membranes is digested.

In addition to the SAPs, other lipid-binding proteins are known (Malinina et al. 2004), including additional proteins of saposin-like structure (Munford et al. 1995); CERT, a protein that transfers ceramide from the ER to Golgi membranes (Hanada et al. 2003); fatty acid–binding proteins (Coe & Bernlohr 1998); and immunoreceptors of the CD1 family (see below). The three-dimensional structure of the complex between a cytoplasmic glycosphingolipid transfer protein, which can transport lactosylceramide, and its ligand has been reported

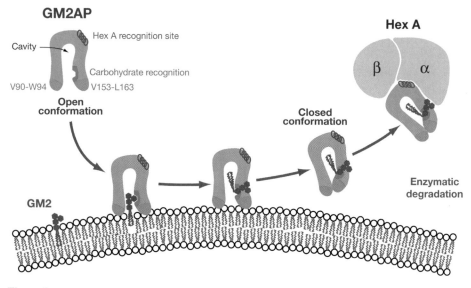

Figure 4

Model for GM2-activator-stimulated hydrolysis of ganglioside GM2 by human β-hexosaminidase A (Wendeler et al. 2004). The GM2AP contains a hydrophobic cavity, with dimensions that can accommodate the ceramide portion of GM2 and other lipids, lined by surface loops and a single short helix. The most flexible of the loops contains the substrate-binding site (V153–L163) and controls the entrance to the cavity, so that two conformations are possible: one open and one closed. The open empty activator binds to the membrane by using the hydrophobic loops and penetrates into the hydrophobic region of the bilayer. Then the lipid recognition site of the activator can interact with the substrate, and its ceramide portion can move inside the hydrophobic cavity. At this point, the conformation of the lipid-loaded activator may change to the closed one, thus the complex becomes more water soluble and leaves the membrane, exposing GM2 to the water-soluble enzyme to be degraded.

(Malinina et al. 2004). Immunoreceptors of the CD1 family present lipid antigens to T-lymphocytes (see below).

Saposins

The Saps or saposins A–D are four acidic, not enzymatically active, heat-stable and protease-resistant glycoproteins of about 8–11 kDa (for review, see Sandhoff et al. 2001). They belong to a family of saposin-like proteins with conserved three-dimensional folds (Munford et al. 1995), of which the solution structures of NK-lysin (Liepinsh et al. 1997) and the pore-forming peptide of *Entamoeba histolytica* (Hecht et al. 2004), Sap-C (de Alba et al. 2003), and the X-ray crystallographic structure of unglycosylated hu-

man recombinant Sap-B (Ahn et al. 2003) are known.

The proteins of this group carry out diverse functions, but share lipid binding- and membrane-perturbing properties. For example, the protozoan parasite *E. histolytica* expresses pore-forming proteins with a saposin-like structure, the amoebapores. Similar to eukaryotic NK-lysin and granulysin, these proteins are able to permeabilize the membranes of target cells, which accounts for their antimicrobial activity (Gutsmann et al. 2003). Although the four Saps share a high degree of homology and similar properties, they act differently and show different specificities.

Sap-A. Sap-A is required for the degradation of galactosylceramide by

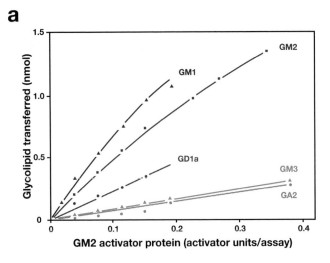

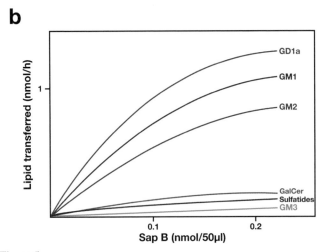

Figure 5

(*a*) Transfer of various glycolipids from donor to acceptor liposomes by the GM2-activator protein. Donor liposomes (250 nmol of lipids) containing 2 mol% of the respective glycolipid (5 nmol) were incubated with an equal amount of acceptor liposomes, 2 μmol of citrate buffer, pH 4.2, 4 μg of bovine serum albumin, and the amount of activator protein indicated in a total volume of 40 μl for 30 min at 37°C. Negatively charged acceptor liposomes were separated from uncharged donor liposomes on DEAE-cellulose columns. Controls were run without activator protein and substracted (Conzelmann et al. 1982). (*b*) Transfer of glycosphingolipids from acceptor to donor liposomes by Sap-B. Donor liposomes (250 nmol lipid) containing 2 mol% of the respective glycosphingolipids were incubated with an equal amount of acceptor liposomes, cytochrome *c* (10 μg) and increasing amounts of activator protein in a total volume of 50 μl, 50 mM citrate, pH 4.0, at 37°C for 1 h. Negatively charged acceptor liposomes were separated from uncharged donor liposomes on small DEAE-cellulose columns. Controls were run without activator proteins and the values subtracted. Modified after (Vogel et al. 1991).

galactosylceramide-β-galactosidase in vivo (**Figure 2**), convincingly demonstrated by the phenotype of mice carrying a mutation in the Sap-A domain of the Sap-precursor. These animals accumulate galactosylceramide and suffer from a late-onset form of Krabbe disease (Matsuda et al. 2001). To date, only one human patient has been reported with an isolated defect of Sap-A (Spiegel et al. 2005).

Sap-B. Sap-B was the first activator protein to be identified (Mehl & Jatzkewitz 1964). It mediates the degradation of sulfatide by arylsulfatase A and of globotriaosylceramide and digalactosylceramide by α-galactosidase A in vivo (**Figure 2**). This has been demonstrated in patients with Sap-B deficiency, where these substrates are found in the urine (Li et al. 1985). Obviously, a small fraction of these amphiphilic and barely water-soluble substances can escape the lysosomes and are detected in the urine after massive accumulation. Sap-B is also required for the degradation of other glycolipids (Li et al. 1988); for example, it cooperates with the GM2-activator protein in the degradation of ganglioside GM1 (Wilkening et al. 2000).

Similar to the GM2-activator, Sap-B acts as a physiological detergent but shows a broader specificity. The crystal structure shows a shell-like homodimer that encloses a large hydrophobic cavity (Ahn et al. 2003); the monomers are composed of four amphipathic α-helices arranged in a long hairpin that is bent into a simple V-shape. As in the GM2-activator, there are two different conformations of the Sap-B dimers, and a similar mechanism for its action has been proposed: The open conformation should interact directly with the membrane, promote a reorganization of the lipid alkyl chains, and extract the lipid substrate accompanied by a change to the closed conformation. Thus the substrate could be exposed to the enzyme in a water-soluble activator-lipid complex (Fischer & Jatzkewitz 1977),

TABLE 1 Lysosomal lipid transfer proteins (modified from Sandhoff et al. 2001)

	GM2-activator	Sap-A	Sap-B	Sap-C	Sap-D	Prosaposin	NPC-2
Subcellular site	le, lys	le, lys	le, lys	le, lys	le, lys	Extracellular, ER, Golgi	le, lys
Acts as lipid transfer protein	Yes	?	Yes	(Yes)	?	?	(Yes)
Stoichiometric complexes	Yes (1:1)	?	Yes	Yes	?	?	?
Solubilizes lipids with BMP and other anionic PL	Yes	?	Yes	Yes	(Yes)	?	?
Defect leads to storage of	GM2, GA$_2$	GalCer (k.o. mouse)	Sulfatides, Gbose3Cer	GlcCer	Hydroxylated Ceramides (k.o. mouse)	Cer, most GSL and inner lysosomal membranes	Chol.
Interaction with exohydrolases	HexA	?	?	Glucosylceramide-β-glucosidase	?	?	?
Lipid antigen transfer to	CD1d?	?	CD1d	CD1b	?	?	?

le, late endosomes; lys, lysosomes.

consistent with the previous observation that Sap-B can act as a lipid-transport protein (**Figure 5***b*; Vogel et al. 1991). The inherited defect of Sap-B leads to an atypical form of metachromatic leukodystrophy, with late infantile or juvenile onset (Kretz et al. 1990). The disease is characterized by accumulation of sulfatides, digalactosylceramide, and globotriaosylceramide (Sandhoff et al. 2001).

Sap-C. Sap-C is a homodimer and was initially isolated from the spleens of patients with Gaucher disease (Ho & O'Brien 1971). It is required for the lysosomal degradation of glucosylceramide by glucosylceramide-β-glucosidase (Ho & O'Brien 1971) (**Figure 2**). In addition, Sap-C renders glucosylceramide-β-glucosidase more protease-resistant inside

the cell (Sun et al. 2003). The solution structure of Sap-C (de Alba et al. 2003) consists of five tightly packed α-helices that form half of a sphere. All charged amino acids are solvent-exposed, whereas the hydrophobic residues are contained within the protein core. In contrast to the mode of action of the GM2-activator and of Sap-B, Sap-C can directly activate glucosylceramide-β-glucosidase in an allosteric manner (Ho & O'Brien 1971, Berent & Radin 1981, Fabbro & Grabowski 1991). Sap-C also supports the interaction of the enzyme with the substrate embedded in vesicles containing anionic phospholipids, and Sap-C is able to destabilize these vesicles (Wilkening et al. 1998). Binding of Sap-C to phospholipid vesicles is a pH-controlled, reversible process (Vaccaro et al. 1995). Sap-C deficiency leads to an abnormal juvenile form of Gaucher disease and an accumulation of glucosylceramide (Christomanou et al. 1986, Schnabel et al. 1991).

Sap-D. Sap-D stimulates lysosomal ceramide degradation by acid ceramidase in cultured cells (Klein et al. 1994) and in vitro (Linke et al. 2001b). Moreover, it stimulates acid sphingomyelinase-catalyzed sphingomyelin hydrolysis, but this seems not to be necessary for the in vivo degradation of sphingomyelin (Morimoto et al. 1988, Linke et al. 2001a). The detailed physiological function and mode of action of Sap-D is unclear. It is able to bind to vesicles containing negatively charged lipids and to solubilize them at an appropriate pH (Ciaffoni et al. 2001). Sap-D-deficient mice accumulate ceramides with hydroxylated fatty acids mainly in the brain and in the kidney (Matsuda et al. 2004).

Prosaposin. All four saposins are derived from a single protein, the Sap-precursor, or prosaposin, which is proteolytically processed to the mature activator proteins in the late endosomes and lysosomes (Fürst et al. 1988, O'Brien et al. 1988, Nakano et al. 1989). Prosaposin is a 70 kDa glycoprotein detected mainly uncleaved in brain, heart, and muscle, whereas mature Saps are found in all organs tested for so far, but mainly in liver, lung, kidney, and spleen. The Sap-precursor also occurs in body fluids such as milk, semen, cerebrospinal fluid, bile, and pancreatic juice. The Sap-precursor is either intracellularly targeted to the lysosomes via mannose-6-phosphate receptors or by sortilin (Lefrancois et al. 2003), or it can be secreted and re-endocytosed by mannose-6-phosphate receptors, low-density, lipoprotein receptor–related protein (LRP), or mannose receptors (Hiesberger et al. 1998).

To date, two different mutations in four human patients have been reported that lead to a complete deficiency of the whole Sap-precursor protein, and consequently of all four Saps. Because the Sap-precursor is proteolytically processed efficiently within the acidic compartments, it can be assumed that the unprocessed protein plays no role in membrane digestion. Inherited deficiency of the protein, however, was an indispensable tool in the elucidation of the specificity of individual saposins (Sandhoff et al. 2001). In human patients with Sap-precursor deficiency, but also in the Sap-precursor knockout mice (Fujita et al. 1996), there is simultaneous storage of many sphingolipids, including ceramide, glucosylceramide, lactosylceramide, ganglioside GM3, galactosylceramide, sulfatides, digalactosylceramide, and globotriaosylceramide, accompanied by a dramatic accumulation of intra-lysosomal membranes. This storage can be completely reversed by exogenous treatment with human Sap-precursor, as demonstrated in prosaposin-deficient fibroblasts (Burkhardt et al. 1997).

Activator Proteins in Lipid Antigen Presentation

In addition to their function as enzyme cofactors, SAPs play an important role in the presentation of lipid and glycolipid antigens. It is now established that CD1 immunoreceptors present lipid antigens to T cells. However, these lipids first must be removed from the membranes in which they are embedded to allow loading of CD1 molecules (**Figure 6**). The human genome encodes four MHC-I-like glycoproteins (CD1a-d) that present lipid antigens to T cells. A fifth gene encodes CD1e, which is synthesized as an integral membrane protein and from which a soluble lipid-binding domain is released by proteolysis within the lysosomes of mature dendritic cells (Angénieux et al. 2005). A possible function of this protein might also be lipid transfer, but this has not been proven to date. The three-dimensional structures of protein-lipid-complexes between human CD1b and two lipids, phosphatidylinositol and ganglioside GM2 (Gadola et al. 2002), and between CD1a

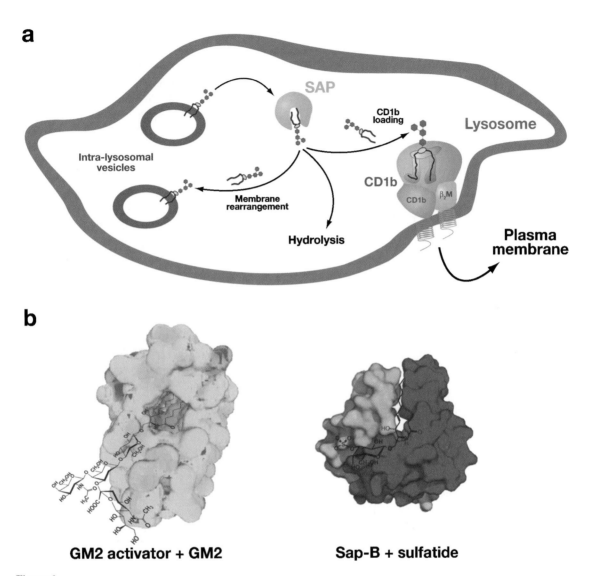

a

SAP

Lysosome

CD1b
loading

Intra-lysosomal
vesicles

Membrane
rearrangement

CD1b

CD1b   β₂M

Hydrolysis

Plasma
membrane

b

GM2 activator + GM2

Sap-B + sulfatide

Figure 6

Model of the presentation of glycolipid antigens to immunoreceptors of the CD1 family by lipid transfer proteins. SAPs, such as Sap-B, or the GM2-activator, extract lipids or glycolipids from inner lysosomal membranes and transfer them to other membranes, to hydrolytic enzymes, or to immunoreceptors of the CD1 family on the perimeter membrane. Lipid-loaded CD1 proteins then travel back to the plasma membrane and activate lipid-antigen-specific T-lymphocytes.

and sulfatide (Zajonc et al. 2003) have been reported. There is evidence that SAPs acting as lipid transfer proteins (**Table 1**) participate in this loading process within the acidic compartments of the cell. Antigen presentation by human CD1b (Winau et al. 2004), as well as human (Kang et al. 2004) and mouse CD1d (Zhou et al. 2004), have been studied: Human CD1b especially requires Sap-C to present different types of glycolipid antigens. In vitro, all saposins can exchange phosphatidylserine bound to murine CD1d against glycosphingolipids, but with different activity. The state of our knowledge in this regard is incomplete. Further research is required to reach a better understanding of the mechanisms involved in this process.

ANIONIC LIPIDS AS ACTIVATORS OF SPHINGOLIPID DEGRADATION

Intra-Endosomal and Intra-Lysosomal Membranes as Degradation Platforms

During maturation of endosomes, the luminal pH value decreases and the composition of the internal membranes changes. Not only the protein composition (Umebayashi 2003) but also the lipid composition is adjusted during this process so that the internal membranes in the acidic compartments of the cell are prepared for degradation. Their lipid composition differs considerably from that of plasma membrane or the limiting membranes of cellular organelles. Membrane-stabilizing cholesterol is continuously removed during this process, and the content of the negatively charged lysosomal lipid, BMP (erroneously also named as lysobisphosphatidic acid), increases (**Figures 1, 7b**). BMP is not present in the limiting lysosomal membrane, but it stimulates sphingolipid degradation on inner membranes of the acidic compartments (see below). Sphingolipid-containing membranes of autophagocytic bodies should contain less cholesterol but should contain metabolic precursors of BMP and might be degraded in a similar manner.

Cholesterol

Cholesterol is enriched in the plasma membrane and in membranes of early endocytic organelles but not in the lysosomes (Umebayashi 2003, Friedland et al. 2003). As indicated by immuno-electronmicroscopical examination of human B-lymphocytes (Möbius et al. 2003), about 80% of the cholesterol detected in the endocytic pathway is present in the recycling compartments and in internal membranes of early and late endosomes. On the other hand, it is nearly completely absent in inner lysosomal membranes (**Figure 7a**).

Data on the differential lipid composition of the internal endo/lysosomal membranes have been obtained with the aid of exogenous addition of ganglioside GM1 derivatives bearing a photoaffinity label and fluorescence- or biotin-labels to cultured cells and subsequent monitoring of endocytosis by fluorescence microscopy (Möbius et al. 1999a,b; von Coburg 2003). Membrane segregation into the intra-endosomal membrane pool has been demonstrated in cultured human fibroblasts, where biotin-labeled ganglioside GM1 derived from the plasma membrane is mainly targeted to intra-lysosomal structures and much less to the lysosomal perimeter membrane (Möbius et al. 1999a).

Analysis of the molecular lipid environment of a short-chain, photoactivatable ganglioside GM1 derivative in the plasma membrane showed an unexpectedly high amount of cholesterol (80%) and comparatively low amounts of phosphatidylcholine (17%) and sphingomyelin (7%). During endocytosis, coupling to cholesterol continuously decreased down to 46%, whereas the amounts of cross-linked phosphatidylcholine increased to 31% and those with sphingomyelin reached up to 23%. Using a long-chain derivative of photoactivatable ganglioside GM1 also enabled examination of its lipid environment in the inner membranes of the lysosomes in cultured human fibroblasts (von Coburg 2003). The amount of cross-linking products with cholesterol dropped down to 1% of all lipid derivatives. Cross-linking with BMP was elevated to 45%, whereas cross-linked phosphatidylcholine and sphingomyelin remained nearly unchanged at values of 31% and 22%, respectively. Similar results were obtained in other cell types, suggesting an efficient sorting of membrane lipids during endocytosis. These results are in agreement with those obtained by electron microscopy (Möbius et al. 1999a, 2003). Other cross-linking experiments using cells derived from patients with NPC1 (see below) yielded 25% of coupled cholesterol in the inner membranes of lysosomes (von Coburg 2003).

Niemann-Pick disease, type C. An inherited disorder, in which intracellular traffic of cholesterol is impaired, is Niemann-Pick disease, type C (NPC) (Patterson et al. 2001, Patterson 2003). Although the molecular details underlying this disorder are far from clear, this disease might shed some light on unexplained aspects of endocytotic lipid sorting. The disease is a neurodegenerative disorder characterized by accumulation of cholesterol and, secondarily also of other membrane components in the lysosomal compartment. Endosomal-lysosomal storage of unesterified cholesterol, neutral glycolipids such as glucosylceramide and lactosylceramide, acidic glycolipids, especially gangliosides GM3 and GM2 (Zervas et al. 2001), sphingomyelin (less than in Niemann-Pick disease, types A and B), BMP, and phospholipids occurs in liver, spleen, brain, and other organs (Patterson et al. 2001).

NPC is an inherited disorder in which mutations in the genes encoding the NPC1 (Carstea et al. 1997) or NPC2 protein (Naureckiene et al. 2000) have been detected. NPC1-deficiency accounts for 95% of NPC cases (Millat et al. 1999), whereas the other cases are from mutations in the HE1 gene encoding the NPC2 protein (Millat et al. 2001). This protein was previously identified as a lysosomal glycoprotein that can bind cholesterol with high affinity in a 1:1 stoichiometry (Okamura et al. 1999). In agreement with results from in vitro experiments (C. Arenz & K. Sandhoff, unpublished results), it probably acts as a cholesterol

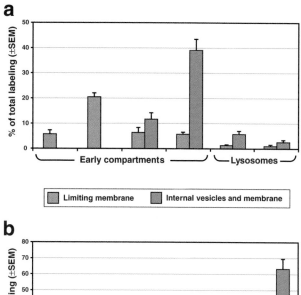

Figure 7

(*a*) Relative distribution of cholesterol in human B-lymphocytes in membranes of different types of endocytic organelles. Cryosections of human B-lymphocytes, which had internalized BSA-gold for different time points, were labeled with biotinylated Θ-toxin and antibiotin antibodies followed by protein A-gold (Möbius et al. 2003). (*b*) Distribution of BMP in human B-lymphocytes in membranes of different types of endocytic organelles. Cryosections of human B-lymphocytes, which had internalized BSA-gold for different time points, were labeled with a monoclonal antibody against BMP followed by rabbit antimouse antibodies and protein A-gold (Möbius et al. 2003).

carrier (**Table 1**) and regulates cholesterol transport from the internal membranes to the NPC1 protein, an integral protein of the perimeter membrane. The NPC1 gene product has 13–16 predicted transmembrane domains and a sterol-sensitive domain with homologies to HMG-CoA-reductase. A strain of mutant BALB/c mice has been used as an animal model for NPC. The mice show storage of unesterified cholesterol and impaired processing of exogenous cholesterol (Pentchev et al. 1984, Liu et al. 2000). Cultured cells derived from these animals show elevated levels of unesterified cholesterol if they are fed with low-density lipoprotein (Pentchev et al. 1986). In skin fibroblasts from NPC patients, the accumulating cholesterol is colocalized with BMP-rich late intra-endosomal membrane structures (Kobayashi et al. 1999). The NPC1 protein is suggested to transport lipophilic molecules through membranes and presumably also out of the endosomal-lysosomal system (Davies et al. 2000).

Ceramide

We assume that during the maturation of intra-endosomal/intra-lysosomal vesicles the levels of ceramide, a cholesterol-competitor (London & London 2004), increase at the expense of sphingomyelin, a cholesterol-binding lipid (see below). Ceramide, the common product of glycosphingolipid and sphingomyelin catabolism, can stabilize lipid phases, i.e., microdomains, more efficiently than cholesterol (Massey 2001, Xu et al. 2001, London & London 2004). Although cholesterol is removed from the internal membranes, the ceramide content of the internal membranes most likely increases. During endosomal maturation, the luminal pH decreases. The enzymes that lead to generation of ceramide have higher pH-optima than acid ceramidase, which converts ceramide into sphingosine that is positively charged at the pH of the lysosol. Glucosylceramide-β-glucosidase, which converts glucosylceramide into ceramide, and acid sphingomyelinase, which converts sphingomyelin into ceramide, have pH optima of about 5.5 (Osiecki-Newman et al. 1988, Goni & Alonso 2002), whereas the pH optimum of acid ceramidase is in the range of 3.8–4.2 in the presence of Sap–D (Bernardo et al. 1995, Linke et al. 2001b). This should lead to a continuous increase of ceramide levels within the intra-lysosomal membranes during the process of endosomal maturation. The replacement of cholesterol by ceramide might facilitate choles-

terol exit out of this membrane population by NPC2.

Bis(monoacylglycero)phosphate

BMP is a characteristic anionic phospholipid of the acidic compartments of the cell. It is biosynthetically formed during the degradation of phosphatidylglycerol and cardiolipin (Amidon et al. 1996, Brotherus et al. 1974), presumably on the surface of intra-lysosomal vesicles. It has an unusual sn1,sn1' configuration, which accounts for its higher resistance to the action of phospholipases than normal phospholipids (Matsuzawa & Hostetler 1979). Other anionic lipids, such as phosphatidylinositol (Kobayashi et al. 1998) and dolichol phosphate (Chojnacki & Dallner 1988), albeit in smaller amounts than BMP, are also found within the lysosomal compartment.

In vivo, enzymatic hydrolysis of most membrane-bound sphingolipids is stimulated not only by sphingolipid activator proteins, but also by BMP. The percentage of BMP increases starting from late endosomes to lysosomes, where the maximal amount is found (**Figure 6***b*). Moreover, BMP was shown to be mainly present in internal membranes; therefore, it distinguishes these membranes from the perimeter membrane (Möbius et al. 2003).

Biophysical Properties

The size, the lateral pressure, and the composition of intra-lysosomal vesicles are contributing factors in the degradation of glycolipids. The diameter of intra-lysosomal vesicles has been determined in tissues from sphingolipid activator protein-deficient patients to be in the range of 50 to 100 nm (Bradova et al. 1993). The lateral surface pressure of most biological membranes is in the range of 30 to 35 mNm^{-1} (Marsh 1996, Maggio et al. 2002). This high lateral pressure seems to contribute to the protection of limiting membranes from inappropriate degradation because in vitro experiments showed that the GM2-activator protein is only able to penetrate into a phospholipid monolayer when the lateral pressure is below a critical value of 15 to 25 mNm^{-1} depending on the lipid composition (Giehl et al. 1999). Even if no data are available on the lateral pressure of intra-lysosomal vesicles, the combination of size and composition can be expected to lower this pressure below this critical value.

Functional Aspects of Membrane Lipid Sorting in the Endocytic Pathway

High amounts of BMP and low amounts of membrane-stabilizing cholesterol in internal lysosomal membranes appear to be required for the degradation of glycosphingolipids. The presence of BMP in these vesicles increases the ability of the GM2-activator to solubilize lipids (Werth et al. 2001). In addition, negatively charged lysosomal lipids drastically stimulate the interfacial hydrolysis of membrane-bound ganglioside GM1 by GM1-β-galactosidase (Wilkening et al. 2000); ganglioside GM2 by β-hexosaminidase A (Werth et al. 2001); the

sulfated gangliotriaosylceramide SM2 by β-hexosaminidases A and S (Hepbildikler et al. 2002) in the presence of the GM2-activator protein; sphingomyelin by acid sphingomyelinase (Linke et al. 2001a); and ceramide by acid ceramidase (Linke et al. 2001b). Furthermore, in the presence of Sap-C, a drastic enhancement of glucosylceramide degradation by glucosylceramide-β-glucosidase is produced by negatively charged model lipids such as phosphatidylserine, phosphatidylglycerol, and phosphatidic acid (Berent & Radin 1981, Sarmientos et al. 1986, Salvioli et al. 2000).

Taken together, degradation of an individual (glyco)sphingolipid (**Figure 2**) embedded in intra-lysosomal membrane structures requires the appropriate combination of hydrolytic enzyme, activator protein, and lysosomal lipid. This ensures that substrates are cleaved only when they are part of membranes destined for degradation.

Inner lysosomal membranes represent the main site of membrane degradation in eukaryotic cells. Their specific lipid profile, low cholesterol, and high BMP content ensure their degradation by water-soluble hydrolases and membrane-perturbing SAPs without affecting the limiting membrane.

ACKNOWLEDGMENTS

We thank Drs. Eskelinen and Saftig (Kiel), and de la Salle (Strasbourg) for helpful discussions and for critically reading the manuscript. Work performed in the laboratory of the authors was supported by the DFG (SFB 284, SFB 400, and SFB 645).

LITERATURE CITED

Ahn VE, Faull KF, Whitelegge JP, Fluharty AL, Prive GG. 2003. Crystal structure of saposin B reveals a dimeric shell for lipid binding. *Proc. Natl. Acad. Sci. USA* 100:38–43

Amidon B, Brown A, Waite M. 1996. Transacylase and phospholipases in the synthesis of bis(monoacylglycero)phosphate. *Biochemistry* 35:13995–4002

Angénieux C, Fraisier V, Maître B, Racine V, van der Wel N, et al. 2005. The cellular pathway of CD1e in immature and maturing dendritic cells. *Traffic* 6:286–302

Berent SL, Radin NS. 1981. Mechanism of activation of glucocerebrosidase by co-β-glucosidase (glucosidase activator protein). *Biochim. Biophys. Acta* 664:572–82

Bernardo K, Hurwitz R, Zenk T, Desnick RJ, Ferlinz K, et al. 1995. Purification, characterization, and biosynthesis of human acid ceramidase. *J. Biol. Chem.* 70:11098–102

Bradova V, Smid F, Ulrich-Bott B, Roggendorf W, Paton BC, Harzer K. 1993. Prosaposin deficiency: further characterization of the sphingolipid activator protein-deficient sibs. Multiple glycolipid elevations (including lactosylceramidosis), partial enzyme deficiencies and ultrastructure of the skin in this generalized sphingolipid storage disease. *Hum. Genet.* 92:143–52

Brotherus J, Renkonen O, Herrmann J, Fischer W, Herrmann, J. 1974. Novel stereoconfiguration in lyso-*bis*-phosphatidic acid of cultured BHK-cells. *Chem. Phys. Lipids* 13:178–82

Brown DA, London E. 2000. Structure and function of sphingolipid- and cholesterol-rich membrane rafts. *J. Biol. Chem.* 275:17221–24

Burkhardt JK, Hüttler S, Klein A, Möbius W, Habermann A, et al. 1997. Accumulation of sphingolipids in SAP-precursor (prosaposin)-deficient fibroblasts occurs as intralysosomal membrane structures and can be completely reversed by treatment with human SAP-precursor. *Eur. J. Cell Biol.* 73:10–18

Carstea ED, Morris JA, Coleman KG, Loftus SK, Zhang D, et al. 1997. Niemann-Pick1 disease gene: homology to mediators of cholesterol homeostasis. *Science* 277:228–31

Chojnacki T, Dallner, G. 1988. The biological role of dolichol. *Biochem. J.* 251:1–9

Christomanou H, Aignesberger A, Linke RP. 1986. Immunochemical characterization of two activator proteins stimulating enzymic sphingomyelin degradation in vitro. Absence of one of them in a human Gaucher disease variant. *Biol. Chem. Hoppe-Seyler* 367:879–90

Ciaffoni F, Salvioli R, Tatti M, Arancia G, Crateri P, Vaccaro AM. 2001. Saposin D solubilizes anionic phospholipid-containing membranes. *J. Biol. Chem.* 276:31583–89

Coe NR, Bernlohr DA. 1998. Physiological properties and functions of intracellular fatty acid-binding proteins. *Biochim. Biophys. Acta* 1391:287–306

Conzelmann E, Burg J, Stephan G, Sandhoff K. 1982. Complexing of glycolipids and their transfer between membranes by the activator protein for degradation of lysosomal ganglioside GM2. *Eur. J. Biochem.* 123:455–64

Conzelmann E, Sandhoff K. 1978. AB variant of infantile GM2 gangliosidosis: deficiency of a factor necessary for stimulation of hexosaminidase A-catalyzed degradation of ganglioside GM2 and glycolipid GA2. *Proc. Natl. Acad. Sci. USA* 75:3979–83

Conzelmann E, Sandhoff K. 1979. Purification and characterization of an activator protein for the degradation of glycolipids GM2 and GA2 by hexosaminidase A. *Hoppe Seylers Z. Physiol. Chem.* 360:1837–49

Davies JP, Chen FW, Ioannou YA. 2000. Transmembrane molecular pump activity of Niemann-Pick C1 protein. *Science* 290:2295–98

de Alba E, Weiler S, Tjandra, N. 2003. Solution structure of human saposin C: pH-dependent interaction with phospholipid vesicles. *Biochemistry* 42:14729–40

Eskelinen E-L, Tanaka Y, Saftig P. 2003. At the acidic edge: emerging functions for lysosomal membrane proteins. *Trends Cell Biol.* 13:137–45

Fabbro D, Grabowski GA. 1991. Human acid β-glucosidase. Use of inhibitory and activating monoclonal antibodies to investigate the enzyme's catalytic mechanism and saposin A and C binding sites. *J. Biol. Chem.* 266:15021–27

Fischer G, Jatzkewitz H. 1977. The activator of cerebroside sulphatase. Binding studies with enzyme and substrate demonstrating the detergent function of the activator protein. *Biochim. Biophys. Acta* 481:561–72

Friedland N, Liou HL, Lobel P, Stock AM. 2003. Structure of a cholesterol-binding protein deficient in Niemann-Pick type C2 disease. *Proc. Natl. Acad. Sci. USA* 100:2512–17

Fürst W, Machleidt W, Sandhoff K. 1988. The precursor of sulfatide activator protein is processed to three different proteins. *Biol. Chem. Hoppe Seyler* 369:317–28

Fürst W, Sandhoff K. 1992. Activator proteins and topology of lysosomal sphingolipid catabolism. *Biochim. Biophys. Acta* 1126:1–16

Fujita N, Suzuki K, Vanier MT, Popko B, Maeda N, et al. 1996. Targeted disruption of the mouse sphingolipid activator protein gene: a complex phenotype, including severe leukodystrophy and wide-spread storage of multiple sphingolipids. *Hum. Mol. Genet.* 5: 711–25

Gadola SD, Zaccai NR, Harlos K, Shepherd D, Castro-Palomino JC, et al. 2002. Structure of human CD1b with bound ligands at 2.3 Å, a maze for alkyl chains. *Nat. Immunol.* 3:721–26

Giehl A, Lemm T, Bartelsen O, Sandhoff K, Blume A. 1999. Interaction of the GM2-activator protein with phospholipid-ganglioside bilayer membranes and with monolayers at the air-water interface. *Eur. J. Biochem.* 261:650–58

Goni FM, Alonso A. 2002. Sphingomyelinases: enzymology and membrane activity. *FEBS Lett.* 531:38–46

Gutsmann T, Riekens B, Bruhn H, Wiese A, Seydel U, Leippe M. 2003. Interaction of amoebapores and NK-lysin with symmetric phospholipid and asymmetric lipopolysaccharide/phospholipid bilayers. *Biochemistry* 42:9804–12

Hanada K, Kumagai K, Yasuda S, Miura Y, Kawano M, et al. 2003. Molecular machinery for non-vesicular trafficking of ceramide. *Nature* 426:803–9

Harzer K, Paton BC, Poulos A. 1989. Sphingolipid activator protein (SAP) deficiency in a 16-week old atypical Gaucher disease patient and his fetal sibling; biochemical signs of combined sphingolipidosis. *Eur. J. Pediatr.* 149:31–39

Hecht O, Van Nuland NA, Schleinkofer K, Dingley AJ, Bruhn H, et al. 2004. Solution structure of the pore forming protein of entamoeba histolytic. *J. Biol. Chem.* 279:17834–41

Heerklotz H. 2002. Triton promotes domain formation in lipid raft mixtures. *Biophys. J.* 83:2693–701

Heerklotz H, Szadkowska H, Anderson T, Seelig J. 2003. The sensitivity of lipid domains to small perturbations demonstrated by the effect of Triton. *J. Mol. Biol.* 329:793–99

Henning R, Stoffel W. 1973. Glycosphingolipids in lysosomal membranes. *Hoppe-Seyler's Z. Physiol. Chem.* 354:760–70

Hepbildikler ST, Sandhoff R, Kölzer M, Proia RL, Sandhoff K. 2002. Physiological substrates for human lysosomal β-hexosaminidase S. *J. Biol. Chem.* 277:2562–72

Hiesberger T, Hüttler S, Rohlmann A, Schneider W, Sandhoff K, Herz J. 1998. Cellular uptake of saposin (SAP) precursor and lysosomal delivery by the low density lipoprotein receptor-related protein (LRP). *EMBO J.* 17:4617–25

Ho MW, O'Brien JS. 1971. Gaucher's disease: deficiency of 'acid' β-glucosidase and reconstitution of enzyme activity in vitro. *Proc. Natl. Acad. Sci. USA* 68:2810–13

Hopkins CR, Gibson A, Shipman M, Miller K. 1990. Movement of internalized ligand-receptor complexes along a continuous endosomal reticulum. *Nature* 346:335–39

Kang SJ, Cresswell P. 2004. Saposins facilitate CD1d-restricted presentation of an exogenous lipid antigen to T cells. *Nature Immunol.* 5:175–81

Katzmann DJ, Babst M, Emr SD. 2001. Ubiquitin-dependent sorting into the multivesicular body pathway requires the function of a conserved endosomal protein sorting complex, ESCRT-I. *Cell* 106:145–55

Klein A, Henseler M, Klein C, Suzuki K, Harzer K, Sandhoff K. 1994. Sphingolipid activator protein D (sap-D) stimulates the lysosomal degradation of ceramide in vivo. *Biochem. Biophys. Res. Commun.* 200:1440–48

Kobayashi T, Beuchat M-H, Lindsay M, Frias S, Palmiter RD. 1999. Late endosomal membranes rich in lysobisphosphatidic acid regulate cholesterol transport. *Nature Cell Biol.* 1:113–18

Kobayashi T, Stang E, Fang KS, de Moerloose P, Parton RG, Gruenberg J. 1998. A lipid associated with the antiphospholipid syndrome regulates endosome structure and function. *Nature* 392:193–97

Kolter T, Proia RL, Sandhoff K. 2002. Minireview: combinatorial ganglioside biosynthesis. *J. Biol. Chem.* 277:25859–62

Kolter T, Sandhoff K. 1998. Recent advances in the biochemistry of sphingolipidoses. *Brain Pathol.* 8:79–100

Kolter T, Sandhoff K. 1999. Sphingolipids—their metabolic pathways and the pathobiochemistry of neurodegenerative diseases. *Angew. Chem. Int. Ed.* 38:1532–68

Kretz KA, Carson GS, Morimoto S, Kishimoto Y, Fluharty AL, O'Brien JS. 1990. Characterization of a mutation in a family with saposin B deficiency: a glycosylation site defect. *Proc. Natl. Acad. Sci. USA* 87:2541–44

Lefrancois S, Zeng J, Hassan AJ, Canuel M, Morales CR. 2003. The lysosomal trafficking of sphingolipid activator proteins (SAPs) is mediated by sortilin. *EMBO J.* 22:6430–37

Li SC, Kihara H, Serizawa S, Li YT, Fluharty AL, Mayes JS, Shapiro LJ. 1985. Activator protein required for the enzymatic hydrolysis of cerebroside sulfate. Deficiency in urine of patients affected with cerebroside sulfatase activator deficiency and identity with activators for the enzymatic hydrolysis of GM1 ganglioside and globotriaosylceramide. *J. Biol. Chem.* 260:1867–71

Li SC, Sonnino S, Tettamanti G, Li YT. 1988. Characterization of a nonspecific activator protein for the enzymatic hydrolysis of glycolipids. *J. Biol. Chem.* 263:6588–91

Liepinsh E, Andersson M, Ruysschaert JM, Otting G. 1997. Saposin fold revealed by the NMR structure of NK-lysin. *Nat. Struct. Biol.* 4:793–95

Linke T, Wilkening G, Lansmann S, Moczall H, Bartelsen O, et al. 2001a. Stimulation of acid sphingomyelinase activity by lysosomal lipids and sphingolipid activator proteins. *Biol. Chem.* 382:283–90

Linke T, Wilkening G, Sadeghlar F, Mozcall H, Bernardo K, et al. 2001b. Interfacial regulation of acid ceramidase activity. Stimulation of ceramide degradation by lysosomal lipids and sphingolipid activator proteins. *J. Biol. Chem.* 276:5760–68

Liu Y, Wu Y-P, Wada R, Neufeld EB, Mullin KA. et al. 2000. Alleviation of neuronal ganglioside storage does not improve the clinical course of the Niemann-Pick C disease mouse. *Hum. Mol. Genet.* 9:1087–92

London M, London E. 2004. Ceramide selectively displaces cholesterol from ordered lipid domains (rafts): implications for lipid raft structure and function. *J. Biol. Chem.* 279:9997–10004

Maggio B, Fanani ML, Oliveira RG. 2002. Biochemical and structural information transduction at the mesoscopic level in biointerfaces containing sphingolipids. *Neurochem. Res.* 27:547–57

Maier T, Strater N, Schuette CG, Klingenstein R, Sandhoff K, Saenger W. 2003. The X-ray crystal structure of human β-hexosaminidase B provides new insights into Sandhoff disease. *J. Mol. Biol.* 328:669–81

Malinina L, Malakhova ML, Teplov A, Brown RE, Patel DJ. 2004. Structural basis for glycosphingolipid transfer specificity. *Nature* 430:1048–53

Mark BL, Mahuran DJ, Cherney MM, Zhao D, Knapp S, James MN. 2003. Crystal structure of human β-hexosaminidase B: understanding the molecular basis of Sandhoff and Tay-Sachs disease. *J. Mol. Biol.* 327:1093–109

Marsh D. 1996. Lateral pressure in membranes. *Biochim. Biophys. Acta* 1286:183–223

Massey JB. 2001. Interaction of ceramides with phosphatidylcholine, sphingomyelin and sphingomyelin/cholesterol bilayers. *Biochim. Biophys. Acta* 1510:167–84

Matsuda J, Vanier MT, Saito Y, Tohyama J, Suzuki K. 2001. A mutation in the saposin A domain of the sphingolipid activator protein (prosaposin) gene results in a late-onset, chronic form of globoid cell leukodystrophy in the mouse. *Hum. Mol. Genet.* 10:1191–99

Matsuda J, Kido M, Tadano-Aritomi K, Ishizuka I, Tominaga K, et al. 2004. Mutation in saposin D domain of sphingolipid activator protein gene causes urinary system defects and cerebellar Purkinje cell degeneration with accumulation of hydroxy fatty acid-containing ceramide in mouse. *Hum. Mol. Genet.* 13:2709–23

Matsuzawa Y, Hostetler KY. 1979. Degradation of bis(monoacylglycero)phosphate by an acid phosphodiesterase in rat liver lysosomes. *J. Biol. Chem.* 254:5997–6001

Maxfield FR, McGraw TE. 2004. Endocytic recycling. *Nat. Rev. Mol. Cell Biol.* 5:121–32

Mehl E, Jatzkewitz H. 1964. A cerebrosidesulfatase from swine kidney. *Hoppe Seylers Z. Physiol. Chem.* 339:260–76

Merrill AH Jr. 2002. De novo sphingolipid biosynthesis: a necessary, but dangerous, pathway. *J. Biol. Chem.* 277:25843–46

Millat G, Marcais C, Rafi MA, Yamamoto T, Morris JA, et al. 1999. Niemann-Pick C1 disease: the I1061T substitution is a frequent mutant allele in patients of Western European descent and correlates with a classic juvenile phenotype. *Am. J. Hum. Genet.* 65:1321–29

Millat G, Chikh K, Naureckiene S, Sleat DE, Fensom AH, et al. 2001. Niemann-Pick disease type C: spectrum of HE1 mutations and genotype/phenotype correlations in the NPC2 group. *Am. J. Hum. Genet.* 69:1013–21

Möbius W, Herzog V, Sandhoff K, Schwarzmann G. 1999a. Intracellular distribution of a biotin-labeled ganglioside GM1 by immunoelectron microscopy after endocytosis in fibroblasts. *J. Histochem. Cytochem.* 47:1005–14

Möbius W, Herzog V, Sandhoff K, Schwarzmann G. 1999b. Gangliosides are transported from the plasma membrane to intralysosomal membranes as revealed by immuno-electron microscopy. *Biosci. Rep.* 19:307–16

Möbius W, van Donselaar E, Ohno-Iwashita Y, Shimada Y, Heijnen HF, et al. 2003. Recycling compartments and the internal vesicles of multivesicular bodies harbor most of the cholesterol found in the endocytic pathway. *Traffic* 4:222–31

Morimoto S, Martin BM, Kishimoto Y, O'Brien JS. 1988. Saposin D: a sphingomyelinase activator. *Biochem. Biophys. Res. Commun.* 156:403–10

Munford RS, Sheppard PO, O'Hara PJ. 1995. Saposin-like proteins (SAPLIP) carry out diverse functions on a common backbone structure. *J. Lipid. Res.* 36:1653–63

Munro S. 2003. Lipid rafts: elusive or illusive? *Cell* 115:377–88

Nakano T, Sandhoff K, Stumper J, Christomanou H, Suzuki K. 1989. Structure of full-length cDNA coding for sulfatide activator, a Co-β-glucosidase and two other homologous proteins: two alternate forms of the sulfatide activator. *J. Biochem.* 105:152–54

Naureckiene S, Sleat DE, Lackland H, Fensom A, Vanier MT, et al. 2000. Identification of HE1 as the second gene of Niemann-Pick C disease. *Science* 290:2298–301

O'Brien JS, Kretz KA, Dewji N, Wenger DA, Esch F, Fluharty AL. 1988. Coding of two sphingolipid activator proteins (SAP-1 and SAP-2) by same genetic locus. *Science* 241:1098–1101

Okamura N, Kiuchi S, Tamba M, Kashima T, Hiramoto S, et al. 1999. A porcine homolog of the major secretory protein of human epididymis, HE1, specifically binds cholesterol. *Biochim. Biophys. Acta* 1438:377–87

Opekarová M, Tanner W. 2003. Specific lipid requirement of membrane proteins—a putative bottleneck in heterologous expression. *Biochim. Biophys. Acta* 1610:11–22

Osiecki-Newman K, Legler G, Grace M, Dinur T, Gatt S, et al. 1988. Human acid β-glucosidase: inhibition studies using glucose analogues and pH variation to characterize the normal and Gaucher disease glycon binding sites. *Enzyme* 40:173–88

Patterson MC. 2003. A riddle wrapped in a mystery: understanding Niemann-Pick disease, type C. *Neurology* 9:301–10

Patterson MC, Vanier MT, Suzuki K, Morris JA, Carstea E, et al. 2001. Niemann-Pick disease type C: a lipid trafficking disorder. In *The Metabolic and MolecularBases of Inherited Disease*, ed. CR Scriver, AL Beaudet, WS Sly, D Valle, III:3611–33. New York: McGraw-Hill. 8th ed.

Pentchev PG, Boothe AD, Kruth HS, Weintroub H, Stivers J, Brady RO. 1984. A genetic storage disorder in BALB/C mice with a metabolic block in esterification of exogenous cholesterol. *J. Biol. Chem.* 259:5784–91

Pentchev PG, Comly ME, Kruth HS, Patel S, Proestel M, Weintroub H. 1986. The cholesterol storage disorder of the mutant BALB/c mouse. *J. Biol. Chem.* 261:2772–77

Platt FM, Walkley SU. 2004. Lysosomal defects and storage. In *Lysosomal Disorders of the Brain*, ed. FM Platt, SU Walkley, pp. 32–49. New York: Oxford Univ. Press

Salvioli R, Tatti M, Ciaffoni F, Vaccaro AM. 2000. Further studies on the reconstitution of glucosylceramidase activity by Sap C and anionic phospholipids. *FEBS Lett.* 472:17–21

Sandhoff K, Kolter T, Harzer K. 2001. Sphingolipid activator proteins. In *The Metabolic and Molecular Bases of Inherited Disease*, ed. CR Scriver, AL Beaudet, WS Sly, D Valle, III:3371–88. New York: McGraw-Hill. 8th ed.

Sandhoff K, Kolter T. 1996. Topology of glycosphingolipid degradation. *Trends Cell Biol.* 6:98–103

Sarmientos F, Schwarzmann G, Sandhoff K. 1986. Specificity of human glucosylceramide β-glucosidase towards synthetic glucosylsphingolipids inserted into liposomes. Kinetic studies in a detergent-free assay system. *Eur. J. Biochem.* 160:527–35

Schnabel D, Schröder M, Fürst W, Klein A, Hurwitz R, et al. 1992. Simultaneous deficiency of sphingolipid activator proteins 1 and 2 is caused by a mutation in the initiation codon of their common gene. *J. Biol. Chem.* 267:3312–15

Schnabel D, Schröder M, Sandhoff K. 1991. Mutation in the sphingolipid activator protein 2 in a patient with a variant of Gaucher disease. *FEBS Lett.* 284:57–59

Simons K, Ikonen E. 1997. Functional rafts in cell membranes. *Nature* 387:569–72

Spiegel R, Bach G, Sury V, Mengistu G, Meidan B, et al. 2005. A mutation in the saposin A coding region of the prosaposin gene in an infant presenting as Krabbe disease: first report of saposin A deficiency in humans. *Mol. Genet. Metab.* 84:160–66

Sun Y, Qi X, Grabowski GA. 2003. Saposin C is required for normal resistance of acid β-glucosidase to proteolytic degradation. *J. Biol. Chem.* 278:31918–23

Suzuki K. 1994. Genetic disorders of lipid, glycoprotein, and mucopolysaccharide metabolism. In *Basic Neurochemistry: Molecular, Cellular, and Medical Aspects*, ed. GJ Siegel, BW Agranoff, RW Albers, PB Molinoff, pp. 793–812. New York: Raven. 5th ed.

Suzuki K, Chen GC. 1968. GM1-gangliosidosis (generalized gangliosidosis). Morphology and chemical pathology. *Pathol. Eur.* 3:389–408

Suzuki K, Vanier MT. 1999. Lysosomal and peroxisomal diseases. In *Basic Neurochemistry—Molecular, Cellular and Medical Aspects*, ed. GJ Siegel, BW Agranoff, RW Albers, SK Fisher, MD Uhler, pp. 821–39. Philadelphia: Lippincott-Raven. 6th ed.

Tettamanti G. 2004. Ganglioside/glycosphingolipid turnover: new concepts. *Glycoconjugate J.* 20:301–17

Umebayashi K. 2003. The roles of ubiquitin and lipids in protein sorting along the endocytic pathway. *Cell. Struct. Funct.* 28:443–53

Vaccaro AM, Ciaffoni F, Tatti M, Salvioli R, Barca A, et al. 1995. pH-dependent conformational properties of saposins and their interactions with phospholipid membranes. *J. Biol. Chem.* 270:30576–80

van der Goot FG, Gruenberg J. 2002. Oiling the wheels of the endocytic pathway. *Trends Cell Biol.* 12:296–99

van Meer G, Lisman Q. 2002. Sphingolipid transport: rafts and translocators. *J. Biol. Chem.* 277:25855–58

van Weely S, Brandsma M, Strijland A, Tager JM, Aerts JM. 1993. Demonstration of the existence of a second, non-lysosomal glucocerebrosidase that is not deficient in Gaucher disease. *Biochim. Biophys. Acta* 1181:55–62

Vogel A, Schwarzmann G, Sandhoff K. 1991. Glycosphingolipid specificity of the human sulfatide activator protein. *Eur. J. Biochem.* 200:591–97

von Coburg A. 2003. *Untersuchung der Lipidnachbarschaft von derivatisiertem GM1 in Modellmembranen und kultivierten Zellen.* PhD thesis. Universität Bonn. 169 pp.

Wendeler M, Hoernschemeyer J, Hoffmann D, Kolter T, Schwarzmann G, Sandhoff K. 2004. Photoaffinity labelling of the human GM2-activator protein. Mechanistic insight into ganglioside GM2 degradation. *Eur. J. Biochem.* 271:614–27

Werth N, Schuette CG, Wilkening G, Lemm T, Sandhoff K. 2001. Degradation of membrane-bound ganglioside GM2 by β-hexosaminidase A. Stimulation by GM2 activator protein and lysosomal lipids. *J. Biol. Chem.* 276:12685–90

Wilkening G, Linke T, Sandhoff K. 1998. Lysosomal degradation on vesicular membrane surfaces. Enhanced glucosylceramide degradation by lysosomal anionic lipids and activators. *J. Biol. Chem.* 273:30271–78

Wilkening G, Linke T, Uhlhorn-Dierks G, Sandhoff K. 2000. Degradation of membrane-bound Ganglioside GM1. *J. Biol. Chem.* 275:35814–19

Winau F, Schwierzeck V, Hurwitz R, Remmel N, Sieling PA, et al. 2004. Saposin C is required for lipid presentation by human CD1b. *Nat. Immunol.* 5:169–74

Winchester BG. 2004. Primary defects in lysosomal enzymes. In *Lysosomal Disorders of the Brain*, ed. FM Platt, SU Walkley, pp. 81–130. New York: Oxford Univ. Press

Wright CS, Li SC, Rastinejad F. 2000. Crystal structure of human GM2-activator protein with a novel β-cup topology. *J. Mol. Biol.* 304:411–22

Wright CS, Zhao Q, Rastinejad F. 2003. Structural analysis of lipid complexes of GM2-activator protein. *J. Mol. Biol.* 331:951–64

Xu X, Bittman R, Duportail G, Heissler D, Vilcheze C, London E. 2001. Effect of the structure of natural sterols and sphingolipids on the formation of ordered sphingolipid/sterol domains (rafts). Comparison of cholesterol to plant, fungal, and disease-associated sterols and comparison of sphingomyelin, cerebrosides, and ceramide. *J. Biol. Chem.* 276:33540–46

Yamashita T, Hashiramoto A, Haluzik M, Mizukami H, Beck S, et al. 2003. Enhanced insulin sensitivity in mice lacking ganglioside GM3. *Proc. Natl. Acad. Sci. USA* 100:3445–49

Zajonc DM, Elsliger MA, Teyton L, Wilson IA. 2003. Crystal structure of CD1a in complex with a sulfatide self antigen at a resolution of 2.15 Å. *Nat. Immunol.* 4:808–15

Zervas M, Dobrenis K, Walkley SU. 2001. Neurons in Niemann-Pick disease Type C accumulate gangliosides as well as unesterified cholesterol and undergo dendritic and axonal alterations. *J. Neuropathol. Exp. Neurol.* 60:49–64

Zhou D, Cantu C 3rd, Sagiv Y, Schrantz N, Kulkarni AB, et al. 2004. Editing of CD1d-bound lipid antigens by endosomal lipid transfer proteins. *Science* 303:523–27

Cajal Bodies: A Long History of Discovery

Mario Cioce[1] and Angus I. Lamond[2]

[1] IRBM (Merck Research Laboratories Rome), Rome, Italy; email: mario_cioce@merck.com

[2] Division of Gene Regulation and Expression, Wellcome Trust Biocentre, University of Dundee, DD15EH Dundee, United Kingdom; email: a.i.lamond@dundee.ac.uk

Annu. Rev. Cell Dev. Biol.
2005. 21:105–31

First published online as a
Review in Advance on
May 16, 2005

The *Annual Review of Cell and Developmental Biology* is online at
http://cellbio.annualreviews.org

doi: 10.1146/
annurev.cellbio.20.010403.103738

Copyright © 2005 by
Annual Reviews. All rights
reserved

1081-0706/05/1110-
0105$20.00

Key Words

coilin, snRNP, stress, nucleus, disease

Abstract

This review surveys what is known about the structure and function of the subnuclear domains called Cajal bodies (CBs). The major focus is on CBs in mammalian cells but we provide an overview of homologous CB structures in other organisms. We discuss the protein and RNA components of CBs, including factors recently found to associate in a cell cycle-dependent fashion or under specific metabolic or stress conditions. We also consider the dynamic properties of both CBs and their molecular components, based largely on recent data obtained thanks to the advent of improved in vivo detection and imaging methods. We discuss how these data contribute to an understanding of CB functions and highlight major questions that remain to be answered. Finally, we consider the interesting links that have emerged between CBs and alterations in nuclear structure apparent in a range of human pathologies, including cancer and inherited neurodegenerative diseases. We speculate on the relationship between CB function and molecular disease.

Contents

approaches have underlined two major concepts in relation to nuclear structure/function: (*a*) the existence and prevalence of specific subnuclear domains or organelles and (*b*) the unexpectedly dynamic behavior of these subnuclear structures and their constituents under both physiological and pathophysiological conditions. In contrast to cytoplasmic organelles, nuclear organelles are characterized by the absence of an outer membrane to separate them from the surrounding nucleoplasm. The fact that all the subnuclear domains are typically enriched for either specific nuclear proteins or RNA-protein complexes has led to the consensus that they can function as sites for coordinating complex molecular assembly and maturation pathways.

This review focuses on the dynamic properties of Cajal bodies (CBs) in mammalian cells. The first section is dedicated to a functional classification of the best characterized CB components, with special attention given to the data that suggest possible roles for these organelles. A short overview of CBs in other organisms follows. Finally, the available data suggesting a possible relationship between CBs and several human diseases are discussed and analyzed.

BACKGROUND AND INTRODUCTION

The cell nucleus has fascinated scientists for over two centuries. The efforts of both morphologists and biochemists have defined some of the basic features of nuclear structure and shown it to be the site of transcription, replication, pre-mRNA splicing, and assembly of ribosomal subunits. Nonetheless, major gaps remain in our understanding of the relationships between nuclear structure, as defined by microscopy, and molecular function, defined largely through the analysis of cell-free in vitro extracts. The advent of detailed fluorescence microscopy studies, made possible by the availability of specific antibody and antisense probes to label individual nuclear proteins and RNA-protein complexes, has helped to identify the wealth of substructure that exists within the nucleus. More recently, the development of fluorescent-protein tagging techniques has also allowed new insights into the dynamic aspects of nuclear structure in living cells. Collectively, these in vivo

CAJAL BODIES IN MAMMALIAN CELLS

CBs are ubiquitous subnuclear organelles found in both plant and animal cells. They can vary in size from less than 0.2 μm up to 2 μm or even larger, depending on cell type and species. As proposed by Gall, they are now called Cajal bodies after their initial discoverer, Ramon y Cajal, who first described them in 1903 as "nucleolar accessory bodies," based on their frequent association with the nucleolar periphery in neurons (Ramon y Cajal 1903). More than sixty years after their first description, the accessory bodies of Ramon y Cajal were rediscovered by electron microscopists and given the name coiled bodies, reflecting their typical appearance in transmission electron micrographs as a tangled ball of fibrillar threads (Hardin et al. 1969, Lafarga et al. 1983, Monneron & Bernhard 1969). By fluorescence microscopy in both fixed and living cells, CBs appear as bright nuclear foci, typically one to six in number, depending on the cell type (**Figure 1**). The size and number of CBs vary during the cell cycle and are maximal at the G1/S-phase (Andrade et al. 1993). Indeed, CBs disassemble during M-phase, and their subsequent reassembly is dependent upon the transcriptional status and growth rate of the cell (Carmo-Fonseca et al. 1993, Fernandez et al. 2002). Interestingly, CBs are also motile. Studies in both plant and animal cells have shown that they can make large movements during interphase, traversing the full diameter of the nucleus in some cases at rates up to ~1 μm min^{-1} (Boudonck et al. 1999, Platani et al. 2000). Individual CBs can show complex dynamic behavior, including moving together and fusing in the nucleoplasm, splitting into two daughter bodies, and reversing movement to and from nucleoli. In cases where CBs split into two smaller bodies, some components, such as fibrillarin, are seen to partition differentially between the daughter structures, whereas other components, such as coilin, are present at similar levels in both (Platani et al. 2000). This is consistent with the view that CBs represent a heterogeneous collection of related structures that can differ in their precise molecular composition and possible biological roles.

Much of the CB movement appears to occur via simple or constrained diffusion, although in some cases it is possible that more active processes are involved (Gorisch et al. 2004, Platani et al. 2000). Not all CBs are mobile at any one time, and in most cases they remain tethered within a confined nuclear volume, probably through interactions with specific regions of chromatin (Platani et al. 2002). This likely includes interactions with gene loci such as snRNA, snoRNA, and histone gene clusters (**Figure 2**) (Frey & Matera 1995, 2001; Gao et al. 1997; Jacobs et al. 1999; Schul et al. 1998; Shopland et al. 2001; Smith et al. 1995; Smith & Lawrence 2000). In the case of

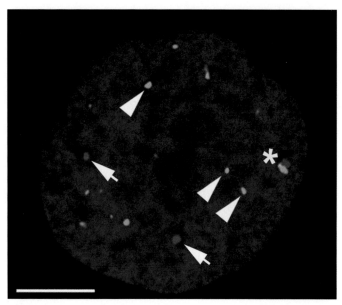

Figure 1

Cajal bodies and promyelocytic leukemia protein (PML) bodies in human cells. Anti-coilin antibodies label CBs (*red*), which appear as spherical, bright foci distributed throughout the nucleoplasm. PML bodies are stained in green, detected here using an anti-PML antibody. Note that some PML bodies and CBs are located in close proximity (*asterisk*). DNA is stained with DAPI (4',6-Diamidino-2-phenylindole: blue). Size bar: 5 μm.

U snRNAs, Matera's group, by studying the in situ localization of repetitive arrays of U2 snRNA–coding cDNAs, has elegantly shown that their association with CBs depends upon the transcriptional status of the locus and requires the presence of nascent U snRNA transcripts and U2 snRNPs (Frey et al. 1999, Frey & Matera 2001). It is possible that CBs play a regulatory role influencing the expression of these genes. In favor of this idea is the accumulation into CBs of NPAT (nuclear protein mapped to the AT locus), a component shown to activate transcription of histone promoters (see below). However, the requirement for U2 snRNPs indicates the non-exclusive possibility that the association of CBs with specific gene loci may also be coupled to the efficient delivery of RNPs required for the processing or maturation of nascent RNA transcripts or in the recycling of RNP components involved in such processes. As well as localizing to specific gene loci, CBs can also be found

Small nuclear RNAs (snRNAs): a class of small RNAs, components of the snRNP particles involved in the pre-mRNA processing

Small nucleolar RNAs (snoRNAs): a class of small RNAs, components of the snoRNP particles involved in the modification and processing of rRNA

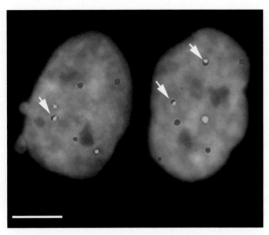

Figure 2

Cajal bodies are spatially associated with U2 snRNA gene loci. Combined detection of CBs (anti-coilin antibody, *red signal*) and U2 snRNA gene loci [fluorescence in situ hybridization (FISH), *green signal*] in HeLa cell nuclei. DNA is visualized by DAPI staining (*blue*). Arrows indicate adjacent domains. Size bar: 10 μm. Courtesy of A.G. Matera, Case Western Reserve University, Cleveland, OH.

sometimes in close proximity to other classes of nuclear bodies. In line with this idea is the observation that CBs can occur close to cleavage bodies, nuclear structures enriched in mRNA 3′ cleavage and processing factors in human bladder carcinoma cells (Schul et al. 1996). CBs can also be found in proximity to PML bodies (**Figure 1**), which are nuclear domains whose complex and heterogeneous protein composition suggests their involvement in many different biological functions, including control of apoptosis, response to DNA damage, antiviral response, and transcriptional modulation (see below). The adjacency of CBs to cleavage bodies and PML bodies possibly reflects the existence of intranuclear trafficking pathways, supporting a continuous and dynamic exchange of components that can ultimately modulate the functional activity of the involved domains.

CAJAL BODIES COMPONENTS

In vivo imaging studies have provided detailed information about the temporal dynamics governing the localization of CB com-

ponents. Indeed, all CB components analyzed so far are dynamic, with relatively fast turnover rates, varying from seconds to minutes (Dundr et al. 2004, Sleeman et al. 2003). As our aim here is to consider the function of CBs as organelles and how this function relates to the localization of specific components, we tentatively subgroup the CB-associated nucleic acid and protein components into potentially functional categories. We believe that this criteria, although not ideal, may help to guide the interpretation of the often preliminary data available. We note that for many CB components, the biological meaning of their localization in CBs is not yet clear, and caution is needed before automatically assuming that their role in CBs must be identical to known roles that they may play in other cell locations. Furthermore, most of the CB-associated factors have been described as such on the basis of morphological evidence and immunolocalization criteria. Little or no biochemical data are available for many of them. Complementing morphological information with biochemical studies is an important future goal that will provide valuable information about targeting mechanisms, post-translational modification(s), and protein-protein interactions likely to play a role in the assembly/integrity of CBs. We consider first what is known about coilin. Interestingly, the available data about this still-enigmatic protein do not allow us to include coilin easily in any of the functional groups that are described below.

Coilin

The human autoantigen p80 coilin was discovered by Tan and coworkers through the analysis of patient autoimmune sera (Andrade et al. 1991, Raska et al. 1991) and has been widely used as a molecular marker for CBs. In physiological conditions, CBs are represented in virtually all cells of fetal tissues (Young et al. 2001). However, not all cell types in adult tissues show CBs, although coilin is still expressed in adult tissues (Andrade et al. 1993,

Young et al. 2000). Interestingly, it is possible to induce the appearance of CBs, even in cells from adult tissues, by increasing the relative abundance of snRNPs; the snRNP concentration itself is a function of the metabolic state of the cell (Sleeman et al. 2001). Indeed, both normal and transformed cells that have high levels of gene expression and metabolic activity often show prominent CBs (Andrade et al. 1993, Carmo-Fonseca et al. 1993, Fernandez et al. 2002, Ochs et al. 1995). Thus coilin may undergo post-translational modification in a way that is controlled by developmental and metabolic stimuli and related to CB formation. A candidate modification is the phosphorylation of coilin. Hyperphosphorylation of coilin has been related to the disassembly of CBs in mitosis (Carmo-Fonseca et al. 1993) and shown to reduce its self-interacting capacity. Indeed, similar to several other components enriched in nuclear organelles, i.e., PML (Fagioli et al. 1998), self-oligomerization of coilin is strictly required for its targeting to CBs (Hebert & Matera 2000).

A mouse knockout model of coilin has been analyzed and shown to exhibit reduced viability, as coilin$^{-/-}$ animals are under-represented in litters. However, surviving coilin$^{-/-}$ animals are viable. Coilin$^{-/-}$ mouse embryonic fibroblasts (MEFs) lack normal CBs: two distinct types of remnant extranucleolar foci can be identified. One form is enriched in snoRNAs (U3), fibrillarin, and NOPP140 (Tucker et al. 2001); another form of remnant CBs accumulates snRNAs (U2 and U5) guide scaRNAs (small Cajal body–specific RNAs) (Jady et al. 2003). Interestingly none of these altered structures in coilin$^{-/-}$ MEFs accumulates either Sm proteins or the survival of motor neurons (SMN) complex, which suggests that coilin is involved either in the targeting or retention of SMN and snRNP complexes in CBs. The symmetrical dimethylation of arginine residues in coilin could play a role in this process. Treatment of cells with either periodate-oxidized adenosine (ADOX) or 5'-deoxy-5'-methylthioadenosine (MTA), which drastically reduces the amount of intracellular sDMA-modified residues, leads to the dissociation of SMN-enriched domains (gems) (see below) and Cajal bodies (Boisvert et al. 2002, Hebert et al. 2002). Additionally, the integrity and the phosphorylation state of the coilin C terminus seems to modulate the number of CBs in mouse and human cells, but this effect does not apparently change the ability of coilin to self-interact (Bohmann et al. 1995, Shpargel et al. 2003).

Coilin could also have other functions. For example, the fact that CB number and size is increased in transformed cells (Spector et al. 1992), that the coilin-containing gene locus (17q22–23) is amplified in anaplastic (but not benign or atypical) meningiomas (Buschges et al. 2002), and that coilin$^{-/-}$ mice are apparently not prone to develop spontaneous tumors, (A.G. Matera, personal communication) provide good reasons to test whether coilin is endowed with oncogenic properties and determine if its forced overexpression can influence the proliferation status and cell cycle behavior of diploid mortal cells. Because in the absence of coilin the intranuclear localization of snoRNAs and snRNAs in the remnant CBs is dissociated, a function for coilin in coupling, (spatially coordinating) the maturation of snRNPs and snoRNPs in a unique organelle could, hypothetically, support its role in cell proliferation.

Cajal Bodies Are Linked with snRNP and snoRNP Biogenesis

RNA/RNP components. The biogenesis and maturation of snRNAs (and snoRNAs) is tightly linked to their subcellular localization. Newly transcribed snRNAs are exported to the cytoplasm, where they bind a set of common proteins (Sm proteins), which triggers their nuclear re-import. In the nucleus, the snRNP particles accumulate in CBs and subsequently in the interchromatin granule clusters (IGCs), which are enriched in splicing factors (Sleeman & Lamond 1999). The snRNAs present in CBs contain a

Small nuclear and nucleolar ribonucleoprotein particles (snRNPs and snoRNPs): nuclear particles composed of a tight complex between a short RNA molecule (snRNA or snoRNA) and proteins

Guide RNA: small RNA molecules that, thanks to complementary base-pairing interactions, guide enzymatic activities to the sequence to be modified or processed

Pseudouridylation: a naturally occurring modification of many stable RNA sequences resulting from the isomerization of uridine residues to pseudouridines

Spliceosome: a high-molecular-weight multiprotein complex that catalyzes the excision of introns from pre-mRNA molecules that leads to the formation of mature mRNAs

hypermethylated 5′ cap, which indicates that they have been re-imported from the cytoplasm and do not correspond to nascent transcripts (**Figure 3*a,b***). In order to give rise to mature, fully functional splicing particles, the RNA component of the snRNPs undergoes several modifications, including 2′-*O*-ribose-methylation and pseudouridylation. At least some of these RNA modifications take place in CBs and are introduced by a recently discovered class of small Cajal body-specific RNAs (Darzacq et al. 2002, Jady et al. 2003). These are members of the so-called guide RNA family, which functions to align RNA modification activities with target sequences through complementary base-pairing interactions. Given that the number of guide RNAs identified to date is still far less than the number of modified nucleotides identified within spliceosomal snRNAs, it is likely that other guide RNAs will be identified. It will be interesting to see how many of these novel guide RNAs will be localized in CBs.

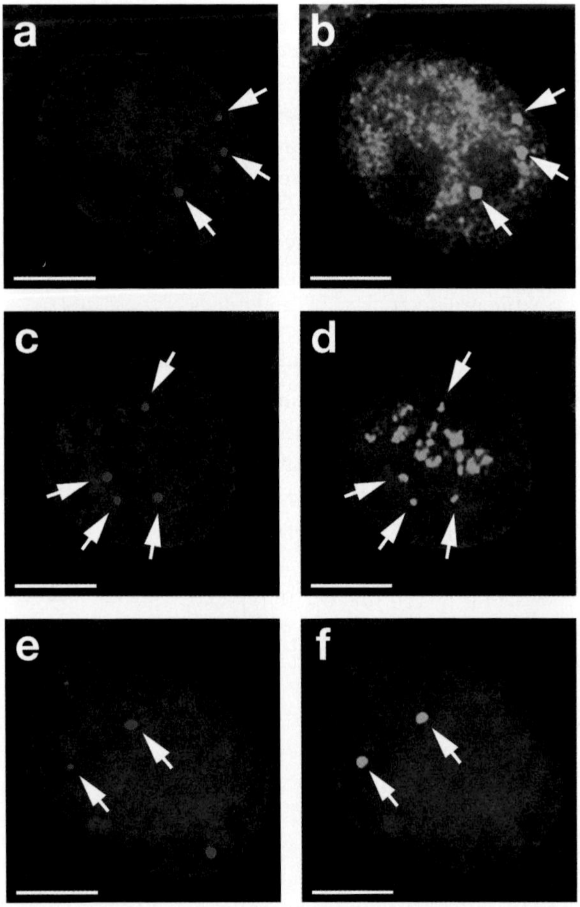

So far, the spliceosome U snRNPs (Carmo-Fonseca et al. 1992, Matera & Ward 1993), the U7 snRNP involved in histone 3′-end processing (Pillai et al. 2001), the U3, and many other snoRNPs (Boulon et al. 2004, Verheggen et al. 2002) involved in processing of pre-rRNA have been definitively shown to be enriched in CBs. Consistent with CBs having a role in snRNP maturation, overexpression of snRNP components appears to promote CB assembly (Sleeman et al. 2001). The RNA component of the telomerase RNP, the telomerase RNA (hTR), is also enriched in CBs (Jady et al. 2004, Zhu et al. 2004), and its targeting could be mediated by SMN (Bachand et al. 2002) and by the presence of sequences homologous to the CB-targeting region of scaRNAs in its 3′-terminal region (Jady et al. 2004) (see below).

NOPP140 and fibrillarin, together with coilin, are among the first CB components identified. Whereas more is known about the nucleolar functions of these proteins, their CB-associated functions are not yet clear. Fibrillarin is an essential (Newton et al. 2003) and highly conserved nucleolar and CB-associated protein (**Figure 3*c,d***). It acts in the nucleolus as a ribose 2′-*O*-methylase targeted to specific sites of RNA modification through its association with a guide snoRNA complementary in sequence to the RNA surrounding the modification site (Feder et al. 2003). NOPP140 is a nucleo-cytoplasmic shuttling phosphoprotein (Meier & Blobel 1992) found in both nucleoli and CBs. It interacts with the largest subunit of RNA PolI and thereby affects rDNA transcription and hence nucleolar biogenesis.

Figure 3

Examples of Cajal bodies components: CBs costained with anti-coilin antibodies (*red signal*: panels *a,c,e*) and in green (*panels b,d,f*).
(*b*) Anti-fibrillarin antibody, which decorates both nucleoli and CBs;
(*d*) anti-TMG CAP antibody; (*e*) anti-NPAT antibody. DNA is stained with DAPI (4′,6-Diamidino-2-phenylindole: blue). Size bar: 5 μm.

In addition to their nucleolar functions, other properties of fibrillarin and NOPP140 link these proteins to CBs. Indeed, NOPP140 has been shown to interact directly with coilin (Isaac et al. 1998), whereas fibrillarin interacts with SMN (Jones et al. 2001, Pellizzoni et al. 2001), and both proteins bind to snoRNPs. Fibrillarin binds to the C/D box snoRNAs; NOPP140 binds to C/D and H/ACA classes of snoRNAs. This suggests that both proteins could act as chaperones for the final stages of maturation of snoRNPs en route to nucleoli. Interestingly, despite the direct interaction with SMN and coilin proteins, both fibrillarin and NOPP140 are still targeted to the remnant CBs in coilin$^{-/-}$ MEFs, which lack coilin and do not accumulate SMN. Although this targeting could be mediated by the binding to snoRNAs that still accumulate in the remnant CBs structures (Jady et al. 2003), it is also possible that the interaction of fibrillarin and NOPP140 with CBs is related to an alternative CB function that has not yet been identified.

Several proteins are involved specifically in the assembly/recycling of U snRNPcomplexes: (*a*) SART3/p110, whose localization to CBs is mediated by the binding to U6 snRNA, requires the presence of coilin and is transcription- and splicing-dependent (Stanek et al. 2003); and (*b*) the U6 snRNP-associated LSm4 and LSm8 proteins (Stanek et al. 2003) as well as the LSm 10, which is a component of U7 snRNPs (Pillai et al. 2001), are present in CBs. The SMN protein is pathogenetically linked to the spinal muscular atrophy (SMA) disease (see below). SMN is present in both the cytoplasm and nucleoplasm where it concentrates in CBs and in the nucleolus. SMN localizes to nuclear structures, called gems, located in close proximity to CBs in fetal tissues and in some cultured cell lines, hence the name meaning gemini or twin of the CBs (Liu & Dreyfuss 1996). However, in adult tissues and in both fetal and adult motoneurons, gems coincide with CBs and are not separate structures (Navascues et al. 2004). SMN plays an important role in the re-import of snRNPs into the nucleus and in the targeting of the U snRNP complexes to CBs (Narayanan et al. 2004). This last function is supported by the existence of an extensive physical and functional crosstalk between SMN and coilin. Thus coilin interacts with SMN through its C terminus (Hebert et al. 2001). Furthermore, de novo imported snRNPs localize almost exclusively to CBs that contain both coilin and SMN and not coilin alone (Sleeman et al. 2001). Moreover, the rearranged CBs in coilin$^{-/-}$ MEFs fail to accumulate both SMN and snRNPs, despite the fact that they still accumulate fibrillarin and that the latter interacts with SMN (Jones et al. 2001, Pellizzoni et al. 2001). Thus the ability of SMN to bind to fibrillarin is not sufficient to either concentrate or retain SMN in CBs in the absence of coilin and/or snRNPs.

Cajal Bodies, Proliferation, and Cell Cycle

CBs change during the cell cycle, and their number and size are maximal at the G1/S-phase transition (Andrade et al. 1993, Fernandez et al. 2002). Furthermore, coilin$^{-/-}$ MEFs show slow growth properties (A.G. Matera, personal communication). Thus it is not entirely surprising that several CB components are functionally related to cell proliferation. Generally, their localization in CBs is increased in highly proliferating cells and, in some cases, their intranuclear accumulation depends on the mitogenic stimulation of the cells, as in the case of both the SMN-interacting ZPR1 putative transcription factor (Galcheva-Gargova et al. 1996, Gangwani et al. 2001) and of an isoform of FGF-2 (fibroblast growth factor-2; 18 kDa) (Claus et al. 2003, Joy et al. 1997). FGF-2 is a member of an expanding class of growth factors whose intranuclear targeting (ignored for years in favor of their better-characterized binding to extracellular membrane receptors) has been shown to be part of an intracrine signaling pathway, which supports their proliferation-

Cyclin-dependent
kinase (cdk): a
protein kinase whose
activity fluctuates
during cell cycle and
that plays
modulatory roles in
cell cycle progression

enhancing abilities (Stachowiak et al. 1997). The presence of FGF-2 and ZPR1 suggests, as observed for PML bodies (Matsuzaki et al. 2003), that mitogenic stimuli can directly signal to CBs.

Pigpen is a member of the TET family of RNA-processing factors (Morohoshi et al. 1998). It is enriched in bovine undifferentiated endothelial cells and retinal pigment epithelial cells and shown to associate with CBs in these cells (Alliegro & Alliegro 1996a). Nuclear microinjection of antipigpen antibodies inhibits the proliferation of endothelial cells, promoting exit from the cell cycle (Alliegro & Alliegro 2002). Pigpen shows 94% identity at the protein level with its human homologue, called TLS, which is involved in the pathogenesis of mixoid liposarcoma (see below). Similar to its human homologue, pigpen can potentially function as both a transcription factor and an RNA-binding protein (Alliegro & Alliegro 1996b).

NPAT has been identified as a cdk2/cyclin E-binding protein (Zhao et al. 1998) and is a large (220 kDa) nucleoplasmic- and CB-associated protein (**Figure 3e,f**). In diploid human fibroblasts, two NPAT-containing CBs can typically be identified and are found adjacent to the histone gene clusters on chromosome 6. In S-phase, two additional NPAT-containing CBs can be observed, both associated with chromosome 1. The number of NPAT-containing CBs increases in transformed, highly proliferating cells (Zhao et al. 2000). Conditional somatic knockout of the protein in HCT116 cells has revealed that its absence blocks the G1/S-phase progression, reduces the levels of histone biosynthesis, and leads to an aberrant localization of coilin when the starved cells traverse the G1/S-phase upon serum induction (Ye et al. 2003). Whether the last phenomenon is a direct consequence of the depletion of NPAT, or an indirect consequence of the cell cycle block, is not yet clear and represents an interesting question to address in the future. In HeLa cells, cyclin E and cdk2 are associated in, or at least co-targeted to, CBs during the G1/S transition (Liu et al. 2000). The role of the CB-associated cdk2/cyclin E complex represents an interesting theater of investigation. Overexpressed NPAT mutants lacking the cdk2/cyclin E phosphorylation sites are still represented in CBs (J.W. Harper, personal communication), suggesting that the phosphorylation of NPAT by the CB-associated cdk2/cyclin E complex does not represent a major targeting signal. Indeed, the CB-associated NPAT is clearly phosphorylated in S-phase, whereas the protein is associated with a subset of CBs throughout the entire cell cycle (Ma et al. 2000). However, DNA damage and overexpression of p21, events that strongly reduce the cdk2/cyclin E kinase activity, trigger dissociation of NPAT (possibly in CBs) from histone loci. This event is concomitant with the inhibition of NPAT phosphorylation and block of histone synthesis and cell cycle progression (Su et al. 2004). Therefore, although the CB-associated cdk2/cyclin E activity seems to be dispensable for the recruitment of NPAT to CBs, it could promote the interaction of NPAT with other partners, an event ultimately relevant to the control of histone gene expression and, consequently, of cell cycle progression. Future studies will clarify this issue.

A further potential connection between the CB localization of NPAT and cell cycle progression comes from the recent demonstration of a physical and functional interaction between NPAT and CBP (CREB-binding protein). NPAT interacts with CBP and the overexpression of the two partners accelerates the entry of cells into S-phase (Wang et al. 2004). CBP is a component of PML bodies that is important for cell cycle progression. However, it is still not clear whether a subset of CBP is localized to CBs in addition to PML bodies.

Cdk7/cyclin H/mat1 is a stable, trimeric complex that has been experimentally proven to be a CAK (cyclin-dependent-activating kinase) complex. It mediates the phosphorylation-dependent activation of cyclin-dependent kinases (CDKs), especially cdk2/

cyclin A and cdk2/cyclin E, thus playing a role in regulating cell cycle progression. Association of the CAK complex with CBs has been demonstrated (Jordan et al. 1997). The role of the CB-associated CAK complex could be more intricate than it appears. Indeed, it has been shown that the presence of the mat1 subunit switches the specificity of the complex toward the p53 protein, which interacts physically with both cyclin H and mat1 (Ko et al. 1997). Conversely, p53 can negatively influence the activity of the CAK complex toward cdk2 and the PolII CTD domains (Schneider et al. 1998). p53 has been shown to accumulate in a large percentage of CBs, probably through interaction with SMN, in conditions of cell stress (Young et al. 2002). In this case, p53 itself could act as a substrate of the associated CAK complex or negatively modulate the activity of the complex toward cdk2/cyclin E. This event can potentially lead to a reduction in the rate of histone biosynthesis (i.e., by reducing the cdk2/cyclin E activity toward NPAT) and to the consequent delay of G1/S progression. Even if it is not possible to exclude that the critical functional interactions take place outside of CBs, the fact that the players are all localized in the same nuclear body is intriguing. This provides a potential conceptual framework for a CB-associated modulation of the cdk2/cyclin E complex in different environmental conditions.

Cajal Bodies and Stress

The previously mentioned interaction of SMN and p53 and the fact that the two proteins colocalize in CBs upon proteasome inhibition and DNA damage is intriguing (Young et al. 2002). p53 is indeed an exquisite sensor of cell stress and plays important roles in cell cycle regulation, DNA repair, control of transcription, and apoptosis (Jin & Levine 2001). Although the meaning of the SMN/P53 interaction is still not clear, the fact that it takes place in a large fraction of these organelles (50–75% of CBs in HeLa cells) (P. Young & C. Lorson, personal communication) suggests

that it likely represents a biologically relevant phenomenon, potentially involving CBs in the cellular response to stress. Interestingly, in addition to SMN, other CB components can interact with p53: For example, the double-strand RNA-activated kinase PKR localizes in CBs and in nucleoli (Jimenez-Garcia et al. 1993) and binds p53 in vivo and can phosphorylate it in vitro (Cuddihy et al. 1999). Furthermore, activation of the PKR (and RNAse L) pathway constitutes the major known mode of intracellular antiviral defense that leads ultimately to the block of host protein synthesis (Flodstrom-Tullberg et al. 2005). Intriguingly, it has been shown that during adenovirus infection CBs become redistributed in hundreds of coilin and fibrillarin-containing microfoci, which are dispersed throughout the nucleoplasm and, even more interestingly, that this redistribution can be recapitulated by the treatment of cells with protein synthesis inhibitors (Rebelo et al. 1996, Rodrigues et al. 1996). These similarities imply a possible involvement of PKR in the adenovirus-induced disruption of CBs. Other evidence also suggests that CBs could represent stress-responsive domains, e.g., the localization of peroxiredoxin V in CBs (Kropotov et al. 2004). As a new entry in the growing collection of Cajal bodies components, peroxiredoxin V is a conserved protein endowed with thioredoxin peroxidase activity and also localized in mitochondria and peroxisomes. Interestingly, peroxiredoxin V has been shown to counteract, at least partially, the increase in intracellular ROS (reactive oxygen species) and prevent the apoptosis induced by p53 overexpression (Zhou et al. 2000). An intriguing possibility is that its presence in CBs is related to a possible local regulation of redox potential. Thus a local redox imbalance could either activate or inactivate stress-responsive CB components, possibly including p53, and be responsible for the observed structural and physical changes of CBs in stress conditions. Ergo, many stress-related CB components seem to be directly or indirectly functionally linked to p53. This raises the interesting

Proteasome: a high-molecular-weight multiprotein complex that catalyzes the degradation of ubiquitinated substrates

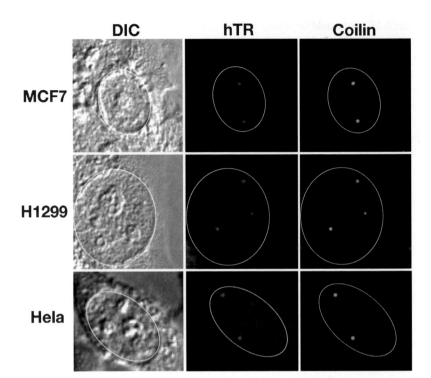

| DIC | hTR | Coilin |

MCF7

H1299

Hela

Figure 4

Telomerase RNA is enriched in CBs. Combined detection of coilin and telomerase RNA (FISH, *red signal*) in CBs (anti-coilin antibody, *green signal*) in three different cancer cell lines. White lines indicate the nuclear perimeter. Size bar: 10 μm. Courtesy of M. Terns, Univ. Georgia, Athens, GA.

question as to whether a CB-associated modulation of p53 takes place in stress conditions. The fact that p53 has been detected in a subset of CBs provides another clue that they represent a heterogeneous family of nuclear bodies.

Cajal Bodies and Aging

Telomerase is an RNP reverse transcriptase that uses an RNA template to synthesize telomeric DNA repeats at the ends of the chromosomes. The enzymatic complex is minimally composed of an internal RNA template (hTR) and a catalytic protein subunit (hTERT). In normal, untransformed cells, the telomeres progressively shorten at each cell division. This contributes to genomic instability (Hande et al. 1999). The progressive shortening of telomeres is an important aspect of the cellular senescence program, because older cells with shorter telomeres will undergo apoptosis and thus be prevented, in physiological conditions, from accumulat-

ing mutations and chromosomal aberrations over time. In cancer, this mechanism of regulation is subverted and there is generally an increase in telomerase activity (Hahn & Meyerson 2001). Consequently, cancer cells do not undergo the physiological shortening of telomeres, which contributes to their immortalization (Satyanarayana et al. 2004). Recent FISH studies from two independent groups revealed that the hTR component is enriched in CBs, especially during the S-phase of the cell cycle (**Figure 4**). Localization of hTR requires the presence of a conserved scaRNA motif in its sequence (Jady et al. 2004) and the expression (endogenous or ectopic) of the catalytic subunit hTERT (Zhu et al. 2004). We note that these data support a possible role for CBs either in the maturation of the telomerase RNA component and/or in the assembly of the active RNP complex. However, several observations warrant further investigation. For example, except for the presence of the hypermethylated 5′ cap structure, no known modifications have been found in mature hTR that could be specifically attributed to its residency in CBs (Jady et al. 2004). Despite the high frequency of colocalization of hTR with CBs, little or no colocalization/adjacency has been reported between CBs and telomeres. This raises the possibility that CBs could act as storage sites and deliver components of the telomerase complex when needed, but it would be expected that a transient colocalization of CBs with telomeres, at least during S-phase, should be observed; so far this has not been reported. Furthermore, the localization of hTR in CBs seems to be specific for cancer cells. In primary cell lines, even in the case of overexpression of hTERT, an event that readily triggers hTR accumulation in their transformed counterparts, the localization of telomerase RNA in CBs is generally not observed (Zhu et al. 2004). One simple explanation could be that CBs are not evident in some primary cell lines (Carmo-Fonseca et al. 1993, Spector et al. 1992). However, it is also possible that in mortal diploid cells, stimuli other than hTERT overexpression could be required for hTR accumulation in CBs. Alternatively, the cancer cell-specific enrichment of hTR in CBs could reflect the high degree of spatial disregulation of nuclear organization in transformed cells (Wong et al. 2002). Notably, the SMN protein, which interacts directly with coilin, binds to components of the telomerase complex and SMN-containing immunoprecipitates show telomerase activity in vitro (Bachand et al. 2002). In coilin$^{-/-}$ MEFs, the residual CB-like structures fail to recruit SMN, although it is not clear whether they can accumulate hTR. Based on what is known about the role of telomerase in aging, we predict that the coilin knockout mice might show signs of premature aging or that the derived coilin$^{-/-}$ MEFs would be more prone to undergo replicative senescence in vitro owing to a dysfunctional telomerase complex. So far there is no clear evidence indicating that this is the case, but it will be interesting to test this prediction in the future.

Cajal Bodies and the Transcription Factor Link

Despite the fact that no nascent RNA has been shown to localize in CBs (Cmarko et al. 1999, Raska 1995), several CB-associated factors either have transcriptional activation domains (**Table 1**) or are involved in active transcription processes, and the CBs themselves can be surrounded by transcription sites (Jordan et al. 1997). The reason why these transcription-related factors are associated with CBs is puzzling. Little is known about either the mechanism of action or the target genes for these factors, except in few cases, such as the factors involved in the transcriptional activation of snRNA genes and ELL/EAF1.

The 45 kDa-subunit of the snRNA gene-specific transcription factor PTFγ has been found enriched in domains partially overlapping with CBs, along with a subset of components of the RNA polymerase II complex, i.e.,TBP, TFIIH, TFIIF, and the hypophosphorylated form of RNA PolII. The hyperphosphorylated form of PolII is instead enriched in dots localized at the periphery of CBs (Schul et al. 1998). ELL is an RNA polymerase II (PolII) transcription elongation factor that interacts with the transcription factor EAF1 (ELL associated factor-1). Both ELL and EAF1 are components of CBs (Polak et al. 2003). Interestingly, ELL has been shown to interact with p53 and to inhibit both sequence-specific transactivation and sequence-independent *trans*-repression by p53 (Shinobu et al. 1999). It is possible that CBs could function as storage sites for transcription factors that could be mobilized when needed to meet the transcriptional requirements of adjacent loci (histone or snRNA loci). However, considering that only a subset of CBs are associated with snRNA and histone loci in steady-state conditions, we consider another possibility, namely that CBs could be involved in the transport of components to chromatin loci during specific conditions, such as stress responses, to promote or

Transcriptional activation domain (TAD): a protein domain that mediates the interaction with the basal transcription machineries, thus promoting an increase in the transcriptional rate of target genes

TABLE 1 Schematic summary of CB components described in this review

	Endogenous	Over-expressed	Morphological evidence	Biochemical interaction	Coilin dep targeting	CB-targeting signal	TAD
Coilin	•	•	•	pos (SMN)		•	
Fibrillarin	•	•	•	pos (SMN)	No		
NOPP140	•		•	pos (coilin)	No		•
SART3/p110	•		•		Yes	•	
Lsm proteins	•	•	•		Yes		
Sm proteins	•	•	•	pos (SMN)	Yes		
SMN	•	•	•	pos (coilin)	Yes	•	
ZPR1	•		•	pos (SMN)			•
FGF-2	•		•				
Pigpen	•		•				•
NPAT	•	•	•				•
Cdk2/cyclin E	•		•				
Cdk7/ cyclinH/mat1	•		•				
TFIIH	•		•				•
EAF1 (EAF2)	•		•	Neg			•
ELL	•	•	•	Neg			
PTFγ	•		•				•
p53	•		•	pos (SMN)			•
Peroxiredoxin	•		•				•
PKR	•		•				
Profilin I	•		•	Neg			
Ataxin-1		•	•	pos (coilin)			
FRG1		•	•				
hTERT		•	•				

TAD, transcriptional activation domain; pos, biochemical data available and interaction demonstrated; Neg, biochemical data available, interaction not demonstrated; Empty space, no data available at present.

enhance the transcription of responsive genes. CBs are known to be motile structures that can undergo dramatic changes in number, size, shape, and intranuclear distribution. Although most of the movements of CBs appear to involve either simple or anomalous diffusion (Platani et al. 2000, 2002), it is still possible that some CB movement may involve forms of active transport. In this regard, it is interesting that profilin I has been localized in CBs. Profilin is an essential and conserved protein whose historically described localization is cytoplasmic and whose functions are linked to the modulation of the nonmuscle actin polymerization, achieved through the formation of a tightly controlled complex with G-actin (Witke 2004). Interestingly, the localization of profilin I in CBs requires active transcription (Skare et al. 2003). Profilin I can possibly bind to nuclear actin and thus influence both the ATP-dependent intranuclear movements and the assembly of transcription complexes, events that require actin (Bettinger et al. 2004). Although a previous study did not observe a change in CB movement after treating cells with actin-depolymerizing agents (Platani et al. 2002), the recent finding that CBs are enriched in actin and that the actin content of CBs is dynamically altered by adenovirus infection adds interest to this hypothesis (Gedge et al. 2005). Future studies are needed to assess whether a complex of profilin I and G-actin exists in CBs and whether the CB-associated pool of actin is related to their intranuclear movements.

CAJAL BODIES IN OTHER ORGANISMS

CB homologues, or similar structures, have been reported in many types of organisms apart from mammalian cells. It is beyond the scope of this review to provide detailed information about the structure and function of CB-like structures in other organisms. However, we do consider some features of non-mammalian CBs to illustrate what could be common principles of nuclear structure and function that have been conserved in eukaryotic cells, for example, the need to coordinate the temporal maturation of nuclear functional complexes with their spatial localization.

Xenopus Cajal bodies.   The homologue of human coilin in *Xenopus* is called SPH-1. It is a component of the so-called sphere organelle (Wu et al. 1994), which is a large structure (>10 μm in diameter) present in variable numbers (50–100) in the germinal vesicle of *Xenopus* oocytes (**Figure 5**). The sphere organelles have been the object of extensive study mainly by Gall's group, often anticipating results that later have been confirmed in mammalian CBs. For example, studies in *Xenopus* provided the first demonstration that U7 snRNAs are concentrated in sphere organelles (Wu & Gall 1993). As in human CBs, *Xenopus* sphere organelles contain, inter alia, coilin (SPH-1), fibrillarin, NOPP140, snoRNAs, and histone mRNA 3′-end-processing factors, such as U7 snRNA and SLBP1. For this reason, Gall proposed that sphere organelles be given the name Cajal bodies.

Despite these striking similarities between mammalian CBs and *Xenopus* CBs/sphere organelles, it is important to note that there are also apparent differences between these struc-

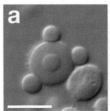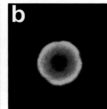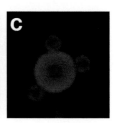

Figure 5

Xenopus CBs. Micrographs representing a (*a*) DIC image of a *Xenopus* CB surrounded by three snurposomes; (*b*) staining of a *Xenopus* CB with an anti-coilin antibody (*green signal*); (*c*) staining (anti-TMG CAP antibody, *red signal*) of TMG-capped RNA-containing structures: the TMG cap is a CB-associated antigen in both amphibian and human cells. Size bar:10 μm. Courtesy of J.G. Gall, Carnegie Institution of Washington, Baltimore, MD.

tures. First of all, *Xenopus* CBs are morphologically distinct with respect to mammalian CBs, being composed of a spherical structure with an internal matrix and often associated with specialized small, rounded substructures enriched in snRNPs, called B snurposomes (**Figure 5**). Interestingly, *Xenopus* CBs contain components of all three types of RNA polymerase complexes and also factors involved in the cleavage and polyadenylation of mRNA transcripts, such as symplekin, cstf 77, and cpsf100 (Hofmann et al. 2002). In mammalian cells, not all of the components of polymerase complexes have been observed to date in CBs, and cleavage and polyadenylation factors, such as cstf 64 and cpsf 100, have been localized instead in cleavage bodies, which are domains sometimes found adjacent to CBs (Schul et al. 1996). Another interesting difference between *Xenopus* and mammalian CBs is represented by the lack of a spatial association of the sphere organelles with the U1- and U2-snRNA gene loci (Abbott et al. 1999). Interestingly, cleavage bodies do not associate with U1- and U2-snRNA gene loci adjacent to CBs (Schul et al. 1999). This similarity raises the possibility that sphere organelles represent prototypical CBs whose components, which in *Xenopus* are grouped in a single type of organelle, have been redistributed, during evolution, to multiple specialized structures: CBs and cleavage bodies. An alternative possibility is that *Xenopus* CBs could instead constitute a specific subset of the CBs observed in mammalian cells, possibly specialized for the unique requirements of the amphibian oocyte.

Cajal body-like structures in *Drosophila*. The characterization of a *Drosophila* CB (similar to the situation in yeast and *Caenorhabditis elegans*) has been limited by the absence of an obvious coilin homologue in these organisms. Surprisingly, the use of an antibody raised against a C-terminal portion of recombinant human coilin (Andrade et al. 1993) has allowed the visualization of CB-like structures in *Drosophila* neurons (Yannoni & White 1997). It is possible that other CB proteins, perhaps sharing conserved structural motifs with coilin, can functionally substitute for the role played by coilin in mammalian CBs. The presence of coilin-like factors in CB-homologous structures in invertebrate species clearly underlines the evolutionary conservation of this nuclear body.

Cajal bodies in plants. CBs have been widely observed in plant cells, and all plant species studied so far have CBs (Shaw & Brown 2004). The advent of suitable probes for detecting plant CBs and of advanced tools for live-cell-imaging studies has shown that, in common with the human CBs, tobacco and *Arabidopsis* CBs are heterogeneous, dynamic structures. They can move through the nucleoplasm, fuse with each other, and move in and out of the nucleolus (Boudonck et al. 1999). Their number changes during plant cell differentiation and during the cell cycle (Boudonck et al. 1998) and, interestingly, with the metabolic activity of the cell (Acevedo et al. 2002). Plant CBs accumulate snRNPs that, strikingly, can be recognized by antibodies produced against their human counterparts, e.g., U2B'' (Beven et al. 1995). The presence of common epitopes and the ability to accumulate similar subsets of U snRNAs clearly suggest that plant and human CBs are related structures that can play similar roles in plant and animal cells.

The yeast nucleolar body. In budding yeast grown on solid media, it is possible to observe an intranucleolar structure, called, for this reason, the nucleolar body (NB). The yeast NB accumulates both mature and precursor forms of U3 snoRNAs and is enriched in TGS-1, a conserved methyltransferase that catalyzes the formation of the 5' terminal tri-methyl-CAP structure, characteristic of most sno- and snRNAs (**Figure 6**) (Verheggen et al. 2002). The same components are enriched in mammalian CBs. Furthermore, mammalian CBs can also be observed within nucleoli under a variety of circumstances. For example, intranucleolar CBs have been reported in human breast carcinoma cells (Ochs et al. 1994) and also in liver cells of hibernating dormice (Malatesta et al. 1994). In addition, treatment of HeLa cells with the phosphatase inhibitor, okadaic acid, or transient expression of a single serine to aspartate mutant of coilin promotes the formation of CB-like structures within nucleoli that contain coilin and snRNPs (Bohmann et al. 1995, Lyon et al. 1997, Sleeman et al. 1998).

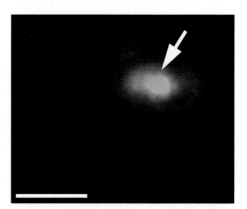

Figure 6

The yeast nucleolar body (NB). A TGS1-GFP fusion protein (*green signal*) is selectively enriched in the yeast NB, a CB-like structure located within the nucleolus (*arrow*). Staining for U3 snoRNA (FISH, *red signal*) is concentrated in the NB as well as in other regions of the nucleolus. The DNA is stained with DAPI (*blue signal*). Size bar: 1 μm. Courtesy of E. Bertrand, Montpellier, France.

LINKS TO DISEASE

There are increasing examples where the detailed organization of the cell nucleus is altered in specific pathologies. Whether these effects are direct or indirect, they likely reflect the disruption of cellular function contributing to the disease state and may provide useful clinical markers as well as possible new insights into the underlying pathogenic mechanisms. In the following discussion we consider examples of human diseases that affect nuclear structure and CB organization.

The expansion of a CAG triplet in an otherwise wild-type protein is a pathological hallmark of a class of hereditary neurodegenerative disorders including at least 14 known syndromes (Costa Lima & Pimentel 2004): for example, DentatoRubral-PallidoLuysian Atrophy (DRPLA), Machado-Joseph disease (MJD), and Spino-Cerebellar-Ataxia type 1 (SCA-1). Severe neuromuscular defects associated with cognitive and behavioral deficits in the DRPLA are a common trait in long-term affected patients. The CAG triplet diseases are characterized by the intranuclear accumulation of neurotoxic cleavage products of the mutant proteins and by a selective loss of neuronal subpopulations (Goti et al. 2004). One of the proposed mechanisms responsible for this phenomenon is the transcriptional deregulation of key survival/antiapoptotic pathways. In DRPLA and MJD the cleavage products of the mutated atrophin-1 and ataxin-3 proteins, respectively, accumulate in membrane-free, spherical aggregates composed of electron-dense material with an amorphous or fibrillar morphology. Interestingly, such structures show striking spatial relationships with CBs in neurons of affected patients. EM analysis has revealed that CBs are either in direct contact with the aggregates or connected to them through filamentous structures (Yamada et al. 2001). The fact that structurally different proteins, such as atrophin-1 and ataxin-3, show similar spatial relationships with CBs strongly suggests that the presence of poly-Q stretches plays

a role in this phenomenon. The biological mechanism and significance of such relationships is not clear. It has been shown that poly-Q protein aggregates are capable of trapping poly-Q-containing transcription factors, even though it is not generally accepted whether this event is directly responsible for transcriptional deregulation. The transient association of transcription factors with CBs could be an explanation for their location adjacent to the poly-Q-containing aggregates. Alternatively, it is also possible that CBs are attracted to the aggregates because of their ability to supply stress-responsive factors. This is in line with the emerging idea that the formation of poly-Q-containing aggregates could represent a cellular stress response mechanism (Arrasate et al. 2004).

A recent report has provided evidence for a physical association between ataxin-1, the protein mutated in SCA-1 (Klement et al. 1998), and coilin. When both proteins are overexpressed, ataxin-1 interacts with and colocalizes with coilin in the ataxin-1-positive inclusions (**Figure 7**) (Hong et al. 2003), and the strength of interaction is not apparently influenced by the number of the ataxin-1 CAG repeats. Thus the interaction between ataxin-1 and coilin does not seem to take place within CBs (Skinner et al. 1997). However, because nucleoplasmatic coilin exchanges quickly and continuously with the CB-associated pool, it cannot be excluded that ataxin-1 can be

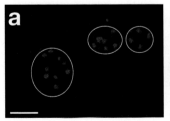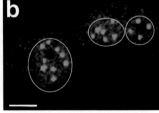

Figure 7

Exogenously expressed ataxin-1 and coilin colocalize. When both coilin (*a*) (*red signal*) and ataxin-1(*b*) (*green signal*) are exogenously overexpressed, they colocalize in ataxin-1-positive inclusions. White lines indicate the nuclear perimeter. Size bar: 10 μm. Courtesy of S. Kang, Korea University, Seoul.

Nuclear speckles:
irregularly shaped
nuclear domains that
likely represent
storage and assembly
sites for most
splicing factors

transiently associated and functionally connected to CBs via its interaction with coilin. In line with this idea, it is noteworthy that ataxin-1 nuclear inclusions recently have been shown to break down upon transcriptional inhibition and to contain RNA-transport factors (Irwin et al. 2005), both features common to CBs (Boulon et al. 2004).

Spinal Muscular Atrophy (SMA) is the most common genetically determined neurodegenerative disease in children in the United States (Pearn 1980). It is caused by either deletion or loss-of-function mutations of the telomeric copy of the duplicated SMN genes that lead, inter alia, to a reduction in the nucleoplasmic levels of SMN (Lefebvre et al. 1997). The clinical manifestations are characterized by a progressive and profound atrophy of the voluntary muscles of the limbs and trunk, which can be variable, depending on the residual levels of protein expression. All the different SMA subtypes are characterized by a loss of motor neurons of the spinal cord. Without forgetting that a pathogenic role could be played by cytoplasmic defects in the axonal transport of RNAs induced by the SMN deficiency (Zhang et al. 2003), it is also possible that CBs play a role in the SMA scenario. It is of note that in SMA there is a defect in the intranuclear targeting of SMN associated with loss of discrete SMN and coilin-containing foci (Frugier et al. 2000): thus less SMN is available for the interaction with CB components. Furthermore, an RNA-binding protein, BRUNOL3, found upregulated in muscle biopsies from SMA patients, has been shown to interact and colocalize with SMN in gems in a motoneuronal cell line (Anderson et al. 2004). Notably, in motoneurons, gems coincide with CBs. On the other hand, the fact that coilin$^{-/-}$ mice apparently do not develop clear phenotypical signs of the SMA disease argues against a dramatic involvement of CBs in the pathogenesis of SMA. Additional data characterizing the localization of BRUNOL3 in CBs will be needed to define the requirement for CBs in the SMN-BRUNOL3 interaction/function. Another interesting experiment would be to cross coilin$^{-/-}$ mice and SMN$^{-/+}$ mice [SMN$^{-/-}$ mice are not viable (Schrank et al. 1997)] in order to unravel any possible contribution of coilin to SMN function, for example, at an embryonal stage.

Recently FRG1, a protein encoded by a candidate gene for the FSHD (facio scapulo humeral distrophy), when overexpressed, has been found to localize to CBs, in addition to nucleoli and ICGs (van Koningsbruggen et al. 2004). The function of FRG1 is still unknown. However it is intriguing to note that it represents another example of a growing group of proteins whose involvement suggests a link between RNA metabolism, neuromuscular disorders, and nuclear organization.

Both the nuclear localization of coilin and the morphology of CBs are altered in cells expressing an MLL-ELL fusion protein, which is the hallmark of a rare form of leukemia (Polak et al. 2003). As previously mentioned, the transcription factor ELL and its associated proteins, EAF-1 and EAF-2, are CB components, and ELL negatively modulates p53 function. Recent work has shown that the leukemic fusion protein is even more powerful than its physiological counterpart in inhibiting p53, by displacing its binding to P300 (Wiederschain et al. 2003).

The TLS/CHOP fusion protein is the product of a translocation found in more than 90% of mixoid and round cell liposarcomas t(12;16) (Aman et al. 1992). The fusion protein has powerful transforming activity and retains the N-terminal region of TLS, which appears essential for its oncogenic properties, and the entire sequence of CHOP, a stress-induced transcription factor. TLS is a member of the TET family of RNA-processing factors (Morohoshi et al. 1998). Thus the fusion protein can potentially affect both RNA processing and transcription processes. An FP-tagged TLS/CHOP fusion protein is localized to discrete nuclear structures that coincide with nuclear speckles (SC35 domains). Interestingly, a fraction of TLS/CHOP-containing domains overlaps with CBs (Goransson et al. 2002). The localization of TLS/CHOP in splicing speckles is temperature sensitive and is lost when cells are exposed to 25°C or lower before fixation. This would indicate the possibility that TLS/CHOP is subject to intranuclear trafficking, involving movements between both CBs and splicing speckles. Furthermore, the bovine homologue of TLS, called Pigpen, is a CB component (Alliegro & Alliegro 1996a), and our preliminary observations indicate that endogenous TLS is present in CBs (M. Cioce,

unpublished results). It is possible that the overlap between TLS/CHOP-containing domains and CBs could be supported by the binding of the TLS wild-type protein and its oncogenic counterpart to a common interactor or protein complex, likely mediated by the N-terminal portion of TLS. It is worth noting that the targeting of a fusion protein to nuclear domains that contain one or both of its physiological counterparts, has been proven to represent an aspect of oncogenesis. For example, the leukemia-specific PML/RARα

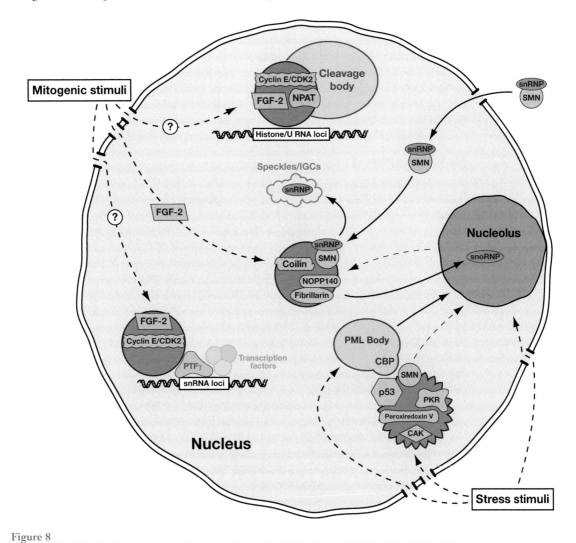

Figure 8

Schematic model illustrating a range of possible functions of CBs discussed in this review. TF: transcription factors, CB: Cajal body.

fusion protein is targeted to PML bodies, leading to their apparent disruption and redistribution in hundreds of nucleoplasmic tiny dots (Dyck et al. 1994). Even more interesting, PML/RARα triggers deacetylation and degradation of p53, which is a component of PML bodies (Insinga et al. 2004), thus contrasting with the stabilizing functions of the wild-type PML protein toward p53 (Bernardi et al. 2004). P53 mutations are rare in leukemia. However, the example of MLL/ELL and of PML/RARα shows that the p53 tumor suppressor protein function can be inactivated in other ways. A possible interference of the MLL/ELL, TLS/CHOP, and, possibly, another TLS-containing leukemic fusion protein, TLS/ERG (Ichikawa et al. 1994) with the function(s) of CB-associated p53, represents an intriguing, if speculative, possibility.

NEW PERSPECTIVES

We have reviewed here recent data about the composition and the behavior of CBs. A significant volume of evidence clearly implicates CBs as having a role in both snRNP and snoRNP maturation, processing of histone mRNA 3′ ends and, probably, the modulation of the expression of snRNA and histone genes. In addition, the presence of new entries in the ever-growing list of CB components warrants further investigation. Indeed, the presence of the telomerase RNA and hTERT in CBs suggests a possible relationship between CBs and cellular aging. The presence of peroxiredoxin V and p53-interacting proteins (SMN, PKR) and of p53 itself in stress conditions opens the possibility that CBs could also be involved in cellular stress response pathways. The fact that the expression of neoplastic fusion proteins can specifically affect the structure of CBs represents an appealing invitation to investigate possible relationships between these conserved nuclear bodies and cancer (**Figure 8**). We look forward to future studies that will shed more light on the function of these old, but always surprising, multifaceted organelles.

SUMMARY POINTS

1. Cajal Bodies are heterogeneous, motile nuclear domains. The number of CBs change during the cell cycle and is influenced by the proliferative and metabolic status of the cell.

2. CBs play a role in the maturation of snRNPs and snoRNPs. CBs are enriched in small guide RNAs that direct the modification of snRNA, a step in the maturation of the spliceosomal snRNP particles.

3. The observation that some components accumulate in CBs preferentially upon mitogenic stimulation of the cells and the presence of both NPAT and of the cdk2/cyclin E complex support a role for CBs in integrating several cell proliferation-related phenomena, such as cell cycle progression and histone biosynthesis.

4. The presence of p53 in CBs during stress conditions and the existence among the CB components of stress-related proteins support a possible involvement of these organelles in the cellular response to stress.

5. CBs of cancer cells are enriched in the human telomerase RNA (hTR) and possibly play a role in the maturation of hTR. This suggests a possible involvement of CBs in cellular replicative senescence.

6. CB-like structures are conserved among mammalian, amphibian, insect, yeast, and plant cells. The fact that these structures accumulate common components (e.g., snRNPs) reveals the existence of evolutionarily conserved functions.

7. The hallmark of an expanding class of neurodegenerative diseases is the accumulation of mutated proteins in intranuclear aggregates, containing abnormally long polyglutamine stretches. CBs are often spatially adjacent to these aggregates. Among the mutated proteins, ataxin-1 interacts with coilin when overexpressed.

8. Expression of the leukemia-specific MLL/ELL fusion protein alters the intranuclear distribution of coilin and the integrity of CBs.

GLOSSARY

Cyclin-dependent kinase (cdk): a protein kinase whose activity depends on the presence of a class of regulatory subunits called cyclins. Cyclins are proteins whose levels of expression fluctuate periodically during the cell cycle. The phosphorylation of target proteins by the cyclin/cdk complexes plays a modulatory role in cell cycle progression

Guide RNA: small RNA molecules that, thanks to complementary base-pairing interactions, guide enzymatic activities to the sequence to be modified or processed

Nuclear speckles: irregularly shaped nuclear domains, typically 20–50 in number per cell nucleus. The consensus is that these domains are storage and assembly sites for most splicing factors

Proteasome: a high-molecular-weight multiprotein complex that catalyzes the degradation of ubiquitinated substrates. Inhibition of the proteasome results in the accumulation of unprocessed substrates and causes cell stress

Pseudouridylation: a naturally occurring modification of many stable RNA sequences resulting from the isomerization of uridine residues to pseudouridines, achieved through the cleavage and reattachment of the base to the sugar

Small nuclear and nucleolar ribonucleoprotein particles (snRNPs and snoRNPs): nuclear particles composed of a tight complex between a short RNA molecule (snRNA or snoRNA) and proteins, involved either in the processing of pre-mRNA (snRNPs) or in the modification and processing of rRNA (snoRNPs)

Small nuclear RNAs (snRNAs): a class of small RNAs, components of the snRNP particles involved in pre-mRNA processing that form base-pairing interactions with the pre-mRNA substrate.

Small nucleolar RNAs (snoRNAs): a class of small RNAs, components of the snoRNP particles involved in the modification and processing of rRNA. SnoRNAs function by coupling the enzymatic activities of the snoRNP complexes to the target sequences, achieved through base-pairing interactions with the substrate RNA

Spliceosome: a high-molecular-weight multiprotein complex that catalyzes the excision of introns from pre-mRNA molecules that leads to the formation of mature mRNAs. It is composed of five snRNPs and many additional protein-splicing factors

Transcriptional activation domain (TAD): a protein domain in a transcription factor responsible for the interaction with the basal transcription machinery that results in an increased transcriptional rate of the target gene

ACKNOWLEDGMENTS

We are indebted to our colleagues Edouard Bertrand, Joseph Gall, Gregory Matera, Seongman Kang, and Michael and Rebecca Terns, who have kindly provided us with the micrographs used in Figures 2, 4, 5, 6, and 7. We are also grateful to Kay Davies, Wade Harper, Chris Lorson, Masato Ota, Jiyong Zhao, and Philip Young for their useful explanations and clear answers to our questions. We thank the members of the Lamond laboratory for their help and encouragement, especially Marco Denegri and David Lleres for their useful comments on this review, and Laura Trinke-Mulcahy for her help with preparing the figures. We apologize to everyone whose work has not been cited for reasons of space and we thank all the authors who have contributed to the Cajal body literature and helped to enhance our understanding of the functions of this fascinating organelle.

LITERATURE CITED

Abbott J, Marzluff WF, Gall JG. 1999. The stem-loop binding protein (SLBP1) is present in coiled bodies of the *Xenopus* germinal vesicle. *Mol. Biol. Cell* 10:487–99

Acevedo R, Samaniego R, Moreno Diaz de la Espina S. 2002. Coiled bodies in nuclei from plant cells evolving from dormancy to proliferation. *Chromosoma* 110:559–69

Alliegro MC, Alliegro MA. 1996a. Identification of a new coiled body component. *Exp. Cell Res.* 227:386–90

Alliegro MC, Alliegro MA. 1996b. A nuclear protein regulated during the transition from active to quiescent phenotype in cultured endothelial cells. *Dev. Biol.* 174:288–97

Alliegro MC, Alliegro MA. 2002. Nuclear injection of anti-pigpen antibodies inhibits endothelial cell division. *J. Biol. Chem.* 277:19037–41

Aman P, Ron D, Mandahl N, Fioretos T, Heim S, et al. 1992. Rearrangement of the transcription factor gene CHOP in myxoid liposarcomas with t(12;16)(q13;p11). *Genes Chromosomes Cancer* 5:278–85

Anderson KN, Baban D, Oliver PL, Potter A, Davies KE. 2004. Expression profiling in spinal muscular atrophy reveals an RNA binding protein deficit. *Neuromuscular Disord.* 14:711–22

Andrade LE, Chan EK, Raska I, Peebles CL, Roos G, Tan EM. 1991. Human autoantibody to a novel protein of the nuclear coiled body: immunological characterization and cDNA cloning of p80-coilin. *J. Exp. Med.* 173:1407–19

Andrade LE, Tan EM, Chan EK. 1993. Immunocytochemical analysis of the coiled body in the cell cycle and during cell proliferation. *Proc. Natl. Acad. Sci. USA* 90:1947–51

Arrasate M, Mitra S, Schweitzer ES, Segal MR, Finkbeiner S. 2004. Inclusion body formation reduces levels of mutant huntingtin and the risk of neuronal death. *Nature* 431:805–10

Bachand F, Boisvert FM, Cote J, Richard S, Autexier C. 2002. The product of the survival of motor neuron (SMN) gene is a human telomerase-associated protein. *Mol. Biol. Cell* 13:3192–202

Bernardi R, Scaglioni PP, Bergmann S, Horn HF, Vousden KH, Pandolfi PP. 2004. PML regulates p53 stability by sequestering Mdm2 to the nucleolus. *Nat. Cell Biol.* 6:665–72

Bettinger BT, Gilbert DM, Amberg DC. 2004. Actin up in the nucleus. *Nat. Rev. Mol. Cell. Biol.* 5:410–15

Beven AF, Simpson GG, Brown JW, Shaw PJ. 1995. The organization of spliceosomal components in the nuclei of higher plants. *J. Cell Sci.* 108(Pt 2):509–18

Bohmann K, Ferreira JA, Lamond AI. 1995. Mutational analysis of p80 coilin indicates a functional interaction between coiled bodies and the nucleolus. *J. Cell Biol.* 131:817–31

Boisvert FM, Cote J, Boulanger MC, Cleroux P, Bachand F, et al. 2002. Symmetrical dimethylarginine methylation is required for the localization of SMN in Cajal bodies and pre-mRNA splicing. *J. Cell Biol.* 159:957–69

Boudonck K, Dolan L, Shaw PJ. 1998. Coiled body numbers in the *Arabidopsis* root epidermis are regulated by cell type, developmental stage and cell cycle parameters. *J. Cell Sci.* 111(Pt 24):3687–94

Boudonck K, Dolan L, Shaw PJ. 1999. The movement of coiled bodies visualized in living plant cells by the green fluorescent protein. *Mol. Biol. Cell* 10:2297–307

Boulon S, Verheggen C, Jady BE, Girard C, Pescia C, et al. 2004. PHAX and CRM1 are required sequentially to transport U3 snoRNA to nucleoli. *Mol. Cell* 16:777–87

Buschges R, Ichimura K, Weber RG, Reifenberger G, Collins VP. 2002. Allelic gain and amplification on the long arm of chromosome 17 in anaplastic meningiomas. *Brain Pathol.* 12:145–53

Carmo-Fonseca M, Ferreira J, Lamond AI. 1993. Assembly of snRNP-containing coiled bodies is regulated in interphase and mitosis—evidence that the coiled body is a kinetic nuclear structure. *J. Cell Biol.* 120:841–52

Carmo-Fonseca M, Pepperkok R, Carvalho MT, Lamond AI. 1992. Transcription-dependent colocalization of the U1, U2, U4/U6, and U5 snRNPs in coiled bodies. *J. Cell Biol.* 117:1–14

Claus P, Doring F, Gringel S, Muller-Ostermeyer F, Fuhlrott J, et al. 2003. Differential intranuclear localization of fibroblast growth factor-2 isoforms and specific interaction with the survival of motoneuron protein. *J. Biol. Chem.* 278:479–85

Cmarko D, Verschure PJ, Martin TE, Dahmus ME, Krause S, et al. 1999. Ultrastructural analysis of transcription and splicing in the cell nucleus after bromo-UTP microinjection. *Mol. Biol. Cell* 10:211–23

Costa Lima MA, Pimentel MM. 2004. Dynamic mutation and human disorders: the spinocerebellar ataxias (review). *Int. J. Mol. Med.* 13:299–302

Cuddihy AR, Wong AH, Tam NW, Li S, Koromilas AE. 1999. The double-stranded RNA activated protein kinase PKR physically associates with the tumor suppressor p53 protein and phosphorylates human p53 on serine 392 in vitro. *Oncogene* 18:2690–702

Darzacq X, Jady BE, Verheggen C, Kiss AM, Bertrand E, Kiss T. 2002. Cajal body-specific small nuclear RNAs: a novel class of 2'-O-methylation and pseudouridylation guide RNAs. *EMBO J.* 21:2746–56

Dundr M, Hebert MD, Karpova TS, Stanek D, Xu H, et al. 2004. In vivo kinetics of Cajal body components. *J. Cell Biol.* 164:831–42

Dyck JA, Maul GG, Miller WH Jr, Chen JD, Kakizuka A, Evans RM. 1994. A novel macromolecular structure is a target of the promyelocyte-retinoic acid receptor oncoprotein. *Cell* 76:333–43

Fagioli M, Alcalay M, Tomassoni L, Ferrucci PF, Mencarelli A, et al. 1998. Cooperation between the RING + B1-B2 and coiled-coil domains of PML is necessary for its effects on cell survival. *Oncogene* 16:2905–13

Feder M, Pas J, Wyrwicz LS, Bujnicki JM. 2003. Molecular phylogenetics of the RrmJ/fibrillarin superfamily of ribose 2'-O-methyltransferases. *Gene* 302:129–38

Fernandez R, Pena E, Navascues J, Casafont I, Lafarga M, Berciano MT. 2002. cAMP-dependent reorganization of the Cajal bodies and splicing machinery in cultured Schwann cells. *Glia* 40:378–88

Flodstrom-Tullberg M, Hultcrantz M, Stotland A, Maday A, Tsai D, et al. 2005. RNase L and double-stranded RNA-dependent protein kinase exert complementary roles in islet cell eefense during Coxsackievirus infection. *J. Immunol.* 174:1171–77

Frey MR, Bailey AD, Weiner AM, Matera AG. 1999. Association of snRNA genes with coiled bodies is mediated by nascent snRNA transcripts. *Curr. Biol.* 9:126–35

Frey MR, Matera AG. 1995. Coiled bodies contain U7 small nuclear RNA and associate with specific DNA sequences in interphase human cells. *Proc. Natl. Acad. Sci. USA* 92:5915–19

Frey MR, Matera AG. 2001. RNA-mediated interaction of Cajal bodies and U2 snRNA genes. *J. Cell Biol.* 154:499–509

Frugier T, Tiziano FD, Cifuentes-Diaz C, Miniou P, Roblot N, et al. 2000. Nuclear targeting defect of SMN lacking the C-terminus in a mouse model of spinal muscular atrophy. *Hum. Mol. Genet.* 9:849–58

Galcheva-Gargova Z, Konstantinov KN, Wu IH, Klier FG, Barrett T, Davis RJ. 1996. Binding of zinc finger protein ZPR1 to the epidermal growth factor receptor. *Science* 272:1797–802

Gangwani L, Mikrut M, Theroux S, Sharma M, Davis RJ. 2001. Spinal muscular atrophy disrupts the interaction of ZPR1 with the SMN protein. *Nat. Cell Biol.* 3:376–83

Gao L, Frey MR, Matera AG. 1997. Human genes encoding U3 snRNA associate with coiled bodies in interphase cells and are clustered on chromosome 17p11.2 in a complex inverted repeat structure. *Nucleic Acids Res.* 25:4740–47

Gedge LJ, Morrison EE, Blair GE, Walker JH. 2005. Nuclear actin is partially associated with Cajal bodies in human cells in culture and relocates to the nuclear periphery after infection of cells by adenovirus 5. *Exp. Cell Res.* 303:229–39

Goransson M, Wedin M, Aman P. 2002. Temperature-dependent localization of TLS-CHOP to splicing factor compartments. *Exp. Cell Res.* 278:125–32

Gorisch SM, Wachsmuth M, Ittrich C, Bacher CP, Rippe K, Lichter P. 2004. Nuclear body movement is determined by chromatin accessibility and dynamics. *Proc. Natl. Acad. Sci. USA* 101:13221–26

Goti D, Katzen SM, Mez J, Kurtis N, Kiluk J, et al. 2004. A mutant ataxin-3 putative-cleavage fragment in brains of Machado-Joseph disease patients and transgenic mice is cytotoxic above a critical concentration. *J. Neurosci.* 24:10266–79

Hahn WC, Meyerson M. 2001. Telomerase activation, cellular immortalization and cancer. *Ann. Med.* 33:123–29

Hande MP, Samper E, Lansdorp P, Blasco MA. 1999. Telomere length dynamics and chromosomal instability in cells derived from telomerase null mice. *J. Cell Biol.* 144:589–601

Hardin JH, Spicer SS, Greene WB. 1969. The paranucleolar structure, accessory body of Cajal, sex chromatin, and related structures in nuclei of rat trigeminal neurons: a cytochemical and ultrastructural study. *Anat. Rec.* 164:403–31

Hebert MD, Matera AG. 2000. Self-association of coilin reveals a common theme in nuclear body localization. *Mol. Biol. Cell* 11:4159–71

Hebert MD, Shpargel KB, Ospina JK, Tucker KE, Matera AG. 2002. Coilin methylation regulates nuclear body formation. *Dev. Cell* 3:329–37

Hebert MD, Szymczyk PW, Shpargel KB, Matera AG. 2001. Coilin forms the bridge between Cajal bodies and SMN, the spinal muscular atrophy protein. *Genes Dev.* 15:2720–29

The first demonstration that, in absence of telomerase activity, cells undergo genomic instability.

Hofmann I, Schnolzer M, Kaufmann I, Franke WW. 2002. Symplekin, a constitutive protein of karyo- and cytoplasmic particles involved in mRNA biogenesis in *Xenopus laevis* oocytes. *Mol. Biol. Cell* 13:1665–76

Hong S, Ka S, Kim S, Park Y, Kang S. 2003. p80 coilin, a coiled body-specific protein, interacts with ataxin-1, the SCA1 gene product. *Biochim. Biophys. Acta* 1638:35–42

Ichikawa H, Shimizu K, Hayashi Y, Ohki M. 1994. An RNA-binding protein gene, TLS/FUS, is fused to ERG in human myeloid leukemia with t(16;21) chromosomal translocation. *Cancer Res.* 54:2865–68

Insinga A, Monestiroli S, Ronzoni S, Carbone R, Pearson M, et al. 2004. Impairment of p53 acetylation, stability and function by an oncogenic transcription factor. *EMBO J.* 23:1144–54

Irwin S, Vandelft M, Pinchev D, Howell JL, Graczyk J, et al. 2005. RNA association and nucleocytoplasmic shuttling by ataxin-1. *J. Cell Sci.* 118:233–42

Isaac C, Yang Y, Meier UT. 1998. Nopp140 functions as a molecular link between the nucleolus and the coiled bodies. *J. Cell Biol.* 142:319–29

Jacobs EY, Frey MR, Wu W, Ingledue TC, Gebuhr TC, et al. 1999. Coiled bodies preferentially associate with U4, U11, and U12 small nuclear RNA genes in interphase HeLa cells but not with U6 and U7 genes. *Mol. Biol. Cell* 10:1653–63

Jady BE, Bertrand E, Kiss T. 2004. Human telomerase RNA and box H/ACA scaRNAs share a common Cajal body-specific localization signal. *J. Cell Biol.* 164:647–52

Jady BE, Darzacq X, Tucker KE, Matera AG, Bertrand E, Kiss T. 2003. Modification of Sm small nuclear RNAs occurs in the nucleoplasmic Cajal body following import from the cytoplasm. *EMBO J.* 22:1878–88

Jimenez-Garcia LF, Green SR, Mathews MB, Spector DL. 1993. Organization of the double-stranded RNA-activated protein kinase DAI and virus-associated VA RNAI in adenovirus-2-infected HeLa cells. *J. Cell Sci.* 106(Pt 1):11–22

Jin S, Levine AJ. 2001. The p53 functional circuit. *J. Cell Sci.* 114:4139–40

Jones KW, Gorzynski K, Hales CM, Fischer U, Badbanchi F, et al. 2001. Direct interaction of the spinal muscular atrophy disease protein SMN with the small nucleolar RNA-associated protein fibrillarin. *J. Biol. Chem.* 276:38645–51

Jordan P, Cunha C, Carmo-Fonseca M. 1997. The cdk7-cyclin H-MAT1 complex associated with TFIIH is localized in coiled bodies. *Mol. Biol. Cell* 8:1207–17

Joy A, Moffett J, Neary K, Mordechai E, Stachowiak EK, et al. 1997. Nuclear accumulation of FGF-2 is associated with proliferation of human astrocytes and glioma cells. *Oncogene* 14:171–83

Klement IA, Skinner PJ, Kaytor MD, Yi H, Hersch SM, et al. 1998. Ataxin-1 nuclear localization and aggregation: role in polyglutamine-induced disease in SCA1 transgenic mice. *Cell* 95:41–53

Ko LJ, Shieh SY, Chen X, Jayaraman L, Tamai K, et al. 1997. p53 is phosphorylated by CDK7-cyclin H in a p36MAT1-dependent manner. *Mol. Cell Biol.* 17:7220–29

Kropotov AV, Grudinkin PS, Pleskach NM, Gavrilov BA, Tomilin NV, Zhivotovsky B. 2004. Downregulation of peroxiredoxin V stimulates formation of etoposide-induced double-strand DNA breaks. *FEBS Lett.* 572:75–79

Lafarga M, Hervas JP, Santa-Cruz MC, Villegas J, Crespo D. 1983. "The accessory body" of Cajal in the neuronal nucleus. A light and electron microscopic approach. *Anat. Embryol. (Berl)* 166:19–30

Lefebvre S, Burlet P, Liu Q, Bertrandy S, Clermont O, et al. 1997. Correlation between severity and SMN protein level in spinal muscular atrophy. *Nat. Genet.* 16:265–69

This study shows that CBs contain the human telomerase RNA (hTR), which shares structural similarities with CB-specific small guide RNA. CBs could play a role in the maturation of hTR.

The residual levels of the SMN protein are strictly related to the phenotype of SMA, thus enforcing the idea that a defect in SMN levels plays a pathogenetic role in this disease.

Liu J, Hebert MD, Ye Y, Templeton DJ, Kung H, Matera AG. 2000. Cell cycle-dependent localization of the CDK2-cyclin E complex in Cajal (coiled) bodies. *J. Cell Sci.* 113(Pt 9):1543–52

Liu Q, Dreyfuss G. 1996. A novel nuclear structure containing the survival of motor neurons protein. *EMBO J.* 15:3555–65

Lyon CE, Bohmann K, Sleeman J, Lamond AI. 1997. Inhibition of protein dephosphorylation results in the accumulation of splicing snRNPs and coiled bodies within the nucleolus. *Exp. Cell Res.* 230:84–93

Ma T, Van Tine BA, Wei Y, Garrett MD, Nelson D, et al. 2000. Cell cycle-regulated phosphorylation of p220(NPAT) by cyclin E/Cdk2 in Cajal bodies promotes histone gene transcription. *Genes Dev.* 14:2298–313

Malatesta M, Zancanaro C, Martin TE, Chan EK, Amalric F, et al. 1994. Cytochemical and immunocytochemical characterization of nuclear bodies during hibernation. *Eur. J. Cell Biol.* 65:82–93

Matera AG, Ward DC. 1993. Nucleoplasmic organization of small nuclear ribonucleoproteins in cultured human cells. *J. Cell Biol.* 121:715–27

Matsuzaki K, Minami T, Tojo M, Honda Y, Saitoh N, et al. 2003. PML-nuclear bodies are involved in cellular serum response. *Genes Cells* 8:275–86

Meier UT, Blobel G. 1992. Nopp140 shuttles on tracks between nucleolus and cytoplasm. *Cell* 70:127–38

Monneron A, Bernhard W. 1969. Fine structural organization of the interphase nucleus in some mammalian cells. *J. Ultrastruct. Res.* 27:266–88

Morohoshi F, Ootsuka Y, Arai K, Ichikawa H, Mitani S, et al. 1998. Genomic structure of the human RBP56/hTAFII68 and FUS/TLS genes. *Gene* 221:191–98

Narayanan U, Achsel T, Luhrmann R, Matera AG. 2004. Coupled in vitro import of U snRNPs and SMN, the spinal muscular atrophy protein. *Mol. Cell* 16:223–34

Navascues J, Berciano MT, Tucker KE, Lafarga M, Matera AG. 2004. Targeting SMN to Cajal bodies and nuclear gems during neuritogenesis. *Chromosoma* 112:398–409

Newton K, Petfalski E, Tollervey D, Caceres JF. 2003. Fibrillarin is essential for early development and required for accumulation of an intron-encoded small nucleolar RNA in the mouse. *Mol. Cell Biol* 23:8519–27

Ochs RL, Stein TW Jr, Andrade LE, Gallo D, Chan EK, et al. 1995. Formation of nuclear bodies in hepatocytes of estrogen-treated roosters. *Mol. Biol. Cell* 6:345–56

Ochs RL, Stein TW Jr, Tan EM. 1994. Coiled bodies in the nucleolus of breast cancer cells. *J. Cell Sci.* 107(Pt 2):385–99

Pearn J. 1980. Classification of spinal muscular atrophies. *Lancet* 1:919–22

Pellizzoni L, Baccon J, Charroux B, Dreyfuss G. 2001. The survival of motor neurons (SMN) protein interacts with the snoRNP proteins fibrillarin and GAR1. *Curr. Biol.* 11:1079–88

Pillai RS, Will CL, Luhrmann R, Schumperli D, Muller B. 2001. Purified U7 snRNPs lack the Sm proteins D1 and D2 but contain Lsm10, a new 14 kDa Sm D1-like protein. *EMBO J.* 20:5470–79

Platani M, Goldberg I, Lamond AI, Swedlow JR. 2002. Cajal body dynamics and association with chromatin are ATP-dependent. *Nat. Cell Biol.* 4:502–8

Platani M, Goldberg I, Swedlow JR, Lamond AI. 2000. In vivo analysis of Cajal body movement, separation, and joining in live human cells. *J. Cell Biol.* 151:1561–74

Polak PE, Simone F, Kaberlein JJ, Luo RT, Thirman MJ. 2003. ELL and EAF1 are Cajal body components that are disrupted in MLL-ELL leukemia. *Mol. Biol. Cell* 14:1517–28

This demonstrates that the expression of a leukemic fusion protein alters the intranuclear distribution of coilin and the integrity of CBs.

Ramon y Cajal S. 1903. Un sencillo metodo de coloracion seletiva del reticulo proto-plasmatico y sus efectos en los diversos organos nerviosos de vertebrados e invertebrados. *Trab. Lab. Invest. Biol.* (*Madrid*) 2:129–221

Raska I. 1995. Nuclear ultrastructures associated with the RNA synthesis and processing. *J. Cell. Biochem.* 59:11–26

Raska I, Andrade LE, Ochs RL, Chan EK, Chang CM, et al. 1991. Immunological and ultrastructural studies of the nuclear coiled body with autoimmune antibodies. *Exp. Cell Res.* 195:27–37

Rebelo L, Almeida F, Ramos C, Bohmann K, Lamond AI, Carmo-Fonseca M. 1996. The dynamics of coiled bodies in the nucleus of adenovirus-infected cells. *Mol. Biol. Cell* 7:1137–51

Rodrigues SH, Silva NP, Delicio LR, Granato C, Andrade LE. 1996. The behavior of the coiled body in cells infected with adenovirus in vitro. *Mol. Biol. Rep.* 23:183–89

Satyanarayana A, Manns MP, Rudolph KL. 2004. Telomeres, telomerase and cancer: an endless search to target the ends. *Cell Cycle* 3:1138–50

Schneider E, Montenarh M, Wagner P. 1998. Regulation of CAK kinase activity by p53. *Oncogene* 17:2733–41

Schrank B, Gotz R, Gunnersen JM, Ure JM, Toyka KV, et al. 1997. Inactivation of the survival motor neuron gene, a candidate gene for human spinal muscular atrophy, leads to massive cell death in early mouse embryos. *Proc. Natl. Acad. Sci. USA* 94:9920–25

Schul W, Groenhout B, Koberna K, Takagaki Y, Jenny A, et al. 1996. The RNA 3′ cleavage factors CstF 64 kDa and CPSF 100 kDa are concentrated in nuclear domains closely associated with coiled bodies and newly synthesized RNA. *EMBO J.* 15:2883–92

Schul W, van Der Kraan I, Matera AG, van Driel R, de Jong L. 1999. Nuclear domains enriched in RNA 3′-processing factors associate with coiled bodies and histone genes in a cell cycle-dependent manner. *Mol. Biol. Cell* 10:3815–24

Schul W, van Driel R, de Jong L. 1998. Coiled bodies and U2 snRNA genes adjacent to coiled bodies are enriched in factors required for snRNA transcription. *Mol. Biol. Cell* 9:1025–36

Shaw PJ, Brown JW. 2004. Plant nuclear bodies. *Curr. Opin. Plant Biol.* 7:614–20

Shinobu N, Maeda T, Aso T, Ito T, Kondo T, et al. 1999. Physical interaction and functional antagonism between the RNA polymerase II elongation factor ELL and p53. *J. Biol. Chem.* 274:17003–10

Shopland LS, Byron M, Stein JL, Lian JB, Stein GS, Lawrence JB. 2001. Replication-dependent histone gene expression is related to Cajal body (CB) association but does not require sustained CB contact. *Mol. Biol. Cell* 12:565–76

Shpargel KB, Ospina JK, Tucker KE, Matera AG, Hebert MD. 2003. Control of Cajal body number is mediated by the coilin C-terminus. *J. Cell Sci.* 116:303–12

Skare P, Kreivi JP, Bergstrom A, Karlsson R. 2003. Profilin I colocalizes with speckles and Cajal bodies: a possible role in pre-mRNA splicing. *Exp. Cell Res.* 286:12–21

Skinner PJ, Koshy BT, Cummings CJ, Klement IA, Helin K, et al. 1997. Ataxin-1 with an expanded glutamine tract alters nuclear matrix-associated structures. *Nature* 389:971–74

Sleeman J, Lyon CE, Platani M, Kreivi JP, Lamond AI. 1998. Dynamic interactions between splicing snRNPs, coiled bodies and nucleoli revealed using snRNP protein fusions to the green fluorescent protein. *Exp. Cell Res.* 243:290–304

Sleeman JE, Ajuh P, Lamond AI. 2001. snRNP protein expression enhances the formation of Cajal bodies containing p80-coilin and SMN. *J. Cell Sci.* 114:4407–19

The first historical description of Cajal Bodies in neuronal cells.

This article demonstrates that the formation of CBs can be stimulated by an increase in the nuclear content of snRNPs.

This demonstrates that CBs specifically accumulate snRNPs newly imported into the nucleus.

Sleeman JE, Lamond AI. 1999. Newly assembled snRNPs associate with coiled bodies before speckles, suggesting a nuclear snRNP maturation pathway. *Curr. Biol.* **9:1065–74**

Sleeman JE, Trinkle-Mulcahy L, Prescott AR, Ogg SC, Lamond AI. 2003. Cajal body proteins SMN and Coilin show differential dynamic behaviour in vivo. *J. Cell Sci.* 116:2039–50

Smith KP, Carter KC, Johnson CV, Lawrence JB. 1995. U2 and U1 snRNA gene loci associate with coiled bodies. *J. Cell Biochem.* 59:473–85

Smith KP, Lawrence JB. 2000. Interactions of U2 gene loci and their nuclear transcripts with Cajal (coiled) bodies: evidence for PreU2 within Cajal bodies. *Mol. Biol. Cell* 11:2987–98

Spector DL, Lark G, Huang S. 1992. Differences in snRNP localization between transformed and nontransformed cells. *Mol. Biol. Cell* 3:555–69

Stachowiak MK, Moffett J, Maher P, Tucholski J, Stachowiak EK. 1997. Growth factor regulation of cell growth and proliferation in the nervous system. A new intracrine nuclear mechanism. *Mol. Neurobiol.* 15:257–83

Stanek D, Rader SD, Klingauf M, Neugebauer KM. 2003. Targeting of U4/U6 small nuclear RNP assembly factor SART3/p110 to Cajal bodies. *J. Cell Biol.* 160:505–16

Su C, Gao G, Schneider S, Helt C, Weiss C, et al. 2004. DNA damage induces downregulation of histone gene expression through the G(1) checkpoint pathway. *EMBO J.* 23:1133–43

Tucker KE, Berciano MT, Jacobs EY, LePage DF, Shpargel KB, et al. 2001. Residual Cajal bodies in coilin knockout mice fail to recruit Sm snRNPs and SMN, the spinal muscular atrophy gene product. *J. Cell Biol.* 154:293–307

van Koningsbruggen S, Dirks RW, Mommaas AM, Onderwater JJ, Deidda G, et al. 2004. FRG1P is localised in the nucleolus, Cajal bodies, and speckles. *J. Med. Genet.* 41:e46

Verheggen C, Lafontaine DL, Samarsky D, Mouaikel J, Blanchard JM, et al. 2002. Mammalian and yeast U3 snoRNPs are matured in specific and related nuclear compartments. *EMBO J.* 21:2736–45

Wang A, Ikura T, Eto K, Ota MS. 2004. Dynamic interaction of p220(NPAT) and CBP/p300 promotes S-phase entry. *Biochem. Biophys. Res. Commun.* 325:1509–16

Wiederschain D, Kawai H, Gu J, Shilatifard A, Yuan ZM. 2003. Molecular basis of p53 functional inactivation by the leukemic protein MLL-ELL. *Mol. Cell Biol.* 23:4230–46

Witke W. 2004. The role of profilin complexes in cell motility and other cellular processes. *Trends Cell Biol.* 14:461–69

Wu CH, Gall JG. 1993. U7 small nuclear RNA in C snurposomes of the *Xenopus* germinal vesicle. *Proc. Natl. Acad. Sci. USA* 90:6257–59

Wu Z, Murphy C, Gall JG. 1994. Human p80-coilin is targeted to sphere organelles in the amphibian germinal vesicle. *Mol. Biol. Cell* 5:1119–27

Yamada M, Sato T, Shimohata T, Hayashi S, Igarashi S, et al. 2001. Interaction between neuronal intranuclear inclusions and promyelocytic leukemia protein nuclear and coiled bodies in CAG repeat diseases. *Am. J. Pathol.* 159:1785–95

Yannoni YM, White K. 1997. Association of the neuron-specific RNA binding domain-containing protein ELAV with the coiled body in *Drosophila* neurons. *Chromosoma* 105:332–41

This article demonstrates that p53 is a component of CBs and potentially suggests a link between CBs and cellular stress response.

Ye X, Wei Y, Nalepa G, Harper JW. 2003. The cyclin E/Cdk2 substrate p220(NPAT) is required for S-phase entry, histone gene expression, and Cajal body maintenance in human somatic cells. *Mol. Cell Biol.* 23:8586–600

Young PJ, Day PM, Zhou J, Androphy EJ, Morris GE, Lorson CL. 2002. A direct interaction between the survival motor neuron protein and p53 and its relationship to spinal muscular atrophy. *J. Biol. Chem.* **277:2852–59**

Young PJ, Le TT, Dunckley M, Nguyen TM, Burghes AH, Morris GE. 2001. Nuclear gems and Cajal (coiled) bodies in fetal tissues: nucleolar distribution of the spinal muscular atrophy protein, SMN. *Exp. Cell Res.* 265:252–61

Young PJ, Le TT, thi Man N, Burghes AH, Morris GE. 2000. The relationship between SMN, the spinal muscular atrophy protein, and nuclear coiled bodies in differentiated tissues and cultured cells. *Exp. Cell Res.* 256:365–74

Zhang HL, Pan F, Hong D, Shenoy SM, Singer RH, Bassell GJ. 2003. Active transport of the survival motor neuron protein and the role of exon-7 in cytoplasmic localization. *J. Neurosci.* 23:6627–37

Zhao J, Dynlacht B, Imai T, Hori T, Harlow E. 1998. Expression of NPAT, a novel substrate of cyclin E-CDK2, promotes S-phase entry. *Genes Dev.* 12:456–61

Zhao J, Kennedy BK, Lawrence BD, Barbie DA, Matera AG, et al. 2000. NPAT links cyclin E-Cdk2 to the regulation of replication-dependent histone gene transcription. *Genes Dev.* 14:2283–97

Zhou Y, Kok KH, Chun AC, Wong CM, Wu HW, et al. 2000. Mouse peroxiredoxin V is a thioredoxin peroxidase that inhibits p53-induced apoptosis. *Biochem. Biophys. Res. Commun.* 268:921–27

Zhu Y, Tomlinson RL, Lukowiak AA, Terns RM, Terns MP. 2004. Telomerase RNA accumulates in Cajal bodies in human cancer cells. *Mol. Biol. Cell* 15:81–90

This work shows that CBs contain components of the human telomerase complex.

Assembly of Variant Histones into Chromatin

Steven Henikoff[1] and Kami Ahmad[2]

[1] Howard Hughes Medical Institute, Fred Hutchinson Cancer Research Center, Seattle, Washington 98109; email: steveh@fhcrc.org

[2] Department of Biological Chemistry and Molecular Pharmacology, Harvard Medical School, Boston, Massachusetts 02115; email: kami_ahmad@hms.harvard.edu

Annu. Rev. Cell Dev. Biol.
2005. 21:133–53

First published online as a Review in Advance on July 18, 2005

The *Annual Review of Cell and Developmental Biology* is online at http://cellbio.annualreviews.org

doi: 10.1146/annurev.cellbio.21.012704.133518

Copyright © 2005 by Annual Reviews. All rights reserved

1081-0706/05/1110-0133$20.00

Key Words

nucleosome, chromatin remodeling, histone replacement, epigenetics, centromeric chromatin

Abstract

Chromatin can be differentiated by the deposition of variant histones at centromeres, active genes, and silent loci. Variant histones are assembled into nucleosomes in a replication-independent manner, in contrast to assembly of bulk chromatin that is coupled to replication. Recent in vitro studies have provided the first glimpses of protein machines dedicated to building and replacing alternative nucleosomes. They deposit variant H2A and H3 histones and are targeted to particular functional sites in the genome. Differences between variant and canonical histones can have profound consequences, either for delivery of the histones to sites of assembly or for their function after incorporation into chromatin. Recent studies have also revealed connections between assembly of variant nucleosomes, chromatin remodeling, and histone post-translational modification. Taken together, these findings indicate that chromosome architecture can be highly dynamic at the most fundamental level, with epigenetic consequences.

Contents

INTRODUCTION

Over the past decade we have witnessed a renaissance of interest in core histones with the general realization that these four simple components of nucleosomal octamers, histones H2A, H2B, H3 and H4, are also key players in basic nuclear processes. This is especially true for post-translational modifications of histones, which have been implicated both in modulating chromatin architecture, and in the regulation of transcription (Brownell et al. 1996, Jenuwein & Allis 2001, Turner et al. 1992). The realization that addition or removal of these modifications can facilitate gene activation or silencing has fueled considerable excitement in the chromatin field.

Much less attention has been paid to the differentiation of chromatin by the incorporation of variant histones (**Table 1**). These are separately encoded forms of canonical histones that are distinguished by sequence differences (Malik & Henikoff 2003). Many variant histones are simply polymorphic versions of the major canonical forms that are assembled into bulk chromatin behind the replication fork. However, other variants are found to have profound differences that distinguish them from canonical forms, either in the way that they are deposited or the way that they function after deposition, or both. In the past few years, we have come to realize that this latter class of histone variants and the special machineries that deposit them play important roles in chromatin differentiation and epigenetic maintenance. The recent study of variants and their assembly has begun to reveal a highly dynamic picture of chromatin, in which processes of post-transcriptional modification appear to be coupled to processes of histone replacement.

In our review, we examine the basis for this more dynamic view of chromatin by considering recent studies on particular core histone variants and chromatin assembly complexes. We explore the possibility that the processes that replace histones at active genes and that propagate chromatin states when DNA replicates involve the concerted action of nucleosome remodeling and histone modification activities. Although the study of chromatin dynamics is technically demanding, we expect that the rapid improvements in molecular biology, cytogenetics, and genomics technologies mean that this area of research is still in its infancy.

Centromeric Chromatin is Identified by a Special Histone H3 Variant

Every eukaryotic chromosome requires a centromere for it to segregate at mitosis, and the uniqueness of this structure has facilitated the cytological identification of centromere-specific protein components. The first such components were identified as epitopes of autoimmune antibodies (Earnshaw & Rothfield 1985), and one of these, CENP-A, was found to be a histone H3 homolog that copurifies with nucleosomes (Palmer et al. 1991, Palmer et al. 1987). Genetic studies

Table 1 Histone variants and associated chromatin assembly complexes

Histones	Features	Assembled by (organism)
Archaeal histones	Ancestral histone fold proteins without tails found in singly wrapped tetrameric units that comprise nucleosome particles.	Unknown
H2A, H2B	Canonical core histones encoded by replication-coupled genes.	FACT (yeast, *Drosophila*)
H2AZ	H2A variant found in nearly all eukaryotes that has a diverged self-interaction domain.	SWR1 (yeast), Tip60 (*Drosophila*)
macroH2A	Vertebrate-specific H2A variant with a C-terminal globular domain. Enriched on the mammalian inactive X-chromosome.	Unknown
H2A-Bbd	Vertebrate-specific H2A variant that is widely distributed. Relatively deficient on the inactive X-chromosome.	Unknown
H2AX	H2A form with an SQ[E/D] Ø (Ø = hydrophobic) C-terminal motif that becomes serine phosphorylated at sites of double-stranded breaks.	INO80 (yeast)
H3, H4	Canonical core histones encoded by replication-coupled genes.	CAF-1 (plants, animals, fungi)
H3.3 (H3.2 in plants)	H3 variant that replaces H3 and differs at position 31 and at a few residues on helix 2 that allow deposition outside of replication.	HIRA (mammals)
Packaging histones	Core and linker histone variants adapted for tight packaging of DNA in sperm and pollen in some organisms.	

reveal that mammalian CENP-A and its counterparts in other eukaryotes (generically referred to as CenH3s) are absolutely required for assembly of the proteinaceous kinetochore to which the spindle microtubules attach at mitosis and meiosis (Blower & Karpen 2001, Buchwitz et al. 1999, Howman et al. 2000, Stoler et al. 1995). Antibodies against CenH3s from both plants and animals have been used to map centromeres (Alonso et al. 2003, Lo et al. 2001, Nagaki et al. 2004), because in these organisms the centromere is not determined by DNA sequence but rather by the presence of centromeric chromatin. In fact, human neocentromeres that show no resemblance in DNA sequence to native alpha-satellite-containing centromeres are nevertheless packaged in CENP-A-containing nucleosomes (Alonso et al. 2003, Amor et al. 2004). These observations imply that the constant location of the centromere in all cells of an organism through millions of years of evolution is maintained by the faithful assembly of CenH3-containing chromatin. Apparently centromeres are maintained indefinitely by the action of a chromatin assembly process.

What features of CenH3s are recognized for assembly into chromatin? Swaps between CENP-A or Cse4p and H3 identified the core region as being crucial (Keith et al. 1999, Shelby et al. 1997), although further inferences were complicated by the possibility that some H3-specific residues might be incompatible with CenH3 function. A more refined approach is to use heterologous CenH3s; in the case of *Drosophila* CenH3 (Cid), this led to the identification of Loop I as being both necessary and sufficient for Cid localization to centromeres (Vermaak et al. 2002) (**Figure 1**). Thus Loop I of human CENP-A is included within a region inferred to be more compact than the corresponding regions of H3 (Black et al. 2004). The possibility that contacts between Loop I and centromeric DNA are important for correct assembly of centromeric nucleosomes is implied by the evidence for adaptive evolution of Loop I in *Drosophila* and *Arabidopsis* (Cooper & Henikoff 2004, Malik & Henikoff 2001). However, mammalian CENP-A is different in that no adaptive evolution is seen (Talbert et al. 2004), and heterologous CenH3s can localize to human centromeres (Henikoff et al. 2000, Wieland et al. 2004). In fact, yeast Cse4p can even functionally replace CENP-A (Wieland et al. 2004), unlike the

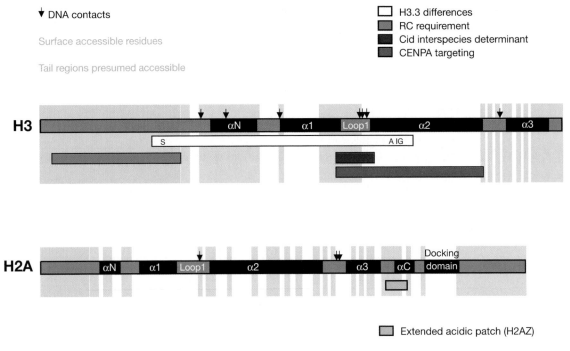

Figure 1

Regions of H3 and H2A variants responsible for chromatin differentiation. Surface-accessible residues (*shaded blue*) were annotated from the 1.9 Å X-ray crystal structure (Davey et al. 2002). N-terminal tails that are not included in nucleosome models but are presumed to be accessible are also indicated (*shaded green*). Unshaded residues are occluded by DNA or by other histones within the octamer. Schematics for the secondary structure of each histone are indicated, and regions of the histones and their variants with the functions described in the text are indicated.

situation for *Drosophila bipectinata* Cid, which requires *Drosophila melanogaster* Loop I to localize to *D. melanogaster* centromeres (Vermaak et al. 2002). These differences between organisms might be attributable to the other centromere-specific DNA-binding protein, CENP-C, which is adaptively evolving in mammals and plants, but which has not been identified in *Drosophila* (Talbert et al. 2004). It has been proposed that CenH3s and CENP-Cs adapt the rapidly evolving centromeric satellites to the conserved kinetochore machinery (Malik & Henikoff 2001), in which case proteins that are not adaptively evolving, such as CENP-A and Cse4p, would not require species-specific interactions to package centromeric chromatin.

Assembly of CenH3-containing nucleosomes is independent of replication (Ahmad & Henikoff 2001a, Shelby et al. 2000). The process presumably initiates with interactions between DNA and CenH3 Loop I of CenH3•H4 units, and when the full core is assembled, the CenH3 N-terminal tail would interact with linker DNA (Vermaak et al. 2002) (**Figure 2**). In the canonical nucleosome, the H3 N-terminal tail exits between the DNA helices and contacts the DNA minor groove where the DNA leaves the nucleosome core (Luger et al. 1997). In contrast to the nearly invariant tail of canonical H3, which is constrained by the density of post-translational modification sites, the CenH3 N-terminal tails are extraordinarily diverse, differing in length and sequence to such an extent that they cannot be aligned between distant species (Malik & Henikoff 2003). Minor groove-binding motifs have been detected

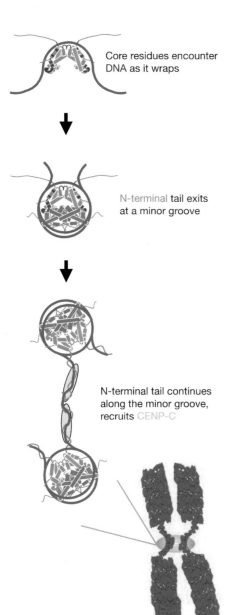

Core residues encounter
DNA as it wraps

N-terminal tail exits
at a minor groove

N-terminal tail continues
along the minor groove,
recruits CENP-C

Figure 2

A model for assembly of centromeric chromatin.

in CenH3 N-terminal tails (Malik et al. 2002), suggesting that contacts with the minor groove stabilize centromeric nucleosome assembly. In support of this possibility, the N-terminal tails of some CenH3s are adaptively evolving (Malik & Henikoff 2001, Talbert

et al. 2002). In *Drosophila* Cid, regions of adaptive evolution are interspersed with patches of sequence conservation that might correspond to sites of interaction with the conserved kinetochore machinery (Keith et al. 1999, Malik et al. 2002). Centromeric chromatin would then mature with the recruitment of CENP-C, a DNA-binding protein that nevertheless depends upon CenH3 for centromeric localization (Moore & Roth 2001, Sugimoto et al. 1994, Yang et al. 1996).

It is striking that centromeric chromatin is organized as interspersed stretches of CenH3- and H3-containing nucleosomes (Blower et al. 2002). It is difficult to envision processive assembly leading to an interspersed organization, which would require switching back and forth between substrates. Alternatively, CenH3- and H3-containing nucleosomes might be assembled at different times in the cell cycle (Ahmad & Henikoff 2002b): Canonical H3 incorporates behind the replication fork, whereas both human and fly CenH3s deposit in an RI manner. Just how CenH3 assembly is maintained at sites of pre-existing CenH3 in the absence of DNA sequence determinants is a major unanswered question. A speculative possibility is that anaphase tension on CenH3-containing chromatin causes adjacent nucleosomes to unravel and subsequent chromatin repair incorporates new CenH3 (Ahmad & Henikoff 2002b, Mellone & Allshire 2003). It is also possible that unique patterns of H3 modifications found in interspersed and flanking stretches of nucleosomes predispose centromeres for deposition of CenH3 (Sullivan & Karpen 2004). How such patterns arise might depend on modes of assembly of H3-containing nucleosomes, which we turn to next.

Active Genes are Sites of H3.3 Replacement

Another universal H3 variant, H3.3, is so similar to canonical H3 that its distinctive properties were not realized until recently. In

RI: replication-independent

RC:
replication-coupled

animals, H3.3 differs from H3 at only four amino acid positions (**Figure 1**), consistent with the view that it is essentially a constitutive version of canonical H3 (Yu & Gorovsky 1997). Indeed, H3.3 is the dominant H3-subtype in nondividing differentiated cells in vertebrates (Pina & Suau 1987, Urban & Zweidler 1983) and in endoreplicating cells during development of a chordate (Chioda et al. 2004). Nevertheless, we found that three of the four differences between H3 and H3.3 determine nucleosome assembly behavior: Changes from the H3 to the H3.3 form allowed RI assembly (Ahmad & Henikoff 2002c). This difference between H3 and H3.3 defines a pathway for RI assembly that is distinct from the RC pathway whereby canonical histones are assembled into bulk DNA. Another difference is that RC assembly of either H3 or H3.3 requires the N-terminal tail, whereas RI does not.

Confirmation that alternative RC and RI assembly pathways exist came from the purification of H3- and H3.3-containing complexes (Tagami et al. 2004). The well-studied RC assembly complex, CAF-1, copurifies with H3, whereas the replication-independent histone chaperone, HIRA, copurifies with H3.3. Both complexes include a common H4-binding component, RbAp48, a homolog of which is also essential for the assembly of *Schizosaccharomyces pombe* CenH3 (Hayashi et al. 2004). Interestingly, the assembly form of histones in both human CAF-1 and HIRA complexes was an H3•H4 (or H3.3•H4) dimer (Tagami et al. 2004), not a tetramer as might have been expected from the existence of $(H3•H4)_2$ tetramers in solution (Wolffe 1992). This implies that histone heterodimers are substrates in all known assembly complexes (**Figure 3**), consistent with a common origin of eukaryotic nucleosome assembly mechanisms. Core eukaryotic histones evolved from structurally similar archaeal histones that assemble to form tetrameric nucleosomes closely resembling $(H3•H4)_2$ tetramers produced in vitro (Pereira & Reeve 1998), and it will be interesting to determine whether there are biochemical similarities in assembly as well.

H3.3 is enriched in active chromatin (Hendzel & Davie 1990), and the basis for enrichment has been elucidated by cytological studies using epitope-tagged versions of H3.3. We showed that RI assembly of H3.3 localizes to active, but not inactive, rDNA arrays and to euchromatin, but not heterochromatin, in *Drosophila* (Ahmad & Henikoff 2002c). In human cells, H3.3 incorporates at a transgene array that has been induced to transcribe (Janicki et al. 2004). The concomitant loss of heterochromatic markers in both *Drosophila* and human cells demonstrates that the process of H3.3 deposition at active genes accompanies replacement of pre-existing histones. The replacement process can be rapid, occurring on the order of an hour at a transgene array observed in living cells (Janicki et al. 2004).

RI assembly of H3.3•H4 provides an attractive mechanism for the resetting and perpetuation of histone modifications (Ahmad & Henikoff 2002c). RC assembly leaves a mixture of old and new nucleosomes on daughter strands and, at active genes, this would be a mixture of H3.3- and H3-containing nucleosomes. If H3.3 is post-translationally modified in a way that suits transcriptional activity, then the mixture of histones after replication would continue to promote transcription (Henikoff et al. 2004). Continued transcriptional activity would then result in nucleosome replacement over the body of genes, leading to differentiation of chromatin whereby active regions are packaged in H3.3-containing nucleosomes. Indeed, there is enough H3.3 in *Drosophila Kc* cell chromatin to densely package transcribed regions (McKittrick et al. 2004). Furthermore, the enrichment in bulk H3.3 of lysine modifications that have been found to correlate with active transcription (McKittrick et al. 2004, Waterborg 1990) provides a connection between histone replacement via the RI pathway and changes in properties of chromatin by post-transcriptional modification.

It remains to be determined whether the form of H3.3•H4 that is assembled at active genes is post-translationally modified prior to or during assembly. Nevertheless, the strong correlations among lysine modifications found in active chromatin of flies and yeast are most easily understood if assembly and modification are concerted processes (Workman & Abmayr 2004). Associations between histone-modifying enzymes and elongating RNA polymerases are also consistent with concerted assembly and modification. In yeast, histone replacement can occur within a minute at sites of active transcription (Schwabish & Struhl 2004), which delimits the processes of histone modification and nucleosome assembly to the same short-time interval.

Nucleosomes are Disassembled at Promoters

Nucleosome replacement during transcriptional elongation might also play a role in gene regulation during development. Transcription within the mouse beta-globin locus control region has been proposed to potentiate activation (Gribnau et al. 2000), and transcription through a *Drosophila* Polycomb response element derepresses expression of genes in *cis* (Bender & Fitzgerald 2002, Drewell et al. 2002, Hogga & Karch 2002, Rank et al. 2002). Inactive chromatin can obstruct the binding of transcription factors to their target sites, but once bound, the result is a heritable state of activity (Ahmad & Henikoff 2001b). Heritable activation of previously silent chromatin by transcription-coupled replacement of histones is one way that the RI assembly process might lead to inheritance of an epigenetic state (Ahmad & Henikoff 2002a).

Although histone modification is the best-established change in chromatin states upon activation of a gene, recent studies have shown that the distribution of nucleosomes can change as well. Activation of the yeast PHO5 gene leads to loss of the nucleosome at the PHO5 promoter (Boeger et al. 2003, Reinke

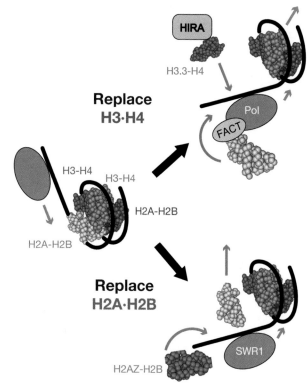

Figure 3

A model for RI replacement with histone variants (Henikoff et al. 2004). Parallels between H2A•H2B and H3•H4 replacement processes suggest a common underlying mechanism, where a large molecular machine (either RNA polymerase or a SWI/SNF remodeler) partially or completely unravels a nucleosome during transit. The result is either retention of heterodimeric subunits, such as the FACT-facilitated transfer of H2A•H2B from in front of RNA polymerase to behind (Belotserkovskaya et al. 2003, Formosa et al. 2002), or loss of a heterodimer. In the latter case, chromatin repair replaces the lost heterodimer with either H3.3•H4 (*top*) or H2AZ•H2B (*bottom*). Failure to repair will result in nucleosome eviction and reduced nucleosome densities, such as has been observed at promoters and in the body of highly transcribed genes in yeast.

& Horz 2003). The nucleosome is not simply moved aside; rather it unravels (**Figure 3**), a process that is facilitated by the ASF1 histone chaperone (Adkins et al. 2004, Boeger et al. 2004). This process is evidently not limited to PHO5 because depletion of nucleosomes at promoters has been found to occur genome-wide in yeast (Bernstein et al. 2004, Lee et al. 2004). In addition, transiting RNA polymerases displace nucleosomes, leading to variation in nucleosome occupancy over the

HP1:
heterochromatin-associated
protein 1

body of genes (Kristjuhan & Svejstrup 2004, Schwabish & Struhl 2004). These findings raise questions about the implicit assumption that the distribution of histone modifications using chromatin immunoprecipitation (ChIP) accurately reflects their relative density along the DNA (Hanlon & Lieb 2004). The action of chromatin remodeling machines, DNA-binding proteins, and RNA polymerases can affect local nucleosome densities such that this assumption is valid, at best, as a first approximation. Thus transcription-coupled replacement or loss of histones could underlie some of the patterns of histone modifications that have been reported. This interpretation might extend to the mapping of histone-modifying enzymes, which are cross-linked to their target sites by formaldehyde, a reagent that primarily cross-links primary amines (Nagy et al. 2003, Solomon & Varshavsky 1985), which are especially abundant and accessible on the histones themselves.

Chromatin Remodeling Machines Replace H2A Variants

Structural alignments of nucleosomal subunits reveal ancestral homology between H3•H4, H2A•H2B, and archaeal dimeric units, where H3 aligns with H2A and H4 with H2B (Pereira & Reeve 1998). This structural equivalence might underlie the fact that H3 and H2A have diverse variant forms, whereas H4 and H2B have (almost) none (Malik & Henikoff 2003). Of the H2A variants, H2AZ is conspicuous in having a single evolutionary origin very early in eukaryotic evolution. Although H2AZ is essential in animals (Faast et al. 2001, van Daal & Elgin 1992), it is nonessential in budding yeast (Dhillon & Kamakaka 2000, Jackson & Gorovsky 2000), and this has facilitated its in vivo study. Thus we know that H2AZ can act as a transcriptional activator and an antisilencer at different loci (Meneghini et al. 2003, Santisteban et al. 2000). Whether these functions of H2AZ generalize to plants and animals is not known. Surprisingly, mammalian H2AZ

shows a heterochromatic distribution and interacts with HP1 (Fan et al. 2004, Rangasamy et al. 2003), making it difficult to draw firm conclusions about conserved roles for H2AZ in chromatin function.

Whereas the mechanism whereby H2AZ affects chromatin remains uncertain, much has been learned about how it is deposited into chromatin, thanks to a combination of in vivo and in vitro studies in yeast. H2AZ is assembled by the SWR1 complex, and mutations in the *swr1* gene, which encodes a key component of this complex, show gene expression phenotypes similar to those of H2AZ (*htz1*) mutants (Kobor et al. 2004, Krogan et al. 2003, Mizuguchi et al. 2004). In vitro, SWR1 replaces H2A•H2B dimers in nucleosomes with H2AZ•H2B dimers (Mizuguchi et al. 2004). The fate of the leaving H2A•H2B dimer is unknown, so that it is premature to refer to this process as an exchange, implying a reciprocal event, as opposed to replacement, which does not. The actual mechanism of H2A•H2B replacement might be very similar to that for H3•H4 replacement, in which chromatin is perhaps repaired after loss of a heterodimeric subunit during transit of a large complex such as RNA polymerase or an ATP-dependent remodeling complex (**Figure 3**).

Replacement of H2A•H2B with H2AZ•H2B requires ATP, as expected from the fact that the SWR1 subunit is a member of the SWI/SNF family of ATP-dependent chromatin remodelers. This finding has important implications both for understanding the assembly of histone variants, and for the understanding of chromatin remodeling. It remains uncertain as to just how actions of various SWI/SNF family members observed in vitro relate to their functions in vivo; but in the case of SWR1, the in vitro replacement of H2A with H2AZ and the supporting in vivo data provide unequivocal evidence for this specific function. We look forward to other examples of specific roles played by chromatin remodelers in nucleosome assembly or disassembly.

The *Drosophila* SWR1 ortholog performs an analogous H2AZ replacement reaction as part of the Tip60 complex (Kusch et al. 2004). An intriguing twist arises from the role of *Drosophila* H2AZ in repair of double-stranded breaks (Madigan et al. 2002). In most eukaryotes, this role is played by the H2AX variant, which is otherwise similar to canonical H2A except for the presence of a four-amino acid C-terminal motif (SQ[D/E]Ø, where Ø represents a hydrophobic amino acid). The *Drosophila* version of H2AZ (H2AvD) has evolved an H2AX-like motif and as a result functions similarly in double-strand break repair. In diverse eukaryotes, H2AX is phosphorylated on the serine of the H2AX C-terminal motif at sites of double-strand breaks, and phosphorylation spreads rapidly to other H2AXs along the chromosome, an event that is important for recruitment of break repair machinery (Fernandez-Capetillo et al. 2004, Rogakou et al. 1998). In vitro, Tip60 specifically binds phosphorylated H2AvD, acetylates it at Lys5 and replaces it with an unphosphorylated H2AvD (Kusch et al. 2004). This combination of activities suggests that the function of Tip60 is to remodel chromatin at sites of double-strand breaks, while restoring the ground state by effectively erasing the phosphorylation mark.

It is likely that a similar process of ATPase-catalyzed replacement occurs in yeast, where the INO80 chromatin remodeling complex is recruited to H2AX when it is phosphorylated following a double-strand break (Morrison et al. 2004, van Attikum et al. 2004). INO80 is an ATPase distinct from SWR1 and is consistent with yeast H2AX, which is actually the canonical version of H2A, distinct from H2AZ. Therefore, two different remodeling machines with distinct H2A substrates appear to have evolved in different organisms to assume similar roles in DNA repair.

Other H2A variants are lineage-specific. The macroH2A histone is a vertebrate-specific variant unique among histones in having an additional globular domain. This C-terminal 200-amino acid domain is homologous to a broad class of polynucleotide and peptide hydrolases, raising the possibility that macroH2A alters chromatin via the action of a tethered enzyme (Allen et al. 2003). macroH2A is enriched in regions of the mammalian inactive X chromosome that are associated with determinants of facultative silencing, including Xist RNA and H3 trimethyl lysine-27 (Chadwick & Willard 2004). In contrast, another vertebrate-specific variant, H2A[Bbd], shows a cytological distribution that indicates an association with active chromatin (Chadwick & Willard 2001). These patterns are suggestive of functional differentiation of variant-containing chromatin.

How Is Variant Structure Related to Function in Chromatin?

As we have seen, histone variants are distinguished from canonical histones both by their mode of assembly into nucleosomes and by their properties in chromatin. Ultimately, processes such as transcription and replication must alter the structure of nucleosomes to expose DNA, and this must involve regulating DNA-histone affinities. The structure of the nucleosome suggests ways that exposure can happen. The protein octamer can be divided into three surfaces, each with distinctive roles: (*a*) a perimeter ramp that underlies the superhelix of wrapped DNA, (*b*) the two exposed faces of the disk, and (*c*) flexible N- and C-terminal tails of the histones that extend out of the nucleosome. The roles of these surfaces are becoming apparent, as are the ways that histone modifications and variants alter accessibility of specific nucleosomes. For example, acetylation of N-terminal lysines may neutralize DNA-histone interactions (Turner et al. 1992). An excellent case has been made that modifications decorating the protein superhelical ramp sterically disrupt DNA-histone contacts (Cosgrove et al. 2004, Freitas et al. 2004). A second mode-of-action for modifications is as binding sites for proteins that remodel or restrict nucleosomes. Modifications

H2A[Bbd]: H2A-Barr-body-deficient

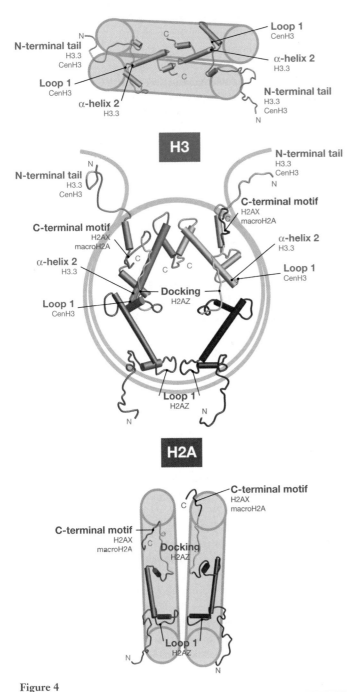

Figure 4

Location of differences between canonical and variant histones H3 (*blue*) and H2A (*brown*) shown on their three-dimensional structures. Segments that differ are highlighted in yellow.

on the histone tails and the exposed disk faces are clearly suitable for display to incoming factors, and both repressive partners (e.g., HP1) and activating ones (e.g., SWI/SNF nucleosome remodelers) are known (Jenuwein & Allis 2001).

Studies of the conserved histone variant H2AZ support these paradigms. The region of *Drosophila* H2AZ homolog (H2AvD) that is essential for development lies in the docking domain (**Figure 4**), where H2A interacts with the H3•H4 dimer within the nucleosome (Clarkson et al. 1999). This specialized docking domain of H2AZ presents a binding site for HP1 exposed on the face of the nucleosome (Fan et al. 2004). Thus the primary sequence differences in the variant create a new binding site in nucleosomes. Additionally, this same region shifts the underlying H3 αN helix (Suto et al. 2000). The H3 αN helix also contacts DNA, and the changes in the H2AZ docking domain appear to alter these DNA contacts, thus subtly destabilizing the nucleosome. Other differences in the Loop 1 region of H2AZ appear to configure the composition of nucleosomes. Comparison of crystal structures shows that the H2A Loop I will clash with H2AZ Loop I within the nucleosome (**Figure 4**). Thus nucleosomes homotypic for H2A or for H2AZ are structurally preferable. This incompatibility between H2A and H2AZ implies that the SWR1-catalyzed replacement of one H2A•H2B dimer by H2AZ•H2B will facilitate replacement of the other H2A•H2B dimer.

Despite these insights gleaned from structures, it is not yet clear how altered structural features of H2AZ-containing nucleosomes lead to their diverse roles inferred from in vivo studies. Comparison of variant and canonical nucleosomes based on their physical properties has led to different conclusions by different laboratories: Some observations are consistent with a destabilizing role for H2AZ (Abbott et al. 2001), whereas others are consistent with greater stabilization (Fan et al. 2002, Park et al. 2004). It is possible that higher order interactions are more important than particle stabilization for in vivo behavior of H2AZ-containing nucleosomes, as suggested by the interaction with HP1 (Fan et al. 2004). Differences in higher order interactions involving H2AZ could help rationalize the different in vivo behaviors implied from studies in yeast and animals.

There is more limited evidence for how other variants of H2A affect function. The H2AX variant is identical to H2A throughout most of the protein, with primary sequence differences in the C-terminal including the defining four-amino acid motif. Thus this variant functions by providing a new phosphorylation site. The macroH2A variant carries a large, globular C-terminal domain that impedes transcription factor binding in vitro, and its histone fold domain interferes with ATP-dependent remodeling (Angelov et al. 2003). Both attributes are consistent with macroH2A playing a role in facultative silencing of the inactivated X chromosome (Costanzi & Pehrson 1998). In contrast, nucleosomes containing the H2A^{Bbd} variant are more accessible than canonical nucleosomes (Angelov et al. 2004), consistent with the striking depletion of H2A^{Bbd} on the inactive X chromosome (Chadwick & Willard 2001).

Some variants of H3 discussed in this review also show evidence of structural differentiation. CenH3s are the most extreme, where positive selection is thought to act as an adaptor between the rapidly evolving centromeric satellite sequences and the conserved kinetochore apparatus (Malik & Henikoff 2001). Rapid substitutions are focused on sites of DNA-histone contacts and may alter the affinity of centromeric nucleosomes for underlying satellite sequences. The Cid Loop 1 that is required for targeting to the centromere is on the disk face, consistent with a role as an exposed binding surface (Vermaak et al. 2002), and the diverged N-terminal tails of all CenH3 histones are thought to be platforms for binding kinetochore components (Malik & Henikoff 2003). In contrast, H3.3 is an example of a variant that appears to be structurally almost identical to its canonical counterpart (**Figure 4**). The cluster of core residues that specify assembly pathway are located behind the sheath of water residues that are structured by the DNA double helices (Davey et al. 2002) and thus are not accessible in the complete nucleosome. The single remaining difference, Ala31 in H3 versus Ser/Thr31 in H3.3, is nearly universal, suggestive of phosphoregulation of H3.3 in as-yet unidentified processes. However, the interchangeability of RC and RI forms in *Tetrahymena* (Yu & Gorovsky 1997), the presence of only H3.3 in yeasts and molds (Malik & Henikoff 2003), and the dominance of H3.3 in certain cell lineages (Chioda et al. 2004, Pina & Suau 1987, Urban & Zweidler 1983) suggest minimal structural differentiation between H3 and H3.3. Rather, a consistent requirement is seen for an RI H3-subtype in all eukaryotes, with the addition of a distinct RC-only form in multicellular eukaryotes. This phyletic pattern—and the evidence that H3.3 assembly is coupled to transcription—suggests that organisms with large, mostly silent genomes have evolved the RC-only form to package silent chromatin (Ahmad & Henikoff 2002c).

However, the functional distinctions between the two subtypes do not seem to be because of alterations of exposed surfaces in a static model of the nucleosome. We will need to consider the dynamics of nucleosome assembly to understand the role of the H3.3 variant.

Functions During Nucleosome Assembly

A number of protein-binding sites on histones are inaccessible in the complete nucleosome structure, implying that the sites are functional only in structural intermediates. One example of this is the RbAp48 subunit of CAF-1, which binds to the αN helix of histone H4 (Vermaak et al. 1999). This interaction probably occurs as histones are being delivered for deposition because the helix will be buried in

CAFs: chromatin
assembly factors

the completely assembled nucleosome. Surprisingly, a number of complexes that regulate transcription and operate on nucleosomal templates also contain RbAp48. This suggests that nucleosomal intermediates with exposed internal portions may be regulatory targets (Vermaak et al. 2003).

Similarly, the cluster of core residues that are not accessible in the complete nucleosome have been proposed to form a binding surface in predeposition complexes that delivers H3.3 to DNA because they are essential for RI deposition of the histone (Ahmad & Henikoff 2002c). However, the inaccessibility of these residues in H3.3 in the nucleosome leaves little to distinguish it from the canonical H3 histone. The observation that H3.3 can deposit in actively transcribed chromatin at any time implies that these regions are structurally and compositionally dynamic, with nucleosomes continually disassembled and then reassembled (**Figure 3**). If this leaves little time to complete nucleosome assembly before a new round of replacement begins, transcribed chromatin would remain in a partially assembled state. Thus the repetition of RI assembly alone would be sufficient to confer high DNA accessibility. This function does not require that the H3.3 variant generates a specialized nucleosome structure. In this view, the lack of structural specialization in H3.3 is a key feature that results from the need for H3.3 to perform the same role as the major histone H3, i.e., to package DNA. The sequence identity between the two H3-subtypes would be especially important as genes become repressed, because transcription will have enriched the H3.3 variant in chromatin that must now be packaged in a silent configuration.

A number of observations suggest that nucleosomal intermediates may be critical for active chromatin. As mentioned above, the very process of targeted RI assembly suggests that intermediates will be common in transcribed chromatin. Structures consistent with split nucleosomes are indeed observed at highly transcribed genes (Lee & Garrard 1991), and in vivo cross-linking studies also imply that

buried nucleosomal surfaces are exposed in active chromatin (Jackson 1978). Finally, a number of transcription-promoting factors have subunits with high histone-binding affinities, which are thought to assist chromatin binding. However, similar histone affinities are found in the protein chaperones that deliver histones for nucleosome assembly (Akey & Luger 2003). Indeed, some transcription factor complexes contain free histones (Keener 1997), and thus do not appear to be using their histone-binding subunits to bind chromatin. Instead, they might act by assisting nucleosome assembly. A nucleosome disassembly role has been reported for one of these factors, ASF1 (Adkins et al. 2004). These observations are consistent with the idea that nucleosomal intermediates are prevalent in active chromatin and distinguish it from inactive regions.

What Are the Functional Consequences of Multiple Nucleosome Assembly Pathways?

The bulk of nucleosome assembly occurs during DNA replication, and experiments with extracts defined a conserved set of CAFs that support histone deposition specifically on replicating DNA (Verreault et al. 1996). The expectation from these studies was that DNA replication without RC nucleosome assembly would be lethal. Thus it was surprising to find that null mutations for CAF components in budding yeast are viable (Enomoto & Berman 1998, Kaufman et al. 1997). CAF mutants show defects in telomere silencing and in DNA damage repair, but have normal packaging of chromatin. Thus other chromatin assembly pathways must compensate when RC assembly is defective. Indeed, other RC and RI nucleosome assembly activities have been identified. For example, the RCAF complex also supports RC assembly and enhances the activity of CAF. The lack of lethal phenotypes for chromatin assembly mutants suggests that any gaps in chromatin are filled by these other pathways. As most of the yeast genome is transcriptionally active, gaps left after replication can be filled by transcription-coupled RI assembly. However, compensation also occurs in organisms with more complex genomes. *FASCIATA* mutations eliminate CAF in *Arabidopsis* and cause some meristematic defects, but plants remain viable, indicating functional redundancy (Kaya et al. 2001). Therefore, even transcriptionally silent heterochromatin is being duplicated by alternative pathways in this mutant. A simple model is that alternative assembly pathways are capable of working on any gapped templates, but normally do not act on regions where CAF rapidly completes nucleosome assembly in S phase. Blocking CAF function in mammalian cells stimulates a DNA damage checkpoint (Hoek & Stillman 2003, Nabatiyan & Krude 2004), consistent with the idea that unpackaged DNA is not tolerated or is easily damaged.

Compensation between nucleosome assembly pathways can also explain why defects in the CAF and HIR nucleosome assembly activities result in mis-targeting of the centromeric Cse4p histone (Sharp et al. 2002). A compensation effect is also consistent with the results of altering the expression of histone variants. Overexpression of centromeric histones in yeast, mammals, or *Drosophila* results in its deposition in euchromatin, as if extra CenH3 fills gaps in transcriptionally active chromatin (Ahmad & Henikoff 2002b, Collins et al. 2004, Shelby et al. 1997). This suggests that there is a balance between all nucleosome assembly pathways in these organisms, even though they normally act on distinct parts of the genome.

CAF mutants in both yeast and *Arabidopsis* package DNA into chromatin and are viable; however, they show defective telomeric silencing and unstable developmental fates. Reduction of an RbAp48 component of CAF in *Arabidopsis* also causes spectacular epigenetic defects (Hennig et al. 2003), although nulls for some of these are lethal (Kohler et al. 2003). These phenotypes point to critical links between the mode of nucleosome assembly and the inheritance of epigenetic states. For example, if each chromatin assembly complex recruits specific chromatin-modifying enzymes, epigenetic patterns would not be preserved when alternate assembly pathways duplicate a chromatin region.

Indeed, there is evidence that the propagation of heterochromatin and DNA replication are linked in this way. Proper heterochromatic localization of HP1 depends in part on binding to CAF-1 at heterochromatic replication forks, as if HP1 is recruited for loading onto H3K9-methylated nucleosomes (Quivy et al. 2004). Moreover, CAF-1 delivers histone H3 pre-methylated at lysine-9 to replication forks at sites of methylated DNA (Sarraf & Stancheva 2004). Both mechanisms for perpetuating heterochromatin require that CAF-1 be used to duplicate the chromatin during replication. These considerations predict that CAF mutations in plants are accumulating replacement H3.2 histones in heterochromatin because RC assembly fails and gap-filling, using RI pathways, compensates. Whereas the phenotypes for elimination of CAF-1 are more severe in mammalian cells, this may reflect its more critical role in epigenetic control of essential processes.

CONCLUSIONS

The differentiation of nucleosomes by incorporation of variant histones must be as ancient as the eukaryotes themselves, insofar as a CenH3-containing centromere is a defining feature of eukaryotic chromosomes. The fact that centromeric chromatin is maintained in the same chromosomal position for millions of years, yet can shift spontaneously to an unrelated DNA sequence, is the most extreme example imaginable of faithful epigenetic inheritance. Nevertheless, the rapid evolution of both the highly repetitive centromeric satellite DNA and the CenH3 variant that evidently adapts to it implies that centromeric stability is dynamically maintained. Yet little is known about how centromeric nucleosomes are assembled and propagated through the cell cycle.

Variant histones, such as CenH3s, are deposited independently of replication, in contrast to the bulk of chromatin, which is assembled by RC assembly of canonical histones. The use of distinct assembly pathways provides a simple means of differentiating chromatin, by targeting replacement of nucleosomes throughout the cell cycle. Transcription itself appears to catalyze the targeting of the H3.3 replacement variant, and the biochemistry of this process has begun to reveal insights into both variant deposition and the possible maintenance of associated active histone modifications. Disruption of nucleosomes is not limited to transcription-coupled replacement, because nucleosomes are "evicted" from promoters upon gene activation.

These dynamic processes of nucleosome replacement and eviction can help account for the abundance and diversity of ATP-dependent chromatin remodeling complexes. Indeed, replacement of the H2AZ variant is catalyzed by one such complex, and it seems likely that other H2A variants are associated with dedicated members of the SWI/SNF family of ATPases. In this way, nucleosome assembly and remodeling, formerly assumed from in vitro work to be distinct processes, are now seen to be aspects of the same in vivo process.

Once incorporated into chromatin, most histone variants have distinct structural properties that are likely to profoundly alter chromatin. Differences between H2A and H2AZ in the docking domain can affect nucleosome integrity, and the large globular domain of macroH2A is suspected to have enzymatic function. H3.3 is the exception, because the only difference from H3 that is exposed in the nucleosome is a single tail residue of uncertain significance; rather, differences in post-translational modifications that are found to distinguish H3 from H3.3 are more likely to affect nucleosome properties.

The study of histone variants and the multiple biochemical processes that deposit them into chromatin has led to new insights into chromatin dynamics. The effects of these processes on gene regulation and chromosome behavior have yet to be elucidated, but the availability of powerful new tools promises to change that. Most importantly, the excitement generated by these new insights has fueled a resurgence of interest in histone variants after decades of relative neglect. We look forward to the deeper insights into eukaryotic biology that now appear to be just around the corner.

SUMMARY POINTS

1. Variants of histones H3 and H2A differentiate chromatin at centromeres, active genes, and heterochromatin.

2. Nucleosomes characterized by a special H3 variant identify the centromeres of every eukaryotic chromosome.

3. The replacement histone, H3.3, marks actively transcribed loci by replication-independent nucleosome assembly.

4. Gene activation is accompanied by disassembly of a nucleosome at the promoter.

5. A chromatin remodeling machine replaces the conserved histone variant, H2AZ.

6. Epigenetically silenced chromatin is enriched or depleted in abundance of diverse H2A variants.

7. Variant structure can affect properties of chromatin.

8. The operation of multiple nucleosome assembly pathways has important implications for nucleosome dynamics.

LITERATURE CITED

Abbott DW, Ivanova VS, Wang X, Bonner WM, Ausio J. 2001. Characterization of the stability and folding of H2A.Z chromatin particles: implications for transcriptional activation. *J. Biol. Chem.* 276:41945–49

Adkins MW, Howar SR, Tyler JK. 2004. Chromatin disassembly mediated by the histone chaperone Asf1 is essential for transcriptional activation of the yeast PHO5 and PHO8 genes. *Mol. Cell.* 14:657–66

Ahmad K, Henikoff S. 2001a. Centromeres are specialized replication domains in heterochromatin. *J. Cell Biol.* 153:101–10

Ahmad K, Henikoff S. 2001b. Modulation of a transcription factor counteracts heterochromatic gene silencing in *Drosophila*. *Cell* 104:839–47

Ahmad K, Henikoff S. 2002a. Epigenetic consequences of nucleosome dynamics. *Cell* 111:281–84

Ahmad K, Henikoff S. 2002b. Histone H3 variants specify modes of chromatin assembly. *Proc. Natl. Acad. Sci. USA* 99(Suppl.) 4:16477–84

Ahmad K, Henikoff S. 2002c. The histone variant H3.3 marks active chromatin by replication-independent nucleosome assembly. *Mol. Cell* 9:1191–200

Akey CW, Luger K. 2003. Histone chaperones and nucleosome assembly. *Curr. Opin. Struct. Biol.* 13:6–14

Allen MD, Buckle AM, Cordell SC, Lowe J, Bycroft M. 2003. The crystal structure of AF1521 a protein from *Archaeoglobus fulgidus* with homology to the non-histone domain of macroH2A. *J. Mol. Biol.* 330:503–11

Alonso A, Mahmood R, Li S, Cheung F, Yoda K, Warburton PE. 2003. Genomic microarray analysis reveals distinct locations for the CENP-A binding domains in three human chromosome 13q32 neocentromeres. *Hum. Mol. Genet.* 12:2711–21

Amor DJ, Bentley K, Ryan J, Perry J, Wong L, et al. 2004. Human centromere repositioning "in progress." *Proc. Natl. Acad. Sci. USA* 101:6542–47

Angelov D, Molla A, Perche PY, Hans F, Cote J, et al. 2003. The histone variant macroH2A interferes with transcription factor binding and SWI/SNF nucleosome remodeling. *Mol. Cell* 11:1033–41

Angelov D, Verdel A, An W, Bondarenko V, Hans F, et al. 2004. SWI/SNF remodeling and p300-dependent transcription of histone variant H2A[Bbd] nucleosomal arrays. *EMBO J.* 23:3815–24

Belotserkovskaya R, Oh S, Bondarenko VA, Orphanides G, Studitsky VM, Reinberg D. 2003. FACT facilitates transcription-dependent nucleosome alteration. *Science* 301:1090–93

Bender W, Fitzgerald DP. 2002. Transcription activates repressed domains in the *Drosophila* bithorax complex. *Development* 129:4923–30

Bernstein BE, Liu CL, Humphrey EL, Perlstein EO, Schreiber SL. 2004. Global nucleosome occupancy in yeast. *Genome Biol.* 5:R62

Black BE, Foltz DR, Chakravarthy S, Luger K, Woods VL Jr, Cleveland DW. 2004. Structural determinants for generating centromeric chromatin. *Nature* 430:578–82

Blower MD, Karpen GH. 2001. The role of *Drosophila* CID in kinetochore formation, cell-cycle progression and heterochromatin interactions. *Nat. Cell Biol.* 3:730–39

Blower MD, Sullivan BA, Karpen GH. 2002. Conserved organization of centromeric chromatin in flies and humans. *Developmental Cell* 2:319–30

Boeger H, Griesenbeck J, Strattan JS, Kornberg RD. 2003. Nucleosomes unfold completely at a transcriptionally active promoter. *Mol. Cell.* 11:1587–98

Boeger H, Griesenbeck J, Strattan JS, Kornberg RD. 2004. Removal of promoter nucleosomes by disassembly rather than sliding in vivo. *Mol. Cell.* 14:667–73

Brownell JE, Zhou J, Ranalli T, Kobayashi R, Edmondson DG, et al. 1996. Tetrahymena histone acetyltransferase A: a homolog to yeast Gcn5p linking histone acetylation to gene activation. *Cell* 84:843–51

Buchwitz BJ, Ahmad K, Moore LL, Roth MB, Henikoff S. 1999. A histone-H3-like protein in *C. elegans*. *Nature* 401:547–48

Chadwick BP, Willard HF. 2001. A novel chromatin protein, distantly related to histone H2A, is largely excluded from the inactive X chromosome. *J. Cell Biol.* 152:375–84

Chadwick BP, Willard HF. 2004. Multiple spatially distinct types of facultative heterochromatin on the human inactive X chromosome. *Proc. Natl. Acad. Sci. USA* 101:17450–55

Chioda M, Spada F, Eskeland R, Thompson EM. 2004. Histone mRNAs do not accumulate during S phase of either mitotic or endoreduplicative cycles in the chordate *Oikopleura dioica*. *Mol. Cell. Biol.* 24:5391–403

Clarkson MJ, Wells JR, Gibson F, Saint R, Tremethick DJ. 1999. Regions of variant histone His2AvD required for *Drosophila* development. *Nature* 399:694–97

Collins KA, Furuyama S, Biggins S. 2004. Proteolysis contributes to the exclusive centromere localization of the yeast Cse4/CENP-A histone H3 variant. *Curr. Biol.* 14:1968–72

Cooper JL, Henikoff S. 2004. Adaptive evolution of the histone fold domain in centromeric histones. *Mol. Biol. Evol.* 21(9):1712–18

Cosgrove MS, Boeke JD, Wolberger C. 2004. Regulated nucleosome mobility and the histone code. *Nat. Struct. Mol. Biol.* 11:1037–43

Costanzi C, Pehrson JR. 1998. Histone macroH2A1 is concentrated in the inactive X chromosome of female mammals. *Nature* 393:599–601

Davey CA, Sargent DF, Luger K, Maeder AW, Richmond TJ. 2002. Solvent mediated interactions in the structure of the nucleosome core particle at 1.9 Å resolution. *J. Mol. Biol.* 319:1097–113

Dhillon N, Kamakaka RT. 2000. A histone variant, Htz1p, and a Sir1p-like protein, Esc2p, mediate silencing at HMR. *Mol. Cell.* 6:769–80

Drewell RA, Bae E, Burr J, Lewis EB. 2002. Transcription defines the embryonic domains of cis-regulatory activity at the *Drosophila* bithorax complex. *Proc. Natl. Acad. Sci. USA* 99:16853–58

Earnshaw WC, Rothfield N. 1985. Identification of a family of human centromere proteins using autoimmune sera from patients with schleroderma. *Chromosoma* 91:313–21

Enomoto S, Berman J. 1998. Chromatin assembly factor I contributes to the maintenance, but not the re-establishment, of silencing at the yeast silent mating loci. *Genes Dev.* 12:219–32

Faast R, Thonglairoam V, Schulz TC, Beall J, Wells JR, et al. 2001. Histone variant H2A.Z is required for early mammalian development. *Curr. Biol.* 11:1183–87

Fan JY, Gordon F, Luger K, Hansen JC, Tremethick DJ. 2002. The essential histone variant H2A.Z regulates the equilibrium between different chromatin conformational states. *Nat. Struct. Biol.* 9:172–76

Fan JY, Rangasamy D, Luger K, Tremethick DJ. 2004. H2A.Z alters the nucleosome surface to promote HP1alpha-mediated chromatin fiber folding. *Mol. Cell.* 16:655–61

Fernandez-Capetillo O, Lee A, Nussenzweig M, Nussenzweig A. 2004. H2AX: the histone guardian of the genome. *DNA Repair* 3:959–67

Formosa T, Ruone S, Adams MD, Olsen AE, Eriksson P, et al. 2002. Defects in SPT16 or POB3 (yFACT) in *Saccharomyces cerevisiae* cause dependence on the Hir/Hpc pathway: polymerase passage may degrade chromatin structure. *Genetics* 162:1557–71

Freitas MA, Sklenar AR, Parthun MR. 2004. Application of mass spectrometry to the identification and quantification of histone post-translational modifications. *J. Cell Biochem.* 92:691–700

Gribnau J, Diderich K, Pruzina S, Calzolari R, Fraser P. 2000. Intergenic transcription and developmental remodeling of chromatin subdomains in the human β-globin locus. *Mol. Cell* 5:377–86

Hanlon SE, Lieb JD. 2004. Progress and challenges in profiling the dynamics of chromatin and transcription factor binding with DNA microarrays. *Curr. Opin. Genet. Dev.* 14:697–705

Hayashi T, Fujita Y, Iwasaki O, Adachi Y, Takahashi K, Yanagida M. 2004. Mis16 and Mis18 are required for CENP-A loading and histone deacetylation at centromeres. *Cell* 118:715–29

Hendzel MJ, Davie JR. 1990. Nucleosomal histones of transcriptionally active/competent chromatin preferentially exchange with newly synthesized histones in quiescent chicken erythrocytes. *Biochem. J.* 271:67–73

Henikoff S, Ahmad K, Platero JS, van Steensel B. 2000. Heterochromatic deposition of centromeric histone H3-like proteins. *Proc. Natl. Acad. Sci. USA* 97:716–21

Henikoff S, Furuyama T, Ahmad A. 2004. Histone variants, nucleosome assembly and epigenetic inheritance. *Trends Genet.* 20:320–26

Hennig L, Taranto P, Walser M, Schonrock N, Gruissem W. 2003. *Arabidopsis* MSI1 is required for epigenetic maintenance of reproductive development. *Development* 130:2555–65

Hoek M, Stillman B. 2003. Chromatin assembly factor 1 is essential and couples chromatin assembly to DNA replication in vivo. *Proc. Natl. Acad. Sci. USA* 100:12183–88

Hogga I, Karch F. 2002. Transcription through the iab-7 *cis*-regulatory domain of the bithorax complex interferes with maintenance of Polycomb-mediated silencing. *Development* 129:4915–22

Howman EV, Fowler KJ, Newson AJ, Redward S, MacDonald AC, et al. 2000. Early disruption of centromeric chromatin organization in centromere protein A (CenpA) null mice. *Proc. Natl. Acad. Sci. USA* 97:1148–53

Jackson JD, Gorovsky MA. 2000. Histone H2A.Z has a conserved function that is distinct from that of the major H2A sequence variants. *Nucleic Acids Res.* 28:3811–16

Jackson V. 1978. Studies on histone organization in the nucleosome using formaldehyde as a reversible cross-linking agent. *Cell* 15:945–54

Janicki SM, Tsukamoto T, Salghetti SE, Tansey WP, Sachidanandam R, et al. 2004. From silencing to gene expression: real-time analysis in single cells. *Cell* 116:683–98

Jenuwein T, Allis CD. 2001. Translating the histone code. *Science* 293:1074–80

Kaufman PD, Kobayashi R, Stillman B. 1997. Ultraviolet radiation sensitivity and reduction of telomeric silencing in *Saccharomyces cerevisiae* cells lacking chromatin assembly factor-I. *Genes Dev.* 11:345–57

Kaya H, Shibahara KI, Taoka KI, Iwabuchi M, Stillman B, Araki T. 2001. FASCIATA genes for chromatin assembly factor-1 in *Arabidopsis* maintain the cellular organization of apical meristems. *Cell* 104:131–42

Keener J, Dodd JA, Lalo D, Nomura M. 1997. Histones H3 and H4 are components of upstream activation factor required for the high-level transcription of yeast rDNA by RNA polymerase I. *Proc. Natl. Acad. Sci. USA* 94:13458–62

Keith KC, Baker RE, Chen Y, Harris K, Stoler S, Fitzgerald-Hayes M. 1999. Analysis of primary structural determinants that distinguish the centromere-specific function of histone variant Cse4p from histone H3. *Mol. Cell. Biol.* 19:6130–39

Kobor MS, Venkatasubrahmanyam S, Meneghini MD, Gin JW, Jennings JL, et al. 2004. A protein complex containing the conserved Swi2/Snf2-related ATPase Swr1p deposits histone variant H2A.Z into euchromatin. *PLoS Biol.* 2:E131

Kohler C, Hennig L, Bouveret R, Gheyselinck J, Grossniklaus U, Gruissem W. 2003. *Arabidopsis* MSI1 is a component of the MEA/FIE Polycomb group complex and required for seed development. *EMBO J.* 22:4804–14

Kristjuhan A, Svejstrup JQ. 2004. Evidence for distinct mechanisms facilitating transcript elongation through chromatin in vivo. *EMBO J.* 23:4243–52

Krogan NJ, Keogh MC, Datta N, Sawa C, Ryan OW, et al. 2003. A Snf2 family ATPase complex required for recruitment of the histone H2A variant Htz1. *Mol. Cell* 12:1565–76

Kusch T, Florens L, Macdonald WH, Swanson SK, Glaser RL, et al. 2004. Acetylation by Tip60 is required for selective histone variant exchange at DNA lesions. *Science* 306:2084–87

Lee CK, Shibata Y, Rao B, Strahl BD, Lieb JD. 2004. Evidence for nucleosome depletion at active regulatory regions genome-wide. *Nat. Genet.* 36:900–5

Lee MS, Garrard WT. 1991. Transcription-induced nucleosome 'splitting': an underlying structure for DNase I sensitive chromatin. *EMBO J.* 10:607–15

Lo AW, Magliano DJ, Sibson MC, Kalitsis P, Craig JM, Choo KHA. 2001. A novel chromatin immunoprecipitation and array (CIA) analysis identifies a 460-kb CENP-A-binding neocentromeric DNA. *Genome Res.* 11:448–57

Luger K, Mader AW, Richmond RK, Sargent DF, Richmond TJ. 1997. Crystal structure of the nucleosome core particle at 2.8 Å resolution. *Nature* 389:251–60

Madigan JP, Chotkowski HL, Glaser RL. 2002. DNA double-strand break-induced phosphorylation of *Drosophila* histone variant H2Av helps prevent radiation-induced apoptosis. *Nucleic Acids Res.* 30:3698–705

Malik HS, Henikoff S. 2001. Adaptive evolution of Cid, a centromere-specific histone in *Drosophila. Genetics* 157:1293–98

Malik HS, Henikoff S. 2003. Phylogenomics of the nucleosome. *Nat. Struct. Biol.* 10:882–91

Malik HS, Vermaak D, Henikoff S. 2002. Recurrent evolution of DNA-binding motifs in the *Drosophila* centromeric histone. *Proc. Natl. Acad. Sci. USA* 99:1449–54

McKittrick E, Gafken PR, Ahmad K, Henikoff S. 2004. Histone H3.3 is enriched in covalent modifications associated with active chromatin. *Proc. Natl. Acad. Sci. USA* 101:1525–30

Mellone BG, Allshire RC. 2003. Stretching it: putting the CEN(P-A) in centromere. *Curr. Opin. Genet. Dev.* 13:191–98

Meneghini MD, Wu M, Madhani HD. 2003. Conserved histone variant H2A.Z protects euchromatin from the ectopic spread of silent chromatin. *Cell* 112:725–36

Mizuguchi G, Shen X, Landry J, Wu WH, Sen S, Wu C. 2004. ATP-driven exchange of histone H2AZ variant catalyzed by SWR1 chromatin remodeling complex. *Science* 303:343–48

Monson EK, de Bruin D, Zakian VA. 1997. The yeast Cac1 protein is required for the stable inheritance of transcriptionally repressed chromatin at telomeres. *Proc. Natl. Acad. Sci. USA* 94:13081–86

Moore LL, Roth MB. 2001. HCP-4, a CENP-C-like protein in *Caenorhabditis elegans*, is required for resolution of sister centromeres. *J. Cell Biol.* 153:1199–208

Morrison AJ, Highland J, Krogan NJ, Arbel-Eden A, Greenblatt JF, et al. 2004. INO80 and gamma-H2AX interaction links ATP-dependent chromatin remodeling to DNA damage repair. *Cell* 119:767–75

Nabatiyan A, Krude T. 2004. Silencing of chromatin assembly factor 1 in human cells leads to cell death and loss of chromatin assembly during DNA synthesis. *Mol. Cell. Biol.* 24:2853–62

Nagaki K, Cheng Z, Ouyang S, Talbert PB, Kim M, et al. 2004. Sequencing of a rice centromere uncovers active genes. *Nat. Genet.* 36:138–45

Nagy PL, Cleary ML, Brown PO, Lieb JD. 2003. Genomewide demarcation of RNA polymerase II transcription units revealed by physical fractionation of chromatin. *Proc. Natl. Acad. Sci. USA* 100:6364–69

Palmer DK, O'Day K, Trong HL, Charbonneau H, Margolis RL. 1991. Purification of the centromere-specific protein CENP-A and demonstration that it is a distinctive histone. *Proc. Natl. Acad. Sci. USA* 88:3734–38

Palmer DK, O'Day K, Wener MH, Andrews BS, Margolis RL. 1987. A 17-kD centromere protein (CENP-A) copurifies with nucleosome core particles and with histones. *J. Cell Biol.* 104:805–15

Park YJ, Dyer PN, Tremethick DJ, Luger K. 2004. A new fluorescence resonance energy transfer approach demonstrates that the histone variant H2AZ stabilizes the histone octamer within the nucleosome. *J. Biol. Chem.* 279:24274–82

Pereira SL, Reeve JN. 1998. Histones and nucleosomes in Archaea and Eukarya: a comparative analysis. *Extremophiles* 2:141–48

Pina B, Suau P. 1987. Changes in histones H2A and H3 variant composition in differentiating and mature rat brain cortical neurons. *Dev. Biol.* 123:51–58

Quivy JP, Roche D, Kirschner D, Tagami H, Nakatani Y, Almouzni G. 2004. A CAF-1 dependent pool of HP1 during heterochromatin duplication. *EMBO J.* 23:3516–26

Rangasamy D, Berven L, Ridgway P, Tremethick DJ. 2003. Pericentric heterochromatin becomes enriched with H2A.Z during early mammalian development. *EMBO. J.* 22:1599–607

Rank G, Prestel M, Paro R. 2002. Transcription through intergenic chromosomal memory elements of the *Drosophila* bithorax complex correlates with an epigenetic switch. *Mol. Cell. Biol.* 22:8026–34

Reinke H, Horz W. 2003. Histones are first hyperacetylated and then lose contact with the activated PHO5 promoter. *Mol. Cell* 23:1599–607

Rogakou EP, Pilch DR, Orr AH, Ivanova VS, Bonner WM. 1998. DNA double-stranded breaks induce histone H2AX phosphorylation on serine 139. *J. Biol. Chem.* 273:5858–68

Santisteban MS, Kalashnikova T, Smith MM. 2000. Histone H2A.Z regulates transcription and is partially redundant with nucleosome remodeling complexes. *Cell* 103:411–22

Sarraf SA, Stancheva I. 2004. Methyl-CpG binding protein MBD1 couples histone H3 methylation at lysine 9 by SETDB1 to DNA replication and chromatin assembly. *Mol. Cell.* 15:595–605

Schwabish MA, Struhl K. 2004. Evidence for eviction and rapid deposition of histones upon transcriptional elongation by RNA polymerase II. *Mol. Cell. Biol.* 24:10111–17

Sharp JA, Franco AA, Osley MA, Kaufman PD. 2002. Chromatin assembly factor I and Hir proteins contribute to building functional kinetochores in *S. cerevisiae*. *Genes Dev.* 16:85–100

Shelby RD, Monier K, Sullivan KF. 2000. Chromatin assembly at kinetochores is uncoupled from DNA replication. *J. Cell Biol.* 151:1113–18

Shelby RD, Vafa O, Sullivan KF. 1997. Assembly of CENP-A into centromeric chromatin requires a cooperative array of nucleosomal DNA contact sites. *J. Cell Biol.* 136:501–13

Solomon MJ, Varshavsky A. 1985. Formaldehyde-mediated DNA-protein crosslinking: a probe for in vivo chromatin structures. *Proc. Natl. Acad. Sci. USA* 82:6470–74

Stoler S, Keith KC, Curnick KE, Fitzgerald-Hayes M. 1995. A mutation in CSE4, an essential gene encoding a novel chromatin-associated protein in yeast, causes chromosome nondisjunction and cell cycle arrest at mitosis. *Genes Dev.* 9:573–86

Sugimoto K, Yata H, Muro Y, Himeno M. 1994. Human centromere protein C (CENP-C) is a DNA-binding protein which possesses a novel DNA-binding motif. *J. Biochem.* 116:877–81

Sullivan BA, Karpen GH. 2004. Centromeric chromatin exhibits a histone modification pattern that is distinct from both euchromatin and heterochromatin. *Nat. Struct. Mol. Biol.* 11:1076–83

Suto RK, Clarkson MJ, Tremethick DJ, Luger K. 2000. Crystal structure of a nucleosome core particle containing the variant histone H2A.Z. *Nat. Struct. Biol.* 7:1121–24

Tagami H, Ray-Gallet D, Almouzni G, Nakatani Y. 2004. Histone H3.1 and H3.3 complexes mediate nucleosome assembly pathways dependent or independent of DNA synthesis. *Cell* 116:51–61

Talbert PB, Bryson TD, Henikoff S. 2004. Adaptive evolution of centromere proteins in plants and animals. *J. Biol.* 3:18

Talbert PB, Masuelli R, Tyagi AP, Comai L, Henikoff S. 2002. Centromeric localization and adaptive evolution of an *Arabidopsis* histone H3 variant. *Plant Cell* 14:1053–66

Turner BM, Birley AJ, Lavendar J. 1992. Histone H4 isoforms acetylated at specific lysine residues define individual chromosomes and chromatin domains in *Drosophila* polytene nuclei. *Cell* 69:375–84

Urban MK, Zweidler A. 1983. Changes in nucleosomal core histone variants during chicken development and maturation. *Dev. Biol.* 95:421–28

van Attikum H, Fritsch O, Hohn B, Gasser SM. 2004. Recruitment of the INO80 complex by H2A phosphorylation links ATP-dependent chromatin remodeling with DNA double-strand break repair. *Cell* 119:777–88

van Daal A, Elgin SC. 1992. A histone variant, H2AvD, is essential in *Drosophila melanogaster*. *Mol. Biol. Cell* 3:593–602

Vermaak D, Ahmad K, Henikoff S. 2003. Maintenance of chromatin states: an open-and-shut case. *Curr. Opin. Cell Biol.* 15:266–74

Vermaak D, Hayden HS, Henikoff S. 2002. Centromere targeting element within the histone fold domain of Cid. *Mol. Cell. Biol.* 22:7553–61

Vermaak D, Wade PA, Jones PL, Shi YB, Wolffe AP. 1999. Functional analysis of the SIN3-histone deacetylase RPD3-RbAp48-histone H4 connection in the *Xenopus* oocyte. *Mol. Cell. Biol.* 19:5847–60

Verreault A, Kaufman PD, Kobayashi R, Stillman B. 1996. Nucleosome assembly by a complex of CAF-1 and acetylated histones H3/H4. *Cell* 87:95–104

Waterborg JH. 1990. Sequence analysis of acetylation and methylation in two histone H3 variants of alfalfa. *J. Biol. Chem.* 265:17157–61

Wieland G, Orthaus S, Ohndorf S, Diekmann S, Hemmerich P. 2004. Functional complementation of human centromere protein A (CENP-A) by Cse4p from *Saccharomyces cerevisiae*. *Mol. Cell. Biol.* 24:6620–30

Wolffe AP. 1992. *Chromatin: Structure and Function*. San Diego: Academic

Workman JL, Abmayr SM. 2004. Histone H3 variants and modifications on transcribed genes. *Proc. Natl. Acad. Sci. USA* 101:1429–30

Yang CH, Tomkiel J, Saitoh H, Johnson DH, Earnshaw WC. 1996. Identification of overlapping DNA-binding and centromere-targeting domains in the human kinetochore protein CENP-C. *Mol. Cell. Biol.* 16:3576–86

Yu L, Gorovsky MA. 1997. Constitutive expression, not a particular primary sequence, is the important feature of the H3 replacement variant hv2 in *Tetrahymena thermophila*. *Mol. Cell. Biol.* 17:6303–10

Planar Cell Polarization: An Emerging Model Points in the Right Direction

Thomas J. Klein and Marek Mlodzik

Mount Sinai School of Medicine, Brookdale Department of Molecular, Cell and Developmental Biology, New York, NY 10029; email: Marek.Mlodzik@mssm.edu, Thomas.Klein@mssm.edu

Annu. Rev. Cell Dev. Biol.
2005. 21:155–76

First published online as a
Review in Advance on
May 23, 2005

The *Annual Review of
Cell and Developmental
Biology* is online at
http://cellbio.annualreviews.org

doi: 10.1146/
annurev.cellbio.21.012704.132806

Copyright © 2005 by
Annual Reviews. All rights
reserved

1081-0706/05/1110-
0155$20.00

Abstract

Polarization is a feature common to many cell types. Epithelial cells, for example, exhibit a characteristic apical-basolateral polarity that is critical for their function. In addition to this ubiquitous form of polarity, whole fields of cells are often polarized in a plane perpendicular to the apical-basal axis. This form of polarity, referred to as planar cell polarity (PCP), exists in all adult *Drosophila* cuticular tissues, as well as in numerous vertebrate tissues, including the mammalian skin and inner ear epithelia. Recent advances in the study of PCP establishment are beginning to unravel the molecular mechanisms underlying this cellular process. This review discusses new developments in the molecular understanding of PCP in *Drosophila* and vertebrates and integrates the current data in a model to illustrate how interactions between PCP factors might function to generate planar polarity.

Contents

INTRODUCTION

PCP: planar cell
polarity

The establishment and maintenance of cellular polarization is an important feature of development and is critical for organ function. Epithelial apical-basolateral polarity enables tissues to perform functions such as the vectorial transport of fluid or the directed secretion of specialized components. In addition to this ubiquitous axis of polarization, many epithelial tissues acquire a second polarity axis within the plane of the epithelium, commonly referred to as epithelial PCP.

PCP was initially studied in *Drosophila*, where all adult cuticular structures show PCP features (Adler 2002, Mlodzik 2002, Strutt 2003). In vertebrates, processes requiring proper PCP signaling include skin development and body hair orientation, polarization of the sensory epithelium in the inner ear, and the directed movement of mesenchymal cell populations during gastrulation (e.g., Dabdoub et al. 2003, Guo et al. 2004, Keller 2002, Montcouquiol et al. 2003, Wallingford et al. 2002). Other vertebrate processes requiring PCP signaling can easily be envisioned and have already been proposed, including the polarization of cilia in the oviduct and respiratory tract (Eaton 1997).

Unraveling the mechanisms of the establishment of PCP is one of the most exciting frontiers in developmental biology. How individual cells hundreds of cell diameters apart acquire the same polarity within the plane of an epithelial field or how mesenchymal cells establish a uniform polarization during their intercalation are fascinating biological problems. In addition, cellular polarization is critical for almost all cell types and is often associated with diseases when disturbed (reviewed in Stein et al. 2002). Although much progress has been made in recent years, the molecular aspects of PCP establishment are still far from being understood.

In this review, we briefly summarize the state of our understanding in the field and then integrate the current data into a working model. On the basis of this model we then discuss unresolved questions that arise from the current state of the field. We apologize to the research areas and viewpoints that we have not been able to include here.

THE HISTORY OF THE STUDY OF PCP

The study of PCP originates from work in the fruit fly, where it was initially referred to as tissue polarity. Elegant work pioneered by the Lawrence and Adler groups put the problem on the map more than 20 years ago (Lawrence & Shelton 1975, Vinson & Adler 1987). This was later accompanied by the genetic analysis of several PCP genes (Adler 1992, Gubb 1993), and the first molecular cloning of a PCP gene from *Drosophila* (Vinson et al. 1989). These initial insights were followed by systematic genetic screens in *Drosophila* by several research teams and the subsequent molecular cloning and analysis of identified PCP factors (**Table 1**) (Adler 2002, Mlodzik 2002, Strutt 2003).

The analysis of PCP generation is now an important feature of developmental studies in many organisms. Starting in *Xenopus* and zebrafish with the analysis of the process of CE during gastrulation and neurulation, PCP-related processes have been discovered and analyzed in vertebrates (reviewed in Keller 2002). Strikingly, the large genome-wide forward genetic screens in zebrafish identified several mutants affecting CE that turned out to be orthologues of *Drosophila* PCP genes (e.g., Jessen et al. 2002, Marlow et al. 2002, Myers et al. 2002).

The parallels and conservation of the PCP gene cassette have recently been extended to mammals as well (**Table 1**). In particular, the mammalian inner ear and epidermis are beautiful examples of epithelia with PCP features. Mammalian homologues of the fly PCP genes have been implicated in the generation of PCP in these tissues (Dabdoub et al. 2003, Guo et al. 2004, Montcouquiol et al. 2003). In addition, some of these homologues are important for both epithelial polarization and CE during gastrulation and neurulation, supporting the commonality of PCP establishment not only across species but also between different polarized cell types and organs.

A recent paper made the striking observation that the hair pattern defects in the mouse Frizzled6 mutant (*mFz6*) are very much reminiscent of the actin-hair pattern defects in the fly *fz* mutant (Guo et al. 2004). This

was exciting not only for the genetic demonstration of the epidermal PCP features in mammals, but also for the discovery that the same evolutionarily conserved protein family regulates this process from flies to mammals and that the defects observed in the respective mutants are virtually identical, suggesting similar principles at work in all contexts. In a way, this finding brought the field full circle—from its beginnings studying hair patterns on flies, to the studies of eye development, CE, and cochlear development—back to studying hair patterns, only this time in mammals. It will be very interesting to see which processes, tissues, and organisms the PCP field will work its way into next.

CE: convergent extension

CONSERVATION OF PLANAR CELL POLARITY SIGNALING

A Conserved Signaling Cassette Gives Varied Readouts

Most of the current evidence indicates that, rather than having many different signaling pathways regulating planar polarity from tissue to tissue and organism to organism, there is a core PCP gene cassette that is conserved across tissues and species. However, as it is used in many different tissues from flies to mammals, the cellular readouts of this conserved signaling cassette must be highly varied. These downstream pathways have been shown to affect cytoskeletal organization, nuclear signaling, orientation of the mitotic spindle, and many other cellular functions.

Cytoskeletal organization is likely the main target of PCP in *Drosophila* cuticular cells and the mammalian inner ear epithelium (Adler 2002, Mlodzik 2002, Montcouquiol et al. 2003, Strutt 2003). A nuclear signaling response is prominent in multicellular units, including the *Drosophila* eye (Mlodzik 1999) and likely in avian feather buds and mammalian hair follicles (Guo et al. 2004, Mlodzik 2002). PCP-mediated regulation of the orientation of mitotic spindles is known

TABLE 1 Primary PCP genes and their relationships

PCP gene	Tissues affected in *Drosophila*	Processes affected in vertebrates[#]	Molecular features	Localization in *Drosophila* wing cells	R3/R4 req. in *Drosophila* eye
frizzled (fz)	All adult tissues	CE[1], inner ear, epidermis	Seven-pass transmembrane receptor, binds Wnt ligands, binds Dsh, recruits Dsh and Dgo to membrane	Distal	R3
dishevelled(dsh)	All adult tissues	CE	Cytoplasmic protein containing DIX, PDZ, DEP domains, recruited to membrane by Fz, binds Fz, Pk, Stbm and Dgo	Distal	R3
prickle (pk)(a.k.a. *prickle-spiny legs*)	All adult tissues	CE	Cytoplasmic protein with 3 LIM domains and PET domain, recruited to membrane by Stbm, physically interacts with Dsh, Stbm and Dgo	Proximal	R4
strabismus (stbm)/Van Gogh (Vang)	All adult tissues	CE, inner ear	Novel 4-pass transmembrane protein, binds Pk, Dsh and Dgo, recruits Pk to membrane	Proximal	R4
flamingo (fmi)/starry night (stan)	All adult tissues	inner ear	Cadherin with seven-pass transmembrane receptor features, homophylic cell adhesions	Proximal + distal	R3 + R4
diego (dgo)	Eye, wing, notum in GOF*	CE	Cytoplasmic Ankyrin repeat protein, recruited to membrane by Fz, binds Dsh, Stbm and Pk	Distal; co-loc. with Fz/Dsh	R3
RhoA	Eye, wing*	CE	Small GTPase, acts downstream of Dsh	n.d.	R3
misshapen (msn)	Eye, wing, notum*	?	STE20-like S/T protein kinase, acts downstream of Dsh	n.d.	R3
Fat (Ft)	All adult tissues	inner ear?	Proto-cadherin, heterophyllic interaction with Ds, binds Atrophin	Distal	R3
dachsous (ds)	All adult tissues	inner ear?	Proto-cadherin, heterophyllic interaction with Fat	Proximal	R4
four jointed (fj)	All adult tissues	n.d.	Type-2 transmembrane or secreted peptide possibly functions in Golgi to modify Ds	n.d.	

[#]only tested tissues mentioned, combination of analysis in Xenopus, zebrafish and mouse.
[1]CE: convergent extension.
*other tissues were not tested.
n.d.: not determined.

in *Drosophila* and *Caenorhabditis elegans* (e.g., Bellaiche et al. 2004, Herman et al. 1995, Sawa et al. 1996) and likely occurs in other animals as well. In addition, there are PCP-controlled processes where the readout is not yet obvious. These include the cellular behavior and movements during CE in gastrulation and neurulation. Although the PCP gene cassette is required in this context, it has not yet been determined what cellular readout it regulates in the mesenchymal cells. It is likely that other shared PCP readouts will emerge in the near future.

Planar Cell Polarity in *Drosophila*

As mentioned above, the *Drosophila* eye and all adult cuticular structures show striking PCP features. In the wing, for example, each cell orients itself with respect to the proximo-distal axis, developing a distally pointing actin-based hair (**Figure 1a**). On the thorax, as in most main body parts, polarization is in the antero-posterior axis. In the eye, PCP is manifest in the regular arrangement of ommatidia with respect to both the antero-posterior and dorso-ventral axes (**Figure 1e**) (Adler 2002, Mlodzik 2002, Strutt 2003). Although these distinct PCP features are regulated by the same set of core PCP genes (**Table 1**), the subsequent mechanistic aspects of polarity are diverse and regulate distinct cellular responses.

In wing cells, the response to the PCP signaling cassette is the formation of a distally pointing actin spike (the wing hair, **Figure 1a**). The main requirement for PCP signaling here is the regulation of cytoskeletal organization, and the asymmetric subcellular distribution of PCP components appears to be sufficient for polarization. In contrast, PCP generation in the eye requires cell-type specification; the precursors of the R3/R4 cell pair need to be differentially specified as either R3 or R4. Thus a nuclear response and the associated cell-type-specific gene expression is a key outcome of PCP signaling in this

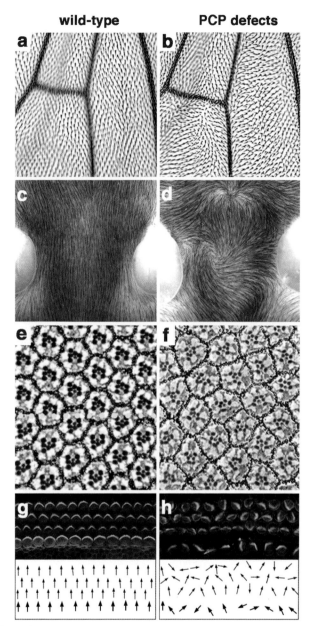

wild-type **PCP defects**

Figure 1

Examples of PCP in *Drosophila* and mammals. Wild-type is shown in panels *a, c, e,* and *g* (note the very regular arrangements in all tissues), and mutant appearance of the same tissues is shown in *b, d, f,* and *h*. (*a,b*) *Drosophila* wing hair patterns; (*c,d*) mouse skin hairs (in dorsal view of neck); (*e,f*) *Drosophila* eye neuroepithelium; (*g,h*) mouse inner ear neuroepithelium. Orientation of inner ear sensory cells is highlighted with arrows in schematic below. A similar schematic could be shown for other tissues reflecting similar underlying defects in all tissues. Panels *c, d* were kindly provided by J. Nathans and panels *g, h* by M. Kelley.

tissue. Detailed descriptions of the tissue-specific PCP responses in *Drosophila* are presented elsewhere (Adler 2002, Bellaiche et al. 2004, Mlodzik 1999, Strutt & Strutt 1999).

Planar Cell Polarity in Vertebrates

PCP establishment and the associated signaling pathways have recently also received attention in vertebrates. The two processes most studied in this context are CE during gastrulation (Keller 2002, Myers et al. 2002, Wallingford et al. 2002) and the polarization of sensory cells in the mammalian cochlea (**Figure 1g**) (Montcouquiol et al. 2003). In addition, the mouse epidermis has also emerged with a beautiful PCP phenotype in *Fz6* mutant mice (Guo et al. 2004) (**Figure 1c,d**). In principle, the same players understood from the analysis of PCP in *Drosophila* play the identical critical roles in vertebrates. Strikingly, homologues of all *Drosophila* PCP genes have been implicated in the CE process, and several also have clear PCP functions in the mammalian inner ear or epidermis (Curtin et al. 2003, Dabdoub et al. 2003, Guo et al. 2004, Montcouquiol et al. 2003). In addition to the core PCP genes, the downstream Fz/PCP signaling components (see below), including the Rho/Rac GTPases, Rho-kinase 2, and JNK, have also been shown to be important for CE in vertebrates (Habas et al. 2001, 2003; Marlow et al. 2002; Yamanaka et al. 2002).

Although there are many similarities between PCP signaling in *Drosophila* and vertebrates, differences do exist. For example, a recent paper identified a murine protein tyrosine kinase as an important PCP signaling factor in a number of different contexts in the mouse, including orientation of inner ear hair cells and CE during neurulation (Lu et al. 2004). To date, no similar factor has been identified in *Drosophila*, despite extensive screening. Thus the possibility remains that there are variations in the PCP signaling cassette across species.

MOLECULAR MECHANISMS OF PLANAR POLARITY

Initiation of PCP Signaling—the Search for a PCP Ligand

The capacity of cells hundreds of cell diameters apart to adopt the same polarity within an epithelial plane poses a fascinating biological problem. Although it has been suggested that long-range signals likely regulate and coordinate this process, the nature of these signals is not known, and this remains a puzzling issue in the study of PCP generation.

As members of the Frizzled (Fz) family of receptors play such an important role in PCP signaling and as they have been shown to function as Wnt receptors during canonical Wnt/β-catenin signaling (Bhanot et al. 1996), Wnt family members seem to be good candidates for the long-range signaling molecules that likely regulate PCP. Evidence for the involvement of Wnts in PCP generation comes from the analysis of CE in vertebrates. In Xenopus and zebrafish, Wnt11/*silberblick* has been identified as a key factor regulating CE (Heisenberg et al. 2000, Tada & Smith 2000). Similarly, Wnt5 has been implicated in *Xenopus* and zebrafish (*pipetail*) gastrulation, suggesting that these Wnt family members are specifically required for CE (Heisenberg et al. 2000, Lele et al. 2001, Westfall et al. 2003). The molecular identification of zebrafish *knypek*, a gene shown to be important for CE (Myers et al. 2002), as a glypican similar to *Drosophila dally* (Topczewski et al. 2001) further supports the importance of Wnts in CE.

However, the existing data in vertebrates argue for a permissive role, rather than an instructive one, for the Wnts during CE. For example, RNA injections of Wnt11 into zebrafish embryos at the one or two cell stage rescue the *silberblick* mutant phenotype, indicating that localized expression may not be required for Wnt11 function (Heisenberg et al. 2000). Moreover, the expression patterns of Wnt11 and Wnt5 do not easily fit a patterning role during CE.

None of the *Drosophila* Wnts is an orthologue of Wnt11 or Wnt5, nor is any Wnt expressed in a pattern that would make it a prime candidate for PCP activation. Moreover, none of the fly Wnt genes has been implicated in PCP in functional analyses. Thus it remains unclear how Fz is activated during PCP signaling in flies, and a requirement for Wnts in PCP establishment in flies is questionable at best.

There is, however, some evidence for upstream regulators of Fz/PCP signaling in *Drosophila*. Recent papers on PCP establishment in the eye and wing suggest that the proto-cadherins Fat and Dachsous (Ds) might play a role in regulating Fz activity (Ma et al. 2003, Rawls et al. 2002, Simon 2004, Yang et al. 2002). Although *fat* and *ds* are not required for Fz activation per se (as Fz/PCP signaling appears largely normal in these mutants), they regulate the Fz signaling bias in the R3/R4 pair in the eye (Yang et al. 2002). Fat acts as a positive regulator of Fz and the related Ds as a Fat antagonist. Fat and Ds interact genetically with *four jointed* (*fj*) (Matakatsu & Blair 2004, Simon 2004, Yang et al. 2002), which has also been suggested to regulate Fz/PCP activity (Zeidler et al. 1999, 2000). It has been proposed that *fj* and *ds* act on *fat*, which in turn positively regulates Fz activity (Matakatsu & Blair 2004, Simon 2004, Yang et al. 2002). Although this model is supported by several experiments, it is not at all clear how and what type of regulation these PCP factors exert on Fz. Thus, as intriguing as these proto-cadherins are as mediators of a long-range patterning element, their role in Fz/PCP regulation remains unclear.

The Core PCP Genes and Their Interactions

As mentioned above, several PCP genes are required for PCP generation in all contexts in flies and most, if not all, PCP aspects in vertebrates. These are generally referred to as primary or core PCP genes and include *frizzled* (*fz*), *dishevelled* (*dsh*), *prickle-spiny legs* (*pk*), *stra-*

bismus/Van Gogh (*stbm/Vang*), *flamingo/starry night* (*fmi/stan*), *diego* (*dgo*), and their respective vertebrate orthologs (reviewed in Adler 2002, Mlodzik 2002, Strutt 2003). In general, Fz and Dsh appear to be the central PCP signaling molecules, with the other core PCP genes serving to regulate Fz/Dsh-PCP signaling activity. For example, Stbm/Vang and Pk appear to inhibit Fz-PCP activity (Jenny et al. 2003, Taylor et al. 1998, Tree et al. 2002, Wolff & Rubin 1998). In particular, Pk has been shown to bind Dsh and to antagonize its Fz-mediated membrane recruitment (Tree et al. 2002). This Pk function is supported by Stbm/Vang, which recruits Pk to the membrane and itself binds and possibly inhibits Dsh function (Bastock et al. 2003, Jenny et al. 2003). Thus the Fz-Dsh and Stbm/Vang-Pk pairs are generally thought to antagonize each other, and their localization domains are resolved to mutually exclusive regions at opposite poles of each cell (**Figure 2**; see below for model).

Although less well defined, Diego appears to promote Fz-Dsh activity and it colocalizes with Fz-Dsh (Das et al. 2004, Feiguin et al. 2001). The role of the atypical cadherin Flamingo (Fmi) is the least understood (Das et al. 2002, Shimada et al. 2001, Usui et al. 1999). Fmi appears to serve a homophylic adhesion function and colocalizes with both Fz-Dsh and Stbm/Vang-Pk complexes (**Figure 2**). Genetically, Fmi also appears to share two functions, promoting and antagonizing Fz-Dsh activity (Das et al. 2002, Usui et al. 1999). The role of Fmi needs to be addressed in more detail. In particular, Fmi still has no known molecular interaction partners.

One of the hallmark features of the core PCP gene products is that prior to PCP signaling these proteins are located uniformly around apical cell membranes but become asymmetrically localized during and after PCP signaling. In the *Drosophila* wing, for example, Fz and Dsh are initially localized in a uniform apical ring but become asymmetrically localized to distal cell membranes (reviewed in Adler 2002, Strutt 2003). Moreover,

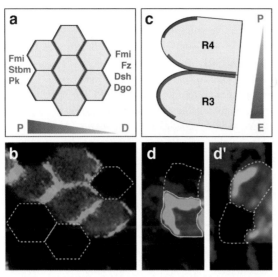

Figure 2

Core PCP protein localization in the Drosophila wing and eye. (*a,b*) In the pupal wing, core PCP proteins become asymmetrically localized to proximal and distal cell membranes. Proximal is left and distal is right. The orientation of the gradient indicates the presumed Fz activity gradient. (*a*) Schematic of the localization of known core PCP proteins. Each hexagon represents a single pupal wing cell. (*b*) Confocal image of a pupal wing demonstrating the distal localization of Diego, as an example. Blue: cells expressing a GFP-Diego construct. Green: GFP-Diego. White outlines: cells not expressing GFP-Diego. (*c,d*) In the third instar eye disc, core PCP proteins become asymmetrically localized to polar and equatorial cell membranes. Equatorial is toward the bottom and polar is toward the top. (*c*) A schematic of the R3 and R4 cells of an ommatidium. Colors are the same as in *a*. Purple: colocalization of both red and blue factors. (*d,d'*) Confocal image of a third instar eye disc demonstrating the localization of Diego, as an example. Blue: cells expressing a GFP-Diego construct. Green: GFP-Diego. White outlines: R3 (*bottom*) and R4 (*top*) precursors.

loss of any of the core PCP genes leads to the random distribution of all the core PCP gene products. This is distinct from components that become asymmetrically localized but are downstream of the core PCP-signaling cassette. For example, *inturned*, a planar polarity effector required for the proper orientation of wing hairs, is initially localized in a uniform apical ring in wing cells and becomes subsequently localized exclusively to proximal cell membranes, but the loss of *inturned* does not affect the localization of any of the core PCP gene products (Adler et al. 2004). The potential mechanism by which the core PCP

proteins generate cellular asymmetry is described in detail in an emerging model (see below).

The Downstream Signaling Cascade

The end result of the positive and negative interactions between the core PCP genes is the localized activation or inhibition of the PCP signaling cascade regulated by Fz and Dsh, the Fz-PCP pathway. Both in flies and vertebrates, a highly related signaling cascade is emerging that is thought to act in many distinct contexts (**Figure 3**).

At the top of the cascade are members of the Fz receptor family. The mechanism of activation of Fz-PCP signaling, however, remains obscure, as it is not clear whether an interaction with Wnt family members (the activating ligands of the Fz-LRP5/6 receptor complex in canonical signaling) is instructive or just permissive (see above). It is quite possible that Fz-PCP activity is regulated largely through the interactions between the core PCP factors as described above. The first known downstream component of Fz-PCP signaling is *dishevelled* (*dsh*). Although Fz and Dsh are both part of canonical Wnt/β-catenin signaling (Boutros & Mlodzik 1999, Cadigan & Nusse 1997), the other known components of canonical Wnt signaling do not appear to be involved in PCP (Axelrod et al. 1998, Boutros et al. 1998), indicating that a distinct pathway functions downstream of Fz/Dsh in PCP establishment.

The specific domain requirements of Dsh in Fz-PCP signaling are the same in flies and vertebrates, but they are distinct from canonical Wnt/β-catenin signaling (e.g., Boutros & Mlodzik 1999, Heisenberg et al. 2000, Wallingford et al. 2000). Although a molecular link between Fz and Dsh is now established (Wong et al. 2003), it remains unclear whether Fz and Dsh share the same type of molecular interaction for PCP and β-catenin signaling. It is likely that there are pathway-specific Fz/Dsh interactions not yet discovered.

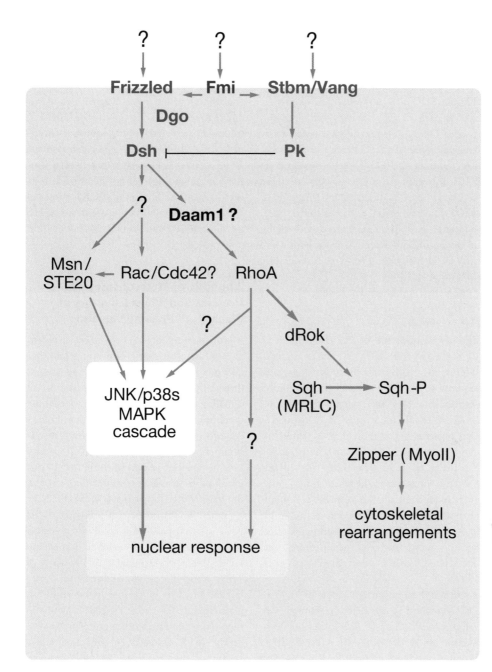

Figure 3

The Frizzled/PCP
pathway. A simplified
schematic view of the
Fz/PCP signaling
cascade(s) is shown.
The nuclear
signaling leading to
transcriptional
activation is observed
in a subset of tissues.
Several members of
the Rho GTPase
family and JNK/p38
MAPK cascade act
largely redundantly.
The effectors of
RhoA in nuclear
signaling are
unknown. The
membrane proximal
components are
color coded, as in
Figures 2 and **4**.
The Stbm/Vang-Pk
complex (*blue*)
antagonizes Fz-Dsh
(*red*) signaling. Fmi
(*purple*) stabilizes
both complexes, and
Dgo (*red*) is thought
to act positively on
Fz-Dsh based on
genetic observations.
See text for details.

The downstream readouts of Fz-PCP signaling differ depending on the context (e.g., cytoskeletal reorganization as the main response in the fly wing versus a transcriptional response in the fly eye) and are discussed in detail elsewhere (e.g., Mlodzik 2002). In brief, a combination of genetic and biochemical studies in flies and vertebrates has demonstrated that the Fz-PCP cascade downstream of Dsh consists of small GTPases of the Rho

GPCRs: G protein coupled receptors

subfamily (Rho, Rac, and cdc42 depending on the context), the Rho-associated kinase (*dROK*), the STE20-like kinase Misshapen (Msn), and the JNK-type MAPK cascade (**Figure 3**). The importance and contribution of each of these factors vary between tissues. There is also significant redundancy between some of the GTPases in this context (Fanto et al. 2000, Hakeda-Suzuki et al. 2002). Interestingly, a Rho-associated kinase has been identified independently in fly PCP (*dROK*) and zebrafish CE (Marlow et al. 2002, Winter et al. 2001). *dROK* affects cytoskeletal aspects of PCP in wing hair formation and links Fz/Dsh signaling to myosin regulation (Winter et al. 2001).

Genetic interactions suggest that JNK/p38-type MAPK modules and Jun-Fos (AP-1) act downstream of Dsh and the Rho GTPases in the fly eye (Boutros et al. 1998, Paricio et al. 1999, Weber et al. 2000) (**Figure 3**). JNK signaling is likely also required in the context of CE in vertebrates (Yamanaka et al. 2002), suggesting a general JNK/p38 requirement in PCP. Consistently, Dsh proteins act as potent JNK activators in biochemical assays (Boutros et al. 1998, Yamanaka et al. 2002). The role of the JNK/p38 kinases in Fz/Dsh signaling is, however, redundant, as the respective mutants do not show strong PCP defects in simple loss-of-function analyses (a phenotypic effect becomes apparent only when more than one related kinase is affected) (Paricio et al. 1999).

Although an important aspect of Fz/Dsh-PCP signaling is the juxta-membrane subcellular localization of Dsh (Axelrod 2001, Axelrod et al. 1998, Boutros et al. 2000, Wong et al. 2003), it is not known how Dsh transduces a signal to downstream effectors. In *Drosophila*, no PCP-specific factors have yet been identified that physically interact with Dsh to provide such a link. In *Xenopus*, the Formin homology domain protein, Daam1, has been proposed as such a bridging factor between Dsh and Rho GTPase (Habas et al. 2001) (**Figure 3**). A Daam1 C-

terminal fragment binds the Dsh PDZ and DEP domains, the N-terminal part of Daam1 binds RhoA, and Daam1 is required for Fz-mediated Rho activation in *Xenopus* (Habas et al. 2001), which makes Daam1 a good candidate to recruit Rho to the sites of Dsh localization. However, the exact sequence of events with Dsh/Daam1/RhoA is not yet resolved, as a C-terminal Daam1 fragment that lacks the RhoA binding region behaves like activated Daam1 (Habas et al. 2001). Although a Daam1 homologue is present in *Drosophila*, its role in PCP establishment has not yet been determined.

The Role of Heterotrimeric G Proteins and Their Effectors in Planar Cell Polarity Signaling

A growing number of recent studies have implicated heterotrimeric G proteins and their effectors in the transduction of Fz signals. Although the role of these molecules in Fz-PCP signaling is still somewhat controversial and much more work needs to be done, there is enough published evidence to merit discussion here.

The simplest reason to think that G proteins and their effectors might play a role in the *Fz-PCP* signaling cascade is that Fz shares a number of features with known GPCRs. Fz is a seven-pass transmembrane receptor with an extracellular N-terminal domain and a cytoplasmic C-terminal tail (Vinson et al. 1989, Wang et al. 1996), and receptor dimerization has been suggested to be important for signaling (Carron et al. 2003). As further support, a chimeric β-adrenergic/Frizzled receptor (with extracellular and membrane-spanning domain from the β-adrenergic receptor and the intracellular loops and cytoplasmic tail of a rat Frizzled) could activate Fz signaling in a α-agonist-specific manner in mouse F9 teratocarcinoma cells (Liu et al. 2001). This indicates that conformational changes that lead to activation of Fz are similar to those in GPCRs. Additionally, members of the β-arrestin family, which are typically involved in the endocytosis of GPCRs, mediate Fz endocytosis (Chen et al. 2003).

Studies of Wnt/Ca^{2+} signaling in zebrafish embryos also indicate that G proteins and their effectors can play a role downstream of Fz. These studies recognized that the activation of calcium flux by Fz is a pertussis toxin-sensitive process (Slusarski et al. 1997a,b). Follow-up studies identified $G_{\beta\gamma}$ subunits, phospholipase C (PLC), and protein kinase C (PKC) as downstream signaling effectors (Penzo-Mendez et al. 2003, Sheldahl et al. 1999). Furthermore, a *Xenopus* PKC was suggested to be both necessary and sufficient for recruitment of Dsh to the plasma membrane (Kinoshita et al. 2003), a process that is critical for Fz-PCP signaling.

The most compelling evidence that G proteins are involved in vivo in PCP signaling comes from recent work out of Tomlinson's group (Katanaev et al. 2005). Using a variety of techniques, Katanaev et al. show that *brokenheart*, the *Drosophila* Gαo gene, is required for PCP signaling. They demonstrate that both gain- and loss-of-function clones in the wing lead to typical PCP defects. They also show that Gαo becomes asymmetrically localized in pupal wing cells at the same time as the core PCP proteins. Lastly, using genetic epistasis experiments, they conclude that Gαo acts downstream of Fz but upstream of Dsh.

Despite this evidence, the actual role of trimeric G proteins and their effectors in PCP signaling remains unclear. Many of the experiments focused on their role in other Fz signaling pathways, including the Wnt/β-catenin and Wnt/Ca^{2+} pathways. Although it seems reasonable that if Fz acts as a GPCR in one context it likely acts as a GPCR in all contexts, the possibility exists that the mechanism of signal transduction is context dependent. Additionally, the requirement for trimeric G proteins may differ from pathway to pathway. For example, the study of *brokenheart* indicated a positive requirement for Gαo in Wnt/β-catenin signaling, but an inhibitory role in Fz-

PCP signaling. In short, more work needs to be done to clarify the role of trimeric G proteins in PCP signal transduction.

AN EMERGING MODEL OF PLANAR CELL POLARITY

A Theoretical Approach to Designing a Model of Planar Cell Polarity

The establishment of PCP is a process that requires the coordination of signals between cells such that these cells are not only precisely aligned relative to one another, but also are precisely aligned relative to a particular body axis. In the *Drosophila* wing, for example, not only are all hairs aligned in parallel to one another, but they are also always aligned in parallel to the proximal-distal axis. Thus any model of PCP signaling must be able to account for both aspects of polarity.

The orientation of cells relative to a body axis is seemingly straightforward. The only requirement for such a process is the presence of a signal at one end of the body axis that is then transmitted toward the other end of the body axis. Using the *Drosophila* wing again as an example, the signal could either be located proximally and transmitted distally or vice versa. The signal itself could be diffusible, transmissible via a cell to cell relay system, or of a completely distinct nature. Regardless of the nature of the signal, some such cue must exist.

The parallel orientation of cells relative to one another is slightly more complicated. One possibility is that no such mechanism exists; cells are simply parallel to one another because they are all parallel to the body axis. If this were true, the orientation of a given cell would not be likely to affect its neighbors because they would all simply orient themselves relative to the body axis. However, because most PCP genes show nonautonomous phenotypes, it is almost certain that communication between neighboring cells exists. The nature of this communication must serve to positively reinforce the initial PCP-establishing signal.

By definition, for a cell to be polarized within the plane of an epithelium, there must be an asymmetric distribution of its subcellular components within the plane. There are at least two ways in which this could feasibly occur. One way in which this might happen is that the cell could relocalize its subcellular components directly in response to the polarizing signal. For example, the presence of more signal on one side of a cell could lead to the preferential activation of the target receptor on that same side of the cell. Alternatively, it is also possible that the graded signal could lead to a higher absolute level of receptor activation in cells that are closer to the source of the signal than those that are farther away. This information could then be transmitted between neighboring cells, which in turn would instruct subcellular components to asymmetrically distribute relative to this information. The initial biasing signals in either of these two systems could potentially be reinforced through positive and negative feedback loops.

Finally, any complete model of the establishment of planar cell polarity will include the following components: (*a*) how the initial axis of planar polarization is established, (*b*) how the cells in the plane respond to this polarizing signal, (*c*) how this signal is communicated within a cell and between neighboring cells, (*d*) what feedback loops exist to reinforce these signals, and (*e*) how these signals are subsequently transmitted to downstream targets. Given the current state of the field, it is impossible to give such a complete model, as insufficient data exist for a number of these components. However, aspects *c* and *d* are really the core of a model for PCP (that is, they describe how cells in the presence of a polarizing gradient become polarized themselves and how they do so in a coordinated manner), and a significant body of evidence exists in regard to them. Thus the model we present below focuses on how cells in a plane interpret a polarity signal, communicate that informa-

tion to one another, and then reinforce that information through feedback loops in order to establish a uniform cellular polarity across an entire tissue.

An Evidence-Based Model of the Establishment of Planar Cell Polarity

In the theoretical approach, we described two possible scenarios for how a cell could become polarized in response to a signal: (*a*) a cell could become directly polarized in response to the signal or (*b*) the absolute level of signal for a given cell could be compared with that of neighboring cells and this would then induce polarization. Given the current evidence, the second possibility is much more likely. One of the clearest pieces of work showing this is a recent paper by Lawrence et al. (2004), which nicely demonstrates the importance of signaling between neighboring cells for the establishment of polarity, and the requirement for a difference in the absolute level of signal between two cells for them to exert an effect on each other.

In addition to the evidence in support of the second possible mechanism, there is also evidence to oppose the first possibility, that a cell becomes polarized directly in response to the polarizing signal. One such piece of evidence comes from protein localization studies. In the *Drosophila* eye, Fz becomes localized to the polar side of the R3 cell, and in the *Drosophila* wing Fz becomes localized to the distal cell membrane. This is interesting because in both of these cases, Fz becomes localized to the side of the cell farthest from the source that is thought to activate Fz signaling. In the eye, the gradient of Fz activity is highest at the equator and lowest at the poles. Likewise, in the wing, the Fz activity gradient is highest proximally and lowest distally. As described above, if the cells in a given tissue are polarized directly by an external signal, then presumably the subcellular components in the cell would migrate to the cell surface closest to the signal source. This is the opposite of what the evidence shows.

Another piece of evidence opposing the possibility that a cell becomes polarized directly in response to the polarizing signal comes from studies looking at the timing of Fz signaling (Strutt 2001). It has been shown in the *Drosophila* wing that Fz signaling acts in at least two distinct phases. First, there is a nonautonomous signaling phase that takes place at approximately 24–30 h after puparium formation (APF). This is then followed by an autonomous signaling phase that takes place approximately 30–36 h APF. The fact that there are these two discrete signaling phases and that the first phase acts to signal between cells while the second acts to signal within a cell clearly opposes the notion that cells become polarized directly in response to an external signal and lends further support to the model that the absolute level of signal for a given cell is first compared with that of neighboring cells and that this comparison is then instructive in polarizing a cell.

There is a wealth of evidence to explain molecularly how such a signaling mechanism likely works. Despite the fact that the nature of the polarity signal and the way in which it interacts with the core PCP gene products are unknown, it is fairly clear that its effect is to establish a Fz activity gradient. In the fly eye this gradient is oriented from the equator to the poles, whereas in the wing it is from the proximal out toward the distal regions. Thus any cell that is closer to the peak of the activity gradient (e.g., the equator in the eye) will have a higher absolute level of Fz activity than will its neighbors that are farther away from the peak (e.g., Wehrli & Tomlinson 1998, Yang et al. 2002, Zheng et al. 1995). This information is then integrated between neighbors via a mechanism that requires Fmi and the nonautonomous signaling function of Frizzled in the signal sending cell (Lawrence et al. 2004, Strutt 2001). The information is received by the neighboring cells in a manner that requires Fmi again, as well as the transmembrane protein Stbm/Vang (Taylor et al. 1998, Wolff & Rubin 1998). In this way,

cells are able to sense their location in relationship to the activity gradient and to their neighbors.

We thus hypothesize the existence of two, distinct Flamingo-containing signaling complexes: one that consists of Fmi, Fz, and Diego, and a second that consists of Fmi, Stbm/Vang, and Pk. Although there is no direct molecular evidence that puts Fmi into either of these complexes, both genetic evidence (described above and see Das et al. 2002, Usui et al. 1999) and protein localization studies point to the existence of two distinct Fmi-containing complexes (Das et al. 2004). These authors showed that Flamingo staining is lost apically only when a tissue is mutant for members of both complexes. For example, in any single mutant staining, either *frizzled, diego, strabismus,* or *prickle,* Flamingo is found apically. Likewise, in double mutants in which both genes are involved in the formation of the same complex, for example a *stbm-pk* double mutant, Flamingo is still found apically. However, in double mutants affecting the formation of the different complexes, for example *stbm-dgo* or *pk-dgo* double mutants, Flamingo staining is no longer maintained apically. This is consistent with the hypothesis that Flamingo exists in two distinct complexes.

If two such Flamingo-containing complexes exist, the model predicts how they probably behave (**Figure 4*a-a''***). The complex consisting of Flamingo, Frizzled, and Diego (Complex A, *red* in all figures) stabilizes the complex consisting of Flamingo, Stbm/Vang, and Pk (Complex B, *blue* in all figures) across cell membranes and vice versa. This likely occurs though the homophylic adhesion properties of Flamingo and is again consistent with a large number of genetic and protein localization studies (e.g., Das et al. 2002, 2004; Shimada et al. 2001; Tree et al. 2002; Usui et al. 1999). Additionally, Complex A is predicted to inhibit Complex A across cell membranes. Likewise, Complex B is predicted to inhibit Complex B across cell membranes (Tree et al. 2002).

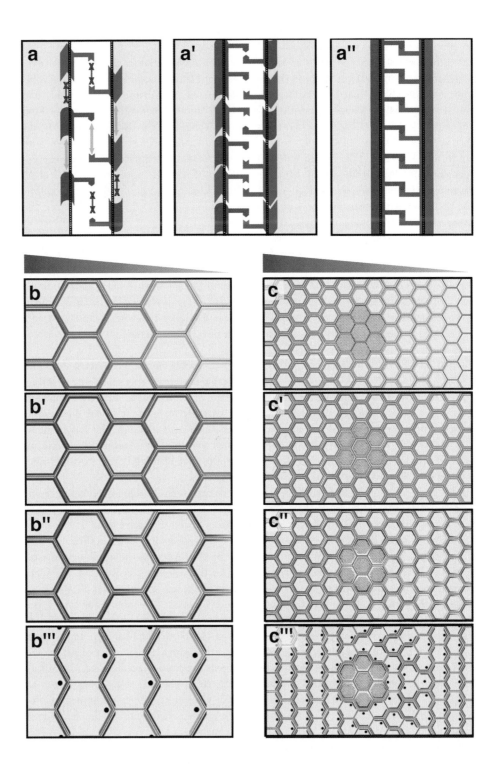

In addition to considering how these complexes interact across cell membranes, we can also consider how they interact within a given cell. Each of these complexes would be predicted to recruit like complexes and inhibit the opposite complexes. For example, the presence of stable Complex B in a given location of a cell would likely recruit more Complex B to that same location in the cell and also inhibit Complex A. Evidence that this, in fact, occurs comes from protein-clustering studies in *Xenopus* animal cap explants (Jenny et al. 2003), where the presence of Pk strongly induced the clustering of GFP-Stbm, indicating that complexes tend to recruit other like complexes.

On the basis of the existing data and concepts described above, we can put together a model of the establishment of PCP. To explain the model, we use the *Drosophila* wing as a system, which can be viewed as a single sheet of cells, all of which become uniformly polarized, because it is the simplest and most straightforward system to conceptualize the establishment of PCP. We will then expand it to make it applicable to more complex systems, including the *Drosophila* eye.

An initial polarizing signal has the early effect of establishing a Fz activity gradient.

Thus cells that are more proximally located will have more Fz activity than those that are more distally located. This, in turn, means that more proximally located cells will have more active Complex A than will more distally located cells. It is important to keep in mind that these complexes are initially located uniformly around the apical circumference of each cell (**Figure 4b**).

Now let us consider what happens to the Complex Bs in each cell. Complex B is stabilized by Complex A across the cell membrane. Because more proximal cells have more active Complex A, more Complex B will be stabilized at the proximal cell border of the adjacent cell (**Figure 4b''**). Although some Complex B will still be located on the anterior, posterior, and distal cell borders, the bias will be toward more Complex B on the proximal side.

This initial bias is then further amplified through the same feedback loops: The presence of more Complex B at the proximal side will cause more clustering of Complex B there, thereby depleting it from other surfaces in the same cell. This, in turn, stabilizes even more Complex A at the distal cell surface of the proximally neighboring cell (**Figure 4b'''**). Moreover, the increased

---

Figure 4

A model of the establishment of PCP in the *Drosophila* wing. Red: Complex A factors, including Fmi, Fz, and Dgo. Blue: Complex B factors, including Fmi, Stbm, and Pk. Pale yellow: intracellular region. In all figures, proximal is left, distal is right, anterior is top, and posterior is bottom. The gradients indicate the presumed Fz activity gradient. (*a-a''*) Rules of interaction. (*a*) In the same cell, like complexes stabilize each other (*green arrows*) and unlike complexes inhibit each other (*red Xs*). Across cell membranes, like complexes inhibit each other and unlike complexes stabilize each other. Using these rules, a slight bias for Complex A at the distal cell membrane (*a'*) leads to a polarization of factors (*a''*). (*b-b'''*) PCP establishment in a wild-type wing. (*b*) An unknown polarity signal establishes a gradient of Fz activity, such that more proximal cells have higher Complex A activity than do more distal cells. (*b'*) Superimposing a uniform level of Complex B on the gradient shown in (*b*) illustrates the initial state in which PCP factors are uniformly distributed around cell membranes. (*b''*) Based on the rules in (*a*), the Complex B in each cell has a tendency to be stabilized proximally and inhibited distally. (*b'''*) Feedback loops based on the rules in (*a*) result in a uniform polarization of all cells, with hairs pointing distally. (*c-c'''*) Distal nonautonomy due to a *fz* mutant clone. The gray hexagons indicate cells in the clone. A localized disruption in the Fz activity gradient (*c*) affects the localization bias in neighboring cells (*c'*). Based on the rules in (*a*), the Complex B in cells adjacent to the clone tends to be inhibited near the clone and stabilized further away from the clone (*c''*), resulting in waves of hairs that point toward the clones (*c'''*). A similar effect is observed in *stbm/vang* mutant clones, but on the proximal side of the clone, as predicted by the model.

presence of Complex B at the proximal surface will inhibit Complex A from stabilizing there. Likewise, the increased presence of Complex A at the distal cell surface of the proximally neighboring cell will inhibit Complex B from stabilizing there. In short, positive and negative feedback loops act to reinforce the initial bias that occurs in response to the Frizzled activity gradient.

The model described here is consistent with all known PCP studies in the *Drosophila* wing. One of the most interesting observations explained by it is that, as PCP signaling occurs, Fz becomes localized to the side of the cell opposite to the peak of its activity (Strutt 2001). Similarly, Dsh and Dgo become localized distally in each cell, despite the fact that the Fz activity gradient is highest proximally (Axelrod 2001, Das et al. 2004). Likewise, Stbm and Pk become localized proximally in each cell despite the fact that the presumed Stbm activity gradient is highest distally (Bastock et al. 2003, Tree et al. 2002). Although these observations are seemingly counter intuitive, they are entirely consistent with the model presented.

Other data supported by this model include a host of protein localization studies looking at PCP gene loss-of-function and overexpression clones in the *Drosophila* wing. In particular, this model explains the phenomenon known as "domineering nonautonomy"; that is, clonal loss or overexpression of many of the core PCP genes leads to a reorganization of the cells around the clone. For example, *fz* clones in the wing cause neighboring wing hairs to point toward them (Vinson & Adler 1987). On a subcellular level, Fz and Dsh in cells neighboring the clone become reoriented such that they are aligned at membranes relative to the clone, not relative to the proximal-distal axis (Strutt 2001, Tree et al. 2002). Such a scenario is explained by the model. For cells within the clone, the neighbor with the highest amount of Complex A will be the closest one outside the clone. Thus the Complex B in a clonal cell on the clone border will be biased toward the cell

membrane on that border (**Figure 4c'**). This, in turn, stabilizes Complex A along the clone border in wild-type cells. Stable Complex A along the clone border will lead to a bias for Complex B at the opposite border, farthest from the clone. It is important to keep in mind that this bias is competing with the bias that is set up by the proximal-distal polarity signal. Thus, close to the clone, the bias will be largely determined by the location of the clone, whereas farther away from the clone, the bias will be largely determined by the proximal-distal signal. This explains why not all hairs in the wing point into a clone, but rather form waves (**Figure 4c'''**).

There are other possible scenarios that could be envisioned and/or have been experimentally tested, including overexpression clones of Complex A genes such as *fz* and *dgo*, and both mutant and overexpression clones of Complex B genes such as *stbm/vang* and *pk*. To the best of our knowledge, all such experimentally tested scenarios, both genetically and by protein localization studies, are in accordance with the model (Axelrod 2001, Bastock et al. 2003, Feiguin et al. 2001, Ma et al. 2003, Strutt 2001, Taylor et al. 1998, Tree et al. 2002, Usui et al. 1999). However, for the sake of brevity, we do not describe all such examples.

Whereas the model explained above makes sense for a uniform field of cells such as in the *Drosophila* wing, it is important to note that it is applicable to other tissue types. The *Drosophila* eye, for example, is fundamentally different from the wing in a number of ways. First, the cells in the eye are not staged uniformly as they are in the wing. Instead, cells that are located posteriorly are more advanced developmentally than those that are located anteriorly. Thus PCP signaling takes place at different times in different parts of the eye. Second, only a select subset of cells is ever responsive to PCP signaling. The developing eye field is composed of ommatidial preclusters consisting of photoreceptor precursors, cone cell precursors, and interommatidial cells. Of these, only two photoreceptor cells per ommatidium, the R3 and R4 precursors, are responsive to PCP signals. Therefore signal fidelity cannot be propagated across a whole field of cells as it can in the wing.

Nevertheless, the model described above is applicable to the eye. For the model to work, only three conditions need to be met: (*a*) at least two cells must be in direct contact with each other, (*b*) these cells must be oriented such that one is closer to the polarity signal than the other, and (*c*) these cells must be staged such that they are responsive to PCP signaling at the same time. In the eye, all three of these conditions are clearly met. First, the R3 and R4 precursors in each ommatidial cluster share a cell boundary. Second, the two cells are oriented parallel to the Frizzled activity gradient, such that the equatorial R3 has a higher Frizzled activity than the polar R4. Third, development in the eye progresses such that cells in a given row, which extends in a stripe from pole to pole, are precisely staged. Thus the model described above works in the *Drosophila* eye, despite the lack of a uniform field of PCP-responsive cells.

The question of how the R3/R4 cells become selectively responsive to PCP signaling has been somewhat of a mystery. In particular, PCP defects in the eye owing to the loss of any core PCP gene can be rescued by expressing the lost gene uniformly in all cells under the control of a ubiquitous promoter (e.g., Strutt et al. 2002, Wu et al. 2004). Immunostaining for the replaced gene product shows that it becomes enriched specifically in R3 and R4, indicating that this enrichment is not controlled by gene expression but rather by affecting protein stability and/or localization. A recent study giving insight into how this might be regulated (Djiane et al. 2005) indicates that the apical-basal determinants aPKC-Bazooka-dPatj regulate Fz activity and allow Fz signaling to take place only in the R3/R4 cell pair. Thus apical-basal polarity is not only a prerequisite for the localization of PCP factors, but it also serves to prime cells

for PCP signaling. It remains to be seen if this paradigm will hold true for all PCP responsive tissues.

Future Directions and Concluding Remarks

Although our understanding of PCP has expanded dramatically in recent years, there is clearly much work that remains to be done. Some of the most exciting work going on right now focuses on the establishment of the initial bias required for PCP to occur. Although genetic experiments have shown a clear requirement for *fat*, *dachsous*, and *four-jointed* in this process, the mechanism by which these genes feed into the core PCP cassette is a complete mystery. Additionally, it has been assumed that the effect of the initial bias is to establish a Fz activity gradient. However, it is possible that the effect is rather to establish a Stbm (or some other) activity gradient. Understanding the mechanistic aspects of signaling upstream of the core PCP cassette should shed light on this subject.

Another area of intense investigation is the study of the molecular interactions within the core PCP signaling cassette. Whereas recent work has demonstrated some of these interactions and hypothesized their likely role in signaling, there are still many connections that have yet to be made. Notably, there is still no concrete molecular evidence to explain how PCP information is transmitted between two cells. The model presented in this review indicates a likely role for Fmi in this process, based on both genetic evidence and protein localization studies, but to date no such molecular evidence exists. In fact, other than the evidence that Fmi can bind to itself across cell membranes as a homophylic adhesion molecule (Usui et al. 1999), no physical interactions between Fmi and other proteins have been shown. Identifying Fmi-binding partners is a critical step in the future understanding of PCP.

Finally, there is still relatively little knowledge about how polarity information from the core PCP factors is transmitted to downstream targets, despite the fact that many of the targets themselves are known. Some of the possible candidates to be of interest in future work include the trimeric G proteins and their effectors, DAAM1, dROK, and many others.

The question of how a field of initially uniform cells becomes polarized is a truly fascinating one. In recent years, our understanding of the mechanisms underlying this process has increased dramatically through genetic and biochemical studies. Still, there are many aspects of planar cell polarity that we do not yet comprehend and, potentially, others that have not even been considered. In this review, we have presented a model of PCP based on the available data that summarizes the initial breaking of symmetry and the subsequent reinforcement of that asymmetry. Future studies should lead to further refinement of this model as well as expansion of it to upstream and downstream signaling events.

ACKNOWLEDGMENTS

We thank all members of the Mlodzik lab for many helpful and stimulating discussions, and all colleagues in the field for sharing ideas and informal discussions. We are grateful to Jeremy Nathans and Matthew Kelley for the pictures shown in **Figure 1c,d** and 1**g,h**. We apologize to all authors of original work in the field that could not be incorporated or cited owing to space limitations. Work in our laboratory is supported by grants from the National Eye Institute and National Institute for General Medical Sciences.

LITERATURE CITED

Adler PN. 1992. The genetic control of tissue polarity in *Drosophila*. *BioEssays* 14:735–41

Adler PN. 2002. Planar signaling and morphogenesis in *Drosophila*. *Dev. Cell* 2:525–35

Adler PN, Zhu C, Stone D. 2004. Inturned localizes to the proximal side of wing cells under the instruction of upstream planar polarity proteins. *Curr. Biol.* 14:2046–51

Axelrod JD. 2001. Unipolar membrane association of Dishevelled mediates Frizzled planar cell polarity signaling. *Genes Dev.* 15:1182–87

Axelrod JD, Miller JR, Shulman JM, Moon RT, Perrimon N. 1998. Differential requirement of Dishevelled provides signaling specificity in the Wingless and planar cell polarity signaling pathways. *Genes Dev.* 12:2610–22

Bastock R, Strutt H, Strutt D. 2003. Strabismus is asymmetrically localised and binds to Prickle and Dishevelled during *Drosophila* planar polarity patterning. *Development* 130:3007–14

Bellaiche Y, Beaudoin-Massiani O, Stuttem I, Schweisguth F. 2004. The planar cell polarity protein Strabismus promotes Pins anterior localization during asymmetric division of sensory organ precursor cells in *Drosophila*. *Development* 131:469–78

Bhanot P, Brink M, Samos CH, Hsieh J-C, Wang Y, et al. 1996. A new member of the *frizzled* family from *Drosophila* functions as a Wingless receptor. *Nature* 382:225–30

Boutros M, Mihaly J, Bouwmeester T, Mlodzik M. 2000. Signaling specificity by Frizzled receptors in *Drosophila*. *Science* 288:1825–28

Boutros M, Mlodzik M. 1999. Dishevelled: at the crossroads of divergent intracellular signaling pathways. *Mech. Dev.* 83:27–37

Boutros M, Paricio N, Strutt DI, Mlodzik M. 1998. Dishevelled activates JNK and discriminates between JNK pathways in planar polarity and *wingless* signaling. *Cell* 94:109–18

Cadigan KM, Nusse R. 1997. Wnt signaling: a common theme in animal development. *Genes Dev.* 11:3286–305

Carron C, Pascal A, Djiane A, Boucaut JC, Shi DL, Umbhauer M. 2003. Frizzled receptor dimerization is sufficient to activate the Wnt/beta-catenin pathway. *J. Cell Sci.* 116:2541–50

Chen W, ten Berge D, Brown J, Ahn S, Hu LA, et al. 2003. Dishevelled 2 recruits beta-arrestin 2 to mediate Wnt5A-stimulated endocytosis of Frizzled 4. *Science* 301:1391–94

Curtin JA, Quint E, Tsipouri V, Arkell RM, Cattanach B, et al. 2003. Mutation of Celsr1 disrupts planar polarity of inner ear hair cells and causes severe neural tube defects in the mouse. *Curr. Biol.* 13:1129–33

Dabdoub A, Donohue MJ, Brennan A, Wolf V, Montcouquiol M, et al. 2003. Wnt signaling mediates reorientation of outer hair cell stereociliary bundles in the mammalian cochlea. *Development* 130:2375–84

Das G, Jenny A, Klein TJ, Eaton S, Mlodzik M. 2004. Diego interacts with Prickle and Strabismus/Van Gogh to localize planar cell polarity complexes. *Development* 131:4467–76

Das G, Reynolds-Kenneally J, Mlodzik M. 2002. The atypical cadherin Flamingo links Frizzled and Notch signaling in planar polarity establishment in the *Drosophila* eye. *Dev. Cell.* 2:655–66

Djiane A, Yogev S, Mlodzik M. 2005. The apical determinants aPKC and dPatj regulate Frizzled-dependent planar cell polarity in the *Drosophila* eye. *Cell.* 121:621–31

Eaton S. 1997. Planar polarity in *Drosophila* and vertebrate epithelia. *Curr. Opin. Cell Biol.* 9:860–66

Fanto M, Weber U, Strutt DI, Mlodzik M. 2000. Nuclear signaling by Rac and Rho GTPases is required in the establishment of epithelial planar polarity in the *Drosophila* eye. *Curr. Biol.* 10:979–88

Feiguin F, Hannus M, Mlodzik M, Eaton S. 2001. The Ankyrin repeat protein Diego mediates Frizzled-dependent planar polarization. *Dev. Cell* 1:93–101

Gubb D. 1993. Genes controlling cellular polarity in *Drosophila*. *Dev. Suppl.* 1993:269–77

Guo N, Hawkins C, Nathans J. 2004. Frizzled6 controls hair patterning in mice. *Proc. Natl. Acad. Sci. USA* 101:9277–81

Habas R, Dawid IB, He X. 2003. Coactivation of Rac and Rho by Wnt/Frizzled signaling is required for vertebrate gastrulation. *Genes Dev.* 17:295–309

Habas R, Kato Y, He X. 2001. Wnt/Frizzled activation of rho regulates vertebrate gastrulation and requires a novel formin homology protein Daam1. *Cell* 107:843–54

Hakeda-Suzuki S, Ng J, Tzu J, Dietzl G, Sun Y, et al. 2002. Rac function and regulation during *Drosophila* development. *Nature* 416:438–42

Heisenberg CP, Tada M, Rauch GJ, Saude L, Concha ML, et al. 2000. Silberblick/Wnt11 mediates convergent extension movements during zebrafish gastrulation. *Nature* 405:76–81

Herman MA, Vassilieva LL, Horvitz HR, Shaw JE, Herman RK. 1995. The *C. elegans* gene *lin-44*, which controls the polarity of certain asymmetric cell divisions, encodes a Wnt protein and acts cell nonautonomously. *Cell* 83:101–10

Jenny A, Darken RS, Wilson PA, Mlodzik M. 2003. Prickle and Strabismus form a functional complex to generate a correct axis during planar cell polarity signaling. *EMBO J.* 22:4409–20

Jessen JR, Topczewski J, Bingham S, Sepich DS, Marlow F, et al. 2002. Zebrafish trilobite identifies new roles for Strabismus in gastrulation and neuronal movements. *Nat. Cell Biol.* 4:610–15

Katanaev VL, Ponzielli R, Semeriva M, Tomlinson A. 2005. Trimeric G protein-dependent frizzled signaling in *Drosophila*. *Cell* 120:111–22

Keller R. 2002. Shaping the vertebrate body plan by polarized embryonic cell movements. *Science* 298:1950–54

Kinoshita N, Iioka H, Miyakoshi A, Ueno N. 2003. PKC delta is essential for Dishevelled function in a noncanonical Wnt pathway that regulates *Xenopus* convergent extension movements. *Genes Dev.* 17:1663–76

Lawrence PA, Casal J, Struhl G. 2004. Cell interactions and planar polarity in the abdominal epidermis of *Drosophila. Development* 131:4651–64

Lawrence PA, Shelton PM. 1975. The determination of polarity in the developing insect retina. *J. Embryol. Exp. Morphol.* 33:471–86

Lele Z, Bakkers J, Hammerschmidt M. 2001. Morpholino phenocopies of the swirl, snailhouse, somitabun, minifin, silberblick, and pipetail mutations. *Genesis* 30:190–94

Liu T, DeCostanzo AJ, Liu X, Wang H, Hallagan S, et al. 2001. G protein signaling from activated rat frizzled-1 to the beta-catenin-Lef-Tcf pathway. *Science* 292:1718–22

Lu X, Borchers AG, Jolicoeur C, Rayburn H, Baker JC, Tessier-Lavigne M. 2004. PTK7/CCK-4 is a novel regulator of planar cell polarity in vertebrates. *Nature* 430:93–98

Ma D, Yang CH, McNeill H, Simon MA, Axelrod JD. 2003. Fidelity in planar cell polarity signalling. *Nature* 421:543–47

Marlow F, Topczewski J, Sepich D, Solnica-Krezel L. 2002. Zebrafish rho kinase 2 acts downstream of wnt11 to mediate cell polarity and effective convergence and extension movements. *Curr. Biol.* 12:876–84

Matakatsu H, Blair SS. 2004. Interactions between Fat and Dachsous and the regulation of planar cell polarity in the *Drosophila* wing. *Development* 131:3785–94

Mlodzik M. 1999. Planar polarity in the *Drosophila* eye: a multifaceted view of signaling specificity and cross-talk. *EMBO J.* 18:6873–79

Mlodzik M. 2002. Planar cell polarization: do the same mechanisms regulate *Drosophila* tissue polarity and vertebrate gastrulation? *Trends Genet.* 18:564–71

Montcouquiol M, Rachel RA, Lanford PJ, Copeland NG, Jenkins NA, Kelley MW. 2003. Identification of Vangl2 and Scrb1 as planar polarity genes in mammals. *Nature* 423:173–77

Myers DC, Sepich DS, Solnica-Krezel L. 2002. Convergence and extension in vertebrate gastrulae: cell movements according to or in search of identity? *Trends Genet.* 18:447–55

Paricio N, Feiguin F, Boutros M, Eaton S, Mlodzik M. 1999. The *Drosophila* STE20-like kinase Misshapen is required downstream of the Frizzled receptor in planar polarity signaling. *EMBO J.* 18:4669–78

Penzo-Mendez A, Umbhauer M, Djiane A, Boucaut JC, Riou JF. 2003. Activation of Gbetagamma signaling downstream of Wnt-11/Xfz7 regulates Cdc42 activity during Xenopus gastrulation. *Dev. Biol.* 257:302–14

Rawls AS, Guinto JB, Wolff T. 2002. The cadherins, Fat and Dachsous, regulate dorsal/ventral signaling in the *Drosophila* eye. *Curr. Biol.* 12:1021–26

Sawa H, Lobel L, Horvitz HR. 1996. The *C. elegans lin-17*, which is required for certain asymmetric cell divisions, encodes a putative seven-transmembrane protein similar to the *Drosophila* Frizzled protein. *Genes Dev.* 10:2189–97

Sheldahl LC, Park M, Malbon CC, Moon RT. 1999. Protein kinase C is differentially stimulated by Wnt and Frizzled homologs in a G-protein-dependent manner. *Curr. Biol.* 9:695–98

Shimada Y, Usui T, Yanagawa S, Takeichi M, Uemura T. 2001. Asymmetric colocalization of Flamingo, a seven-pass transmembrane cadherin, and Dishevelled in planar cell polarization. *Curr. Biol.* 11:859–63

Simon MA. 2004. Planar cell polarity in the *Drosophila* eye is directed by graded Four-jointed and Dachsous expression. *Development* 131:6175–84

Slusarski DC, Corces VG, Moon RT. 1997a. Interaction of Wnt and a Frizzled homologue triggers G-protein-linked phosphatidylinositol signalling. *Nature* 390:410–13

Slusarski DC, Yang-Snyder J, Busa WB, Moon RT. 1997b. Modulation of embryonic intracellular Ca^{2+} signaling by Wnt-5A. *Dev. Biol.* 182:114–20

Stein M, Wandinger-Ness A, Roitbak T. 2002. Altered trafficking and epithelial cell polarity in disease. *Trends Cell Biol.* 12:374–81

Strutt D. 2003. Frizzled signalling and cell polarisation in *Drosophila* and vertebrates. *Development* 130:4501–13

Strutt D, Johnson R, Cooper K, Bray S. 2002. Asymmetric localization of frizzled and the determination of notch-dependent cell fate in the *Drosophila* eye. *Curr. Biol.* 12:813–24

Strutt DI. 2001. Asymmetric localization of frizzled and the establishment of cell polarity in the *Drosophila* wing. *Mol. Cell* 7:367–75

Strutt H, Strutt D. 1999. Polarity determination in the *Drosophila* eye. *Curr. Opin. Genet. Dev.* 9:442–46

Tada M, Smith JC. 2000. Xwnt11 is a target of *Xenopus* Brachyury: regulation of gastrulation movements via Dishevelled, but not through the canonical Wnt pathway. *Development* 127:2227–38

Taylor J, Abramova N, Charlton J, Adler PN. 1998. Van Gogh: a new *Drosophila* tissue polarity gene. *Genetics* 150:199–210

Topczewski J, Sepich DS, Myers DC, Walker C, Amores A, et al. 2001. The zebrafish glypican knypek controls cell polarity during gastrulation movements of convergent extension. *Dev. Cell* 2:251–64

Tree DRP, Shulman JM, Rousset R, Scott MP, Gubb D, Axelrod JD. 2002. Prickle mediates feedback amplification to generate asymmetric planar cell polarity signaling. *Cell* 109:371–81

Usui T, Shima Y, Shimada Y, Hirano S, Burgess RW, et al. 1999. Flamingo, a seven-pass transmembrane cadherin, regulates planar cell polarity under the control of Frizzled. *Cell* 98:585–95

Vinson CR, Adler PN. 1987. Directional non-cell autonomy and the transmission of polarity information by the *frizzled* gene of *Drosophila*. *Nature* 329:549–51

Vinson CR, Conover S, Adler PN. 1989. A *Drosophila* tissue polarity locus encodes a protein containing seven potential transmembrane domains. *Nature* 338:263–64

Wallingford JB, Fraser SE, Harland RM. 2002. Convergent extension: the molecular control of polarized cell movement during embryonic development. *Dev. Cell* 2:695–706

Wallingford JB, Rowning BA, Vogeli KM, Rothbacher U, Fraser SE, Harland RM. 2000. Dishevelled controls cell polarity during *Xenopus* gastrulation. *Nature* 405:81–85

Wang Y, Macke JP, Abella BS, Andreasson K, Worley P, et al. 1996. A large family of putative transmembrane receptors homologous to the product of the *Drosophila* tissue polarity gene frizzled. *J. Biol. Chem.* 271:4468–76

Weber U, Paricio N, Mlodzik M. 2000. Jun mediates Frizzled induced R3/R4 cell fate distinction and planar polarity determination in the *Drosophila* eye. *Development* 127:3619–29

Wehrli M, Tomlinson A. 1998. Independent regulation of anterior/posterior and equatorial/polar polarity in the *Drosophila* eye; evidence for the involvement of Wnt signaling in the equatorial/polar axis. *Development* 125:1421–32

Westfall TA, Brimeyer R, Twedt J, Gladon J, Olberding A, et al. 2003. Wnt-5/pipetail functions in vertebrate axis formation as a negative regulator of Wnt/beta-catenin activity. *J. Cell Biol.* 162:889–98

Winter CG, Wang B, Ballew A, Royou A, Karess R, et al. 2001. *Drosophila* Rho-associated kinase (Drok) links Frizzled-mediated planar cell polarity signaling to the actin cytoskeleton. *Cell* 105:81–91

Wolff T, Rubin GM. 1998. *strabismus*, a novel gene that regulates tissue polarity and cell fate decisions in *Drosophila*. *Development* 125:1149–59

Wong HC, Bourdelas A, Krauss A, Lee HJ, Shao Y, et al. 2003. Direct binding of the PDZ domain of Dishevelled to a conserved internal sequence in the C-terminal region of Frizzled. *Mol. Cell* 12:1251–60

Wu J, Klein TJ, Mlodzik M. 2004. Subcellular localization of frizzled receptors, mediated by their cytoplasmic tails, regulates signaling pathway specificity. *PLoS Biol.* 2:E158

Yamanaka II, Moriguchi T, Masuyama N, M. K, Hanafusa H, et al. 2002. JNK functions in the non-canonical Wnt pathway to regulate convergent extension movements in vertebrates. *EMBO Rep.* 3:69–75

Yang C, Axelrod JD, Simon MA. 2002. Regulation of Frizzled by Fat-like cadherins during planar polarity signaling in the *Drosophila* compound eye. *Cell* 108:675–88

Zeidler MP, Perrimon N, Strutt DI. 1999. The four-jointed gene is required in the *Drosophila* eye for ommatidial polarity specification. *Curr. Biol.* 9:1363–72

Zeidler MP, Perrimon N, Strutt DI. 2000. Multiple roles for four-jointed in planar polarity and limb patterning. *Dev. Biol.* 228:181–96

Zheng L, Zhang J, Carthew RW. 1995. *frizzled* regulates mirror-symmetric pattern formation in the *Drosophila* eye. *Development* 121:3045–55

Molecular Mechanisms of Steroid Hormone Signaling in Plants

Grégory Vert,* Jennifer L. Nemhauser,*
Niko Geldner,* Fangxin Hong, and Joanne Chory

Plant Biology Laboratory and Howard Hughes Medical Institute, The Salk Institute
for Biological Studies, La Jolla California 92037; email: vert@salk.edu,
nemhauser@salk.edu, geldner@salk.edu, fhong@salk.edu, chory@salk.edu

Annu. Rev. Cell Dev. Biol.
2005. 21:177–201

First published online as a
Review in Advance on
June 14, 2005

The *Annual Review of
Cell and Developmental
Biology* is online at
http://cellbio.annualreviews.org

doi: 10.1146/
annurev.cellbio.21.090704.151241

Copyright © 2005 by
Annual Reviews. All rights
reserved

*These authors
contributed equally

1081-0706/05/1110-
0177$20.00

Key Words

Arabidopsis, brassinosteroids, receptor, signal transduction, gene
expression

Abstract

Brassinosteroids (BRs), the polyhydroxylated steroid hormones of
plants, regulate the growth and differentiation of plants throughout
their life cycle. Over the past several years, genetic and biochemical approaches have yielded great progress in understanding BR
signaling. Unlike their animal counterparts, BRs are perceived at
the plasma membrane by direct binding to the extracellular domain
of the BRI1 receptor S/T kinase. BR perception initiates a signaling cascade, acting through a GSK3 kinase, BIN2, and the BSU1
phosphatase, which in turn modulates the phosphorylation state and
stability of the nuclear transcription factors BES1 and BZR1. Microarray technology has been used extensively to provide a global
view of BR genomic effects, as well as a specific set of target genes
for BES1 and BZR1. These gene products thus provide a framework
for how BRs regulate the growth of plants.

Contents

belong to plant-specific families, suggesting that the role of steroids as signaling molecules may have arisen multiple times on the road to multicellularity.

The BRs are important regulators of growth and differentiation in plants. BR biosynthesis is fairly well understood as a result of the identification of many BR-deficient dwarf mutants and numerous feeding experiments in cultured cells (Fujioka & Yokota 2003). In the past few years, tremendous progress has been made in *Arabidopsis* in understanding how BRs are perceived and how the information is transduced to promote genomic responses (Clouse 2002, Peng & Li 2003). In this review, we present a critical analysis of currently available data on BR signaling pathway components, highlighting the latest findings on the cell surface-localized BR receptor and on the specific control of gene expression by a novel family of transcription factors.

LIGAND PERCEPTION AND RECEPTOR ACTIVATION

Brassinosteroids are Perceived by a Receptor Serine/Threonine Kinase

In contrast to animal steroid signals, BRs are perceived by a plasma membrane-localized receptor kinase. This kinase is encoded by the *BRI1* gene, which was initially identified as a BL-insensitive mutant (Clouse et al. 1996) and is defined by a large number of recessive mutations (**Figure 1**). *bri1* mutants display a light-grown morphology in the dark, show extremely dwarfed growth in the light, and have numerous other phenotypes, all of which are also seen in strong BR biosynthetic mutants.

BRI1 is part of a large, plant-specific family of S/T LRR-RLKs, consisting of more than 200 members in *Arabidopsis* (Shiu & Bleecker 2001). The BRI1 extracellular region consists of more than 20 LRRs, interrupted by a stretch of amino acids termed the island domain. Initial annotations predicted a putative N-terminal leucine-zipper followed

BRs:
brassinosteroids

BL: brassinolide

S/T:
serine/threonine

LRR-RLK:
leucine-rich repeat
receptor-like kinase

INTRODUCTION

Polyhydroxylated steroid hormones are widely distributed in nature. They have been identified in fungi, plants, and animals. The likelihood of an ancient origin for these molecules is underscored by the remarkable conservation in activity between plant and human forms of at least one key biosynthetic enzyme (Li et al. 1997). In recent years, many of the proteins required for steroid response in plants have been identified. Strikingly, almost every protein in the pathway appears to

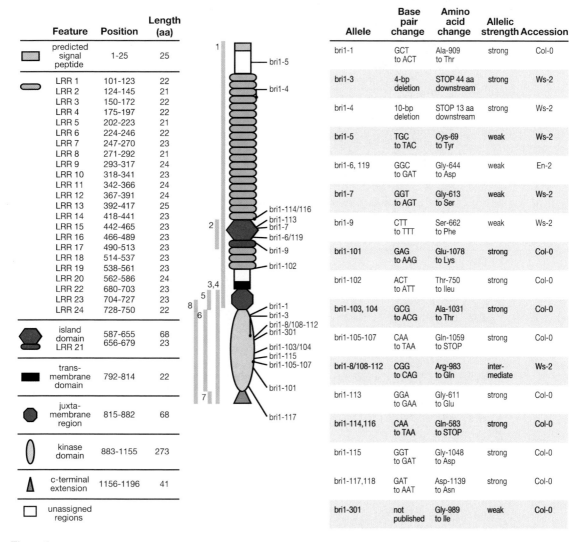

Feature	Position	Length (aa)
predicted signal peptide	1-25	25
LRR 1	101-123	22
LRR 2	124-145	21
LRR 3	150-172	22
LRR 4	175-197	22
LRR 5	202-223	21
LRR 6	224-246	22
LRR 7	247-270	23
LRR 8	271-292	21
LRR 9	293-317	24
LRR 10	318-341	23
LRR 11	342-366	24
LRR 12	367-391	24
LRR 13	392-417	25
LRR 14	418-441	23
LRR 15	442-465	23
LRR 16	466-489	23
LRR 17	490-513	23
LRR 18	514-537	23
LRR 19	538-561	23
LRR 20	562-586	24
LRR 22	680-703	23
LRR 23	704-727	23
LRR 24	728-750	22
island domain	587-655	68
LRR 21	656-679	23
trans-membrane domain	792-814	22
juxta-membrane region	815-882	68
kinase domain	883-1155	273
c-terminal extension	1156-1196	41
unassigned regions		

Allele	Base pair change	Amino acid change	Allelic strength	Accession
bri1-1	GCT to ACT	Ala-909 to Thr	strong	Col-0
bri1-3	4-bp deletion	STOP 44 aa downstream	strong	Ws-2
bri1-4	10-bp deletion	STOP 13 aa downstream	strong	Ws-2
bri1-5	TGC to TAC	Cys-69 to Tyr	weak	Ws-2
bri1-6, 119	GGC to GAT	Gly-644 to Asp	weak	En-2
bri1-7	GGT to AGT	Gly-613 to Ser	weak	Ws-2
bri1-9	CTT to TTT	Ser-662 to Phe	weak	Ws-2
bri1-101	GAG to AAG	Glu-1078 to Lys	strong	Col-0
bri1-102	ACT to ATT	Thr-750 to Ileu	strong	Col-0
bri1-103, 104	GCG to ACG	Ala-1031 to Thr	strong	Col-0
bri1-105-107	CAA to TAA	Gln-1059 to STOP	strong	Col-0
bri1-8/108-112	CGG to CAG	Arg-983 to Gln	inter-mediate	Ws-2
bri1-113	GGA to GAA	Gly-611 to Glu	strong	Col-0
bri1-114,116	CAA to TAA	Gln-583 to STOP	strong	Col-0
bri1-115	GGT to GAT	Gly-1048 to Asp	strong	Col-0
bri1-117,118	GAT to AAT	Asp-1139 to Asn	strong	Col-0
bri1-301	not published	Gly-989 to Ile	weak	Col-0

Figure 1

(Continued on next page)

by 25 LRRs, with the island domain residing between repeats 21 and 22 (Li & Chory 1997). For this review, we have reannotated BRI1; the new annotation no longer predicts a leucine zipper. Furthermore, it now appears that BRI1 has 24 rather than 25 LRRs, with LRR21 (formerly LRR22) being an unusual methionine-rich repeat (**Figure 1**). The intracellular region can be subdivided into a JM, followed by a canonical S/T kinase and a short C-terminal extension (**Figure 1**). Thus

by its overall structure, BRI1 is an archetypal receptor kinase (Li & Chory 1997), and several lines of evidence established BRI1 as a critical and limiting component for BR binding and perception. *BRI1* overexpression increases the number of BL binding sites, and this binding activity can be precipitated using specific antibodies (Wang et al. 2001). In competition experiments, binding affinities of these sites correlate with the bioactivity of the respective compounds. The

JM: juxtamembrane region

	Peptide aa positions	Number of sites	Possible positions
1 ▬	825-841	1	S-838
2 ▬	842-854	2	T-842, T-846
3 ▬	855-869	1	S-858
4 ▬	870-874	1	T-872
5 ▬	886-895	1	S-887, S-891
6 ▬	978-983	1	S-981, T-982
7 ▬	1038-1062	3	T-1039, S-1042, S-1044, T-1045, T-1049, S-1060
8 ▬	1165-1171	1	S-1166, S-1168, T-1169
9 ▬	1172-1109	1	S-1172, S-1179, T-1180
10 ●	n.a.	1	S-1162
11 ●	n.a.	1	T-1180

Kinase sub-domains and functional regions	aa position	Length
I	883-903	21
II	904-918	15
III	919-934	16
IV	935-949	15
V	950-980	30
VIa	981-1003	23
VIb	1004-1020	17
VII	1021-1037	17
VIII	1038-1058	21
IX	1059-1085	27
X	1086-1106	21
XI	1107-1149	43
ATP-binding signature	889-912	24
catalytic loop	1007-1014	8
activation loop	1027-1056	30

Number	Position	Length (aa)	Description	Reference
1	1-879	879	extracellular/ transmembrane/ juxtamembrane region (NGT-1)	He et al., 2000
2	580-673	94	BL-binding region (Island+LRR22)	Kinoshita et al. 2005
3	815-1196	382	region described as BRI1-KD or JKC	Oh et al., 2000 Wang et al. 2005
4	814-1196	383	region used for kinase and interaction assays	Li et al., 2002
5	815-882	68	juxtamembrane region	Wang et al. 2005
6	883-1155	273	kinase domain	Wang et al. 2005
7	1156-1196	41	c-terminal region	Wang et al. 2005
8	847-1196	350	region used for yeast-two-hybrid	Nam and Li, 2002

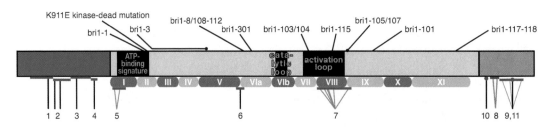

Figure 1

Compilation of BRI1 data. Features were predicted using the following web resources. Signal peptide and transmembrane domain: **www.cbs.dtu.dk/services/**; LRR repeats and kinase regions: **www.ebi.ac.uk/InterProScan/**. LRR repeats were refined manually. Kinase subdomains are based on Hanks & Hunter (1995). Alignments with cAPK (NP_00,2721), resources at **http://pkr.sdsc.edu/** and MSAs at **www.botany.wisc.edu/prkr/** were used to define kinase subdomain borders. Phosphorylation data are from Oh et al. (2000) and Wang et al. 2005. BRI1 mutation data are from Friedrichsen et al. (2000), Noguchi et al. (1999) and Nam & Li (2002). (Tables can be downloaded from Table S1. Follow the Supplemental Material link from the Annual Reviews home page at **http://www.annualreviews.org**).

BRI1 N-terminal region, consisting of the extracellular domain, the transmembrane pass, and the JM (**Figure 1**), was shown to function as a BR signal-transducing module (He et al. 2000). By fusing this region to the kinase domain of a rice LRR-RLK involved in pathogen defense, BL-inducible defense responses could be transferred to transformed cells. However, because these experiments were done in rice, the presence of additional plant-specific factors involved in BR binding could not be excluded.

Recently, direct binding of BL to BRI1 was demonstrated with native and recombinant BRI1 proteins (Kinoshita et al. 2005). It was shown that a BR analog could be cross-linked to BRI1, both in microsomal preparations and in pull-down fractions highly enriched for BRI1-GFP, indicating that BRs and BRI1 directly interact. Moreover, recombinant proteins consisting of the island domain and the neighboring C-terminal LRR repeat (**Figure 1**) were sufficient to bind radioactive BL with an affinity comparable to that observed for full-length BRI1 from plants. In addition to BRI1, three highly similar homologs have been characterized (Cano-Delgado et al. 2004, Clay & Nelson 2002, Zhou et al. 2004). Two display high BL-binding affinity. Genetic analysis suggests that these receptors play a restricted and partially redundant role in BR signaling. Thus BRI1 apparently represents the single most important BR binding activity in *Arabidopsis*. Recently, the *BRI1* ortholog in tomato was shown to act as the receptor for systemin as well as for BRs (reviewed in Wang & He 2004). However, systemin, a small peptide signal involved in plant defense, is present only in a subgroup of higher plants, not including *Arabidopsis*. Why BRI1 was co-opted for this dual role is not known.

Possible Mechanisms of BRI1 Activation

How is ligand binding transduced across the membrane? Since the cloning of *BRI1*, numerous analogies to animal receptor pathways have been drawn (Peng & Li 2003, Wang & He 2004, Yin et al. 2002c). Such comparisons are inevitable and potentially useful. However, many receptor pathways have developed during the independent acquisition of multi-cellularity in plants and animals, and it is possible that mechanistic similarities between the BR and animal receptor pathways might merely represent random evolutionary convergences.

In animals, ligand-induced activation of single-pass transmembrane receptors is often associated with dimerization or multimerization of the receptor with itself and/or coreceptors. In mammalian cells, ligand-induced oligomerization was proposed to initiate downstream signaling by bringing intracellular kinase domains together and allowing their *trans*-phosphorylation (Schlessinger 2000). In many cases, however, this simple "induction by dimerization" model does not appear to apply. The insulin receptor, for example, exists as a constitutive, ligand-independent dimer (Jiang & Hunter 1999). Pre-formed dimers of epidermal growth factor (EGF) receptors were shown to exist in vivo (Gadella & Jovin 1995) and structures of the receptor's extracellular domain suggest a model whereby EGF binds with high affinity to dimeric receptor forms, leading to their stabilization rather than inducing their formation (Ferguson et al. 2003, Garrett et al. 2002, Ogiso et al. 2002). Current models suggest that ligand binding induces a reorientation of subunits with respect to each other. For the EGF receptor, some evidence supports a rotational rearrangement of subunits (Moriki et al. 2001). For Epo, another receptor tyrosine kinase, a scissor-like activation mechanism has been put forward (Jiang & Hunter 1999). In summary, it appears that dimerization is required but not sufficient for activation of single-pass transmembrane receptor kinases.

BAK1, an LRR-RLK with five extracellular LRRs, is a candidate for BRI1's coreceptor. BAK1 was independently found as a

FRET: fluorescence resonance energy transfer

gain-of-function suppressor of a weak allele of *bri1*, as well as a BRI1 yeast-two-hybrid interactor (Li et al. 2002, Nam & Li 2002). BRI1 and BAK1 expressed in yeast interact with each other and are able to mutually *trans*-phosphorylate. The phenotypes of BAK1 knockouts and kinase-dead, dominant-negative variants are consistent with its role as a positive component of BR-signaling. However, knockout phenotypes of *BAK1* are rather subtle compared with *BRI1* knockouts, indicating that BAK1 is either not strictly required or functions redundantly with the four other members of its subfamily (Hecht et al. 2001). It will be important to determine if multiple knockouts will eventually give rise to a *bri1*-like phenotype. Neither knockout nor over-expression of *BAK1* influences ligand binding to BRI1 (Kinoshita et al. 2005, Wang et al. 2005). Thus current data suggests that BAK1 is a coreceptor and/or downstream target of BRI1.

Recently, self-interaction of BRI1 was demonstrated by FRET using cell-culture transfection assays and by pull-down experiments in transgenic plants (Russinova et al. 2004, Wang et al. 2005). BR effects on interaction were not tested in the FRET experiments, but the pull-down experiments showed that BRI1 interaction increases upon BL treatment. In the future, it will be critical to address the oligomerization status of BRI1 and BAK1 together, in a functional but non-stimulated plant cell system. This can be done in the background of strong biosynthetic mutants or in the presence of high concentrations of BR biosynthetic inhibitors.

Thus initial BL binding to the island-LRR domain of BRI1 may occur on BRI1 monomers or with a preformed homo-oligomer. Current data cannot exclude the possibility that BL binds to a BRI1, which is part of a BRI1-BAK1 hetero-dimer or hetero-tetramer. The fact that BRI1 and BAK1 interact in yeast in the absence of ligand suggests that there may be pre-existing hetero-dimers or tetramers. For the animal BMP-receptors, all possible modes of receptor/coreceptor in-

teraction states have been demonstrated in the absence of ligand (Gilboa et al. 2000). BL is a relatively small molecule compared with the ligands of most animal receptor kinases, and it is hard to imagine how BL could bridge two receptor molecules via bivalent interaction, as has been shown for ligands in animals (Schlessinger 2000, Wiesmann & de Vos 1999). To our knowledge, the family of TOLL-like receptors (TLRs) (but not TOLL itself) is the only example where smaller molecules (bacterial components) activate a single-pass transmembrane receptor in animals (Akira & Takeda 2004). Unfortunately, not much is known about the activation mechanism of TLRs. Rather than bridging subunits, BL could induce a conformational change that stabilizes a pre-existing dimer, as discussed for the EGF receptor. A consequent conformational change would then reorient the kinase subunits and allow for initial *trans*-phosphorylation, either between BRI1 subunits or between BRI1 and BAK1. *Trans*-phosphorylation is considered to be the critical initial event in receptor kinase activation, releasing the kinases from an auto-inhibited state of low activity (Hubbard 2004).

What is known about the regulation of BRI1 kinase activity? It has been demonstrated that BL treatment leads to BRI1 phosphorylation in planta (Wang et al. 2001, Wang et al. 2005). In yeast, BAK1/BRI1 *trans*-phosphorylation activity is apparently interdependent since neither of the two proteins can be phosphorylated if one is in its kinase-dead form (Nam & Li 2002). Slightly different results were obtained using recombinant BRI1 and BAK1 kinases (Li et al. 2002). The cytosolic parts were shown to interact and *trans*-phosphorylate each other, even if one partner was inactive, although not as efficiently. Thus isolated intracellular domains seem to be less dependent on each other than are full-length proteins. This might be explained by some topological restraints imposed on the full-length proteins. Therefore, back-and-forth signaling between BRI1 and

BAK1 is possibly needed for full activation of both.

In order to be activated, many kinases require phosphorylation in their activation loop, which increases kinase activity by a number of mechanisms (Johnson et al. 1996). The BRI1 kinase contains all the signatures of an activation-loop-dependent kinase, and it was shown that S/T residues in the activation loop are subject to auto-phosphorylation (Oh et al. 2000; **Figure 1**), suggesting that this might be an initial activation event for BRI1. Additional mechanisms of receptor auto-inhibition have been described, namely inhibition by C-terminal extensions or JM regions of the cytosolic domain (Hubbard 2004). Small insertion/deletions in the JM region of the RTK KIT, for example, lead to ligand-independent receptor activity (Hirota et al. 1998). Deletion of the BRI1-JM results in an inactive receptor, precluding conclusions about a possible role in auto-inhibition (Wang et al. 2005). Nonetheless, the JM domain is subject to BRI1 auto-phosphorylation in vitro (**Figure 1**), and it will be interesting to see if an in vivo function can be assigned to these phosphorylation sites.

The C-terminal extension of BRI1 appears to have an auto-inhibitory function (Wang et al. 2005). A BRI1 C-terminal deletion construct is functional and slightly hyperactive in vivo. Moreover, the deletion variant is less dependent on ligand, and a kinase domain lacking the C terminus displays increased kinase activity in vitro. The C terminus is phosphorylated at multiple sites (**Figure 1**), and "phosphorylation-mimic" mutations have similar effects as deleting the domain. Taken together, these results provide a first clue of how BRI1 kinase is auto-inhibited and activated by phosphorylation. However, because there is still a clear ligand dependency of C-terminally deleted BRI1, other mechanisms must provide additional layers of regulation. A fully phosphorylated receptor kinase will either directly phosphorylate downstream targets or simply interact with them, thereby recruiting them to their site of action.

Downstream Targets of BRI1

The direct targets of BRI1 in vivo are not known, but several candidates exist. As discussed, BAK1 and its homologs may be the main direct targets that initiate signaling events that ultimately inactivate the downstream kinase BIN2 (see below). Therefore, identifying BAK1 interaction partners promises to further our understanding of BR signaling. The second candidate for a direct BRI1 target, transthyretin-like protein (TTL), was identified in a yeast-two-hybrid with BRI1 (Nam & Li 2004). The interaction depends on BRI1 kinase activity, and TTL is phosphorylated by BRI1 in vitro. Genetic analysis, however, suggests that TTL is a negative modulator of BRI1 signaling. TTL is largely or completely associated with membranes. Therefore, TTL could be involved in recruitment of deactivating phosphatases or be necessary for receptor down-regulation.

BRI1 Deactivation

Understanding receptor deactivation is as important as understanding its activation because speed and mode of inactivation will determine the amplitude and duration of ligand-induced signaling. Virtually nothing is known about how the activated BRI1 receptor is turned off. Co-overexpression of BRI1, together with BAK1, in cowpea protoplasts leads to dramatic shifts of BRI1 localization toward endosomal compartments, and FRET between BRI1/BAK1 preferentially occurs in endosomes and at restricted plasma membrane sites. This suggests that BAK1 might somehow regulate BRI1 endocytosis (Russinova et al. 2004). It remains to be seen how this finding relates to the mechanism of BR signaling in planta and to BRI1 deactivation. Enzymes catalyzing inactivating hydroxylation reactions on BRs have been identified and shown to be important in BR homeostasis in vivo (Neff et al. 1999). If and how these enzymes act in deactivating receptor-bound BL is unknown. In animals, receptors can be

GSK3: glycogen synthase kinase-3

inactivated by pH-dependent ligand separation in the acidic endosomal compartments (Rudenko et al. 2002). This is unlikely to occur in plants, however, as the extracellular space already has a low pH. Therefore, a possible ligand/receptor separation in endosomes would have to occur by a different mechanism. Studies of BRI1 endocytosis and its turn-over rates upon ligand binding will help us to understand how BRI1 deactivation is achieved.

SIGNAL TRANSDUCTION

BIN2, a GSK3 Kinase Critical for BR Signaling

Downstream from BRI1/BAK1, a major signaling component in the BR pathway is defined by semidominant *bin2* gain-of-function mutations. These mutants are allelic to *dwarf12* (Choe et al. 2002) and *ucu1* (Perez-Perez et al. 2002), uncovered in genetic screens for BR-related dwarfism and altered leaf morphology, respectively. *bin2* mutants resemble *bri1* mutants, but are distinguished from *bri1* mutants by an extreme downward curling of the leaves. As in *bri1* mutants, the feedback down-regulation of the BR-biosynthetic gene *CPD* is lost in *bin2* (Choe et al. 2002, Li et al. 2001), accounting for the higher accumulation of BL and its precursors (Choe et al. 2002).

BIN2 encodes a protein kinase, 70% similar in its catalytic domain to the mammalian GSK3 (Choe et al. 2002, Li & Nam 2002, Perez-Perez et al. 2002). GSK3s are a group of highly conserved constitutively active S/T kinases implicated in numerous signaling pathways and controlling metabolism, cell fate determination, and tissue patterning in various organisms.

BIN2 is a negative regulator of the BR pathway.
With a dominant mutant, unambiguous assignment of the affected gene to a given pathway is more difficult than with loss-of-function alleles. Indeed, as animal GSK3s are known to be fairly promiscuous in their substrates, a gain-of-function mutation in one family member could interfere with substrates of other GSK3s or unrelated kinases. Gene dosage analyses revealed that the *bin2-1* mutation was either hypermorphic or neomorphic (Li et al. 2001), whereas the *ucu1* mutation was likely to be antimorphic (Perez-Perez et al. 2002), although several studies argue in favor of the first hypothesis.

Three lines of evidence suggest that increased activity of BIN2 negatively regulates BR signal transduction. First, treatment of plants with Li$^+$, a known inhibitor of GSK3 (Klein & Melton 1996), provokes cell elongation and shows the typical BR-feedback down-regulation of *CPD* expression (J. Li, unpublished results), as well as dephosphorylation of a BIN2 substrate, BES1 (S. Mora-Garcia, unpublished results). This clearly indicates that one physiological function of GSK3s is to negatively regulate BR signaling. Second, BIN2 protein carrying the original *bin2-1* mutation displays a higher kinase activity in vitro toward both a GSK3-peptide substrate (Li & Nam 2002) and its substrate BES1 (Zhao et al. 2002), compared with activity of the wild-type BIN2 protein. Finally, overexpression of *BIN2* in the sensitized genetic background of a weak *bri1* mutant leads to either (*a*) severe dwarfing in plants with increased levels of *BIN2* or (*b*) wild-type-like plants resulting from co-suppression of endogenous *BIN2* (Li & Nam 2002). Though these observations point to a negative role of GSK3s in the BR signaling pathway, the function of BIN2 itself remains somewhat unresolved and will await the identification of a loss-of-function mutant for *BIN2*.

Are other GSK3s involved?
Although well-characterized in animals, very little is known about plant GSK3s. In *Arabidopsis*, *BIN2* belongs to a 10-member family organized in four phylogenetic subclasses (Jonak & Hirt 2002). Plant GSK3s show a highly conserved S/T kinase domain, but divergent N- and C termini. The function of most

GSK3s remains largely unknown and may not be restricted to specific pathways. In mammals, GSK3β is indeed involved in diverse cellular processes such as phosphorylation of glycogen synthase and β-catenin in the insulin and Wnt signaling pathways, respectively, yet no cross-talk is observed between the two pathways.

Several lines of evidence suggest that plant GSK3s are involved in stress responses and developmental processes (Jonak & Hirt 2002). Interestingly, *BIN2* has been shown to be expressed and restricted to the suspensor cells and excluded from the hypophysis (Dornelas et al. 1999). Whether this specific expression pattern carries a BR-related function is unknown. Genetic evidence suggests a stress involvement for *ASK*τ (a close *BIN2* relative) that is ABA- and salt-induced and whose overexpression in plants enhances salt tolerance (Piao et al. 1999). It is not yet known if *ASK*τ acts in BR signaling, but it could represent a molecular link between BRs and their reported role in salt-stress tolerance (Anuradha & Rao 2001). Uncovering the degree of redundancy and specialization within the plant GSK3s awaits in-depth genetic and biochemical investigation.

Atypical regulation of BIN2 activity. In animals, GSK3s are usually constitutively active enzymes, tightly regulated by two major mechanisms: phosphorylation and protein-protein interactions.

Phosphorylation. Many GSK3 substrates need to be prime-phosphorylated by a different kinase at position $n + 4$ before being phosphorylated at position n by GSK3s. Also, GSK3s themselves are regulated by phosphorylation. For example, upon insulin binding to its receptor, protein kinase B (PKB)/AKT phosphorylates GSK3s at a highly conserved N-terminal serine residue (Cross et al. 1995). This mimicks a prime phosphorylation and therefore turns the GSK3 N terminus into a pseudosubstrate, blocking access to its catalytic site.

Multiprotein complex. The best characterized example is the canonical Wnt pathway, where GSK3β-binding proteins control access to its substrate β-catenin, generating a high degree of specificity in regulating GSK3β. In the absence of stimulus, the scaffold protein axin binds GSK3β and β-catenin, triggering the phosphorylation of β-catenin and thereby promoting its ubiquination and subsequent degradation by the proteasome (Aberle et al. 1997). Upon Wnt binding by the Frizzled family receptor, the GSK3-binding protein FRAT facilitates the disruption of the GSK3β-containing complex. This decreases the phosphorylation of β-catenin, which results in β-catenin accumulation and activation.

At present, the biochemical characterization of plant GSK3s is scarce. The absence of both plant PKB and of the highly conserved N-terminal serine residue in BIN2 suggests that BIN2 is regulated by a different mechanism than the one seen for insulin. Moreover, BIN2 activity has been shown to act following a new docking mechanism independently of prime phosphorylation and of a multiprotein complex formation (Zhao et al. 2002).

Neither BRI1 nor BAK1 physically interacts with or phosphorylates BIN2 (Li & Nam 2002, Peng & Li 2003), suggesting additional steps in the pathway. Out of seven alleles of *bin2/ucu1/dwf12* identified, six are gain-of-function mutations that cluster in the four-residue threonine-arginine-glutamic acid-glutamic acid (TREE) domain, highlighting its importance in BIN2 function (Choe et al. 2002, Li & Nam 2002, Perez-Perez et al. 2002). The TREE domain is part of a short α-helix at the surface of the protein (Peng & Li 2003) and could be part of a phosphorylation site for CK2. CK2 indeed phosphorylates a S/T residue in an environment of acidic residues (Meggio & Pinna 2003). In this sense, the different *bin2/ucu1/dwf12* mutations would affect either the target residue or its environment by substituting basic residues for acidic ones.

BRZ: brassinazole

Investigating a possible role for CK2 in the BR-signaling pathway may shed some light on how this key kinase is regulated by BRs.

BES1/BZR1, Two Nuclear Downstream Components of BR Signaling

Two independent genetic screens identified homologous proteins acting as positive regulators of the BR signaling pathway. The *bzr1* mutant was identified as resistant to the BR-biosynthesis inhibitor BRZ in the dark (Wang et al. 2002). A suppressor screen of a weak *bri1* allele identified the *bes1* mutant, which not only suppresses the dwarf phenotype of *bri1* but also leads to constitutive BR responses (Yin et al. 2002b). *BES1* and *BZR1* encode plant-specific proteins that are 88% identical at the amino acid level. BES1 and BZR1 belong to a family of six closely related members with unknown function in *Arabidopsis*. All contain a bipartite nuclear localization signal, a central region rich in S/T, including many consensus phosphorylation sites for GSK3s, and a proteolysis-related PEST domain that encompasses the same P to L substitution in both mutants.

BES1 and BZR1 are positive regulators. BES1 and BZR1 proteins exist as two different forms, visualized as a slow- and a fast-migrating band on a Western blot, corresponding to a difference in the phosphorylation status of the two proteins (He et al. 2002, Yin et al. 2002b). Following BL treatment, only the hypophosphorylated form of both proteins is detected, accumulating to higher level compared with that in non-treated plants. This post-transcriptional regulation by BR was recently shown for four other members of the BES1 family (Yin et al. 2005). Fusion of BES1 and BZR1 to fluorescent proteins indicates that the accumulation of the hypophosphorylated form of both proteins following BL treatment correlates with their accumulation in the nucleus (Wang

et al. 2002, Yin et al. 2002b). The hyperphosphorylated form of BZR1 is stabilized in the presence of the proteasome inhibitor MG132, suggesting that the phosphorylation of BZR1 increases its degradation by the proteasome (He et al. 2002). In this sense, the respective mutations would uncouple the phosphorylation of both BES1 and BZR1 from their degradation. Phosphorylation appears necessary, but not sufficient, for the degradation of both proteins as both forms are detected in the cell under normal conditions. An additional modification of BES1 and BZR1, which could be from additional phosphorylations, may be required to efficiently target them for degradation. Consistent with this, the hyperphosphorylated form of the mutated BZR1 protein migrates as a faster band compared with that of the wild-type hyperphosphorylated BZR1 (He et al. 2002). These findings support a model where the BL-dependent accumulation of BES1 and BZR1 in their hypophosphorylated forms is regulated by a negatively acting kinase via proteasome degradation.

Although *bes1* and *bzr1* are gain-of-function mutations, several results argue for their specific involvement as positive regulators in the BR-pathway (He et al. 2002, Wang et al. 2002, Yin et al. 2002b, Zhao et al. 2002). Recently a loss-of-function dwarf phenotype was reported from RNAi knock-down plants for *BES1* and its relatives, further supporting the redundant role of these proteins in BR signaling (Yin et al. 2005).

One key question is why *bes1* and *bzr1* mutants, which share the same lesion in virtually identical proteins and result in similar BRZ-resistant phenotypes in the dark, exhibit opposite phenotypes in the light. For example, in the light, *bes1* displays constitutive BR responses, including long, bending petioles and pale green leaves reminiscent of *DWF4* or *BRI1* overexpressing plants (Choe et al. 2001, Wang et al. 2001). In contrast, *bzr1* displays a semidwarf phenotype and increased sensitivity to BRZ. *bzr1* shows reduced expression of the biosynthetic gene *CPD* (Wang et al. 2002), a difference that may account for such phenotypic observations.

BES1 and BZR1: actual substrates of BIN2?

BES1 and BZR1 proteins were shown to exist as two different forms and to specifically accumulate the hypophosphorylated form as early as 10 min after BL treatment (He et al. 2002). *bes1* and *bzr1* gain-of-function mutations, as well as *BES1* and *BZR1* overexpression, suppress the *bin2* dwarf phenotype, suggesting that BES1 and BZR1 act downstream from BIN2 (He et al. 2002, Yin et al. 2002b, Zhao et al. 2002). BIN2 was shown in vitro to interact with and to phosphorylate BES1 and BZR1 (He et al. 2002, Yin et al. 2002b, Zhao et al. 2002). Moreover, BES1 and BZR1 protein levels are low in the *bin2* gain-of-function background (Wang et al. 2002, Yin et al. 2002b). Finally, the drastic deletion of the central region of BES1, which contains the putative GSK3 phosphorylation sites, gives rise to constitutive BR responses (Yin et al. 2005). The next challenge will be the identification of the precise sites in BES1 and BZR1 that are phosphorylated by BIN2 in vivo and determining how this correlates with their biological activity.

BSU1, A Nuclear Phosphatase Promoting BES1 Dephosphorylation

A *bri1* suppressor screen by activation tagging led to the identification of the *bsu1-1D* mutant (Mora-Garcia et al. 2004). *BSU1* encodes a plant-specific protein with a long, Kelch-repeat-containing N-terminal region hooked up to a C-terminal S/T phosphatase domain. *bsu1* partially suppresses the dwarf phenotype of the *bin2* mutant. In addition, BES1 accumulates in its hypophosphorylated form in *bsu1* mutants, and in vitro BIN2-phosphorylated BES1 is dephosphorylated in the presence of BSU1 protein. Finally, RNAi knock-down plants show a compact phenotype resembling weak *bri1* alleles, providing additional support for a model where BSU1 directly counters the

effects of BIN2 on BES1, and likely BZR1 (**Figure 2**).

Cracking the Code of the BES1/BZR1 Signaling Mechanism

Three distinct BR effects have been described for BES1/BZR1.

Dephosphorylation. The rapid conversion of the pool of BES1 to its hypophosphorylated form correlates with the first measurable changes in transcription of BR-responsive genes. The robustness of this response makes the disappearance of the hyperphosphorylated form of BES1 the best marker for BR signaling.

Accumulation. In some cases, a clear overall increase in BES1 protein levels can be observed, whereas in other experiments the total amount of protein appears unchanged, although shifted to the hypophosphorylated form. BES1 accumulation may reflect conversion of the BES1 pool to the more stable hypophosphorylated BES1 rather than to active stabilization, suggesting a minor role for protein accumulation in BR signaling.

Nuclear translocation. Nuclear accumulation of a BES1-GFP fusion protein was reported following BR treatment (Yin et al. 2002b). These data were interpreted as evidence of a nuclear translocation correlating with a shift from hypo- to hyperphosphorylated form, by analogy with what is known for β-catenin in the canonical Wnt signaling pathway (**Figure 3a**). However, the data would also be consistent with stabilization of a constitutively nuclear protein. In agreement with this idea, mBES1 and mBZR1 mutant proteins, known to accumulate high levels of both hypo- and hyperphosphorylated forms, are detected exclusively in the nucleus (Wang et al. 2002, Yin et al. 2002b). One report described BES1 and BZR1 as constitutively nuclear proteins (Zhao et al. 2002), which also correlates with the nuclear localization

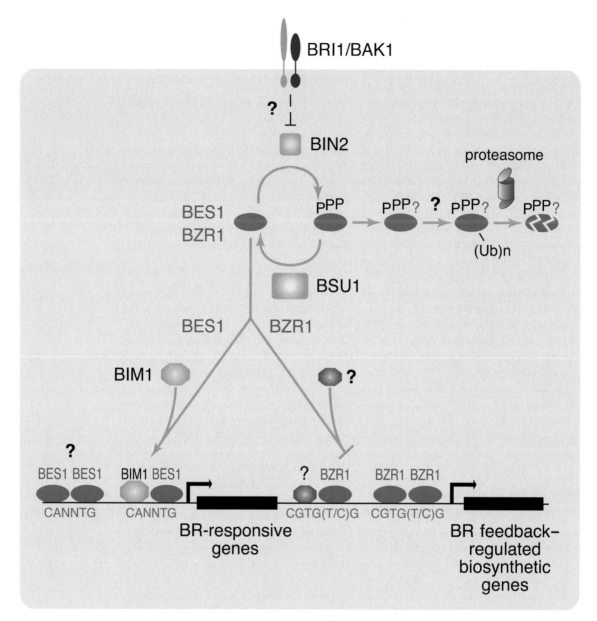

Figure 2

A model for downstream events in the BR signal transduction pathway. In resting cells, the BIN2 GSK3 kinase is active and phosphorylates the transcription factors BES1 and BZR1, targeting them for ubiquitination and subsequent proteasome-dependent degradation. In BR-stimulated cells, BRI1/BAK1 inhibits BIN2 and/or activates BSU1 activities by a yet unknown mechanism, leading to the conversion of the BES1/BZR1 pool to the hypophosphorylated form. BES1, in association with the bHLH transcription factor BIM1, promotes transcription of a subset of BR-regulated genes by binding to E-box motifs, CANNTG. BZR1 directly represses the transcription of BR feedback–regulated genes such as *CPD* to adjust BR homeostasis by binding to CGTG(T/C)G elements.

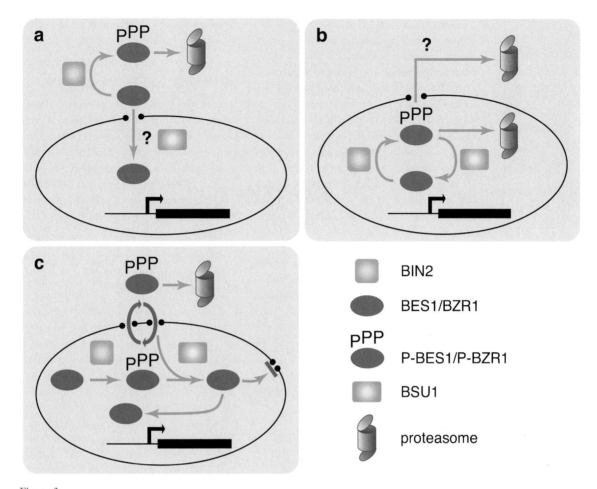

Figure 3

Models for downstream signaling. (*a*) Original nuclear translocation model. BIN2 phosphorylates BES1 and BZR1 in the cytosol. Upon BR stimulation, hypophosphorylated BES1 and BZR1 shuttle to the nucleus to promote BR responses (*b*) Nuclear model. BIN2, BES1, and BZR1 are constitutively in the nucleus; the activity of BES1 and BZR1 is primarily regulated by their phosphorylation status (*c*) Alternative nucleocytoplasmic model. Hyperphosphorylated BES1 and BZR1 are constantly cycling between the cytosol and the nucleus. Stimulation by BR triggers conversion of BES1 and BZR1 to their hypophosphorylated forms. This may lead to a greater affinity to DNA and/or prevent the two transcription factors from exiting the nucleus, thereby accounting for their accumulation in the nucleus.

of BSU1 (Mora-Garcia et al. 2004). Importantly, the subcellular localization of BIN2 is unknown to date and could help to solve this issue. Its placement in the cytosol was based on analogy with the Wnt pathway without supporting evidence, although there is evidence that plant GSK3s can be localized in the nucleus (Tavares et al. 2002).

These observations raise important questions about the overall design of the pathway and suggest that phosphorylation is the primary mode of regulation of BES1 protein activity. We therefore present a second model where BES1 phosphorylation and dephosphorylation events would occur exclusively in the nucleus, assuming that BIN2 could be localized in this compartment (**Figure 3b**). Whether the degradation of BES1 happens in the nucleus or in the cytosol is unknown, but nucleocytoplasmic transport and subsequent degradation in the cytosol could be involved, as described for p53, for instance (Liang & Clarke 2001). Alternatively, BES1 and BZR1 could undergo a rapid nucleocytoplasmic cycling between the two compartments even though the steady state of both proteins is in the nucleus (**Figure 3c**). This phenomenon has been described for many transcription regulators such as ERF, SMADs, and STATs in the Ras/Erk, TGF-β, and JAK/STAT signaling pathways, respectively (Le Gallic et al. 2004, Marg et al. 2004, Nicolas et al. 2004, Pranada et al. 2004). Unraveling the relationship between BR-induced dephosphorylation of BES1 and BZR1 and their localization, uncovering the subcellular localization of the BIN2 protein, identifying the compartment where BIN2 interacts with BES1 and BZR1, as well as monitoring the possible dynamic distribution of all the players will be essential for determining the true architecture of the BR signaling pathway.

GENOMIC EFFECTS OF BRS

Studies have linked BRs to several nongenomic effects, including changes in wall extensibility (Zurek et al. 1994), osmotic permeability (Morillon et al. 2001), vacuolar function (Schumacher et al. 1999), and intracellular calcium fluxes (Allen et al. 2000). The best characterized direct effects, however, are the early transcriptional responses to BR treatment.

A High Confidence List of BR-Regulated Genes

Several recent reviews have described historical approaches to measuring BR responses (i.e., Mussig & Altmann 2003). The focus of this section is on the application of genome-scale tools to the question of the BR genomic response. In 2002, three groups published reports on short-term effects of BR treatment on gene expression, using Affymetrix chips representing approximately one third of the genome (Goda et al. 2002, Mussig et al. 2002, Yin et al. 2002b). Surprisingly, the findings from these groups showed little overlap in the genes identified (**Figure 4**), although similarities in the broad functional categories represented by each group's gene list could be observed. One important result common to all three reports was the modest nature of the BR response. Whereas studies on other plant hormones, such as auxin, have reported transcript-induction in excess of 10-fold (Zhao et al. 2003), few BR-regulated genes were shown to be induced by more than 2-fold. This is an interesting result from a biological perspective but also presents a challenge for current analysis methods. In an attempt to resolve the question of whether the results reported from each study reflected differences in experimental design or were largely attributable to varying analytical methods, we initiated a combined

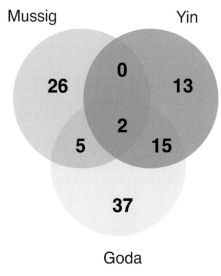

Figure 4

Early studies of BR-regulated gene expression have little overlap. Genes shown to be induced by BR treatment of seedlings are shown from three studies: Mussig et al. (2002), cyan; Yin et al. (2002b), magenta; Goda et al. (2002), yellow.

analysis with new data from two of the original groups.

Both the Chory and Shimada groups had applied their original experimental conditions to the Affymetrix ATH1 microarray, representing approximately 22,000 genes. Importantly, to perform a joint analysis of the data from both laboratories, all differences, including experimenter, treatment, growth conditions, and age, were combined into a factor called lab effect (for details of analysis see Appendix A. Follow the Supplemental Material link from the Annual Reviews home page at **http://www.annualreviews.org**). Three replicates were available from the Chory laboratory, where 10-day-old seedlings grown on plates were submersed in 1 μM BL or mock treatments for 2.5 h (Nemhauser et al. 2004). Two replicates were available from the Shimada laboratory, where 7-day-old liquid-culture-grown seedlings were exposed to 10 nM BL or mock treatments for 3 h (**http://web.unifrankfurt.de/fb15/botanik/mcb/AFGN/atgenex.html**). To establish a high confidence list of BR-regulated genes, two diverse approaches were taken. In the first, linear models were used (Gentleman et al. 2004; limma library). Very few genes were found to be differentially expressed by linear models unless a term for lab effect was included. This lab effect was found to be significant for over half of the genes. Linear models identified 480 genes whose transcript levels increased following BR treatment and 386 genes whose transcript levels decreased at a false discovery rate (FDR) = 0.05.

A description of an alternative, nonparametric approach, called Rank Product, was recently published (Breitling et al. 2004; Gentleman et al. 2004; RankProd library of bioconductor). This approach was proposed to offer several advantages over linear modeling, including fewer assumptions under the model, no requirement to normalize all data together, and increased performance with noisy data and/or low numbers of replicates. At an FDR = 0.05, 681 transcripts

increased following BR treatment, and 558 transcripts decreased. The overlap between the gene lists identified in these approaches is substantial (424 up-regulated genes and 332 down-regulated genes; Table S2, S3. Follow the Supplemental Material link from the Annual Reviews home page at **http://www.annualreviews.org**).

With this high confidence list in hand, we returned to the original microarray data performed with the first-generation Affymetrix microarrays. All seedling data from these studies were used, including data from biosynthetic and signaling mutants. From the Altmann experiments, 20-day-old wild-type and weak BR-deficient *dwf1* seedlings were exposed to 300 nM epi-BL or mock treatments (Mussig et al. 2002). In the Shimada experiments, in addition to wild-type seedlings, weak mutants from either BR signaling (*bri1–5*) or biosynthesis (*det2*) pathways were exposed to 10 nM BL or mock treatments (Goda et al. 2002). Also, seedlings were exposed to BRZ. The Chory group published two papers on using BL treatment. One included arrays representing BL and mock treatments of BL-insensitive mutants *bin3* and *bin5*, subunits of topoisomerase VI (Yin et al. 2002a). In a second paper, strong BR-insensitive *bri1-116* mutants and *bes1*-hypersensitive mutants were exposed to 1 μM BL or mock treatments (Yin et al. 2002b). Both papers also had wild-type seedlings exposed to both treatments. All data were quantile-normalized within experiment (Gentleman et al. 2004; rma library of bioconductor), and then ratios were taken between important contrasts (i.e., WT + BL/WT + mock; mutant + BL/mutant + mock; mutant + BL/WT + BL; mutant + mock/WT + mock). The resulting 30 ratios were then clustered on the basis of the correlated expression of the 282 genes from the high confidence list, represented on the earlier version of the microarray (Table S4, S5. Follow the Supplemental Material link from the Annual Reviews home page at **http://www.annualreviews.org**). Very clear clusters emerged, distinguishing up- and down-regulated genes, and clustering together experiments from different laboratories expected to have similar results (**Figure 5**). This analysis provides strong evidence that while BR genomic effects are undoubtedly affected by the various factors confounded in the lab effect, there are many genes with robustly detectable BR effects regardless of these factors.

One important result from this analysis is that more than 80% of consistently detected BR-regulated genes show estimated expression changes of less than twofold. All three original analyses used an arbitrary twofold cut-off in identifying differentially expressed genes. Determining whether such modest effects are biologically relevant will be a critical question for future studies of the BR response. Several alternative explanations have been proposed, including larger changes in a small subset of cells, highly responsive pathways, and the coupling of modest expression changes with large changes in protein stability or activity.

Biological Implications of BR-Regulated Gene Expression

Which pathways are clearly affected by BRs, as assayed by the genomic response? First, it should be stated that a large proportion of the genes identified by the analysis described above have no known function or only a vague hint without specific assignment to a biological process (e.g., DNA-binding domains). However, a few conclusions can be drawn with confidence. In support of decades of physiological data, BRs clearly initiate loosening of the cell wall and biogenesis of new cell wall material (Table S6. Follow the Supplemental Material link from the Annual Reviews home page at **http://www.annualreviews.org**). The strength of primary cell walls depends upon steel-like cables of cellulose microfibrils reinforced with cross-linking glycans (Reiter 2002). A gel-like pectin matrix surrounding this framework regulates porosity and other physiological properties. Structural proteins, such as the hydroxyproline-rich glycoprotein, extensin, and arabinogalactan proteins, contribute in largely undefined ways to cell wall architecture. One of the first genes identified as BR induced was *BRU1* in soybean, encoding a xyloglucan endotransglusylases/hydrolases (XTHs-formerly known as XETs) (Zurek & Clouse 1994). Consistent with their role in cell growth, many cell wall components and the enzymes that produce them are BR regulated, including extensins, arabinogalactans, and cellulose synthase subunits. Endo-glucanases and expansins are also up-regulated. Decreases in expression of several genes involved in cell division, including two cyclins, are also observed.

Interestingly, a number of genes involved in the production and secretion of very-long-chain fatty acids are also up-regulated following BR treatment (Table S6). This may reflect an increased requirement for waxy cuticle to cover rapidly elongating epidermal cells and could contribute to the biotic and abiotic stress protective effects of BR treatment (Krishna 2003). The cytoskeleton is also a target of BR regulation. In particular, two tubulin-encoding genes, *TUB1* and *TUB8*, are up-regulated by BRs (Table S6). Studies in the *bul1/dwf7-3* mutant suggest that one aspect of the dwarfing phenotype observed in BR mutants results from a defect in microtubule organization and concomitant loss of cellulose microfibrils (Catterou et al. 2001). BR treatment of the BR-deficient mutant induces correct orientation of cortical microtubules.

Connections with other hormones are plentiful, including components of both biosynthesis and signaling pathways (Table S7. Follow the Supplemental Material link from the Annual Reviews home page at **http://www.annualreviews.org**). A large number of genes previously identified as auxin-responsive has been noted by many groups, which reflects the close association of the BR and auxin genomic responses

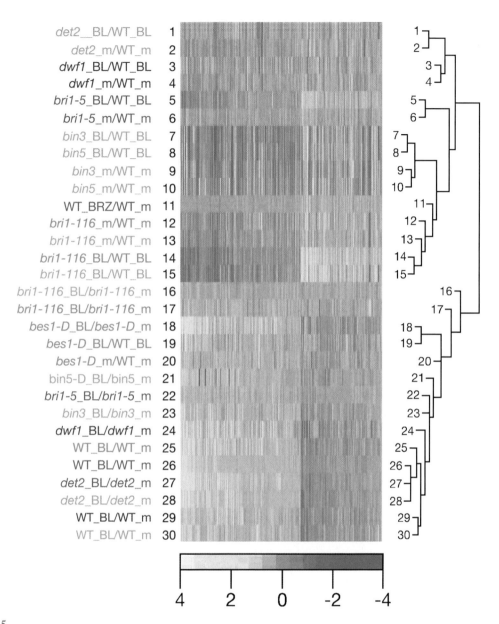

Figure 5

Re-analysis of microarray results. There are now many genes that behave consistently across BR microarray experiments. Log ratios were taken for relevant contrasts in four separate microarray experiments (e.g., WT treated with BL divided by WT mock-treated). These 30 ratios were clustered on the basis of correlated distance among the 282 genes identified as BR-responsive in our previous analysis (for details of analysis see Appendix A). A heat map is shown with each ratio represented as a row (described on the left) and each gene as a column. Columns represent up-regulated genes on the left and down-regulated genes on the right. The exact identity and order of these genes can be found in Tables S4 and S5. Experiments from Yin et al. (2002a) are shown in orange, from Mussig et al. (2002) in purple, from Goda et al. (2002) in blue, and from Yin et al. (2002b) in green. The results of the clustering analysis are shown at the right. Note that experiments from different groups are clustered together, suggesting that many genes behave reproducibly across laboratories.

(Goda et al. 2004; Nakamura et al. 2003a,b; Nemhauser et al. 2004). Several Aux/IAA transcriptional repressors are up-regulated by BRs and three ARF transcription factors are down-regulated. Multiple studies have demonstrated that both gene expression and growth effects of these two hormone pathways are interdependent (Bao et al. 2004; Nakamura et al. 2003a,b; Nemhauser et al. 2004). Genes involved in IAA homeostasis have also been found, and a number of genes involved in auxin transport are down-regulated, including members of the AUX1, PIN, and MDR families. Together, these effects might serve to reinforce local peaks in auxin concentration, perhaps as part of a canalization process. Effects of BRs on the ethylene biosynthesis enzymes ACC synthase (ACS) were observed many years ago in mung bean (*Vigna radiata*) (Yi et al. 1999). At least three ACS genes are up-regulated by BRs in our survey. Interestingly, one gene encoding an ACS was down-regulated by BRs. The *HOOKLESS1* (*HLS1*) gene, which encodes an N-acetyltransferase, is also down-regulated by BR treatment. HLS1 was recently shown to promote turnover of ARF2 protein in response to ethylene or light stimuli, perhaps providing another mechanism for regulation of the shared auxin:BR pathway (Li et al. 2004). Three type A response regulators, ARR3, ARR5, and ARR6, are down-regulated by BR treatment. These genes encode transcriptional repressors induced by cytokinin treatment and are thought to function as part of a negative feedback loop in that pathway (Suzuki et al. 2004).

A clear antagonistic relationship with the light response is also apparent in BR genomic responses (Table S8. Follow the Supplemental Material link from the Annual Reviews home page at **http://www.annualreviews.org**). Three photoreceptors, phototropin1 and phytochromes B and E, are down-regulated by BRs. Three other proteins connected with the light response, CIP7, DRT100, and an NPH3-like gene, are up-regulated by BRs, although their precise biological roles are not well established. Several papers have suggested that BR levels might be regulated by light, either through direct regulation of the DDWF1 BR biosynthetic enzyme or through BAS1-mediated hydroxylation/deactivation (Kang et al. 2001, Neff et al. 1999). Together, these findings suggest a complex web of interactions among both phytohormones and the light response modulating development and physiology.

A large number of transcription factors are regulated by BRs, including more than 10% of the BR down-regulated genes (Table S9. Follow the Supplemental Material link from the Annual Reviews home page at **http://www.annualreviews.org**). Among the 41 down-regulated genes predicted to encode transcription factors, 5 contain AP2 domains, 7 contain homeobox domains, and another 7 are predicted to contain Zn finger domains. Interestingly, several of the homeobox genes are expressed in vascular tissue, a developmental fate closely associated with BR response (Cano-Delgado et al. 2004). A major shift in transcriptional programs is likely to precede and support the significant changes in seedling morphology observed with changes in BR response.

From the Signal to Specific Target Gene Expression

Because BES1 and BZR1 share no significant homology to any known protein, the mechanism by which they control transcription was obscure until recently. DNA-binding activity and regulation of transcription were uncovered for both BZR1 and BES1.

A yeast two-hybrid approach using BES1 as a bait identified the bHLH transcription factor BIM1 (Yin et al. 2005). Gel shift experiments showed that both BIM1 and BES1 are able to bind CANNTG E-box motifs in the promoter of a *SAUR-AC1* BR-responsive gene and likely to form a heterodimer in a cooperative manner. Those E-box motifs are known binding sites for many bHLH transcription factors (Toledo-Ortiz et al. 2003) and are also overrepresented in the promoter of BR-induced genes (Nemhauser et al. 2004). BES1 binds DNA through its N terminus, which contains a highly basic domain as well as certain key residues of bHLHs predicted to form a helix-loop-helix type structure. BES1 appears to activate *SAUR-AC1* expression, although this remains to be directly shown using a heterologous system.

The actual involvement of BIM1 and its paralogs in the BR-signaling pathway comes from both gain- and loss-of-function mutants, although the mutant phenotypes obtained are weak. This suggests that other transcription factors likely act redundantly in the pathway or that the response mediated by BIMs affects only a subset of BR-responsive genes. The first candidates potentially acting in concert with BIMs are BEE1, BEE2, and BEE3, distantly related bHLHs previously shown to be positive regulators of the BR response (Friedrichsen et al. 2002). Importantly, direct binding of BES1 was shown only for the promoters of two SAUR-like genes and could not be detected on the promoter of a XET and BEE1, which are known to be BR responsive (Yin et al. 2005). This suggests that BES1 binding is not required on all BR-responsive promoters or that levels of BES1 protein on such promoters are quite low.

BZR1 was shown to act as a transcriptional repressor through direct binding to CGTG(T/C)G elements in the promoter of the BR biosynthetic gene *CPD*, a motif also conserved in the promoter of other biosynthetic genes such as *DWF4*, *ROT3*, and *BR6OX* (He et al. 2005). These observations strengthen the role of BZR1 in the control of BR homeostasis through the direct repression of BR feedback-regulated BR biosynthetic genes. In addition to its repressor role, a positive role of BZR1 on gene expression has been observed and is therefore likely dependent on its interaction with different partners that could switch BZR1 from a repressor to an activator depending on the context or, conversely, could simply be mediated by the repression of a BR-regulated transcriptional repressor.

These studies clearly argue for a direct role of BZR1 and BES1 in the repression of biosynthetic genes and promotion of BR responses, respectively, but do not completely explain the differences seen between *bes1* and *bzr1* gain-of-function phenotypes. Surprisingly, *DWF4* promoter activity, another target of the feedback regulation of BR biosynthesis by signaling, is down-regulated in both *bes1* and *bzr1* mutants (M. Lee, unpublished results). Consistent with this observation, the in-depth analysis done in the present study of previously published *bes1* microarray experiments (Yin et al. 2002b) indicates that several biosynthetic genes are down-regulated in *bes1*. This brings up the question about the opposite phenotype of *bes1* and *bzr1* in the light and gives rise to new questions at the molecular level. Does BES1 binding to the same promoter element that BZR1 is binding to directly repress *CPD* expression? Is BZR1 acting like BES1 to positively regulate BR response genes through E-box motifs? How do almost identical proteins act differently in the pathway? A comparative analysis of BES1 and BZR1 transcriptional activity should therefore be carried out using the same target promoters from both biosynthetic and other BR-regulated genes to determine their target specificity. Also, microarray analysis reveals that BZR1 transcripts are moderately induced by BR treatment and that two other family members, BEH1 and BEH2, show reduced transcript levels following BR treatment (Table S2, S3. Follow the Supplemental Material link from the Annual Reviews home page at **http://www.annualreviews.org**). This may reflect a more complex relationship among family members in promoting BR responses. A detailed analysis of spatial and temporal expression pattern of the entire family will also help clarify the apparent paradox of *bes1* and *bzr1* phenotypes.

CONCLUDING REMARKS

Despite significant progress in understanding the mechanisms of BR signaling, several fundamental questions remain unsolved. A major question is how the activity of BIN2 is regulated and whether this regulation involves BRI1/BAK1 directly. How BES1 and BZR1 and perhaps other family members coordinately regulate the large number of target genes is also unknown. In order to truly understand the role of BRs as developmental signals, we need to unravel the determinants of BR homeostasis: where and when BRs are synthesized and degraded, how they are transported out of the cell, and to what extent they are distributed in the plant. Finally, integration of BRs with other key signals, such as auxin and light, must be understood to gain further insight into the complexity of plant development.

SUMMARY POINTS

1. Brassinosteroids are perceived at the plasma membrane by direct binding to the extracellular domain of the BRI1 receptor. How ligand binding transduces the information across the membrane and activates BRI1, as well as the mechanism of receptor deactivation, is unknown.

2. BR-induced changes in gene expression are mainly achieved through the control of the phosphorylation state of the transcription factors, BES1 and BZR1.

3. The specific contribution of transcription factors in BR responses is emerging. Whereas the transcription factor BES1 is involved in the promotion of BR responses, BZR1 represses BR-biosynthetic genes.

4. The genomic response to BRs gives a good picture of their direct effects on growth and differentiation, which is correlated with physiological observations.

ACKNOWLEDGMENTS

We thank former and present members of the Chory laboratory for stimulating discussions, Dr. Stéphane Richard (SBL, Salk Institute) for reannotation of LRR repeats, and Drs. Carsten Mussig, Thomas Altmann, Hideki Goda, and Yukihisa Shimada for sharing microarray data. The work was supported by grants from the USDA and NSF to J.C.; by long-term fellowships from EMBO and HFSP to both G.V. and N.G.; and by grants from the NIH to J.L.N. J.C is an investigator of the Howard Hughes Medical Institute.

LITERATURE CITED

Aberle H, Bauer A, Stappert J, Kispert A, Kemler R. 1997. Beta-catenin is a target for the ubiquitin-proteasome pathway. *EMBO J.* 16:3797–804

Akira S, Takeda K. 2004. Toll-like receptor signalling. *Nat. Rev. Immunol.* 4:499–511

Allen GJ, Chu SP, Schumacher K, Shimazaki CT, Vafeados D, et al. 2000. Alteration of stimulus-specific guard cell calcium oscillations and stomatal closing in *Arabidopsis det3* mutant. *Science* 289:2338–42

Anuradha S, Rao S. 2001. Effect of brassinosteroids on salinity stress induced inhibition of seed germination and seedling growth of rice (*Oryza sativa* L.). *Plant Growth Regul.* 33:151–53

Bao F, Shen J, Brady SR, Muday GK, Asami T, Yang Z. 2004. Brassinosteroids interact with auxin to promote lateral root development in *Arabidopsis*. *Plant Physiol.* 134:1624–31

Breitling R, Armengaud P, Amtmann A, Herzyk P. 2004. Rank products: a simple, yet powerful, new method to detect differentially regulated genes in replicated microarray experiments. *FEBS Lett.* 573:83–92

Cano-Delgado A, Yin Y, Yu C, Vafeados D, Mora-Garcia S, et al. 2004. BRL1 and BRL3 are novel brassinosteroid receptors that function in vascular differentiation in *Arabidopsis*. *Development* 131:5341–51

Catterou M, Dubois F, Schaller H, Aubanelle L, Vilcot B, et al. 2001. Brassinosteroids, microtubules and cell elongation in *Arabidopsis thaliana*. II. Effects of brassinosteroids on microtubules and cell elongation in the *bul1* mutant. *Planta* 212:673–83

Choe S, Fujioka S, Noguchi T, Takatsuto S, Yoshida S, Feldmann KA. 2001. Overexpression of *DWARF4* in the brassinosteroid biosynthetic pathway results in increased vegetative growth and seed yield in *Arabidopsis*. *Plant J.* 26:573–82

Choe S, Schmitz RJ, Fujioka S, Takatsuto S, Lee MO, et al. 2002. *Arabidopsis* brassinosteroid-insensitive *dwarf12* mutants are semidominant and defective in a glycogen synthase kinase 3beta-like kinase. *Plant Physiol.* 130:1506–15

Clay NK, Nelson T. 2002. VH1, a provascular cell-specific receptor kinase that influences leaf cell patterns in *Arabidopsis*. *Plant Cell* 14:2707–22

Clouse SD. 2002. Brassinosteroid signal transduction: clarifying the pathway from ligand perception to gene expression. *Mol. Cell* 10:973–82

Clouse SD, Langford M, McMorris TC. 1996. A brassinosteroid-insensitive mutant in *Arabidopsis thaliana* exhibits multiple defects in growth and development. *Plant Physiol.* 111:671–78

Cross DA, Alessi DR, Cohen P, Andjelkovich M, Hemmings BA. 1995. Inhibition of glycogen synthase kinase-3 by insulin mediated by protein kinase B. *Nature* 378:785–89

Dornelas MC, Wittich P, von Recklinghausen I, van Lammeren A, Kreis M. 1999. Characterization of three novel members of the *Arabidopsis* SHAGGY-related protein kinase (ASK) multigene family. *Plant Mol. Biol.* 39:137–47

Ferguson KM, Berger MB, Mendrola JM, Cho HS, Leahy DJ, Lemmon MA. 2003. EGF activates its receptor by removing interactions that autoinhibit ectodomain dimerization. *Mol. Cell* 11:507–17

Friedrichsen DM, Joazeiro CA, Li J, Hunter T, Chory J. 2000. Brassinosteroid-insensitive-1 is a ubiquitously expressed leucine-rich repeat receptor serine/threonine kinase. *Plant Physiol.* 123:1247–56

Friedrichsen DM, Nemhauser J, Muramitsu T, Maloof JN, Alonso J, et al. 2002. Three redundant brassinosteroid early response genes encode putative bHLH transcription factors required for normal growth. *Genetics* 162:1445–56

Fujioka S, Yokota T. 2003. Biosynthesis and metabolism of brassinosteroids. *Annu. Rev. Plant Biol.* 54:137–64

Gadella TW Jr, Jovin TM. 1995. Oligomerization of epidermal growth factor receptors on A431 cells studied by time-resolved fluorescence imaging microscopy. A stereochemical model for tyrosine kinase receptor activation. *J. Cell Biol.* 129:1543–58

Garrett TP, McKern NM, Lou M, Elleman TC, Adams TE, et al. 2002. Crystal structure of a truncated epidermal growth factor receptor extracellular domain bound to transforming growth factor alpha. *Cell* 110:763–73

Gentleman R, Carey V, Bates D, Bolstad B, Dettling M, et al. 2004. Bioconductor: Open software development for computational biology and bioinformatics. *Genome Biol.* 5:R80

Gilboa L, Nohe A, Geissendorfer T, Sebald W, Henis YI, Knaus P. 2000. Bone morphogenetic protein receptor complexes on the surface of live cells: a new oligomerization mode for serine/threonine kinase receptors. *Mol. Biol. Cell* 11:1023–35

Goda H, Sawa S, Asami T, Fujioka S, Shimada Y, Yoshida S. 2004. Comprehensive comparison of auxin-regulated and brassinosteroid-regulated genes in *Arabidopsis*. *Plant Physiol.* 134:1555–73

Goda H, Shimada Y, Asami T, Fujioka S, Yoshida S. 2002. Microarray analysis of brassinosteroid-regulated genes in *Arabidopsis*. *Plant Physiol.* 130:1319–34

Hanks SK, Hunter T. 1995. Protein kinases 6. The eukaryotic protein kinase superfamily: kinase (catalytic) domain structure and classification. *FASEB J.* 9:576–96

He JX, Gendron JM, Sun Y, Gampala SS, Gendron N, et al. 2005. BZR1 is a transcriptional repressor with dual roles in brassinosteroid homeostasis and growth response. *Science* 307:1634–38

He JX, Gendron JM, Yang Y, Li J, Wang ZY. 2002. The GSK3-like kinase BIN2 phosphorylates and destabilizes BZR1, a positive regulator of the brassinosteroid signaling pathway in *Arabidopsis*. *Proc. Natl. Acad. Sci. USA* 99:10185–90

He Z, Wang ZY, Li J, Zhu Q, Lamb C, et al. 2000. Perception of brassinosteroids by the extracellular domain of the receptor kinase BRI1. *Science* 288:2360–63

Hecht V, Vielle-Calzada JP, Hartog MV, Schmidt ED, Boutilier K, et al. 2001. The *Arabidopsis SOMATIC EMBRYOGENESIS RECEPTOR KINASE* 1 gene is expressed in developing ovules and embryos and enhances embryogenic competence in culture. *Plant Physiol.* 127:803–16

Hirota S, Isozaki K, Moriyama Y, Hashimoto K, Nishida T, et al. 1998. Gain-of-function mutations of c-kit in human gastrointestinal stromal tumors. *Science* 279:577–80

Hubbard SR. 2004. Juxtamembrane autoinhibition in receptor tyrosine kinases. *Nat. Rev. Mol. Cell. Biol.* 5:464–71

Jiang G, Hunter T. 1999. Receptor signaling: when dimerization is not enough. *Curr. Biol.* 9R:568–71

Johnson LN, Noble ME, Owen DJ. 1996. Active and inactive protein kinases: structural basis for regulation. *Cell* 85:149–58

Jonak C, Hirt H. 2002. Glycogen synthase kinase 3/SHAGGY-like kinases in plants: an emerging family with novel functions. *Trends Plant Sci.* 7:457–61

Kang JG, Yun J, Kim DH, Chung KS, Fujioka S, et al. 2001. Light and brassinosteroid signals are integrated via a dark-induced small G protein in etiolated seedling growth. *Cell* 105:625–36

Kinoshita T, Cano-Delgado A, Seto H, Hiranuma S, Fujioka S, et al. 2005. Binding of brassinosteroids to the extracellular domain of plant receptor kinase BRI1. *Nature* 433:167–71

Klein PS, Melton DA. 1996. A molecular mechanism for the effect of lithium on development. *Proc. Natl. Acad. Sci. USA* 93:8455–59

Krishna P. 2003. Brassinosteroid-mediated stress responses. *J. Plant Growth Regul.* 22:289–97

Le Gallic L, Virgilio L, Cohen P, Biteau B, Mavrothalassitis G. 2004. ERF nuclear shuttling, a continuous monitor of Erk activity that links it to cell cycle progression. *Mol. Cell Biol.* 24:1206–18

Li H, Johnson P, Stepanova A, Alonso JM, Ecker JR. 2004. Convergence of signaling pathways in the control of differential cell growth in *Arabidopsis*. *Dev. Cell* 7:193–204

Li J, Biswas MG, Chao A, Russell DW, Chory J. 1997. Conservation of function between mammalian and plant steroid 5alpha-reductases. *Proc. Natl. Acad. Sci. USA* 94:3554–59

Li J, Chory J. 1997. A putative leucine-rich repeat receptor kinase involved in brassinosteroid signal transduction. *Cell* 90:929–38

Li J, Nam KH. 2002. Regulation of brassinosteroid signaling by a GSK3/SHAGGY-like kinase. *Science* 295:1299–301

This study shows that BZR1 is a transcriptional repressor of BR biosynthetic genes under feedback control by signaling.

This publication finally establishes that BRI1 directly binds BL and defines the island domain plus a neighboring downstream LRR as the BR binding domain.

Li J, Nam KH, Vafeados D, Chory J. 2001. BIN2, a new brassinosteroid-insensitive locus in *Arabidopsis*. *Plant Physiol.* 127:14–22

Li J, Wen J, Lease KA, Doke JT, Tax FE, Walker JC. 2002. BAK1, an *Arabidopsis* LRR receptor-like protein kinase, interacts with BRI1 and modulates brassinosteroid signaling. *Cell* 110:213–22

Liang SH, Clarke MF. 2001. Regulation of p53 localization. *Eur. J. Biochem.* 268:2779–83

Marg A, Shan Y, Meyer T, Meissner T, Brandenburg M, Vinkemeier U. 2004. Nucleocytoplasmic shuttling by nucleoporins Nup153 and Nup214 and CRM1-dependent nuclear export control the subcellular distribution of latent Stat1. *J. Cell Biol.* 165:823–33

Meggio F, Pinna LA. 2003. One-thousand-and-one substrates of protein kinase CK2? *FASEB J.* 17:349–68

Mora-Garcia S, Vert G, Yin Y, Cano-Delgado A, Cheong H, Chory J. 2004. Nuclear protein phosphatases with Kelch-repeat domains modulate the response to brassinosteroids in *Arabidopsis*. *Genes Dev.* 18:448–60

Moriki T, Maruyama H, Maruyama IN. 2001. Activation of preformed EGF receptor dimers by ligand-induced rotation of the transmembrane domain. *J. Mol. Biol.* 311:1011–26

Morillon R, Catterou M, Sangwan RS, Sangwan BS, Lassalles JP. 2001. Brassinolide may control aquaporin activities in *Arabidopsis thaliana*. *Planta* 212:199–204

Mussig C, Altmann T. 2003. Genomic brassinosteroid effects. *J. Plant Growth Regul.* 22:313–24

Mussig C, Fischer S, Altmann T. 2002. Brassinosteroid-regulated gene expression. *Plant Physiol.* 129:1241–51

Nakamura A, Higuchi K, Goda H, Fujiwara MT, Sawa S, et al. 2003a. Brassinolide induces IAA5, IAA19, and DR5, a synthetic auxin response element in *Arabidopsis*, implying a cross talk point of brassinosteroid and auxin signaling. *Plant Physiol.* 133:1843–53

Nakamura A, Shimada Y, Goda H, Fujiwara MT, Asami T, Yoshida S. 2003b. AXR1 is involved in BR-mediated elongation and *SAUR-AC1* gene expression in *Arabidopsis*. *FEBS Lett.* 553:28–32

Nam KH, Li J. 2002. BRI1/BAK1, a receptor kinase pair mediating brassinosteroid signaling. *Cell* 110:203–12

Nam KH, Li J. 2004. The *Arabidopsis* transthyretin-like protein is a potential substrate of BRASSINOSTEROID-INSENSITIVE 1. *Plant Cell* 16:2406–17

Neff MM, Nguyen SM, Malancharuvil EJ, Fujioka S, Noguchi T, et al. 1999. *BAS1*: a gene regulating brassinosteroid levels and light responsiveness in *Arabidopsis*. *Proc. Natl. Acad. Sci. USA* 96:15316–23

Nemhauser JL, Mockler TC, Chory J. 2004. Interdependency of brassinosteroid and auxin signaling in *Arabidopsis*. *PLoS Biol.* 2:E258

Nicolas FJ, De Bosscher K, Schmierer B, Hill CS. 2004. Analysis of Smad nucleocytoplasmic shuttling in living cells. *J. Cell Sci.* 117:4113–25

Noguchi T, Fujioka S, Choe S, Takatsuto S, Yoshida S, et al. 1999. Brassinosteroid-insensitive dwarf mutants of *Arabidopsis* accumulate brassinosteroids. *Plant Physiol.* 121:743–52

Ogiso H, Ishitani R, Nureki O, Fukai S, Yamanaka M, et al. 2002. Crystal structure of the complex of human epidermal growth factor and receptor extracellular domains. *Cell* 110:775–87

Oh MH, Ray WK, Huber SC, Asara JM, Gage DA, Clouse SD. 2000. Recombinant brassinosteroid insensitive 1 receptor-like kinase autophosphorylates on serine and threonine residues and phosphorylates a conserved peptide motif in vitro. *Plant Physiol.* 124:751–66

Peng P, Li J. 2003. Brassinosteroid signal transduction: a mix of conservation and novelty. *J. Plant Growth Regul.* 22:298–312

Perez-Perez JM, Ponce MR, Micol JL. 2002. The *UCU1 Arabidopsis* gene encodes a SHAGGY/GSK3-like kinase required for cell expansion along the proximodistal axis. *Dev. Biol.* 242:161–73

Piao HL, Pih KT, Lim JH, Kang SG, Jin JB, et al. 1999. An *Arabidopsis* GSK3/shaggy-like gene that complements yeast salt stress-sensitive mutants is induced by NaCl and abscisic acid. *Plant Physiol.* 119:1527–34

Pranada AL, Metz S, Herrmann A, Heinrich PC, Muller-Newen G. 2004. Real time analysis of STAT3 nucleocytoplasmic shuttling. *J. Biol. Chem.* 279:15114–23

Reiter WD. 2002. Biosynthesis and properties of the plant cell wall. *Curr. Opin. Plant Biol.* 5:536–42

Rudenko G, Henry L, Henderson K, Ichtchenko K, Brown MS, et al. 2002. Structure of the LDL receptor extracellular domain at endosomal pH. *Science* 298:2353–58

Russinova E, Borst JW, Kwaaitaal M, Cano-Delgado A, Yin Y, et al. 2004. Heterodimerization and endocytosis of *Arabidopsis* brassinosteroid receptors BRI1 and AtSERK3 (BAK1). *Plant Cell* 16:3216–29

Schlessinger J. 2000. Cell signaling by receptor tyrosine kinases. *Cell* 103:211–25

Schumacher K, Vafeados D, McCarthy M, Sze H, Wilkins T, Chory J. 1999. The *Arabidopsis det3* mutant reveals a central role for the vacuolar H$^+$-ATPase in plant growth and development. *Genes Dev.* 13:3259–70

Shiu SH, Bleecker AB. 2001. Receptor-like kinases from *Arabidopsis* form a monophyletic gene family related to animal receptor kinases. *Proc. Natl. Acad. Sci. USA* 98:10763–68

Suzuki M, Kamide Y, Nagata N, Seki H, Ohyama K, et al. 2004. Loss of function of 3-hydroxy-3-methylglutaryl coenzyme A reductase 1 (HMG1) in *Arabidopsis* leads to dwarfing, early senescence and male sterility, and reduced sterol levels. *Plant J.* 37:750–61

Tavares R, Vidal J, van Lammeren A, Kreis M. 2002. AtSKtheta, a plant homologue of SGG/GSK-3 marks developing tissues in *Arabidopsis thaliana*. *Plant Mol. Biol.* 50:261–71

Toledo-Ortiz G, Huq E, Quail PH. 2003. The *Arabidopsis* basic/helix-loop-helix transcription factor family. *Plant Cell* 15:1749–70

Wang X, Li X, Meisenhelder J, Hunter T, Yoshida S, et al. 2005. Autoregulation and homodimerization are involved in the activation of the plant steroid receptor BRI1. *Dev. Cell* 8:855–65

Wang ZY, He JX. 2004. Brassinosteroid signal transduction–choices of signals and receptors. *Trends Plant Sci.* 9:91–96

Wang ZY, Nakano T, Gendron J, He J, Chen M, et al. 2002. Nuclear-localized BZR1 mediates brassinosteroid-induced growth and feedback suppression of brassinosteroid biosynthesis. *Dev. Cell* 2:505–13

Wang ZY, Seto H, Fujioka S, Yoshida S, Chory J. 2001. BRI1 is a critical component of a plasma-membrane receptor for plant steroids. *Nature* 410:380–83

Wiesmann C, de Vos AM. 1999. Putting two and two together: crystal structure of the FGF-receptor complex. *Structure Fold. Des.* 7:R251–55

Yi HC, Joo S, Nam KH, Lee JS, Kang BG, Kim WT. 1999. Auxin and brassinosteroid differentially regulate the expression of three members of the 1-aminocyclopropane-1-carboxylate synthase gene family in mung bean (*Vigna radiata* L.). *Plant Mol. Biol.* 41:443–54

Yin Y, Cheong H, Friedrichsen D, Zhao Y, Hu J, et al. 2002a. A crucial role for the putative *Arabidopsis* topoisomerase VI in plant growth and development. *Proc. Natl. Acad. Sci. USA* 99:10191–96

Yin Y, Vafeados D, Tao Y, Yoshida S, Asami T, Chory J. 2005. A new class of transcription factors mediates brassinosteroid-regulated gene expression in *Arabidopsis*. *Cell* 120:249–59

BES1 acts in a cooperative manner with a bHLH transcription factor, BIM1, to induce the expression of a subset of BR-regulated genes.

Yin Y, Wang ZY, Mora-Garcia S, Li J, Yoshida S, et al. 2002b. BES1 accumulates in the nucleus in response to brassinosteroids to regulate gene expression and promote stem elongation. *Cell* 109:181–91

Yin Y, Wu D, Chory J. 2002c. Plant receptor kinases: systemin receptor identified. *Proc. Natl. Acad. Sci. USA* 99:9090–92

Zhao J, Peng P, Schmitz RJ, Decker AD, Tax FE, Li J. 2002. Two putative BIN2 substrates are nuclear components of brassinosteroid signaling. *Plant Physiol.* 130:1221–29

Zhao Y, Dai X, Blackwell HE, Schreiber SL, Chory J. 2003. SIR1, an upstream component in auxin signaling identified by chemical genetics. *Science* 301:1107–10

Zhou A, Wang H, Walker JC, Li J. 2004. BRL1, a leucine-rich repeat receptor-like protein kinase, is functionally redundant with BRI1 in regulating *Arabidopsis* brassinosteroid signaling. *Plant J.* 40:399–409

Zurek DM, Clouse SD. 1994. Molecular cloning and characterization of a brassinosteroid-regulated gene from elongating soybean (*Glycine max* L.) epicotyls. *Plant Physiol.* 104:161–70

Zurek DM, Rayle DL, McMorris TC, Clouse SD. 1994. Investigation of gene expression, growth kinetics, and wall extensibility during brassinosteroid-regulated stem elongation. *Plant Physiol.* 104:505–13

Anisotropic Expansion of the Plant Cell Wall

Tobias I. Baskin

Biology Department, University of Massachusetts, Amherst, Massachusetts 01003;
email: baskin@bio.umass.edu

Annu. Rev. Cell Dev. Biol.
2005. 21:203–22

The *Annual Review of
Cell and Developmental
Biology* is online at
http://cellbio.annualreviews.org

doi: 10.1146/
annurev.cellbio.20.082503.103053

Copyright © 2005 by
Annual Reviews. All rights
reserved

1081-0706/05/1110-
0203$20.00

Key Words

cellulose microfibrils, cortical microtubules, morphogenesis,
elongation, radial expansion

Abstract

Plants shape their organs with a precision demanded by optimal
function; organ shaping requires control over cell wall expansion
anisotropy. Focusing on multicellular organs, I survey the occurrence
of expansion anisotropy and discuss its causes and proposed controls.
Expansion anisotropy of a unit area of cell wall is characterized by
the direction and degree of anisotropy. The direction of maximal
expansion rate is usually regulated by the direction of net alignment
among cellulose microfibrils, which overcomes the prevailing stress
anisotropy. In some stems, the directionality of expansion of epidermal cells is controlled by that of the inner tissue. The degree of
anisotropy can vary widely as a function of position and of treatment. The degree of anisotropy is probably controlled by factors in
addition to the direction of microfibril alignment. I hypothesize that
rates of expansion in maximal and minimal directions are regulated
by distinct molecular mechanisms that regulate interactions between
matrix and microfibrils.

Contents

INTRODUCTION

Plants shape their organs with a precision demanded by optimal function, from the thin, flat solar panels of leaves to the coiled grappling hooks of tendrils. Thompson (1917) realized that adaptive advantage is insufficient to explain form; he argued that additionally the process of construction plays a role. The only construction process used to shape plant organs is expansion of cell walls. The cell wall surrounds neighboring cells in a continuous sheet and stretching it requires a large force. These features preclude plants from driving morphogenesis with cell migration or motility, processes used routinely by animals to build organs. The cell wall also seems to preclude plants from using programmed cell death in morphogenesis, as used, for example, to shape the human hand, because when a plant cell dies, programmatically or otherwise, the cell wall remains. The shape of a plant organ thus reflects the history of the expansion undergone by its cell walls.

When an area of cell wall expands at the same rate in all directions, expansion rate is isotropic, whereas when the rate in one direction differs from the rate in another, expansion rate is anisotropic. Integration of all the local expansion behavior throughout the growing regions gives the organ its shape and for this reason understanding the anisotropic expansion of the cell wall is pivotal for understanding plant development.

I review the patterns of anisotropic expansion that have been documented in plants and discuss the causes and controls of such expansion. My review focuses on multicellular organs. To my knowledge, anisotropic expansion per se has not been reviewed, but certain aspects are treated by Green (1980) and Taiz (1984). Cosgrove (1999) provides an authoritative treatment of cell wall yielding. Readers interested in understanding how anisotropic growth fits into the overall problem of morphogenesis will enjoy reading the paper by Coen and colleagues (2004). Finally, Harold (1990, 2002) masterfully explores shape generation in single cells of all kingdoms.

Defining Terms

Two-dimensional expansion of a unit area of cell wall. This review focuses on a unit area of cell wall and its expansion. Cell walls are thin, and changes in thickness do not contribute directly to changes in cell or organ size; therefore, I treat cell wall expansion as a two-dimensional problem. Even though the third dimension (i.e., thickness) is of undoubted relevance to the behavior of the wall (e.g., Dumais et al. 2004), this simplification is unavoidable because there are almost no data from which this dimension can be assessed. Except where noted, I refer to expansion in the maximal direction as elongation and in the minimal direction as radial.

Strain rate. This review focuses on anisotropy of the rates of expansion for a unit area of cell wall, and not on the anisotropy of shape. These expansion rates are best treated relatively because cell walls expand throughout their area, making the absolute amount of expansion proportional to area. Engineers refer to relative expansion rates as strain rates; botanists often refer to them as relative elemental expansion rates. For brevity, I use engineering nomenclature.

Direction and degree of anisotropy. For any material that expands anisotropically, fundamental laws of mechanics dictate that the direction in which the maximal rate occurs is perpendicular to the direction of the minimal rate (**Figure 1**). Therefore, anisotropy is characterized by two parameters: direction and degree. Direction specifies the direction in which the maximal strain rate occurs, and degree specifies the relationship between the maximal and minimal strain rates. A widespread convention for showing anisotropy diagrammatically is the so-called strain ellipse in which the major and minor axes represent the magnitudes of maximal and minimal expansion rate, respectively.

The orientation of the ellipse shows the direction and the ellipticity shows the degree. The greater the difference between maximal and minimal strain rates, the closer the ellipse is to a line and the greater the degree of anisotropy. Mathematically, there are various ways to represent the degree of anisotropy; the simplest of these is the ratio of maximal to minimal rates, as used here. There are more complex representations; for example, one may calculate the eccentricity of the ellipse or use the difference between maximal and minimal rates divided by the sum.

Historical Foundation

The foundation for the modern understanding of anisotropic expansion was built on studies of the Brobdingnagian internodal cells of *Nitella*. Three influential results emerged from these studies. First, the degree of growth anisotropy is constant; the rate of elongation is invariably four to five times greater than the rate of radial expansion (Probine & Preston 1961, Green 1965). Second, the anisotropic expansion of the cell wall originates from its anisotropic mechanical construction, specifically in the deposition of aligned cellulose microfibrils (Probine & Preston 1961, 1962). Third, the ability of a cell to expand anisotropically requires an array of microtubules called cortical microtubules, which are beneath the plasma membrane and thought to determine the deposition direction of the microfibrils

Anisotropy: the state where a property differs as a function of direction, in contrast to isotropy where the property is the same in all directions

Strain rate: Strain is a relative deformation, typically defined as the ratio of final size to initial size; strain rate is the temporal rate change of this quantity.

Microfibril: the most pronounced structural unit of the cell wall; formed by the lateral noncovalent association of many $1 \rightarrow 4$ ß-linked glucose chains.

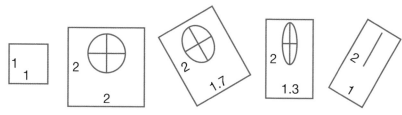

Figure 1

A unit area of cell wall (*left-most box*) expands, doubling in length (*right-hand boxes*). The numbers denote the length and width of the cell wall area, and the ellipses represent the anisotropy, which can vary in direction and degree. The direction of anisotropy is defined by the angle formed by the major ellipse axis and a given reference and the degree of anisotropy is defined conveniently by the length ratio between the major and minor axes of the ellipse.

(Green 1962). These three results permeate all subsequent experiments but, as discussed below, need modification as applied to multicellular organs.

QUANTIFICATION AND OCCURRENCE OF EXPANSION ANISOTROPY

Before considering the causes and controls of anisotropic expansion in multicellular organs, one needs to know what the patterns of anisotropic cellular expansion actually are. I describe these patterns next, starting with a description of how to measure them.

Measuring Expansion Anisotropy

Theory. In essence, plant growth is the movement of water, and to a first approximation, is described appropriately via the formalism of continuum mechanics (Silk & Erickson 1979, Silk 1992). In a moving continuum, each element has a velocity, which is a vector quantity with a magnitude and direction. For a moving body, the set of velocities of all elements fully characterizes the movement, including growth. If the velocities among a group of elements are all the same, then that group is moving but not otherwise changing. But if the velocities diverge, then new material has to be added, and if the velocities converge, then material has to be removed; otherwise the continuum would fail. Diverging velocities mean growth, whereas converging velocities mean shrinkage. Shrinkage is rare in plants and is not considered further in this review.

Quantitatively, the divergence of the velocity field represents the strain rates within the material. The problem of finding those strain rates amounts to quantifying the rates at which elements within a material move and then differentiating along a given reference to obtain the divergence. One may use either a spatial or a temporal reference (Silk 1984). A spatial reference, sometimes called a Eulerian reference, characterizes the behavior of

the movement for a set of spatial coordinates; in contrast, a temporal reference, sometimes called a material or Lagrangian reference, follows one element as it moves over time. The difference can be appreciated from a waterfall: Euler would measure the velocity at which water droplets are moving at various places in the fall, whereas Lagrange would choose a drop at the top and measure its velocity at different times as it fell.

Two complexities arise: time-dependent behavior and organ geometry. Only if the pattern of movement is constant in time is there a straightforward mapping between time and position. Movements due to growth are three dimensional but methods for observing growth, such as photography, are inherently two dimensional. Furthermore, growth within a volume can be spatially heterogeneous, but usually only the surface is accessible. Thus the velocities at which elements inside the organ move are obscure. Both of these problems are illustrated by a leaf: A photographic record of the lamina over time could provide information about the velocity vectors in the plane of the leaf but not about those in the perpendicular plane. But even were an accompanying map of changes in leaf thickness to be obtained, it would not be possible to know whether the observed increase in thickness occurred uniformly among leaf tissues or reflected instead a thickness increase in only a single tissue (e.g., palisade mesophyll). To assess how the mechanical properties of the cell wall regulate expansion anisotropy, the measured rates of expansion must be related to specific cell walls (Liang et al. 1997). In the case of a leaf, measurements of the lamina could be coupled to structural data for the outer epidermal wall, but leaf thickness measurements could not be coupled to data for the anticlinal walls unless the spatial distribution of expansion in thickness were resolved.

Practice. Embodying the essence of most later approaches, often termed kinematic, a method for handling one-dimensional velocity fields was worked out in the 1950s for the plant root (Silk 1992). The root (or other organ whose expansion in one direction is of interest) is photographed as it grows, and from the photographs, the trajectories of marks on the surface are plotted and used to calculate velocities. The marks may be exogenous, such as ink or beads, or endogenous, such as cross walls. The plot of velocity versus distance from a suitable reference, for example the root tip, is differentiated, and this derivative plot gives the strain rate as a function of distance from the reference point. This approach can be extended to two dimensions although the mathematics becomes more complicated (Silk 1984). Recently, researchers have developed software that can recover velocity fields from image sequences algorithmically without requiring marking (Schmundt et al. 1998, van der Weele et al. 2003), greatly increasing measurement resolution and processing speed compared to manual methods.

An alternative approach for characterizing the velocity field was developed by Hejnowicz & Romberger (1984). The approach, called the growth tensor, is based on the strain rate tensor, which Silk & Erickson (1979), using the tools of continuum mechanics, introduced in their treatment of plant growth. A line can move through a growing continuum and retain its orientation only if it is parallel to one of the three principal directions of expansion rate. In a median longitudinal section of a growing plant organ, many such lines are present in the form of periclinal cell walls forming cell files, maintained through the growth zone. Starting with the assumption that these periclinal cell wall lines are parallel to one of the principal growth directions (likewise, the associated cross walls are parallel to another of the principal directions), researchers derived a complete velocity field for root apices (Hejnowicz 1989). Although these calculated velocity fields reproduce the main features of apices qualitatively, the fit has not been assessed quantitatively, nor is it clear that the solution is unique, meaning that there may be a family of velocity fields that can give rise to the observed cell files. Perhaps the growth-tensor approach could be enhanced if it were constrained by direct velocity data for the surface.

In the growth-tensor approach, cell boundaries are used to derive an overarching function for the entire growing object, but cell boundaries can also be used to infer the local field (Hejnowicz & Brodzki 1960, Silk et al. 1989). For example, a cell in the elongation zone of a root moves away from the tip at a speed proportional to its length divided by its length at maturity, provided that elongation rate is constant over time and there is no cell division in the elongation zone. The proportionality constant is the total root elongation rate. This approach is ideal for situations where the organ cannot be photographed over time, such as when a root is growing in compact soil, but suffers from the facts that cell lengths are extremely variable and that cell division or time-dependent changes invalidate the results.

Because cell boundaries are customarily viewed in sections, they are difficult to use for organs with complex geometry, which are also nearly impossible to photograph. However, cell boundaries can be revealed when a cell undergoes a mutation, such as in a pigment biosynthetic pathway, that alters the cell's appearance compared with that of its neighbors. If the cell is dividing, then its progeny are likewise marked. Groups of marked cells are called clones; these have been used for many years in studies of fate determination. The shape of a clone carries information about the directional growth experienced by that boundary. Recently Rolland-Lagan et al. (2005) developed a painstaking method to extract that information: By following the shapes of clones induced at different times, these authors constructed the growth history of the organ. Although the method yielded measurements with relatively low precision, it provided the first assessment of local expansion anisotropy for an inaccessible organ (Rolland-Lagan et al. 2003).

Tip-Growing Cells

Stress: the force on a body divided by the cross-sectional area across which the force acts.

Pollen tubes, root hairs, and fungal hyphae, as well as apical cells in some filamentous algae and lower plants, grow by tip growth. In this mode, expansion is confined to one end of the cell, the tip, in contrast with diffuse growth in which growth occurs throughout one or more faces of the cell. For a single tip-growing cell, the stress distribution can be calculated from geometric considerations (Hejnowicz et al. 1977). Unfortunately, because shape and growth rate change steeply as a function of position within the tip, observations are prone to measurement error.

Over the past half century, anisotropic expansion rates have been quantified for a small number of tip-growing cells, including the young sporangiophore of the fungus, *Phycomyces blakesleeanus* (Castle 1958); the shoot and leaf apical cells (Green 1965) and rhizoid (Chen 1973) of the characean green alga, *Nitella*, as well as the rhizoid of *Chara* (Hejnowicz et al. 1977); the shoot apical cell of the xanthophycean green alga, *Vaucheria geminata* (Kataoka 1982); and the root hair of the angiosperm, alfalfa (Dumais et al. 2004). Despite this taxonomic diversity and the somewhat different approaches used for calculations, a key regularity emerges: Near the apex, expansion is isotropic, but within the basal part of the growth zone, expansion rate in circumference is greater than the longitudinal rate.

Taking advantage of the well-understood mechanical behavior of a thin-walled, pressurized shell with rotational symmetry, Dumais and colleagues (2004) calculated the distribution of longitudinal and circumferential stresses as a function of position in the dome. Then, they modeled how those anisotropic stresses would deform the cell wall under three alternative cell wall reinforcements: complete isotropy (uniform material in three dimensions), transverse isotropy (a wall made in layers but with each layer isotropic), and full anisotropy (e.g., a wall with aligned microfibrils). A wall with full anisotropy matches the observed expansion anisotropy, but so does a wall with transverse anisotropy, which indicates that directional wall reinforcement is not required to generate anisotropic expansion patterns in a tip-growing cell; instead, the anisotropic distribution of stresses suffices. Consistently, cell wall layers at the tips of tip-growing cells appear to be transversely isotropic when analyzed structurally (references in Dumais et al. 2004). Thus, tip growth strikingly contrasts diffuse growth, for which the degree of anisotropy is usually high, the cell wall layers are anisotropic structurally, and the maximal stress is probably parallel to the minimal expansion rate.

Multicellular Cylindrical Organs: Stems and Roots

Cylindrical organs, such as stems and roots, are figures of revolution and thus share the rotational symmetry of single cells. I note in passing that whereas roots generally are circular in cross section, stems frequently have more complicated shapes, indicating that expansion around the stem circumference is not uniform. To my knowledge, the only attempt to consider this has been a theoretical derivation of growth fields required to sustain the helical form of twining vines (Silk 1989).

Only a handful of papers have measured, at an elemental level, expansion anisotropy of stems and roots, despite their relatively straightforward geometry. Silk & Abou Haidar (1986) measured longitudinal and circumferential strain for the stem of morning glory (*Pharbitis nil*). Strain rates in each direction are roughly constant through the growth zone, indicating that the degree of anisotropy is constant and the longitudinal rate exceeds the tangential rate by approximately a factor of two. The long growth zone (more than 10 cm), small strain rates ($\sim$2% h^{-1}), and twining habit all hindered the analysis; finer scale patterns may have been missed. On the other hand, Cavalieri & Boyer (1982) found that when the dark-grown soybean hypocotyl acclimates to water deficit, elongation and radial expansion vary independently throughout the growth zone, although the authors did not calculate strain rates in this study.

The sedate strain rates of the morning glory stem contrast with those of the root, where maximal longitudinal strain rates can reach 50% h^{-1}. Data from roots confirm that longitudinal and radial strain rates can vary independently. This can be inferred from cell-length data for the maize root swelling under the influence of a microtubule inhibitor (Bystrova 1984), although rates were not quantified explicitly. Likewise, tomato roots, both wild type and gibberellin deficient, have bell-shaped spatial profiles of elongation and circumferential strain rates that peak at different locations (Barlow et al. 1991).

Expansion rates in length and width have been thoroughly quantified for maize roots growing at low water potential (Liang et al. 1997). The roots of many species, when exposed to water deficit in a substrate that avoids compaction, acclimate by thinning. In maize roots, circumferential strain in the stele and cortex and radial strain in the cortex vary independently of longitudinal strain. In particular, water stress decreases longitudinal strain in the basal half of the growth zone while leaving radial and circumferential strain rates unaffected, but in the apical half, water stress inhibits radial and circumferential strain rates without affecting the longitudinal rate. The degree of anisotropy varies from 3 to more than 30. Similarly, longitudinal and circumferential strain rates vary independently in arabidopsis roots exposed to low concentrations of the microtubule inhibitor oryzalin (Baskin et al. 2004). These roots become thicker, and because the oryzalin concentration is low enough to permit cell division to continue, they grow at approximately steady state.

In sum, whereas the morning glory stem grows with a constant degree of anisotropy in accord with the classic picture from *Nitella*, the soybean hypocotyl, as well as the maize and arabidopsis roots, grow with a variable degree of expansion anisotropy. This suggests that distinct mechanisms exist to control expansion rates in different directions.

Multicellular Laminar Organs: Leaves and Petals

Although leaves are found in many shapes and sizes, there have been few attempts to quantify the anisotropy of local expansion rates. In contrast, leaf shape has long been characterized allometrically (Tsukaya 2003), with logarithmic plots of length and width during development often remaining constant (e.g., Haber & Foard 1963). For *Nitella* internodal cells, the linearity of logarithmic plots of cell length versus width allows one to conclude that the cell wall expands with a constant degree of anisotropy, but a log-linear relation for the length and width of leaves allows no similar conclusion. This is because the leaf has thousands of growing cells in each dimension and the allometric relation pertains to the integrated output of them all. Allometric constancy requires only that the total growth in each direction stay steady in time (or change in both directions proportionally) and does not preclude cell walls at different leaf regions from expanding with various degrees of anisotropy. Indeed, anisotropic leaf shape can arise from purely isotropic local expansion rates when the magnitude of area expansion rate differs in different regions of the leaf (Green 1965).

Perhaps the earliest quantification of expansion anisotropy of any material was made for the tobacco leaf (Richards & Kavanagh 1943); later, similar measurements were made for cocklebur (*Xanthium pensylvanicum*) (Erickson 1966). Both data sets show that expansion in the laminar plane of these broad-leaf plants is essentially isotropic. There are small changes in the degree of anisotropy, but these may reflect measurement errors, although a few marginal regions appear to have significant expansion anisotropy. These photographic records were made rather late in development when the leaf could be reasonably flat for photography, so earlier patterns are unknown. To my knowledge, spatial profiles of thickness have never been mapped for the dicot leaf. Plots of average thickness over time for the cocklebur leaf suggest that growth in thickness follows a trajectory that is independent of growth in area (Maksymowych 1990), but spatially resolved measurements are required before conclusions can be extended to the level of cell wall expansion anisotropy.

Despite major interest in the growth of grass leaves, there is apparently only one published characterization of local expansion anisotropy. Maurice et al. (1997) combined spatial and material methods to quantify the spatial profiles of strain in length, width, and thickness for tall fescue (*Festuca arundinacea*). Leaves continue to thicken for several centimeters past the terminus of the elongation zone; this behavior may relate to the differentiation of epidermal cell types or to de novo cell thickening (Macadam & Nelson 2002). In the elongation zone, longitudinal strain has a bell-shaped profile, typical of grass leaves, which remains steady over time with a peak ($6\%\ h^{-1}$) about one third of the way through the growth zone. In contrast, strain rates in width and thickness are maximal ($6\%\ h^{-1}$) at the very base of the growth zone, decrease steeply with position, and change over time. Although strain rates in width and thickness were obtained for the entire leaf width or thickness and hence do not apply directly at the cellular scale, it is clear that cell walls in the grass leaf expand with considerable anisotropy, where the degree and even sign change during development.

In the papers above on leaves, researchers used material that is far past the primordium stage because of the requirements for marking and of photography. But many of the growth transformations required to bring about the complex forms of plant organs presumably happen soon after the primordium bulges up and away from the shoot meristem. Recently, Rolland-Lagan et al. (2003, 2005) used clonal analysis to analyze the anisotropic expansion in the laminar plane of the snapdragon flower's dorsal petal. Like the dicot leaf, expansion in the petal lamina is only slightly anisotropic and variations in the degree of anisotropy in different regions of the petal were not large enough to be resolved. Interestingly, the direction of maximal expansion rate did change in a programmatic way, which the authors argued is essential for the development of the petal's shape.

Multicellular Organs: Shoot Apical Meristem

Attempts to characterize the expansion of the shoot apical meristem (Kwiatkowska 2004) either have been highly theoretical (e.g., Nakielski 1987, Hejnowicz et al. 1988) or have failed to account for the curvature of the surface, which limits their precision (Hernández et al. 1991). In a tour de force, Kwiatkowska & Dumais (2003) quantified the expansion behavior of the surface of the shoot apical meristem (including early primordia) of *Anagallis arvensis*. This was accomplished by viewing successive replicas of the same meristem in the scanning electron microscope at two angles and then calculating the elevations of points on the surface (Dumais & Kwiatkowska 2001). As may be expected for a meristem initiating leaves with spiral phyllotaxis, expansion patterns are complex. Cell walls at the center of the dome expand more or less isotropically as do cells at the distal tips of primordia, but cell walls at other locations can expand with considerable anisotropy, especially walls at the meristem flank or those forming an axil. These data for the meristem further confirm that plants readily modify both the direction and the degree of cell wall expansion anisotropy.

CAUSES AND CONTROLS OF ANISOTROPIC EXPANSION RATES

A unit area of material may expand anisotropically because of two (not exclusive) causes: It may be acted on by forces whose distribution is anisotropic, or its ability to resist force may be anisotropic (Harold 2002).

Causes of Growth Anisotropy: Force and Resistance

Force. The forces acting on the plant cell wall originate from hydrostatic pressure, which is strictly isotropic. The force must act over a cross section of cell wall, thus creating a stress. The shape of the cell impinges on the distribution of stresses so that, for example in a single cylindrical cell, the stress acting circumferentially is twice that acting longitudinally (Probine & Preston 1961). Given that single cylindrical cells expand faster longitudinally than radially, the resistance of the cell wall to stress must be anisotropic.

However, even for a single cell with an anisotropic cell wall, the distribution of stresses and its relation to wall reinforcement is complex and can give rise to unexpected behavior, such as helical twisting (Sellen 1983). For a cell in a tissue, and especially for growing tissues, the theoretical problems are formidable (Bruce 2003). A major complexity arises for shoots from the thick inextensible epidermal layer, which sheathes more compliant inner tissue (Hejnowicz & Sievers 1996a, Nicklas & Paolillo 1998). The epidermis and inner layers are connected so that forces generated in one may be borne by the other. This force-sharing is called tissue tension (Hejnowicz & Sievers 1995a, Passioura & Boyer 2003).

In the seedling stems of sunflower and tulip, which are subject to tissue tensions, the longitudinal stress was reported to be larger than the transverse stress by threefold for the inner tissue and by sixfold for the epidermis (Hejnowicz & Sievers 1995b, Hejnowicz et al. 2000). The direction of this anisotropy is opposite to the classic twofold excess of transverse stress for single, cylindrical cells and implies that anisotropic expansion rates in a stem could occur without anisotropic reinforcement of the cell wall. This surprising stress anisotropy was discovered by means of measuring the mechanical properties of mechanically isolated inner and outer tissue layers; excising tissue alters hydraulic relations and may confound analysis (Peters & Tomos 2000). Furthermore, when stems are treated so as to remove or reduce the anisotropic mechanical reinforcement of the cell wall, they often expand faster radially than longitudinally. For example, in maize coleoptiles, colchicine treatment converts a sixfold strain rate anisotropy favoring length to a fourfold anisotropy favoring girth (Schopfer 2000), a result hardly in accord with a longitudinally dominant stress anisotropy and underscoring the importance of mechanical resistance of the cell walls.

An unexpected stress anisotropy has also been reported for the flattened dome of the sunflower capitulum, only in this case the model agreed with the qualitative stress distribution obtained experimentally from the gaping of cuts made on the surface (Dumais & Steele 2000). It would be worthwhile to obtain directly the stress distribution for growing stems.

Resistance. For multicellular organs, the mechanical anisotropy of the cell wall, as compared with the anisotropy of stress, is well described. Principally, cell walls are reinforced anisotropically by cellulose microfibrils (Brett 2000), which are made from long polymers of $1 \rightarrow 4$, β-linked glucose that associate laterally to form a partly crystalline lattice (Doblin et al. 2002). The word microfibril applies to any group of glucose chains, whether it be the "elementary" microfibril whose number of chains is a matter of debate or to higher-order assemblies of already crystalline domains. Neighboring microfibrils tend to be roughly parallel, giving the cell wall a mat-like appearance and a distinct structural anisotropy (**Figure 2**).

Microfibril alignment is sometimes equated to resistance. However, resistance emerges from the way in which microfibrils interact with each other and with other wall components; all these interactions sum

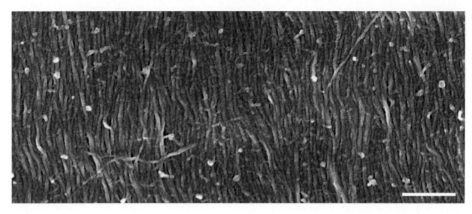

Figure 2

Micrograph of the innermost cell wall layer of a cucumber hypocotyl imaged with field-emission scanning electron microscopy (Marga et al. 2005). The long axis of the stem runs left to right. The cell wall is structurally anisotropic. Scale bar = 250 nm.

Polarized-light microscopy: an optical technique whereby contrast is generated from molecular alignment, as in crystals. Modes can use absorption, florescence, or phase. Useful for cell walls because of the crystallinity of microfibrils.

to give the resistance of the wall to stress. Maximal resistance in a composite material with anisotropic fibrillar reinforcement is predictably parallel to the long axis of the fibers (Sellen 1983); accordingly, the alignment of microfibrils commonly does reflect the direction of strain rate anisotropy. However, the magnitude of the resistances and hence the degree of anisotropy depend on the interactions between microfibrils.

Controls on Growth Anisotropy: Microfibril Synthesis

Cellulose microfibrils are probably necessary for anisotropic expansion to take place. Several herbicides target the synthesis of cellulose and reduce the anisotropy of expansion (Sabba & Vaughn 1999, Scheible et al. 2003). Similarly, numerous mutants that either directly or indirectly reduce the rate of cellulose synthesis are known, and organ growth in these mutants is invariably less anisotropic (Robert et al. 2004). Cellulose synthesis continues, albeit at a lower rate, in these treatments and mutants, and the new microfibrils are poorly aligned in some cases (Wasteneys 2004) but not in others (Refrégier et al. 2004). Surprisingly, several laboratories have selected tissue culture cell lines that grow well despite hav-

ing scant microfibrillar cellulose (Shedletzky et al. 1992, Sabba et al. 1999); also, pollen tubes have little if any cellulose, particularly at the tip (Ferguson et al. 1998). However, these cellulose-deficient cells grow isotropically: The herbicide-adapted cultures grow as cell clumps without a major axis of expansion, and any anisotropy in the expansion of the pollen tube tip, as in root hairs, probably occurs without mechanical anisotropy within the plane of the cell wall (Dumais et al. 2004).

Controls on Growth Anisotropy: The Direction of Anisotropy

The paradoxical behavior of the stem epidermis. For multicellular organs, the direction of maximal expansion rate is generally perpendicular to the net orientation among microfibrils. This generalization applies to microfibril order throughout the cell wall, as assayed with polarized-light microscopy, and has been reported widely (Roelofsen 1965). Recent examples include roots (Green 1984, Baskin et al. 1999), rice internodal parenchyma (Sauter et al. 1993), grass leaves for both cortical parenchyma and epidermis (Hogetsu 1989, Paolillo 1995), and a variety of expansion planes and tissues within the shoot apical meristem (Green 1988, Sakaguchi et al. 1990). Additionally, the outer epidermis in dicot leaves is isotropic in both microfibril alignment and expansion (Hogetsu 1989, Kerstens et al. 2001). The generalization has been strengthened experimentally by finding that, when the direction of maximal expansion changes from longitudinal to radial, the net orientation of microfibrils in parenchyma also changes (e.g., pea epicotyls exposed to horomones) (Probine 1965, Veen 1970, Ridge 1973). However, as noted by Roelofsen (1965) and recently confirmed by others for many species (Iwata & Hogetsu 1989, Paolillo 2000, Verbelen & Karstens 2000), in stems (and coleoptiles) the net alignment of microfibrils in the epidermis is longitudinal, even in rapidly elongating regions. Note that an exception is the rice coleoptile, which has transverse microfibrils throughout its epidermis (Paolillo 2000).

Microfibril alignment is also widely assessed with electron microscopy. Cell walls viewed in cross sections often have many layers of microfibrils, taking on complex three-dimensional patterns (Roland et al. 1987). Furthermore, cellulose in conventional ultra-thin sections is erratically stained by heavy metals, and researchers have resorted to extraction regimes to enhance microfibrillar contrast (Roland et al. 1982). However, extraction even with relatively gentle reagents can disorganize microfibrils (Crow & Murphy 2000). For these reasons, assessing the direction of microfibril alignment in electron microscopy is problematic.

A partial solution is offered by the innermost cell wall layer. This layer can be imaged reliably with scanning or transmission electron microscope images of the cell wall surface and is the layer of microfibrils whose orientation is directly controlled by the cell. Some researchers have further justified examining the innermost wall layer by arguing that the innermost layer in a multicellular organ bears most of the load. This argument is based on results from *Nitella* indicating that

only the inner quarter of the wall is load bearing (Richmond et al. 1980) but this extrapolation is probably invalid (Baskin et al. 1999, Hejnowicz & Borowska-Wykręt 2005).

Studies on the innermost cell wall layer support the generalization that microfibrils are perpendicular to the direction of maximal expansion, for example, in cortical parenchyma of roots (Hogetsu 1986, Baskin et al. 1999), bamboo shoots (Crow & Murphy 2000), and the cambium of a conifer (Abe et al. 1995). In elongating stems with polylamellate cell walls, inner microfibril orientation can be longitudinal, transverse, or oblique, but it is transverse in a majority of parenchymatous cells (Takeda & Shibaoka 1981a,b, Iwata & Hogetsu 1989).

Again, electron microscopy confirms that stem (and coleoptile) epidermis is exceptional. Microfibrils of the innermost epidermal wall in lettuce hypocotyls elongating anisotropically are consistently longitudinal (Sawhney & Srivastiva 1975), as in epidermis of intact, elongating maize coleoptiles (Bergfeld et al. 1988). Furthermore, microfibril alignment in the innermost layer of the azuki bean epidermal cell wall is predominantly transverse only in short, apical cells, whereas in longer but still anisotropically elongating cells, alignments among cells are mixed between transverse, oblique, and longitudinal, mirroring alignments in cortical parenchyma of epicotyls treated to induce more or less isotropic expansion (Takeda & Shibaoka 1981a,b).

If microfibril alignment determines the directionality of expansion, then why do stem epidermal cells not swell? The most likely answer is that epidermal cells in stems require tissue tensions to expand. Apparently, epidermal cell turgor is weaker than the stiffness of the outer cell wall and cannot drive expansion without help from the inner tissues (e.g., Hejnowicz & Sievers 1996b). That epidermal reinforcement direction can be overridden is supported by findings that developmental swelling in onion bulbs occurs despite transverse epidermal microtubules

(Mita & Shibaoka 1983) and microfibrils (Verbelen & Kerstens 2000). If the coherent anisotropic mechanical reinforcement in cortical parenchyma transmits to the epidermis a large axial force but a small transverse force, then this could be sufficient to determine the yielding behavior of the epidermis. The possibility that the inner tissue plays a dominant role in regulating the directionality of organ expansion suggests that this neglected tissue requires further study.

Controls on Growth Anisotropy: The Degree of Anisotropy

Green's degree of alignment hypothesis. Green (1964) noticed that *Nitella* internodal cell walls have better-aligned microfibrils than those of another, green algal species, which expand less anisotropically. He hypothesized that the degree of expansion anisotropy is proportional to the degree of alignment among microfibrils (**Figure 3a**).

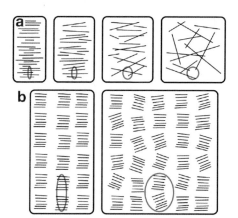

Figure 3

Models for controlling the degree of anisotropy by microfibril alignment. The degree of anisotropy is shown by the blue ellipses. (*a*) Green's (1964) hypothesis, where the degree of anisotropy is proportional to the degree of local alignment among microfibrils. (*b*) Modified hypothesis for a tissue (*black rectangle with rounded corners*) in which the degree of anisotropy is proportional to the uniformity of alignment in different cells or cell regions (cells are not shown). Figure redrawn from Baskin et al. (2004).

Green's hypothesis is economical because it allows the microfibrils to control the direction of anisotropy by means of their net orientation and the degree of anisotropy by means of how well they are aligned.

To test this hypothesis, my collaborators and I quantified the alignment of microfibrils throughout the cell wall with polarized-light microscopy and on the innermost layer with surface views for the maize root growing more anisotropically in response to water stress (Baskin et al. 1999) as well as for the arabidopsis root growing less anisotropically in response to low doses of a microtubule inhibitor (Baskin et al. 2004). In both systems, we found that, contrary to Green's hypothesis, the degree of orientation among microfibrils is uncorrelated to the degree of anisotropic expansion experienced by the wall. However, the imaging method used for arabidopsis allowed the angle of net orientation of each wall area's microfibrils to be measured, and this revealed that, where the degree of expansion anisotropy is less, the alignment among neighboring cell walls is less uniform. Recently, polarized-light microscopy revealed that a similar patchiness in microfibril alignment accompanies a decreased degree of expansion anisotropy in the root of the arabidopsis *cobra-1* mutant (Roudier et al. 2005). Green's hypothesis could be revised by invoking microfibrillar alignment among cells (global order) rather than among individual microfibrils (local order) in the regulation of the degree of expansion anisotropy (**Figure 3b**).

In this view, a uniform microfibril alignment in cells across a tissue promotes a high degree of expansion anisotropy, whereas the tissue expands less anisotropically when the mechanical reinforcement within the tissue becomes patchy, even though local regions of cell wall contain well-aligned cellulose. This view is consistent with experiments where the induction of more or less isotropic expansion is associated with organs losing a preferential alignment direction among constituent cells but with each cell having perfectly well-aligned microfibrils (Itoh 1976, Takeda & Shibaoka 1981b, Iwata & Hogetsu 1989). It is also consistent with models in which the loss of uniform reinforcement around the periphery of the shoot apical meristem is associated with bulge formation to begin a leaf primordium (Green 1988).

Generalizing Green's hypothesis. The economy of invoking microfibril alignments to explain both the direction and the degree of anisotropy is no guarantee of validity. The uniformity of microfibril alignment may be one mechanism among many that the plant uses to regulate radial expansion. Several lines of evidence suggest that microfibril alignment, although necessary, is insufficient to explain strain rate anisotropy. Studies have shown that several root morphology mutants of arabidopsis have stimulated radial expansion without any worsening of the alignment among microfibrils (Wiedemeier et al. 2002, Sugimoto et al. 2003) and that no change in microfibril alignment occurs to explain the increased strain rate anisotropy of maize roots growing at low water potential (Baskin et al. 1999), although none of these studies checked explicitly for patchiness of reinforcement direction among cells. In *Nitella*, Richmond et al. (1980), who prepared cell wall ghosts from growing cells and pressurized them to mimic turgor-driven expansion, found that the ghosts expanded more anisotropically than did the living cells, an observation that implies there is more to anisotropic yielding than the structural disposition of the cell wall.

I propose as a general hypothesis that expansion rates in length and width are regulated by independent molecular mechanisms. A similar proposal has been made for leaves (Tsuge et al. 1996, Tsukaya 2003), although it was based on leaf allometry rather than cellular expansion rates. Recently, Hejnowicz and Borowska-Wykręt (2005) obtained evidence that the longitudinal stress in stems is borne principally by the outer layers of the outer epidermal cell wall, and they proposed that the outer layers of the cell wall limit elongation while the inner layers of the wall limit radial expansion.

Whether controls on expansion in different directions reside in different layers of the cell wall, in different tissues, or within the same cell wall, reactions needed to shear microfibrils parallel to their lengths can be expected to differ from reactions needed to separate them. Elongation rate has been modeled as depending on the tension developed in a system of xyloglucan tethers between parallel microfibrils (Passioura 1994, Veytsman & Cosgrove 1998), and experimental evidence suggests that elongation involves the separation of parallel microfibrils (Marga et al. 2005); however, in principle, longitudinally taut tethers weakly resist shear between microfibrils. Biochemical influences on transverse expansion do differ from those on longitudinal expansion in *Nitella* cell walls assayed in vitro—for example, the threshold for acid-induced deformation is more than a whole pH unit more acidic for radial expansion compared with elongation (Métraux & Taiz 1979, Richmond et al. 1980)—but few comparable data exist for higher plants. A more explicit hypothesis must await improved understanding of the molecular interactions among cell wall polymers and the development of engineering models that handle anisotropic deformation.

Controls on Growth Anisotropy: Microtubules

That microtubules participate in the regulation of anisotropic expansion is indisputable. Whether microtubules are inhibited chemically (Vaughn & Lehnen 1991) or genetically (Wasteneys 2004), growing organs become swollen, indicating the loss of anisotropy. This has been interpreted as indicating that microtubules are required to control microfibril alignment (Green 1980, Baskin 2001). In the absence of microtubules, expansion tends toward isotropy and cellulose alignment often becomes more isotropic in the statistical sense (averaged among cells in a tissue or through several cell wall layers or regions) (Takeda & Shibaoka 1981b, Iwata & Hogetsu 1989, Hogetsu 1989, Inada et al. 2002, Sugimoto et al. 2003).

This interpretation has been questioned pointedly on the basis of studies of the *mor1* mutant, which has disrupted cortical microtubules, reduced expansion anisotropy, and well-aligned microfibrils (Himmelspach et al. 2003, Sugimoto et al. 2003). The *mor1* mutant retains some cortical microtubules, which may be sufficient to align microfibrils insofar as a depleted cortical array is sufficient to align microfibrils and prevent swelling in other cell types (Baskin et al. 1994, Inada et al. 2002). But because the *MOR1* gene encodes a microtubule-associated protein, the swelling in the mutant, despite aligned microfibrils, plausibly results from the loss of some microtubule function other than that of aligning cellulose. Wasteneys (2004) has suggested the microtubules govern the effective functional length of the microfibril: When microtubules are disrupted, were the average microfibril length to decrease, then adjacent microfibrils would become easier to shear and radial expansion would be stimulated. This mechanism offers a straightforward way for the cell to regulate rates of radial expansion independently of elongation, as hypothesized above.

SUMMARY AND PERSPECTIVES FOR FUTURE RESEARCH

Anisotropic expansion in plant cells depends on the alignment among cellulose microfibrils, but this dependence is incomplete. In general, in multicellular organs, expansion rates in maximal and minimal directions change independently, and a means for controlling expansion rate in the minimal direction needs to be discovered. Cortical microtubules may be part of a cellular mechanism that conditions cell wall resistance in the direction parallel to the microfibrils. This role could be in addition to that of specifying the direction of anisotropy by aligning microfibrils. Knowledge is limited as there have been few attempts to quantify strain rate anisotropy at a cell wall level, even in studies where cell wall or cellular ultrastructure is examined. Future research should involve testing the biochemical properties of radial expansion, as compared with those of elongation, and developing a mechanical analysis of stress and compliance anisotropy in growing, multicellular plant organs. Then we can begin to comprehend how the shape of the organ and its cells, the movement of water between cells, the architecture of the constituent cell walls, and the biochemical reactions modifying those cell walls are all united to generate the predictable and beautiful forms of higher plant organs.

SUMMARY POINTS

1. The expansion anisotropy of a unit area of cell wall is characterized by the direction and degree of anisotropy; the plant must control both direction and degree to build organs with specific and heritable shapes.

2. The anisotropy of expansion rate throughout a growth zone has been quantified for only a few multicellular organs.

3. In multicellular organs, the degree of anisotropy can vary widely across a growth zone and as a function of treatment.

4. The direction of maximal expansion rate is usually regulated by the direction of net alignment among cellulose microfibrils.

5. In stems, the directionality of expansion of epidermal cells is apparently controlled by that of the inner tissue.

6. The degree of anisotropy is probably controlled by factors in addition to the direction of microfibril alignment.

7. The rates of expansion in maximal and minimal directions are hypothesized to be regulated by distinct molecular mechanisms.

ACKNOWLEDGMENTS

I am indebted to Winfried Peters (University of Frankfurt) and Jacques Dumais (Harvard University) for enlightening discussions, and to Wendy Silk (University of California, Davis) for reading the manuscript critically. Work in my lab on morphogenesis is supported by the U.S. Department of Energy (award number 03ER15421), which does not constitute endorsement by that Department of views expressed herein.

LITERATURE CITED

Abe H, Funada R, Ohtani J, Fukazawa K. 1995. Changes in the arrangement of microtubules and microfibrils in differentiating conifer tracheids during the expansion of cells. *Ann. Bot.* 75:305–10

Barlow PW, Brain P, Parker JS. 1991. Cellular growth in roots of a gibberellin-deficient mutant of tomato (*Lycopersicon esculentum* Mill.) and its wild-type. *J. Exp. Bot.* 42:339–51

Baskin TI. 2001. On the alignment of cellulose microfibrils by cortical microtubules: A review and a model. *Protoplasma* 215:150–71

Baskin TI, Beemster GTS, Judy-March JE, Marga F. 2004. Disorganization of cortical microtubules stimulates tangential expansion and reduces the uniformity of cellulose microfibril alignment among cells in the root of *Arabidopsis thaliana*. *Plant Physiol.* 135:2279–90

Baskin TI, Meekes HTHM, Liang BM, Sharp RE. 1999. Regulation of growth anisotropy in well watered and water-stressed maize roots. II. Role of cortical microtubules and cellulose microfibrils. *Plant Physiol.* 119:681–92

Baskin TI, Wilson JE, Cork A, Williamson RE. 1994. Morphology and microtubule organization in *Arabidopsis* roots exposed to oryzalin or taxol. *Plant Cell Physiol.* 35:935–44

Bergfeld R, Speth V, Schopfer P. 1988. Reorientation of microfibrils and microtubules at the outer epidermal wall of maize coleoptiles during auxin-mediated growth. *Bot. Acta* 101: 57–67

Brett CT. 2000. Cellulose microfibrils in plants: Biosynthesis, deposition, and integration into the cell wall. *Int. Rev. Cytol.* 199:161–99

Bruce DM. 2003. Mathematical modelling of the cellular mechanics of plants. *Philos. Trans. R. Soc. London Ser. B* 358:1437–44

Bystrova EI. 1984. Analysis of swelling formation on roots under the influence of isopropyl N-(3-chlorophenyl) carbamate. *Russ. J. Plant Physiol.* 31:720–25

Castle ES. 1958. The topography of tip growth in a plant cell. *J. Gen. Physiol.* 41:913–26

Cavalieri AJ, Boyer JS. 1982. Water potentials induced by growth in soybean hypocotyls. *Plant Physiol.* 69:492–96

This is a readable and authoritative summary of mathematical models for cell wall biomechanics, including growing tissue.

Chen JCW. 1973. The kinetics of tip growth in the *Nitella* rhizoid. *Plant Cell Physiol.* 14:631–40

Coen E, Rolland-Lagan A-G, Matthews M, Bangham JA, Prusinkiewicz P. 2004. The genetics of geometry. *Proc. Natl. Acad. Sci. USA* 101:4728–35

Cosgrove DJ. 1999. Enzymes and other agents that enhance cell wall extensibility. *Annu. Rev. Plant Physiol. Plant Mol. Biol.* 50:391–417

Crow E, Murphy RJ. 2000. Microfibril orientation in differentiating and maturing fibre and parenchyma cell walls in culms of bamboo (*Phyllostachys viridiglaucescens* (Carr.) Riv. & Riv.). *Bot. J. Linn. Soc.* 134:339–59

Doblin MS, Kurek I, Jacob-Wilk D, Delmer DP. 2002. Cellulose biosynthesis in plants: from genes to rosettes. *Plant Cell Physiol.* 43:1407–20

Dumais J, Kwiatkowska D. 2001. Analysis of surface growth in shoot apices. *Plant J.* 31:229–41

Dumais J, Long SR, Shaw SL. 2004. The mechanics of surface expansion anisotropy in *Medicago tarantula* root hairs. *Plant Physiol.* 136:3266–75

Dumais J, Steele C. 2000. New evidence for the role of mechanical forces in the shoot apical meristem. *J. Plant Growth Regul.* 19:7–18

Erickson RO. 1966. Relative elemental rates and anisotropy of growth in area: a computer programme. *J. Exp. Bot.* 17:390–403

Ferguson C, Teeri TT, Siika-aho M, Read SM, Bacic A. 1998. Location of cellulose and callose in pollen tubes and grains of *Nicotiana tabacum*. *Planta* 206:452–60

Green PB. 1962. Mechanism for plant cellular morphogenesis. *Science* 138:1404–5

Green PB. 1964. Cell walls and the geometry of plant growth. *Brookhaven Symp. Biol.* 16:203–17

Green PB. 1965. Pathways of cellular morphogenesis. A diversity in *Nitella*. *J. Cell Biol.* 27:343–63

Green PB. 1980. Organogenesis—a biophysical view. *Annu. Rev. Plant Physiol.* 31:51–82

Green PB. 1984. Shifts in plant cell axiality: Histogenetic influences on cellulose orientation in the succulent, *Graptopetalum*. *Dev. Biol.* 103:18–27

Green PB. 1988. A theory for inflorescence development and flower formation based on morphological and biophysical analysis in *Echeveria*. *Planta* 175:153–69

Haber AH, Foard DE. 1963. Nonessentiality of concurrent cell divisions for degree of polarization of leaf growth. II. Evidence from untreated plants and from chemically induced changes of the degree of polarization. *Am. J. Bot.* 50:937–44

Harold FM. 1990. To shape a cell: an inquiry into the causes of morphogenesis of microorganisms. *Microbiol. Rev.* 54:381–431

Harold FM. 2002. Force and compliance: rethinking morphogenesis in walled cells. *Fungal Genet. Biol.* 37:271–82

Hejnowicz Z. 1989. Differential growth resulting in the specification of different types of cellular architecture in root meristems. *Environ. Exp. Bot.* 29:85–93

Hejnowicz Z, Borowska-Wykręt D. 2005. Buckling of inner cell wall layers after manipulations to reduce tensile stress: observations and interpretations for stress transmission. *Planta* 220:465–73

Hejnowicz Z, Brodzki P. 1960. The growth of root cells as the function of time and their position in the root. *Acta Soc. Bot. Pol.* 29:625–44

Hejnowicz Z, Heinmann B, Sievers A. 1977. Tip growth: Patterns of growth rate and stress in the *Chara* rhizoid. *Z. Pflanzenphysiol.* 81:409–24

Hejnowicz Z, Nakielski J, Wloch W, Beltowski M. 1988. Growth and development of shoot apex in barley. III. Study of growth rate variation by means of the growth tensor. *Acta Soc. Bot. Pol.* 57:31–50

This is a clear and cogent essay on the interplay between molecular biology and mechanics.

This study explains the observed anisotropic growth field of the root hair in terms of the stresses and mechanical properties of the cell wall.

This is perhaps the only study for a growing organ where the calculated stress distribution (itself a challenge) was confirmed experimentally.

This is a Michelin *Red Book* for cellular morphogenesis.

Hejnowicz Z, Romberger JA. 1984. Growth tensor of plant organs. *J. Theor. Biol.* 110:93–114

Hejnowicz Z, Rusin A, Rusin T. 2000. Tensile tissue stress affects the orientation of cortical microtubules in the epidermis of sunflower hypocotyl. *J. Plant Growth Regul.* 19:31–44

Hejnowicz Z, Sievers A. 1995a. Tissue stresses in organs of herbaceous plants. I. Poisson ratios of tissues and their role in determination of the stresses. *J. Exp. Bot.* 46:1035–43

Hejnowicz Z, Sievers A. 1995b. Tissue stresses in organs of herbaceous plants. II. Determination in three dimensions in the hypocotyl of sunflower. *J. Exp. Bot.* 46:1045–53

Hejnowicz Z, Sievers A. 1996a. Tissue stresses in organs of herbaceous plants. III. Elastic properties of the tissues of sunflower hypocotyl and origin of tissue stresses. *J. Exp. Bot.* 47:519–28

Hejnowicz Z, Sievers A. 1996b. Acid-induced elongation of *Reynoutria* stems requires tissue stresses. *Physiol. Plant.* 98:345–48

Hernández LF, Havelange A, Bernier G, Green PB. 1991. Growth behavior of single epidermal cells during flower formation: sequential scanning electron micrographs provide kinematic patterns for Anagallis. *Planta* 185:139–47

Himmelspach R, Williamson RE, Wasteneys GO. 2003. Cellulose microfibril alignment recovers from DCB-induced disruption despite microtubule disorganization. *Plant J.* 36:565–75

Hogetsu T. 1986. Orientation of wall microfibril deposition in root cells of *Pisum sativum* L. var. Alaska. *Plant Cell Physiol.* 27:947–51

Hogetsu T. 1989. The arrangement of microtubules in leaves of monocotyledonous and dicotyledonous plants. *Can. J. Bot.* 67:3506–12

Inada S, Sonobe S, Shimmen T. 2002. Regulation of directional expansion by the cortical microtubule array in roots of Lemna minor. *Funk. Plant Biol.* 29:1273–78

Itoh T. 1976. Microfibrillar orientation of radially enlarged cells of coumarin- and colchicine-treated pine seedlings. *Plant Cell Physiol.* 17:385–98

Iwata K, Hogetsu T. 1989. Orientation of wall microfibrils in *Avena* coleoptiles and mesocotyls and in *Pisum* epicotyls. *Plant Cell Physiol.* 30:749–57

Kataoka H. 1982. Colchicine-induced expansion of Vaucheria cell apex alteration from isotropic to transversally anisotropic growth. *Bot. Mag.* 95:317–30

Kerstens S, Decraemer WF, Verbelen J-P. 2001. Cell walls at the plant surface behave mechanically like fiber-reinforced composite materials. *Plant Physiol.* 127:381–85

Kwiatkowska D. 2004. Structural integration at the shoot apical meristem: Models, measurements, and experiments. *Am. J. Bot.* 91:1277–93

Kwiatkowska D, Dumais J. 2003. Growth and morphogenesis at the vegetative shoot apex of *Anagallis arvensis*. *J. Exp. Bot.* 54:1585–95

Liang BM, Sharp RE, Baskin TI. 1997. Regulation of growth anisotropy in well watered and water-stressed maize roots. I. Spatial distribution of longitudinal, radial and tangential expansion rates. *Plant Physiol.* 115:101–11

Macadam JW, Nelson CJ. 2002. Secondary cell wall deposition causes radial growth of fibre cells in the maturation zone of elongating tall fescue leaf blades. *Ann. Bot.* 89:89–96

Maksymowych R. 1990. *Analysis of Growth and Development of Xanthium*, pp. 28–30. Cambridge, UK: Cambridge Univ. Press. 220 pp.

Marga F, Grandbois M, Cosgrove DJ, Baskin TI. 2005. Cell wall extension results in the coordinate separation of parallel microfibrils: Evidence from scanning electron microscopy and atomic force microscopy. *Plant J.* 43:181–90

Maurice I, Gastal F, Durand J-L. 1997. Generation of form and associated mass deposition during leaf development in grasses: a kinematic approach for non-steady growth. *Ann. Bot.* 80:673–83

These three papers (Hejnowitz & Sievers 1995b, 1996a,b) represent a pioneering theoretical and experimental analysis of the anisotropy of stress and compliance for a growing stem.

This study quantifies the local expansion rates of a grass leaf in all three dimensions.

Métraux J-P, Taiz L. 1979. Transverse viscoelastic extension in Nitella. II. Effects of acid and ions. *Plant Physiol.* 63:657–59

Mita T, Shibaoka H. 1983. Changes in microtubules in onion leaf sheath cells during bulb development. *Plant Cell Physiol.* 24:109–17

Nakielski J. 1987. Variations of growth in shoot apical domes of spruce seedlings: A study using the growth tensor. *Acta Soc. Bot. Pol.* 56:625–43

Nicklas KJ, Paolillo DJ Jr. 1998. Preferential states of longitudinal tension in the outer tissues of *Taraxacum officinale* (Asteraceae) peduncles. *Am. J. Bot.* 85:1068–81

Paolillo DJ Jr. 1995. The net orientation of wall microfibrils in the outer periclinal epidermal walls of seedling leaves of wheat. *Ann. Bot.* 76:589–96

Paolillo DJ Jr. 2000. Axis elongation can occur with net longitudinal orientation of wall microfibrils. *New Phytol.* 145:449–55

Passioura JB. 1994. The physical chemistry of the primary cell wall: implications for the control of expansion rate. *J. Exp. Bot.* 45:1675–82

Passioura JB, Boyer JS. 2003. Tissue stresses and resistance to water flow conspire to uncouple the water potential of the epidermis from that of the xylem in elongating plant stems. *Funct. Plant Biol.* 30:325–34

Peters WS, Tomos AD. 2000. The mechanic state of "inner tissue" in the growing zone of sunflower hypocotyls and the regulation of its growth rate following excision. *Plant Physiol.* 123:605–12

Probine MC. 1965. Chemical control of plant cell wall structure and of cell shape. *Proc. R. Soc. London Ser. B* 161:526–37

Probine MC, Preston RD. 1961. Cell growth and the structure and mechanical properties of the wall in internodal cells of *Nitella opaca*. I. Wall structure and growth. *J. Exp. Bot.* 12:261–82

Probine MC, Preston RD. 1962. Cell growth and the structure and mechanical properties of the wall in internodal cells of *Nitella opaca*. II. Mechanical properties of the walls. *J. Exp. Bot.* 13:111–27

Refregier G, Pelletier S, Jaillard D, Höfte H. 2004. Interaction between wall deposition and cell elongation in dark-grown hypocotyl cells in arabidopsis. *Plant Physiol.* 135:959–68

Richards OW, Kavanagh AJ. 1943. The analysis of the relative growth gradients and changing form of growing organisms: Illustrated by the tobacco leaf. *Am. Nat.* 77:385–99

Richmond PA, Métraux J-P, Taiz L. 1980. Cell expansion patterns and directionality of wall mechanical properties in Nitella. *Plant Physiol.* 65:211–17

Ridge I. 1973. The control of cell shape and rate of cell expansion by ethylene: effects on microfibril orientation and cell wall extensibility in etiolated peas. *Acta Bot. Neerl.* 22:144–58

Robert S, Mouille G, Hofte H. 2004. The mechanism and regulation of cellulose synthesis in primary walls: lessons from cellulose-deficient Arabidopsis mutants. *Cellulose* 11:351–64

Roelofsen PA. 1965. Ultrastructure of the wall in growing cells and its relation to the direction of the growth. *Adv. Bot. Res.* 2:69–149

Roland JC, Reis D, Mosiniak M, Vian B. 1982. Cell wall texture along the growth gradient of the Mung bean hypocotyl: ordered assembly and dissipative processes. *J. Cell Sci.* 56:303–18

Roland JC, Reis D, Vian B, Satiat-Jeunemaitre B, Mosiniak M. 1987. Morphogenesis of plant cell walls at the supramolecular level: internal geometry and versatility of helicoidal expression. *Protoplasma* 140:75–91

Rolland-Lagan A-G, Bangham JA, Coen E. 2003. Growth dynamics underlying petal shape and asymmetry. *Nature* 422:161–63

This is an essential read for anyone who thinks the epidermis "controls" growth.

Rolland-Lagan A-G, Coen E, Impey SJ, Bangham JA. 2005. A computational method for inferring growth parameters and shape changes during development based on clonal analysis. *J. Theor. Biol.* 232:157–77

Roudier F, Fernandez AG, Fujita M, Himmelspach R, Borner GH, et al. 2005. COBRA, an extracellular glycosyl-phosphatidyl inositol-anchored protein, specifically controls highly anisotropic expansion through its involvement in cellulose microfibril orientation. *Plant Cell* 17:1749–63

Sabba RP, Durso NA, Vaughn KC. 1999. Structural and immunocytochemical characterization of the walls of dichlobenil-habituated BY-2 tobacco cells. *Int. J. Plant Sci.* 160:275–90

Sabba RP, Vaughn KC. 1999. Herbicides that inhibit cellulose biosynthesis. *Weed Sci.* 47:757–63

Sakaguchi S, Hogetsu T, Hara N. 1990. Specific arrangements of cortical microtubules are correlated with the architecture of meristems in shoot apices of angiosperms and gymnosperms. *Bot. Mag.* 103:143–63

Sauter M, Seagull RW, Kende H. 1993. Internodal elongation and orientation of cellulose microfibrils and microtubules in deepwater rice. *Planta* 190:354–62

Sawhney VK, Srivastava LM. 1975. Wall fibrils and microtubules in normal and gibberellic-acid-induced growth of lettuce hypocotyl cells. *Can. J. Bot.* 53:824–35

Scheible WR, Fry B, Kochevenko A, Schindelasch D, Zimmerli L, et al. 2003. An Arabidopsis mutant resistant to thaxtomin A, a cellulose synthesis inhibitor from *Streptomyces* species. *Plant Cell* 15:1781–94

Schmundt D, Stitt M, Jähne B, Schurr U. 1998. Quantitative analysis of the local rates of growth of dicot leaves at high temporal and spatial resolution, using image sequence analysis. *Plant J.* 16:505–14

Schopfer P. 2000. Cell-wall mechanics and extension growth. In *Plant Biomechanics 2000*, ed. H-C Spatz, T Speck, pp. 218–28. Stuttgart: Georg Thieme. 681 pp.

Sellen DB. 1983. The response of mechanically anisotropic cylindrical cells to multiaxial stress. *J. Exp. Bot.* 34:681–87

Shedletzky E, Shmuel M, Trainin T, Kalman S, Delmer D. 1992. Cell wall structure in cells adapted to growth on the cellulose-synthesis inhibitor 2,6-dichlorobenzonitrile. *Plant Physiol.* 100:120–30

Silk WK. 1984. Quantitative descriptions of development. *Annu. Rev. Plant Physiol.* 35:479–518

Silk WK. 1989. Growth rate patterns which maintain a helical tissue tube. *J. Theor. Biol.* 138:311–27

Silk WK. 1992. Steady form from changing cells. *Int. J. Plant Sci.* 153:S49–58

Silk WK, Abou Haidar S. 1986. Growth of the stem of *Pharbitis nil*: Analysis of longitudinal and radial components. *Physiol. Vég.* 24:109–16

Silk WK, Erickson RO. 1979. Kinematics of plant growth. *J. Theor. Biol.* 76:481–501

Silk WK, Lord EM, Eckard KJ. 1989. Growth patterns inferred from anatomical records. Emperical tests using longisections of roots of *Zea mays* L. *Plant Physiol.* 90:708–13

Sugimoto K, Himmelspach R, Williamson RE, Wasteneys GO. 2003. Mutation or drug-dependent microtubule disruption causes radial swelling without altering parallel cellulose microfibril deposition in Arabidopsis root cells. *Plant Cell* 15:1414–29

Taiz L. 1984. Plant cell expansion: regulation of cell wall mechanical properties. *Annu. Rev. Plant Physiol.* 35:585–657

Takeda K, Shibaoka H. 1981a. Changes in microfibril arrangement on the inner surface of the epidermal cell walls in the epicotyl of *Vigna angularis* Ohwi et Ohashi during cell growth. *Planta* 151:385–92

Takeda K, Shibaoka H. 1981b. Effects of gibberellin and colchicine on microfibril arrangement in epidermal cell walls of *Vigna angularis* Ohwi et Ohashi epicotyls. *Planta* 151:393–98

Thompson DW. 1917. *On Growth and Form*. Cambridge, UK: Cambridge Univ. Press

Still essential reading, this is the cornerstone of quantitative and biophysical approaches to development.

Tsuge T, Tsukaya H, Uchiyama H. 1996. Two independent and polarized processes of cell elongation regulate leaf blade expansion in *Arabidopsis thaliana*. *Development* 122:1589–1600

Tsukaya H. 2003. Organ shape and size: a lesson from studies of leaf morphogenesis. *Curr. Opin. Plant Biol.* 6:57–62

van der Weele CM, Jiang H, Palaniappan KK, Ivanov VB, Palaniappan K, Baskin TI. 2003. A new algorithm for computational image analysis of deformable motion at high spatial and temporal resolution applied to root growth. Roughly uniform elongation in the meristem and also, after an abrupt acceleration, in the elongation zone. *Plant Physiol.* 132:1138–48

Vaughn KC, Lehnen LP Jr. 1991. Mitotic disrupter herbicides. *Weed Sci.* 39:450–57

Veen BW. 1970. Control of plant cell shape by cell wall structure. *Proc. Kon. Ned. Akad. Wet. C* 73:118–21

Verbelen J-P, Kerstens S. 2000. Polarization confocal microscopy and Congo red fluorescence: a simple and rapid method to determine the mean cellulose fibril orientation in plants. *J. Microsc.* 198:101–7

Veytsman BA, Cosgrove DJ. 1998. A model of cell wall expansion based on thermodynamics of polymer networks. *Biophys. J.* 75:2240–50

Wasteneys GO. 2004. Progress in understanding the role of microtubules in plant cells. *Curr. Opin. Plant Biol.* 7:651–60

Wiedemeier AMD, Judy-March JE, Hocart CH, Wasteneys GO, Williamson RE, Baskin TI. 2002. Mutant alleles of arabidopsis *RADIALLY SWOLLEN 4* and *RSW7* reduce growth anisotropy without altering the transverse orientation of cortical microtubules or cellulose microfibrils. *Development* 129:4821–30

RNA Transport and Local Control of Translation

Stefan Kindler,[1] Huidong Wang,[2] Dietmar Richter,[1] and Henri Tiedge[3]

[1]Institute for Cell Biochemistry and Clinical Neurobiology, University Hospital Hamburg-Eppendorf, University of Hamburg, D-20246 Hamburg, Germany; [2]Laboratory of Molecular Neuro-Oncology, The Rockefeller University, New York, New York 10021; [3]Department of Physiology and Pharmacology, Department of Neurology, State University of New York, Health Science Center at Brooklyn, Brooklyn, New York 11203; email: kindler@uke.uni-hamburg.de; hwang@mail.rockefeller.edu; richter@uke.uni-hamburg.de; htiedge@downstate.edu

Annu. Rev. Cell Dev. Biol.
2005. 21:223–45

The *Annual Review of Cell and Developmental Biology* is online at
http://cellbio.annualreviews.org

doi: 10.1146/
annurev.cellbio.21.122303.120653

Copyright © 2005 by
Annual Reviews. All rights
reserved

1081-0706/05/1110-
0223$20.00

Key Words

cis-acting targeting element, *trans*-acting factor, molecular motor, activity-dependent translation

Abstract

In eukaryotes, the entwined pathways of RNA transport and local translational regulation are key determinants in the spatio-temporal articulation of gene expression. One of the main advantages of this mechanism over transcriptional control in the nucleus lies in the fact that it endows local sites with independent decision-making authority, a consideration that is of particular relevance in cells with complex cellular architecture such as neurons. Localized RNAs typically contain codes, expressed within *cis*-acting elements, that specify subcellular targeting. Such codes are recognized by *trans*-acting factors, adaptors that mediate translocation along cytoskeletal elements by molecular motors. Most transported mRNAs are assumed translationally dormant while en route. In some cell types, especially in neurons, it is considered crucial that translation remains repressed after arrival at the destination site (e.g., a postsynaptic microdomain) until an appropriate activation signal is received. Several candidate mechanisms have been suggested to participate in the local implementation of translational repression and activation, and such mechanisms may target translation at the level of initiation and/or elongation. Recent data indicate that untranslated RNAs may play important roles in the local control of translation.

Contents

INTRODUCTION

It has been 40 years since publication of the first reports of RNA localization in eukaryotic cells (Bodian 1965, Koenig 1965a,b). Curiously, these early data were initially given little attention. Even after evidence for localized RNAs and ribosomes had emerged from several laboratories in the early 1980s (Colman et al. 1982, Jeffery et al. 1983, Palacios-Prü et al. 1981, Steward & Levy 1982), RNA localization continued to be considered something of an obscure oddity in cell biology. The notion was greeted with skepticism.

In the late 1980s and early 1990s, however, it became increasingly clear that protein synthetic machinery and individual RNAs can be targeted to distant sites, sometimes at considerable distances from perikaryal somatic regions. Because evidence supporting this notion was emerging from diverse cell types including *Xenopus* oocytes, mammalian neurons, and glial cells, and from *Drosophila* embryos, among others, researchers began to realize that a common principle might be at work. Today we understand that mRNAs are transported, localized, and locally translated in many eukaryotic cell types. These mechanisms are of fundamental importance in the regulation of gene expression as they allow cells to delegate control to autonomously acting local sites. In the past several years we have witnessed dynamic progress in our understanding of mechanisms that govern RNA localization and local translational control, mechanisms that underlie the spatiotemporal modulation of local protein repertoires. Hand-in-hand with this development has come the realization that small untranslated RNAs, many hitherto unappreciated (Couzin 2002), play important roles as mediators of translational control.

Why do cells localize RNAs? RNA transport and local translation have now been documented in vertebrates, invertebrates, and unicellular organisms, and it therefore appears that RNA localization is an ancient cellular mechanism. Today this mechanism is used by diverse cell types for various purposes, but one of the main benefits that cells—especially complex and highly polarized cells—derive from it is the ability to control gene expression locally. This consideration is particularly relevant in neurons where the plastic modulation of synaptic connections forms the cellular substratum for higher brain functions such as learning and memory. Long-term, experience-dependent, and input-specific synaptic remodeling requires changes of gene expression and de novo protein synthesis; given the fact that a typical neuron makes several thousand synaptic connections on its dendritic arbor, the advantages of local translational control are obvious.

The dynamic expansion of the field in recent years dictates that we focus on selected model systems. Our main emphasis here is on mammalian cells, but we refer to non-mammalian cells whenever pertinent. For the same reason, we restrict this review to literature published within the last four years. The reader is referred to a number of excellent review articles that cover earlier periods or subject areas that are outside the scope of this review. Several of these articles have appeared in the Annual Reviews series (Chartrand et al. 2001, Goldstein & Yang 2000, Kapp & Lorsch

---

Translational control: the regulation of gene expression at the level of translation

Untranslated RNA: an RNA that, in contrast to an mRNA, does not encode a polypetide sequence. It typically carries codes to specify functionality, e.g., catalytic action or subcellular targeting

2004, Palacios & St. Johnston 2001, Piper & Holt 2004, Steward & Schuman 2001). Piper & Holt (2004) provide a current account of local translation in axons, a topic not covered here. Other relevant reviews have appeared elsewhere (Bassell & Kelic 2004, Carson et al. 2001, Darnell 2003, Giuditta et al. 2002, Job & Eberwine 2001, Kloc & Etkin 2005, Kloc et al. 2002, Martin 2004, Smith 2004, Steward & Schuman 2003, Wang & Tiedge 2004).

RNA TRANSPORT AND LOCALIZATION

RNA localization is a widespread phenomenon that has been observed in many eukaryotic cell types, including unicellular organisms (e.g., yeast), developing germ cells and embryos (e.g., in *Drosophila* and *Xenopus*), and plant and other animal cells (e.g., mammalian neurons). Active targeting involves recognition of *cis*-acting RNA elements, i.e., RNA segments that contain codes to specify targeting, by *trans*-acting RNA-binding factors. This interaction results in the formation of ribonucleotide protein (RNP) complexes, which travel along cytoskeletal filaments with the help of motor proteins. At their destination sites, delivered transcripts are anchored and are now ready for translation.

Cis-Acting Targeting Elements

Location-coding, *cis*-acting targeting elements have been identified in various localized transcripts of higher eukaryotes (**Table 1**). In *Xenopus* oocytes, more than 20 RNAs are transported via three different pathways to either the vegetal pole or the animal hemisphere (King et al. 1999). Vegetal localization occurs during an early and a late phase of oogenesis. RNAs of the early METRO (messenger transport organizer) pathway (Kloc & Etkin 2005), such as Xcat-2, Xdaz1, Xpat, Xwnt-11, and Xlsirts, initially associate with the mitochondrial cloud (stages I–II) and subsequently migrate to the vegetal cortex between late stage II and early stage III. Transcripts of the late

pathways, including Vg1 (Melton 1987) and VegT (Zhang & King 1996), are transported during late stage III and early stage IV and remain anchored to the vegetal cortex until the end of oogenesis. This pathway localizes germ layer determinants to the vegetal hemisphere. In Vg1 mRNAs, a 340-nucleotide (nt) motif, termed Vg1 localization element (VLE), mediates vegetal targeting (Mowry & Melton 1992). The VLE and the *cis*-elements of virtually all other vegetally localized transcripts contain clusters of short CAC-containing motifs (Betley et al. 2002). Slight sequence variations between these motifs appear to correlate with distinct functions during the localization process (Zhou & King 2004). However, Xvelo1 transcripts of the late vegetal pathway use a 75-nt stem-loop localization element (Claussen & Pieler 2004). Thus targeting of different mRNAs to identical cellular sites may involve diverse as well as overlapping molecular mechanisms.

Numerous RNAs are targeted to dendrites in mammalian neurons (Eberwine et al. 2002). The dendritic delivery of some of these mRNAs is dependent on neuronal activity. Synaptic activation in vivo strongly upregulates Arc/arg3.1 gene expression (Link et al. 1995, Lyford et al. 1995) and results in the selective, *N*-methyl-D-aspartate (NMDA) receptor (NMDAR)–dependent recruitment of corresponding transcripts to dendritic segments in which synapses had been stimulated (Steward & Worley 2001). Similarly, epileptogenic stimuli in vivo produce increased brain-derived neurotrophic factor (BDNF) mRNA levels, coupled with an NMDAR-dependent transcript accumulation in proximal dendritic segments in hippocampus (Tongiorgi et al. 2004). A common requirement in these pathways would be retrograde synapse-to-nucleus signaling, and mechanisms underlying such signaling are now beginning to be addressed (Thompson et al. 2004).

Dendritic targeting elements (DTEs) have been identified in a number of neuronal mRNAs. Such DTEs appear to be quite diverse in length and sequence, and they may

Cis-acting element: a segment within an RNA that contains a code to specify functionality (e.g., subcellular targeting)

Dendritic targeting element (DTE): a *cis*-acting element that directs a neuronal RNA to or along dendrites

nt: nucleotide

RNP: ribonucleoprotein

Targeting code: information contained within an RNA, expressed by an RNA motif within a *cis*-acting element, that specifies subcellular targeting

Table 1 *cis*-acting elements and *trans*-acting factors in cell types of vertebrate organisms. *cis*-acting elements have often been only broadly defined. Position range and length therefore represent maxima that in many cases are likely to be further narrowed down in the future

Species, cell type, RNA	Position	Name, length (nucleotides)	*trans*-acting factor	References
Xenopus oocytes				
Vg1	3′ UTR	VLE, 340	Vg1RBP/VERA, Prrp, VgRBP78, VgRBP69, VgRBP60, VgRBP40, VgRBP36, and VgRBP33	(Deshler et al. 1997, Mowry & Cote 1999, Mowry & Melton 1992, Schwartz et al. 1992, Zhao et al. 2001)
VegT		?, ?	?	(Betley et al. 2002)
Xvelo1		LE, 75	UV-crosslinking pattern similar to VLE	(Claussen & Pieler 2004)
Chicken fibroblasts				
β-actin	3′ UTR	zipcode, 54	ZBP1, ZBP2	(Kislauskis et al. 1994, Ross et al. 1997)
Chicken neurons (dendrites)				
β-actin	3′ UTR	zipcode, 54	ZBP1	(Zhang et al. 2001)
Mammalian oligodendrocytes				
MBP	3′ UTR	A2RE, 11	hnRNP A2	(Ainger et al. 1997, Hoek et al. 1998, Munro et al. 1999, Shan et al. 2003)
Mammalian neurons (dendrites)				
β-actin	3′ UTR	zipcode, 54	ZBP1	(Eom et al. 2003, Tiruchinapalli et al. 2003)
BC1	5′ domain	BC1 DTE, 65	?	(Muslimov et al. 1997)
CaMKIIα, ligatin	ORF, 3′ UTR	Y element, ~14	translin	(Severt et al. 1999)
CaMKIIα, neurogranin	3′ UTR	CNDLE, 28–30	?	(Mori et al. 2000)
CaMKIIα	3′ UTR	CaMKIIα DTE, ~1200	?	(Blichenberg et al. 2001)
CaMKIIα	3′ UTR	CPE, 6	CPEB	(Huang et al. 2003)
GluR2	3′ UTR	?	?	(Ju et al. 2004)
MAP2	3′ UTR	MAP2 DTE, 640	MARTA1, MARTA2	(Blichenberg et al. 1999; Rehbein et al. 2000, 2002)
MAP2	3′ UTR	CPE, 6	CPEB	(Huang et al. 2003)
MAP2	ORF	A2RE, 11	hnRNP A2	(Shan et al. 2003)
PKMζ	5′ UTR, ORF	Mζ DTE1, 499	?	(Muslimov et al. 2004)
PKMζ	3′ UTR	Mζ DTE2, 42	?	(Muslimov et al. 2004)
Vasopressin	ORF, 3′ UTR	DLS, 395	PABP	(Mohr et al. 2002, Mohr & Richter 2001)
Mammalian neurons (axon hillock)				
tau	3′ UTR	fragment H, 240	HuD	(Aranda-Abreu et al. 1999)

also encode different destination sites within dendritic arborizations. Protein repertoires in dendritic domains are typically mosaic and diverse. Synapses, in particular, feature highly specialized complements of macromolecular components. It has now become apparent that part of this complexity is implemented and maintained by on-site translation of locally available mRNAs. The complexity of neuronal DTEs may therefore be a reflection of the diversity of intracellular target sites, as is illustrated by the following examples.

Calcium/calmodulin-dependent protein kinase II (CaMKII) is enriched in postsynaptic microdomains. The 3' untranslated region (UTR) of CaMKIIα mRNA has been reported to contain two distinct nonoverlapping DTEs, one of about 30 nt (Mori et al. 2000), the other one located within about 1200 nt (Blichenberg et al. 2001). In mouse brain in vivo, however, the smaller element alone is not sufficient to mediate dendritic targeting (Miller et al. 2002). In addition, the 3' UTR of CaMKIIα mRNA contains two copies of a hexanucleotide motif called cytoplasmic polyadenylation element (CPE), which facilitate dendritic translocation (Huang et al. 2003).

In shank1 transcripts, a DTE is contained within 200 nt of the 3' UTR (Böckers et al. 2004), whereas in vasopressin transcripts, a DTE resides within a 395-nt segment that spans part of the open reading frame (ORF) and the 3' UTR (Prakash et al. 1997). In protein kinase Mζ (PKMζ) mRNA, one DTE is positioned within 499 nt at the interface of the 5' UTR and the ORF (499 nt) and mediates somatic export, whereas a second one, a stem-loop structure of 42 nt in the 3' UTR, is responsible for distal dendritic mRNA targeting (Muslimov et al. 2004). The element encoding distal targeting features a GA kink-turn (K-turn) motif. K-turns have been identified as sites for RNA-protein interactions in various RNAs (Klein et al. 2001), and the question is thus raised whether such or similar motifs play a more widespread role in RNA localization. In small untranslated BC1 RNA, a 65-nt

segment in the 5' stem-loop domain is sufficient for dendritic targeting (Muslimov et al. 1997). (Because all localized RNAs, regardless of whether they are translated, carry codes to specify their destination sites, the term noncoding should be avoided; see Brosius & Tiedge 2004.) In various other dendritic mRNAs, DTEs have not yet been mapped or identified. Taken together, however, dendritic targeting signals of different mRNAs appear to be quite diverse, differing in length, sequence, relative position, and number per RNA, and encoding differential destination sites. Therefore, one is left to wonder whether the apparent complexity of DTEs is indeed dictated solely by biological necessity or if underlying principles have been slow to emerge.

Mammalian oligodendrocytes are highly elongated glial cells that extend extensively branched processes to form insulating myelin sheaths around axons. Several mRNAs, including the one encoding myelin basic protein (MBP), are transported into these processes (Barbarese et al. 1999, Boccaccio 2000). Targeting of MBP mRNA is mediated by an 11-nt cis-acting element, referred to as the heterogeneous ribonucleoprotein (hnRNP) A2 response element (A2RE), that has also been implicated in RNP granule formation and cap-dependent translation (Carson et al. 2001). A2REs in three different transcripts of the human immunodeficiency virus 1 (HIV-1) also mediate extrasomatic targeting of microinjected mRNAs in cultured oligodendrocytes, albeit with distinct efficiencies (Mouland et al. 2001). Flanking sequences thus seem to modulate the targeting competence of A2RE. A2RE has also been shown to support dendritic targeting in neurons (Shan et al. 2003, Smith 2004). An A2RE-like element was identified in the ORF of dendritic MAP2 mRNAs; in this case, however, a DTE is located in the 3' UTR, whereas the ORF was found targeting incompetent (Blichenberg et al. 1999).

Actin polymerization at the leading edge of moving cells regulates the extension of lamellipodia, filopodia, or pseudopodia (Rafelski

CPE: cytoplasmic polyadenylation element

ORF: open reading frame

RNA motif: a three-dimensional structural design that is contained, in identical or similar form, in a number of RNAs. Such motifs may represent codes that are recognized by *trans*-acting factors

UTR: untranslated region

KH domain:
ribonucleoprotein K
homology domain

RRM: RNA
recognition motif

Trans-acting factor
(TAF): a factor,
typically a protein,
that mediates RNA
functionality (e.g.,
subcellular targeting)
by recognizing a
code within a
cis-acting element

& Theriot 2004). In chicken embryonic fibroblasts and myoblasts, β-actin transcripts are concentrated just proximal to the leading edge (Lawrence & Singer 1986). A 54-nt "zipcode" in the 3′ UTR mediates mRNA sorting to the leading edge (Kislauskis et al. 1994). The same zipcode also mediates targeting of β-actin messages into both growth cones of immature chicken neurons (Zhang et al. 2001) and dendrites of mature rat hippocampal neurons (Eom et al. 2003). Thus the functional role of the β-actin *cis*-acting element appears to be conserved in avian and mammalian species. On the other hand, restriction of avian MAP2 mRNA to neuronal somata in the chicken retina (in contrast to a somatodendritic distribution in the mammalian retina) and the absence of a DTE-like sequence in the 3′ UTR of the avian mRNA suggest that acquisition of dendritic targeting competence has been a relatively recent event in evolution (Cristofanilli et al. 2004). Taken together, the currently available data indicate that localized mRNAs use both shared and unique *cis*-acting elements that may have been acquired at distinct time points during evolution. Long *cis*-acting elements often appear to consist of multiple short motifs, sometimes cooperating, sometimes partially redundant. The physical description of such motifs will be an important future challenge.

Trans-Acting Factors

By definition, *trans*-acting factors (TAFs) are proteins that support RNA transport and/or localization by binding to *cis*-acting elements (**Figure 1**). In this context, TAFs are decoding devices, proteins that recognize location-specifying codes in RNAs.

Vg1RBP/VERA and ZBP1, TAFs that interact with the Vg1-VLE and the β-actin *cis*-acting element, respectively, represent two members of a family of RNA-binding proteins that contain two RNA recognition motifs (RRMs) and four hnRNP K homology (KH) domains (Bassell & Kelic 2004, Yaniv & Yisraeli 2002). Zipcode mutations reduce ZBP1 association with β-actin mRNA and transcript localization to the leading edge of fibroblasts (Ross et al. 1997). In developing neurons, disruption of the interaction between the zipcode and ZBP1 inhibits neurotrophin-induced β-actin mRNA localization into growth cones and reduces growth cone motility (Zhang et al. 1999). In mature neurons, ZBP1 and β-actin mRNA granules reside in dendritic shafts and spines (Tiruchinapalli et al. 2003). Reduced ZBP1 levels diminish dendritic β-actin mRNA concentrations and impair the growth of dendritic filopodia after neurotrophin stimulation, whereas overexpression of zipcode-containing mRNAs increases the density of dendritic filopodia (Eom et al. 2003). Members of the Vg1RBP/ZBP1 family of RNA-binding proteins are thus mediators of cytoplasmic mRNA targeting in different species and cell systems.

Two TAFs that interact with dendritic MAP2 mRNA, MARTA1 and MARTA2, belong to the FUSE-binding protein (FBP) subfamily of KH-domain proteins (Duncan et al. 1994, Rehbein et al. 2002; K. Zivraj, M. Rehbein, F. Buck, M. Schweizer, D. Richter, S. Kindler, manuscript in preparation). MARTA1 is the rodent ortholog of human KH-type splicing regulatory protein (KSRP/FBP2), a component of a multiprotein complex that has been implicated in neuron-specific splicing (Min et al. 1997). ZBP2, the chicken ortholog of MARTA1, interacts with the zipcode of β-actin mRNA (Gu et al. 2002). Located predominantly in the nucleus, ZBP2 undergoes nucleo-cytoplasmic shuttling, and overexpression of truncated ZBP2 partially disrupts β-actin mRNA targeting to both lamellipodia of polarized chicken fibroblasts and growth cones of neurons in culture.

In contrast to MARTA1/ZBP2, MARTA2/FBP3 predominantly resides in the somatodendritic compartment of neurons where it was found associated with polysomes. When expressed in cultured neurons, exogenous

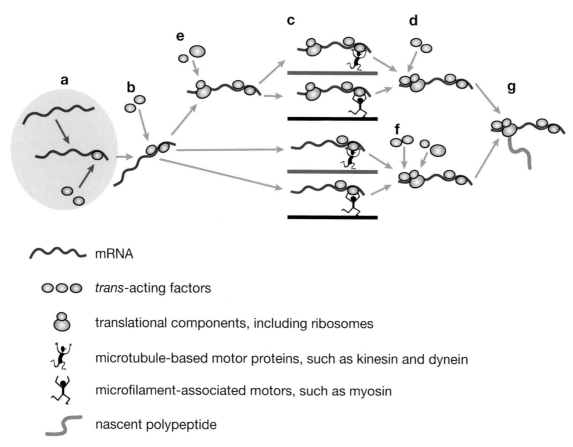

mRNA

trans-acting factors

translational components, including ribosomes

microtubule-based motor proteins, such as kinesin and dynein

microfilament-associated motors, such as myosin

nascent polypeptide

Figure 1

Active transport of mRNAs from the nucleus (*left, light blue*) to peripheral cytoplasmic destination sites in eukaryotic cells. Transport proceeds through a number of sequential phases as follows: (*a*) Recognition of *cis*-acting elements by nuclear *trans*-acting factors (TAFs) (*purple ellipses*) and formation of ribonucleotide protein (RNP) complexes; (*b*) nuclear export of RNP particles and recruitment of cytoplasmic TAFs (*yellow ellipses*); (*c*) motor-based active transport of cytoplasmic RNP particles along cytoskeletal filaments (*green and blue lines*); (*d, f*) association with additional cytoplasmic TAFs, such as anchoring proteins, at destination sites (*light green ellipses*); (*e, f*) recruitment of ribosomes and other translational components (not shown) either before (*e*) or after (*f*) cytoplasmic translocation; (*g*) locally controlled translation.

MARTA2 partially colocalizes with recombinant DTE-containing mRNAs in granules along dendrites. Overexpression of a truncated MARTA2 version completely disrupts dendritic targeting of endogenous MAP2 mRNA granules. Thus two members of the FBP family appear to participate in RNA localization in different cell types.

Testis/brain RNA-binding protein (TB-RBP; known as translin in primates) binds to conserved Y and H elements of several testis and brain mRNAs and represses their translation (Han et al. 1995, Kwon & Hecht 1993). TB-RBP attaches translationally repressed mRNAs to the microtubule cytoskeleton (Han et al. 1995). Its presence in intercellular cytoplasmic bridges between spermatids suggests that TB-RBP may mediate mRNA transport between male germ cells (Morales et al. 1998). These data indicate an interplay between RNA transport and translational repression.

Translational control element (TCE): a segment within an RNA that contains a code to instruct regulation of its translation

Other proteins may also play dual roles in RNA targeting and translational control. One example is provided by the Staufen family of double-stranded (ds) RNA-binding proteins (Roegiers & Jan 2000). In *Drosophila* oocytes and early embryos, Staufen is essential for localization and translational control of different maternal transcripts and for the establishment of the antero-posterior body pattern. Two Staufen orthologs, Stau1 and Stau2, have been described in mammals (Duchaine et al. 2002, Monshausen et al. 2001, Tang et al. 2001). Stau1 is found in most tissues, whereas Stau2 is preferentially present in brain (Duchaine et al. 2002, Monshausen et al. 2001). In neurons, both proteins are located in dendrites, associate with dendritic microtubules, and reside in granules (Duchaine et al. 2002, Monshausen et al. 2001, Tang et al. 2001). At the same time, enrichment in polysome fractions may suggest an additional role in translation (see below). Similarly, the cytoplasmic polyadenylation element (CPE), contained in a subset of dendritic mRNAs, performs the dual functions as a translational control element (TCE) and as a DTE. Its binding protein, CPEB, mediates both cytoplasmic polyadenylation-induced translation (see below) and transport of CPE-containing mRNAs to dendrites (Huang et al. 2003). In summary, it appears that localized RNAs may recruit binding proteins for the dual purpose of RNA transport and local control of translation, underscoring once again that the two mechanisms are functionally interdependent.

Cytoplasmic mRNA targeting may already be initiated in the nucleus, as newly accumulating evidence from various cell types has been suggesting. Several TAFs are nucleo-cytoplasmic shuttling proteins that first appear to associate with transcript in the nucleus and subsequently direct mRNA targeting in the cytoplasm (Farina & Singer 2002). In *Drosophila* oocytes, the nuclear processing history of oskar transcripts is a determinant of their cytoplasmic fate (Hachet & Ephrussi 2004). Similarly, transcript recognition and formation of a specific RNP complex in the nucleus is an early event in Vg1 mRNA localization in *Xenopus* oocytes (Kress et al. 2004). The RNP is remodeled after its export to the cytoplasm, and additional transport factors are recruited into the complex. Several other TAFs also seem to shuttle between the nucleus and the cytoplasm (Bassell & Kelic 2004). The notion thus emerges that RNA targeting in eukaryotic cells is often a multi-step process in which an RNA, from the time of its transcription in the nucleus to the anchoring at its final destination site, enlists a set of TAFs in a sequentially orchestrated manner. Relay-type targeting, with multiple TAFs cooperating in the delivery of the cargo, may be advantageous especially when long and complex transport routes have to be negotiated, e.g., in dendrites with thousands of potential synaptic destination sites.

Cytoskeletal Elements and Molecular Motors

Different cytoskeletal elements have been shown to support RNA transport and local anchoring (**Figure 1**). In fibroblasts, microfilaments are used to localize particles containing β-actin transcripts and ZBP1. In neurons, on the other hand, ZBP1 and its β-actin mRNA target seem to move predominantly along microtubules (Bassell & Kelic 2004). Thus ZBP1 may serve as an adaptor between mRNA and either microfilament- or microtubule-based molecular motors. It is plausible that in neurons, long-range RNA transport is mediated by microtubules, whereas localization in the destination microdomain is supported by actin filaments (Muslimov et al. 2002). Neuronal CPEB granules contain both kinesin and dynein motors, and their bidirectional movement in dendrites is microtubule dependent (Huang et al. 2003). In addition, kinesin 1 complexes from mammalian brain were shown to contain several of the aforementioned TAFs (Kanai et al. 2004). In the axon-like processes of P19 embryonic carcinoma cells, knock-down of another kinesin family member, KIF3A, impairs the sorting of tau mRNA (Aronov et al. 2002). In oligo-dendrocytes, inhibition of kinesin disrupts targeting of MBP mRNAs (Carson et al. 1997); in testis, kinesin KIF17b associates with TB-RBP, suggesting that a microtubule-dependent RNA transport system operates in mammalian male germ cells (Chennathukuzhi et al. 2003). Evidence is thus accumulating that long-range RNA transport in various mammalian cell types is mediated by microtubule-based kinesin- and dynein-type molecular motors.

LOCAL CONTROL OF TRANSLATION

In this section, we discuss mechanisms of local regulation of translation, using selected examples to highlight common principles. We probe the significance of translational repression en route and at destination sites. A brief synopsis of the translational pathway in eukaryotes serves as an introduction.

Mechanisms of Local Translational Control

The translation of an mRNA into a cognate protein proceeds in the three sequential steps of initiation, elongation, and termination (reviewed in Sonenberg et al. 2000). Regulation can occur at any of these steps, but initiation is typically rate-limiting and thus often a target for regulation (Gingras et al. 1999, Kapp & Lorsch 2004).

During initiation, a 43S preinitiation complex is formed by the binding of an eIF2•GTP•Met-tRNAi ternary complex to the 40S ribosomal subunit (**Figure 2**). The 43S complex is then recruited to the initiator codon of the mRNA to form a stable 48S complex. This step requires participation of a set of factors from the eIF4 family. These factors subsequently dissociate, the 60S ribosomal subunit joins the 40S subunit to form an 80S complex, and elongation ensues (**Figure 2**; Hershey & Merrick 2000). During elongation, eEF1A guides aminoacyl-tRNAs to the A site on the ribosome; following peptide bond formation, the ribosome is translocated by one codon along the mRNA, a step that is catalyzed by eEF2. Finally, termination at a stop codon is mediated by a set of release factors.

Because of its relevance for the long-term, experience-dependent modulation of synaptic strength (and thus for neuronal plasticity), translational control of gene expression in synapto-dendritic domains has increasingly become a subject of general interest. Therefore, we use this system as a case in point to exemplify some of the emerging general principles of local translational control (**Figure 3**).

Translation initiation factor eIF4E, a subunit of the eIF4F complex, binds to the 5′ cap of mRNAs and promotes recruitment of the 43S preinitiation complex (**Figure 2**; Gingras et al. 1999, Kapp & Lorsch 2004). Several signaling pathways have been reported in neurons that modulate translation by targeting eIF4E. One involves BDNF and the mammalian target of rapamycin (mTOR), a serine/threonine kinase. Activated mTOR phosphorylates eIF4E-binding proteins (eIF4E-BPs), resulting in their dissociation from eIF4E and consequently in the activation of translation initiation (Raught et al. 2000). This pathway has been implicated in translational regulation in mammalian neurons (Takei et al. 2001, 2004; Tang et al. 2002). Key components of the mTOR pathway, such as mTOR, eIF4E, and eIF4E-BP, have been identified in postsynaptic domains (Tang et al. 2002). BDNF, a member of the neurotrophin family, has been shown to be involved in long-term synaptic potentiation (Korte et al. 1995). Application of rapamycin, an inhibitor of the mTOR pathway, to hippocampal slices prevents BDNF-induced synaptic potentiation (Tang et al. 2002). BDNF induces phosphorylation of eIF4E-BP, eIF4E, and mTOR in cortical neurons, as well as in isolated dendrites (Takei et al. 2001, 2004), indicating that the mTOR pathway is essential for BDNF to activate translation in dendrites. A subset of mRNAs has been identified as targets

Eukaryotic initiation factor (eIF): a factor that promotes translation initiation, i.e., the sequence of events that results in the formation of an 80S ribosomal complex at the AUG initiator codon

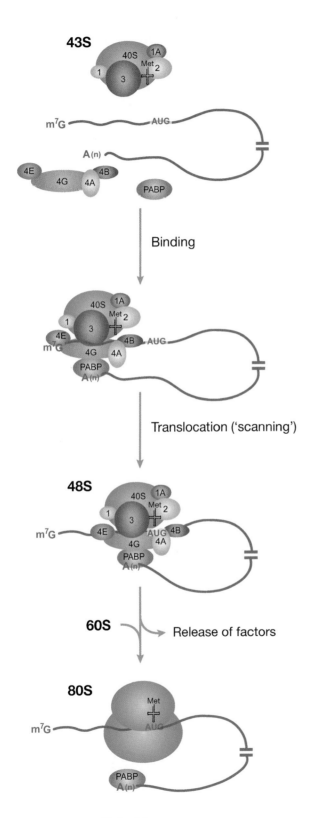

of the BDNF-activated mTOR pathway (Schratt et al. 2004).

The mitogen-activated protein kinase (MAPK) signaling pathway has also been reported to activate local translation through its downstream target eIF4E (Kelleher et al. 2004). Upon phosphorylation by MAPK/ERK kinase (MEK), extracellular signal-regulated kinase (ERK), a member of the MAPK family, activates its downstream substrate MAP kinase-interacting kinase 1 (Mnk1) (Fukunaga & Hunter 1997, Waskiewicz et al. 1997). Activated Mnk subsequently phosphorylates eIF4E, thus promoting translation initiation (Pyronnet et al. 1999, Waskiewicz et al. 1999). These combined findings suggest a coregulation of local translation by targeting eIF4E through the MEK and mTOR signaling pathways (Gingras et al. 1999; **Figure 3**).

Work with *Drosophila* has recently indicated that expression of eIF4E itself is subject to regulation. Pumilio (Pum), a translational repressor, plays an important role in determining the *Drosophila* anterior-posterior body pattern during early embryogenesis (Johnstone & Lasko 2001). Pum has been found to control dendritic morphogenesis (Ye et al. 2004) and to be involved in long-term memory formation (Dubnau et al. 2003). At

Figure 2

Cap-dependent translation initiation in eukaryotes. Recruitment of the 43S preinitiation complex to the mRNA, mediated by factors of the eIF4 family and poly(A) binding protein (PABP), results in the formation of a 48S complex at the AUG initiator codon. Alternatively to the scenario shown here, eIF4E may remain bound to the cap during scanning (resulting in a looping-out of the 5′ UTR; Jackson 2000); in general, it remains to be established at which point after initial cap-binding the eIF4 factors dissociate from the mRNA and from PABP. Translation may also be initiated in cap-independent fashion, for instance by binding of the 43S complex to an internal ribosome entry site (IRES), which is typically located directly upstream of the AUG start codon (not shown; see Hellen & Sarnow 2001). Some factors (e.g., eIF5) have been omitted for clarity.

the *Drosophila* neuromuscular junction, postsynaptic Pum downregulates eIF4E expression (Menon et al. 2004). Pum selectively interacts with the 3′ UTR of eIF4E mRNA, indicating that Pum may modulate synaptic function through direct control of eIF4E expression (Menon et al. 2004).

eIF4F, the heterotrimeric protein that mediates recruitment of the 43S complex to the mRNA, is composed of eIF4E, the cap-binding protein; eIF4G, a large scaffolding protein; and eIF4A, an ATP-dependent RNA helicase that unwinds secondary structure elements in the 5′ UTR prior to recruitment of the 43S complex (**Figure 2**; Gingras et al. 1999, Kapp & Lorsch 2004). The activity of eIF4F is significantly enhanced by its interaction with PABP, a protein that binds to the poly(A) tails of mRNAs. This interaction has been shown to be the target of dendritic BC1 RNA (Wang et al. 2002). BC1 RNA is a small, untranslated RNA that is selectively transported to dendrites (Muslimov et al. 1997). It represses translation initiation by inhibiting formation of the 48S preinitiation complex (Wang et al. 2002). BC1-mediated repression is effective in cap-dependent initiation and in one subtype of IRES-mediated initiation. Because IRES-mediated initiation may be the preferred mode at the synapse, particularly in response to stimulation (Dyer et al. 2003, Pinkstaff et al. 2001), BC1 RNA is well positioned for a modulatory role in synaptic translation.

Also targeting translation initiation, the CPE pathway has first been analyzed in *Xenopus* oocytes. Some mRNAs remain translationally dormant until their short poly(A) tails are extended in the cytoplasm (Mendez & Richter 2001). Cytoplasmic polyadenylation is dependent on the presence of a CPE in the 3′ UTR of the mRNA. Upon phosphorylation of CPEB by the kinase aurora, CPEB binds to CPE and recruits poly(A) polymerase to polyadenylate the mRNA. CPEB is in turn recognized by two CPEB-binding proteins, Symplekin, a scaffold protein, and xGLD-2, a poly(A) polymerase (Barnard et al. 2004).

Polyadenylation promotes the dissociation of maskin from eIF4E, thus "demasking" eIF4E and allowing it to participate in translation initiation (Richter 2000).

Dendritic CaMKIIα mRNA contains two CPE-like sequences in its 3′ UTR (Wu et al. 1998). Translation of CaMKIIα mRNA depends on the same polyadenylation mechanism that has previously been described in *Xenopus* oocytes, and factors such as aurora, CPEB, poly(A) polymerase, and maskin have been identified at synaptic sites of hippocampal neurons (Huang et al. 2002, Wells et al. 2001, Wu et al. 1998). NMDAR activation was found essential for CPE-dependent synthesis of CaMKIIα (Wells et al. 2001). NMDAR stimulation induced phosphorylation of aurora, thus activating local translation of CaMKIIα mRNA (Huang et al. 2002). Dendritic plasminogen activator (tPA) mRNA has recently also been shown to interact with CPEB; it was rapidly polyadenylated and translated into cognate protein after metabotropic glutamate receptor (mGluR) activation (Shin et al. 2004). Cytoplasmic polyadenylation thus provides a mechanism of translational control that is selective for CPE-bearing mRNAs.

Elongation may also be a target for local translational control, if not as frequently as initiation. One such target is eEF2 (Scheetz et al. 1997, 2000). In the amphibian tectum, phosphorylation of eEF2 is dependent on the activation of NMDARs and can be induced by visual stimulation (Scheetz et al. 1997). Phosphorylation of eEF2 results in reduced overall protein synthesis in eukaryotes (Nairn et al. 2001) but may enhance translation of some mRNAs in developing neurons (Scheetz et al. 2000). This mechanism may thus provide an alternative to the typically initiation-targeted local translational control (**Figure 3**).

FMRP has been implicated in the regulation of dendritic translation, but the exact mode of action has not yet been resolved (Antar & Bassell 2003, O'Donnell & Warren 2002). The protein inhibits translation whereas a disease-causing mutant FMRP does

IRES: internal ribosome entry site

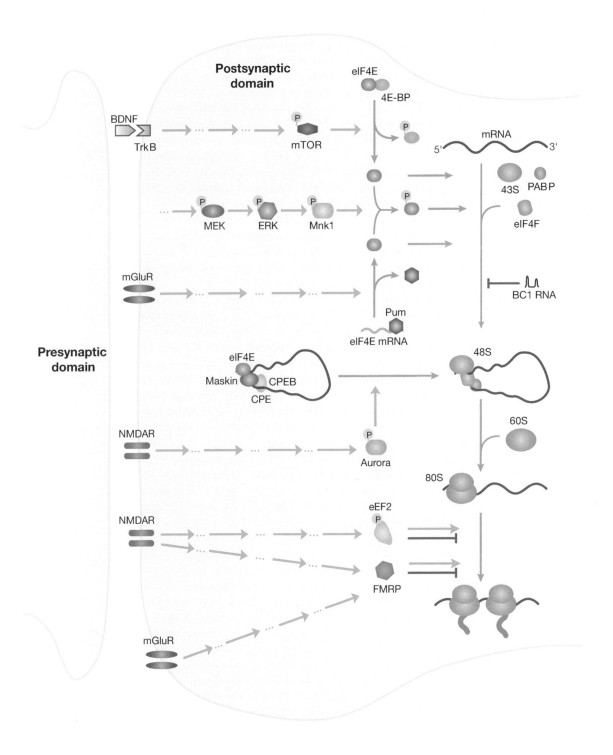

not (Laggerbauer et al. 2001). FMRP has been suggested to inhibit translation through interaction with mRNAs (Li et al. 2001). At the synapse, NMDAR, as well as mGluR activation, has been implicated in the FMRP pathway (Antar et al. 2004, Gabel et al. 2004, Weiler et al. 1997). FMRP has also been suggested to interact with untranslated BC1 RNA to repress translation of such mRNAs that base-pair with the 5′ BC1 domain (Zalfa et al. 2003). However, because FMRP associates with polyribosomes (Corbin et al. 1997; Eberhart et al. 1996; Feng et al. 1997; Khandjian et al. 1996, 2004; Stefani et al. 2004; Weiler et al. 2004), whereas BC1 RNA is enriched in the lighter RNP fractions (Krichevsky & Kosik 2001), it appears more likely that FMRP and BC1 RNA operate in translational control without directly interacting with each other. Furthermore, the functional role of FMRP in translation may be more complex in vivo—inhibitory or stimulatory, depending on the target mRNA— as polysome loading was found differentially altered in the absence of FMRP (Brown et al. 2001). The phosphorylation status of FMRP may be another determinant of translational regulation: Whereas nonphosphorylated FMRP was found associated with actively translating polyribosomes, phosphorylated FMRP tended to be associated with apparently stalled polyribosomes (Ceman et al. 2003). In this scenario, FMRP would modulate translation at the level of elongation (**Figure 3**). Another twist is added by the fact that FMRP binds eEF1A mRNA and represses its translation in vivo (Sung et al. 2003). Another proposed mechanism links FMRP with microRNAs (miRNAs) (Jin et al.

2004). miRNAs are small RNAs (21–24 nt) that function as translation repressors through partial base-pairing with their target mRNAs (Ambros 2003, Ruvkun 2001). miRNAs have been identified from mammalian brain preparations where they were found enriched in polyribosome fractions (Kim et al. 2004). dFXR, the *Drosophila* ortholog of FMRP, has also been suggested to interact in the RNAi pathway (Caudy et al. 2002, Ishizuka et al. 2002), supporting the hypothesis that FMRP regulates translation of target mRNAs through miRNAs and/or small interfering RNAs (siRNAs).

Translational Repression During Transport

It is generally assumed that mRNAs will have to be translationally repressed while en route to prevent ectopic expression (Kwon et al. 1999). This principle has been confirmed in several cell types.

During oogenesis and early embryogenesis, *Drosophila* oskar mRNA is guided by its 3′ UTR from nurse cells into the anterior margin of the oocytes to be subsequently anchored at the posterior pole (Johnstone & Lasko 2001). During transport, the translation of oskar mRNA is repressed by several *trans*-acting factors, including Bruno and Apontic proteins (Lie & MacDonald 1999, Webster et al. 1997). oskar mRNA is translated into cognate protein only after it reaches the posterior pole; in fact, ectopic expression of oskar mRNA during transport results in severe developmental defects (Kim-Ha et al. 1995). Support for this model comes from recent work with Cup, a newly identified

---

Figure 3

Many roads lead to Rome: possible pathways of translational control at the synapse. Translational control can be implemented by activation of various receptor systems (*left*) and can be mediated at the level of initiation and elongation (*right*). Mechanisms in addition to those summarized here have been discussed; for example, FMRP has also been reported to repress translation at the level of initiation (Laggerbauer et al. 2001). Green arrows, stimulation; red arrows, inhibition. Aspects of this diagram remain speculative; note that not all factors shown are currently known to be active at the synapse. Factors not drawn to scale.

component of the oskar RNP complex (Wilhelm et al. 2003). Cup, an eIF4E-binding protein, is required for both transport and translational repression of oskar mRNA, suggesting again a close coordination between the two processes (Wilhelm et al. 2003).

A similar coordination is provided by CPEB in neurons. As discussed above, CPEB has been shown to facilitate dendritic transport of CPE-containing mRNAs (Huang et al. 2003). During transport, CPEB is colocalized with the translational repressor maskin in transport particles. An interplay between both proteins may therefore play a role in the transport of translationally dormant mRNAs in dendrites (Huang et al. 2003).

In many cell types, RNP complexes have been identified as the actual motile transport units. These macromolecular complexes appear to contain multiple RNAs, associated TAFs, and/or components of the translational machinery. Work with MBP mRNA showed that transport granules were formed within minutes after microinjection of the RNA into oligodendrocytes (Ainger et al. 1997). Translation of MBP mRNA can be activated only when it arrives within the myelinating processes, thereby avoiding inappropriate localization of MBP (Boccaccio 2000). RNA transport granules have been reported to lack essential translational components and to be unable to incorporate radioactive amino acids, indicating that they are translationally incompetent (Krichevsky & Kosik 2001). The translational repressor FMRP has recently been found to interact with the human ZBP1 ortholog IMP1 in granules (Rackham & Brown 2004). Similarly, the RNA-binding protein RNG105 is a translational repressor that has been found associated with neuronal RNA granules (Shiina et al 2005). These data suggest further functional links between RNA transport and translational repression.

In summary, translational repression seems to be important during RNA transport to prevent ectopic expression, and—at least in neurons—also after arrival at the destination site. Inappropriately controlled protein synthesis at the synapse may upset a delicate neuronal excitation-inhibition balance, and therefore, it may be essential to have multiple systems in place to keep local translation in the "off" state until a valid "go" signal is received. The nature of such a signal, and the mechanism by which it is transduced to the local translational machinery, will be a major challenge for future work.

PERSPECTIVE

Impressive as it may be, recent progress in the field of RNA transport and local translational control can not distract from the fact that we remain ignorant of some very fundamental aspects. Various *cis*-acting RNA elements have been described that encode subcellular destination sites, but we do not understand how such codes are expressed in the structural design of RNA motifs. We are therefore not yet able to read such codes, and we do not know how they are decoded by TAFs. What is urgently needed is a physical description—in conjunction with a functional dissection—of code-carrying RNA motifs that specify targeting. The number of such motifs may in fact be limited (Moore 1999). In evolutionary terms, RNA has predated proteins, and at least some functional RNA motifs may therefore have evolved early. It will be interesting to see how such motifs have been recruited and adapted by RNAs to encode subcellular destination sites, and how they have evolved to differ from one another in the various aspects of RNA targeting. A physical description of RNA target-encoding motifs will also be essential for us to understand how transport RNPs are assembled around *cis*-acting elements, how such RNP formation shapes the molecular architecture of transport-competent particles that have been described in various systems, and how these particles are delivered to their final destination sites.

Similarly, although much has been learned about translational control mechanisms, our understanding of how translation is repressed during transport and at the destination site remains rudimentary. It is poorly understood how repression is implemented in transport granules, and how mRNAs are released from granules. When and how are released RNAs activated (i.e., derepressed) for translation, and what are the roles of untranslated RNAs in these mechanisms? These questions are particularly relevant at synaptic sites in neurons where translational repression/activation is assumed to be controlled by the local activity status (Smith et al. 2005).

Activity-dependent local protein synthesis is considered one of the key mechanisms in the long-term modulation of synaptic connections, and the pathway is thus seen underlying higher brain functions, including learning and memory. This area will certainly attract more attention, in part because it appears that neurological diseases such as the Fragile X Syndrome are causally related to RNA localization and/or translation at synaptic sites. However, we are still far from arriving at a clear picture of the underlying molecular mechanism and the way it is disrupted in the disease. To understand the disease, it appears, it will be necessary that we first understand the codes.

SUMMARY POINTS

1. Both mRNAs and untranslated RNAs are localized in various eukaryotic cell types.

2. *Cis*-acting elements contain codes, likely to be expressed as RNA motifs, that specify RNA targeting.

3. Targeting codes are recognized by *trans*-acting factors, proteins that bind *cis*-acting elements and act as adaptors in the assembly of larger ribonucleoprotein (RNP) particles.

4. RNPs are translocated along cytoskeletal elements by molecular motors.

5. Microtubules and microfilaments have been implicated in subcellular RNA localization; long-range transport in neurons is typically mediated by microtubules.

6. Most transported mRNAs appear to be translationally dormant while en route.

7. In neurons, mRNAs at their destination sites will often have to remain translationally repressed until an activation signal is received.

8. Untranslated RNAs may participate in the local control of translation.

ACKNOWLEDGMENTS

Work in the authors' laboratories has been supported by the Deutsche Forschungsgemeinschaft (Ki488/2-6 to S.K., Ri192/24-1 to D.R.); the Human Frontier Science Program Organization (RG0120/1999-B to S.K.); the Volkswagenstiftung (i/74,041 to D.R.); the European Commission (QLG3-CT-1999-00,908 to D.R.); the U.S. Department of Defense (DAMD17-02-1-0520 to H.T.); and the National Institutes of Health (NS34158 and NS46769 to H.T.). H.T. thanks Jun Zhong for comments on the manuscript, and Jürgen Brosius and Christopher Hellen for stimulating discussions.

The authors declare that they have no conflicting financial interests.

LITERATURE CITED

Ainger K, Avossa D, Diana AS, Barry C, Barbarese E, Carson JH. 1997. Transport and localization elements in myelin basic protein mRNA. *J. Cell Biol.* 138:1077–87

Ambros V. 2003. MicroRNA pathways in flies and worms: growth, death, fat, stress, and timing. *Cell* 113:673–76

Antar LN, Afroz R, Dictenberg JB, Carroll RC, Bassell GJ. 2004. Metabotropic glutamate receptor activation regulates fragile X mental retardation protein and FMR1 mRNA localization differentially in dendrites and at synapses. *J. Neurosci.* 24:2648–55

Antar LN, Bassell GJ. 2003. Sunrise at the synapse: the FMRP mRNP shaping the synaptic interface. *Neuron* 37:555–58

Aranda-Abreu GE, Behar L, Chung S, Furneaux H, Ginzburg I. 1999. Embryonic lethal abnormal vision-like RNA-binding proteins regulate neurite outgrowth and tau expression in PC12 cells. *J. Neurosci.* 19:6907–17

Aronov S, Aranda G, Behar L, Ginzburg I. 2002. Visualization of translated tau protein in the axons of neuronal P19 cells and characterization of tau RNP granules. *J. Cell Sci.* 115:3817–27

Barbarese E, Brumwell C, Kwon S, Cui H, Carson JH. 1999. RNA on the road to myelin. *J. Neurocytol.* 28:263–70

Barnard DC, Ryan K, Manley JL, Richter JD. 2004. Symplekin and xGLD-2 are required for CPEB-mediated cytoplasmic polyadenylation. *Cell* 119:641–51

Bassell GJ, Kelic S. 2004. Binding proteins for mRNA localization and local translation, and their dysfunction in genetic neurological disease. *Curr. Opin. Neurobiol.* 14:574–81

Betley JN, Frith MC, Graber JH, Choo S, Deshler JO. 2002. A ubiquitous and conserved signal for RNA localization in chordates. *Curr. Biol.* 12:1756–61

Blichenberg A, Rehbein M, Muller R, Garner CC, Richter D, Kindler S. 2001. Identification of a *cis*-acting dendritic targeting element in the mRNA encoding the alpha subunit of Ca^{2+}/calmodulin-dependent protein kinase II. *Eur. J. Neurosci.* 13:1881–88

Blichenberg A, Schwanke B, Rehbein M, Garner CC, Richter D, Kindler S. 1999. Identification of a *cis*-acting dendritic targeting element in MAP2 mRNAs. *J. Neurosci.* 19:8818–29

Boccaccio GL. 2000. Targeting of mRNAs within the glial cell cytoplasm: how to hide the message along the journey. *J. Neurosci. Res.* 62:473–79

Böckers TM, Segger-Junius M, Iglauer P, Bockmann J, Gundelfinger ED, et al. 2004. Differential expression and dendritic transcript localization of Shank family members: identification of a dendritic targeting element in the 3′ untranslated region of Shank1 mRNA. *Mol. Cell. Neurosci.* 26:182–90

Bodian D. 1965. A suggestive relationship of nerve cell RNA with specific synaptic sites. *Proc. Natl. Acad. Sci. USA* 53:418–25

Brosius J, Tiedge H. 2004. RNomenclature. *RNA Biol.* 1:81–83

Brown V, Jin P, Ceman S, Darnell JC, O'Donnell WT, et al. 2001. Microarray identification of FMRP-associated brain mRNAs and altered mRNA translational profiles in fragile X syndrome. *Cell* 107:477–87

Carson JH, Cui H, Krueger W, Schlepchenko B, Brumwell C, Barbarese E. 2001. RNA trafficking in oligodendrocytes. *Results Probl. Cell Differ.* 34:69–81

Carson JH, Worboys K, Ainger K, Barbarese E. 1997. Translocation of myelin basic protein mRNA in oligodendrocytes requires microtubules and kinesin. *Cell Motil. Cytoskelet.* 38:318–28

Caudy AA, Myers M, Hannon GJ, Hammond SM. 2002. Fragile X-related protein and VIG associate with the RNA interference machinery. *Genes Dev.* 16:2491–96

Ceman S, O'Donnell WT, Reed M, Patton S, Pohl J, Warren ST. 2003. Phosphorylation influences the translation state of FMRP-associated polyribosomes. *Hum. Mol. Genet.* 12:3295–305

Chartrand P, Singer RH, Long RM. 2001. RNP localization and transport in yeast. *Annu. Rev. Cell Dev. Biol.* 17:297–310

Chennathukuzhi V, Morales CR, Fl-Alfy M, Hecht NB. 2003. The kinesin KIF17b and RNA-binding protein TB-RBP transport specific cAMP-responsive element modulator-regulated mRNAs in male germ cells. *Proc. Natl. Acad. Sci. USA* 100:15566–71

Claussen M, Pieler T. 2004. Xvelo1 uses a novel 75-nucleotide signal sequence that drives vegetal localization along the late pathway in *Xenopus* oocytes. *Dev. Biol.* 266:270–84

Colman DR, Kreibich G, Frey AB, Sabatini DD. 1982. Synthesis and incorporation of myelin polypeptides into CNS myelin. *J. Cell Biol.* 95:598–608

Corbin F, Bouillon M, Fortin A, Morin S, Rousseau F, Khandjian EW. 1997. The fragile X mental retardation protein is associated with poly(A)$^+$ mRNA in actively translating polyribosomes. *Hum. Mol. Genet.* 6:1465–72

Couzin J. 2002. Small RNAs make big splash. *Science* 298:2296–97

Cristofanilli M, Thanas S, Brosius J, Kindler S, Tiedge H. 2004. Neuronal MAP2 mRNA: species-dependent differential dendritic targeting competence. *J. Mol. Biol.* 341:927–34

Darnell RB. 2003. Memory, synaptic translation, and...prions? *Cell* 115:767–68

Deshler JO, Highett MI, Schnapp BJ. 1997. Localization of *Xenopus* Vg1 mRNA by Vera protein and the endoplasmic reticulum. *Science* 276:1128–31

Dubnau J, Chiang AS, Grady L, Barditch J, Gossweiler S, et al. 2003. The staufen/pumilio pathway is involved in *Drosophila* long-term memory. *Curr. Biol.* 13:286–96

Duchaine TF, Hemraj I, Furic L, Deitinghoff A, Kiebler MA, DesGroseillers L. 2002. Staufen2 isoforms localize to the somatodendritic domain of neurons and interact with different organelles. *J. Cell Sci.* 115:3285–95

Duncan R, Bazar L, Michelotti G, Tomonaga T, Krutzsch H, et al. 1994. A sequence-specific, single-strand binding protein activates the far upstream element of c-myc and defines a new DNA-binding motif. *Genes Dev.* 8:465–80

Dyer JR, Michel S, Lee W, Castellucci VF, Wayne NL, Sossin WS. 2003. An activity-dependent switch to cap-independent translation triggered by eIF4E dephosphorylation. *Nat. Neurosci.* 6:219–20

Eberhart DE, Malter HE, Feng Y, Warren ST. 1996. The fragile X mental retardation protein is a ribonucleoprotein containing both nuclear localization and nuclear export signals. *Hum. Mol. Genet.* 5:1083–91

Eberwine J, Belt B, Kacharmina JE, Miyashiro K. 2002. Analysis of subcellularly localized mRNAs using in situ hybridization, mRNA amplification, and expression profiling. *Neurochem. Res.* 27:1065–77

Eom T, Antar LN, Singer RH, Bassell GJ. 2003. Localization of a beta-actin messenger ribonucleoprotein complex with zipcode-binding protein modulates the density of dendritic filopodia and filopodial synapses. *J. Neurosci.* 23:10433–44

Farina KL, Singer RH. 2002. The nuclear connection in RNA transport and localization. *Trends Cell Biol.* 12:466–72

Feng Y, Absher D, Eberhart DE, Brown V, Malter HE, Warren ST. 1997. FMRP associates with polyribosomes as an mRNP, and the I304N mutation of severe fragile X syndrome abolishes this association. *Mol. Cell* 1:109–18

Fukunaga R, Hunter T. 1997. MNK1, a new MAP kinase-activated protein kinase, isolated by a novel expression screening method for identifying protein kinase substrates. *EMBO J.* 16:1921–33

Gabel LA, Won S, Kawai H, McKinney M, Tartakoff AM, Fallon JR. 2004. Visual experience regulates transient expression and dendritic localization of fragile X mental retardation protein. *J. Neurosci.* 24:10579–83

Gingras AC, Raught B, Sonenberg N. 1999. eIF4 initiation factors: effectors of mRNA recruitment to ribosomes and regulators of translation. *Annu. Rev. Biochem.* 68:913–63

Giuditta A, Kaplan BB, van Minnen J, Alvarez J, Koenig E. 2002. Axonal and presynaptic protein synthesis: new insights into the biology of the neuron. *Trends Neurosci.* 25:400–4

Goldstein LS, Yang Z. 2000. Microtubule-based transport systems in neurons: the roles of kinesins and dyneins. *Annu. Rev. Neurosci.* 23:39–71

Gu W, Pan F, Zhang H, Bassell GJ, Singer RH. 2002. A predominantly nuclear protein affecting cytoplasmic localization of beta-actin mRNA in fibroblasts and neurons. *J. Cell Biol.* 156:41–51

Hachet O, Ephrussi A. 2004. Splicing of oskar RNA in the nucleus is coupled to its cytoplasmic localization. *Nature* 428:959–63

Han JR, Yiu GK, Hecht NB. 1995. Testis/brain RNA-binding protein attaches translationally repressed and transported mRNAs to microtubules. *Proc. Natl. Acad. Sci. USA* 92:9550–54

Hellen CU, Sarnow P. 2001. Internal ribosome entry sites in eukaryotic mRNA molecules. *Genes Dev.* 15:1593–612

Hershey JWB, Merrick WC. 2000. The pathway and mechanism of initiation of protein synthesis. See Sonenberg et al. 2000, pp. 33–88

Hoek KS, Kidd GJ, Carson JH, Smith R. 1998. hnRNP A2 selectively binds the cytoplasmic transport sequence of myelin basic protein mRNA. *Biochemistry* 37:7021–29

Huang YS, Carson JH, Barbarese E, Richter JD. 2003. Facilitation of dendritic mRNA transport by CPEB. *Genes Dev.* 17:638–53

Huang YS, Jung MY, Sarkissian M, Richter JD. 2002. N-methyl-D-aspartate receptor signaling results in Aurora kinase-catalyzed CPEB phosphorylation and alpha CaMKII mRNA polyadenylation at synapses. *EMBO J.* 21:2139–48

Ishizuka A, Siomi MC, Siomi H. 2002. A *Drosophila* fragile X protein interacts with components of RNAi and ribosomal proteins. *Genes Dev.* 16:2497–508

Jackson RJ. 2000. A comparative view of initiation site selection mechanisms. See Sonenberg et al. 2000, pp. 127–83

Jeffery WR, Tomlinson CR, Brodeur RD. 1983. Localization of actin messenger RNA during early ascidian development. *Dev. Biol.* 99:408–17

Jin P, Zarnescu DC, Ceman S, Nakamoto M, Mowrey J, et al. 2004. Biochemical and genetic interaction between the fragile X mental retardation protein and the microRNA pathway. *Nat. Neurosci.* 7:113–17

Job C, Eberwine J. 2001. Localization and translation of mRNA in dendrites and axons. *Nat. Rev. Neurosci.* 2:889–98

Johnstone O, Lasko P. 2001. Translational regulation and RNA localization in *Drosophila* oocytes and embryos. *Annu. Rev. Genet.* 35:365–406

Ju W, Morishita W, Tsui J, Gaietta G, Deerinck TJ, et al. 2004. Activity-dependent regulation of dendritic synthesis and trafficking of AMPA receptors. *Nat. Neurosci.* 7:244–53

Kanai Y, Dohmae N, Hirokawa N. 2004. Kinesin transports RNA: isolation and characterization of an RNA-transporting granule. *Neuron* 43:513–25

Kapp LD, Lorsch JR. 2004. The molecular mechanics of eukaryotic translation. *Annu. Rev. Biochem.* 73:657–704

Kelleher RJ, Govindarajan A, Jung HY, Kang H, Tonegawa S. 2004. Translational control by MAPK signaling in long-term synaptic plasticity and memory. *Cell* 116:467–79

Khandjian EW, Corbin F, Woerly S, Rousseau F. 1996. The fragile X mental retardation protein is associated with ribosomes. *Nat. Genet.* 12:91–93

This study demonstrates that nuclear RNA processing and RNP assembly regulate cytoplasmic mRNA localization events.

Ligand binding to NMDARs triggers activation of Aurora kinase, resulting in the phosphorylation of CPEB, which in turn is required for the polyadenylation of CPE-containing dormant mRNAs and thus for their translational activation.

This paper and the related paper by Caudy et al. (2002) show that the *Drosophila* fragile X protein is associated with components of the RNAi pathway. FMRP may also interact with components of the miRNA pathway (Jin et al. 2004).

Khandjian EW, Huot ME, Tremblay S, Davidovic L, Mazroui R, Bardoni B. 2004. Biochemical evidence for the association of fragile X mental retardation protein with brain polyribosomal ribonucleoparticles. *Proc. Natl. Acad. Sci. USA* 101:13357–62

Kim J, Krichevsky A, Grad Y, Hayes GD, Kosik KS, et al. 2004. Identification of many microRNAs that copurify with polyribosomes in mammalian neurons. *Proc. Natl. Acad. Sci. USA* 101:360–65

Kim-Ha J, Kerr K, MacDonald PM. 1995. Translational regulation of oskar mRNA by Bruno, an ovarian RNA-binding protein, is essential. *Cell* 81:403–12

King ML, Zhou Y, Bubunenko M. 1999. Polarizing genetic information in the egg: RNA localization in the frog oocyte. *Bioessays* 21:546–57

Kislauskis EH, Zhu X, Singer RH. 1994. Sequences responsible for intracellular localization of beta-actin messenger RNA also affect cell phenotype. *J. Cell Biol.* 127:441–51

Klein DJ, Schmeing TM, Moore PB, Steitz TA. 2001. The kink-turn: a new RNA secondary structure motif. *EMBO J.* 20:4214–21

Kloc M, Etkin LD. 2005. RNA localization mechanisms in oocytes. *J. Cell Sci.* 118:269–82

Kloc M, Zearfoss NR, Etkin LD. 2002. Mechanisms of subcellular mRNA localization. *Cell* 108:533–44

Koenig E. 1965a. Synthetic mechanisms in the axon. I. Local axonal synthesis of acetylcholinesterase. *J. Neurochem.* 12:343–55

Koenig E. 1965b. Synthetic mechanisms in the axon. II. RNA in myelin-free axons of the cat. *J. Neurochem.* 12:357–61

Korte M, Carroll P, Wolf E, Brem G, Thoenen H, Bonhöffer T. 1995. Hippocampal long-term potentiation is impaired in mice lacking brain-derived neurotrophic factor. *Proc. Natl. Acad. Sci. USA* 92:8856–60

Kress TL, Yoon YJ, Mowry KL. 2004. Nuclear RNP complex assembly initiates cytoplasmic RNA localization. *J. Cell Biol.* 165:203–11

Krichevsky AM, Kosik KS. 2001. Neuronal RNA granules: a link between RNA localization and stimulation-dependent translation. *Neuron* 32:683–96

Kwon S, Barbarese E, Carson JH. 1999. The *cis*-acting RNA trafficking signal from myelin basic protein mRNA and its cognate *trans*-acting ligand hnRNP A2 enhance cap-dependent translation. *J. Cell Biol.* 147:247–56

Kwon YK, Hecht NB. 1993. Binding of a phosphoprotein to the 3′ untranslated region of the mouse protamine 2 mRNA temporally represses its translation. *Mol. Cell. Biol.* 13:6547–57

Laggerbauer B, Ostareck D, Keidel EM, Ostareck-Lederer A, Fischer U. 2001. Evidence that fragile X mental retardation protein is a negative regulator of translation. *Hum. Mol. Genet.* 10:329–38

Lawrence JB, Singer RH. 1986. Intracellular localization of messenger RNAs for cytoskeletal proteins. *Cell* 45:407–15

Li Z, Zhang Y, Ku L, Wilkinson KD, Warren ST, Feng Y. 2001. The fragile X mental retardation protein inhibits translation via interacting with mRNA. *Nucleic Acids Res.* 29:2276–83

Lie YS, MacDonald PM. 1999. Apontic binds the translational repressor Bruno and is implicated in regulation of oskar mRNA translation. *Development* 126:1129–38

Link W, Konietzko U, Kauselmann G, Krug M, Schwanke B, et al. 1995. Somatodendritic expression of an immediate early gene is regulated by synaptic activity. *Proc. Natl. Acad. Sci. USA* 92:5734–38

Lyford GL, Yamagata K, Kaufmann WE, Barnes CA, Sanders LK, et al. 1995. Arc, a growth factor and activity-regulated gene, encodes a novel cytoskeleton-associated protein that is enriched in neuronal dendrites. *Neuron* 14:433–45

Martin KC. 2004. Local protein synthesis during axon guidance and synaptic plasticity. *Curr. Opin. Neurobiol.* 14:305–10

Melton DA. 1987. Translocation of a localized maternal mRNA to the vegetal pole of *Xenopus* oocytes. *Nature* 328:80–82

Mendez R, Richter JD. 2001. Translational control by CPEB: a means to the end. *Nat. Rev. Mol. Cell Biol.* 2:521–29

Menon KP, Sanyal S, Habara Y, Sanchez R, Wharton RP, et al. 2004. The translational repressor Pumilio regulates presynaptic morphology and controls postsynaptic accumulation of translation factor eIF-4E. *Neuron* 44:663–76

Miller S, Yasuda M, Coats JK, Jones Y, Martone ME, Mayford M. 2002. Disruption of dendritic translation of CaMKIIalpha impairs stabilization of synaptic plasticity and memory consolidation. *Neuron* 36:507–19

Min H, Turck CW, Nikolic JM, Black DL. 1997. A new regulatory protein, KSRP, mediates exon inclusion through an intronic splicing enhancer. *Genes Dev.* 11:1023–36

Mohr E, Kachele I, Mullin C, Richter D. 2002. Rat vasopressin mRNA: a model system to characterize *cis*-acting elements and *trans*-acting factors involved in dendritic mRNA sorting. *Prog. Brain Res.* 139:211–24

Mohr E, Richter D. 2001. Messenger RNA on the move: implications for cell polarity. *Int. J. Biochem. Cell Biol.* 33:669–79

Monshausen M, Putz U, Rehbein M, Schweizer M, DesGroseillers L, et al. 2001. Two rat brain Staufen isoforms differentially bind RNA. *J. Neurochem.* 76:155–65

Moore PB. 1999. Structural motifs in RNA. *Annu. Rev. Biochem.* 68:287–300

Morales CR, Wu XQ, Hecht NB. 1998. The DNA/RNA-binding protein, TB-RBP, moves from the nucleus to the cytoplasm and through intercellular bridges in male germ cells. *Dev. Biol.* 201:113–23

Mori Y, Imaizumi K, Katayama T, Yoneda T, Tohyama M. 2000. Two *cis*-acting elements in the 3′ untranslated region of alpha-CaMKII regulate its dendritic targeting. *Nat. Neurosci.* 3:1079–84

Mouland AJ, Xu H, Cui H, Krueger W, Munro TP, et al. 2001. RNA trafficking signals in human immunodeficiency virus type 1. *Mol. Cell. Biol.* 21:2133–43

Mowry KL, Cote CA. 1999. RNA sorting in *Xenopus* oocytes and embryos. *FASEB J.* 13:435–45

Mowry KL, Melton DA. 1992. Vegetal messenger RNA localization directed by a 340-nt RNA sequence element in *Xenopus* oocytes. *Science* 255:991–94

Munro TP, Magee RJ, Kidd GJ, Carson JH, Barbarese E, et al. 1999. Mutational analysis of a heterogeneous nuclear ribonucleoprotein A2 response element for RNA trafficking. *J. Biol. Chem.* 274:34389–95

Muslimov IA, Nimmrich V, Hernandez AI, Tcherepanov A, Sacktor TC, Tiedge H. 2004. Dendritic transport and localization of protein kinase Mζ mRNA: implications for molecular memory consolidation. *J. Biol. Chem.* 279:52613–22

Muslimov IA, Santi E, Homel P, Perini S, Higgins D, Tiedge H. 1997. RNA transport in dendrites: a *cis*-acting targeting element is contained within neuronal BC1 RNA. *J. Neurosci.* 17:4722–33

Muslimov IA, Titmus M, Koenig E, Tiedge H. 2002. Transport of neuronal BC1 RNA in Mauthner axons. *J. Neurosci.* 22:4293–301

Nairn AC, Matsushita M, Nastiuk K, Horiuchi A, Mitsui K, et al. 2001. Elongation factor-2 phosphorylation and the regulation of protein synthesis by calcium. *Prog. Mol. Subcell. Biol.* 27:91–129

Via mutation of the 3′ UTR of the endogenous CaMKIIα gene, this study showed that local synthesis of CaMKIIα contributes to synaptic and behavioral plasticity.

O'Donnell WT, Warren ST. 2002. A decade of molecular studies of fragile X syndrome. *Annu. Rev. Neurosci.* 25:315–38

Palacios IM, St Johnston D. 2001. Getting the message across: the intracellular localization of mRNAs in higher eukaryotes. *Annu. Rev. Cell Dev. Biol.* 17:569–614

Palacios-Prü EL, Palacios L, Mendoza RV. 1981. Synaptogenetic mechanisms during chick cerebellar cortex development. *Submicrosc. Cytol.* 13:145–67

Pinkstaff J, Chappell SA, Mauro VP, Edelman G, Krushel LA. 2001. Internal initiation of translation of five dendritically localized neuronal mRNAs. *Proc. Natl. Acad. Sci. USA* 98:2770–75

Piper M, Holt C. 2004. RNA translation in axons. *Annu. Rev. Cell Dev. Biol.* 20:505–23

Prakash N, Fehr S, Mohr E, Richter D. 1997. Dendritic localization of rat vasopressin mRNA: ultrastructural analysis and mapping of targeting elements. *Eur. J. Neurosci.* 9:523–32

Pyronnet S, Imataka H, Gingras AC, Fukunaga R, Hunter T, Sonenberg N. 1999. Human eukaryotic translation initiation factor 4G (eIF4G) recruits mnk1 to phosphorylate eIF4E. *EMBO J.* 18:270–79

Rackham O, Brown CM. 2004. Visualization of RNA-protein interactions in living cells: FMRP and IMP1 interact on mRNAs. *EMBO J.* 23:3346–55

Rafelski SM, Theriot JA. 2004. Crawling toward a unified model of cell mobility: spatial and temporal regulation of actin dynamics. *Annu. Rev. Biochem.* 73:209–39

Raught B, Gingras AC, Sonenberg N. 2000. Regulation of ribosomal recruitment in eukaryotes. See Sonenberg et al. 2000, pp. 245–93

Rehbein M, Kindler S, Horke S, Richter D. 2000. Two *trans*-acting rat-brain proteins, MARTA1 and MARTA2, interact specifically with the dendritic targeting element in MAP2 mRNAs. *Brain Res. Mol. Brain Res.* 79:192–201

Rehbein M, Wege K, Buck F, Schweizer M, Richter D, Kindler S. 2002. Molecular characterization of MARTA1, a protein interacting with the dendritic targeting element of MAP2 mRNAs. *J. Neurochem.* 82:1039–46

Richter JD. 2000. Influence of polyadenylation-induced translation on metazoan development and neuronal synaptic function. See Sonenberg et al. 2000, pp. 785–805

Roegiers F, Jan YN. 2000. Staufen: a common component of mRNA transport in oocytes and neurons? *Trends Cell Biol.* 10:220–24

Ross AF, Oleynikov Y, Kislauskis EH, Taneja KL, Singer RH. 1997. Characterization of a beta-actin mRNA zipcode-binding protein. *Mol. Cell. Biol.* 17:2158–65

Ruvkun G. 2001. Glimpses of a tiny RNA world. *Science* 294:797–99

Scheetz AJ, Nairn AC, Constantine-Paton M. 1997. *N*-methyl-D-aspartate receptor activation and visual activity induce elongation factor-2 phosphorylation in amphibian tecta: a role for *N*-methyl-D-aspartate receptors in controlling protein synthesis. *Proc. Natl. Acad. Sci. USA* 94:14770–75

Scheetz AJ, Nairn AC, Constantine-Paton M. 2000. NMDA receptor-mediated control of protein synthesis at developing synapses. *Nat. Neurosci.* 3:211–16

Schratt GM, Nigh EA, Chen WG, Hu L, Greenberg ME. 2004. BDNF regulates the translation of a select group of mRNAs by a mammalian target of rapamycin-phosphatidylinositol 3-kinase-dependent pathway during neuronal development. *J. Neurosci.* 24:9366–77

Schwartz SP, Aisenthal L, Elisha Z, Oberman F, Yisraeli JK. 1992. A 69-kDa RNA-binding protein from *Xenopus* oocytes recognizes a common motif in two vegetally localized maternal mRNAs. *Proc. Natl. Acad. Sci. USA* 89:11895–99

Severt WL, Biber TU, Wu X, Hecht NB, DeLorenzo RJ, Jakoi ER. 1999. The suppression of testis-brain RNA binding protein and kinesin heavy chain disrupts mRNA sorting in dendrites. *J. Cell Sci.* 112:3691–702

Using a novel approach, this study found that the TAF ZBP1/IMP1 and the translational inhibitor FMRP interacted in an RNA-independent way, which suggests a functional link between mRNA localization and translational repression.

Shan J, Munro TP, Barbarese E, Carson JH, Smith R. 2003. A molecular mechanism for mRNA trafficking in neuronal dendrites. *J. Neurosci.* 23:8859–66

Shiina N, Shinkura K, Tokunaga M. 2005. A novel RNA-binding protein in neuronal RNA granules: regulatory machinery for local translation. *J. Neurosci.* 25:4420–34

Shin CY, Kundel M, Wells DG. 2004. Rapid, activity-induced increase in tissue plasminogen activator is mediated by metabotropic glutamate receptor-dependent mRNA translation. *J. Neurosci.* 24:9425–33

Smith R. 2004. Moving molecules: mRNA trafficking in mammalian oligodendrocytes and neurons. *Neuroscientist* 10:495–500

Smith WB, Starck SR, Roberts RW, Schuman EM. 2005. Dopaminergic stimulation of local protein synthesis enhances surface expression of GluR1 and synaptic transmission in hippocampal neurons. *Neuron* 45:765–79

Sonenberg N, Hershey JWB, Mathews MB, eds. 2000. *Translational Control of Gene Expression.* Cold Spring Harbor, NY: Cold Spring Harbor Lab. Press. 1020 pp.

Stefani G, Fraser CE, Darnell JC, Darnell RB. 2004. Fragile X mental retardation protein is associated with translating polyribosomes in neuronal cells. *J. Neurosci.* 24:9272–76

Steward O, Levy WB. 1982. Preferential localization of polyribosomes under the base of dendritic spines in granule cells of the dentate gyrus. *J. Neurosci.* 2:284–91

Steward O, Schuman EM. 2001. Protein synthesis at synaptic sites on dendrites. *Annu. Rev. Neurosci.* 24:299–325

Steward O, Schuman EM. 2003. Compartmentalized synthesis and degradation of proteins in neurons. *Neuron* 40:347–59

Steward O, Worley PF. 2001. Selective targeting of newly synthesized Arc mRNA to active synapses requires NMDA receptor activation. *Neuron* 30:227–40

Sung YJ, Dolzhanskaya N, Nolin SL, Brown T, Currie JR, Denman RB. 2003. The fragile X mental retardation protein FMRP binds elongation factor 1A mRNA and negatively regulates its translation in vivo. *J. Biol. Chem.* 278:15669–78

Takei N, Inamura N, Kawamura M, Namba H, Hara K, et al. 2004. Brain-derived neurotrophic factor induces mammalian target of rapamycin-dependent local activation of translation machinery and protein synthesis in neuronal dendrites. *J. Neurosci.* 24:9760–69

Takei N, Kawamura M, Hara K, Yonezawa K, Nawa H. 2001. Brain-derived neurotrophic factor enhances neuronal translation by activating multiple initiation processes: comparison with the effects of insulin. *J. Biol. Chem.* 276:42818–25

Tang SJ, Meulemans D, Vazquez L, Colaco N, Schuman E. 2001. A role for a rat homolog of staufen in the transport of RNA to neuronal dendrites. *Neuron* 32:463–75

Tang SJ, Reis G, Kang H, Gingras AC, Sonenberg N, Schuman EM. 2002. A rapamycin-sensitive signaling pathway contributes to long-term synaptic plasticity in the hippocampus. *Proc. Natl. Acad. Sci. USA* 99:467–72

Thompson KR, Otis KO, Chen DY, Zhao Y, O'Dell TJ, Martin KC. 2004. Synapse to nucleus signaling during long-term synaptic plasticity; a role for the classical active nuclear import pathway. *Neuron* 44:997–1009

Tiruchinapalli DM, Oleynikov Y, Kelic S, Shenoy SM, Hartley A, et al. 2003. Activity-dependent trafficking and dynamic localization of zipcode binding protein 1 and beta-actin mRNA in dendrites and spines of hippocampal neurons. *J. Neurosci.* 23:3251–61

Tongiorgi E, Armellin M, Giulianini PG, Bregola G, Zucchini S, et al. 2004. Brain-derived neurotrophic factor mRNA and protein are targeted to discrete dendritic laminas by events that trigger epileptogenesis. *J. Neurosci.* 24:6842–52

Wang H, Iacoangeli A, Popp S, Muslimov IA, Imataka H, et al. 2002. Dendritic BC1 RNA: functional role in regulation of translation initiation. *J. Neurosci.* 22:10232–41

This study shows that the RNA-binding protein CPEB appears to regulate tPA mRNA translation via polyadenylation. Local synthesis and release of tPA at synapses may therefore be effectors of synaptic plasticity.

In dentate gyrus neurons, newly synthesized Arc mRNA is selectively delivered to synapses that have recently been activated. NMDAR activation is required for Arc mRNA targeting.

Wang H, Tiedge H. 2004. Translational control at the synapse. *Neuroscientist* 10:456–66

Waskiewicz A, Flynn A, Proud CG, Cooper JA. 1997. Mitogen-activated protein kinases activate the serine/threonine kinases Mnk1 and Mnk2. *EMBO J.* 16:1909–20

Waskiewicz A, Johnson JC, Penn B, Mahalingam M, Kimball SR, Cooper JA. 1999. Phosphorylation of the cap-binding protein eukaryotic translation initiation factor 4E by protein kinase Mnk1 in vivo. *Mol. Cell. Biol.* 19:1871–80

Webster PJ, Liang L, Berg CA, Lasko P, MacDonald PM. 1997. Translational repressor bruno plays multiple roles in development and is widely conserved. *Genes. Dev.* 11:2510–21

Weiler IJ, Irwin SA, Klintsova AY, Spencer CM, Brazelton AD, et al. 1997. Fragile X mental retardation protein is translated near synapses in response to neurotransmitter activation. *Proc. Natl. Acad. Sci. USA* 94:5395–400

Weiler IJ, Spangler CC, Klintsova AY, Grossman AW, Kim SH, et al. 2004. Fragile X mental retardation protein is necessary for neurotransmitter-activated protein translation at synapses. *Proc. Natl. Acad. Sci. USA* 101:17504–9

Wells DG, Dong X, Quinlan EM, Huang YS, Bear MF, et al. 2001. A role for the cytoplasmic polyadenylation element in NMDA receptor-regulated mRNA translation in neurons. *J. Neurosci.* 21:9541–48

Wilhelm JE, Hilton M, Amos Q, Henzel WJ. 2003. Cup is an eIF4E binding protein required for both the translational repression of oskar and the recruitment of Barentsz. *J. Cell Biol.* 163:1197–204

Wu L, Wells D, Tay J, Mendis D, Abbott MA, et al. 1998. CPEB-mediated cytoplasmic polyadenylation and the regulation of experience-dependent translation of α-CaMKII mRNA at synapses. *Neuron* 21:1129–39

Yaniv K, Yisraeli JK. 2002. The involvement of a conserved family of RNA binding proteins in embryonic development and carcinogenesis. *Gene* 287:49–54

Ye B, Petritsch C, Clark IE, Gavis ER, Jan LY, Jan YN. 2004. Nanos and Pumilio are essential for dendrite morphogenesis in *Drosophila* peripheral neurons. *Curr. Biol.* 14:314–21

Zalfa F, Giorgi M, Primerano B, Moro A, Di Penta A, et al. 2003. The fragile X syndrome protein FMRP associates with BC1 RNA and regulates the translation of specific mRNAs at synapses. *Cell* 112:317–27

Zhang HL, Eom T, Oleynikov Y, Shenoy SM, Liebelt DA, et al. 2001. Neurotrophin-induced transport of a beta-actin mRNP complex increases beta-actin levels and stimulates growth cone motility. *Neuron* 31:261–75

Zhang HL, Singer RH, Bassell GJ. 1999. Neurotrophin regulation of beta-actin mRNA and protein localization within growth cones. *J. Cell Biol.* 147:59–70

Zhang J, King ML. 1996. *Xenopus* VegT RNA is localized to the vegetal cortex during oogenesis and encodes a novel T-box transcription factor involved in mesodermal patterning. *Development* 122:4119–29

Zhao WM, Jiang C, Kroll TT, Huber PW. 2001. A proline-rich protein binds to the localization element of *Xenopus* Vg1 mRNA and to ligands involved in actin polymerization. *EMBO J.* 20:2315–25

Zhou Y, King ML. 2004. Sending RNAs into the future: RNA localization and germ cell fate. *IUBMB Life* 56:19–27

Rho GTPases: Biochemistry and Biology

Aron B. Jaffe and Alan Hall

MRC Laboratory for Molecular Cell Biology, Cancer Research UK, Oncogene and
Signal Transduction Group, and Department of Biochemistry and Molecular Biology,
University College London, London WC1E 6BT, United Kingdom;
email: a.jaffe@ucl.ac.uk; alan.hall@ucl.ac.uk

Annu. Rev. Cell Dev. Biol.
2005. 21:247–69

First published online as a
Review in Advance on
June 28, 2005

The *Annual Review of
Cell and Developmental
Biology* is online at
http://cellbio.annualreviews.org

doi: 10.1146/
annurev.cellbio.21.020604.150721

Copyright © 2005 by
Annual Reviews. All rights
reserved

1081-0706/05/1110-
0247$20.00

Key Words

cytoskeleton, cell cycle, morphogenesis, migration

Abstract

Approximately one percent of the human genome encodes proteins that either regulate or are regulated by direct interaction with
members of the Rho family of small GTPases. Through a series
of complex biochemical networks, these highly conserved molecular switches control some of the most fundamental processes of cell
biology common to all eukaryotes, including morphogenesis, polarity, movement, and cell division. In the first part of this review, we
present the best characterized of these biochemical pathways; in the
second part, we attempt to integrate these molecular details into a
biological context.

Contents

INTRODUCTION

Rho GTPases constitute a distinct family within the superfamily of Ras-related small GTPases and are found in all eukaryotic cells. Twenty-two mammalian genes encoding Rho GTPases have been described—three Rho isoforms A, B, and C; three Rac isoforms 1, 2, and 3; Cdc42, RhoD, Rnd1, Rnd2, RhoE/Rnd3, RhoG, TC10, and TCL; RhoH/TTF; Chp and Wrch-1; Rif, RhoBTB1, and 2; and Miro-1 and 2 (Aspenstrom et al. 2004). The yeast *Saccharomyces cerevisiae* has 5 Rho proteins (Rho 1, 2, 3, and 4 and Cdc42), whereas *Caenorhabditis elegans* and *Drosophila melanogaster* are predicted to have 10 and 11, respectively. Similar to other regulatory GTPases, they act as molecular switches cycling between an active GTP-bound state and an inactive GDP-bound state. This activity is controlled by (*a*) guanine nucleotide exchange factors (GEFs) that catalyze exchange of GDP for GTP to activate the switch (Schmidt & Hall 2002); (*b*) GTPase-activating proteins (GAPs) that stimulate the intrinsic GTPase activity to inactivate the switch (Bernards 2003); and (*c*) guanine nucleotide dissociation inhibitors (GDIs), whose role appears to be to block spontaneous activation (Olofsson 1999) (**Figure 1**). Interestingly, the three Rnd proteins and RhoH lack any de-tectable GTPase activity and so it is not clear that they should be regarded as molecular switches, at least in the conventional sense. Finally, Rho GTPases can be regulated through direct phosphorylation or ubiquitination (Lang et al. 1996, Wang et al. 2003), but the extent to which these covalent modifications play a role in normal physiology is unclear.

It is in the active GTP-bound state that Rho GTPases perform their regulatory function through a conformation-specific interaction with target (effector) proteins. Over 50 effectors have been identified so far for Rho, Rac, and Cdc42 that include serine/threonine kinases, tyrosine kinases, lipid kinases, lipases, oxidases, and scaffold proteins. For the handful of targets that have been examined structurally, it appears that they exist in a closed inactive conformation that is relieved through GTPase binding (Bishop & Hall 2000). However, it is possible that GTPases might also serve to recruit targets to specific locations or complexes.

BIOCHEMICAL FUNCTIONS

Actin Cytoskeleton

The activation of Rho, Rac, or Cdc42 leads to the assembly of contractile actin:myosin filaments, protrusive actin-rich lamellipodia, and protrusive actin-rich filopodia, respectively (Etienne-Manneville & Hall 2002). These highly specific effects on the actin cytoskeleton point to a series of well-defined signal transduction pathways controlled by each GTPase leading to both the formation (actin polymerization) and the organization (filament bundling) of actin filaments.

Actin polymerization. Actin polymerization in eukaryotic cells occurs through the coordinated activities of filament severing and capping proteins and the two major actin polymerization factors Arp2/3 and Formin.

Arp2/3. Although Rac and Cdc42 lead to morphologically distinct protrusions at the

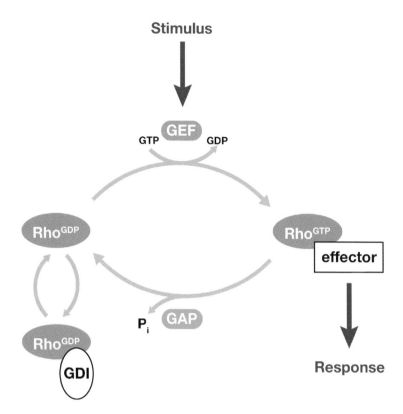

Stimulus

GTP **GEF** GDP

RhoGDP

RhoGTP

effector

RhoGDP

P_i **GAP**

GDI

Response

Figure 1

The GTPase cycle. Rho GTPases cycle between an inactive GDP-bound form and an active GTP-bound form. In mammalian cells, their activity is regulated by a large family of 85 GEFs, an equally large family of 80 GAPs, and 3 GDIs. Active GTPases interact with effector proteins to mediate a response.

plasma membrane (i.e., lamellipodia and filopodia), they both initiate peripheral actin polymerization through the Arp2/3 complex. This heptameric, actin-nucleation machine is found in all eukaryotic cells; it associates with the sides and, perhaps, also the ends of existing actin filaments to initiate new branched filaments (Millard et al. 2004) (**Figure 2**). Both GTPases activate Arp2/3 indirectly through members of the Wiskott-Aldrich syndrome protein (WASP) family. In vitro, Cdc42.GTP binds directly to N-WASP, or the closely related, hemopoietic-specific WASP, to relieve an intra-molecular, auto-inhibitory interaction and expose a C-terminal Arp2/3 binding/activation site. However, recent work has shown that the situation may be more complicated, with the majority of cellular N-WASP bound to a protein, WIP or CR16, which suppresses activation by Cdc42 (Ho et al. 2001, Martinez-Quiles et al. 2001). This suggests that in vivo N-WASP may be *trans*-inhibited

rather than auto-inhibited, and indeed a new Cdc42 target, Toca-1 (transducer of Cdc42-dependent actin assembly), has been identified that is required for activation of the N-WASP/WIP complex (Ho et al. 2004). The significance of having two direct targets of Cdc42 (i.e., N-WASP and Toca-1) in the signal transduction pathway to Arp2/3 is unclear; perhaps it allows more finely regulated control.

Activation of Arp2/3 by Rac is mediated by WAVE family proteins (**Figure 2**), which although structurally related to N-WASP, do not interact directly with the GTPase. Progress in understanding this signal transduction pathway has also come from a biochemical approach, which has led to the isolation of a complex containing WAVE1, along with three other proteins, HSPC300, Nap125, and PIR121 (Eden et al. 2002). Nap125 and PIR121 are both direct Rac targets, and it has been proposed that Rac

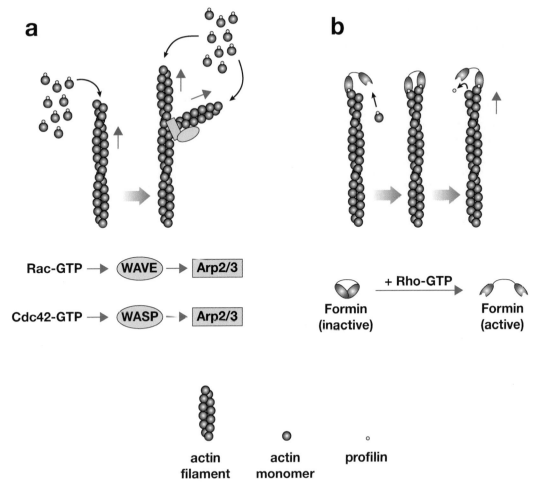

a

Rac-GTP ➡ (WAVE) ➡ [Arp2/3]

Cdc42-GTP ➡ (WASP) ⇢ [Arp2/3]

b

+ Rho-GTP

Formin
(inactive)

Formin
(active)

actin
filament

actin
monomer

profilin

Figure 2

Rho GTPases regulate the two major modes of actin polymerization. (*a*) Rac and Cdc42 activate Arp2/3 via WAVE (a WASP family protein) and WASP, respectively, to initiate a branched filament network. (*b*) Rho activates formins to promote linear elongation of filaments at barbed ends.

promotes the disassembly of this inactive complex, allowing WAVE to interact directly with Arp2/3. A second group has isolated a similar complex, containing WAVE2, but this has an additional component, Abi1, and is active in actin polymerization assays (Innocenti et al. 2004). This group argues that the role of Rac is to localize this complex to the cell periphery to promote actin nucleation. Further analysis is required to resolve the issue.

Formins. The other major mechanism for inducing actin polymerization in eukaryotic cells is through the formin family of proteins. In particular, Rho stimulates actin polymerization in mammalian cells through the diaphanous-related formin (DRF), mDia1 (and possibly mDia2), and in *S. cerevisiae* through Bnr1 and Bni1, the only two formins in this organism. mDia1 is a direct target of Rho. GTP and binding of the GTPase relieves an auto-inhibitory interaction, exposing an FH2 domain that then binds to the barbed end of an actin filament (Zigmond 2004). mDia1 also contains an essential FH1 domain, which interacts with a profilin/actin complex and delivers it to the filament end. The remarkable thing here is that once mDia1 has added an actin monomer, it remains bound to the barbed end ready to add another actin monomer. This processive mechanism of filament elongation has been described as a leaky cap and is schematically depicted in **Figure 2**. Exactly how monomer assembly occurs at the barbed end while the formin is still bound is unclear; most other barbed end–binding proteins inhibit filament elongation. Given the complexity found with Arp2/3 activation, it would not be surprising to find that mDia is part of a larger complex and subject to further regulation.

Cofilin. ADF/cofilin severs actin filaments, leading to an increase in uncapped barbed ends that serve as sites for actin polymerization and filament elongation (Ghosh et al. 2004). Cofilin also participates in filament disassembly by promoting actin monomer dissociation from the pointed end, and both activities seem to be important for productive membrane protrusions (Dawe et al. 2003, Desmarais et al. 2005, Pollard & Borisy 2003). Cofilin is tightly regulated, and its activity is affected by phosphorylation, PIP$_2$ binding, changes in intracellular pH, and protein-protein interactions. Phosphorylation of cofilin leads to inactivation and occurs primarily through LIM kinases (LIMK), which in turn are activated by the PAK family of Rac/Cdc42-dependent kinases. How to reconcile the requirement for both active Rac and active (i.e., non-phosphorylated) cofilin at sites of membrane protrusion is not yet clear, but may involve spatially distinct compartments (Dawe et al. 2003, Svitkina & Borisy 1999). LIMK-dependent phosphorylation of cofilin can also be induced by Rho acting through its target Rho kinase (ROCK), and this may be an important event in the stabilization of actin:myosin filaments (Ohashi et al. 2000).

Actin filament organization. In addition to elongation, the discrete changes to the actin cytoskeleton induced by Rho, Rac, or Cdc42 require the correct spatial organization of filaments. This has been best characterized for Rho-induced assembly of contractile actin:myosin filaments, which is mediated by ROCK. Although this serine/threonine kinase has many substrates, the key event appears to be phosphorylation-induced inactivation of myosin light chain (MLC) phosphatase (Riento & Ridley 2003). This in turn leads to increased phosphorylation of MLC, which promotes the actin filament cross-linking activity of myosin II.

Less is known about how Rac and Cdc42 organize actin filaments into branched and unbranched filaments, respectively. Recent studies suggest that the formation of unbranched bundles of actin filaments found in filopodia originate from a branched network initiated by Cdc42 activation of Arp2/3 (Svitkina et al. 2003). Remodeling then occurs through a combination of (*a*) inhibiting barbed end capping to allow filament elongation and (*b*) promoting filament cross-linking through actin-bundling proteins such as fascin (Vignjevic et al. 2003).

Microtubule Cytoskeleton

Similar to actin filaments, microtubules have an intrinsic polarity, with a minus end (usually, but not always, anchored at the centrosome) and a dynamic plus end (usually at the cell periphery). The change from growth to shrinking (catastrophe) and from shrinking to growth (rescue) at the plus end is referred to as dynamic instability. In addition, the intracellular organization of microtubules makes a major contribution to cell polarity and to the distribution of intracellular organelles, such as the Golgi and mitotic spindle.

Microtubule dynamics. Microtubule plus end–binding proteins profoundly influence microtubule dynamics, and Rho GTPases can regulate this in different ways. The Op18/stathmin family, for example, interacts both with microtubule plus ends to promote catastrophic disassembly and with tubulin dimers to inhibit polymerization (Cassimeris 2002). Op18/stathmin can be phosphorylated at four key residues, any one of which leads to its inactivation, thereby resulting in net elongation of microtubule ends. Phosphorylation at Ser16 is mediated by Cdc42/Rac-dependent activation of PAK, which occurs in response to a number of extracellular stimuli (Daub et al. 2001).

The effect of Rho on microtubule dynamics is likely to be context dependent. In neurons, collapsin response mediator protein-2 (CRMP-2) binds tubulin heterodimers and promotes microtubule assembly, perhaps by delivering dimers to the plus ends of growing microtubules (Y. Fukata et al. 2002). CRMP-2 is phosphorylated and inactivated by ROCK (at Thr555), which correlates well with growth cone collapse induced by LPA, although not by semaphorin 3A (Arimura et al. 2000). In migrating fibroblasts, on the other hand, Rho promotes the formation of stabilized microtubules, as visualized by an increase in detyrosinated tubulin. It is not clear how stabilization occurs; it appears to be mediated by mDia but does not involve changes to the actin cytoskeleton (Palazzo et al. 2001).

Microtubule plus end capture. Microtubules play a major role in defining cell shape and polarity through the specific interaction of their plus ends with proteins at the cell cortex. This plus end capture of microtubules has been attributed to a number of plus end–binding proteins, whose activities are influenced by Rho GTPases. CLIP-170, for example, can simultaneously bind to microtubules and to the scaffold protein IQGAP, a Rac/Cdc42 effector that is enriched at the leading edge of migrating cells. Expression of constitutively active Rac or Cdc42 enhances the ability of CLIP-170 to bind to IQGAP, thereby promoting plus end capture (M. Fukata et al. 2002). IQGAP was originally identified as a Rac/Cdc42 effector that can influence the actin cytoskele-

ton, and these observations therefore provide a potential biochemical link between the actin and microtubule cytoskeletons (Bashour et al. 1997).

Another plus end–binding protein, EB1, interacts with the adenomatous polyposis coli (APC) tumor suppressor protein; this not only stabilizes microtubules but also can facilitate interactions with proteins at the cell cortex. Two distinct pathways have been reported to regulate the APC/EB1 interaction and thereby promote microtubule capture. In fibroblasts, both EB1 and APC bind to mDia, suggesting a role for Rho in microtubule capture (Wen et al. 2004). In migrating astrocytes, the interaction of APC with EB1 at microtubule plus ends is regulated by Cdc42, acting through the Par6/PKCζ effector complex (Etienne-Manneville & Hall 2003). Interestingly, APC was recently found to bind to IQGAP, and both proteins are required for the proper maintenance of CLIP-170 at the leading edge of migrating cells (Watanabe et al. 2004).

Gene Expression

In addition to their cytoskeletal effects, Rho GTPases regulate several signal transduction pathways that lead to alterations in gene expression.

SRF—actin sensor. The serum response element (SRE) is found in many promoters, including those of genes encoding components of the cytoskeleton, most notably actin. Two transcription factors act at the SRE: (*a*) the ternary complex factor (TCF) regulated by the Ras/MAP kinase pathway and (*b*) the serum response factor (SRF) regulated by Rho. Recent work has established that SRF requires a co-activator, MAL, which translocates from the cytoplasm to the nucleus in response to Rho activation (Miralles et al. 2003). Furthermore, Rho-mediated changes to the actin cytoskeleton promote this translocation, although precisely how is not clear. MAL binds to monomeric G-actin and actin polymerization results in MAL nuclear translocation, suggesting that MAL is sensitive to a decrease in cellular G-actin levels. However, it must be more complicated than this because nuclear translocation occurs even when the total levels of G-actin remain unchanged, and Rac and Cdc42, which are also strong inducers of actin polymerization, activate SRF poorly compared with activation by Rho.

Actin-independent pathways. Rho, Rac, and Cdc42 also affect gene transcription through signal transduction pathways not involving the actin cytoskeleton. All three GTPases are capable of activating the JNK and p38 MAP kinase pathway, although this is dependent on cell context, and there are many examples where they seem not to be involved (Coso et al. 1995, Minden et al. 1995, Puls et al. 1999). At least four MAP kinase kinase kinases (MAPKKKs) are direct targets of Rho GTPases: MLK2, MLK3, and MEKK4 interact with Rac/Cdc42, whereas MEKK1 interacts with Rho and with Rac/Cdc42, although through different sites (Burbelo et al. 1995, Gallagher et al. 2004, Teramoto et al. 1996).

Scaffold proteins play a major role in controlling the activation and specificity of MAP kinase pathways in vivo and, interestingly, at least three Rho GTPase targets, POSH, CNK, and MEKK1, act as scaffold proteins (Morrison & Davis 2003). POSH interacts with MLK2/3 (MAPKKK), MKK4/7 (MAPKK), and JNK (MAPK) and is required

for Rac-dependent activation of JNK in neurons upon NGF withdrawal, whereas CNK interacts with MLK2/3 and MKK7 and is required for Rho-dependent JNK activation upon serum addition to HeLa cells (Jaffe et al. 2005, Xu et al. 2003).

Rho, Rac, and Cdc42 have been reported to activate NFκB in response to a variety of stimuli, particularly inflammatory cytokines (Perona et al. 1997). A potential complication in elucidating the mechanism of GTPase activation of NFκB is that Rac and Cdc42 stimulate the production of reactive oxygen species (ROS) (see below) and inflammatory cytokines, both of which are potent activators of NFκB (Joneson & Bar-Sagi 1998, Kheradmand et al. 1998, Tapon et al. 1998). Therefore, it is still not entirely clear whether NFκB activation by Rho GTPases is direct or indirect.

Regulation of Enzymatic Activities

A number of additional enzymatic activities (**Table 1**) are influenced by Rho GTPases. Many are involved in lipid metabolism and some have been implicated in GTPase-mediated changes to the actin cytoskeleton (Tolias et al. 2000, Wang et al. 2002). One of the first targets of Rac to be identified was p67phox, an essential structural component of the NADPH oxidase complex found in phagocytic cells (Diekmann et al. 1994). Since then, Rac has been reported to promote ROS production in many cells, and a new family of widely expressed oxidases (Nox family) may mediate this activity (Lambeth 2002, Takeya et al. 2003).

Table 1   Additional enzymatic activities regulated by Rho GTPases

Enzymatic activity	Effector protein	GTPase	References
Lipid metabolism	PI4P 5-kinase	Rho, Rac	(Weernink et al. 2004)
	PI-3-kinase	Rac, Cdc42	(Zheng et al. 1994)
	DAG kinase	Rho, Rac	(Houssa et al. 1999)
	PLD	Rho, Rac, Cdc42	(Hess et al. 1997)
	PLC	Rac, Cdc42	(Illenberger et al. 1998)
ROS generation	NADPH oxidases	Rac	(Takeya & Sumimoto 2003)

BIOLOGICAL FUNCTIONS

Cell Cycle

The eukaryotic cell cycle consists of a DNA replication phase (S) and a nuclear/cell division phase (M) separated by two gap phases (G_1 and G_2). Rho GTPases influence the activity of cyclin-dependent kinases during G_1 and the organization of the microtubule and actin cytoskeletons during M (**Figure 3**).

G_1 progression. Inhibition of Rho, Rac, or Cdc42 blocks G_1 progression in a variety of mammalian cell types, but the mechanisms are cell-type dependent and have proven difficult to elucidate (Olson et al. 1995, Yamamoto et al. 1993). G_1 progression is controlled by two types of cyclin-dependent kinases (Cdks), Cdk4/Cdk6 and Cdk2, which are activated by binding to cyclin D and cyclin E, respectively, and inhibited by binding to the INK4A and

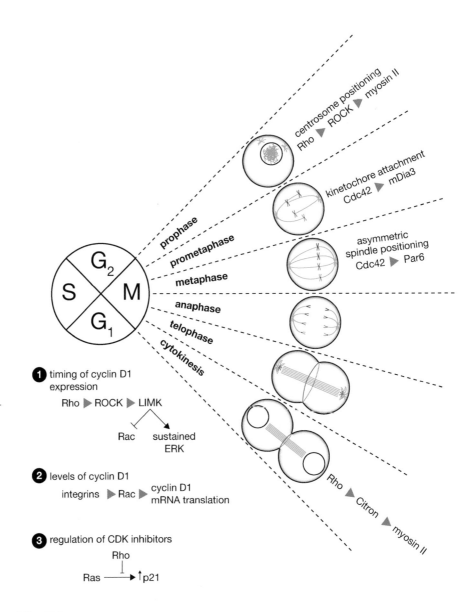

Figure 3

Rho GTPases and the cell cycle. Rho GTPases control multiple aspects of M phase and G_1 progression. The signaling pathways are shown for the relevant stages. Microtubules, green; actin, red; condensed chromosomes, blue.

Cip/Kip family of proteins. The key to G_1 progression lies in the cellular levels of cyclins and the Cdk inhibitors, and these are controlled by two major extrinsic factors: growth factors and extracellular matrix proteins.

The best understood route for affecting cyclin D levels is through growth factor–induced activation of transcription by the Ras/ERK pathway (Coleman et al. 2004). However, to achieve the correct cyclin D levels and at the right time (i.e., in mid G_1), ERK activation must be sustained. In normal cells this requires additional signal inputs from adhesion to the extracellular matrix. This phenomenon of adhesion- or anchorage-dependent proliferation is likely to account for many of the effects of Rho GTPases on G_1 progression.

There have been several reports that Rac and Cdc42 (but not Rho) stimulate cyclin D1 transcription when ectopically expressed in cells, and in one case at least, this was mediated by NFκB (Joyce et al. 1999, Westwick et al. 1997). More detailed analysis of endogenous Rho GTPases in NIH 3T3 fibroblasts has, however, painted a more complicated picture. In these cells, Rho is required for normal cyclin D expression, but Rho inhibition leads to premature cyclin D expression, which is Rac dependent (Welsh et al. 2001). The authors of this work argue that during the normal cell cycle, Rho acts to suppress Rac and to promote sustained activation of ERK, both of which are required for correct temporal control of cyclin D levels. Sustained ERK activity involves the Rho target kinase ROCK and its substrate LIMK, as well as the induction of stress fibers, whereas suppression of Rac is independent of actin and involves nuclear translocation of LIMK (Roovers & Assoian 2003, Roovers et al. 2003). How stress fibers influence ERK activity is unknown, and because cofilin is the only known substrate for LIMK, it is unclear how this actin-independent function of LIMK is mediated. A rather different story has emerged through analyzing G_1 progression in endothelial cells. Here too, matrix-dependent integrin activation is essential, but

this triggers a Rac pathway that controls cyclin D1 mRNA translation (Mettouchi et al. 2001). Finally, there has been one report that expression of cyclin E, which occurs in late G_1, can be stimulated by Cdc42 through its effector p70 S6 kinase (Chou et al. 2003). Unlike cyclin D1, the timing of cyclin E expression is independent of Rho kinase activity (Roovers & Assoian 2003).

Rho GTPases also regulate the levels of the Cdk2 inhibitors, p21^{cip1} and p27^{kip1}. The first indication of this came from studies showing that activated Ras is unable to stimulate G_1 progression in the absence of Rho. It appears that Ras promotes the accumulation of high levels of p21^{cip1} and this is attenuated by Rho either through inhibition of p21^{cip1} transcription or, possibly, through promotion of protein degradation (Olson et al. 1998, Weber et al. 1997). In fibroblasts and colon carcinoma cell lines, the effect is ROCK independent, but in other cell types, ROCK seems to be involved (Lai et al. 2002, Sahai et al. 2001). Rho regulation of p27^{kip1} is post-transcriptional, although there are examples of it acting though protein degradation or mRNA translation (Hu et al. 1999, Vidal et al. 2002).

Many of the effects of Rho GTPases on G_1 progression are thought to reflect the crucial role of anchorage- or adhesion-dependent signals for cell proliferation. The loss of anchorage dependence is one of the hallmarks of cancer, which has sparked much interest in the possible contribution of deregulated Rho GTPase pathways to tumor progression (Jaffe & Hall 2002). In this respect, it is interesting to note that many members of the Rho GEF family can induce a transformed phenotype when expressed in immortalized cells.

Mitosis. The alignment of chromosomes during prophase and metaphase is driven primarily by microtubules emanating from the two centrosomes: Astral microtubules interact with the cell cortex, whereas spindle microtubules interact with the kinetochore. A major role of astral microtubules is to align the spindle along an axis perpendicular to the future cell division plane, and it was recently shown that actin:myosin filaments, under the control of ROCK, are required at the cortex to allow positioning of the centrosomes (Rosenblatt et al. 2004). It is too early to say how myosin II promotes cortical movement of astral microtubules, but microtubules have been reported to influence Rho GTPases so localized inhibition of Rho (and therefore contractility) might be involved (Wittmann & Waterman-Storer 2001).

Another role of astral microtubules is to determine the position of the spindle. Although in cell culture this is symmetrical, producing two equal daughter cells after division, the asymmetric positioning of the spindle, giving rise to daughter cells of different size and specification, is crucially important both in development and in the adult (e.g., during stem cell division). Much of our understanding of this process has come from genetic screens in *C. elegans* and in particular, in the discovery of six Par proteins (Par1-6), along with atypical protein kinase C (aPKC), that are essential for asymmetric cell division in the zygote. These proteins are themselves asymmetrically distributed, with Par6/aPKC/Par3 at the anterior and Par1/Par2 at the posterior (the external cue here being the site of sperm entry), and they determine the positioning of the spindle so as to produce a large and a small daughter cell. The major interest here is that Par6 is a target for Cdc42, and this GTPase is also essential for asymmetric cell division (Gotta et al. 2001). Cdc42 induces a conformational change in Par6 and can activate aPKC, but precisely how this influences spindle positioning is still being elucidated (Ahringer 2003, Etienne-Manneville & Hall 2001, Peterson et al. 2004).

Spindle microtubules interact with chromosomes at the kinetochore, a complex of at least 50 proteins that includes the Cdc42-specific effector mDia3. Inhibition of Cdc42 or depletion of mDia3 causes a mitotic arrest in which many chromosomes are not properly attached to microtubules (Yasuda et al. 2004).

The GEF and GAP that control Cdc42 activity during this stage of mitosis have been identified as Ect2 and MgcRacGAP, respectively (Oceguera-Yanez et al. 2005). Interestingly, Cdc42-null ES cells proliferate normally, and although this could mean that spindle attachment here is GTPase independent, it more likely reflects some redundancy with the other close relatives of Cdc42, such as TCL or TC10 (Chen et al. 2000). The Cdc42-dependent attachment of microtubules to a kinetochore shows some similarities to the capture of microtubules at the cell cortex, and it will be interesting to see whether Cdc42 regulates the assembly of proteins at the tip of spindle microtubules as well as at the chromosomal attachment site (Narumiya et al. 2004). Finally, the mDia proteins are best known for their role in promoting actin polymerization (see above) leading to speculation that actin might play a role in spindle attachment.

Cytokinesis. Cell division is initiated at the end of mitosis through the assembly of a cleavage furrow and an associated contractile ring consisting of actin and myosin II filaments. Rho plays a crucial role in contractile ring function and localizes to the cleavage furrow along with at least three known effectors, ROCK, Citron kinase, and mDia (Glotzer 2001). Although the majority of actin filaments at the cleavage furrow originates from pre-existing filaments, it is likely that some de novo actin polymerization is required and perhaps this is the role of mDia (Kato et al. 2001). The respective roles of the two Rho-dependent kinases have been difficult to tease out. Myosin is activated by phosphorylation of its light chain (MLC) at Thr18 and Ser19, and this drives contraction of the actin filament ring (Komatsu et al. 2000, Matsumura et al. 1998). Citron kinase phosphorylates MLC directly at both sites, whereas ROCK affects these sites indirectly through phosphorylation and inhibition of MLC phosphatase (Yamashiro et al. 2003). Inhibition of Citron kinase, but not ROCK, blocks cytokinesis in HeLa cells; however, in other cell types

ROCK is also important, and the phenotype of the Citron kinase knockout mice, although severe, is not lethal (Di Cunto et al. 2000, Kosako et al. 2000, Madaule et al. 1998). The dynamic changes in contractile forces during the division process are likely to be complex and perhaps both kinases play distinct, but partially overlapping roles. In this respect it is interesting that when ectopically expressed, Citron kinase can replace ROCK in stress fiber assembly (Yamashiro et al. 2003). Finally, the GEF and GAP responsible for regulating Rho activity during cytokinesis are Ect2 and MgcRacGAP (Lee et al. 2004, Prokopenko et al. 1999, Tatsumoto et al. 1999). This raises some very interesting questions concerning regulation and specificity, since earlier in mitosis the same GEF and GAP apparently con-trol Cdc42 activity to promote kinetochore attachment (Oceguera-Yanez et al. 2005).

Cell Morphogenesis

Cell morphology is intimately linked to function. In response to external cues, Rho GTPases contribute to morphogenesis by regulating both the actin and microtubule cytoskeletons and the core machinery involved in establishing polarity (**Figure 4**).

Cell-cell interactions. The assembly of molecularly distinct cell-cell adhesion complexes and the concomitant establishment of polarity drive the morphogenesis of many cell types. In epithelial cells, E-cadherin,

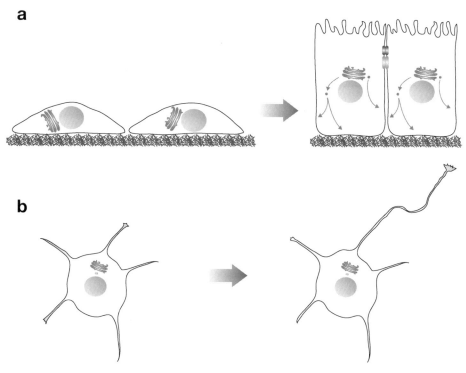

Figure 4

Rho GTPases and cell morphogenesis. (*a*) Epithelial morphogenesis involves cell-cell interactions leading to the formation of apical and basolateral domains, separated by tight junctions (*blue*) and adherens junctions (*orange*). Cell polarity is reinforced through the directed trafficking of vesicles from the Golgi (*brown*). (*b*) Neuronal morphogensis begins with the establishment of axonal and somato-dendritic compartments.

a member of a family of Ca^{2+}-dependent *trans*-membrane proteins, participates in homophilic interactions to produce stable adherens junctions through the subsequent recruitment of intracellular proteins (such as α- and β-catenin) and the actin cytoskeleton. Rho, Rac, and Cdc42 have each been implicated in adherens junction assembly.

The inhibition of Rho or Rac in keratinocytes prevents the formation of adherens junctions, whereas Tiam-1, a GEF for Rac, is essential for the formation and maintenance of junctions in Madin-Darby canine kidney (MDCK) cells plated on fibronectin (Braga et al. 1997, Malliri et al. 2004). Cadherin ligation leads to the recruitment of activated Rac to adhesion sites, where its major role is to stabilize the junction through the assembly of actin filaments (Hordijk et al. 1997, Takaishi et al. 1997). The Rac target IQGAP has been implicated in actin assembly at junctions; however, it has been suggested that IQGAP may also sequester β-catenin and that Rac and Cdc42 can disrupt this complex allowing β-catenin to participate in junction assembly (Kuroda et al. 1999, Noritake et al. 2004). An additional complication to the story is that hyperactivation of Rac in keratinocytes leads to junction disassembly, and activation of Rac in MDCK cells plated on collagen promotes migration rather than cell-cell adhesion (Braga et al. 2000, Sander et al. 1998). Because Rac is involved in two seemingly opposing activities, namely cell-cell junction assembly and cell migration, it is likely that its effects will be greatly influenced by environmental factors and cell type. A major role for Rho at junctions is probably through its effector ROCK and F-actin assembly, although it too has been shown to interact directly with junctional proteins (α-catenin and p120ctn), at least in *Drosophila* (Magie et al. 2002, Vaezi et al. 2002).

Numerous reports indicate that adherens junction assembly is preceded by the localized induction of filopodia and/or lamellipodia and that these protrusive structures drive intimate membrane contact between adjacent cells (Ehrlich et al. 2002, Jacinto et al. 2000, Vasioukhin et al. 2000). The formation of localized protrusions may be initiated by early cadherin interactions that activate Rac, and perhaps Cdc42, which could then generate a positive feedback loop (Noritake et al. 2004). However, an alternative possibility has emerged from work on a relatively new family of cell-cell adhesion proteins, the nectins. These Ca^{2+}-independent, immunoglobulin-like *trans*-membrane proteins form homo- and heteromolecular interactions, which suggests that this is required for subsequent cadherin-based junctional assembly (Irie et al. 2004, Kawakatsu et al. 2002). Nectin-nectin interactions activate both Cdc42 and Rac, and this could provide the localized protrusive activity that facilitates subsequent cadherin-based adhesion (Fukuhara et al. 2004).

Cell polarity. The morphogenesis of epithelial cells requires the establishment of cell polarity to form apical and basolateral domains, which involves the assembly of tight junctions formed by homophilic interactions involving another family of integral membrane proteins, the claudins. Although the details of how tight junctions are assembled apically to adherens junctions are far from well understood, genetic analyses in flies and worms have identified some key players in this process. Specifically, three protein complexes have been implicated in epithelial morphogenesis: Par6/aPKC/Par3, Dlg/Lgl/scribble, and Crumbs/PALS1/PATJ (Gibson & Perrimon 2003).

As described above, Par6 is a direct target of Cdc42 and is required for asymmetric cell division. Although morphogenesis is a very different biological process, it also involves the establishment of asymmetry through the specification of distinct domains around the cell periphery. Work with cultured mammalian epithelial cells has shown that the Par6/aPKC/Par3 complex localizes apically with tight junctions, whereas Par1 localizes laterally (Izumi et al. 1998). Par3 binds directly to JAM, a *trans*-membrane adhesion protein that interacts directly with a core, tight junctional component, ZO-1. When overexpressed, Par3 promotes the formation of tight junctions (Hirose et al. 2002, Itoh et al. 2001). Furthermore, the Cdc42-dependent activation of aPKC is required for junction assembly (Suzuki et al. 2001, Yamanaka et al. 2001). One of the substrates of aPKC is Lgl, and recently Cdc42 and aPKC have been shown to induce localization of Dlg to the front of migrating cells (Plant et al. 2003, Yamanaka et al. 2003, S. Etienne-Manneville & A. Hall, unpublished results). These observations suggest that in some situations Cdc42 may signal via the Par6/aPKC to regulate the Dlg/Lgl/scribble complex. Other observations point to a more complex relationship between the Par complex and tight junction assembly. For example, Par3 inhibits aPKC activity in vitro, and in one report Par6 negatively regulates tight junction formation (Gao et al. 2002, Lin et al. 2000).

Epithelial morphogenesis in vivo requires additional signals to establish polarity within the context of a tissue (Zegers et al. 2003). Extracellular matrix provides a likely signal for defining the appropriate orientation of the apical surface and in vitro three-dimensional morphogenesis assays have uncovered a potential role for Rac in this process (Yu et al. 2005). Finally, polarization may also take place across tissues along the proximal-distal axis referred to as planar polarity (Fanto & McNeill 2004). Work in the *Drosophila* eye and wing have shed most light on this process, and it appears that Rac and Rho, but not Cdc42, acting downstream of the Dishevelled protein and the family of Frizzled receptors, play a key role. The effects of Rac and Rho are complex, but it interesting to note that they likely involve both the actin cytoskeleton and JNK-dependent gene transcription.

Although the initiating signals are different, neuronal morphogenesis also involves Cdc42 and the Par proteins. When hippocampal neurons are plated in culture, they spontaneously polarize, in the absence of cell-cell contacts, to form a single axon and a somato-dendritic compartment (extracellular cues presumably control this response in vivo) (**Figure 4**). Cdc42 and the Par3/Par6/aPKC complex are specifically enriched in the developing axon and disruption of their activity results in polarity defects, leading to zero or multiple axons (Schwamborn & Puschel 2004, Shi et al. 2003). Localized inhibition of GSK-3 is required in the presumptive axon, and this promotes the polarized localization of Par3 and two microtubule plus end–binding proteins, APC and CRMP-2 (Jiang et al. 2005, Shi et al. 2004, Yoshimura et al. 2005). It is interesting to compare this effect with polarity establishment in migrating cells (see below), where the localized accumulation of APC requires Par/aPKC-dependent inhibition of GSK-3 (Etienne-Manneville & Hall 2003).

Concomitant with the establishment of polarity, the morphogenetic program in neurons and epithelia is reinforced by the polarized trafficking of vesicles. In MDCK cells, for example, Cdc42 is required for the sorting of proteins to the basolateral surface, although not to the apical surface (Kroschewski et al. 1999). The small GTPase Ral and its effector Sec5, a component of the exocyst complex, have been implicated in Cdc42-induced filopodia formation in fibroblasts (Sugihara et al. 2002), raising the possibility that Cdc42 regulates the exocyst in mammalian cells, as it does in yeast, to influence basolateral trafficking (Sugihara et al. 2002, Zhang et al. 2001). Finally, Rho GTPases may regulate vesicular trafficking by regulating the microtubule cytoskeleton, which is dramatically reorganized during both epithelial and neuronal morphogenesis (Musch 2004).

Cell Migration

Actin polymerization and filament elongation at the front, coupled to actin:myosin filament contraction at the rear, are thought to provide the major driving forces for migration in animal cells. Polarization of these two activities can occur spontaneously and is often seen with single cells in culture, but this is usually short-lived and leads to random migration.

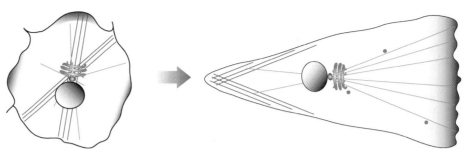

Figure 5

Rho GTPases and cell migration. Cell migration requires actin-dependent protrusions at the front (*red*) and contractile actin:myosin filaments (*red*) at the rear. In addition, microtubules (*green*) originating from the centrosome (*purple*) are preferentially stabilized in the direction of migration allowing targeted vesicle trafficking from the Golgi (*brown*) to the leading edge.

In vivo, cell migration is directed and extracellular cues polarize the actin cytoskeleton accordingly. Persistent and efficient directed migration requires additional cellular changes involving polarization of the microtubule cytoskeleton and the secretory pathway (Ridley et al. 2003) (**Figure 5**).

Movement. Rac, acting through WAVE and Arp2/3, is required to promote actin polymerization at the front of migrating cells, and this pushes forward the leading edge membrane. GTP-bound active Rac accumulates at the front of migrating cells, and in chemotaxing neutrophils this is sustained through a positive feedback loop with PI 3-kinase and its lipid product PIP_3 (Gardiner et al. 2002, Itoh et al. 2002, Kraynov et al. 2000). Work in *Dictyostelium* and neutrophils has shown that this polarized accumulation of PIP_3 is reinforced through localization of the lipid phosphatase PTEN to the sides and rear of the migrating cell (Li et al. 2003, Merlot & Firtel 2003).

This scenario of Rac activation at the front guiding movement is almost certainly an oversimplification. For example, fluorescence resonance energy transfer (FRET) -based assays have revealed the presence of Rac.GTP at the rear as well as the front of migrating neutrophils (Gardiner et al. 2002). Because protrusions are not found at the rear, it must be assumed that Rac is playing a different role in the two locations. One clear example where

Rac is known to make a contribution to cell migration other than through actin is during *Drosophila* dorsal closure. Here, Rac is required in leading edge cells not only for actin polymerization but also for JNK MAP kinase activation (Ricos et al. 1999). JNK signaling promotes transcription of the TGFβ–family member, Dpp, which when secreted acts on cells behind the leading edge to coordinate movement of the whole epithelial sheet (Glise & Noselli 1997, Hou et al. 1997). JNK may make other contributions to cell migration; it can, for example, phosphorylate paxillin, which leads to focal adhesion turnover, a prerequisite at the rear of cells for efficient movement (Huang et al. 2003).

Rho acts at the rear of the cell to generate contractile forces through ROCK-mediated MLC phosphorylation, which move the cell body forward (Riento & Ridley 2003). In addition, ROCK may inhibit inappropriate lateral protrusions, perhaps by restricting the formation of new integrin adhesion complexes (Worthylake & Burridge 2003). However, again things may not be so simple, and in some situations inhibition of ROCK stimulates cell migration (Nobes & Hall 1999). Furthermore, an E3 ubiquitin ligase, Smurf1, which ubiquitinates Rho, has been localized at the front of migrating cells (Wang et al. 2003). The significance of this is not clear—it would seem more economical not to activate Rho there in the first place, but perhaps it has a role to play at the front and there is more to Rho ubiquitination than simply degradation.

More recently, cell migration studies using tumor cells in three-dimensional matrices have revealed striking differences from migration on two-dimensional tissue culture plates, which could have important implications in metastasis. Some tumor cells generate leading edge protrusive structures as might be expected, but they do not seem to require Rho or ROCK (Sahai & Marshall 2003). Presumably an alternative mechanism exists to promote actin:myosin contraction in the rear. However, another class of tumor cell moves with a rounded, blebbing morphology and here Rho and ROCK are essential. Rac is required for both types of migration (Sahai & Marshall 2003).

Directional sensing. The mechanisms by which external cues direct Rac and the activation of Arp2/3 to the leading edge and thereby determine the direction of migration (directional sensing) are less well characterized, but in some cases, at least, this involves Cdc42. The first indication that Cdc42 links extracellular cues to intracellular polarity came through studies on the pheromone-mating response in yeast, and this was later confirmed by work in animal cells using either single cells moving in a chemotactic gradient or sheets of cells moving in an in vitro scratch or wound assay (Allen et al. 1998, Nobes & Hall 1999, Simon et al. 1995). For example, Cdc42 is essential for the directional migration of a macrophage cell in a gradient of the chemoattractant M-CSF1, and when this GTPase is inhibited, cells still move, but do so randomly (Allen et al. 1998). Inhibition of Rac in the same cells blocks migration totally. Similarly, Cdc42 is essential for restricting Rac-dependent actin polymerization to the front of fibroblasts induced to migrate by scratching a monolayer (Cau & Hall 2005). It appears, therefore, that Cdc42 is locally activated by external cues (chemoattractant or loss of cell-cell contact) and then activates a pathway that determines the spatial localization of active

Rac. Recent work in neutrophils suggests that this involves the localization of α-PIX, a Rac-specific GEF, through an unexpected pathway in which the Cdc42 target, PAK, acts as a scaffold rather than a kinase (Li et al. 2003).

Directed migration also involves polarization of the microtubule cytoskeleton, readily visualized in most, although not all, cell types through reorientation of the centrosome to the side of the nucleus facing the direction of migration. Cdc42 also regulates this aspect of cell polarity through a mechanism that involves some interesting parallels with polarity establishment during morphogenesis and asymmetric cell division. In particular, active Cdc42 at the front of the migrating cell leads to localized activation of the atypical PKC in the Par6/aPKC complex (Etienne-Manneville & Hall 2001). This has at least two consequences important for establishing microtubule polarity: (*a*) inactivation of the serine/threonine kinase GSK3 to promote association of APC with microtubule plus ends preferentially at the leading edge (Etienne-Manneville & Hall 2003) and (*b*) association of Dlg with the leading edge cortex (S. Etienne-Manneville & A. Hall, unpublished results). The interaction between microtubule-bound APC and cortex-bound Dlg is required for microtubule polarization (S. Etienne-Manneville & A. Hall, unpublished results).

CONCLUSIONS

It is impossible to present the whole range of biochemical pathways and biological processes that are influenced by Rho GTPases in a single review. Instead, we have tried to focus on areas that are better characterized and that provide more general insights into cellular functions. It has been disappointing not to be able to do justice to neuronal morphogenesis or to cover host-pathogen interactions and developmental processes, but perhaps this is better left to experts in those areas; the interested reader will certainly have no problem in finding suitable reviews.

ACKNOWLEDGMENTS

We thank members of the Hall laboratory for helpful discussions and Christian Dillon for invaluable advice on the use of Adobe Illustrator. Our research is generously supported by a program grant from Cancer Research UK.

LITERATURE CITED

Ahringer J. 2003. Control of cell polarity and mitotic spindle positioning in animal cells. *Curr. Opin. Cell Biol.* 15:73–81

Allen WE, Zicha D, Ridley AJ, Jones GE. 1998. A role for Cdc42 in macrophage chemotaxis. *J. Cell Biol.* 141:1147–57

Arimura N, Inagaki N, Chihara K, Menager C, Nakamura N, et al. 2000. Phosphorylation of collapsin response mediator protein-2 by Rho-kinase. Evidence for two separate signaling pathways for growth cone collapse. *J. Biol. Chem.* 275:23973–80

Aspenstrom P, Fransson A, Saras J. 2004. Rho GTPases have diverse effects on the organization of the actin filament system. *Biochem. J.* 377:327–37

Bashour AM, Fullerton AT, Hart MJ, Bloom GS. 1997. IQGAP1, a Rac- and Cdc42-binding protein, directly binds and cross-links microfilaments. *J. Cell Biol.* 137:1555–66

Bernards A. 2003. GAPs galore! A survey of putative Ras superfamily GTPase activating proteins in man and *Drosophila*. *Biochim. Biophys. Acta* 1603:47–82

Bishop AL, Hall A. 2000. Rho GTPases and their effector proteins. *Biochem. J.* 348(Pt 2):241–55

Braga VM, Betson M, Li X, Lamarche-Vane N. 2000. Activation of the small GTPase Rac is sufficient to disrupt cadherin-dependent cell-cell adhesion in normal human keratinocytes. *Mol. Biol. Cell* 11:3703–21

Braga VM, Machesky LM, Hall A, Hotchin NA. 1997. The small GTPases Rho and Rac are required for the establishment of cadherin-dependent cell-cell contacts. *J. Cell Biol.* 137:1421–31

Burbelo PD, Drechsel D, Hall A. 1995. A conserved binding motif defines numerous candidate target proteins for both Cdc42 and Rac GTPases. *J. Biol. Chem.* 270:29071–74

Cau J, Hall A. 2005. Cdc42 controls the polarity of the actin and microtubule cytoskeletons through two distinct signal transduction pathways. *J. Cell Sci.* 118(Pt. 12):2579–87

Cassimeris L. 2002. The oncoprotein 18/stathmin family of microtubule destabilizers. *Curr. Opin. Cell Biol.* 14:18–24

Chen F, Ma L, Parrini MC, Mao X, Lopez M, et al. 2000. Cdc42 is required for PIP_2-induced actin polymerization and early development but not for cell viability. *Curr. Biol.* 10:758–65

Chou MM, Masuda-Robens JM, Gupta ML. 2003. Cdc42 promotes G1 progression through p70 S6 kinase-mediated induction of cyclin E expression. *J. Biol. Chem.* 278:35241–47

Coleman ML, Marshall CJ, Olson MF. 2004. RAS and RHO GTPases in G1-phase cell-cycle regulation. *Nat. Rev. Mol. Cell Biol.* 5:355–66

Coso OA, Chiariello M, Yu JC, Teramoto H, Crespo P, et al. 1995. The small GTP-binding proteins Rac1 and Cdc42 regulate the activity of the JNK/SAPK signaling pathway. *Cell* 81:1137–46

Daub H, Gevaert K, Vandekerckhove J, Sobel A, Hall A. 2001. Rac/Cdc42 and p65PAK regulate the microtubule-destabilizing protein stathmin through phosphorylation at serine 16. *J. Biol. Chem.* 276:1677–80

Dawe HR, Minamide LS, Bamburg JR, Cramer LP. 2003. ADF/cofilin controls cell polarity during fibroblast migration. *Curr. Biol.* 13:252–57

Desmarais V, Ghosh M, Eddy R, Condeelis J. 2005. Cofilin takes the lead. *J. Cell Sci.* 118:19–26

Di Cunto F, Imarisio S, Hirsch E, Broccoli V, Bulfone A, et al. 2000. Defective neurogenesis in citron kinase knockout mice by altered cytokinesis and massive apoptosis. *Neuron* 28:115–27

Diekmann D, Abo A, Johnston C, Segal AW, Hall A. 1994. Interaction of Rac with p67phox and regulation of phagocytic NADPH oxidase activity. *Science* 265:531–33

Eden S, Rohatgi R, Podtelejnikov AV, Mann M, Kirschner MW. 2002. Mechanism of regulation of WAVE1-induced actin nucleation by Rac1 and Nck. *Nature* 418:790–93

Ehrlich JS, Hansen MD, Nelson WJ. 2002. Spatio-temporal regulation of Rac1 localization and lamellipodia dynamics during epithelial cell-cell adhesion. *Dev. Cell* 3:259–70

Etienne-Manneville S, Hall A. 2001. Integrin-mediated activation of Cdc42 controls cell polarity in migrating astrocytes through PKCzeta. *Cell* 106:489–98

Etienne-Manneville S, Hall A. 2002. Rho GTPases in cell biology. *Nature* 420:629–35

Etienne-Manneville S, Hall A. 2003. Cdc42 regulates GSK-3beta and adenomatous polyposis coli to control cell polarity. *Nature* 421:753–56

Fanto M, McNeill H. 2004. Planar polarity from flies to vertebrates. *J. Cell Sci.* 117:527–33

Fukata M, Watanabe T, Noritake J, Nakagawa M, Yamaga M, et al. 2002. Rac1 and Cdc42 capture microtubules through IQGAP1 and CLIP-170. *Cell* 109:873–85

Fukata Y, Itoh TJ, Kimura T, Menager C, Nishimura T, et al. 2002. CRMP-2 binds to tubulin heterodimers to promote microtubule assembly. *Nat. Cell Biol.* 4:583–91

Fukuhara T, Shimizu K, Kawakatsu T, Fukuyama T, Minami Y, et al. 2004. Activation of Cdc42 by *trans* interactions of the cell adhesion molecules nectins through c-Src and Cdc42-GEF FRG. *J. Cell Biol.* 166:393–405

Gallagher ED, Gutowski S, Sternweis PC, Cobb MH. 2004. RhoA binds to the amino terminus of MEKK1 and regulates its kinase activity. *J. Biol. Chem.* 279:1872–77

Gao L, Joberty G, Macara IG. 2002. Assembly of epithelial tight junctions is negatively regulated by Par6. *Curr. Biol.* 12:221–25

Gardiner EM, Pestonjamasp KN, Bohl BP, Chamberlain C, Hahn KM, Bokoch GM. 2002. Spatial and temporal analysis of Rac activation during live neutrophil chemotaxis. *Curr. Biol.* 12:2029–34

Ghosh M, Song X, Mouneimne G, Sidani M, Lawrence DS, Condeelis JS. 2004. Cofilin promotes actin polymerization and defines the direction of cell motility. *Science* 304:743–46

Gibson MC, Perrimon N. 2003. Apicobasal polarization: epithelial form and function. *Curr. Opin. Cell Biol.* 15:747–52

Glise B, Noselli S. 1997. Coupling of Jun amino-terminal kinase and Decapentaplegic signaling pathways in *Drosophila* morphogenesis. *Genes Dev.* 11:1738–47

Glotzer M. 2001. Animal cell cytokinesis. *Annu. Rev. Cell Dev. Biol.* 17:351–86

Gotta M, Abraham MC, Ahringer J. 2001. CDC-42 controls early cell polarity and spindle orientation in *C. elegans*. *Curr. Biol.* 11:482–88

Hess JA, Ross AH, Qiu RG, Symons M, Exton JH. 1997. Role of Rho family proteins in phospholipase D activation by growth factors. *J. Biol. Chem.* 272:1615–20

Hirose T, Izumi Y, Nagashima Y, Tamai-Nagai Y, Kurihara H, et al. 2002. Involvement of ASIP/PAR-3 in the promotion of epithelial tight junction formation. *J. Cell Sci.* 115:2485–95

Ho HY, Rohatgi R, Lebensohn AM, Le M, Li J, et al. 2004. Toca-1 mediates Cdc42-dependent actin nucleation by activating the N-WASP-WIP complex. *Cell* 118:203–16

Ho HY, Rohatgi R, Ma L, Kirschner MW. 2001. CR16 forms a complex with N-WASP in brain and is a novel member of a conserved proline-rich actin-binding protein family. *Proc. Natl. Acad. Sci. USA* 98:11306–11

Hordijk PL, ten Klooster JP, van der Kammen RA, Michiels F, Oomen LC, Collard JG. 1997. Inhibition of invasion of epithelial cells by Tiam1-Rac signaling. *Science* 278:1464–66

Hou XS, Goldstein ES, Perrimon N. 1997. *Drosophila* Jun relays the Jun amino-terminal kinase signal transduction pathway to the Decapentaplegic signal transduction pathway in regulating epithelial cell sheet movement. *Genes Dev.* 11:1728–37

Houssa B, de Widt J, Kranenburg O, Moolenaar WH, van Blitterswijk WJ. 1999. Diacylglycerol kinase theta binds to and is negatively regulated by active RhoA. *J. Biol. Chem.* 274:6820–22

Hu W, Bellone CJ, Baldassare JJ. 1999. RhoA stimulates p27(Kip) degradation through its regulation of cyclin E/CDK2 activity. *J. Biol. Chem.* 274:3396–401

Huang C, Rajfur Z, Borchers C, Schaller MD, Jacobson K. 2003. JNK phosphorylates paxillin and regulates cell migration. *Nature* 424:219–23

Illenberger D, Schwald F, Pimmer D, Binder W, Maier G, et al. 1998. Stimulation of phospholipase C-beta2 by the Rho GTPases Cdc42Hs and Rac1. *EMBO J.* 17:6241–49

Innocenti M, Zucconi A, Disanza A, Frittoli E, Areces LB, et al. 2004. Abi1 is essential for the formation and activation of a WAVE2 signalling complex. *Nat. Cell Biol.* 6:319–27

Irie K, Shimizu K, Sakisaka T, Ikeda W, Takai Y. 2004. Roles and modes of action of nectins in cell-cell adhesion. *Semin. Cell Dev. Biol.* 15:643–56

Itoh M, Sasaki H, Furuse M, Ozaki H, Kita T, Tsukita S. 2001. Junctional adhesion molecule (JAM) binds to PAR-3: a possible mechanism for the recruitment of PAR-3 to tight junctions. *J. Cell Biol.* 154:491–97

Itoh RE, Kurokawa K, Ohba Y, Yoshizaki H, Mochizuki N, Matsuda M. 2002. Activation of rac and cdc42 video imaged by fluorescent resonance energy transfer-based single-molecule probes in the membrane of living cells. *Mol. Cell Biol.* 22:6582–91

Izumi Y, Hirose T, Tamai Y, Hirai S, Nagashima Y, et al. 1998. An atypical PKC directly associates and colocalizes at the epithelial tight junction with ASIP, a mammalian homologue of *Caenorhabditis elegans* polarity protein PAR-3. *J. Cell Biol.* 143:95–106

Jacinto A, Wood W, Balayo T, Turmaine M, Martinez-Arias A, Martin P. 2000. Dynamic actin-based epithelial adhesion and cell matching during *Drosophila* dorsal closure. *Curr. Biol.* 10:1420–26

Jaffe AB, Hall A. 2002. Rho GTPases in transformation and metastasis. *Adv. Cancer Res.* 84:57–80

Jaffe AB, Hall A, Schmidt A. 2005. Association of CNK1 with Rho guanine nucleotide exchange factors controls signaling specificity downstream of Rho. *Curr. Biol.* 15:405–12

Jiang H, Guo W, Liang X, Rao Y. 2005. Both the establishment and the maintenance of neuronal polarity require active mechanisms: critical roles of GSK-3beta and its upstream regulators. *Cell* 120:123–35

Joneson T, Bar-Sagi D. 1998. A Rac1 effector site controlling mitogenesis through superoxide production. *J. Biol. Chem.* 273:17991–94

Joyce D, Bouzahzah B, Fu M, Albanese C, D'Amico M, et al. 1999. Integration of Rac-dependent regulation of cyclin D1 transcription through a nuclear factor-kappaB-dependent pathway. *J. Biol. Chem.* 274:25245–49

Kato T, Watanabe N, Morishima Y, Fujita A, Ishizaki T, Narumiya S. 2001. Localization of a mammalian homolog of diaphanous, mDia1, to the mitotic spindle in HeLa cells. *J. Cell Sci.* 114:775–84

Kawakatsu T, Shimizu K, Honda T, Fukuhara T, Hoshino T, Takai Y. 2002. *Trans*-interactions of nectins induce formation of filopodia and lamellipodia through the respective activation of Cdc42 and Rac small G proteins. *J. Biol. Chem.* 277:50749–55

Kheradmand F, Werner E, Tremble P, Symons M, Werb Z. 1998. Role of Rac1 and oxygen radicals in collagenase-1 expression induced by cell shape change. *Science* 280:898–902

Komatsu S, Yano T, Shibata M, Tuft RA, Ikebe M. 2000. Effects of the regulatory light chain phosphorylation of myosin II on mitosis and cytokinesis of mammalian cells. *J. Biol. Chem.* 275:34512–20

Kosako H, Yoshida T, Matsumura F, Ishizaki T, Narumiya S, Inagaki M. 2000. Rho-kinase/ROCK is involved in cytokinesis through the phosphorylation of myosin light chain and not ezrin/radixin/moesin proteins at the cleavage furrow. *Oncogene* 19:6059–64

Kraynov VS, Chamberlain C, Bokoch GM, Schwartz MA, Slabaugh S, Hahn KM. 2000. Localized Rac activation dynamics visualized in living cells. *Science* 290:333–37

Kroschewski R, Hall A, Mellman I. 1999. Cdc42 controls secretory and endocytic transport to the basolateral plasma membrane of MDCK cells. *Nat. Cell Biol.* 1:8–13

Kuroda S, Fukata M, Nakagawa M, Kaibuchi K. 1999. Cdc42, Rac1, and their effector IQ-GAP1 as molecular switches for cadherin-mediated cell-cell adhesion. *Biochem. Biophys. Res. Commun.* 262:1–6

Lai JM, Wu S, Huang DY, Chang ZF. 2002. Cytosolic retention of phosphorylated extracellular signal-regulated kinase and a Rho-associated kinase-mediated signal impair expression of p21(Cip1/Waf1) in phorbol 12-myristate-13-acetate-induced apoptotic cells. *Mol. Cell Biol.* 22:7581–92

Lambeth JD. 2002. Nox/Duox family of nicotinamide adenine dinucleotide (phosphate) oxidases. *Curr. Opin. Hematol.* 9:11–17

Lang P, Gesbert F, Delespine-Carmagnat M, Stancou R, Pouchelet M, Bertoglio J. 1996. Protein kinase A phosphorylation of RhoA mediates the morphological and functional effects of cyclic AMP in cytotoxic lymphocytes. *EMBO J.* 15:510–19

Lee JS, Kamijo K, Ohara N, Kitamura T, Miki T. 2004. MgcRacGAP regulates cortical activity through RhoA during cytokinesis. *Exp. Cell Res.* 293:275–82

Li Z, Hannigan M, Mo Z, Liu B, Lu W, et al. 2003. Directional sensing requires G beta gamma-mediated PAK1 and PIX alpha-dependent activation of Cdc42. *Cell* 114:215–27

Lin D, Edwards AS, Fawcett JP, Mbamalu G, Scott JD, Pawson T. 2000. A mammalian PAR-3-PAR-6 complex implicated in Cdc42/Rac1 and aPKC signalling and cell polarity. *Nat. Cell Biol.* 2:540–47

Madaule P, Eda M, Watanabe N, Fujisawa K, Matsuoka T, et al. 1998. Role of citron kinase as a target of the small GTPase Rho in cytokinesis. *Nature* 394:491–4

Magie CR, Pinto-Santini D, Parkhurst SM. 2002. Rho1 interacts with p120ctn and alpha-catenin, and regulates cadherin-based adherens junction components in *Drosophila*. *Development* 129:3771–82

Malliri A, van Es S, Huveneers S, Collard JG. 2004. The Rac exchange factor Tiam1 is required for the establishment and maintenance of cadherin-based adhesions. *J. Biol. Chem.* 279:30092–98

Martinez-Quiles N, Rohatgi R, Anton IM, Medina M, Saville SP, et al. 2001. WIP regulates N-WASP-mediated actin polymerization and filopodium formation. *Nat. Cell Biol.* 3:484–91

Matsumura F, Ono S, Yamakita Y, Totsukawa G, Yamashiro S. 1998. Specific localization of serine 19 phosphorylated myosin II during cell locomotion and mitosis of cultured cells. *J. Cell Biol.* 140:119–29

Merlot S, Firtel RA. 2003. Leading the way: directional sensing through phosphatidylinositol 3-kinase and other signaling pathways. *J. Cell Sci.* 116:3471–18

Mettouchi A, Klein S, Guo W, Lopez-Lago M, Lemichez E, et al. 2001. Integrin-specific activation of Rac controls progression through the G_1 phase of the cell cycle. *Mol. Cell* 8:115–27

Millard TH, Sharp SJ, Machesky LM. 2004. Signalling to actin assembly via the WASP (Wiskott-Aldrich syndrome protein)-family proteins and the Arp2/3 complex. *Biochem. J.* 380:1–17

Minden A, Lin A, Claret FX, Abo A, Karin M. 1995. Selective activation of the JNK signaling cascade and c-Jun transcriptional activity by the small GTPases Rac and Cdc42Hs. *Cell* 81:1147–57

Miralles F, Posern G, Zaromytidou AI, Treisman R. 2003. Actin dynamics control SRF activity by regulation of its coactivator MAL. *Cell* 113:329–42

Morrison DK, Davis RJ. 2003. Regulation of MAP kinase signaling modules by scaffold proteins in mammals. *Annu. Rev. Cell Dev. Biol.* 19:91–118

Musch A. 2004. Microtubule organization and function in epithelial cells. *Traffic* 5:1–9

Narumiya S, Oceguera-Yanez F, Yasuda S. 2004. A new look at Rho GTPases in cell cycle: role in kinetochore-microtubule attachment. *Cell Cycle* 3:855–57

Nobes CD, Hall A. 1999. Rho GTPases control polarity, protrusion, and adhesion during cell movement. *J. Cell Biol.* 144:1235–44

Noritake J, Fukata M, Sato K, Nakagawa M, Watanabe T, et al. 2004. Positive role of IQGAP1, an effector of Rac1, in actin-meshwork formation at sites of cell-cell contact. *Mol. Biol. Cell* 15:1065–76

Oceguera-Yanez F, Kimura K, Yasuda S, Higashida C, Kitamura T, et al. 2005. Ect2 and MgcRacGAP regulate the activation and function of Cdc42 in mitosis. *J. Cell Biol.* 168:221–32

Ohashi K, Nagata K, Maekawa M, Ishizaki T, Narumiya S, Mizuno K. 2000. Rho-associated kinase ROCK activates LIM-kinase 1 by phosphorylation at threonine 508 within the activation loop. *J. Biol. Chem.* 275:3577–82

Olofsson B. 1999. Rho guanine dissociation inhibitors: pivotal molecules in cellular signalling. *Cell Signal* 11:545–54

Olson MF, Ashworth A, Hall A. 1995. An essential role for Rho, Rac, and Cdc42 GTPases in cell cycle progression through G1. *Science* 269:1270–72

Olson MF, Paterson HF, Marshall CJ. 1998. Signals from Ras and Rho GTPases interact to regulate expression of p21Waf1/Cip1. *Nature* 394:295–99

Palazzo AF, Cook TA, Alberts AS, Gundersen GG. 2001. mDia mediates Rho-regulated formation and orientation of stable microtubules. *Nat. Cell Biol.* 3:723–29

Perona R, Montaner S, Saniger L, Sanchez-Perez I, Bravo R, Lacal JC. 1997. Activation of the nuclear factor-kappaB by Rho, CDC42, and Rac-1 proteins. *Genes Dev.* 11:463–75

Peterson FC, Penkert RR, Volkman BF, Prehoda KE. 2004. Cdc42 regulates the Par-6 PDZ domain through an allosteric CRIB-PDZ transition. *Mol. Cell* 13:665–76

Plant PJ, Fawcett JP, Lin DC, Holdorf AD, Binns K, et al. 2003. A polarity complex of mPar-6 and atypical PKC binds, phosphorylates and regulates mammalian Lgl. *Nat. Cell Biol.* 5:301–8

Pollard TD, Borisy GG. 2003. Cellular motility driven by assembly and disassembly of actin filaments. *Cell* 112:453–65

Prokopenko SN, Brumby A, O'Keefe L, Prior L, He Y, et al. 1999. A putative exchange factor for Rho1 GTPase is required for initiation of cytokinesis in *Drosophila*. *Genes Dev.* 13:2301–14

Puls A, Eliopoulos AG, Nobes CD, Bridges T, Young LS, Hall A. 1999. Activation of the small GTPase Cdc42 by the inflammatory cytokines TNF-α and IL-1, and by the Epstein-Barr virus transforming protein LMP1. *J. Cell Sci.* 112:2983–92

Ricos MG, Harden N, Sem KP, Lim L, Chia W. 1999. Dcdc42 acts in TGF-beta signaling during *Drosophila* morphogenesis: distinct roles for the Drac1/JNK and Dcdc42/TGF-beta cascades in cytoskeletal regulation. *J. Cell Sci.* 112(Pt 8):1225–35

Ridley AJ, Schwartz MA, Burridge K, Firtel RA, Ginsberg MH, et al. 2003. Cell migration: integrating signals from front to back. *Science* 302:1704–9

Riento K, Ridley AJ. 2003. Rocks: multifunctional kinases in cell behaviour. *Nat. Rev. Mol. Cell Biol.* 4:446–56

Roovers K, Assoian RK. 2003. Effects of rho kinase and actin stress fibers on sustained extracellular signal-regulated kinase activity and activation of G_1 phase cyclin-dependent kinases. *Mol. Cell Biol.* 23:4283–94

Roovers K, Klein EA, Castagnino P, Assoian RK. 2003. Nuclear translocation of LIM kinase mediates Rho-Rho kinase regulation of cyclin D1 expression. *Dev. Cell* 5:273–84

Rosenblatt J, Cramer LP, Baum B, McGee KM. 2004. Myosin II-dependent cortical movement is required for centrosome separation and positioning during mitotic spindle assembly. *Cell* 117:361–72

Sahai E, Marshall CJ. 2003. Differing modes of tumour cell invasion have distinct requirements for Rho/ROCK signalling and extracellular proteolysis. *Nat. Cell Biol.* 5:711–19

Sahai E, Olson MF, Marshall CJ. 2001. Cross-talk between Ras and Rho signalling pathways in transformation favours proliferation and increased motility. *EMBO J.* 20:755–66

Sander EE, van Delft S, ten Klooster JP, Reid T, van der Kammen RA, et al. 1998. Matrix-dependent Tiam1/Rac signaling in epithelial cells promotes either cell-cell adhesion or cell migration and is regulated by phosphatidylinositol 3-kinase. *J. Cell Biol.* 143:1385–98

Schmidt A, Hall A. 2002. Guanine nucleotide exchange factors for Rho GTPases: turning on the switch. *Genes Dev.* 16:1587–609

Schwamborn JC, Puschel AW. 2004. The sequential activity of the GTPases Rap1B and Cdc42 determines neuronal polarity. *Nat. Neurosci.* 7:923–29

Shi SH, Cheng T, Jan LY, Jan YN. 2004. APC and GSK-3beta are involved in mPar3 targeting to the nascent axon and establishment of neuronal polarity. *Curr. Biol.* 14:2025–32

Shi SH, Jan LY, Jan YN. 2003. Hippocampal neuronal polarity specified by spatially localized mPar3/mPar6 and PI 3-kinase activity. *Cell* 112:63–75

Simon MN, De Virgilio C, Souza B, Pringle JR, Abo A, Reed SI. 1995. Role for the Rho-family GTPase Cdc42 in yeast mating-pheromone signal pathway. *Nature* 376:702–5

Sugihara K, Asano S, Tanaka K, Iwamatsu A, Okawa K, Ohta Y. 2002. The exocyst complex binds the small GTPase RalA to mediate filopodia formation. *Nat. Cell Biol.* 4:73–78

Suzuki A, Yamanaka T, Hirose T, Manabe N, Mizuno K, et al. 2001. Atypical protein kinase C is involved in the evolutionarily conserved par protein complex and plays a critical role in establishing epithelia-specific junctional structures. *J. Cell Biol.* 152:1183–96

Svitkina TM, Borisy GG. 1999. Arp2/3 complex and actin depolymerizing factor/cofilin in dendritic organization and treadmilling of actin filament array in lamellipodia. *J. Cell Biol.* 145:1009–26

Svitkina TM, Bulanova EA, Chaga OY, Vignjevic DM, Kojima S, et al. 2003. Mechanism of filopodia initiation by reorganization of a dendritic network. *J. Cell Biol.* 160:409–21

Takaishi K, Sasaki T, Kotani H, Nishioka H, Takai Y. 1997. Regulation of cell-cell adhesion by rac and rho small G proteins in MDCK cells. *J. Cell Biol.* 139:1047–59

Takeya R, Sumimoto H. 2003. Molecular mechanism for activation of superoxide-producing NADPH oxidases. *Mol. Cell.* 16:271–77

Takeya R, Ueno N, Kami K, Taura M, Kohjima M, et al. 2003. Novel human homologues of p47phox and p67phox participate in activation of superoxide-producing NADPH oxidases. *J. Biol. Chem.* 278:25234–46

Tapon N, Nagata K, Lamarche N, Hall A. 1998. A new rac target POSH is an SH3-containing scaffold protein involved in the JNK and NF-kappaB signalling pathways. *EMBO J.* 17:1395–404

Tatsumoto T, Xie X, Blumenthal R, Okamoto I, Miki T. 1999. Human ECT2 is an exchange factor for Rho GTPases, phosphorylated in G2/M phases, and involved in cytokinesis. *J. Cell Biol.* 147:921–28

Teramoto H, Coso OA, Miyata H, Igishi T, Miki T, Gutkind JS. 1996. Signaling from the small GTP-binding proteins Rac1 and Cdc42 to the c-Jun N-terminal kinase/stress-activated protein kinase pathway. A role for mixed lineage kinase 3/protein-tyrosine kinase 1, a novel member of the mixed lineage kinase family. *J. Biol. Chem.* 271:27225–28

Tolias KF, Hartwig JH, Ishihara H, Shibasaki Y, Cantley LC, Carpenter CL. 2000. Type Iα phosphatidylinositol-4-phosphate 5-kinase mediates Rac-dependent actin assembly. *Curr. Biol.* 10:153–56

Vaezi A, Bauer C, Vasioukhin V, Fuchs E. 2002. Actin cable dynamics and Rho/Rock orchestrate a polarized cytoskeletal architecture in the early steps of assembling a stratified epithelium. *Dev. Cell* 3:367–81

Vasioukhin V, Bauer C, Yin M, Fuchs E. 2000. Directed actin polymerization is the driving force for epithelial cell-cell adhesion. *Cell* 100:209–19

Vidal A, Millard SS, Miller JP, Koff A. 2002. Rho activity can alter the translation of p27 mRNA and is important for RasV12-induced transformation in a manner dependent on p27 status. *J. Biol. Chem.* 277:16433–40

Vignjevic D, Yarar D, Welch MD, Peloquin J, Svitkina T, Borisy GG. 2003. Formation of filopodia-like bundles in vitro from a dendritic network. *J. Cell Biol.* 160:951–62

Wang F, Herzmark P, Weiner OD, Srinivasan S, Servant G, Bourne HR. 2002. Lipid products of PI$_3$Ks maintain persistent cell polarity and directed motility in neutrophils. *Nat. Cell Biol.* 4:513–8

Wang HR, Zhang Y, Ozdamar B, Ogunjimi AA, Alexandrova E, et al. 2003. Regulation of cell polarity and protrusion formation by targeting RhoA for degradation. *Science* 302:1775–79

Watanabe T, Wang S, Noritake J, Sato K, Fukata M, et al. 2004. Interaction with IQGAP1 links APC to Rac1, Cdc42, and actin filaments during cell polarization and migration. *Dev. Cell* 7:871–83

Weber JD, Hu W, Jefcoat SC Jr, Raben DM, Baldassare JJ. 1997. Ras-stimulated extracellular signal-related kinase 1 and RhoA activities coordinate platelet-derived growth factor-induced G1 progression through the independent regulation of cyclin D1 and p27. *J. Biol. Chem.* 272:32966–71

Weernink PA, Meletiadis K, Hommeltenberg S, Hinz M, Ishihara H, et al. 2004. Activation of type I phosphatidylinositol 4-phosphate 5-kinase isoforms by the Rho GTPases, RhoA, Rac1, and Cdc42. *J. Biol. Chem.* 279:7840–49

Welsh CF, Roovers K, Villanueva J, Liu Y, Schwartz MA, Assoian RK. 2001. Timing of cyclin D1 expression within G1 phase is controlled by Rho. *Nat. Cell Biol.* 3:950–57

Wen Y, Eng CH, Schmoranzer J, Cabrera-Poch N, Morris EJ, et al. 2004. EB1 and APC bind to mDia to stabilize microtubules downstream of Rho and promote cell migration. *Nat. Cell Biol.* 6:820–30

Westwick JK, Lambert QT, Clark GJ, Symons M, Van Aelst L, et al. 1997. Rac regulation of transformation, gene expression, and actin organization by multiple, PAK-independent pathways. *Mol. Cell Biol.* 17:1324–35

Wittmann T, Waterman-Storer CM. 2001. Cell motility: Can Rho GTPases and microtubules point the way? *J. Cell Sci.* 114:3795–803

Worthylake RA, Burridge K. 2003. RhoA and ROCK promote migration by limiting membrane protrusions. *J. Biol. Chem.* 278:13578–84

Xu Z, Kukekov NV, Greene LA. 2003. POSH acts as a scaffold for a multiprotein complex that mediates JNK activation in apoptosis. *EMBO J.* 22:252–61

Yamamoto M, Marui N, Sakai T, Morii N, Kozaki S, et al. 1993. ADP-ribosylation of the rhoA gene product by botulinum C3 exoenzyme causes Swiss 3T3 cells to accumulate in the G1 phase of the cell cycle. *Oncogene* 8:1449–55

Yamanaka T, Horikoshi Y, Sugiyama Y, Ishiyama C, Suzuki A, et al. 2003. Mammalian Lgl forms a protein complex with PAR-6 and aPKC independently of PAR-3 to regulate epithelial cell polarity. *Curr. Biol.* 13:734–43

Yamanaka T, Horikoshi Y, Suzuki A, Sugiyama Y, Kitamura K, et al. 2001. PAR-6 regulates aPKC activity in a novel way and mediates cell-cell contact-induced formation of the epithelial junctional complex. *Genes Cells* 6:721–31

Yamashiro S, Totsukawa G, Yamakita Y, Sasaki Y, Madaule P, et al. 2003. Citron kinase, a Rho-dependent kinase, induces di-phosphorylation of regulatory light chain of myosin II. *Mol. Biol. Cell* 14:1745–56

Yasuda S, Oceguera-Yanez F, Kato T, Okamoto M, Yonemura S, et al. 2004. Cdc42 and mDia3 regulate microtubule attachment to kinetochores. *Nature* 428:767–71

Yoshimura T, Kawano Y, Arimura N, Kawabata S, Kikuchi A, Kaibuchi K. 2005. GSK-3beta regulates phosphorylation of CRMP-2 and neuronal polarity. *Cell* 120:137–49

Yu W, Datta A, Leroy P, O'Brien LE, Mak G, et al. 2005. β1-integrin orients epithelial polarity via Rac1 and laminin. *Mol. Biol. Cell* 16:433–45

Zegers MM, O'Brien LE, Yu W, Datta A, Mostov KE. 2003. Epithelial polarity and tubulo-genesis in vitro. *Trends Cell Biol.* 13:169–76

Zhang X, Bi E, Novick P, Du L, Kozminski KG, et al. 2001. Cdc42 interacts with the exocyst and regulates polarized secretion. *J. Biol. Chem.* 276:46745–50

Zheng Y, Bagrodia S, Cerione RA. 1994. Activation of phosphoinositide 3-kinase activity by Cdc42Hs binding to p85. *J. Biol. Chem.* 269:18727–30

Zigmond SH. 2004. Formin-induced nucleation of actin filaments. *Curr. Opin. Cell Biol.* 16:99–105

Spatial Control of Cell Expansion by the Plant Cytoskeleton

Laurie G. Smith[1] and David G. Oppenheimer[2]

[1] Section of Cell and Developmental Biology, University of California, San Diego, La Jolla, California 92093-0116; email: lsmith@biomail.ucsd.edu

[2] Department of Botany and UF Genetics Institute, University of Florida, Gainesville, Florida 32611-8526; email: doppen@botany.ufl.edu

Annu. Rev. Cell Dev. Biol.
2005. 21:271–95

First published online as a
Review in Advance on
June 28, 2005

The *Annual Review of
Cell and Developmental
Biology* is online at
http://cellbio.annualreviews.org

doi: 10.1146/
annurev.cellbio.21.122303.114901

Copyright © 2005 by
Annual Reviews. All rights
reserved

1081-0706/05/1110-
0271$20.00

Key Words

morphogenesis, actin, microtubules, tip growth, trichomes

Abstract

The cytoskeleton plays important roles in plant cell shape determination by influencing the patterns in which cell wall materials are deposited. Cortical microtubules are thought to orient the direction of cell expansion primarily via their influence on the deposition of cellulose into the wall, although the precise nature of the microtubule-cellulose relationship remains unclear. In both tip-growing and diffusely growing cell types, F-actin promotes growth and also contributes to the spatial regulation of growth. F-actin has been proposed to play a variety of roles in the regulation of secretion in expanding cells, but its functions in cell growth control are not well understood. Recent work highlighted in this review on the morphogenesis of selected cell types has yielded substantial new insights into mechanisms governing the dynamics and organization of cytoskeletal filaments in expanding plant cells and how microtubules and F-actin interact to direct patterns of cell growth. Nevertheless, many important questions remain to be answered.

Contents

INTRODUCTION

Plant cells exhibit a wide variety of shapes that make important contributions to cell function. A plant cell's shape is determined by its wall and is acquired during development according to the direction and pattern in which the wall extends as the cell expands under the force of turgor pressure. By directing the deposition of cell wall materials, the cytoskeleton plays a central role in the control of plant cell growth and its spatial regulation. Cellulose, the principle load-bearing structural element of expanding cell walls, is synthesized by enzyme complexes in the plasma membrane. The arrangement of cellulose microfibrils in the wall is a key determinant of cell expansion pattern and is clearly related to the arrangement of cortical microtubules in expanding cells. Cell wall components other than cellulose and the related polymer callose are introduced into the wall via secretion. This includes a variety of carbohydrates that form cross-links between cellulose microfibrils, as well as enzymes that catalyze the breakage and reformation of these cross-links. Thus, spatial regulation of secretion by the cytoskeleton is crucial for determining patterns of wall extensibility and thus patterns of growth.

In this review, we discuss the roles of microtubules and actin filaments in plant cell growth and its spatial regulation, as well as a rapidly growing body of knowledge about how the dynamics and organization of both classes of filaments are controlled in expanding cells. Owing to the availability of several reviews on similar topics published in the last several years (e.g., Hepler et al. 2001, Smith 2003, Wasteneys & Galway 2003, Lloyd & Chan 2004, Mathur 2004) and to space limitations, this review does not comprehensively discuss all the relevant literature. Instead, we emphasize recent work leading to important new insights and ideas about regulation of the cytoskeleton and its contributions to patterning of plant cell growth. After a general discussion about the contributions of microtubules and actin, we discuss recent advances gained from studies on the morphogenesis of three cell types: pollen tubes, trichomes (epidermal hairs), and leaf epidermal pavement cells.

MICROTUBULES AND GROWTH DIRECTION: OLD AND NEW MODELS

The importance of both microtubules and cellulose in determining the direction of cell expansion has been clear for many years (reviewed in Baskin 2001, Wasteneys & Yang 2004, Lloyd & Chan 2004). The observation that the cellulose deposition pattern normally mirrors the pattern of cortical microtubules in expanding cells led to formulation of the cortical microtubule/cellulose microfibril co-alignment hypothesis, hereafter referred to as the co-alignment hypothesis, which states

Cellulose microfibril: a bundle of cellulose polymers each consisting of covalently linked glucose subunits

Microtubule: a 25-nm wide hollow tube that is a polymer of subunits consisting of α-tubulin/β-tubulin heterodimers

that movement of cellulose synthase enzyme complexes in the plasma membrane is constrained by interactions with the cortical microtubules (Giddings & Staehelin 1991). The energy of cellulose polymerization supplies the force needed to move the enzyme complex through the membrane, and it only needs to be guided by direct or indirect interactions with the cortical microtubules.

Unfortunately, this straightforward co-alignment hypothesis is inconsistent with observations of continued synthesis of organized cellulose microfibrils following cortical microtubule disruption (reviewed in Baskin 2001, Sugimoto et al. 2003). In addition, the model fails to account for the inability of cortical microtubules to form ordered arrays when cellulose synthesis is inhibited (Fisher & Cyr 1998). To accommodate these observations, Baskin (2001) extended the co-alignment hypothesis with the templated incorporation model. In this model, bi-functional scaffold factors bind to existing microfibrils and newly synthesized microfibrils, facilitating local order. The model also posits the existence of integral membrane components linking the scaffold factors to the cortical microtubules. The persistence of a membrane-based scaffold following cortical microtubule depolymerization can explain how ordered deposition of nascent microfibrils can occur in the absence of cortical microtubule organization. Although the templated incorporation model goes a long way toward accounting for observations apparently contradicting the co-alignment hypothesis, it suffers from a lack of direct evidence as to either the scaffold or the membrane-based components that link the scaffold to cortical microtubules.

However, one of the predictions of the templated incorporation model is that mutants that fail to properly construct the putative membrane-based scaffold should exhibit normal cortical microtubule organization but show altered patterns of cellulose microfibril organization resulting in cell expansion defects. Recently, a mutant that displays this phenotype has been identified. The

fragile fiber 1 (*fra1*) mutation was identified in a screen for mutants that showed reduced mechanical strength of the inflorescence stem due to defects in interfascicular fiber cell differentiation (Zhong et al. 2002). In addition to weak inflorescences, *fra1* mutants showed moderate cell expansion defects that led to a general shortening of plant organs. Analysis of the cell wall composition of *fra1* mutants showed no significant difference compared with wild type, but examination of the cellulose microfibrils in fiber cell walls revealed that *fra1* mutants lack the densely packed, parallel arrangement of microfibrils observed in wild type. Instead, the cellulose microfibrils in *fra1* mutants were more loosely arranged and oriented in different directions. On the basis of the co-alignment hypothesis, one might expect that the cortical microtubules would show similar disorganization. Interestingly, the arrangement of cortical microtubules in *fra1* mutants is indistinguishable from wild type. Positional cloning of the *FRA1* gene showed that it encodes a member of the KIF4 family of kinesin motor proteins. Localization of the FRA1 protein to the cell cortex in expanding cells provides support for its role in microfibril orientation. If the templated incorporation model (Baskin 2001) is correct, then this gene product may be involved in organizing the membrane-based elements that link the putative scaffold to the cortical microtubules. Identification of the cargo of this kinesin will most likely provide important insight into its role in cellulose microfibril orientation.

Another contradiction with the co-alignment hypothesis arises from the study of the *microtubule organization1-1* (*mor1-1*) mutant in *Arabidopsis* (Whittington et al. 2001). This temperature-sensitive mutant appears completely normal at permissive temperatures but rapidly loses cortical microtubule organization at restrictive temperatures, resulting in growth isotropy (loss of growth directionality). Surprisingly, mutant cells undergoing isotropic expansion still maintain transversely aligned cellulose microfibrils.

Actin filament (filamentous or F-actin): an 8-nm wide, helical polymer of globular actin (G-actin) subunits

Dynamics (cytoskeletal/microtubule/actin dynamics): refers to the turnover of filaments via lengthening (polymerization) and shortening (depolymerization)

Trichome: an epidermal hair; in *Arabidopsis*, trichomes have a branched morphology

Pavement cell: an unspecialized epidermal cell; in the leaves of most flowering plant species they have a lobed morphology

Motor protein: a protein that associates with a cytoskeletal filament and uses energy derived from ATP hydrolysis to move its cargo along the filament in a particular direction; kinesin motors move cargoes along microtubules; myosin motors move cargoes along actin filaments

FRA1: Fragile Fiber 1

Cortex: a thin shell of cytoplasm lining the inner surface of the plasma membrane

Isotropic expansion: growth that is oriented uniformly in all directions

mor1: microtubule organization1

Tip growth: an extremely polarized mode of plant cell growth exhibited by pollen tubes and root hairs in which wall extension and incorporation of new wall material occurs at a single site on the cell surface (the tip)

This led Wasteneys (2004) to propose the microfibril length regulation hypothesis, in which cortical microtubules participate in regulating the length of cellulose microfibrils rather than their alignment. As the major load-bearing structural element in plant cell walls, the arrangement of cellulose microfibrils is thought to resist expansion in the direction parallel to the orientation of the microfibrils much like a coil spring resists expansion in the radial direction. Loosening of the matrix polysaccharides linking adjacent microfibrils allows expansion perpendicular to the orientation of the microfibrils. But unless individual microfibrils extend for a considerable distance around the cell, loosening of the matrix polysaccharides would also allow adjacent microfibrils to slip past one another, resulting in radial expansion as well as elongation (**Figure 1**). The microfibril length regulation hypothesis thus explains how radial expansion could occur in cells with transversely aligned microfibrils while still providing a role for cortical microtubules in guiding directional cell expansion. A test of this model would be to measure cellulose microfibril lengths in root cells from *mor1-1* mutants before and after growth at the restrictive temperature. Additionally, radial expansion of wild-type roots treated with short pulses of the cellulose synthase inhibitor, isoxaben, might provide additional support for this hypothesis.

ROLES FOR ACTIN IN CELL GROWTH AND ITS SPATIAL REGULATION

Tip Growth

Essential functions for F-actin in cell growth have long been recognized because of the growth-arresting effects of actin depolymerizing drugs (cytochalasins and latrunculins). The dependence of cell growth on F-actin was first recognized in tip-growing cells in which growth and extension of the cell wall is focused at a single site on the cell surface resulting in production of a cylindrical shape (**Figure 2a**). Thus, much work has been devoted to understanding how F-actin contributes to tip growth. Studies of F-actin organization in tip-growing cells have produced rather different results depending on the localization method being used, and no one method is ideal in every respect. However, as illustrated in **Figure 2a**, there is now widespread agreement that longitudinally oriented actin cables run along the length of tip-growing cells, and that a dense meshwork of fine actin filaments occupies an area near the tip but does not extend to the extreme apex (e.g., Miller et al. 1996, Kost et al. 1998, Miller et al. 1999, Ketelaar et al. 2002, Y.-S. Wang et al. 2004), although highly dynamic actin filaments have been reported to penetrate transiently into the apex (Fu et al. 2001). This fine F-actin network, occupying both the cell cortex and

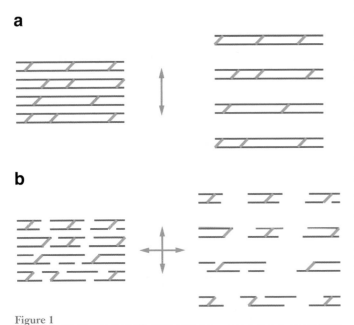

a

b

Figure 1

Microfibril length regulation model for control of anisotropic cell expansion proposed by Wasteneys (2004). (*a*) Long cellulose microfibrils (*blue*) are cross-linked by a given number of matrix polysaccharides (*orange*) that prevent sliding of adjacent microfibrils. Only expansion in the longitudinal direction is allowed (*green arrows*). (*b*) Short cellulose microfibrils cross-linked by the same number of matrix polysaccharides as in (*a*) cannot prevent slippage of some adjacent microfibrils. Expansion in the radial as well as the longitudinal direction is allowed.

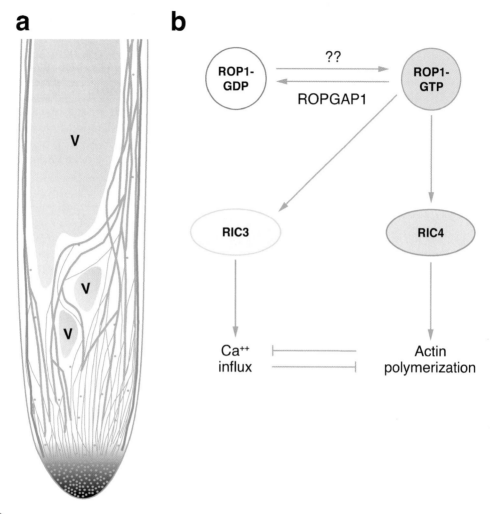

a

V

V

V

b

ROP1-GDP

?? → ROP1-GTP

ROPGAP1

RIC3

RIC4

Ca⁺⁺ influx

Actin polymerization

Figure 2

Cytoskeletal regulation of pollen tube growth. (*a*) Schematic, cross-sectional view of an elongating pollen tube. Cortical microtubules are shown in green, actin filaments and cables in orange (V, vacuole). Tip high cytoplasmic Ca^{2+} gradient is represented by gray shading in tip region with black representing the highest Ca^{2+} concentration. Blue dots represent vesicles (adapted from Ketelaar & Emons 2001). (*b*) Model schematically illustrating interactions occurring at the growing pollen tube tip between ROP1, RIC3, and RIC4 demonstrated by Gu et al. (2005).

cytoplasm, is often referred to as a subapical F-actin fringe or collar.

In tip-growing cells, myosins drive the movement of Golgi-derived vesicles and other cell components along longitudinally oriented actin cables toward the growth site via cytoplasmic streaming (Hepler et al. 2001). Thus, one important role for actin in promoting tip growth is to drive the long-range movement of vesicles, ferrying the raw materials for growth (plasma membrane and cell wall materials) toward the tip. However, actin clearly plays additional roles important for growth and its spatial regulation in tip-growing cells. Treatment of growing pollen tubes and root hairs with low

Actin cable: a thick bundle of actin filaments

Golgi: a cellular
compartment
consisting of a stack
of flattened,
membrane-bounded
sacs where proteins
destined for
secretion are
modified and sorted
as they pass through
it on their way from
the endoplasmic
reticulum to the cell
surface

Vesicle: a small
membrane-bounded,
spherical organelle
inside which proteins
and other materials
are transported
within the cell from
one location to
another

Cytoplasmic
streaming: a
directed, continuous
flow of vesicles and
other organelles that
is driven by the
actomyosin system in
plant cells

Actin assembly/
polymerization:
elongation of an
actin filament

Diffuse growth: the
mode of growth
exhibited by most
plant cells in which
extension of the wall
and incorporation of
new wall material is
broadly distributed
across the cell surface

ADF: actin-
depolymerizing
factor

concentrations of actin depolymerizing drugs can inhibit growth without blocking cytoplasmic streaming (Miller et al. 1999, Gibbon et al. 1999, Vidali et al. 2001). These treatments eliminate the subapical fine F-actin fringe without disrupting longitudinal F-actin cables and also result in loss of the vesicle-rich body of cytoplasm normally found at the apex. The sensitivity of subapical actin filaments to drug treatment suggests that they are more dynamic than the longitudinal F-actin cables. The effects of selective depletion of the subapical F-actin fringe show that it functions in some way to promote the accumulation and/or retention of vesicles at the growth site. Many ideas have been proposed to explain how subapical F-actin might accomplish these functions (reviewed in Geitmann & Emons 2000). A simple idea based on the observation that cytoplasmic streaming stops short of the apex and then reverses direction to produce a reverse fountain pattern of motility is that subapical F-actin traps vesicles, preventing them from leaving the tip region via cytoplasmic streaming, and actively transports them to the tip. This could be achieved via myosin-dependent transport through the actin mesh and/or via actin polymerization–driven propulsion of vesicles toward the apex. The subapical F-actin fringe may also serve as a barrier to prevent the loss of vesicles from the apex.

Interestingly, at concentrations even lower than those that inhibit tip growth in the presence of continued cytoplasmic streaming, actin depolymerizing drugs cause a slight depolarization of growth in the tip region, producing a swelling of the tip (e.g., Gibbon et al. 1999, Ketelaar et al. 2003). This observation indicates a role for subapical F-actin in fine-tuning the spatial distribution of growth in the tip region. One possible explanation for this finding is that subapical F-actin functions as a physical barrier to inhibit vesicle fusion in the subapical region. However, an alternative explanation is suggested by recent observations regarding the role of F-actin in Ca^{2+} channel regulation, to be discussed below in con-

nection with recent discoveries regarding the roles of ROP GTPases in regulation of pollen tube growth.

Diffuse Growth

Until relatively recently, studies on the role of F-actin in cell growth have focused mainly on tip growth. However, it is now clear from both pharmacological and genetic studies that F-actin also plays important roles in diffuse growth, the mode of growth exhibited by most plant cells, in which wall extension and incorporation of new wall material are distributed across the cell surface. Treatment with actin-depolymerizing drugs inhibits diffuse growth, causing an overall dwarfing effect (Thimann et al. 1992, Baluska et al. 2001). Moreover, although mutations knocking out individual actin isoforms have no obvious effects on diffuse growth, double mutants lacking both *ACTIN2* and *ACTIN7* function are severely growth-inhibited (Gilliland et al. 2002). Diffuse growth and tip growth have classically been viewed as mechanistically distinct growth processes, but it is now clear that they have much in common with regard to the regulation of F-actin dynamics and probably also with regard to the functions of F-actin in promotion of growth. For example, a variety of actin-binding proteins have similar roles in diffuse cell expansion and tip growth in vivo. Overexpression of actin-depolymerizing factor (ADF) disrupts cytoplasmic F-actin cables and reduces the expansion of diffusely growing cells and root hairs, whereas antisense inhibition of ADF produces an increase in the density of cytoplasmic F-actin and excess expansion of both diffusely growing cells and root hairs (Dong et al. 2001). RNAi-mediated downregulation of actin-interacting protein 1 (AIP1), which can modulate F-actin dynamics by capping actin filaments and enhancing the activity of ADF, causes excessive bundling of cytoplasmic F-actin and reduces the expansion of diffusely growing cells and root hairs alike (Ketelaar et al. 2004). Modulation of profilin levels achieved via overexpression and gene knockout approaches also has similar effects on root hair elongation and diffuse cell expansion (Ramachandran et al. 2000, McKinney et al. 2001). These perturbations show that F-actin promotes both diffuse growth and tip growth. However, other lesions in the F-actin cytoskeleton reveal roles for F-actin in the spatial regulation of diffuse growth, as discussed below in connection with trichome and pavement cell morphogenesis. In diffusely growing cells, F-actin cables permeate the cytoplasm, promoting cytoplasmic streaming as in tip-growing cells and driving the motility of a variety of organelles. Several investigators have proposed that cytoplasmic F-actin is important for properly targeted delivery of Golgi-derived vesicles to diffusely growing cell walls, just as in tip-growing cells (Baskin & Bivens 1995, Miller et al. 1999, Dong et al. 2001).

POLLEN TUBE GROWTH: A CASE IN POINT

As discussed above, a variety of actin-binding proteins are implicated in regulation of actin dynamics in tip-growing cells such as pollen tubes, but there is little information at present regarding the specific roles of most of these proteins in the spatial regulation of growth in the tip region. However, members of a plant-specific family of Rho-related GTPases (called ROPs for Rho of plants) play key roles in the polarization of tip growth owing in part to their impact on F-actin dynamics. Analyses of ROP function in pollen tube growth have focused primarily on ROP1, but ROP3 and ROP5 are also expressed in growing pollen tubes and are nearly 100% identical to ROP1, so they are probably functionally equivalent (Gu et al. 2004). ROP1 is localized to the plasma membrane at the tips of elongating pollen tubes (Lin et al. 1996). Inhibition of ROP1 function results in loss of F-actin in the tip region (but not of longitudinal actin cables) and growth arrest (Lin & Yang 1997, Li et al. 1999). Conversely, overexpression of wild-type ROP1, or expression of a con-

stitutively active form of ROP1 (ROP1-CA), causes depolarization of growth, which is associated with excess/ectopic F-actin polymerization (Li et al. 1999). Analysis of cytoplasmic Ca^{2+} distribution in pollen tubes with inhibited or excess ROP1 function showed that ROP1 also promotes an influx of extracellular Ca^{2+} needed to maintain the tip-high Ca^{2+} gradient characteristic of tip-growing cells (Li et al. 1999), which is essential for tip growth and is thought to promote vesicle fusion at the growth site (reviewed in Hepler et al. 2001).

Recent work indicates that stimulation of F-actin assembly and extracellular Ca^{2+} influx are separate functions of ROP1 that are mediated by different RIC (ROP-interacting CRIB domain) effector proteins (Gu et al. 2005). Overexpression of either RIC3 or RIC4 causes ROP1-dependent delocalization of growth in pollen tubes via distinct mechanisms: RIC4 stimulates actin assembly at the tip, whereas RIC3 stimulates Ca^{2+} influx at the tip (Gu et al. 2005) (**Figure 2b**). Other than its CRIB domain, the sequence of RIC4 is novel, so it remains to be determined how it functions to stimulate actin assembly. One possibility is that it positively regulates the activity of formins, a class of actin-nucleating proteins recently implicated in actin nucleation in elongating pollen tubes (Cheung & Wu 2004). Another possibility is that RIC4 stimulates actin assembly by mediating ROP1-dependent downregulation of ADF. This possibility is suggested by recent evidence that the tobacco homolog of ROP1 (NtRAC1) inactivates ADF in elongating pollen tubes (Chen et al. 2003). In both plant and animal cells, Rho-family GTPases regulate production of reactive oxygen species (ROS) by NADPH oxidases (Gu et al. 2004), and NADPH oxidase-dependent ROS production has been shown to play an essential role in root hair elongation by stimulating an influx of extracellular Ca^{2+} at the tip (Foreman et al. 2003). Thus, RIC3 may stimulate Ca^{2+} influx at the tip by mediating ROP1 regulation of NADPH oxidases. This

GTPase: guanine nucleotide triphosphatase

ROP: Rho of plants

RIC: ROP-interacting CRIB domain protein

Actin nucleation: the initial assembly of actin monomers to form a seed for further actin polymerization; occurs inefficiently in living cells except in the presence of a nucleator such as the Arp2/3 complex

notion is consistent with the finding that plant NADPH oxidases lack the regulatory subunit that mediates Rac activation of mammalian NADPH oxidases (Gu et al. 2004).

Interestingly, although analysis of *ric3* and *ric4* loss-of-function mutants indicates that both RIC3 and RIC4 positively regulate tip growth, various observations demonstrate that these two proteins antagonize each other's functions in a manner that depends on their downstream effectors (F-actin and Ca^{2+}) (Gu et al. 2005). Thus, ROP1/RIC4-mediated stimulation of F-actin assembly at the tip negatively regulates Ca^{2+} influx, whereas ROP1/RIC3-mediated Ca^{2+} influx negatively regulates F-actin assembly at the tip (**Figure 2b**). Consistent with these findings, negative regulation by F-actin of inward Ca^{2+} channel activity has recently been demonstrated in pollen protoplasts and pollen tubes (Y.-F. Wang et al. 2004). In turn, RIC3-induced Ca^{2+} influx may suppress actin assembly at the tip by stimulating the Ca^{2+}-dependent actin depolymerizing activities of profilins, gelsolins, and/or perhaps villins as well (Kovar et al. 2000, Holdaway-Clarke & Hepler 2003, Huang et al. 2004). But how can mutual antagonism between RIC3 and RIC4 be reconciled with the ability of both of these proteins to positively regulate tip growth? One possibility is that RIC3/RIC4 antagonism is at least partially responsible for the temporal growth oscillations that have been observed in elongating pollen tubes and root hairs, in which pulses of Ca^{2+} influx and growth (reviewed in Hepler et al. 2001) alternate with pulses of F-actin polymerization (Fu et al. 2001). Consistent with this possibility, treatment of elongating pollen tubes with very low doses of actin depolymerizing drugs showed that their growth oscillations (and thus presumably pulses of Ca^{2+} influx) are F-actin dependent (Geitmann et al. 1996). Another possibility, not mutually exclusive with the first, is suggested by the observation mentioned above that pollen tubes and root hairs elongating in the presence of very low doses of actin depolymerizing drugs exhibit a slight depolarization of growth in the tip region (tip swelling; Gibbon et al. 1999, Ketelaar et al. 2003). Subapical F-actin may negatively regulate Ca^{2+} influx in the subapical region, sharpening the Ca^{2+} gradient and thereby focusing vesicle fusion at the apex.

TRICHOME MORPHOGENESIS

The development of epidermal hairs (trichomes) in *Arabidopsis* has long been recognized as an excellent model for the study of cell shape control (Marks et al. 1991). *Arabidopsis* trichomes are single, epidermally derived cells that consist of three or four branches of equal length symmetrically arranged on top of a stalk. Their development begins with expansion of the trichome precursor as a domed cylinder out of the plane of the leaf blade, followed by two or three successive branching events. Further global expansion by diffuse growth (Hülskamp 2000) takes the 20- to 30-μm diameter incipient trichome to the final 300- to 500-μm tall mature trichome. The trichome's large size, distinctive shape, and dispensability makes them an ideal target for mutational studies of cell morphogenesis. Indeed, at least 20 genes have been identified that control trichome branching, and at least that many control the enlargement/branch elongation phase of trichome development (Mathur 2004). Because trichome development has been reviewed recently (Mathur 2004, Szymanski 2005), here we focus on new developments that link the cytoskeleton to specific events in trichome morphogenesis. The important role of the plant cytoskeleton in defining complex cell shapes was highlighted by the effects of actin and microtubule inhibitors on trichome development (Mathur et al. 1999, Szymanski et al. 1999, Mathur & Chua 2000). These studies initially defined separate roles for the microtubule and actin cytoskeletons in polar growth/branch initiation and extension growth, respectively. Genetic and molecular analyses support this view: Mutations in genes encoding microtubule-related proteins affect branch number, and mutations in genes encoding actin polymerization regulators affect extension growth (reviewed in Mathur 2004). However, synergistic effects between mutations that affect branch number and those that affect branch elongation suggest that the same gene products are involved in both processes (Zhang et al. 2005a). In addition, recent work discussed below suggests that both cytoskeletal systems cooperate throughout trichome morphogenesis.

Trichome Branch Initiation: Three Important Points

Genetic studies have led to the identification of a variety of proteins involved in the initiation of trichome branches. The functional relationships between these proteins have been murky, but recent work discussed here suggests that Golgi stacks may be the key to understanding how the cytoskeleton directs trichome branch initiation.

Zwichel and Angustifolia: seemingly unlikely partners? The *zwichel* (*zwi*) mutant was first identified by Hülskamp et al. (1994) in a screen for mutants that affected trichome development. Mutations in *zwi* lead to trichomes with a reduced stalk and fewer than normal branches (Hülskamp et al. 1994, Folkers et al. 1997, Oppenheimer et al. 1997). In addition, one of the branches on *zwi* mutants often fails to elongate properly (Folkers et al. 1997, Oppenheimer et al. 1997, Luo & Oppenheimer 1999). Thus, this mutation provides the first evidence that branch initiation and elongation are related processes. Also, the site of branch initiation is altered in *zwi* mutants, resulting in reduced stalk height (Oppenheimer et al. 1997). This phenotype suggests that ZWI is involved in proper localization of a necessary branch initiation factor. The *ZWI* gene was cloned by T-DNA tagging (Oppenheimer et al. 1997) and shown to encode a novel kinesin previously identified in a screen for calmodulin-binding proteins [kinesin-like calmodulin-binding protein

(KCBP) (Reddy et al. 1996a,b)]. Hereafter, we refer to the *ZWI* product as KCBP. Although the *Arabidopsis* genome encodes 61 kinesins (Reddy & Day 2001), *ZWI* exists as a single copy gene. A unique feature of KCBP among plant kinesins is that its presumed cargo-binding tail contains a domain also found in unconventional myosins and talin (Oppenheimer et al. 1997, Reddy & Reddy 1999). This suggests that KCBP plays a role in one or more actin-dependent processes in addition to its function as a microtubule motor. The function of KCBP in trichome development has been recently reviewed (Reddy & Day 2000). Here, we focus on its interaction with AN.

Mutations in *ANGUSTIFOLIA* (*AN*) cause cell expansion defects that result in narrow cotyledons and leaves, bent and/or twisted seed pods, and reduced branching in trichomes (Rédei 1962, Hülskamp et al. 1994, Tsuge et al. 1996). At the cellular level, *an* mutants show altered microtubule organization in trichomes and leaf cells. The *AN* gene encodes a distantly related member of the C-terminal binding protein/brefeldin A-ADP ribosylated substrate (CtBP/BARS) family (Folkers et al. 2002, Kim et al. 2002). In animal cells, members of this family function as transcriptional co-repressors and as regulators of vesicle budding from Golgi stacks (De Matteis et al. 1994, Schaeper et al. 1998, Spano et al. 1999, Weigert et al. 1999, Nardini et al. 2003). The two functions are controlled by allosteric changes induced by the binding of the molecules NAD(H) and acyl-CoA. Binding of NAD(H) to CtBP/BARS induces a conformational change that allows it to function as a co-repressor, whereas binding to acyl-CoA induces an alternative conformation that activates the membrane fission function of the protein (Nardini et al. 2003). Currently, there is no direct evidence that AN possesses either of these functions. However, consistent with a role for AN as a transcriptional co-repressor are the observations that AN-GFP fusions are detected in the nucleus and that at least

ZWI: zwichel

KCBP: kinesin-like calmodulin-binding protein

AN: *ANGUSTIFOLIA*

CtBP/BARS: C-terminal binding protein/brefeldin A-ADP ribosylated substrate

GFP: green fluorescent protein

Trafficking (membrane/vesicle/golgi trafficking): directed movement of organelles within the cell, generally involving the cytoskeleton

XTH: xyloglucan endotransglucosylase/hydrolase

Exocytosis: fusion of secretory vesicles with the plasma membrane to release vesicle contents into the extracellular environment

eight genes are modestly (about three-fold) upregulated in *an* mutants (Kim et al. 2002). Moreover, the finding that AN-GFP fusions are also found in the cytoplasm and that AN and KCBP interact genetically and biochemically support a cytoplasmic role for AN in trichome branching (Folkers et al. 2002, Kim et al. 2002). In either case, AN's effect on microtubule organization is likely to be indirect.

Kinesin-13a: the golgi connection. The recent discovery of a Golgi-localized kinesin provides a key link between Golgi dynamics and localized cell expansion (Lu et al. 2005). The GhKinesin-13A protein was identified as a kinesin that was abundantly expressed during cotton fiber development. This kinesin has an internal motor domain and is most closely related to the mammalian mitotic centromere-associated kinesin (MCAK)/kinesin-13 subfamily. A role in trichome branching was uncovered by analysis of *Arabidopsis kinesin-13A* mutants; trichomes on the mutants had more branches than wild type. Co-immunolocalization of KINESIN-13A and Golgi markers showed that this kinesin is associated with Golgi stacks. Additional studies using GFP fusions to Golgi markers revealed association of some of the Golgi stacks with cortical microtubules. These observations were surprising because previous work had suggested that only the acto-myosin system is responsible for the intracellular transport of Golgi stacks (Nebenführ et al. 1999). However, Lu and co-workers (2005) also found that Golgi stacks were more clustered in trichomes and other cell types of *kinesin-13A* mutants than in wild type, supporting a role for KINESIN-13A in distribution of Golgi stacks at the cell periphery. Lu and co-workers (2005) present a model in which KINESIN-13A-associated Golgi stacks are transported from the perinuclear region along actin filaments to the cell cortex. The recent demonstration that the myosin XI MYA2 is required for trichome branch formation suggests that this myosin may be responsible for actin-based movement of Golgi stacks to the cell periphery (Holweg & Nick 2004). According to the model of Lu et al. (2005), once Golgi come into contact with cortical microtubules, KINESIN-13A takes over transport of the Golgi stacks.

Golgi delivery model for trichome branch initiation. An intriguing model emerges from consideration of the work of Lu et al. (2005) on KINESIN-13A in relation to earlier work on ZWI and AN, which can explain the ZWI-AN interaction and the connection between Golgi trafficking and trichome branch initiation. As illustrated in **Figure 3**, this Golgi delivery model proposes that KINESIN-13A-directed transport of Golgi stacks along cortical microtubules brings them into contact with AN, which is localized to branch initiation sites by its interaction with KCBP. The postulated Golgi membrane fission-promoting activity of AN then facilitates delivery of Golgi-derived vesicles to the plasma membrane at branch initiation sites. The plant Golgi apparatus is not only the site of synthesis of matrix polysaccharides, but is also the site of post-translational modifications of the key cell wall–loosening enzyme, xyloglucan endotransglucosylase/hydrolase (XTH) (Campbell & Braam 1998, Rose et al. 2002). Localization of XTH activity at branch initiation sites in the cell wall could lead to localized weakening and bulge formation. In addition, once XTHs are released into the cell wall by exocytosis, they appear to be targeted to cellulose microfibrils, and this localization pattern relies on an intact cortical microtubule array (Vissenberg et al. 2005), providing another link between Golgi localization and cortical microtubule organization. Although increased activity of XTHs at trichome branch initiation sites has not yet been demonstrated, there is a precedent for this mechanism. A localized region of high XTH activity in root trichoblasts presages the site of bulge formation that signals root hair initiation (Vissenberg et al. 2001). In addition to XTHs, other cell wall–loosening enzymes are likely

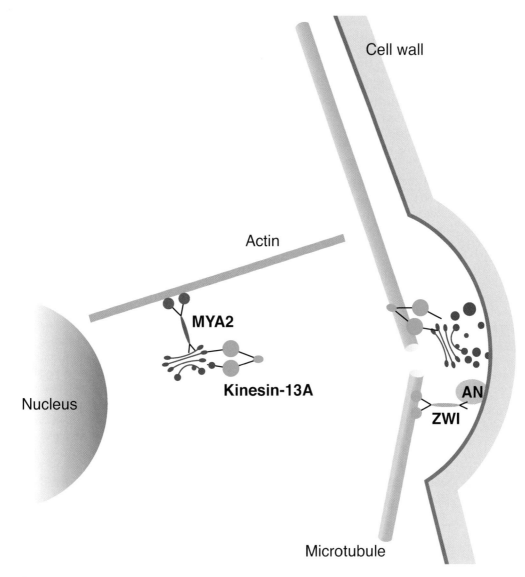

Figure 3

Golgi delivery model for trichome branch initiation. Golgi stacks (*blue*) are transported along actin filaments by the MYA2 myosin to the cell cortex where the Golgi-associated KINESIN-13A contacts cortical microtubules. KINESIN-13A transports Golgi stacks toward the minus ends of cortical microtubules (*yellow*). Localization of AN to the minus ends of microtubules is facilitated by its interaction with the minus-end-directed kinesin, ZWI. Interaction of AN with Golgi stacks promotes membrane fission and delivery of cell wall–loosening enzymes to the cell wall via exocytosis.

to be involved in the initial bulge formation as well.

The Golgi delivery model for trichome branch initiation is consistent with the phenotypes of several branching mutants. In *kinesin-*

13A mutants, Golgi stacks are transported to the cell cortex but apparently are not efficiently transported along the cortical microtubules because of the absence of KINESIN-13A. The resulting clustering of Golgi stacks

Arp2/3 complex
(actin-related
proteins 2 and 3):
a highly conserved
complex of seven
subunits, including
two actin-related
proteins (Arp2 and
Arp3); functions in
cells to nucleate
actin polymerization
in a temporally and
spatially controlled
manner

at random locations in the cell periphery could lead to cell wall loosening at ectopic sites owing to aberrant localization of XTH and other wall-loosening enzymes. Localized cell wall expansion could then set in motion the full process of branch initiation and expansion analogous to the initiation of leaf primordia by the localized application of the cell wall–loosening enzyme expansin (Fleming et al. 1997, Pien et al. 2001), thus producing an ectopic trichome branch. Also, according to this model, *an* mutants would have fewer branches than wild type because of the lack of the Golgi membrane fission-promoting activity. Similarly, lack of proper localization of AN in *zwi* mutants would lead to fewer branches as well. Even though AN and ZWI are single-copy genes in *Arabidopsis*, loss of either function does not completely block branching. This suggests that Golgi localization is unlikely to be the sole mechanism regulating branch initiation.

Trichome Branch Elongation

Role of the Arp2/3 complex. Drug treatments and mutations that interfere with actin polymerization demonstrate that F-actin plays a critical role in trichome branch elongation. Similar to the effects of near-complete depolymerization of F-actin by treatment with cytochalasin D or latrunculin B (Szymanski et al. 1999, Mathur et al. 1999), mutations disrupting four different subunits of the putative actin-nucleating Arp2/3 complex in *Arabidopsis* produce a distorted trichome phenotype characterized by a lack of trichome branch elongation, accompanied by swelling of trichome stalks (Le et al. 2003; Li et al. 2003; Mathur et al. 2003a,b; Saedler et al. 2004; El-Assal et al. 2004b). Because the volumes of distorted trichomes are similar to wild type, these findings indicate a critical role for Arp2/3-dependent actin polymerization in the spatial regulation of trichome growth (Szymanski 2005). Surprisingly, however, mutations disrupting the putative Arp2/3 complex have relatively subtle effects on the ap-

pearance of F-actin in expanding trichomes. In wild-type trichomes, longitudinally oriented F-actin cables extend throughout the length of elongating branches, whereas a fine network of cortical actin filaments (and microtubules) is aligned transversely to the branch axis (**Figure 4a**). In Arp2/3 complex subunit mutants, the most conspicuous difference in the F-actin cytoskeleton is a disorganization of cytoplasmic F-actin bundles (Le et al. 2003; Li et al. 2003; Mathur et al. 2003a,b; Saedler et al. 2004; El-Assal et al. 2004b). Thus, the putative Arp2/3 complex apparently influences actin dynamics in a manner that is vital for proper growth distribution in expanding trichomes, even though an extensive F-actin network can be maintained without it. Another surprising feature of Arp2/3 complex subunit mutants is that the shapes of other cell types are affected very little or not at all, even though the subunit genes are widely expressed. Why is the putative Arp2/3 complex so important for trichomes in particular? Compared to other leaf cell types, trichomes expand very rapidly, which might in part explain why trichomes are so sensitive to loss of Arp2/3 complex function. However, pollen tubes and root hairs also grow extremely rapidly and are affected very little by disruption of the putative Arp2/3 complex, even though their growth is clearly actin-dependent. Thus, although many proteins involved in the regulation of F-actin dynamics appear to have similar roles in diffusely growing and tip-growing cells, the putative Arp2/3 complex has a role in diffusely growing trichomes that is not critical for tip growth.

Several ideas have been proposed to explain the role of Arp2/3-dependent actin polymerization in trichome morphogenesis that are not mutually exclusive. If cytoplasmic F-actin bundles play important roles in targeted vesicle delivery to the cell surface, then disorganization of those bundles observed in Arp2/3 complex mutants could alter the spatial distribution of growth owing to vesicle mistargeting. The Golgi delivery model for

a

b

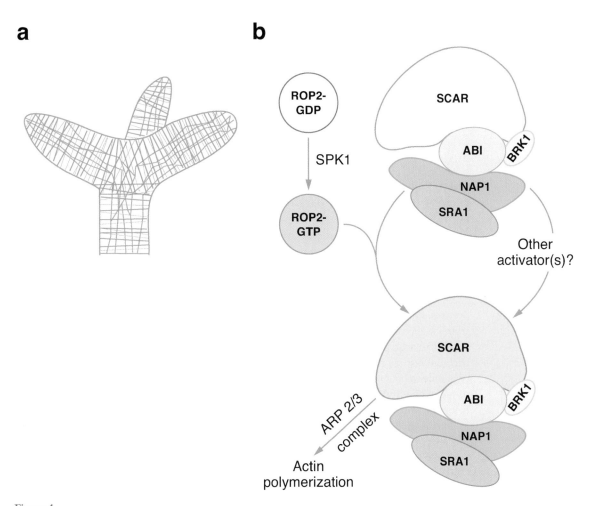

Figure 4

Cytoskeletal regulation of trichome morphogenesis. (*a*) Schematic illustration representing a two dimensional projection of the outer half of an expanding trichome showing transversely aligned microtubules (*green*) and fine actin filaments (*orange*) in the cell cortex, and longitudinally oriented actin cables (*orange*) in the central cytoplasm (based on data shown in Szymanski et al. 1999; Basu et al. 2005; S.N. Djakovic, M.P. Burke, M.J. Frank, L.G. Smith, manuscript submitted). (*b*) Model schematically illustrating activation of ARP2/3 complex–dependent actin polymerization in expanding trichomes by a plant Scar complex as proposed in the text. This pathway is also likely to contribute to actin polymerization in epidermal pavement cells. Note that there is controversy as to whether the WAVE complex remains intact upon Rac activation as shown here (Innocenti et al. 2004) or dissociates into two subcomplexes (Eden et al. 2002).

trichome branch initiation discussed above suggests a related possibility. If the distribution of Golgi stacks at the cell periphery is crucial for proper distribution of trichome growth, and if trafficking of Golgi stacks from the central cytoplasm to the cell periphery is actin-dependent (as shown for other cell types; Nebenführ et al. 1999), then the disorganization of cytoplasmic actin bundles could alter the distribution of growth by disrupting Golgi trafficking. Consistent with this idea, aberrant motility and clustering of Golgi was observed in expanding trichomes of *Arabidopsis crooked* mutants lacking the ARPC5 subunit

Scar: suppressor of cAMP receptor

WAVE: WASP (Wiscott-Aldrich Syndrome protein) family verprolin-homologous protein

Sra1: specifically Rac-associated 1

Nap1: Nck-associated protein 1

Abi: Abelson-interacting protein

HSPC300: heat shock protein C300

of the putative *Arabidopsis* Arp2/3 complex (Mathur et al. 2003b). Another possibility stems from the observation that expanding trichomes of Arp2/3 complex subunit mutants show alterations in the organization of cortical microtubules as well as of F-actin bundles (Schwab et al. 2003, Saedler et al. 2004). Consistent with a variety of observations made in the past demonstrating interdependence between microtubule and F-actin organization, this observation suggests that aberrant growth patterns in mutant trichomes may be due at least in part to defects in microtubule organization, presumably leading to defects in the arrangement of cellulose microfibrils (Saedler et al. 2004). Another possibility is that potentially actin-dependent membrane trafficking events critical for growth (e.g., exocytosis, endocytosis, and release of vesicles from Golgi stacks) are facilitated by Arp2/3-dependent actin polymerization in expanding trichomes. If so, then disorganization of cytoplasmic actin bundles could be the result of membrane trafficking defects rather than being the primary cause of growth distortion in mutant trichomes (Szymanski 2005). Further insight into the role of Arp2/3-dependent actin polymerization in the spatial regulation of trichome expansion will likely emerge from identification of the intracellular sites where this polymerization occurs through localization of the putative Arp2/3 complex.

Regulation of the Arp2/3 complex by the Scar/WAVE complex. Although many questions remain regarding the role of Arp2/3-dependent actin polymerization in the spatial regulation of trichome growth, rapid advances have been made recently in understanding its regulation. Although the plant Arp2/3 complex has not yet been purified or reconstituted for use in biochemical studies, it has long been known that efficient nucleation of F-actin in vitro by the mammalian or yeast Arp2/3 complex depends on the presence of an activator, such as a member of the WASP or Scar/WAVE family (Pollard & Borisy 2003). In mammalian cells, WAVE

proteins are regulated by the small GTPase Rac and the SH domain–containing adaptor protein, Nck (Higgs & Pollard 2001). Regulation of WAVE by Rac and Nck is mediated by their interaction with a five-protein complex consisting of WAVE, the Rac-binding protein Sra1, Nck-associated protein 1 (Nap1), Abelson-interacting protein (Abi)1 or Abi2, which is critical for complex assembly, and a small protein of unknown function called HSPC300 (Eden et al. 2002, Innocenti et al. 2004, Steffen et al. 2004, Gautreau et al. 2004). Although plants were thought initially not to have Arp2/3 complex activators of the WASP/Scar/WAVE family, a family of four *Arabidopsis* proteins distantly related at their amino and carboxy termini to Scar/WAVE proteins has recently been shown to activate the mammalian Arp2/3 complex in vitro (Frank et al. 2004, Basu et al. 2005). *Arabidopsis distorted3/irregular trichome branch1* mutations disrupt one member of this family, *SCAR2*, producing trichome morphology defects and associated alterations in the F-actin cytoskeleton similar to (although less severe than) those observed in Arp2/3 complex subunit mutants (Basu et al. 2005, Zhang et al. 2005b). Thus, SCAR2 is implicated to play an important role in activation of the Arp2/3 complex in expanding trichomes. However, since trichome morphology defects in *scar2* mutants are less severe than those of Arp2/3 complex subunit mutants, other activators must also be involved, perhaps including other members of the *Arabidopsis* SCAR family.

Homologs of Sra1, Nap1, Abi, and HSPC300 are also present in *Arabidopsis*; recent analyses of these proteins and the corresponding genes strongly suggest that, as in mammalian cells, these proteins form a complex that is essential for SCAR-mediated activation of the putative *Arabidopsis* Arp2/3 complex (**Figure 4b**). *Arabidopsis pirogi* and *gnarled* mutations *affecting SRA1* and *NAP1, respectively*, produce trichome morphology defects and alterations in the F-actin cytoskeleton of expanding trichomes similar to those observed in Arp2/3 complex subunit mutants

(Brembu et al. 2004, Basu et al. 2004, El-Assal et al. 2004a, Li et al. 2004, Zimmerman et al. 2004). HSPC300 is the mammalian homolog of *BRICK1* (*BRK1*), originally discovered because of its essential role in formation of localized cortical F-actin enrichments and epidermal pavement cell lobe formation in maize (Frank & Smith 2002). Like mutations in *SCAR2*, *NAP1*, and *SRA1*, mutations disrupting *Arabidopsis BRK1* result in trichome morphology defects and alterations in the F-actin cytoskeleton similar to those of Arp2/3 complex subunit mutants (L.G. Smith, unpublished observation; Szymanski 2005). Analyses of double mutants lacking both an Arp2/3 complex subunit and NAP1, BRK1, or SCAR2 provide genetic evidence that all three of these putative SCAR complex components act in the same pathway with the putative Arp2/3 complex (Deeks et al. 2004; Basu et al. 2005; L.G. Smith, unpublished observation). Moreover, *Arabidopsis* SRA1 and NAP1 interact with each other in the yeast two-hybrid system (Basu et al. 2004, El Assal et al. 2004b), and *Arabidopsis* BRK1 binds directly to the N-terminal Scar homology domains of *Arabidopsis* SCAR1, SCAR2, and SCAR3 (Frank et al. 2004, Zhang et al. 2005b). Although no genetic evidence has yet been reported supporting the expectation that *Arabidopsis* homologs of Abi proteins also function to activate the putative Arp2/3 complex, one member of the family of four predicted Abi-related proteins in *Arabidopsis* has recently been shown to interact with the Scar homology domain of *Arabidopsis* SCAR2 in the yeast two-hybrid system (Basu et al. 2005).

Given that the mammalian WAVE complex is activated by Rac (Eden et al. 2002, Innocenti et al. 2004, Steffen et al. 2004), an obvious question is whether ROPs function to activate the putative *Arabidopsis* SCAR complex. ROP2 is a member of the ROP family that is expressed in developing leaves and as discussed in detail below, it plays a critical role in the spatial regulation of epidermal pavement cell expansion owing in part to its ability to stimulate localized actin polymerization.

The idea that ROP2 is similarly involved in trichome morphogenesis is supported by the finding that expression of a constitutively active version of ROP2 causes a mild distorted trichome phenotype (Fu et al. 2002). Supporting the notion that ROP2 activates the putative *Arabidopsis* SCAR complex, ROP2 interacts with *Arabidopsis* SRA1 (homologous to the Rac-binding component of the mammalian WAVE complex) in the yeast two-hybrid system (Basu et al. 2004). Interestingly, the *Arabidopsis* SPIKE1 (SPK1) protein, which is required for the formation of trichome branches as well as for normal epidermal pavement cell morphogenesis, contains a domain found in a class of unconventional guanine nucleotide exchange factors that stimulate the GTPase activity of Rho family GTPases in animal cells (Qiu et al. 2002, Brugnera et al. 2002, Meller et al. 2002). Thus, SPK1 may function to activate ROP2 in developing trichomes and pavement cells. Further work will be needed to establish whether, as illustrated in **Figure 4b**, ROP2 activates the putative *Arabidopsis* SCAR complex, as well as to elucidate the potential role of SPK1 in that activation. In any case, other proteins are likely to be involved in regulation of the putative *Arabidopsis* SCAR complex. For example, the mammalian WAVE complex is activated by Nck as well as by Rac, and this activation is thought to involve binding of Nck to Nap1 (Eden et al. 2002). Thus, an as yet unidentified *Arabidopsis* Nck ortholog may interact with NAP1 to activate the putative SCAR complex (**Figure 4b**).

PAVEMENT CELL MORPHOGENESIS: BATTLE OF THE BULGES

Lobe Formation: A Collaboration Between Microtubules and F-Actin

Unspecialized leaf epidermal cells (so-called pavement cells) are an interesting case study in cytoskeletal regulation of cell growth pattern. As illustrated in **Figure 5a** for *Arabidopsis*,

BRK1: BRICK1

WAVE complex: a complex of five proteins, including the Arp2/3 activator WAVE; functions in mammalian cells to regulate the activity and localization of WAVE

SPK1: SPIKE1

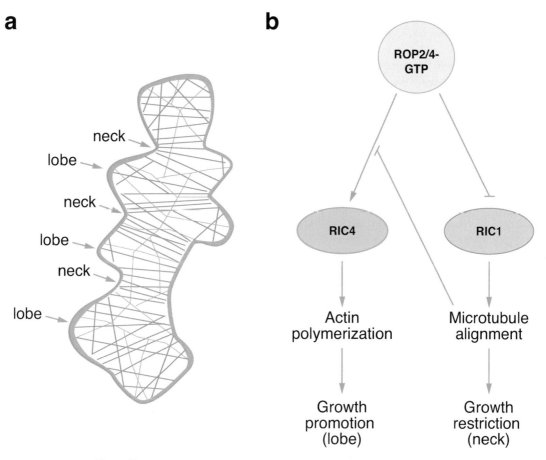

a

neck

lobe

neck

lobe

neck

lobe

b

ROP2/4-GTP

RIC4

RIC1

Actin polymerization

Microtubule alignment

Growth promotion (lobe)

Growth restriction (neck)

Figure 5

Cytoskeletal regulation of epidermal pavement cell morphogenesis. (*a*) Schematic illustration representing a two-dimensional projection of the outer half of an expanding pavement cell. Actin filaments are shown in orange, with cortical actin around the cell periphery and cytoplasmic actin cables permeating the cytoplasm; cortical microtubules are shown in green. For the sake of clarity, cortical actin on the outer face of the cell is not shown. (*b*) Model schematically illustrating regulation of actin polymerization and microtubule organization in expanding pavement cells by ROPs and two ROP-interacting RIC proteins (based on data presented in Fu et al. 2005).

epidermal pavement cells of most flowering plant species have lobed morphologies. The lobes of each pavement cell interdigitate with those of its nearest neighbors to form an interlocking cellular array. Thus, pavement cells not only acquire complex shapes, but they do so in a manner involving coordination of growth patterns between adjacent cells. Studies on the cytoskeletal basis of pavement cell morphogenesis have shown that microtubules are required for lobe formation and that they tend to be organized into parallel bundles in areas of the cell periphery where lobes are not emerging (**Figure 5a**; reviewed in Smith 2003). Thus, microtubules have been thought to contribute to lobe formation by constraining cell expansion between lobes. Recent observations indicate that actin also plays a critical role in the spatial regulation of pavement cell growth. In expanding leaf epidermal pavement cells, localized accumulations of dense, fine cortical F-actin are found at sites of lobe outgrowth in both maize and *Arabidopsis* (**Figure 5a**). In maize, *brk* mutations eliminate the formation of these localized F-actin enrichments and also eliminate the formation of lobes (Frank & Smith 2002, Frank et al. 2003). As discussed above, BRK1 is the plant homolog of a mammalian WAVE complex component and is thereby implicated as a regulator of Arp2/3 complex–dependent actin polymerization, which therefore appears to be essential for pavement cell lobe formation in maize.

The presence of dense, fine F-actin networks at sites of lobe outgrowth in pavement cells presents an intriguing parallel with tip-growing cells, where a related actin configuration is observed near the growth site, as discussed above. In a further parallel with tip growth, two closely related members of the ROP GTPase family (ROP2 and ROP4) contribute to lobe outgrowth in part by stimulating localized F-actin assembly. Plasma membrane–localized ROP2-GFP is concentrated at sites of lobe outgrowth. Localized cortical F-actin accumulation and lobe outgrowth are both reduced in plants with impaired ROP2 and ROP4 function, although cytoplasmic F-actin density and organization is normal (Fu et al. 2002, 2005). Conversely, expression of a constitutively active version of ROP2 results in delocalization of cortical F-actin accumulation and causes growth to be uniformly distributed as well. Interestingly, ROP2 and ROP4 have also been implicated in polarization of actin polymerization and growth in tip-growing root hairs (Molendijk et al. 2001, Jones et al. 2002). Moreover, like ROP1 in pollen tubes, ROP2-dependent localization of F-actin polymerization in expanding pavement cells involves an interaction with RIC4 (this interaction is discussed below in more detail). Thus, polarization of diffuse growth in pavement cells is mechanistically related to that in tip-growing cells. However, whether the contribution of ROP-dependent F-actin polymerization to growth polarization is the same in pavement cells and tip-growing cells is unclear. As discussed

above, in pollen tubes exocytosis appears to be concentrated in the apical-most area of the tip where little or no F-actin is present. The subapical F-actin fringe, although important for vesicle delivery to and/or retention at the apex, has also been implicated in suppression of exocytosis in the subapical area (e.g., Ketelaar et al. 2003) (**Figure 2a**). In contrast, cortical F-actin is enriched at sites where exocytosis rates are presumably highest in expanding pavement cells, although patterns of exocytosis have not been directly examined. Although these findings may seem contradictory, there is evidence that cortical F-actin plays both inhibitory and stimulatory roles in exocytosis in neuroendocrine PC-12 cells (Lang et al. 2000). Thus, ROP-dependent actin polymerization may locally facilitate exocytosis in expanding pavement cells, and locally inhibit exocytosis in tip-growing cells.

In any case, the effects of ROP2 and ROP4 on pavement cell morphogenesis are not limited to their influence on F-actin polymerization. In plants with reduced *ROP2* and *ROP4* gene function, parallel bundles of transversely aligned microtubules are more broadly distributed throughout the cell cortex than they are in wild type (Fu et al. 2005). Conversely, expression of constitutively active ROP2 inhibits the formation of well-ordered arrays of cortical microtubules (Fu et al. 2002). Thus, ROP2 and ROP4 appear to have dual roles in promoting pavement cell lobe outgrowth: They locally activate F-actin polymerization at sites of lobe outgrowth and also suppress the formation of ordered arrays of transversely aligned cortical microtubule bundles in these areas.

Coordination of Microtubules and Actin by ROPs and RICs

As discussed above, the observed interaction between ROP2 and *Arabidopsis* SRA1 suggests that ROP2 can activate the putative SCAR complex. However, pavement cell lobe outgrowth and localized F-actin accumulation are reduced considerably less in Arp2/3 complex subunit mutants (Li et al. 2003) than they are in plants with reduced ROP2/4 function (Fu et al. 2002, 2005). Thus, in expanding pavement cells, activation of the putative SCAR complex apparently cannot be the only pathway through which ROP2 acts to stimulate F-actin assembly. Indeed, recent work analyzing interactions between ROP2 and CRIB-domain-containing RIC proteins has shown that, as for ROP1 in pollen tubes, ROP2/4 stimulates cortical F-actin assembly in expanding pavement cells via interaction with RIC4 (Fu et al. 2005). GFP-RIC4 is localized to sites of incipient lobe formation and lobe tips in young pavement cells, and this localization pattern is dependent upon ROP2/4 activity. Moreover, loss of RIC4 activity results in reduced accumulation of fine cortical F-actin and reduced outgrowth of lobes.

Interestingly, ROP2/4-mediated suppression of the formation of well-ordered cortical microtubule bundles also involves another CRIB-domain-containing protein, RIC1 (Fu et al. 2005). Similar to what was observed in plants with reduced ROP2/4 function, RIC1 overexpression causes transversely aligned microtubule bundles to form along the entire length of expanding pavement cells and reduces lobe outgrowth. Conversely, in *ric1* loss-of-function mutants, cortical microtubules are fewer, less bundled, and less well-ordered than they are in the neck regions of expanding wild-type pavement cells, and excess expansion of neck regions occurs. RIC1-GFP colocalizes with cortical microtubules; this localization is inhibited by expression of constitutively active ROP2, but is increased in mutants with reduced ROP2/4 function. Thus, RIC1 appears to mediate the formation of ordered arrays of transversely aligned cortical microtubules via a direct association with microtubules, and ROP2/4 activity inhibits this function of RIC1. RIC1 activity, in turn, suppresses cortical F-actin accumulation by inhibiting the interaction between ROP2 and RIC4. This effect of RIC1 is likely to be mediated by microtubules themselves because depolymerization of microtubules by oryzalin

treatment or by shifting *mor1-1* mutants to restrictive temperature enhances the RIC4-ROP2 interaction and increases cortical F-actin accumulation (Fu et al. 2005).

Together, these observations support the following model to explain the patterning of pavement cell growth via ROP-dependent activities of RIC1 and RIC4 (**Figure 5b**). Local enrichment of ROP2/4 activity at sites of lobe emergence promotes RIC4-dependent activation of cortical F-actin assembly and simultaneously suppresses RIC1-dependent formation of well-ordered cortical microtubule arrays in these areas. These effects of ROP2/4 cooperatively promote lobe outgrowth. Between sites of lobe emergence where ROP2/4 and RIC4 are less abundant, RIC1-dependent formation of transversely aligned cortical microtubule bundles can take place. Promotion of cortical microtubule alignment by RIC1 is self-reinforcing because the resulting microtubule arrays inhibit the ROP2/RIC4 interaction, further reducing the inhibition of RIC1 activity in neck regions. RIC1-dependent cortical microtubule arrays restrict cell expansion between lobes, amplifying the difference in growth rates between areas of the cell surface where lobes are emerging and neck regions between these lobes. This model goes a long way toward explaining cytoskeletal regulation of pavement cell growth pattern, but the question remains open as to what initially determines the sites where ROP2/4 will become enriched. Because growth patterns of adjacent cells must be coordinated, it seems likely that the initial localization of ROP2/4 enrichment sites depends on some form of cell-cell communication. Thus, important questions remain to be answered regarding the coordination of growth patterns among neighboring pavement cells.

CONCLUDING PERSPECTIVES

In recent years, dramatic advances have been made in our understanding of mechanisms regulating cytoskeletal dynamics and organization that are important for plant cell shape determination. These advances have come from studies combining tools of genetics, genomics, molecular biology, cell biology, and biochemistry. However, much remains to be learned. For example, studies of the putative plant Arp2/3 complex have made it clear that the majority of F-actin in plant cells is nucleated in an Arp2/3-independent manner. Formins, which constitute a family of 21 predicted proteins in *Arabidopsis* (Deeks et al. 2002), are likely to serve as the primary F-actin nucleators in plant cells, but have only begun to be studied. A multitude of microtubule and actin-binding proteins are known to be important for cell growth and its spatial regulation in plants, but their precise roles remain to be elucidated. Another area still awaiting major breakthroughs is that of understanding how cytoskeletal filaments promote or spa-

tially regulate growth. More than a decade after the initial formulation of the cortical microtubule/cellulose microfibril co-alignment hypothesis, basic questions are still being asked about the microtubule/microfibril relationship, and the molecular nature of postulated linkages between microtubules and cellulose microfibrils remains a mystery. The precise mechanisms by which actin filaments promote and spatially regulate growth are also largely obscure, and there may be many of them. Elucidating the mechanisms by which cytoskeletal filaments control the spatial distribution of growth will most likely require innovative approaches employing tools not yet widely used to date, such as ultrastructural and biophysical analyses. Thus, many important challenges lie ahead in our progress toward understanding plant cell shape determination.

SUMMARY POINTS

1. The cytoskeleton plays key roles in the spatial regulation of plant cell growth primarily by influencing the pattern in which cell wall materials are deposited.

2. Microtubules are thought to control growth direction by influencing the pattern of cellulose deposition into the cell wall. A variety of models have been proposed to explain the precise nature of the microtubule/cellulose microfibril relationship.

3. Actin is essential for plant cell growth and also participates in the spatial regulation of growth, but the roles played by actin in these processes are only partially understood.

4. Polarization of pollen tube growth depends on spatially localized activities of ROP GTPases, which act through RIC effector proteins to stimulate actin polymerization and Ca^{2+} influx at the tip.

5. Trichome branch initiation is a microtubule- and actin-dependent process that may involve motor-driven transport of Golgi stacks to sites of branch initiation.

6. The putative actin-nucleating Arp2/3 complex is required for proper spatial distribution of trichome expansion and appears to be regulated by a complex of five proteins, including the Arp2/3 activator SCAR.

7. Lobe formation in epidermal pavement cells depends on the coordinated activities of microtubules and actin filaments, both of which are controlled by ROP GTPases. ROP-interacting RIC effector proteins promote the outgrowth of lobes by locally stimulating actin polymerization and restrict growth between lobes by stimulating microtubule alignment.

LITERATURE CITED

Baluska F, Jasik J, Edelmann HG, Salajova T, Volkmann D. 2001. Latrunculin B-induced dwarfism: plant cell elongation is F-actin-dependent. *Dev. Biol.* 231:113–24

Baskin TI. 2001. On the alignment of cellulose microfibrils by cortical microtubules: a review and a model. *Protoplasma* 215:150–71

Baskin TI, Bivens NJ. 1995. Stimulation of radial expansion in *Arabidopsis* roots by inhibitors of actomyosin and vesicle secretion but not by various inhibitors of metabolism. *Planta* 197:514–21

Basu D, El-Assal SE, Le J, Mallery EL, Szymanski DB. 2004. Interchangeable functions of *Arabidopsis PIROGI* and the human WAVE complex subunit SRA1 during leaf epidermal development. *Development* 131:4345–55

Basu D, Le J, El-Assal SE, Huang S, Zhang C, et al. 2005. DISTORTED3/SCAR2 is a putative *Arabidopsis* WAVE complex subunit that activates the Arp2/3 complex and is required for epidermal morphogenesis. *Plant Cell* 17:502–24

Brembu T, Winge P, Seem M, Bones AM. 2004. NAPP and PIRP encode subunits of a putative WAVE regulatory protein complex involved in plant cell morphogenesis. *Plant Cell* 16:2335–49

Brugnera E, Haney L, Grimsley C, Lu M, Walk SF, et al. 2002. Unconventional Rac-GEF activity is mediated through the Dock180-ELMO complex. *Nat. Cell Biol.* 4:574–82

Campbell P, Braam J. 1998. Co- and/or post-translational modifications are critical for TCH4 XET activity. *Plant J.* 15:553–61

Chen CY, Cheung AY, Wu HM. 2003. Actin-depolymerizing factor mediates Rac/Rop GTPase-regulated pollen tube growth. *Plant Cell* 15:237–49

Cheung AY, Wu HM. 2004. Overexpression of an *Arabidopsis* formin stimulates actin cable formation from pollen tube cell membrane. *Plant Cell* 16:257–69

Deeks MJ, Hussey PJ, Davies B. 2002. Formins: intermediates in signal-transduction cascades that affect cytoskeletal reorganization. *Trends Plant Sci.* 7:492–98

Deeks MJ, Kaloriti D, Davies B, Malho R, Hussey PJ. 2004. *Arabidopsis* NAP1 is essential for Arp2/3-dependent trichome morphogenesis. *Curr. Biol.* 14:1410–14

De Matteis MA, Di Girolamo M, Colanzi A, Pallas M, Di Tullio G, et al. 1994. Stimulation of endogenous ADP-ribosylation by brefeldin A. *Proc. Natl. Acad. Sci. USA* 91:1114–18

Dong C-H, Xia G-X, Hong Y, Ramachandran S, Kost B, Chua N-H. 2001. ADF proteins are involved in the control of flowering and regulate F-actin organization, cell expansion, and organ growth in *Arabidopsis*. *Plant Cell* 13:1333–46

Eden S, Rohtagi R, Podtelejnikov AV, Mann M, Kirschner M. 2002. Mechanism of regulation of WAVE1-induced actin nucleation by Rac1 and Nck. *Nature* 418:790–93

El-Assal SE, Le J, Basu D, Mallery EL, Szymanski DB. 2004a. *Arabidopsis GNARLED* encodes a NAP125 homolog that positively regulates Arp2/3. *Curr. Biol.* 14:1405–9

El-Assal SE, Le J, Basu D, Mallery EL, Szymanski DB. 2004b. *DISTORTED2* encodes an ARPC2 subunit of the putative *Arabidopsis* ARP2/3 complex. *Plant J.* 38:526–38

Fisher DD, Cyr RJ. 1998. Extending the microtubule/microfibril paradigm: cellulose synthesis is required for normal cortical microtubule alignment in elongating cells. *Plant Physiol.* 116:1043–51

Fleming AJ, McQueen-Mason S, Mandel T, Kuhlemeier C. 1997. Induction of leaf primordia by the cell wall protein expansin. *Science* 276:1415–18

Folkers U, Berger J, Hülskamp M. 1997. Cell morphogenesis of trichomes in *Arabidopsis*: differential control of primary and secondary branching by branch initiation regulators and cell growth. *Development* 124:3779–86

Folkers U, Kirik V, Schobinger U, Falk S, Krishnakumar S, et al. 2002. The cell morphogenesis gene ANGUSTIFOLIA encodes a CtBP/BARS-like protein and is involved in the control of the microtubule cytoskeleton. *EMBO J.* 21:1280–88

Foreman J, Demidchik V, Bothwell JHF, Mylona P, Miedema H, et al. 2003. Reactive oxygen species produced by NADPH oxidase regulate plant cell growth. *Nature* 422:442–46

Frank M, Egile C, Dyachok J, Djakovic S, Nolasco M, et al. 2004. Activation of Arp2/3 complex-dependent actin polymerization by plant proteins distantly related to Scar/WAVE. *Proc. Nat. Acad. Sci. USA* 101:16379–84

Frank MJ, Cartwright HN, Smith LG. 2003. Three *Brick* genes have distinct functions in a common pathway promoting polarized cell division and cell morphogenesis in the maize leaf epidermis. *Development* 130:753–62

Frank MJ, Smith LG. 2002. A small, novel protein highly conserved in plants and animals promotes the polarized growth and division of maize leaf epidermal cells. *Curr. Biol.* 12:849–53

Fu Y, Gu Y, Zheng Z, Wasteneys G, Yang Z. 2005. *Arabidopsis* interdigitating cell growth requires two antagonistic pathways with opposing action on cell morphogenesis. *Cell* 120:687–700

Fu Y, Li H, Yang Z. 2002. The ROP2 GTPase controls the formation of cortical fine F-actin and the early phase of directional cell expansion during *Arabidopsis* organogenesis. *Plant Cell* 14:777–94

Fu Y, Wu G, Yang Z. 2001. Rop GTPase-dependent dynamics of tip-localized F-actin controls tip growth in pollen tubes. *J. Cell Biol.* 152:1019–32

Gautreau A, Ho HY, Steen H, Gygi SP, Kirschner MW. 2004. Purification and architecture of the ubiquitous WAVE complex. *Proc. Natl. Acad. Sci. USA* 101:4379–83

Geitmann A, Emons AMC. 2000. The cytoskeleton in plant and fungal cell tip growth. *J. Microsc.* 198:218–45

Geitmann A, Li YQ, Cresti M. 1996. The role of the cytoskeleton and dictyosome activity in the pulsatory growth of *Nicotiana tabacum* and *Petunia hybrida* pollen tubes. *Bot. Acta* 109:102–9

Gibbon BC, Kovar DR, Staiger CJ. 1999. Latrunculin B has different effects on pollen germination and tube growth. *Plant Cell* 11:2349–63

Giddings THJ, Staehelin LA. 1991. Microtubule-mediated control of microfibril deposition: a re-examination of the hypothesis. In *The Cytoskeletal Basis of Plant Growth and Form*, ed. CW Lloyd, pp. 85–99. New York: Academic

Gilliland LU, Kandasamy MK, Pawloski LC, Meagher RB. 2002. Both vegetative and reproductive actin isovariants complement the stunted root hair phenotype of the *Arabidopsis act 2–1* mutation. *Plant Physiol.* 130:2199–209

Gu Y, Fu Y, Dowd P, Li S, Vernoud V, et al. 2005. A Rho-family GTPase controls actin dynamics and tip growth via two counteracting downstream pathways in pollen tubes. *J. Cell Biol.* 169:127–38

Gu Y, Wang Z, Yang Z. 2004. ROP/RAC GTPase: an old new master regulator for plant signaling. *Curr. Opin. Plant Biol.* 7:527–36

Hepler PK, Vidali L, Cheung AY. 2001. Polarized cell growth in higher plants. *Annu. Rev. Cell Dev. Biol.* 17:159–87

Higgs HN, Pollard TD. 2001. Regulation of actin filament network formation through the Arp2/3 complex: activation by a diverse array of proteins. *Annu. Rev. Biochem.* 70:649–76

Holdaway-Clarke TL, Hepler PK. 2003. Control of pollen tube growth: role of ion gradients and fluxes. *New Phytol.* 159:539–63

A groundbreaking study demonstrating that ROP GTPases interact with two different RIC effector proteins to promote formation of a lobed morphology in epidermal pavement cells.

An important paper showing that ROP GTPases interact with one RIC protein to promote actin assembly and another to promote influx of Ca^{2+} at the tip.

Holweg C, Nick P. 2004. *Arabidopsis* myosin XI mutant is defective in organelle movement and polar auxin transport. *Proc. Nat. Acad. Sci. USA* 101:10488–93

Huang S, Blanchoin L, Chaudhry F, Franklin-Tong VE, Staiger CJ. 2004. A gelsolin-like protein from *Papaver rhoeas* pollen (PrABP80) stimulates calcium-regulated severing and depolymerization of actin filaments. *J. Biol. Chem.* 279:23364–75

Hülskamp M. 2000. How plants split hairs. *Curr. Biol.* 10:R308–10

Hülskamp M, Miséra S, Jürgens G. 1994. Genetic dissection of trichome cell development in *Arabidopsis*. *Cell* 76:555–66

Innocenti M, Zucconi A, Disanza A, Frittoli E, Areces L, et al. 2004. Abi1 is essential for the formation and activation of a WAVE2 signaling complex mediating Rac-dependent actin remodeling. *Nat. Cell Biol.* 6:319–27

Jones MA, Shen J-J, Fu Y, Li H, Yang Z, Grierson CS. 2002. The *Arabidopsis* Rop2 GTPase is a positive regulator of both root hair initiation and tip growth. *Plant Cell* 14:763–76

Ketelaar T, Allwood EG, Anthony R, Voigt B, Menzel D, Hussey PJ. 2004. The actin-interacting protein AIP1 is essential for actin organization and plant development. *Curr. Biol.* 14:145–49

Ketelaar T, de Ruijter NCA, Emons AMC. 2003. Unstable F-actin specifies the area and microtubule direction of cell expansion in *Arabidopsis* root hairs. *Plant Cell* 15:285–92

Ketelaar T, Emons AMC. 2001. The cytoskeleton in plant cell growth: lessons from root hairs. *New Phytol.* 152:409–18

Ketelaar T, Faivre-Moskalenko C, Esseling JJ, de Ruijter NCA, Grierson CS, et al. 2002. Positioning of nuclei in *Arabidopsis* root hairs: an actin-regulated process of tip growth. *Plant Cell* 14:2941–55

Kim GT, Shoda K, Tsuge T, Cho KH, Uchimiya H, et al. 2002. The *ANGUSTIFOLIA* gene of *Arabidopsis*, a plant CtBP gene, regulates leaf-cell expansion, the arrangement of cortical microtubules in leaf cells and expression of a gene involved in cell-wall formation. *EMBO J.* 21:1267–79

Kost B, Spielhofer P, Chua NH. 1998. A GFP-mouse talin fusion protein labels plant actin filaments in vivo and visualizes the actin cytoskeleton in growing pollen tubes. *Plant J.* 16:393–401

Kovar DR, Drøbak BK, Staiger CJ. 2000. Maize profilin isoforms are functionally distinct. *Plant Cell* 12:583–98

Lang T, Wacker I, Wunderlich I, Rohrbach A, Giese G, et al. 2000. Role of actin cortex in the subplasmalemmal transport of secretory granules in PC-12 cells. *Biophys. J.* 78:2863–77

Le J, El-Assal SE, Basu D, Saad ME, Szymanski DB. 2003. Requirements for *Arabidopsis* *ATARP2* and *ATARP3* during epidermal development. *Curr. Biol.* 13:1341–47

Li H, Lin Y, Heath RM, Zhu MX, Yang Z. 1999. Control of pollen tube tip growth by a Rop GTPase-dependent pathway that leads to tip localized calcium influx. *Plant Cell* 11:1731–42

Li S, Blanchoin L, Yang Z, Lord EM. 2003. The putative *Arabidopsis* Arp2/3 complex controls leaf cell morphogenesis. *Plant Physiol.* 132:2034–44

Li Y, Sorefan K, Hemmann G, Bevan MW. 2004. *Arabidopsis* NAP and PIR regulate actin-based cell morphogenesis and multiple developmental processes. *Plant Physiol.* 136:3616–27

Lin Y, Wang Y, Zhu J-K, Yang Z. 1996. Localization of a Rho GTPase implies a role in tip growth and movement of the generative cell in pollen tubes. *Plant Cell* 8:293–303

Lin Y, Yang Z. 1997. Inhibition of pollen tube elongation by microinjected anti-Rop1 antibodies suggests a crucial role for Rho-type GTPases in the control of tip growth. *Plant Cell* 9:1647–59

Interestingly, results in this paper suggest that actin-based processes are important for trichome branch initiation, contradicting results from earlier pharmacological studies using actin inhibitors.

Lloyd CW, Chan J. 2004. Microtubules and the shape of plants to come. *Nat. Rev. Mol. Cell Biol.* 5:13–22

Lu L, Lee YR, Pan R, Maloof JN, Liu B. 2005. An internal motor kinesin is associated with the Golgi apparatus and plays a role in trichome morphogenesis in *Arabidopsis*. *Mol. Biol. Cell* 16:811–23

Luo D, Oppenheimer DG. 1999. Genetic control of trichome branch number in *Arabidopsis*: the roles of the *FURCA* loci. *Development* 126:5547–57

Marks MD, Esch J, Herman P, Sivakumaran S, Oppenheimer D. 1991. A model for cell-type determination and differentiation in plants. In *Molecular Biology of Plant Development*, ed. G Jenkins, W Schuch, pp. 259–75. Cambridge, UK: Co. Biol. Ltd.

Mathur J. 2004. Cell shape development in plants. *Trends Plant Sci.* 9:583–90

Mathur J, Chua N-H. 2000. Microtubule stabilization leads to growth reorientation in *Arabidopsis* trichomes. *Plant Cell* 12:465–77

Mathur J, Spielhofer P, Kost B, Chua N. 1999. The actin cytoskeleton is required to elaborate and maintain spatial patterning during trichome cell morphogenesis in *Arabidopsis thaliana*. *Development* 126:5559–68

Mathur J, Mathur N, Kernebeck B, Hülskamp M. 2003a. Mutations in actin-related proteins 2 and 3 affect cell shape development in *Arabidopsis*. *Plant Cell* 15:1632–45

Mathur J, Mathur N, Kirik V, Kernebeck B, Srinivas BP, Hülskamp M. 2003b. *Arabidopsis CROOKED* encodes for the smallest subunit of the ARP2/3 complex and controls cell shape by region specific fine F-actin formation. *Development* 130:3137–46

McKinney EC, Kandasamy MK, Meagher RB. 2001. Small changes in the regulation of one *Arabidopsis* profilin isovariant, PRF1, alter seedling development. *Plant Cell* 13:1179–91

Meller N, Irani-Tehrani M, Kiosses WB, Del Pozo MA, Schwartz MA. 2002. Zizimin1, a novel Cdc42 activator, reveals a new GEF domain for Rho proteins. *Nat. Cell Biol.* 4:639–47

Miller DD, de Ruijter NCA, Bisseling T, Emons AMC. 1999. The role of actin in root hair morphogenesis: studies with lipochito-oligosaccharide as a growth stimulator and cytochalasin as an actin perturbing drug. *Plant J.* 17:141–54

Miller DD, Lancelle SA, Hepler PK. 1996. Actin microfilaments do not form a dense meshwork in *Lilium longiflorum* pollen tube tips. *Protoplasma* 195:123–32

Molendijk A, Bischoff F, Rajendrakumar CSV, Friml J, Braun M, et al. 2001. *Arabidopsis thaliana* Rop GTPases are localized to tips of root hairs and control polar growth. *EMBO J.* 20:2779–88

Nardini M, Spano S, Cericola C, Pesce A, Massaro A, et al. 2003. CtBP/BARS: a dual-function protein involved in transcription co-repression and Golgi membrane fission. *EMBO J.* 22:3122–30

Nebenführ A, Gallagher LA, Dunahay TG, Frohlick JA, Mazurkiewicz AM, et al. 1999. Stop-and-go movements of plant Golgi stacks are mediated by the acto-myosin system. *Plant Physiol.* 121:1127–42

Oppenheimer DG, Pollock MA, Vacik J, Szymanski DB, Ericson B, et al. 1997. Essential role of a kinesin-like protein in *Arabidopsis* trichome morphogenesis. *Proc. Nat. Acad. Sci. USA* 94:6261–66

Qiu JL, Jilk R, Marks MD, Szymanski DB. 2002. The *Arabidopsis SPIKE1* gene is required for normal cell shape control and tissue development. *Plant Cell* 14:101–18

Pien S, Wyrzykowska J, McQueen-Mason S, Smart C, Fleming A. 2001. Local expression of expansin induces the entire process of leaf development and modifies leaf shape. *Proc. Nat. Acad. Sci. USA* 98:11812–17

Pollard TD, Borisy GG. 2003. Cellular motility driven by assembly and disassembly of actin filaments. *Cell* 112:453–65

This report provides direct evidence that microtubule-dependent Golgi transport is critical for trichome morphogenesis in *Arabidopsis*.

Ramachandran S, Christensen HEM, Ishimaru Y, Dong C-H, Chao-Ming W, et al. 2000. Profilin plays a role in cell elongation, cell shape maintenance, and flowering in *Arabidopsis*. *Plant Physiol.* 124:1637–47

Reddy AS, Day IS. 2000. The role of the cytoskeleton and a molecular motor in trichome morphogenesis. *Trends Plant Sci.* 5:503–5

Reddy AS, Day IS. 2001. Kinesins in the *Arabidopsis* genome: a comparative analysis among eukaryotes. *BMC Genomics* 2:2

Reddy ASN, Narasimhulu SB, Safadi F, Golovkin M. 1996a. A plant kinesin heavy chain-like protein is a calmodulin-binding protein. *Plant J.* 10:9–21

Reddy ASN, Safadi F, Narasimhulu SB, Golovkin M, Hu X. 1996b. A novel plant calmodulin-binding protein with a kinesin heavy chain motor domain. *J. Biol. Chem.* 271:7052–60

Reddy VS, Reddy AS. 1999. A plant calmodulin-binding motor is part kinesin and part myosin. *Bioinformatics* 15:1055–57

Rédei GP. 1962. Single locus heterosis. *Z. Vererbungsl.* 93:164–70

Rose JK, Braam J, Fry SC, Nishitani K. 2002. The XTH family of enzymes involved in xyloglucan endotransglucosylation and endohydrolysis: current perspectives and a new unifying nomenclature. *Plant Cell Physiol.* 43:1421–35

Saedler R, Mathur N, Srinivas BP, Kernebeck B, Hülskamp M, Mathur J. 2004. Actin control over microtubules suggested by *DISTORTED2* encoding the *Arabidopsis* ARPC2 subunit homolog. *Plant Cell Physiol.* 45:813–22

Schaeper U, Subramanian T, Lim L, Boyd JM, Chinnadurai G. 1998. Interaction between a cellular protein that binds to the C-terminal region of adenovirus E1A (CtBP) and a novel cellular protein is disrupted by E1A through a conserved PLDLS motif. *J. Biol. Chem.* 273:8549–52

Schwab B, Mathur J, Saedler R, Schwarz H, Frey B, et al. 2003. Regulation of cell expansion by the *DISTORTED* genes in *Arabidopsis thaliana*: actin controls the spatial organization of microtubules. *Mol. Genet. Genomics* 269:350–60

Smith LG. 2003. Cytoskeletal control of plant cell shape: getting the fine points. *Curr. Opin. Plant Biol.* 6:63–73

Spano S, Silletta MG, Colanzi A, Alberti S, Fiucci G, et al. 1999. Molecular cloning and functional characterization of brefeldin A-ADP-ribosylated substrate. A novel protein involved in the maintenance of the Golgi structure. *J. Biol. Chem.* 274:17705–10

Steffen A, Rottner K, Ehinger J, Innocenti M, Scita G, et al. 2004. Sra-1 and Nap1 link Rac to actin assembly driving lamellipodia formation. *EMBO J.* 23:749–59

Sugimoto K, Himmelspach R, Williamson RE, Wasteneys GO. 2003. Mutation or drug-dependent microtubule disruption causes radial swelling without altering parallel cellulose microfibril deposition in *Arabidopsis* root cells. *Plant Cell* 15:1414–29

Szymanski DB. 2005. Breaking the WAVE complex: the point of *Arabidopsis* trichomes. *Curr. Opin. Plant Biol.* 8:103–112

Szymanski DB, Marks MD, Wick SM. 1999. Organized F-actin is essential for normal trichome morphogenesis in *Arabidopsis*. *Plant Cell* 11:2331–47

Thimann KV, Reese K, Nachmias VT. 1992. Actin and the elongation of plant cells. *Protoplasma* 171:153–66

Tsuge T, Tsukaya H, Uchimiya H. 1996. Two independent and polarized processes of cell elongation regulate leaf blade expansion in *Arabidopsis thaliana* (L.) Heynh. *Development* 122:1589–600

Vidali L, McKenna ST, Hepler PK. 2001. Actin polymerization is essential for pollen tube growth. *Mol. Biol. Cell* 12:2534–45

An in-depth review of the role of the putative Arp2/3 complex in *Arabidopsis* trichome morphogenesis and its regulation by a putative SCAR complex.

Vissenberg K, Fry SC, Pauly M, Höfte H, Verbelen JP. 2005. XTH acts at the microfibril-matrix interface during cell elongation. *J. Exp. Bot.* 56:673–83

Vissenberg K, Fry SC, Verbelen JP. 2001. Root hair initiation is coupled to a highly localized increase of xyloglucan endotransglycosylase action in *Arabidopsis* roots. *Plant Physiol.* 127:1125–35

Wang Y-F, Fan L-M, Zhang W-Z, Zhang W, Wu W-H. 2004. Ca^{2+}-permeable channels in the plasma membrane of *Arabidopsis* pollen are regulated by actin microfilaments. *Plant Physiol.* 136:3892–904

Wang Y-S, Motes CM, Mohamalawari DR, Blancaflor EB. 2004. Green fluorescent protein fusions to *Arabidopsis* Fimbrin 1 for spatio-temporal imaging of F-actin dynamics in roots. *Cell Motil. Cytoskel.* 59:79–93

Wasteneys GO. 2004. Progress in understanding the role of microtubules in plant cells. *Curr. Opin. Plant Biol.* 7:651–60

Wasteneys GO, Galway ME. 2003. Remodeling the cytoskeleton for growth and form: an overview with some new views. *Annu. Rev. Plant Biol.* 54:691–722

Wasteneys GO, Yang Z. 2004. New views on the plant cytoskeleton. *Plant Physiol.* 136:3884–91

Weigert R, Silletta MG, Spano S, Turacchio G, Cericola C, et al. 1999. CtBP/BARS induces fission of Golgi membranes by acylating lysophosphatidic acid. *Nature* 402:429–33

Whittington AT, Vugrek O, Wei KJ, Hasenbein NG, Sugimoto K, et al. 2001. MOR1 is essential for organizing cortical microtubules in plants. *Nature* 411:610–13

Zhang X, Dyachok J, Krishnakumar S, Smith LG, Oppenheimer DG. 2005b. The *Arabidopsis IRREGULAR TRICHOME BRANCH 1 (ITB1)* gene encodes a plant homolog of the Arp2/3 complex activator Scar/WAVE that regulates actin and microtubule organization. *Plant Cell.* In press

Zhang X, Grey PH, Krishnakumar S, Oppenheimer DG. 2005a. The *IRREGULAR TRICHOME BRANCH* loci regulate trichome elongation in *Arabidopsis. Plant Cell Physiol.* 17:1–13

Zhong R, Burk DH, Morrison WH 3rd , Ye ZH. 2002. A kinesin-like protein is essential for oriented deposition of cellulose microfibrils and cell wall strength. *Plant Cell* 14:3101–17

Zimmermann I, Saedler R, Mutondo M, Hülskamp M. 2004. The *Arabidopsis GNARLED* gene encodes the NAP125 homolog and controls several actin-based cell shape changes. *Mol. Genet. Genomics* 272:290–96

The microfibril length regulation hypothesis presented in this review accounts for how radial expansion can occur in the *mor 1-1* mutant at restrictive temperature before changes in cellulose microfibril orientation.

This paper analyzes the *fra1* mutant in *Arabidopsis*, showing normal cell wall polymer composition and cortical microtubule organization but altered patterns of cellulose microfibril deposition.

RNA Silencing Systems and Their Relevance to Plant Development

Frederick Meins, Jr., Azeddine Si-Ammour, and Todd Blevins

Friedrich Miescher Institute for Biomedical Research, Maulbeerstrasse 66, CH-4058, Basel, Switzerland; email: meins@fmi.ch, azeddine.si-ammour@fmi.ch, todd.blevins@fmi.ch

Annu. Rev. Cell Dev. Biol. 2005. 21:297–318

First published online as a Review in Advance on June 28, 2005

The *Annual Review of Cell and Developmental Biology* is online at http://cellbio.annualreviews.org

doi: 10.1146/ annurev.cellbio.21.122303.114706

Copyright © 2005 by Annual Reviews. All rights reserved

1081-0706/05/1110-0297$20.00

Key Words

microRNAs, RNA interference, posttranscriptional gene silencing, RNA-directed DNA methylation, regulatory networks

Abstract

RNA silencing refers to a broad range of phenomena sharing the common feature that large, double-stranded RNAs or stem-loop precursors are processed to ca. 21–26 nucleotide small RNAs, which then guide the cleavage of cognate RNAs, block productive translation of these RNAs, or induce methylation of specific target DNAs. Although the core mechanisms are evolutionarily conserved, epigenetic maintenance of silencing by amplification of small RNAs and the elaboration of mobile, RNA-based silencing signals occur predominantly in plants. Plant RNA silencing systems are organized into a network with shared components and overlapping functions. MicroRNAs, and probably *trans*-acting small RNAs, help regulate development at the posttranscriptional level. Small interfering RNAs associated with transgene- and virus-induced silencing function primarily in defending against foreign nucleic acids. Another system, which is concerned with RNA-directed methylation of DNA repeats, seems to have roles in epigenetic silencing of certain transposable elements and genes under their control.

Contents

PGTS:
posttranscriptional
gene silencing

RNAi: RNA
interference

> If things do not turn out as we wish, we
> should wish for them as they turn out.
>
> Aristotle

INTRODUCTION

RNA silencing, also known as posttranscriptional gene silencing (PGTS) or RNA interference (RNAi), is a topic of enormous current interest. Impressive advances over the past few years have highlighted the importance of RNA silencing in developmental regulation, maintenance of genome integrity, and defense against foreign nucleic acids. Because specific RNAs are targeted for degradation, RNA silencing has also provided powerful new technologies for characterizing gene function (Novina & Sharp 2004).

RNA silencing was first described for transgenic plants, in which it was shown that *trans*-interactions between multiple copies of entirely foreign transgenes and between transgenes and homologous host genes led to the sequence-specific degradation of mRNAs (Meyer & Saedler 1996). Later, VIGS was shown to share this RNA-degradation pathway and to be a major mechanism for the protection of plants against virus infection (Ruiz et al. 1998). VIGS may function in animals as well (Li et al. 2002, Andino 2004). These early studies established the epigenetic nature of RNA silencing, sequence specificity, cell-to-cell transmission, links to DNA methylation, and TGS. They also implicated double-stranded RNA (dsRNA) and ca. 21–26 nucleotide (nt) smRNA as components of the RNA-degradation pathway (Kooter et al. 1999).

The turning point in the field came in the late 1990s with the discovery of RNAi—namely through studies showing the silencing of genes by dsRNA in *Caenorhabditis elegans* and biochemical studies, mainly of *Drosophila* and mammals, showing that Dicer RNase III processes large dsRNAs to ca. 21 nt-long, double-stranded siRNAs with 2–3 nt 3′ overhangs. These RNAs mark complementary regions of cognate RNAs for a single endonucleolytic cleavage (Hannon 2002). Cloning experiments led to the discovery that plants and animals have another class of ca. 21 nt smRNAs, the miRNAs, which also depend on Dicer activity for their biogenesis. Several of these miRNAs have now been shown to regulate gene expression at the level of RNA degradation and productive translation (Bartel 2004).

RNA silencing is a large, rapidly moving field that has been frequently reviewed. Our review focuses on how plant silencing systems are organized and the relevance of these systems to development. The thesis we develop is that plants have evolved a complex network of pathways with shared as well as unique components leading to smRNAs exhibiting diverse but overlapping functions. To emphasize this point, we define RNA silencing to encompass a broad range of phenomena sharing the common feature that large dsRNAs or stem-loop precursors are processed to ca. 21–26 nt smRNAs, which can then guide the cleavage of cognate RNAs, block productive translation of these RNAs, or induce the methylation of cytosines in specific target DNAs. Because smRNA-related nomenclature is often ambiguous or implies an underlying mechanism, we follow the recommendation of Steimer et al. (2004) and use the collective term smRNA, except for well-annotated miRNAs and siRNAs (Ambros et al. 2003).

RNA SILENCING: AN OVERVIEW

Transgene-Induced RNA Silencing and VIGS

Initiation, maintenance, and systemic spread. Plants carrying sense transgenes, antisense transgenes, inverted repeats, transcriptionally silenced transgenes, promoterless transgenes, and even DNA fragments can exhibit spontaneous silencing (Fagard & Vaucheret 2000a, Meins 2000). This often occurs stochastically, i.e., the frequency of individual plants showing silencing in a genetically identical population is variable and individual plants can show variegation in the silent phenotype. RNA silencing can also be initiated locally by introducing additional gene copies by virus infection (i.e., by VIGS), by infiltration of leaves with *Agrobacterium tumefaciens* that deliver T-DNAs to plant cells, and by biolistic delivery of DNA and RNA molecules (Mlotshwa et al. 2002).

Once initiated, the silent state is usually stable and maintained during vegetative growth and in cuttings. The few cases studied in detail have revealed that high-level expression of silenced genes is restored early in embryonic development (Dehio & Schell 1994, Kunz et al. 1996, Balandin & Castresana 1997, Mitsuhara et al. 2002), which indicates that silencing is a meiotically inherited form of epigenetic regulation (Meins 1996).

Transgene-induced RNA silencing and VIGS can spread both locally and systemically (Meins 2000, Mlotshwa et al. 2002). Local silencing induced by introduction of additional gene copies, combined with grafting experiments, have shown that silencing signals exist that are capable of cell-to-cell movement via plasmodesmata and systemic spread via phloem (Palauqui et al. 1997, Voinnet et al. 1998).

All aspects of silencing—its incidence, timing, site of initiation, maintenance, and systemic spread—depend on the nature, structure, and copy number of the transgene insert (Meins 2000, Fagard & Vaucheret 2000b, Mlotshwa et al. 2002). Silencing is also strongly influenced by developmental and environmental factors, which implies that the basic mechanism is closely linked to a variety of pathways for sensing stress and developmental cues. These cues may act in part on the target gene, as silencing requires that target genes be actively transcribed.

The RNA-degradation pathway. Run-on transcription experiments and measurements of mRNA turnover indicate that transgene-induced silencing and VIGS result primarily from increased degradation of mRNA. This occurs in the cytoplasm because nuclear and unspliced transcripts are not decreased by silencing and nuclear genes can be silenced by RNA viruses with an exclusively cytoplasmic

Virus-induced gene silencing (VIGS): RNA silencing triggered by a virus infection that degrades viral transcripts and transcripts of cognate nuclear genes

Transcriptional gene silencing (TGS): chromatin-associated, epigenetic inhibition of transcription

dsRNA: double-stranded RNA

nt: nucleotide

small RNA (smRNA): low-molecular-weight RNAs associated with RNA silencing that are defined operationally by their size: as short (ca. 21 nt) or long (24–26 nt)

small interfering RNAs (siRNAs): ca. 21 nt smRNAs that are processed by RNase III-like activity from large dsRNA into duplexes with 2–3 nt 3′ overhangs and that guide the cleavage of RNA at a single site in regions of complementarity

microRNAs
(miRNAs): ca.
21–24 nt regulatory
smRNAs that are
derived from hairpin
RNAs and that
target mRNAs for
degradation or block
productive
translation of these
mRNAs

DCL: Dicer like

life cycle (Kooter et al. 1999). Biochemical studies and the conservation of silencing-related genes suggest that the core mechanism for silencing is similar in plants and animals (Tijsterman et al. 2002, Matzke et al. 2004, Bartel 2004) (**Figure 1**). In plants, short (21–22 nt) and long (24–26 nt) smRNAs are generated from dsRNAs by Dicer-like (DCL) activities, which have been found in wheat-germ extracts (Hamilton et al. 2002, Tang et al. 2003). The antisense strand of siRNA duplexes, in association with the RISC, then guides the cleavage of cognate RNA at a single site in the region of complementarity (Tijsterman et al. 2002, Hannon 2002). In mammals, this cleavage is catalyzed by the Argonaute (AGO)2 protein "slicer" (Liu et al. 2004).

miRNA and *trans*-Acting smRNAs

Identification of miRNAs. miRNAs are regulatory, ca. 21–24 nt smRNAs found in plants and animals (Ambros et al. 2003, Bartel 2004). First identified in *C. elegans*, miRNAs have been cloned on numerous occasions from *Arabidopsis*, rice, and several other plant species by ligation of RNA adaptors to endogenous small RNAs and concatemerization of the ligated segments (Llave et al. 2002a, Reinhard et al. 2002, Park et al. 2002, Sunkar & Zhu 2004). Additional miRNAs

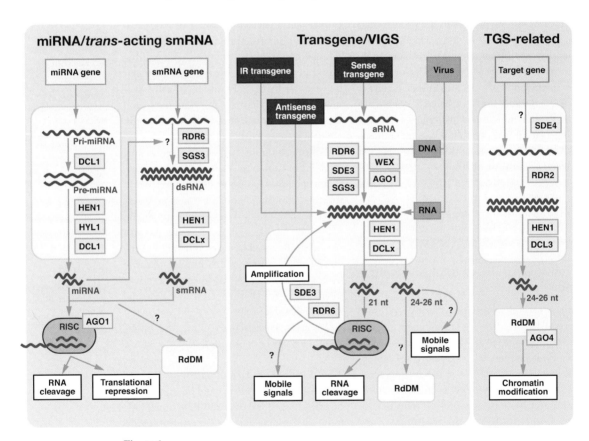

Figure 1

A scheme summarizing current models for RNA silencing in plants. To facilitate discussion, silencing pathways are organized into three systems with dedicated functions. Functional modules are indicated in yellow and components identified by genetic analysis in pale blue. DCLx refers to a postulated Dicer-like activity.

have been identified by using algorithms to detect miRNA-like hairpins conserved in rice and *Arabidopsis* (Jones-Rhoades & Bartel 2004, Bonnet et al. 2004, Wang et al. 2004). To date, 112 *Arabidopsis thaliana* miRNA genes representing 29 families, among them 9 families conserved in rice and 29 conserved in maize, are known (Griffiths-Jones 2004).

The miRNA Pathway. The current view is that the miRNA pathway for plants (**Figure 1**) is very similar to that described for animals (Bartel 2004, Kidner & Martienssen 2005). Most miRNAs arise from larger precursors transcribed from smRNA/miRNA genes that do not encode proteins. Others appear to arise from introns present in transcripts encoding proteins. Primary transcripts of miRNA genes are often spliced, capped, and polyadenylated; this suggests that, as in humans, they have their own promoters and are transcribed by RNA polymerase II. These primary transcripts are processed to smaller RNAs with self-complementary structures, which are then cleaved by DCL activities to miRNAs. In animals, miRNA duplexes are transported by the nuclear export receptor Exportin-5 to the cytoplasm. In Arabidopsis, single-stranded miRNAs are transported to the cytoplasm by the Exportin-5 homologue, HASTY (Bollman et al. 2003, Park et al. 2005).

After it is transported to the cytoplasm, miRNA is assumed, as in animals, to enter the RISC complex, where it guides the cleavage of the mRNA target by a "slicer" endonuclease (Liu et al. 2004). miRNAs pair with near-perfect complementarity to their targets, mainly in exons; in contrast to animal miRNAs, plant miRNAs repress the expression of targeted genes primarily by RNA cleavage at the site where the miRNA pairs (Llave et al. 2002b, Bartel 2004). Nevertheless, translational repression has also been reported for the *Arabidopsis AP2* gene, which suggests that dual mechanisms for the same target may exist in plants (Chen 2003). For some *Arabidopsis* miRNA targets, the 3′ end of the cleaved

mRNA is degraded by the 5′–3′ exoribonuclease AtXRN4, while the 5′ fragment is digested by a 3′–5′ exonuclease-containing exosome (Souret et al. 2004). The 5′ fragment generated from many miRNA targets is marked by addition of a 3′-oligouridine signature that could enhance its decay (Shen & Goodman 2004).

***trans*-acting smRNAs.** *trans*-acting smRNAs comprise a subset of endogeneous smRNAs (Vazquez et al. 2004, Peragine et al. 2004, Allen et al. 2005). These smRNAs correspond to the sense and antisense strands of a noncoding miRNA/smRNA gene. Transcripts of this gene appear to be converted into a dsRNA intermediate, which is then processed into 21 nt increments by DCL activity. Although their biogenesis is similar to that of siRNAs, *trans*-smRNAs resemble miRNAs in that they act *in trans* to guide the cleavage of RNAs generated from another gene.

RNA-Directed DNA Methylation

RNA-directed DNA methylation (RdDM) in plants is associated with all known RNA silencing pathways (**Figure 1**) (Matzke & Birchler 2005). In the cases of transgene RNA silencing and VIGS, de novo methylation of cytosines occurs in all sequence contexts and methylation is largely confined to regions of homology in the transcribed region of target genes. Although RdDM is not required for degradation of target RNAs, target RNA levels decrease with increasing DNA methylation, which suggests that transcription may be affected (Depicker & Van Montagu 1997, van Houdt et al. 1997). The promoter region of genes can also be targeted for RdDM by inverted-repeat transgenes encoding hairpin RNAs and results in TGS (Matzke et al. 2004). A related form of RdDM that depends on production of long smRNAs is required for de novo methylation and transcriptional silencing of several endogeneous repeat DNAs (Chan et al. 2004).

RNA-induced silencing complex (RISC): a protein complex in which the antisense strand of siRNA duplexes guides endonucleolytic cleavage of target RNAs

trans-acting small RNAs: endogenous small RNAs derived from noncoding genes that resemble siRNAs in their biogenesis but act in trans-like miRNAs to guide cleavage of target RNAs

RdDM: RNA-directed DNA methylation

RDR:
RNA-dependent
RNA polymerase

AGO: Argonaute

A GENETIC ANALYSIS OF RNA SILENCING

EGS1, SDE3, SDE4/RPD1, RPD2, SGS3, and WEX

Arabidopsis genes important for RNA silencing have been identified by screening for mutants altered in transgene-mediated silencing, by similarity in developmental phenotype to other silencing-deficient mutants, or by their homology to genes important for silencing in other organisms. The first silencing gene described, *EGS1*, appears to regulate sensitivity to transgene silencing (Dehio & Schell 1994). *SDE3* is required for efficient transgene silencing, but not for VIGS, and encodes a putative RNA-helicase (Dalmay et al. 2001). *SDE4/RPD1* and *RPD2*, which encode subunits of a putative RNA polymerase IV, influence the onset time of RNA silencing and are required for the accumulation of long smRNAs and RdDM of direct repeats (Hamilton et al. 2002, Chan et al. 2004, Herr et al. 2005, Onodera et al. 2005). *SGS3* is required for transgene silencing and encodes a protein of unknown function with a coiled-coil domain thought to function in protein-protein interactions (Mourrain et al. 2000). *WEX* (*Werner Syndrome-like exonuclease*) is required for sense-transgene silencing but not for VIGS (Glazov et al. 2003). *WEX* encodes a functional RNase D exonuclease domain (Plchova et al. 2003) similar to that in human Werner Syndrome protein and in MUT7, which is essential for RNA silencing and transposon activity in *C. elegans*.

The RNA-Dependent RNA Polymerase Family

Proteins encoded by six *Arabidopsis* genes, *RDR1-RDR6* (Yu et al. 2003), show homology to the tomato RNA-dependent RNA polymerase (RDR) LeRDR1 (Schiebel et al. 1998) and the *Neurospora crassa* RDR QDE-1 (Makeyer & Bamford 2002) with proven RDR activity. Members of the *RDR* gene family are important for RNA silencing in several organisms, including *C. elegans*, *Dictyostelium discoideum*, *Schizosaccharomyces pombe*, and *N. crassa* (Tijsterman et al. 2002), and also have been implicated in germ line development in *C. elegans* (Smardon et al. 2000) and centromere function in *S. pombe* (Volpe et al. 2002).

Deficiency mutants of *RDR6*, also known as *SDE1/SGS2*, show developmental abnormalities, such as leaf curling, associated with deficiencies in several other silencing-related genes (Peragine et al. 2004, Vazquez et al. 2004). *RDR6* is necessary for sense-transgene-mediated silencing but not inverted-repeat-mediated silencing and is important for silencing-based resistance to certain viruses (Dalmay et al. 2000, Mourrain et al. 2000, Beclin et al. 2002).

RDR2 has been implicated in the methylation of histones, long smRNA production, and silencing of certain repetitive DNAs such as the SINE-like retroelement *AtSN1*. It is expressed in inflorescences but not leaves, appears to have a role in the timing of flowering, and is required for the de novo methylation of the epigenetically regulated FWA locus (Xie et al. 2004, Chan et al. 2004).

RDR1, like its tobacco homologue *NtRDR1*, appears to be important for silencing-based resistance to virus infection. Both of these genes are induced as part of the virus defense response, and *rdr1* loss-of-function mutants exhibit increased accumulation of *Potato virus X* and *Tobacco mosaic virus* RNAs (Xie et al. 2001, Yu et al. 2003).

Although *RDR6*, *RDR1*, and *RDR2* are not required for production of miRNAs from stem-loop precursors (Xie et al. 2004, Vazquez et al. 2004), *RDR6* is required for production of *trans*-acting smRNAs (Vazquez et al. 2004, Peragine et al. 2004).

The *Argonaute* Family

The large *Argonaute* (*AGO*) gene family encodes proteins with characteristic N-terminal

PAZ and C-terminal PIWI domains (Carmell et al. 2002, Hunter et al. 2003). The 10 *Arabidopsis AGO* genes are members of the *AGO1* subfamily, which includes several genes implicated in RNA silencing or developmental regulation. For example, the mouse AGO2 "slicer" protein is essential for early development (Liu et al. 2004, Song et al. 2004). The *S. pombe AGO1* gene is required for formation of heterochromatin associated with the centromere, accurate segregation of chromosomes, and epigenetic silencing (Volpe et al. 2002, Hall et al. 2003). The *N. crassa QDE-2* and *C. elegans RDE-1* genes, which are similar in sequence to *Arabidopsis AGO1*, are required for RNA silencing (Tijsterman et al. 2002).

Deficiency *ago1* mutants show severe pleiotropic effects on morphogenesis (Bohmert et al. 1998). Whereas the shoot axis and phyllotaxy are normal, axillary meristems are rare, and leaves are filamentous in appearance and lack a blade. Flowers form filamentous organs and are infertile. These mutants are deficient in sense but not inverted-repeat silencing, which suggests that AGO1 acts upstream of dsRNA (Fagard et al. 2000, Boutet et al. 2003). AGO1 appears to function at different sites in transgene silencing and miRNA regulation, as *ago1* mutants show increased accumulation of miRNA but decreased target cleavage, which may indicate that AGO1 acts downstream of miRNA formation (Kidner & Martienssen 2004, Vaucheret et al. 2004). Analysis of hypomorphic *ago1* alleles suggests that the developmental abnormalities observed are not completely linked to the deficiency in transgene silencing. The *ago1* mutants also show hypersensitivity to *Cucumber mosaic virus* infection and increased viral RNA accumulation; this suggests that *AGO1* may also be important for silencing-based virus resistance (Morel et al. 2002).

AGO4 functions in smRNA pathways important for DNA methylation and epigenetic regulation. It is required for TGS of the SUPERMAN gene, locus-specific histone methylation, non-CpG DNA methyla-tion, and long smRNA production from certain repetitive loci, but it is not required for RNA silencing induced by inverted repeats (Zilberman et al. 2003, 2004).

AGO7/ZIPPY functions in developmental timing of *Arabidopsis* (Hunter et al. 2003). As judged from leaf shape and from the placement of trichomes on leaves, loss-of-function *ago7* mutants show a precocious juvenile-to-adult transition. AGO7 is not required for either RNA silencing of sense transgenes or production of miRNAs and *trans*-acting smRNAs (Peragine et al. 2004, Vazquez et al. 2004).

The *Dicer-Like* Family

The *Dicer* gene family encodes proteins with a similar N-terminal DExH-box, a DUF283 domain (where DUF means "domain of unknown function"), a PAZ domain, two RNase III motifs, and at least one C-terminal dsRNA-binding domain (Schauer et al. 2002). A single Dicer in mouse and humans processes miRNAs from stem-loop precursor RNAs and siRNAs from dsRNAs (Hannon 2002), whereas these functions are mediated by two different Dicers in *Drosophila* (Lee et al. 2004). On the basis of their domain structure, four *DCL* genes have been identified in *Arabidopsis* (Schauer et al. 2002). Studies with GFP (green fluorescent protein)–tagged proteins suggest that DCL1, DCL2, and DCL3 are localized primarily in the nucleus (Papp et al. 2003, Xie et al. 2004). The presence of a putative nuclear localization signal indicates that DCL4 may be a nuclear protein as well (Papp et al. 2003).

Deficiency mutants of *Arabidopsis DCL1* were first identified on the basis of developmental phenotypes (Schauer et al. 2002). Several mutants of *SUSPENSOR1* are arrested at the heart stage of embryogenesis but show continued proliferation of the suspensor. Mutants *dcl1-7* and *dcl1-8* of *SHORT INTEGU-MENTS1* form abnormal ovules that fail to expand their integuments, are late flowering, and show enhanced conversion of floral

HEN: *Hua Enhancer*

HYL: *hyponastic leaves*

meristems to an indeterminate state. The mutant *dcl1-9* of *CARPEL FACTORY* forms flowers in which carpels fail to fuse. There is excessive proliferation of floral whorls W3 and W4 in this mutant. *DCL1* RNA is expressed in all cells of the apical and floral meristems and in flowers, cauline leaves, and stems (Jacobsen et al. 1999).

DCL1 is required for the production of at least 14 different miRNAs (Reinhart et al. 2002, Park et al. 2002) and *trans*-acting smRNAs (Peragine et al. 2004, Vazquez et al. 2004, Allen et al. 2005). Interestingly, miR162 targets *DCL1* mRNA for degradation (Xie et al. 2003). The mRNA of another component of the miRNA pathway, AGO1, is targeted by miR168 (Vaucheret et al. 2004), which suggests that the miRNA pathway is negatively regulated by a miRNA-based mechanism. Unlike animal Dicers, DCL1 appears to have both Dicer and Drosha functions in miRNA biogenesis (Kurihara & Watanabe 2004). High-level DCL1 expression is not required for the generation of smRNAs associated with silencing triggered by sense or inverted-repeat transgenes (Finnegan et al. 2003), the generation of long smRNAs associated with repetitive DNA and retrotransposable elements, or RNA silencing–based virus resistance (Xie et al. 2004, Chan et al. 2004).

DCL2, unlike DCL1, is not required for production of the two miRNAs tested or of long smRNAs associated with DNA repeats (Xie et al. 2004). DCL2 is involved in the RNA silencing–based defense against a specific virus. Measurements of virus susceptibility and viral smRNA production have shown that the loss-of-function mutant *dcl2-1* is increased in susceptibility to *Turnip crinkle virus* but not to a GFP-tagged version of *Turnip mosaic virus* or *Cucumber mosaic virus*. The function of DCL2 in virus infection is unclear, as *DCL2* RNA is expressed in inflorescences but not in leaves.

DCL3, which also is expressed in inflorescences but not in leaves, has functions in DNA methylation and epigenetic regulation,

but is not required for either miRNA production or virus resistance (Xie et al. 2004, Chan et al. 2004). The deficiency mutant *dcl3-1* shows a late flowering phenotype and reduced methylation of the epigenetically regulated loci FWA and MEA-ISR. DCL3 is required for production of long smRNAs representing several different DNA repeats and is important for histone and DNA methylation at those loci.

Hua Enhancer 1

The *Arabidopsis Hua Enhancer* (*HEN1*) gene, which is expressed in roots, stems, leaves, and inflorescences (Chen et al. 2002), encodes a methyltransferase required for O-methylation of the 3'-ribose of miRNAs (Yu et al. 2005a). The EMS deficiency mutants *hen1-1* and *hen1-2* show pronounced pleiotropic effects on development (Chen et al. 2002). *HEN1* has functions in flower development similar to those of the type C gene *AGAMOUS*: It specifies floral organ identity, represses type A gene function, and controls determinacy. Deficiency mutants show pleiotropic effects, including infertility, late flowering, reduced organ and cell size, shortened internodes, and alterations in leaf number and morphology.

HEN1 functions in several RNA silencing pathways. It is required for the accumulation of at least 11 *Arabidopsis* miRNAs (Park et al. 2002), for production of *trans*-acting smRNAs (Vazquez et al. 2004) and smRNAs associated with sense-transgene silencing, but not for smRNAs associated with inverted-repeat silencing (Boutet et al. 2003). HEN1 also contributes to resistance to infection by CMV. Single point mutations uncouple the effects on miRNA and transgene silencing, which suggests that HEN1 has distinct functions in these pathways.

Hyponastic Leaves 1

The *Arabidopsis hyponastic leaves* (*HYL1*) gene encodes a dsRNA-binding protein with two dsRNA-binding motifs, a putative nuclear localization signal, and a C-terminal repeat structure that may be a protein-protein interaction domain (Lu & Fedoroff 2000). Evidence suggests that the protein is present in cells predominantly as a macromolecular complex associated with bodies and ring-like structures in the nucleus (Han et al. 2004). Although reporter gene experiments show that the *HYL1* promoter is active primarily in vascular tissues of petioles and mid-veins of rosette leaves (Yu et al. 2005b), *HYL1* mRNA accumulates in essentially all tissues and organs (Lu & Fedoroff 2000).

Transposon insertion *hyl1* mutants show narrowing and curling of leaves, reduced organ size, and late flowering reminiscent of *hen1* mutants (Lu & Fedoroff 2000) and *ago1* mutants (Vaucheret et al. 2004). In addition, *hyl1* shows abnormalities suggestive of altered auxin regulation, including increased lateral branching, decreased sensitivity of roots to the effects of auxins and cytokinin, reduced root gravitropic response, and plagiotropic growth.

HYL1 functions primarily in miRNA-related pathways. It was shown to be required for the accumulation of several miRNAs and *trans*-acting smRNAs, but is not required for transgene silencing (Han et al. 2004, Vazquez et al. 2004).

EPIGENETIC MAINTENANCE: SECONDARY smRNA PRODUCTION AND TRANSITIVITY

Although the core mechanism for silencing has been highly conserved in evolution, epigenetic maintenance of silencing and systemic silencing have only been reported for plants and *C. elegans* (Tijsterman et al. 2002). Early threshold models proposed that epigenetic maintenance in plants involves a self-sustaining, positive feedback loop, with siRNAs and target RNAs as

components, that is triggered when one of the components exceeds a critical threshold concentration (Meins 2000). The resultant bistable, reversible switch nicely explains the requirement for target-gene transcription, the stochastic all-or-none nature of silencing, and the sensitivity of silencing to developmental and environmental cues.

The present view is that dsRNAs are produced, using siRNAs as primers, from a target RNA template (Tijsterman et al. 2002). In plants and *C. elegans*, this results in transitivity (Sonoda & Nishiguchi 2000, Sijen et al. 2001, Klahre et al. 2002, Vaistij et al. 2002, Van Houdt et al. 2003). Transitivity can spread along the length of the target RNA, generating secondary siRNAs able to trigger degradation of additional target RNAs in a self-amplifying fashion (Lipardi et al. 2001, Bergstrom et al. 2003). Transitivity and RNA silencing are blocked in deficiency mutants of *RDR6* in *Arabidopsis* (Vaistij et al. 2002) and *RRF-1* in *C. elegans* (Sijen et al. 2001). These genes are homologous to *N. crassa QDE-1*, which is known to be required for RNA silencing and to encode an RDR that catalyzes primed and unprimed production of dsRNAs in vitro (Makeyer & Bamford 2002). Tobacco secondary siRNAs have been shown to be of the short 21 nt type (Garcia-Perez et al. 2004); in *Arabidopsis*, they depend on the SDE3 RNA helicase-like protein for their production (Himber et al. 2003). Together, these results support the hypothesis that the silent state is maintained by RDR- and RNA helicase-dependent amplification of short siRNAs. The detailed mechanism for this is not known (Baulcombe 2004). For example, it is unclear why transitivity proceeds mainly bidirectionally in plants but only 5′ to 3′ in *C. elegans*, or why some endogenous RNAs are targets of transitivity whereas others are not. The possibility that single-stranded transcripts derived from silencing loci are also part of the amplification loop has not been ruled out.

Transitivity: production of secondary smRNAs from regions outside of the sequence initially targeted for silencing by smRNAs

THE CHEMICAL NATURE OF MOBILE SILENCING SIGNALS

aRNA: aberrant RNA

Current studies have focused on RNA-based signals (Mlotshwa et al. 2002). Biolistic delivery of large sense-, antisense-, and dsRNAs as well as double-stranded siRNAs can trigger systemic silencing, which indicates that any or all of these could be signals (Klahre et al. 2002). Attempts to correlate specific RNAs with systemic silencing have led to conflicting conclusions (Mlotshwa et al. 2002). Experiments in which smRNA accumulation was inhibited by expressing the viral suppressor protein P1/HC-Pro suggest that systemic silencing in tobacco can occur without detectable smRNA accumulation in graft recipients (Mallory et al. 2002). Although this suggests that mobile signals arise upstream of smRNAs, the results are difficult to interpret because the block of smRNA accumulation by P1/HC-Pro is often incomplete and depends on the transgene construct (Schöb 2001, Mallory et al. 2003, Dunoyer et al. 2004). Alternatively, signals may arise downstream of dsRNA. On the one hand, in *Nicotiana benthamiana* and *Arabidopsis*, systemic silencing is correlated with accumulation of long smRNAs but not short smRNAs (Hamilton et al. 2002). On the other hand, systemic silencing in tobacco is correlated with secondary short smRNAs generated from a target transgene (Garcia-Perez et al. 2004).

Grafting experiments have shown that smRNAs derived from a silent transgene can be transported via phloem under conditions that rule out contamination from surrounding tissues (Yoo et al. 2004). Moreover, a pumpkin phloem protein PSRP1, which specifically binds smRNAs, can promote intercellular trafficking of smRNAs in a heterologous system. Although consistent with the hypothesis that mobile silencing signals are smRNAs bound to specific proteins, full-length transgene transcripts also accumulate in phloem. Evidence that the transport of smRNAs is necessary and sufficient for systemic silencing is lacking.

Multiple silencing signals may exist. It has been proposed that primary short smRNAs are signals for short-range spread, which is limited to a region of ca. 15 cells surrounding the initiation site and does not depend on the presence of cognate transcripts. Secondary short smRNAs are signals for progressive local spread, which, in *Arabidopsis*, depends on RDR6 and SDE3. According to this model, these smRNAs trigger production of long smRNAs, which then act as systemic signals (Himber et al. 2003).

A NETWORK MODEL FOR RNA SILENCING

Evidence for Three Distinct Systems

Genetic and molecular analyses provide reasonably strong evidence for at least three distinct RNA silencing systems in plants (**Figure 1**). The transgene/VIGS system consists of a branched pathway which converges on dsRNA (Meins 2000, Beclin et al. 2002). How this key intermediate is generated depends on the nature of the transgene and type of virus. Sense transgenes form dsRNA indirectly by an unknown mechanism postulated to involve ill-defined aberrant RNAs (aRNA), such as mRNAs lacking a cap; such aRNAs could be substrates for dsRNA formation (Gazzani et al. 2004). This pathway requires a putative RDR (RDR6), a putative RNA helicase (SDE3), an endonuclease (WEX), AGO1, and SGS3 (Beclin et al. 2002, Himber et al. 2003, Glazov et al. 2003). Geminiviruses, which are bipartite, single-stranded DNAs, probably enter the same pathway, as their silencing requires RDR6, SDE3, AGO1, and SGS3 (Muangsan et al. 2004). Inverted-repeat-encoded dsRNAs, antisense transgenes capable of forming duplexes with transcripts of endogeneous target genes, and viral RNAs, which depend on viral RDR for dsRNA formation, appear to enter the pathway directly at the dsRNA step (Waterhouse et al. 2001, Baulcombe 2004). Processing of dsRNAs to short and long smRNAs probably depends on distinct DCL activities (Tang et al. 2003). In *Arabidopsis*, this requires HEN1 (Boutet et al. 2003), but the gene(s) for the two putative DCL activities have not been identified. Studies with deficient mutants suggest that high-level DCL1 expression is not required (Finnegan et al. 2003). Loss-of-function mutants of DCL2 and DCL3 have not been tested and these proteins are still likely candidates (Xie et al. 2004). The function of DCL4 is not known.

The miRNA/*trans*-acting smRNA system is also branched. (**Figure 1**). The miRNA-specific pathway processes miRNAs from primary miRNA transcripts and requires HEN1, HYL1, and DCL1 (Kidner & Martienssen 2005). In contrast, *trans*-acting smRNAs are generated via a pathway that requires RDR6, SGS3, HEN1, HYL1, and DCL1, combining features of both miRNA processing and sense-transgene silencing (Vazquez et al. 2004, Peragine et al. 2004). Recent studies have shown that miR173 and miR390 set the phase for processing primary *trans*-acting smRNA transcripts to 21 nt-long smRNAs. This supports a model in which miRNAs guide the RDR6-dependent formation of a dsRNA, which is then cleaved by a DCL activity to give phased *trans*-acting smRNAs (Allen et al. 2005). After transport to the cytoplasm, both types of smRNA have been proposed to associate with a miRNA-specific RISC, with AGO1 as a component (Vaucheret et al. 2004).

The TGS-related system (**Figure 1**) methylates DNA and histones associated primarily with genes containing direct repeats (Zilberman et al. 2004). In *Arabidopsis*, this process requires RDR2, DCL3, AGO4, SDE4/RPD1, RPD2, and, for some target genes, HEN1 (Xie et al. 2004, Chan et al. 2004). According to a recent model (Herr et al. 2005), RNA polymerase IV-dependent and -independent pathways generate substrates for RDR-mediated dsRNA formation. dsRNA is processed by DCL3 to long smRNAs, which guide AGO4-dependent RdDM (Zilberman et al. 2004).

Cross-Talk, Shared Components, and Overlapping Functions

The available evidence suggests that RNA silencing systems constitute a network with cross-talk, shared components, and overlapping functions. The nature of input genes, of exclusive protein components of functional modules, and of smRNAs determine specificity within each system. Upstream modules in the nucleus link input genes to smRNA production. Specific interactions of these modules with chromatin and the sequence of input transcripts probably establish the source of smRNAs. How cytoplasmic RISCs distinguish between RNA cleavage and translational repression is not known (Bartel 2004). Nuclear RdDM modules, such as the RITS complex concerned with TGS in *S. pombe* (Verdel et al. 2004), interact with target-gene chromatin and depend on smRNAs or other RNA signals for specificity (Matzke & Birchler 2005). We propose that smRNAs are intracellular signals linking upstream modules to RISC and RdDM. A major question is how smRNAs are specifically channeled to downstream modules. The associations of short smRNAs with RISC and of long smRNAs with RdDM suggest that the length of the smRNA may determine its transport and affinity for a particular module.

Shared components and smRNAs provide possible links for cross-talk between systems. System-specific functions of a common RdDM module, i.e., transcribed-sequence methylation and TGS-related promoter methylation, plausibly may be determined by the sequence of smRNAs. Studies with viral suppressor proteins suggest that transgene silencing and miRNA pathways share a common RISC module (Chapman et al. 2004). Moreover, there are hints that translational repression associated with the miRNA RISC (Chen 2003) is also a feature of transgene silencing. Sense-transgene silencing can lead to decreased protein-to-mRNA ratios indicative of translational regulation (van Houdt et al. 1997). Antisense silencing, which shares the sense-silencing pathway (Stam et al. 2000, Di Serio et al. 2001), reduces both mRNA accumulation and the efficiency of translation (Cornelissen & Vandewiele 1989).

Viral suppressor proteins block transgene silencing, VIGS, and miRNA pathways, but not the production of RdDM-associated smRNAs (Silhavy & Burgyan 2004, Chapman et al. 2004, Dunoyer et al. 2004). One of these viral proteins, P1/HC-Pro, which acts on RISC, interacts with a calmodulin-like protein in tobacco that blocks transgene silencing and produces developmental abnormalities typical of mutations in the miRNA pathway (Anandalakshmi et al. 2000). Thus, the two systems are linked at the level of RISC by an endogenous regulatory pathway. There is also cross-talk between the transgene/VIGS and TSG-related systems. *Arabidopsis* mutants deficient in DDM1, MET1, SDE4/RPD1, and RPD2, which are required for the RdDM pathway and for TGS, are impaired in sense-transgene RNA silencing (Morel et al. 2000, Hamilton et al. 2002, Herr et al. 2005).

RELEVANCE TO DEVELOPMENT

miRNAs and Other smRNAs

miRNAs have important roles in developmental regulation (Bartel 2004, Dugas & Bartel 2004). Plants expressing viral suppressor proteins and deficiency mutants with impaired miRNA functions show characteristic developmental abnormalities. Approximately 50% of validated miRNA targets in *Arabidopsis* are transcription factors with known functions in development (Jones-Rhoades & Bartel 2004). Several studies have linked specific *Arabidopsis* miRNAs to organ identity, morphogenesis, and polarity. The first of these studies showed that the characteristic crinkly-leaf phenotype of the *jaw* mutant results from miR-JAW-guided cleavage of transcripts encoding TCP transcription factors (Palatnik et al. 2003). The same study showed that miR159a, which differs

from miR-JAW at three positions, downregulates AtMYB33, AtMYB65, and AtMYB101 targets. These GAMYB-related transcription factors regulate short-day (SD) floral transition by activating the floral-meristem identity gene LEAFY via a gibberellin (GA)-dependent GA-DELLA signaling pathway. Transgenic lines with increased levels of miR159a are affected in anther development and delayed in SD floral transition. Moreover, GA regulates miR159a by opposing DELLA function in both *Arabidopsis* and barley in which miR159 is conserved (Achard et al. 2004). Using a similar forward-genetic approach, miR172 was shown to target the *AP2* and the *AP2*-like *TOE1* and *TOE2* genes belonging to the class A floral homeotic genes (Aukerman & Sakai 2003, Chen 2003). Overexpression of miR172 resulted in floral abnormalities typical of the strong *ap2–9* mutant allele and repressed *TOE1* and *TOE2*, known to suppress flowering; this suggests that miR172 controls both flowering time and floral whorl identity.

The *CUP-SHAPED COTYLEDON* genes *CUC1*, *CUC2*, and *CUC3*, which act in the cells forming the boundary domain around primordia at the shoot apical and floral meristems, are also miRNA targets (Laufs et al. 2004, Mallory et al. 2004). Constitutive expression of miR164 phenocopies the *cuc1cuc2* double mutant, which exhibits fused cotyledons, sepals, and stamens, as well as defective shoot-meristem formation. Expression of miR164-resistant CUC1 and CUC2 transcripts lead to reduced organ size and to distinctive abnormalities in leaf and flower morphology. *REVOLUTA* (*REV*), *PHABULOSA*, *PHAVOLUTA*, *ATHB8*, and *ATHB15* are transcription factors important for the determination of the adaxial-abaxial polarity of *Arabidopsis* organs. A considerable body of evidence suggests that targeting of these genes by miRNA165/miR166 regulates leaf polarity and the arrangement of xylem and phloem (Kidner & Martienssen 2004). For example, plants expressing a mutated *REV* resistant to cleavage by miR165/166 and regulated by the endogenous *REV* promoter pheno-copy the mutant *rev-10d*, which shows ectopic leaf tissue fused to the stem, abnormal stem vascular bundles with an xylem surrounding the phloem, and adaxialized trumpet-shaped leaves (Emery et al. 2003). These gain-of-function phenotypes are due to an accumulation of the miRNA-resistant *REV* transcript in the abaxial domain of lateral organs and in the phloem. The maize mutant *Rolled leaf1-Original* shows the *rev-10d* phenotype and is located at a site with miR166 complementarity (Juarez et al. 2004). This suggests that this miRNA pathway has been highly conserved in Monocot and Eudicot evolution.

Aside from the examples cited above, few developmental functions have been assigned to specific smRNAs. Genetic analyses of double and triple mutants have led to the proposal that AGO7, RDR6, and HASTY are part of a novel smRNA pathway concerned with regulating juvenile-to-adult transition in *Arabidopsis* (Peragine et al. 2004). The related *trans*-acting smRNA pathway also targets several genes implicated in developmental regulation; mutants impaired in the biogenesis of these RNAs show characteristic developmental abnormalities (Peragine et al. 2004, Vazquez et al. 2004). To date, only one loss-of-function mutant in a specific *trans*-acting smRNA locus has been tested, and it did not show either developmental or other obvious abnormalities. Thus, the functional significance of the *trans*-acting smRNA pathway is still unclear.

In principle, smRNAs could also function in epigenetic regulation at the chromatin level (Steimer et al. 2004, Matzke & Birchler 2005). RdDM is important for TGS of some transposable elements in plants (Lippman et al. 2003, Lippman et al. 2004). Evidence for a similar role of RdDM in development is scant. A SINE-type retroelement provides the promoter and first two exons of the *Arabidopsis FWA* gene (Lippman et al. 2004). Expression of *FWA* is normally imprinted in the endosperm by a DNA methylation-dependent mechanism. *FWA* is silenced in most tissues by methylation of two direct repeats in the promoter region; these repeats also give rise to smRNAs. Overexpression of *FWA* in hypomethylated epimutants results in a late-flowering phenotype. Genetic analyses have shown that the initiation and maintenance of DNA methylation required for epigenetic regulation of *FWA* depend on long smRNA production and the AGO4-RDR2-DCL3-SDE4/RPD1 RdDM system (Chan et al. 2004).

Intra- and Intercellular Regulatory Networks

miRNAs and possibly other smRNAs are likely important for establishing regulatory networks at the cellular and supracellular level. Complex networks showing positive and negative feedback are believed to contribute to the stability and progressive nature of development (Freeman 2000, Oliveri & Davidson 2004). These networks also have the potential to generate switch-like behavior (Becskei & Serrano 2000), to buffer the deleterious effects of mutations (Wagner 2000), and to ensure the fidelity of developmental pathways (Paulsson 2004). Transcriptional networks are important for meristem identify and patterning in plants (Benfrey & Weigel 2001). miRNA pathways could provide additional posttranscriptional networks. Remarkably, sets of miRNA genes can generate miRNAs with the same or very similar sequence that then target the same gene (Dugas & Bartel 2004). Moreover, most individual miRNAs can target up to 12 different genes, and some miRNAs exhibit negative autoregulation (**Figure 2**). This suggests that miRNA pathways may be network-building blocks capable of integrating many inputs and controlling sets of genes in a hierarchical fashion. Epigenetic stability could result from network architecture, positive autoregulation based on RNA replication, RdDM, or a combination of these mechanisms. Although miRNAs can trigger RdDM methylation (Bao et al. 2004), it is not known if stable miRNA states exist or

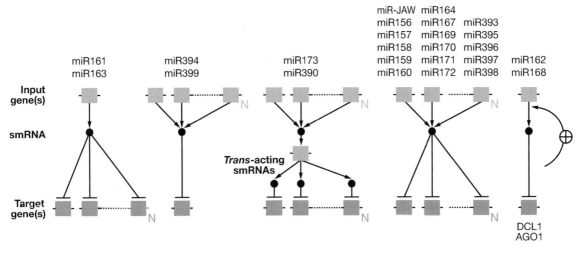

Figure 2

Input-output relationships for validated miRNA and *trans*-acting smRNA pathways. Input smRNA genes (*blue boxes*) and target genes (*orange boxes*) are indicated. The number (N) varies from 1 to 12 for input genes and 1 to 11 for target genes. Based on data from Dugas & Bartel (2004), Vazquez et al. (2004), and Allen et al. (2005).

if these states depend on the amplification of secondary miRNAs (Peragine et al. 2004).

Determination and pattern formation in plants occurs predominantly at the supracellular level; this implies that communication between cells is involved (Meins 1996). Certain forms of cell differentiation are self-inducing (Lang 1965b). In some species, initiation of flowering is regulated by graft-transmittable, systemic signals that can be self-sustaining (Lang 1965a, Bernier et al. 1981). Mechanisms exist for the transport of RNA in phloem and systemically transported RNAs can regulate the development of target cells (Wu et al. 2002). These similarities to systemic silencing, which is also developmentally regulated (Meins 2000), lead us to speculate that one mechanism for supracellular regulation is the transport of endogeneous smRNAs. Evidence for this hypothesis is still indirect. In maize, miR166, which regulates development of leaf primordia, appears to be transported via the phloem (Juarez et al. 2004). This miRNA, which controls leaf polarity in maize by targeting *rolled leaf1*, accumulates in phloem but not surrounding tissues, and spreads progressively as leaf development proceeds. Numerous smRNAs have been cloned from the phloem sap of squash (Yoo et al. 2004). Many of these smRNAs appear to be orthologs of miRNAs that target *Arabidopsis* genes with developmental functions; other smRNAs appear to target transposon-like sequences. Although contamination from surrounding tissues has not been ruled out, the similarity of the phloem smRNAs to Arabidopsis smRNAs suggests that miRNAs as well as smRNAs concerned with transposon regulation may be transported via the phloem. These findings, while fairly preliminary, implicate shared miRNAs in linking distant intracellular networks. That similar coordination may exist for RdDM is problematic because systemic RdDM could not be demonstrated by grafting (Mlotshwa et al. 2002) and RdDM generates clonal patterns suggestive of cell-autonomous silencing.

CONCLUDING REMARKS

Rapid progress in research over the past 10 years has established the basic molecular mechanisms for RNA silencing in plants and other organisms. This has generated renewed interest in regulatory RNAs and has provided powerful tools for examining gene function. Our understanding of how silencing systems are organized and of their developmental relevance is still at an early stage. To facilitate discussion, we, like others, have tried to organize silencing pathways into systems with dedicated functions. Such an analysis has led to several conclusions. Functional diversity depends primarily on system-specific, upstream events leading to smRNA production. Downstream functions such as RNA cleavage, translational repression, and RdDM are likely to be shared by several systems, although modulation by specific components is also possible. Generation of systemic signals does not seem to be a feature of TGS-related RdDM but could provide mechanisms for supracellular regulation at the posttranscriptional level. The miRNA pathway is important for developmental regulation; it seems likely that the *trans*-acting smRNA pathway and other related pathways will prove to have similar functions. Because they involve multiple inputs and outputs, these pathways are potential building blocks in regulatory networks. In contrast, the transgene-silencing/VIGS system seems to function primarily in defending against foreign nucleic acids (Baulcombe 2004). TGS-related RdDM in plants may regulate the activity of certain transposons and may play a role in transposon-associated regu-

lation of developmental genes (Lippman et al. 2004).

There are many unanswered questions. Complete data sets for the same mutant alleles and comparable input genes are needed to assign components and functions to particular silencing systems. This also requires additional and better-characterized loss-of-function mutants. Of particular value would be mutants affecting smRNA/miRNA genes and genes functioning further upstream as tools for linking regulation of specific smRNAs to specific developmental pathways. Studies of epistatic relationships in multiple mutants could help establish where genes act in a pathway. In animals, spectacular progress has been made by combining molecular genetics with biochemical approaches based on in vitro systems and proteomics. Our present knowledge of how silencing-related genes function biochemically in plants is very limited: Only two of these genes, *WEX* (Glazov et al. 2003, Plchova et al. 2003) and *HEN1* (Yu et al. 2005a), have been shown to encode proteins with the predicted enzymatic activity. Nevertheless, the future looks promising. Plant in vitro systems have been established and collections of well-characterized mutants are publicly available. The identification of additional mutants and targets for microRNAs is proceeding at a rapid pace, and pathways regulating smRNA biogenesis are being elucidated.

SUMMARY POINTS

1. Analysis of informative mutants and molecular studies suggests that three RNA silencing systems with dedicated functions exist in plants. These systems depend on production of regulatory smRNAs from precursors catalyzed by DCL enzymes, which then guide the cleavage of RNA targets, repress productive translation, and trigger methylation of cytosines in DNA.

2. RNA silencing systems are organized into a network with shared components that may facilitate cross-talk and with partially overlapping functions.

3. The miRNA/*trans*-acting smRNA system is concerned with developmental regulation. miRNAs help regulate organ identity, morphogenesis, and polarity.

4. miRNA pathways often link multiple miRNA genes with multiple targets. This suggests that miRNAs may be building blocks in posttranscriptional regulatory networks.

5. The transgene RNA silencing/VIGS has a key role in defending against foreign nucleic acids. This system has two special features: Silencing states are maintained by amplification of smRNAs, and mobile signals—probably smRNA-protein complexes—that transmit silencing via an intercellular pathway are produced.

6. A third system is concerned with RNA-directed methylation of DNA repeats and seems to have roles in epigenetic silencing of certain transposable elements and of genes under their control.

7. Complete data sets for the same mutant alleles and comparable input genes, additional mutants affecting specific smRNA genes, and biochemical studies are needed to better understand the organization and functions of silencing systems.

ACKNOWLEDGMENTS

We thank our FMI colleagues for many interesting, provocative discussions; Franck Vazquez for his critical comments; and David Baulcombe for communicating unpublished information. Work from our laboratory was supported in part by Grants Nos. 96.0250 and 00.0224 from the Swiss Office for Education and Science as part of the European Union Gene Silencing in Transgenic Plants Projects, and by the Novartis Research Foundation.

LITERATURE CITED

Achard P, Herr A, Baulcombe DC, Harberd NP. 2004. Modulation of floral development by a gibberellin-regulated microRNA. *Development* 131:3357–65

Allen E, Xie Z, Gustafson AM, Carrington JC. 2005. microRNA-directed phasing during trans-Acting siRNA biogenesis in plants. *Cell* 121:207–21

Ambros V, Bartel B, Bartel DP, Burge CB, Carrington JC, et al. 2003. A uniform system for microRNA annotation. *RNA* 9:277–79

Anandalakshmi R, Marathe R, Ge X, Herr JM Jr, Mau C, et al. 2000. A calmodulin-related protein that suppresses posttranscriptional gene silencing in plants. *Science* 290:142–44

Andino R. 2004. RNAi puts a lid on virus replication. *Nat. Biotechnol.* 21:629–30

Aukerman MJ, Sakai H. 2003. Regulation of flowering time and floral organ identity by a microRNA and its APETALA2-like target genes. *Plant Cell* 15:2730–41

Balandin T, Castresana C. 1997. Silencing of a β-1,3-glucanase transgene is overcome during seed formation. *Plant Mol. Biol.* 34:125–37

Bao N, Lye KW, Barton MK. 2004. MicroRNA binding sites in Arabidopsis class III HD-ZIP mRNAs are required for methylation of the template chromosome. *Dev. Cell* 7:653–62

Bartel DP. 2004. MicroRNAs: genomics, biogenesis, mechanism, and function. *Cell* 116:281–97

Baulcombe D. 2004. RNA silencing in plants. *Nature* 431:356–63

Beclin C, Boutet S, Waterhouse P, Vaucheret H. 2002. A branched pathway for transgene-induced RNA silencing in plants. *Curr. Biol.* 12:684–88

Becskei A, Serrano L. 2000. Engineering stability in gene networks by autoregulation. *Nature* 405:590–93

In this paper, the authors show that plant miRNAs, like animal miRNAs, also can repress translation.

Benfrey PN, Weigel D. 2001. Transcriptional networks controlling plant development. *Plant Physiol.* 125:109–11

Bergstrom CT, McKittrick E, Antia R. 2003. Mathematical models of RNA silencing: unidirectional amplification limits accidental self-directed reactions. *Proc. Natl. Acad. Sci. USA* 100:11511–16

Bernier G, Kinet J-M, Sachs RM. 1981. *The Physiology of Flowering.* Boca Raton, FL: CRC Press

Bohmert K, Camus I, Bellini C, Bouchez D, Caboche M, Benning C. 1998. AGO1 defines a novel locus of Arabidopsis controlling leaf development. *EMBO J.* 17:170–80

Bollman KM, Aukerman MJ, Park MY, Hunter C, Berardini TZ, Poethig RS. 2003. HASTY, the Arabidopsis ortholog of exportin 5/MSN5, regulates phase change and morphogenesis. *Development* 130:1493–504

Bonnet E, Wuyts J, Rouze P, Van de Peer Y. 2004. Detection of 91 potential conserved plant microRNAs in *Arabidopsis thaliana* and *Oryza sativa* identifies important target genes. *Proc. Natl. Acad. Sci. USA* 101:11511–16

Boutet S, Vazquez F, Liu J, Beclin C, Fagard M, et al. 2003. Arabidopsis HEN1: a genetic link between endogenous miRNA controlling development and siRNA controlling transgene silencing and virus resistance. *Curr. Biol.* 13:843–48

Carmell MA, Xuan Z, Zhang MQ, Hannon GJ. 2002. The Argonaute family: tentacles that reach into RNAi, developmental control, stem cell maintenance, and tumorigenesis. *Genes Dev.* 16:2733–42

Chan SWL, Zilberman D, Xie Z, Johansen LK, Carrington JC, Jacobsen SE. 2004. RNA silencing genes control de novo DNA methylation. *Science* 303:1336

> This paper reports that RDR2, DCL3, SDE4/RPD1, and AGO4 are required for the siRNA-dependent methylation of direct repeats and perpetuation of non–CG DNA methylation.

Chapman EJ, Prokhnevsky AI, Gopinath K, Dolja VV, Carrington JC. 2004. Viral RNA silencing suppressors inhibit the microRNA pathway at an intermediate step. *Genes Dev.* 18:1179–86

Chen X. 2003. A microRNA as a translational repressor of APETALA2 in Arabidopsis flower development. *Science* 303:2022–25

Chen X, Liu J, Cheng Y, Jia D. 2002. HEN1 functions pleiotropically in Arabidopsis development and acts in C function in the flower. *Development* 129:1085–94

Cornelissen M, Vandewiele M. 1989. Both RNA level and translation efficiency are reduced by anti-sense RNA in transgenic tobacco. *Nucl. Acids Res.* 17:833–43

Dalmay T, Hamilton A, Rudd S, Angell S, Baulcombe DC. 2000. An RNA-dependent RNA polymerase gene in Arabidopsis is required for posttranscriptional gene silencing mediated by a transgene but not by a virus. *Cell* 101:543–53

Dalmay T, Horsefield R, Braunstein TH, Baulcombe DC. 2001. SDE3 encodes an RNA helicase required for post-transcriptional gene silencing in Arabidopsis. *EMBO J.* 20:2069–78

Dehio C, Schell J. 1994. Identification of plant genetic loci involved in a posttranscriptional mechanism for meiotically reversible transgene silencing. *Proc. Natl. Acad. Sci. USA* 91:5538–42

Depicker A, Van Montagu M. 1997. Post-transcriptional gene silencing in plants. *Curr. Opin. Cell Biol.* 9:373–82

Di Serio F, Schöb H, Iglesias A, Tarina C, Bouldoires E, Meins F Jr. 2001. Sense- and antisense-mediated gene silencing in tobacco is inhibited by the same viral suppressors and is associated with accumulation of small RNAs. *Proc. Natl. Acad. Sci. USA* 98:6506–10

Dugas DV, Bartel B. 2004. MicroRNA regulation of gene expression in plants. *Curr. Opin. Plant Biol.* 7:512–20

Dunoyer P, Lecellier CH, Parizotto EA, Himber C, Voinnet O. 2004. Probing the microRNA and small interfering RNA pathways with virus-encoded suppressors of RNA silencing. *Plant Cell* 16: 235–50

Emery JF, Floyd SK, Alvarez J, Eshed Y, Hawker NP, et al. 2003. Radial patterning of Arabidopsis shoots by class III HD-ZIP and KANADI genes. *Curr. Biol.* 13:1768–74

Fagard M, Boutet S, Morel J-B, Bellini C, Vaucheret H. 2000. AGO1, QDE-2, and RDE-1 are related proteins required for post-transcriptional gene silencing in plants, quelling in fungi, and RNA interference in animals. *Proc. Natl. Acad. Sci. USA* 97:11650–54

Fagard M, Vaucheret H. 2000a. (*Trans*)gene silencing in plants: how many mechanisms? *Annu. Rev. Plant Physiol. Plant Mol. Biol.* 51:168–221

Fagard M, Vaucheret H. 2000b. Systemic silencing signal(s). *Plant Mol. Biol.* 43:285–93

Finnegan EJ, Margis R, Waterhouse PM. 2003. Posttranscriptional gene silencing is not compromised in the Arabidopsis *CARPEL FACTORY* (*DICER-Like1*) mutant, a homolog of Dicer-1 from Drosophila. *Curr. Biol.* 13:236–40

Freeman M. 2000. Feedback control of intercellular signalling in development. *Nature* 408:313–19

Garcia-Perez RD, Houdt HV, Depicker A. 2004. Spreading of post-transcriptional gene silencing along the target gene promotes systemic silencing. *Plant J.* 38:594–602

Gazzani S, Lawrenson T, Woodward C, Headon D, Sablowski R. 2004. A link between mRNA turnover and RNA interference in Arabidopsis. *Science* 306:1046–48

Glazov E, Phillips K, Budziszewski GJ, Schöb H, Meins FJ, Levin JZ. 2003. A gene encoding an RNase D exonuclease-like protein is required for posttranscriptional silencing in Arabidopsis. *Plant J.* 35:342–49

Griffiths-Jones S. 2004. The microRNA registry. *Nucl. Acids Res.* 32:D109–D111

Hall IM, Noma Ki, Grewal SIS. 2003. RNA interference machinery regulates chromosome dynamics during mitosis and meiosis in fission yeast. *Proc. Natl. Acad. Sci. USA* 100(1):193–98

Hamilton A, Voinnet O, Chappell L, Baulcombe D. 2002. Two classes of short interfering RNA in RNA silencing. *EMBO J.* 21(17):4671–79

Han MH, Goud S, Song L, Fedoroff N. 2004. The Arabidopsis double-stranded RNA-binding protein HYL1 plays a role in microRNA-mediated gene regulation. *Proc. Natl. Acad. Sci. USA* 101:1093–98

Hannon GJ. 2002. RNA interference. *Nature* 418:244–51

Herr AJ, Jensen MB, Dalmay T, Baulcombe DC. 2005. RNA polymerase IV directs silencing of endogenous DNA. *Science* 308:118–20

Himber C, Dunoyer P, Moissiard G, Ritzenthaler C, Voinnet O. 2003. Transitivity-dependent and -independent cell-to-cell movement of RNA silencing. *EMBO J.* 22:4523–33

Hunter C, Sun H, Poethig RS. 2003. The Arabidopsis heterochronic gene ZIPPY is an ARGONAUTE family member. *Curr. Biol.* 13:1734–39

Jacobsen SE, Running MP, Meyerowitz EM. 1999. Disruption of an RNA helicase/RNAse III gene in Arabidopsis causes unregulated cell division in floral meristems. *Development* 126:5231–43

Jones-Rhoades MW, Bartel DP. 2004. Computational identification of plant microRNAs and their targets, including a stress-induced miRNA. *Mol. Cell* 14:787–99

Juarez MT, Kui JS, Thomas J, Heller BA, Timmermans MCP. 2004. microRNA-mediated repression of rolled leaf1 specifies maize leaf polarity. *Nature* 428:84–88

Kidner CA, Martienssen RA. 2004. Spatially restricted microRNA directs leaf polarity through ARGONAUTE1. *Nature* 428:81–84

Kidner CA, Martienssen RA. 2005. The developmental role of microRNA in plants. *Curr. Opin. Plant Biol.* 8:38–44

Klahre U, Crété P, Leuenberger SA, Iglesias VA, Meins FJ. 2002. High molecular weight RNAs and small interfering RNAs induce systemic posttranscriptional gene silencing in plants. *Proc. Natl. Acad. Sci. USA* 99:11981–86

Kooter JM, Matzke MA, Meyer P. 1999. Listening to the silent genes: transgene silencing, gene regulation and pathogen control. *Trends Plant Sci.* 4:340–47

Kunz C, Schöb H, Stam M, Kooter JM, Meins F Jr. 1996. Developmentally regulated silencing and reactivation of tobacco chitinase transgene expression. *Plant J.* 10:437–50

Kurihara Y, Watanabe Y. 2004. From the cover: Arabidopsis micro-RNA biogenesis through Dicer-like 1 protein functions. *Proc. Natl. Acad. Sci. USA* 101:12753–58

Lang A. 1965a. Physiology of flower initiation. *Encycl. Plant Physiol.* 15(1):1380

Lang A. 1965b. Progressiveness and contagiousness in plant differentiation and development. *Encycl. Plant Physiol.* 15(1):409–23

Laufs P, Peaucelle A, Morin H, Traas J. 2004. MicroRNA regulation of the CUC genes is required for boundary size control in Arabidopsis meristems. *Development* 131:4311–22

Lee YS, Nakahara K, Pham JW, Kim K, He Z, et al. 2004. Distinct roles for Drosophila Dicer-1 and Dicer-2 in the siRNA/miRNA silencing pathways. *Cell* 117:69–81

Li H, Li WX, Ding SW. 2002. Induction and suppression of RNA silencing by an animal virus. *Science* 296(5571):1319–21

Lipardi C, Wei Q, Paterson BM. 2001. RNAi as random degradative PCR: siRNA primers convert mRNA into dsRNAs that are degraded to generate new siRNAs. *Cell* 107:297–307

Lippman Z, May B, Yordan C, Singer T, Martienssen R. 2003. Distinct mechanisms determine transposon inheritance and methylation via small interfering RNA and histone modification. *PLOS Biol.* 1:420–28

Lippman Z, Gendrel AV, Black M, Vaughn MW, Dedhia N, et al. 2004. Role of transposable elements in heterochromatin and epigenetic control. *Nature* 430:471–76

Liu J, Carmell MA, Rivas FV, Marsden CG, Thomson JM, et al. 2004. Argonaute2 is the catalytic engine of mammalian RNAi. *Science* 305:1437–41

Llave C, Kasschau KD, Rector MA, Carrington JC. 2002a. Endogenous and silencing-associated small RNAs in plants. *Plant Cell* 14:1605–19

Llave C, Xie Z, Kasschau KD, Carrington JC. 2002b. Cleavage of scarecrow-like mRNA targets directed by a class of Arabidopsis miRNA. *Science* 297:2053–56

This paper demonstrates, for the first time, that mRNAs can be targeted for degradation by an miRNA.

Lu C, Fedoroff N. 2000. A mutation in the Arabidopsis HYL1 gene encoding a dsRNA binding protein affects responses to abscisic acid, auxin, and cytokinin. *Plant Cell* 12:2351–66

Makeyer EV, Bamford DH. 2002. Cellular RNA-dependent RNA polymerase involved in posttranscriptional gene silencing has two distinct activity modes. *Mol. Cell* 10:1417–27

Mallory AC, Dugas DV, Bartel DP, Bartel B. 2004. MicroRNA regulation of NAC-domain targets is required for proper formation and separation of adjacent embryonic, vegetative, and floral organs. *Curr. Biol.* 14:1035–46

Mallory AC, Mlotshwa S, Bowman LH, Vance VB. 2003. The capacity of transgenic tobacco to send a systemic RNA silencing signal depends on the nature of the inducing transgene locus. *Plant J.* 35:82–92

Mallory AC, Reinhart BJ, Bartel D, Vance VB, Bowman LH. 2002. A viral suppressor of RNA silencing differentially regulates the accumulation of short interfering RNAs and micro-RNAs in tobacco. *Proc. Natl. Acad. Sci. USA* 99(23):15228–33

Matzke MA, Birchler JA. 2005. RNAi-mediated pathways in the nucleus. *Nat. Rev. Genet.* 6:24–34

Matzke M, Aufsatz W, Kanno T, Daxinger L, Papp I, et al. 2004. Genetic analysis of RNA-mediated transcriptional gene silencing. *Biochim. Biophys. Acta* 1677:129–41

Meins F Jr. 1996. Epigenetic modifications and gene silencing in plants. In *Epigenetic Mechanisms of Gene Regulation*, ed. VEA Russo, RA Martienssen, AD Riggs, pp. 415–42. Cold Spring Harbor, NY: Cold Spring Harbor Lab. Press. 692 pp.

Meins F Jr. 2000. RNA degradation and models for posttranscriptional gene silencing. *Plant Mol. Biol.* 43:261–73

Meyer P, Saedler H. 1996. Homology-dependent gene silencing in plants. *Annu. Rev. Plant Physiol. Plant Mol. Biol.* 47:23–48

Mitsuhara I, Shirasawa-Seo N, Iwai T, Nakamura S, Honkura R, Ohashi Y. 2002. Release from post-transcriptional gene silencing by cell proliferation in transgenic tobacco plants. Possible mechanism for noninheritance of the silencing. *Genetics* 160:343–52

Mlotshwa S, Voinnet O, Mette MF, Matzke M, Vaucheret H, et al. 2002. RNA silencing and the mobile silencing signal. *Plant Cell* 14(90001):S289–S301

Morel JB, Godon C, Mourrain P, Beclin C, Boutet S, et al. 2002. Fertile hypomorphic ARGONAUTE (ago1) mutants impaired in post-transcriptional gene silencing and virus resistance. *Plant Cell* 14:629–39

Morel JB, Mourrain P, Beclin C, Vaucheret H. 2000. DNA methylation and chromatin structure affect transcriptional and post-transcriptional transgene silencing in Arabidopsis. *Curr. Biol.* 10:1591–94

Mourrain P, Béclin C, Elmayan T, Feuerbach F, Godon C, et al. 2000. Arabidopsis *SGS2* and *SGS3* genes are required for posttranscriptional gene silencing and natural virus resistance. *Cell* 101:533–42

Muangsan N, Beclin C, Vaucheret H, Robertson D. 2004. Geminivirus VIGS of endogenous genes requires SGS2/SDE1 and SGS3 and defines a new branch in the genetic pathway for silencing in plants. *Plant J.* 38:1004–14

Novina CD, Sharp PA. 2004. The RNAi revolution. *Nature* 430:161–64

Oliveri P, Davidson EH. 2004. Gene regulatory network controlling embryonic specification in the sea urchin. *Curr. Opin. Genet. Dev.* 14:351–60

Onodera Y, Haag JR, Ream T, Nunes PC, Pontes O, Pikaard CS. 2005. Plant nuclear RNA polymerase IV mediates siRNA and DNA methylation-dependent heterochromatin formation. *Cell* 120:613–22

Palatnik JF, Allen E, Wu X, Schommer C, Schwab R, et al. 2003. Control of leaf morphogenesis by microRNAs. *Nature* 425:257–63

Palauqui J, Elmayan T, Pollien J, Vaucheret H. 1997. Systemic acquired silencing: transgene-specific post-transcriptional silencing is transmitted by grafting from silenced stocks to non-silenced scions. *EMBO J.* 16:4738–45

Papp I, Mette MF, Aufsatz W, Daxinger L, Schauer SE, et al. 2003. Evidence for nuclear processing of plant micro RNA and short interfering RNA precursors. *Plant Physiol.* 132:1382–90

Park MY, Wu G, Gonzalez-Sulser A, Vaucheret H, Poethig RS. 2005. Nuclear processing and export of microRNAs in Arabidopsis. *Proc. Natl. Acad. Sci. USA* 102:3691–96

Park W, Li J, Song R, Messing J, Chen X. 2002. CARPEL FACTORY, a Dicer homolog, and HEN1, a novel protein, act in microRNA metabolism in *Arabidopsis thaliana*. *Curr. Biol.* 12:1484–95

Paulsson J. 2004. Summing up the noise in gene networks. *Nature* 427:415–18

This is the first report identifying a developmental defect resulting from altered production of a specific miRNA.

Peragine A, Yoshikawa M, Wu G, Albrecht HL, Poethig RS. 2004. SGS3 and SGS2/SDE1/RDR6 are required for juvenile development and the production of trans-acting siRNAs in Arabidopsis. *Genes Dev.* 18:2368–79

Plchova H, Hartung F, Puchta H. 2003. Biochemical characterization of an exonuclease from *Arabidopsis thaliana* reveals similarities to the DNA exonuclease of the human Werner Syndrome protein. *J. Biol. Chem.* 278:44128–38

Reinhard BJ, Weinstein EG, Rhoades MW, Bartel B, Bartel DP. 2002. MicroRNAs in plants. *Genes Dev.* 16:1616–26

Ruiz MT, Voinnet O, Baulcombe DC. 1998. Initiation and maintenance of virus-induced gene silencing. *Plant Cell* 10:937–46

Schauer SE, Jacobsen SE, Meinke D, Ray A. 2002. DICER-LIKE1: blind men and elephants in *Arabidopsis* development. *Trends Plant Sci.* 7:487–91

Schiebel W, Pélissier T, Riedel L, Thalmeir S, Schiebel R, et al. 1998. Isolation of an RNA-directed RNA polymerase-specific cDNA clone from tomato. *Plant Cell* 10:1–16

Schöb H. 2001. *Sense- and antisense gene silencing in tobacco are mechanistically related.* Ph.D. Thesis. Univ. Basel. 128 pp.

Shen B, Goodman HM. 2004. Uridine addition after microRNA-directed cleavage. *Science* 306:997

Sijen T, Fleenor J, Simmer F, Thijssen KL, Parrish S, et al. 2001. On the role of RNA amplification in dsRNA-triggered gene silencing. *Cell* 107:465–76

Silhavy D, Burgyan J. 2004. Effects and side-effects of viral RNA silencing suppressors on short RNAs. *Trends Plant Sci.* 9:76–83

Smardon A, Spoerke JM, Stacey SC, Klein ME, Maine EM. 2000. EGO-1 is related to RNA-directed/RNA polymerase and functions in germ-line development and RNA interference in *C. elegans. Curr. Biol.* 10:169–78

Song JJ, Smith SK, Hannon GJ, Joshua-Tor L. 2004. Crystal structure of Argonaute and its implications for RISC slicer activity. *Science* 304:1409–10

Sonoda S, Nishiguchi M. 2000. Graft transmission of post-transcriptional gene silencing: Target specificity for RNA degradation is transmissible between silenced and non-silenced plants, but not between silenced plants. *Plant J.* 21:1–8

Souret FF, Kastenmayer JP, Green PJ. 2004. AtXRN4 degrades mRNA in *Arabidopsis* and its substrates include selected miRNA targets. *Mol. Cell* 15:173–83

Stam M, de Bruin R, van Blokland R, Van der Hoorn RAL, Mol JNM, Kooter JM. 2000. Distinct features of post-transcriptional gene silencing by antisense transgenes in single copy and inverted T-DNA repeat loci. *Plant J.* 21:27–42

Steimer A, Schob H, Grossniklaus U. 2004. Epigenetic control of plant development: new layers of complexity. *Curr. Opin. Plant Biol.* 7:11–19

Sunkar R, Zhu JK. 2004. Novel and stress-regulated microRNAs and other small RNAs from Arabidopsis. *Plant Cell* 16(8):2001–19

Tang G, Reinhard BJ, Bartel DP, Zamore PD. 2003. A biochemical framework for RNA silencing in plants. *Genes Dev.* 17:49–63

Tijsterman M, Ketting RF, Plasterk RHA. 2002. The genetics of RNA silencing. *Annu. Rev. Genet.* 36(1):489–519

Vaistij FE, Jones L, Baulcombe DC. 2002. Spreading of RNA targeting and DNA methylation in RNA silencing requires transcription of the target gene and a putative RNA-dependent RNA polymerase. *Plant Cell* 14(4):857–67

van Houdt H, Ingelbrecht I, Van Montagu M, Depicker A. 1997. Post-transcriptional silencing of a neomycin phosphotransferase II transgene correlates with the accumulation of

This paper provides evidence for a *trans*-acting smRNA pathway and a novel AGO7-dependent smRNA pathway.

unproductive RNAs and with increased cytosine methylation of 3' flanking regions. *Plant J.* 12:379–92

van Houdt H, Bleys A, Depicker A. 2003. RNA target sequences promote spreading of RNA silencing. *Plant Physiol.* 131:245–53

Vaucheret H, Vazquez F, Crete P, Bartel DP. 2004. The action of ARGONAUTE1 in the miRNA pathway and its regulation by the miRNA pathway are crucial for plant development. *Genes Dev.* 18:1187–97

Vazquez F, Vaucheret H, Rajagopalan R, Lepers C, Gasciolli V, et al. 2004. Endogenous *trans*-acting siRNAs regulate the accumulation of Arabidopsis mRNAs. *Mol. Cell* 16:69–79

Verdel A, Jia S, Gerber S, Sugiyama T, Gygi S, Grewal SIS, Moazed D. 2004. RNAi-mediated targeting of heterochromatin by the RITS complex. *Science* 303:672–76

Voinnet O, Vain P, Angell S, Baulcombe DC. 1998. Systemic spread of sequence-specific transgene RNA degradation in plants is initiated by localized introduction of ectopic promoterless DNA. *Cell* 95:177–87

Volpe TA, Kidner C, Hall IM, Teng G, Grewal SIS, Martienssen RA. 2002. Regulation of heterochromatic silencing and histone H3 lysine-9 methylation by RNAi. *Science* 297:1833–37

Wagner A. 2000. Robustness against mutations in genetic networks of yeast. *Nat. Genet.* 24:355–61

Wang XJ, Reyes J, Chua NH, Gaasterland T. 2004. Prediction and identification of *Arabidopsis thaliana* microRNAs and their mRNA targets. *Genome Biol.* 5:R65

Waterhouse PM, Wang M-B, Finnegan EJ. 2001. Role of short RNAs in gene silencing. *Trends Plant Sci.* 6:297–301

Wu X, Weigel D, Wigge PA. 2002. Signaling in plants by intercellular RNA and protein movement. *Genes Dev.* 16:151–58

Xie Z, Fan B, Chen C, Chen Z. 2001. An important role of an inducible RNA-dependent RNA polymerase in plant antiviral defense. *Proc. Natl. Acad. Sci. USA* 98:6516–21

Xie Z, Johansen LK, Gustafson AM, Kasschau KD, Lellis AD, et al. 2004. Genetic and functional diversification of small RNA pathways in plants. *PLOS Biol.* 2:1–11

Xie Z, Kasschau KD, Carrington JC. 2003. Negative feedback regulation of *Dicer-like1* in *Arabidopsis* by microRNA-guided mRNA degradation. *Curr. Biol.* 13:784–89

Yoo BC, Kragler F, Varkonyi-Gasic E, Haywood V, Archer-Evans S, et al. 2004. A systemic small RNA signaling system in plants. *Plant Cell* 16(8):1979–2000

Yu D, Fan B, MacFarlane SA, Chen Z. 2003. Analysis of the involvement of an inducible *Arabidopsis* RNA-dependent RNA polymerase in antiviral defense. *Mol. Plant Microbe Interact.* 16:206–16

Yu B, Yang Z, Li J, Minakhina S, Yang M, et al. 2005a. Methylation as a crucial step in plant microRNA biogenesis. *Science* 307:932–35

Yu L, Yu X, Shen R, He Y. 2005b. HYL1 gene maintains venation and polarity of leaves. *Planta* 221:231–42

Zilberman D, Cao X, Jacobsen SE. 2003. ARGONAUTE4 control of locus-specific siRNA accumulation and DNA and histone methylation. *Science* 299(5607):716–19

Zilberman D, Cao X, Johansen LK, Xie Z, Carrington JC, Jacobsen SE. 2004. Role of Arabidopsis ARGONAUTE4 in RNA-directed DNA methylation triggered by inverted repeats. *Curr. Biol.* 14:1214–20

In this study, the authors identified a new class of small RNAs called *trans*-acting smRNAs.

The authors in this paper provide evidence for the functional diversity of DCL and RDR gene family members.

In this study, the authors demonstrate that smRNAs can be transported systemically and provide evidence that trafficking of smRNAs requires specific proteins.

Quorum Sensing: Cell-to-Cell Communication in Bacteria

Christopher M. Waters and Bonnie L. Bassler

Department of Molecular Biology, Princeton University, Princeton, New Jersey 08544-1014; emails: cwaters@molbio.princeton.edu, bbassler@molbio.princeton.edu

Annu. Rev. Cell Dev. Biol. 2005. 21:319–46

First published online as a Review in Advance on June 28, 2005

The *Annual Review of Cell and Developmental Biology* is online at http://cellbio.annualreviews.org

doi: 10.1146/ annurev.cellbio.21.012704.131001

Copyright © 2005 by Annual Reviews. All rights reserved

1081-0706/05/1110-0319$20.00

Key Words

autoinducer, quorum quenching, regulon

Abstract

Bacteria communicate with one another using chemical signal molecules. As in higher organisms, the information supplied by these molecules is critical for synchronizing the activities of large groups of cells. In bacteria, chemical communication involves producing, releasing, detecting, and responding to small hormone-like molecules termed autoinducers. This process, termed quorum sensing, allows bacteria to monitor the environment for other bacteria and to alter behavior on a population-wide scale in response to changes in the number and/or species present in a community. Most quorum-sensing-controlled processes are unproductive when undertaken by an individual bacterium acting alone but become beneficial when carried out simultaneously by a large number of cells. Thus, quorum sensing confuses the distinction between prokaryotes and eukaryotes because it enables bacteria to act as multicellular organisms. This review focuses on the architectures of bacterial chemical communication networks; how chemical information is integrated, processed, and transduced to control gene expression; how intra- and inter-species cell-cell communication is accomplished; and the intriguing possibility of prokaryote-eukaryote cross-communication.

Contents

QUORUM SENSING

Quorum-sensing bacteria produce and release chemical signal molecules termed autoinducers whose external concentration increases as a function of increasing cell-population density. Bacteria detect the accumulation of a minimal threshold stimulatory concentration of these autoinducers and alter gene expression, and therefore behavior, in response. Using these signal-response systems, bacteria synchronize particular behaviors on a population-wide scale and thus function as multicellular organisms. Here, we describe some well-characterized quorum-sensing systems with the aim of illustrating their similarities and differences. We presume similarities in these systems exist because the ability to communicate is fundamental to bacteria. Differences in the systems likely exist because each system has been optimized to promote survival in the specialized niche in which a particular species of bacteria resides. Thus, the types of signals, receptors, mechanisms of signal transduction, and target outputs of each quorum-sensing system reflect the unique biology carried out by a particular bacterial species.

Quorum Sensing in Gram-Negative Bacteria

The first described quorum-sensing system is that of the bioluminescent marine bacterium *Vibrio fischeri*, and it is considered the paradigm for quorum sensing in most gram-negative bacteria (Nealson & Hastings 1979). *V. fischeri* colonizes the light organ of the Hawaiian squid *Euprymna scolopes*. In this organ, the bacteria grow to high cell density and induce the expression of genes required for bioluminescence. The squid uses the light provided by the bacteria for counterillumination to mask its shadow and avoid predation (Visick et al. 2000). The bacteria benefit because the light organ is rich in nutrients and allows proliferation in numbers unachievable in seawater. Two

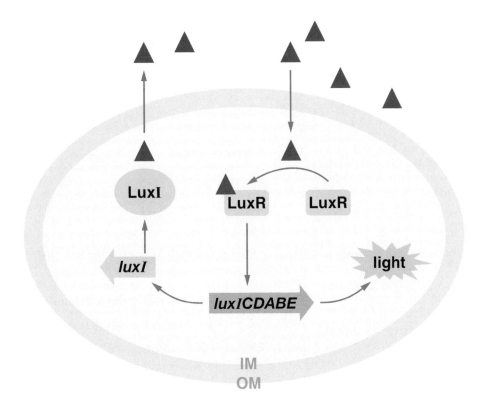

Figure 1

Quorum sensing in
Vibrio fischeri; a
LuxIR signaling
circuit. Red triangles
indicate the
autoinducer that is
produced by LuxI.
OM, outer
membrane; IM,
inner membrane.

proteins, LuxI and LuxR, control expression of the luciferase operon (*luxICDABE*) required for light production (**Figure 1**). LuxI is the autoinducer synthase, which produces the acyl-homoserine lactone (AHL) autoinducer 3OC6-homoserine lactone (**Figure 2a** and Eberhard et al. 1981, Engebrecht & Silverman 1984), and LuxR is the cytoplasmic autoinducer receptor/DNA-binding transcriptional activator (Engebrecht et al. 1983). Following production, the AHL freely diffuses in and out of the cell and increases in concentration with increasing cell density (Kaplan & Greenberg 1985). When the signal reaches a critical, threshold concentration, it is bound by LuxR and this complex activates transcription of the operon encoding luciferase (Stevens et al. 1994). Importantly, the LuxR-AHL complex also induces expression of *luxI* because it is encoded in the luciferase operon. This regulatory configuration floods the environment with

the signal. This creates a positive feedback loop that causes the entire population to switch into "quorum-sensing mode" and produce light.

A large number of other gram-negative proteobacteria possess LuxIR-type proteins and communicate with AHL signals (Manefield & Turner 2002). These systems are used predominantly for intraspecies communication as extreme specificity exists between the LuxR proteins and their cognate AHL signals. LuxI-type proteins link and lactonize the methionine moiety from *S*-adenosylmethionine (SAM) to particular fatty acyl chains carried on acyl-acyl carrier proteins (More et al. 1996, Parsek et al. 1999). A diverse set of fatty acyl side chains of varying length, backbone saturation, and side-chain substitutions are incorporated into AHL signals; these differences are crucial for signaling specificity (**Figure 2a** and Fuqua 1999). Structural studies of LuxI-type

Quorum sensing: a
process of cell-cell
communication in
bacteria

Autoinducers:
small molecules
secreted by bacteria
that are used to
measure population
density

AHL:
acyl-homoserine
lactone

SAM: *S*-adenosyl-
methionine

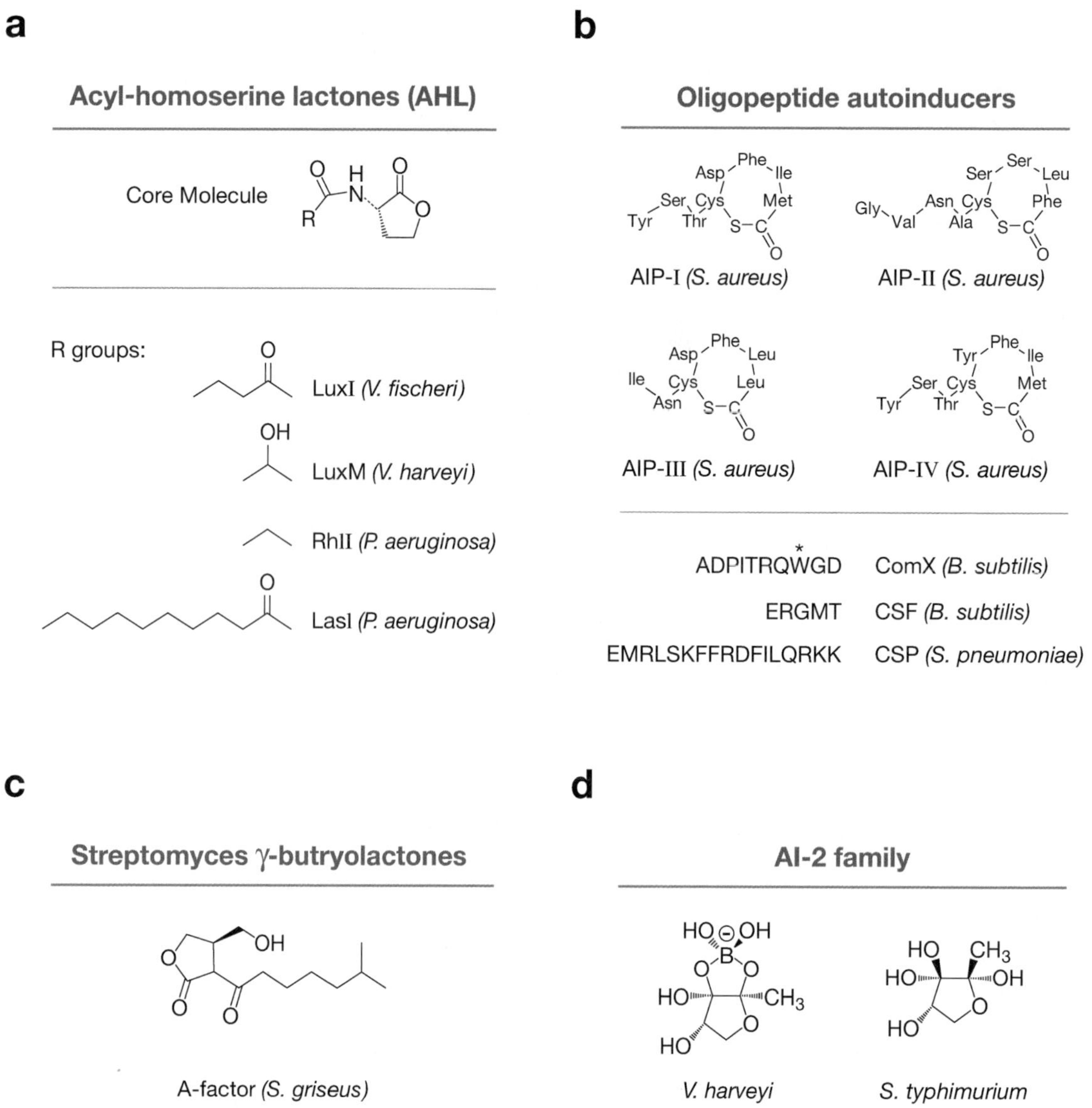

Figure 2

Representative bacterial autoinducers. The asterisk above the tryptophan in ComX represents an isoprenyl modification.

proteins indicate that each possesses an acyl-binding pocket that precisely fits a particular side-chain moiety (Gould et al. 2004, Watson et al. 2002). This structural feature apparently confers specificity in signal production. Thus, each LuxI protein produces the correct signal molecule with high fidelity. There are some LuxI-type proteins that produce multiple AHLs, although it is not clear if all are biologically relevant (Marketon et al. 2002). The structures of LuxR proteins suggest that LuxR proteins also possess specific acyl-binding pockets that allow each LuxR to bind and be activated only by its cognate signal (Vannini et al. 2002, Zhang et al. 2002b). Hence, it appears that in mixed-species environments in which multiple AHL signals are present, each species can distinguish, measure, and respond only to the buildup of its own signal. Importantly, bacteria rarely rely exclusively on one LuxIR quorum-sensing system. Rather, bacteria use one or more LuxIR systems, often in conjunction with other types of quorum-sensing circuits.

Mechanisms must exist to prevent premature activation of LuxIR-type quorum-sensing circuits because both the signal and the detector are synthesized and interact in the cytoplasm (**Figure 1**). One such mechanism, evidenced by the LuxR homologue TraR in the plant pathogen *Agrobacterium tumefaciens*, is the stability of LuxR-type proteins increases upon AI binding. In the absence of autoinducer, TraR has a half-life of a few minutes. However, in the presence of AHL, the half-life of TraR increases to over 30 minutes (Zhu & Winans 1999). The crystal structure of TraR predicts that AHL binding is required for folding of the nascent polypeptide (Zhang et al. 2002b), and indeed radiolabeled TraR was stabilized only when its cognate AHL was added prior to labeling of the protein (Zhu & Winans 2001). Hence, only when AHL accumulates to a significant concentration (both outside and inside the cell) can TraR bind it, fold, and initiate the quorum-sensing cascade. Another mechanism that prevents "short-circuiting" of LuxIR systems is active export of AHL signals (Pearson et al. 1999). When a significant concentration of signal has accumulated, which is indicative of high cell density, diffusion into the cell overwhelms export and thus engages the circuit. AHLs with long acyl side chains are thought to require active export to transverse the bacterial membrane (Pearson et al. 1999).

Quorum Sensing in Gram-Positive Bacteria

Gram-positive bacteria communicate using modified oligopeptides as signals and "two-component"-type membrane-bound sensor histidine kinases as receptors. Signaling is mediated by a phosphorylation cascade that influences the activity of a DNA-binding transcriptional regulatory protein termed a response regulator. Similar to the mechanisms by which gram-negative bacteria use LuxIR quorum-sensing systems, each gram-positive bacterium uses a signal different from that used by other bacteria and the cognate receptors are exquisitely sensitive to the signals' structures. Thus, as in LuxIR systems, peptide quorum-sensing circuits are understood to confer intraspecies communication. Peptide signals are not diffusible across the membrane, hence signal release is mediated by dedicated oligopeptide exporters. In most cases, concomitant with signal release is signal processing and modification. While the biochemistry underlying these events is poorly defined, it is known that most peptide quorum-sensing signals are cleaved from larger precursor peptides, which then are modified to contain lactone and thiolactone rings, lanthionines, and isoprenyl groups (Ansaldi et al. 2002, Booth et al. 1996, Mayville et al. 1999, Nakayama et al. 2001). Many gram-positive bacteria communicate with multiple peptides in combination with other types of quorum-sensing signals.

A fascinating example of peptide quorum sensing exists in *Staphylococcus aureus*, which is normally a benign human commensal but becomes a deadly pathogen upon penetration into host tissues (reviewed in Tenover & Gaynes 2000). *S. aureus* uses a biphasic strategy to cause disease: At low cell density, the bacteria express protein factors that promote attachment and colonization, whereas at high cell density, the bacteria repress these traits and initiate secretion of toxins and proteases that are presumably required for dissemination (reviewed in Lyon & Novick 2004). This switch in gene expression programs is regulated by the Agr quorum-sensing system (**Figure 3**). The system consists of an autoinducing peptide of *Staphylococcus aureus* (AIP) (**Figure 2b**) encoded by *agrD* (Ji et al. 1995) and a two-component sensor kinase-response regulator pair, AgrC and AgrA, respectively

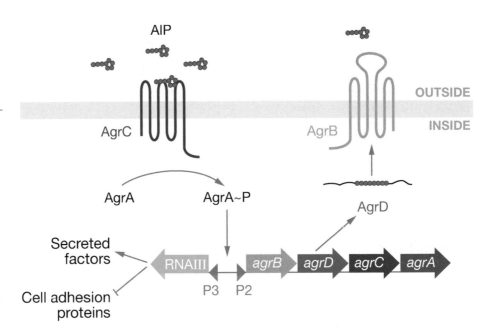

Figure 3

Using a two-component response regulatory system, *Staphylococcus aureus* detects and responds to an extracellular peptide. Small red circles indicate the AIP. P2 and P3 designate the promoters for *agrBDCA* and RNAIII, respectively.

AIP: autoinducing peptide of *Staphylococcus aureus*

(Novick et al. 1995). The AgrB protein exports and adds the thiolactone ring modification to *S. aureus* AIPs (Saenz et al. 2000). Binding of the AIP to AgrC leads to phosphorylation of AgrA. Phospho-AgrA induces the expression of a regulatory RNA termed RNAIII, which represses expression of cell-adhesion factors while inducing expression of secreted factors (Novick et al. 1993). Activated AgrA also induces expression of the *agrBDCA*. This results in increased AIP levels, which ensures that the entire population switches from the low-cell-density to the high-cell-density state (Novick et al. 1995).

S. aureus strains are classified on the basis of the sequence of their thiolactone-containing AIP. At present, four different AIPs (**Figure 2b** and Dufour et al. 2002), and thus four different groups of *S. aureus*, are known. Surprisingly, each AIP specifically activates its cognate AgrC receptor but inhibits activation of all others by competitive binding to the non-cognate receptors (Lyon et al. 2002b). Thus, each AIP inhibits activation of the virulence cascade in the other three groups of *S. aureus* while not affecting the other groups' growth. Coinfection with two different *S. au-*reus groups results in intraspecies competition; the *S. aureus* group that first establishes its quorum-sensing cascade outcompetes the other group. Consistent with this idea, purified AIP II attenuates virulence of a Group I *S. aureus* in a mouse infection model (Mayville et al. 1999). Thus, in *S. aureus*, quorum sensing allows dissemination of closely related progeny while inhibiting the spread of non-kin. Clinical analyses show that each *S. aureus* group is the primary causative agent of a specific type of *S. aureus* disease. This suggests that cell-cell communication has been instrumental in establishing a specific niche for each "strain" (Novick 2003). The codivergence of the signal-receptor pairs occurring in these bacteria may be one molecular mechanism underlying the evolution of new bacterial species.

Streptomycetes are a diverse family of gram-positive soil-dwelling bacteria that are of clinical relevance because they are a major biological reservoir of secondary metabolites, many of which are used as antibiotics (reviewed in Chater & Horinouchi 2003). Streptomycetes use γ-butyrolactones (**Figure 2c**) as autoinducers and control morphological differentiation and secondary metabolite production via quorum sensing. Their signals are intriguing because they are structurally related to AHL autoinducers. However, there has not yet been any report describing either cross-communication between or cross-inhibition of streptomycetes and Gram-negative bacteria that communicate with AHLs.

QUORUM-SENSING NETWORK ARCHITECTURE

Identification of the chemical signals, receptors, target genes, and mechanisms of signal transduction involved in quorum sensing is leading to a comprehensive understanding of cell-cell communication in bacteria. This research is providing insight into the variety of molecular arrangements that enable communication between cells as well as the unique characteristics that the various signaling architectures provide in terms of information dissemination, detection, relay, and response. Below we highlight a few quorum-sensing systems and discuss how each particular network arrangement leads to distinct signaling features.

Parallel Quorum-Sensing Circuits

The first observation that bacteria could communicate with multiple quorum-sensing signals was in the quorum-sensing system of the Gram-negative, bioluminescent marine bacterium *Vibrio harveyi* (**Figure 4**). The *V. harveyi* quorum-sensing system consists of three autoinducers and three cognate receptors functioning in parallel to channel information into a shared regulatory pathway. Similar to other Gram-negative bacteria, *V. harveyi* produces an AHL signal termed HAI-1 (3OHC4-homoserine lactone; **Figure 2a** and Cao & Meighen 1989). Its synthase, LuxM, shares no homology to LuxI-type enzymes but catalyzes the identical biochemical reaction to generate a specific AHL (Bassler et al. 1993, Hanzelka et al. 1999). HAI-1 binds to a membrane-bound sensor histi-

CHEMICAL COMPLEXITY IN BACTERIAL AUTOINDUCERS

Recent research shows that a rich diversity of chemical molecules is used for communication in the bacterial world. New genetic, biochemical, and imaging techniques have enhanced our ability to identify and measure the readouts of cell-cell communication. These tools have led to the identification of several novel molecules and classes of molecules that are clearly bona fide autoinducers mediating cell-cell communication. A few examples are

PQS The molecule 3,4-dihydroxy-2-heptylquinoline, termed PQS, is a signal that is integral to the *P. aeruginosa* quorum-sensing cascade (Pesci et al. 1999). This signal acts as an additional regulatory link between the Las and Rhl quorum-sensing circuits.

3OH PAME 3OH palmitic acid methyl ester (3OH PAME) transmits information via the two-component sensor histidine kinase-response regulator pair, PhcS-PhcR, to cause the plant pathogen *Ralstonia solanacearum* to switch from a motile to an infective state (Flavier et al. 1997).

CYCLIC DIPEPTIDES Newly described in a number of gram-negative bacteria, at high concentrations, cyclic-dipeptides antagonize AHL binding to cognate receptors (Holden et al. 1999).

dine kinase (LuxN) similar to sensors in Gram-positive quorum-sensing signaling circuits (Bassler et al. 1993, Freeman et al. 2000). The second *V. harveyi* signal is a furanosyl borate diester known as AI-2 (**Figure 2d** and Bassler et al. 1994a, Chen et al. 2002), production of which requires the LuxS enzyme (Surette et al. 1999, Xavier & Bassler 2003). AI-2 is bound in the periplasm by the protein LuxP; the LuxP-AI-2 complex interacts with another membrane-bound sensor histidine kinase, LuxQ (Bassler et al. 1994a). The third *V. harveyi* signal, an unidentified molecule termed CAI-1, is produced by the CqsA enzyme, and again, this signal interacts with a membrane-bound sensor histidine kinase, CqsS (Henke & Bassler 2004b).

At low cell density, in the absence of appreciable amounts of autoinducers, the three sensors—LuxN, LuxQ, and CqsA—act as

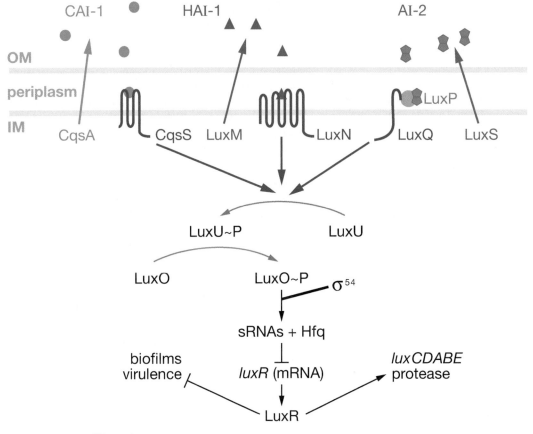

Figure 4

Vibrio harveyi produces and responds to three distinct autoinducers. The sensory information is fed into a shared two-component response regulatory pathway. The arrows indicate the direction of phosphate flow in the low-cell-density state. CAI-1, HAI-1, and AI-2 are respectively represented by green circles, red triangles, and blue double pentagons. OM, outer membrane; IM, inner membrane.

sRNAs: small RNAs

kinases, autophosphorylate, and subsequently transfer the phosphate to the cytoplasmic protein LuxU (**Figure 4**). LuxU passes the phosphate to the DNA-binding response regulator protein LuxO (Bassler et al. 1994b, Freeman & Bassler 1999a,b, Freeman et al. 2000). Phospho-LuxO, in conjunction with a transcription factor termed σ^{54}, activates transcription of the genes encoding five regulatory small RNAs (sRNAs) termed Qrr1–5 (for Quorum Regulatory RNA) (Lilley & Bassler 2000, Lenz et al. 2004). The Qrr sRNAs interact with an RNA chaperone termed Hfq, which is a member of the Sm family of eukaryotic RNA chaperones in-

volved in mRNA splicing (Carrington & Ambros 2003). The sRNAs, together with Hfq, bind to and destabilize the mRNA encoding the transcriptional activator termed LuxR (not similar to LuxR of *V. fischeri*) (Lenz et al. 2004). LuxR is required to activate transcription of the luciferase operon *luxCD-ABE* (Swartzman et al. 1992). Thus, at low cell density, because the *luxR* mRNA is degraded, the bacteria do not express bioluminescence. At high cell density, when the autoinducers accumulate to the level required for detection, the three sensors switch from being kinases to being phosphatases and drain phosphate from LuxO via LuxU. Unphosphorylated LuxO cannot induce expression of the sRNAs. This allows translation of *luxR* mRNA, production of LuxR, and expression of bioluminescence. This pathway controls many genes in addition to those encoding luciferase (Henke & Bassler 2004a, Mok et al. 2003).

The human pathogen *Vibrio cholerae*, the causative agent of the endemic diarrheal disease cholera, possesses a quorum-sensing network similar to that of *V. harveyi* (Miller et al. 2002). *V. cholerae* has no equivalent to the AI-1/LuxN branch of the system. However, this bacterium does possess the AI-2/LuxPQ and CAI-1/CqsS branches as well as LuxU, LuxO, four Qrr sRNAs, and a *V. harveyi* LuxR-like protein termed HapR. The *V. cholerae* systems function analogously to those of *V. harveyi* but control virulence instead of regulating bioluminescence (Miller et al. 2002, Zhu et al. 2002). Surprisingly, quorum sensing promotes *V. cholerae* virulence factor expression and biofilm formation at low cell density and represses these traits at high cell density (Hammer & Bassler 2003). Quorum sensing commonly controls bacterial virulence factor expression, but typically, induction occurs at high cell density. This opposite regulatory pattern exhibited by *V. cholerae* can be understood in terms of the specific disease that the bacterium causes. Following a successful *V. cholerae* infection, the ensuing diarrhea wash huge numbers of bacteria from the human intestine into the environment. Repression of virulence factor production and biofilm formation genes at high cell density may promote dissemination of *V. cholerae*.

Upon the recent completion of the *V. fischeri* genome sequence, it was revealed that, in addition to LuxIR, homologues of two of the *V. harveyi* quorum-sensing circuits and the shared downstream components are present: LuxMN, LuxSPQ, LuxU, LuxO, and LuxR (referred to as LitR in *V. fischeri*) (Fidopiastis et al. 2002, Lupp & Ruby 2004, Lupp et al. 2003, Miyamoto et al. 2003). In *V. fischeri*, the *V. harveyi*-like quorum-sensing systems activate expression of *litR* at low cell densities. LitR induces expression of *luxR*, which in turn

promotes light production, as described above (Fidopiastis et al. 2002). This latter event occurs at relatively high cell densities, which presumably can be achieved only in the squid and cannot be achieved in the open ocean.

These three vibrio quorum-sensing systems underscore the way in which a common quorum-sensing network can be modified to fit the unique biology of the bacteria. Whereas the two-tiered *V. fischeri* circuit is adapted for two disparate lifestyles, inside and outside of the squid, *V. harveyi* and *V. cholerae* do not possess LuxIR homologues and they are not known to exist in symbiotic relationships. Although *V. harveyi* and *V. cholerae* share many of the same signals and receptors, the relative input from each signal is different in the two species. CqsA/CqsS is the dominant signaling-circuit in *V. cholerae* whereas it is the weakest in *V. harveyi* (Henke & Bassler 2004b). These signaling variations, coupled with their regulation of distinct downstream virulence factors, may be determining factors that allow *V. cholerae*, but not *V. harveyi*, to infect humans.

In each vibrio circuit, all signal-receptor pairs channel phosphate to LuxO in the absence of a signal and remove phosphate from LuxO in the presence of a signal. Thus, because all signals lead to a reduction in the level of LuxO-phosphate, each signal reinforces the information encoded in the other signals. This arrangement may allow the network to function as a coincidence detector that significantly activates or represses gene expression only when all signals are simultaneously present or absent (Mok et al. 2003). This signaling architecture may be critical for filtering out noise from molecules in the environment that are related to the true signals and/or noise from signal mimics produced by other bacteria in the vicinity.

Quorum-Sensing Circuits Arranged in Series

As in the vibrios, the *Pseudomonas aeruginosa* quorum-sensing circuit is responsive to

Figure 5

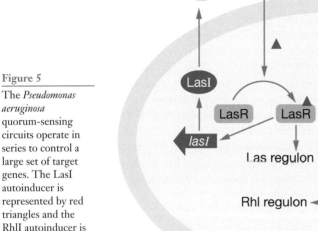

Figure 5
The *Pseudomonas aeruginosa* quorum-sensing circuits operate in series to control a large set of target genes. The LasI autoinducer is represented by red triangles and the RhlI autoinducer is shown as blue triangles. OM, outer membrane; IM, inner membrane.

CF: cystic fibrosis

multiple autoinducers, however, unlike those in vibrios, the *P. aeruginosa* regulatory systems are arranged in series rather than in parallel. *P. aeruginosa*, a common soil organism, is also an opportunistic pathogen most notorious for its devastating effects on cystic fibrosis (CF) patients (Eberl & Tummler 2004). Quorum sensing is essential for chronic *P. aeruginosa* respiratory infection because it controls adhesion, biofilm formation, and virulence factor expression, all of which allow persistence in the lung and are required for disease progression (Smith & Iglewski 2003).

The *P. aeruginosa* quorum-sensing network consists of two LuxIR circuits, termed LasIR and RhlIR (**Figure 5** and Gambello & Iglewski 1991, Ochsner et al. 1994). LasI, a LuxI homologue, produces an AHL autoinducer (3OC12-homoserine lactone; **Figure 2a** and Pearson et al. 1994) that binds to LasR. The LasR-autoinducer complex activates a variety of target genes including *lasI*, which sets up the characteristic

positive feedback loop that further activates the system (**Figure 5** and Seed et al. 1995). The LasR-autoinducer complex also activates the expression of *rhlR* and *rhlI* encoding another quorum-sensing circuit. RhlI produces the AHL C4-homoserine lactone (**Figure 2a** and Pearson et al. 1995). Following accumulation, RhlR binds the RhlI-directed signal; this complex activates its own set of target genes. Importantly, because the LasIR system induces both *rhlI* and *rhlR*, induction of the genes under RhlIR control occurs subsequent to induction of genes under LasIR control.

Microarray analyses of *P. aeruginosa* quorum-sensing-controlled gene expression revealed three classes of genes: genes that respond to only one autoinducer, genes that respond to either autoinducer, and genes that require both autoinducers simultaneously for activation (Hentzer et al. 2003; Schuster et al. 2003; Wagner et al. 2003, 2004). Furthermore, transcriptome analyses showed that these classes of genes are expressed at different times over the growth cycle. This indicates that the tandem network architecture indeed produces a temporally ordered sequence of gene expression that may be critical for the ordering of early and late events in a successful infection (Schuster et al. 2003, Whiteley et al. 1999).

Competitive Quorum-Sensing Circuits

The above quorum-sensing networks rely on multiple signals acting synergistically. Other quorum-sensing networks are arranged such that the signals antagonize one another. For example, *Bacillus subtilis* has two autoinducing peptides functioning in a network arrange-ment that allows *B. subtilis* to commit to one of two mutually exclusive lifestyles: competence (the ability to take up exogenous DNA) and sporulation (**Figure 6**). ComX, a 10-amino acid peptide (**Figure 2b** and Magnuson et al. 1994, Solomon et al. 1996) that is processed and secreted by ComQ (Bacon Schneider et al. 2002), is detected by the membrane-bound histidine sensor kinase ComP. ComX binding stimulates ComP to autophosphorylate and transfer phosphate to the DNA-binding response regulator ComA (Solomon et al. 1995). Phosphorylated ComA regulates transcription of a variety of genes encoding factors required for competence development (Nakano & Zuber 1991). A second oligopeptide autoinducer, competence and sporulation

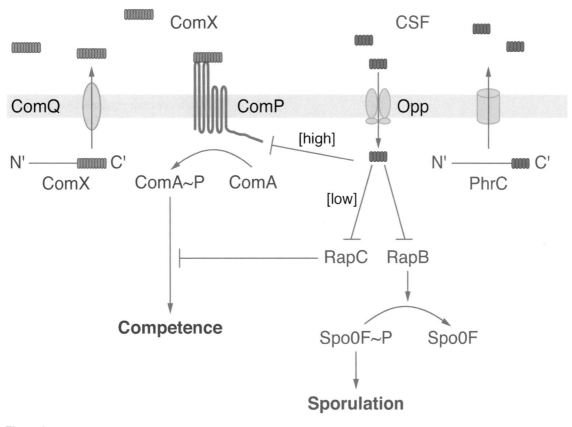

Figure 6

Bacillus subtilis produces two autoinducing peptides that regulate two different developmental pathways: competence and sporulation. ComX is represented as a chain of purple ovals and CSF is shown as a chain of red ovals.

CSF: competence and sporulation factor of *B. subtilis*

CSP: competence-stimulating peptide of *S. pneumoniae*

factor of *B. subtilis* (CSF; encoded by the gene *phrC*), is released via the general secretory apparatus, is re-internalized through the Opp peptide transporter, and acts in the cytoplasm (**Figure 2b**; **Figure 6**; and Lazazzera et al. 1997, Solomon et al. 1996). At low internal concentrations, CSF binds to a protein named RapC and disrupts RapC binding to ComA (Perego 1997, Solomon et al. 1996). RapC binding to ComA inhibits competence development because DNA binding by ComA is prevented. Thus, CSF binding to RapC promotes competence development (Core & Perego 2003). However, at high concentrations, internalized CSF inhibits the ComP-ComA signaling cascade through an unknown mechanism, decreasing competence development and favoring sporulation (Lazazzera et al. 1997, Solomon et al. 1996). CSF also directly promotes sporulation by inhibiting RapB-mediated dephosphorylation of a response regulator named Spo0F, which, in its phosphorylated state, indirectly activates genes required for sporulation (Grossman 1995, Perego 1997).

Quorum-Sensing Circuits with On-Off Switches

The above quorum-sensing circuits allow bacteria to transition from a set of low cell density behaviors to a different set of high cell density behaviors. There are, however, quorum-sensing circuits that promote transient expression of particular traits followed by reversion to the original set of behaviors. Such an on-off switch controls competence development in the Gram-positive bacterium *Streptococcus pneumoniae*, which uses an oligopeptide autoinducer named competence-stimulating peptide (CSP) to monitor cell density (**Figure 2b**). CSP is encoded by *comC* (Havarstein et al. 1995, Tomasz & Hotchkiss 1964). The transporter ComAB exports and modifies CSP (Hui et al. 1995). CSP is detected by the membrane-bound sensor histidine kinase ComD, which transfers phosphate to the cytoplasmic response regulator ComE (Pestova et al. 1996). This circuit controls the transcription of gene subsets in a precise temporal order. Early genes are expressed maximally 6–7 min after CSP accumulation; late genes are maximally induced at 9–10 min (Peterson et al. 2000). ComE directly activates transcription of early genes that include *comAB* and *comCDE*; this causes increased signal production and detection (Pestova et al. 1996). This positive feedback loop results in a dramatic, population-wide spike in competence when the bacteria reach the critical cell density (**Figure 7**). ComE also activates transcription of *comX*, a gene encoding an alternate sigma factor (Lee & Morrison 1999), and *comW*, which is required for transcription of late genes encoding proteins essential for DNA uptake (Luo et al. 2004).

A novel feature of *S. pneumoniae* quorum-sensing circuit is the rapidity with which the process of competence development initiates and terminates (**Figure 7** and Tomasz & Hotchkiss 1964). Importantly, competent

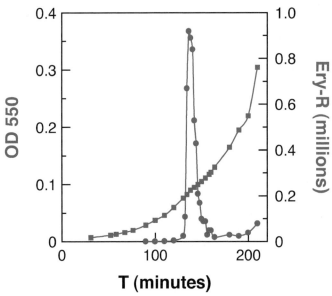

Figure 7

Competence in *Streptococcus pneumoniae* (*red line*) rises sharply at a specific growth stage (*blue line*), followed by a rapid decline. Optical Density (OD550) was used to measure cell number and resistance to erythromycin (Ery-R) was used to assess competence. Figure courtesy of D. Morrison.

S. pneumoniae cells are more prone to autolysis than are noncompetent cells (Dagkessamanskaia et al. 2004, Morrison & Baker 1979, Seto & Tomasz 1975, Steinmoen et al. 2003). Thus, the benefit gained from acquiring DNA that can be used as a repository of new genes is maximized by efficiently activating and terminating the process to minimize lethality by autolysis. The events leading to termination have not been defined. It is known that ComX rapidly disappears when competence is terminated, which suggests that regulated proteolysis occurs (Luo et al. 2004).

Quorum-Sensing Systems Responsive to Host Cues

Agrobacterium tumefaciens induces crown gall tumors in plants through transfer and integration of a tumor-inducing (TI) plasmid into plant cells (Zhu et al. 2000). The tumors produce molecules termed opines, which the bacteria use as nutrients (Dessaux et al. 1992). The quorum-sensing circuit of *A. tumefaciens* is especially interesting because it is only activated at the host-bacterial interface owing to a requirement for both plant- and bacteria-produced signals. Mobilization of the TI plasmid is responsive to proximity to the plant because it requires detection of opines by a cytoplasmic receptor termed AccR or OccR (Beck von Bodman et al. 1992, Fuqua & Winans 1994). AccR/OccR-opine binding induces expression of the *V. fischeri*-like *luxR* homologue, *traR* (Fuqua & Winans 1994). TraR responds to an AHL autoinducer produced by the *V. fischeri* LuxI-type enzyme, TraI (Hwang et al. 1994, Zhang et al. 1993). Hence, bacterial number controls TI transfer because TraR bound to its autoinducer induces TI plasmid replication and bacterial-bacterial conjugation, which lead to increased infectivity of the population (Zhu & Winans 1999). Anti-TraR activators exist that limit TraR activity and thus presumably optimize the ratio of bacteria-bacteria to bacterial-plant TI transfer (Chai et al. 2001, Fuqua et al. 1995, Hwang et al. 1995).

GLOBAL CONTROL: QUORUM-SENSING REGULONS

The advent of genomic profiling has shown that quorum sensing, in many bacteria, controls gene expression in a global manner. Two transcription profiling studies identified over 150 competence-regulated genes in *S. pneumoniae* that were categorized as early, late, delayed-induction, and repressed (Dagkessamanskaia et al. 2004, Peterson et al. 2004). As previously mentioned, early genes are required for signal production, export, and detection whereas some late genes are necessary for DNA internalization. Many of the delayed genes are involved in bacterial stress responses (Dagkessamanskaia et al. 2004, Peterson et al. 2004). Gene-disruption experiments analyzing 124 quorum-sensing-controlled genes found that only 23 are required for competence (Peterson et al. 2004). Quorum-sensing mutants of *S. pneumoniae* and related streptococci show defects in multiple pathways, including biofilm formation, acid tolerance, bacteriocin production, and virulence (reviewed in Suntharalingam & Cvitkovitch 2005). Together, these results suggest that quorum sensing in streptococcus controls the initiation of a global developmental program in which competence development represents only one aspect.

Further evidence that quorum sensing coordinates the control of a large subset of genes comes from transcriptome analyses of *P. aeruginosa* that identify 616 genes as part of the regulon. In one study, addition of autoinducers repressed 222 genes (Wagner et al. 2003). A concurrent study identified 315 quorum-sensing-controlled targets, of which only 38 were repressed (Schuster et al. 2003). Although the two experiments were performed under different growth and autoinducer conditions, the reasons for the discrepancies remain unclear. Importantly, prior to these profiling analyses, quorum-sensing-repressed targets had not been identified in *P. aeruginosa*. Similarly, transcriptional analysis of *V. cholerae* quorum-sensing mutants shows that

TI: tumor-inducing

Regulon: a set of genes under common regulatory control

SAH: *S*-adenosyl-homocysteine

DPD:
4,5-dihydroxy-2,3-pentonedione, a molecule generated by LuxS

the entire virulence regulon (>70 genes) is repressed by quorum sensing (Zhu et al. 2002).

These recent whole-genome quorum-sensing studies highlight two important ideas. First, quorum sensing allows bacteria to alternate between distinct genome-wide programs. These findings, along with an enhanced appreciation of the complexity of the quorum-sensing network architectures, have fundamentally changed the perception of bacteria as primitive single-celled organisms. Bacteria now are understood to undergo complex programs of development similar in many respects to eukaryotic organisms. Second, large groups of genes are repressed by quorum sensing. This finding challenges the notion that the primary function of quorum sensing is to initiate activities that are only beneficial to bacterial participation in group activities. Rather, an equally important function of quorum sensing may be to terminate processes that are only beneficial to bacteria living in relative isolation outside of a community structure.

INTERSPECIES COMMUNICATION AMONG BACTERIA

Beyond controlling gene expression on a global scale, quorum sensing allows bacteria to communicate within and between species. This notion arose with the discovery and study of the autoinducer AI-2, which is one of several signals used by *V. harveyi* in quorum sensing (**Figure 2d** and **Figure 4**). Specifically, *luxS* encoding the AI-2 synthase is present in roughly half of all sequenced bacterial genomes, AI-2 production has been verified in a large number of these species, and AI-2 controls gene expression in a variety of bacteria. Together, these findings have led to the hypothesis that bacteria use AI-2 to communicate between species (reviewed in Xavier & Bassler 2003).

LuxS functions in the pathway for metabolism of SAM, the major cellular methyl donor. Transfer of the methyl moiety to various substrates produces the toxic byproduct *S*-adenosylhomocysteine (SAH) (Schauder et al. 2001). In non-LuxS-containing bacteria and eukaryotes, the enzyme SAH hydrolase metabolizes SAH to adenosine and homocysteine. However, in bacteria containing LuxS, two enzymes, Pfs and LuxS, act sequentially to convert SAH to adenine, homocysteine, and the signaling molecule DPD (**Figure 8** and Schauder et al. 2001). DPD is a highly reactive product that can rearrange and undergo additional reactions, which suggests that distinct but related molecules derived from DPD may be the signals that different bacterial species recognize as AI-2. Two distinct DPD-derived signals were identified in *V. harveyi* and *Salmonella typhimurium* by trapping the active molecules in their respective receptors (LuxP for *V. harveyi* and LsrB for *S. typhimurium*), crystallizing the complexes, and solving their structures (Chen et al. 2002, Miller et al. 2004, Taga et al. 2001). In *V. harveyi*, AI-2 is (2*S*,4*S*)-2-methyl-2,3,3,4-tetrahydroxytetrahydrofuran-borate (*S*-THMF borate); in *S. typhimurium*, AI-2 is (2*R*,4*S*)-2-methyl-2,3,3,4-tetrahydroxytetrahydrofuran (*R*-THMF) (**Figure 2d** and **Figure 8**). Straightforward chemistry links these two molecules, as DPD can cyclize with two equally feasible stereochemistries. Following hydration and borate addition, the upper cyclization pathway in **Figure 8** yields the *V. harveyi* AI-2 and the lower cyclization and hydration pathway yields the *S. typhimurium* AI-2.

Identification of boron in *V. harveyi* AI-2 is surprising, as few biological roles for boron are known. However, boron is present in high concentrations (~0.4 mM) in the marine environment, which makes it a reasonable element in the *V. harveyi* AI-2 signal (Bowen 1966). Significantly lower boron concentration is found in terrestrial environments. This makes boron an unlikely component of the *S. typhimurium* AI-2 signal (Fresenius 1988). Importantly, all of the chemical species shown in **Figure 8** exist in equilibrium and rapidly

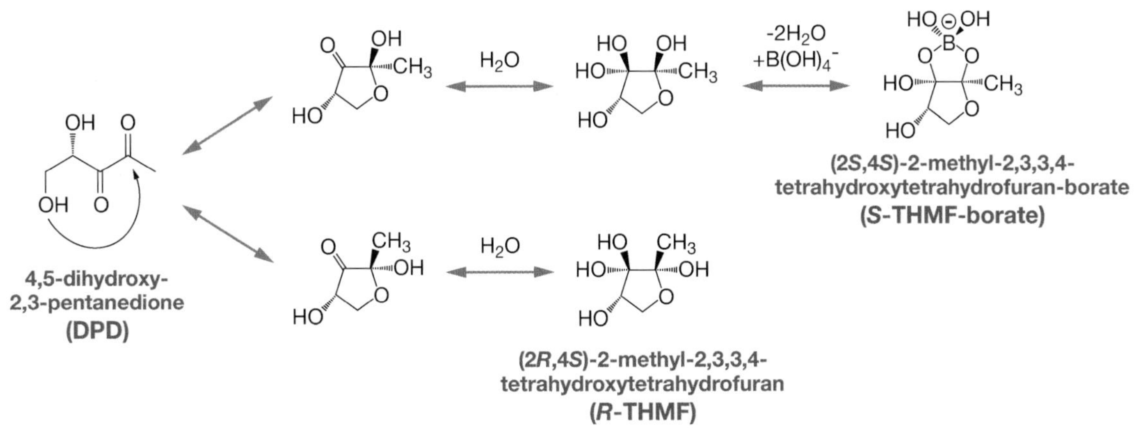

Figure 8

AI-2 is a family of interconverting molecules derived from DPD. *Vibrio harveyi* AI-2 is *S*-THMF-borate and *Salmonella typhimurium* AI-2 is *R*-THMF. Figure from Miller et al. 2004.

interconvert. Moreover, the concentrations of each molecule can be altered by manipulating the boron concentration. For example, addition of boron to DPD preparations promotes formation of the *V. harveyi* AI-2 molecule at the expense of the *S. typhimurium* signal. This shift is biologically relevant because DPD supplemented with boron causes *V. harveyi* to produce maximal bioluminescence whereas the same mixture inhibits the AI-2 response from *S. typhimurium*. Conversely, DPD preparations depleted for boron promote formation of the *S. typhimurium* signal with the concomitant loss of the *V. harveyi* AI-2. Again, this chemistry is borne out in the effect on AI-2-responsive gene expression in the two bacterial species (Miller et al. 2004).

These initial AI-2 investigations show that bacteria employ a conserved biosynthetic pathway to synthesize chemical signal intermediates whose fates are ultimately defined by the chemistry of the particular environment. Other DPD derivatives may exist and be biologically active. Additionally, some bacteria may possess two or more AI-2 receptors for recognition of different derivatives of DPD and alter particular behaviors in response to the information conveyed by each signal. Because only one enzyme (LuxS) is required to synthesize this family of interconverting signal molecules, this pathway may represent an especially economical method for evolving a complex bacterial lexicon.

QUORUM QUENCHING

The fundamental role of quorum sensing appears to be global control of the physiology of bacterial populations. This control is often exerted at the interface of different bacterial populations or at the bacterial-host margin. In niches in which bacterial populations compete for limited resources, the ability to disrupt quorum sensing may give one bacterial species an advantage over another that relies on quorum sensing. Likewise, a host's ability to interfere with bacterial cell-cell communication may be crucial in preventing colonization by pathogenic bacteria that use quorum sensing to coordinate virulence. Thus, it is not surprising that mechanisms have evolved to interfere with bacterial cell-cell communication in processes termed quorum quenching. Analogous mechanisms presumably exist for promoting quorum-sensing-controlled behaviors when such behaviors provide benefits to organisms cohabitating with quorum-sensing bacteria. These latter processes are not yet well defined, so we will focus our discussion primarily on mechanisms of quorum quenching.

Prokaryote-to-Prokaryote Quorum Quenching

As mentioned, cross-inhibition of AIP-mediated signaling in *S. aureus* represents a clear example of a quorum-quenching mechanism because each of the four AIPs specifically inhibits quorum sensing in competing *S. aureus* groups while not disrupting growth and other cellular functions (Lyon et al. 2002a). Many *Bacillus* species secrete an enzyme, AiiA, that cleaves the lactone rings from the acyl moieties of AHLs and renders the AHLs inactive in signal transduction (Dong et al. 2000). AiiA is extremely nonspecific with regard to the AHL acyl side chain, which suggests that this strategy interferes generically with AHL-mediated communication between gram-negative bacteria (Dong et al. 2001). Significantly, as previously mentioned, *Bacillus* relies on oligopeptide-mediated quorum sensing. Therefore, this tactic, while disrupting gram-negative bacterial communication, leaves *Bacillus* cell-cell communication unperturbed.

The soil bacterium *Variovorax paradoxus* uses a different generalized anti-AHL quorum-quenching tactic (Leadbetter & Greenberg 2000). Like *Bacillus*, *V. paradoxus* also degrades AHLs. However, in this case, AHL destruction occurs via an acylase-mediated lactone ring opening. *V. paradoxus* uses the linearized product of the reaction as a source of carbon and nitrogen. This strategy provides *V. paradoxus* with a double benefit: It terminates competitors' group behaviors and simultaneously increases its own growth potential (Leadbetter & Greenberg 2000). Particular *Ralstonia* isolates contain an AHL acylase encoded by *aiiD*, which suggests a similar anti-quorum-sensing mechanism to that of *V. paradoxus* (Lin et al. 2003). The *Ralstonia* quorum-sensing system is immune because the *Ralstonia* autoinducer, 3OH-PAME (Flavier et al. 1997), is not affected by AiiD activity (described in the sidebar).

In some cases, bacteria may degrade their own autoinducers, which presumably terminates quorum-sensing activities. For example, in stationary phase, *A. tumefaciens* produces the AttM AHL lactonase, which can degrade the *A. tumefaciens* autoinducer (Zhang et al. 2002a). It is hypothesized that it is disadvantageous for *A. tumefaciens* to continue to participate in group activities at this late growth stage and that AttM halts these processes. *Erwinia carotovora* and *Xanthomonas campestris* show a similar loss of AHL in stationary phase growth, which suggests an autoinducer degradative activity (Barber et al. 1997, Holden et al. 1998). *P. aeruginosa* degrades long, but not short, chain AHLs through an AiiD-type acylase named PvdQ (Huang et al. 2003). In this case, the RhlI autoinducer, C4-homoserine lactone, is immune, and the LasI autoinducer, 3OC12-homoserine lactone, can be destroyed. Interestingly, *pvdQ* is a member of the LasIR regulon and it is thus under 3OC12-homoserine control (Huang et al. 2003, Whiteley et al. 1999).

Some enteric bacteria, including *S. typhimurium* and *Escherichia coli*, import AI-2 with an AI-2-specific transporter (Surette et al. 1999; Taga et al 2001, 2003; Xavier & Bassler 2005). Once AI-2 is in the cytoplasm, a series of enzymatic reactions inactivates its signaling activity. This process reduces extracellular AI-2 concentrations to levels indicative of low cell density and—because AI-2 is used for interspecies communication—indicative of monospecies environments. AI-2 internalization is suspected to be another mechanism for interference with chemical communication among bacteria. Because *E. coli* and *S. typhimurium* also produce AI-2 and respond to this signal (Taga et al. 2001, 2003), it is not clear how these bacteria regulate AI-2 import while protecting the fidelity of their own AI-2 signaling cascades.

Eukaryote-to-Prokaryote Quorum Quenching

Several eukaryotic mechanisms that counteract bacterial quorum sensing have recently been discovered. The Australian red macro-agla *Delisea pulchra* coats its surface with a mixture of halogenated furanones that bear structural similarity to AHLs (Givskov et al. 1996). The furanones are internalized by bacteria, bind to LuxR-type proteins, and cause the degradation of these proteins (Manefield et al. 2002). This strategy prevents bacterial colonization of the algal surface by inhibiting quorum-sensing-controlled biofilm formation.

The legume *Medicago truncatula* controls over 150 proteins in response to AHLs produced by two model quorum-sensing bacteria; *Sinorhizobium meliloti* and *P. aeruginosa* (Mathesius et al. 2003). The plant secretes compounds in response to AHLs. These factors inhibit AI-2 signaling and stimulate AHL signaling in quorum-sensing reporter strains. The plant presumably encourages signaling between AHL-producing bacteria but not AI-2-producing bacteria because only the former are beneficial to the plant. Similarly, *Pisum sativum* (pea) produces AHL mimics that both positively and negatively affect AHL-regulated behaviors in a number of bacterial reporter strains (Teplitski et al. 2000).

Reactive oxygen and nitrogen intermediates generated by NADPH oxidase inactivate the *S. aureus* autoinducing peptide in a mouse air pouch skin model (Rothfork et al. 2004). These findings indicate a novel role for NADPH oxidase, an important component of innate immunity, in protection from bacterial infections. Consistent with this, mice deficient in NADPH oxidase have reduced resistance to infection by *S. aureus*, whereas infection from quorum-sensing mutant *S. aureus* remains unaffected by the loss of NAPH oxidase (Rothfork et al. 2004). This latter result suggests that reactive oxygen species influence infectivity only through quorum quenching. The authors of this study speculate that oxidation of other kinds of quorum-sensing molecules by NADPH oxidase is likely (Rothfork et al. 2004).

Human cells also have quorum-quenching activity. Analysis of primary and immortal-ized human epithelial cell lines show specific inactivation of the *P. aeruginosa* 3OC12-homoserine lactone autoinducer (the product of LasI) but not of the C4-homoserine lactone autoinducer (the product of RhlI) (Chun et al. 2004). Although presently uncharacterized, the quenching activity is membrane associated and heat liable, which suggests that it is a protein. This activity is intriguing in terms of the development of anti-*P. aeruginosa* therapies for treatment of CF.

Biotechnological Applications of Quorum Quenching

Naturally occurring quorum-quenching processes are being tested as novel antimicrobial therapies. Overexpression of *aiiA* in tobacco and potato plants confers resistance to *E. carotovora*, which requires AHL-controlled virulence factor expression to cause disease (Dong et al. 2001). Likewise, coculture of *Bacillus thuringiensis* decreased *E. carotovora*–mediated plant disease in an *aiiA*-dependent manner (Dong et al. 2004). Mice treated with synthetic antagonists of *S. aureus* AIP show resistance to infection (Mayville et al. 1999). Similarly, purified halogenated furanones appear to attenuate virulence of bacteria in mouse models (Hentzer et al. 2003, Wu et al. 2004). These and other examples predict that inhibition of quorum sensing offers an attractive alternative to traditional antibiotics because these strategies are not bactericidal and the occurrence of bacterial resistance therefore could be reduced. Likewise, approaches aimed at promoting beneficial quorum-sensing associations may enhance industrial-scale production of natural or engineered bacterial products.

EVOLUTION AND MAINTENANCE OF QUORUM SENSING IN BACTERIA

Quorum sensing presumably provides bacteria benefits from group activities that may be unattainable to an individual bacterium acting alone. For example, the benefit derived from secretion of antibiotics or proteases may only occur when these exoproducts exceed a particular extracellular concentration, and achieving this concentration is only possible through the synchronous activity of a group of cells. The idea that bacteria cooperate has led to new questions regarding the evolution of cell-cell communication in bacteria, the cost bacteria pay for communicating, how fidelity is maintained in quorum-sensing systems, how cheating is controlled, and if and how "eavesdropping" occurs.

Although these evolutionary questions are new in the context of the molecular mechanisms underlying quorum-sensing-controlled behaviors, there exists an extensive literature dealing with these topics in other social organisms (Bourke 2001, Bradley 1999, Korb & Heinze 2004). For example, some social insects (e.g., ants and bees) have sterile worker castes that promote colony fitness even though the workers have no chance at reproduction (Queller & Strassmann 2002). The predominant explanation for these behaviors rests on Hamilton's kin-selection theory that predicts that even without directly contributing to reproduction, organisms belonging to a multimember group promote the inheritance of their own genes by increasing the fitness of closely related kin (Hamilton 1964a,b). A key component of kin selection is the ability to recognize another individual as kin. The ability to distinguish between and communicate with specific chemical signal molecules may enable a type of "kin selection" in bacteria. Consistent with this, many higher social organisms rely heavily on chemical signaling to maintain the integrity of the social order (Breed et al. 2004, Holldobler 1995, Queller & Strassmann 2002).

Cases of sacrifice of the individual for the group benefit also exist in microorganisms. For example, both the soil-dwelling bacterium *Myxococcus xanthus* and the slime mold *Dictyostelium discoidium*, in the absence of nutrients, produce resistant spores that survive nonvegetatively for long periods and can be dispersed to new environments (Dao et al. 2000, Strassmann et al. 2000). Spore development requires a large percentage of the population to undergo a lethal differentiation event that leads to structures whose function is to promote spore generation and dispersal. Chemical communication is required for these developmental events in both *D. discoidium* and *M. xanthus*: cAMP and Differentiation-Inducing Factor initiates development of fruiting bodies in *D. discoidium* (Konijn et al. 1969, Town & Stanford 1979, Town et al. 1976), whereas quorum-sensing communication controls the process in *M. xanthus* (Shimkets 1999).

Two examples exist to date that illustrate a selection for maintenance of quorum sensing. In *V. fischeri*, mutants incapable of luciferase production are outcompeted by the wild-type bacteria in the squid host. This indicates that the squid may possess a policing mechanism to eliminate cheater cells. Interestingly, the defect in these mutants was in luciferase itself, which suggests that the squid somehow distinguishes between cells that can and cannot contribute to light production (Visick et al. 2000). *A. tumefaciens*–induced plant tumors contain a large percentage of plasmid-free bacterial cells (Belanger et al. 1995). These cells have a faster growth rate and may therefore more efficiently grow on the opine nutrients produced in the plant tumors. However, plasmid-free bacteria are unable to initiate new tumor formation. Interestingly, as bacterial density increases and nutrients become limiting, increased bacterial-bacterial conjugation occurs and the TI plasmid is replicated to a higher copy number. Both of these events require quorum sensing and ensure that most of the bacteria acquire copies of the plasmid before they disseminate to a new location. This elegant strategy optimizes growth inside the tumor while maintaining the population's virulence and at least partially explains why quorum sensing is maintained.

Mechanisms for eavesdropping apparently also exist in quorum-sensing systems.

P. aeruginosa does not have *luxS* and therefore does not produce AI-2. However, *P. aeruginosa* detects AI-2 produced by the indigenous nonpathogenic microflora present in CF sputum samples (Duan et al. 2003). In the CF lung, *P. aeruginosa* exists in a complex microbial community composed of a variety of pathogenic and nonpathogenic bacteria. The detection of AI-2 may alert *P. aeruginosa* that *P. aeruginosa* is in the lung and that a program of gene expression that enhances persistence/virulence in the host is required. Consistent with this idea, CF sputum contains high concentrations of AI-2 and AI-2 induces *P. aeruginosa* virulence factor expression (Duan et al. 2003). In another example, *Salmonella enterica*, which has a *V. fischeri*–type LuxR-type protein (SdiA) but no LuxI-type enzyme, intercepts AHLs produced by other LuxI-containing gram-negative bacteria. In response to these signals, *S. enterica* expresses the *rck* operon and other genes that protect *S. enterica* from host defenses in the intestine (Ahmer et al. 1998). This result is interpreted to mean that the AHL signals signify that *S. enterica* is in a dense population of bacteria, which can presumably be attained only inside a host.

The ecological and evolutionary implications of quorum sensing in bacteria are only beginning to be addressed (Travisano & Velicer 2004). However, continued study of such questions hopefully will provide insight into the evolution and maintenance of group dynamics and behavior.

RHOMBOID: A SHARED PROKARYOTIC AND EUKARYOTIC CHEMICAL COMMUNICATION MECHANISM

New data suggest that some bacterial and eukaryotic signaling mechanisms have a common evolutionary origin. The inner membrane protein AarA of *Providencia stuartii* is required for the release of an extracellular quorum-sensing signal whose structure has not been defined (Rather et al. 1999). AarA has homology to the *Drosophila melanogaster* RHO (Gallio et al. 2002), which is a serine protease required for intramembrane cleavage, release, and activation of Epidermal Growth Factor receptor ligands (Klambt 2000, 2002). RHO is essential for many developmental processes in *D. melanogaster*, including proper wing vein development and organization of the fly eye (Schweitzer & Shilo 1997). Consistent with the idea that AarA and RHO have a common signaling function, expression of *P. stuartii aarA* in a *D. melanogaster rho* mutant rescued wing vein development. Likewise, expression of *rho* in a *P. stuartii aarA* mutant complemented the quorum-sensing signaling defect (Gallio et al. 2002). Homologues of RHO/AarA are nearly ubiquitous in all three kingdoms of life: bacteria, archea, and eukaryotes (Koonin et al. 2003). Five of eight tested bacterial Aar/RHO orthologues specifically cleaved RHO substrates, which suggests a widespread conservation of the mechanism of RHO with its bacterial homologues (Urban et al. 2002). These fascinating results show that bacteria and higher eukaryotes share a common cell-cell communication system; however, it has not been determined if any cross-kingdom communication can be mediated by RHO or its homologues.

A recent bioinformatics study suggests that the RHO/AarA finding is not an anomaly but rather that many signaling mechanisms may be shared by prokaryotes and eukaryotes. Enzymes involved in the production of cell-cell signaling molecules in vertebrates have homologues in bacteria but are absent from plants and archea (Iyer et al. 2004). A few of numerous examples are the enzymes phenylethanolamine N-methyltransferase (which catalyzes the conversion of norepinephrine to epinephrine), histidine decarboxylase (which catalyzes histidine to histamine), and glutamate decarboxylase (which catalyzes glutamate to γ-aminobutyric acid). It is hypothesized that eukaryotes acquired these genes from bacteria through a series of horizontal gene transfer

RHO: rhomboid protein

events (Iyer et al. 2004). These findings suggest that bacteria and eukaryotes share enzymes responsible for many cell-cell signaling pathways. This points to the exciting possibility that prokaryotic-to-eukaryotic cross-kingdom communication may be more prevalent than is currently appreciated.

CONCLUSIONS

It is now clear that cell-cell communication is the norm in the bacterial world and that understanding this process is fundamental to all of microbiology, including industrial and clinical microbiology. Our knowledge of quorum sensing may ultimately affect our understanding of higher-organism development. Quorum sensing was, until recently, considered to promote exclusively intraspecies communication and thus enable clonal populations of bacteria to count their cell numbers and alter gene expression in unison. While some autoinducers indeed appear to be extremely species-specific, new research shows that others are either genus-specific or promote intergenera communication. Further, hints that interkingdom communication occurs are becoming increasingly prevalent. Coincident with these findings are the beginnings of an understanding that prokaryotic and eukaryotic mechanisms that enhance and interfere with bacterial chemical communication also exist in nature. Bacterial quorum-sensing signal detection and relay apparatuses are complex and often consist of multiple circuits organized in a variety of configurations. Because bacteria routinely exist in fluctuating environments containing complex mixtures of chemicals, some of which are signals and some of which presumably do not convey meaningful information, we hypothesize that each quorum-sensing network organization evolved to solve the particular set of communication needs a particular species of bacteria encounters. Elements of these elegant solutions for deciphering complex chemical vocabularies appear to be conserved and used for analogous purposes in eukaryotes.

ACKNOWLEDGMENTS

This work was supported by NSF grant MCB-0343821, NIH grants 5R01 GM065859 and 1R01 AI054442, ONR grant N00014-03-0183, and an NIH Postdoctoral Fellowship (C.M.W.). We are grateful to the members of the Bassler lab for insightful discussions. We thank Dr. D. Morrison for **Figure 7** and Dr. S. Miller for **Figure 8**.

LITERATURE CITED

Ahmer BM, van Reeuwijk J, Timmers CD, Valentine PJ, Heffron F. 1998. *Salmonella typhimurium* encodes an SdiA homolog, a putative quorum sensor of the LuxR family, that regulates genes on the virulence plasmid. *J. Bacteriol.* 180:1185–93

Ansaldi M, Marolt D, Stebe T, Mandic-Mulec I, Dubnau D. 2002. Specific activation of the *Bacillus* quorum-sensing systems by isoprenylated pheromone variants. *Mol. Microbiol.* 44:1561–73

Bacon Schneider K, Palmer TM, Grossman AD. 2002. Characterization of *comQ* and *comX*, two genes required for production of ComX pheromone in *Bacillus subtilis*. *J. Bacteriol.* 184:410–19

Barber CE, Tang JL, Feng JX, Pan MQ, Wilson TJ, et al. 1997. A novel regulatory system required for pathogenicity of *Xanthomonas campestris* is mediated by a small diffusible signal molecule. *Mol. Microbiol.* 24:555–66

Bassler BL, Wright M, Showalter RE, Silverman MR. 1993. Intercellular signalling in *Vibrio harveyi*: sequence and function of genes regulating expression of luminescence. *Mol. Microbiol.* 9:773–86

Bassler BL, Wright M, Silverman MR. 1994a. Multiple signalling systems controlling expression of luminescence in *Vibrio harveyi*: sequence and function of genes encoding a second sensory pathway. *Mol. Microbiol.* 13:273–86

Bassler BL, Wright M, Silverman MR. 1994b. Sequence and function of LuxO, a negative regulator of luminescence in *Vibrio harveyi*. *Mol. Microbiol.* 12:403–12

Beck von Bodman S, Hayman GT, Farrand SK. 1992. Opine catabolism and conjugal transfer of the nopaline Ti plasmid pTiC58 are coordinately regulated by a single repressor. *Proc. Natl. Acad. Sci. USA* 89:643–47

Belanger C, Canfield ML, Moore LW, Dion P. 1995. Genetic analysis of nonpathogenic Agrobacterium tumefaciens mutants arising in crown gall tumors. *J. Bacteriol.* 177:3752–57

Booth MC, Bogie CP, Sahl HG, Siezen RJ, Hatter KL, Gilmore MS. 1996. Structural analysis and proteolytic activation of *Enterococcus faecalis* cytolysin, a novel lantibiotic. *Mol. Microbiol.* 21:1175–84

Bourke AF. 2001. Social insects and selfish genes. *Biologist* 48:205–8

Bowen HJM. 1966. *Trace Elements in Biochemistry*, pp. 9–19. London: Academic

Bradley BJ. 1999. Levels of selection, altruism, and primate behavior. *Q. Rev. Biol.* 74:171–94

Breed MD, Guzman-Novoa E, Hunt GJ. 2004. Defensive behavior of honey bees: organization, genetics, and comparisons with other bees. *Annu. Rev. Entomol.* 49:271–98

Cao JG, Meighen EA. 1989. Purification and structural identification of an autoinducer for the luminescence system of *Vibrio harveyi*. *J. Biol. Chem.* 264:21670–76

Carrington JC, Ambros V. 2003. Role of microRNAs in plant and animal development. *Science* 301:336–38

Chai Y, Zhu J, Winans SC. 2001. TrlR, a defective TraR-like protein of Agrobacterium tumefaciens, blocks TraR function in vitro by forming inactive TrlR:TraR dimers. *Mol. Microbiol.* 40:414–21

Chater KF, Horinouchi S. 2003. Signalling early developmental events in two highly diverged Streptomyces species. *Mol. Microbiol.* 48:9–15

Chen X, Schauder S, Potier N, Van Dorsselaer A, Pelczer I, et al. 2002. Structural identification of a bacterial quorum-sensing signal containing boron. *Nature* 415:545–49

Chun CK, Ozer EA, Welsh MJ, Zabner J, Greenberg EP. 2004. Inactivation of a *Pseudomonas aeruginosa* quorum-sensing signal by human airway epithelia. *Proc. Natl. Acad. Sci. USA* 101:3587–90

Core L, Perego M. 2003. TPR-mediated interaction of RapC with ComA inhibits response regulator-DNA binding for competence development in *Bacillus subtilis*. *Mol. Microbiol.* 49:1509–22

Dagkessamanskaia A, Moscoso M, Henard V, Guiral S, Overweg K, et al. 2004. Interconnection of competence, stress and CiaR regulons in *Streptococcus pneumoniae*: competence triggers stationary phase autolysis of ciaR mutant cells. *Mol. Microbiol.* 51:1071–86

Dao DN, Kessin RH, Ennis HL. 2000. Developmental cheating and the evolutionary biology of Dictyostelium and Myxococcus. *Microbiology* 146(Pt. 7):1505–12

Dessaux Y, Petit AS, Tempe J. 1992. Opines in *Agrobacterium* biology. In *The Molecular Signals in Plant-Microbe Interaction*, ed. DPS Verma, pp. 109–36. Boca Raton, FL: CRC Press

Dong YH, Wang LH, Xu JL, Zhang HB, Zhang XF, Zhang LH. 2001. Quenching quorum-sensing-dependent bacterial infection by an N-acyl homoserine lactonase. *Nature* 411:813–17

This is the first report identifying a small molecule by X-ray crystallography of a ligand bound to its receptor.

This work highlights the potential for targeting quorum sensing in the development of novel antibacterial therapeutics.

Dong YH, Xu JL, Li XZ, Zhang LH. 2000. AiiA, an enzyme that inactivates the acylhomoserine lactone quorum-sensing signal and attenuates the virulence of *Erwinia carotovora*. *Proc. Natl. Acad. Sci. USA* 97:3526–31

Dong YH, Zhang XF, Xu JL, Zhang LH. 2004. Insecticidal *Bacillus thuringiensis* silences *Erwinia carotovora* virulence by a new form of microbial antagonism, signal interference. *Appl. Environ. Microbiol.* 70:954–60

Duan K, Dammel C, Stein J, Rabin H, Surette MG. 2003. Modulation of *Pseudomonas aeruginosa* gene expression by host microflora through interspecies communication. *Mol. Microbiol.* 50:1477–91

Dufour P, Jarraud S, Vandenesch F, Greenland T, Novick RP, et al. 2002. High genetic variability of the agr locus in Staphylococcus species. *J. Bacteriol.* 184:1180–86

Eberhard A, Burlingame AL, Eberhard C, Kenyon GL, Nealson KH, Oppenheimer NJ. 1981. Structural identification of autoinducer of *Photobacterium fischeri* luciferase. *Biochemistry* 20:2444–49

Eberl L, Tummler B. 2004. *Pseudomonas aeruginosa* and *Burkholderia cepacia* in cystic fibrosis: genome evolution, interactions and adaptation. *Int. J. Med. Microbiol.* 294:123–31

Engebrecht J, Nealson K, Silverman M. 1983. Bacterial bioluminescence: isolation and genetic analysis of functions from *Vibrio fischeri*. *Cell* 32:773–81

Engebrecht J, Silverman M. 1984. Identification of genes and gene products necessary for bacterial bioluminescence. *Proc. Natl. Acad. Sci. USA* 81:4154–58

Fidopiastis PM, Miyamoto CM, Jobling MG, Meighen EA, Ruby EG. 2002. LitR, a new transcriptional activator in Vibrio fischeri, regulates luminescence and symbiotic light organ colonization. *Mol. Microbiol.* 45:131–43

Flavier AB, Clough SJ, Schell MA, Denny TP. 1997. Identification of 3-hydroxypalmitic acid methyl ester as a novel autoregulator controlling virulence in *Ralstonia solanacearum*. *Mol. Microbiol.* 26:251–59

Freeman JA, Bassler BL. 1999a. A genetic analysis of the function of LuxO, a two-component response regulator involved in quorum sensing in *Vibrio harveyi*. *Mol. Microbiol.* 31:665–77

Freeman JA, Bassler BL. 1999b. Sequence and function of LuxU: a two-component phosphorelay protein that regulates quorum sensing in *Vibrio harveyi*. *J. Bacteriol.* 181(3):899–906

Freeman JA, Lilley BN, Bassler BL. 2000. A genetic analysis of the functions of LuxN: a two-component hybrid sensor kinase that regulates quorum sensing in *Vibrio harveyi*. *Mol. Microbiol.* 35:139–49

Fresenius W, Quentin KE, Schneider W, eds. 1988. *Water Analysis: A Practical Guide to Physicochemical, Chemical and Microbiological Water Examination and Quality Assurance*, pp. 420–21. New York: Springer-Verlag. 804 pp.

Fuqua C, Burbea M, Winans SC. 1995. Activity of the Agrobacterium Ti plasmid conjugal transfer regulator TraR is inhibited by the product of the traM gene. *J. Bacteriol.* 177:1367–73

Fuqua C, Eberhard A. 1999. Signal generation in autoinduction systems: synthesis of acylated homoserine lactones by LuxI-type proteins. In *Cell-Cell Signaling in Bacteria*, ed. GM Dunny, SC Winans, pp. 211–30. Washington, DC: ASM Press

Fuqua WC, Winans SC. 1994. A LuxR-LuxI type regulatory system activates Agrobacterium Ti plasmid conjugal transfer in the presence of a plant tumor metabolite. *J. Bacteriol.* 176:2796–806

Gallio M, Sturgill G, Rather P, Kylsten P. 2002. A conserved mechanism for extracellular signaling in eukaryotes and prokaryotes. *Proc. Natl. Acad. Sci. USA* 99:12208–13

This study's results suggest a conserved Rhomboid-signaling mechanism in eukaryotes and bacteria.

Gambello MJ, Iglewski BH. 1991. Cloning and characterization of the *Pseudomonas aeruginosa* lasR gene, a transcriptional activator of elastase expression. *J. Bacteriol.* 173:3000–9

Givskov M, de Nys R, Manefield M, Gram L, Maximilien R, et al. 1996. Eukaryotic interference with homoserine lactone-mediated prokaryotic signalling. *J. Bacteriol.* 178:6618–22

Gould TA, Schweizer HP, Churchill ME. 2004. Structure of the *Pseudomonas aeruginosa* acyl-homoserinelactone synthase LasI. *Mol. Microbiol.* 53:1135–46

Grossman AD. 1995. Genetic networks controlling the initiation of sporulation and the development of genetic competence in *Bacillus subtilis*. *Annu. Rev. Genet.* 29:477–508

Hamilton WD. 1964a. The genetical evolution of social behaviour. I. *J. Theor. Biol.* 7:1–16

Hamilton WD. 1964b. The genetical evolution of social behaviour. II. *J. Theor. Biol.* 7:17–52

Hammer BK, Bassler BL. 2003. Quorum sensing controls biofilm formation in *Vibrio cholerae*. *Mol. Microbiol.* 50:101–4

Hanzelka BL, Parsek MR, Val DL, Dunlap PV, Cronan JE Jr, Greenber EP. 1999. Acylhomoserine lactone synthase activity of the *Vibrio fischeri* AinS protein. *J. Bacteriol.* 181:5766–70

Havarstein LS, Coomaraswamy G, Morrison DA. 1995. An unmodified heptadecapeptide pheromone induces competence for genetic transformation in *Streptococcus pneumoniae*. *Proc. Natl. Acad. Sci. USA* 92:11140–44

Henke JM, Bassler BL. 2004a. Quorum sensing regulates type III secretion in *Vibrio harveyi* and *Vibrio parahaemolyticus*. *J. Bacteriol.* 186:3794–805

Henke JM, Bassler BL. 2004b. Three parallel quorum-sensing systems regulate gene expression in *Vibrio harveyi*. *J. Bacteriol.* 186:6902–14

Hentzer M, Wu H, Andersen JB, Riedel K, Rasmussen TB, et al. 2003. Attenuation of *Pseudomonas aeruginosa* virulence by quorum sensing inhibitors. *EMBO J.* 22:3803–15

Holden MT, McGowan SJ, Bycroft BW, Stewart GS, Williams P, Salmond GP. 1998. Cryptic carbapenem antibiotic production genes are widespread in *Erwinia carotovora*: facile trans activation by the carR transcriptional regulator. *Microbiology* 144(Pt. 6):1495–508

Holden MT, Ram Chhabra S, de Nys R, Stead P, Bainton NJ, et al. 1999. Quorum-sensing cross talk: isolation and chemical characterization of cyclic dipeptides from *Pseudomonas aeruginosa* and other gram-negative bacteria. *Mol. Microbiol.* 33:1254–66

Holldobler B. 1995. The chemistry of social regulation: multicomponent signals in ant societies. *Proc. Natl. Acad. Sci. USA* 92:19–22

Huang JJ, Han JI, Zhang LH, Leadbetter JR. 2003. Utilization of acyl-homoserine lactone quorum signals for growth by a soil pseudomonad and *Pseudomonas aeruginosa* PAO1. *Appl. Environ. Microbiol.* 69:5941–49

Hui FM, Zhou L, Morrison DA. 1995. Competence for genetic transformation in *Streptococcus pneumoniae*: organization of a regulatory locus with homology to two lactococcin A secretion genes. *Gene* 153:25–31

Hwang I, Cook DM, Farrand SK. 1995. A new regulatory element modulates homoserine lactone-mediated autoinduction of Ti plasmid conjugal transfer. *J. Bacteriol.* 177:449–58

Hwang I, Li PL, Zhang L, Piper KR, Cook DM, et al. 1994. TraI, a LuxI homologue, is responsible for production of conjugation factor, the Ti plasmid N-acylhomoserine lactone autoinducer. *Proc. Natl. Acad. Sci. USA* 91:4639–43

Iyer LM, Aravind L, Coon SL, Klein DC, Koonin EV. 2004. Evolution of cell-cell signaling in animals: did late horizontal gene transfer from bacteria have a role? *Trends Genet.* 20:292–99

Ji G, Beavis RC, Novick RP. 1995. Cell density control of staphylococcal virulence mediated by an octapeptide pheromone. *Proc. Natl. Acad. Sci. USA* 92:12055–59

Kaplan HB, Greenberg EP. 1985. Diffusion of autoinducer is involved in regulation of the *Vibrio fischeri* luminescence system. *J. Bacteriol.* 163:1210–14

Klambt C. 2000. EGF receptor signalling: the importance of presentation. *Curr. Biol.* 10:R 388–91

Klambt C. 2002. EGF receptor signalling: roles of star and rhomboid revealed. *Curr. Biol.* 12:R21–23

Konijn TM, van deMeene JG, Chang YY, Barkley DS, Bonner JT. 1969. Identification of adenosine-3′,5′-monophosphate as the bacterial attractant for myxamoebae of *Dictyostelium discoideum*. *J. Bacteriol.* 99:510–12

Koonin EV, Makarova KS, Rogozin IB, Davidovic L, Letellier MC, Pellegrini L. 2003. The rhomboids: a nearly ubiquitous family of intramembrane serine proteases that probably evolved by multiple ancient horizontal gene transfers. *Genome Biol.* 4:R19

Korb J, Heinze J. 2004. Multilevel selection and social evolution of insect societies. *Naturwissenschaften* 91:291–304

Lazazzera BA, Solomon JM, Grossman AD. 1997. An exported peptide functions intracellularly to contribute to cell density signaling in *B. subtilis*. *Cell* 89:917–25

Leadbetter JR, Greenberg EP. 2000. Metabolism of acyl-homoserine lactone quorum-sensing signals by *Variovorax paradoxus*. *J. Bacteriol.* 182:6921–26

Lee MS, Morrison DA. 1999. Identification of a new regulator in *Streptococcus pneumoniae* linking quorum sensing to competence for genetic transformation. *J. Bacteriol.* 181:5004–16

Lenz DH, Mok KC, Lilley BN, Kulkarni RV, Wingreen NS, Bassler BL. 2004. The small RNA chaperone Hfq and multiple small RNAs control quorum sensing in *Vibrio harveyi* and *Vibrio cholerae*. *Cell* 118:69–82

Lilley BN, Bassler BL. 2000. Regulation of quorum sensing in *Vibrio harveyi* by LuxO and sigma-54. *Mol. Microbiol.* 36:940–54

Lin YH, Xu JL, Hu J, Wang LH, Ong SL, et al. 2003. Acyl-homoserine lactone acylase from *Ralstonia* strain XJ12B represents a novel and potent class of quorum-quenching enzymes. *Mol. Microbiol.* 47:849–60

Luo P, Li H, Morrison DA. 2004. Identification of ComW as a new component in the regulation of genetic transformation in *Streptococcus pneumoniae*. *Mol. Microbiol.* 54:172–83

Lupp C, Ruby EG. 2004. *Vibrio fischeri* LuxS and AinS: comparative study of two signal synthases. *J. Bacteriol.* 186:3873–81

Lupp C, Urbanowski M, Greenberg EP, Ruby EG. 2003. The *Vibrio fischeri* quorum-sensing systems ain and lux sequentially induce luminescence gene expression and are important for persistence in the squid host. *Mol. Microbiol.* 50:319–31

Lyon GJ, Novick RP. 2004. Peptide signaling in *Staphylococcus aureus* and other Gram-positive bacteria. *Peptides* 25:1389–403

Lyon GJ, Wright JS, Christopoulos A, Novick RP, Muir TW. 2002a. Reversible and specific extracellular antagonism of receptor-histidine kinase signaling. *J. Biol. Chem.* 277:6247–53

Lyon GJ, Wright JS, Muir TW, Novick RP. 2002b. Key determinants of receptor activation in the *agr* autoinducing peptides of *Staphylococcus aureus*. *Biochemistry* 41:10095–104

Magnuson R, Solomon J, Grossman AD. 1994. Biochemical and genetic characterization of a competence pheromone from *B. subtilis*. *Cell* 77:207–16

Manefield M, Rasmussen TB, Henzter M, Andersen JB, Steinberg P, et al. 2002. Halogenated furanones inhibit quorum sensing through accelerated LuxR turnover. *Microbiology* 148:1119–27

This study shows that, unlike most gram-positive, bacterial autoinducing peptides that act from the extracellular environment, CSF, a peptide signal in *B. subtilis*, is imported into the cytoplasm.

The regulatory link connecting the quorum-sensing machinery and the genes required for competence is established in this study.

A genetic screen combined with bioinformatics reveals that Hfq and multiple redundant sRNAs control the translation of the master regulators of quorum sensing, LuxR (*V. harveyi*) and HapR (*V. cholerae*).

Manefield M, Turner SL. 2002. Quorum sensing in context: out of molecular biology and into microbial ecology. *Microbiology* 148:3762–64

Marketon MM, Gronquist MR, Eberhard A, Gonzalez JE. 2002. Characterization of the *Sinorhizobium meliloti* sinR/sinI locus and the production of novel N-acyl homoserine lactones. *J. Bacteriol.* 184:5686–95

Mathesius U, Mulers S, Gao M, Teplitski M, Caetano-Anolles G, et al. 2003. Extensive and specific responses of a eukaryote to bacterial quorum-sensing signals. *Proc. Natl. Acad. Sci. USA.* 100:1444–49

Mayville P, Ji G, Beavis R, Yang H, Goger M, et al. 1999. Structure-activity analysis of synthetic autoinducing thiolactone peptides from *Staphylococcus aureus* responsible for virulence. *Proc. Natl. Acad. Sci. USA* 96:1218–23

Miller MB, Skorupski K, Lenz DH, Taylor RK, Bassler BL. 2002. Parallel quorum sensing systems converge to regulate virulence in *Vibrio cholerae. Cell* 110:303–14

Miller ST, Xavier KB, Campagna SR, Taga ME, Semmelhack MF, et al. 2004. A novel form of the bacterial quorum sensing signal AI-2 recognized by the *Salmonella typhimurium* receptor LsrB. *Mol. Cell.* 15:677–87

Miyamoto CM, Dunlap PV, Ruby EG, Meighen EA. 2003. LuxO controls *luxR* expression in *Vibrio harveyi*: evidence for a common regulatory mechanism in *Vibrio. Mol. Microbiol.* 48:537–48

Mok KC, Wingreen NS, Bassler BL. 2003. *Vibrio harveyi* quorum sensing: a coincidence detector for two autoinducers controls gene expression. *EMBO J.* 22:870–81

More MI, Finger LD, Stryker JL, Fuqua C, Eberhard A, Winans SC. 1996. Enzymatic synthesis of a quorum-sensing autoinducer through use of defined substrates. *Science* 272:1655–58

Morrison DA, Baker MF. 1979. Competence for genetic transformation in pneumococcus depends on synthesis of a small set of proteins. *Nature* 282:215–17

Nakano MM, Zuber P. 1991. The primary role of comA in establishment of the competent state in *Bacillus subtilis* is to activate expression of srfA. *J. Bacteriol.* 173:7269–74

Nakayama J, Cao Y, Horii T, Sakuda S, Akkermans AD, et al. 2001. Gelatinase biosynthesis-activating pheromone: a peptide lactone that mediates a quorum sensing in *Enterococcus faecalis. Mol. Microbiol.* 41:145–54

Nealson KH, Hastings JW. 1979. Bacterial bioluminescence: its control and ecological significance. *Microbiol. Rev.* 43:496–518

Novick RP. 2003. Autoinduction and signal transduction in the regulation of staphylococcal virulence. *Mol. Microbiol.* 48:1429–49

Novick RP, Projan SJ, Kornblum J, Ross HF, Ji G, et al. 1995. The agr P2 operon: an autocatalytic sensory transduction system in *Staphylococcus aureus. Mol. Gen. Genet.* 248:446–58

Novick RP, Ross HF, Projan SJ, Kornblum J, Kreiswirth B, Moghazeh S. 1993. Synthesis of staphylococcal virulence factors is controlled by a regulatory RNA molecule. *EMBO J.* 12:3967–75

Ochsner UA, Koch AK, Fiechter A, Reiser J. 1994. Isolation and characterization of a regulatory gene affecting rhamnolipid biosurfactant synthesis in *Pseudomonas aeruginosa. J. Bacteriol.* 176:2044–54

Parsek MR, Val DL, Hanzelka BL, Cronan JE Jr, Greenberg EP. 1999. Acyl homoserine-lactone quorum-sensing signal generation. *Proc. Natl. Acad. Sci. USA* 96:4360–65

Pearson JP, Gray KM, Passador L, Tucker KD, Eberhard A, et al. 1994. Structure of the autoinducer required for expression of *Pseudomonas aeruginosa* virulence genes. *Proc. Natl. Acad. Sci. USA* 91:197–201

Pearson JP, Passador L, Iglewski BH, Greenberg EP. 1995. A second N-acylhomoserine lactone signal produced by *Pseudomonas aeruginosa. Proc. Natl. Acad. Sci. USA* 92:1490–94

Pearson JP, Van Delden C, Iglewski BH. 1999. Active efflux and diffusion are involved in transport of *Pseudomonas aeruginosa* cell-to-cell signals. *J. Bacteriol.* 181:1203–10

Perego M. 1997. A peptide export-import control circuit modulating bacterial development regulates protein phosphatases of the phosphorelay. *Proc. Natl. Acad. Sci. USA* 94:8612–17

Pesci EC, Milbank JB, Pearson JP, McKnight S, Kende AS, et al. 1999. Quinolone signaling in the cell-to-cell communication system of *Pseudomonas aeruginosa. Proc. Natl. Acad. Sci. USA* 96:11229–34

Pestova EV, Havarstein LS, Morrison DA. 1996. Regulation of competence for genetic transformation in *Streptococcus pneumoniae* by an auto-induced peptide pheromone and a two-component regulatory system. *Mol. Microbiol.* 21:853–62

Peterson S, Cline RT, Tettelin H, Sharov V, Morrison DA. 2000. Gene expression analysis of the *Streptococcus pneumoniae* competence regulons by use of DNA microarrays. *J. Bacteriol.* 182:6192–202

Peterson SN, Sung CK, Cline R, Desai BV, Snesrud EC, et al. 2004. Identification of competence pheromone responsive genes in *Streptococcus pneumoniae* by use of DNA microarrays. *Mol. Microbiol.* 51:1051–70

Queller DC, Strassmann JE. 2002. The many selves of social insects. *Science* 296:311–13

Rather PN, Ding X, Baca-DeLancey RR, Siddiqui S. 1999. *Providencia stuartii* genes activated by cell-to-cell signaling and identification of a gene required for production or activity of an extracellular factor. *J. Bacteriol.* 181:7185–91

Rothfork JM, Timmins GS, Harris MN, Chen X, Lusis AJ, et al. 2004. Inactivation of a bacterial virulence pheromone by phagocyte-derived oxidants: new role for the NADPH oxidase in host defense. *Proc. Natl. Acad. Sci. USA* 101:13867–72

Saenz HL, Augsburger V, Vuong C, Jack RW, Gotz F, Otto M. 2000. Inducible expression and cellular location of AgrB, a protein involved in the maturation of the staphylococcal quorum-sensing pheromone. *Arch. Microbiol.* 174:452–55

Schauder S, Shokat K, Surette MG, Bassler BL. 2001. The LuxS family of bacterial autoinducers: biosynthesis of a novel quorum-sensing signal molecule. *Mol. Microbiol.* 41:463–76

Schuster M, Lostroh CP, Ogi T, Greenberg EP. 2003. Identification, timing, and signal specificity of *Pseudomonas aeruginosa* quorum-controlled genes: a transcriptome analysis. *J. Bacteriol.* 185:2066–79

Schweitzer R, Shilo BZ. 1997. A thousand and one roles for the Drosophila EGF receptor. *Trends Genet.* 13:191–96

Seed PC, Passador L, Iglewski BH. 1995. Activation of the *Pseudomonas aeruginosa lasI* gene by LasR and the *Pseudomonas* autoinducer PAI: an autoinduction regulatory hierarchy. *J. Bacteriol.* 177:654–59

Seto H, Tomasz A. 1975. Protoplast formation and leakage of intramembrane cell components: induction by the competence activator substance of pneumococci. *J. Bacteriol.* 121:344–53

Shimkets LJ. 1999. Intercellular signaling during fruiting-body development of *Myxococcus xanthus. Annu. Rev. Microbiol.* 53:525–49

Smith RS, Iglewski BH. 2003. *P. aeruginosa* quorum-sensing systems and virulence. *Curr. Opin. Microbiol.* 6:56–60

Solomon JM, Lazazzera BA, Grossman AD. 1996. Purification and characterization of an extracellular peptide factor that affects two different developmental pathways in *Bacillus subtilis. Genes Dev.* 10:2014–24

This study, using microarray analysis, shows that the *Pseudomonas aeruginosa* quorum-sensing regulon consists of a large number of genes and defines the temporal order of their expression.

Solomon JM, Magnuson R, Srivastava A, Grossman AD. 1995. Convergent sensing pathways mediate response to two extracellular competence factors in *Bacillus subtilis*. *Genes Dev.* 9:547–58

Steinmoen H, Teigen A, Havarstein LS. 2003. Competence-induced cells of *Streptococcus pneumoniae* lyse competence-deficient cells of the same strain during cocultivation. *J. Bacteriol.* 185:7176–83

Stevens AM, Dolan KM, Greenberg EP. 1994. Synergistic binding of the *Vibrio fischeri* LuxR transcriptional activator domain and RNA polymerase to the lux promoter region. *Proc. Natl. Acad. Sci. USA* 91:12619–23

Strassmann JE, Zhu Y, Queller DC. 2000. Altruism and social cheating in the social amoeba *Dictyostelium discoideum*. *Nature* 408:965–67

Suntharalingam P, Cvitkovitch DG. 2005. Quorum sensing in streptococcal biofilm formation. *Trends Microbiol.* 13:3–6

Surette MG, Miller MB, Bassler BL. 1999. Quorum sensing in *Escherichia coli*, *Salmonella typhimurium*, and *Vibrio harveyi*: a new family of genes responsible for autoinducer production. *Proc. Natl. Acad. Sci. USA* 96:1639–44

Swartzman E, Silverman M, Meighen EA. 1992. The *luxR* gene product of *Vibrio harveyi* is a transcriptional activator of the lux promoter. *J. Bacteriol.* 174:7490–93

Taga ME, Miller ST, Bassler BL. 2003. Lsr-mediated transport and processing of AI-2 in *Salmonella typhimurium*. *Mol. Microbiol.* 50:1411–27

Taga ME, Semmelhack JL, Bassler BL. 2001. The LuxS-dependent autoinducer AI-2 controls the expression of an ABC transporter that functions in AI-2 uptake in *Salmonella typhimurium*. *Mol. Microbiol.* 42:777–93

Tenover FC, Gaynes RP. 2000. The epidemiology of *Staphylococcus aureus* infections. In *Gram-Positive Pathogens*, ed. VA Fischetti, RP Novick, JJ Ferretti, DA Portnoy, JI Rood, pp. 414–21. Washington, DC: ASM Press

Teplitski M, Robinson JB, Bauer WD. 2000. Plants secrete substances that mimic bacterial N-acyl homoserine lactone signal activities and affect population density-dependent behaviors in associated bacteria. *Mol. Plant Microbe Interact.* 13:637–48

Tomasz A, Hotchkiss RD. 1964. Regulation of the transformability of pneumococcal cultures by macromolecular cell products. *Proc. Natl. Acad. Sci. USA* 51:480–87

Town C, Stanford E. 1979. An oligosaccharide-containing factor that induces cell differentiation in *Dictyostelium discoideum*. *Proc. Natl. Acad. Sci. USA* 76:308–12

Town CD, Gross JD, Kay RR. 1976. Cell differentiation without morphogenesis in *Dictyostelium discoideum*. *Nature* 262:717–19

Travisano M, Velicer GJ. 2004. Strategies of microbial cheater control. *Trends Microbiol.* 12:72–78

Urban S, Schlieper D, Freeman M. 2002. Conservation of intramembrane proteolytic activity and substrate specificity in prokaryotic and eukaryotic rhomboids. *Curr. Biol.* 12:1507–12

Vannini A, Volpari C, Gargioli C, Muraglia E, Cortese R, et al. 2002. The crystal structure of the quorum sensing protein TraR bound to its autoinducer and target DNA. *EMBO J.* 21:4393–401

Visick KL, Foster J, Doino J, McFall-Ngai M, Ruby EG. 2000. *Vibrio fischeri* lux genes play an important role in colonization and development of the host light organ. *J. Bacteriol.* 182:4578–86

Wagner VE, Bushnell D, Passador L, Brooks AI, Iglewski BH. 2003. Microarray analysis of *Pseudomonas aeruginosa* quorum-sensing regulons: effects of growth phase and environment. *J. Bacteriol.* 185:2080–95

Wagner VE, Gillis RJ, Iglewski BH. 2004. Transcriptome analysis of quorum-sensing regulation and virulence factor expression in *Pseudomonas aeruginosa*. *Vaccine* 22(Suppl. 1):S15–20

Watson WT, Minogue TD, Val DL, von Bodman SB, Churchill ME. 2002. Structural basis and specificity of acyl-homoserine lactone signal production in bacterial quorum sensing. *Mol. Cell* 9:685–94

Whiteley M, Lee KM, Greenberg EP. 1999. Identification of genes controlled by quorum sensing in *Pseudomonas aeruginosa*. *Proc. Natl. Acad. Sci. USA* 96:13904–9

Wu H, Song Z, Hentzer M, Andersen JB, Molin S, et al. 2004. Synthetic furanones inhibit quorum-sensing and enhance bacterial clearance in *Pseudomonas aeruginosa* lung infection in mice. *J. Antimicrob. Chemother.* 53:1054–61

Xavier KB, Bassler BL. 2003. LuxS quorum sensing: more than just a numbers game. *Curr. Opin. Microbiol.* 6:191–17

Xavier KB, Bassler BL. 2005. Regulation of uptake and processing of the quorum-sensing autoinducer AI-2 in *Escherichia coli*. *J. Bacteriol.* 187:238–48

Zhang HB, Wang LH, Zhang LH. 2002a. Genetic control of quorum-sensing signal turnover in *Agrobacterium tumefaciens*. *Proc. Natl. Acad. Sci. USA* 99:4638–43

Zhang L, Murphy PJ, Kerr A, Tate ME. 1993. Agrobacterium conjugation and gene regulation by N-acyl-L-homoserine lactones. *Nature* 362:446–48

Zhang RG, Pappas T, Brace JL, Miller PC, Oulmassov T, et al. 2002b. Structure of a bacterial quorum-sensing transcription factor complexed with pheromone and DNA. *Nature* 417:971–74

Zhu J, Miller MB, Vance RE, Dziejman M, Bassler BL, Mekalanos JJ. 2002. Quorum-sensing regulators control virulence gene expression in *Vibrio cholerae*. *Proc. Natl. Acad. Sci. USA* 99:3129–34

Zhu J, Oger PM, Schrammeijer B, Hooykaas PJ, Farrand SK, Winans SC. 2000. The bases of crown gall tumorigenesis. *J. Bacteriol.* 182:3885–95

Zhu J, Winans SC. 1999. Autoinducer binding by the quorum-sensing regulator TraR increases affinity for target promoters in vitro and decreases TraR turnover rates in whole cells. *Proc. Natl. Acad. Sci. USA* 96:4832–37

Zhu J, Winans SC. 2001. The quorum-sensing transcriptional regulator TraR requires its cognate signaling ligand for protein folding, protease resistance, and dimerization. *Proc. Natl. Acad. Sci. USA* 98:1507–12

This study supports the hypothesis that autoinducer binding is required for protein stability of LuxR-like AHL receptors and suggests a mechanism for how bacteria avoid premature activation of their quorum-sensing cascades.

Pushing the Envelope: Structure, Function, and Dynamics of the Nuclear Periphery

Martin W. Hetzer,[1] Tobias C. Walther,[2] and Iain W. Mattaj[3]

[1]Molecular and Cell Biology Laboratory, Salk Institute for Biological Studies, La Jolla, California 92037; email: hetzer@salk.edu

[2]Department of Biochemistry and Biophysics, University of California, San Francisco, California 94143-2200; email: twalther@itsa.ucsf.edu

[3]Gene Expression Program, European Molecular Biology Laboratory, 69115 Heidelberg, Germany; email: mattaj@embl-heidelberg.de

Annu. Rev. Cell Dev. Biol.
2005. 21:347–80

First published online as a
Review in Advance on
June 28, 2005

The *Annual Review of
Cell and Developmental
Biology* is online at
http://cellbio.annualreviews.org

doi: 10.1146/
annurev.cellbio.21.090704.151152

Copyright © 2005 by
Annual Reviews. All rights
reserved

1081-0706/05/1110-
0347$20.00

Key Words

nuclear envelope, chromatin, nuclear pore complex, lamina, membrane fusion, Ran GTPase

Abstract

The nuclear envelope (NE) is a highly specialized membrane that delineates the eukaryotic cell nucleus. It is composed of the inner and outer nuclear membranes, nuclear pore complexes (NPCs) and, in metazoa, the lamina. The NE not only regulates the trafficking of macromolecules between nucleoplasm and cytosol but also provides anchoring sites for chromatin and the cytoskeleton. Through these interactions, the NE helps position the nucleus within the cell and chromosomes within the nucleus, thereby regulating the expression of certain genes. The NE is not static, rather it is continuously remodeled during cell division. The most dramatic example of NE reorganization occurs during mitosis in metazoa when the NE undergoes a complete cycle of disassembly and reformation. Despite the importance of the NE for eukaryotic cell life, relatively little is known about its biogenesis or many of its functions. We thus are far from understanding the molecular etiology of a diverse group of NE-associated diseases.

Contents

INTRODUCTION

More than 50 years ago, electron microscopy showed that the NEs of amphibian oocytes consist of a double lipid bilayer (Callan & Tomlin 1950). In the following decades, a variety of techniques revealed the complex and unique structure of this membrane system. The NE is composed of two concentric membranes joined with one another at the sites of NPCs. The ONM is not only continuous with the rER, but is thought, with a small number of exceptions discussed below, to harbor an identical set of integral proteins (Gerace & Burke 1988, Newport & Forbes 1987). The INM hosts a unique set of integral and associated proteins that provide attachment sites for the lamina and chromatin (Burke & Stewart 2002). NPCs are very large protein complexes that span the ONM and INM and form gated channels through which nucleocytoplasmic transport occurs (Görlich & Kutay 1999, Vasu & Forbes 2001, Wente 2000) (**Figure 1**).

Textbook illustrations of the NE often imply a static architecture. Whereas some components are indeed stably associated with the NE, others—for example, some nucleoporins

NE: nuclear envelope

NPC: nuclear pore complex

ONM: outer nuclear membrane

Rough ER (rER): the presence of ribosomes bound to the cytosolic side of the endoplasmic reticulum membrane gives the rER its "rough" appearance and indicates that it is involved in protein synthesis

INM: inner nuclear membrane

MDa: megadalton

STEM: scanning transmission electron microscope

EM: electron microscope

(NPC proteins)—are dynamic and can bind to and dissociate from the NE within seconds (Daigle et al. 2001, Griffis et al. 2003, Rabut et al. 2004). Certain proteins associate with the NE only at specific times. For example, mitotic checkpoint control proteins move between kinetochores and the NPC, and nucleoporins physically interact with both kinetochores and the spindle during mitosis (Belgareh et al. 2001, Campbell et al. 2001, Iouk et al. 2002, Joseph et al. 2004, Loiodice et al. 2004, Salina et al. 2003).

The NE has become a major area of research activity and recent proteomics studies have provided a long list of potential NE proteins (Dreger et al. 2001, Schirmer et al. 2003), several of which, when mutated, lead to pathological manifestations (Burke & Stewart 2002, Worman & Courvalin 2003). Analysis of the molecular functions of NE components, however, barely has begun. We briefly discuss interphase NE structure and function before concentrating on NE dynamics.

The Nuclear Pore Complex

The NPC is a very large protein assembly with an estimated mass of between 60 and 125 MDa in vertebrates (Cronshaw et al. 2002, Reichelt et al. 1990) and 44 and 66 MDa in *Saccharomyces cerevisiae* (Rout & Blobel 1993, Rout et al. 2000, Yang et al. 1998). The higher and lower estimates are based on STEM and biochemical purification, respectively. The striking differences in these estimates may be explained by inclusion of soluble transport receptors and their cargo in the EM NPC preparations or by the loss of nucleoporins during subcellular fractionation.

Structural analyses of the NPC (**Figure 1**) have been continuously refined over the last decade using transmission and scanning electron microscopy, atomic force microscopy, and most recently, cryoelectron tomography (Akey 1989, Beck et al. 2004, Goldberg & Allen 1993, Stoffler et al. 2003). Three-dimensional reconstructions have been calculated to ~8 nm resolution from isolated NPCs

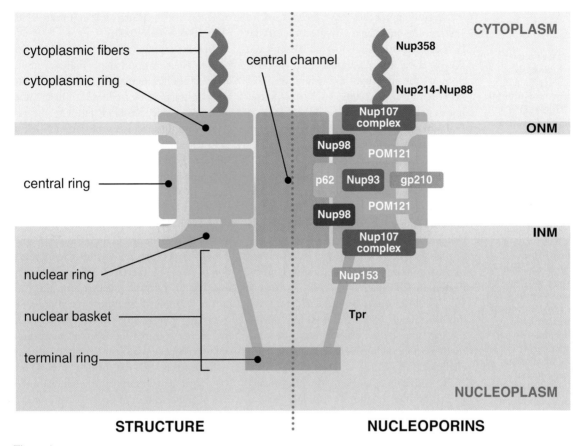

Figure 1

Schematic cross-section of a vertebrate NPC (*left*). Model of NPC architecture. Shown are the cytoplasmic, central and nuclear ring (*green*). INM and ONM (*yellow*) are joined at the sites of NPC insertion. The lumenal space between INM and ONM is continuous with the ER (not shown). Cytoplasmic and nuclear filaments are attached; the latter form a basket structure closed by a distal ring (*right*). Nucleoporins are shown within the ring, filament, and basket structures. The color code represents the three dynamic classes of nucleoporins (Rabut et al. 2004): structural scaffold (*blue*), mobile (*purple*), and highly dynamic (*orange*).

or intact nuclei (Beck et al. 2004) and have been aided by the conservation of the overall structure of NPCs between species. Early estimates suggested that NPCs contain up to 100 different proteins; although the overall number of individual protein subunits in a single pore is indeed several hundred, this is due to the presence of multiple copies of roughly 30 distinct nucleoporins per NPC (Rout et al. 2000, Cronshaw et al. 2002). Because the NPC is eightfold radially symmetrical (Maul et al. 1971), each protein is thought to be present in at least eight copies; many proteins, however, are present at higher copy number (Cronshaw et al. 2002, Rabut et al. 2004, Rout et al. 2000). Although the composition of the NPC is expected to be similar in all tissues, there appear to be subtle differences between cell types. For instance, the transmembrane nucleoporin gp210 is undetectable in many mouse cell types (Olsson et al. 2004). A cell type-specific role also has been proposed for the human nucleoporin ALADIN (Cronshaw & Matunis 2003, Cronshaw et al. 2002).

WD-40 repeat
proteins: short ~40
amino acid motifs,
often terminating in
a Trp-Asp (W-D)
dipeptide; have four
to 16 repeating units,
all of which are
thought to form a
circularized
beta-propeller
structure. A large
family found in all
eukaryotes, these
proteins coordinate
multiprotein
complex assemblies,
where the repeating
units serve as a rigid
scaffold for protein
interactions in
functions ranging
from signal
transduction and
transcription
regulation to cell
cycle control and
apoptosis

RanGAP: Ran
GTPase-activating
protein

ATP: adenosine
triphosphate

Several approaches have been used to analyze the molecular architecture of the NPC. Systematic immuno-gold labeling and EM analysis have assigned relative positions to the nucleoporins within the yeast NPC (Rout et al. 2000); a subset of vertebrate nucleoporins has been similarly mapped (Fahrenkrog et al. 2000). Analysis of the in vitro binding partners of recombinant nucleoporins, coupled with previous analyses of individual nucleoporins and of genetic interactions between nucleoporins, have led to an interaction map of the NPC (Allen et al. 2002, Huang et al. 2002). The yeast Nup84 complex, which consists of seven nucleoporins, has been successfully reconstituted in vitro (Siniossoglou et al. 2000). Higher-resolution information on NPC structure will assist enormously in elucidating the mechanism of NPC translocation during nuclear import or export, but obtaining such information represents a huge technical challenge.

By convention, nucleoporins are usually named by their predicted molecular weight, as in the case of yeast and vertebrates (e.g., yNup84, Nup107). However, there are many exceptions (**Table 1**). Based on their primary structure, nucleoporins can be classified into FxFG-, GLGF-, and WD-40 (β-propeller) repeat–containing and nonrepeat nucleoporins. The first two classes are thought to be generally involved in NPC translocation (Bayliss et al. 2000, 2002, Iovine et al. 1995, Strawn et al. 2004), whereas the others are more likely to play an architectural role in NPC structure (Allen et al. 2001, Rabut et al. 2004).

NPCs are the exclusive site of nucleocytoplasmic transport (Görlich & Kutay 1999, Wente 2000). Although transport is a major NE-related function, we do not discuss it in detail here. A model for most types of nuclear transport has emerged, and the role of transport receptors of the Importin β family and the Ran GTPase with its regulators (**Figures 2** and **3**) is generally accepted and has been the subject of major reviews (Görlich & Kutay 1999, Mattaj & Englmeier 1998, Weis 2003, Wente 2000).

In this simplified model, import receptors bind to their cargo in the cytoplasm, interact specifically with the NPC, and translocate through the pore by facilitated diffusion. On the nuclear side, RanGTP binds to the receptor and thus displaces the cargo. Nuclear export is almost the mirror image of import (**Figure 3**). The localization of RanGTP is high only in the nucleus because its exchange factor RCC1, which allows replacement of GDP for GTP on Ran, is a chromatin protein. In addition, GTP is hydrolyzed by Ran only with help from the cytoplasmic RanBP1/2 and RanGAP1 (**Figure 2**). Asymmetric localization of RanGTP therefore defines the directionality of transport mediated by the Importin β family of transport receptors and, by making translocation irreversible, provides the energy required for cargo accumulation in the target compartment.

In contrast to the overall model for Ran-dependent transport, the actual process of translocation of cargo-receptor complexes through the NPC is still subject to lively debate. Unlike many other intracellular transport processes, translocation of transport receptor-cargo complexes through the NPC occurs without ATP or GTP hydrolysis (Englmeier et al. 1999, Ribbeck et al. 1999). In one model for the NPC translocation process, increasing affinities of nucleoporins for receptor/cargo complexes along the axis of the NPC contribute to directionality (Ben-Efraim & Gerace 2001). Because the direction of transport can be reversed, and because NPC translocation of the same transport complex in the forward and reverse directions can occur at comparable rates in vivo under normal conditions, this model can, at best, account for only a minor contribution to directionality (Becskei & Mattaj 2003, Nachury & Weis 1999, Zeitler & Weis 2004).

Other models assume random diffusion as the basis of translocation. In one, Brownian motion of the cytoplasmic filaments

Table 1   Vertebrate, yeast, worm and fly nucleoporins are listed[*]

Nucleoporins in different species			
Vertebrate	*S. cerevisiae*	*C. elegans*	*D. melanogaster*
Nup358		npp-9	Nup358
Nup214	yNup159	npp-14	DNup214
gp210	—	npp-12	Gp210
Nup205	yNup192	npp-3	CG11953
Tpr	Mlp1/Mlp2	R07G3.3	—
Nup188	yNup188	—	—
Nup160	yNup120	npp-6	CG4738
Nup155	yNup170/yNup157	npp-8	CG4579
Nup153	yNup1/2, yNup60	npp-7	Nup153
Nup133	yNup133	npp-15	CG6958
POM121	—	—	—
Nup107	yNup84	npp-5	CG6743
Nup98	yNup145-N	npp-10	Nup98
Nup96	yNup145-C	npp-10	Nup96
Nup93	yNic96	npp-13	CG11092
Nup88	yNup82	—	Mbo
Nup85/75	—	npp-2	—
p62/Nup62	yNsp1	npp-11	Nup62
Nup58	yNup49	—	Nup58
Nup54	yNup57	npp-1	Nup54
Nup50	—	npp-16	—
Nup45	—	npp-4	—
Nup43	—	C09G9.2	CG7671
Nup37	—	—	CG11875
Nup35	—	npp-19	—
Sec13	ySec13	npp-20	CG6773
Seh1	ySeh1	npp-18	CG8722
RAE1	—	npp-17	—
Gle1	yGle1	—	—
Gle2	yGle2 (Rae1)	—	—
hCG1	yNup42 (Rip)ddNdc1	—	—
—	POM152, Nup60, POM34	—	—
—	Ndc1, yNup85, yNup53, yNup59	—	—
—	yNup59, cdc31	—	—

[*]The vertebrate nucleoporins are listed according to their molecular weight. The *S. cerevisiae, C. elegans, and D. melanogaster* homologs are listed. Drosophila nucleoporins that have not been annotated but show clear homology are listed with flybase ORF gene nomenclature. A Tpr homolog was identified in *C. elegans* (R07G3.3).

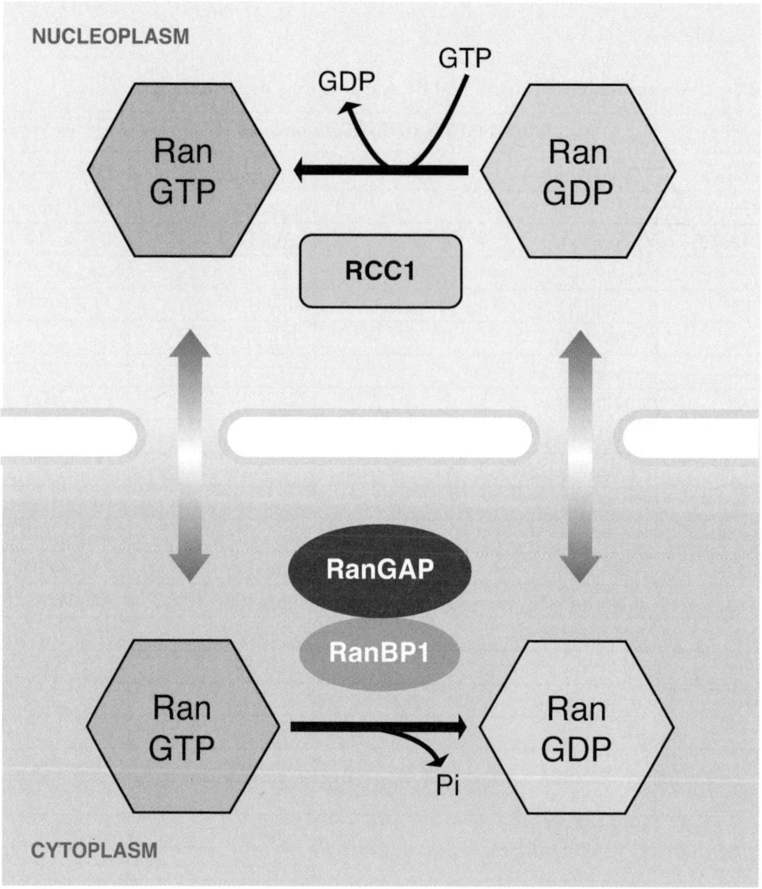

Figure 2

The Ran GTPase cycle. The asymmetric, compartmentalized distribution of the regulators of Ran (*red*) confines RanGTP mainly to the nucleus and RanGDP mainly to the cytoplasm. Ran's exchange factor, RCC1 (*pink*), is chromatin bound. When Ran leaves the nucleus, two Ran-binding proteins (RanBP1 and RanBP2 = Nup358) cooperate with the Ran GTPase-activating protein (RanGAP) to induce GTP hydrolysis by Ran.

or the FG–repeat containing segments of nucleoporins is proposed to exclude non-transported proteins from the central NPC channel by forming an entropic barrier, whereas weak interactions with nucleoporins favor translocation of proteins that bind to them (Rout et al. 2003). Once proteins reach the central channel, they can move through it by diffusion to the opposite face of the NPC, where either RanGTP binding, for import, or RanGTP removal, for export, terminates the transport event (**Figure 3**).

Another suggestion is that the NPC works by hydrophobic exclusion of soluble proteins. Nucleoporins are proposed to form a mesh-work via weak interactions between their FG-repeats. Hydrophilic proteins are excluded from this network, whereas transport receptors dissolve into and sever it by interacting with the FG-repeats. Predictions from this model have been simulated and agree with experimental data for simple cargoes (Gorlich et al. 2003, Ribbeck & Gorlich 2002). Support for this idea also comes from single molecule analyses of translocation, which conclude that a cargo spends most of its 10 ms interaction time with the NPC in a random walk in the central channel (Kubitscheck et al. 2005, Yang et al. 2004), and from findings that asymmetrically localized nucleoporins are dispensable for nucleoplasmic transport in vitro and in vivo (Strawn et al. 2004, Zeitler & Weis 2004). Comparison of the measured rate of NPC translocation for single molecules with the saturated transport

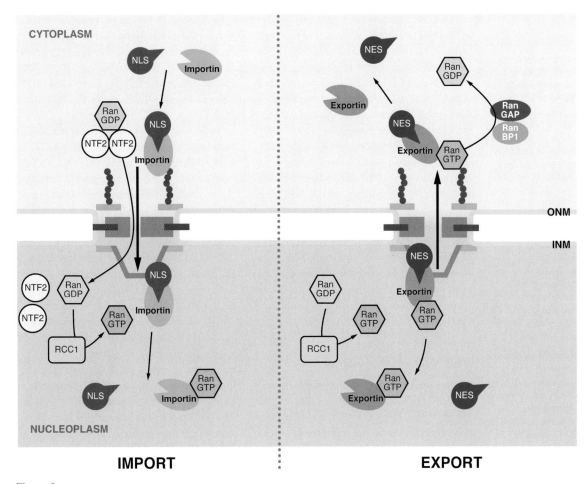

CYTOPLASM

NUCLEOPLASM

IMPORT   **EXPORT**

ONM

INM

Figure 3

Import and export cycles mediated by Importins (*right*) and Exportins (*left*). Importins (*green*) bind to cargo molecules (NLS for nuclear localization signal) in the cytoplasm and mediate interactions with the NPC to translocate into the nucleus (*left*). RanGTP in the nucleus binds to Importin and induces cargo release. The Importin-RanGTP complex is then recycled to the cytoplasm. The export cycle is similar but with the crucial difference that RanGTP induces cargo binding to an export receptor in the nucleus (*right*).

rate suggests that as many as 10 molecules simultaneously can traverse an NPC (Yang et al. 2004).

Some observations are difficult to explain by any current translocation model. The observed specificity of inhibition of particular transport pathways after deletion or depletion of individual nucleoporins is one (Nehrbass et al. 1993, Strawn et al. 2004). Another is the recent finding that overlapping sets of FG-repeats from different nu-

cleoporins, up to and including over 50% of the mass of the FG-repeats in the yeast NPC, can be deleted without any apparent effect on the NPC exclusion diameter and without causing major defects in nucleocytoplasmic transport (Strawn et al. 2004). The second observation represents a challenge both to understanding how the NPC functions during exclusion of nontransported proteins and during translocation of transport substrates.

Fluorescence
recovery after
photobleaching
(FRAP): a
technique in which a
pool of fluorescent
molecules is
destroyed locally by
high-intensity laser
irradiation. After this
"photobleach," the
exchange of the
nonfluorescent
molecules with the
surrounding
fluorescent
molecules is
monitored and
measured as
"recovery" of
fluorescence in the
bleached area

Green fluorescent
protein (GFP): a
spontaneously
fluorescent protein
isolated from
coelenterates such as
the Pacific jellyfish,
Aequoria victoria

Centrosome: the
microtubule-
organizing center; it
contains the
centrioles, which
anchor the minus
ends of microtubules

The NPC traditionally has been perceived as a static channel, at least during interphase. This view has had to be revised following several analyses of nucleoporin dynamics employing FRAP. In a systematic study, Rabut et al. (2004) demonstrated that 19 GFP-tagged nucleoporins may be divided into roughly three categories based on the stability of their association with the NPC. The most stable nucleoporins (with a residence time of >35 h) were suggested to be architectural components of the NPC scaffold; those of intermediate stability (with residence times between 2.5 and 30 h) may be involved in interactions with complexes undergoing transport, whereas the most dynamic (with residence times of seconds or minutes) may have functions distinct from their roles at the NPC (Rabut et al. 2004).

The Nuclear Membranes

As mentioned above, the protein composition of the ONM is very similar to that of the rER. Recent evidence nevertheless demonstrates that some proteins specifically locate to the ONM and, via interactions bridged by ONM protein-INM protein interactions, can connect the peripheral nucleoskeleton to the cytoskeleton. Good examples of such proteins are UNC-83 and ZYG-12 from *Caenorhabditis elegans*, which connect to the microtubule cytoskeleton and centrosomes, respectively (Lee et al. 2002, Malone et al. 2003). UNC-83 interacts with the SUN family protein UNC-84, and ZYG-12 with SUN1 (or matefin). Both SUN proteins are proposed to be located in the INM. All these proteins are required for nuclear positioning and nuclear movement during early *C. elegans* development (Fridkin et al. 2004). The SUN family of proteins is conserved in metazoa, and, aside from interactions with microtubule-binding ONM proteins, may also help anchor actin filament-binding proteins of the nesprin family (Zhang et al. 2001). In this way, a nuclear periphery–INM-

ONM–cytoskeleton connection is created that is responsible for nuclear positioning, nuclear movement, and possibly additional functions (Apel et al. 2000, Malone et al. 2003, Padmakumar et al. 2004, Starr et al. 2001, Starr & Han 2002, Zhang et al. 2001, Zhen et al. 2002).

Interations involving the nesprin family members Syne, Myne, and NUANCE illustrate that NE proteins also fulfill a specialized role in establishing and maintaining the three-dimensional architecture of certain differentiated cells (Mislow et al. 2002a,b). These large integral NE proteins contain several spectrin repeats, a C-terminal transmembrane domain, and an actin-binding motif. Some nesprins bind both lamin A and emerin in vitro and are proposed to locate to the INM (Apel et al. 2000, Zhang et al. 2001). In skeletal and cardiac smooth muscle, the binding of nesprin-1 to both the nuclear lamina and cytoplasmic actin is required for correct, stable nuclear positioning within the cell. Loss of the *Drosophila* or *C. elegans* nesprin homologues (MSP-300 and ANC-1, respectively) also affects nuclear positioning (Starr & Han 2002, Zhang et al. 2002). The *Drosophila* nesprin family member Klarsicht, whose activity is required for eye development, does not have a cytoplasmic actin-binding domain. Instead, like ZYG-12 or UNC83, Klarsicht regulates the movement of photoreceptor nuclei by linking the microtubule organizing center to the nuclear lamina (Fischer-Vize & Mosley 1994, Patterson et al. 2004). Connection between the NE and the cytoskeleton is the subject of an excellent recent review (Gruenbaum et al. 2005). Proteins of the INM (**Figure 4**) are discussed below.

MITOSIS

Nuclear Envelope Breakdown

The NE's dynamic nature is particularly obvious in metazoa, in which the whole nucleus breaks down and reforms during cell division.

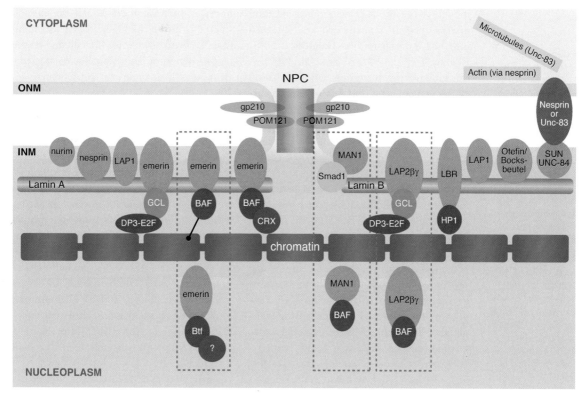

Figure 4

Schematic illustration of the NE. INM proteins are grouped into Lamin A and Lamin B binding proteins, and interactions with transcription and chromatin regulators are indicated. Chromatin is shown in blue. Proteins are grouped as INM proteins (*orange*), lamins (*gray*), chromatin (*blue*), and transcription regulators (as indicated). See text for details.

In vivo and in vitro systems, most of the latter recapitulating pronuclear formation, have been used to study nuclear disassembly and assembly. At the onset of mitosis, during the prophase-prometaphase transition, the NE breaks down (reviewed in (Burke & Ellenberg 2002)). Open mitosis has dramatic consequences for the dividing cell. The components of the nucleoplasm and the cytoplasm are mixed as the interphase compartmentalization disappears in order to allow the establishment of a mitotic spindle apparatus sufficiently large to capture and segregate all the duplicated chromosomes.

NEBD has been investigated by transmission electron microscopy (Roos 1973, Zatsepina et al. 1976, Zeligs & Wollman 1979) and fluorescence microscopy

(Beaudouin et al. 2002, Ellenberg et al. 1997). Based on these studies, two very different models for NEBD have been proposed. According to the first, the NE membranes break down into vesicles distinct from the intact mitotic ER network. This vesiculation model is supported by findings that membrane vesicles enriched in NE proteins can be isolated from embryonic extracts, such as those of *Xenopus* or sea urchin egg, that are widely used to study NE assembly (Collas & Courvalin 2000, Sasagawa et al. 1999, Vigers & Lohka 1991). Studies of the movement of transmembrane INM and ER proteins in intact mitotic mammalian cells, however, have revealed that these two protein classes are located in the same mitotic tubular network, leading to a second

NEBD: nuclear envelope breakdown

Endoplasmic
reticulum (ER): an
interconnected
network of tubules,
vesicles, and sacs that
serves specialized
cellular functions,
including protein
synthesis,
sequestration of
calcium, production
of steroids, storage
and production of
glycogen, and
insertion of
membrane proteins

COPI AND COPII
vesicles: traffic
between the
endoplasmic
reticulum (ER) and
the Golgi complex,
COPI primarily
from the Golgi to
the ER and between
Golgi cisternae, and
COPII from the ER
to the Golgi

model for NEBD: NE membranes and their integral protein constituents are absorbed into the ER during mitosis (Daigle et al. 2001, Ellenberg et al. 1997, Yang et al. 1997). A third model, which attempts a reconciliation of the two concepts and sets of data, is that NE proteins are indeed redistributed into the ER, but remain segregated from bulk ER proteins in microdomains in the reticular network (Collas & Courvalin 2000, Mattaj 2004). Although the last model is speculative, it provides a reasonable reconciliation of most of the published literature on both NE membrane vesicle fraction and live cell imaging of NE proteins.

The disassembly of the NE membrane, with the dispersal of nucleoporins, lamins, and INM proteins, is a key event for mitotic progression in metazoa. Although most NE components are dispersed throughout the cytoplasm or ER, there are interesting exceptions. Certain nucleoporins, such as the Nup107-160 complex, associate with kinetochores during cell division (Belgareh et al. 2001, Enninga et al. 2003, Loiodice et al. 2004, Salina et al. 2003). Nup358, in association with RanGAP, relocates to kinetochores and specific regions of the mitotic spindle (Joseph et al. 2002). siRNA-mediated depletion of Nup358 was recently shown to inhibit kinetochore assembly and chromosome segregation, which suggests that Nup358 plays an essential role in kinetochore function and maturation (Joseph et al. 2004, Salina et al. 2003). A model that can be formulated based on these results is that nucleoporin relocalization to kinetochores is part of a signal that "reports" on NPC breakdown and is also essential for mitotic maturation of kinetochores. If kinetochore assembly is dependent on the mitotic recruitment of otherwise stable NPC components, then the formation of this structure will be temporally restricted to phases of the cell cycle in which NPCs disassemble. The targeting of the kinetochore and spindle checkpoint proteins Mad1 and Mad2 to the NPC in interphase may serve an analogous function (Campbell et al. 2001, Iouk et al. 2002). The

interplay between the NE and the mitotic spindle is further highlighted by the finding that NE breakdown is itself facilitated by the tearing action of centrosome-associated microtubules (Beaudouin et al. 2002, Salina et al. 2002).

Another suggested link of potential interest for NEBD is that between Nup153 and components of the COPI coatomer complex (Liu et al. 2003b), which mediates retrograde transport from the Golgi to the ER (Lee et al. 2004). Nup153 may recruit the COPI complex to the NE membrane, where it participates in NE vesiculation. Antibodies against β-COP, one of the COPI subunits, also inhibited NEBD. However, because Nup153 is on the nuclear face of the NPC, whether Nup153 can interact with COPI before NEBD occurs is uncertain (Liu et al. 2003b). Furthermore, as discussed above, it is not clear that vesiculation is required for NEBD. How Nup153 affects NEBD and what role COPI plays in the process is a subject of further study.

Analysis of NEBD in starfish oocytes has led to identification of two phases of NE permeabilization: first, a relatively slow progressive dismantling of the NPC that leaves the NE largely intact but allows passive access to the nucleoplasm of large cytoplasmic proteins, followed by a rapid breakdown of the NE. The latter phase possibly is caused by the fenestration of the nuclear membrane induced by the complete removal of one or a few NPCs (Lenart et al. 2003). Similarly, the cytoplasmic fibers of the NPC are removed while the NE remains intact in *Drosophila* (Kiseleva et al. 2001), which suggests that NPC destabilization generally may occur early during NEBD and could trigger later steps in membrane disassembly.

Even some yeasts that were generally thought to maintain an intact NE throughout their so-called closed mitosis exhibit at least partial mitotic NPC disassembly, potentially analogous to the first stage of metazoan NEBD. In the filamentous fungus, *A. nidulans*, the NPC undergoes significant disassembly under the influence of two mitotic kinases, NIMA and Cdk1 (De Souza et al. 2004). Several normally cytoplasmic proteins, including tubulin, gain access to the nucleoplasm and chromosomes during mitosis, owing to this partial NPC disassembly (De Souza et al. 2004). It will be interesting to examine other single-celled eukaryotes to determine the number of distinct types of "closed" mitosis, as well as the similarity of the mechanisms employed in mitotic NPC remodeling in fungi to those used during NE disassembly in metazoa.

NE Reformation at the End of Mitosis

A key event at the end of mitosis is the molecular "unmixing" of nucleoplasmic and cytoplasmic components and the re-formation of the nuclear compartment around the segregated chromosomes. This process involves the coordination of membrane fusion and the assembly of massive protein complexes (NPCs and the nuclear lamina). The process of NE formation is regulated, at least in part, by proteins that have well-defined functions in interphase. The GTPase Ran, the major regulator of nucleocytoplasmic transport, is required both for NE membrane fusion and NPC assembly steps [reviewed in (Hetzer et al. 2002)]. Importantly, Ran executes its role using the same mechanism and the same molecular players with which it interacts during nucleocytoplasmic transport: the import and export receptors of the Importin β family (**Figure 3**) (Harel et al. 2003a, Walther et al. 2003b). Ran's nucleotide exchange factor, RCC1 (**Figure 2**) (Bischoff & Ponstingl 1991), which allows conversion of Ran to its active, GTP-bound form, is a chromatin-bound protein (Nemergut et al. 2001). This has led to the model that local generation of RanGTP around chromatin acts as a spatial signal that directs the mitotic spindle and the NE to form in the correct place (Hetzer et al. 2002). This concept is supported further by the finding that Ran can bind chromatin in vitro and in vivo during mitosis

(Bilbao-Cortes et al. 2002, Hinkle et al. 2000, 2002) and by the visualization of a region of high RanGTP concentration surrounding chromatin in egg extract (Kalab et al. 2002).

This form of RanGTP's spatial regulation of NE assembly is supplemented by temporal regulation, which ultimately stems from the cyclin-dependent kinases that drive progression through mitosis. Many NE proteins, including nucleoporins, lamins, and INM proteins, are phosphorylated by p34/cdc2 and other downstream kinases (Courvalin et al. 1992, Enoch et al. 1991, Macaulay et al. 1995, Peter et al. 1990). Phosphorylation is assumed to participate in the disruption of protein-protein interactions that hold the NE together, although as yet there is sparse evidence that phosphorylation is a cause, rather than a consequence, of NE disassembly. Among the best examples are the *Aspergillus* studies in which the NIMA and cdk1 kinases have been clearly implicated in causing mitotic NPC rearrangement (De Souza et al. 2004). It is assumed that NE assembly will require removal of the mitotic phosphates, but there is currently no detailed information on this, nor is there identification of phosphatases that function in NE assembly.

NPC Assembly

The organization of 30 different proteins into an eight-fold symmetrical structure that is embedded in a double membrane is a fascinating example of self-assembly. In multicellular eukaryotes, NPC assembly occurs during two different cell cycle phases. First, NPCs are reassembled at the end of mitosis (e.g., Bodoor et al. 1999, Ellenberg et al. 1997). Second, newly synthesized nucleoporins are inserted into the intact NE during S-phase, which results in a doubling of the number of NPCs prior to the next cell division (e.g., Maul et al. 1972). In *S. cerivisiae*, where the NE remains intact during mitosis, NPC assembly occurs during the entire cell cycle, with the highest assembly activity observed at G1/S and G2/M (Winey et al. 1997,

Mutvei et al. 1992). It should be noted that much of our current mechanistic information comes from analysis of NPC reassembly during anaphase/telophase, and it is not clear to what extent interphase NPC assembly is similar.

Theoretically, several mechanisms can be envisioned for NPC formation in a closed membrane sheet. NPCs may be generated and inserted de novo, or existing NPCs may grow and divide. Both mechanisms require membrane fusion (**Figure 5**). In the first case, the fusion has to occur between the INM and ONM, presumably requiring a mechanism of communication between the two and their associated proteins, as well as a mechanism to prevent the hole formed by fusion of the INM and ONM expanding catastrophically. The lumenal leaflets of the INM and ONM would touch first. One might imagine that the INM/ONM fusion events are strictly coupled to NPC assembly and insertion. Alternatively, INM/ONM fusion may occur spontaneously and either reverse or, if an NPC precursor (see below) is present, be coupled

to NPC insertion. In the second case of NPC division, membrane fusion occurs at the pore membrane domain of the NPC, essentially dividing one pore into two, and first contact is via the outer, cytoplasmic leaflets of the NE membranes (**Figure 5**). To date, there have been no experiments that distinguish between these possibilities. Interestingly, NPCs with more than eightfold symmetry are observed (Hinshaw & Milligan 2003). These may represent intermediates of NPC growth and division but may also result from rare assembly mistakes.

Yeast, where NPCs always insert into an existing NE, and whose genetic manipulability is legendary, in principle should represent the system of choice to study interphase NPC assembly. It has proven difficult to identify an assay where phenotypes definitively can be assigned to defects in NPC assembly, but some progress has been made. Cells depleted of the nucleoporin Nsp1, or containing a nonfunctional *nic96* allele, have significantly reduced pore number, which is suggestive of a defect in NPC assembly (Mutvei et al. 1992, Zabel et al.

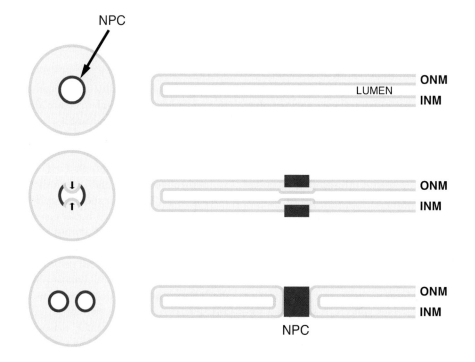

Figure 5

Two models for NPC assembly into an existing NE. New NPCs may be generated by splitting of a preexisting NPC (*purple circle*). Fusion occurs between the outer, cytoplasmic leaflets of the NE membranes (*left*). NPCs may insert into the NE and require fusion of ONM and INM (*right*).

1996). Yeast genetics has also led to the identification of a group of genes that encode components of the secretory pathway–including *SEC1*, *SEC18*, *SEC13*, *SEC23*, *SEC27*, and *BET*–that are required for the maintenance of normal NE morphology and NPC distribution (Nanduri et al. 1999, Ryan et al. 2003). This common requirement, and the connection between the secretory pathway and the NPC, the latter phenomenon represented by the fact that Sec13 is a component of both the NPC and the COPII complex, have prompted the suggestion that secretory vesicles might be involved in NPC assembly into the NE. However, a recent study, while confirming that mutations that disrupt transport between the ER and Golgi lead to a severe distortion of the NE, suggested that defects in the secretory pathway had no effect on NPC assembly per se. This suggests that ER organization is the primary defect observed in this and earlier studies, while the NE and NPC defects are secondary (A.S. Madrid, J.M. McCaffery & K. Weis, private communication).

Postmitotic assembly of NPCs at the end of mitosis has been studied in more detail. Starting at the end of anaphase II, membranes are recruited to chromatin (Daigle et al. 2001, Haraguchi et al. 2000). Time course studies using immunofluorescence or GFP-tagged nucleoporins have revealed that the recruitment of nucleoporins to the reforming nucleus occurs sequentially in vertebrate cells (Belgareh et al. 2001, Bodoor et al. 1999, Daigle et al. 2001, Haraguchi et al. 2000). Among the earliest nucleoporins found at the chromatin surface are the members of the Nup107-160 complex, as well as Nup358, Nup153, and Pom121. These are followed by the Nup62 complex and the cytoplasmically oriented Nup214 and Nup88. The second transmembrane nucleoporin, gp210, arrives at the NPC very late in assembly, at about the same time as the nuclear basket constituent Tpr. This time course suggests that the NPC is built in stages, ending with the cytoplasmic face and the nuclear basket. The bulk of the nuclear lamins arrive only after the establishment of a closed NE and lamins can be removed or their assembly prevented without preventing NE or NPC assembly (Daigle et al. 2001, Newport et al. 1990, Steen & Collas 2001). Nuclei that lack lamins are functionally defective, however, and have an abnormal NE composition.

Recruitment of nucleoporins to chromatin in the *Xenopus* egg extract nuclear assembly system also occurs sequentially. In particular, the Nup107-160 complex associates with chromatin early, and can do so in the absence of membranes (Walther et al. 2003a). This association is required for the recruitment of certain additional nucleoporins to the chromatin surface. Depletion of the Nup107-160 complex results in the formation of nuclei with a closed envelope, but without NPCs (Harel et al. 2003b, Walther et al. 2003a). Consistent with these results, siRNA-mediated depletion of components of the Nup107-160 complex from mammalian cultured cells leads to a reduction in NPC number (Boehmer et al. 2003, Loiodice et al. 2004, Walther et al. 2003a).

When extracts containing the Nup107-160 complex were added to preassembled nuclei devoid of NPCs, no incorporation of NPCs was observed, which shows that the complex needs to have access to the inner, chromatin-associated, face of the NE in order to promote NPC assembly (Walther et al. 2003a). This data led to the proposal of a model in which the Nup107-160 complex binds to chromatin independent of membranes and forms a "pre-pore" there. This, in turn, recruits other nucleoporins and permits further steps of NPC assembly. The pore membrane domain is suggested to form around the pre-pore structure, obviating the requirement for NPC-specific NE membrane fusion events (Walther et al. 2003a). This model follows an earlier suggestion that pre-pore structures form on chromatin before membrane assembly is based on the observation, by electron microscopy, of ring-like structures on sperm chromatin during nuclear assembly (Maul 1977, Sheehan et al. 1988).

Nuclear basket: a fishtrap-like structure on the nuclear side of the nuclear pore that is made up of eight fibrils joined by a distal ring

Regulation of NPC assembly. The mechanism of Nup107-160 targeting to chromatin is currently unknown. However, some insight has been obtained into the regulation of early steps of NPC assembly. The activated, GTP-bound form of Ran stimulates the recruitment of nucleoporins to chromatin in the absence of membranes. Specifically, RanGTP increases the recruitment of Nup107-160 complex and nucleoporins Nup153 and Nup358/RanBP2 was increased by RanGTP. In addition, in the presence of membranes and absence of chromatin, RanGTP uncouples the formation of NPCs from chromatin and induces the formation of arrays of NPC-dense membrane stacks in the ER (Walther et al. 2003b). This latter effect of RanGTP is mediated by the transport receptor Importin β, which acts as a negative regulator of NPC assembly and whose inhibitory activity can be reversed by RanGTP (Harel et al. 2003a, Walther et al. 2003b).

From these experiments, Importin β can be placed into the model for the early steps of NPC formation described above. Removal of bound Importin β from the complex may facilitate targeting of the Nup107-160 complex to chromatin as well as regulate the recruitment of additional FxFG-containing nucleoporins. By covering and preventing the aggregation of nucleoporin interaction surfaces before the appropriate time and place for NPC assembly is reached, Importin β thus may play a role in NPC assembly analogous to its function as a chaperone for some newly translated proteins during interphase (Jakel et al. 2002).

In vivo data consistent with the roles of RanGTP and Importin β in NE assembly exist (see below), but are more difficult to interpret due to the many functions of both Ran and Importin β. Nevertheless, dsRNA-mediated depletion of the Ran system components or of Importin β leads to defects, including formation of extranuclear clusters of nucleoporins and NE abnormalities (Askjaer et al. 2002, Bamba et al. 2002, Walther et al. 2003b). Screens for mutants that affect the incorporation of GFP-tagged nucleoporins into the NE in yeast have yielded the genes encoding Ran, RanGAP, NTF2, and RCC1, all components of the Ran cycle (Ryan & Wente 2002, Ryan et al. 2003).

Other nucleocytoplasmic receptors probably have roles in NPC assembly, as exemplified by study of the yeast nucleoporin, Nup53p. When Nup53p is overexpressed, intranuclear double membrane lamellae form that are fenestrated by what look like empty NPCs (Marelli et al. 2001). These structures are coated with Nup53p, suggesting that it may have a role in inducing pore membrane domain formation. Nup53p binds to the Importin β family member Kap121p, and localization of Nup53p to the nucleus and NPC is dependent on Kap121p (Marelli et al. 2001). Once in the NPC, Nup53p does not interact with Kap121p during interphase but forms a mutually exclusive interaction with Nup170p (Marelli et al. 2001, Lusk et al. 2002, Makhnevych et al. 2003). During mitosis, the Nup170p-Nup53p interaction is disrupted, allowing Nup53p to interact with Kap121p at the NPC. This interaction has a mitosis-specific inhibitory effect on Kap121p movement through the NPC and thus on Kap121p-mediated protein import (Makhnevych et al. 2003). The rearrangement of the yeast NPC is probably more extensive, because Nup53p also makes a mitotic-specific interaction with Nic96p, another nucleoporin (Makhnevych et al. 2003). Nevertheless, unlike the situation in metazoa and *Aspergillus* described above, the NPC in *S. cerevisiae* is thought to remain largely intact during mitosis, with the maintenance of at least some aspects of active nucleocytoplasmic transport.

NPC-membrane interaction during NE assembly. How are early steps of NPC assembly coordinated with the formation of the nuclear membranes and the recruitment of the transmembrane nucleoporins? Early models (Cohen et al. 2003, Drummond & Wilson 2002, Greber et al. 1990, Wozniak & Blobel 1992) proposed that the transmembrane nucleoporins are critical as anchoring points in the membrane and as nucleating centers on which the rest of the NPC is constructed. An apparent setback for this model was the realization that transmembrane nucleoporins are very poorly conserved in evolution (Mans et al. 2004, Vasu & Forbes 2001). One of the two characterized vertebrate proteins, gp210, has recognizable homologues in animals and plants, but it appears to have been secondarily lost in fungi (Mans et al. 2004). The second, Pom121, is even more restricted; it is detectable only in vertebrates. In yeast, there are two transmembrane nucleoporins, Pom34p and Pom152p, that are restricted to NPCs, and a third, Ndc1p, that is a component not only of NPCs but also the SPB—the yeast centrosome—which is embedded in the NE. Of these, only Ndc1p is essential, and Ndc1p requirement likely is related to the SPB rather than to NPC assembly (Chial et al. 1998, Lau et al. 2004, Winey et al. 1993). Recent studies have shown that even Pom34p and Pom152p double knockout strains are viable (A.S. Madrid, J.M. McCaffery & K. Weis, private communication; S.R. Wente, private communication). This does not mean that neither plays a critical role in NPC assembly or function, as Ndc1p and Pom152p appear to carry out a redundant function that is required for maintenance of NPCs in the NE (A.S. Madrid, J.M. McCaffery & K. Weis, private communication).

Studies of gp210 in vertebrate systems have been inconclusive about its detailed functional role. A highly mobile nucleoporin, gp210 is only detected at the NPC very late in NE assembly in comparison with other nucleoporins (Antonin et al. 2005, Bodoor et al. 1999, Daigle et al. 2001), making it unlikely that it plays an essential role in NPC assembly. Nevertheless, RNAi-mediated depletion of gp210 from both human cultured cells and *C. elegans* embryos affects cell viability (Cohen et al. 2003). Although some mammalian cells lack gp210 (Olsson et al. 2004), the situation is complicated by the presence of a gp210-related gene in some mammalian genomes, including those of human and rat (E. Hartmann,

private communication). A recent study concluded that depletion of gp210 has no apparent effect on NE or NPC assembly in the *Xenopus* in vitro system. Nuclei lacking gp210 also has normal size, indicating that nuclear protein import operates (Antonin et al. 2005). The function of gp210 therefore remains to be discovered. In sharp contrast, no in vitro nuclei forms in the absence of Pom121, and NE formation is arrested at a very early step (Antonin et al. 2005). The importance of Pom121 for NE and NPC formation is consistent with the observation that Pom121 assembles early into NPCs and is a very stable constituent of NPCs once they form (Daigle et al. 2001, Rabut et al. 2004). However, a lack of any major phenotype in HeLa cells treated with anti-Pom121 siRNAs, and the relatively minor effect of similar treatment of rat NRK cells (Antonin et al. 2005), suggests that there is more to learn about the functions of transmembrane nucleoporins in NPC assembly.

Perhaps the most interesting aspect of the Pom121 depletion analysis was that it revealed the existence of cross-regulation between NE and NPC assembly. In the absence of Pom121, no NE assembly was observed. However, if the Nup107-160 complex was co-depleted with Pom121, a closed NE was formed. As is expected from the absence of Nup107-160, these NEs lacked NPCs. Nevertheless, removal of Nup107-160 suppressed the effect of Pom121 depletion on NE membrane fusion events (Antonin et al. 2005). Formally, this means that the Nup107-160 complex acts to inhibit NE membrane fusion in the absence of Pom121, and the entire regulatory phenomenon can be described, by analogy to mitotic controls, as an NE/NPC assembly checkpoint. A closed NE was formed even when both gp210 and pom121 were co-depleted with Nup107-160 complex, demonstrating that neither of the transmembrane nucleoporins is required for NE membrane fusion. The assembly of NPCs is closely linked to the formation of the NE. When NE membrane fusion is blocked, many steps of NPC assembly are inhibited (Hetzer et al.

SPB: spindle pole body

2000, Macaulay & Forbes 1996). In the following section, we discuss recent progress in the understanding of NE membrane fusion.

The Formation of a Highly Organized Membrane Barrier Around Chromatin

At the end of anaphase, the chromatin is packaged into a nucleus. This normally occurs via formation of one single nucleus, but this is not always the case. In *Xenopus* and other species that exhibit very rapid early embryonic cell division cycles, single chromosomes form individual karyomeres, each with its own NE, before karyomere fusion forms a single nucleus (e.g., Lemaitre et al. 1998). At first sight, this looks like a recipe for disaster, but cleavage-stage nuclei do not have to transcribe their genes. Thus even if the process of formation of a single nucleus from the karyomeres is not highly efficient, the errant chromosomes that fail to incorporate into a large nucleus at one cell division may be re-collected into a complete metaphase plate at the next mitosis.

The formation of a double membrane around daughter nuclei requires extensive membrane fusion (Collas & Poccia 2000). This reaction may be related to the fusion of nuclei after yeast cell mating (karyogamy). Two classes of genes that affect karyogamy have been identified. Whereas members of the first class show a defect in nuclear migration, the class II mutants *kar2*, *kar5*, *kar7*, *kar8*, and *prm3* (Rose 1996) are defective in fusion of the two NEs. All of these proteins are located in the ER/NE lumen or the ER membrane. Kar2p is the yeast homologue of BiP, a ubiquitous chaperone involved in protein translocation into the ER. This suggests either that a fusogen must be translocated into the ER/NE membrane to promote nuclear fusion or that the ER translocation apparatus has a direct role in fusion. The isolation of mutations in the ER translocon components *sec63*, *sec71*, and *sec72*, which

have no detected translocation defect but block karyogamy, also points in this direction (Brizzio et al. 1999, Rose 1996). *Kar2*, together with ATP, is also required for ER/NE vesicle fusion in vitro (Kurihara et al. 1994, Latterich & Schekman 1994, Latterich et al. 1995). However, these studies have not separated requirements for constitutive ER fusion from the fusion events occurring during karyogamy.

Our knowledge of the mechanisms of vertebrate NE membrane fusion is almost exclusively derived from cell-free systems (Lohka & Masui 1983, Newport & Dunphy 1992, Burke & Gerace 1986). These have allowed the experimental separation of NE assembly into distinct steps (see **Figure 6**): (*a*) targeting of NE precursor membranes to the chromatin surface (Newport & Dunphy 1992, Sheehan et al. 1988, Vigers & Lohka 1991), (*b*) fusion of these vesicles to form a tubular network on the chromatin surface (Hetzer et al. 2001), (*c*) further fusion of membranes to form closed INM and ONM, (*d*) the assembly of NPC subunits and integral membrane proteins of the pore membrane domains to produce individual NPCs, and (*e*) import of nuclear components such as lamins and expansion of the NE (Hetzer et al. 2000). More recently, several in vivo studies, notably in *C. elegans* and cultured mammalian cells, have provided complementary insights into the molecular mechanisms of NE formation, mostly using RNAi as a tool (e.g., Liu et al. 2003a).

NE vesicle populations: poorly understood diversity. NE formation in vitro is initiated by the binding of a subset of membrane vesicles to decondensed chromatin (Collas et al. 1996). The population of membrane vesicles purified from egg extracts is largely composed of membranes and proteins that will give rise to the ER (Dreier & Rapoport 2000, Hetzer et al. 2001). Only a minor component of the membrane fraction actually can bind to chromatin (Sasagawa et al. 1999, Wilson & Newport 1988). Chromatin can be used as an affinity matrix to purify

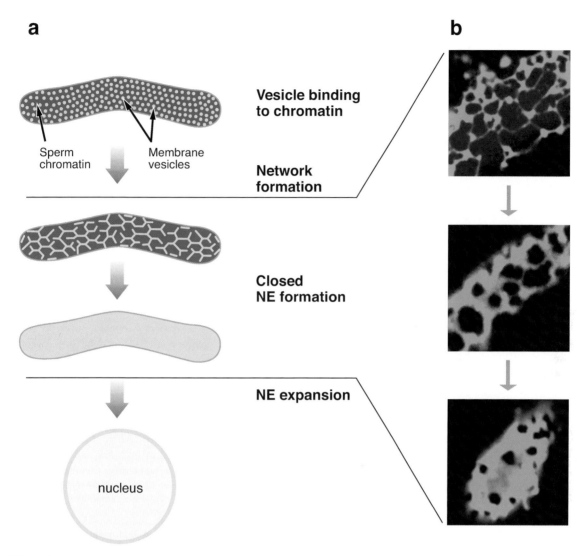

Figure 6

Steps of NE formation in vitro. (*a*) A subset of membrane vesicles (*yellow*) binds to decondensed sperm chromatin (*blue*). In the presence of Ran, a membrane network forms on the chromatin surface. This network is continuous with the ER. After closed NE formation, nuclear growth occurs. (*b*) NE networks seal to form a closed NE. Different steps of an in vitro assembly reaction are shown. Fluorescently labeled membranes (*green*) form a network on the surface of chromatin and progressively form membrane patches that close the remaining holes.

vesicles that are required for NE formation (Sasagawa et al. 1999). Various strategies of membrane fractionation have been employed in the study of NE assembly, leading to the conclusion that minimally two, and possibly more, distinct fractions of membrane vesicles are separable and required for NE assembly (Collas & Poccia 1996a, Vigers & Lohka 1991). The fractions display distinct biochemical, chromatin-binding, and fusion properties. In *Xenopus*, two distinct populations have been separated: one that binds to chromatin and another that associates with the chromatin-bound fraction (Sasagawa et al.

HP1:
heterochromatin
protein;
heterochromatin is
highly compacted,
transcriptionally
inactive chromatin
and includes
gene-free regions of
chromosomes (e.g.,
centrosomes) as well
as silenced genes

BAF: barrier-to-
autointegration
factor

1999). Two morphologically distinct *Xenopus* vesicle populations have also been observed by EM (Wiese et al. 1997): a ribosome-containing vesicle fraction that binds to chromatin, and smooth vesicles that in turn bind to the chromatin-bound vesicles; these may correspond to the biochemical classes.

Recently, a different fractionation procedure separated two vesicle fractions from *Xenopus* egg extracts that contained either the Pom121 or gp210 transmembrane nucleoporins as molecular markers (Antonin et al. 2005). As mentioned above, only Pom121-containing vesicles were required for NE formation and this marker will provide a tool to allow characterization of the protein composition of the vesicle fraction. If, as suggested earlier, vesicles derived from meiotic or mitotic cell extracts are derived from a tubular ER network, the number of different vesicle populations present may depend on and vary with the experimental conditions applied to prepare the extracts (Collas & Courvalin 2000).

So what distinguishes the membranes that will give rise to the NE from bulk ER-derived membranes? Clearly, NE vesicles can attach to chromosomes and therefore must be associated with proteins that bind to chromatin. Several INM proteins bind to chromatin constituents such as HP1 or BAF, and these interactions have been indirectly implicated in NE assembly. For example, two studies have demonstrated that the integral INM protein LBR, which binds to HP1, is required for targeting and anchoring NE vesicles to chromatin in vitro (Collas et al. 1996, Pyrpasopoulou et al. 1996). Whether these interactions play a role in the early steps of NE assembly in vivo remains to be seen. Different NPC components are recruited to the chromatin and the assembling NE membranes in a spatially ordered process. INM and chromatin proteins, e.g., LBR, emerin, and BAF, exhibit distinct localizations on the decondensing chromatin surface during late anaphase and early telophase relative to the axes of the mitotic spindle (Dechat et al. 2004, Haraguchi

et al. 2000). The molecular distinctions between these regions of the chromatin surface are completely unknown.

NE fusion in vitro. NE vesicles incorporate into ER in the absence of chromatin, whereas ER-derived membranes are incapable of forming an NE even at high concentrations (Dreier & Rapoport 2000, Hetzer et al. 2001). The capacity to form ER is therefore an intrinsic feature of both ER and NE vesicles. This suggests that they might share similar fusion requirements and machineries. Similarities between NE and ER are further substantiated by the finding that an early step of in vitro NE formation occurs via a chromatin-associated membrane network that is morphologically similar to ER (**Figure 6**) (Hetzer et al. 2001). Also, ER and NE vesicles seem to share at least one common GTP-dependent "basic fusion" step (Newport & Dunphy 1992, Dreier & Rapoport 2000). Ran is not detected on NE membranes prior to NPC assembly nor is it required for ER formation, so this step likely requires a different GTPase whose identification will help reveal whether this step is normally involved in both ER and NE formation. It is important to note that the formation of ER tubules, in contrast to large vesicles, is strictly dependent on the addition of cytosol (Dreier & Rapoport 2000), indicating that membrane tubule formation is not an intrinsic property of the membrane fractions involved in NE and ER formation.

Experiments in the *Xenopus* in vitro NE assembly system, however, suggest that NE and ER tubule formation also have distinct requirements. A member of the AAA-ATPase family, p97/Cdc48, and its adaptors p47 and Ufd1/Npl4, are required at two separate stages of NE formation (Hetzer et al. 2001). Inhibition or removal of p97 or Ufd1/Npl4 from *Xenopus* egg extracts or autoantibodies from biliary cirrhosis patients that bind to p97 (Miyachi et al. 2004) causes a block at an early step in NE membrane fusion (Hetzer et al. 2001). In contrast, p97/p47 is not required for the formation of a closed NE, but it is needed for the NE growth that occurs following NE formation. The lack of p47 thus leads to the formation of small nuclei that resemble those formed in assembly reactions in which nucleocytoplasmic transport has been inhibited (Hetzer et al. 2001). p97 and its yeast homologue Cdc48p are involved in a considerable variety of cellular processes (Woodman 2003), and the mechanism by which p97/Ufd1/Npl4 functions in NE assembly is presently unclear. The growth of NE membranes also may be regulated by controlling the quantity of NE-associated proteins (see, e.g., Prufert et al. 2004).

Ran is required for NE assembly in vivo and in vitro (Askjaer et al. 2002, Bamba et al. 2002, Hetzer et al. 2000, Zhang & Clarke 2000, 2001). Removal of either Ran or Ran's nucleotide exchange factor RCC1 from *Xenopus* egg extracts completely blocks NE assembly from proceeding beyond the energy-independent step of vesicle binding to chromatin (Hetzer et al. 2000). These results suggest that Ran either is required directly to allow NE vesicle fusion or to modify chromatin, such that the chromatin can stimulate fusion of bound vesicles.

Experimental support for the first model came from the demonstration that Sepharose beads coupled to wildtype Ran can nucleate the assembly of functional, NPC-containing NEs in extracts of either *Xenopus* eggs or mammalian cultured cells (Zhang & Clarke 2000). Recent data has shown that Ran associates with meiotic or mitotic chromatin in both the in vitro nuclear assembly extracts and in vivo in a variety of cell types (Bilbao-Cortes et al. 2002, Hinkle et al 2000, 2002). Importin β was shown to block NE membrane fusion when added in excess to in vitro assembly reactions (Harel et al. 2003a), but how it does so is unclear. In addition to Importin β, Importin α, the protein that acts as an adaptor between Importin β and NLS-containing proteins during nuclear import, functions in NE assembly. Importin α seems to play two roles in the process. One requires interaction with NLS-containing proteins in the cytosol. The second is more unusual and involves membrane-associated Importin α (Hachet et al. 2004). Membrane association of Importin α does not require NLS binding or interaction with Importin β, CAS (the Importin α export receptor), or RanGTP; but is regulated by phosphorylation and is needed for NE assembly (Hachet et al. 2004).

In order to determine whether the differences between the NE and ER assembly processes are fundamental rather than superficial (e.g., through the addition of supplementary regulatory mechanisms to the same basic fusion process), it will be necessary to identify the targets of Ran and p97/Ufd1/Npl4 regulation, and also to obtain a better understanding of the relationship between these two different regulators, which may interact with one another either in series on the same NE formation reactions or in parallel on distinct processes.

Formation of a double membrane around chromatin. After NE vesicles have fused to form a chromatin-associated network, the NE seals (**Figure 6**) (Dreier & Rapoport 2000, Hetzer et al. 2001). This fusion process eventually leads to the formation of two flat membrane sheets perforated by and joined at the sites of NPCs. This step does not necessarily require chromatin because AL, stacked membrane sheets densely packed with NPCs, also can form in the cytoplasm (Dabauvalle et al. 1991, Walther et al. 2003b). AL are found in embryonic and cancer cells (Kessel 1992) and were thought to be a storage form of NPCs mobilized in fast-growing cells during cancer or embryonic cleavage stages, but a study in *Drosophila* embryos surprisingly demonstrated that AL are not consumed during embryogenesis (Onischenko et al. 2004). AL can be induced to assemble in egg extracts in the absence of chromatin, and excess RanGTP can greatly increase the efficiency of this process (Dabauvalle et al. 1991, Harel et al. 2003a, Meier et al. 1995, Walther et al. 2003b).

AL: annulate lamellae

Although the signals that mediate cytoplasmic AL formation have not been elucidated, in vitro studies suggest that the balance between Importin β–bound and free nucleoporins is important for AL assembly (Harel et al. 2003a, Walther et al. 2003b).

To form a sealed NE, the final steps must include the closure of holes and gaps in the expanded tubular network (**Figure 6**). It has been suggested that closure of these holes requires a molecular drawstring similar to the GTPase dynamin (Burke 2001). One model for dynamin action is that it constricts the necks of budding vesicles after association with their concave cytoplasmic surface, which leads to fusion of the neck-forming membranes and vesicle release from the donor membrane (Hinshaw 2000). In the case of the NE, a functionally equivalent molecule may be involved, but it has to reside in the lumenal space between the ONM and INM (Burke 2001). Whether the final steps of NE membrane assembly, like many other characterized membrane fusion events, involve SNARE-like proteins (Jahn & Grubmuller 2002) remains to be seen. It is attractive to imagine constriction occurring at the sites where NPCs form; this obviates the need for the final fusion steps, as the remaining hole is stabilized around an NPC. However, because closed NEs can form in the absence of NPCs (Harel & Forbes 2004, Macaulay & Forbes 1996, Walther et al. 2003a), closure events must be able to occur during NE assembly.

The INM proteins (**Figure 4**) studied to date accumulate after NE and NPC formation by a passive mechanism (Burke & Stewart 2002). They are inserted into the rER and diffuse in the plane of the membrane through the ER, ONM, and NPC pore membrane domain to the INM. INM proteins harbor binding sites for lamins, chromatin proteins, or for each other, and the interactions mediated by these sites cause the retention and accumulation of INM proteins in the INM.

The Nuclear Periphery, Disease, and Gene Expression

In the last few years, a variety of human diseases have been linked to mutations in components of the NE (Burke & Stewart 2002, Worman & Courvalin 2004). The rapidly growing list of diseases ranges from cancers to tissue-specific inherited disorders, and can be roughly categorized into three groups based on the NE structure that is involved, namely diseases associated with mutations in proteins of (*a*) the NPC, (*b*) the INM, and (*c*) the nuclear lamina (**Figures 1** and **4**). Two basic models to explain NE-related disease have been proposed. The first suggests that they are due to fragility of the nuclear periphery and are caused by physical damage to the NE. Although such fragility has been observed (Sullivan et al. 1999), it is not known whether it is causative for disease. The second model suggests that NE components may regulate gene expression and that disease results when this regulation is disrupted. We will discuss some of the data pertinent to the second model in this section.

In principle, chromatin or gene localization at the NE could regulate transcription either positively or negatively. The majority of actively transcribed genes are not found at the nuclear periphery and gene-poor chromosomes with lower levels of transcription are more likely to be associated with the NE than gene-rich chromosomes (for review see Taddei et al. 2004b). These results suggest that chromatin-NE interactions may play a negative role in gene expression.

The capacity of lamins to influence RNA polymerase (pol II) activity has been examined. Microinjection of a C-terminal fragment of lamin A leads to a dramatic decrease in pol II–mediated transcription (Spann et al. 2002). Conversely, inhibition of pol II causes a coordinated reorganization of lamins A/C and splicing factors. The mechanistic link between these events currently is not understood. Lamins also might regulate transcription through association with transcriptional regulators. For example, the retinoblastoma gene product Rb interacts with lamin A and LAP2 in vivo (Markiewicz et al. 2002). Because Rb binds to E2F, thereby repressing E2F-dependent transcription, this interaction might influence transcription of E2F-responsive genes. The conserved and ubiquitously expressed transcription regulator GCL also binds a member of the E2F family, E2F-DP3, and thereby represses E2F-dependent gene transcription (for detailed discussion see Taddei et al. 2004b). GCL interacts with LEM-domain INM proteins and localizes together with E2F-DP3 to the NE (Holaska et al. 2003, Kimura et al. 2003).

Several negative regulators of transcription localize to the NE and it is thought that this may favor the incorporation of their target genes into a repressive environment. Examples include SREBP 1, a key regulator of cholesterol metabolism (Lloyd et al. 2002), transcription factor Oct-1 (Imai et al. 1997), and the cell death–promoting repressor Btf (Haraguchi et al. 2000). During *Xenopus* embryogenesis, the NE protein MAN1 antagonizes bone morphogenic protein signaling and neural induction by interacting with the coactivator Smad1 (Osada et al. 2003). Taken together, these accumulating results provide some support for the idea that the NE may participate generally in the establishment and maintenance of transcriptional repression.

As stated above, chromosomes are not randomly oriented in the nucleus. For example, in some cells, both telomeres and centromeres are found in close proximity to the NE. Telomeric and centromeric regions of chromosomes are organized into heterochromatin and are known to contain mainly repressed genes. In yeast, telomere-NE association is regulated by the Ku heterodimer and Sir4 (Luo et al. 2002, Roy et al. 2004, Taddei et al. 2004a).

However, there seem to be no rules in biology without exceptions, and the notion

that the nuclear periphery is a repressive domain is also not universally applicable. First, normally peripheral, repressed domains remain silent even when relocated within the nucleus (Gartenberg et al. 2004). Second, *S. cerevisiae* chromatin boundary activities, i.e., factors that bind boundary elements and prevent the spread of repressive chromatin structures into transcriptionally active areas include several exportins: Cse1p, Mex67p, and Los1p. These all exhibit robust boundary activity when artificially attached to a boundary element and block spreading of heterochromatin by physical tethering of chromosome sites to Nup2, a nucleoporin, suggesting that association of genomic loci to the NPC can alter positively their transcriptional state (Ishii et al. 2002).

Perspectives and Less Well-Studied Problems

It is evident that the study of many aspects of NE dynamics and functions is in its infancy. There are a considerable number of open or partially answered questions about NE composition, structure, dynamics, and function during the cell cycle (**Figure 7**). In this section, we will mention a few interesting but understudied areas of NE research.

We know very little about what activities mediate NE membrane assembly, or how the roughly spherical double bilayer of the NE is shaped and separated from the ER. A single study of this topic in yeast (Siniossoglou et al. 1998) suggested that integral membrane proteins of the NE and ER influenced yeast NE shape, but did not elucidate how. In metazoa, interphase nuclear shape varies considerably between differentiated cell types. In one study of this, expression of a spermatocyte-specific B-type lamin in somatic cells was found to alter nuclear shape to one resembling the meiotic cells from which the lamin was derived (Furukawa & Hotta 1993). But how or why different lamins assemble into distinct three-dimensional forms is unknown.

E2F transcription factor family: The transcription factor E2F1 was identified as a DNA-binding protein required for transcription of the adenovirus E2A promoter. The E2F binding site is also present in many growth-responsive and growth-promoting genes

GCL: Germcell-less

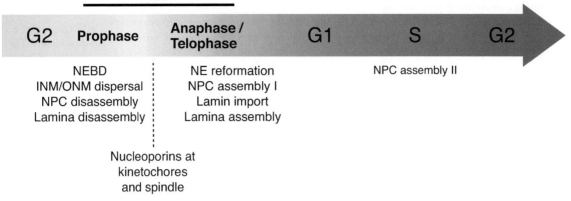

Figure 7

Schematic illustrations of events that are associated with NE dynamics during the cell cycle. See text for details.

A related topic is that of NPC number. The density of NPCs in the NE is known to vary in a manner that roughly correlates with cellular metabolic activity. Metabolically active cells have a high NPC density and a total of roughly 3000 NPCs, whereas the dormant nuclei of avian erythroblasts have only a handful of NPCs (Maul et al. 1971, Maul et al. 1972). But how metabolic activity can be signaled to the machinery responsible for NPC synthesis, assembly, and turnover is a mystery.

The lipid composition of the NE, and the possible role of lipids in distinguishing the NE from other endomembranes, has been studied in sea urchin, where phosphatidylinositol and kinases that modify the NE are re-quired for its assembly (Byrne et al. 2004, Larijani et al. 2001). However, the sea urchin pronuclear NE may be unusual (Mattaj 2004), and study of the role of lipids in other NEs is required. The role of peripheral nuclear structures in gene expression is another un-derresearched area. In this case, the fact that various disease states may be explicable by changes in gene expression caused by muta-tion in peripheral NE components lends ur-gency to the acquisition of more information. Further challenges include understanding the checkpoint that links NPC and NE assem-bly. There is no doubt that research on the NE will be a growth area for the foreseeable future.

ACKNOWLEDGMENTS

We thank Wolfram Antonin, Maximiliano D'Angelo, Jan Ellenberg, Cerstin Franz, Vincent Galy, Matyas Gorjanacz, Melpomeni Platani, Katharina Ribbeck, and Sebastian Ulbert for critically reading the manuscript. We are grateful to many colleagues for sharing their unpub-lished information. An extended version of this review, with many more citations, is available from Martin Hetzer.

LITERATURE CITED

Akey CW. 1989. Interactions and structure of the nuclear pore complex revealed by cryo-electron microscopy. *J. Cell Biol.* 109:955–70

Allen NP, Huang L, Burlingame A, Rexach M. 2001. Proteomic analysis of nucleoporin inter-acting proteins. *J. Biol. Chem.* 276:29268–74

Allen NP, Patel SS, Huang L, Chalkley RJ, Burlingame A, et al. 2002. Deciphering networks of protein interactions at the nuclear pore complex. *Mol. Cell Proteomics* 1:930–46

Antonin W, Franz C, Haselmann U, Antony C, Mattaj IW. 2005. The integral membrane nucleoporin pom121 functionally links nuclear pore complex assembly and nuclear envelope formation. *Mol. Cell* 17:83–92

Apel ED, Lewis RM, Grady RM, Sanes JR. 2000. Syne-1, a dystrophin- and Klarsicht-related protein associated with synaptic nuclei at the neuromuscular junction. *J. Biol. Chem.* 275:31986–95

Askjaer P, Galy V, Hannak E, Mattaj IW. 2002. Ran GTPase cycle and importins alpha and beta are essential for spindle formation and nuclear envelope assembly in living *Caenorhabditis elegans* embryos. *Mol. Biol. Cell* 13:4355–70

Bai SW, Rouquette J, Umeda M, Faigle W, Loew D, et al. 2004. The fission yeast Nup107–120 complex functionally interacts with the small GTPase Ran/Spi1 and is required for mRNA export, nuclear pore distribution, and proper cell division. *Mol. Cell Biol.* 24:6379–92

Bamba C, Bobinnec Y, Fukuda M, Nishida E. 2002. The GTPase Ran regulates chromosome positioning and nuclear envelope assembly in vivo. *Curr. Biol.* 12:503–7

Bayliss R, Littlewood T, Stewart M. 2000. Structural basis for the interaction between FxFG nucleoporin repeats and importin-beta in nuclear trafficking. *Cell* 102:99–108

Bayliss R, Littlewood T, Strawn LA, Wente SR, Stewart M. 2002. GLFG and FxFG nucleoporins bind to overlapping sites on importin-beta. *J. Biol. Chem.* 277:50597–606

Beaudouin J, Gerlich D, Daigle N, Eils R, Ellenberg J. 2002. Nuclear envelope breakdown proceeds by microtubule-induced tearing of the lamina. *Cell* 108:83–96

Beck M, Forster F, Ecke M, Plitzko JM, Melchior F, et al. 2004. Nuclear pore complex structure and dynamics revealed by cryoelectron tomography. *Science* 306:1387–90

Becskei A, Mattaj IW. 2003. The strategy for coupling the RanGTP gradient to nuclear protein export. *Proc. Natl. Acad. Sci. USA* 100:1717–22

Belgareh N, Rabut G, Bai SW, van Overbeek M, Beaudouin J, et al. 2001. An evolutionarily conserved NPC subcomplex, which redistributes in part to kinetochores in mammalian cells. *J. Cell Biol.* 154:1147–60

Ben-Efraim I, Gerace L. 2001. Gradient of increasing affinity of importin beta for nucleoporins along the pathway of nuclear import. *J. Cell Biol.* 152:411–17

Bernad R, van der Velde H, Fornerod M, Pickersgill H. 2004. Nup358/RanBP2 attaches to the nuclear pore complex via association with Nup88 and Nup214/CAN and plays a supporting role in CRM1-mediated nuclear protein export. *Mol. Cell Biol.* 24:2373–84

Bilbao-Cortes D, Hetzer M, Langst G, Becker PB, Mattaj IW. 2002. Ran binds to chromatin by two distinct mechanisms. *Curr. Biol.* 12:1151–56

Bischoff FR, Ponstingl H. 1991. Catalysis of guanine nucleotide exchange on Ran by the mitotic regulator RCC1. *Nature* 354:80–82

Bodoor K, Shaikh S, Salina D, Raharjo WH, Bastos R, et al. 1999. Sequential recruitment of NPC proteins to the nuclear periphery at the end of mitosis. *J. Cell Sci.* 112(Pt. 13):2253–64

Boehmer T, Enninga J, Dales S, Blobel G, Zhong H. 2003. Depletion of a single nucleoporin, Nup107, prevents the assembly of a subset of nucleoporins into the nuclear pore complex. *Proc. Natl. Acad. Sci. USA* 100:981–85

Brizzio V, Khalfan W, Huddler D, Beh CT, Andersen SS, et al. 1999. Genetic interactions between KAR7/SEC71, KAR8/JEM1, KAR5, and KAR2 during nuclear fusion in Saccharomyces cerevisiae. *Mol. Biol. Cell* 10:609–26

Burke B. 2001. The nuclear envelope: filling in gaps. *Nat. Cell Biol.* 3:E273–74

Burke B, Ellenberg J. 2002. Remodelling the walls of the nucleus. *Nat. Rev. Mol. Cell Biol.* 3:487–97

Burke B, Gerace L. 1986. A cell free system to study reassembly of the nuclear envelope at the end of mitosis. *Cell* 44:639–52

Burke B, Stewart CL. 2002. Life at the edge: the nuclear envelope and human disease. *Nat. Rev. Mol. Cell. Biol.* 3:575–85

Byrne RD, Barona TM, Garnier M, Koster G, Katan M, et al. 2004. Nuclear envelope assembly is promoted by phosphoinositide-specific PLC with selective recruitment of phosphatidylinositol enriched membranes. *Biochem. J.* 387:393–400

Callan HG, Tomlin SG. 1950. Experimental studies on amphibian oocyte nuclei. I. Investigation of the structure of the nuclear membrane by means of the electron microscope. *Proc. R. Soc. London Ser. B.* 137:367–78

Campbell MS, Chan GK, Yen TJ. 2001. Mitotic checkpoint proteins HsMAD1 and HsMAD2 are associated with nuclear pore complexes in interphase. *J. Cell Sci.* 114:953–63

Carazo-Salas RE, Gruss OJ, Mattaj IW, Karsenti E. 2001. Ran-GTP coordinates regulation of microtubule nucleation and dynamics during mitotic-spindle assembly. *Nat. Cell Biol.* 3:228–34

Chaudhary N, Courvalin JC. 1993. Stepwise reassembly of the nuclear envelope at the end of mitosis. *J. Cell Biol.* 122:295–306

Chial HJ, Rout MP, Giddings TH, Winey M. 1998. Saccharomyces cerevisiae Ndc1p is a shared component of nuclear pore complexes and spindle pole bodies. *J. Cell Biol.* 143:1789–800

Cohen M, Feinstein N, Wilson KL, Gruenbaum Y. 2003. Nuclear pore protein gp210 is essential for viability in HeLa cells and *Caenorhabditis elegans*. *Mol. Biol. Cell* 14:4230–37

Collas I, Courvalin JC. 2000. Sorting nuclear membrane proteins at mitosis. *Trends Cell Biol.* 10:5–8

Collas P, Courvalin JC, Poccia D. 1996. Targeting of membranes to sea urchin sperm chromatin is mediated by a lamin B receptor-like integral membrane protein. *J. Cell Biol.* 135:1715–25

Collas P, Poccia D. 1996a. Distinct egg membrane vesicles differing in binding and fusion properties contribute to sea urchin male pronuclear envelopes formed in vitro. *J. Cell Sci.* 109(Pt. 6):1275–83

Collas P, Poccia D. 1998. Methods for studying in vitro assembly of male pronuclei using oocyte extracts from marine invertebrates: sea urchins and surf clams. *Methods Cell Biol.* 53:417–52

Collas P, Poccia D. 2000. Membrane fusion events during nuclear envelope assembly. *Subcell. Biochem.* 34:273–302

Courvalin JC, Segil N, Blobel G, Worman HJ. 1992. The lamin B receptor of the inner nuclear membrane undergoes mitosis-specific phosphorylation and is a substrate for p34cdc2-type protein kinase. *J. Biol. Chem.* 267:19035–38

Cronshaw JM, Krutchinsky AN, Zhang W, Chait BT, Matunis MJ. 2002. Proteomic analysis of the mammalian nuclear pore complex. *J. Cell Biol.* 158:915–27

Cronshaw JM, Matunis MJ. 2003. The nuclear pore complex protein ALADIN is mislocalized in triple A syndrome. *Proc. Natl. Acad. Sci. USA* 100:5823–27

Dabauvalle MC, Loos K, Merkert H, Scheer U. 1991. Spontaneous assembly of pore complex-containing membranes ("annulate lamellae") in Xenopus egg extract in the absence of chromatin. *J. Cell Biol.* 112:1073–82

This study is an important step toward a detailed molecular description of NPC organization.

Daigle N, Beaudouin J, Hartnell L, Imreh G, Hallberg E, et al. 2001. Nuclear pore complexes form immobile networks and have a very low turnover in live mammalian cells. *J. Cell Biol.* 154:71–84

Danker T, Oberleithner H. 2000. Nuclear pore function viewed with atomic force microscopy. *Pflugers Arch.* 439:671–81

Dasso M. 2002. The Ran GTPase: theme and variations. *Curr. Biol.* 12:R502–8

Dechat T, Gajewski A, Korbei B, Gerlich D, Daigle N, et al. 2004. LAP2α and BAF transiently localize to telomeres and specific regions on chromatin during nuclear assembly. *J. Cell Sci.* 117:6117–28

Delphin C, Guan T, Melchior F, Gerace L. 1997. RanGTP targets p97 to RanBP2, a filamentous protein localized at the cytoplasmic periphery of the nuclear pore complex. *Mol. Biol. Cell* 8:2379–90

Denning DP, Patel SS, Uversky V, Fink AL, Rexach M. 2003. Disorder in the nuclear pore complex: the FG repeat regions of nucleoporins are natively unfolded. *Proc. Natl. Acad. Sci. USA* 100:2450–55

De Souza CP, Osmani AH, Hashmi SB, Osmani SA. 2004. Partial nuclear pore complex disassembly during closed mitosis in Aspergillus nidulans. *Curr. Biol.* 14:1973–84

Devos D, Dokudovskaya S, Alber F, Williams R, Chait BT, et al. 2004. Components of coated vesicles and nuclear pore complexes share a common molecular architecture. *PLoS Biol.* 2:e380

Dilworth DJ, Suprapto A, Padovan JC, Chait BT, Wozniak RW, et al. 2001. Nup2p dynamically associates with the distal regions of the yeast nuclear pore complex. *J. Cell Biol.* 153:1465–78

Doye V, Hurt E. 1997. From nucleoporins to nuclear pore complexes. *Curr. Opin. Cell Biol.* 9:401–11

Dreger M, Bengtsson L, Schoneberg T, Otto H, Hucho F. 2001. Nuclear envelope proteomics: novel integral membrane proteins of the inner nuclear membrane. *Proc. Natl. Acad. Sci. USA* 98:11943–48

Dreier L, Rapoport TA. 2000. In vitro formation of the endoplasmic reticulum occurs independently of microtubules by a controlled fusion reaction. *J. Cell Biol.* 148:883–98

Drummond S, Ferrigno P, Lyon C, Murphy J, Goldberg M, et al. 1999. Temporal differences in the appearance of NEP-B78 and an LBR-like protein during Xenopus nuclear envelope reassembly reflect the ordered recruitment of functionally discrete vesicle types. *J. Cell Biol.* 144:225–40

Drummond SP, Wilson KL. 2002. Interference with the cytoplasmic tail of gp210 disrupts "close apposition" of nuclear membranes and blocks nuclear pore dilation. *J. Cell Biol.* 158:53–62

Ellenberg J, Siggia ED, Moreira JE, Smith CL, Presley JF, et al. 1997. Nuclear membrane dynamics and reassembly in living cells: targeting of an inner nuclear membrane protein in interphase and mitosis. *J. Cell Biol.* 138:1193–206

Englmeier L, Olivo JC, Mattaj IW. 1999. Receptor-mediated substrate translocation through the nuclear pore complex without nucleotide triphosphate hydrolysis. *Curr. Biol.* 9:30–41

Enninga J, Levay A, Fontoura BM. 2003. Sec13 shuttles between the nucleus and the cytoplasm and stably interacts with Nup96 at the nuclear pore complex. *Mol. Cell Biol.* 23:7271–84

Enoch T, Peter M, Nurse P, Nigg EA. 1991. p34cdc2 acts as a lamin kinase in fission yeast. *J. Cell Biol.* 112:797–807

Fahrenkrog B, Aris JP, Hurt EC, Pante N, Aebi U. 2000. Comparative spatial localization of protein-A-tagged and authentic yeast nuclear pore complex proteins by immunogold electron microscopy. *J. Struct. Biol.* 129:295–305

Favreau C, Bastos R, Cartaud J, Courvalin JC, Mustonen P. 2001. Biochemical characterization of nuclear pore complex protein gp210 oligomers. *Eur. J. Biochem.* 268:3883–89

Fischer-Vize JA, Mosley KL. 1994. Marbles mutants: uncoupling cell determination and nuclear migration in the developing Drosophila eye. *Development* 120:2609–18

Fontoura BM, Blobel G, Matunis MJ. 1999. A conserved biogenesis pathway for nucleoporins: proteolytic processing of a 186-kilodalton precursor generates Nup98 and the novel nucleoporin, Nup96. *J. Cell Biol.* 144:1097–112

Forbes DJ, Kirschner MW, Newport JW. 1983. Spontaneous formation of nucleus-like structures around bacteriophage DNA microinjected into Xenopus eggs. *Cell* 34:13–23

Franke WW, Scheer U, Krohne G, Jarasch ED. 1981. The nuclear envelope and the architecture of the nuclear periphery. *J. Cell Biol.* 91:S39–50

Fridkin A, Mills E, Margalit A, Neufeld E, Lee KK, et al. 2004. Matefin, a *Caenorhabditis elegans* germ line-specific SUN-domain nuclear membrane protein, is essential for early embryonic and germ cell development. *Proc. Natl. Acad. Sci. USA* 101:6987–92

Frosst P, Guan T, Subauste C, Hahn K, Gerace L. 2002. Tpr is localized within the nuclear basket of the pore complex and has a role in nuclear protein export. *J. Cell Biol.* 156:617–30

Furukawa K, Hotta Y. 1993. cDNA cloning of a germ cell specific lamin B3 from mouse spermatocytes and analysis of its function by ectopic expression in somatic cells. *EMBO J.* 12: 97–106

Gartenberg MR, Neumann FR, Laroche T, Blaszczyk M, Gasser SM. 2004. Sir-mediated repression can occur independently of chromosomal and subnuclear contexts. *Cell* 119:955–67

Gerace L, Burke B. 1988. Functional organization of the nuclear envelope. *Annu. Rev. Cell Biol.* 4:335–74

Goldberg MW, Allen TD. 1993. The nuclear pore complex: three-dimensional surface structure revealed by field emission, in-lens scanning electron microscopy, with underlying structure uncovered by proteolysis. *J. Cell Sci.* 106(Pt. 1):261–74

Goldberg MW, Wiese C, Allen TD, Wilson KL. 1997. Dimples, pores, star-rings, and thin rings on growing nuclear envelopes: evidence for structural intermediates in nuclear pore complex assembly. *J. Cell Sci.* 110(Pt. 4):409–20

Gorlich D, Kutay U. 1999. Transport between the cell nucleus and the cytoplasm. *Annu. Rev. Cell Dev. Biol.* 15:607–60

Gorlich D, Seewald MJ, Ribbeck K. 2003. Characterization of Ran-driven cargo transport and the RanGTPase system by kinetic measurements and computer simulation. *EMBO J.* 22:1088–100

Greber UF, Senior A, Gerace L. 1990. A major glycoprotein of the nuclear pore complex is a membrane-spanning polypeptide with a large lumenal domain and a small cytoplasmic tail. *EMBO J.* 9:1495–502

Griffis ER, Xu S, Powers MA. 2003. Nup98 localizes to both nuclear and cytoplasmic sides of the nuclear pore and binds to two distinct nucleoporin subcomplexes. *Mol. Biol. Cell* 14:600–10

Gruenbaum Y, Margalit A, Goldman RD, Shumaker DK, Wilson KL. 2005. The nuclear lamina comes of age. *Nat. Rev. Mol. Cell. Biol.* 6:21–31

Hachet V, Kocher T, Wilm M, Mattaj IW. 2004. Importin alpha associates with membranes and participates in nuclear envelope assembly in vitro. *EMBO J.* 23:1526–35

Hallberg E, Wozniak RW, Blobel G. 1993. An integral membrane protein of the pore membrane domain of the nuclear envelope contains a nucleoporin-like region. *J. Cell Biol.* 122:513–21

Haraguchi T, Koujin T, Hayakawa T, Kaneda T, Tsutsumi C, et al. 2000. Live fluorescence imaging reveals early recruitment of emerin, LBR, RanBP2, and Nup153 to reforming functional nuclear envelopes. *J. Cell Sci.* 113(Pt. 5):779–94

Harel A, Chan RC, Lachish-Zalait A, Zimmerman E, Elbaum M, Forbes DJ. 2003a. Importin beta negatively regulates nuclear membrane fusion and nuclear pore complex assembly. *Mol. Biol. Cell* 14:4387–96

Harel A, Forbes DJ. 2004. Importin beta: conducting a much larger cellular symphony. *Mol. Cell* 16:319–30

Harel A, Orjalo AV, Vincent T, Lachish-Zalait A, Vasu S, et al. 2003b. Removal of a single pore subcomplex results in vertebrate nuclei devoid of nuclear pores. *Mol. Cell* 11:853–64

Hetzer M, Bilbao-Cortes D, Walther TC, Gruss OJ, Mattaj IW. 2000. GTP hydrolysis by Ran is required for nuclear envelope assembly. *Mol. Cell* 5:1013–24

Hetzer M, Gruss OJ, Mattaj IW. 2002. The Ran GTPase as a marker of chromosome position in spindle formation and nuclear envelope assembly. *Nat. Cell Biol.* 4:E177–84

Hetzer M, Meyer HH, Walther TC, Bilbao-Cortes D, Warren G, Mattaj IW. 2001. Distinct AAA-ATPase p97 complexes function in discrete steps of nuclear assembly. *Nat. Cell Biol.* 3:1086–91

Hinkle B, Rolls MM, Stein P, Rapoport T, Terasaki M. 2000. Ran is associated with chromosomes during starfish oocyte meiosis and embryonic mitoses. *Zygote* 8(Suppl. 1):S91

Hinkle B, Slepchenko B, Rolls MM, Walther TC, Stein PA, et al. 2002. Chromosomal association of Ran during meiotic and mitotic divisions. *J. Cell Sci.* 115:4685–93

Hinshaw JE. 2000. Dynamin and its role in membrane fission. *Annu. Rev. Cell Dev. Biol.* 16:483–519

Hinshaw JE, Milligan RA. 2003. Nuclear pore complexes exceeding eightfold rotational symmetry. *J. Struct. Biol.* 141:259–68

Holaska JM, Lee KK, Kowalski AK, Wilson KL. 2003. Transcriptional repressor germ cell-less (GCL) and barrier to autointegration factor (BAF) compete for binding to emerin in vitro. *J. Biol. Chem.* 278:6969–75

Holmer L, Worman HJ. 2001. Inner nuclear membrane proteins: functions and targeting. *Cell Mol. Life Sci.* 58:1741–47

Huang L, Baldwin MA, Maltby DA, Medzihradszky KF, Baker PR, et al. 2002. The identification of protein-protein interactions of the nuclear pore complex of Saccharomyces cerevisiae using high throughput matrix-assisted laser desorption ionization time-of-flight tandem mass spectrometry. *Mol. Cell Proteomics* 1:434–50

Hutchison CJ, Alvarez-Reyes M, Vaughan OA. 2001. Lamins in disease: why do ubiquitously expressed nuclear envelope proteins give rise to tissue-specific disease phenotypes? *J. Cell Sci.* 114:9–19

Imai S, Nishibayashi S, Takao K, Tomifuji M, Fujino T, et al. 1997. Dissociation of Oct-1 from the nuclear peripheral structure induces the cellular aging-associated collagenase gene expression. *Mol. Biol. Cell* 8: 2407–19

Iouk T, Kerscher O, Scott RJ, Basrai MA, Wozniak RW. 2002. The yeast nuclear pore complex functionally interacts with components of the spindle assembly checkpoint. *J. Cell Biol.* 159:807–19

Iovine MK, Watkins JL, Wente SR. 1995. The GLFG repetitive region of the nucleoporin Nup116p interacts with Kap95p, an essential yeast nuclear import factor. *J. Cell Biol.* 131:1699–713

Ishii K, Arib G, Lin C, Van Houwe G, Laemmli UK. 2002. Chromatin boundaries in budding yeast: the nuclear pore connection. *Cell* 109: 551–62

This report explained, for the first time, that the previously described GTPase sensitive step in nuclear assembly reflects a requirement for Ran.

Jahn R, Grubmuller H. 2002. Membrane fusion. *Curr. Opin. Cell Biol.* 14:488–95

Jakel S, Mingot JM, Schwarzmaier P, Hartmann E, Gorlich D. 2002. Importins fulfill a dual function as nuclear import receptors and cytoplasmic chaperones for exposed basic domains. *EMBO J.* 21:377–86

Joseph J, Liu ST, Jablonski SA, Yen TJ, Dasso M. 2004. The RanGAP1-RanBP2 complex is essential for microtubule-kinetochore interactions in vivo. *Curr. Biol.* 14:611–17

Joseph J, Tan SH, Karpova TS, McNally JG, Dasso M. 2002. SUMO-1 targets RanGAP1 to kinetochores and mitotic spindles. *J. Cell Biol.* 156:595–602

Kalab P, Pu RT, Dasso M. 1999. The ran GTPase regulates mitotic spindle assembly. *Curr. Biol.* 9:481–84

Kalab P, Weis K, Heald R. 2002. Visualization of a Ran-GTP gradient in interphase and mitotic Xenopus egg extracts. *Science* 295:2452–56

Kau TR, Way JC, Silver PA. 2004. Nuclear transport and cancer: from mechanism to intervention. *Nat. Rev. Cancer* 4:106–17

Kessel RG. 1992. Annulate lamellae: a last frontier in cellular organelles. *Int. Rev. Cytol.* 133:43–120

Kimura T, Ito C, Watanabe S, Takahashi T, Ikawa M, et al. 2003. Mouse germ cell-less as an essential component for nuclear integrity. *Mol. Cell Biol.* 23:1304–15

Kiseleva E, Allen TD, Rutherford S, Bucci M, Wente SR, Goldberg MW. 2004. Yeast nuclear pore complexes have a cytoplasmic ring and internal filaments. *J. Struct. Biol.* 145:272–88

Kiseleva E, Rutherford S, Cotter LM, Allen TD, Goldberg MW. 2001. Steps of nuclear pore complex disassembly and reassembly during mitosis in early Drosophila embryos. *J. Cell Sci.* 114:3607–18

Krull S, Thyberg J, Bjorkroth B, Rackwitz HR, Cordes VC. 2004. Nucleoporins as components of the nuclear pore complex core structure and tpr as the architectural element of the nuclear basket. *Mol. Biol. Cell* 15:4261–77

Kubitscheck U, Grunwald D, Hoekstra A, Rohleder D, Kues T, et al. 2005. Nuclear transport of single molecules: dwell times at the nuclear pore complex. *J. Cell Biol.* 168:233–43

Kurihara LJ, Beh CT, Latterich M, Schekman R, Rose MD. 1994. Nuclear congression and membrane fusion: two distinct events in the yeast karyogamy pathway. *J. Cell Biol.* 126:911–23

Larijani B, Barona TM, Poccia DL. 2001. Role for phosphatidylinositol in nuclear envelope formation. *Biochem. J.* 356:495–501

Latterich M, Frohlich KU, Schekman R. 1995. Membrane fusion and the cell cycle: Cdc48p participates in the fusion of ER membranes. *Cell* 82:885–93

Latterich M, Schekman R. 1994. The karyogamy gene KAR2 and novel proteins are required for ER-membrane fusion. *Cell* 78:87–98

Lau CK, Giddings TH Jr, Winey M. 2004. A novel allele of Saccharomyces cerevisiae NDC1 reveals a potential role for the spindle pole body component Ndc1p in nuclear pore assembly. *Eukaryot. Cell* 3:447–58

Lee KK, Starr D, Cohen M, Liu J, Han M, et al. 2002. Lamin-dependent localization of UNC-84, a protein required for nuclear migration in *Caenorhabditis elegans*. *Mol. Biol. Cell* 13:892–901

Lee MC, Miller EA, Goldberg J, Orci L, Schekman R. 2004. Bi-directional protein transport between the ER and Golgi. *Annu. Rev. Cell Dev. Biol.* 20:87–123

Lemaitre JM, Geraud G, Mechali M. 1998. Dynamics of the genome during early Xenopus laevis development: karyomeres as independent units of replication. *J. Cell Biol.* 142:1159–66

This study is the first experimental evidence for the existence of the Ran-GTP gradient.

Lenart P, Rabut G, Daigle N, Hand AR, Terasaki M, Ellenberg J. 2003. Nuclear envelope breakdown in starfish oocytes proceeds by partial NPC disassembly followed by a rapidly spreading fenestration of nuclear membranes. *J. Cell Biol.* 160:1055–68

Liu J, Lee KK, Segura-Totten M, Neufeld E, Wilson KL, Gruenbaum Y. 2003a. MAN1 and emerin have overlapping function(s) essential for chromosome segregation and cell division in *Caenorhabditis elegans*. *Proc. Natl. Acad. Sci. USA* 100:4598–603

Liu J, Prunuske AJ, Fager AM, Ullman KS. 2003b. The COPI complex functions in nuclear envelope breakdown and is recruited by the nucleoporin Nup153. *Dev. Cell* 5:487–98

Lohka MJ, Masui Y. 1983. Formation in vitro of sperm pronuclei and mitotic chromosomes induced by amphibian ooplasmic components. *Science* 220:719–21

Loiodice I, Alves A, Rabut G, VanOverbeek M, Ellenberg J, et al. 2004. The entire Nup107-160 complex, including three new members, is targeted as one entity to kinetochores in mitosis. *Mol. Biol. Cell* 15:3333–44

Luo K, Vega-Palas MA, Grunstein M. 2002. Rap1-Sir4 binding independent of other Sir, yKu, or histone interactions initiates the assembly of telomeric heterochromatin in yeast. *Genes Dev.* 16: 1528–39

Lusk CP, Makhnevych T, Marelli M, Aitchison JD, Wozniak RW. 2002. Karyopherins in nuclear pore biogenesis: a role for Kap121p in the assembly of Nup53p into nuclear pore complexes. *J. Cell Biol.* 159:267–78

Lutzmann M, Kunze R, Buerer A, Aebi U, Hurt E. 2002. Modular self-assembly of a Y-shaped multiprotein complex from seven nucleoporins. *EMBO J.* 21:387–97

Macaulay C, Forbes DJ. 1996. Assembly of the nuclear pore: biochemically distinct steps revealed with NEM, GTP gamma S, and BAPTA. *J. Cell Biol.* 132:5–20

Macaulay C, Meier E, Forbes DJ. 1995. Differential mitotic phosphorylation of proteins of the nuclear pore complex. *J. Biol. Chem.* 270:254–62

Makhnevych T, Lusk CP, Anderson AM, Aitchison JD, Wozniak RW. 2003. Cell cycle regulated transport controlled by alterations in the nuclear pore complex. *Cell* 115:813–23

Malone CJ, Misner L, Le Bot N, Tsai MC, Campbell JM, et al. 2003. The *C. elegans* hook protein, ZYG-12, mediates the essential attachment between the centrosome and nucleus. *Cell* 115:825–36

Mans BJ, Anantharaman V, Aravind L, Koonin EV. 2004. Comparative genomics, evolution and origins of the nuclear envelope and nuclear pore complex. *Cell Cycle* 3:1612–37

Maraldi NM, Lattanzi G, Sabatelli P, Ognibene A, Squarzoni S. 2002. Functional domains of the nucleus: implications for Emery-Dreifuss muscular dystrophy. *Neuromuscul. Disord.* 12:815–23

Marelli M, Lusk CP, Chan H, Aitchison JD, Wozniak RW. 2001. A link between the synthesis of nucleoporins and the biogenesis of the nuclear envelope. *J. Cell Biol.* 153:709–24

Markiewicz E, Dechat T, Foisner R, Quinlan RA, Hutchison CJ. 2002. Lamin A/C binding protein LAP2alpha is required for nuclear anchorage of retinoblastoma protein. *Mol. Biol. Cell* 13:4401–13

Mattaj IW. 2004. Sorting out the nuclear envelope from the endoplasmic reticulum. *Nat. Rev. Mol. Cell. Biol.* 5:65–69

Mattaj IW, Englmeier L. 1998. Nucleocytoplasmic transport: the soluble phase. *Annu. Rev. Biochem.* 67:265–306

Maul GG. 1977. Nuclear pore complexes. Elimination and reconstruction during mitosis. *J. Cell Biol.* 74:492–500

This is a landmark paper in cell biology; it laid the foundations for many of the in vitro studies during the past two decades that focused on nuclear re-formation.

The authors present evidence for mitotic NPC rearrangements in fission yeast.

This is the most comprehensive study on NPC density.

Maul GG, Maul HM, Scogna JE, Lieberman MW, Stein GS, et al. 1972. Time sequence of nuclear pore formation in phytohemagglutinin-stimulated lymphocytes and in HeLa cells during the cell cycle. *J. Cell Biol.* 55:433–47

Maul GG, Price JW, Lieberman MW. 1971. Formation and distribution of nuclear pore complexes in interphase. *J. Cell Biol.* 51:405–18

Meier E, Miller BR, Forbes DJ. 1995. Nuclear pore complex assembly studied with a biochemical assay for annulate lamellae formation. *J. Cell Biol.* 129:1459–72

Mislow JM, Holaska JM, Kim MS, Lee KK, Segura-Totten M, et al. 2002a. Nesprin-1alpha self-associates and binds directly to emerin and lamin A in vitro. *FEBS Lett.* 525:135–40

Mislow JM, Kim MS, Davis DB, McNally EM. 2002b. Myne-1, a spectrin repeat transmembrane protein of the myocyte inner nuclear membrane, interacts with lamin A/C. *J. Cell Sci.* 115:61–70

Miyachi K, Hirano Y, Horigome T, Mimori T, Miyakawa H, et al. 2004. Autoantibodies from primary biliary cirrhosis patients with anti-p95c antibodies bind to recombinant p97/VCP and inhibit in vitro nuclear envelope assembly. *Clin. Exp. Immunol.* 136:568–73

Mounkes L, Kozlov S, Burke B, Stewart CL. 2003. The laminopathies: nuclear structure meets disease. *Curr. Opin. Genet. Dev.* 13:223–30

Mutvei A, Dihlmann S, Herth W, Hurt F.C. 1992. NSP1 depletion in yeast affects nuclear pore formation and nuclear accumulation. *Eur. J. Cell Biol.* 59:280–95

Nachury MV, Weis K. 1999. The direction of transport through the nuclear pore can be inverted. *Proc. Natl. Acad. Sci. USA* 96:9622–27

Nanduri J, Mitra S, Andrei C, Liu Y, Yu Y, et al. 1999. An unexpected link between the secretory path and the organization of the nucleus. *J. Biol. Chem.* 274:33785–89

Nehrbass U, Fabre E, Dihlmann S, Herth W, Hurt EC. 1993. Analysis of nucleo-cytoplasmic transport in a thermosensitive mutant of nuclear pore protein NSP1. *Eur. J. Cell Biol.* 62:1–12

Nemergut ME, Mizzen CA, Stukenberg T, Allis CD, Macara IG. 2001. Chromatin docking and exchange activity enhancement of RCC1 by histones H2A and H2B. *Science* 292:1540–43

Newport J, Dunphy W. 1992. Characterization of the membrane binding and fusion events during nuclear envelope assembly using purified components. *J. Cell Biol.* 116:295–306

Newport JW, Forbes DJ. 1987. The nucleus: structure, function, and dynamics. *Annu. Rev. Biochem.* 56:535–65

Newport JW, Wilson KL, Dunphy WG. 1990. A lamin-independent pathway for nuclear envelope assembly. *J. Cell Biol.* 111:2247–59

Ohtsubo M, Okazaki H, Nishimoto T. 1989. The RCC1 protein, a regulator for the onset of chromosome condensation locates in the nucleus and binds to DNA. *J. Cell Biol.* 109:1389–97

Olsson M, Scheele S, Ekblom P. 2004. Limited expression of nuclear pore membrane glycoprotein 210 in cell lines and tissues suggests cell-type specific nuclear pores in metazoans. *Exp. Cell Res.* 292:359–70

Onischenko EA, Gubanova NV, Kieselbach T, Kiseleva EV, Hallberg E. 2004. Annulate lamellae play only a minor role in the storage of excess nucleoporins in Drosophila embryos. *Traffic* 5:152–64

Osada S, Ohmori SY, Taira M. 2003. XMAN1, an inner nuclear membrane protein, antagonizes BMP signaling by interacting with Smad1 in Xenopus embryos. *Development* 130:1783–94

Padmakumar VC, Abraham S, Braune S, Noegel AA, Tunggal B, et al. 2004. Enaptin, a giant actin-binding protein, is an element of the nuclear membrane and the actin cytoskeleton. *Exp. Cell Res.* 295:330–39

Patterson K, Molofsky AB, Robinson C, Acosta S, Cater C, Fischer JA. 2004. The functions of Klarsicht and nuclear lamin in developmentally regulated nuclear migrations of photoreceptor cells in the Drosophila eye. *Mol. Biol. Cell* 15:600–10

Peter M, Nakagawa J, Doree M, Labbe JC, Nigg EA. 1990. In vitro disassembly of the nuclear lamina and M phase-specific phosphorylation of lamins by cdc2 kinase. *Cell* 61:591–602

Prufert K, Vogel A, Krohne G. 2004. The lamin CxxM motif promotes nuclear membrane growth. *J. Cell Sci.* 117:6105–16

Pyrpasopoulou A, Meier J, Maison C, Simos G, Georgatos SD. 1996. The lamin B receptor (LBR) provides essential chromatin docking sites at the nuclear envelope. *EMBO J.* 15:7108–19

Rabut G, Doye V, Ellenberg J. 2004. Mapping the dynamic organization of the nuclear pore complex inside single living cells. *Nat. Cell Biol.* 6:1114–21

Reichelt R, Holzenburg A, Buhle EL Jr, Jarnik M, Engel A, Aebi U. 1990. Correlation between structure and mass distribution of the nuclear pore complex and of distinct pore complex components. *J. Cell Biol.* 110:883–94

Ribbeck K, Gorlich D. 2002. The permeability barrier of nuclear pore complexes appears to operate via hydrophobic exclusion. *EMBO J.* 21:2664–71

Ribbeck K, Kutay U, Paraskeva E, Gorlich D. 1999. The translocation of transportin-cargo complexes through nuclear pores is independent of both Ran and energy. *Curr. Biol.* 9:47–50

Roos UP. 1973. Light and electron microscopy of rat kangaroo cells in mitosis. II. Kinetochore structure and function. *Chromosoma* 41:195–220

Rose M. 1996. Nuclear fusion in the yeast Saccharomyces cerevisiae. *Annu. Rev. Biochem.* 12:663–95

Rout MP, Aitchison JD. 2001. The nuclear pore complex as a transport machine. *J. Biol. Chem.* 276:16593–96

Rout MP, Aitchison JD, Magnasco MO, Chait BT. 2003. Virtual gating and nuclear transport: the hole picture. *Trends Cell Biol.* 13:622–28

Rout MP, Aitchison JD, Suprapto A, Hjertaas K, Zhao Y, Chait BT. 2000. The yeast nuclear pore complex: composition, architecture, and transport mechanism. *J. Cell Biol.* 148:635–51

Rout MP, Blobel G. 1993. Isolation of the yeast nuclear pore complex. *J. Cell Biol.* 123:771–83

Roy R, Meier B, McAinsh AD, Feldmann HM, Jackson SP. 2004. Separation-of-function mutants of yeast Ku80 reveal a Yku80p-Sir4p interaction involved in telomeric silencing. *J. Biol. Chem.* 279:86–94

Ryan KJ, McCaffery JM, Wente SR. 2003. The Ran GTPase cycle is required for yeast nuclear pore complex assembly. *J. Cell Biol.* 160:1041–53

Ryan KJ, Wente SR. 2002. Isolation and characterization of new Saccharomyces cerevisiae mutants perturbed in nuclear pore complex assembly. *BMC Genet.* 3:17

Saitoh H, Sparrow DB, Shiomi T, Pu RT, Nishimoto T, et al. 1998. Ubc9p and the conjugation of SUMO-1 to RanGAP1 and RanBP2. *Curr. Biol.* 8:121–24

Salina D, Bodoor K, Eckley DM, Schroer TA, Rattner JB, Burke B. 2002. Cytoplasmic dynein as a facilitator of nuclear envelope breakdown. *Cell* 108:97–107

Salina D, Enarson P, Rattner JB, Burke B. 2003. Nup358 integrates nuclear envelope breakdown with kinetochore assembly. *J. Cell Biol.* 162:991–1001

Sasagawa S, Yamamoto A, Ichimura T, Omata S, Horigome T. 1999. In vitro nuclear assembly with affinity-purified nuclear envelope precursor vesicle fractions, PV1 and PV2. *Eur. J. Cell Biol.* 78:593–600

The authors demonstrate that the NPC is preferentially permeable to hydrophobic proteins and formulate the hydrophobic exclusion model, and that certain aliphatic alcohols act as permeation enhancers.

In this study, proteins from the purified yeast pore were subjected to mass spectroscopy sequencing. A "Brownian ratchet" model was proposed to explain nuclear translocation.

Schirmer EC, Florens L, Guan T, Yates JR 3rd, Gerace L. 2003. Nuclear membrane proteins with potential disease links found by subtractive proteomics. *Science* 301:1380–82

Sheehan MA, Mills AD, Sleeman AM, Laskey RA, Blow JJ. 1988. Steps in the assembly of replication-competent nuclei in a cell-free system from Xenopus eggs. *J. Cell Biol.* 106:1–12

Siniossoglou S, Lutzmann M, Santos-Rosa H, Leonard K, Mueller S, et al. 2000. Structure and assembly of the Nup84p complex. ***J. Cell Biol.*** **149:41–54**

Siniossoglou S, Santos-Rosa H, Rappsilber J, Mann M, Hurt E. 1998. A novel complex of membrane proteins required for formation of a spherical nucleus. *EMBO J.* 17:6449–64

Soderqvist H, Hallberg E. 1994. The large C-terminal region of the integral pore membrane protein, POM121, is facing the nuclear pore complex. *Eur. J. Cell Biol.* 64:186–91

Spann TP, Goldman AE, Wang C, Huang S, Goldman RD. 2002. Alteration of nuclear lamin organization inhibits RNA polymerase II-dependent transcription. *J. Cell Biol.* 156:603–8

Starr DA, Han M. 2002. Role of ANC-1 in tethering nuclei to the actin cytoskeleton. *Science* 298:406–9

Steen RL, Collas P. 2001. Mistargeting of B-type lamins at the end of mitosis: implications on cell survival and regulation of lamins A/C expression. *J. Cell Biol.* 153:621–26

Stoffler D, Feja B, Fahrenkrog B, Walz J, Typke D, Aebi U. 2003. Cryo-electron tomography provides novel insights into nuclear pore architecture: implications for nucleocytoplasmic transport. *J. Mol. Biol.* 328:119–30

Stoffler D, Goldie KN, Feja B, Aebi U. 1999. Calcium-mediated structural changes of native nuclear pore complexes monitored by time-lapse atomic force microscopy. *J. Mol. Biol.* 287:741–52

Strambio-de-Castillia C, Blobel G, Rout MP. 1999. Proteins connecting the nuclear pore complex with the nuclear interior. *J. Cell Biol.* 144:839–55

Strawn LA, Shen T, Shulga N, Goldfarb DS, Wente SR. 2004. Minimal nuclear pore complexes define FG repeat domains essential for transport. *Nat. Cell Biol.* 6:197–206

Sullivan T, Escalante-Alcalde D, Bhatt H, Anver M, Bhat N, et al. 1999. Loss of A-type lamin expression compromises nuclear envelope integrity leading to muscular dystrophy. *J. Cell Biol.* 147:913–20

Suntharalingam M, Wente SR. 2003. Peering through the pore: nuclear pore complex structure, assembly, and function. *Dev. Cell* 4:775–89

Taddei A, Hediger F, Neumann FR, Bauer C, Gasser SM. 2004a. Separation of silencing from perinuclear anchoring functions in yeast Ku80, Sir4 and Esc1 proteins. *EMBO J.* 23:1301–12

Taddei A, Hediger F, Neumann FR, Gasser SM. 2004b. The function of nuclear architecture: a genetic approach. *Annu. Rev. Genet.* 38:305–45

Tcheperegine SE, Marelli M, Wozniak RW. 1999. Topology and functional domains of the yeast pore membrane protein Pom152p. *J. Biol. Chem.* 274:5252–58

Thomas JH, Botstein D. 1986. A gene required for the separation of chromosomes on the spindle apparatus in yeast. *Cell* 44:65–76

Ulitzur N, Harel A, Goldberg M, Feinstein N, Gruenbaum Y. 1997. Nuclear membrane vesicle targeting to chromatin in a Drosophila embryo cell-free system. *Mol. Biol. Cell* 8:1439–48

Vasu S, Shah S, Orjalo A, Park M, Fischer WH, Forbes DJ. 2001. Novel vertebrate nucleoporins Nup133 and Nup160 play a role in mRNA export. *J. Cell Biol.* 155:339–54

Vasu SK, Forbes DJ. 2001. Nuclear pores and nuclear assembly. *Curr. Opin. Cell Biol.* 13:363–75

Vigers GP, Lohka MJ. 1991. A distinct vesicle population targets membranes and pore complexes to the nuclear envelope in Xenopus eggs. *J. Cell Biol.* 112:545–56

This is a detailed study of the Nup84 subcomplex of the yeast nuclear pore, demonstrating that sec13 plays a significant role in the nuclear pore.

Walther TC, Alves A, Pickersgill H, Loiodice I, Hetzer M, et al. 2003a. The conserved Nup107-160 complex is critical for nuclear pore complex assembly. *Cell* 113:195–206

Walther TC, Askjaer P, Gentzel M, Habermann A, Griffiths G, et al. 2003b. RanGTP mediates nuclear pore complex assembly. *Nature* 424:689–94

Walther TC, Pickersgill HS, Cordes VC, Goldberg MW, Allen TD, et al. 2002. The cytoplasmic filaments of the nuclear pore complex are dispensable for selective nuclear protein import. *J. Cell Biol.* 158:63–77

Weirich CS, Erzberger JP, Berger JM, Weis K. 2004. The N-terminal domain of Nup159 forms a beta-propeller that functions in mRNA export by tethering the helicase Dbp5 to the nuclear pore. *Mol. Cell* 16:749–60

Weis K. 2002. Nucleocytoplasmic transport: cargo trafficking across the border. *Curr. Opin. Cell Biol.* 14:328–35

Weis K. 2003. Regulating access to the genome: nucleocytoplasmic transport throughout the cell cycle. *Cell* 112:441–51

Wente SR. 2000. Gatekeepers of the nucleus. *Science* 288:1374–77

Wiese C, Goldberg MW, Allen TD, Wilson KL. 1997. Nuclear envelope assembly in Xenopus extracts visualized by scanning EM reveals a transport-dependent 'envelope smoothing' event. *J. Cell Sci.* 110(Pt. 13):1489–502

Wilson KL, Newport J. 1988. A trypsin-sensitive receptor on membrane vesicles is required for nuclear envelope formation in vitro. *J. Cell Biol.* 107:57–68

Winey M, Hoyt MA, Chan C, Goetsch L, Botstein D, Byers B. 1993. NDC1: a nuclear periphery component required for yeast spindle pole body duplication. *J. Cell Biol.* 122:743–51

Winey M, Yarar D, Giddings TH Jr, Mastronarde DN. 1997. Nuclear pore complex number and distribution throughout the Saccharomyces cerevisiae cell cycle by three-dimensional reconstruction from electron micrographs of nuclear envelopes. *Mol. Biol. Cell* 8:2119–32

Woodman PG. 2003. p97, a protein coping with multiple identities. *J. Cell Sci.* 116:4283–90

Worman HJ, Courvalin JC. 2000. The inner nuclear membrane. *J. Membr. Biol.* 177:1–11

Worman HJ, Courvalin JC. 2004. How do mutations in lamins A and C cause disease? *J. Clin. Invest.* 113:349–51

Wozniak RW, Blobel G. 1992. The single transmembrane segment of gp210 is sufficient for sorting to the pore membrane domain of the nuclear envelope. *J. Cell Biol.* 119:1441–49

Wozniak RW, Lusk CP. 2003. Nuclear pore complexes. *Curr. Biol.* 13:R169

Yang L, Guan T, Gerace L. 1997. Integral membrane proteins of the nuclear envelope are dispersed throughout the endoplasmic reticulum during mitosis. *J. Cell Biol.* 137:1199–210

Yang Q, Rout MP, Akey CW. 1998. Three-dimensional architecture of the isolated yeast nuclear pore complex: functional and evolutionary implications. *Mol. Cell* 1:223–34

Yaseen NR, Blobel G. 1999. Two distinct classes of Ran-binding sites on the nucleoporin Nup-358. *Proc. Natl. Acad. Sci. USA* 96:5516–21

Yokoyama N, Hayashi N, Seki T, Pante N, Ohba T, et al. 1995. A giant nucleopore protein that binds Ran/TC4. *Nature* 376:184–88

Zabel U, Doye V, Tekotte H, Wepf R, Grandi P, Hurt EC. 1996. Nic96 is required for nuclear pore formation and functionally interacts with a novel nucleoporin, Nup188p. *J. Cell Biol.* 133:1141–52

Zatsepina OV, Poliakov V, Chentsov IuS.1976. Electron-microscopic study of the nuclear membrane of the PKEV cells during mitosis. 1. The breakdown of the nuclear membrane (prophase, prometaphase, metaphase). *Tsitologiia* 18:1299–304

Zeitler B, Weis K. 2004. The FG-repeat asymmetry of the nuclear pore complex is dispensable for bulk nucleocytoplasmic transport in vivo. *J. Cell Biol.* 167:583–90

Zeligs JD, Wollman SH. 1979. Mitosis in rat thyroid epithelial cells in vivo. I. Ultrastructural changes in cytoplasmic organelles during the mitotic cycle. *J. Ultrastruct. Res.* 66:53–77

Zhang C, Clarke PR. 2000. Chromatin-independent nuclear envelope assembly induced by Ran GTPase in Xenopus egg extracts. *Science* 288:1429–32

Zhang C, Clarke PR. 2001. Roles of Ran-GTP and Ran-GDP in precursor vesicle recruitment and fusion during nuclear envelope assembly in a human cell-free system. *Curr. Biol.* 11:208–12

Zhang Q, Ragnauth C, Greener MJ, Shanahan CM, Roberts RG. 2002. The nesprins are giant actin-binding proteins, orthologous to Drosophila melanogaster muscle protein MSP-300. *Genomics* 80:473–81

Zhang Q, Skepper JN, Yang F, Davies JD, Hegyi L, et al. 2001. Nesprins: a novel family of spectrin-repeat-containing proteins that localize to the nuclear membrane in multiple tissues. *J. Cell Sci.* 114:4485–98

Integrin Structure, Allostery, and Bidirectional Signaling

M.A. Arnaout, B. Mahalingam, and J.-P. Xiong

Structural Biology Program, Leukocyte Biology and Inflammation Program, Nephrology Division, Department of Medicine, Massachusetts General Hospital and Harvard Medical School, Charlestown, Massachussetts 02129; email: arnaout@receptor.mgh.harvard.edu, bmahalingam@partners.org, xiong@helix.mgh.harvard.edu

Annu. Rev. Cell Dev. Biol. 2005. 21:381–410

First published online as a Review in Advance on June 28, 2005

The *Annual Review of Cell and Developmental Biology* is online at http://cellbio.annualreviews.org

doi: 10.1146/ annurev.cellbio.21.090704.151217

Copyright © 2005 by Annual Reviews. All rights reserved

1081-0706/05/1110-0381$20.00

Key Words

cell adhesion, inflammation, cancer, hemostasis, therapeutics

Abstract

$\alpha\beta$ heterodimeric integrins mediate dynamic adhesive cell-cell and cell-extracellular matrix (ECM) interactions in metazoa that are critical in growth and development, hemostasis, and host defense. A central feature of these receptors is their capacity to change rapidly and reversibly their adhesive functions by modulating their ligand-binding affinity. This is normally achieved through interactions of the short cytoplasmic integrin tails with intracellular proteins, which trigger restructuring of the ligand-binding site through long-range conformational changes in the ectodomain. Ligand binding in turn elicits conformational changes that are transmitted back to the cell to regulate diverse responses. The publication of the integrin $\alpha V\beta 3$ crystal structure has provided the context for interpreting decades-old biochemical studies. Newer NMR, crystallographic, and EM data, reviewed here, are providing a better picture of the dynamic integrin structure and the allosteric changes that guide its diverse functions.

Contents

INTRODUCTION

Cells exist in a highly dynamic extracellular milieu consisting of complex chemicals and mechanical stress signals to which cells must continuously adjust and in turn modulate. In metazoa, the task of integrating these mechanochemical cues across the plasma membrane is mediated mainly by integrins: heterodimeric cell surface glycoproteins comprised of α- and β-subunits each spanning the plasma membrane once. Since the identification of this receptor family more than two decades ago, intensive efforts have been made to understand its complex functions as mechanochemical sensors and transducers. More recent studies have focused on elucidating the underlying structure-activity relationships in order to understand the regulatory roles of these receptors in cellular function. Since integrins also participate in injury and disease, a potential outcome is the discovery of small molecule antagonists to treat common diseases linked to integrin dysfunction such as cancer, thrombosis, and chronic inflammation. Here, we review these new structural advances, placing them in the biologic context.

THE INTEGRIN FAMILY

In mammals, 24 integrins have been identified to date, resulting from different pairings among 18 α- and 8 β-subunits. The extracellular segment of the α- and β-subunits has up to 1104 residues and 778 residues respectively, with the N-terminal portions of each subunit combining to form a globular ligand-binding "head" connected to the membrane by a long (~170 Å) stalk. One-half of the mammalian α-subunits contain an additional ~190 amino acid vWFA domain (αA, also known as I domain); nine integrins belong to this αA-containing subclass.

Integrins recognize a large number of physiologic ligands, including soluble and surface-bound proteins. Unlike binding to soluble ligands, integrin binding to ligands on rigid surfaces results in force generation at the focal contacts (meeting points of the integrin with ligand), which directly affects the cell's contractile apparatus and is a key element in the assembly and remodeling of the extracellular matrix (Ingber 2003, Ridley 2004). Integrin binding to ligands requires divalent cations and invariably involves a

vWFA: von Willebrand Factor type A

solvent-exposed aspartate or a glutamate (in the case of the αA-containing subclass) from the ligand. High-affinity binding of αA-lacking or -containing integrins to physiologic ligands is not constitutive but requires a conformational switch of the ligand-binding site. This regulatory mechanism allows leukocytes and platelets, for example, to circulate in the blood without pathologically adhering to each other or to the vascular wall. The switch to the high-affinity state is rapid, with a subsecond time frame, is reversible in less than a minute, and is triggered from within the cell (hence the term inside-out activation) in response to extracellular chemical (Constantin et al. 2000, Lollo et al. 1993) and/or mechanical stress signals (Zwartz et al. 2004b). The intracellular activation signals are channeled through the short integrin cytoplasmic and transmembrane segments to the extracellular ligand-binding pocket. Many integrins [e.g., αMβ2 (CD11b/CD18) or αIIbβ3] exist in two activation states (active and inactive) (Litvinov et al. 2002, Xiong et al. 2000), while some [e.g., α4β1 (Chigaev et al. 2001)] may also display intermediate-affinity states. Integrins bound to soluble or immobilized physiologic ligands form micro- (tens of angstroms scale) or macro- (tens of nanometer scale) clusters (Kim et al. 2004) and transmit mechanochemical signals inwards (outside-in signaling) that reorganize the cytoskeleton; in this way, integrins modulate much of the cell's metabolic and signal transduction machinery (Ingber 2003, Schwartz & Ginsberg 2002).

The inside-out–driven change in integrin affinity occurs primarily through a 30–100-fold reduction in the dissociation rate, with a smaller contribution from the increased association rate, implying that conformation of the ligand-binding pocket rather than its accessibility on the cell surface is the main contributor to stimulus-induced affinity switching (Zwartz et al. 2004a). The dissociation rates of high-affinity integrins (0.02–0.2 s^{-1}) responsible for firm adhesion are typically an order of magnitude less than those of selectins (Bhatia et al. 2003). The high

dissociation rates ($\sim$2.6–4.6 s^{-1}) (Shimaoka et al. 2003b, Smith et al. 1999) of the low- and intermediate-affinity states allow some integrins (e.g., α4 integrins) to mediate cell rolling, like selectins. Thus depending on their affinity state, such integrins may mediate rolling or firm adhesion.

MIDAS: metal ion-dependent adhesion site

THE ALPHA SUBUNIT vWFA DOMAIN (αA)

Conformational States of αA

In 1995, the first crystal structure of the integrin αA domain was published (Lee et al. 1995b), soon after it was demonstrated that this domain is the major ligand-binding site in αA-containing integrins (Michishita et al. 1993). The globular αA is composed of a GTPase-like domain, in which the catalytic center on the domain's upper surface is replaced with an invariant ligand-binding site named MIDAS. Isolated αA domains exist in two conformations— closed and open (**Figure 1**)—which equate respectively with the low- and high-affinity ligand-binding states of the domain (Emsley et al. 2000, Lee et al. 1995a,b, Li et al. 1998, Shimaoka et al. 2003b, Xiong et al. 2000). In the closed state, a metal ion (Mn^{2+}) is coordinated by a D140xSxS MIDAS motif from the βA-α1 loop (loop 1, x for any amino acid, residue numbering is from CD11b), a T209 from the α3-α4 loop (loop 2) that hydrogen bonds to a metal-bound water, and a D242 from the βD-α5 loop (loop 3) that directly coordinates the metal ion producing the Thr-water-metal-Asp closed MIDAS configuration. A water molecule completes the octahedral coordination sphere around the metal. In all closed αA structures, the βF-α7 (F-α7) loop occupies an identical position, stabilized by hydrophobic contacts with the central β-sheet, but the C-terminal α7 helix is not always packed against the central strand. In the αA structure from CD11a, this helix is flexible following its first helical turn (Qu &

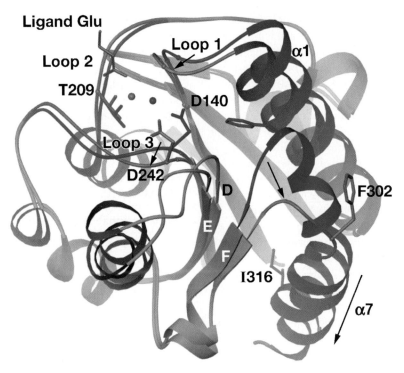

Figure 1

Structures of closed and open αA from integrin CD11b. Ribbon diagram of the crystal structures of open and closed αA from integrin CD11b. Major conformational differences are shown in red (*open*) and blue (*closed*), with the MIDAS metal shown in the respective color. The ligand Glu is in red and the MIDAS residues T209, D242, and the D140xSxS motif are indicated (for clarity, only the side chains for the first two are shown). Arrows indicate direction of major movements. This figure and the subsequent figures are made using Ribbons (Carson 1987).

Leahy 1995, 1996), perhaps as a function of the construct and/or crystal packing.

All three MIDAS loops rearrange as the domain switches to the open state (**Figure 1**). This appears to be driven by a major restructuring of the F-α7 loop and is associated with a downward two-turn displacement of the α7 helix. The rearranged contacts of the F-α7 loop with the hydrophobic core lead to a 2 Å inward pull of the α1 helix, resulting in the open MIDAS configuration Thr-metal-water-Asp. This change is linked to a 2 Å inward movement loop 1 (bearing the MIDAS motif) and in position of loop 3 bearing D242 that allows T209 to coordinate directly the metal ion (Mg^{2+} or Mn^{2+}) while D242 hydrogen bonds to a metal-bound water molecule. The resulting increased elec-

trophilicity of MIDAS permits a glutamate from the ligand to provide the sixth metal coordination, replacing a water molecule. This restructuring of MIDAS also results in a ten fold enthalpy-driven increase (from mM to ~0.1 mM) in affinity of MIDAS to the proadhesive physiologic cation Mg^{2+} approaching that of Mn^{2+} (~0.03 mM), whose affinity remains unchanged in both conformations. Activation of αA does not alter the low (>1 mM) affinity of this domain to the generally inhibitory Ca^{2+} (Ajroud et al. 2004). Atomic force microscopy (AFM) measurements of the rupture forces of the cell-bound αA-integrin CD11a/CD18 from its surface-immobilized ligand were 20–320 pN at a loading rate of 0.02–50 nN s^{-1} (Zhang et al. 2002), revealing inner and outer activation barriers. The inner steep barrier probably reflects the ionic interaction between the ligand glutamate and the MIDAS ion; this interaction allows the complex to resist large pulling forces. The outer barrier is likely explained by the high-affinity determining tertiary changes in the structure.

Engineering Open Forms of αA

Stable open forms of αA have been produced by locking the F-α7 segment into the open conformation. A first approach was to break the α7 helix by substituting an invariant I316 (in CD11b, **Figure 1**) with glycine (Xiong et al. 2000). In the closed state, Ile316 is buried into a hydrophobic socket (socket for isoleucine or SILEN), linking the lower part of α7 to the central sheet of the CD11b and CD11c αA domains, but is displaced from this socket by the downward displacement of α7 (Vorup-Jensen et al. 2003, Xiong et al. 2000). The expressed domain assumed the open conformation, was of high affinity even in the context of the whole integrin, and exhibited a distinctive mAb-binding profile in solution (Xiong et al. 2000) and a 2 Å increase in its hydrodynamic Stokes radius (Vorup-Jensen et al. 2003). Engineered disulfides (based on the open structure of CD11b) targeting the N-terminal region of the α7 helix (Shimaoka et al. 2003b) or Lys315 of CD11b (McCleverty & Liddington 2003) yielded open but unliganded high-affinity αA domains (K_d ∼200–600 nM) that mirror the high-affinity state of the respective active parent integrin.

Mapping studies have identified the residues on the MIDAS face involved in αA binding to physiologic ligands (Li et al. 1998, Zhang & Plow 1999). These residues derive from the three conformationally sensitive MIDAS loops and encircle the MIDAS metal. More recently, crystal structure determination of αA domains in complex with certain physiologic ligands, i.e., a collagen peptide and CD54 (ICAM-1), defined the integrin-ligand interface in atomic detail and identified some of the very same MIDAS loop residues

identified earlier as directly contacting the ligand (Emsley et al. 2000, Shimaoka et al. 2003b). The F-α7 loop does not contribute directly to the αA-ligand interface but affects it allosterically. In liganded αA structures, the position of the F-α7 loop is almost superimposable, despite the C-terminal α7 helix's highly variable conformation, which is dictated by the disulfides used to lock open the domain (**Figure 2**). These findings suggest that rearrangement of the F-α7 loop in αA domains, normally through the axial downward

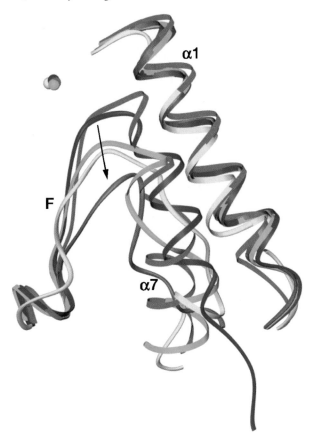

Figure 2

Structures of wild-type and mutant forms of closed and open αA domains in the F-α7 region. Ribbon diagrams of the F-α7 and α1 helix and MIDAS metal were superimposed using TOP (Lu 2000) from the crystal structures of the closed wild-type αA of CD11b (*magenta*) and the closed wild-type (*blue*), unliganded (*green*), and liganded (*yellow*) intermediate-affinity forms and the unliganded high-affinity form (*red*), all from CD11a. The MIDAS ion is colored accordingly, except for the unliganded high-affinity form, where no MIDAS metal is present in the structure (Shimaoka et al. 2003b). The arrow indicates direction of major movements in the F-α7 loop.

PSI domain: plexin-semaphorin-integrin domain

displacement of the $\alpha7$ helix or artificially by mutations, is the critical component in induction of the high-affinity state.

αA Antagonists

Although the high-affinity binding of physiologic ligands to αA-integrins is activation-dependent (i.e., requires the open state), nonphysiologic ligands that function as competitive antagonists bind with nM affinity to closed αA domains expressed either alone or in the context of the whole integrin. Examples include the hookworm-derived neutrophil inhibitory factor (Rieu et al. 1996), mAb 107 (Li et al. 2002b), and mAb AQC2 (Karpusas et al. 2003). A recent crystal structure of αA from integrin $\alpha1\beta1$ in complex with AQC2 Fab (Karpusas et al. 2003) revealed that a ligand Asp (instead of the Glu provided by physiologic ligands) completes the MIDAS coordination sphere, which assumes the open MIDAS configuration, yet the structure adopts the closed state (i.e., it lacks the associated protein movements in the F-$\alpha7$ loop and $\alpha7$ helix) (**Figure 3a**). Thus, with ligand-mimetic antagonists, induction of the open MIDAS configuration does not trigger the conformational changes in the F-$\alpha7$ axis at the base of the structure. This may be explained by a tighter interface or by additional contacts that prevent such movements, "freezing" the domain in the closed state.

A second class of αA antagonists acts noncompetitively to lock the domain in the inactive state. These antagonists act by stabilizing the hydrophobic contact between the $\alpha7$ helix and the central β-sheet (**Figure 3b**). The prototype of this group is lovastatin, identified using a chemical library screening approach (Kallen et al. 1999). Lovastatin wedges itself in a large hydrophobic pocket, IDAS (I domain allosteric site), between the central sheet and the $\alpha7$ helix of CD11a (Kallen et al. 1999); IDAS is equivalent to SILEN in CD11b (Xiong et al. 2000). The lovastatin binding pocket is present in one structure of isolated

closed αA of CD11a (Qu & Leahy 1995) but not another (generated in the absence of metal ions (Qu & Leahy 1996), but is readily accessible in native CD11a/CD18 (Kallen et al. 1999, Weitz-Schmidt et al. 2004). Additionally, mutations of the invariant I316 equivalent in CD11a are activating in the native integrin but not in isolated αA (Lupher et al. 2001; Li and Arnaout, unpublished data). These findings again suggest that flexibility of the CD11a $\alpha7$ helix seen in closed recombinant αA is not reflective of the normal position of this helix in the native inactive form of this integrin.

STRUCTURE OF THE INTEGRIN ECTODOMAIN

Integrin Domains and Quaternary Structure

The crystal structure of the unliganded ectodomain from the αA-lacking integrin αVβ3 was determined in the presence of Ca^{2+} and Mn^{2+} (Xiong et al. 2001, 2002, 2004). It revealed 12 domains: four in the αV-subunit and eight in the β3-subunit (**Figure 4a,b**). The two subunits assemble into a globular head built by two predicted domains: an N-terminal seven-bladed β-propeller domain of αV (Springer 1997) and an αA-like domain (βA) from β3 (Lee et al. 1995b). βA loops out from an Ig-like "hybrid" domain (β3 residues 55–108 and 353–432), which itself is inserted in the N-terminal PSI domain (residues 1–54 and residues 433–435) of β3. The PSI domain is followed by four EGF-like domains (of which EGF1 and 2 are poorly visible in the electron density map) and a novel membrane-proximal β-tail domain (βTD), all together forming the β3 leg. The corresponding αV leg is formed of an Ig-like thigh domain followed by two large β-sandwich domains, calf-1 and calf-2, the latter containing the proteolytic cleavage site that results in the heavy and light chains of αV upon reduction. An unexpected feature of the structure (Xiong et al. 2001) is that the legs are bent at the "knees" and folded back against the head of the same

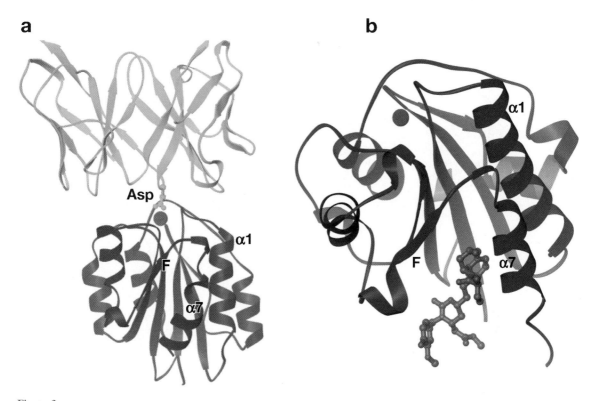

Figure 3

Structures of competitive and allosteric antagonists of αA. (*a*) Ribbon diagram of αA-Fab complex. αA is from integrin α1 (*blue*), and the antibody's variable heavy (*green*) and light (*cyan*) chains are shown. The MIDAS Mn^{2+} is the purple sphere. The sidechain of the ligand-mimetic D101, which coordinates MIDAS, is labeled, as is the F-α7 segment. (*b*) Structure of αA-lovastatin complex. Ribbon diagram drawn though the Cα positions of αA from integrin CD11a. Lovastatin (*red*) binds at a crevice between the F-strand and α7 helix (*labeled*). The MIDAS metal is the purple sphere.

molecule. This sharp bending occurs between the thigh and calf-1 domains of αV(α-genu) and approximately between EGF domains 1 and 2 of β3 (β-genu). Extension at the knees produces an integrin (**Figure 4***b*), which resembles the conventional shape of integrins derived from previous electron microscopy images, showing an elongated molecule with an oval ligand-binding head sitting on top of two extended legs (Nermut et al. 1988). A metal ion (Ca^{2+} or Mn^{2+}) occupies the α-genu in both the unliganded and liganded structures. At the base of the propeller, blades 4–7 each contain a metal ion coordinated in a β-hairpin loop, which can be occupied with Ca^{2+} or Mn^{2+}. These cations may help rigidify the propeller's interface with the thigh domain.

Heterodimer Formation

The major continuous intersubunit contact is present at the βA/propeller interface. At this interface's core is an Arg621 (a lysine in all other integrins except β8); the Arg621 protrudes from a 3_{10} helix of βA into the center of the propeller channel, and is caged into place by two rings of predominantly aromatic amino acids. The striking similarity of this interface to that which is between the Gα (an αA-like fold) and Gβ (a seven-bladed β-propeller) of G proteins suggests that this interface is likewise dynamic (Xiong et al. 2001). This core arrangement is duplicated in the binding interface of the βA domain with the propeller from integrin αIIb (Xiao et al. 2004).

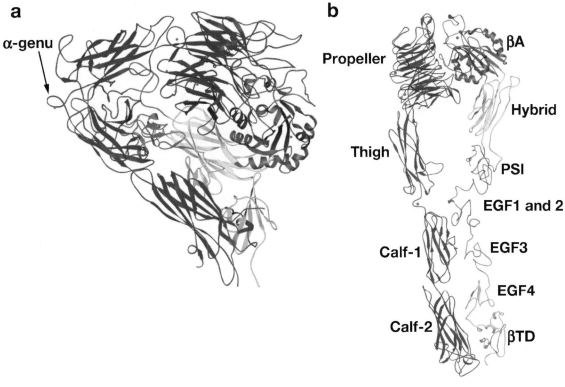

Figure 4

Structure of the integrin $\alpha V\beta 3$ ectodomain. (*a*) Ribbon diagram of the structure of the unliganded $\alpha V\beta 3$. The protein is bent at a flexible region (*α-genu, arrow*), occupied by a metal ion (*purple sphere*). (*b*) A genu-extended model clearly revealing the various integrin domains (*labeled*). EGF1 and EGF2 are not visible in the structure; their approximate location is indicated in gray. The four metal ions at the bottom of the propeller and the ADMIDAS ion (*all in purple*) are shown.

Additional contacts bury a surface area of about 1500 Å² in each integrin. Eight residues from βA [L258, D259, R261, L262 (from the 3_{10} helix), S291 and Y321, D217, and T254] contact (≤ 3.5 Å) the αV propeller. The first six also contact αIIb in integrin $\alpha IIb\beta 3$; in addition, R216, Lys253, S162, S168, P169, D179 (the last four from the specificity-determining loop, SDL), and E297 (from the $\alpha 5$ helix) contribute selectively to the βA interface with αIIb. The SDL segment may also contribute to the heterodimer interface in other integrins such as $\alpha 6\beta 1$, $\alpha V\beta 1$, and $\alpha L\beta 2$ but not $\alpha 5\beta 1$, $\alpha 4\beta 1$, $\alpha V\beta 3$, or $\alpha 6\beta 4$ (Takagi et al. 2002a). Inclusion of the $\beta 3$ SDL region in the interface with αIIb (but not αV) is accounted for by the longer B2-C2 (of blade 2) and D2-A3

loops (between blades 2 and 3) of the αIIb propeller. These findings indicate that usage of subunit interface residues varies among integrins, with a variable contribution of the SDL loop to this interface.

Structures of the Unliganded and Liganded βA Domain

The structure of βA is largely superimposable onto that of αA (Xiong et al. 2003). One distinguishing feature is the presence of two insertions in βA: one that forms the core of the interface with the α-subunit's propeller domain and the second, the SDL loop, which is involved in ligand binding and also contributes to the intersubunit interface in some integrins. A second distinguishing feature is the presence of a new cation site, ADMIDAS, which is adjacent to the MIDAS cation. The Ca^{2+} at this site is coordinated by the side chains of D126 and D127 (both from the α1 helix) and the carbonyl oxygens of MIDAS S123 (from loop 1) and of M335 (from the activation regulatory F-α7 loop) (**Figure 5a**). The stable occupation by Ca^{2+} is in large part explained by the much higher propensity of Ca^{2+} for carbonyl coordination (Harding 2001). Thus, a novel ionic contact stabilizes the F-α7 loop in βA through linkage to the activation-sensitive top of the α1 helix and thereby replaces the largely hydrophobic con-

tact at this site in αA domains. A third feature of βA, not found in αA, is that the unliganded αVβ3 structure does not have a MIDAS cation, as the side chain of E220 (equivalent to T209 from loop 2) blocks access to MIDAS. An Mn^{2+} replaces the Ca^{2+} at AD-MIDAS when the ectodomain is crystallized in the presence of this cation; this replacement was associated with very small movements in the F-α7 loop and the top of the α7 helix, which are similar in direction to those found in the liganded structure (see below). Although these movements did not alter the pattern of ADMIDAS coordination, they may have a priming effect by lowering the energy

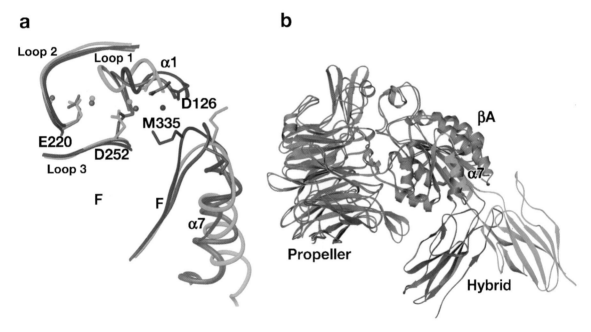

Figure 5

Structures of unliganded and liganded βA domains in the bent and legless integrins. (*a*) Ribbon diagram showing the three MIDAS loops and the side chains for key MIDAS coordinating residues in the unliganded (*blue*) and liganded (*red*) bent integrin from αVβ3 and the liganded legless integrin (*green*) from αIIbβ3. The side chains of two of the six ADMIDAS metal-coordinating residues M335 (from the F-α7 loop) and D126 (from α1 helix) in the three structures are shown in their respective colors, but labeled only in the unliganded one. Note that the MIDAS coordinating residues are superimposible in the two liganded structures. The unliganded structure only has the ADMIDAS metal (*blue*). All three metal ions (from left to right: LIMBS, MIDAS, ADMIDAS) are bound in both liganded structures and colored accordingly (for detailed metal coordination spheres, see Xiong et al. 2002, Xiao et al. 2004). (*b*) Ribbon diagram of the propeller, βA, and hybrid domains from the liganded bent and legless integrin forms. Non-moving parts of the Cα traces are in gray. All major moving segments from liganded bent (*red*) and liganded legless (*green*) are shown.

LIMBS cation: a
cation occupying a
ligand-associated
metal-binding site

barrier for activation (Xiong et al. 2002). Consistently, Mn^{2+} was found to enhance ligand-binding function of an inactive form of full-length $\alpha IIb\beta 3$ but was unable to convert this form into the active state (Yan et al. 2000).

The Integrin Ligand-Binding Site and Ligand-Associated Structural Changes

When a cyclic pentapeptide containing the prototypical ligand Arg-Gly-Asp is soaked into the Mn^{2+}-bound structure, it binds at a crevice between the propeller and the βA domains (Xiong et al. 2002). Its binding is associated with both tertiary and quaternary changes in the integrin. In the ligand-bound integrin, the ionic contact of the F-$\alpha 7$ loop with the $\alpha 1$ helix through M335 is broken, which leads to a ~2 Å inward pull of this helix as occurs with liganded αA domains, along with a large inward displacement of the ADMIDAS Mn^{2+} with the MIDAS carboxyl oxygen of D251 now replacing the carbonyl oxygen of M335. This new coordination favors the occupation of ADMIDAS with Mn^{2+} versus Ca^{2+} (**Figure 5a**). MIDAS is now occupied by Mn^{2+}, which is coordinated in an open configuration through the D119xSxS of loop 1, directly through contact by E220 (from loop 2), and indirectly through D252 of loop 3. As in αA, the sixth metal coordination sphere is now provided by the ligand Asp, while the ligand Arg occupies a pocket in the propeller, thus bringing the propeller and βA domains closer together. The position of MIDAS is stabilized further through the repositioned ADMIDAS and by a third LIMBS cation. The LIMBS cation is coordinated by the carboxylate oxygen of E220; the side chains of D158, N215, and D217; and the carbonyl oxygens of D217 and P219 (Xiong et al. 2002). Consistent with its predicted stabilizing effect on the MIDAS metal, LIMBS has been shown to act as a positive regulator of high-affinity ligand binding (Chen et al. 2003). It has been shown (Hu et al.

1999) that $\beta 3$ integrins contain two classes of ion-binding sites: a ligand-competent high-affinity site that must be occupied for the ligand to bind, most likely MIDAS; and a low-affinity Ca^{2+}-binding inhibitory (I) site, likely to be ADMIDAS, allosterically linked to the ligand-binding pocket, which increases the rate of ligand dissociation. Mn^{2+} activates integrins by competing with Ca^{2+} at the ADMIDAS. Thus, ADMIDAS appears to have a dual role in integrin function stabilizing the unliganded state (in the Ca^{2+}-bound form) as well as the liganded state (by stabilizing the MIDAS metal). This dual role may explain the discrepant data in the literature assessing the role of ADMIDAS in adhesion. In one study, disruption of ADMIDAS from integrin $\alpha 4\beta 7$ switched adhesion from low-affinity rolling to high-affinity firm adhesion (Chen et al. 2003). A second study, carried out in $\alpha 5\beta 1$, found that disrupting ADMIDAS impaired ligand binding and outside-in signaling (Mould et al. 2003a).

Four ligand- or pseudoligand-occupied crystal structures (Xiao et al. 2004) of a legless version of integrin $\alpha IIb\beta 3$ (lacking the three leg domains of αIIb and the five leg domains of $\beta 3$) have been recently determined in the absence (one form) or presence of the complex-specific and function-blocking mAb 10E5 Fab, which binds the inactive integrin (Coller et al. 1983). All four structures are stabilized by major crystal contacts at the F-$\alpha 7$ loop from symmetry-related molecules. There is minimal movement of the F-$\alpha 7$ loop but a greater inward movement of the $\alpha 1$ helix when compared to those of the liganded $\alpha V\beta 3$ ectodomain structure (**Figure 5a,b**). These movements had no impact on the already open MIDAS configuration found in the liganded full-length $\alpha V\beta 3$ ectodomain (**Figure 5a,b**). In the legless structure, a one-turn downward movement of the $\alpha 7$ helix, which also pivots laterally, is observed, in association with large swing-out motions of the hybrid domain at the hybrid-βA interface (69°, 57°, 59°, and 61° degrees in the four structures) (**Figure 5b**). These hybrid swings, which in the crystal structures are stabilized by a non-physiologic salt bridge between the PSI and propeller domains of the same molecule, are associated with a smaller βA/hybrid interface ($\sim$350 Å^2 or half that in the bent state) that is contributed mainly by Y110 and Y348, and that varies slightly in the four structures. If, as it is in the αA domains, the F-α7 loop is the primary trigger in the transition from the closed to open MIDAS configuration in βA, the change in this loop seen in the liganded full-length αVβ3 ectodomain appears to have achieved this goal. The small additional movement in this loop seen in the legless form adds little to this functionality. It has been proposed (see below) that the one-turn downward and lateral movement of the α7 helix seen in the legless crystal form is necessary to transition βA to the high-affinity state and that it did not occur in the liganded full-length ectodomain of αVβ3 because of unfavorable intramolecular contacts from the leg segments. An alternative explanation is that the α7 movement in the legless structure may be induced by ligand, and mimicked in the legless structure by the introduced truncation; as noted above, the α7 helix of CD11a is highly flexible in the isolated domain but apparently not in the intact integrin.

The αA Domain Is an Endogenous Ligand of βA

Integrins are activated from inside-out whether or not they have the ligand-binding αA domain, which is a recent addition to the integrin during evolution (Whittaker & Hynes 2002). How is this achieved? αA loops out from the D3A3 loop of the propeller (between its second and third blades). In αA-lacking integrins, this loop forms part of the ligand-binding interface. αA is linked to the D3A3 loop by three residues N-terminally and 15 residues C-terminally. Function-blocking mAbs to the linker regions have been described (reviewed in Alonso et al. 2002).

Alanine or glutamine substitution of an invariant E320 (in CD11b) at the very end of the α7 helix of closed αA (which becomes disordered in the open structure) inactivates the integrin even in the presence of Mn^{2+}, whereas an aspartate substitution of this residue is better tolerated (Alonso et al. 2002). These findings indicate that the length and charge of the residue at position 320 are both important in activation of this integrin. This leads us to propose that E320 acts as an endogenous ligand linking αA to the activation-sensitive βA domain (Alonso et al. 2002). We envision that the open and closed conformations of αA exist in an equilibrium that heavily favors the closed state in the low-affinity integrin (Li et al. 1998). Inside-out activation of the βA domain allows E320 of open αA to coordinate directly the βA MIDAS metal and thus to stabilize the open state (**Figure 6**). Consistently, it was recently shown that mutating the equivalent glutamate residue in CD11a produces the same outcome (Shimaoka et al. 2003a).

A new class of antagonists that block ligand binding by interfering with this αA/βA interface have been identified (Gadek et al. 2002, Welzenbach et al. 2002). The non-statin-derived XVA143 acts by directly coordinating the βA MIDAS metal, thus severing the αA link to βA. Binding of XVA143 induces the activation epitopes 24, KIM127 and NKI-L16; this indicates that this class acts in fact as pseudoligands, competitively inhibiting E320 or its equivalent from ligating βA and hence preventing αA from stably assuming the open state (Shimaoka et al. 2003a).

THE INTEGRIN TRANSMEMBRANE AND CYTOPLASMIC SEGMENTS

Structures of the Cytoplasmic Tails

Several NMR structures of integrin α- and/or β-subunit cytoplasmic tails were reported (Li et al. 2001, Ulmer et al. 2001, Vinogradova et al. 2002, 2004, Weljie et al. 2002). The initial studies did not agree on

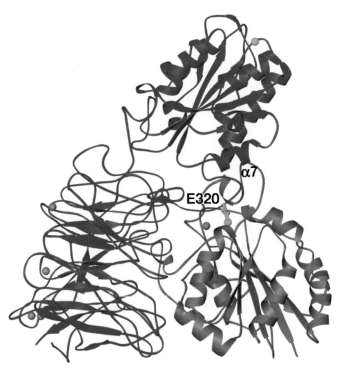

Figure 6

αA is an endogenous integrin ligand. Hypothetical model of the CD11b/CD18 head segment based on the αVβ3 and the CD11b αA crystal structures. The invariant E320 (*green side chain*) at the C terminus of the α7 helix coordinates the MIDAS ion (*cyan sphere*) of active βA, but only when the α7 helix has moved downward in open αA. The four metal ions (*orange spheres*) at the bottom of the propeller are also shown. The ADMIDAS and LIMBS ions are in magenta and gray, respectively. The αA MIDAS ion is in cyan.

contacts (with V990; F992 of αIIb; and L718, I719, I721, and H722 of β3 being interfacial residues). The αIIb helix terminates at P998, followed by a turn that allows the C-terminal loop to fold back and interact with the membrane-proximal region. The final 25 residues following the β3 membrane-proximal helix are disordered. Disruption of either the hydrophobic or electrostatic interaction destabilized the cytoplasmic complex, as did the addition of the cytoskeletal protein talin, a known integrin activator (Garcia-Alvarez et al. 2003). Using the synthetic αIIb and β3 peptides M987-R995 and K716-D740, Weljie et al. (2002) found evidence for two conformations. In both NMR structures, the β3 helices of each are situated on the opposite side of the αIIb N-terminal helix found by Vinogradova et al.; in this position, these helices would clash with the αIIb C-terminal segment. In one conformation, the main interhelical contact involves V990 and F993 of αIIb and I719 of β3 in addition to the R995-D723 salt bridge. In the second conformation, a bend between D723 and A728 of β3 is found, which prevents formation of the salt bridge. Weljie et al. (2002) proposed that the cytoplasmic tails remain complexed upon activation, with the β3-hinge at D723-A728 playing a central role in the switch between the low- and high-affinity states.

Significant changes in the structure of the full-length α and β cytoplasmic tails are seen in the presence of the membrane-mimetic dodecylphosphocholine (DPC) (Vinogradova et al. 2004) (**Figure 7b**). These changes reconcile some of the structural differences observed in the two studies discussed above (Vinogradova et al. 2002 and Weljie et al. 2002). The membrane-proximal α and β segments no longer associate, as each segment is now embedded in DPC; as a result, the αIIb tail does not fold back to interact with the α-helical membrane-proximal region; thus, the predicted clash of the αIIb tail with the β3 helices in Weljie et al. (2002)'s structures may be avoided. Significantly, the disordered C-terminal segment of β3 tail is

whether the α and β cytoplasmic tails interacted at all. Two subsequent studies then demonstrated that such an interaction took place in the aqueous environment but disagreed on the nature of the αβ interface. Vinogradova et al. (2002), using full-length cytoplasmic tails of αIIb (K989-E1008) and β3 (K716-T762) (**Figure 7a**), found a single population of tails thought to represent the integrin tails in the low-affinity state, where the conserved membrane-proximal regions of αIIb (K989xGFFKR995) and β3 (K716LLxxIHD723) are α-helical and interact via a combination of electrostatic (involving a salt bridge between the cytoplasmic residues R995 and D723) and hydrophobic

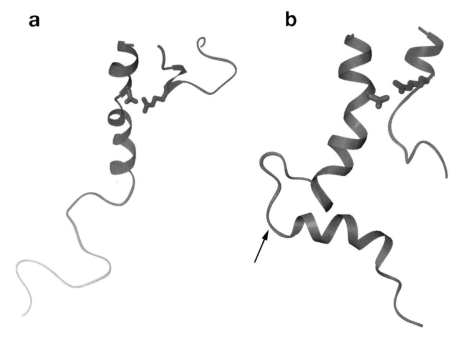

a

b

Figure 7

Structures of the αIIbβ3 cytoplasmic tails. (*a*) Ribbon diagram showing the NMR structure of the αIIb (*blue*) and β3 (*red*) tail complex. The side chains of D723 (from β3) and Arg995 (αIIb) form a salt bridge. (*b*) Structures of αIIb (*blue*) and β3 (*red*) cytoplasmic tails in detergent, where no association was detected by NMR. Note that the C-terminal portion of the β3 is now structured, with a membrane-anchoring NPxY site indicated by an arrow. We have oriented the αIIb structure so as to form the salt bridge with β3.

now ordered: the membrane-proximal α-helix (K716-R734) is followed by a flexible NPxY-containing loop and a second short helix at Y747-T755. A kink at D723 bends the long helix slightly and thus brings the flexible loop into possible contact with the membrane surface. A similar bend at this position observed in one of the two conformations in Weljie et al. (2002) may thus be reflective of the conformational space the TM segments can occupy under different conditions in the absence of the extracellular domain, which probably restricts this space. Vinogradova et al. (2004) proposed a model in which the interacting α (K989xGFF) and β (K716LLxxI) membrane-proximal regions that form in solution (and naturally within the cytoplasm) represent the low-affinity state of the integrin and the noninteracting tail structures in detergent represent the high-affinity integrin, in which these membrane-proximal regions are now intramembranous. An alternative model, however, is that the interacting α (K989xGFF) and β (K716LLxxI) membrane-proximal regions are embedded in the membrane in the low-affinity state, stabilized by the cytoplasmic R995-D723 salt bridge at their C termini and by the membrane-associated N744PxY loop. Binding of the known integrin-activator talin (Ulmer et al. 2003) detaches this loop from the membrane, breaks the salt bridge, and dissociates the membrane-proximal αβ interface. This model is more consistent with recent data showing that the TM segment is shorter in the high-affinity state (see below). The mechanism of the talin-β3 interaction has been determined from crystallographic analysis of a chimeric protein containing the integrin-binding portion of talin linked to a β-cytoplasmic domain

TM:
transmembrane

GpA: glycophorin A

peptide (Garcia-Alvarez et al. 2003). In the structure, the phosphotyrosine-binding (PTB) domain from the talin FERM domain binds the NPxY helical region of $\beta 3$. Talin has also been shown to bind the membrane-proximal segment of $\beta 3$, but the structural basis of this interaction remains to be determined.

Structure Models of the TM Segments

If the conserved α and β membrane-proximal segments K989xGFF and K716LLxxI are embedded in the plasma membrane, then the TM segments will be 25–29 amino acids long—three to five amino acids longer than the typical TM segment of type I membrane proteins. Direct support that membrane-proximal segments can in fact be membrane embedded comes from glycosylation mapping studies, which suggested that, given their length, the TM segments are likely tilted or coiled inside the membrane (Armulik et al. 1999, Stefansson et al. 2004). Previous studies have also shown that either deletions involving the membrane-proximal segments [both inclusive of or limited to the R995-D723 salt bridge (Hughes et al. 1996)] or mutations that shorten the TM domains will switch integrins into high affinity (Partridge et al. 2004). Armulik et al. (1999) suggested that in response to inside-out activation, the conserved intramembranous K989 of αIIb and K716 of $\beta 3$ become cytosolic, probably in response to talin binding to the now cytoplasmic membrane-proximal K716LLxI segment. Given the hydrophobic nature of the membrane-proximal region, this transition would not be energetically costly. Since the TM borders at the outer membrane leaflet do not change in this process (Stefansson et al. 2004), the α and β TM segments shorten by three to five amino acids and may become less tilted or coiled, but do not slide past each other in a piston-like movement.

Although there is general agreement that modulation of the interhelical TM interface is necessary for inside-out activation, there is no consensus on the precise nature of this intramembranous interface or how the interface is modulated to effect activation. Several structural models of $\alpha\beta$ TM segments have been recently proposed, some based on experimental data. In a cryo-EM study of the full-length αIIb$\beta 3$, Adair and Yeager (2002) proposed that in the low-affinity state, αIIb and $\beta 3$ TM segments associate in an α-helical coiled coil (**Figure 8a**). The αIIb$\beta 3$ TM segments were aligned using the R995-D723 salt bridge and mapped onto a right-handed coiled coil or a left-handed leucine zipper. Computational search of the conformational space of the TM domains suggested that the α and β TM segments each interact through a conserved GxxxG-like motif (Gottschalk et al. 2002) in the midsection of the TM segments, This interaction is used in GpA homodimer formation. A GpA-like crossing angle of 40° between the energetically favored right-handed helices (Li et al. 2005) brings αIIb Gly972 and Gly976 of the GxxxG motif into intimate contact with the $\beta 3$ TM domain, but does not allow formation of the membrane-proximal salt bridge; this implied to the authors that this form represented the high-affinity state. A second right-handed coiled-coil conformation inclusive of the membrane-proximal segment and associated with ~100° rotation of the α helix permitted the formation of the salt bridge and was thought to represent the low-affinity state (Gottschalk et al. 2002). Thus activation in these models involves interhelical rotation motion without complete dissociation of the $\alpha\beta$ TM complex. Using asparagine scanning throughout the TM helix of the $\beta 3$-subunit, Li et al. (2003) found that G708N and M701N mutations, both along one face of the helix, enable integrin αIIb$\beta 3$ to become active and constitutively cluster, and enhance the tendency of the TM helix to form homotrimers. This led Li et al. to conclude that homo-oligomerization plays a key role in integrin activation. Asparagine scanning of helices has the potential, however, to weaken some helix-helix associations while

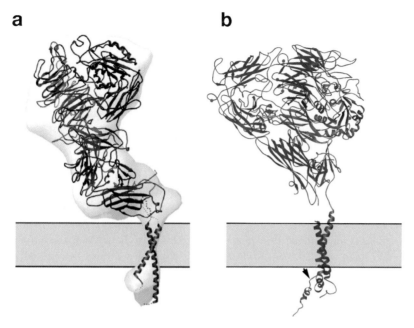

a

b

Figure 8

A hypothetical model of a membrane-bound integrin. (*a*) Ribbon model of the X-ray crystal structure of the ectodomain fitted within the map derived by electron cryomicroscopy of the full-length αIIbβ3. Hinge movements at three pivot points of the αVβ3 crystal structure (Xiong et al. 2001) were introduced to fit the map. αV- and β3-subunits are in blue and red, respectively. The map shows that the transmembrane domains are associated as a single rod of density, which has been modeled as a right-handed, parallel, α-helical coiled coil. This figure is taken from Adair & Yeager 2002 with permission. (*b*) An atomic model of a full-length β3 subunit of αVβ3. The predicted α-helical TM segments of the α- and β-subunits (Rost et al. 2004) begin with residues I966 of αIIb, V964 of αV, and I693 of β3 (based on glycation mapping) (Stefansson et al. 2004). The crystal structure of the β3-subunit from the αVβ3 ectodomain ends with G690. Further analysis of the electron density map has extended the structure to P691 (unpublished observations), which we place in the model as the first residue of the TM α helix. This placement is in view of the proline residue's propensity to stabilize strongly an α-helical conformation when placed in the first position of the helix (Senes et al. 2004). The model places the TM segment at ∼45° with respect to the long axis of calf-2 domain, compared to the ∼90° angle in the EM structure in *a*. The TM helix is continuous with the α-helical membrane-proximal and cytoplasmic tail of the NMR-derived structure of β3 (Vinogradova et al. 2004). Note that the proximal NPxY segment faces the membrane (*arrow*). Positioning of the αIIb TM helix in relation to that of β3 was done according to Partridge et al. (2004); this positioning allowed formation of the R995-D723 salt bridge without further modifications. The NMR structure of cytoplasmic αIIb is also from Vinogradova et al. (2004).

strengthening others (DeGrado et al. 2003); the activating effect of the G708N mutation found by Li et al. (Li et al. 2003) could thus act primarily by disrupting the αβ TM helical packing and favoring homo-dimerization. In this view, homo-oligomerization, which may be favored by the GxxxG-like motifs, may be a postactivation event.

In a fourth study (Partridge et al. 2004), an unbiased random mutagenesis of the β3 TM and cytoplasmic tail identified activating point mutations in the TM segment and cytoplasmic tail but none were C-terminal to the salt bridge forming D723. A Monte Carlo simulation (Kim et al. 2003b) produced two alternative right-handed helical structures for the low-affinity state, with the TM helices packing at crossing angles either near their C or N termini. The former structure was favored by the random mutagenesis data, which

FRET: fluorescent resonance energy transfer

suggested a specific TM helix-helix packing interface in which the C-terminal $\beta3$ TM residues G708, A711, and I714 are helical contacts, with $\beta3$ G708 packing against the TM residue Thr981 of αIIb. The GxxxG-like sequences of αIIb (V971 and L974) and $\beta3$ (V700) may make some contribution to this interface but are not at its core. Finally, an in vivo cysteine cross-linking study placed the interhelical interface of the low-affinity state in the outer leaflet of the membrane (Luo et al. 2004a), in contrast to the study by Partridge et al. (2004). Treating cells with the oxidation catalyst Cu-phenanthroline improved efficiency of disulfide formation up to αIIb G972 and $\beta3$ V700 but no deeper, perhaps as a result of this catalyst's limited diffusion into the inner membrane leaflet. The proposed interhelical contact involved N-terminal residues at positions I966, W968, and G972 of αIIb; and I693, L694, V696, L697, and V700 of $\beta3$; and placed the $\beta3$ GxxxG-like (S699xxxA) motif on the opposite face of this interface, i.e., not facing the corresponding G972xxxG motif of αIIb. This interface was lost and no alternative $\alpha\beta$ disulfides or homomeric associations were detected in an activating truncating mutation of αIIb (after G991 of the KGFFKR motif or by replacing this motif with KGAAKR). The interface proposed here is largely incompatible with the one proposed by Partridge et al. (2004), since the respective core contact residues W968 and T981 are not on the same side of the helix. This may be a reflection of the limitation of such models and/or of the techniques used; for example, in vivo intramembranous cysteine cross-linking does not invariably imply that the respective cross-linked positions constitute a contact site (Homma et al. 2004). The findings from the cysteine scanning study also do not support the NMR (Weljie et al. 2002) and computational (Gottschalk et al. 2002) models in which the TM regions remain complexed in the active state. In fact, a FRET-based study, in which yellow and cyan fluorescent protein tags were fused to the C termini of

the integrin α- and β-subunit cytoplasmic domains, concluded that the TM segments separate laterally in the lipid bilayer by >100 Å (Kim et al. 2003a). A lateral separation of the α- and β-transmembrane helices upon integrin activation would not likely be a fast and reversible process but energetically costly and probably slow owing to the viscosity of the plasma membrane. Although the evidence for interacting TM segments in the high-affinity state is weak at present, a reorganization of the interhelical interface in the active state through a change in tilt or rotation may be more energetically favorable and faster. It may be relevant that an inactivating disulfide is formed between αIIb G972 and $\beta3$ L697 in the active KGAAKR mutant αIIb$\beta3$ integrin (Luo et al. 2004a); since disulfide bond formation implies a proximity of 1.9 Å or less, some components of the active integrin's TM segments appear to remain spatially close. Furthermore, if the noted disorder of the $\beta3$ C-terminal segment in the aqueous NMR structure is activation-induced, the decreased FRET observed in the GFP-tagged C termini may in part reflect this flexibility.

The Integrin Ectodomain-TM Connection: A Model Structure of the Full-Length $\beta3$-Subunit

The crystal structure of bent αV$\beta3$ lacked the TM and cytoplasmic tails and therefore provided no information on how integrins are positioned relative to the plasma membrane. The transmembrane helices are commonly depicted as parallel to the legs; in this orientation, the ligand-binding pocket would be facing the plasma membrane and thus inaccessible to the ligand. Three-dimensional reconstruction of EM images of full-length αIIb$\beta3$ in the low-affinity state showed, however, that the TM helices are not parallel but almost perpendicular to the long axis of the legs (**Figure 8a**). This orientation can be readily accommodated in an atomic model of the $\beta3$-subunit inclusive of its TM and cytoplasmic segments—where the model incorporates the NMR structure of the membrane-proximal and cytoplasmic tails of β3 in detergent (Vinogradova et al. 2004)—and the data that suggest that a long TM segment is a feature of the inactive state (Partridge et al. 2004) (**Figure 8b**). The modeled β3 TM helix includes the exofacial residue D692 and extends through the membrane-proximal segment in the low-affinity state; the helix is tilted by ∼35–40° to fit into the 30 Å membrane thickness. The calf-2-TM tilt angle may vary among integrins as a function of the length of their TM segments [e.g., it is 26, 22, 29, and 25 amino acids long in β1, β8, α10, and α2, respectively (Stefansson et al. 2004)]; the longer the TM segment, the greater the tilt angle. On the inner side of the β3 TM helix, i.e., facing the α-subunit, are G708, A711, I714, and cytoplasmic D723, the last of which forms the salt bridge with R995. Mutations of any of these residues have been shown to activate the integrin (Partridge et al. 2004). In our model, the α-helical structure of the αIIb TM segment and its cytoplasmic tail can be accommodated in coiled-coil (Adair & Yeager 2002) or N- (Luo et al. 2004a) or C-terminal (Partridge et al. 2004) TM crossed-angle packing patterns. A C-terminal crossing based on the recent structure model of Partridge et al. (2004) is displayed in **Figure 8b**. All three TM crossing-angles models are allowed by the presence of a five amino acid sequence linking P957, the last residue in the crystal structure of the αV ectodomain, with V964 the first TM residue.

MODELS OF INTEGRIN ACTIVATION AND SIGNALING

Inside-Out Integrin Activation

The most surprising feature of the αVβ3 integrin ectodomain's crystal structure is its knee-bent conformation. Subsequent studies have confirmed that this conformation exists naturally in αVβ3 as well as in other integrins. First, EM images of recombinant integrins in the presence of Ca^{2+} (Adair et al. 2005, Takagi et al. 2002b) or Mn^{2+} (Adair et al. 2005) revealed a bent form, consistent with the bent structure. Second, engineering a head-to-leg disulfide bond between the αV propeller and EGF-4 of β3, which should lock the integrin in the bent form, led to a cell surface-expressed heterodimer (Takagi et al. 2002b). Third, clusters of monoclonal antibodies that bind to the head or membrane-proximal leg segments of native αIIbβ3 on intact platelets cross-compete, consistent with the bent state of the integrin (Calzada et al. 2002). Fourth, FRET is observed between cell-bound integrins bearing fluorescent ligands and lipophilic probes, which suggests that the ligand-binding site is within ∼100 Å of the lipid bilayer, consistent with a bent-integrin conformation (Chigaev et al. 2003).

It is currently thought that affinity switching is triggered intracellularly by the binding of one of two NPxY/F motifs of the integrin β cytoplasmic tail to the PTB-like F3 FERM subdomain of talin (Garcia-Alvarez et al. 2003, Yan et al. 2001): expressing talin shRNA specifically blocks binding of the activation-sensitive mAb PAC-1 to αIIbβ3 in response to ADP or thrombin (Tadokoro et al. 2003). Talin shRNA-induced suppression is not reversed by activated R-Ras or the CD98 heavy chain, two previously known positive regulators of integrin activation (Tadokoro et al. 2003). Binding of talin induces spectral perturbation in the membrane-proximal region of β3 (Li et al. 2002a, Vinogradova et al. 2002). These perturbations are prevented by the presence of the αIIb cytoplasmic tail (Ulmer et al. 2003). The role of serine/threonine and tyrosine phosphorylation of the α- and β-subunit cytoplasmic tails in this context is unclear, as is the role of RhoA, which regulates integrin affinity through a region distinct from the one that regulates avidity (Giagulli et al. 2004). Other data suggest, however, that talin-mediated activation may not be prototypic of inside-out activation of all integrins. First, down-regulation of talin by antisense mRNA or shRNA

ILK: integrin-linked kinase

reduces integrin processing and transport to the cell surface (Albiges-Rizo et al. 1995, Martel et al. 2000, 2001). Second, integrins in talin-deficient *Drosophila melanogaster* can still bind to physiologic ligands but fail to connect to the actin cytoskeleton (Brown et al. 2002, von Wichert et al. 2003). Third, $\beta 2$ integrins in resting neutrophils are linked constitutively to talin and switch to α-actinin upon activation (Sampath et al. 1998), which suggests that talin stabilizes the inactive state in this case. Fourth, the $\beta 7$ integrin binds talin poorly but binds tightly to filamin through a phosphorylation-modified ITTTI motif found between the two NPxY motifs (Calderwood et al. 2001). Even if talin is a final common pathway for activation in some integrins, its regulation and recruitment to the cell surface remains poorly understood. Currently, the only known physiologic activator of talin is PtdIns(4,5)P2 (Martel et al. 2001) produced by the action of type I PtdIns(4)P 5-kinases and regulated by PtdIns(4,5)P2 phosphatases. The type I PtdIns(4)P 5-kinases are themselves regulated by phosphatidic acid, the small G proteins Arf1 and Rho, and Src. The transience of the integrin activation response indicates the presence of downregulatory molecules as well. Independently of ERK/MAPK, H-ras suppresses integrin activation via Raf-1 (Hughes et al. 1997); this is reversed by PEA15 (phosphoprotein-enriched in astrocytes 15) (Kinbara et al. 2003). H-ras does not regulate chemokine-induced integrin activation in other cells, which suggests that the suppressive effect may be integrin-specific (Weber et al. 2001a). Other potential negative regulators of chemokine-induced activation include the serine-threonine kinase ILK (Friedrich et al. 2002) and Rac1.

The switchblade model. How are the activation events triggered inside the cell transmitted across the plasma membrane to the integrin ligand-binding site some 200 Å away? An early study, reporting the NMR structure of the recombinant EGF3 domain and a modeled EGF2 domain from the $\beta 2$-subunit, suggested that the epitope for the activation-sensitive mAb KIM127 found in EGF2/3 (Tan et al. 2000) is buried in the bent integrin structure but exposed in the knee-extended model (Beglova et al. 2002). Further, the assumption that the long axis of calf-2 is perpendicular to the plane of the plasma membrane results in an integrin in which access to the ligand-binding pocket is blocked by the membrane. The so-called switchblade model of inside-out integrin activation emerged on the basis of (*a*) additional data showing that a head-to-leg disulfide bonded integrin is inactive, and (*b*) EM two-dimensional images of the recombinant ectodomain. The latter showed mainly bent forms in 1mM Ca^{2+} changing to predominantly knee-extended, leg-separated, and wide-angled βA/hybrid forms in the presence of the activating cation Mn^{2+} with or without RGD (Takagi et al. 2002b). In the switchblade model, inside-out activation is driven by separation of the cytoplasmic and TM segments, causing a jackknife-like knee-extension of the inactive bent integrin. This extension releases the hybrid domain from its restraining interface with the leg domains, which allows the domain to swing outward and pull down the $\alpha 7$ helix of βA, thus switching βA to high affinity. Consistent with such a global conformational change, the calculated Stokes radius, based on gel elution profiles, increases from 56 Å (in 1mM Ca^{2+}) to 60 Å (in Mn^{2+}), with a further increase to 63 Å upon addition of cyclic RGD (Takagi et al. 2002b). As noted above, spatial separation of the cytoplasmic and TM segments has been demonstrated (Kim et al. 2003a). N-glycosylation sites introduced in the $\beta 1$-subunit through a P333N mutation (Luo et al. 2003a) or the better-expressed G429N (G420 in $\beta 3$) mutations (Luo et al. 2004b) are activating from a proposed wedging open of the βA/hybrid. In addition, mAb SG/19, which maintains $\alpha 5 \beta 1$ in the low-affinity state, binds in 1 mM Mn^{2+} to recombinant legless integrins at the outer surface of the βA/hybrid junction (Luo et al. 2004b); it was argued that mAb

SG/19 inactivates the integrin by preventing the hybrid out-swing. Three-dimensional EM imaging of the head segment of recombinant legless $\alpha 5\beta 1$ in 1mM Ca^{2+} and 1mM Mn^{2+} yielded the acute βA/hybrid-angle form seen in the bent X-ray structure of $\alpha V\beta 3$ (Takagi et al. 2003); the absence of the wide-angled βA/hybrid form in Mn^{2+}, predominant in the two-dimensional EM structure of the complete ectodomain, is notable (Takagi et al. 2002b). A large outward swing of the hybrid domain was seen in only ~25% of the legless integrin molecules upon addition of saturating amounts of GRGDNP peptide ligand in 1 mM Ca^{2+} and in ~20% of the molecules when the physiologic ligand Fn7-10 was added in 1mM Mn^{2+} (Takagi et al. 2003); in the latter case, most FN-bound integrins displayed the wide βA/hybrid angle. Low-angle X-ray solution scattering (Mould et al. 2003c) and crystal structures of legless $\alpha IIb\beta 3$ with small RGD-like ligands or pseudoligands (Xiao et al. 2004) all showed an outward swing of the hybrid domain of wide-ranging angles (45–80°), and the crystal structures (Xiao et al. 2004) revealed a rearranged F-$\alpha 7$ loop and a one-turn downward and outward displacement of the $\alpha 7$ helix. Disulfides introduced in the F-$\alpha 7$ region of βA of the $\beta 3$-subunit produced high- (F-strand V332C/F-$\alpha 7$ loop M335C) or low-affinity (F-strand T329C/$\alpha 7$ helix A347C) (Luo et al. 2003b, Tng et al. 2004) forms, and one-turn- ($\beta 2$-subunit residues 336–339 or 340–343) or two-turn ($\beta 2$-residues 337–343) deletions of $\alpha 7$ were also activating (Yang et al. 2004) (**Figure 9**). Furthermore, a N329S mutant in the F-$\alpha 7$ loop (equivalent to N339 in $\beta 3$), an L358A mutation in the $\alpha 7$ helix of the $\beta 1$-subunit (equivalent to I351 in the $\beta 3$-subunit and to the conserved SILEN residue I316 in αA) (Barton et al. 2004, Mould et al. 2003b) and a natural V409D of the $\alpha 7$ helix in the insect's βPS-subunit (equivalent to L341 in $\beta 3$) (Jannuzi et al. 2004) (**Figure 9**) were also activating, suggesting involvement of the F-$\alpha 7$ region in integrin activation.

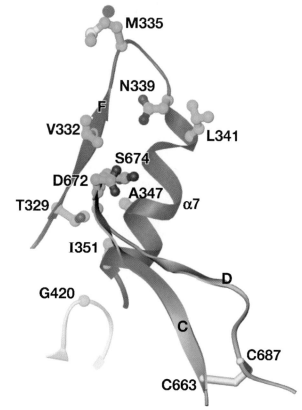

Figure 9

Activating mutations in the F-$\alpha 7$ region of βA: the deadbolt model. Ribbon diagram of the F-$\alpha 7$ region of βA (*red*). Side chains of amino acids in the F-strand, F-$\alpha 7$ loop, $\alpha 7$ helix [including the deleted segment (in *gray*)], and CD loop of the βTD are labeled; these amino acids activate the integrin when mutated. Position of the activating G420N glycan wedge mutation in the hybrid domain (*yellow*) is also shown. The CD loop interfaces with βA in a similar way to that of the competitive inhibitor lovastatin (see **Figure 3b**), inviting the deadbolt analogy (see text for details). The cation at ADMIDAS is in magenta.

A number of other observations, however, are not consistent with knee-extension as a prerequisite for inside-out integrin activation. First, the cross-competition between clusters of monoclonal antibodies directed to the head and leg segments is retained when platelets are treated with activating agents such as ADP and thrombin (Calzada et al. 2002); this is not possible if knee-extension has taken place. Second, the complete $\alpha V\beta 3$ ectodomain, which was crystallized in 1mM Mn^{2+} in the absence (Xiong et al. 2002) or presence of cyclic RGD, assumes the bent

FN: fibronectin

uPAR: urokinase
plasminogen
activator receptor

conformation. Third, transmission electron microscopy and single-particle image analysis were recently used to determine the three-dimensional structure of a stable complex of the complete $\alpha V\beta 3$ ectodomain with the soluble physiologic ligand FN; this complex contains type III domains 7 to 10 and the EDB domain (FN7-EDB-10) (Adair et al. 2005). Most Mn^{2+}-bound integrin particles, whether unliganded or FN-bound, assume the compact form of the bent X-ray structure; only a rotation of the hybrid domain by ~10° with respect to that of the $\alpha V\beta 3$ X-ray structure was observed. Thus, the large outward swing of the hybrid domain is not a necessary feature of the liganded state in the complete ectodomain. Furthermore, stable binding to ligand is not associated with detectable separation of the αV- and $\beta 3$-subunit leg domains. Difference maps comparing the three-dimensional reconstructed images of the ligand-free and FN-bound $\alpha V\beta 3$ reveal density that could accommodate the RGD-containing FN10 in proximity to the ligand-binding site of $\beta 3$, with FN9 just adjacent to the synergy site–binding region of αV (**Figure 10a**). The discrepancy of these results with those obtained earlier (Takagi et al. 2002b) may reflect in part the unbiased automatic selection routine used in the more recent study and the limitation of two-dimensional versus three-dimensional image analysis of large multidomain structures such as integrins. Fourth, the binding of the GPI-linked uPAR to cell-bound $\alpha 5\beta 1$ has been shown to stabilize interaction of the integrin with FN (Wei et al. 2005). uPAR is a small three-domain structure that binds through its N-terminal domain to blade 4 of the propeller (Simon et al. 2000); this interaction is not possible in a genuextended integrin (**Figure 10b**). This finding also suggests that the ligand-binding site in the cell-bound bent integrin is accessible to large ligands, consistent with the modeled orientation of the integrin ectodomain described above (**Figure 8b**). The ability of the bent integrin to bind ligands is also consistent with a number of additional observations. If the leg domains restrict the ability of the integrin to bind physiologic ligands, then removal of these domains should be activating. However, truncating the leg domains does not result in a constitutively active integrin unless induced by Mn^{2+} (Mould et al. 2003a,b). Mn^{2+}-induced activation of the legless integrin takes place in the absence of the outward swing of the hybrid domain (Takagi et al. 2003), and does not require the physical separation of the leg domains or TM segments in cell-bound integrins (Kim et al. 2003a).

How can these data be reconciled? One consideration is that the bent state is of intermediate affinity, as observed in αA domains, with high affinity achieved only in the knee-straightened conformation. However, binding to soluble physiologic ligands is a characteristic of the integrin's high- and not intermediate-affinity state (Ma et al. 2004), and the bent $\alpha V\beta 3$ ectodomain binds soluble FN. Second, the ability of the bent integrin to bind soluble FN is induced by a nonphysiologic stimulus Mn^{2+} independent of the normal inside-out activation pathway, which arguably requires the hybrid swing movement. However, the stable binding of the complete $\alpha V\beta 3$ ectodomain to soluble FN is also supported by the physiologic cation Mg^{2+} (Adair et al. 2005), consistent with the activating effect of this metal ion in various integrins (Dransfield et al. 1992). Further, 1 mM Mn^{2+} did not induce the outward hybrid swing in legless integrins unless a ligand was also present (Takagi et al. 2003). A third possibility is that the hybrid swing seen in legless integrins is an artifact resulting from truncations of the leg domains; these truncations may also result in the unstable and variable binding of FN to the recombinant legless integrin forms (Mould et al. 2003c, Takagi et al. 2003). A fourth and most likely explanation in our view is that the hybrid swing may not be a prerequisite of inside-out activation, but is ligand induced, i.e., it is a feature of outside-in signaling. The absence of the large hybrid swing in the FN-bound complete ectodomain, otherwise seen in EM and crystal

a

b

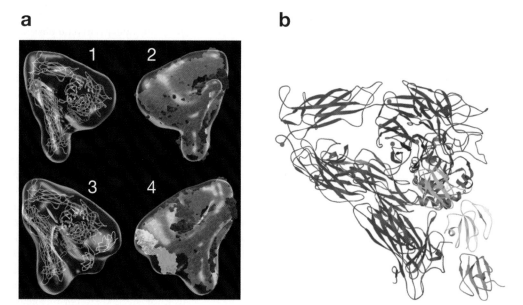

Figure 10

Ligand-binding by bent αVβ3. (*a*) Pseudoatomic models of unliganded and FN-bound αVβ3. αV- and β3-subunits are in blue and red, respectively. The isosurface levels for the maps have been set to 2.6σ for each. (1) A ribbon diagram of the αVβ3 crystal structure (Xiong et al. 2001) including the PSI domain (Xiong et al. 2004), fitted within the uncomplexed map. (2) The accessible surface of αVβ3 (1.4 Å probe radius) fitted in the map in the same view as (1). The molecules in (2) are colored as in the ribbon models. (3) Ribbon diagram of the αVβ3/FN9-10 complex fitted within the αVβ3/FN7-EDB-10 map in the same orientations as (1). (4) The accessible surface model within the map viewed in the same orientation as (3). FN9 (*yellow*) and FN10 (*green*) domains are based on the X-ray structure of FN7-10 (Leahy et al. 1996) and positioned with the RGD loop of FN10 superimposed onto the cyclic RGD peptide in αVβ3 (Xiong et al. 2002). This figure is modified from Adair et al. 2005 with permission. (*b*) Ribbon diagram of a hypothetical model αVβ3 (colors as in *a*) in complex of the GPI-linked uPAR. The three domains of uPAR (*green*, *yellow*, and *gray*, respectively) are modeled based on the homologous CD59 domain (PDB 1CDR) (Fletcher et al. 1994). Positioning of domain 1 (*green*) of uPAR is based on mapping studies showing that this domain interacts with a short sequence in blade 4 of the propeller in proximity to the RGD-binding site. The metal ions of ADMIDAS, genu, and propeller are in magenta.

structures of the legless integrin, may reflect normal constraints by the leg domains that are overcome by binding of ligand to the intact cell-bound integrin.

The deadbolt model. So what are other potential means of achieving inside-out activation? As noted earlier, the bulk of the activating mutations in βA described to date are clustered in the F-α7 region (**Figure 9**). Such mutations include swapping the F-α7 loop residues E322LSED from the β2-subunit with T339LSAN from β1 (V332LSMD in β3) (Ehirchiou et al. 2005); this mutation may

act in part by disrupting the direct coordination of the ADMIDAS metal found in the unliganded state. In the crystal structure of the unliganded αVβ3 ectodomain, V332 of the F-strand contacts Ser674 (∼60 Å² contact area) from the CD loop (S673SG) of the membrane-proximal βTD. The swap of this loop with that of β2 (G659MD) is also activating in integrin CD11b/CD18 (unpublished data). Of significance, a recent study, which systematically mutated each of the 56 cysteines in the β3-subunit to serines, found outside the EGF domains only two among thirteen pairs that are activating (Kamata et al.

2004); one pair (C663–C687) stabilizes the CD strands of βTD (**Figure 9**). This CD strand-loop also contains epitopes for activating mAbs (Wilkins et al. 1996). Furthermore, of all the known β-subunits from sponges to humans, the βTD CD strands, inclusive of their disulfide bridge, are naturally missing in vertebrate β8. The functional implications of this deletion, if any, so far remain unexplained. This constellation of observations, unexplained by the switchblade model, has invited an alternate minimalist "deadbolt" model for inside-out activation (Xiong et al. 2003). The deadbolt model is inspired by the fact that the allosteric inhibitor lovastatin acts by binding at the F-α7 interface of the αA domain, irreversibly locking the domain in its inactive state (**Figure 3b**). It is proposed that the novel βTD domain, through its CD loop, likewise locks βA in the low-affinity state. The deletion of the integrin TM and cytoplasmic segments (reviewed in Humphries 2000) lead to the partial disengagement of the deadbolt. This creates the priming effect that is likely the source for the weak contact observed in the ectodomain crystal structure. It is suggested that the CD loop is more engaged in the native integrin. Modifying the TM tilt angle by inside-out signals may readily and perhaps directly disengage the deadbolt (**Figure 8b**), switching βA into the high-affinity state through restructuring of the F-α7 loop, disengaging this loop from the ADMIDAS metal and triggering the inward movement of the α1 helix. The activating mutations in the F-α7 region and the glycan wedge mutation G420N (**Figure 9**) can be reinterpreted as acting mainly by disengaging the deadbolt. The Mn^{2+}-dependent ability of the bent integrin to bind ligand in solution takes place not by leg straightening (Mould et al. 2003a) but through an Mn^{2+}-induced shift of the ADMIDAS metal coordination to that of the liganded state. This shift triggers the tertiary changes seen in the active state. The β8-subunit, with its short TM segment and a cytoplasmic tail that lacks the talin-binding site, may require the CD deletion

to allow the β8 integrin heterodimer to become active, perhaps constitutively so. Transmission of inside-out activation signals across a single domain is more energetically favored, especially for a rapidly reversible reaction, in comparison with that proposed across at least five domains in the switchblade model.

Integrin Avidity

Changes in integrin avidity (clustering) complement the changes in affinity and appear critical for linking the liganded integrin to the cytoskeleton to initiate cell signaling (Hato et al. 1998). Clustering can take place in seconds when induced by chemokines and can be detected by certain mAbs such as NKI-L16 (van Kooyk et al. 1994). Integrin clustering is RhoA-dependent (Giagulli et al. 2004), and proceeds via recruitment of ζPKC to the plasma membrane, in combination with PI3K, Rap1 (Kinashi et al. 2004, Shimonaka et al. 2003), and cytohesin-1 (Weber et al. 2001b). Changes in integrin affinity are not necessarily accompanied by changes in avidity, and vice versa. For example, in T cells derived from activated Rap1 transgenic mice, integrin CD11a clusters spontaneously, but with no increase in affinity (Sebzda et al. 2002). Blockade of Src kinases or of actin polymerization prevents β2-integrin clustering without changing binding of the affinity-sensitive KIM127 mAb (Piccardoni et al. 2004). Also, inhibition of the calcium-dependent neutral protease calpain or ζPKC in leukocytes blocks clustering but not affinity modulation (Constantin et al. 2000, Giagulli et al. 2004). These findings suggest that although affinity and avidity regulation may coexist or occur sequentially, they may be driven by independent, although perhaps intersecting, pathways.

The structural basis of integrin clustering (avidity modulation) is unknown. A recent study showed that the minimal complex necessary for formation of an integrin-cytoskeleton complex is a trimer of the fibronectin's integrin-binding domain

FNIII7-10 (Jiang et al. 2003). Another study proposed that integrin clustering is the result of the TM domain–mediated homo-oligomerization (Li et al. 2003). This structure model is based on observations made with integrin TM and cytoplasmic fragments that formed dimers and trimers in sodium dodecyl sulfate or dodecylphosphocholine detergent micelles (Li et al. 2003). Full-length integrin heterodimers do not form in these same detergents, however, which suggests that alternative models of integrin clustering likely exist and remain to be discovered.

Outside-In Signaling

Extracellular ligand-bound integrins transduce a variety of signals that modulate many cellular processes, including adhesion, migration, and differentiation. Integrin-mediated signaling induces dramatic changes in the organization of the cytoskeleton (see Ridley et al. 2003 for a recent review). How are these signals transduced from the outside-in? Here, the switchblade model may provide one answer. Ligand binding to the cell-bound integrin may drive the outward hybrid swing, leg domain separation, and separation of the TM and cytoplasmic segments, inviting new interactions with cytoplasmic tails that result in the assembly of cytoskeletal-based signaling complexes. A second, although not mutually exclusive, possibility is that extracellular ligand binding triggers outside-in signaling through *cis*-interactions between integrin ectodomains and/or with other cell-surface moieties such as members of the tetraspanin protein family (Hemler 2003). As integrins transmit not only chemical but also mechanical stress signals, the flexible integrin knees may be necessary for adhesion strengthening, especially in contractile tissues.

CONCLUDING REMARKS

Spurred by the atomic resolution of the integrin ectodomain, valuable insight into the inner working of integrins has been provided to date. The new structures have explained volumes of biochemical observations but, as expected, they also raise new questions. As shown here, such questions are many. How can integrins rapidly and reversibly become activated? Do their TM segments become shorter or longer with activation? What is the nature of the TM interface and does activation require physical separation or just realignment of the TM interface? What is the structural basis of clustering? What is the nature of the moieties that maintain the cytoplasmic tails in the inactive state and how do talin and likely additional proteins modulate the structure of the C-tails to activate? How is outside-in signaling channeled and how is it translated into different cellular responses? Last but not least, what is the nature of multimolecular assemblies that mediate adhesion signaling? Answering even some of these questions will keep many busy for a long while.

ACKNOWLEDGMENTS

We thank the National Institutes of Health for grant support (DK48549, DK50305, and HL70219).

LITERATURE CITED

Adair BD, Xiong J-P, Maddock C, Goodman SL, Arnaout MA, Yeager M. 2005. Three-dimensional EM structure of the ectodomain of integrin $\alpha V\beta3$ in a complex with fibronectin. *J. Cell Biol.* 168:1109–18

Adair BD, Yeager M. 2002. Three-dimensional model of the human platelet integrin alpha IIbbeta 3 based on electron cryomicroscopy and x-ray crystallography. *Proc. Natl. Acad. Sci. USA* 99:14059–64

Ajroud K, Sugimori T, Goldmann WH, Fathallah DF, Xiong JP, Arnaout MA. 2004. Binding affinity of metal ions to the CD11b A-domain is regulated by integrin activation and ligand. *J. Biol. Chem.* 279:25483–88

Albiges-Rizo C, Frachet P, Block MR. 1995. Down regulation of talin alters cell adhesion and the processing of the alpha 5 beta 1 integrin. *J. Cell Sci.* 108(Pt. 10):3317–29

Alonso JL, Essafi M, Xiong JP, Stehle T, Arnaout MA. 2002. Does the integrin alphaA domain act as a ligand for its betaA domain? *Curr. Biol.* 12:R340–42

Armulik A, Nilsson I, von Heijne G, Johansson S. 1999. Determination of the border between the transmembrane and cytoplasmic domains of human integrin subunits. *J. Biol. Chem.* 274:37030–34

Barton SJ, Travis MA, Askari JA, Buckley PA, Craig SE, et al. 2004. Novel activating and inactivating mutations in the integrin beta1 subunit A domain. *Biochem. J.* 380:401–7

Bhatia SK, King MR, Hammer DA. 2003. The state diagram for cell adhesion mediated by two receptors. *Biophys. J.* 84:2671–90

Beglova N, Blacklow SC, Takagi J, Springer TA. 2002. Cysteine-rich module structure reveals a fulcrum for integrin rearrangement upon activation. *Nat. Struct. Biol.* 9:282–87

Brown NH, Gregory SL, Rickoll WL, Fessler LI, Prout M, et al. 2002. Talin is essential for integrin function in Drosophila. *Dev. Cell* 3:569–79

Calderwood DA, Huttenlocher A, Kiosses WB, Rose DM, Woodside DG, et al. 2001. Increased filamin binding to beta-integrin cytoplasmic domains inhibits cell migration. *Nat. Cell Biol.* 3:1060–68

Calzada MJ, Alvarez MV, Gonzalez-Rodriguez J. 2002. Agonist-specific structural rearrangements of integrin alpha IIbbeta 3. Confirmation of the bent conformation in platelets at rest and after activation. *J. Biol. Chem.* 277:39899–908

Carson M. 1987. Ribbon models of macromolecules. *J. Mol. Graph.* 5:103–6

Chen J, Salas A, Springer TA. 2003. Bistable regulation of integrin adhesiveness by a bipolar metal ion cluster. *Nat. Struct. Biol.* 10:995–1001

Chigaev A, Blenc AM, Braaten JV, Kumaraswamy N, Kepley CL, et al. 2001. Real time analysis of the affinity regulation of alpha 4-integrin. The physiologically activated receptor is intermediate in affinity between resting and Mn(2+) or antibody activation. *J. Biol. Chem.* 276:48670–78

Chigaev A, Buranda T, Dwyer DC, Prossnitz ER, Sklar LA. 2003. FRET detection of cellular alpha4-integrin conformational activation. *Biophys. J.* 85:3951–62

Coller BS, Peerschke EI, Scudder LE, Sullivan CA. 1983. A murine monoclonal antibody that completely blocks the binding of fibrinogen to platelets produces a thrombasthenic-like state in normal platelets and binds to glycoproteins IIb and/or IIIa. *J. Clin. Invest.* 72:325–38

Constantin G, Majeed M, Giagulli C, Piccio L, Kim JY, et al. 2000. Chemokines trigger immediate beta2 integrin affinity and mobility changes: differential regulation and roles in lymphocyte arrest under flow. *Immunity* 13:759–69

DeGrado WF, Gratkowski H, Lear JD. 2003. How do helix-helix interactions help determine the folds of membrane proteins? Perspectives from the study of homo-oligomeric helical bundles. *Protein Sci.* 12:647–65

Dransfield I, Cabanas C, Craig A, Hogg N. 1992. Divalent cation regulation of the function of the leukocyte integrin LFA-1. *J. Cell Biol.* 116:219–26

Ehirchiou D, Xiong YM, Li Y, Brew S, Zhang L. 2005. Dual function for a unique site within the {beta2}I domain of integrin {alpha}M{beta}2. *J. Biol. Chem.* 280:8324–31

Emsley J, Knight CG, Farndale RW, Barnes MJ, Liddington RC. 2000. Structural basis of collagen recognition by integrin alpha2beta1. *Cell* 101:47–56

Fletcher CM, Harrison RA, Lachmann PJ, Neuhaus D. 1994. Structure of a soluble, glycosylated form of the human complement regulatory protein CD59. *Structure* 2:185–99

Friedrich EB, Sinha S, Li L, Dedhar S, Force T, et al. 2002. Role of integrin-linked kinase in leukocyte recruitment. *J. Biol. Chem.* 277:16371–75

Gadek TR, Burdick DJ, McDowell RS, Stanley MS, Marsters JC Jr, et al. 2002. Generation of an LFA-1 antagonist by the transfer of the ICAM-1 immunoregulatory epitope to a small molecule. *Science* 295:1086–89

Garcia-Alvarez B, de Pereda JM, Calderwood DA, Ulmer TS, Critchley D, et al. 2003. Structural determinants of integrin recognition by talin. *Mol. Cell* 11:49–58

Giagulli C, Scarpini E, Ottoboni L, Narumiya S, Butcher EC, et al. 2004. RhoA and zeta PKC control distinct modalities of LFA-1 activation by chemokines: critical role of LFA-1 affinity triggering in lymphocyte in vivo homing. *Immunity* 20:25–35

Gottschalk KE, Adams PD, Brunger AT, Kessler H. 2002. Transmembrane signal transduction of the alpha(IIb)beta(3) integrin. *Protein Sci.* 11:1800–12

Harding MM. 2001. Geometry of metal-ligand interactions in proteins. *Acta Crystallogr. D* 57:401–11

Hato T, Pampori N, Shattil SJ. 1998. Complementary roles for receptor clustering and conformational change in the adhesive and signaling functions of integrin alphaIIb beta3. *J. Cell Biol.* 141:1685–95

Hemler ME. 2003. Tetraspanin proteins mediate cellular penetration, invasion, and fusion events and define a novel type of membrane microdomain. *Annu. Rev. Cell Dev. Biol.* 19:397–422

Homma M, Shiomi D, Kawagishi I. 2004. Attractant binding alters arrangement of chemoreceptor dimers within its cluster at a cell pole. *Proc. Natl. Acad. Sci. USA* 101:3462–67

Hu DD, White CA, Panzer-Knodle S, Page JD, Nicholson N, Smith JW. 1999. A new model of dual interacting ligand binding sites on integrin alphaIIbbeta3. *J. Biol. Chem.* 274:4633–39

Hughes PE, Diaz-Gonzalez F, Leong L, Wu C, McDonald JA, et al. 1996. Breaking the integrin hinge. *J. Biol. Chem.* 271:6571–74

Hughes PE, Renshaw MW, Pfaff M, Forsyth J, Keivens KM, et al. 1997. Suppression of integrin activation: A novel function of a Ras/Raf-initiated MAP kinase pathway. *Cell* 88:521–30

Humphries MJ. 2000. Integrin structure. *Biochem. Soc. Trans.* 28:311–39

Ingber DE. 2003. Mechanosensation through integrins: cells act locally but think globally. *Proc. Natl. Acad. Sci. USA* 100:1472–74

Jannuzi AL, Bunch TA, West RF, Brower DL. 2004. Identification of integrin beta subunit mutations that alter heterodimer function in situ. *Mol. Biol. Cell* 15:3829–40

Jiang G, Giannone G, Critchley DR, Fukumoto E, Sheetz MP. 2003. Two-piconewton slip bond between fibronectin and the cytoskeleton depends on talin. *Nature* 424:334–37

Kallen J, Welzenbach K, Ramage P, Geyl D, Kriwacki R, et al. 1999. Structural basis for LFA-1 inhibition upon lovastatin binding to the CD11a I-domain. *J. Mol. Biol.* 292:1–9

Kamata T, Ambo H, Puzon-McLaughlin W, Tieu KK, Handa M, et al. 2004. Critical cysteine residues for regulation of integrin alphaIIbbeta3 are clustered in the epidermal growth factor domains of the beta3 subunit. *Biochem. J.* 378:1079–82

Karpusas M, Ferrant J, Weinreb PH, Carmillo A, Taylor FR, Garber EA. 2003. Crystal structure of the alpha1beta1 integrin I domain in complex with an antibody Fab fragment. *J. Mol. Biol.* 327:1031–41

Kim M, Carman CV, Springer TA. 2003a. Bidirectional transmembrane signaling by cytoplasmic domain separation in integrins. *Science* 301:1720–25

Kim M, Carman CV, Yang W, Salas A, Springer TA. 2004. The primacy of affinity over clustering in regulation of adhesiveness of the integrin {alpha}L{beta}2. *J. Cell Biol.* 167:1241–53

Kim S, Chamberlain AK, Bowie JU. 2003b. A simple method for modeling transmembrane helix oligomers. *J. Mol. Biol.* 329:831–40

Kinashi T, Aker M, Sokolovsky-Eisenberg M, Grabovsky V, Tanaka C, et al. 2004. LAD-III, a leukocyte adhesion deficiency syndrome associated with defective Rap1 activation and impaired stabilization of integrin bonds. *Blood* 103:1033–36

Kinbara K, Goldfinger LE, Hansen M, Chou FL, Ginsberg MH. 2003. Ras GTPases: integrins' friends or foes? *Nat. Rev. Mol. Cell Biol.* 4:767–76

Leahy DJ, Aukhil I, Erickson HP. 1996. 2.0 A crystal structure of a four-domain segment of human fibronectin encompassing the RGD loop and synergy region. *Cell* 84:155–64

Lee J-O, Bankston LA, Arnaout MA, Liddington RC. 1995a. Two conformations of the integrin A-domain (I-domain): a pathway for activation? *Structure* 3:1333–40

Lee J-O, Rieu P, Arnaout MA, Liddington R. 1995b. Crystal structure of the A-domain from the α-subunit of $\beta2$ integrin complement receptor type 3 (CR3, CD11b/CD18). *Cell* 80:631–38

Li R, Babu CR, Lear JD, Wand AJ, Bennett JS, DeGrado WF. 2001. Oligomerization of the integrin alphaIIbbeta3: roles of the transmembrane and cytoplasmic domains. *Proc. Natl. Acad. Sci. USA* 98:12462–67

Li R, Babu CR, Valentine K, Lear JD, Wand AJ, et al. 2002a. Characterization of the monomeric form of the transmembrane and cytoplasmic domains of the integrin beta 3 subunit by NMR spectroscopy. *Biochemistry* 41:15618–24

Li R, Haruta I, Rieu P, Sugimori T, Xiong JP, Arnaout MA. 2002b. Characterization of a conformationally sensitive murine monoclonal antibody directed to the metal ion-dependent adhesion site face of integrin CD11b. *J. Immunol.* 168:1219–25

Li R, Mitra N, Gratkowski H, Vilaire G, Litvinov R, et al. 2003. Activation of integrin alphaI-Ibbeta3 by modulation of transmembrane helix associations. *Science* 300:795–98

Li R, Rieu P, Griffith DL, Scott D, Arnaout MA. 1998. Two functional states of the CD11b A-domain: correlations with key features of two Mn2+-complexed crystal structures. *J. Cell Biol.* 143:1523–34

Li W, Metcalf DG, Gorelik R, Li R, Mitra N, et al. 2005. A push-pull mechanism for regulating integrin function. *Proc. Natl. Acad. Sci. USA* 102:1424–29

Litvinov RI, Shuman H, Bennett JS, Weisel JW. 2002. Binding strength and activation state of single fibrinogen-integrin pairs on living cells. *Proc. Natl. Acad. Sci. USA* 99:7426–31

Lollo BA, Chan KWH, Hanson EM, Moy VT, Brian AA. 1993. Direct evidence for two affinity states for lymphocyte function-associated antigen 1 on activated T cells. *J. Biol. Chem.* 268:21693–700

Lu G. 2000. TOP: A new method for protein structure comparisons and similarity searches. *J. Appl. Cryst.* 33:176–83

Lupher ML Jr, Harris EA, Beals CR, Sui LM, Liddington RC, Staunton DE. 2001. Cellular activation of leukocyte function-associated antigen-1 and its affinity are regulated at the I domain allosteric site. *J. Immunol.* 167:1431–39

Luo BH, Springer TA, Takagi J. 2003a. Stabilizing the open conformation of the integrin headpiece with a glycan wedge increases affinity for ligand. *Proc. Natl. Acad. Sci. USA* 100:2403–8

Luo BH, Springer TA, Takagi J. 2004a. A specific interface between integrin transmembrane helices and affinity for ligand. *PLoS Biol.* 2:e153

Luo BH, Strokovich K, Walz T, Springer TA, Takagi J. 2004b. Allosteric beta1 integrin antibodies that stabilize the low affinity state by preventing the swing-out of the hybrid domain. *J. Biol. Chem.* 279:27466–71

Luo BH, Takagi J, Springer TA. 2003b. Locking the beta 3 integrin I-like domain into high and low affinity conformations with disulfides. *J. Biol. Chem.* 279:10215–21

Ma YQ, Plow EF, Geng JG. 2004. P-selectin binding to P-selectin glycoprotein ligand-1 induces an intermediate state of alphaMbeta2 activation and acts cooperatively with extracellular stimuli to support maximal adhesion of human neutrophils. *Blood* 104:2549–56

Martel V, Racaud-Sultan C, Dupe S, Marie C, Paulhe F, et al. 2001. Conformation, localization, and integrin binding of talin depend on its interaction with phosphoinositides. *J. Biol. Chem.* 276:21217–27

Martel V, Vignoud L, Dupe S, Frachet P, Block MR, Albiges-Rizo C. 2000. Talin controls the exit of the integrin alpha 5 beta 1 from an early compartment of the secretory pathway. *J. Cell Sci.* 113(Pt. 11):1951–61

McCleverty CJ, Liddington RC. 2003. Engineered allosteric mutants of the integrin alphaM-beta2 I domain: structural and functional studies. *Biochem. J.* 372:121–27

Michishita M, Videm V, Arnaout MA. 1993. A novel divalent cation-binding site in the A domain of the beta 2 integrin CR3 (CD11b/CD18) is essential for ligand binding. *Cell* 72:857–67

Mould AP, Barton SJ, Askari JA, Craig SE, Humphries MJ. 2003a. Role of ADMIDAS cation-binding site in ligand recognition by integrin {alpha}5{beta}1. *J. Biol. Chem.* 278:51622–29

Mould AP, Barton SJ, Askari JA, McEwan PA, Buckley PA, et al. 2003b. Conformational changes in the integrin beta A domain provide a mechanism for signal transduction via hybrid domain movement. *J. Biol. Chem.* 278:17028–35

Mould AP, Symonds EJ, Buckley PA, Grossmann JG, McEwan PA, et al. 2003c. Structure of an integrin-ligand complex deduced from solution x-ray scattering and site-directed mutagenesis. *J. Biol. Chem.* 278:39993–99

Nermut MV, Green NM, Eason P, Yamada SS, Yamada KM. 1988. Electron microscopy and structural model of human fibronectin receptor. *EMBO J.* 7:4093–99

Partridge AW, Liu S, Kim S, Bowie JU, Ginsberg MH. 2004. Transmembrane domain helix packing stabilizes integrin alpha IIbbeta 3 in the low affinity state. *J. Biol. Chem.* 280:7294–300

Piccardoni P, Manarini S, Federico L, Bagoly Z, Pecce R, et al. 2004. SRC-dependent outside-in signaling is a key step in the process of auto-regulation of beta2 integrins in polymorphonuclear leukocytes. *Biochem J.* 380(Pt. 1):57–65

Qu A, Leahy DJ. 1995. Crystal structure of the I-domain from the CD11a/CD18 (LFA-1, $\alpha L\beta 2$) integrin. *Proc. Natl. Acad. Sci.* 92:10277–81

Qu A, Leahy DJ. 1996. The role of the divalent cation in the structure of the I domain from the CD11a/CD18 integrin. *Structure* 4:931–42

Ridley AJ. 2004. Pulling back to move forward. *Cell* 116:357–58

Ridley AJ, Schwartz MA, Burridge K, Firtel RA, Ginsberg MH, et al. 2003. Cell migration: integrating signals from front to back. *Science* 302:1704–9

Rieu P, Sugimori T, Griffith DL, Arnaout MA. 1996. Solvent accessible residues on the MIDAS face of integrin CR3 mediate its binding to the neutrophil adhesion inhibitor NIF. *J. Biol. Chem.* 271:15858–61

Rost B, Yachdav G, Liu J. 2004. The PredictProtein server. *Nucleic Acids Res.* 32:W321–26

Sampath R, Gallagher PJ, Pavalko FM. 1998. Cytoskeletal interactions with the leukocyte integrin beta2 cytoplasmic tail. Activation-dependent regulation of associations with talin and alpha-actinin. *J. Biol. Chem.* 273:33588–94

Schwartz MA, Ginsberg MH. 2002. Networks and crosstalk: integrin signalling spreads. *Nat. Cell Biol.* 4:E65–68

Sebzda E, Bracke M, Tugal T, Hogg N, Cantrell DA. 2002. Rap1A positively regulates T cells via integrin activation rather than inhibiting lymphocyte signaling. *Nat. Immunol.* 3:251–58

Senes A, Engel DE, DeGrado WF. 2004. Folding of helical membrane proteins: the role of polar, GxxxG-like and proline motifs. *Curr. Opin. Struct. Biol.* 14:465–79

Shimaoka M, Salas A, Yang W, Weitz-Schmidt G, Springer TA. 2003a. Small molecule integrin antagonists that bind to the beta2 subunit I-like domain and activate signals in one direction and block them in the other. *Immunity* 19:391–402

Shimaoka M, Xiao T, Liu JH, Yang Y, Dong Y, et al. 2003b. Structures of the alphaL I domain and its complex with ICAM-1 reveal a shape-shifting pathway for integrin regulation. *Cell* 112:99–111

Shimonaka M, Katagiri K, Nakayama T, Fujita N, Tsuruo T, et al. 2003. Rap1 translates chemokine signals to integrin activation, cell polarization, and motility across vascular endothelium under flow. *J. Cell Biol.* 161:417–27

Simon DI, Wei Y, Zhang L, Rao NK, Xu H, et al. 2000. Identification of a urokinase receptor-integrin interaction site. Promiscuous regulator of integrin function. *J. Biol. Chem.* 275:10228–34

Smith MJ, Berg EL, Lawrence MB. 1999. A direct comparison of selectin-mediated transient, adhesive events using high temporal resolution. *Biophys. J.* 77:3371–83

Springer TA. 1997. Folding of the N terminal, ligand binding region of integrin alpha-subunits into a beta-propeller domain. *Proc. Natl. Acad. Sci. USA* 94:65–72

Stefansson A, Armulik A, Nilsson I, von Heijne G, Johansson S. 2004. Determination of N- and C-terminal borders of the transmembrane domain of integrin subunits. *J. Biol. Chem.* 279:21200–5

Tadokoro S, Shattil SJ, Eto K, Tai V, Liddington RC, et al. 2003. Talin binding to integrin beta tails: a final common step in integrin activation. *Science* 302:103–6

Takagi J, DeBottis DP, Erickson HP, Springer TA. 2002a. The role of the specificity-determining loop of the integrin beta subunit I-like domain in autonomous expression, association with the alpha subunit, and ligand binding. *Biochemistry* 41:4339–47

Takagi J, Petre BM, Walz T, Springer TA. 2002b. Global conformational rearrangements in integrin extracellular domains in outside-in and inside-out signaling. *Cell* 110:599–11

Takagi J, Strokovich K, Springer TA, Walz T. 2003. Structure of integrin alpha5beta1 in complex with fibronectin. *EMBO J.* 22:4607–15

Tan SM, Hyland RH, Al-Shamkhani A, Douglass WA, Shaw JM, Law SK. 2000. Effect of integrin beta 2 subunit truncations on LFA-1 (CD11a/CD18) and Mac-1 (CD11b/CD18) assembly, surface expression, and function. *J. Immunol.* 165:2574–81

Tng E, Tan SM, Ranganathan S, Cheng M, Law SK. 2004. The integrin alpha L beta 2 hybrid domain serves as a link for the propagation of activation signal from its stalk regions to the I-like domain. *J. Biol. Chem.* 279:54334–39

Ulmer TS, Calderwood DA, Ginsberg MH, Campbell ID. 2003. Domain-specific interactions of talin with the membrane-proximal region of the integrin beta3 subunit. *Biochemistry* 42:8307–12

Ulmer TS, Yaspan B, Ginsberg MH, Campbell ID. 2001. NMR analysis of structure and dynamics of the cytosolic tails of integrin alpha IIb beta 3 in aqueous solution. *Biochemistry* 40:7498–508

van Kooyk Y, Weder P, Heiji K, Figdor CG. 1994. Extracellular Ca^{2+} modulates leukocyte function-associated antigen-1 cell surface distribution on T lymphocytes and consequently affects cell adhesion. *J. Cell Biol.* 124:1061–70

Vinogradova O, Vaynberg J, Kong X, Haas TA, Plow EF, Qin J. 2004. Membrane-mediated structural transitions at the cytoplasmic face during integrin activation. *Proc. Natl. Acad. Sci. USA* 101:4094–99

Vinogradova O, Velyvis A, Velyviene A, Hu B, Haas T, et al. 2002. A structural mechanism of integrin alpha(IIb)beta(3) "inside-out" activation as regulated by its cytoplasmic face. *Cell* 110:587–97

von Wichert G, Haimovich B, Feng GS, Sheetz MP. 2003. Force-dependent integrin-cytoskeleton linkage formation requires downregulation of focal complex dynamics by Shp2. *EMBO J.* 22:5023–35

Vorup-Jensen T, Ostermeier C, Shimaoka M, Hommel U, Springer TA. 2003. Structure and allosteric regulation of the alpha X beta 2 integrin I domain. *Proc. Natl. Acad. Sci. USA* 100:1873–78

Weber KS, Ostermann G, Zernecke A, Schroder A, Klickstein LB, Weber C. 2001a. Dual role of H-Ras in regulation of lymphocyte function antigen-1 activity by stromal cell-derived factor-1alpha: implications for leukocyte transmigration. *Mol. Biol. Cell* 12:3074–86

Weber KS, Weber C, Ostermann G, Dierks H, Nagel W, Kolanus W. 2001b. Cytohesin-1 is a dynamic regulator of distinct LFA-1 functions in leukocyte arrest and transmigration triggered by chemokines. *Curr. Biol.* 11:1969–74

Wei Y, Czekay RP, Robillard L, Kugler MC, Zhang F, et al. 2005. Regulation of alpha5beta1 integrin conformation and function by urokinase receptor binding. *J. Cell Biol.* 168:501–11

Weitz-Schmidt G, Welzenbach K, Dawson J, Kallen J. 2004. Improved lymphocyte function-associated antigen-1 (LFA-1) inhibition by statin derivatives: molecular basis determined by x-ray analysis and monitoring of LFA-1 conformational changes in vitro and ex vivo. *J. Biol. Chem.* 279:46764–71

Weljie AM, Hwang PM, Vogel HJ. 2002. Solution structures of the cytoplasmic tail complex from platelet integrin alpha IIb- and beta 3-subunits. *Proc. Natl. Acad. Sci. USA* 99:5878–83

Welzenbach K, Hommel U, Weitz-Schmidt G. 2002. Small molecule inhibitors induce conformational changes in the I domain and the I-like domain of lymphocyte function-associated antigen-1. Molecular insights into integrin inhibition. *J. Biol. Chem.* 277:10590–98

Whittaker CA, Hynes RO. 2002. Distribution and evolution of von Willebrand/integrin A domains: widely dispersed domains with roles in cell adhesion and elsewhere. *Mol. Biol. Cell* 13:3369–87

Wilkins JA, Li A, Ni H, Stupack DG, Shen C. 1996. Control of beta1 integrin function. Localization of stimulatory epitopes. *J. Biol. Chem.* 271:3046–51

Xiao T, Takagi J, Coller BS, Wang JH, Springer TA. 2004. Structural basis for allostery in integrins and binding to fibrinogen-mimetic therapeutics. *Nature* 432:59–67

Xiong JP, Li R, Essafi M, Stehle T, Arnaout MA. 2000. An isoleucine-based allosteric switch controls affinity and shape shifting in integrin CD11b A-domain. *J. Biol. Chem.* 275:38762–67

Xiong JP, Stehle T, Diefenbach B, Zhang R, Dunker R, et al. 2001. Crystal structure of the extracellular segment of integrin alpha Vbeta3. *Science* 294:339–45

Xiong JP, Stehle T, Goodman SL, Arnaout MA. 2003. New insights into the structural basis of integrin activation. *Blood* 102:1155–59

Xiong JP, Stehle T, Goodman SL, Arnaout MA. 2004. A novel adaptation of the integrin PSI domain revealed from its crystal structure. *J. Biol. Chem.* 279:40252–54

Xiong JP, Stehle T, Zhang R, Joachimiak A, Frech M, et al. 2002. Crystal structure of the extracellular segment of integrin alpha Vbeta3 in complex with an Arg-Gly-Asp ligand. *Science* 296:151–55

Yan B, Calderwood DA, Yaspan B, Ginsberg MH. 2001. Calpain cleavage promotes talin binding to the beta 3 integrin cytoplasmic domain. *J. Biol. Chem.* 276:28164–70

Yan B, Hu DD, Knowles SK, Smith JW. 2000. Probing chemical and conformational differences in the resting and active conformers of platelet integrin alpha(IIb)beta(3). *J. Biol. Chem.* 275:7249–60

Yang W, Shimaoka M, Chen J, Springer TA. 2004. Activation of integrin beta-subunit I-like domains by one-turn C-terminal alpha-helix deletions. *Proc. Natl. Acad. Sci. USA* 101:2333–38

Zhang L, Plow EF. 1999. Amino acid sequences within the alpha subunit of integrin alpha M beta 2 (Mac-1) critical for specific recognition of C3bi. *Biochemistry* 38:8064–71

Zhang X, Wojcikiewicz E, Moy VT. 2002. Force spectroscopy of the leukocyte function-associated antigen-1/intercellular adhesion molecule-1 interaction. *Biophys. J.* 83:2270–79

Zwartz G, Chigaev A, Foutz T, Larson RS, Posner R, Sklar LA. 2004a. Relationship between molecular and cellular dissociation rates for VLA-4/VCAM-1 interaction in the absence of shear stress. *Biophys. J.* 86:1243–52

Zwartz GJ, Chigaev A, Dwyer DC, Foutz TD, Edwards BS, Sklar LA. 2004b. Real-time analysis of very late antigen-4 affinity modulation by shear. *J. Biol. Chem.* 279:38277–86

Centrosomes in Cellular Regulation

Stephen Doxsey,[1] Dannel McCollum,[2]
and William Theurkauf[1]

[1]Program in Molecular Medicine and [2]Molecular Genetics and Microbiology,
University of Massachusetts Medical School, Worcester, Massachusetts 01605;
email: stephen.doxsey@umassmed.edu; dannel.mccollum@umassmed.edu;
william.theurkauf@umassmed.edu

Annu. Rev. Cell Dev. Biol.
2005. 21:411–34

First published online as a
Review in Advance on
June 28, 2005

The *Annual Review of
Cell and Developmental
Biology* is online at
http://cellbio.annualreviews.org

doi: 10.1146/
annurev.cellbio.21.122303.120418

Copyright © 2005 by
Annual Reviews. All rights
reserved

1081-0706/05/1110-
0411$20.00

Key Words

spindle pole body, centrosome inactivation, checkpoints,
cytokinesis, cell cycle, mitotic exit network, septation initiation
network

Abstract

Centrosomes, spindle pole bodies, and related structures in other
organisms are a morphologically diverse group of organelles that
share a common ability to nucleate and organize microtubules and
are thus referred to as microtubule organizing centers or MTOCs.
Features associated with MTOCs include organization of mitotic
spindles, formation of primary cilia, progression through cytokinesis, and self-duplication once per cell cycle. Centrosomes bind
more than 100 regulatory proteins, whose identities suggest roles in
a multitude of cellular functions. In fact, recent work has shown that
MTOCs are required for several regulatory functions including cell
cycle transitions, cellular responses to stress, and organization of signal transduction pathways. These new liaisons between MTOCs and
cellular regulation are the focus of this review. Elucidation of these
and other previously unappreciated centrosome functions promises
to yield exciting scientific discovery for some time to come.

Contents

INTRODUCTION

Microtubule organizing center (MTOC): structures of diverse morphology that duplicate every cell cycle and nucleate the growth of microtubules

Spindle pole body (SPB): yeast equivalent of the centrosome

Microtubule organizing centers (MTOCs) assume various shapes and sizes and include centrosomes of vertebrate cells and spindle pole bodies (SPB) of yeasts. Despite their diversity of form, all MTOCs share a common function in the nucleation and organization of microtubules. MTOC-organized cytoplasmic microtubules perform a variety of functions, whereas those in mitotic cells play a central role in the organization of mitotic spindles. For more information on MTOC structure, function, and re-lated diseases see Doxsey 2001, Khodjakov & Rieder 2001, Nigg 2002, and Sobel 1997. More recent studies demonstrate that MTOCs play new and unexpected roles in several other processes including cell cycle control, cytokinesis, and responses to cellular stress. These burgeoning new areas of centrosome biology are the focus of this review.

CENTROSOMES AND CELL CYCLE CONTROL

Centrosomes (and other MTOCs) are made up of numerous proteins whose amino acid sequence suggests a coiled coil tertiary structure. Increasing evidence indicates that this molecular structure may be well designed for the organization of multiprotein scaffolds that can anchor a diversity of activities ranging from protein complexes involved in microtubule nucleation (Diviani et al. 2000, Diviani & Scott 2001) to multicomponent pathways for cellular regulation (Elliott et al. 1999). By physically linking components of a common pathway, molecular scaffolds can increase the local concentration of components, limit nonspecific interactions, and provide spatial control for regulatory pathways by positioning them at specific sites in proximity to downstream targets or upstream modulators. On the basis of the increasing number of regulatory molecules anchored at the centrosome/MTOC (>100), it is likely that this organelle serves as a centralized control center for regulating a diversity of cellular activities. Recent studies have provided some of the first functional links between centrosomes and regulatory networks. In this section, we focus on work that provides the most direct links between centrosomes and cell cycle progression. We discuss the role of the centrosome in cell cycle transitions from G_1 to S-phase, G_2 to M-phase and metaphase to anaphase. The role of centrosomes in progression through cytokinesis is addressed in the next section.

Centrosomes and the G_2 to M Transition

Centrosomal Cdk1 activation during the G_2/M transition. Early work on the G_2/M transition showed that centrosomes could induce progression into mitosis following injection into G_2-arrested starfish oocytes (Picard et al. 1987). Centrosomes were also able to activate maturation promoting factor [MPF, now known as cyclin-dependent kinase1-cyclinB complexes (Cdk1-cycB)] and induce premature mitotic entry in *Xenopus* eggs (Perez-Mongiovi et al. 2000). Other studies showed that mitotic kinases and cyclins were present at centrosomes. More recent work indicates that centrosomes are involved in the G_2 to M transition in mammalian cells. Mitosis is initiated in part by activation of Cdk1-cycB. CycB1 is present throughout the cytoplasm prior to prophase. However, active Cdk1-cycB1 is first detected at the centrosome during prophase and prior to the Cdk1-cycB1-dependent phosphorylation of histone H3 in the nucleus (Jackman et al. 2003). This observation suggested that centrosomes might function as sites of integration and activation of proteins that trigger mitosis.

A clever centrosome targeting strategy was recently used to provide evidence that mitotic entry requires centrosome localization of Cdk1 and its modulators (Kramer et al. 2004). The authors demonstrated that the checkpoint kinase 1 (Chk1) associated with centrosomes in nonmitotic cells and inhibited Cdk1 activity. Inhibition of Chk1 induced premature activation of Cdk1 at centrosomes and premature mitotic entry. When the centrosome localization sequence (CLS) of the centrosome protein AKAP450 was fused to Chk1, the kinase was immobilized at the centrosome where it was unable to phosphorylate nuclear substrates. Chk1 lacking the CLS did not localize to centrosomes and phosphorylated substrates normally. Centrosome immobilization of a kinase-inactive form of Chk1 induced premature Cdk1 activation

and premature mitotic entry, whereas centrosome immobilization of wild-type Chk1 prevented Cdk1 activation at centrosomes and induced mitotic failure, polyploidy, and multiple centrosomes.

The Chk1-mediated inhibition of Cdk1 activity was not direct but linked to inhibition of the Cdk1 activating phosphatase, Cdc25B (Kramer et al. 2004). Cdc25-mediated activation of Cdk1 seemed to occur through centrosome-localized Chk1 (Forrest et al. 1999, Kramer et al. 2004). In addition, aurora-A kinase was required for recruitment of Cdk1-cycB to centrosomes and thus, for activation of the kinase (Hirota et al. 2003). Centrosome-associated Polo kinase (Polo) is also involved in mitotic entry, although it does not appear to be directly linked to Cdk1 activation. Localization of Polo to centrosomes during G2/M depends upon the polo box (Jang et al. 2002, Lee et al. 1998, Reynolds & Ohkura 2003), which prevents mitotic entry and arrests cells with 4N DNA content when overexpressed. In contrast, polo box mutants do not localize to centrosomes or inhibit mitotic progression (Jang et al. 2002).

Taken together, these results suggest that there is a cell cycle regulatory module at the centrosome that integrates positive and negative pathways to control mitotic entry. However, removal of centrosomes/centrioles does not prevent entry into mitosis (Hinchcliffe et al. 2001, Khodjakov & Rieder 2001), suggesting that centrosomal regulation of Cdk1 activation by Chk1 may not be required for mitotic entry. On the other hand, the pericentriolar material (PCM) and other material remaining in the acentriolar MTOC after extraction of the centrioles may serve this function (Hinchcliffe et al. 2001, Khodjakov & Rieder 2001). Thus it seems that centrosome-associated regulatory pathways may be dominant over centrosome-independent pathways for mitotic entry (Kramer et al. 2004).

Centrosomal γ tubulin and the G_2/M transition. Centrosomal microtubule nucleation is mediated in part by γ tubulin ring complexes

Cdk: cyclin-dependent kinase

Cyc: cyclin

Centrioles: microtubule-based structures comprised of α/β tubulin and other proteins surrounded by pericentriolar material

Pericentriolar material (PCM): fibrillar material surrounding centrioles in the centrosome that nucleates the growth of new microtubules

(γTuRCs). Pericentrin is a coiled coil scaffolding protein that anchors γTuRCs at centrosomes (Zimmerman et al. 2004). Uncoupling of the pericentrin-γTuRC interaction by peptides encoding pericentrin's γTuRC-binding domain, or by siRNA (small interfering RNA)–mediated pericentrin depletion, induced arrest at G_2/M followed by apoptosis. Some cell lines appeared insensitive to the G_2/M arrest, continued to cycle, and revealed a reduction in centrosomal γtubulin and astral microtubules in mitosis. It will be interesting to determine if centrosome-associated regulatory molecules involved in cell cycle progression are mislocalized from centrosomes under these conditions (see above).

γTuRC: gamma tubulin ring complex

The Centrosome/Spindle Pole in the Metaphase to Anaphase Transition

Although the kinetochore is best known for sensing and regulating the metaphase to anaphase transition (Maiato et al. 2004), the centrosome/spindle pole also appears to play a role in this process. The was first suggested by data showing that destruction of GFP-tagged and endogenous cycB at the M-A transition in *Drosophila* was initiated at the spindle pole, which then proceeded up the spindle (Huang & Raff 1999). When centrosomes were lost from spindles in the *Drosophila* mutant *cfo*, cycB was lost from centrosomes but not spindles, and cells arrested in anaphase (Wakefield et al. 2000). Recent work in cellularized *Drosophila* embryos showed that Fzy/Cdc20 was responsible for the spindle-associated wave of cycB destruction, whereas Fzr/Cdh1 was required for destruction of cytoplasmic cycB (Raff et al. 2002).

More recent studies have implicated the centrosome proteins γtubulin and the pericentrin B homologue, Pcp1p, in regulation of the M-A transition. In *Aspergillus nidulans*, a cold-sensitive γtubulin allele did not inhibit spindle formation at the restrictive temperature but delayed the M-A transition and induced cytokinesis failure (Prigozhina et al. 2004). Likewise, a mutation in *Schizosaccha-*

romyces pombe, Pcp1p, inhibited the M-A transition without disrupting bipolar spindle assembly (Rajagopalan et al. 2004). Thus it appears that centrosomal pericentrin homologs and γtubulin complexes may be involved in regulation of multiple cell cycle transitions (G_2/M and M-A).

The Centrosome in the G_1 to S-phase Transition

Removal of centrioles and other core centrosome components induces G_1 arrest. Recent data indicate that centrosomes can affect progression through the cell cycle (**Figure 1**). Studies designed to remove centrosomes/centrioles from cells by microsurgical cutting (Hinchcliffe et al. 2001) or laser ablation (Khodjakov & Rieder 2001) have provided direct evidence for centrosomes in cell cycle progression. Under both conditions, the centriole pair and associated proteins were removed, yet it is important to note that these cells formed acentriolar MTOCs containing proteins of the PCM and perhaps other centrosome structures (Khodjakov & Rieder 2001, Khodjakov et al. 2002). Similar acentriolar MTOCs are functional in higher plants and some animal meiotic systems (Shimamura et al. 2004, Theurkauf & Hawley 1992). In mammlian cells with acentriolar MTOCs, early mitotic events occurred normally but many cells exhibited cytokinesis defects or failure. All cells with acentriolar MTOCs generated by microsurgery failed to initiate DNA replication (BrdU-negative) regardless of whether they completed cytokinesis. Moreover, ablation of one of two centrosomes in prometaphase cells produced a centrosome-containing daughter that continued to cycle and a daughter cell with an acentriolar MTOC that arrested in G_1 (BrdU-negative). Both strategies used to remove core centrosome structures and components showed that acentriolar MTOCs experience problems during cytokinesis and subsequently undergo G_1 arrest. However, extra centrosomes created by cell fusions or inhibition of

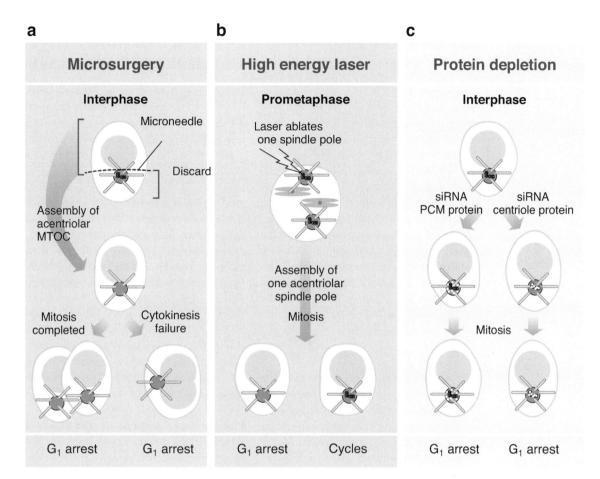

a

Microsurgery

Interphase

Microneedle

Discard

Assembly of acentriolar MTOC

Mitosis completed

Cytokinesis failure

G_1 arrest G_1 arrest

b

High energy laser

Prometaphase

Laser ablates one spindle pole

Assembly of one acentriolar spindle pole

Mitosis

G_1 arrest   Cycles

c

Protein depletion

Interphase

siRNA PCM protein   siRNA centriole protein

Mitosis

G_1 arrest G_1 arrest

Figure 1

Centrosomes are required for G_1 to S-phase progression. Removal of centrosome (*blue/red*) by (*a*) microsurgery or (*b*) laser ablation produces cells with loosely focused microtubule arrays (*bottom row*, *yellow*) organized by an MTOC-containing PCM proteins (*blue*) but lacking centrioles (*red*). Acentriolar MTOC-containing cells undergo G_1 arrest. (*b*) Cells with intact centrosomes cycle (*right*). (*c*) Depletion of centrosome proteins of the centrioles (*red turned white*), PCM (*blue turned white*), or other structures by small interfering RNAs (siRNA) results in G_1 arrest (*bottom row*). Nucleus, gray (adapted from *Trends in Cell Biology*, in press).

cytokinesis do not inhibit cell cycle progression (Uetake & Sluder 2004, Wong & Stearns 2003). Similarly, centrosome-associated microtubules do not appear to play an essential role in cell cycle progression. Addition of the microtubule depolymerizing drug nocodazole to normal cycling diploid cells (2N) did not induce arrest in G_1 of cells with diploid genomes (2N), even after they were first synchronized in G_1/G_0 by serum starvation. Instead, nocodazole-treated cells continued to cycle, delayed in mitosis, experienced mitotic failure, and then arrested as tetraploid cells (4N) in a G_1-like state (Lanni & Jacks 1998, Trielli et al. 1996). Because G_1 arrest was observed after mitotic failure in tetraploid cells with multiple centrosomes, it will be important to confirm this result under physiological conditions using normal cycling diploid cells.

Changes in centrosome protein levels or localization induce G$_1$ arrest. Recent studies have identified centrosome proteins that function in cytokinesis and cell cycle progression (**Figure 1**). Centriolin is a centrosome protein that shares homology with yeast proteins involved in cytokinesis and mitotic exit (Nud1p and Cdc11p) (Gromley et al. 2003, Guertin et al. 2002). Overexpression of the Nud1p/Cdc11p homology domain or centriolin depletion induced cytokinesis defects followed by cell cycle arrest in G$_1$ as shown by flow cytometry and BrdU staining. A similar cell cycle arrest in G$_1$ was observed when the centrosome protein AKAP450 and PKA were mislocalized from centrosomes (Gillingham & Munro 2000, Keryer et al. 2003). The data thus far show that removal of entire centrioles or changes in individual centrosome proteins induce cytokinesis defects that appear to lead to G$_1$ arrest.

SiRNA-mediated depletion of several centrosome proteins induces G$_1$ arrest. As discussed above, centrosomes play key roles in spindle function and cytokinesis. This suggests that the G$_1$ arrest observed after perturbation of centrosomes or centrosome proteins may result from mitotic dysfunction. Recent data provided evidence for a mitosis-independent cell cycle arrest. Antibodies to the centrosome protein PCM-1 prevented entry into S-phase when microinjected into early interphase mouse zygotes (Balczon et al. 2002). In another study, RNA interference was used to individually deplete more than 20 centrosome proteins that localized to several independent centrosome sites (e.g., PCM, centriole) (**Figure 1**). Depletion of nearly all these proteins induced G$_1$ arrest as shown by accumulation in the 2N peak by flow cytometry, lack of BrdU incorporation, and reduction in the activity of cyclin-dependent kinase2-cyclinA/E complexes (Cdk2-cycA/E) (Mikule et al. 2003). G$_1$ arrest could be rescued by returning targeted centrosome proteins to normal levels. No common functional change was observed for proteins that induced cell cycle

arrest. These results suggested that cell cycle arrest could be induced through discrete alterations in centrosome composition.

G$_1$ arrest is induced in postmitotic cells. The role of mitotic dysfunction in G$_1$ arrest was examined in more detail using postmitotic cells (early G$_1$). Cells that had recently completed cytokinesis were microinjected with a plasmid encoding the centrosome-targeting region of pericentrin/AKAP450 that mislocalizes both proteins from centrosomes (Gillingham & Munro 2000). This reduced the centrosome-bound fraction of endogenous pericentrin and prevented cells from entering S-phase and incorporating BrdU (K. Mikule & S. Doxsey, unpublished observations). Another recent study showed that ablation of centrosomes in postmitotic cells inhibited progression through the cell cycle (14/16 cells; A. Khodjakov, personal communication). Taken together, results from multiple approaches show that G$_1$ arrest can be induced from within G$_1$.

Possible mechanisms of centrosome-associated G$_1$ arrest. How centrosome loss or alteration leads to G$_1$ arrest is currently unclear. Regulation of centrosome duplication and entry into S-phase are similar in that both require Cdk2-cycE/A complexes (Lacey et al. 1999). As described above, ectopic expression of cycE accelerated entry into S-phase whereas expression of the cycE centrosome localization domain (CLD) disrupted centrosome binding of endogenous cycE (and cycA) and prevented entry into S-phase (Matsumoto & Maller 2004). We currently do not know if centrosome targeting of cycE is lost during centrosome removal or centrosome protein depletion, and if this contributes to the centrosome-induced G$_1$ arrest. In this regard, the centrosome could control S-phase entry and therefore the nuclear replication cycle through mechanisms such as cycE binding that are independent of the cell cycle because cell cycle arrest by cycE CLD does not require Cdk2 binding.

A centrosome-induced checkpoint? Studies on centrosome protein depletion suggest that G_1 arrest involves activation of a cell cycle checkpoint. For example, G_1 arrest was suppressed in human tumor cells with abrogated p53 function and in cells acutely depleted of p53 following centrosome protein depletion (K. Mikule & S. Doxsey, unpublished observation). In addition, the p38 stress-activated signal transduction pathway was shown to be involved in the G_1 arrest. These observations suggest that the inability of some cells to arrest in the absence of centrioles may be related to loss or abrogation of p53 or p38 or related regulatory molecules (Bobinnec et al. 1998, Piel et al. 2001). It is also possible that checkpoint signaling may occur at the centrosome given the localization of both p53 and p38 to this site (Ciciarello et al. 2001; Liu et al. 2004; K. Mikule & S. Doxsey, unpublished observations). Tumor cells with abrogated p53 or p38 function may avoid checkpoint activation, continue to cycle and propagate centrosome defects, mitotic dysfunction, and genetic instability.

ROLE OF CENTROSOMES AND SPINDLE POLE BODIES IN CYTOKINESIS AND MITOTIC EXIT

Recent studies have implicated SPBs and centrosomes in numerous aspects of cell cycle progression including mitotic exit and cytokinesis. Studies in both the fission yeast *S. pombe* and the budding yeast *Saccharomyces cerevisiae* have delineated two conserved signaling pathways termed the septation initiation network (SIN) and the mitotic exit network (MEN), respectively, which localize to the SPBs and regulate cytokinesis and mitotic exit. Both of these pathways have been covered extensively in recent reviews (Seshan & Amon 2004, Simanis 2003) and are not be reviewed in detail here. Instead, we focus on more recent studies, novel functions for these pathways, and potential conserved functions in mammalian cells.

The SIN Pathway in *S. pombe*

As mentioned above, the SIN pathway is essential for cytokinesis in fission yeast (**Figure 2**). The SIN is a SPB localized-GTPase-regulated protein kinase cascade (for a list of SIN homologs in *S. pombe*, *S. cerevisiae*, and mammalian cells, see **Table 1**). SIN mutants proceed normally through the cell cycle and mitosis and can form cytokinetic actomyosin contractile rings, a key structure required for cytokinesis analogous to the cleavage furrow in animal cells. However, SIN mutants fail to initiate constriction of the ring and the rings fall apart causing the cells to fail cytokinesis and become multinucleate (Balasubramanian et al. 1998, Fankhauser et al. 1995). Inappropriate activation of the SIN can also drive exit from mitosis, suggesting that the SIN may play a nonessential role in exit from mitosis (Fankhauser et al. 1993, Guertin et al. 2002).

Asymmetry of SIN signaling. One curious feature of the SIN is the asymmetric pattern of loca lization of some components to the SPBs (**Figure 2**). It has been unclear both how the asymmetry of the SIN pathway is generated as well as why SIN signaling is asymmetric. Although there is still little known about the functional significance of asymmetry for SIN signaling, a recent study revealed the basis of the asymmetry of the SIN pathway (Grallert et al. 2004). This study found that the asymmetry of the SIN reflected underlying asymmetry of the spindle poles. After SPB duplication, each cell has an old SPB that was inherited from the previous cell cycle and the new SPB that forms upon SPB duplication. It was discovered that in anaphase, when the asymmetry of the SIN arises, the SIN inhibitors Cdc16p-Byr4p localize to the old SPB, whereas the SIN activator Cdc7p and presumably Sid1p-Cdc14p localize to the new SPB. It is not known why SIN signaling is asymmetric, but it may have to do with downregulating the SIN because most of the known mutations that activate SIN signaling

Septation initiation network (SIN): signaling network in *S. pombe*, analogous to MEN in *S. cerevisiae*, required for cytokinesis

Mitotic exit network (MEN): signaling network required for mitotic exit in the budding yeast *S. cerevisiae*

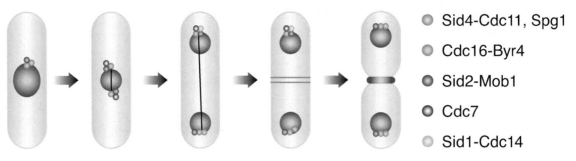

○ Sid4-Cdc11, Spg1

○ Cdc16-Byr4

● Sid2-Mob1

○ Cdc7

○ Sid1-Cdc14

Figure 2

Cell cycle–dependent localization of SIN components. All SIN components localize to the spindle pole body (SPB). Sid4p and Cdc11p form a complex at the SPB that functions as a scaffold required for localization of all known SIN components to the SPB (Chang & Gould 2000; Guertin et al. 2000; Hou et al. 2000; Krapp et al. 2001, 2004; C. Li et al. 2000; Morrell et al. 2004; Sparks et al. 1999). Sid4p-Cdc11p and Spg1 (*green*), and to some extent Sid2p-Mob1p (*blue*), localize to the SPB throughout the cell cycle. SIN signaling is negatively regulated by Cdc16p and Byr4p (*orange*), which function as part of a two-component GTPase activating protein (GAP) for Spg1p (Furge et al. 1998). A guanine-nucleotide exchange factor (GEF) for Spg1p has not been identified. Both the activity and localization of SIN proteins are regulated through the cell cycle. In interphase the Cdc16p-Byr4p GAP complex localizes to the SPB (Cerutti & Simanis 1999, C. Li et al. 2000) and, consistent with this, Spg1p is at the pole but in the inactive GDP-bound state (Sohrmann et al. 1998). As the mitotic spindle forms during metaphase, Spg1p becomes activated at both SPBs (GTP-bound form), and Cdc16p-Byr4p leaves both SPBs (Cerutti & Simanis 1999, C. Li et al. 2000). Cdc7p (*red*) is recruited to the SPB by the GTP-bound form of Spg1p to which it binds directly (Sohrmann et al. 1998). However, the SIN does not become activated at this time. SIN activation in metaphase is restrained by Cdk activity. During anaphase B, after Cdk inactivation, Cdc16p-Byr4p returns to one SPB (Cerutti & Simanis 1999, C. Li et al. 2000). Spg1p is inactivated, and Cdc7p becomes delocalized at that SPB. Also at this time Sid1p-Cdc14p (*aqua*) localizes to the single Cdc7p containing SPB in anaphase (Guertin et al. 2000). Once Sid1p-Cdc14p localizes to the SPB, it and possibly Cdc7p are then presumed to activate Sid2p-Mob1p and cause them to translocate to the actomyosin ring to trigger ring constriction and septation (Guertin et al. 2000, Sparks et al. 1999). Once septum formation is complete, the SIN becomes inactive and returns to its interphase configuration.

result in localization of SIN activators to both SPBs.

The SIN, CDK regulation, and the cytokinesis checkpoint. The SIN is kept inactive in early mitosis by Cdk activity (Chang et al. 2001, Guertin et al. 2000). Cdk inhibition of the SIN may be from direct phosphorylation of SIN components by Cdk1, since Cdk1p-Cdc13p binds directly to the SIN scaffold protein Cdc11p, positioning it to phosphorylate SIN proteins (Morrell et al. 2004). The targets of Cdk phosphorylation in the SIN are not known. Because, in *S. pombe* and most other organisms, Cdk1p inactivation occurs coincident with chromosome segregation, coupling initiation of cytokinesis to Cdk1p inactivation ensures that cell division does not initiate before chromosomes have been segregated. However, this mechanism renders cytokinesis sensitive to Cdk activity, which begins to rise shortly after completion of cytokinesis, as the next cell cycle initiates. If cytokinesis is delayed, the rising Cdk1p activity could inhibit the SIN and cytokinesis unless the cell has a way to inhibit Cdk activity until cytokinesis is complete. Clp1p/Flp1p, the *S. pombe* homolog of the budding yeast Cdc14 phosphatase homolog Clp1p/Flp1p (hereinafter referred to as Clp1p) plays a crucial role in maintaining SIN activity if cytokinesis is delayed (Mishra et al. 2004, Trautmann et al. 2001). Because Cdc14-family phosphatases dephosphorylate sites phosphorylated by Cdks (Esteban et al. 2004, Kaiser et al. 2002, L. Li et al. 2000, Visintin et al. 1998, Wolfe & Gould 2004),

Clp1p presumably maintains SIN signaling when cytokinesis is delayed by antagonizing the inhibitory effects of Cdk phosphorylation on the SIN. Clp1p also inhibits Cdk activity by dephosphorylating and destabilizing Cdc25p (Esteban et al. 2004, Wolfe & Gould 2004). Through this mechanism Clp1p antagonizes Cdk1 activity by promoting inhibitory tyrosine phosphorylation on Cdk1.

Clp1p localizes to the nucleolus in interphase where it is thought to be sequestered and inactive (Cueille et al. 2001, Trautmann et al. 2001). Clp1p is released from the nucleolus in mitosis and the SIN acts to keep Clp1p out of the nucleolus until cytokinesis is complete. Thus Clp1p and the SIN seem to function together, each acting to maintain the others activity until cytokinesis is complete. This suggests that the SIN inhibits cell cycle progression through Clp1p. However, recent evidence showed that hyperactivation of the SIN can block cell cycle progression independently of Clp1p, suggesting that the SIN can antagonize Cdk activity through another mechanism (Mishra et al. 2004). This activity becomes crucial when cytokinesis is delayed. When cytokinesis is delayed in wild-type cells, the SIN remains active, Clp1p stays out of the nucleolus, the cytokinetic apparatus is maintained, and the cells arrest further nuclear division until cytokinesis is complete. This ensures that the cell does not become multinucleate and polyploid if cytokinesis is delayed. Complete inhibition of cytokinesis results in a prolonged delay in nuclear division (Cueille et al. 2001, Trautmann et al. 2001). In contrast, cells with weakened SIN signaling, or deletion of *clp1*, are not able to maintain the cytokinetic apparatus in response to delays in cytokinesis, and the cytoskeleton returns to the interphase configuration. These cells then proceed with further rounds of nuclear division resulting in multinucleate, polyploid cells. This is reminiscent of the delay or block in cell cycle progression observed in mammalian cells after failure to complete mitosis or cytokinesis (for review, see Stukenberg 2004).

Table 1 SIN and MEN proteins and potential mammalian homologs

S. pombe	S. cerevisiae	Mammals	Protein function
plo1	CDC5	Polo	Kinase
sid4	Unknown	?	SPB scaffold
cdc11	NUD1	centriolin	SPB scaffold
spg1	TEM1	?	GTPase
cdc7	CDC15	Mst2?	Kinase
sid1	Unknown	Mst2?	Kinase
cdc14	Unknown	?	Sid1 binding
sid2	DBF2	Warts1/Lats1, Lats2	Kinase
mob1	MOB1	Mob1	Sid2/Dbf2 binding
clp1	CDC14	Cdc14A/B	Phosphatase
cdc16	BUB2	?	Part of GAP
byr4	BFA1/BYR4	?	Part of GAP

The SIN consists of two scaffolding proteins (Sid4p, Cdc11p) (Krapp et al. 2004, Morrell et al. 2004), four protein kinases (Plo1p, Cdc7p, Sid1p, and Sid2p) (Fankhauser & Simanis 1994, Guertin et al. 2000, Ohkura et al. 1995, Sparks et al. 1999) and one small GTPase, Spg1p (Schmidt et al. 1997). Additionally, the Cdc14p (Fankhauser & Simanis 1993) and Mob1p (Hou et al. 2000, Salimova et al. 2000) proteins function as subunits of the Sid1p and Sid2p kinases, respectively. The MEN consists of a scaffolding protein (Nud1p) (Gruneberg et al. 2000), four protein kinases (Cdc5p, Cdc15p, Dbf2p, and Dbf20p), a GTPase (Tem1p), an exchange factor (Lte1p), a protein phosphatase (Cdc14p), and a Dbf2p-binding protein (Mob1p) (Johnston et al. 1990; Kitada et al. 1993; Luca & Winey 1998; Schweitzer & Philippsen 1991; Shirayama et al. 1994a,b; Toyn et al. 1991; Wan et al. 1992). Potential homologs of several SIN/MEN components have been identified, including centriolin (Gromley et al. 2003), Polo kinase (Golsteyn et al. 1994), Cdc14A/B (Li et al. 1997), Mst2 kinase (Hay & Guo 2003), Mob1 (Luca & Winey 1998), Warts/Lats1 (Nishiyama et al. 1999, Tao et al. 1999), and Lats2 (Hori et al. 2000, Yabuta et al. 2000) kinases.

See below for speculation about whether homologs of the SIN/MEN proteins y function in the mitosis to interphase transition in mammalian cells to deal with mitotic or cytokinetic failures.

The MEN Pathway in *S. cerevisiae*

In *S. cerevisiae* the pathway analogous to the SIN is termed the MEN. Like the SIN, the MEN is a GTPase-regulated protein kinase cascade, whose components localize to both the SPB and the bud neck (Seshan & Amon 2004, Simanis 2003). Mutants in the MEN pathway arrest at the end of anaphase with elongated spindles and high Cdk activity. The MEN is required to antagonize Cdk activity and bring about mitotic exit. Similar to the SIN, the MEN also plays a role, albeit nonessential, in cytokinesis. The MEN functions in anaphase to inhibit Cdk activity and cause mitotic exit by promoting release of the Cdc14p phosphatase from the nucleolus by an unknown mechanism (Shou et al. 1999, Visintin et al. 1999). Cdc14p then dephosphorylates a number of Cdk substrates including Cdh1p/Hct1p and the Cdk inhibitor Sic1p (Visintin et al. 1998). Dephosphorylation by Cdc14p stabilizes Sic1p and also causes activation of Cdh1p/Hct1p, which promotes cyclin B proteolysis. Together, these events result in Cdk inactivation and exit from mitosis.

Regulation of MEN activity. An elegant model has been proposed for MEN activation in anaphase, where elongation of the spindle into the bud would bring the SPB localized Tem1p GTPase into contact with its bud localized activator Tem1p (Bardin et al. 2000, Pereira et al. 2000) (**Figure 3**). This model would allow for temporal and spatial coupling of mitotic exit and chromosome segregation, such that exit from mitosis occurs only after chromosomes have segregated to each daughter cell. Some recent evidence suggests that this model may not be quite so straightforward. Lte1p has not been shown to directly activate Tem1p in vitro, and further experiments suggest that its exchange activity may not be

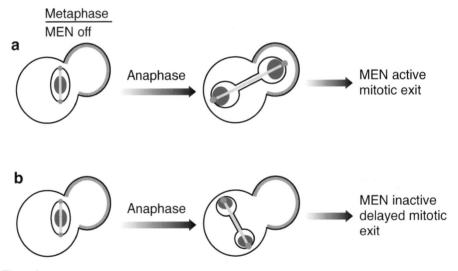

Figure 3

Regulation of MEN activity. (*a*) Several mechanisms function to keep the MEN inactive in metaphase. Premature activation of the MEN in metaphase is prevented by the Bub2p-Bfa1p, which serves as a GAP for the Tem1p GTPase (Alexandru et al. 1999, Fesquet et al. 1999, Fraschini et al. 1999, Li 1999). Additionally, Tem1p is spatially separated from its proposed GTP exchange factor Lte1p, which localizes to the bud cortex (*orange*). Cdk phosphorylation of Cdc15p is also thought to antagonize MEN signaling in metaphase (Jaspersen & Morgan 2000, Menssen et al. 2001). Similarly, activation of the MEN in anaphase is brought about by multiple mechanisms. Activation of the MEN depends on the passage of one of the spindle poles through the bud neck during anaphase spindle elongation (Molk et al. 2004, Yeh et al. 1995). Once this has occurred, it brings that pole in proximity to the putative exchange factor Lte1p, which localizes specifically to the bud cortex (Bardin et al. 2000, Pereira et al. 2000), presumably allowing activation of the Tem1p GTPase. Although some recent results raise questions about certain aspects of this model (see text), it provides a nice explanation for how spindle orientation in anaphase is coupled to mitotic exit. (*b*) Cells with defects in spindle orientation often undergo anaphase chromosome separation in the mother cell. These cells delay MEN activation and exit from mitosis until correct spindle orientation is achieved. Correct spindle orientation results in exit from mitosis, presumably in part by bringing SPB localized Tem1p into contact with bud localized Tem1p (*orange*).

required for its role in mitotic exit (Yoshida et al. 2003), raising questions about whether Lte1p is the GEF for Tem1. Also, Lte1p is essential only for mitotic exit at low temperatures, indicating that Tem1p can become activated in the absence of Lte1p. Thus other mechanisms may contribute to activation of the MEN, including reduced Cdk inhibition of the MEN that occurs when Cdk activity decreases in anaphase through partial proteolysis of B-type cyclins and early anaphase release of the Cdc14p phosphatase triggered by the FEAR network (see D'Amours & Amon 2004 for review). It has also been proposed that the loss of astral microtubules from the bud neck that occurs when the spindle elongates through the bud neck functions independently from Lte1p to promote mitotic exit (Castillon et al. 2003). In addition, Cdc5p, the *S. cerevisiae* polo homolog, acts upstream to antagonize Bub2p/Byr4p by phosphorylating Byr4p directly facilitating activation of the MEN (Hu et al. 2001). Together, these studies suggest that regulation of the MEN is complex and controlled by numerous inputs.

Asymmetry at SPBs and MEN regulation. Although not all reports are in agreement, the MEN components Bub2p and Tem1p seem to preferentially localize to the SPB that enters the bud (Bardin et al. 2000, Pereira et al. 2000). Recent analysis (Molk et al. 2004) of localization of GFP fusions in live cells shows that Bub2p, Tem1p, and Cdc15p show asymmetric localization to the SPB in the bud. As the old SPB enters the bud, Tem1p localization increases at that site. This coincides with Cdc15 localization to the same SPB, which is presumably recruited by Tem1p. Surprisingly, localization of the Tem1p inhibitor Bub2p also increases at the old SPB as it passes into the bud. The reason for the asymmetry is not clear. It is presumed that asymmetry is important for MEN signaling. In dynein mutants, where anaphase takes place in the bud, Tem1p localizes in a symmetric manner to both SPBs, and the MEN remains inactive. It is curious that both MEN and the SIN proteins local-ize asymmetrically on the SPBs, but the active SIN components are on the new SPB, whereas the MEN components are on the old SPB. However, as with the SIN, proof of the importance of asymmetry in MEN signaling is lacking.

MEN and cytokinesis. Recent evidence suggests that the MEN, similar to the SIN, also plays a more direct role in cytokinesis, although this function does not appear essential. Cells in which the requirement for the MEN in mitotic exit has been relieved are viable, but they still have defects in cytokinesis (Lippincott et al. 2001, Luca et al. 2001, Menssen et al. 2001). Moreover, certain alleles of *cdc15* and *mob1* have been identified that are specifically defective for cytokinesis (Jimenez et al. 1998, Luca et al. 2001, Menssen et al. 2001). The intracellular localization of some MEN components are also consistent with a role in cytokinesis as Cdc15p (Xu et al. 2000), Cdc5p (Cheng et al. 1998, Song et al. 2000), and Dbf2p-Mob1p (Frenz et al. 2000, Luca et al. 2001, Yoshida & Toh-e 2001) have been observed to localize at the bud neck. The MEN is required to promote splitting of the septin ring and constriction of the actomyosin ring in telophase (Cid et al. 2001, Lippincott et al. 2001); however, as with the SIN, it is not yet clear at a molecular level how the MEN triggers cytokinesis.

There is also evidence that MEN function in cytokinesis is inhibited by Cdk activity, similar to the SIN in *S. pombe*. For instance, the MEN promotes release of Cdc14p from the nucleolus; however, Cdc14p also seems to promote the cytokinesis function of the MEN. In early mitosis, Cdc15p is phosphorylated by Cdk1, and in late mitosis, as Cdk activity drops, Cdc15p becomes dephosphorylated by Cdc14p (Menssen et al. 2001, Xu et al. 2000). Dephosphorylation of Cdc15p seems to be important for its ability to promote cytokinesis but not mitotic exit (Menssen et al. 2001). Cdc14p may also be important for regulating Dbf2p-Mob1p localization, because Dbf2p becomes activated in *cdc14* mutants

(Lee et al. 2001, Mah et al. 2001), but does not localize to the bud neck (Frenz et al. 2000, Yoshida & Toh-e 2001). This is reminiscent of observations in *S. pombe* showing that the SIN is inhibited by Cdk activity to ensure that cytokinesis does not occur until exit from mitosis is complete.

Centrosomes and Cytokinesis in Animal Cells

Although the centrosome has not been implicated in exit from mitosis in animal cells as in budding yeast, several studies suggest a role for the centrosome in cytokinesis. The most direct evidence for a role for centrosomes in cytokinesis has come from observations of cell division in mammalian tissue culture cells in which the centrosome was surgically removed, or in an acentriolar *Drosophila* cell line, which showed that a high frequency of these cells specifically failed to complete cytokinesis (Piel et al. 2001). These cells were able to complete cleavage furrow ingression and often remained connected by a narrow cytoplasmic bridge for an extended period of time. Unlike normal cells, these cells were not able to undergo abscission and separate this bridge and complete cytokinesis. As discussed earlier, these cells also failed to enter the next cell cycle. Observational studies are consistent with the centrosome playing an important role in cytokinesis. Using time-lapse video microscopy of cells stably expressing the centrosomal protein centrin fused to GFP as a centriole marker, Piel et al. observed that immediately before abscission, the mother centriole transiently and quickly migrates to the intracellular bridge near the midbody. Each centrosome is comprised of two distinguishable centrioles, a mother and daughter. Only when the mother centriole moved back from the bridge to the center of the cell did cytokinesis (abscission) finish. This is reminiscent of previous observations in which centrosomes in association with the Golgi complex, appear to shift localization from poles to the intracellular bridge and back at the end of mitosis (Mack & Rattner 1993, Moskalewski & Thyberg 1992). Similarly, the recycling endosome-associated protein FIP3 localizes first to the centrosomes during anaphase, and then localizes to the midbody in telophase, where it is required for cytokinesis/abscission (Wilson et al. 2004). These observations suggest that centrosomes and associated organelles must temporally and spatially come in close contact with the midbody for completion of cytokinesis.

What function might this centrosome repositioning serve? Because Golgi complex associates with migrating centrosomes, this transient movement may deliver membrane and secretory vesicles required for complete cell separation. However another exciting possibility is that centrosomes harbor regulatory components required for mitotic exit and cytokinesis, as in budding and fission yeast. Recent studies are beginning to suggest that both ideas may be correct. Potential homologs of several SIN/MEN components have been identified (see Table 1). Several have been characterized (centriolin, Polo kinase, Cdc14A, Lats1,2 kinases) and shown to localize to centrosomes and play a role in cytokinesis (Carmena et al. 1998, Gromley et al. 2003, Kaiser et al. 2002, Mailand et al. 2002, McPherson et al. 2004, Yang et al. 2004). Centriolin, which is related to *S. pombe* Cdc11p and *S. cerevisiae* Nud1p, localizes to maternal centrioles and to the midbody (Gromley et al. 2003). Interestingly, these authors showed that centriolin could bind yeast Bub2p. Loss of function of Centriolin causes cells to have defects in abscission and remain connected for extended periods of time in a manner very similar to the phenotype of cells lacking centrosomes (Gromley et al. 2003). Centriolin depletion using RNAi also causes a G_1 arrest, as described above. A recent study showed that Centriolin may function to recruit factors required for targeted secretion to the midbody, which is required for abscission (Gromley et al. 2004). Interestingly, depletion of the centrosomal protein Pericentrin/Kendrin (Gromley et al.

2003) or deletion of the Sid2p/Dbf2p homolog Lats2 (McPherson et al. 2004) also causes a similar cytokinesis defect. Together, these studies suggest that there may be a SIN/MEN pathway in mammalian cells that regulates cytokinesis; however, it has not yet been shown that these proteins function together as part of a signaling network as in yeast.

SIN/MEN HOMOLOGS AND CELL CYCLE CHECKPOINTS

Several laboratories have shown that cytokinesis failures induced by a number of different treatments cause cells to arrest in the G_1 phase of the following cell cycle as tetraploid cells (see Stukenberg 2004 for review). The arrest depends on p53 (Andreassen et al. 2001b). This arrest was initially termed a tetraploidy checkpoint (Andreassen et al.2001b), although recent results indicate that it may not be tetraploidy or cleavage failure that triggers the checkpoint (Uetake & Sluder 2004). What actually is monitored remains a mystery. Regardless of what triggers the arrest, it bears some interesting similarities to the cytokinesis checkpoint in fission yeast, which arrests cells as binucleate cells for a prolonged period following cleavage failure. It would be interesting to determine if any of the mammalian homologs of components required for the checkpoint in fission yeast function in this arrest in mammalian cells. Recent results suggest that this may be the case. Normally, cells treated with microtubule-depolymerizing drugs arrest in mitosis because of the spindle checkpoint. These cells eventually leak past this checkpoint, exit mitosis without undergoing cytokinesis, and arrest in G_1 phase of the next cell cycle in a p53-dependent manner. One report showed that after this treatment, the arrested cells express high levels of p53 consistent with it being required for the arrest (Iida et al. 2004). However, if the same treatment was done to cells overexpressing a dominant-negative form of the Sid2

[SIN pathway)/Dbf2 (MEN pathway)] homolog Lats1 (Warts), these cells failed to arrest in G_1 and induce p53. It has also been noticed that the mammalian Cdc14 homologs Cdc14A and Cdc14B bind p53 and dephosphorylate it at Ser315 (L. Li et al. 2000). This would likely stabilize p53 because phosphorylation at this site has been shown to promote degradation of p53 (Katayama et al. 2004). It will be interesting to see if Lats1/Warts kinase acts through Cdc14 to promote p53 stability and G_1 arrest following cleavage failure.

Studies in *Drosophila melanogaster* have shown that Warts functions together with a second kinase hippo, probably a homolog of mammalian Mst2 kinase, to prevent tumor formation (Harvey et al. 2003, Hay & Guo 2003, Pantalacci et al. 2003). Mst2 is related to the SIN kinases Sid1p and Cdc7p, as well as to budding yeast Cdc15p. Warts/Lats1 kinase has been shown to function as a tumor suppressor in mice as well (St. John et al. 1999). It will be interesting to determine whether Warts kinase and the Mst2 (hippo) kinase function as tumor suppressors by inhibiting cell cycle progression following cleavage failure. At present, these results suggest that SIN/MEN homologs may be required to promote p53-dependent arrest following cleavage failure, reminiscent of the cytokinesis checkpoint in *S. pombe*. This arrest also bears similarity to the p53-dependent G_1 arrest following depletion of a number of centrosome proteins (see above) and perhaps the G_1 arrest induced by surgical removal of the centrosome (Hinchcliffe et al. 2001). Whether SIN/MEN homologs are involved in this arrest is unknown.

The Centrosome and Genotoxic Stress

Maintenance of genomic integrity is critical to normal development and disease prevention, and conserved DNA damage and replication checkpoints delay the cell cycle to allow repair of genetic lesions or completion of

Cytokinesis checkpoint: halts further rounds of nuclear division if cytokinesis is delayed

DNA replication. In systems ranging from mammalian tumors to early *Drosophila* embryos, checkpoint failures that allow DNA damage or incomplete replication to persist into mitosis triggers "mitotic catastrophe," a poorly understood process characterized by delays in metaphase followed by chromosome segregation and cytokinesis failures. The resulting cells then arrest in G_0 or die by apoptotic or nonapoptotic mechanisms (Andreassen et al. 2001a, Canman 2001, Roninson et al. 2001, Sibon et al. 2000). (The cellular and molecular basis of mitotic catastrophe is poorly understood, but this process appears to be a significant cause of chemotherapy-induced cell death in tumors and may serve an important genome maintenance function (reviewed in Roninson et al. 2001). Recent studies in *Drosophila* embryos and mammalian cultured cells indicate that mitotic catastrophe is linked to centrosome disruption, which may contribute to subsequent chromosome segregation and cytokinesis failures (Roninson et al. 2001). Recent studies in flies and human cells indicate that the Chk2 kinase is required for centrosome disruption on checkpoint failure, indicating that mitotic catastrophe is a genetically programmed response to genotoxic lesions (Takada et al. 2003). Chk2 is a human tumor suppressor, which raises the possibility that defects in damage-dependent centrosome disruption contribute to genomic instability and cancer progression.

Centrosome Inactivation in Early Embryos

Drosophila embryogenesis is initiated by 13 very rapid mitotic divisions that proceed without cytokinesis. These syncytial divisions, similar to the cleavage stage divisions in other embryos, are characterized by alternating S and M phases without intervening gap phases (Foe & Alberts 1983). The first 9 divisions take place in the interior of the embryo, but the majority of nuclei migrate to the cortex and form a monolayer by interphase of division 10. The final 4 syncytial blastoderm stage nuclear divisions (mitosis 10–13) take place in a cortical monolayer, and during these divisions the length of S phase progressively increases whereas M phase remains relatively constant. The DNA replication checkpoint is required to delay mitosis as S phase slows during these final syncytial blastoderm divisions (Sibon et al. 1997, 1999). As a result, embryos mutant for the replication checkpoint spontaneously initiate mitosis before S phase is completed, triggering mitotic catastrophe. Time-lapse confocal microscopic analyses show that checkpoint failure triggers mitosis-specific centrosome inactivation, anastral spindle assembly, and delays in mitosis, and centrosome inactivation correlate with loss of multiple components of the γTuRC from a core centrosome structure (Sibon et al. 2000). In wild-type embryos, identical mitotic defects are triggered by DNA replication inhibitors, a wide range of DNA damaging agents, and direct injection of restriction enzyme-digested DNA. Centrosome disruption and mitotic division failure thus appear to be a normal response to genotoxic lesions at the onset of mitosis (Sibon et al. 2000, Takada et al. 2003). Following division failure, the resulting nuclei drop into the interior of the embryo and are degraded. In syncytial embryos, mitotic catastrophe eliminates nuclei carrying DNA damage and thus serves a genome maintenance function analogous to apoptosis.

Mutations in the *Drosophila* homolog of Checkpoint kinase 2 (Chk2), encoded by the *mnk* gene, block all aspects of mitotic catastrophe in early embryos (Takada et al. 2003). The response is restored by a wild-type Chk2 transgene or injection of GST-Chk2 fusion protein, demonstrating that mitotic catastrophe, at least in fly embryos, is a genetically programmed response to genotoxic stress.

Centrosome Inactivation in Mammalian Cells

Mitotic catastrophe in mammalian cells is also triggered by G_2/M checkpoint failures and is characterized by delays in metaphase, chromosome segregation and cytokinesis failures, and cell death. In some cells, mitotic division failure is followed by apoptosis. However, cells more commonly arrest in G_1 or die by a nonapoptotic mechanism (Roninson et al. 2001). Cytologically similar mitotic catastrophe responses have been described in diverse systems, including primary mouse embryo fibroblasts, *Drosophila* embryos, and a number of cultured cells (Brown & Baltimore 2000, Bunz et al. 1998, Chan et al. 1999, Liu et al. 2000, Sibon et al. 2000) Significantly, mitotic catastrophe may be the primary mechanism of cell death in a number of tumor cell lines following treatment with chemotherapeutic agents (Roninson et al. 2001).

A hallmark of damage-induced mitotic catastrophe is accumulation of cells with large polyploid nuclei or multiple nuclei. In vivo studies in human colorectal tumor cells demonstrate that these cells can be formed by mitotic division failure (Bunz et al. 1998). Following ionizing radiation, HCT116 cells progress into mitosis and chromosomes align at the metaphase plate. However, anaphase chromosome segregation and cytokinesis fail, producing polyploid cells that contain bi-lobed nuclei (Bunz et al. 1998). Following division failure, nuclei fragment into compact masses that resemble clusters of grapes. Similar nuclear morphology is observed during apoptosis, and conventional apoptosis is sometimes observed following damage-induced division failure. However, cells produced by damage-induced mitotic failure are often TUNEL negative, and DNA isolated from these cells does not show laddering characteristic of apoptosis (Lock & Stribinskiene 1996, Nabha et al. 2002). In addition, in some cases cell death is not blocked by apoptotic inhibitors, cells do not contract or bleb, and apoptotic bodies are not formed (reviewed by Roninson et al. 2001). Cell death by mitotic catastrophe thus appears to be distinct from apoptosis in both cell cycle phase and mechanism of execution.

Several recent studies have linked mitotic catastrophe to Chk2-dependent centrosome disruption, suggesting that a conserved signaling mechanism triggers this response. Hut et al. analyzed centrosomes as hamster cells progress into mitosis prior to completion of DNA replication (Hut et al. 2003), whereas Castedo et al. (Castedo et al. 2004b) analyzed γ tubulin distribution when interphase and mitotic cells are fused, driving the interphase nucleus into mitosis and bypassing the G_2/M checkpoint. Using GFP-γ tubulin as a centrosome marker, Hut et al. showed that centrosomes frequently fragment when checkpoint control is disrupted and mitosis is initiated before S phase is completed. These cells often assemble multi-polar spindles and progress through an aborted mitotic division to produce a single polypoid cell. These authors also found that cells carrying a mutation that disrupts DNA damage repair spontaneously show similar mitotic defects, indicating that centrosome fragmentation is not due to the caffeine treatments used in the replication studies. γ tubulin localization is also disrupted when G_2/M phase checkpoint control is bypassed by cell fusion (Castedo et al. 2004a,b). Significantly, this response requires Chk2 kinase (Castedo et al. 2004a,b), suggesting that active disruption of centrosome function in response to checkpoint failure is triggered by a conserved kinase pathway.

The timing of DNA damage may be critical to the mitotic response to DNA damage. Mikhailov et al. (2002) used laser light to induce DNA damage during prometaphase and found that these cells do not show centrosome defects. However, mitosis was delayed and H2Ax histone was phosphorylated, indicating that damage was present and detected by the cellular machinery. We have found that inducing DNA damage during mitosis has no clear effect on centrosome structure in early *Drosophila* embryos (S. Takada & W. Theurkauf, unpublished data), suggesting that centrosome disruption may require transit through the G_2/M transition with DNA lesions. Mitotic catastrophe signaling may therefore require association of proteins with DNA lesions prior to mitotic chromosome condensation.

As discussed elsewhere in this review, a growing body of evidence links centrosome function to mitotic exit and cytokinesis. The observations on checkpoint-defective cells and embryos described above suggest that DNA lesions lead to centrosome inactivation, which lead to chromosome segregation and cytokinesis failures on mitotic exit. This mitotic catastrophe response eliminates damaged cells from the population. Inhibition of centrosome function by the Chk2 tumor suppressor, leading to mitotic exit defects, may therefore serve a function analogous to apoptosis in maintaining genome integrity. This pathway could have important implications for tumor suppression and chemotherapeutic treatment of cancer.

CONCLUSIONS

A growing body of evidence demonstrates that centrosomes and SPBs are involved in an increasing number of regulatory processes in cells. Centrosomes and SPBs provide a scaffold for binding numerous regulatory molecules. Among these are proteins involved in cell cycle progression and checkpoint control. Recent work shows that centrosomes are crucial for several cell cycle transitions, including entry into mitosis and progression from S phase to G_1. In budding and fission yeast, SPBs control exit from mitosis and progression through cytokinesis. Proteins involved in mitotic exit are sometimes positioned asymmetrically on the two SPBs. Centrosomes can also respond to cellular changes. For example, centrosomes lose their microtubule organizing activity and prevent mitosis when cells are exposed to genotoxic and other stresses. The ensuing failure in mitosis or cytokinesis results in polyploid cells that usually result in cell death, thus eliminating damaged cells from the population. The multitude of regulatory proteins that associate with centrosomes suggests that the number of regulatory processes in which centrosomes participate is only beginning to be revealed.

ACKNOWLEDGMENTS

We acknowledge support from the National Institutes of Health and the National Cancer Institute.

LITERATURE CITED

Alexandru G, Zachariae W, Schleiffer A, Nasmyth K. 1999. Sister chromatid separation and chromosome re-duplication are regulated by different mechanisms in response to spindle damage. *EMBO J.* 18:2707–21

Andreassen PR, Lacroix FB, Lohez OD, Margolis RL. 2001a. Neither p21WAF1 nor 14-3-3sigma prevents G2 progression to mitotic catastrophe in human colon carcinoma cells after DNA damage, but p21WAF1 induces stable G1 arrest in resulting tetraploid cells. *Cancer Res.* 61:7660–68

Andreassen PR, Lohez OD, Lacroix FB, Margolis RL. 2001b. Tetraploid state induces p53-dependent arrest of nontransformed mammalian cells in G1. *Mol. Biol. Cell* 12:1315–28

Balasubramanian MK, McCollum D, Chang L, Wong KC, Naqvi NI, et al. 1998. Isolation and characterization of new fission yeast cytokinesis mutants. *Genetics* 149:1265–75

Balczon R, Simerly C, Takahashi D, Schatten G. 2002. Arrest of cell cycle progression during first interphase in murine zygotes microinjected with anti-PCM-1 antibodies. *Cell Motil. Cytoskelet.* 52:183–92

Bardin AJ, Visintin R, Amon A. 2000. A mechanism for coupling exit from mitosis to partitioning of the nucleus. *Cell* 102:21–31

Bobinnec Y, Khodjakov A, Mir LM, Rieder CL, Edde CL, Bornens M. 1998. Centriole disassembly in vivo and its effect on centrosome structure and function in vertebrate cells. *J. Cell Biol.* 143:1575–89

Brown EJ, Baltimore D. 2000. ATR disruption leads to chromosomal fragmentation and early embryonic lethality. *Genes Dev.* 14:397–402

Bunz F, Dutriaux A, Lengauer C, Waldman T, Zhou S, et al. 1998. Requirement for p53 and p21 to sustain G2 arrest after DNA damage. *Science* 282:1497–501

Canman CE. 2001. Replication checkpoint: preventing mitotic catastrophe. *Curr. Biol.* 11: R121–24

Carmena M, Riparbelli MG, Minestrini G, Tavares AM, Adams R, et al. 1998. *Drosophila* polo kinase is required for cytokinesis. *J. Cell. Biol.* 143:659–71

Castedo M, Perfettini JL, Roumier T, Valent A, Raslova H, et al. 2004a. Mitotic catastrophe constitutes a special case of apoptosis whose suppression entails aneuploidy. *Oncogene* 23:4362–70

Castedo M, Perfettini JL, Roumier T, Yakushijin K, Horne D, et al. 2004b. The cell cycle checkpoint kinase Chk2 is a negative regulator of mitotic catastrophe. *Oncogene* 23:4353–61

Castillon GA, Adames NR, Rosello CH, Seidel HS, Longtine MS, et al. 2003. Septins have a dual role in controlling mitotic exit in budding yeast. *Curr. Biol.* 13:654–58

Cerutti L, Simanis V. 1999. Asymmetry of the spindle pole bodies and spg1p GAP segregation during mitosis in fission yeast. *J. Cell Sci.* 112:2313–21

Chan GK, Jablonski SA, Sudakin V, Hittle JC, Yen TJ. 1999. Human BUBR1 is a mitotic checkpoint kinase that monitors CENP-E functions at kinetochores and binds the cyclosome/APC. *J. Cell. Biol.* 146:941–54

Chang L, Gould KL. 2000. Sid4p is required to localize components of the septation initiation pathway to the spindle pole body in fission yeast. *Proc. Natl. Acad. Sci. USA* 97:5249–54

Chang L, Morrell JL, Feoktistova A, Gould KL. 2001. Study of cyclin proteolysis in anaphase-promoting complex (apc) mutant cells reveals the requirement for apc function in the final steps of the fission yeast septation initiation network. *Mol. Cell Biol.* 21:6681–94

Cheng L, Hunke L, Hardy CFJ. 1998. Cell cycle regulation of the *Saccharomyces cerevisiae* polo-like kinase cdc5p. *Mol. Cell Biol.* 18:7360–70

Ciciarello M, Mangiacasale R, Casenghi M, Zaira Limongi M, D'Angelo M, et al. 2001. p53 displacement from centrosomes and p53-mediated G1 arrest following transient inhibition of the mitotic spindle. *J. Biol. Chem.* 276:19205–13

Cid VJ, Adamikova L, Sanchez M, Molina M, Nombela C. 2001. Cell cycle control of septin ring dynamics in the budding yeast. *Microbiology* 147:1437–50

Cueille N, Salimova E, Esteban V, Blanco M, Moreno S, et al. 2001. Flp1, a fission yeast orthologue of the *S. cerevisiae* CDC14 gene, is not required for cyclin degradation or rum1p stabilisation at the end of mitosis. *J. Cell Sci.* 114:2649–64

D'Amours D, Amon A. 2004. At the interface between signaling and executing anaphase–Cdc14 and the FEAR network. *Genes Dev.* 18:2581–95

Diviani D, Langeberg LK, Doxsey SJ, Scott JD. 2000. Pericentrin anchors protein kinase A at the centrosome through a newly identified RII-binding domain. *Curr. Biol.* 10:417–20

Diviani D, Scott JD. 2001. AKAP signaling complexes at the cytoskeleton. *J. Cell Sci.* 114:1431–37

Doxsey SJ. 2001. Re-evaluating centrosome function. *Nat. Rev. Mol. Biol.* 2:688–99

This paper proposes an explanation of why MEN activation and mitotic exit in budding yeast only occurs after the spindle is properly oriented and has passed through the bud neck.

This paper shows that colorectal cancer cells undergo aborted mitotic division when DNA damage persists into mitosis.

This paper suggests that Chk2 has a conserved role in regulating centrosome function in response to G_2/M phase checkpoint failure.

Elliott S, Knop M, Schlenstedt G, Schiebel E. 1999. Spc29p is a component of the Spc110p subcomplex and is essential for spindle pole body duplication. *Proc. Natl. Acad. Sci. USA* 96:6205–10

Esteban V, Blanco M, Cueille N, Simanis V, Moreno S, Bueno A. 2004. A role for the Cdc14-family phosphatase Flp1p at the end of the cell cycle in controlling the rapid degradation of the mitotic inducer Cdc25p in fission yeast. *J. Cell Sci.* 117:2461–68

Fankhauser C, Marks J, Reymond A, Simanis V. 1993. The *S. pombe cdc16* gene is required both for maintenance of p34 Cdc2 kinase activity and regulation of septum formation: a link between mitosis and cytokinesis? *EMBO J.* 12:2697–704

Fankhauser C, Reymond A, Cerutti L, Utzig S, Hofmann K, Simanis V. 1995. The *S. pombe cdc15* gene is a key element in the reorganization of F-actin at mitosis. *Cell* 82:435–44. Erratum. 1997. *Cell* 89(7):1185

Fankhauser C, Simanis V. 1993. The *Schizosaccharomyces pombe cdc14* gene is required for septum formation and can also inhibit nuclear division. *Mol. Biol. Cell* 4:531–39

Fankhauser C, Simanis V. 1994. The cdc7 protein kinase is a dosage dependent regulator of septum formation in fission yeast. *EMBO J.* 13:3011–19

Fesquet D, Fitzpatrick PJ, Johnson AL, Kramer KM, Toyn JH, Johnston LH. 1999. A Bub2p-dependent spindle checkpoint pathway regulates the Dbf2p kinase in budding yeast. *EMBO J.* 18:2424–34

Foe VE, Alberts BM. 1983. Studies of nuclear and cytoplasmic behaviour during the five mitotic cycles that precede gastrulation in *Drosophila* embryogenesis. *J. Cell Sci.* 61:31–70

Forrest AR, McCormack AK, DeSouza CP, Sinnamon JM, Tonks ID, et al. 1999. Multiple splicing variants of cdc25B regulate G2/M progression. *Biochem. Biophys. Res. Commun.* 260:510–15

Fraschini R, Formenti E, Lucchini G, Piatti S. 1999. Budding yeast Bub2 is localized at spindle pole bodies and activates the mitotic checkpoint via a different pathway from Mad2. *J. Cell. Biol.* 145:979–91

Frenz LM, Lee SE, Fesquet D, Johnston LH. 2000. The budding yeast Dbf2 protein kinase localises to the centrosome and moves to the bud neck in late mitosis. *J. Cell Sci.* 113:3399–408

Furge KA, Wong K, Armstrong J, Balasubramanian M, Albright CF. 1998. Byr4 and Cdc16 form a two-component GTPase-activating protein for the Spg1 GTPase that controls septation in fission yeast. *Curr. Biol.* 8:947–54

Gillingham AK, Munro S. 2000. The PACT domain, a conserved centrosomal targeting motif in the coiled-coil proteins AKAP450 and pericentrin. *EMBO Rep.* 1:524–29

Golsteyn RM, Schultz SJ, Bartek J, Ziemiecki A, Ried T, Nigg EA. 1994. Cell cycle analysis and chromosomal localization of human Plk1, a putative homologue of the mitotic kinases *Drosophila* polo and *Saccharomyces cerevisiae* Cdc5. *J. Cell Sci.* 107(Pt 6):1509–17

Grallert A, Krapp A, Bagley S, Simanis V, Hagan IM. 2004. Recruitment of NIMA kinase shows that maturation of the *S. pombe* spindle-pole body occurs over consecutive cell cycles and reveals a role for NIMA in modulating SIN activity. *Genes Dev.* 18:1007–21

Gromley A, Jurczyk A, Mikule K, Doxsey S. 2004. Centriolin-anchoring of exocyst and SNARE complexes at the midbody is required for localized secretion and abscission during cytokinesis. *Mol. Biol. Cell* 14(Suppl.):141a

Gromley A, Jurczyk A, Sillibourne J, Halilovic E, Mogensen M, et al. 2003. A novel human protein of the maternal centriole is required for the final stages of cytokinesis and entry into S phase. *J. Cell. Biol.* 161:535–45

This paper shows asymmetric localization of SIN inhibitors Cdc16p-Byr4p to "old" SPBs, and SIN activators Cdc7p (and Sid1p-Cdc14p) to "new" SPBs.

Gruneberg U, Campbell K, Simpson C, Grindlay J, Schiebel E. 2000. Nud1p links astral microtubule organization and the control of exit from mitosis. *EMBO J.* 19:6475–88

Guertin DA, Chang L, Irshad F, Gould KL, McCollum D. 2000. The role of the sid1p kinase and cdc14p in regulating the onset of cytokinesis in fission yeast. *EMBO J.* 19:1803–15

Guertin DA, Trautmann S, McCollum D. 2002. Cytokinesis in eukaryotes. *Microbiol. Mol. Biol. Rev.* 66:155–78

Harvey KF, Pfleger CM, Hariharan IK. 2003. The *Drosophila* Mst ortholog, hippo, restricts growth and cell proliferation and promotes apoptosis. *Cell* 114:457–67

Hay BA, Guo M. 2003. Coupling cell growth, proliferation, and death. Hippo weighs in. *Dev. Cell* 5:361–63

Hinchcliffe EH, Miller FJ, Cham M, Khodjakov A, Sluder G. 2001. Requirement of a centrosomal activity for cell cycle progression through G1 into S phase. *Science* 291:1547–50

Hirota T, Kunitoku N, Sasayama T, Marumoto T, Zhang D, et al. 2003. Aurora-A and an interacting activator, the LIM protein Ajuba, are required for mitotic commitment in human cells. *Cell* 114:585–98

Hori T, Takaori-Kondo A, Kamikubo Y, Uchiyama T. 2000. Molecular cloning of a novel human protein kinase, kpm, that is homologous to warts/lats, a *Drosophila* tumor suppressor. *Oncogene* 19:3101–9

Hou MC, Salek J, McCollum D. 2000. Mob1p interacts with the Sid2p kinase and is required for cytokinesis in fission yeast. *Curr. Biol.* 10:619–22

Hu F, Wang Y, Liu D, Li Y, Qin J, Elledge SJ. 2001. Regulation of the Bub2/Bfa1 GAP complex by Cdc5 and cell cycle checkpoints. *Cell* 107:655–65

Huang J, Raff JW. 1999. The disappearance of cyclin B at the end of mitosis is regulated spatially in *Drosophila* cells. *EMBO J.* 18:2184–95

Hut HM, Lemstra W, Blaauw EH, Van Cappellen GW, Kampinga HH, Sibon OC. 2003. Centrosomes split in the presence of impaired DNA integrity during mitosis. *Mol. Biol. Cell* 14:1993–2004

Iida S, Hirota T, Morisaki T, Marumoto T, Hara T, et al. 2004. Tumor suppressor WARTS ensures genomic integrity by regulating both mitotic progression and G1 tetraploidy checkpoint function. *Oncogene* 23:5266–74

Jackman M, Lindon C, Nigg EA, Pines J. 2003. Active cyclin B1-Cdk1 first appears on centrosomes in prophase. *Nat. Cell Biol.* 5:143–48

Jang YJ, Ma S, Terada Y, Erikson RL. 2002. Phosphorylation of threonine 210 and the role of serine 137 in the regulation of mammalian polo-like kinase. *J. Biol. Chem.* 277:44115–20

Jaspersen SL, Morgan DO. 2000. Cdc14 activates cdc15 to promote mitotic exit in budding yeast. *Curr. Biol.* 10:615–18

Jimenez J, Cid VJ, Cenamor R, Yuste M, Molero G, et al. 1998. Morphogenesis beyond cytokinetic arrest in *Saccharomyces cerevisiae*. *J. Cell. Biol.* 143:1617–34

Johnston LH, Eberly SL, Chapman JW, Araki H, Sugino A. 1990. The product of the *Saccharomyces cerevisiae* cell cycle gene DBF2 has homology with protein kinases and is periodically expressed in the cell cycle. *Mol. Cell. Biol.* 10:1358–66

Kaiser BK, Zimmerman ZA, Charbonneau H, Jackson PK. 2002. Disruption of centrosome structure, chromosome segregation, and cytokinesis by misexpression of human Cdc14A phosphatase. *Mol. Biol. Cell* 13:2289–300

Katayama H, Sasai K, Kawai H, Yuan ZM, Bondaruk J, et al. 2004. Phosphorylation by aurora kinase A induces Mdm2-mediated destabilization and inhibition of p53. *Nat. Genet.* 36:55–62

This paper shows that removal of centrosomes by microsurgery prevented entry into S phase consistent with arrest in the G_1 stage of the cell cycle.

In vivo studies show that DNA lesions trigger mitotic centrosome fragmentation and division defects.

Keryer G, Witczak O, Delouvee A, Kemmner WA, Rouillard D, et al. 2003. Dissociating the centrosomal matrix protein AKAP450 from centrioles impairs centriole duplication and cell cycle progression. *Mol. Biol. Cell* 14:2436–46

Khodjakov A, Rieder CL. 2001. Centrosomes enhance the fidelity of cytokinesis in vertebrates and are required for cell cycle progression. *J. Cell. Biol.* 153:237–42

Khodjakov A, Rieder CL, Sluder G, Cassels G, Sibon O, Wang CL. 2002. De novo formation of centrosomes in vertebrate cells arrested during S phase. *J. Cell. Biol.* 158:1171–81

Kitada K, Johnson AL, Johnston LH, Sugino A. 1993. A multicopy suppressor gene of the *Saccharomyces cerevisiae* G1 cell cycle mutant gene *dbf4* encodes a protein kinase and is identified as CDC5. *Mol. Cell. Biol.* 13:4445–57

Kramer A, Mailand N, Lukas C, Syljuasen RG, Wilkinson CJ, et al. 2004. Centrosome-associated Chk1 prevents premature activation of cyclin-B-Cdk1 kinase. *Nat. Cell Biol.* 6(9):884–91

Krapp A, Cano E, Simanis V. 2004. Analysis of the *S. pombe* signalling scaffold protein Cdc11p reveals an essential role for the N-terminal domain in SIN signalling. *FEBS Lett.* 565:176–80

Krapp A, Schmidt S, Cano E, Simanis V. 2001. *S. pombe cdc11p*, together with *sid4p*, provides an anchor for septation initiation network proteins on the spindle pole body. *Curr. Biol.* 11:1559–68

Lacey KR, Jackson PK, Stearns T. 1999. Cyclin-dependent kinase control of centrosome duplication. *Proc. Natl. Acad. Sci. USA* 96:2817–22

Lanni JS, Jacks T. 1998. Characterization of the p53-dependent postmitotic checkpoint following spindle disruption. *Mol. Cell Biol.* 18:1055–64

Lee KS, Grenfell TZ, Yarm FR, Erikson RL. 1998. Mutation of the polo-box disrupts localization and mitotic functions of the mammalian polo kinase Plk. *Proc. Natl. Acad. Sci. USA* 95:9301–6

Lee SE, Frenz LM, Wells NJ, Johnson AL, Johnston LH. 2001. Order of function of the budding-yeast mitotic exit-network proteins Tem1, Cdc15, Mob1, Dbf2, and Cdc5. *Curr. Biol.* 11:784–88

Li C, Furge KA, Cheng QC, Albright CF. 2000. Byr4 localizes to spindle-pole bodies in a cell cycle-regulated manner to control Cdc7 localization and septation in fission yeast. *J. Biol. Chem.* 275:14381–87

Li L, Ernsting BR, Wishart MJ, Lohse DL, Dixon JE. 1997. A family of putative tumor suppressors is structurally and functionally conserved in humans and yeast. *J. Biol. Chem.* 272:29403–66

Li L, Ljungman M, Dixon JE. 2000. The human Cdc14 phosphatases interact with and dephosphorylate the tumor suppressor protein p53. *J. Biol. Chem.* 275:2410–14

Li R. 1999. Bifurcation of the mitotic checkpoint pathway in budding yeast. *Proc. Natl. Acad. Sci. USA* 96:4989–94

Lippincott J, Shannon KB, Shou W, Deshaies RJ, Li R. 2001. The Tem1 small GTPase controls actomyosin and septin dynamics during cytokinesis. *J. Cell Sci.* 114:1379–86

Liu J, Puscheck EE, Wang F, Trostinskaia A, Barisic D, et al. 2004. Serine-threonine kinases and transcription factors active in signal transduction are detected at high levels of phosphorylation during mitosis in preimplantation embryos and trophoblast stem cells. *Reproduction* 128:643–54

Liu Q, Guntuku S, Cui XS, Matsuoka S, Cortez D, et al. 2000. Chk1 is an essential kinase that is regulated by Atr and required for the G(2)/M DNA damage checkpoint. *Genes Dev.* 14:1448–59

This paper shows that laser ablation of centrosomes prevents entry into S phase consistent with arrest in the G_1 stage of the cell cycle.

Ectopic localization of Chk1 (Cdk1 inhibitor) to centrosomes induces premature mitotic entry; centrosome-targeting of wild-type Chk1 prevents Cdk1 activation at centrosomes, inducing mitotic defects.

Lock RB, Stribinskiene L. 1996. Dual modes of death induced by etoposide in human epithelial tumor cells allow Bcl-2 to inhibit apoptosis without affecting clonogenic survival. *Cancer Res.* 56:4006–12

Luca FC, Mody M, Kurischko C, Roof DM, Giddings TH, Winey M. 2001. *Saccharomyces cerevisiae* Mob1p is required for cytokinesis and mitotic exit. *Mol. Cell Biol.* 21:6972–83

Luca FC, Winey M. 1998. *MOB1*, an essential yeast gene required for completion of mitosis and maintenance of ploidy. *Mol. Biol. Cell* 9:29–46

Mack G, Rattner JB. 1993. Centrosome repositioning immediately following karyokinesis and prior to cytokinesis. *Cell Motil. Cytoskel.* 26:239–47

Mah AS, Jang J, Deshaies RJ. 2001. Protein kinase Cdc15 activates the Dbf2-Mob1 kinase complex. *Proc. Natl. Acad. Sci. USA* 98:7325–30

Maiato H, DeLuca J, Salmon ED, Earnshaw WC. 2004. The dynamic kinetochore-microtubule interface. *J. Cell Sci.* 117:5461–77

Mailand N, Lukas C, Kaiser BK, Jackson PK, Bartek J, Lukas J. 2002. Deregulated human Cdc14A phosphatase disrupts centrosome separation and chromosome segregation. *Nat. Cell Biol.* 4:318–22

Matsumoto Y, Maller JL. 2004. A centrosomal localization signal in cyclin E required for Cdk2-independent S phase entry. *Science* 306:885–88

McPherson JP, Tamblyn L, Elia A, Migon E, Shehabeldin A, et al. 2004. Lats2/Kpm is required for embryonic development, proliferation control and genomic integrity. *EMBO J.* 23:3677–88

Menssen R, Neutzner A, Seufert W. 2001. Asymmetric spindle pole localization of yeast Cdc15 kinase links mitotic exit and cytokinesis. *Curr. Biol.* 11:345–50

Mikhailov A, Cole RW, Rieder CL. 2002. DNA damage during mitosis in human cells delays the metaphase/anaphase transition via the spindle-assembly checkpoint. *Curr. Biol.* 12:1797–806

Mikule K, Jurczyk A, Gromley A, Doxsey SJ. 2003. siRNA-mediated centrosome damage activates a G1 checkpoint. *Mol. Biol. Cell* 14(Suppl.):136a

Mishra M, Karagiannis J, Trautmann S, Wang H, McCollum D, Balasubramanian MK. 2004. The Clp1p/Flp1p phosphatase ensures completion of cytokinesis in response to minor perturbation of the cell division machinery in *Schizosaccharomyces pombe*. *J. Cell Sci.* 117:3897–910

Molk JN, Schuyler SC, Liu JY, Evans JG, Salmon ED, et al. 2004. The differential roles of budding yeast Tem1p, Cdc15p, and Bub2p protein dynamics in mitotic exit. *Mol. Biol. Cell* 15:1519–32

Morrell JL, Tomlin GC, Rajagopalan S, Venkatram S, Feoktistova AS, et al. 2004. Sid4p-Cdc11p assembles the septation initiation network and its regulators at the *S. pombe* SPB. *Curr. Biol.* 14:579–84

Moskalewski S, Thyberg J. 1992. Synchronized shift in localization of the Golgi complex and the microtubule organizing center in the terminal phase of cytokinesis. *J. Submicrosc. Cytol. Pathol.* 24:359–70

Nabha SM, Mohammad RM, Dandashi MH, Coupaye-Gerard B, Aboukameel A, et al. 2002. Combretastatin-A4 prodrug induces mitotic catastrophe in chronic lymphocytic leukemia cell line independent of caspase activation and poly(ADP-ribose) polymerase cleavage. *Clin. Cancer Res.* 8:2735–41

Nigg EA. 2002. Centrosome aberrations: cause or consequence of cancer progression? *Nat. Rev. Cancer* 2:815–25

This paper shows that cycE binding to centrosomes is required for entry into S-phase.

Nishiyama Y, Hirota T, Morisaki T, Hara T, Marumoto T, et al. 1999. A human homolog of *Drosophila warts tumor suppressor, h-warts*, localized to mitotic apparatus and specifically phosphorylated during mitosis. *FEBS Lett.* 459:159–65

Ohkura H, Hagan IM, Glover DM. 1995. The conserved *Schizosaccharomyces pombe* kinase plo1, required to form a bipolar spindle, the actin ring, and septum, can drive septum formation in G1 and G2 cells. *Genes Dev.* 9:1059–73

Pantalacci S, Tapon N, Leopold P. 2003. The Salvador partner Hippo promotes apoptosis and cell-cycle exit in *Drosophila*. *Nat. Cell Biol.* 5:921–27

Pereira G, Hofken T, Grindlay J, Manson C, Schiebel E. 2000. The Bub2p spindle checkpoint links nuclear migration with mitotic exit. *Mol. Cell* 6:1–10

Perez-Mongiovi D, Beckhelling C, Chang P, Ford CC, Houliston E. 2000. Nuclei and microtubule asters stimulate maturation/M phase promoting factor (MPF) activation in *Xenopus* eggs and egg cytoplasmic extracts. *J. Cell. Biol.* 150:963–74

Picard A, Karsenti E, Dabauvalle MC, Doree M. 1987. Release of mature starfish oocytes from interphase arrest by microinjection of human centrosomes. *Nature* 327:170–72

Piel M, Nordberg J, Euteneuer U, Bornens M. 2001. Centrosome-dependent exit of cytokinesis in animal cells. *Science* 291:1550–53

Prigozhina NL, Oakley CE, Lewis AM, Nayak T, Osmani SA, Oakley BR. 2004. Gamma-tubulin plays an essential role in the coordination of mitotic events. *Mol. Biol. Cell* 15:1374–86

Raff JW, Jeffers K, Huang JY. 2002. The roles of Fzy/Cdc20 and Fzr/Cdh1 in regulating the destruction of cyclin B in space and time. *J. Cell. Biol.* 157:1139–49

Rajagopalan S, Bimbo A, Balasubramanian MK, Oliferenko S. 2004. A potential tension-sensing mechanism that ensures timely anaphase onset upon metaphase spindle orientation. *Curr. Biol.* 14:69–74

Reynolds N, Ohkura H. 2003. Polo boxes form a single functional domain that mediates interactions with multiple proteins in fission yeast polo kinase. *J. Cell Sci.* 116:1377–87

Roninson IB, Broude EV, Chang BD. 2001. If not apoptosis, then what? Treatment-induced senescence and mitotic catastrophe in tumor cells. *Drug Resist. Update* 4:303–13

Salimova E, Sohrmann M, Fournier N, Simanis V. 2000. The *S. pombe* orthologue of the *S. cerevisiae mob1* gene is essential and functions in signalling the onset of septum formation. *J. Cell Sci.* 113:1695–704

Schmidt S, Sohrmann M, Hofmann K, Woollard A, Simanis V. 1997. The Spg1p GTPase is an essential, dosage-dependent inducer of septum formation in *Schizosaccharomyces pombe*. *Genes Dev.* 11:1519–34

Schweitzer B, Philippsen P. 1991. *CDC15*, an essential cell cycle gene in *Saccharomyces cerevisiae*, encodes a protein kinase domain. *Yeast* 7:265–73

Seshan A, Amon A. 2004. Linked for life: temporal and spatial coordination of late mitotic events. *Curr. Opin. Cell Biol.* 16:41–48

Shimamura M, Brown RC, Lemmon BE, Akashi T, Mizuno K, et al. 2004. Gamma-tubulin in basal land plants: characterization, localization, and implication in the evolution of acentriolar microtubule organizing centers. *Plant Cell* 16:45–59

Shirayama M, Matsui Y, Tanaka K, Toh-e A. 1994a. Isolation of a CDC25 family gene, MSI2/LTE1, as a multicopy suppressor of ira1. *Yeast* 10:451–61

Shirayama M, Matsui Y, Toh EA. 1994b. The yeast *TEM1* gene, which encodes a GTP-binding protein, is involved in termination of M phase. *Mol. Cell Biol.* 14:7476–82

Shou W, Seol JH, Shevchenko A, Baskerville C, Moazed D, et al. 1999. Exit from mitosis is triggered by Tem1-dependent release of the protein phosphatase Cdc14 from nucleolar RENT complex. *Cell* 97:233–44

This paper also proposes an explanation of why MEN activation and mitotic exit in budding yeast only occurs after the spindle is properly oriented and has passed through the bud neck.

Sibon OC, Kelkar A, Lemstra W, Theurkauf WE. 2000. DNA-replication/DNA-damage-dependent centrosome inactivation in *Drosophila* embryos. *Nat. Cell Biol.* 2:90–95

Sibon OC, Laurencon A, Hawley R, Theurkauf WE. 1999. The *Drosophila* ATM homologue Mei-41 has an essential checkpoint function at the midblastula transition. *Curr. Biol.* 9:302–12

Sibon OC, Stevenson VA, Theurkauf WE. 1997. DNA-replication checkpoint control at the *Drosophila* midblastula transition. *Nature* 388:93–97

Simanis V. 2003. Events at the end of mitosis in the budding and fission yeasts. *J. Cell Sci.* 116:4263–75

Sobel SG. 1997. Mini review: mitosis and the spindle pole body in *Saccharomyces cerevisiae*. *J. Exp. Zool.* 277:120–38

Sohrmann M, Schmidt S, Hagan I, Simanis V. 1998. Asymmetric segregation on spindle poles of the *Schizosaccharomyces pombe* septum-inducing protein kinase Cdc7p. *Genes Dev.* 12:84–94

Song S, Grenfell TZ, Garfield S, Erikson RL, Lee KS. 2000. Essential function of the polo box of Cdc5 in subcellular localization and induction of cytokinetic structures. *Mol. Cell Biol.* 20:286–98

Sparks CA, Morphew M, McCollum D. 1999. Sid2p, a spindle pole body kinase that regulates the onset of cytokinesis. *J. Cell. Biol.* 146:777–90

St John MA, Tao W, Fei X, Fukumoto R, Carcangiu ML, et al. 1999. Mice deficient of Lats1 develop soft-tissue sarcomas, ovarian tumours and pituitary dysfunction. *Nat. Genet.* 21:182–86

Stukenberg PT. 2004. Triggering p53 after cytokinesis failure. *J. Cell. Biol.* 165:607–8

Takada S, Kelkar A, Theurkauf WE. 2003. *Drosophila* checkpoint kinase 2 couples centrosome function and spindle assembly to genomic integrity. *Cell* 113:87–99

Tao W, Zhang S, Turenchalk GS, Stewart RA, St John MA, et al. 1999. Human homologue of the *Drosophila melanogaster* lats tumour suppressor modulates CDC2 activity. *Nat. Genet.* 21:177–81

Theurkauf WE, Hawley RS. 1992. Meiotic spindle assembly in *Drosophila* females: behavior of nonexchange chromosomes and the effects of mutations in the nod kinesin-like protein. *J. Cell. Biol.* 116:1167–80

Toyn JH, Araki H, Sugino A, Johnston LH. 1991. The cell-cycle-regulated budding yeast gene *DBF2*, encoding a putative protein kinase, has a homologue that is not under cell-cycle control. *Gene* 104:63–70

Trautmann S, Wolfe BA, Jorgensen P, Tyers M, Gould KL, McCollum D. 2001. Fission yeast Clp1p phosphatase regulates G2/M transition and coordination of cytokinesis with cell cycle progression. *Curr. Biol.* 11:931–40

Trielli MO, Andreassen PR, Lacroix FB, Margolis RL. 1996. Differential Taxol-dependent arrest of transformed and nontransformed cells in the G1 phase of the cell cycle, and specific-related mortality of transformed cells. *J. Cell. Biol.* 135:689–700

Uetake Y, Sluder G. 2004. Cell cycle progression after cleavage failure: mammalian somatic cells do not possess a "tetraploidy checkpoint." *J. Cell. Biol.* 165:609–15

Visintin R, Craig K, Hwang ES, Prinz S, Tyers M, Amon A. 1998. The phosphatase Cdc14 triggers mitotic exit by reversal of Cdk- dependent phosphorylation. *Mol. Cell.* 2:709–18

Visintin R, Hwang ES, Amon A. 1999. Cfi1 prevents premature exit from mitosis by anchoring Cdc14 phosphatase in the nucleolus. *Nature* 398:818–23

Wakefield JG, Huang JY, Raff JW. 2000. Centrosomes have a role in regulating the destruction of cyclin B in early *Drosophila* embryos. *Curr. Biol.* 10:1367–70

This paper shows that the *S. pombe* Cdc14 phosphatase homolog Clp1p/Flp1p functions together with the SIN in a positive feedback loop important for delaying further rounds of nuclear division if cytokinesis is delayed.

Wan J, Xu H, Grunstein M. 1992. CDC14 of *Saccharomyces cerevisiae*. Cloning, sequence analysis, and transcription during the cell cycle. *J. Biol. Chem.* 267:11274–80

Wilson GM, Fielding AB, Simon GC, Yu X, Andrews PD, et al. 2004. The FIP3-Rab11 protein complex regulates recycling endosome targeting to the cleavage furrow during late cytokinesis. *Mol. Biol. Cell* 6(2):849–60

Wolfe BA, Gould KL. 2004. Fission yeast Clp1p phosphatase affects G_2/M transition and mitotic exit through Cdc25p inactivation. *EMBO J.* 23:919–29

Wong C, Stearns T. 2003. Centrosome number is controlled by a centrosome-intrinsic block to reduplication. *Nat. Cell Biol.* 5:539–44

Xu S, Huang HK, Kaiser P, Latterich M, Hunter T. 2000. Phosphorylation and spindle pole body localization of the Cdc15p mitotic regulatory protein kinase in budding yeast. *Curr. Biol.* 10:329–32

Yabuta N, Fujii T, Copeland NG, Gilbert DJ, Jenkins NA, et al. 2000. Structure, expression, and chromosome mapping of LATS2, a mammalian homologue of the *Drosophila* tumor suppressor gene *lats/warts*. *Genomics* 63:263–70

Yang X, Yu K, Hao Y, Li DM, Stewart R, et al. 2004. LATS1 tumour suppressor affects cytokinesis by inhibiting LIMK1. *Nat. Cell Biol.* 6:609–17

Yeh E, Skibbens RV, Cheng JW, Salmon ED, Bloom K. 1995. Spindle dynamics and cell cycle regulation of dynein in the budding yeast, *Saccharomyces cerevisiae*. *J. Cell. Biol.* 130:687–700

Yoshida S, Ichihashi R, Toh-e A. 2003. Ras recruits mitotic exit regulator Lte1 to the bud cortex in budding yeast. *J. Cell. Biol.* 161:889–97

Yoshida S, Toh-e A. 2001. Regulation of the localization of Dbf2 and mob1 during cell division of *Saccharomyces cerevisiae*. *Genes Genet. Syst.* 76:141–47

Zimmerman WC, Sillibourne J, Rosa J, Doxsey SJ. 2004. Mitosis-specific anchoring of gamma tubulin complexes by pericentrin controls spindle organization and mitotic entry. *Mol. Biol. Cell* 15:3642–57

Endoplasmic Reticulum–Associated Degradation

Karin Römisch

University of Cambridge, Cambridge Institute for Medical Research, Hills Road, Cambridge CB2 2XY, United Kingdom; email: kbr20@cam.ac.uk

Annu. Rev. Cell Dev. Biol.
2005. 21:435–56

First published online as a
Review in Advance on
June 28, 2005

The *Annual Review of
Cell and Developmental
Biology* is online at
http://cellbio.annualreviews.org

doi: 10.1146/
annurev.cellbio.21.012704.133250

Copyright © 2005 by
Annual Reviews. All rights
reserved

1081-0706/05/1110-
0435$20.00

Key Words

protein secretion, protein misfolding, Sec61 channel, retrotranslocation, proteasome

Abstract

Secretory and transmembrane proteins enter the secretory pathway through the protein-conducting Sec61 channel in the membrane of the endoplasmic reticulum. In the endoplasmic reticulum, proteins fold, are frequently covalently modified, and oligomerize before they are packaged into transport vesicles that shuttle them to the Golgi complex. Proteins that misfold in the endoplasmic reticulum are selectively transported back across the endoplasmic reticulum membrane to the cytosol for degradation by proteasomes. Depending on the topology of the defect in the protein, cytosolic or lumenal chaperones are involved in its targeting to degradation. The export channel for misfolded proteins is likely also formed by Sec61p. Export may be powered by AAA-ATPases of the proteasome 19S regulatory particle or Cdc48p/p97. Exported proteins are frequently ubiquitylated prior to degradation and are escorted to the proteasome by polyubiquitin-binding proteins.

Contents

INTRODUCTION

In eukaryotic cells an estimated 30% of all newly synthesized proteins misfold early during their biogenesis and are degraded (Schubert et al. 2000). Mutations can further increase the proportion of protein turnover during biosynthesis (Lomas & Parfrey 2004). Subunits that fail to assemble into oligomeric complexes frequently have short half lives, and postsynthetic damage, e.g., by oxidation, glycation, and deamidation, also results in protein turnover (Goldberg 2003). Efficient dis-

ER: endoplasmic reticulum

posal of defective proteins is essential because these proteins may compete with their functional counterparts for substrate binding or for complex formation with interaction partners; they also may form toxic aggregates (Goldberg 2003, Römisch 2004). The toxicity of damaged cytosolic proteins is usually restricted to the cell in which they reside. Defective secretory and transmembrane proteins present a much higher inherent risk because, when released from the cell or exposed at the cell surface, they have the potential to interfere with cell-cell communication, with negative consequences for development, for example, or for the recognition of self by the immune system (Römisch 2004). Eukaryotic secretory proteins are therefore subject to stringent quality control, which begins at their site of biogenesis, the endoplasmic reticulum (ER) (Trombetta & Parodi 2003).

Targeting to the ER is mediated by a hydrophobic signal peptide at the N terminus of secretory proteins or by a transmembrane domain (Johnson & van Waes 1999). Proteins are translocated into the ER lumen or integrated into the ER membrane through a channel formed by the heterotrimeric Sec61 complex (Sec61α, Sec61β, Sec61γ in mammals; Sec61p, Sbh1p, Sss1p in yeast) (**Figure 1**) (Johnson & van Waes 1999). During and after translocation into the ER, secretory proteins fold with the help of chaperones (Johnson & van Waes 1999). Depending on the protein, they may acquire disulfide bonds and N-linked glycans and oligomerize into higher-order complexes before they are deemed fully matured, which results in their packaging into ER-to-Golgi transport vesicles and shuttling to the next compartment of the secretory pathway (**Figure 1**) (Johnson & van Waes 1999). Secretory proteins that fail to fold or are not appropriately modified or oligomerized are either directly retained in the ER or recycled to the ER from downstream compartments of the secretory pathway and are often subsequently degraded (Johnson & van Waes 1999, Trombetta & Parodi 2003).

Initially it was assumed that degradation of misfolded secretory proteins took place in the ER lumen (Klausner & Sitia 1990). The discovery in the mid-1990s that defective secretory proteins were degraded by proteasomes made it obvious that such proteins are transported back across the ER membrane into the cytosol (**Figure 1**) (Jensen et al. 1995, Ward et al. 1995, Werner et al. 1996, Wiertz et al. 1996a). The process of selective protein export from the ER to the cytosol followed by proteasomal degradation is now known as ER-associated degradation (McCracken & Brodsky 1996). The investigations are far from complete, but most available evidence suggests that misfolded proteins are retrotranslocated through the Sec61 channel (**Figure 1**), whereas we know little about protein targeting to the channel in the ER lumen and about channel opening from the lumenal side (Wiertz et al. 1996b, Pilon et al. 1997, Schmitz et al. 2000).

In this review, I summarize current data on recognition of ERAD substrates in the ER and prerequisites for export to the cytosol. I discuss the nature of the export channel and the cytosolic protein complexes proposed to promote export through the channel, most likely by a molecular ratcheting mechanism. Finally, I describe the cytosolic factors involved in post-export events that ultimately lead to protein degradation.

RECOGNITION OF ERAD SUBSTRATES

Checkpoints and Topology of Degradation Signals

All secretory pathway cargo has the potential to misfold in the ER and thus become a substrate for ERAD. The machinery responsible for substrate recognition must therefore be able to recognize defects in molecules as diverse as the cystic fibrosis transmembrane conductance regulator (CFTR), a large polytopic transmembrane glycoprotein, and mutant alpha-factor precur-

sor (pαF), a small soluble yeast pheromone precursor that is not glycosylated and has little inherent structure (Ward et al. 1995, McCracken & Brodsky 1996). Because chaperones recognize and bind to incompletely folded proteins, it seems likely that chaperones are also involved in distinguishing folding intermediates from terminally defective proteins and in the initiation of targeting to degradation (**Figure 1**). By comparing observations from a number of laboratories, Brodsky & McCracken (1999) found that all yeast ERAD substrates required proteasomes for their degradation but strikingly differed in their requirement for the ER-lumenal Hsp70 BiP. Whereas degradation of soluble substrates such as pαF and a mutant form of the vacuolar protease carboxypeptidase Y (CPY*) were dependent on BiP, degradation of transmembrane proteins Pdr5*p, Ste6-166p, Sec61-2p, and Hmg2p was not impaired in yeast strains with mutations in *KAR2*, the gene encoding BiP. Degradation of transmembrane proteins with large cytosolic domains requires the cytosolic Hsp70, Ssa1p, cytosolic Hsp40s, Ydj1p and Hlj1p, and in one instance cytosolic Hsp104, but as one would expect, selection of soluble secretory proteins for degradation is independent of cytosolic chaperones (Zhang et al. 2001, Taxis et al. 2003, Huyer et al. 2004b). In an attempt to address the chaperone requirements for soluble and transmembrane proteins, Taxis and colleagues (Taxis et al. 2003) generated derivatives of CPY* in which the soluble protein was fused to a transmembrane domain, or to a transmembrane domain with green flourescent protein (GFP) attached as a tightly folded cytosolic domain. The authors found that the chaperone requirements for CPY*-GFP degradation were different from CPY* degradation but were similar to the chaperone requirements for the cytosolic proteasome substrate Deg1-GFP, suggesting that the cytosolic chaperones were primarily required to unfold GFP and make it available to the proteasome (Taxis et al. 2003). Rather surprisingly, soluble substrates (but

ER-associated degradation (ERAD): recognition of defective proteins in the ER, their targeting for export, and degradation in the cytoplasm

Sec61 channel: consists of one or 3–4 copies of the Sec61 complex; mediates protein import into the ER and perhaps also protein export

CFTR: cystic fibrosis transmembrane conductance regulator

pαF: mutant alpha-factor precursor

CPY*: mutant carboxypeptidase Y

GFP: green fluorescent protein

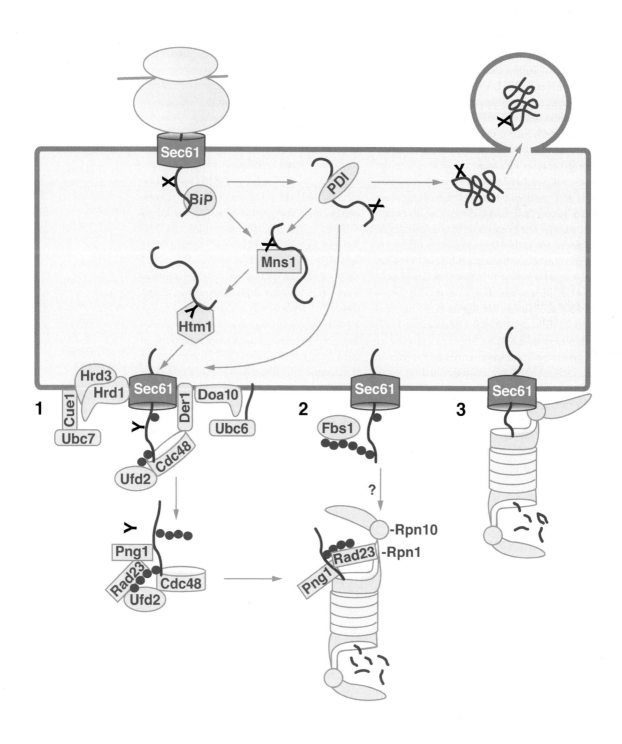

not transmembrane ERAD substrates) also rely on genes required for vesicle budding from the ER and transport to the Golgi complex (*SEC12*, *SEC18*, *ERV29*, *ERV14*, *SEC23*, *SEC13*, *UFE1*, *SED5*) (Caldwell et al. 2001, Vashist et al. 2001, Taxis et al. 2002, Fu & Sztul 2003).

Vashist & Ng (2004) systematically investigated the basis for the differences in ERAD requirements between soluble and transmembrane proteins by constructing chimeras between the soluble degradation substrate KHN and the polytopic transmembrane substrate Ste6-166p. They found that the decisive factor was not the presence of a transmembrane domain, but rather the location of the defect in the protein: Substrates with defects in a cytosolic domain such as Ste6-166p were degraded BiP-independently, whereas proteins with defects in an ER-lumenal domain required BiP and the ER-to-Golgi transport machinery for ERAD whether or not the substrates were membrane anchored (Vashist & Ng 2004). Using chimeras between KHN and Ste6-166p, Vashist & Ng also demonstrated that ER quality control uses sequential checkpoints (Vashist & Ng 2004). Cytosolic domains are examined first, and when a defect is found the protein is degraded by the BiP-independent pathway, even if it also has a lesion in a lumenal domain (Vashist & Ng 2004).

Roles of Cytosolic and ER-Lumenal Chaperones

The mechanism by which cytosolic chaperones target defective ER transmembrane proteins with cytosolic lesions to degradation is not understood. Whereas prolonged association with chaperones involved in protein folding may be the first indicator of a problem, the actual targeting into the ERAD pathway may be mediated by chaperones that do not interact with normal folding intermediates. For CFTR, for example, Hsp90 is required for the folding of its large cytoplasmic domains, but Hsp70 and Hsp40 are required for its targeting to degradation (Loo et al. 1998, Zhang et al. 2001, Youker et al. 2004). What is recognized by cytosolic chaperones as misfolded remains unidentified, but it is likely a global structural feature rather than an exposed sequence motif. CFTR contains an unstable

KHN: Kar2 signal peptide fused to lumenal domain of simian virus 5 hemagglutinin neuraminidase

---

Figure 1

ER-associated degradation. Secretory proteins enter the ER through the Sec61 channel cotranslationally (depicted here) or posttranslationally. Proteins carrying appropriate acceptor sites are N-glycosylated (*black X*). Proteins fold with the help of lumenal chaperones such as BiP and PDI. Fully folded proteins are packaged into ER-to-Golgi transport vesicles. Misfolding leads to prolonged association with chaperones, trimming of a mannose residue from N-glycans by ER-mannosidase I (Mns1p) and recognition of the trimmed N-glycan (*black Y*) by Htm1p, which targets misfolded glycoproteins for degradation. Targeting for export of nonglycosylated proteins may be mediated directly by PDI. Misfolded proteins are exported to the cytosol through a channel most likely formed by Sec61p and multiple accessory proteins. During export, misfolded proteins are oligo-ubiquitylated (*blue circles*) by either Hrd1p (E3) and Ubc7p (E2) or Doa10p (E3) and Ubc6p (E2). (1) Der1p-anchored Cdc48p at the ER membrane may recognize oligo-ubiquitylated substrates and assist their size-restricted polyubiquitylation by Ufd2p (E4). After substrate deglycosylation by Png1p, Rad23p (or Dsk2p) can bind to the Ufd2p-associated polyubiquitin and target the substrate to the proteasome by binding the 19S RP subunit Rpn1p. Some or all of the factors involved may be bound to the ER membrane. (2) Alternatively, polyubiquitylation may be mediated by the N-glycan-specific SCF[Fbs1] complex; the resulting longer polyubiquitin chains may be recognized by the 19S RP subunit Rpn10p. (3) In a third scenario, the 19S RP binds directly to the Sec61 channel and extracts misfolded proteins from the ER. If 20S core particles associate with these 19S RP complexes, as shown here, export and degradation can be coupled. Alternatively, 19S RPs could function on their own in driving export similar to Cdc48p in scenario (1). In this case, some or all of the steps depicted in (1) and (2) may also precede proteolysis when the driving force for export is provided by the 19S RP.

HMG-CoA:
hydroxymethyl-
glutaryl-coenzyme A

PDI: protein
disulfide isomerase

membrane segment whose maintenance in the transmembrane topology is dependent on the CFTR cytosolic domains; thus the ΔF508 mutation in one of the cytosolic domains leads to significant changes in the packing of the transmembrane domains of the entire protein (Tector & Hartl 1999, Chen et al. 2004). Similarly, degradation of the polytopic transmembrane hydroxymethylglutaryl-coenzyme A (HMG-CoA) reductase by ERAD is triggered by a regulated structural change of the entire 523 amino acid N-terminal domain of the protein (Gardner & Hampton 1999).

In the ER lumen, misfolded nonglycosylated proteins are first recognized by BiP, but subsequent interaction with protein disulfide isomerase (PDI) or with Eps1p, a membrane-anchored member of the PDI family is a prerequisite for export even for substrates lacking disulfide bridges (**Figure 1**) (Gillece et al. 1999, Wang & Chang 2003). Transport of cholera toxin from the ER lumen to cytosol is also preceded by interaction with PDI (Tsai et al. 2001). The disulfide isomerase function of PDI is required for export of disulfide-bonded substrates to the cytosol, which confirms data from experiments with chemical reductants and oxidants that disulfide bonds need to be reduced prior to translocation across the ER membrane (Tortorella et al. 1998, Gillece et al. 1999, Fagioli et al. 2001, Tsai et al. 2001). The active site cysteines of Eps1p are necessary for formation of mixed disulfides with the nondisulfide-bonded but cysteine-containing ERAD substrate, Pma1-D378N, and its targeting to degradation (Wang & Chang 2003). For ERAD substrates without cysteines, integrity of the protein-binding site of PDI is required for ERAD, but it remains unclear which substrate feature this domain recognizes (Gillece et al. 1999). No ternary complexes have been identified so far between PDI, ERAD substrates, and the protein translocation channel, which suggests either that another step exists between substrate interaction with PDI and substrate interaction

with the channel or, perhaps, that the ternary complex is too transient to be captured by the methods employed.

The ER-lumenal Hsp70 BiP is required for posttranslational protein import into the ER, protein folding in the ER lumen, and ERAD of proteins with lumenal lesions (**Figure 1**) (Kabani et al. 2003, Vashist & Ng 2004). BiP binds to exposed hydrophobic regions of folding intermediates and misfolded proteins and prevents their aggregation in the ER lumen (Flynn et al. 1991, Kabani et al. 2003). The rate of proteolysis of ERAD substrates is determined by the rate of their release from BiP (Chillaron & Haas 2000). Brodsky and colleagues identified two mutants in the gene encoding yeast BiP, *kar2-1* and *kar2-133*, which result in a reduced affinity of BiP for its substrates and in ERAD defects, but have no effect on posttranslational protein import (Kabani et al. 2003). In *kar2-1* and *kar2-133* yeast, ERAD substrates aggregate in the ER lumen; aggregation, but not the ERAD defect, can be rescued by coexpression of wild-type BiP (Kabani et al. 2003). This suggests that solubility is not sufficient for export and that BiP has an additional function in targeting substrate proteins to degradation. The formation of ER chaperone complexes that are important for organized protein folding may also be disrupted in cells expressing Kar2-1p and Kar2-133p (Gillece et al. 1999, Meunier et al. 2002, Molinari et al. 2002). Modest overexpression of *KAR2* and *PDI1* even in a wild-type background also has negative effects on ERAD, which indicates that the ratio of chaperones in the ER lumen is important for ER functions (Kabani et al. 2003; K. Römisch, unpublished data).

N-Glycans

N-glycan structure contributes to the recognition of ERAD substrates in the ER (**Figure 1**). Because this topic has been discussed in detail in recent excellent reviews

(Trombetta & Parodi 2003, Helenius & Aebi 2004), I summarize here only the salient points: N-glycosylated proteins in the ER enter the calnexin/calreticulin cycle (Helenius & Aebi 2004). These lectins recognize the terminal glucose residue of the N-glycan, which can be removed by glucosidase II; deglucosylation allows fully folded substrates to exit the cycle, whereas folding intermediates are reglucosylated and rebind calnexin/calreticulin (Helenius & Aebi 2004). ER-resident mannosidase I removes one terminal mannose from N-glycans of proteins which remain in the ER for a prolonged period (**Figure 1**) (Herscovics 2001). This trimming makes N-glycans poorer substrates for reglucosylation and results in exit of proteins from the calnexin cycle (Molinari et al. 2003). De-mannosylation is required for degradation of misfolded N-glycosylated proteins, and overexpression of ER mannosidase I accelerates turnover of glycoproteins (Helenius & Aebi 2004, Wu et al. 2003). The resulting Man8-GlcNAc2 structure is the substrate for a mannnosidase-related lectin in the ER [Htm1p in yeast; ER-degradation enhancing α-mannosidase-like protein (EDEM) in mammalian cells] (Helenius & Aebi 2004). Deletion of *HTM1* in yeast specifically interferes with ERAD of glycoproteins, and overexpression of EDEM enhances degradation of glycosylated ERAD substrates in mammalian cells, suggesting that this lectin is involved in targeting of misfolded glycoproteins to ERAD (**Figure 1**) (Jakob et al. 2001, Molinari et al. 2003).

Calcium

Treatment of cells expressing mutant ΔF508 CFTR with the calcium pump inhibitor thapsigargin increases transport of the mutant protein to the cell surface, but at the concentrations used this did not trigger release of ER-resident proteins or induce the unfolded protein response (UPR) (Egan et al. 2002). One of the genes identified in a screen for yeast mutants defective in ERAD encodes the

calcium pump Pmr1p; deletion of *PMR1* leads to stabilization of CPY* (Durr et al. 1998). Deletion of a gene encoding a structurally related P-type ATPase, *COD1*, abolishes the regulation of HMG-CoA reductase degradation in yeast, but $\Delta pmr1$ has no effect (Cronin et al. 2000). Cod1p is localized in the ER, whereas Pmr1p is a Golgi-resident protein (Durr et al. 1998, Cronin et al. 2002, Vashist et al. 2002), but as these organelles are connected by vesicular transport in both directions, calcium homeostasis in both compartments is altered by individual deletions of *PMR1* or *COD1*. *COD1* was also identified in a screen for mutants synthetically lethal with defects in the UPR, and in $\Delta cod1$ cells the UPR is constitutively activated (Vashist et al. 2002); in contrast, UPR-defective yeast are viable when *PMR1* is deleted (Cronin et al. 2002, Vashist et al. 2002). Pmr1p can pump both calcium and magnesium, whereas direct transport of calcium by Cod1p has not been observed so far, and its ATPase activity is not stimulated by calcium, but rather by high concentrations of magnesium (Durr et al. 1998, Cronin et al. 2002). Nevertheless, deletion of *COD1* and *PMR1* causes a synergistic increase in cellular calcium, and deletion of *COD1* alone induces expression of calcium-responsive genes, suggesting a role for Cod1p in cellular calcium homeostasis (Cronin et al. 2002). Although many of the chaperones in the ER are calcium dependent, deletion of *PMR1* and *COD1* does not affect ERAD in general, and Ste6-166p and Hmg2p are efficiently degraded (Cronin et al. 2002, Vashist et al. 2002). Degradation of CPY*, on the other hand, is significantly retarded in individual mutants and strongly defective in $\Delta pmr1$ $\Delta cod1$ yeast (Vashist et al. 2002). The most likely explanation for this turnover defect is the impaired trimming of the N-glycans from Man9 to Man8 in the ER of $\Delta cod1 \Delta pmr1$ cells, which would explain why the turnover of the glycoprotein CPY* is affected, whereas the nonglycosylated Ste6-166p and Hmg2p are still degraded (Vashist et al. 2002).

EDEM: ER-degradation enhancing α-mannosidase-like protein

UPR: unfolded protein response

Role of the ER-to-Golgi Transport Machinery

Based on the observation that components of the COPII coat, which drives vesicle formation from the ER, are required for degradation of proteins with lesions in the ER lumen, Vashist et al. and Caldwell and colleagues proposed that such proteins were subject to one round of transport to the Golgi complex followed by recycling to the ER (Caldwell et al. 2001, Vashist et al. 2001). Taxis and colleagues made similar observations but in addition investigated the role of COPI coat components, which are required for recycling to the ER, in ERAD (Taxis et al. 2002). Similar to data by Vashist and Ng, Taxis et al. found a modest ERAD defect in *sec21-1* cells under restrictive conditions, but a second COPI mutant, *sec27-1*, had no effect on CPY* turnover (Taxis et al. 2002). Because both COPI mutants were profoundly defective in recycling to the ER under the conditions used, Taxis et al. concluded that the ERAD defects in ER-to-Golgi vesicle trafficking mutants could not reflect transport of ERAD substrates to the Golgi and recycling to the ER, but rather that the ERAD defects were indirectly caused by changes in ER structure and function in these mutants (Taxis et al. 2002). At the restrictive temperature, Prinz and colleagues observed a dramatic collapse of peripheral ER structures in all ER-to-Golgi vesicle trafficking mutants (Prinz et al. 2000). In addition, the distribution of BiP, and to a lesser extent PDI, in the ER changes from homogenous to punctate by immunofluorescence (Nishikawa et al. 1994); electron micrographs of these structures show clusters of short tubules and cisternae connected to each other and the nuclear envelope, which may represent ER exit sites (Nishikawa et al. 1994, Vashist et al. 2001). The sequestration of ER lumenal chaperones into these sites may reduce their availability for their functions in ERAD and elicit the observed defects in turnover of substrates dependent on lumenal chaperones.

Heterologously expressed CFTR misfolds and is subject to ERAD in yeast, independently of whether it carries the disease-relevant mutation ΔF508 in its cytoplasmic domain (Gnann et al. 2004). Fu & Sztul (2003) observed defects in CFTR turnover in yeast with mutations in *sec12*, *sec13*, *sec23*, and, to a lesser degree, in *sec18*. Sec18p is required for vesicle fusion with the Golgi complex. Fu & Sztul took the limited effect of the *sec18* mutant on CFTR turnover as an indicator that vesicle formation and transport to the Golgi were not required for this process and that the role of the COPII coat here was to sequester CFTR into an ER subcompartment from which it could subsequently be degraded (Fu & Sztul 2003).

Although there was a morphological change in the CFTR-containing ER subcompartments in a *sec23* mutant from small compact puncta to larger more diffuse structures, CFTR in these cells was still not distributed homogeneously throughout the ER (Fu & Sztul 2003). Huyer and colleagues subsequently showed that expression of CFTR and of some mutant alleles encoding the structurally related yeast Ste6 protein triggers formation of tubuloreticular membrane clusters attached to the nucleus, which are morphologically reminiscent of the structures formed in ER exit mutants under restrictive conditions (Nishikawa et al. 2001, Fu & Sztul 2003, Huyer et al. 2004a). Formation of these clusters, however, is solely a function of the presence of the overexpressed transmembrane proteins and does not require any additional defects in the ER-to-Golgi transport machinery (Huyer et al. 2004a). In this respect, formation of these ER-associated compartments resembles HMG-CoA reductase-induced karmellae, but physically, the two compartments are distinct (Wright et al. 1988, Huyer et al. 2004a). Huyer and colleagues showed that overexpression of proteins inducing the ER-associated compartments does not trigger the UPR and proposed that these membranes serve as a holding compartment for defective transmembrane proteins, which prevents them from interfering with normal ER functions, a hypothesis that is hard to test in the absence of means to disrupt formation of these structures (Huyer et al. 2004a).

PREREQUISITES FOR EXPORT FROM THE ER TO THE CYTOSOL

Solubility

Misfolded secretory proteins have a propensity to form aggregates in the ER (Nishikawa et al. 2001, Kabani et al. 2003). BiP activity and the J-domain proteins Jem1p and Scj1p are required to keep ERAD substrates soluble in the ER (Nishikawa et al. 2001, Kabani et al. 2003), but BiP may have an additional role in ERAD because coexpression of wild-type BiP in an ERAD-defective *kar2-1* strain alleviates the substrate aggregation but not the ERAD phenotype (Kabani et al. 2003). Solubility of aberrant proteins in the ER can also be enhanced by O-mannosylation, but this does not necessarily improve turnover (Harty et al. 2001, Nakatsukasa et al. 2004). A number of mutant serine protease inhibitors (serpins) cause human genetic diseases; these proteins form polymers in the ER in vivo and are turned over inefficiently, which triggers both the UPR and the ER overload reponse and results in inflammation and cell death and in severe diseases in the patients (Lomas & Parfrey 2004, Miranda et al. 2004). Strategies to disassemble these polymers are being discussed, but it remains to be seen whether turnover of aberrant serpins can be enhanced solely by maintaining mutant serpins in the monomeric form (Lomas & Parfrey 2004, Römisch 2004).

Unfolding?

In order to be translocated across the ER membrane to the cytosol, ERAD substrates need to be soluble and dissociated from oligomeric complexes, and their disulfide bonds need to be reduced (Tortorella et al. 1998, Gillece et al. 1999, Fagioli et al. 2001,

Nishikawa et al. 2001, Sevilla et al. 2004). Protein export from the ER of many misfolded proteins is dependent on Sec61p, and during protein import into the ER through the Sec61 channel, proteins have to be fully unfolded (**Figure 1**) (Johnson & van Waes 1999). The need for unfolding during retrograde transport, however, has been questioned recently: Fiebiger and colleagues studied the dislocation of the type I transmembrane protein major histocompatibility complex (MHC) class I heavy chain (HC) by fusing GFP to its ER-lumenal N terminus (Fiebiger et al. 2002). Upon proteasome inhibition, the authors observed accumulation of fluorescent GFP-HC in the cytosol of transfected cells (Fiebiger et al. 2002). Because GFP-HC was unable to refold into a fluorescent molecule after chemical denaturation in vitro, the authors concluded that dislocation of the GFP to the cytosol had taken place without unfolding the GFP moiety (Fiebiger et al. 2002). Intracellularly, however, cytosolic GFP-HC is exposed to high concentrations of chaperones that may be able to refold the GFP domain (Zietkiewicz et al. 2004).

In a second attempt to address the unfolding question, Tirosh and colleagues fused dihydrofolate reductase (DHFR) to the HC N terminus and found that a DHFR substrate analogue, trimetraxate, which induces tight folding of the enyzme, does not inhibit dislocation or degradation of the DHFR-HC chimera (Tirosh et al. 2003). Moreover, the authors observed that DHFR-HC molecules, which bound trimetraxate during a 15 min pulse, retained the substrate throughout a 60 min chase. Tirosh and colleagues concluded that DHFR-HC molecules were dislocated tightly folded around the DHFR substrate. Dislocation of HC in cells, however, is extremely efficient with a t1/2 of 1–2 min (Wiertz et al. 1996a); another possible interpretation of the results is that because of the high concentration of chaperones in the ER lumen, DHFR was unfolded for retrotranslocation even in the presence of

MHC: major histocompatibility complex

HC: heavy chain

DHFR: dihydrofolate reductase

trimetraxate, and DHFR rebound the trimetraxate during the long pulse after dislocation into the cytosol. Whether unfolding is required for retrotranslocation could perhaps be addressed more conclusively by attaching ERAD substrates to inert molecules of specific sizes; alternatively full characterization of the export channel in the ER membrane may provide the answer to this question.

THE EXPORT CHANNEL

Sec61 Complex

Wiertz and colleagues were the first to find two transmembrane ERAD substrates, HC and US2, associated with the heterotrimeric Sec61 complex prior to degradation (Wiertz et al. 1996b). Mutations in the core component of the channel, Sec61p, have profound defects on ERAD of mutant pαF in a cell-free system and cause prolonged association of the substrate with Sec61p and prolonged association of the backed-up substrate in the ER lumen with PDI (**Figure 1**) (Pilon et al. 1997, 1998; Gillece et al. 1999). Mutant *sec61* alleles also reduce ERAD of many substrates in pulse-chase experiments, but these data are more difficult to evaluate because of the potential pleiotropic effects of *sec61* mutations on ER protein composition and ER function (Plemper et al. 1997, Biederer et al. 1996). Direct blockade of the Sec61 channels of mammalian microsomes with ribosome-nascent chain complexes prevents transport of cholera toxin and amyloid beta-peptide from the ER lumen into the cytosol (Schmitz et al. 2000, 2004). In dendritic cells, phagocytosed antigens gain access to the ER and are transported to the cytosol for degradation by proteasomes and subsequent presentation of the resulting peptides on MHC class I molecules (cross-presentation) (Imai et al. 2005). Antigen degradation and cross-presentation are strongly inhibited if Sec61α expression is reduced by RNAi, suggesting that the Sec61 channel is responsible for transport of antigens to the cytosol (Imai et al. 2005). Yeast

strains that are unable to induce the expression of *SEC61* upon overexpression of CPY* develop an ER protein import defect that is alleviated upon overexpression of *SEC61*, suggesting that the protein translocation channel becomes limiting for import into the ER if there is a high demand for export (Ng et al. 2000). HC destined for degradation associates with complexes containing both Sec61γ and Sec61β (Wiertz et al. 1996b). The two genes encoding the beta-subunit, *SBH1* and *SBH2*, are not essential in yeast, and deletion of both causes a protein import defect into the ER only at elevated temperatures (Finke et al. 1996). Microsomes from *sbh1 sbh2* yeast are export competent for pαF in vitro (K. Römisch, unpublished data). Conditional alleles in the gene encoding the essential gamma-subunit of the protein translocation channel, *SSS1*, are not available, so the requirement for Sss1p for export cannot not be tested. Taken together, the data strongly suggest that protein export from the ER is mediated by a channel containing Sec61p (**Figure 1**).

Sec63 Complex

In yeast the Sec61 channel formed by Sec61p, Sbh1p, and Sss1p can associate with the heterotetrameric Sec63 complex (Sec63p, Sec62p, Sec71p, Sec72) to mediate posttranslational import of specific substrates into the ER (Johnson & van Waes 1999). Although there is virtually no posttranslational import into mammalian ER, both Sec62p and Sec63p have homologs of thus far unknown function in mammals that are associated with Sec61p (Meyer et al. 2000a, Tyedmers et al. 2000). *SEC71*, *SEC72*, and *SEC62* are dispensable for pαF ERAD in vitro and CPY* turnover in vivo, suggesting that the Sec63 complex is not required for ERAD (Pilon et al. 1997, Plemper et al. 1997). The role of Sec63p itself is controversial: both CPY* and pαF ERAD are delayed in *sec63-1* mutants, but because the allele is leaky, it remains unclear whether Sec63p is required for ERAD or whether ER function in general is compromised in this mutant (Pilon et al. 1997, Plemper et al. 1997).

Sec61p-Independent Export?

The effects of different *sec61* alleles on ERAD vary widely from a modest delay in export in *sec61-2* membranes to an almost complete block in *sec61-32* membranes (Pilon et al. 1997). In general, ER import and export defects in *sec61* mutants are correlated (Pilon et al. 1998, Wilkinson et al. 2000), but degradation of a number of transmembrane ERAD substrates seems less dependent on Sec61p than does degradation of soluble substrates: Two sec61 mutations that strongly affect soluble substrate turnover, *sec61-32* and *sec61-41*, do not affect turnover of the tail-anchored membrane protein Ubc6p (Pilon et al. 1997, Walter et al. 2001); *sec61-2*, which has a modest effect on turnover of soluble substrates, does not affect degradation of transmembrane polytopic Ste6-166p, or the unstable Sec61-2 protein itself, but partially stabilizes Pdr5*p, and causes dramatic accumulation of CFTR in yeast (Biederer et al. 1996, Pilon et al. 1998, Plemper et al. 1998, Kiser et al. 2001, Huyer et al. 2004b). Because Ste6-166p, Pdr5*p, and CFTR are closely related structurally, these results are unlikely to reflect substrate specificity, but may result from different strain backgrounds or different substrate expression levels. The protein translocation channel has to open laterally in order to accommodate transmembrane proteins, and transversely to allow passage of soluble proteins; certain *sec61* mutant alleles may specifically affect one mode of opening and thus differentially affect dislocation of soluble versus transmembrane proteins, but a systematic comparison of the effects of *sec61* alleles on different dislocation substrates is still missing.

A further complication is the existence of a Sec61p homolog in yeast, Ssh1p (Finke et al. 1996). Ssh1p, which is nonessential, forms channels with Sbh2p and Sss1p and enhances the cotranslational protein import ca-

pacity of the ER (Finke et al. 1996, Wilkinson et al. 2001). Δ*ssh1* strains with *sec61* mutations have a tendency to become respiration deficient, which in turn affects protein turnover (Wilkinson et al. 2001). The loss of respiration competence may vary in different strain backgrounds and with experimental protocols, which may explain the lack of congruent results about the role of Ssh1p in retrograde protein transport from the ER (Plemper et al. 1997, Wilkinson et al. 2001, Huyer et al. 2004b).

Two recent publications question the contribution of Sec61p to protein export from the ER and propose instead that a protein with four transmembrane domains, Derlin-1 (Der1p in yeast), forms the export channel (Lilley & Ploegh 2004, Ye et al. 2004). Lilley and colleagues isolated dislocation intermediates of HC from cells expressing a viral protein, US11, which promotes HC ERAD, and from cells expressing a mutant form of US11, which still forms complexes with HC, but no longer triggers its dislocation (Lilley & Ploegh 2004). Wild-type but not mutant US11-HC complexes contain Derlin-1, and expression of dominant-negative Derlin-1 interferes with US11-mediated HC degradation (Lilley & Ploegh 2004). In a quest for a receptor of the hexameric AAA-ATPase chaperone complex formed by p97 (Cdc48p in yeast), which is required for ERAD of many substrates, Ye and colleagues (Ye et al. 2004) found p97 associated with two transmembrane proteins, Derlin-1 and VCP-interacting membrane protein (VIMP), in dog pancreas microsomes (**Figure 1**). Depletion of Derlin-1 by RNAi in *Caenorhabditis elegans* induced the UPR (Ye et al. 2004). The gene encoding the yeast homologue of Derlin-1, *DER1*, was identified in a screen for mutants that stabilize CPY* (Knop et al. 1996). Deletion of *DER1* also increases the half life of another mutant vacuolar protease, PrA*, about fourfold, but for other substrates there is only a modest effect (a twofold increase in the half life of KHN, KWW, pαF) or none at all (Pdr5*, Sec61-2p) (Knop et al. 1996, Vashist

AAA-ATPase: ATPase associated with a variety of cellular functions

& Ng 2004, Plemper et al. 1998; K. Römisch, unpublished data). Whereas Der1p/Derlin-1 may be associated with the export channel (**Figure 1**), the lack of a general stringent requirement for its presence makes it unlikely that it is the only, or even the major, export channel.

EXPORT MECHANISM

Cdc48p/p97

Cdc48p (p97 in mammals) is a member of the AAA-ATPase family of chaperones involved in many cellular functions (Woodman 2003). Two cofactors, p47 and Ufd1p/Npl4p, which bind mutually exclusively to the N-terminal domain of p97 determine functions of the resulting complexes in membrane fusion and ubiquitin-dependent protein degradation, respectively (Meyer et al. 2000b). The Cdc48p N-terminal domain also binds ubiquitin with a preference for polyubiquitin with a K48 linkage (Dai & Li 2001). Two not necessarily mutually exclusive roles have been proposed for the Cdc48p/Ufd1p/Npl4p complex in ERAD: Jentsch and colleagues suggest that the complex acts as a "segregase," which recognizes and binds to ubiquitinated proteins at the ER membranes (**Figure 1**), separates them from associated nonubiquitinated proteins, and thus makes substrates available to the proteasome for degradation (Rape et al. 2001, Braun et al. 2002). This is a role similar to that of Cdc48p/Ufd1p/Npl4p in mobilization of the membrane-tethered transcription factor Spt23p after proteolytic processing. Support for this view comes from the observation that polyubiquitinated proteins remain membrane-associated in *ufd1-1* mutants, but accumulate in the soluble fraction if the Cdc48 complex is active and protein turnover is inhibited (Rape et al. 2001, Jarosch et al. 2002). Rapoport and colleagues propose that the Cdc48p/Ufd1p/Npl4p complex is a "dislocase" that directly exerts force on the substrate protein and extracts it from the protein translocation channel (**Figure 1**) (Ye

et al. 2001, 2003). Membrane-anchored AAA-ATPases can indeed extract defective proteins from bacterial and mitochondrial membranes for degradation (Langer 2000).

The difficulty is to distinguish between the two models of Cdc48p action experimentally. The decisive issue is whether the substrates, which remain membrane-associated in mutants of the Cdc48p/Ufd1p/Npl4p complex, are associated with the cytoplasmic face of the ER membrane or reside in the ER lumen. Typically, protease resistance is used as a measure for sequestration inside the ER, but most experiments performed to address the role of Cdc48p do not include protease protection assays. And is protease resistance a reliable measure for location inside the ER lumen? In cell-free ER-import assays, for example, the relatively hydrophobic pαF has a propensity to aggregate on the cytoplasmic face of the ER membrane and is therefore partially protease resistant in the absence of detergent; solubilization of the membranes, however, also dissolves the aggregates and makes the protein protease-susceptible (Pilon et al. 1997). The in vitro dislocation assay that Rapoport and colleagues use is based on a permeabilized mammalian cell system that expresses HC and one of two viral proteins, US2 or US11, which trigger dislocation of the HC to the cytosol with a t1/2 of less than 1 min (Wiertz et al. 1996a). In the 3–5 min pulse that is typically applied prior to cell permeabilization, a significant fraction of the HCs may have already been extracted from the ER membrane, and the subsequent release and degradation assays in the permeabilized cells may measure disaggregation of HCs on the cytoplasmic face of the ER by Cdc48p rather than extraction from the ER membrane (Shamu et al. 1999; Ye et al. 2001, 2003). It also remains to be seen how representative US2/US11-mediated dislocation of HC is for ERAD in general, because Sommer and colleagues recently demonstrated that dislocation mediated by another viral protein, Vpu, differs significantly from generic ERAD (Meusser & Sommer 2004).

Proteasome 19S Regulatory Particle

Cdc48p can direct substrates to proteasomes (Verma et al. 2004). Cdc48p binding to dislocation intermediates in the Sec61 channel could therefore recruit proteasomes and thus couple extraction from the ER and degradation (Braun et al. 2002). Export into the cytosol is proteolysis independent for most substrates, but the six nonequivalent AAA-ATPases of the proteasome 19S regulatory particle (19S RP), which have the capacity to unfold substrates and feed them into the proteolytic proteasome core particle, may contribute to export (Wiertz et al. 1996a, Glickman et al. 1998, Rubin et al. 1998, Lee et al. 2004). For the nonubiquitinated ERAD substrate pαF, the 19S RP is indeed the only cytosolic factor required for export from the ER in vitro (Lee et al. 2004). A significant fraction of proteasomes in the cell is associated with the ER (Enenkel et al. 1998, Rivett 1998), and we recently found a direct interaction of the AAA-ATPase containing base of the 19S RP with Sec61 channels (**Figure 1**) (Kalies et al. 2005). These Sec61 channel/19S RP complexes contained sub-stoichiometric amounts of Cdc48p, suggesting that whereas Cdc48p can interact with the 19S RP bound to Sec61 channels, it is not required for channel binding (K. Römisch & K. Kalies, unpublished data). Individual mutations in some of the proteasome AAA-ATPases (*RPT1-6*) reduce but do not abolish export of CPY* from the ER (Jarosch et al. 2002). However, a systematic investigation of the effects of mutations in 19S RP subunits on export from the ER has not yet been done, and the relative contributions of Cdc48p and the 19S RP to this process remain unclear.

POST-EXPORT EVENTS

Ubiquitylation

Ubiquitylation takes place in the cytosol and therefore cannot be involved in initiating the export of soluble proteins from the ER (Hampton 2002). For the transmembrane HC, removal of all lysines in the cytosolic tail does not abrogate export, suggesting that ubiquitin modification takes place after the ER-lumenal N terminus of the protein has been made accessible to the ubiquitylation machinery on the cytoplasmic face of the ER membrane (Shamu et al. 1999). Both HC and the soluble CPY* are initially oligo-ubiquitylated, and these forms of the proteins remain membrane associated, but it is not clear whether oligo-ubiquitylated CPY* and HC are export intermediates (**Figure 1**) or are fully exported but tightly associated with the cytoplasmic face of the ER membrane (Shamu et al. 2001, Jarosch et al. 2002, Flierman et al. 2003). Release from the membrane requires lysine48-linked polyubiquitylation and ATP hydrolysis by factors such as Cdc48/p97 or 19S RP that presumably bind to and segregate the ERAD substrates from the ER membrane (Shamu et al. 2001, Jarosch et al. 2002, Flierman et al. 2003).

The best-characterized ubiquitylation of ERAD substrates is mediated by ER membrane-resident RING-motif ubiquitin ligases (E3 enzymes), Hrd1p and Doa10p (Hampton 2002). Hrd1p forms a complex with Hrd3p, a transmembrane protein with a large ER-lumenal domain that stabilizes Hrd1p (**Figure 1**). The complex, together with the ubiquitin-conjugating enzyme (E2) Ubc7p, which binds to the ER via Cue1p (**Figure 1**), catalyzes ubiquitylation of ERAD-substrates with lumenal lesions (Biederer et al. 1997, Hampton 2002, Vashist & Ng 2004). The Hrd1/3p complex can also cooperate with the E2 Ubc1p (Hampton 2002). Doa10p together with Ubc7p/Cue1p or the tail-anchored E2 Ubc6p (**Figure 1**) ubiquitylates primarily transmembrane ERAD substrates with lesions in cytosolic domains or unassembled subunits of oligomeric complexes (Hampton 2002, Vashist & Ng 2004). ERAD of the transmembrane protein Ole1p is independent of both Hrd1p and Doa10p but relies on Ubc6p and Ubc7p, which suggests that there may be additional E3s involved in ERAD (Braun et al. 2002).

19S RP:
proteasome 19S
regulatory particle

SCF: Skp1,
Cdc53/Cullin, F box
receptor

FBS: F-box protein
for sugar recognition

TCR: T cell
receptor

26S proteasome:
abundant cytosolic
protease complex;
consists of a 20S core
particle containing
proteolytic activities,
and one or two 19S
RP

Yoshida and colleagues discovered two members of the mammalian SCF ubiquitin ligase family, SCF[FBS1] and SCF[FBS2], which specifically recognize and ubiquitylate N-glycosylated proteins that have been transported to the cytosol (**Figure 1**) (Yoshida et al. 2002, 2003). SCF ubiquitin ligases are cytosolic protein complexes with four subunits; the F-box- containing subunit recognizes and binds the substrate. The F-box proteins Fbs1 and Fbs2 recognize mannose-terminated N-glycans (**Figure 1**) (Yoshida et al. 2003). Fbs1 is expressed mostly in neurons and binds strongly to the chitobiose moiety of denatured glycoproteins, whereas Fbs2, which is ubiquitously expressed, also recognizes the mannosyl residues of the oligosaccharide (Yoshida et al. 2003). Deletion of the F-box of Fbs1 reduces turnover of CFTR and T cell receptor (TCR)α in mammalian cells, but ΔF-box Fbs2 has a much stronger stabilizing effect on TCRα, suggesting that the two F-box proteins have different substrate specificities (Yoshida et al. 2002, 2003). These sugar-recognizing F-box proteins are members of a family of mammalian proteins with several more uncharacterized members; there is no obvious orthologue in the yeast genome (Yoshida et al. 2003).

Targeting to Proteasomes

Hrd1p and Doa10p mediate only the initial oligo-ubiquitylation of ERAD substrates; polyubiquitylation, which is required for degradation by proteasomes, is catalyzed by so-called E4 enzymes such as Ufd2p (**Figure 1**), which was discovered in a screen for genes required for degradation of monoubiquitin-fusion proteins (Koegl et al. 1999, Richly et al. 2005). Ufd2p, Cdc48p, the 19S RP subunits Rpn10p and Rpt5p, and two other proteins that can associate with 19S RPs, Rad23p and Dsk2p, all bind ubiquitin conjugates and may transport substrates to the proteasome proteolytic core, but their respective roles are unclear (**Figure 1**) (Koegl et al. 1999, Lam et al. 2002, Medicherla

et al. 2004, Verma et al. 2004). An elegant study by Richly and colleagues demonstrated recently that Cdc48p binds to both oligo-ubiquitylated substrates and Ufd2p (**Figure 1**) (Richly et al. 2005). This interaction stimulates the E4 activity of Udf2 but also restricts the size of the multiubiquitin chains assembled to 3–6 ubiquitin moieties (Richly et al. 2005). This chain size, however, is sufficient for the recognition by the multi-ubiquitin chain receptors Rad23p and Dsk2p, which directly interact with the E4 Ufd2p (**Figure 1**) in a manner that is dependent on the Cdc48/Npl4/Ufd1p complex, although there is no direct interaction of the complex with Rad23p or Dsk2p (Richly et al. 2005). Rad23p and Dsk2p can also bind directly to the 19S RP subunit Rpn1p, and Rad23p binding to the proteasome and Ufd2p is mutually exclusive, suggesting that after binding to multiubiquitylated proteins, Rad23p and Dsk2p dissociate from Ufd2p and ferry the substrates to 26S proteasomes (**Figure 1**) (Elsasser et al. 2002, Kim et al. 2004, Richly et al. 2005). Dsk2p and Rad23p seem to have overlapping functions in turnover of ERAD substrates because only the $\Delta rad23 \ \Delta dsk2$ double mutant displays significant ERAD defects (Medicherla et al. 2004). Ufd2p and the 19S RP subunit Rpn10p, which binds to longer multiubiquitin chains, function in parallel pathways that lead to proteasomal turnover (**Figure 1**): Whereas individual deletions of *ufd2* and *rpn10* result in moderate or no effects on turnover of a variety of ERAD substrates, the $\Delta ufd2 \ \Delta rpn10$ double mutant has a much stronger phenotype (Richly et al. 2005). The authors propose that ubiquitinated substrates are escorted to proteasomes by ubiquitin-binding proteins that may prevent deubiquitination and seem to use increasing multiubiquitin chain lengths to ensure directionality (Richly et al. 2005). Although the 19S RP subunit Rpt5p can also recognize tetra-ubiquitin chains, it is not sufficient to target a ubiquitinated substrate to degradation in a defined cell-free system (Lam et al. 2002, Verma et al. 2004). Verma and colleagues suggested that Rpt5p may act downstream of Rad23p, Dsk2p, and Rpn10p and position the delivered substrates for de-ubiquitination and delivery into the proteolytic channel of the proteasome core particle (Verma et al. 2004).

Deglycosylation

Wiertz and colleagues observed that in the presence of proteasome inhibitors, dislocated deglycosylated HC accumulates in the cytosol (Wiertz et al. 1996a). N-glycanase activity has also been detected in mammalian cytosol in vitro (Suzuki et al. 1994, Römisch & Ali 1997). Yeast peptide N-glycanase encoded by *PNG1* is a soluble 42.5-kDa protein with orthologues in all eukaryotes (Suzuki et al. 2000). The enzyme is expressed at low levels, and deletion of *PNG1* results in merely a doubling of the CPY* half life, suggesting that deglycosylation is not essential for proteasomal degradation (Suzuki et al. 2000). Png1p binds to Rad23p, which appears to mediate its interaction with proteasomes (**Figure 1**) (Katiyar et al. 2004). Because mammalian Rad23p can bind only to deglycosylated proteins, protein N-glycanase (PNGase) must act upstream of Rad23p, but it remains unclear whether deglycosylation is required before or after substrate interaction with Cdc48p and Ufd2p (Katiyar et al. 2004).

Yeast and mammalian PNGase recognize high-mannose N-glycans on denatured proteins (Hirsch et al. 2003, 2004). Soluble N-glycanase can be inhited in vitro and in cells by Z-VAD-fmk, which results in the presence of N-glycosylated dislocated HC in the cytosol; degradation, however, is not affected by the presence of the N-glycans (Misaghi et al. 2004). Deglycosylation of TCRα, which is dislocated with slower kinetics from the ER membrane, cannot be inhibited by Z-VAD-fmk, suggesting perhaps that the active site of the membrane-associated fraction of PN-Gase is not accessible to the inhibitor and that dislocation and deglycosylation may be temporally coupled (Hirsch et al. 2003, Misaghi

et al. 2004). The presence of N-glycans does not pose a principal obstacle to proteasomal turnover, but because glycans often stabilize protein structures, unfolding of specific domains by the proteasome 19S RP may be more efficient after deglycosylation.

PNGase: protein N-glycanase

ERAD INHIBITORS

Compounds that interfere with ERAD include proteasome inhibitors, ER-mannosidase I inhibitors, and agents that change the redox potential in the ER (Fiebiger et al. 2004). By screening a chemical library for compounds that stabilize HC in cells coexpressing US11, Fiebiger and colleagues identified two large, hydrophobic, and structurally related ERAD inhibitors, Eeyarestatin I and II (Fiebiger et al. 2004). Eeyarestatin I/II cause stabilization of the glycosylated HC and also of TCRα in the membrane fraction after relatively long incubation times (8–16 h), suggesting that the inhibitors are metabolized to produce the active compound (Fiebiger et al. 2004). Eeyarestatin I/II do not inhibit PNGase and have no effect on the proteolytic activity of the proteasome, but it is not clear which stage in ERAD they block (Fiebiger et al. 2004). Their effect on ER-mannosidase I has not been studied, and although the inhibitors seem to reduce the biosynthesis of HC, no experiments were done to address whether they interfere with the activity of the ER protein translocation channel that is responsible for membrane integration of newly synthesized HC and most likely also for its dislocation (Fiebiger et al. 2004). It would also be interesting to know if Eeyarestatins specifically inhibit ERAD of transmembrane proteins or proteins with lesions in specific domains.

PERSPECTIVES

The discovery that the majority of mutations in transmembrane and secretory proteins leading to human genetic diseases simply results in retention of these proteins in the

ER and subsequent ERAD made it possible for human geneticists to begin to develop therapeutics that promote folding and transport through the secretory pathway of the respective mutant proteins (Römisch 2004). Inefficient ERAD results in accumulation of misfolded proteins in the ER and triggers cell death, which is involved in diabetes and common neurodegenerative diseases such as Alzheimer's and Parkinson's diseases. Developing strategies to increase our capacity to prevent or remove protein aggregates by boosting ERAD may therefore significantly improve our health in old age. Insight into ERAD-related processes has been growing exponentially in the last few years, and although many questions remain open, the hope that we will understand at least the basic ERAD mechanism by the time the next large review is written is quite realistic.

SUMMARY POINTS

1. Topology of the lesion in a protein determines the targeting pathway for degradation.

2. Solubility is required, but not sufficient, for export of misfolded proteins from the ER lumen to the cytosol.

3. Direct blockade of Sec61 channels with translating ribosomes and RNAi of Sec61α reduces protein transport from the ER lumen to the cytosol.

4. 19S RP is sufficient for ERAD in vitro and binds directly to the Sec61 channel.

5. Cdc48p binds to Der1p in the ER membrane.

6. Cdc48p coordinates substrate recruitment for polyubiquitylation by Ufd2 and subsequent proteasomal targeting.

ACKNOWLEDGMENTS

K.R. was funded by a Senior Fellowship from The Wellcome Trust (042216). Apologies to the many colleagues whose research could not be included here; there are currently over 26,000 ERAD-relevant papers in PubMed, 80% of which were published in the past two years. I thank the members of my extended lab for input, Jaime Gallo for a last-minute boost of motivation, and Randy Schekman for critically reading the manuscript.

LITERATURE CITED

Biederer T, Volkwein C, Sommer T. 1996. Degradation of subunits of the Sec61p complex, an integral component of the ER membrane, by the ubiquitin-proteasome pathway. *EMBO J.* 15:2069–76

Biederer T, Volkwein C, Sommer T. 1997. Role of Cue1p in ubiquitination and degradation at the ER surface. *Science* 278:1806–9

Braun S, Matuschewski K, Rape M, Thoms S, Jentsch S. 2002. Role of the ubiquitin-selective CDC48 chaperone in ERAD of OLE1 and other substrates. *EMBO J.* 21:615–21

Brodsky JL, McCracken AA. 1999. ER protein quality control and proteasome-mediated protein degradation. *Semin. Cell Dev. Biol.* 10:507–13

Caldwell SR, Hill KJ, Cooper AA. 2001. Degradation of endoplasmic reticulum (ER) quality control substrates requires transport between the ER and Golgi. *J. Biol. Chem.* 276:23296–303

Chen EY, Bartlett MC, Loo TW, Clarke DM. 2004. The DeltaF508 mutation disrupts packing of the transmembrane segments of the cystic fibrosis transmembrane conductance regulator. *J. Biol. Chem.* 279:39620–27

Chillaron J, Haas IG. 2000. Dissociation from BiP and retrotranslocation of unassembled immunoglobulin light chains are tightly coupled to proteasome activity. *Mol. Biol. Cell.* 11:217–26

Cronin SR, Khoury A, Ferry DK, Hampton RY. 2000. Regulation of HMG-CoA reductase degradation requires the P-type ATPase Cod1p/Spf1p. *J. Cell. Biol.* 148:915–24

Cronin SR, Rao R, Hampton RY. 2002. Cod1p/Spf1p is a P-type ATPase involved in ER function and Ca^{2+} homeostasis. *J. Cell. Biol.* 157:1017–28

Dai RM, Li CC. 2001. Valosin-containing protein is a multi-ubiquitin chain-targeting factor required in ubiquitin-proteasome degradation. *Nat. Cell. Biol.* 3:740–74

Durr G, Strayle J, Plemper R, Elbs S, Klee SK, et al. 1998. The medial-Golgi ion pump Pmr1 supplies the yeast secretory pathway with Ca^{2+} and Mn^{2+} required for glycosylation, sorting, and endoplasmic reticulum-associated protein degradation. *Mol. Biol. Cell.* 9:1149–62

Egan ME, Glockner-Pagel J, Ambrose C, Cahill PA, Pappoe L, et al. 2002. Calcium-pump inhibitors induce functional surface expression of delta F508-CFTR protein in cystic fibrosis epithelial cells. *Nat. Med.* 8:485–92

Elsasser S, Gali RR, Schwickart M, Larsen CN, Leggett DS, et al. 2002. Proteasome subunit Rpn1 binds ubiquitin-like domains. *Nat. Cell. Biol.* 4:725–30

Enenkel C, Lehmann A, Kloetzel PM. 1998. Subcellular distribution of proteasomes implicates a major location of protein degradation in the nuclear envelope-ER network in yeast. *EMBO J.* 17:6144–54

Fagioli C, Mezghrani A, Sitia R. 2001. Reduction of interchain disulfide bonds precedes the dislocation of Ig-mu chains from the endoplasmic reticulum to the cytosol for proteasomal degradation. *J. Biol. Chem.* 276:40962–67

Fiebiger E, Hirsch C, Vyas JM, Gordon E, Ploegh HL, Tortorella D. 2004. Dissection of the dislocation pathway for type I membrane proteins with a new small molecule inhibitor, eeyarestatin. *Mol. Biol. Cell.* 15:1635–46

Fiebiger E, Story C, Ploegh HL, Tortorella D. 2002. Visualization of the ER-to-cytosol dislocation reaction of a type I membrane protein. *EMBO J.* 21:1041–53

Finke K, Plath K, Panzner S, Prehn S, Rapoport TA, et al. 1996. A second trimeric complex containing homologs of the Sec61p complex functions in protein transport across the ER membrane of *S. cerevisiae*. *EMBO J.* 15:1482–94

Flierman D, Ye Y, Dai M, Chau V, Rapoport TA. 2003. Polyubiquitin serves as a recognition signal, rather than a ratcheting molecule, during retrotranslocation of proteins across the endoplasmic reticulum membrane. *J. Biol. Chem.* 278:34774–82

Flynn GC, Pohl J, Flocco MT, Rothman JE. 1991. Peptide-binding specificity of the molecular chaperone BiP. *Nature* 353:726–30

Fu L, Sztul E. 2003. Traffic-independent function of the Sar1p/COPII machinery in proteasomal sorting of the cystic fibrosis transmembrane conductance regulator. *J. Cell. Biol.* 160:157–63

Gardner RG, Hampton RY. 1999. A 'distributed degron' allows regulated entry into the ER degradation pathway. *EMBO J.* 18:5994–6004

This study describes a chemical inhibitor of dislocation of transmembrane ERAD substrates; the target of the inhibitor is not known.

Gillece P, Luz JM, Lennarz WJ, de La Cruz FJ, Römisch K. 1999. Export of a cysteine-free misfolded secretory protein from the endoplasmic reticulum for degradation requires interaction with protein disulfide isomerase. *J. Cell. Biol.* 147:1443–56

Glickman MH, Rubin DM, Fried VA, Finley D. 1998. The regulatory particle of the *Saccharomyces cerevisiae* proteasome. *Mol. Cell. Biol.* 18:3149–62

Gnann A, Riordan JR, Wolf DH. 2004. Cystic fibrosis transmembrane conductance regulator degradation depends on the lectins Htm1p/EDEM and the Cdc48 protein complex in yeast. *Mol. Biol. Cell.* 15:4125–35

Goldberg AL. 2003. Protein degradation and protection against misfolded or damaged proteins. *Nature* 426:895–99

Hampton RY. 2002. ER-associated degradation in protein quality control and cellular regulation. *Curr. Opin. Cell. Biol.* 14:476–82

Harty C, Strahl S, Römisch K. 2001. O-mannosylation protects mutant alpha-factor precursor from endoplasmic reticulum-associated degradation. *Mol. Biol. Cell.* 12:1093–101

Helenius A, Aebi M. 2004. Roles of N-linked glycans in the endoplasmic reticulum. *Annu. Rev. Biochem.* 73:1019–49

Herscovics A. 2001. Structure and function of Class I alpha 1,2-mannosidases involved in glycoprotein synthesis and endoplasmic reticulum quality control. *Biochimie* 83:757–62

Hirsch C, Blom D, Ploegh HL. 2003. A role for N-glycanase in the cytosolic turnover of glycoproteins. *EMBO J.* 22:1036–46

Hirsch C, Misaghi S, Blom D, Pacold ME, Ploegh HL. 2004. Yeast N-glycanase distinguishes between native and non-native glycoproteins. *EMBO Rep.* 5:201–6

Huyer G, Longsworth GL, Mason DL, Mallampalli MP, McCaffery JM, et al. 2004a. A striking quality control subcompartment in *Saccharomyces cerevisiae*: the endoplasmic reticulum-associated compartment. *Mol. Biol. Cell.* 15:908–21

Huyer G, Piluek WF, Fansler Z, Kreft SG, Hochstrasser M, et al. 2004b. Distinct machinery is required in *Saccharomyces cerevisiae* for the endoplasmic reticulum-associated degradation of a multispanning membrane protein and a soluble luminal protein. *J. Biol. Chem.* 279:38369–78

Imai J, Hasegawa H, Maruya M, Koyasu S, Yahara I. 2005. Exogenous antigens are processed through the endoplasmic reticulum-associated degradation (ERAD) in cross-presentation by dendritic cells. *Int. Immunol.* 17:45–53

Jakob CA, Bodmer D, Spirig U, Battig P, Marcil A, et al. 2001. Htm1p, a mannosidase-like protein, is involved in glycoprotein degradation in yeast. *EMBO Rep.* 2:423–30

Jarosch E, Taxis C, Volkwein C, Bordallo J, Finley D, et al. 2002. Protein dislocation from the ER requires polyubiquitination and the AAA-ATPase Cdc48. *Nat. Cell. Biol.* 4:134–39

Jensen TJ, Loo MA, Pind S, Williams DB, Goldberg AL, Riordan JR. 1995. Multiple proteolytic systems, including the proteasome, contribute to CFTR processing. *Cell* 83:129–35

Johnson AE, van Waes MA. 1999. The translocon: a dynamic gateway at the ER membrane. *Annu. Rev. Cell. Dev. Biol.* 15:799–842

Kabani M, Kelley SS, Morrow MW, Montgomery DL, Sivendran R, et al. 2003. Dependence of endoplasmic reticulum-associated degradation on the peptide binding domain and concentration of BiP. *Mol. Biol. Cell.* 14:3437–48

Kalies KU, Allan S, Sergeyenko T, Kroger H, Römish K. 2005. The protein translocation channel binds proteasomes to the endoplasmic reticulum membrane. *EMBO J.* 24:2284–93

This paper shows how RNAi of Sec61α inhibits transport of exogenous antigens from phagosomes into the cytosol.

This paper demonstrates that Cdc48p/Ufd1p/Npl4p are involved in release of ERAD substrates from the ER membrane.

Katiyar S, Li G, Lennarz WJ. 2004. A complex between peptide:N-glycanase and two proteasome-linked proteins suggests a mechanism for the degradation of misfolded glycoproteins. *Proc. Natl. Acad. Sci. USA* 101:13774–79

Kim I, Mi K, Rao H. 2004. Multiple interactions of rad23 suggest a mechanism for ubiquitylated substrate delivery important in proteolysis. *Mol. Biol. Cell.* 15:3357–65

Kiser GL, Gentzsch M, Kloser AK, Balzi E, Wolf DH, et al. 2001. Expression and degradation of the cystic fibrosis transmembrane conductance regulator in *Saccharomyces cerevisiae*. *Arch. Biochem. Biophys.* 390:195–205

Klausner RD, Sitia R. 1990. Protein degradation in the endoplasmic reticulum. *Cell* 62:611–14

Knop M, Finger A, Braun T, Hellmuth K, Wolf DH. 1996. Der1, a novel protein specifically required for endoplasmic reticulum degradation in yeast. *EMBO J.* 15:753–63

Koegl M, Hoppe T, Schlenker S, Ulrich HD, Mayer TU, Jentsch S. 1999. A novel ubiquitination factor, E4, is involved in multiubiquitin chain assembly. *Cell* 96:635–44

Lam YA, Lawson TG, Velayutham M, Zweier JL, Pickart CM. 2002. A proteasomal ATPase subunit recognizes the polyubiquitin degradation signal. *Nature* 416:763–67

Langer T. 2000. AAA proteases: cellular machines for degrading membrane proteins. *Trends Biochem. Sci.* 25:247–51

Lee RJ, Liu C, Harty C, McCracken AA, Romisch K, et al. 2004. The 19S cap of the 26S proteasome is sufficient to retro-translocate and deliver a soluble polypeptide for ER-associated degradation. *EMBO J.* 23:2206–15

Lilley BN, Ploegh HL. 2004. A membrane protein required for dislocation of misfolded proteins from the ER. *Nature* 429:834–40

Lomas DA, Parfrey H. 2004. Alpha1-antitrypsin deficiency: molecular pathophysiology. *Thorax* 59:529–35

Loo MA, Jensen TJ, Cui L, Hou Y, Chang XB, Riordan JR. 1998. Perturbation of Hsp90 interaction with nascent CFTR prevents its maturation and accelerates its degradation by the proteasome. *EMBO J.* 17:6879–87

McCracken AA, Brodsky JL. 1996. Assembly of ER-associated protein degradation in vitro: dependence on cytosol, calnexin, and ATP. *J. Cell. Biol.* 132:291–98

Medicherla B, Kostova Z, Schaefer A, Wolf DH. 2004. A genomic screen identifies Dsk2p and Rad23p as essential components of ER-associated degradation. *EMBO Rep.* 5:692–97

Meunier L, Usherwood YK, Chung KT, Hendershot LM. 2002. A subset of chaperones and folding enzymes form multiprotein complexes in endoplasmic reticulum to bind nascent proteins. *Mol. Biol. Cell.* 13:4456–69

Meusser B, Sommer T. 2004. Vpu-mediated degradation of CD4 reconstituted in yeast reveals mechanistic differences to cellular ER-associated protein degradation. *Mol. Cell* 14:247–58

Meyer HA, Grau H, Kraft R, Kostka S, Prehn S, et al. 2000a. Mammalian Sec61 is associated with Sec62 and Sec63. *J. Biol. Chem.* 275:14550–57

Meyer HH, Shorter JG, Seemann J, Pappin D, Warren G. 2000b. A complex of mammalian ufd1 and npl4 links the AAA-ATPase, p97, to ubiquitin and nuclear transport pathways. *EMBO J.* 19:2181–92

Miranda E, Römisch K, Lomas DA. 2004. Mutants of neuroserpin that cause dementia accumulate as polymers within the endoplasmic reticulum. *J. Biol. Chem.* 279:28283–91

Misaghi S, Pacold ME, Blom D, Ploegh HL, Korbel GA. 2004. Using a small molecule inhibitor of peptide: N-glycanase to probe its role in glycoprotein turnover. *Chem. Biol.* 11:1677–87

Molinari M, Calanca V, Galli C, Lucca P, Paganetti P. 2003. Role of EDEM in the release of misfolded glycoproteins from the calnexin cycle. *Science* 299:1397–400

This study shows that the proteasome 19S regulatory particle can, on its own, promote export of a misfolded protein from yeast microsomes.

This paper demonstrates that Derlin-1 (Der1p) interacts with export intermediates and p97 in the ER membrane.

Molinari M, Galli C, Piccaluga V, Pieren M, Paganetti P. 2002. Sequential assistance of molecular chaperones and transient formation of covalent complexes during protein degradation from the ER. *J. Cell. Biol.* 158:247–57

Nakatsukasa K, Okada S, Umebayashi K, Fukuda R, Nishikawa S, Endo T. 2004. Roles of O-mannosylation of aberrant proteins in reduction of the load for endoplasmic reticulum chaperones in yeast. *J. Biol. Chem.* 279:49762–72

Ng DT, Spear ED, Walter P. 2000. The unfolded protein response regulates multiple aspects of secretory and membrane protein biogenesis and endoplasmic reticulum quality control. *J. Cell. Biol.* 150:77–88

Nishikawa S, Fewell SW, Kato Y, Brodsky JL, Endo T. 2001. Molecular chaperones in the yeast endoplasmic reticulum maintain the solubility of proteins for retrotranslocation and degradation. *J. Cell Biol.* 153:1061–70

Nishikawa S, Hirata A, Nakano A. 1994. Inhibition of endoplasmic reticulum (ER)-to-Golgi transport induces relocalization of binding protein (BiP) within the ER to form the BiP bodies. *Mol. Biol. Cell.* 5:1129–43

Pilon M, Römisch K, Quach D, Schekman R. 1998. Sec61p serves multiple roles in secretory precursor binding and translocation into the endoplasmic reticulum membrane. *Mol. Biol. Cell.* 9:3455–73

Pilon M, Schekman R, Römisch K. 1997. Sec61p mediates export of a misfolded secretory protein from the endoplasmic reticulum to the cytosol for degradation. *EMBO J.* 16:4540–48

Plemper RK, Bohmler S, Bordallo J, Sommer T, Wolf DH. 1997. Mutant analysis links the translocon and BiP to retrograde protein transport for ER degradation. *Nature* 388:891–95

Plemper RK, Egner R, Kuchler K, Wolf DH. 1998. Endoplasmic reticulum degradation of a mutated ATP-binding cassette transporter Pdr5 proceeds in a concerted action of Sec61 and the proteasome. *J. Biol. Chem.* 273:32848–56

Prinz WA, Grzyb L, Veenhuis M, Kahana JA, Silver PA, Rapoport TA. 2000. Mutants affecting the structure of the cortical endoplasmic reticulum in *Saccharomyces cerevisiae*. *J. Cell. Biol.* 150:461–74

Rape M, Hoppe T, Gorr I, Kalocay M, Richly H, Jentsch S. 2001. Mobilization of processed, membrane-tethered SPT23 transcription factor by CDC48 (UFD1/NPL4), a ubiquitin-selective chaperone. *Cell* 107:667–77

Richly J, Rape M, Braun S, Rumpf S, Hoege C, Jentsch S. 2005. A series of ubiquitin binding factors connects CDC48/p97 to substrate multiubiquitylation and proteasomal targeting. *Cell* 120:73–84

Rivett AJ. 1998. Intracellular distribution of proteasomes. *Curr. Opin. Immunol.* 10:110–14

Römisch K. 2004. A cure for traffic jams: small molecule chaperones in the endoplasmic reticulum. *Traffic* 5:815–20

Römisch K, Ali BRS. 1997. Similar processes mediate glycopeptide export from the endoplasmic reticulum in mammalian cells and *Saccharomyces cerevisiae*. *Proc. Natl. Acad. Sci. USA* 94:6730–34

Rubin DM, Glickman MH, Larsen CN, Dhruvakamar S, Finley D. 1998. Active site mutants in the six regulatory particle ATPases reveal multiple roles for ATP in the proteasome. *EMBO J.* 17:4909–19

Schmitz A, Herrgen H, Winkeler A, Herzog V. 2000. Cholera toxin is exported from microsomes by the Sec61p complex. *J. Cell. Biol.* 148:1203–12

This paper shows that Cdc48p coordinates limited polyubiquitination of degradation substrates by sequential interactions with substrate, E4 enzyme, and polyubiquitin-binding proteins.

Schmitz A, Schneider A, Kummer MP, Herzog V. 2004. Endoplasmic reticulum-localized amyloid beta-peptide is degraded in the cytosol by two distinct degradation pathways. *Traffic* 5:89–101

Schubert U, Anton LC, Gibbs J, Norbury CC, Yewdell JW, Bennink JR. 2000. Rapid degradation of a large fraction of newly synthesized proteins by proteasomes. *Nature* 404:770–74

Sevilla LM, Comstock SS, Swier K, Miller J. 2004. Endoplasmic reticulum-associated degradation-induced dissociation of class II invariant chain complexes containing a glycosylation-deficient form of p41. *J. Immunol.* 173:2586–93

Shamu CE, Flierman D, Ploegh HL, Rapoport TA, Chau V. 2001. Polyubiquitination is required for US11-dependent movement of MHC class I heavy chain from endoplasmic reticulum into cytosol. *Mol. Biol. Cell.* 12:2546–55

Shamu CE, Story CM, Rapoport TA, Ploegh HL. 1999. The pathway of US11-dependent degradation of MHC class I heavy chains involves a ubiquitin-conjugated intermediate. *J. Cell. Biol.* 147:45–58

Suzuki T, Park H, Hollingsworth NM, Sternglanz R, Lennarz WJ. 2000. *PNG1*, a yeast gene encoding a highly conserved peptide:N-glycanase. *J. Cell. Biol.* 149:1039–52

Suzuki T, Seko A, Kitajima K, Inoue Y, Inoue S. 1994. Purification and enzymatic properties of peptide: N-glycanase from C3H mouse-derived L-929 fibroblast cells. Possible widespread occurrence of post-translational remodification of proteins by N-deglycosylation. *J. Biol. Chem.* 269:17611–18

Taxis C, Hitt R, Park SH, Deak PM, Kostova Z, Wolf DH. 2003. Use of modular substrates demonstrates mechanistic diversity and reveals differences in chaperone requirement of ERAD. *J. Biol. Chem.* 278:35903–13

Taxis C, Vogel F, Wolf DH. 2002. ER-Golgi traffic is a prerequisite for efficient ER degradation. *Mol. Biol. Cell.* 13:1806–18

Tector M, Hartl FU. 1999. An unstable transmembrane segment in the cystic fibrosis transmembrane conductance regulator. *EMBO J.* 18:6290–98

Tirosh B, Furman MH, Tortorella D, Ploegh HL. 2003. Protein unfolding is not a prerequisite for endoplasmic reticulum-to-cytosol dislocation. *J. Biol. Chem.* 278:6664–72

Tortorella D, Story CM, Huppa JB, Wiertz EJ, Jones TR, et al. 1998. Dislocation of type I membrane proteins from the ER to the cytosol is sensitive to changes in redox potential. *J. Cell. Biol.* 142:365–76. Erratum. 1999. *J. Cell. Biol.* 145(3):642

Trombetta ES, Parodi AJ. 2003. Quality control and protein folding in the secretory pathway. *Annu. Rev. Cell. Dev. Biol.* 19:649–76

Tsai B, Rodighiero C, Lencer WI, Rapoport TA. 2001. Protein disulfide isomerase acts as a redox-dependent chaperone to unfold cholera toxin. *Cell* 104:937–48

Tyedmers J, Lerner M, Bies C, Dudek J, Skowronek MH, et al. 2000. Homologs of the yeast Sec complex subunits Sec62p and Sec63p are abundant proteins in dog pancreas microsomes. *Proc. Natl. Acad. Sci. USA* 97:7214–19

Vashist S, Frank CG, Jakob CA, Ng DT. 2002. Two distinctly localized p-type ATPases collaborate to maintain organelle homeostasis required for glycoprotein processing and quality control. *Mol. Biol. Cell.* 13:3955–66

Vashist S, Kim W, Belden WJ, Spear ED, Barlowe C, Ng DT. 2001. Distinct retrieval and retention mechanisms are required for the quality control of endoplasmic reticulum protein folding. *J. Cell. Biol.* 155:355–68

Vashist S, Ng DT. 2004. Misfolded proteins are sorted by a sequential checkpoint mechanism of ER quality control. *J. Cell. Biol.* 165:41–52

Verma R, Oania R, Graumann J, Deshaies RJ. 2004. Multiubiquitin chain receptors define a layer of substrate selectivity in the ubiquitin proteasome system. *Cell* 118:99–110

This study demonstrates that topology of the lesion in a misfolded protein determines the mode of targeting to degradation.

Walter J, Urban J, Volkwein C, Sommer T. 2001. Sec61p-independent degradation of the tail-anchored ER membrane protein Ubc6p. *EMBO J.* 20:3124–31

Wang Q, Chang A. 2003. Substrate recognition in ER-associated degradation mediated by Eps1, a member of the protein disulfide isomerase family. *EMBO J.* 22:3792–802

Ward CL, Omura S, Kopito RR. 1995. Degradation of CFTR by the ubiquitin-proteasome pathway. *Cell* 83:121–27

Werner ED, Brodsky JL, McCracken AA. 1996. Proteasome-dependent endoplasmic reticulum-associated protein degradation: an unconventional route to a familiar fate. *Proc. Natl. Acad. Sci. USA* 93:13797–801

Wiertz EJ, Jones TR, Sun L, Bogyo M, Geuze HJ, Ploegh HL. 1996a. The human cytomegalovirus US11 gene product dislocates MHC class I heavy chains from the endoplasmic reticulum to the cytosol. *Cell* 84:769–79

Wiertz EJ, Tortorella D, Bogyo M, Yu J, Mothes W, et al. 1996b. Sec61-mediated transfer of a membrane protein from the endoplasmic reticulum to the proteasome for destruction. *Nature* 384:432–38

Wilkinson BM, Tyson JR, Reid PJ, Stirling CJ. 2000. Distinct domains within yeast Sec61p involved in post-translational translocation and protein dislocation. *J. Biol. Chem.* 275:521–29

Wilkinson BM, Tyson JR, Stirling CJ. 2001. Ssh1p determines the translocation and dislocation capacities of the yeast endoplasmic reticulum. *Dev. Cell* 1:401–9

Woodman PG. 2003. p97, a protein coping with multiple identities. *J. Cell. Sci.* 116:4283–90

Wright R, Basson M, D'Ari L, Rine J. 1988. Increased amounts of HMG-CoA reductase induce "karmellae": a proliferation of stacked membrane pairs surrounding the yeast nucleus. *J. Cell. Biol.* 107:101–14

Wu Y, Swulius MT, Moremen KW, Sifers RN. 2003. Elucidation of the molecular logic by which misfolded alpha 1-antitrypsin is preferentially selected for degradation. *Proc. Natl. Acad. Sci. USA* 100:8229–34

Ye Y, Meyer HH, Rapoport TA. 2001. The AAA ATPase Cdc48/p97 and its partners transport proteins from the ER to the cytosol. *Nature* 414:652–56

Ye Y, Meyer HH, Rapoport TA. 2003. Function of the p97-Ufd1-Npl4 complex in retrotranslocation from the ER to the cytosol: dual recognition of non-ubiquitinated polypeptide segments and polyubiquitin chains. *J. Cell. Biol.* 162:71–84

Ye Y, Shibata Y, Yun C, Ron D, Rapoport TA. 2004. A membrane protein complex mediates retro-translocation from the ER lumen into the cytosol. *Nature* 429:841–47

Yoshida Y, Chiba T, Tokunaga F, Kawasaki H, Iwai K, et al. 2002. E3 ubiquitin ligase that recognizes sugar chains. *Nature* 418:438–42

Yoshida Y, Tokunaga F, Chiba T, Iwai K, Tanaka K, Tai T. 2003. Fbs2 is a new member of the E3 ubiquitin ligase family that recognizes sugar chains. *J. Biol. Chem.* 278:43877–84

Youker RT, Walsh P, Beilharz T, Lithgow T, Brodsky JL. 2004. Distinct roles for the Hsp40 and Hsp90 molecular chaperones during cystic fibrosis transmembrane conductance regulator degradation in yeast. *Mol. Biol. Cell.* 15:4787–97

Zhang Y, Nijbroek G, Sullivan ML, McCracken AA, Watkins SC, et al. 2001. Hsp70 molecular chaperone facilitates endoplasmic reticulum-associated protein degradation of cystic fibrosis transmembrane conductance regulator in yeast. *Mol. Biol. Cell.* 12:1303–14

Zietkiewicz S, Krzewska J, Liberek K. 2004. Successive and synergistic action of the Hsp70 and Hsp100 chaperones in protein disaggregation. *J. Biol. Chem.* 279:44376–83

This study demonstrates that Cdc48p/Ufd1p/Npl4p are involved in release of ERAD substrates from the ER membrane.

This paper identifies Derlin1 as a p97 receptor in the ER membrane.

The Lymphatic Vasculature: Recent Progress and Paradigms

Guillermo Oliver[1] and Kari Alitalo[2]

[1]Department of Genetics and Tumor Cell Biology, St. Jude Children's Research Hospital, Memphis, Tennessee 38105; email: guillermo.oliver@stjude.org

[2]Molecular/Cancer Biology Laboratory and Ludwig Institute for Cancer Research, Biomedicum Helsinki, University of Helsinki, FIN-00014 Helsinki, Finland; email: Kari.Alitalo@Helsinki.FI

Annu. Rev. Cell Dev. Biol.
2005. 21:457–83

The *Annual Review of
Cell and Developmental
Biology* is online at
http://cellbio.annualreviews.org

doi: 10.1146/
annurev.cellbio.21.012704.132338

Copyright © 2005 by
Annual Reviews. All rights
reserved

1081-0706/05/1110-
0457$20.00

Key Words

lymphatic development, lymphangiogenesis, lymphedema, tumor metastasis

Abstract

The field of lymphatic research has been recently invigorated by the identification of genes and mechanisms that control various aspects of lymphatic development. We are beginning to understand how, starting from a subgroup of embryonic venous endothelial cells, the whole lymphatic system forms in a stepwise manner. The generation of genetically engineered mice with defects in different steps of the lymphangiogenic program has provided models that are increasing our understanding of the lymphatic system in health and disease. This knowledge, in turn, should lead to the development of better diagnostic methods and treatments of lymphatic disorders and tumor metastasis.

Contents

CHARACTERISTICS OF THE LYMPHATIC VASCULATURE

Evolutionary Aspects

In vertebrates, two specialized vascular systems have evolved for effective circulation: the blood vasculature, which delivers oxygen and nutrients and carries away waste products for detoxification and replenishment, and the lymphatic vasculature, which returns the interstitial protein-rich exudate fluid (lymph) to the bloodstream. Although G. Aselli first described the lymphatic vessels in 1627 (Aselli 1627), most of the original morphologic descriptions of the lymphatic system in different organisms were performed during the 19th and the beginning of the 20th centuries by anatomic criteria. The recent identification of molecular markers specific for the lymphatic vasculature now allows a better identification and characterization of the evolution of the lymphatic system among metazoans. As the complexity of the blood circulatory system increased among higher species in the phylogenetic tree, an additional mechanism to clear the tissues of substances not absorbed by the blood vascular system became necessary (Yoffey & Courtice 1970). In teleost fishes, two groups of lymphatic vessels can be identified; however, these vessels only occasionally have lymph hearts and lack valves. Lymph hearts, which are enlarged portions of the lymph vessels, have contractile walls and pulsatile function. The lymph vessels are flanked by valves to prevent the backflow of fluid and to force the lymph into the veins. Similar to fishes, amphibians and reptiles exhibit a pattern of longitudinal lymphatic vessels. Most primitive amphibians also have a pair of lymph hearts per body segment. Advanced amphibians such as frogs have fewer lymph hearts, and amniotes (i.e., reptiles, birds, and mammals) have a dramatically decreased number of these structures (Yoffey & Courtice 1970). Transient rudimentary lymph hearts are present in a few species of birds, but they are absent in most birds and mammals. Mammals and some birds that are mainly aquatic have lymph nodes (Jeltsch et al. 2003).

Anatomic Features

The lymphatic system is composed of a vascular network of thin-walled collecting vessels that drain lymph from the extracellular spaces within most organs into larger thicker-walled collecting trunks. At the base of the neck, lymph joins the venous circulation via the larger lymphatic collecting vessels (thoracic duct and the right lymphatic duct). In contrast to the blood vascular system, the lymphatic vasculature does not form a circulatory system; instead, the lymph is transported unidirectionally from tissues back to the blood circulation (Witte et al. 2001).

The smaller absorbing lymphatic capillaries are different from blood capillaries in that they are blind-ended structures that lack fenestrations, a continuous basal lamina, and pericytes (which cover ECs of blood capillaries). Lymphatic capillaries are composed of a single-cell layer of overlapping ECs that contains few intercellular tight junctions or adherens junctions. These ECs form loose intercellular valve-like junctions and exhibit large interendothelial pores and anchoring filaments that connect the vessels to the ECM. All of these features make the lymphatic capillaries highly permeable to large macromolecules, pathogens, and migrating cells. In contrast to the smaller collecting vessels, the larger lymphatic collecting vessels contain a continuous muscular layer, an adventitial layer, a basement membrane, and valves to prevent retrograde flow. These larger vessels drain into the thoracic duct and the right lymphatic duct, which in turn, discharge lymph into the large veins at the base of the neck.

The lymphatic system also includes lymphoid organs such as the lymph nodes, tonsils, Peyer's patches, spleen, and thymus, all of which play an important role in the immune responses. Lymphatic vessels are not normally present in avascular structures such as the epidermis, hair, nails, cartilage, and cornea, nor are they present in some vascularized organs such as the brain and retina.

The lymphatic system develops in parallel, but secondarily to, the blood vascular system through a process known as lymphangiogenesis. In birds and most mammals, the initiation of lymphangiogenesis is marked by the formation of primitive lymph sacs as the result of the fusion of lymphatic capillaries (jugular, subclavian, posterior, retroperitoneal, and the cisterna chyli). In humans, the first lymph sacs have been found in six- to seven-week-old embryos, nearly one month after the development of the first blood vessels (van der Putte 1975). In mice, the lymphatic system starts to develop at around embryonic day (E) 10.5, after the cardiovascular system is already functional (Oliver & Detmar 2002).

Lymphatic Function

Lymphatic vessels are essential for transporting tissue fluid, extravasated plasma proteins, and cells back to the blood circulation. The lymphatic system also contributes to the immune surveillance of the body and absorbs lipids from the intestinal tract. Lymphangiogenesis is an essential feature of tissue repair and inflammatory reactions in most organs, and congenital or acquired dysfunctions of the lymphatic system resulting in the formation of lymphedema (a disfiguring, disabling, and occasionally life-threatening disorder) are often associated with impaired immune function (Witte et al. 2001).

The lymphatic system is essential for the immune response to infectious agents. For example, during exposure to an inflammatory agent, afferent lymphatic vessels are the route through which, after antigen uptake, dendritic cells migrate to the lymph nodes and lymphoid organs. Recent studies have suggested that spindle cells and cells lining the irregular spaces in Kaposi sarcoma have a lymphatic endothelial origin (Hong et al. 2004, Jussila et al. 1998, Skobe et al. 1999, Wang et al. 2004).

To metastasize, a tumor cell must be able to pass through various barriers and finally attach, grow, and proliferate in the target tissue. Tumor cell dissemination can occur by invasion, by spreading via blood or lymphatic vessels, or by direct seeding of body surfaces or cavities. In general, tumor cells more easily penetrate lymphatic vessels than blood vessels, because the former have looser junctions between the ECs, and their basement membrane is discontinuous (Alitalo & Carmeliet 2002). Therefore, lymphatic vessels are a primary route for malignant tumor dissemination to the regional lymph nodes and possibly via the thoracic duct and the blood circulation, to distant organs. Moreover, tumor cell metastasis to lymph nodes represents a major

Pericyte: a cell that is similar to smooth-muscle cells and covers the outer surface of the endothelial cells of blood vessels

BEC: blood vascular endothelial cell

Endothelial cell (EC): an angioblast-derived cell that lines the inner surface of blood or lymphatic vessels

ECM: extracellular matrix

Lymphangiogenesis: the growth of lymphatic vessels

Lymphedema: failure of lymphatic drainage followed by the abnormal accumulation of interstitial fluid and swelling in the affected body part(s)

Vasculogenesis: multiplication and differentiation of angioblasts into endothelial cells

Angioblasts: the progenitor cells of the endothelial layer of blood cells

Angiogenesis: the process by which the primary capillary plexus grows and is remodeled and pruned into a capillary bed, arteries, and veins

SMC: smooth muscle cell

VEGF: vascular endothelial growth factor

VEGFR: vascular endothelial growth factor receptor

criterion for evaluating the prognosis of patients with cancer and for choosing additional chemotherapy, radiation therapy, or both after excision of primary tumors.

GENES AND MECHANISMS INVOLVED IN LYMPHATIC VASCULATURE DEVELOPMENT AND PATTERNING

Development of the Blood Vasculature Mediates that of the Lymphatic Vasculature

The blood vascular system is the first functional system formed during embryonic development. In vertebrates, the lymphatic vasculature appears once the blood vasculature has formed, a finding that supports the proposal that lymphatics have a blood vasculature origin (Sabin 1902, 1904). During embryonic vasculogenesis, angioblasts, the precursors of the ECs that will form the lining of the blood vessels, originate from mesoderm-derived progenitor cells called hemangioblasts (Risau & Flamme 1995). Hemangioblasts aggregate to form the primary blood islands; cells in the interior of the island differentiate into hematopoietic cells, and those in the periphery differentiate into angioblasts. Angioblasts also differentiate in the developing head mesenchyme. They then cluster and reorganize to form a network of primitive capillary-like plexus. The primary capillary plexus is remodeled by a process called angiogenesis, whereby new arteries and veins form by sprouting or splitting from preexisting capillaries or from postcapillary venules (Risau 1997). Angiogenesis by sprouting consists of EC disruption of the ECM, allowing EC migration and proliferation into the perivascular space (Folkman & Shing 1992). The newly formed blood vessels are stabilized by pericytes, SMCs, and a basal lamina. Circulating EC progenitors have been identified in bone marrow and in peripheral blood of adult tissues, and some data suggest that these cells can participate in postnatal formation of new blood vessels (Asahara et al. 1997, Shi et al. 1998).

The process of vasculogenesis is restricted to early stages of embryonic development, whereas postnatal angiogenesis occurs during physiologic and pathologic neovascularization. In adults, physiologic angiogenesis occurs during the female reproductive cycle and in wound healing (Ferrara 2001). Pathologic angiogenesis, on the other hand, is involved in a variety of disorders such as tumor growth and metastasis, diabetic retinopathy, rheumatoid arthritis, cardiovascular diseases, and psoriasis (Folkman 1995).

A number of signaling pathways crucial for vasculogenesis and angiogenesis have been identified. Members of the VEGF family, angiopoietin/Tie, Notch, and ephrin/Eph pathways play major roles in these processes. The different members of the VEGF family interact with a family of three closely related tyrosine kinase receptors: VEGFR1–3. VEGF-A promotes vascular EC proliferation and angioblast differentiation (Carmeliet et al. 1996, Ferrara 1999, Ferrara et al. 1996, Shalaby et al. 1995). Removal of both *Vegfa* alleles in mice resulted in an almost complete lack of vasculature (Carmeliet et al. 1996, Ferrara et al. 1996). Furthermore, removal of a single *Vegfa* allele leads to arrested vasculogenesis and embryonic lethality (Carmeliet et al. 1996, Ferrara et al. 1996), a finding that indicates that *Vegfa* expression is required for vascular patterning and assembly. Similar to *Vegfa* gene ablation, ablation of the gene that encodes the *Vegf* receptor Vegfr-2 in mice results in the absence of a functional vasculature as a consequence of the arrest in the differentiation of hematopoietic cells and ECs in embryos. VEGFR-2 is required for the migration of the mesodermal precursor cells to the extraembryonic region (Shalaby et al. 1995). In the embryo, FGF-1 may be involved in the induction of angioblasts from the mesoderm (Cox & Poole 2000, Flamme & Risau 1992).

In addition to the three VEGF receptors, the neuropilin family of coreceptors modulates binding to the main receptors and probably contributes to VEGF-mediated vasculogenic events, as demonstrated by the finding that embryos null for both neuropilin genes display arrested vasculogenesis, similar to that observed in *Vegfa* and *Vegfr2* knockout mice (Takashima et al. 2002).

The four angiopoietins (Ang1–Ang4) and their two-Tie (Tie1 and Tie2) family of tyrosine kinase receptors are also key players during vasculogenesis. The Tie receptors are expressed in the vascular endothelium and in the hematopoietic lineage (Iwama et al. 1993, Partanen et al. 1992, Sato et al. 1993). Although mice lacking Ang1, Tie1, or Tie-2 develop a rather normal primary vasculature, the process of remodeling is defective (Dumont et al. 1994, Puri et al. 1995). Functional inactivation of *Ang2* (*Agpt2*) indicated that this molecule is required for postnatal blood vascular remodeling and proper development of the lymphatic vasculature (Gale et al. 2002). Ang2 also activates the EC-specific Tie2 receptor on some cells and blocks it on others. Besides assuming a functional role during embryonic lymphangiogenesis (see below), Ang2 regulates vascular remodeling by participating in vessel sprouting and regression (Gale et al. 2002). Members of the VEGF and Ang families are partners during development of the blood vasculature (Suri et al. 1996). A schematic view of blood and lymphatic capillaries and the VEGF and Ang growth factors is shown in **Figure 1**.

Members of the Notch family are required to mediate the choice of fate between arterial and venous ECs (Lawson et al. 2001, Zhong et al. 2001). Notch activity promotes arterial cell fate, at least partially, via the activity of *gridlock*, a transcriptional repressor in zebrafish that is a member of the Hes/Hey gene family. Gridlock negatively regulates venous cell identity (Lawson et al. 2001, Zhong et al. 2001). The ephrins and their Eph receptor tyrosine kinases were initially identified in the nervous system, but recently, ephrinB2 and its EphB4 receptor also have been implicated in determining arterial and venous cell identities, probably via mediating a repulsive signal that separates arterial and venous ECs (Adams et al. 1999, Wang et al. 1998).

The current model indicates that venous identity is the default status of ECs, and that upon VEGF and Notch signaling events, arterial fate is gained. This results in the separation of arterial and venous ECs via ephrinB2/EphB4 signaling (Adams et al. 1999, Harvey & Oliver 2004).

Lymphatic Vasculature Development

Historically, Sabin proposed the most widely accepted model of lymphatic development: She suggested that early in development, isolated primitive lymph sacs originate from ECs that bud from the veins (Sabin 1902, 1904). From these primary lymph sacs, the peripheral lymphatic system spreads by endothelial sprouting into the surrounding tissues and organs, where local capillaries form (Sabin 1902, 1904). In 1910, Huntington and McClure proposed an alternative model, suggesting that the initial lymph sacs arise in mesenchyme, independent of the veins, and subsequently establish venous connections (Huntington & McClure 1910).

Recently, a number of detailed review articles describing some of the recent advances in the lymphatic field have been published (Harvey & Oliver 2004, Jeltsch et al. 2003, Oliver 2004, Saharinen et al. 2004, Scavelli et al. 2004, Stacker et al. 2004); therefore, here we focus only on a detailed description of those genes whose function is crucial in the control of different steps of embryonic and adult lymphangiogenesis. Some of the genes whose functional activity has been described in the lymphatic vasculature are listed in **Table 1**.

Lymphatic endothelial cell–fate specification. Competence is the capacity of a cell to respond to an initial inducing signal. The presence of the blood vasculature system alone is not sufficient to initiate lymphatic

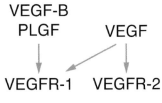

SIGNALING MECHANISMS

Blood vessels

VEGF-B
PLGF   VEGF

VEGFR-1   VEGFR-2

Dysfunction:
tumor angiogenesis
tissue ischemia

Lymphatic vessels

VEGF-C, VEGF-D

VEGFR-3

Dysfunction:
lymphatic metastasis
lymphedema

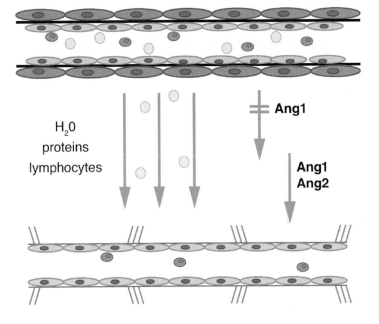

H$_2$0
proteins
lymphocytes

Ang1

Ang1
Ang2

Figure 1

Functions and growth factor regulation of the lymphatic vessels. Fluid and macromolecules extravasated from blood vessels are taken up by lymphatic vessels. In most blood vessels, the endothelial cells (*yellow*) are surrounded by a continuous basement membrane (*black line*) and pericytes/smooth muscle cells (*orange*). Blood vessel permeability is regulated via VEGF signaling, and Ang1 can block the VEGF-induced increase of vessel permeability. The protein-rich fluid from the interstitial tissues enters the lymphatic capillaries (*bottom*). Lymphatic vessels are characterized by an incomplete basement membrane; anchoring filaments; and, in the collecting lymphatic vessels, by luminal valves, which prevent lymph backflow, and SMCs, which have intrinsic contractility for lymph propulsion. VEGFR-3 is expressed on the surface of lymphatic endothelial cells; VEGF-C as well as VEGF-D are important mediators of lymphatic endothelial cell growth via this receptor. Tumor angiogenesis can be inhibited by blocking VEGF, and tissue ischemia may be relieved by an increase of VEGF growth factor activity. In an analogous fashion, inhibition or stimulation of VEGF-C and VEGF-D activity may be beneficial in lymphatic metastasis and lymphedema, respectively. Figure adapted from Karkkainen et al. 2001a.

vasculature development; some additional gene product(s) must begin to be expressed by venous ECs to become progressively competent to respond to a local lymphatic-inducing signal. Although the molecular factor(s) that regulates this initial stage of lymphatic competence remains unknown, the expression of the lymphatic vessel endothelial hyaluronan receptor 1 (Lyve1) (Banerji et al. 1999) by a few of the ECs that line the anterior

Table 1   Genes that mediate lymphatic vasculature formation. Except when indicated, all names and symbols refer to the mouse gene

Gene name	Gene symbol	Expression profile	Functional role
Angiopoietin-2	*Agpt2*	SMCs (smooth muscle cells) of large arteries, large veins, and venules	Postnatal angiogenesis and lymphatic patterning. $Agpt2^{-/-}$ mice died within 2 weeks of birth and exhibited chylous ascites and peripheral edema.
ephrin B2	*Efnb2*	Arterial ECs (endothelial cells), SMCs, and LECs (lymphatic endothelial cells) in collecting vessels	Remodeling of primary lymphatic plexus. $ephrinB2^{\Delta V/\Delta V}$ mice exhibit hyperplasia of collecting lymphatic vessels that lack luminal valves, and have chylothorax.
Forkhead-box-C2	*Foxc2*	LEC progenitors in the anterior cardinal vein and surrounding mesenchyme, jugular lymph sacs, lymphatic collectors, and capillaries during development	Lymphatic vasculature patterning, formation of lymphatic valves. *Foxc2*-null embryos exhibit mispatterning of lymphatics, increased pericyte investment, agenesis of valves, and lymphatic dysfuntion. Human *FOXC2* mutations cause lymphedema-distichasis.
Integrin α9β1	*Integrin α9β1*	Epithelial cells, smooth and skeletal muscle, neutrophils, and a subset of endothelial cells	*Integrin α9β1*-null mice die at birth. They exhibit chylothorax and structurally defective thoracic duct and peripheral lymphatic vessels. VEGF-C and VEGF-D are ligands for Integrin α9β1.
Neuropilin-2	*Nrp2*	Anterior cardinal vein, budding LECs and lymphatic vessels, adult LECs	Selectively required for the formation of small lymphatic vessels and capillaries. *Nrp2*-null mice exhibit transient absence or reduction of small lymphatic vessels and capillaries.
Podoplanin	*T1a*	Budding embryonic and adult LECs	LEC adhesion and migration, formation of connecting lymphatics between superficial and deep lymphatic plexuses. *T1a*-null mice have congenital lymphedema and dilation of lymphatic vessels.
Prospero-related homeobox 1	*Prox1*	Embryonic and adult LECs	Lymphatic endothelial cell type specification. *Prox1*-null embryos are lethal and are devoid of lymphatic vasculature.
SRC homology 2 domain-containing leukocyte protein of 76 kDa; spleen tyrosine kinase Syk	*Lcp2 and Syk*	Circulating hematopoietic cells	*Syk*-null mouse embryos exhibit abnormal blood-lymphatic connections, hemorrhage, and arterio-venous shunting
Vascular endothelial growth factor receptor-3	*Vegfr3*	Embryonic and adult LECs, embryonic blood vascular endothelial cells	Blood vessel development. *Vegfr3*-null embryos die of cardiovascular defects. Mutations of human VEGFR3 result in Milroy disease.
Vascular endothelial growth factor-C	*Vegfc*	Mesenchymal cells around embryonic veins	Budding and survival of embryonic LEC progenitors. *Vegfc*-mutant embryos lack lymphatic vasculature.

LEC: lymphatic endothelial cell

cardinal vein of mice at E9.0–9.5 could be considered the first morphologic indication that venous ECs are already competent to respond to a lymphatic-inducing signal (Oliver 2004).

Following the initial stage of lymphatic competence, cells acquire the ability to give rise to a particular cell type. In mice, a few hours after Lyve1 expression is initiated in the anterior cardinal vein (around E9.5), expression of the transcription factor prospero-related homeobox 1 (Prox1) is also observed in that tissue (**Figure 2**, Wigle & Oliver 1999). Interestingly, expression of Prox1 is restricted to a subpopulation of venous ECs located on one side of the vein (Wigle & Oliver 1999). As a first step in the process leading to LEC specification, the Prox1-expressing LEC progenitors start to bud from the veins in a polarized manner, a finding that suggests the presence of some type of guidance mechanism (Wigle & Oliver 1999). Subsequently, these budding LEC progenitors proliferate and migrate to form the embryonic lymph sacs and lymphatic vascular network (**Figure 2**). Localization of Prox1 expression in the venous ECs provides strong support for Sabin's original model of a venous origin of the lymphatic vasculature (Harvey & Oliver 2004).

One of the first indications of the crucial role played by Prox1 in the development of the lymphatic vasculature was provided by the finding that $Prox1^{-/-}$ mouse embryos display multiple phenotypic abnormalities, including the complete absence of lymphatic vessels, and die around midgestation (Wigle & Oliver 1999). Furthermore, the absence of lymphatic vasculature in $Prox1^{-/-}$ embryos was not the consequence of an arrest in LEC budding, but rather of a failure in lymphatic cell–type specification (Wigle et al. 2002). The detailed characterization of the $Prox1^{-/-}$ embryos demonstrated that the endothelial cells that budded from the anterior cardinal vein at E11.0–11.5 did not express LEC markers; instead, they expressed blood vascular markers (Wigle et al. 2002). Based on these findings, it was proposed that Prox1 activity is required

and sufficient to confer a LEC phenotype (or bias) on LEC progenitors located the embryonic veins that would otherwise remain as blood vascular ECs (Oliver 2004, Wigle et al. 2002). In brief, embryonic venous ECs are the default ground state, and following expression of Prox1, these LEC progenitors adopt a lymphatic cell–type phenotype (Oliver & Detmar 2002, Wigle et al. 2002). At the same time, the isolation and cell culture studies of BECs and LECs showed that overexpression of Prox1 is sufficient to reprogram cultured BECs into LECs (Hong et al. 2002, Petrova et al. 2002). Prox1 repressed about 40% of the BEC-specific markers and induced the expression of about 20% of the LEC-specific genes. Together, these data indicate that $Prox1$ is a master control gene whose activity is necessary and sufficient to initiate the process of lymphatic cell–type specification in a subpopulation of competent venous ECs. During this specification stage, which can be morphologically identified by the budding of LEC progenitors from the veins, the expression of other lymphatic markers (e.g., neuropilin-2, podoplanin, and Vegfr3) (see **Table 1**) is initiated in a stepwise manner in the budding Prox1-expressing LEC progenitors, and expression of some blood vascular markers (e.g., CD34 and laminin) progressively decreases (Oliver 2004).

Lymphatic endothelial cell budding. The expression of some receptor-ligand signaling system appears required to promote and guide the polarized budding of LECs toward the receptor-ligand source, thereby ensuring the precise location of the primary lymph sacs (Oliver 2004). VEGFR-3 (also known as Fms-like tyrosine kinase 4, Flt4) (**Table 1**) was one of the first markers associated with lymphatic differentiation (Kaipainen et al. 1995). VEGFR-3 was soon shown to bind VEGF-C and VEGF-D (Joukov et al. 1996, 1997; Achen et al. 1998). Early during mouse development, $Vegfr-3$ is widely expressed by the lymphatic endothelium and the blood vascular endothelium, including that of the anterior

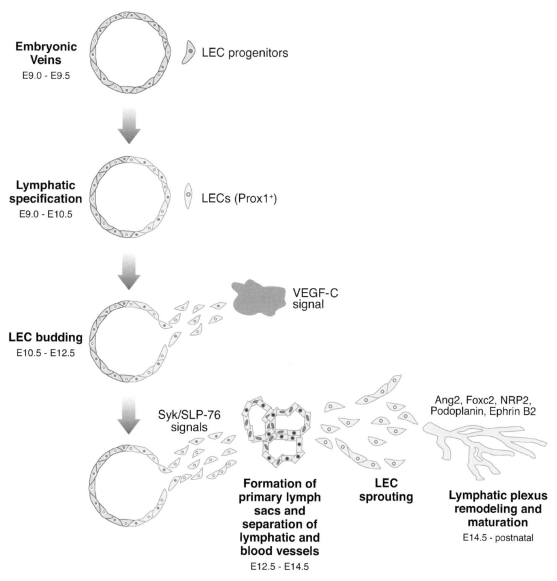

Embryonic Veins
E9.0 - E9.5

LEC progenitors

Lymphatic specification
E9.0 - E10.5

LECs (Prox1⁺)

LEC budding
E10.5 - E12.5

VEGF-C signal

Syk/SLP-76 signals

Ang2, Foxc2, NRP2, Podoplanin, Ephrin B2

Formation of primary lymph sacs and separation of lymphatic and blood vessels
E12.5 - E14.5

LEC sprouting

Lymphatic plexus remodeling and maturation
E14.5 - postnatal

Figure 2

Stepwise model for the development of the mammalian lymphatic vasculature. Early during mouse embryonic development (E9.0–9.5), Prox1 begins to be expressed in LEC progenitors located in one side of the anterior cardinal veins. Following this expression, those progenitors acquire a LEC phenotype (LEC bias) and bud in a polarized manner. This budding, which is dependent on a Vegf-c signal in the surrounding mesenchyme, gives rise to the primary lymph sacs. Sometime during this process, the lymphatic and venous systems become separated in a process that involves the Syk/SLP-76 signaling pathway, and the sprouting of LECs from those sacs is initiated. Finally, lymphatic vessel maturation and remodeling occurs in a stepwise manner, leading to the formation of the complete lymphatic network.

cardinal vein (Dumont et al. 1998, Kaipainen et al. 1995). This expression is subsequently downregulated in blood vasculature and becomes restricted mostly to the lymphatic endothelium during late development and adulthood (Kaipainen et al. 1995, Wigle et al. 2002).

Contact with SMCs downregulates *Vegfr3* in ECs (Veikkola et al. 2003). This finding suggests that *Vegfr3* signaling is important in nascent blood vessels and that *Vegfr3* becomes redundant as the vessels mature. Indeed, *Vegfr-3* gene–targeted mice exhibit a dramatic blood vascular phenotype, with embryonic lethality at E9.5 due to defective remodeling of the primary vascular plexus and disturbed hematopoiesis (Dumont et al. 1998, Hamada et al. 2000). Transgenic mice overexpressing a soluble VEGFR-3-Ig fusion protein in the skin lack dermal lymphatic vessels and have hypoplastic deeper lymphatic vessels, though the phenotype is less pronounced as the mice age (Mäkinen et al. 2001a). The early lethality of *Vegfr-3*-null embryos has hampered the analysis of the role of this growth factor receptor in lymphatic system development. Inactivation of *Vegfr-3* causes cardiovascular failure and death of the embryo before the emergence of lymphatic vessels (Dumont et al. 1998). However, the identification of missense mutations in *VEGFR3* in patients with hereditary lymphedema (Karkkainen et al. 2000) has provided support for an important role of this gene in lymphatic development.

The VEGFR-3 ligand VEGF-C (**Table 1**) was the first lymphangiogenic growth factor identified (Joukov et al. 1996, Kukk et al. 1996), and studies of VEGF-C have provided further support for the importance of the VEGFR-3 signaling pathway during lymphatic development. VEGF-C is produced as a precursor protein, which is activated by intracellular secretory proprotein convertases (Joukov et al. 1997, Siegfried et al. 2003). The secreted VEGF-C 31/29 kD subunits are bound by disulphide bonds (**Figure 3**). This form binds VEGFR-3, but the factor is fur-

ther proteolyzed in the extracellular environment by plasmin and other proteases to generate a 21 kD nondisulfide-linked homodimeric protein with high affinity for both VEGFR-2 and VEGFR-3. The mature form of VEGF-C induces mitogenesis, migration, and survival of ECs (Mäkinen et al. 2001b). VEGF-D is similarly processed at its N-terminal and C-terminal ends. Fully processed human VEGF-D activates VEGFR-2, and VEGFR-3 is mitogenic for ECs and angiogenic (Stacker et al. 1999). In contrast, mouse Vegf-d binds only Vegfr-3 (Baldwin et al. 2001).

After embryonic development, Vegf-c expression decreases in most tissues, remaining high in the lymph nodes (Lymboussaki et al. 1999). Vegf-d is also expressed most abundantly in the lung and skin during embryogenesis, and downregulated thereafter (Farnebo et al. 1999).

VEGF-C has been identified as a potent inducer of lymphatic sprouting (Karkkainen et al. 2004, Oh et al. 1997, Saaristo et al. 2002) and is expressed mainly in the mesenchyme of areas where embryonic lymphatics are developing (Karkkainen et al. 2004, Kukk et al. 1996). Overexpression of VEGF-C in the skin of transgenic mice results in hyperplasia of cutaneous lymphatic vessels (Jeltsch et al. 1997), and overexpression of VEGF-C through the use of a recombinant adenovirus promotes lymphangiogenesis in adult mouse skin (Enholm et al. 2001). VEGF-C gene transfer to the skin of mice with lymphedema induces regeneration of the network of cutaneous lymphatic vessels (Karkkainen et al. 2001b). Whereas VEGF-C is predominantly lymphangiogenic, the fully processed form of VEGF-D is able to induce angiogenesis in addition to lymphangiogenesis upon gene transduction into rabbit hindlimb muscle (Byzova et al. 2002, Rissanen et al. 2003).

The recent inactivation of the *Vegfc* gene in mice has provided additional information regarding its role during embryonic lymphangiogenesis (Karkkainen et al. 2004). The

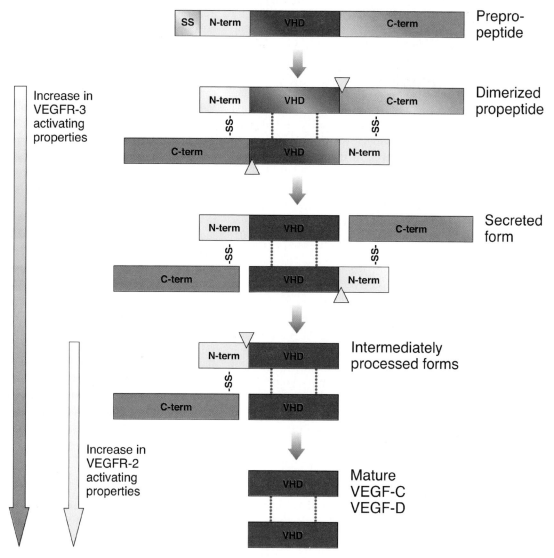

Figure 3

Stepwise proteolytic processing regulates VEGF-C and VEGF-D activity. Both factors are synthesized as precursor proteins. The C-terminal cleavage occurs in all producing cells and activates binding to VEGFR-3 (lymphangiogenic receptor), which is the only signal needed for lymphangiogenic activity in vivo (Veikkola et al. 2001). After the N-terminus is fully processed, the short form binds to and activates VEGFR-2 (angiogenic receptor) as well, resulting in angiogenic activity (Cao et al. 1998). Thus, both angiogenic and lymphangiogenic signals can be generated from a single molecule depending on the degree of processing and the relative expression of the receptors. VHD, VEGF homology domain.

analysis of the generated mutant embryos demonstrated that Vegf-c activity is essential during the earliest events in lymphatic development; its functional inactivation results in embryonic lethality. In these mutant embryos, the lymphatic vasculature does not form; therefore, lymph fails to drain and the animal dies. However, blood vessel development is not affected (Karkkainen et al. 2004). Vegf-c activity is essential for promoting the budding and proliferation of the Prox1-expressing LECs located in the embryonic veins (**Figure 2**) (Karkkainen et al. 2004). In *Vegfc*-null embryos, budding of the Prox1-expressing venous LEC progenitors is arrested, and those cells remain abnormally confined to the wall of the cardinal veins (Karkkainen et al. 2004). However, in the absence of Vegf-c, lymphatic specification still takes place, as indicated by the expression of Prox1 by the venous LEC progenitors (Karkkainen et al. 2004). These studies provide further support to Sabin's hypothesis (Sabin 1902, 1904).

Interestingly, *Vegfc* haploinsufficiency results in delayed lymphatic vascular development and lymphedema. In $Vegfc^{+/-}$ pups, accumulation of chyle (milky fluid) in the peritoneal cavity was observed, which suggests defects in intestinal lymphatic vessel function (Karkkainen et al. 2004). Although in most organs of the $Vegfc^{+/-}$ mice the lymphatic vessels eventually grow postnatally, the cutaneous lymphatic vessels remain hypoplastic, and adult *Vegfc*-heterozygous mice exhibit swelling of the paws. These results indicate that Vegf-c is essential to promote and/or maintain the budding of the LEC progenitors from the veins and that *Vegfc* expression levels are crucial for normal lymphatic vascular development (Karkkainen et al. 2004). Sprouting of endothelial cells committed to the lymphatic endothelial lineage in *Vegfc* gene–targeted mice can be rescued by application of recombinant VEGF-C and to a lesser degree by VEGF-D, but not by VEGF-A (Karkkainen et al. 2004). This finding indicates that the lymphatic vessel development depends on VEGFR-3 and VEGF-C in a nonredundant fashion.

Finally, by using morpholino antisense oligonucleotides to reduce Vegfc levels in zebrafish, Ober et al. (2004) recently demonstrated that in fish, *Vegfc* activity is required for vascular development and endoderm morphogenesis.

Separation of the blood and lymphatic vasculature systems. Although derived from the embryonic veins, the lymphatic and blood networks normally remain separated at all but one connection point: the junction between the thoracic duct and the left subclavian vein. While the lymphatic vessels and capillaries start to spread along the embryo, they must be separated from the veins. Two of the key players controlling this critical step are the adaptor protein Slp76 (SRC homology 2-domain–containing leukocyte protein of 76 kDa) and the tyrosine kinase Syk (**Figure 2**); these two molecules are expressed mainly by circulating hematopoietic cells (Abtahian et al. 2003) (**Table 1**).

Mice in which any of these molecules have been functionally inactivated exhibit abnormal blood-lymphatic connections, embryonic hemorrhage, and arterio-venous shunting (Abtahian et al. 2003). In addition, blood is present in the mesenteric lymph vessels of these mutant mice, a defect that can be phenocopied if bone marrow from *Syk-* or *Slp76*-deficient mice is transplanted into lethally irradiated wild-type mice (Abtahian et al. 2003). The mechanisms by which these signaling pathways mediate the separation of the blood and lymphatic vascular systems remain unknown.

Maturation and remodeling of the lymphatic vasculature. By E14.5, the lymphatic vasculature has spread throughout most of the developing mouse embryo by budding and sprouting from the primary lymph sacs; however, these growing vessels have not yet become fully mature or terminally differentiated (Oliver 2004). This finds support in the fact that the terminal differentiation and maturation of the lymphatic vasculature appear to occur in a stepwise manner, i.e., the forming lymphatic vessels and capillaries start expressing other gene products as embryonic development progresses (Harvey & Oliver 2004, Oliver & Detmar 2002). Therefore, the detection and levels of expression of some gene products on the differentiating lymphatics will depend on the developmental stage of the embryo, the type of tissue being analyzed, or both. Only near the time of birth do lymphatic vessels express the complete profile of markers found in mature, terminally differentiated lymphatics (Oliver 2004, Saharinen et al. 2004). Therefore, the function of at least some of these gene products is most likely to affect later aspects of lymphatic development such as terminal differentiation/maturation and remodeling (**Figure 2**, **Table 1**).

Vascular remodeling is a necessary, well-characterized step during the formation of the blood vasculature (Risau 1997). A recent report (Makinen et al. 2005) has provided strong support to the idea that this process is also required during the maturation of the lymphatic system. Below we describe the genes and mechanisms that have been identified as controls for different aspects of lymphatic differentiation/maturation and remodeling.

Angiopoietin-2. Angiopoietin-2, or Ang2 (Agpt2), is a ligand for the endothelial Tie2 receptor tyrosine kinase (**Table 1**) and has a dual function in postnatal angiogenesis and developmental lymphangiogenesis (Veikkola & Alitalo 2002). Ang2 expression is restricted to sites of physiologic angiogenesis in which vascular remodeling occurs (Maisonpierre et al. 1997). Transgenic mice that overexpress Ang2 under the *Tie2* promoter show disrupted blood vessel formation during embryogenesis, a phenotype similar to that of *Ang1*- or *Tie2*-knockout embryos (Maisonpierre et al. 1997, Suri et al. 1996). Functional inactivation of Ang2 has highlighted a previously unknown role for this molecule during embryonic lymphangiogenesis (Gale et al. 2002). *Ang2*-null mice die within two weeks of birth, and these animals exhibit chylous ascites, subcutaneous edema, a disorganized and leaky lymphatic vasculature, and defects in the postnatal vascular remodeling in the retina—all conditions that indicate lymphatic dysfunction (**Table 1**, Gale et al. 2002). These de-

fects indicate that normal lymphatic development and aspects of postnatal blood vascular remodeling require Ang2 activity (**Figure 2**, Gale et al. 2002). Interestingly, the lymphatic defects of *Ang2*-null mice were rescued by *Ang1* (Gale et al. 2002). Ang1, which induces both Tie1 and Tie2 phosphorylation, also promotes LEC sprouting and proliferation in adult tissues (Saharinen et al. 2005, Tammela et al. 2005). These data suggest that both Ang1 and Ang2 provide agonistic signals via the Tie receptors in the development of the lymphatic network.

Neuropilin-2. VEGF-C also binds to the non-receptor tyrosine kinase neuropilin-2 (Nrp2), which is involved in axon guidance as a semaphorin receptor (Chen et al. 1997, Karkkainen et al. 2001b). Nrp2 appears to play a selective role in the development of small lymphatic vessels (Yuan et al. 2002). Functional inactivation of this gene product reveal that homozygous *Nrp2*-mutant mice exhibit a transient absence or severe reduction of small lymphatic vessels and capillaries during development (**Table 1**, Yuan et al. 2002). The lymphatic vessels of *Nrp2*-mutant mice regenerate postnatally; however, the arteries, veins, and larger collecting lymphatic vessels develop normally. This finding suggests a differential control of the formation of large- and small-caliber lymphatic networks and a selective requirement of Nrp2 for the formation of small lymphatic vessels and capillaries (Yuan et al. 2002).

Podoplanin. T1a/podoplanin, a transmembrane protein, is predominantly expressed by the lymphatic endothelium, and in mouse it is detected as early as E11.0 in the budding LECs (**Table 1**, Breiteneder-Geleff et al. 1999, Oliver 2004, Schacht et al. 2003). At least in some genetic backgrounds, the *podoplanin*-null mice die at birth due to respiratory failure. These animals display defects in lymphatic vessel structure and function as well as obvious dilations of the cutaneous and submucosal intestinal lymphatic vasculature

Chylous ascites: accumulation of lymphatic fluid in the abdominal cavity caused by defects of lymphatic vessels

that lead to lymphedema (Schacht et al. 2003). Podoplanin may contribute to LEC adhesion and migration and to the formation of connecting lymphatics between superficial and deep lymphatic plexuses (**Figure 2**, Schacht et al. 2003).

Foxc2. In the lymphatic syndrome lymphedema-distichiasis, mutations in the forkhead transcription factor FOXC2 have been identified as responsible for the malfunction of the lymphatic vessels (Fang et al. 2000). In addition to its expression in the paraxial mesoderm and somites, *Foxc2* is also normally expressed in LEC progenitors located in the embryonic cardinal veins and their surrounding mesenchyme as well as in jugular lymph sacs, lymphatic collectors and capillaries, podocytes, developing eyelids, and other tissues associated with abnormalities in lymphedema-distichiasis syndrome (**Table 1**, Dagenais et al. 2004, Petrova et al. 2004).

As previously discussed, lymphatic capillaries lack associated mural cells, and collecting lymphatic vessels have valves, which prevent lymph backflow. Petrova et al. (2004) recently reported that *Foxc2*-null mice exhibit abnormal lymphatic vascular patterning, increased pericyte investment of lymphatic vessels, lymphatic dysfunction, and agenesis of valves. Loss of *Foxc2* also results in increased levels of expression of *Pdgfb*, *endoglin*, and *collagen IV* in the lymphatic vessels (Petrova et al. 2004). Furthermore, *Foxc2*-heterozygous mice exhibit an abnormally large proportion of skin lymphatic vessels associated with smooth muscle actin–positive periendothelial cells (Petrova et al. 2004).

Foxc2 activity most likely is required for the establishment of a pericyte- and SMC-free lymphatic capillary network through the specific repression of *Pdgfb* expression by the lymphatic vessels (Petrova et al. 2004). In addition, partial recapitulation of a default blood vascular endothelial phenotype or a shift toward a mature collecting lymphatic vessel phenotype could cause the abnormalities identified in the basement membrane, *Pdgfb* expression, and pericyte and SMC coverage exhibited by the *Foxc2*-null lymphatics; these characteristics are typical of blood vessels (Petrova et al. 2004). Foxc2 activity is also essential for morphogenesis of the lymphatic valves, and Foxc2 cooperates with Vegfr-3 during lymphatic vascular patterning (Petrova et al. 2004).

EphrinB2. The transmembrane ligand ephrinB2 and its Eph receptor tyrosine kinases are important regulators of embryonic blood vascular morphogenesis. EphrinB2 is expressed in arterial ECs and SMCs, and its activity is required for the remodeling of the primary blood capillary plexus and the formation of major vessels (Adams et al. 1999, Wang et al. 1998).

EphrinB2 is also expressed in the LECs of collecting lymphatic vessels (**Table 1**, Makinen et al. 2005). Homozygous mutant mice generated by expressing a mutated form of ephrinB2 (*ephrinB2$^{\Delta V/\Delta V}$*) lacking the carboxy-terminal interaction site for PDZ domain–containing proteins do not need ephrinB2 in embryonic vascular remodeling. Interestingly, in those mutant mice, the early development of the dermal lymphatic vessels is not affected; however, these mice exhibit major lymphatic defects, including hyperplasia of the collecting lymphatic vessels, lack of luminal valve formation, failure to remodel their primary lymphatic capillary plexus, and chylothorax (Makinen et al. 2005). The dermal lymphatic vasculature of these mutant mice fails to mature and resembles a primitive capillary network (Makinen et al. 2005).

Normally, the ephrinB2- and ephrinB4-expressing dermal lymphatic vasculature develops first as a primary capillary plexus, and the subsequent remodeling involves sprouting of new lymphatic capillaries from the initial plexus (Makinen et al. 2005). The ephrinB2- and ephrinB4-expressing vessels remaining in the deeper dermal layers become transformed into smaller precollecting lymphatic vessels that contain valves, acquire SMC coverage, and downregulate Lyve-1 expression. Instead, the newly formed, Lyve-1-positive lymphatic capillaries express ephrinB4 but not ephrinB2, suggesting a molecular distinction between collecting vessels and capillary lymphatic vessels (Makinen et al. 2005). These data not only identify ephrinB2 as an important player during the remodeling of the primary lymphatic capillary plexus but also demonstrate the importance of this process during normal lymphatic vasculature formation (**Figure 2**). In addition, they suggest that downstream PDZ effectors of Ephrin signaling must play key roles in this process.

The data summarized in this section highlight the major progress made in the last few years toward identifying and characterizing the genes and mechanisms involved in normal and pathologic lymphangiogenesis. However, many important questions remain unanswered. Sabin's original model of lymphatic vasculature development has received important support in recent years. Nevertheless, it is not yet clear whether the alternative Huntington-McClure model suggesting that primary lymph sacs arise in the mesenchyme independent of the veins and then secondarily establish venous connections may also apply to the development of some subpopulation of lymphatic vessels in some mammals and birds. Furthermore, whether there are other sources of lymphatic progenitor cells (e.g., circulating lymphangioblasts) remains to be determined. What is the origin and identity of the initial signal that specifies lymphatic endothelial cell fate? Are the veins a constant source of lymphatic progenitors at later stages of embryonic development or in the adult? Are the mechanisms controlling embryonic lymphangiogenesis similar to those required for the growth of lymphatic vessels during wound healing and inflammation?

Although investigators have identified a number of key players whose activity is necessary during different stages of lymphangio-genesis, very little, if any, information about the regulatory pathways of these players is yet available.

PROPERTIES OF THE LYMPHATIC ENDOTHELIUM IN ADULT TISSUES

In the past few years, genes expressed specifically in LECs have been identified. Specific cell-surface markers have been used in immunomagnetic purification of LECs and BECs (Hirakawa et al. 2003, Kriehuber et al. 2001, Mäkinen et al. 2001b, Petrova et al. 2002, Podgrabinska et al. 2002).

The detailed comparison of genes expressed by these two cell populations has revealed that some of the most prominent differences are in genes that encode proinflammatory cytokines and chemokines and in those involved in cytoskeletal or cell matrix interactions. Interestingly, genes such as *ß-catenin* and *plakoglobin*, which are known to be involved in connecting cadherins to the actin cytoskeleton, show specificity for BECs and LECs, respectively. Furthermore, the whole actin cytoskeleton is organized differently in the two cell types: Integrin $\alpha 5$ is expressed by BECs, and integrins $\alpha 1$ and $\alpha 9$ are expressed by LECs (Petrova et al. 2002). Integrin $\alpha 5$ is part of the fibronectin receptor, integrin $\alpha 1$ is a subunit of the laminin and collagen receptors, and integrin $\alpha 9$ is a subunit of the osteopontin and tenascin receptors. Integrin $\alpha 9$ also binds VEGF-C, which suggests that integrin $\alpha 9$ behaves as a VEGFR-3 coreceptor (Vlahakis et al. 2004).

Isolated LECs are heterogenous for several lymphatic markers (Mäkinen et al. 2001b, Petrova et al. 2002). It is as yet unclear whether the different LEC populations, which are defined by such expression patterns, represent functionally different cell types. The pattern of gene expression in ECs is highly dependent on the EC microenvironment. Similar plasticity in phenotype has been observed in gene profiling studies of LECs maintained in culture; a comparison of these cells with LECs directly isolated from in vivo material will be required for a better evaluation of these results.

PATHOLOGIES OF THE LYMPHATIC VASCULATURE

Lymphatic Dysfunction, Lymphedema

In lymphedema, the transport capacity of lymphatic vessels is decreased and fluid accumulates in tissues. This leads to chronic and disabling swelling, tissue fibrosis, adipose degeneration, poor immune function, susceptibility to infections, and impaired wound healing (Rockson 2001). Primary lymphedemas are rare genetic developmental disorders. The symptoms are apparent from birth (Milroy disease) or at puberty (Meige disease) (Witte et al. 1998). Primary lymphedema develops in approximately one of 6000 newborns. The superficial or subcutaneous lymphatic vessels are usually aplastic or hypoplastic in Milroy disease, whereas the microlymphatic network is dysfuctional and may be hyperplastic in late-onset lymphedema. Secondary lymphedema results from various types of damage to the lymphatic vessels caused by, for example, radiation therapy, surgery, or infections. This form of lymphedema is much more common; approximately three to five million people in the U.S. suffer from the acquired form of the disease. However, the most common form of lymphedema is caused by filariasis; an estimated 100 million patients worldwide have this disease. Filariasis can lead to permanent disability due to massive edema and subsequent deformation of the limbs or genitals (Rockson 2001).

The genetic defects underlying several primary lymphedemas have recently been characterized. In Milroy disease, several heterozygous *VEGFR3* missense mutations result in the expression of an inactive tyrosine kinase (Irrthum et al. 2000, Karkkainen et al. 2000). Heterozygous *Vegfr-3* inactivation causes *Chy* mice to develop chylous ascites and lymphedema (Karkkainen et al. 2001b). Mutations in the *FOXC2* gene cause the hereditary lymphedema-distichiasis syndrome, and as previously indicated, *Foxc2*-targeted mice have lymphatic abnormalities (Fang et al. 2000, Kriederman et al. 2003, Petrova et al. 2004). Interestingly, patients with *FOXC2* mutations display abnormal mural cell coating of their lymphatic vessels and lack lymphatic valves (Petrova et al. 2004). Hypotrichosis-lymphedema-telangiectasia syndrome in humans has been associated with mutations of the transcription factor *SOX18* (Irrthum et al. 2003). Congenital lymphedema and accumulation of chylous ascites has also been reported in patients with mutations of the *REELIN* gene, which encodes an ECM protein guiding neuronal cell migration (Hong et al. 2000). Other human diseases such as Turner syndrome and cholestasis-lymphedema involve lymphedema, but the molecular mechanisms of that involvement remain unknown.

Lymphatic Vessels in Inflammation and Immune Responses

Several lines of evidence indicate that inflammation triggers lymphangiogenic signals that may regulate lymphatic vessel function. *VEGFC* mRNA transcription is induced in ECs in response to proinflammatory cytokines, possibly via sites that bind the transcription factor NF-κB identified in the *VEGFC* promoter (Chilov et al. 1997, Ristimaki et al. 1998). VEGF-C and VEGF-D are also highly expressed in activated macrophages (Schoppmann et al. 2002, Skobe et al. 2001a). Prostaglandin E2–mediated activation of cyclo-oxygenase-2 increases *VEGFC* mRNA transcription, which suggests that prostanoids induce VEGF-C-driven lymphangiogenesis (Su et al. 2004). Rejection of human kidney transplants is often accompanied by inflammatory lymphangiogenesis (Kerjaschki et al. 2004). VEGF-C is also highly expressed in the arthritic joint synovium in patients with rheumatoid arthritis (Paavonen et al. 2002). Furthermore, stimulation of the VEGFR-3 pathway has been associated with reactive lymphadenitis (Baluk et al. 2005), and inhibition of this pathway in growing lymphatic vasculature leads to hypoplasia of the lymph nodes and lymphatic vessels (Baluk et al. 2005, Mäkinen et al. 2001a). Lymphangiogenesis is also associated with inflammatory angiogenesis in the rabbit cornea (Cursiefen et al. 2004). The angiogenic and lymphangiogenic responses are blocked by depletion of macrophages, which suggests that inflammatory cells mediate the formation of lymphatic vessels (Cursiefen et al. 2004). Such findings suggest that growth of the lymphatic vasculature is intimately coupled to the control of immune function.

Lymphatic Involvement in Tumor Metastasis

In most cancers, the metastatic spread of tumor cells to distant organs results in patient mortality. Tumor metastasis to lymph nodes and distant organs occurs via lymphatic or blood vessels (Nathanson 2003). The patterns of metastasis vary from tumor to tumor. Regional lymph nodes are often the first site of metastasis in carcinomas. Metastatic nodes are important in the staging, treatment, and follow up of many solid tumors. Regional lymph node metastasis is often the most important prognostic factor for patients with malignant tumors of epithelial origin (Pepper et al. 2001). The sentinel lymph node is the first regional lymph node to which tumor cells metastasize; further metastasis then may occur to other nodes and into the systemic circulation. The recent introduction of sentinel lymph node biopsy and its replacement by regional lymph node excision in patients with melanoma or breast cancer reflect our current understanding of the organized nature of lymph node metastases.

The process of metastasis is complex and involves changes in the expression of many genes (Ramaswamy et al. 2003). It is not known whether lymphatic metastasis selects tumor cells with increased potential for dis-

tant organ metastasis or whether the appearance of lymphatic metastasis simply indicates that the tumor has entered a metastatic phase in general. The anatomic pathway starts with the primary tumor and involves a complex interaction among several factors: intratumoral and peritumoral interstitial pressures, connective tissue channels, cell adhesion molecules, cytokines and chemokines, chemokine receptors, specific microanatomic structures in lymphatic capillaries, interendothelial gaps, tumor growth within the lymphatic vessels, and the flow of lymph carrying tumor cells in the direction of regional draining lymph nodes (He et al. 2004, Jain 2002).

Growth factor–mediated stimulation of lymphatic vessels appears to be required for lymphatic metastasis, but many questions about the mechanisms of lymphatic dissemination remain unanswered. Accumulating data from clinicopathologic studies suggest that the spread of cancer cells to regional lymph nodes is an early event for many solid tumors and that lymphatic vessels are the primary route of this spread. Although the cellular mechanisms that induce VEGF-C expression in tumors are still unclear, a link between VEGF-C and VEGF-D expression and metastatic spread has been confirmed in a number of studies.

Preclinical investigations have shown that VEGF-C and VEGF-D can enhance lymphatic metastasis. In an elegant study, Mandriota et al. (2001) showed that VEGF-C transgenic mice that overexpressed VEGF-C in pancreatic β-cell tumors had well-developed peritumoral lymphatic vessels and a higher rate of regional lymph node metastases than did wild-type mice with the same tumors. Increased tumor lymphangiogenesis and angiogenesis was also demonstrated in VEGF-C-transfected human melanoma and breast carcinoma xenografts in nude mice (Karpanen et al. 2001, Mattila et al. 2002, Skobe et al. 2001b). Expression of a soluble VEGFR-3 fusion protein via adenovirus delivery into mouse blood circulation inhibited these

Lymphatic metastasis: invasion of detached primary tumor cells through the lymphatic system to lymph nodes effects (Karpanen et al. 2001). Thus, there is a direct link between VEGF-C expression and metastatic progression.

In a model of human lung carcinoma, lymph node metastases produced by VEGF-C-expressing tumor cells occurred early and were associated with a high density of lymphatic vessels (He et al. 2002). Lymphatic metastases, but not lung metastases, were blocked with a soluble VEGFR-3 fusion protein (He et al. 2002). The lung metastases probably resulted from tumor cells spreading directly via the blood vasculature rather than the lymphatic vessels. However, in another experiment using an immunocompetent rat mammary tumor model, VEGF-C expression promoted metastases to the lymph nodes and lungs (Krishnan et al. 2003). In this tumor model, soluble VEGFR-3 inhibited both lung and lymph node metastases. This work also suggested that interactions between the tumor cells and host cells stimulate the production of VEGF-C and VEGF-D.

VEGF-D introduced via gene transfer promoted lymphangiogenesis and lymphatic metastases in a tumor cell model in which VEGF had failed to produce the same effects. The role of VEGF-D in the lymphatic spread of the tumor was confirmed when metastases was blocked by anti-VEGF-D antibodies (Stacker et al. 2001).

Together, these results provide strong support for the contribution of VEGF-C, VEGF-D, and their receptor, VEGFR-3, in lymphatic spread in malignancy. Furthermore, the ability to inhibit lymphangiogenesis and lymphatic metastases with soluble VEGFR-3 (He et al. 2005) or antibodies against VEGF-C, VEGF-D, or VEGFR-3 suggests potential therapeutic approaches.

Clinical and Therapeutic Perspectives

Currently, lymphedema is treated by manual lymph drainage and compression bandages and stockings. *VEGFC* gene therapy induced the growth of new lymphatic vessels in the skin of the *Chy* mutant mice whose disease resembles human lymphedema associated with Milroy's disease. This suggests that some forms of human lymphedema might be treated via similar approaches (Karkkainen et al. 2001a,b). In addition, recombinant VEGF-C protein induced lymphangiogenesis in a surgical model of lymphedema in the rabbit ear; VEGF-C treatment improved lymphatic function and reversed the abnormalities in tissue architecture resulting from chronic lymphatic insufficiency (Szuba et al. 2002). The use of a VEGF-C plasmid in rabbit ear and mouse tail lymphedema models also resulted in improved lymphatic function and alleviated lymphedema (Yoon et al. 2003).

VEGFR-3 is upregulated on blood vascular ECs in pathologic conditions, such as vascular tumors including Kaposi's sarcoma, and in the periphery of solid tumors (Jussila et al. 1998, Kaipainen et al. 1995, Valtola et al. 1999). As stated previously, monoclonal antibodies that target VEGF-C, VEGF-D, or their receptor(s) and that antagonize the ligand binding or dimerization of the receptor(s) may prove useful as inhibitors of lymphatic metastasis. Alternatively, a soluble version of the extracellular domain of VEGFR-3 to sequester VEGF-C and VEGF-D may prove clinically useful. Another approach might be to use orally administered small molecules that may inhibit the tyrosine kinase catalytic domain of the VEGFR-3 (**Figure 4**). Interestingly, some of the small-molecule inhibitors of the catalytic domain of VEGFR-2 inhibit all three VEGF receptors (Eskens 2004). The success of a number of these strategies has been confirmed in animal models, but the evaluation of these compounds' effects on lymphatic vessels in human patients has not been reported.

These results from preclinical models are encouraging and suggest that regimens using lymphangiogenic growth factors can be developed as therapeutics for lymphedema. Furthermore, characterization of other genes involved in the regulation of lymphatic vessel growth or the pathogenesis of lymphatic

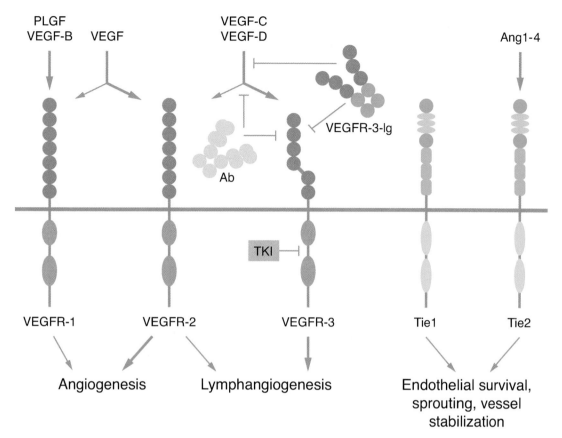

Figure 4

Schematic drawing of the stimulators and inhibitors of lymphangiogenesis and their interactions with the VEGF and Ang growth factors and receptors. VEGF-A, VEGF-B, VEGF-C, VEGF-D, VEGF-E, PlGF, and Ang1–4 bind to cell-surface tyrosine kinase receptors. The binding induces receptor dimerization and activation, which lead to intracellular transduction of signals that direct cellular functions. VEGFR-2 is essential for vascular development and is a high-affinity receptor for VEGF and the processed forms of VEGF-C and VEGF-D. VEGFR-1 is a high-affinity receptor for VEGF, VEGF-B, and PLGF. VEGFR-3 is a high-affinity receptor for VEGF-C and VEGF-D and the major signaling pathway for lymphangiogenesis. Although VEGF-C and VEGF-D appear to be the primary lymphangiogenic factors, they also link angiogenesis with lymphangiogenesis, as they can stimulate both processes. Recent data show that Ang1 can also stimulate Tie1 and angiogenesis (Saharinen et al. 2005, Tammela et al. 2005). Soluble versions of the extracellular domain of VEGFR-3 (VEGFR-3-Ig) provide one form of lymphangiogenesis inhibitor: They sequester VEGF-C and VEGF-D, blocking the binding of VEGF-C and VEGF-D to VEGFR-3 on the lymphatic endothelial cell surface. Another strategy is to use antibodies (Ab) that bind VEGF-C and VEGF-D or their receptor(s) and antagonize interaction of VEGC/VEGF-D or receptor activation by dimerization. A third strategy is to generate small orally active VEGFR-3 tyrosine kinase inhibitors (TKIs) that prevent the transduction of signals stimulating lymphangiogenesis.

dysfunction should give us more insight into the molecular mechanisms of lymphedema. Although progress has been made toward understanding lymphangiogenesis and its importance in the metastatic spread of tumors, much remains to be learned about the molecular mechanisms of lymphangiogenesis, the specificity of growth factors and lymphatic markers, and possible organ-specific lymphangiogenic molecules that might contribute to tumor-associated lymphangiogenesis. What host and tumor factors contribute to lymphangiogenesis, lymphatic dissemination of tumor cells, and subsequent arrival and growth of these cells at regional and distant sites? Given the importance of lymphangiogenesis and its regulators in both lymphatic as well as hematogenous spread, a number of strategies aimed at inhibition of these pathways are under investigation (He et al. 2004, Stacker et al. 2002).

SUMMARY POINTS

1. The recent identification of genes involved in different aspects of lymphatic vasculature formation has paved the way to a better understanding of this process in health and disease.

2. Blood endothelial cells are the default state from which a lymphatic vasculature phenotype is specified in a stepwise process.

3. Developmental defects of lymphatic vasculature lead to primary lymphedema.

4. Inflammation stimulates lymphangiogenesis and lymphangiogenesis promotes tumor metastasis.

5. Stimulation of lymphatic growth is an attainable goal in the future treatment of lymphatic diseases.

6. Inhibition of lymphangiogenesis via blocking the VEGF-C/VEGF-D signal transduction pathway inhibits lymphatic metastasis.

LITERATURE CITED

This is an intriguing first report describing some of the genes and possible mechanisms involved in the physical separation of the blood and lymphatic systems.

Abtahian F, Guerriero A, Sebzda E, Lu MM, Zhou R, et al. 2003. Regulation of blood and lymphatic vascular separation by signaling proteins SLP-76 and Syk. *Science* 299:247–51

Achen MG, Jeltsch M, Kukk E, Makinen T, Vitali A, et al. 1998. Vascular endothelial growth factor D (VEGF-D) is a ligand for the tyrosine kinases VEGF receptor 2 (Flk1) and VEGF receptor 3 (Flt4). *Proc. Natl. Acad. Sci. USA.* 95:548–53

Adams RH, Wilkinson GA, Weiss C, Diella F, Gale NW, et al. 1999. Roles of ephrinB ligands and EphB receptors in cardiovascular development: demarcation of arterial/venous domains, vascular morphogenesis, and sprouting angiogenesis. *Genes Dev.* 13:295–306

Alitalo K, Carmeliet P. 2002. Molecular mechanisms of lymphangiogenesis in health and disease. *Cancer Cell* 1:219–27

Asahara T, Murohara T, Sullivan A, Silver M, van der Zee R, et al. 1997. Isolation of putative progenitor endothelial cells for angiogenesis. *Science* 275:964–67

Aselli G. 1627. *De Lacteibus sive Lacteis Venis, Quarto Vasorum Mesarai corum Genere novo invento.* Milan: Mediolani

Baldwin ME, Catimel B, Nice EC, Roufail S, Hall NE, et al. 2001. The specificity of receptor binding by vascular endothelial growth factor-d is different in mouse and man. *J. Biol. Chem.* 276:19166–71

Baluk P, Tammela T, Ator E, Lyubynska N, Achen MG, et al. 2005. Pathogenesis of persistent lymphatic vessel hyperplasia in chronic airway inflammation. *J. Clin. Invest.* 115:247–57

Banerji S, Ni J, Wang S-X, Clasper S, Su J, et al. 1999. LYVE-1, a new homologue of the CD44 glycoprotein, is a lymph-specific receptor for hyaluronan. *J. Cell Biol.* 144:789–801

Breiteneder-Geleff S, Soleiman A, Kowalski H, Horvat R, Amann G, et al. 1999. Angiosarcomas express mixed endothelial phenotypes of blood and lymphatic capillaries: podoplanin as a specific marker for lymphatic endothelium. *Am. J. Pathol.* 154:385–94

Byzova TV, Goldman CK, Jankau J, Chen J, Cabrera G, et al. 2002. Adenovirus encoding vascular endothelial growth factor-D induces tissue-specific vascular patterns in vivo. *Blood* 99(12):4434–42

Cao Y, Linden P, Farnebo J, Cao R, Eriksson A, et al. 1998. Vascular endothelial growth factor C induces angiogenesis in vivo. *Proc. Natl. Acad. Sci. USA* 95:14389–94

Carmeliet P, Ferreira V, Breier G, Pollefeyt S, Kieckens L, et al. 1996. Abnormal blood vessel development and lethality in embryos lacking a single VEGF allele. *Nature* 380:435–39

Chen H, Chedotal A, He Z, Goodman CS, Tessier-Lavigne M. 1997. Neuropilin-2, a novel member of the neuropilin family, is a high affinity receptor for the semaphorins Sema E and Sema IV but not Sema III. *Neuron* 19:547–59

Chilov D, Kukk E, Taira S, Jeltsch M, Kaukonen J, et al. 1997. Genomic organization of human and mouse genes for vascular endothelial growth factor C. *J. Biol. Chem.* 272:25176–83

Cox CM, Poole TJ. 2000. Angioblast differentiation is influenced by the local environment: FGF-2 induces angioblasts and patterns vessel formation in the quail embryo. *Dev. Dyn.* 218:371–82

Cursiefen C, Chen L, Borges LP, Jackson D, Cao J, et al. 2004. VEGF-A stimulates lymphangiogenesis and hemangiogenesis in inflammatory neovascularization via macrophage recruitment. *J. Clin. Invest.* 113:1040–50

Dagenais SL, Hartsough RL, Erickson RP, Witte MH, Butler MG, Glover TW. 2004. Foxc2 is expressed in developing lymphatic vessels and other tissues associated with lymphedema-distichiasis syndrome. *Gene Expr. Patterns* 6:611–19

Dumont DJ, Gradwohl G, Fong GH, Puri MC, Gertsenstein M. 1994. Dominant-negative and targeted null mutations in the endothelial receptor tyrosine kinase, tek, reveal a critical role in vasculogenesis of the embryo. *Genes Dev.* 16:1897–909

Dumont DJ, Jussila L, Taipale J, Lymboussaki A, Mustonen T, et al. 1998. Cardiovascular failure in mouse embryos deficient in VEGF receptor-3. *Science* 282:946–49

Enholm B, Karpanen T, Jeltsch M, Kubo H, Stenback F, et al. 2001. Adenoviral expression of vascular endothelial growth factor-C induces lymphangiogenesis in the skin. *Circ. Res.* 88:623–29

Eskens FA. 2004. Angiogenesis inhibitors in clinical development; where are we now and where are we going? *Br. J. Cancer* 90:1–7

Fang J, Dagenais SL, Erickson RP, Arlt MF, Glynn MW, et al. 2000. Mutations in FOXC2 (MFH-1), a forkhead family transcription factor, are responsible for the hereditary lymphedema-distichiasis syndrome. *Am. J. Hum. Genet.* 67:1382–88

Farnebo F, Piehl F, Lagercrantz J. 1999. Restricted expression pattern of vegf-d in the adult and fetal mouse: high expression in the embryonic lung. *Biochem. Biophys. Res. Commun.* 257(3):891–94

Ferrara N. 1999. Vascular endothelial growth factor: molecular and biological aspects. *Curr. Top. Microbiol. Immunol.* 237:1–30

Ferrara N. 2001. Role of vascular endothelial growth factor in regulation of physiological angiogenesis. *Am. J. Physiol. Cell Physiol.* 280:1358–66

Ferrara N, Carver-Moore K, Chen H, Dowd M, Lu L, O'Shea KS, et al. 1996. Heterozygous embryonic lethality induced by targeted inactivation of the VEGF gene. *Nature* 380:439–42

Flamme I, Risau W. 1992. Induction of vasculogenesis and hematopoiesis in vitro. *Development* 116:435–39

Folkman J. 1995. Angiogenesis in cancer, vascular, rheumatoid and other disease. *Nat. Med.* 1:27–31

Folkman J, Shing Y. 1992. Angiogenesis. *J. Biol. Chem.* 267:10931–34

Gale NW, Thurston G, Hackett SF, Renard R, Wang Q, et al. 2002. Angiopoietin-2 is required for postnatal angiogenesis and lymphatic patterning, and only the latter role is rescued by Angiopoietin-1. *Dev. Cell* 3:411–23

Hamada K, Oike Y, Takakura N, Ito Y, Jussila L, et al. 2000. VEGF-C signaling pathways through VEGFR-2 and VEGFR-3 in vasculoangiogenesis and hematopoiesis. *Blood* 96:3793–800

Harvey NL, Oliver G. 2004. Choose your fate: artery, vein or lymphatic vessel? *Curr. Opin. Genet. Dev.* 5:499–505

He Y, Karpanen T, Alitalo K. 2004. Role of lymphangiogenic factors in tumor metastasis. *Biochim. Biophys. Acta* 1654:3–12

He Y, Kozaki K, Karpanen T, Koshikawa K, Yla-Herttuala S, et al. 2002. Suppression of tumor lymphangiogenesis and lymph node metastasis by blocking vascular endothelial growth factor receptor 3 signaling. *J. Natl. Cancer Inst.* 94:785–87

He Y, Rajantie I, Pajusola K, Yla-Herttuala S, Jooss K, et al. 2005. VEGFR-3 mediated activation of lymphatic endothelium is crucial for tumor cell entry and spread via lymphatic vessels. *Cancer Res.* 65:4739–46

Hirakawa S, Hong YK, Harvey N, Schacht V, Matsuda K, et al. 2003. Identification of vascular lineage-specific genes by transcriptional profiling of isolated blood vascular and lymphatic endothelial cells. *Am. J. Pathol.* 162:575–86

Hong SE, Shugart YY, Huang DT, Shahwan SA, Grant PE, et al. 2000. Autosomal recessive lissencephaly with cerebellar hypoplasia is associated with human RELN mutations. *Nat. Genet.* 26:93–96

Hong YK, Harvey N, Noh YH, Schacht V, Hirakawa S, et al. 2002. Prox1 is a master control gene in the program specifying lymphatic endothelial cell fate. *Dev. Dyn.* 225:351–57

Hong YK, Foreman K, Shin JW, Hirakawa S, Curry CL, et al. 2004. Lymphatic reprogramming of blood vascular endothelium by Kaposi sarcoma-associated herpesvirus. *Nat. Genet.* 36(7):683–85

Huntington GS, McClure CFW. 1910. The anatomy and development of the jugular lymph sac in the domestic cat (*Felis domestica*). *Am. J. Anat.* 10:177–311

Irrthum A, Devriendt K, Chitayat D, Matthijs G, Glade C, et al. 2003. Mutations in the transcription factor gene SOX18 underlie recessive and dominant forms of hypotrichosis-lymphedema-telangiectasia. *Am. J. Hum. Genet.* 72(6):1470–78

Irrthum A, Karkkainen MJ, Devriendt K, Alitalo K, Vikkula M. 2000. Congenital hereditary lymphedema caused by a mutation that inactivates VEGFR3 tyrosine kinase. *Am. J. Hum. Genet.* 7:295–301

Iwama A, Hamaguchi I, Hashiyama M, Murayama Y, Yasunaga K, et al. 1993. Molecular cloning and characterization of mouse TIE and TEK receptor tyrosine kinase genes and their expression in hematopoietic stem cells. *Biochem. Biophys. Res. Commun.* 195(1):301–9

Jain RK. 2002. Angiogenesis and lymphangiogenesis in tumors: insights from intravital microscopy. *Cold Spring Harbor Symp. Quant. Biol.* 67:239–34

Jeltsch M, Kaipainen A, Joukov V, Meng X, Lakso M, et al. 1997. Hyperplasia of lymphatic vessels in VEGF-C transgenic mice. *Science* 276:1423–25

Jeltsch M, Tammela T, Alitalo K, Wilting J. 2003. Genesis and pathogenesis of lymphatic vessels. *Cell Tissue Res.* 314:69–84

This is the first demonstration that VEGF-C is lymphangiogenic in vivo.

Joukov V, Kumar V, Sorsa T, Arighi E, Weich H, et al. 1998. A recombinant mutant vascular endothelial growth factor-C that has lost vascular endothelial growth factor receptor-2 binding, activation, and vascular permeability activities. *J. Biol. Chem.* 273(12):6599–602

Joukov V, Pajusola K, Kaipainen A, Chilov DA, Lahtinen I, et al. 1996. A novel vascular endothelial growth factor, VEGF-C, is a ligand for the Flt4 (VEGFR-3) and KDR (VEGFR-2) receptor tyrosine kinases. *EMBO J.* 15(2):290–98. Erratum. 1996. *EMBO J.* 15(7):1751

Joukov V, Sorsa T, Kumar V, Jeltsch M, Claesson-Welsh L, et al. 1997. Proteolytic processing regulates receptor specificity and activity of VEGF-C. *EMBO J.* 16(13):3898–911

Jussila L, Valtola R, Partanen TA, Salven P, Heikkila P, et al. 1998. Lymphatic endothelium and Kaposi's sarcoma spindle cells detected by antibodies against the vascular endothelial growth factor receptor-3. *Cancer Res.* 58:1599–604

Kaipainen A, Korhonen J, Mustonen T, van Hinsbergh VWM, Fang GH, et al. 1995. Expression of the *fms*-like tyrosine kinase FLT4 gene becomes restricted to lymphatic endothelium during development. *Proc. Natl. Acad. Sci. USA* 92:3566–70

Karkkainen MJ, Ferrell RE, Lawrence EC, Kimak MA, Levinson KL, et al. 2000. Missense mutations interfere with VEGFR-3 signaling in primary lymphoedema. *Nat. Genet.* 25:153–59

Karkkainen MJ, Haiko P, Sainio K, Partanen J, Taipale J, et al. 2004. Vascular endothelial growth factor C is required for sprouting of the first lymphatic vessels from embryonic veins. *Nat. Immunol.* 5:74–80

Karkkainen MJ, Jussila L, Ferrell RE, Finegold DN, Alitalo K. 2001a. Molecular regulation of lymphangiogenesis and targets for tissue oedema. *Trends Mol. Med.* 7:18–22

Karkkainen MJ, Saaristo A, Jussila L, Karila KA, Lawrence EC, et al. 2001b. A model for gene therapy of human hereditary lymphedema. *Proc. Natl. Acad. Sci. USA* 98:12677–82

Karpanen T, Egeblad M, Karkkainen MJ, Kubo H, Jackson DG, et al. 2001. Vascular endothelial growth factor C promotes tumor lymphangiogenesis and intralymphatic tumor growth. *Cancer Res.* 61:1786–90

Kerjaschki D, Regele HM, Moosberger I, Nagy-Bojarski K, Watschinger B, et al. 2004. Lymphatic neoangiogenesis in human kidney transplants is associated with immunologically active lymphocytic infiltrates. *J. Am. Soc. Nephrol.* 15:603–12

Kriederman BM, Myloyde TL, Witte MH, Dagenais SL, Witte CL, et al. 2003. FOXC2 haploinsufficient mice are a model for human autosomal dominant lymphedema-distichiasis syndrome. *Hum. Mol. Genet.* 12:1179–85

Kriehuber E, Breiteneder-Geleff S, Groeger M, Soleiman A, Schoppmann SF, et al. 2001. Isolation and characterization of dermal lymphatic and blood endothelial cells reveal stable and functionally specialized cell lineages. *J. Exp. Med.* 194:797–808

Krishnan J, Kirkin V, Steffen A, Hegen M, Weih D, et al. 2003. Differential in vivo and in vitro expression of vascular endothelial growth factor (VEGF)-C and VEGF-D in tumors and its relationship to lymphatic metastasis in immunocompetent rats. *Cancer Res.* 63(3):713–22

Kukk E, Lymboussaki A, Taira S, Kaipainen A, Jeltsch M, et al. 1996. VEGF-C receptor binding and pattern of expression with VEGFR-3 suggests a role in lymphatic vascular development. *Development* 122:3829–37

Lawson ND, Scheer N, Pham VN, Kim CH, Chitnis AB, et al. 2001. Notch signaling is required for arterial-venous differentiation during embryonic vascular development. *Development* 128:3675–83

This study reported the first identification, purification, and cloning of the VEGFR-3 ligand, VEGF-C, the first growth factor for lymphatic endothelia.

This is the first identification of a gene product with restricted expression in the lymphatic system.

This study identified mutations in the vascular endothelial growth-factor receptor 3 (*VEGFR3*) responsible for certain forms of hereditary lymphedema.

This study demonstrated that VEGF-C is essential for the migration and proliferation of lymphatic progenitor cells from the cardinal veins during the earliest stages of lymphatic development.

This work demonstrated that gene therapy eventually could be achieved in patients with lymphedema.

Lymboussaki A, Olofsson B, Eriksson U, Alitalo K. 1999. Vascular endothelial growth factor (VEGF) and VEGF-C show overlapping binding sites in embryonic endothelia and distinct sites in differentiated adult endothelia. *Circ. Res.* 85(11):992–99

Maisonpierre PC, Suri C, Jones PF, Bartunkova S, Wiegand SJ, et al. 1997. Angiopoietin-2, a natural antagonist for Tie2 that disrupts in vivo angiogenesis. *Science* 277:55–60

Makinen A, Adams RH, Bailey J, Lu Q, Ziemiecki A, et al. 2005. PDZ interaction site in ephrinB2 is required for the remodeling of lymphatic vasculature. *Genes Dev.* 19:397–410

Mäkinen T, Jussila L, Veikkola T, Karpanen T, Kettunen MI, et al. 2001a. Inhibition of lymphangiogenesis with resulting lymphedema in transgenic mice expressing soluble VEGF receptor-3. *Nat. Med.* 7:199–205

Mäkinen T, Veikkola T, Mustjoki S, Karpanen T, Catimel B, et al. 2001b. Isolated lymphatic endothelial cells transduce growth, survival and migratory signals via the VEGF-C/D receptor VEGFR-3. *EMBO J.* 20:4762–73

Mandriota SJ, Jussila L, Jeltsch M, Compagni A, Baetens D, et al. 2001. Vascular endothelial growth factor-C-mediated lymphangiogenesis promotes tumour metastasis. *EMBO J.* 20:672–82

Mattila MM, Ruohola JK, Karpanen T, Jackson DG, Alitalo K, et al. 2002. VEGF-C induced lymphangiogenesis is associated with lymph node metastasis in orthotopic MCF-7 tumors. *Int. J. Cancer* 98(6):946–51

Nathanson SD. 2003. Insights into the mechanisms of lymph node metastasis. *Cancer* 98:413–23

Ober EA, Olofsson B, Makinen T, Jin SW, Shoji W. 2004. Vegfc is required for vascular development and endoderm morphogenesis in zebrafish. *EMBO Rep.* 5:78–84

Oh SJ, Jeltsch MM, Birkenhager R, McCarthy JE, Weich HA, et al. 1997. VEGF and VEGF-C: specific induction of angiogenesis and lymphangiogenesis in the differentiated avian chorioallantoic membrane. *Dev. Biol.* 188(1):96–109

Oliver G. 2004. Lymphatic vasculature development. *Nat. Rev. Immunol.* 4:35–45

Oliver G, Detmar M. 2002. The rediscovery of the lymphatic system: old and new insights into the development and biological function of the lymphatic vasculature. *Genes Dev.* 16:773–83

Paavonen K, Mandelin J, Partanen T, Jussila L, Li TF, et al. 2002. Vascular endothelial growth factors C and D and their VEGFR-2 and 3 receptors in blood and lymphatic vessels in healthy and arthritic synovium. *J. Rheumatol.* 29:39–45

Partanen J, Armstrong E, Makela TP, Korhonen J, Sandberg M, et al. 1992. A novel endothelial cell surface receptor tyrosine kinase with extracellular epidermal growth factor homology domains. *Mol. Cell. Biol.* 12(4):1698–707

Pepper MS. 2001. Lymphangiogenesis and tumor metastasis: myth or reality? *Clin. Cancer Res.* 7:462–68

Petrova TV, Karpanen T, Norrmen C, Mellor R, Tamakoshi T, et al. 2004. Defective valves and abnormal mural cell recruitment underlie lymphatic vascular failure in lymphedema distichiasis. *Nat. Med.* 10:974–81

Petrova TV, Makinen T, Makela TP, Saarela J, Virtanen I, et al. 2002. Lymphatic endothelial reprogramming of vascular endothelial cells by the Prox-1 homeobox transcription factor. *EMBO J.* 21:4593–45

Podgrabinska S, Braun P, Velasco P, Kloos B, Pepper MS, Skobe M. 2002. Molecular characterization of lymphatic endothelial cells. *Proc. Natl. Acad. Sci. USA* 99:16069–74

This study demonstrates that ephrinB2 is an essential regulator of lymphatic development and gives a detailed description of the process of lymphatic maturation and remodeling.

Puri MC, Rossant J, Alitalo K, Bernstein A, Partanen J. 1995. The receptor tyrosine kinase TIE is required for integrity and survival of vascular endothelial cells. *EMBO J.* 14(23):5884–91

Ramaswamy S, Ross KN, Lander ES, Golub TR. 2003. A molecular signature of metastasis in primary solid tumors. *Nat. Genet.* 33(1):49–54

Risau W. 1997. Mechanisms of angiogenesis. *Nature* 386:671–74

Risau W, Flamme I. 1995. Vasculogenesis. *Annu. Rev. Cell Dev. Biol.* 11:73–91

Rissanen TT, Markkanen JE, Gruchala M, Heikura T, Puranen A, et al. 2003. VEGF-D is the strongest angiogenic and lymphangiogenic effector among VEGFs delivered into skeletal muscle via adenoviruses. *Circ. Res.* 92(10):1098–106

Ristimaki A, Narko K, Enholm B, Joukov V, Alitalo K. 1998. Proinflammatory cytokines regulate expression of the lymphatic endothelial mitogen vascular endothelial growth factor-C. *J. Biol. Chem.* 273:8413–18

Rockson SG. 2001. Lymphedema. *Am. J. Med.* 110:288–95

Saaristo A, Veikkola T, Enholm B, Hytönen M, Arola J, et al. 2002. Adenoviral VEGF-C over-expression induces blood vessel enlargement, tortuosity, and leakiness but no sprouting angiogenesis in the skin or mucous membranes. *FASEB J.* 16:1041–49

Sabin FR. 1902. On the origin of the lymphatic system from the veins, and the development of the lymph hearts and thoracic duct in the pig. *Am. J. Anat.* 1:367–89

Sabin FR. 1904. On the development of the superficial lymphatics in the skin of the pig. *Am. J. Anat.* 3:183–95

Saharinen P, Kerkelä K, Ekman N, Marron M, Brindle N, et al. 2005. Multiple angiopoietin recombinant proteins activate the Tie1 receptor tyrosine kinase and promote its interaction with Tie2. *J. Cell Biol.* 169:239–43

Saharinen P, Tammela T, Karkkainen MJ, Alitalo K. 2004. Lymphatic vasculature: development, molecular regulation and role in tumor metastasis and inflammation. *Trends Immunol.* 25:387–95

Sato TN, Qin Y, Kozak CA, Audus KL. 1993. Tie-1 and tie-2 define another class of putative receptor tyrosine kinase genes expressed in early embryonic vascular system. *Proc. Natl. Acad. Sci. USA* 90:9355–58

Scavelli C, Weber E, Agliano M, Cirulli T, Nico B, et al. 2004. Lymphatics at the crossroads of angiogenesis and lymphangiogenesis. *J. Anat.* 204:433–34

Schacht V, Ramirez MI, Hong YK, Hirakawa S, Feng D, et al. 2003. T1alpha/podoplanin deficiency disrupts normal lymphatic vasculature formation and causes lymphedema. *EMBO J.* 22:3546–56

Schoppmann SF, Birner P, Stockl J, Kalt R, Ullrich R, et al. 2002. Tumor-associated macrophages express lymphatic endothelial growth factors and are related to peritumoral lymphangiogenesis. *Am. J. Pathol.* 161:947–56

Shalaby F, Rossant J, Yamaguchi TP, Gertsenstein M, Wu XF, et al. 1995. Failure of blood island formation and vasculogenesis in Flk-1-deficient mice. *Nature* 376:62–66

Shi Q, Rafii S, Wu MH, Wijelath ES, Yu C, et al. 1998. Evidence for circulating bone marrow-derived endothelial cells. *Blood* 92:362–67

Siegfried G, Basak A, Cromlish JA, Benjannet S, Marcinkiewicz J, et al. 2003. The secretory proprotein convertases furin, PC5, and PC7 activate VEGF-C to induce tumorigenesis. *J. Clin. Invest.* 111(11):1723–32

Skobe M, Hamberg LM, Hawighorst T, Schirner M, Wolf GL, et al. 2001a. Concurrent induction of lymphangiogenesis, angiogenesis, and macrophage recruitment by vascular endothelial growth factor-C in melanoma. *Am. J. Pathol.* 159:893–903

Skobe M, Hawighorst T, Jackson DG, Prevo R, Janes L, et al. 2001b. Induction of tumor lymphangiogenesis by VEGF-C promotes breast cancer metastasis. *Nat. Med.* 7:192–98

Stacker SA, Achen MG, Jussila L, Baldwin ME, Alitalo K. 2002. Lymphangiogenesis and cancer metastasis. *Nat. Rev. Cancer.* 2(8):573–83

Stacker SA, Caesar C, Baldwin ME, Thornton GE, Williams RA, et al. 2001. VEGF-D promotes the metastatic spread of tumor cells via the lymphatics. *Nat. Med.* 7:186–91

Stacker SA, Stenvers K, Caesar C, Vitali A, Domagala T, et al. 1999. Biosynthesis of vascular endothelial growth factor-D involves proteolytic processing which generates non-covalent homodimers. *J. Biol. Chem.* 274(45):32127–36

Stacker SA, Williams RA, Achen MG. 2004. Lymphangiogenic growth factors as markers of tumor metastasis. *APMIS* 112:539–49

Su JL, Shih JY, Yen ML, Jeng YM, Chang CC, et al. 2004. Cyclooxygenase-2 induces EP1- and HER-2/Neu-dependent vascular endothelial growth factor-C up-regulation: a novel mechanism of lymphangiogenesis in lung adenocarcinoma. *Cancer Res.* 64:554–64

Suri C, Jones PF, Patan S, Bartunkova S, Maisonpierre PC, et al. 1996. Requisite role of Angiopoietin-1, a ligand for the TIE2 receptor, during embryonic angiogenesis. *Cell* 87:1161–69

Szuba A, Skobe M, Karkkainen MJ, Shin WS, Beynet DP, et al. 2002. Therapeutic lymphangiogenesis with human recombinant VEGF-C. *FASEB J.* 16:1985–87

Takashima S, Kitakaze M, Asakura M, Asanuma H, Sanada S, et al. 2002. Targeting of both mouse neuropilin-1 and neuropilin-2 genes severely impairs developmental yolk sac and embryonic angiogenesis. *Proc. Natl. Acad. Sci. USA* 99:3657–62

Tammela T, Saaristo A, Lohela M, Morisada T, Tornberg J, et al. 2005. Angiopoietin-1 promotes lymphatic sprouting and hyperplasia. *Blood.* 105:4642–48

Valtola R, Salven P, Heikkila P, Taipale J, Joensuu H, et al. 1999. VEGFR-3 and its ligand VEGF-C are associated with angiogenesis in breast cancer. *Am. J. Pathol.* 154(5):1381–90

van der Putte SCJ. 1975. The development of the lymphatic system in man. *Adv. Anat. Embryol. Cell. Biol.* 51:3–60

Veikkola T, Alitalo K. 2002. Dual role of Ang2 in postnatal angiogenesis and lymphangiogenesis. *Dev. Cell* 3:302–4

Veikkola T, Jussila L, Makinen T, Karpanen T, Jeltsch M, et al. 2001. Signaling via vascular endothelial growth factor receptor-3 is sufficient for lymphangiogenesis in transgenic mice. *EMBO J.* 6:1223–31

Veikkola T, Lohela M, Ikenberg K, Makinen T, Korff T, et al. 2003. Intrinsic versus microenvironmental regulation of lymphatic endothelial cell phenotype and function. *FASEB J.* 17:2006–13

Vlahakis NE, Young BA, Atakilit A, Sheppard D. 2004. The lymphangiogenic growth factors VEGF-C and D are ligands for the integrin alpha 9beta 1. *J. Biol. Chem.* 280(6):4544–52

Wang HU, Chen ZF, Anderson DJ. 1998. Molecular distinction and angiogenic interaction between embryonic arteries and veins revealed by ephrin-B2 and its receptor Eph-B4. *Cell* 93:741–53

Wang HW, Trotter MW, Lagos D, Bourboulia D, Henderson S, et al. 2004. Kaposi sarcoma herpesvirus-induced cellular reprogramming contributes to the lymphatic endothelial gene expression in Kaposi sarcoma. *Nat. Genet.* 36(7):687–93

Wigle JT, Harvey N, Detmar M, Lagutina I, Grosveld G, et al. 2002. An essential role for Prox1 in the induction of the lymphatic endothelial cell phenotype. *EMBO J.* 21:1505–13

Wigle JT, Oliver G. 1999. Prox1 function is required for the development of the murine lymphatic system. *Cell* 98:769–78

Witte MH, Bernas MJ, Martin CP, Witte CL. 2001. Lymphangiogenesis and lymphangiodysplasia: from molecular to clinical lymphology. *Microsc. Res. Tech.* 55:122–45

Witte MH, Erickson R, Bernas M, Andrade M, Reiser F, et al. 1998. Phenotypic and genotypic heterogeneity in familial Milroy lymphedema. *Lymphology* 31:145–55

Yoffey JM, Courtice FC. 1970. *Lymphatics, Lymph and the Lymphomyeloid Complex.* London: Academic

Yoon YS, Murayama T, Gravereaux E, Tkebuchava T, Silver M, et al. 2003. VEGF-C gene therapy augments postnatal lymphangiogenesis and ameliorates secondary lymphedema. *J. Clin. Invest.* 111:717–25

Yuan L, Moyon D, Pardanaud L, Breant C, Karkkainen MJ, et al. 2002. Abnormal lymphatic vessel development in neuropilin 2 mutant mice. *Development* 129:4797–806

Zhong TP, Childs S, Leu JP, Fishman MC. 2001. Gridlock signalling pathway fashions the first embryonic artery. *Nature* 414:216–20

This is the first demonstration that venous endothelial cells are the default ground state of LECs and that Prox1 activity is necessary to specify the lymphatic phenotype of venous endothelial cells.

This is the first report of a gene product, the targeted inactivation of which leads to embryos without any lymphatic vasculature.

Regulation of Root Apical Meristem Development

Keni Jiang and Lewis J. Feldman

Department of Plant and Microbial Biology, University of California, Berkeley, California 94720; email: kenij@nature.berkeley.edu; feldman@nature.berkeley.edu

Annu. Rev. Cell Dev. Biol.
2005. 21:485–509

First published online as a
Review in Advance on
June 28, 2005

The *Annual Review of
Cell and Developmental
Biology* is online at
http://cellbio.annualreviews.org

doi: 10.1146/
annurev.cellbio.21.122303.114753

Copyright © 2005 by
Annual Reviews. All rights
reserved

1081-0706/05/1110-
0485$20.00

Key Words

quiescent center, auxin, redox

Abstract

The establishment of the Angiosperm root apical meristem is dependent on the specification of a stem cell niche and the subsequent development of the quiescent center at the presumptive root pole. Distribution of auxin and the establishment of auxin maxima are early formative steps in niche specification that depend on the expression and distribution of auxin carriers. Auxin specifies stem cell niche formation by directly and indirectly affecting gene activities. Part of the indirect regulation by auxin may involve changes in redox, favoring local, oxidized microenvironments. Formation of a QC is required for root meristem development and elaboration. Many signals likely pass between the QC and the adjacent root meristem tissues. Disappearance of the QC is associated with roots becoming determinate. Given the many auxin feedback loops, we hypothesize that roots evolved as part of an auxin homeostasis mechanism.

Contents

INTRODUCTION

In this review we focus on the developmental biology of roots, with an emphasis on the root apical meristem of Angiosperms. Several excellent reviews (Aeschbacher et al. 1994; Casson & Lindsay 2003; Jiang & Feldman 2003; Jürgens 2001; Scheres & Wolkenfelt 1998; Scheres et al. 1994, 1996, 2002) and articles (Aida et al. 2004, Benfey et al. 1993, Dolan et al. 1993, Friml et al. 2003) documenting

RAM: root apical meristem

our increased understanding of the molecular controls of root development, particularly with regard to auxin controls of root development in *Arabidopsis*, have recently appeared. Here we build on those reviews, expand the number of species considered, and propose a model that integrates many of these observations. In the development of this model, we believe it instructive to consider current thinking on the evolutionary origin of roots.

THE EVOLUTIONARY ORIGIN OF ROOTS

The earliest evidence for roots associated with vascular plants comes from fossil lycopods, which were extant in the early- to mid-Devonian period, about 400 mya. In these seedless plants, the rooting structures appear little different morphologically from the shoot system, reflecting the hypothesized origin of the root as a product of a embryonic shoot dichotomy. In this scenario, the root meristem was transformed from a shoot meristem (Gensel & Edwards 2001). Thus, precocious occurrence of the root-shoot dichotomy may be viewed as a mechanism for inserting root formation into embryogeny and thereby suggests that there exists a correlation between the morphology of the mature root system and embryo morphology, as postulated by Goebel (1928). In another part of this scenario, transformation of a shoot apex into a root apex was associated with the shoot system's adaptive requirements, most probably anchorage and the absorption of water and nutrients. Current molecular efforts may eventually elucidate the mechanisms underlying this hypothesized evolutionary push for shoot meristem transformation. Recent investigations of mutations affecting both root and shoot meristems would support this transformation theory (Ueda et al. 2004), as do the similarities in the expression of *SCR* and *WOX* genes in both root and shoot meristems (Haecker et al. 2004, Wysocka-Diller et al. 2000). On the other hand, molecular data may also suggest a de novo origin of the embryonic root without the need for any dichotomy. The ability to profile and compare global gene expression patterns in root and shoot meristems may help in deciding whether roots originated de novo or whether root and shoot systems differentiated from a single, homologous organ system.

MORPHOLOGICAL AND ANATOMICAL LANDMARKS

Much of the recent and most exciting data on embryonic primary-root development come from *Arabidopsis thaliana*. As a consequence, the root anatomy of this species is well-characterized both in relation to mutants and with regard to cell- and tissue-specific expression of many genes (Casson & Lindsey 2003 and references therein). But findings from other species, especially maize and pea, have also impacted greatly our understanding of root development (Clowes 1978, Jiang & Feldman 2003).

At the very tip of most roots is the RC (**Figure 1**), which serves to protect the underlying root meristem as the root advances through the soil. With the exception of a few species in highly specialized families, such as the Podostemaceae—in which the RC is reduced in size or absent during some portion of root ontogeny (Rutishauser 1997, Suzuki et al. 2002)—a RC is otherwise always present (von Guttenberg 1968), thereby pointing to nontrivial, central roles for it. Just proximal to the RC is the RAM. In many species such as *Arabidopsis* and maize, a discrete boundary, the RC junction, is recognized between the RC and the RAM (**Figure 1**). Roots maintaining this type of architecture are said to show "closed" organization, as contrasted with those in species such as pea, in which no sharp boundary is discerned between the RC and the root proper. This latter type of root is described as having "open" organization (Clowes 1981, 1982). Recent work suggests that an organizational state that is intermediate between open and closed, termed "intermediate open," was the ancestral orga-

nizational state of Angiosperm RAMs (Groot et al. 2004). But no matter what the architecture (open, closed, or intermediate), all Angiosperm RAMs consist of files of cells that converge to a small group of cells located just proximal to the RC proper (**Figure 1**). Because of their location, cells situated at the point of file convergence were long considered to function as the initial cells for the root. Another important conclusion, derived from lineage analyses, was that in roots showing closed organization, maintenance of the RC and formation of new RC cells could be traced to a set of initials that functioned solely in the production of new RC derivatives. This contrasted with RAMs showing open organization, where files of several different cell types converged to a common group of initials; this suggested that more than one cell/tissue type is produced from a common initial cell(s) (Cutter 1971). Much effort has been directed at determining the minimum number of initial cells, which taken together are said to form the promeristem (Clowes 1954). Based on lineage analyses, early researchers concluded that the promeristem was quite small, perhaps consisting of 20 cells or less (**Figure 1c**) (Clowes 1954, 1961).

This notion of a small number of initials was challenged by Clowes (1954), who showed that the promeristem was broad and consisted of a relatively large number of cells. Using radiolabeled DNA precursors and autoradiography, Clowes demonstrated that cells positioned at the point of lineage convergence were relatively inactive mitotically (**Figure 2**). He termed this region of slowly dividing cells the QC (Clowes 1956, 1975). We now know that the QC is a ubiquitous feature of all Angiosperm RAMs for at least part or all of their ontogeny. Clowes's proposal of a group of mitotically inactive cells at the point of lineage convergence completely upset all previous ideas of the promeristem (Clowes 1956). The presence of a QC meant that the functioning initials in a growing root were not located at the point of lineage convergence, and hence necessitated a

Auxin: from the very first mitotic division of the zygote, gradients of auxin, a plant hormone, guide patterning of the embryo into the parts that will become the plant's organs. The most important auxin produced by plants is indole-3-acetic acid (IAA)

RC: root cap

QC: quiescent center. Located just proximal to the RC, the QC is a collection of slowly dividing stem cells that spend long periods in G1

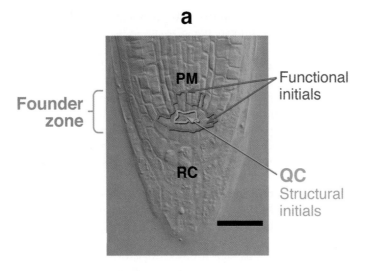

a

PM

Founder zone

Functional initials

RC

QC
Structural initials

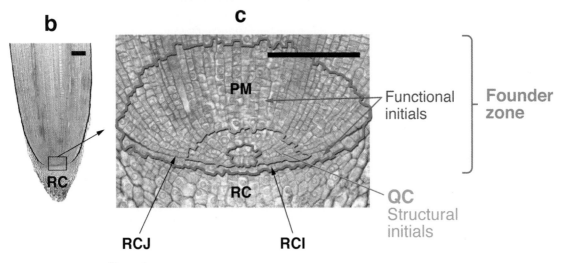

b

RC

c

PM

Functional initials

Founder zone

RC

QC
Structural initials

RCJ

RCI

Figure 1

(*a*) RAMs of *Arabidopsis thaliana* and (*b,c*) *Zea mays* showing the location of various cell populations. For *b* and *c*, note the convergence of cell files to a small number of cells, circumscribed in green (the originally designated promeristem). QC, quiescent center; PM, proximal meristem; RC, root cap; RCI, root cap initials; RCJ, root cap junction. Scale bar = 100 μm.

reconsideration of these initials' location. From the autoradiographs, Clowes also deduced that the mitotically active initials comprising the promeristem circumscribed the QC and that the location of these initials could shift as the QC enlarged or diminished in size (**Figure 2**). In this sense, a cell is an initial not because of any inherent properties but because of its position within the RAM and, more specifically, because of its position in relation to the QC. Hence, if the promeristem is considered to be the total collection of initials encircling the QC, then the promeristem is not permanent.

Although the QC lately has attracted much attention (Aida et al. 2004, Haecker et al. 2004, Ueda et al. 2004), its role in root development remains obscure. The development

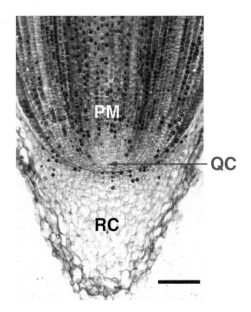

Figure 2

Autoradiograph of a *Zea mays* root apex supplied with ^{3}H-thymidine to indicate which cells are synthesizing DNA and thus able to divide. Black "dots" demarcate cells incorporating the radiolabel. Note the absence of labeled cells at the point of lineage convergence, in the region designated the QC. Most new derivatives in a growing root occur just proximal to the QC in the proximal meristem. QC, quiescent center; PM, proximal meristem; RC: root cap. Scale bar = 100 μm.

and interpolation of a zone of low mitotic activity within a region of meristematic activity raises a number of questions, including questions as to the reason for and consequences of rapidly dividing and quiescent cells being located adjacent to each other. Here, upon reviewing the literature on the QC, we conclude that the QC is central to root development and that it serves as an "integrator" for many processes and events requisite for meristem establishment and maintenance. Hence, the QC will be a focus for this review.

CONCEPT OF STEM CELLS

A number of researchers suggest that QC cells should be viewed as stem cells because of their apparent ability for unlimited proliferation,

self-maintenance, and self-renewal (Barlow 1997, Ivanov 2004). Barlow (1997) hypothesizes that in roots with large QCs, such as maize, the QC is composed of cells showing varying degrees of "stemness" (Cai et al. 2004), including variations in their ability to divide (or not divide) as well as their states of differentiation and self-renewal. Hence, he suggests that we consider QC cells as constituting part of "founder zones" within which exist cells displaying gradations or gradients in the traits that define a cell as a stem cell. Within the founder zone, lineages converge to "initials" (**Figure 1**), which, because they divide infrequently, appear compromised as to the active providing of new cells. To reconcile this apparent contradiction, Barlow distinguishes between two types of initials: "structural" and "functional." Functional initials are considered the rapidly dividing cells encircling and abutting directly onto the surfaces of the QC. Functional initials comprise the promeristem and because of their rapid rates of division, they serve as the source for most new derivatives in a growing root. Structural initials, on the other hand, are viewed as cells which divide relatively infrequently. Such divisions, when they do occur, result in additions to, and often replacement of, the functional initials (Barlow 1997). How the balance of structural and functional initials is maintained is not known. However, the recent application of the stem cell niche concept (Aida et al. 2004, Laux 2003)—the notion that stem cells are controlled by local environments known as niches—may offer some insight. This concept has already been usefully applied to animals (Spradling et al. 2001). When applied to the QC, this concept suggests that the cells comprising the QC acquire or express stem cell features not because of any preexisting properties but rather because of the particular microenvironment that defines the stem cell niche. Thus, for root meristems it is probably correct to consider founder zones and stem cell niches as either partly or completely overlapping (**Figure 1**).

Stem cells: undifferentiated cells with the capacity for unlimited proliferation, self-maintenance, and self-renewal

Stem cell niche: a microenvironment in which cells are maintained as stem cells

IAA: indole-3-acetic acid

Polar auxin transport: the unidirectional, energy-requiring movement of auxin from the shoot to the root tip

As Spradling et al. (2001) note, a niche can exist in the absence of resident stem cells. Clearly, understanding the developmental mechanisms that activate or silence niches is an important challenge, and with regard to root development, we have begun to make some progress. In maize roots from which the terminal half-millimeter is surgically excised (i.e., both the RC and QC are removed), a new root meristem will re-form, but only after the development of a new QC (Feldman 1976). Insight into QC redevelopment has come from work showing that stem cell niche reestablishment is auxin-IAA dependent. Inhibiting polar auxin transport prevents both stem cell niche reestablishment and reformation of the QC (Sabatini et al. 1999). More recent work using molecular markers specific to the QC (Friml et al. 2002, Sabatini et al. 1999) has shown that long-term exposure of *Arabidopsis* roots to auxin transport inhibitors induces ectopic QC formation (Sabatini et al. 1999), which further supports the notion that auxin is central to stem cell niche development. Linking the establishment of this niche with the formation of the QC is and will continue to be an active research thrust.

Interestingly, in applying the niche concept to *Arabidopsis*, Aida et al. (2004) do not believe the four-celled *Arabidopsis* QC is composed of stem cells. Rather, they view the QC as an organizing center that, through the production and local distribution of a signal, maintains as stem cells the adjacent, rapidly dividing initial cells. If we accept that a stem cell must divide, at least occasionally, then under the experimental conditions used by most researchers, the *Arabidopsis* QC indeed does not appear to be composed of stem cells. There are no reports of the QC cells dividing under the conditions under which *Arabidopsis* plants are normally grown (Fujie et al. 1993). However, if *Arabidopsis* plants are exposed to elevated temperatures (30–42°C) for varying periods (5–180 min), QC cells activate and replace, at different rates, the surrounding stem cells (Kidner et al. 2000). More recently Baum et al. (2002) concluded that the four

cells comprising the *Arabidopsis* QC "do divide initially after germination and apparently continue to divide during subsequent root growth until the primary root reaches its final length." They further noted that whereas divisions within the QC may be rare during early stages of seedling root growth (stages used by most investigators), they are relatively more frequent in older roots. Thus, given the criteria by which stem cells are typically defined, including the capacity for occasional, asymmetric divisions—where one daughter cell may differentiate—and the capacity of QC cells to respond (activate) under certain conditions in which neighboring cells are lost or destroyed (Barlow 1997, Cai et al. 2004, Kidner et al. 2000), we conclude that the four cells comprising the *Arabidopsis* QC are most certainly stem cells and should be regarded as structural initials as opposed to the surrounding functional initials (which Aida et al. regard as *the* stem cells) (Aida et al. 2004, Sablowski 2004). Significantly, this conclusion means that *Arabidopsis* root development is not so different from that of larger roots in other members of the Cruciferae, which have much larger QCs (Cutter 1971). As well, this conclusion is central to understanding determinate growth in roots, both in *Arabidopsis* and in other species, as we discuss below.

TOWARD AN UNDERSTANDING OF QC FORMATION

Cells occupying the QC niche begin acquiring QC-specific molecular characters very early in embryogenesis (Friml et al. 2002). In *Arabidopsis*, the QC's origins can be traced to divisions of the upper hypophysis derivative, which generates four nascent QC cells located just proximal to the RC junction, at the pole of the stele (**Figure 3**). It is postulated that sometime thereafter, the cell cycles of these cells lengthen, mostly because G1 is prolonged. Quiescence is, significantly, not an all-or-nothing state but rather a continuum. In roots with small QCs, such as *Arabidopsis*, each of the four QC cells can be at the same state

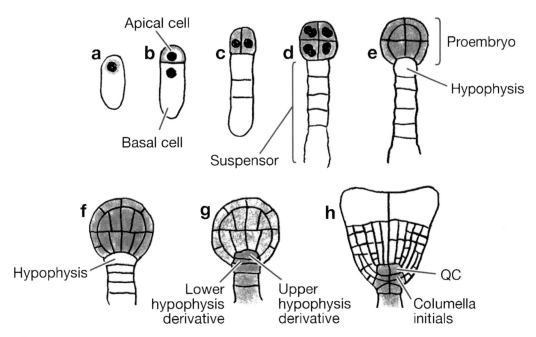

Figure 3

Schematic diagram of the early stages of *Arabidopsis* embryogenesis [after Hanstein (1870)]. Green shading indicates the distribution of auxin at various stages of embryo development; the darker the green, the more auxin [after Friml et al. (2003)].

of quiescence. However, in roots with considerably larger QCs, such as maize (with 600–1000 QC cells) (Clowes 1961), the degree of quiescence probably varies with position in the QC: Cells located at the point of lineage convergence are at the "deepest" level of quiescence and divide the least frequently, whereas cells located toward the proximal (basal) face of the QC divide more frequently (Barlow & MacDonald 1973). These differences in the state of quiescence again emphasize the fact that root meristem cells express certain characteristics because of their position in the meristem and not because these cells are inherently different from other cells in the root meristem.

Definitive data are now lacking on the timing of formation of the stereotypical (functional) QC. However, expression of *WOX5* (encoding a putative homeodomain transcription factor) in early globular (16-cell) embryos led Haecker et al. (2004) to conclude that QC identity is specified early in hypoph-

ysis formation. As well, other putative transcription factors, including *SHORT-ROOT, SCARECROW,* and *PLETHORA* (*PLT1* and *PLT2*) appear to be required for QC identity and specification and are expressed early in root development (Aida et al. 2004, Di Laurenzio et al. 1996, Helariutta et al. 2000, Sabatini et al. 2003, Wysocka-Diller 2000). *PLT1,* which is transcribed in response to auxin, is expressed as early as the eight-cell globular embryo stage, whereas *PLT2* is expressed in slightly older globular embryos. Through the generation of double and triple mutants, Aida et al. (2004) were able to show that *PLT* and *SCR/SHR* work via parallel pathways that appear to converge to a subset of target QC-specific promoters. *PLT* genes are not involved with auxin accumulation but rather act downstream of auxin, as indicated by the fact that auxins "cannot bypass the requirement for the *PLT* genes" (Aida et al. 2004). Additionally, the expression of a number of so-far uncharacterized QC-specific markers (QC25,

CLV: CLAVATA
WUS: WUSCHEL

QC46) has been detected in early heart-stage embryos (Aida et al. 2004, Friml et al. 2002). In summary, we can now conclude that the beginnings of QC specification can be traced to the 8–16-cell embryo.

MAINTENANCE OF THE QC AND STEM CELL NICHE

Once the stem cell niche is determined and the QC consequently developed, how is the niche (QC) maintained? Is the QC solely dependent on auxin? As Cai et al. (2004) note, a common problem faced by all stem cell populations is how they balance the number of differentiated cells with the maintenance of an adequate pool of self-renewing stem cells. Drawing from parallels with animals, Barlow (1997) suggests that the most straightforward way of viewing this balance in the QC (both structural and functional initials) is to consider that as cells leave the stem cell population, they in turn regulate the proliferative rate of the "true" stem cells. Such interactions between the stem cell niche and derivative, proliferating cells imply that communication exists between these various regions and thus point to future research directions.

In this regard, recent work demonstrating molecular controls for the location and number of stem cells in the shoot meristem of *Arabidopsis* may offer insights about similar controls in root meristems. In *Arabidopsis* shoot meristems, control of the size and positioning of the stem cell niche involves two-way signaling between the stem cells and an adjacent, underlying population of cells comprising the organizing center (Sablowski 2004). To maintain this balance, the CLV pathway, which restricts the number of stem cells, is balanced by WUS, a homeodomain transcription factor secreted by cells that constitute the underlying organizing center (Clark et al. 1997, Mayer et al. 1998). A feedback loop between these two proteins is part of the mechanism maintaining shoot meristem homeostatis (Brand et al. 2000, Carles & Fletcher 2003). In roots of *Arabidopsis*, overexpression of *CLE19*,

which encodes a CLV3 homolog, causes a diminution in the size of the root meristem (Casamitjana-Martínez et al. 2003). This suggests that CLE19 functions by overactivating an endogenous CLV-like pathway involved in root meristem maintenance. Interestingly, overexpression of *CLE19* in the root meristem via the RCH1 promoter reduces meristem size without affecting either QC specification (QC-specific markers are still correctly expressed) or stem cell maintenance (Casamitjana-Martínez et al. 2003). However, when *CLE19* is placed under regulation of the CaMV35S promoter, the QC in the primary root disappears and the root meristem is fully differentiated (Fiers et al. 2004). Thus, although there is evidence of a role for *CLV3*-like genes operating in the root meristem, a pathway similar to the CLV/WUS feedback loop has not been found to date in *Arabidopsis* roots (Sablowski 2004). Although the CLV pathway has not yet been shown to operate in roots, it is clear that, as in shoots, many factors are likely involved in regulating root meristem organization (Sabatini et al. 2003).

THE CENTRAL ROLE OF AUXIN

One of the most satisfying conclusions to come from work with *Arabidopsis* root mutants is the central role of the plant hormone, auxin (IAA), in root development. Recent work demonstrates that auxin influences root development and specification from the earliest stages of embryogenesis (Hamann et al. 1999, 2002). In a truly landmark paper, Friml and others (2003) report that after division of the zygote of *Arabidopsis*, auxin accumulates in the smaller apical derivative (**Figure 3**). The auxin maximum remains apically positioned throughout the development of the preglobular embryo. However, by the 32-cell stage and apparently coincident with the initial expression of certain QC-specific markers, the auxin maximum shifts basally to the most distal (uppermost) cells of the suspensor, which includes the hypophysis, the progenitor of the QC. Some time after the QC is established, the auxin maximum moves yet again to pre-cursors of the columella initials (Sabatini et al. 1999). Changes in location of the auxin maximum are correlated with the distribution and redistribution of members of the PIN protein family, putative auxin efflux carriers of which at least four are expressed during *Arabidopsis* embryogenesis (Aida et al. 2002). Particular attention was drawn to the cellular distribution of PIN1, PIN4, and PIN7, whose locations shift during the early stages of embryo development. This led Friml et al. (2003) to postulate that the repositioning of the efflux carriers underlies and accounts for changes in auxin distribution and in the relocation of the auxin maximum. Further, Friml et al. (2003) suggest that auxin gradients may be reversed and that maxima may shift, not only because of changes in the location of the auxin transporters, but also because rates of auxin efflux may differ among the transporters.

In this model, therefore, specification of the root occurs as a consequence of a PIN1- and PIN4-dependent accumulation of auxin. In evaluating this model, it is worthwhile to consider the origins of this auxin. Friml et al. (2003) suggest that the auxin in preglobular embryos originates either in the suspensor or externally, perhaps in the embryo sac. However, by the globular stage and coincident with a change in the expression pattern of the PIN-dependent transporters, the embryo most likely begins auxin production in the apical region (Friml et al. 2003). Moreover, as the embryo matures, auxin production likely continues and increases at the apical end. Genes mediating the auxin response, including *BO-DENLOS* (which affects sensitivity to auxin) (Hamann 2001, Hamann et al. 1999, 2002), *MONOPTEROS* (which encodes an auxin response factor affecting polar auxin transport) (Aida et al. 2002, Berleth & Jürgens 1993, Hardtke & Berleth 1998), and *AUXIN RESIS-TANT1* (which affects the auxin response by mediating ubiquitination) (Leyser et al. 1993), have mutants that do not form an embryonic root. For two of these mutants, *monopterous* (*mp*) and *bodenlos* (*bdl*), initial regulation by

auxin of root meristem development may be direct. For *mp* mutants, the first recognizable morphological effects of the mutation are seen at the octant stage of embryogenesis, in which four rather than two tiers of embryonic cells are formed. For *bdl* mutants, defects are first observed at the two-cell proembryo stage, in which the apical cell undergoes a transverse rather than longitudinal division.

SPECIFICATION AND DIFFERENTIATION OF THE HYPOPHYSIS

The specific controls in *Arabidopsis* for the division of the upper hypophysis derivative into the four-celled QC are not known (**Figure 3**). However, an early role for auxin is suggested. Mutations of the *HOBBIT* gene cause the hypophysis to develop incorrectly, which disrupts QC development (Willemsen et al. 1998). Moreover, *hobbit* mutants show a reduction in auxin reporter gene expression and accumulate the AXR3/IAA17 repressor of auxin responses, which supports a role for auxin in the earliest stages of root meristem specification (Blilou et al. 2002). Additionally, the *pin*7-mutant *Arabidopsis* lines show abnormal, premature divisions of the hypophysis. Finally, both the shift of PIN1 to the basal face of the hypophysis and the repositioning of the auxin maximum from the QC point to a role for auxin in the control of hypophysis mitoses. But why establish such high levels of auxin in these cells? Is it possible that a certain auxin threshold must be reached before the hypophysis can develop further and form the QC? And if so, are these auxin-regulated events dependent mainly on the positioning (and repositioning) of the PIN proteins (specifically PIN4)? If yes, what then regulates both these efflux carriers' positioning and subsequent shift in position in the hypophysis and its descendants? Possibly auxin directs or canalizes its own movement by regulating the repositioning of its own carriers. The canalization theory (Berleth & Sachs 2001, Sachs 2000) considers this mechanism and is based

Auxin efflux carriers: transmembrane proteins inserted in localized portions of the plasma membrane that preferentially move auxin out of the cell

upon the view that small, local differences in auxin concentration can be amplified by a self-reinforcing accumulation mechanism, which results in local auxin elevation and depletion in surrounding tissue. Can we use this theme of self-reinforcement to understand changes in the positioning of the auxin maximum at different stages of embryogenesis? Is auxin canalization linked with a repositioning of the efflux carriers? Friml et al. (2004) suggest that the localization of PIN proteins toward either the apical or basal end of cells depends, in part, on the level of expression of PINOID (PID), a serine-threonine kinase. Overexpression of PID causes a basal to apical relocalization of PIN and a consequent reduction of auxin in primary root tips. Moreover, if the *PID* gene is attached to a promoter primarily active in young embryos, globular- and heart-stage embryos are unable to establish an auxin maximum in the hypophysis. This results in a misspecification of the hypophysis. Most intriguing is that auxin itself controls cellular PID levels (Benjamins et al. 2001), which thereby suggests a feedback mechanism by which auxin regulates its own distribution. In the context of embryogenic root meristem organization, the accumulation of auxin in the apical cell following the first zygotic division (Friml et al. 2003) may be in response to an auxin feedback loop. If correct, the challenge is to discover the earliest steps of this auxin self-regulation. The elucidation of the role of PID is an important step toward realizing this goal (Friml et al. 2004). As well, the demonstration that other proteins, such as GNOM, influence localization of auxin efflux carriers (Steinmann et al. 1999), suggests the existence of other targets (proteins or genes) for auxin feedback and self-regulation. Taken together, substantial evidence now points to the likelihood that root specification is an auxin-dependent/linked process that begins early in the chain of events leading to QC formation. Significantly, however, high auxin levels do not always translate into quiescence. For example, the columella initials are the most rapidly cycling cells in the mature root yet

have the highest auxin levels. Thus, high auxin per se does not cause cells to divide more slowly. Rather, auxin regulation of cell division likely depends both on auxin levels in a cell or group of cells and these cells' position in the root, especially position in relation to the QC and stem cell niche. In this regard, Blilou et al. (2005) have recently proposed a way in which auxin accumulation is linked to root specification; PIN proteins restrict *PLT* expression and in turn *PLT* genes maintain *PIN* transcription.

While specification of the uppermost suspensor cell as the hypophysis appears required for *Arabidopsis* embryonic root meristem initiation and organization (Friml et al. 2003), there are a large number of species in which the root meristem organizes without forming a hypophysis (von Guttenberg 1968). Examples include pea (*Pisum sativum*) (Tiegs 1912) and many grasses (e.g., maize) in which root meristem formation occurs from an irregular mass of embryonic parenchyma tissue within which the root pole organizes (von Guttenberg 1968). It is unknown whether the fact that a hypophysis is not required for embryonic root meristem development implies that a polar auxin gradient is also absent in embryos lacking a hypophysis. However, we consider this unlikely and rather posit that apical-basal auxin gradients are central to root meristem organization in all Angiosperms whether or not a hypophysis develops in the embryonic root.

A ROLE FOR THE QC IN THE ESTABLISHMENT, MAINTENANCE, AND ELABORATION OF PATTERNS

What does the QC do? The origins of the QC, or at least the temporal expression of some QC-specific genes early in root embryogenesis, suggests a role for the QC in root meristem establishment, but there is not yet much evidence to support this view. The evidence is mostly correlative: In embryos in which a QC fails to form, patterning of the primary root meristem is disrupted (Friml et al. 2002). There is considerable evidence, however, that in mature roots a QC is essential for normal root growth (Feldman & Torrey 1976). By exposing roots to environmental extremes such as cold (Barlow & Adam 1989) or radiation (Clowes 1959), or by culturing roots in media supplemented with varying levels of nutrients (Feldman & Torrey 1976), QC size can be altered and meristem architecture affected. Exactly how these treatments lead to a change in QC size, that is, induce some QC cells to divide, is not known. In some cases it seems likely that QC cells are activated indirectly in response to environmental damage to the adjacent, rapidly dividing initials, which again suggests that a balance exists between the stem cells comprising the QC and the rapidly dividing initials. Damage to the initials thus shifts this balance, which is restored by activation of the QC and replacement of the damaged initials. Other evidence pointing to the necessity of the QC in the maintenance of root meristem organization comes from surgical approaches, including laser ablation. Destruction of one or more of the four *Arabidopsis* QC cells alters the developmental fate of contacting initial cells (van den Berg et al. 1997). In maize, excising the QC (and the RC) leads to a reorganization of the remaining root meristem tissue (Feldman 1976). A new root meristem will re-form, but only *after* a new QC begins to re-form (Feldman 1976). This suggests that it is not the cells occupying the QC niche that are special, but rather the niche that is special. Interestingly, if additional proximal root tissue (>0.5 mm in maize) is also excised with the QC, a new QC niche cannot re-form. Hence, the damaged root meristem is not replaced, which suggests that the "new" distalmost cells may have lost their competence to respond to auxin (Sabatini et al. 1999).

Changes in QC size and shape correlate with changes in tissue (vascular) patterns. Feldman & Torrey (1976) reported that as a consequence of shifting the location of the basal (proximal) edge of the QC, the PM is moved into a wider or narrower region of the root, further from and closer to the RC, respectively. These researchers showed that maize roots with small QCs had simpler vascular patterns and that an increase in QC size (upon moving the edge of the PM to a wider region of the root) corresponded to an increase in vascular complexity. They thus proposed that the QC influences root tissue patterns through alterations in its size.

ROOT DETERMINATION AND THE QC AND STEM CELL NICHE

Although we tend to think of individual roots (and their root meristems) as having the potential for unlimited growth, most roots in fact have a limited lifetime (Chapman et al. 2003, Varney & McCulley 1991 and references cited therein). In some special instances, however, individual roots can be quite long-lived, such as those of certain Gymnosperms, which can live for more than one year (Wilcox 1962), and for roots of certain species that can be maintained for long periods in tissue culture (Torrey 1958). But this behavior is the exception. Roots of most species, including those of *Arabidopsis*, become determinate when they are four to five weeks old (Barlow 1997, Baum et al. 2002). In wildtype *Arabidopsis*, root determination is preceded by an activation of the structural initials (the QC) resulting in a change from closed to open in apical meristem architecture. Ultimately, the four cells comprising the QC are no longer distinct and found in their usual position, just proximal to the RC, are "disorganized," vacuolated cells (Baum et al. 2002). The stem cell niche has been lost; having become determinate, the root ceases to grow. Barlow (1997) suggests that we may view this determination essentially as a loss of the QC's ability to divide and thereby replenish the functional initials. How might we view the underlying causes of determination? If, as suggested, the niche is primarily dependent on auxin, then perhaps insufficient auxin accumulates to maintain the niche? Or perhaps in older root meristems,

PM: proximal meristem. In an actively growing root, the collection of dividing cells overlaying the basal (proximal) face of the QC. Through the proliferative activity of the proximal meristem, new root derivatives form

there is a lessening of the inhibitory effects of the QC cells on the differentiation of the neighboring functional initials (van den Berg et al. 1997), which thereby upsets the balance between functional and structural initials and leads to determination. Perhaps, too, one might consider possible parallels with the model of stem cell termination proposed for the conversion of cells in a vegetative shoot meristem to part of a floral meristem (Lenhard et al. 2001).

An extreme example of root determination is found in species of the Podostemaceae, in which roots either never form or abort very early during embryogenesis (Rutishauser 1997, Suzuki et al. 2002). A number of known mutations that underlie similar phenoypes in *Arabidopsis* [e.g., *axr6* (Hobbie et al. 2000), *short-root* (Helariutta et al. 2000), *monopterous* (Hardtke & Berleth 1998), *rfc3* (Horiguchi et al. 2003), and *root meristemless* (Cheng et al. 1995)] show premature root determination, which directs us to an examination of the underlying genetic controls of root determination. In particular, in the *rml* mutant of *Arabidopsis*, the root meristem forms normally during embryogenesis only to disorganize shortly after germination. Because the *RML* gene encodes the first enzyme of glutathione (GSH) biosynthesis, gamma-glutamylcysteine synthetase, it has been suggested that this mutation causes meristem disorganization by interfering with a redox-regulated G1/S transition of the cell cycle. This would lead to an imbalance in the stem cell population (Vernoux et al. 2000). These efforts thus point to candidate genes whose homologues may underlie the determinate root phenotypes, as in the Podostemaceae.

Although determination of roots is generally viewed as an "ending" of root function, in some instances, the determination process can also be considered a "beginning." For example, in some palms (e.g., *Cryosophila*), the first-order roots produced on the trunk become determinate and are converted into protective spines (Tomlinson 1990). Other examples of determinate spine roots occur at the nodes of some bamboos. How this transformation and determination occur, especially in regard to already well-characterized meristem landmarks such as the QC, has recently been explored in roots of cacti (e.g., *Pachycereus* and *Stenocereus*) by Rodríguez-Rodríguez et al. (2003). These researchers showed that shortly after germination, the primary root becomes determined, which leads to the formation and outgrowth of many branch roots requisite for seedling establishment (Dubrovsky 1997). In this determination process, a QC disappears at or shortly after germination. Rodríguez-Rodríguez et al. (2003) thus concluded that maintenance of typical meristem architecture and indeterminate growth are impossible in the absence of a QC. Toward an understanding of the mechanism of determination, Rodríguez-Rodríguez et al. (2003) suggest similarities between their observations and those regarding defects in *Arabidopsis* roots caused by mutations of either *HOBBIT* (Blilou et al. 2002, Willemsen et al. 1998) or *PINOID* (Bennett et al. 1995, Christensen et al. 2000, Friml et al. 2004). Because a mutation of *PINOID* results in increased auxin efflux in *Arabidopsis* roots, Rodríguez-Rodríguez et al. (2003) treated the developing (not-yet-determined) cacti roots with the auxin transport inhibitor NPA to ascertain whether they could rescue the cacti roots and maintain the indeterminate state. They were unsuccessful. This may be explained by the recent finding that PINOID overexpression leads to meristem disorganization and auxin reduction at the root tip, which suggests that cacti roots may become determinate because of too little and not too much auxin (Friml et al. 2004). Supporting this suggestion are reports that the terminal differentiation of primary-root meristems of 35S::*PID* seedlings is preceded by a reduction in the expression of auxin-responsive reporter genes (Benjamins et al. 2001). Although little is known of the molecular mechanisms leading to QC disappearance and root determination, that *PLT* expression is required for maintenance of both the QC and normal meristem architecture may mean that molecular events underlie the indeterminate state of roots. Aida et al. (2004) explored this possibility by making double mutants in *Arabidopsis* of this two-gene family (*plt1 plt2*), which showed that lateral root meristems disorganized and terminally differentiated shortly after their initiation.

THE ROLE OF THE RC IN MERISTEM ESTABLISHMENT AND MAINTENANCE

Because division of the hypophysis simultaneously produces derivatives that are the progenitors for both the QC and RC, the question arises as to the requirement of a RC for meristem establishment. In the absence of a RC, would a root meristem organize? Two types of indirect evidence suggest that a RC is central to primary root meristem establishment and maintenance. First, no capless primary (embronic-in-origin) roots exist in nature (von Guttenberg 1968). While it is true that lateral roots of some highly derived species lack a RC (Hiyama et al. 2002, Suzuki et al. 2002), there are no reports of embryonic roots developing without a RC. Second, no capless, embryonically formed roots have been reported from *Arabidopsis* mutant screens. This may simply be coincidental since in *Arabidopsis*, any mutation affecting the hypophysis would be expected to impact development of both the RC and root meristem (Willemsen et al. 1998). To the contrary, we conclude that the simultaneous origin of the progenitors of both the RC and QC is not coincidental and points to a developmental linkage between the RC and QC that is established early in embryogenesis. Additional evidence of the interdependence of the RC and QC comes from experiments in which the RC can be excised (via a tear in the columella initials) (Feldman 1977, Lim et al. 2000). As a consequence of RC removal, the QC activates and meristem architecture is dramatically altered (Feldman 1976). In this regard, one would predict that destruction of RCIs,

employing the same approach used to ablate QC and other root tip cells (van den Berg et al. 1995, 1997), would also activate the QC. The recent report that transgenic toxin expression in RCs leads to alterations in root meristem activity reinforces the notion of the RC's importance in overall meristem function (Tsugeki & Federoff 1999), in part by affecting lateral auxin redistribution (Blilou et al. 2005). As well, reports that RC border cells in pea affect meristem activity and gene expression provide additional evidence for a regulatory role for the RC in root meristem function (Woo et al. 1999).

Experimental data thus show that the RC is required not only for meristem establishment but also for meristem maintenance. While at this time we can only speculate about the nature of this regulation, the shift of the auxin maximum during early embryogenesis to the columella initials suggests that the nascent RC may function as a positional marker that defines where the new QC will re-form. The recent report that RCs are short-lived in determinate roots of certain cacti (Rodríguez-Rodríguez et al. 2003) supports this suggestion. It also raises the intriguing possibility that root determination may have its origins in changes in the RC that could translate into changes in the positioning and degree of the auxin maximum. Given the fact that a small amount of auxin can create a sink and cause more auxin to move toward this sink (Feldman 1981), the shift in the auxin maximum to the columella initials may be an outcome of the development of an auxin sink in the RC. In this regard, studies on the timing and distribution (and redistribution) in the RCs of PIN and PINOID should be very informative (Friml et al. 2004).

COMMUNICATION IN THE ROOT MERISTEM

The coordinated activities and developmental links between the RC and the QC point to a two-way communication between these two populations. Not only does the RC

RCI: root cap initial

ROS: reactive oxygen species. ROS are molecules or ions formed by the incomplete one-electron reduction of oxygen. Reactive oxygen intermediates include singlet oxygen, superoxides, peroxides, and hydroxyl radicals. ROS participate in the regulation of signal transduction and gene expression as well as in the process of oxidative damage to nuclei acids, proteins, and lipids

communicate with the QC (as shown by decapping experiments) (Feldman 1976, Lim et al. 2000) but, as suggested from laser-ablation experiments, the QC communicates with the RC (van den Berg et al. 1995, 1997). In *Arabidopsis*, ablation of one or more QC cells causes the adjacent, rapidly cycling columella cells to cease dividing and instead undergo differentiation (van den Berg et al. 1995). Although the nature of this hypothesized two-way communication is not known, we have a few clues. One of the more intriguing comes from maize roots, which increase slightly their rate of growth following cap excision (L.J. Feldman, personal observations). As well, the increase in mitoses in the QC and the differentiation of membrane-bound starch grains in the terminal-most QC cells following cap excision (Barlow 1974) suggest that signals from and processes in the RC are important in the maintenance of both quiescence and the undifferentiated status of QC cells. We can thus conclude that these signals, whether originating in or redistributed by the RC, play important roles in balancing cell division and differentiation in the root meristem. At this point, while we can only speculate as to the nature of such communication and positional information, the communication mechanism surely involves auxin in part. Upsetting auxin transport perturbs the balance between the QC and RC and can result in the QC penetrating and displacing the original cap (Jiang et al. 2003). Also, given the high levels of oxidative stress in the QC (Jiang et al. 2003), one could hypothesize a role for free radicals in short-range communication as has already been shown for plant responses to pathogens (Laloi et al. 2004, Wojtaszek 1997). In addition, the recent demonstration of the requirement of SCARECROW (SCR) for QC specification suggests that interactions such as those of WUS/CLV in the shoot may operate in root meristems in balancing proliferation and differentiation (Doerner 2003, Sabatini et al. 2003). Sabatini et al. (2003) postulate that SCR activity in the QC is important in maintaining function in the surrounding

stem cells (functional initials). *SCR* activity is regulated by SHORT-ROOT, a transcription factor that originates in the stele (Benfey et al. 1993, Helariutta et al. 2000). Thus, taken together, these observations provide specific evidence for short-range communication feedback loops between the various cell types comprising the root meristem. Additionally, this work (Sabatini et al. 2003) focuses attention on communication between the QC and the cells bordering on the proximal (basal) face of the QC, the proximal meristem (PM) (**Figures 1, 2**). Damaging the PM by exposing roots to environmental extremes activates the QC, which results in a replacement of the damaged PM cells and a re-formation of the QC (Barlow & Adam 1989). It would be interesting to view this damage and activation in *Arabidopsis* in relation to *SCR* and *SHR* activity. In this regard, activation of the QC might provide insight into the unidentified signals believed to move from the QC as a result of *SCR* action (Sabatini et al. 2003). The challenge now is to determine the interplay, or feedback, between QC-, RC-, and PM-derived signals and to show which of these act as positional signals and thereby promote or inhibit cell-type specific differentiation.

REDOX REGULATION OF ROOT DEVELOPMENT

We next consider recent views on the mechanisms linking auxin to the development of the stem cell niche, the QC, and root meristem establishment. An important insight into these links has come from recent work showing that the redox status of the QC is different from that in adjacent, rapidly dividing cells: The QC has a more oxidizing environment (Jiang et al. 2003, Kerk & Feldman 1995, Kerk et al. 2000, Liso et al. 2004, Sánchez-Fernández et al. 1997). Consequently, Jiang et al. (2003) have proposed that auxin affects the cell cycle in the QC via changes in redox.

Overall cell redox is determined by the net contribution of different redox couples and ROS. In biological systems, the major

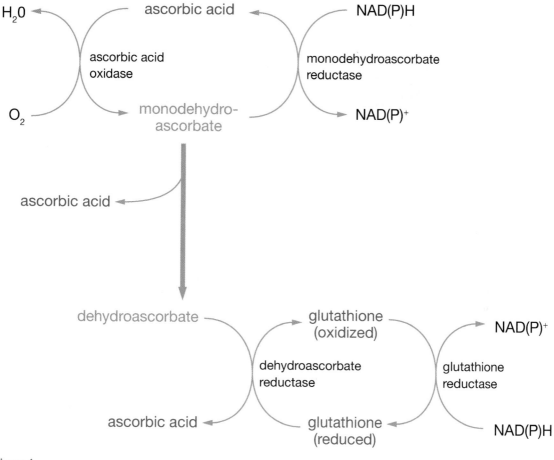

Figure 4

Ascorbate/gluthatione cycle.

(and usually most abundant) redox couples are GSH/GSSG and AA/DHA (Arrigo 1999; Arrigoni & De Tullio 2002; Banhegyi et al. 1997; Conklin & Barth 2004; De Tullio & Arrigoni 2003, 2004; Filomeni et al. 2002; May et al. 1998; Noctor et al. 2002; Potters et al. 2002, 2004; Schafer & Buettner 2001), which are biochemically interconnected via the ascorbate/glutathione cycle (May et al. 1998) (**Figure 4**). The levels and ratios of the reduced and oxidized forms of these two couples are a direct indicator of a tissue's overall redox status.

Many lines of evidence suggest that cellular redox status plays a critical role in regulating cell proliferation (Barzilai &

Yamamoto 2004, den Boer & Murray 2000, Reichheld et al. 1999, Shackelford et al. 2000). In Chinese hamster ovary fibroblasts, GSH concentration has been related to the cell-cycle phase, with the lowest GSH concentration at G1, a higher concentration at S, and the greatest concentration at G2/M (Conour et al. 2004). In human colon cancer CaCo-2 cells, a change in the ratio of GSH to GSSH to favor GSSH, the oxidized form, resulted in cells arresting at G1/S (Noda et al. 2001). In plants too, changes in the absolute amounts and/or ratios of reductants/oxidants affect cell proliferation, as shown by the *rml1* mutant, in which less glutathione is synthesized because of a mutation

GSH/GSSG:
glutathione[reduced]/
glutathione[oxidized]

AA/DHA:
ascorbate/
dehydroascorbate

AAO: ascorbic acid oxidase

GR: glutathione reductase

in gamma-glutamylcysteine synthetase, the first enzyme in glutathione biosynthesis. In this mutant, the root meristem forms normally but ceases producing new derivatives shortly after germination and disorganizes (Vernoux et al. 2000). Similarly, treating cultured plant cells with a specific inhibitor of gamma-glutamylcysteine synthetase can block the plant cell cycle (Potters et al. 2004, Sánchez-Fernández et al. 1997). Addition of AA increases the rates of cell proliferation in plant cells (Kerk & Feldman 1995, Liso et al. 1988, 2004), whereas addition of the oxidized form of AA, DHA, delays cell-cycle progression (Potters et al. 2002, 2004). Reichheld et al. (1999) showed that tobacco cells in G1 were more sensitive to oxidative stress than cells in S phase; this accords similar findings with regard to mammalian cells, in which G1 progression is influenced to a large extent by extracellular signals (Hulleman & Boonstra 2001, Juliano 2003). Reichheld et al. (1999) thus proposed that redox sensing could be central to the control of cell-cycle progression under conditions of environmental stress. These observations, plus reports of relatively low levels of GSH (Jiang et al. 2003, May et al. 1998) and AA (Kerk & Feldman 1995, Liso et al. 2004) in the QC, and the observation that AA, or its immediate precursor, L-galactono-gamma-lactone, activates the QC (Kerk & Feldman 1995, Liso et al. 2004), has led to the proposal of a linkage between redox status and the arrest of QC cells in G1 (Jiang et al. 2003).

Fundamental cell cycle mechanisms are conserved in eukaryotes and involve as core regulators cyclins, cyclin-dependent kinases, cyclin-dependent kinase inhibitors, and the protein retinoblastoma (reviewed in De Veylder et al. 2003, Dewitte & Murray 2003, Massague 2004, Schafer 1998). Reports that oxidative stress and redox status affect the expression and/or activities of these and other cell-cycle regulators and thus lead to cell-cycle arrest highlight possible mechanisms underlying direct redox control of the cell cycle (reviewed in Barzilai & Yamamoto 2004,

Boonstra & Post 2004, den Boer & Murray 2000). The general characteristics of these controls include downregulation of G1 D-type cyclins, inhibition of the ubiquitin pathway by reducing the activities of proteasome or the enzymes involved in the ubiqitination process, upregulation of CKIs, and suppresion of retinoblastoma phosphorylation (Esposito et al. 2000, Hosako et al. 2004). In plants, A-type cyclins are involved in S-phase progression (Dewitte & Murray 2003). Under oxidative stress, the expression of two A-type cyclins in BY-2 tobacco cells is downregulated and results in cell-cycle arrest at G1/S (Reichheld et al. 1999). Intriguingly, Burssens et al. (2000) showed that an A-type cyclin is not detected in the QC of the *Arabidopsis* primary root; this correlates well with the more-oxidized status of QC and the arrest of QC in G1. We thus conclude that redox directly regulates the plant cell cycle.

In plants, auxins have been linked to changes in redox (Jiang et al. 2003, Takahama 1996). High levels of auxin are correlated with the generation of H_2O_2, $O_2 \cdot^-$, and other ROS (Joo et al. 2001, Pfeiffer & Höftberger 2001, Schopfer 2001, Schopfer et al. 2002). The mechanisms leading to ROS generation are not known but may involve free-radical generation via peroxidase-catalyzed oxidation of IAA (reviewed in Kawano 2003). ROS generation may also occur as a consequence of auxin affecting the activities of redox-associated systems (Jiang et al. 2003, Kisu et al. 1997, Pignocchi & Foyer 2003, Pignocchi et al. 2003, Takahama 1996). In this regard, upregulation by auxin of AAO in the QC (Jiang et al. 2003, Kerk & Feldman 1995) and concomitant downregulation of GR activity (Jiang et al. 2003) might possibly underlie auxin-induced changes in redox in the QC. In the QC, more AAO would lead to a rapid conversion of AA to DHA (Pignocchi & Foyer 2003, Pignocchi et al. 2003) and thereby diminish the cells' capacity to neutralize newly formed ROS (**Figure 4**) (Conklin et al. 1996, Pastori et al. 2003). As well, the concomitant, IAA-regulated reduction in GR in the QC would reduce even further the capacity to buffer redox changes. This is because a lowering in GR activity results in less AA being regenerated (**Figure 4**). In addition, auxin may influence tissue redox status through the induction of NADH oxidase leading to the production of H_2O_2 (Liszkay et al. 2003, Morre et al. 2003). In summary, limited but convincing evidence points to ROS generation as part of the mechanism of auxin action in the QC.

Given the G1 state of most QC cells, the demonstrable oxidized redox status of the QC, and the central role for auxin in root development, a model linking auxin to quiescence and to root meristem establishment has been proposed (**Figure 5**) (Jiang et al. 2003, Kerk & Feldman 1995, Kerk et al. 2000). The key elements of this pathway are initiated early in embryogenesis with the polar accumulation of auxin at the presumptive root pole. As a consequence, mechanisms are developed to regulate auxin levels, which results not only in auxin homeostasis but the formation of a region with a relatively oxidized redox status. This leads to the accumulation of cells at the G1/S checkpoint and the origin of the QC. That a QC is specified before a root meristem organizes (Feldman 1976) and that a QC must be maintained for root meristem integrity suggests that the QC has a central role in integrating and translating auxin's biochemical and molecular effects in root organogenesis.

SUMMARY

The delimitation of the stem cell niche, containing both the structural initials (the QC) and the functional initials, is a defining step in root meristem organization. Upon such delimitation, a "reference point" is established and the root meristem organizes. Central to this view is the supposition that specific patterns and timing of auxin distribution are requisite for normal root development. We propose that specification involves direct and indirect interactions of auxin with essential genes (Aida et al. 2004, Reed 2001) and, as we previously have suggested, auxin-regulated,

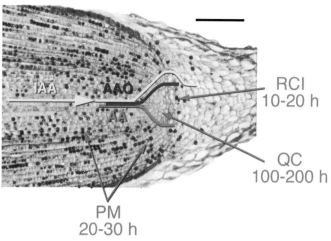

Figure 5

Schematic view of the distribution in the root apex (location of maxima and minima) of auxin (IAA), ascorbic acid oxidase (AAO), and ascorbic acid in relation to the QC. Average cell-cycle times (in hours) are indicated for various regions. PM, proximal meristem; RCI, root cap initials. Scale bar = 100 μm.

localized changes in redox (Jiang et al. 2003). Specification likely also involves an overlaying of the auxin maximum onto other morphogen gradients (Berleth 2001).

Underlying this specification, and essential for root meristem establishment, is the self-regulation by auxin of its own distribution and redistribution (Benjamins et al. 2001, Feldman 1981). We suggest that auxin effects its own transport by regulating the location and directionality of its own carriers, and thereby determines the size and position of the auxin maximum. A change in the maximum may also be accounted for by local differences in auxin synthesis and/or metabolism (Zazimalova & Napier 2003). In this regard, evidence linking auxin in the QC to high levels of auxin-metabolizing enzymes (Kerk & Feldman 1995) suggests that the "shift" of the auxin maximum from the QC to the RCI may in part be explained by auxin catabolism in the QC. Understanding why (and how) the auxin maximum is positioned and why it "pauses" in certain cells of the embryo are crucial to an appreciation of auxin-regulated meristem initiation (Uggla et al. 1996).

Finally, based on the fact that root meristems are sensitive to and very effective in metabolizing auxin (Feldman 1981), we propose that specification of the QC and stem cell niche, and subsequent root meristem organization, are inevitable developmental outcomes of the necessity to regulate auxin levels. In other words, root meristems that formed in response to auxin in turn become the mechanism for modulating auxin levels and thereby maintain auxin homeostasis. This may be a perfect feedback loop! Carrying this view to an extreme, one might even speculate that roots evolved as a consequence of plants' "need" to respond to increased levels of polarly transported auxin.

SUMMARY POINTS

1. The QC is composed of stem cells.

2. The stem cell niche is, in part, specified by auxin.

3. The stem cell niche and the QC are specified early in embryogenesis.

4. The formation of a QC is necessary for and precedes root meristem organization.

5. A QC must be maintained in order for the root meristem to remain organized and indeterminate.

6. Auxin specifies the stem cell niche and the QC by mediating localized changes in redox, preferentially favoring a more oxidizing environment.

7. The origin of roots may be linked to auxin-regulated development of oxidizing microenvironments, which culminates in root meristem establishment.

LITERATURE CITED

Aeschbacher RA, Schiefelbein KW, Benfey PN. 1994. The genetic and molecular basis of root development. *Annu. Rev. Plant Physiol. Plant Mol. Biol.* 45:25–45

Aida M, Beis D, Heidstra R, Willemsen V, Blilou I, et al. 2004. The *PLETHORA* genes mediate patterning of the *Arabidopsis* root stem cell niche. *Cell* 119:109–20

Aida M, Vernoux T, Furutani M, Traas J, Tasaka M. 2002. Roles of *PIN-FORMED1* and *MONOPTEROS* in pattern formation of the apical region of the *Arabidopsis* embryo. *Development* 129:3965–74

Arrigo AP. 1999. Gene expression and the thiol redox state. *Free Radic. Biol. Med.* 27:936–44

Arrigoni O, De Tullio MC. 2002. Ascorbic acid: Much more than just an antioxidant. *Biochim. Biophys. Acta* 1569:1–9

Banhegyi G, Braun L, Csala M, Puskas F, Mandl J. 1997. Ascorbate metabolism and its regulation in animals. *Free Radic. Biol. Med.* 23:793–803

Barlow P. 1974. Regeneration of the cap of primary roots of *Zea Mays*. *New Phytol.* 73:937–54

Barlow PW. 1997. Stem cells and founder zones in plants, particularly their roots. In *Stem Cells*, ed. CS Potten, pp. 29–57. London: Academic. 474 pp.

Barlow PW, Adam JS. 1989. The response of the primary root meristem of *Zea mays* L. to various periods of cold. *J. Exp. Bot.* 40:81–88

Barlow PW, MacDonald PDM. 1973. An analysis of the mitotic cycle in the root meristem of *Zea mays*. *Proc. R. Soc. London Ser. B* 183:385–98

This paper links auxin to *PLT* transcription and specification of the stem cell niche.

Barzilai A, Yamamoto K-I. 2004. DNA damage responses to oxidative stress. *DNA Repair* 3:1109–15

Baum S, Dubrovsky JF, Rost TL. 2002. Apical organization and maturation of the cortex and vascular tissue in *Arabidopsis thaliana* (Brassicaceae) roots. *Am. J. Bot.* 89:908–20

Benfey PN, Linstead PJ, Roberts K, Schiefelbein JW, Hauser M-T, et al. 1993. Root development in *Arabidopsis*: Four mutants with dramatically altered root morphogenesis. *Development* 199:57–70

Benjamins R, Quint A, Weijers D, Hooykaas P, Offringa R. 2001. The PINOID protein kinase regulates organ development in *Arabidopsis* by enhancing polar auxin transport. *Development* 128:4057–67

Bennett SRM, Alvarez J, Bossinger G, Smyth DR. 1995. Morphogenesis in *pinoid* mutants of *Arabidopsis thaliana*. *Plant J.* 8:505–20

Berleth T. 2001. Top-down and inside-out: Directionality of signaling in vascular and embryo development. *J. Plant Growth Regul.* 20:14–21

Berleth T, Jürgens G. 1993. The role of the *monopteros* gene in organising the basal body region of the *Arabidopsis* embryo. *Development* 118:575–87

Berleth T, Sachs T. 2001. Plant morphogenesis: long-distance coordination and local patterning. *Curr. Opin. Plant Biol.* 4:57–62

Blilou I, Frugier F, Folmer S, Serralbo O, Willemsen V, et al. 2002. The *Arabidopsis HOBBIT* gene encodes a CDC27 homolog that links the plant cell cycle to progression of cell differentiation. *Genes Dev.* 16:2566–75

Blilou I, Wildwater M, Willemsen V, Paponov I, Friml J, et al. 2005. The PIN auxin efflux facilitator network controls growth and patterning in Arabidopsis roots. *Nature* 43:39–44

Boonstra J, Post JA. 2004. Molecular events associated with reactive oxygen species and cell cycle progression in mammalian cells. *Gene* 337:1–13

Brand UC, Fletcher J, Hobe H, Meyerowitz EM, Simon R. 2000. Dependence of stem cell fate in *Arabidopsis* on a feedback loop regulated by *CLV3* activity. *Science* 289:617–19

Burssens S, de Almeida Engler J, Beeckman T, Richard C, Shaul O, et al. 2000. Developmental expression of the *Arabidopsis thaliana CycA2;1* gene. *Planta* 211:623–31

Cai J, Weiss ML, Rao HS. 2004. In search of "stemness." *Exp. Hematol.* 32:585–98

Carles CC, Fletcher JC. 2003. Shoot apical meristem maintenance: the art of a dynamic balance. *Trends Plant Sci.* 8:394–401

Casamitjana-Martínez E, Hofhuis HF, Xu J, Liu C-M, Heidstra R, et al. 2003. Root-specific *CLE19* overexpression and the *sol1/2* suppressors implicate a CLV-like pathway in the control of *Arabidopsis* root meristem maintenance. *Curr. Biol.* 13:1435–41

Casson SA, Lindsey K. 2003. Genes and signaling in root development. *New Phytol.* 158:11–38

Chapman K, Groot EP, Nichol S, Rost TL. 2003. The pattern of root apical meristem organization and primary root determinate growth are coupled. *J. Plant Growth Regul.* 21:287–95

Cheng JC, Seeley KA, Sung ZR. 1995. *RML1* and *RML2, Arabidopsis* genes required for cell proliferation at the root tip. *Plant Physiol.* 107:365–76

Christensen SK, Dagenais N, Chory J, Weigel D. 2000. Regulation of the auxin response by the protein kinase PINOID. *Cell* 100:469–78

Clark SE, Williams RW, Meyerowitz EM. 1997. The *CLAVATA1* gene encodes a putative receptor kinase that controls shoot and floral meristem size in *Arabidopsis*. *Cell* 89:575–85

Clowes FAL. 1954. The promeristem and the minimal constructional center in grass root apices. *New Phytol.* 53:108–16

Clowes FAL. 1956. Nucleic acids in root apical meristems of *Zea*. *New Phytol.* 55:29–35

Clowes FAL. 1959. Reorganization of root apices after irradiation. *Ann. Bot. (N.S.)* 23:205–10

Clowes FAL. 1961. *Apical Meristems*. Oxford: Blackwell. 217 pp.

Clowes FAL. 1975. The quiescent centre. In *The Development and Function of Roots*, ed. JG Torrey, DT Clarkson, pp. 3–19. London: Academic. 618 pp.

Clowes FAL. 1978. Origin of quiescence at the root pole of pea embryos. *Ann. Bot.* 42:1237–39

Clowes FAL. 1981. The difference between open and closed meristems. *Ann. Bot.* 48:761–67

Clowes FAL. 1982. Changes in cell population kinetics in an open meristem during root growth. *New Phytol.* 91:741–48

Conklin PL, Barth C. 2004. Ascorbic acid, a familiar small molecule intertwined in the response of plants to ozone, pathogens, and the onset of senescence. *Plant Cell Environ.* 27:959–70

Conklin PL, Williams EH, Last RL. 1996. Environmental stress sensitivity of an ascorbic acid-deficient Arabidopsis mutant. *Proc. Natl. Acad. Sci. USA* 93:9970–74

Conour JE, Graham WV, Gaskins HR. 2004. A combined in vitro/bioinformatic investigation of redox regulatory mechanisms governing cell cycle progression. *Physiol. Genomics* 18:196–205

Cutter EG. 1971. *Plant Anatomy: Experiment and Interpretation. Part II. Organs*. London: Edward Arnold. 343 pp.

den Boer BGW, Murray JAH. 2000. Triggering the cell cycle in plants. *Trends Cell Biol.* 10:245–50

De Tullio MC, Arrigoni O. 2003. The ascorbic acid system in seeds: to protect and to serve. *Seed Sci. Res.* 13:249–60

De Tullio MC, Arrigoni O. 2004. Hopes, disillusions and more hopes from vitamin C. *Cell. Mol. Life Sci.* 61:209–19

De Veylder L, Joubes J, Inze D. 2003. Plant cell cycle transitions. *Curr. Opin. Plant Biol.* 6:536–43

Dewitte W, Murray JAH. 2003. The plant cell cycle. *Annu. Rev. Plant Biol.* 54:235–64

Di Laurenzio L, Wysocka-Diller J, Malamy J, Pysh E, Helariutta Y, et al. 1996. The *SCARECROW* gene regulates an asymmetric cell division that is essential for generating the radial organization of the *Arabidopsis* root. *Cell* 86:423–33

Doerner P. 2003. Plant meristems: A merry-go-round of signals. *Curr. Biol.* 13:R368–74

Dolan L, Janmaat K, Willemsen V, Linstead P, Poethig S, et al. 1993. Cellular organisation of the *Arabidopsis thaliana* root. *Development* 119:71–84

Dubrovsky JG. 1997. Determinate primary-root growth in seedlings of Sonoran Desert Cactaceae; its organization, cellular basis, and ecological significance. *Planta* 203:85–92

Esposito F, Russo L, Russo T, Cimino F. 2000. Retinoblastoma protein dephosphorylation is an early event of cellular response to prooxidant conditions. *FEBS Lett.* 470:211–15

Feldman LJ. 1976. The *de novo* origin of the quiescent center in regenerating root apices of *Zea mays*. *Planta* 128:207–12

Feldman LJ. 1977. The generation and elaboration of primary vascular tissue patterns in roots of *Zea*. *Bot. Gaz.* 138:393–401

Feldman LJ. 1981. Effect of auxin on acropetal auxin transport in roots of corn. *Plant Physiol.* 67:278–81

Feldman LJ, Torrey JG. 1976. The quiescent center and primary vascular tissue pattern formation in cultured roots of *Zea*. *Can. J. Bot.* 53:2796–803

Fiers M, Hause G, Boutilier K, Casamitjana-Martinez E, Weijers D, et al. 2004. Misexpression of the *CLV3/ESR*-like gene *CLE19* in *Arabidopsis* leads to a consumption of root meristem. *Gene* 327:37–49

Filomeni G, Rotilio G, Ciriolo MR. 2002. Cell signalling and the glutathione redox system. *Biochem. Pharmacol.* 64:1057–64

Friml J, Benková E, Blilou I, Wisniewska J, Hamann T, et al. 2002. AtPIN4 mediates sink-driven auxin gradients and root patterning in *Arabidopsis*. *Cell* 108:661–73

Friml J, Vieten A, Sauer M, Weijers D, Schwarz H, et al. 2003. Efflux-dependent auxin gradients establish the apical-basal axis of *Arabidopsis*. *Nature* 4261:47–153

Friml J, Yang X, Michniewicz M, Weijers D, Quint A, et al. 2004. A PINOID-dependent binary switch in apical-basal PIN polar targeting directs auxin efflux. *Science* 306:862–65

Fujie M, Kurowia H, Suzuki T, Kwano S, Kuroiwa T. 1993. Organelle DNA synthesis in the quiescent centre of *Arabidopsis thaliana* (Col.). *J. Exp. Bot.* 44:689–93

Gensel PG, Edwards D. 2001. *Plants Invade the Land*. New York: Columbia Univ. Press. 304 pp.

Goebel K. 1928–1933. *Organographie der Pflanzen*, Vols. 1–3. Jena, Ger.: Gustav Fisher

Groot EP, Doyle JA, Nichol SA, Rost TL. 2004. Phylogenetic distribution and evolution of root apical meristem organization in dicotyledonous angiosperms. *Int. J. Plant Sci.* 165:97–105

Haecker A, Gross-Hardt R, Geiges B, Sarkar A, Breuninger H, et al. 2004. Expression dynamics of *WOX* genes mark cell fate decisions during early embryonic patterning in *Arabidopsis thaliana*. *Development* 131:657–68

Hamann T. 2001. The role of auxin in apical-basal pattern formation during *Arabidopsis* embryogenesis. *J. Plant Growth Regul.* 20:292–99

Hamann T, Benkova E, Baürle I, Kientz M, Jürgens G. 2002. The *Arabidopsis BODENLOS* gene encodes an auxin response protein inhibiting MONOPTEROS-mediated embryo patterning. *Genes Dev.* 16:1610–15

Hamann T, Mayer U, Jürgens G. 1999. The auxin-insensitive *bodenlos* mutation affects primary root formation and apical-basal patterning in the *Arabidopsis* embryo. *Development* 126:1387–95

Hanstein J. 1870. Die Entwicklung des Keimes der Monokotylen und der Dikotylen. *Bot. Abh.* 1:1–112

Hardtke CS, Berleth T. 1998. The *Arabidopsis* gene *MONOPTEROS* encodes a transcription factor mediating embryo axis formation and vascular development. *EMBO J.* 17:1405–11

Helariutta Y, Fukaki H, Wysocka-Diller J, Nakajima K, Jung J. 2000. The *SHORT-ROOT* gene controls radial patterning of the *Arabidopsis* root through radial signaling. *Cell* 101:555–67

Hiyama Y, Tsukamoto I, Imaichi R, Kato M. 2002. Developmental anatomy and branching of roots of four *Zeylanidium* species (Podostemaceae), with implications for evolution of foliose roots. *Ann. Bot.* 90:735–44

Hobbie L, McGovern M, Hurwitz LR, Pierro A, Liu NY, et al. 2000. The *axr6* mutants of *Arabidopsis thaliana* define a gene involved in auxin response and early development. *Development* 127:23–32

Horiguchi G, Kodama H, Iba K. 2003. Mutations in a gene for plastid ribosomal protein S6-like protein reveal novel developmental processes required for the correct organization of lateral root meristem in *Arabidopsis*. *Plant J.* 33:521–29

Hosako M, Ogino T, Omori M, Okada S. 2004. Cell cycle arrest by monochloramine through the oxidation of retinoblastoma protein. *Free Radic. Biol. Med.* 36:112–22

Hulleman E, Boonstra J. 2001. Regulation of G1 phase progression by growth factors and the extracellular matrix. *Cell. Mol. Life Sci.* 58:80–93

Ivanov VB. 2004. Meristem as a self-renewal system: maintenance and cessation of cell proliferation (a review). *Russ. J. Plant Physiol.* 51:834–47

This paper describes and links the distribution of AtPIN4 to auxin distribution and patterning in the root meristem.

This paper shows that specific patterns in auxin accumulation and distribution occur as early as the two-cell stage of the *Arabidopsis* embryo.

This paper shows that PINOID controls PIN polarity and thereby mediates the directionality of auxin flow to establish auxin gradients and maxima.

Jiang K, Feldman LJ. 2003. Root meristem establishment and maintenance: The role of auxin. *J. Plant Growth Regul.* 21:432–40

Jiang K, Meng YL, Feldman LJ. 2003. Quiescent center formation in maize roots is associated with an auxin-regulated oxidizing environment. *Development* **130:1429–38**

Joo JH, Bae YS, Lee JS. 2001. Role of auxin-induced reactive oxygen species in root gravitropism. *Plant Physiol.* **126:1055–60**

Juliano RL. 2003. Regulation of signaling and the cell cycle by cell intractions with the extracellular matrix. In *G1 Phase Progression*, ed. J Boonstra, pp. 110–19. New York: Kluwer Acad./Plenum. 267 pp.

Jürgens G. 2001. Apical-basal pattern formation in *Arabidopsis* embryogenesis. *EMBO J.* 20:3609–16

Kawano T. 2003. Roles of the reactive oxygen species-generating peroxidase reactions in plant defense and growth induction. *Plant Cell Rep.* 21:829–37

Kerk NM, Feldman LJ. 1995. A biochemical model for the initiation and maintenance of the quiescent center: implications for organization of root meristems. *Development* 121:2825–33

Kerk NM, Jiang K, Feldman LJ. 2000. Auxin metabolism in the root apical meristem. *Plant Physiol.* 122:925–32

Kidner C, Sundaresan V, Roberts K, Dolan L. 2000. Clonal analysis of the *Arabidopsis* root confirms that position, not lineage, determines cell fate. *Planta* 211:191–99

Kisu Y, Harada Y, Goto M, Esaka M. 1997. Cloning of the pumpkin ascorbate oxidase gene and analysis of a cis-acting region involved in induction by auxin. *Plant Cell Physiol.* 38:631–37

Laloi C, Apel K, Danon A. 2004. Reactive oxygen signaling: the latest news. *Curr. Opin. Plant Biol.* 7:323–28

Laux T. 2003. The stem cell concept in plants: a matter of debate. *Cell* 113:281–83

Lenhard M, Bohnert A, Jürgens G, Laux T. 2001. Termination of stem cell maintenance in *Arabidopsis* floral meristems by interactions between *WUSCHEL* and *AGAMOUS*. *Cell* 105:805–14

Leyser O, Lincoln CA, Timpte C, Lammer D, Turner J, et al. 1993. *Arabidopsis* auxin-resistance gene *AXR1* encodes a protein related to ubiquitin-activating enzyme E1. *Nature* 364:161–64

Lim J, Helariutta Y, Specht CD, Jung J, Sims L, et al. 2000. Molecular analysis of the *SCARECROW* gene in maize reveals a common basis for radial patterning in diverse meristems. *Plant Cell* 12:1307–18

Liso R, De Tullio MC, Ciraci S, Balestrini R, La Rocca N, et al. 2004. Localization of ascorbic acid, ascorbic acid oxidase, and glutathione in roots of *Cucurbita maxima* L. *J. Exp. Bot.* 55:2589–97

Liso R, Innocenti AM, Bitonti MB, Arrigoni O. 1988. Ascorbic acid-induced progression of quiescent centre cells from G1 to S phase. *New Phytol.* 110:469–71

Liszkay A, Kenk B, Schopfer P. 2003. Evidence for the involvement of cell wall peroxidase in the generation of hydroxyl radicals mediating extension growth. *Planta* 217:658–67

Massague J. 2004. G1 cell-cycle control and cancer. *Nature* 432:298–306

May MJ, Vernoux T, Leaver C, Van Montagu M, Inze D. 1998. Glutathione homeostasis in plants: Implications for environmental sensing and plant development. *J. Exp. Bot.* 49:649–67

Mayer KF, Schoof H, Haecker A, Lenhard M, Jürgens G, et al. 1998. Role of *WUSCHEL* in regulating stem cell fate in the *Arabidopsis* shoot meristem. *Cell* 95:805–15

This paper demonstrates the oxidized redox status of the QC and proposes a linkage between auxin, an oxidizing redox, and QC establishment.

This paper reports that auxin can induce ROS in roots and that ROS plays a role as a downstream component in auxin-mediated signaling pathways.

Morre DJ, Morre DM, Ternes P. 2003. Auxin-activated NADH oxidase activity of soybean plasma membranes is distinct from the constitutive plasma membrane NADH oxidase and exhibits prion-like properties. *In Vitro Cell. Dev. Biol. Plan.* 39:368–76

Noctor G, Gomez L, Vanacker H, Foyer CH. 2002. Interactions between biosynthesis, compartmentation and transport in the control of glutathione homeostasis and signalling. *J. Exp. Bot.* 53:1283–304

Noda T, Iwakiri R, Fujimoto K, Aw TY. 2001. Induction of mild intracellular redox imbalance inhibits proliferation of CaCo-2 cells. *FASEB J.* 15:2131–39

Pastori GM, Kiddle G, Antoniw J, Bernard S, Veljovic-Jovanovic S, et al. 2003. Leaf vitamin C contents modulate plant defense transcripts and regulate genes that control development through hormone signaling. *Plant Cell* 15:939–51

Pfeiffer W, Höftberger M. 2001. Oxidative burst in *Chenopodium rubrum* suspension cells: Induction by auxin and osmotic changes. *Physiol. Plant.* 111:144–50

Pignocchi C, Fletcher JM, Wilkinson JE, Barnes JD, Foyer CH. 2003. The function of ascorbate oxidase in tobacco. *Plant Physiol.* 132:1631–41

Pignocchi C, Foyer CH. 2003. Apoplastic ascorbate metabolism and its role in the regulation of cell signalling. *Curr. Opin. Plant Biol.* 6:379–89

Potters G, De Gara L, Asard H, Horemans N. 2002. Ascorbate and glutathione: Guardians of the cell cycle, partners in crime? *Plant Physiol. Biochem.* 40:537–48

Potters G, Horemans N, Bellone S, Caubergs RJ, Trost P, et al. 2004. Dehydroascorbate influences the plant cell cycle through a glutathione-independent reduction mechanism. *Plant Physiol.* 134:1479–87

Reed JW 2001. Roles and activities of AUX/IAA proteins in Arabidopsis. *Trends Plant Sci.* 6:420–25

Reichheld JP, Vernoux T, Lardon F, Van Montagu M, Inze D. 1999. Specific checkpoints regulate plant cell cycle progression in response to oxidative stress. *Plant J.* 17:647–56

Rodríguez-Rodríguez JF, Shishkova S, Napsucialy-Mendivil S, Dubrovsky JG. 2003. Apical meristem organization and lack of establishment of the quiescent center in Cactaceae roots with determinate growth. ***Planta*** **271:849–57**

Rutishauser R.1997. Structural and developmental diversity in Podostemaceae (river-weeds). *Aquat. Bot.* 57:29–70

Sabatini S, Beis D, Wolkenfelt H, Murfett, J, Guilfoyle T, et al. 1999. An auxin-dependent distal organizer of pattern and polarity in the *Arabidopsis* root. *Cell* 99:463–72

Sabatini S, Heidstra R,Wildwater M, Scheres B. 2003. SCARECROW is involved in positioning the stem cell niche in the Arabidopsis root meristem. *Genes Dev.* 17:354–58

Sablowski R. 2004. Plant and animal stem cells: conceptually similar, molecularly distinct? *Trends Cell Biol.* 14:605–11

Sachs T. 2000. Integrating cellular and organismic aspects of vascular differentiation. *Plant Cell Physiol.* 41:649–56

Sánchez-Fernández R, Fricker M, Corben LB, White NS, Sheard N, et al. 1997. Cell proliferation and hair tip growth in the Arabidopsis root are under mechanistically different forms of redox control. *Proc. Natl. Acad. Sci. USA* 94:2745–50

Schafer FQ, Buettner G. 2001. Redox environment of the cell as viewed through the redox state of the glutathione disulfide/glutathione couple. *Free Radic. Biol. Med.* 30:1191–12

Schafer KA. 1998. The cell cycle: a review. *Vet. Pathol.* 35:461–78

Scheres B, Benfey P, Dolan L. 2002. Root development. In *The Arabidopsis Book*, ed. C Somerville, E Meyerowitz, pp. 1–18. Rockville, MD: Am. Soc. Plant Biol.

Scheres B, McKhann HI, van den Berg C. 1996. Roots redefined: Anatomical and genetic analysis of root development. *Plant Physiol.* 111:959–64

This paper links the disappearance/absence of a QC with determinate growth in roots.

Scheres B, Wolkenfelt H. 1998. The Arabidopsis root as a model to study plant development. *Plant Physiol. Biochem.* 36:21–32

Scheres B, Wolkenfelt H, Willemsen V, Terlouw M, Lawson E, et al. 1994. Embryonic origin of the *Arabidopsis* primary root and root meristem initials. *Development* 120:2475–87

Schopfer P. 2001. Hydroxyl radical-induced cell-wall loosening in vitro and in vivo: Implications for the control of elongation growth. *Plant J.* 28:679–88

Schopfer P, Liszkay A, Bechtold M, Frahry G, Wagner A. 2002. Evidence that hydroxyl radicals mediate auxin-induced extension growth. *Planta* 214:821–28

Shackelford RE, Kaufmann WK, Paules RS. 2000. Oxidative stress and cell cycle checkpoint function. *Free Radic. Biol. Med.* 28:1387–404

Spradling A, Drummond-Barbosa D, Kai T. 2001. Stem cells find their niche. *Nature* 414:98–104

Steinmann T, Geldner N, Grebe M, Mangold S, Jackson CL, et al. 1999. Coordinated polar localization of auxin efflux carrier PIN1 by GNOM ARF GEF. *Science* 286:316–18

Suzuki K, Kita Y, Kato M. 2002. Comparative developmental anatomy of seedlings in nine species of Podostemaceae (Subfamily Podostemoideae). *Ann. Bot.* 89:755–65

Takahama U. 1996. Effects of fusicoccin and indole-3-acetic acid on the levels of ascorbic acid and dehydroascorbic acid in the apoplast during elongation of epicotyl segments of *Vigna angularis*. *Physiol. Plant.* 98:731–36

Tiegs E. 1912. Beiträge zur Kenntnis der Entstehung und des Wachstums der Wurzelhaube einiger Leguminosen. *Jahrb. Wiss Bot.* 52:622–46

Tomlinson PB. 1990. *The Structural Biology of Palms.* Oxford: Clarendon Press/Oxford Press. 477 pp.

Torrey JG. 1958. Endogenous bud and root formation by isolated roots of Convolvulus grown in vitro. *Plant Physiol.* 33:258–63

Tsugeki R, Federoff NV. 1999. Genetic ablation of root cap cells in *Arabidopsis*. *Proc. Natl. Acad. Sci. USA* 96:12941–46

Ueda M, Matsui K, Ishiguro S, Sano R, Wada T, et al. 2004. The *HALTED ROOT* gene encoding the 26S proteasome subunit RPT2a is essential for the maintenance of *Arabidopsis* meristems. *Development* 131:2101–11

Uggla C, Moritz T, Sandberg G, Sundberg B. 1996. Auxin as a positional signal in pattern formation in plants. *Proc. Natl. Acad. Sci. USA* 93:9282–86

van den Berg C, Willemsen V, Hage WP, Weisbeek P, Scheres B. 1995. Cell fate in the Arabidopsis root meristem determined by directional signaling. *Nature* 378:62–65

van den Berg C, Willemsen V, Hendriks G, Weisbeek P, Scheres B. 1997. Short-range control of cell differentiation in the *Arabidopsis* root meristem. *Nature* 390:287–89

Varney GT, McCully ME. 1991. The branch roots of Zea. II. Developmental loss of the apical meristem in field-growth roots. *New Phytol.* 118:535–46

Vernoux T, Wilson RC, Seeley KA, Reichheld J-P, Muroy S, et al. 2000. The *ROOT MERISTEMLESS1/CADMIUM SENSITIVE2* gene defines a glutathione-dependent pathway involved in initiation and maintenance of cell division during postembryonic root development. *Plant Cell* 12:97–109

von Guttenberg H. 1968. *Der Primäre Bau der Angiospermenwurzel.* Berlin: Gebrüder Borntraeger. 472 pp.

Wilcox H. 1962. Growth studies of the root incense cedar, *Libocedrus decurrens*. The origin and development of primary tissue. *Am. J. Bot.* 49:221–36

Willemsen V, Wolkenfelt H, de Vrieze G, Weisbeek P, Scheres B. 1998. The *HOBBIT* gene is required for formation of the root meristem in the *Arabidopsis* embryo. *Development* 125:521–31

Wojtaszek P. 1997. Oxidative burst: an early plant response to pathogen infection. *Biochem. J.* 322:681–92

Woo H-H, Orbach MJ, Hirsch AM, Hawes MC. 1999. Meristem-localized inducible expression of a UDP-glycosyltransferase gene is essential for growth and development in pea and alfalfa. *Plant Cell* 11:2303–15

Wysocka-Diller JW, Helariutta Y, Fukaki H, Malamy JE, Benfey PN. 2000. Molecular analysis of SCARECROW function reveals a radial patterning mechanism common to root and shoot. *Development* 127:595–603

Zazimalova E, Napier RM. 2003. Points of regulation for auxin action. *Plant Cell Rep.* 21:625–34

Phagocytosis: At the Crossroads of Innate and Adaptive Immunity

Isabelle Jutras and Michel Desjardins

Département de pathologie et biologie cellulaire, Université de Montréal,
Montréal, Quebec H3C 3J7, Canada; email: isabelle.jutras@umontreal.ca,
michel.desjardins@umontreal.ca

Annu. Rev. Cell Dev. Biol.
2005. 21:511–27

First published online as a
Review in Advance on
August 3, 2005

The *Annual Review of
Cell and Developmental
Biology* is online at
http://cellbio.annualreviews.org

doi: 10.1146/
annurev.cellbio.20.010403.102755

Copyright © 2005 by
Annual Reviews. All rights
reserved

1081-0706/05/1110-
0511$20.00

Key Words

phagosomes, phagolysosome biogenesis, antigen
cross-presentation

Abstract

Phagocytosis, the process by which cells engulf large particles, requires a substantial contribution of membranes. Recent studies have
revealed that intracellular compartments, including endocytic organelles and the endoplasmic reticulum (ER), can engage in fusion
events with the plasma membrane at the sites of nascent phagosomes.
The finding that ER proteins are delivered to phagosomes, where degraded peptides are loaded onto major histocompatibility complex
(MHC) class II molecules, has significantly enhanced our understanding of the immune functions associated with these organelles.
Although it is well known that pathogens are killed in phagosomes,
the contribution of ER proteins to phagosomes has provided a novel
pathway for the loading of exogenous peptides onto MHC class I
molecules, a process known as cross-presentation. Thus, phagocytosis has evolved from a nutritional function in unicellular organisms
to play key roles in both innate and adaptive immunity in vertebrates.

Contents

INTRODUCTION

Eukaryotic cells continuously internalize constituents of their environment. The nature of the internalized constituents, either fluids or particles of different sizes, delineates the various internalization processes (Conner & Schmid 2003). Phagocytosis involves the entry of large particles, typically 1 μm or more, including particles as diverse as inert beads, apoptotic cells, and microbes. Phagocytosis is used for different purposes by a wide range of organisms from amoeba to vertebrates. The capacity to engulf large particles likely appeared as a nutritional function in unicellular organisms, exemplified by the amoeba *Dictyostelium discoideum*, which feed on bacteria. More complex organisms have taken advantage of the phagocytic machinery to fulfill additional functions, such as the clearance of apoptotic cells during embryonic development. The molecular framework of phagocytosis appears to have been conserved throughout evolution, hence enabling the use of genetically tractable model organisms, such as the aforementioned *Dictyostelium* and the nematode *Caenorhabditis elegans*, to decipher the mechanisms underlying phagocytosis. For example, the study of phagocytosis in *C. elegans* has provided a better understanding of the molecular mechanisms associated with the engulfment of apoptotic cells; these mechanisms are often conserved in mammals (Reddien & Horvitz 2004).

Although it is easy to imagine how, in theory, cells could invaginate their cell surface to form a phagosome, the early events leading to particle engulfment during phagocytosis are, in fact, extremely complex. Recognition of specific ligands on the particle by cell-surface receptors initiates the phagocytic process. Carbohydrate binding to lectin receptors, used by a variety of cell types and organisms for phagocytosis, is probably the most widespread and ancient ligand-receptor combination (McGreal et al. 2004). This binding motif on bacteria, for instance, contributes to their engulfment by both amoeba and mammalian immune cells. Via phagocytosis, amoeba derive nutritional benefits whereas mammals may avoid infection (Cardelli 2001, Underhill & Ozinsky 2002). Mammalian professional phagocytes, such as neutrophils, dendritic cells, and macrophages, display a more substantial array of phagocytic receptors, coupled to distinct signal transduction pathways, than do unicellular organisms. In mammalian phagocytes, the study of phagocytic pathways has been focused mainly on those triggered by the Fc receptors (see

Swanson & Hoppe 2004) and complement receptors, which mediate, respectively, phagocytosis of antibody-opsonized and complement-opsonized particles. As various sets of receptors can be engaged by a single particle, a process referred to as crosstalk, phagocytosis is more likely to turn on a complex network of signaling pathways (Underhill & Ozinsky 2002). Integration of these signaling events triggers a dynamic process of cytoskeletal rearrangement accompanied by membrane remodeling at the cell surface that leads to engulfment. The complexity of this process has been discussed in detail elsewhere (Brumell & Grinstein 2003). Membrane remodeling leads to the complete wrapping of the particles and their release in the cytoplasm as novel membrane-bound organelles: the phagosomes. Newly formed phagosomes are unsuited to perform their basic tasks, the killing and degradation of their content, and acquire these properties by a complex maturation process characterized by the sequential fusion with endosomes of increasing age and ultimately with lysosomes. In addition to acidification, one of the hallmarks of phagosome maturation is the acquisition of hydrolytic enzymes, which digest and degrade the content of the internalized particle.

The "digestive" properties of phagosomes, which were the basis of Metchnikoff's cellular theory of immunity (Silverstein 2003), have had a major influence on the ability of complex organisms to adapt to their environment and have allowed these organelles to play a pivotal role in both innate immunity and adaptive immunity. Indeed, phagocytosis and the subsequent killing of microbes in phagosomes form the basis of an organism's innate defense against intracellular pathogens. Furthermore, the degradation of pathogens in the phagosome lumen and the production of antigenic peptides, which are presented by phagocytic cells to activate specific lymphocytes, also link phagocytosis to adaptive immunity. In this review, we focus on recent developments that underscore the importance of phagocytosis in both the innate and adaptive immune functions.

PHAGOLYSOSOME BIOGENESIS: A TALE OF MULTIPLE MEMBRANES

Contribution of the Plasma Membrane to Nascent Phagosomes

In the classical view of phagocytosis, which had prevailed since the original description of phagocytes by Metchnikoff, the PM extended around the particle and formed the phagosome membrane (for a review on the early history of phagocytosis, see Stossel 1999). Several observations have since led to a reconsideration of the sole involvement of the PM in phagocytosis. The membrane surface required to enclose the multiple particles that professional phagocytes can phagocytose may represent an area equivalent to the entire cell surface (Werb & Cohn 1972). This is well illustrated by the fact that macrophages display a gargantuan appetite and can internalize beads larger than themselves without impairing their basic functions (Cannon & Swanson 1992). Nevertheless, no substantial decrease in the cell surface occurs following phagocytosis (Werb & Cohn 1972). The phagocytic rate observed in professional phagocytes may also imply an important waste of PM, a crucial organelle in monitoring the extracellular milieu. Thus, it appeared highly improbable that the PM would serve as the only source of membrane involved in phagocytosis, because the replacement of that membrane by neosynthesis may be expected to occur only within hours (Werb & Cohn 1972). Capacitance measurements made during the internalization of latex beads in J774 macrophages showed that phagocytosis is accompanied by an increase, rather than a decrease, in the cell surface area (Holevinsky & Nelson 1998). These observations were the first to suggest that endomembranes are recruited at the cell surface for the formation of phagosomes in a process referred to as focal exocytosis (Bajno et al. 2000).

Innate immunity: the initial defensive response against encountered microbes. This immune response involves the recognition of microbes by germ line–encoded receptors and relies mainly on phagocytic cells

Adaptive immunity: immune response that occurs as an adaptation to the microbes encountered during the organism's lifetime and that confers protection against these microbes. This response involves the recognition of peptide antigens by specific T cells and the production of antibodies

PM: plasma membrane

Focal Exocytosis of Endosomes and Lysosomes

Recycling endosomes: intracellular compartments, composed of vesicles and tubular structures, where proteins internalized by endocytosis are sorted and transported back to the plasma membrane or to other organelles

Syt VII: synaptotagmin VII

Phagolysosomes: organelles resulting from extensive fusion events between phagosomes and lysosomes; these organelles are the sites where phagocytosed material is degraded

The observation that the cell surface area initially rises prior to the engulfment of particles raises several questions. What is the source of the intracellular membranes contributing to this increase? Where does exocytosis occur at the PM? How is exocytosis synchronized with phagocytosis? Although still incomplete, some answers to these questions have started to emerge. Focal exocytosis of VAMP-3-containing vesicles, presumably originating from recycling endosomes, was shown to occur in the vicinity of nascent phagosomes, and insertion of these membranes was suggested to account for the growth of pseudopods (Bajno et al. 2000). The small GTP-binding protein ARF6, which is activated during Fc receptor–mediated phagocytosis, appears to regulate the recruitment of these vesicles at phagocytic sites (Niedergang et al. 2003). The exocytosis of late endosomes at sites of phagocytosis was also recently demonstrated (Braun et al. 2004). This process appears to be regulated by VAMP-7, as pseudopod extension was inhibited in the absence of VAMP-7 (Braun et al. 2004).

Lysosomes are also capable of fusing with the PM, a process particularly important in pathogen phagocytosis. The secretory properties of specialized lysosomes, such as azurophil granules in neutrophils, are well known (Henson et al. 1992). Fusion of azurophil granules occurs with nascent unclosed phagosomes (Tapper & Grinstein 1997), leading to the release of azurophil granule enzymes in the extracellular medium. The release of these enzymes, which is likely to depend on the type of receptors triggered by phagocytosis, can be inhibited by certain pathogens, including mycobacteria (Cougoule et al. 2002). More recently, neutrophils have also been shown to secrete chromatin, which, together with the degradative content of the granules, forms an extracellular trap capable of retaining and killing bacteria (Brinkmann et al. 2004). The exocyto-

sis of conventional lysosomes during phagocytosis was first demonstrated with the parasite *Trypanosoma cruzi* (Tardieux et al. 1992). Host-cell lysosomes were observed to aggregate at sites of parasite attachment and appeared to fuse with nascent phagosomes (Tardieux et al. 1992). This process was later shown to depend on the elevation of intracellular free Ca^{2+} concentrations, which are triggered upon the attachment of the parasite to the host cell (Andrews 2002). Syt VII, a member of the synaptotagmin family of Ca^{2+}-binding proteins, regulates lysosome exocytosis and is required for cell invasion by *T. cruzi* (Caler et al. 2001). Importantly, Ca^{2+}-regulated exocytosis appears to be a general property of lysosomes and does not occur exclusively during parasite invasion. Lysosome exocytosis has been suggested, for instance, to play a key role in the repair of PM injuries, a process also regulated by Syt VII (Reddy et al. 2001). Consistent with these findings, fibroblasts from Syt VII-deficient mice are less susceptible to *T. cruzi* invasion and are defective in membrane repair (Chakrabarti et al. 2003). Altogether, these studies clearly indicate that, in addition to the PM, membranes of various origins along the endosomal/lysosomal pathway may contribute directly to the formation of nascent phagosomes (**Figure 1**).

Phagosome Maturation

Although generally referred to as phagosomes, the compartments formed by phagocytosis can be widely heterogeneous depending on the type of particles internalized and the type of receptors used at the cell surface. Nevertheless, regardless of their content, phagosomes are extremely dynamic organelles that are continuously modified upon their formation. The conditions prevailing in the phagosome lumen, as well as their overall protein composition, change significantly from when phagosomes are formed at the cell surface to when phagolysosomes, the mature compartment where killing and degradation of pathogens occur, are generated. The

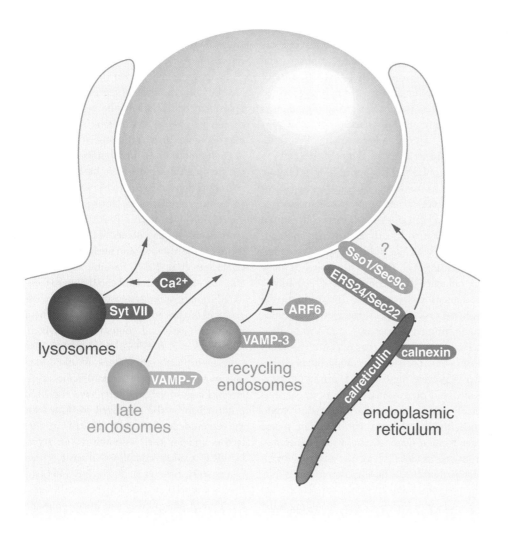

Figure 1

Multiple origins of the phagosome membrane. Phagocytosis of large particles involves the extension of plasma membrane pseudopods around the particle. However, the plasma membrane alone is insufficient to engulf the particle; phagocytosis is a process that requires the contribution of intracellular membranes. Several intracellular compartments—including the endoplasmic reticulum (ER), recycling endosomes (VAMP-3-containing vesicles), late endosomes (VAMP-7-containing vesicles), and lysosomes—may donate membranes to nascent phagosomes. Fusion of ER and plasma membranes involves the ER-resident SNARE protein ERS24, the mammalian homolog of yeast Sec22. Sec22 is capable of pairing with the plasma membrane t-SNARE Sso1/Sec9c to mediate fusion, a mechanism that may also take place in mammalian cells. Focal exocytosis of VAMP-3 vesicles is regulated by the small GTP-binding protein ARF6. The secretion of lysosomes is triggered by an increase in intracellular free Ca^{2+} concentration and is regulated by the Ca^{2+}-binding protein Syt VII.

BSA: bovine serum albumin

Kiss and run fusion: transient and incomplete fusion of organelles that allows limited exchange of their respective contents

RILP: Rab7-interacting lysosomal protein

multiple ways by which phagosome composition can be modulated include direct exchange of molecules with the cytoplasmic pool and transient interactions with various cell organelles. Early kinetic studies of phagosome maturation have indicated that cell surface markers rapidly disappear from newly formed phagosomes, and that maturing phagosomes sequentially display markers of early endosomes, late endosomes, and lysosomes (Desjardins et al. 1994, Pitt et al. 1992).

By monitoring the transfer of BSA-gold particles from endocytic organelles to nascent phagosomes at the electron microscope level, our group showed that phagosomes rapidly fuse with early endosomes and subsequently with late endosomes and lysosomes (Desjardins et al. 1997). This process is likely to influence the acquisition of proteolytic and microbicidal activities. Interestingly, the exchange of BSA-gold particles from endosomes to phagosomes was shown to be size-selective, suggesting that fusing membranes form a pore between the compartments in order to prevent the complete mixing of endosomal and phagosomal contents (Desjardins et al. 1997). Phagosome maturation could thus involve multiple but transient membrane fusion/fission events with endosomes, a process referred to as kiss and run fusion (Desjardins 1995). Fusion/fission events during phagolysosome biogenesis are controlled by a subset of Rab GTPases, including Rab5 and Rab7, which are thought to regulate the transient nature of the fusion events with early endosomes and late endosomes, respectively (Duclos et al. 2000). The expression of Rab5 molecules blocked in their active GTP-bound form impairs the transient nature of these fusion events and results in the formation of giant phagosomes that are unable to kill the intracellular parasite *Leishmania donovani* (Duclos et al. 2000). An effector of Rab7 termed RILP (Cantalupo et al. 2001) has been shown to bind dynein-dynactin, a microtubule-associated motor complex (Harrison et al. 2003). As maturing phagosomes are known to move along microtubules

from the cell surface to a perinuclear region (Blocker et al. 1997), RILP may bind Rab7 on phagosomes and hence function in the attachment of phagosomes to microtubules (Harrison et al. 2003).

Phagosome maturation is also accompanied by the acquisition of sets of proteins and lipids that contribute to the segregation of certain phagosome constituents in membrane microdomains. Lipid raft-associated proteins, including flotillin-1 and subunits of the proton pump ATPase, appear to be absent from nascent phagosomes and to be recruited from endocytic compartments during phagosome maturation (Dermine et al. 2001). Lipid microdomains on phagosomes may serve as platforms for the assembly and nucleation of actin (Defacque et al. 2002). Because actin nucleation on phagosomes has been shown to facilitate fusion between phagosomes and late endosomes but has no significant involvement in fusion with early endosomes (Defacque et al. 2000, Kjeken et al. 2004), lipid rafts may play key roles in phagosome maturation and the acquisition of a phagosome's microbicidal properties. Phagosomal lipid rafts have also been proposed to be the privileged platforms for the recruitment of cytosolic NADPH oxidase factors and for their assembly in an active NADPH oxidase complex (Vilhardt & van Deurs 2004, Shao et al. 2003); NADPH oxidase is a crucial enzyme for the microbicidal function of phagosomes and innate immune defense against infections.

How changes in the composition of phagosome proteins and lipids are modulated and coordinated in time represents a complex issue. Although substantial progress has been made, via a proteomics approach, in characterizing the protein composition of phagosome populations (Garin et al. 2001), the dynamics of association and dissociation of proteins on maturing phagosomes are largely unknown. To make things even more complex, a recent study indicated that rather than being regulated at the cell level, some features of phagosome maturation, such as the pattern of phospholipid association, may vary between individual phagosomes in a given cell (Henry et al. 2004).

Toll-Like Receptors and Phagosome Maturation

Phagocytic cells express a large number of receptors that bind specifically to various microbial molecules, allowing the innate immune system to respond to different types of infection in a process referred to as microbial sensing (Janeway & Medzhitov 2002, Gordon 2002). Among these receptors, recent studies have focused on the family of TLRs, which may suffice to recognize all types of microbes (Beutler 2004). In mammals, 13 TLRs have been identified; these directly sense microbial molecules (Beutler 2004). For instance, bacterial peptidoglycan and lipopolysaccharide are recognized by TLR2 and TLR4, respectively. The observed recruitment of TLR2 to the phagosome membrane suggests that TLRs may sense the microbial content of phagosomes to elicit an appropriate inflammatory response (Underhill et al. 1999). The effect of TLR activation on phagocytosis has been recently investigated. Pretreatment of macrophages with various TLR ligands, especially with the ligand of TLR9, apparently increased the number of *Escherichia coli* taken up by macrophages (Doyle et al. 2004). In agreement with this finding, one study showed that phagocytosis of various bacteria was impaired in macrophages lacking both TLR2 and TLR4 signaling (Blander & Medzhitov 2004). This study also highlighted an unexpected function for TLRs, as TLR activity was shown to stimulate fusion between endocytic organelles and phagosomes containing bacteria (Blander & Medzhitov 2004). This study found that the maturation of phagosomes containing apoptotic cells was not affected by TLR signaling (Blander & Medzhitov 2004); however, a concomitant study suggests that TLR4 has an inhibitory effect on the fusion of endocytic organelles with phagosomes containing apoptotic cells in macrophages (Shiratsuchi et al. 2004). Clearly, more stud-

ies are needed to decipher the role that TLRs might play in membrane fusion and phagosome maturation.

FUNCTION OF PHAGOSOMES IN ANTIGEN PRESENTATION

The killing of pathogens in the phagosome lumen after phagocytosis is a simple defense mechanism that is derived from the fact that phagosomes specialize in the degradation of complex molecules for recycling or nutritional purposes. This feature is sufficient for organisms such as amoebae to survive in their challenging environment and feed on bacteria. The degradative properties of phagosomes were exploited, through evolution, to improve the ability of complex organisms to monitor their environment and build more efficient defense mechanisms. In jawed vertebrates, peptides generated through the degradation of exogenous proteins can be loaded in the phagosome lumen onto MHC molecules and presented at the cell surface to initiate an adaptive immune response. This response involves recognition of MHC-associated peptides by TCR molecules, which have coevolved with the MHC (Du Pasquier 2004). Phagosomes in jawed vertebrates thus represent a compartment that has evolved to integrate functions of the innate and adaptive immune system.

Antigen Presentation by MHC Class II Molecules

The proteins present on engulfed particles encounter an array of degrading proteases in phagosomes. Yet, this destructive environment generates peptides that are capable of binding to MHC class II molecules. Interestingly, evidence indicates that hydrolases are not acquired by phagosomes through a simple "one-step" fusion event with lysosomes; as implied in the term "phagolysosomes," such a process would transfer the bulk of hydrolases all at once. Instead, as indicated by the protein profile of phagosomes

TLR: Toll-like receptor, a family of receptors that recognizes microbial molecules and induces the expression of inflammatory cytokines upon being activated, hence contributing to innate immunity

MHC: major histocompatibility complex

TCR: T cell receptor

ER: endoplasmic reticulum

Ii: invariant chain

CLIP: class II–associated Ii peptide

AEP: asparagine endopeptidase

GFP: green fluorescent protein

obtained by two-dimensional gel electrophoresis, the subsets of hydrolases acquired by phagosomes vary throughout the maturation process whereby newly formed phagosomes become phagolysosomes (Garin et al. 2001). The kinetics of protease acquisition may also vary between phagocytes; for instance, it is faster in macrophages than in dendritic cells (Lennon-Duménil et al. 2002). These conditions may limit the proteolysis of proteins in dendritic cells and favor the generation of peptide antigens of appropriate length for loading onto MHC class II molecules.

In the ER, newly synthesized MHC class II α and β chains associate with the Ii, which occupies the peptide-binding groove of the MHC class II molecules and targets them to the endocytic pathway (Watts 2004). There, Ii undergoes a series of proteolytic cleavages involving intermediates known as p22 and p10 and culminating in a fragment referred to as the CLIP. The processing of Ii can be initiated by AEP, which generates p22 and p10. However, other unidentified proteases can also mediate Ii degradation (Manoury et al. 2003). Importantly, humans and mice may exhibit differences in these proteases; for instance, AEP does not generate p10 in humans (Manoury et al. 2003). Although several cysteine proteases are capable of converting p10 into CLIP, this step is mainly dependent on cathepsin S in antigen-presenting cells (Watts 2004). The protease responsible for processing p10 into CLIP may also vary between species. In murine thymic epithelial cells, cathepsin L appears to be the main p10-cleaving protease (Nakagawa et al. 1998), whereas in humans p10 may be processed only by cathepsin S (Bania et al. 2003). CLIP remains bound to the peptide-binding site of the MHC class II molecules until it is replaced by a peptide antigen, in a process catalyzed by the DM protein. DM favors the dissociation of weakly bound peptides, such as CLIP, and the binding of high-affinity antigens (Watts 2004).

Antigen loading on MHC class II molecules takes place in various compartments, from early endosomes to lysosomes, of the endocytic pathway (Hiltbold & Roche 2002). Phagosomes containing latex beads have been shown to acquire MHC class II molecules both from a recycling pool on the PM and from a newly synthesized pool (Ramachandra & Harding 2000). The capacity of pathogens to prevent phagosome maturation is likely to affect antigen processing and presentation. For instance, in phagosomes containing *Mycobacterium tuberculosis*, the formation of bacterial antigen-MHC class II complexes is decreased when live bacteria, rather than heat-killed bacteria, are phagocytosed (Ramachandra et al. 2001). Likewise, phagocytosed *Legionella*, which reside in an ER-derived vacuole, may be expected to avoid the MHC class II presentation pathway. On the contrary, however, *Legionella* peptide antigens access compartments competent for peptide loading on MHC class II molecules (Neild & Roy 2003). The pathway by which these ER-derived antigens are trafficked to endocytic compartments remains to be defined.

Newly formed antigen-MHC class II complexes are delivered to the cell surface for presentation to CD4[+] T cells. Recent results using GFP-tagged MHC molecules in dendritic cells suggest that peptide-MHC complexes are transported to the cell surface on lysosome-derived tubules, which extend toward the interface between dendritic cells and T cells (Boes et al. 2002, Chow et al. 2002). These lysosomal tubular structures, or vesicles derived from them, were also observed to fuse with the PM (Chow et al. 2002, Kleijmeer et al. 2001). Although mostly examined in dendritic cells, the trafficking of peptide-MHC class II complexes is likely to follow a similar pathway in other types of antigen-presenting cells.

ER-Mediated Phagocytosis and Antigen Cross-Presentation

A large body of evidence indicates that, in addition to the clear involvement of the

MHC class II pathway in the immune response against phagocytosed pathogens, antigens from pathogens, including mycobacteria, *Salmonella*, *Brucella*, and *Leishmania*, can elicit an MHC class I–dependent response promoting the proliferation of CD8[+] cytotoxic T cells (**Figure 2**) (Ackerman & Cresswell 2004, Kaufmann & Schaible 2005). Yet the common view has long been that exogenous proteins, internalized by endocytosis or phagocytosis, are presented by the MHC class II pathway. In contrast, the MHC class I presentation pathway was considered restricted to endogenously synthesized proteins, including self proteins and those resulting from viral infections. The presentation of exogenous proteins on MHC class I molecules was thus referred to as cross-presentation, but the molecular basis of the mechanisms that might explain cross-presentation remained elusive. Clearly, the proliferation of CD8[+] T cells is usually triggered by proteins that are from endogenous sources and processed in the cytoplasm by proteasomes to generate peptides that are loaded onto MHC class I molecules in the ER. Insights into the mechanisms enabling the cross-presentation of exogenous antigens came from observations, made by the Harding group, that phagocytosis of bacteria, with no known mechanism for cytosolic penetration, resulted in the presentation of bacterial antigens by MHC class I molecules in a process insensitive to Brefeldin A treatment (Pfeifer et al. 1993). These results indicated that because this drug disrupts the biosynthetic apparatus and blocks the classical class I processing pathway, a novel vacuolar pathway, competent for the processing of exogenous peptides and not linked to transport steps through the Golgi apparatus, also existed (Pfeifer et al. 1993).

Further insights into the molecular mechanisms linked to cross-presentation came from proteomics analysis of latex bead-containing phagosomes indicating that several ER proteins known to play a role in MHC class I-mediated antigen presentation, but no Golgi-resident protein, were present on phagosomes

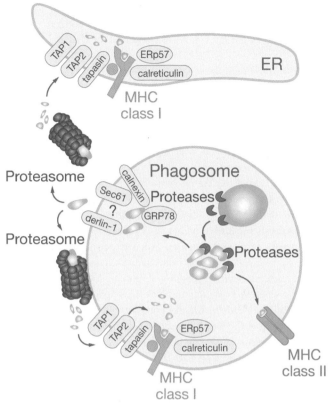

Figure 2

Antigen processing and loading on MHC molecules in phagosomes. Phagocytosed particles are degraded by proteases acquired during phagolysosome biogenesis. The degradation of foreign proteins generates peptide antigens that are loaded onto MHC class II molecules. Through ER-mediated phagocytosis and the delivery of ER proteins to phagosomes, cross-presentation of exogenous proteins on MHC class I molecules can be initiated in these organelles. Antigen cross-presentation involves the retro-translocation of exogenous peptides from the phagosome lumen to the cytosol, either by the Sec61 translocon or by the putative channel formed by the recently identified Derlin-1. Retro-translocated peptides are polyubiquitinated, processed by the proteasome, transported back to the phagosome lumen through TAP, and loaded onto MHC class I molecules. An alternative pathway for cross-presentation involves the transport of processed peptides from the cytosol to the ER lumen, where they can be loaded onto MHC class I molecules. Some of the chaperones involved in retro-translocation (calnexin and GRP78/Bip) and in MHC class I antigen loading (calreticulin and Erp57) are shown.

(Garin et al. 2001). On the basis of these observations, it was proposed that ER membranes could be recruited to phagosomes at some point during phagosome formation or maturation into phagolysosomes (Garin et al. 2001). Analysis of phagosome formation by

WHAT INTRACELLULAR PATHOGENS TELL US ABOUT CROSS-PRESENTATION

Antigen presentation by MHC class I and class II molecules has generally been considered to involve two segregated pathways and to take place in distinct organelles. Namely, endogenous proteins processed in the cytoplasm are loaded onto MHC class I molecules in the ER, whereas exogenous proteins are processed and loaded onto MHC class II molecules in endocytic compartments. However, this model is difficult to reconcile with our comprehension of the immune response to pathogens. Intracellular pathogens have evolved ways to survive in their host by subverting host-membrane trafficking to establish replicative niches in compartments lacking the properties of phagolysosomes. For instance, *Salmonella* and *Mycobacteria* species reside in immature phagosomes lacking the typical markers of lysosomes (Holden 2002, Vergne et al. 2004), whereas *Legionella* and *Brucella* bypass the endocytic pathway and reside in ER-like compartments (Roy & Tilney 2002, Celli & Gorvel 2004). Hence, the segregated presentation pathways imply that peptides derived from the former species are presented on class II molecules, whereas those derived from the latter are presented on class I molecules. In fact, all of these pathogens can elicit responses from both the MHC class I and II presentation pathways. Recent findings in the trafficking of intracellular organelles during phagocytosis indicate that all types of phagosomes interact, at least to some level, with the endocytic and biosynthetic pathways to build mixed compartments in which, in principle, both class I and class II processing can take place. Thus, the concept of cross-presentation may have arisen from a misunderstanding of the complex trafficking events occurring in antigen-presenting cells.

the PM at the cell surface was unexpected, as no molecular basis could explain the fusion of these two compartments. Interestingly, the Gerisch group showed that the ER was likely to play a direct role in the formation of phagosomes, because a double knockout for the ER proteins calnexin and calreticulin in *Dictyostelium* abolished phagocytosis in this cell (Müller-Taubenberger et al. 2001).

Membrane fusion requires the formation of a SNARE complex, which is composed of SNAREs located on the vesicle and the target membranes, termed respectively the v-SNARE and the t-SNARE. The appropriate pairing of v-SNAREs and t-SNAREs ensures the specificity of these fusion events. Using an in vitro assay that measures the fusion of reconstituted liposomes, Rothman's group tested different combinations of yeast SNAREs for their capacity to sustain fusion (McNew et al. 2000). This study revealed that liposomes containing the PM t-SNARE Sso1/Sec9c could fuse with liposomes bearing the ER-resident SNARE Sec22—a finding that was, at the time, difficult to reconcile with the SNARE hypothesis as ER-PM were not expected to be fusion partners. The significance of this finding was recently investigated in the context of phagocytosis using ERS24, the mammalian homolog of Sec22. It was shown that ERS24 is recruited to nascent phagosomes and that its inhibition impairs the phagocytosis of particles larger than 0.8 μm (Becker et al. 2005).

Cytotoxic T cells: a class of T lymphocytes that kills infected target cells displaying pathogens' peptide fragments bound to MHC class I molecules

various biochemical and morphological methods led to the surprising observation that during phagocytosis, the ER is recruited to the cell surface, where it appears to fuse underneath phagocytic cups to provide some of the membrane required for the formation of phagosomes, a process referred to as ER-mediated phagocytosis (Gagnon et al. 2002). Although several types of organelles were shown to be able to contribute membrane to forming phagosomes (see above and **Figure 1**), the direct fusion of the ER with

The Phagosome is a Competent Organelle for Cross-Presentation

Arguments in favor of the proposal that the ER contributes a significant part of the phagosome membrane came from morphological and biochemical studies showing that calnexin was present in its proper orientation in the membrane of isolated phagosomes, whereas calreticulin, a lumenal protein, was protected from hydrolysis by proteases, as expected from a direct fusion event between the ER and the phagosomes (Gagnon et al. 2002).

Calnexin was also observed on membranes surrounding latex particles in dendritic cells (Guermonprez et al. 2003). In addition to calnexin, most of the molecules known to participate in the processing and presentation of peptides on MHC class I molecules were shown to be present on phagosomes and to contribute to defining a compartment competent for the processing of exogenous proteins for MHC class I presentation (Ackerman et al. 2003, Guermonprez et al. 2003, Houde et al. 2003).

The ability of phagosomes to degrade microorganisms and generate derived peptides is well established. Moreover, the sequential acquisition of hydrolases by phagosomes (Garin et al. 2001) may favor regulated patterns of peptide degradation suitable for presentation (Lennon-Duménil et al. 2002). Recently, cathepsin S-deficient phagosomes were shown to generate poorly presented class I peptides (Shen et al. 2004). However, the overall efficiency of intraphagosomal degradation to generate appropriate peptides for direct loading onto MHC class I molecules remains to be established. The 8- to 10-amino acid peptides preferentially loaded on MHC class I molecules are best produced by proteasomes; this implies that peptides partially degraded in the phagosome lumen would be better suited for presentation if they underwent further processing by proteasomes. This step would require the translocation of peptides from the phagosome lumen to the cytoplasm. This process has been shown to occur and be part of a "phagosome-to-cytosol" pathway that feeds exogenous peptides to the "classical" MHC class I pathway (Kovacsovics-Bankowski & Rock 1995). Swanson and colleagues tested various carriers as a source of exogenous proteins by phagocytosis and found that translocation of ovalbumin to the cytoplasm was more efficient when delivered by polystyrene particles or biodegradable particles rather than by sheep red blood cells (Oh et al. 1997).

Although these translocated peptides could be targeted effectively to the classical MHC class I pathway, recent findings suggest that phagosomes from macrophages possess a machinery of their own for antigen processing (**Figure 2**). Proteomics analyses have revealed that proteins involved in each steps of the process required for MHC class I antigen processing and presentation, including chaperones, the ubiquitination machinery, proteasome subunits, and elements of the loading complex, were present on purified latex bead–containing phagosomes (Houde et al. 2003). In functional assays, peptides from fluorescently labeled ovalbumin that were loaded onto latex beads were shown to be efficiently translocated to the cytoplasm, where they were polyubiquitinated and observed to associate with proteasomes on the phagosome membrane (Houde et al. 2003). It is still unclear which molecule is responsible for the translocation of peptides from the phagosome lumen to the cytoplasm. Because CTA1 can also be translocated to the cytoplasm after phagocytosis, it was proposed that Sec61, the ER translocon also detected on phagosomes, might be involved in this process (Houde et al. 2003). Indeed, Sec61 was shown previously to be involved in the retro-translocation of CTA1 from the ER lumen to the cytoplasm (Schmitz et al. 2000). Recently, a novel membrane protein, Derlin1, was shown to mediate retro-translocation from the ER lumen to the cytosol (Lilley & Ploegh 2004, Ye et al. 2004). Whether Derlin 1, which also localizes to phagosomes (P. Cresswell, personal communication), can also function on this organelle remains to be established. Using in vitro assays, two groups have also demonstrated that functional TAP transporters on dendritic cell phagosomes are able to translocate added peptides to the phagosome lumen (Ackerman et al. 2003, Guermonprez et al. 2003). Intraphagosomal loading of peptides on MHC class I molecules was also observed (Ackerman et al. 2003, Guermonprez et al. 2003, Houde et al. 2003). Finally, it was shown that following phagocytosis, the MHC class I ovalbumin–derived peptide SIINFEKL could trigger the proliferation of

Phagocytic cups: extensions of plasma membrane around a particle during the initial steps of phagocytosis

SNARE: soluble N-ethylmaleimide-sensitive factor attachment protein receptor

CTA1: A1 subunit of cholera toxin

TAP: transporter for antigen processing

CD8[+] T cells by a mechanism that was only partially inhibited by Brefeldin A (Houde et al. 2003).

Taken together, these results indicate that the contribution of the ER to phagosome formation and/or maturation is not only meant to spare the use of the PM but also brings additional functional properties to the phagosome. The advantage of using the ER as a source of membrane for phagosome formation has been discussed in previous reviews (Aderem 2002, Desjardins 2003). Recent evidence further extends the possible use of the ER as a source of membrane for the formation of macropinosomes, where it might also contribute to the ability of this compartment to process soluble exogenous material for cross-presentation (Ackerman et al. 2003). Moreover, a pathway that brings exogenous molecules to the ER via endocytosis has also been described (Ackerman & Cresswell 2004). This pathway might also be used to allow contact between the ER-localized TLR9 and its endocytosed ligand consisting of bacterial unmethylated CpG DNA (Latz et al. 2004). ER-mediated phagocytosis would indeed favor the natural encounter of bacterial DNA with TLR9 after the killing and degradation of pathogens in phagosomes.

CONCLUSION

The contribution of both endocytic organelles and the ER to phagosome formation and maturation obviously allows for the building of a cellular compartment where killing, degradation, processing, and presentation of exogenous peptides can take place in an integrated way. Mammalian cell phagosomes have clearly inherited part of their innate immune function from the "digestive" properties of their amoeba counterparts. However, the functions related to antigen presentation have appeared only recently during evolution with jawed vertebrates, in which the essential elements of the adaptive immune system have co-evolved in the immunoglobulin-TCR-MHC unit (Du Pasquier 2004). Because several of the molecules involved in MHC class I antigen presentation encoded in the MHC are expressed in the ER—e.g., TAP—this raises the question of whether the contribution of the ER to phagosome formation is restricted to vertebrates.

Interestingly, the few pieces of evidence available at the moment indicate that this is not the case. As mentioned before, ER-mediated phagocytosis is a likely process in *Dictyostelium* because phagosomes formed in these cells appear to be surrounded by membrane-containing ER molecules, as shown by confocal microscopy (Müller-Taubenberger et al. 2001). More importantly, a double knockout of calnexin and calreticulin inhibits phagocytosis (Müller-Taubenberger et al. 2001). These results are supported by a recent study showing that the uptake of *Legionella* in *Dictyostelium* is accompanied by the recruitment of GFP-tagged calnexin and calreticulin to phagocytic cups (Fajardo et al. 2004). This contrasts with current models in mammals suggesting that ER is recruited to phagosomes at later points during infection (Roy & Tilney 2002). Furthermore, preliminary results using a proteomics approach indicate that latex bead–containing phagosomes purified from *Dictyostelium* display a significant number of ER proteins, whereas such phagosomes are mostly devoid of contaminants from other intracellular organelles (T. Soldati & M. Desjardins, unpublished results). Similar results were obtained from phagosomes isolated from Schneider S2 cells derived from *Drosophila* (J. Boulais & M. Desjardins, unpublished results). Although preliminary, these results suggest that the ER contributed to the formation and/or maturation of phagosomes from their earliest evolutionary origins. This argues for the interesting concept that because phagosome formation involved the ER in species with no adaptive immunity, novel properties beneficial to the immune system appeared on phagosomes concomitant with the appearance of the MHC locus and the expression of some of its gene products in the ER. This enabled the evolution or refinement of phagosomes from a lytic compartment capable of playing a role in innate immunity by killing microorganisms into a well-integrated organelle linking innate and adaptive immunity. Not surprisingly, understanding of the fine molecular mechanisms supporting a central role for phagosomes in immunity awaits further study.

SUMMARY POINTS

1. Phagocytosis, the internalization of large particles by cells, has evolved from a nutritional process in amoeba to an immune process in complex organisms.

2. The phagosome membrane is composed of membranes of multiple cellular origins, including the plasma membrane, endocytic organelles, and the ER.

3. The protein composition of phagosomes undergoes modifications through transient fusion/fission events with compartments of both the endocytic and biosynthetic pathways.

4. Phagosome fusion with late endosomes and lysosomes leads to the formation of phagolysosomes.

5. Phagolysosomes contain proteolytic and microbicidal activities, which are essential features of the innate immune system.

6. Phagosomes are the site of peptide antigen processing and loading onto MHC class II molecules.

7. The fusion of ER membranes with phagosomes allows the delivery of the machinery required for the processing and loading of peptides onto MHC class I molecules.

8. Phagosomes are competent organelles for antigen cross-presentation.

9. The contribution of ER to phagosome formation may be a conserved process from amoeba to vertebrates.

LITERATURE CITED

Ackerman AL, Cresswell P. 2004. Cellular mechanisms governing cross-presentation of exogenous antigens. *Nat. Immunol.* 5:678–84

Ackerman AL, Kyritsis C, Tampe R, Cresswell P. 2003. Early phagosomes in dendritic cells form a cellular compartment sufficient for cross presentation of exogenous antigens. *Proc. Natl. Acad. Sci. USA* 100:12889–94

Aderem A. 2002. How to eat something bigger than your head. *Cell* 110:5–8

Andrews NW. 2002. Lysosomes and the plasma membrane: trypanosomes reveal a secret relationship. *J. Cell Biol.* 158:389–94

Bajno L, Peng X-R, Schreiber AD, Moore H-P, Trimble WS, et al. 2000. Focal exocytosis of VAMP3-containing vesicles at sites of phagosome formation. *J. Cell Biol.* 149:697–705

Bania J, Gatti E, Lelouard H, David A, Cappello F, et al. 2003. Human cathepsin S, but not cathepsin L, degrades efficiently MHC class II-associated invariant chain in nonprofessional APCs. *Proc. Natl. Acad. Sci. USA* 100:6664–69

Becker T, Volchuk A, Rothman JE. 2005. Differential use of ER membrane for phagocytosis in J774 macrophages. *Proc. Natl. Acad. Sci. USA* 102:4022–26

Beutler B. 2004. Innate immunity: an overview. *Mol. Immunol.* 40:845–59

Blander JM, Medzhitov R. 2004. Regulation of phagosome maturation by signals from toll-like receptors. *Science* 304:1014–18

Blocker A, Severin FF, Burkhardt JK, Bingham JB, Yu H, et al. 1997. Molecular requirements for bi-directional movement of phagosomes along microtubules. *J. Cell Biol.* 137:113–29

Boes M, Cerny J, Massol R, Op den Brouw M, Kirchhausen T, et al. 2002. T-cell engagement of dendritic cells rapidly rearranges MHC class II transport. *Nature* 418:983–88

Braun V, Fraisier V, Raposo G, Hurbain I, Sibarita JB, et al. 2004. TI-VAMP/VAMP7 is required for optimal phagocytosis of opsonised particles in macrophages. *EMBO J.* 23:4166–76

Brinkmann V, Reichard U, Goosmann C, Fauler B, Uhlemann Y, et al. 2004. Neutrophil extracellular traps kill bacteria. *Science* 303:1532–35

Brumell JH, Grinstein S. 2003. Role of lipid-mediated signal transduction in bacterial internalization. *Cell. Microbiol.* 5:287–97

Caler EV, Chakrabarti S, Fowler KT, Rao S, Andrews NW. 2001. The exocytosis-regulatory protein synaptotagmin VII mediates cell invasion by *Trypanosoma cruzi*. *J. Exp. Med.* 193:1097–104

Cannon GJ, Swanson JA. 1992. The macrophage capacity for phagocytosis. *J. Cell Sci.* 101:907–13

Cantalupo G, Alifano P, Roberti V, Bruni CB, Bucci C. 2001. Rab-interacting lysosomal protein (RILP): the Rab7 effector required for transport to lysosomes. *EMBO J.* 20:683–93

Cardelli J. 2001. Phagocytosis and macropinocytosis in *Dictyostelium*: phosphoinositide-based processes, biochemically distinct. *Traffic* 2:311–20

Celli J, Gorvel JP. 2004. Organelle robbery: *Brucella* interactions with the endoplasmic reticulum. *Curr. Opin. Microbiol.* 7:93–97

Chakrabarti S, Kobayashi KS, Flavell RA, Marks CB, Miyake K, et al. 2003. Impaired membrane resealing and autoimmune myositis in synaptotagmin VII-deficient mice. *J. Cell Biol.* 162:543–49

Chow A, Toomre D, Garrett W, Mellman I. 2002. Dendritic cell maturation triggers retrograde MHC class II transport from lysosomes to the plasma membrane. *Nature* 418:988–94

Conner SD, Schmid SL. 2003. Regulated portals of entry into the cell. *Nature* 422:37–44

Cougoule C, Constant P, Etienne G, Daffe M, Maridonneau-Parini I. 2002. Lack of fusion of azurophil granules with phagosomes during phagocytosis of *Mycobacterium smegmatis* by human neutrophils is not actively controlled by the bacterium. *Infect. Immun.* 70:1591–98

Defacque H, Bos E, Garvalov B, Barret C, Roy C, et al. 2002. Phosphoinositides regulate membrane-dependent actin assembly by latex bead phagosomes. *Mol. Biol. Cell* 4:1190–202

Defacque H, Egeberg M, Habermann A, Diakonova M, Roy C, et al. 2000. Involvement of ezrin/moesin in *de novo* actin assembly on phagosomal membranes. *EMBO J.* 19:199–212

Dermine JF, Duclos S, Garin J, St-Louis F, Rea S, et al. 2001. Flotillin-1-enriched lipid raft domains accumulate on maturing phagosomes. *J. Biol. Chem.* 276:18507–12

Desjardins M. 1995. Biogenesis of phagolysosomes: the 'kiss and run' hypothesis. *Trends Cell Biol.* 5:183–86

Desjardins M. 2003. ER-mediated phagocytosis: a new membrane for new functions. *Nat. Rev. Immunol.* 3:280–91

Desjardins M, Huber LA, Parton RG, Griffiths G. 1994. Biogenesis of phagolysosomes proceeds through a sequential series of interactions with the endocytic apparatus. *J. Cell Biol.* 124:677–88

Desjardins M, Nzala NN, Corsini R, Rondeau C. 1997. Maturation of phagosomes is accompanied by changes in their fusion properties and size-selective acquisition of solute materials from endosomes. *J. Cell Sci.* 110:2303–14

Doyle SE, O'Connell RM, Miranda GA, Vaidya SA, Chow EK, et al. 2004. Toll-like receptors induce a phagocytic gene program through p38. *J. Exp. Med.* 199:81–90

Duclos S, Diez R, Garin J, Papadopoulou B, Descoteaux A, et al. 2000. Rab5 regulates the kiss and run fusion between phagosomes and endosomes and the acquisition of phagosome leishmanicidal properties in RAW 264.7 macrophages. *J. Cell Sci.* 113:3531–41

Du Pasquier L. 2004. Innate immunity in early chordates and the appearance of adaptive immunity. *C. R. Biol.* 327:591–601

Fajardo M, Schleicher M, Noegel A, Bozzaro S, Killinger S, et al. 2004. Calnexin, calreticulin and cytoskeleton-associated proteins modulate uptake and growth of *Legionella pneumophila* in *Dictyostelium discoideum*. *Microbiology* 150:2825–35

Gagnon E, Duclos S, Rondeau C, Chevet E, Cameron PH, et al. 2002. Endoplasmic reticulum-mediated phagocytosis is a mechanism of entry into macrophages. *Cell* 110:119–31

Garin J, Diez R, Kieffer S, Dermine JF, Duclos S, et al. 2001. The phagosome proteome: insight into phagosome functions. *J. Cell Biol.* 152:165–80

Gordon S. 2002. Pattern recognition receptors: doubling up for the innate immune response. *Cell* 111:927–30

Guermonprez P, Saveanu L, Kleijmeer M, Davoust J, van Endert P, et al. 2003. ER-phagosome fusion defines an MHC class I cross-presentation compartment in dendritic cells. *Nature* 425:397–402

Harrison RE, Bucci C, Vieira OV, Schroer TA, Grinstein S. 2003. Phagosomes fuse with late endosomes and/or lysosomes by extension of membrane protrusions along microtubules: role of Rab7 and RILP. *Mol. Cell. Biol.* 23:6494–506

Henry RM, Hoppe AD, Joshi N, Swanson JA. 2004. The uniformity of phagosome maturation in macrophages. *J. Cell Biol.* 164:185–94

Henson PM, Henson IE, Fittschen C, Bratton DL, Riches DWH. 1992. Degranulation and secretion by phagocytic cells. In *Inflammation: Basic Principles and Clinical Correlates*, ed. JI Gallin, IM Goldstein, R Snyderman, pp. 511–39. New York: Raven. 1186 pp.

Hiltbold EM, Roche PA. 2002. Trafficking of MH class II molecules in the late secretory pathway. *Curr. Opin. Immunol.* 14:30–35

Holden DW. 2002. Trafficking of the *Salmonella* vacuole in macrophages. *Traffic* 3:161–69

Holevinsky KO, Nelson DJ. 1998. Membrane capacitance changes associated with particle uptake during phagocytosis in macrophages. *Biophys. J.* 75:2577–86

Houde M, Bertholet S, Gagnon E, Brunet S, Goyette G, et al. 2003. Phagosomes are competent organelles for antigen cross-presentation. *Nature* 425:402–6

Janeway CA Jr, Medzhitov R. 2002. Innate immune recognition. *Annu. Rev. Immunol.* 20:197–216

Kaufmann SH, Schaible UE. 2005. Antigen presentation and recognition in bacterial infections. *Curr. Opin. Immunol.* 17:79–87

Kjeken R, Egeberg M, Habermann A, Kuehnel M, Peyron P, et al. 2004. Fusion between phagosomes, early and late endosomes: a role for actin in fusion between late, but not early endocytic organelles. *Mol. Biol. Cell* 1:345–58

Kleijmeer M, Ramm G, Schuurhuis D, Griffith J, Rescigno M, et al. 2001. Reorganization of multivesicular bodies regulates MHC class II antigen presentation by dendritic cells. *J. Cell Biol.* 155:53–63

This was the first demonstration that direct fusion events between the endoplasmic reticulum and the plasma membrane occur at the site of nascent phagosomes.

This paper questions the idea that phagosomes containing a single type of particle mature in a uniform manner and shows differences in phosphoinositide labeling between individual phagosomes.

Together with Ackerman et al. (2003) and Guermonprez et al. (2003), this paper reveals that phagosomes are self-sufficient organelles for antigen cross-presentation. Phagosome-associated proteins undergo retro-translocation to the cytosol, ubiquitination, proteasome-dependent degradation, and loading onto MHC class I molecules.

Kovacsovics-Bankowski M, Rock KL. 1995. A phagosome-to-cytosol pathway for exogenous antigens presented on MHC class I molecules. *Science* 267:243–46

Latz E, Schoenemeyer A, Visintin A, Fitzgerald KA, Monks BG, et al. 2004. TLR9 signals after translocating from the ER to CpG DNA in the lysosome. *Nat. Immunol.* 5:190–98

Lennon-Duménil AM, Bakker AH, Maehr R, Fiebiger E, Overkleeft HS, et al. 2002. Analysis of protease activity in live antigen-presenting cells shows regulation of the phagosomal proteolytic contents during dendritic cell activation. *J. Exp. Med.* 196:529–40

Lilley BN, Ploegh HL. 2004. A membrane protein required for dislocation of misfolded proteins from the ER. *Nature* 429:834–40

Manoury B, Mazzeo D, Li DN, Billson J, Loak K, et al. 2003. Asparagine endopeptidase can initiate the removal of the MHC class II invariant chain chaperone. *Immunity* 18:489–98

McGreal EP, Martinez-Pomares L, Gordon S. 2004. Divergent roles for C-type lectins expressed by cells of the innate immune system. *Mol. Immunol.* 11:1109–21

McNew JA, Parlati F, Fukuda R, Johnston RJ, Paz K, et al. 2000. Compartmental specificity of cellular membrane fusion encoded in SNARE proteins. *Nature* 407:153–59

Müller-Taubenberger A, Lupas AN, Li H, Ecke M, Simmeth E, et al. 2001. Calreticulin and calnexin in the endoplasmic reticulum are important for phagocytosis. *EMBO J.* 20:6772–82

Nakagawa T, Roth W, Wong P, Nelson A, Farr A, et al. 1998. Cathepsin L: critical role in Ii degradation and CD4 T cell selection in the thymus. *Science* 280:450–53

Neild AL, Roy CR. 2003. *Legionella* reveal dendritic cell functions that facilitate selection of antigens for MHC class II presentation. *Immunity* 18:813–23

Niedergang F, Colucci-Guyon E, Dubois T, Raposo G, Chavrier P. 2003. ADP ribosylation factor 6 is activated and controls membrane delivery during phagocytosis in macrophages. *J. Cell Biol.* 161:1143–50

Oh YK, Harding CV, Swanson JA. 1997. The efficiency of antigen delivery from macrophage phagosomes into cytoplasm for MHC class I-restricted antigen presentation. *Vaccine* 15:511–18

Pfeifer JD, Wick MJ, Roberts RL, Findlay K, Normark SJ, et al. 1993. Phagocytic processing of bacterial antigens for class I MHC presentation to T cells. *Nature* 361:359–62

Pitt A, Mayorga LS, Stahl PD, Schwartz AL. 1992. Alterations in the protein composition of maturing phagosomes. *J. Clin. Invest.* 90:1978–83

Ramachandra L, Harding CV. 2000. Phagosomes acquire nascent and recycling class II MHC molecules but primarily use nascent molecules in phagocytic antigen processing. *J. Immunol.* 164:5103–12

Ramachandra L, Noss E, Boom WH, Harding CV. 2001. Processing of *Mycobacterium tuberculosis* antigen 85B involves intraphagosomal formation of peptide-major histocompatibility complex II complexes and is inhibited by live bacilli that decrease phagosome maturation. *J. Exp. Med.* 194:1421–32

Reddien PW, Horvitz HR. 2004. The engulfment process of programmed cell death in *Caenorhabditis elegans. Annu. Rev. Cell Dev. Biol.* 20:193–221

Reddy A, Caler EV, Andrews NW. 2001. Plasma membrane repair is mediated by Ca^{2+}-regulated exocytosis of lysosomes. *Cell* 106:157–69

Roy CR, Tilney LG. 2002. The road less traveled: transport of *Legionella* to the endoplasmic reticulum. *J. Cell Biol.* 158:415–19

Schmitz A, Herrgen H, Winkeler A, Herzog V. 2000. Cholera toxin is exported from microsomes by the Sec61p complex. *J. Cell Biol.* 148:1203–12

Shao D, Segal AW, Dekker LV. 2003. Lipid rafts determine efficiency of NADPH oxidase activation in neutrophils. *FEBS Lett.* 550:101–6

In this systematic characterization of the possible combinations of SNARE molecules, an ER SNARE was shown to pair with a PM SNARE to mediate fusion.

This study shows the possible role of two ER proteins, calnexin and calreticulin, in phagocytosis.

This study characterizes the role of the small GTP-binding protein ARF6 in regulating focal membrane delivery at the site of nascent phagosomes.

Shen L, Sigal LJ, Boes M, Rock KL. 2004. Important role of cathepsin S in generating peptides for TAP-independent MHC class I crosspresentation in vivo. *Immunity* 21:155–65

Shiratsuchi A, Watanabe I, Takeuchi O, Akira S, Nakanishi Y. 2004. Inhibitory effect of Toll-like receptor 4 on fusion between phagosomes and endosomes/lysosomes in macrophages. *J. Immunol.* 172:2039–47

Silverstein AM. 2003. Darwinism and immunology: from Metchnikoff to Burnet. *Nat. Immunol.* 4:3–6

Stossel TP. 1999. The early history of phagocytosis. In *Advances in Cell and Molecular Biology of Membranes and Organelles, Volume 5, Phagocytosis: The Host*, ed. S Gordon, 5:3–18. JAI. 536 pp.

Swanson JA, Hoppe AD. 2004. The coordination of signaling during Fc receptor-mediated phagocytosis. *J. Leukoc. Biol.* 76:1093–103

Tapper H, Grinstein S. 1997. Fc receptor-triggered insertion of secretory granules into the plasma membrane of human neutrophils. *J. Immunol.* 159:409–18

Tardieux I, Webster P, Ravesloot J, Boron W, Lunn JA, et al. 1992. Lysosome recruitment and fusion are early events required for trypanosome invasion of mammalian cells. *Cell* 71:1117–30

Underhill DM, Ozinsky A. 2002. Phagocytosis of microbes: complexity in action. *Annu. Rev. Immunol.* 20:825–52

Underhill DM, Ozinsky A, Hajjar AM, Stevens A, Wilson CB, et al. 1999. The Toll-like receptor 2 is recruited to macrophage phagosomes and discriminates between pathogens. *Nature* 401:811–15

Vergne I, Chua J, Singh SB, Deretic V. 2004. Cell biology of mycobacterium tuberculosis phagosome. *Annu. Rev. Cell Dev. Biol.* 20:367–94

Vilhardt F, van Deurs B. 2004. The phagocyte NADPH oxidase depends on cholesterol-enriched membrane microdomains for assembly. *EMBO J.* 23:739–48

Watts C. 2004. The exogenous pathway for antigen presentation on major histocompatibility complex class II and CD1 molecules. *Nat. Immunol.* 5:685–92

Werb Z, Cohn ZA. 1972. Plasma membrane synthesis in the macrophage following phagocytosis of polystyrene latex particles. *J. Biol. Chem.* 247:2439–46

Ye Y, Shibata Y, Yun C, Ron D, Rapoport TA. 2004. A membrane protein complex mediates retro-translocation from the ER lumen into the cytosol. *Nature* 429:841–47

Protein Translocation by the Sec61/SecY Channel

Andrew R. Osborne,[1] Tom A. Rapoport,[1]
and Bert van den Berg[2]

[1] Howard Hughes Medical Institute and Department of Cell Biology, Harvard Medical
School, Boston, Massachusetts 02115; email: Andrew_Osborne@hms.harvard.edu,
tom_rapoport@hms.harvard.edu

[2] University of Massachusetts Medical School, Program in Molecular Medicine,
Biotech II, Worcester, Massachusetts 01605; email: bert.vandenberg@umassmed.edu

Annu. Rev. Cell Dev. Biol.
2005. 21:529–50

First published online as a
Review in Advance on
June 29, 2005

The *Annual Review of
Cell and Developmental
Biology* is online at
http://cellbio.annualreviews.org

doi: 10.1146/
annurev.cellbio.21.012704.133214

Copyright © 2005 by
Annual Reviews. All rights
reserved

1081-0706/05/1110-
0529$20.00

Key Words

membrane protein integration, ribosome-channel complex

Abstract

The conserved protein-conducting channel, referred to as the Sec61
channel in eukaryotes or the SecY channel in eubacteria and archaea,
translocates proteins across cellular membranes and integrates proteins containing hydrophobic transmembrane segments into lipid
bilayers. Structural studies illustrate how the protein-conducting
channel accomplishes these tasks. Three different mechanisms, each
requiring a different set of channel binding partners, are employed
to move polypeptide substrates: The ribosome feeds the polypeptide
chain directly into the channel, a ratcheting mechanism is used by
the eukaryotic endoplasmic reticulum chaperone BiP, and a pushing
mechanism is utilized by the bacterial ATPase SecA. We review these
translocation mechanisms, relating biochemical and genetic observations to the structures of the protein-conducting channel and its
binding partners.

Contents

INTRODUCTION

Protein transport across the ER membrane in eukaryotes is an early and decisive step in the biosynthesis of many proteins (for earlier reviews, see Hegde & Lingappa 1997, Johnson & van Waes 1999, Matlack et al. 1998). These proteins can be divided into two groups: soluble proteins, such as those ultimately secreted from the cell or localized to the ER lumen, and membrane proteins, such as those in the plasma membrane or in other organelles of the secretory pathway. In eubacteria and archaea, protein transport occurs directly through the plasma membrane and is also an important step in the biosynthesis of secreted and membrane proteins. Soluble proteins cross the membrane completely and usually have N-terminal cleavable signal sequences, whose major feature is a short hydrophobic segment (typically 7–12 amino acid residues). Membrane proteins have different topologies, with one or more TM segments, each containing about 20 hydrophobic residues. Membrane proteins have soluble domains that are translocated through

ER: endoplasmic reticulum

TM: transmembrane

the membrane as well as soluble domains that remain in the cytosol. Both types of proteins use the same machinery for translocation across the membrane: a protein-conducting channel with a hydrophilic interior (Crowley et al. 1993, Simon & Blobel 1991). This channel, in contrast to those channels that transport ions and small molecules, has the unusual property of being able to open in two directions: perpendicular to the plane of the membrane to allow a polypeptide segment across and within the membrane to allow a hydrophobic TM segment of a membrane protein to exit laterally into the lipid phase. The protein-conducting channel is formed by an evolutionarily conserved heterotrimeric membrane protein complex termed the Sec61 complex in eukaryotes and the SecY complex in eubacteria and archaea. In this review, we summarize our current understanding of how the channel functions in protein translocation, with special reference to its recently determined X-ray structure (van den Berg et al. 2004).

THE Sec61/SecY COMPLEX

The largest subunit of the heterotrimeric Sec61/SecY complex is the α-subunit, termed Sec61α in mammals, Sec61p in *Saccharomyces cerevisiae*, and SecY in eubacteria and archaea (for a review, see Rapoport et al. 1996). This subunit spans the membrane ten times, with both the N- and C termini in the cytosol. The β-subunit is termed Sec61β in mammals, Sbh1p in *S. cerevisiae*, SecG in eubacteria, and Secβ in archaea. In eukaryotes and archaea, this subunit spans the membrane once with the N- terminus in the cytosol. SecG in eubacteria spans the membrane twice. The γ-subunit is termed Sec61γ in mammals, Sss1p in *S. cerevisiae*, and SecE in eubacteria and archaea. In most species, this subunit is a single-spanning protein with its N terminus in the cytosol. In some eubacteria, e.g., *Escherichia coli*, the γ-subunit has two additional N-terminal TM segments that are not essential for its function. The α- and γ-subunits of the

Sec61/SecY complex are found in all organisms and show low, but significant, sequence conservation. The β-subunits are homologous among eukaryotes and archaea but have no obvious sequence similarity to the eubacterial SecG. The α- and γ-subunits are essential for viability of yeast and eubacteria, whereas the β-subunit is not. Together, these observations indicate that the α- and γ-subunits constitute the core of the channel-forming complex. Several organisms have two copies of Sec61 or SecY (Bensing & Sullam 2002, Rapoport et al. 1996), and in some cases the second copy may transport specific substrates (Bensing & Sullam 2002). In S. cerevisiae, there is a second Sec61 complex (Ssh1 complex), which is not essential for viability and seems to function exclusively in cotranslational translocation (Finke et al. 1996).

The initial evidence that the Sec61/SecY complex forms a protein-conducting channel came from systematic cross-linking experiments in which photoreactive probes were placed at different positions of a polypeptide substrate (Mothes et al. 1994). Substrates with probes at positions predicted to be within the membrane could be cross-linked to the α-subunit of the Sec61 complex, but not to other membrane proteins. These data indicated that the α-subunit surrounds the polypeptide chain as the chain passes through the membrane. Strong support for the notion that the Sec61/SecY complex forms a channel came from experiments in which the purified complex was reconstituted into proteoliposomes and shown to be the essential membrane component for protein translocation (Akimaru et al. 1991, Brundage et al. 1990, Gorlich & Rapoport 1993).

THREE DIFFERENT MODES OF TRANSLOCATION

The protein-conducting channel formed by the Sec61/SecY complex is a passive pore that allows a polypeptide chain to slide back and forth. The channel therefore needs to associate with partners that provide a driving force

for translocation. Depending on the partner, the channel can function in three different translocation modes.

The first mode, cotranslational translocation, involves the ribosome as the major channel partner (**Figure 1**). This is a general translocation mechanism found in all organisms and cells, and it is responsible for the integration of most membrane proteins. Cotranslational translocation begins with a targeting phase during which a ribosome-nascent chain complex is directed to the membrane by the signal recognition particle (SRP) and its membrane receptor (SRP receptor) (for review, see Halic & Beckmann 2005, Luirink & Sinning 2004). Once the ribosome is bound to the protein-conducting channel, the elongating polypeptide chain is moved from the ribosome to the membrane channel; GTP hydrolysis during translation provides the energy for translocation (**Figure 1**). When the ribosome synthesizes a cytosolic domain of a membrane protein, the polypeptide chain emerges from the ribosome-channel junction sideways into the cytosol (Mothes et al. 1997). In a later section we discuss cotranslational translocation in more detail.

In eukaryotes, there is a second mode of translocation by which proteins are transported after completion of their synthesis (posttranslational translocation). Proteins that use this mode have a less hydrophobic signal sequence and may therefore escape interaction with SRP during their synthesis (Ng et al. 1996). The mechanism of posttranslational translocation has been determined in S. cerevisiae (Matlack et al. 1999), and it is likely to be the same in higher eukaryotes. In this mode of translocation, the channel partners are another membrane protein complex (the Sec62/63 complex) and the lumenal protein BiP, a member of the Hsp70 family of ATPases. In yeast, the Sec62/63 complex is a tetramer that, together with the Sec61 complex, forms a seven-component Sec complex (Deshaies et al. 1991, Panzner et al. 1995). In addition to the essential proteins Sec62p and Sec63p, this complex contains the

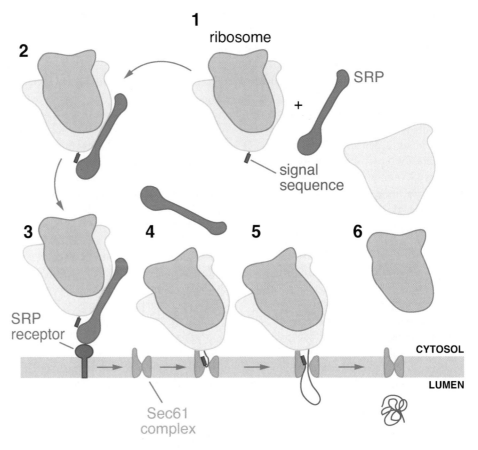

Figure 1

Cotranslational translocation of a secretory protein. The scheme shows different steps in the translocation of a eukaryotic secretory protein. (*1* and *2*) The signal recognition particle (SRP) binds to the signal sequence in a growing polypeptide chain as well as to the ribosome (large subunit, *light blue*; small subunit, *pink*). (*3*) The entire complex is targeted to the membrane by an interaction of the SRP with the SRP receptor. (*4* and *5*) The SRP is released, and the ribosome binds to the protein-conducting channel formed by the Sec61 complex. The polypeptide inserts into the channel as a loop, with the N and C termini in the cytosol. The signal sequence is intercalated into the wall of the channel, and the following polypeptide segment is located in the pore proper. (*6*) The remainder of the polypeptide chain moves from the ribosome tunnel, through the channel, and to the other side of the membrane. The signal sequence is cleaved at some point during translocation.

nonessential components Sec71p and Sec72p. Mammalian cells have Sec62p and Sec63p but lack the other two proteins (Meyer et al. 2000, Tyedmers et al. 2000).

The driving force for posttranslational translocation is generated by a ratcheting mechanism (**Figure 2**) (Matlack et al. 1999). A polypeptide in the channel can slide in either direction, but its binding to BiP inside the ER lumen prevents movement back into the cytosol, resulting in net forward translocation. ATP-bound BiP, with an open peptide-binding pocket, interacts with a lumenal domain of Sec63p, termed the J domain. This interaction stimulates rapid ATP hydrolysis and closure of the peptide-binding pocket around the incoming polypeptide chain. When the polypeptide has moved a sufficient distance in the forward direction, another BiP molecule can bind to it; this process

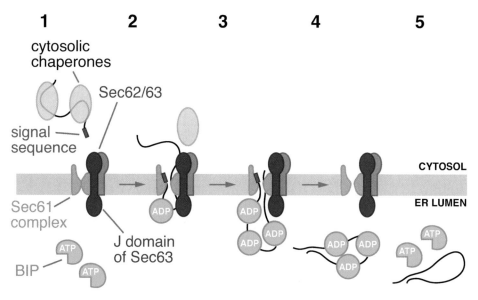

Figure 2

Posttranslational translocation in eukaryotes. (*1*) After it is synthesized in the cytosol, an unfolded polypeptide is kept in solution by cytosolic chaperones. (*2*) It is targeted by its signal sequence to the translocation channel, comprised of the Sec61 complex and the Sec62/63 complex, and the cytosolic chaperones are released. The J domain of Sec63 stimulates ATP hydrolysis by BiP, and ADP bound BiP binds to the polypeptide chain emerging into the ER lumen. (*3*) When the polypeptide has moved a sufficient distance into the ER lumen, another BiP molecule can bind to it. (*4*) This process is repeated until the polypeptide chain has completely traversed the channel. (*5*) BiP is released upon exchange of ADP for ATP; this exchange opens the peptide-binding pocket.

is repeated until the polypeptide chain has completely traversed the channel. When ADP is exchanged for ATP, the peptide-binding pocket opens, and BiP is released.

Several aspects of the ratcheting mechanism deserve comment. First, prior to translocation, a polypeptide substrate loses all bound cytosolic chaperones, facilitating its passive forward movement. Cross-linking experiments show that several different chaperones bind to the completed polypeptide and probably cycle on and off (Plath & Rapoport 2000). However, once a polypeptide chain has bound to the Sec complex through its N-terminal signal sequence, even chaperones that interact with the C terminus are released. The Sec complex does not stimulate the dissociation of chaperones, but rather prevents their rebinding, perhaps through the sizable cytosolic domains of Sec62p and Sec63p. A specific targeting molecule, similar to SRP in cotransla-

tional translocation, has not been found. Second, BiP binds to a diverse set of substrates and, within each polypeptide, to different segments. Although BiP preferentially binds hydrophobic peptides under equilibrium conditions, it shows little sequence specificity when activated by the J domain of Sec63p (Misselwitz et al. 1998). Under such nonequilibrium conditions, even segments that do not fit perfectly into the peptide-binding pocket can bind. Third, the location of the J domain ensures that BiP activation only occurs close to the channel, where BiP binding to the polypeptide chain is most effective in preventing its backsliding (**Figure 2**). Once a polypeptide segment has moved away from the channel, new BiP molecules do not bind, whereas those that bound previously can dissociate. Fourth, forward movement of the polypeptide chain is likely by Brownian motion. This is supported by the observation that

EM: electron
microscopy

in proteoliposomes containing the Sec complex, ATP-independent translocation occurs if BiP is replaced by antibodies to the substrate (Matlack et al. 1999). Mathematical modeling also shows that a Brownian ratcheting mechanism is sufficient to explain the kinetics of translocation (Liebermeister et al. 2001).

A third mode of translocation, found only in eubacteria, also occurs posttranslationally; it is used by most secretory proteins (for review, see Mori & Ito 2001). In this case, the channel partner is a cytosolic ATPase, termed SecA. SecA likely undergoes conformational changes coupled to its ATPase cycle and pushes polypeptides through the SecY channel in a stepwise manner (**Figure 3**) (Economou & Wickner 1994). We discuss the mechanism of SecA-mediated translocation in more detail later.

Archaea probably have both co- and posttranslational translocation (Irihimovitch & Eichler 2003, Ortenberg & Mevarech 2000), but it is unclear how they perform the latter, as they lack both SecA and the Sec62/63 complex.

THE X-RAY STRUCTURE OF THE SecY COMPLEX AND ITS IMPLICATIONS

Significant insight into the function of the protein-conducting channel is provided by the 3.2 Å resolution X-ray structure of the detergent-solubilized SecY complex from the archaebacterium *Methanococcus jannaschii* (van den Berg et al. 2004). Given the sequence similarities mentioned above, it is likely that the structure is representative of all species. In addition, the structure of the *E. coli* SecY complex, determined by EM of two-dimensional crystals (Breyton et al. 2002), shows that all TM segments are virtually superimposable onto those of the archaeal complex (van den Berg et al. 2004). This observation also means that the structure of the SecY complex in detergent is very similar to that in a lipid bilayer.

In the X-ray structure, the SecY complex contains one copy of each of the three subunits (**Figure 4**). Viewed from the cytosol, the complex has an approximately square shape. The two small subunits (SecE and Secβ) are

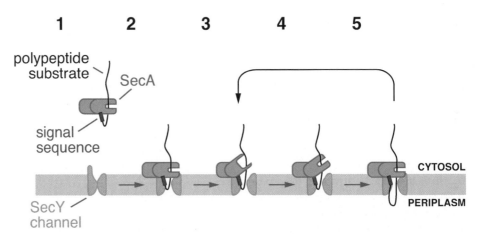

Figure 3

SecA-mediated posttranslational translocation in eubacteria. The scheme shows a model for the different steps in translocation. (*1*) SecA binds to a polypeptide substrate bearing an N-terminal signal sequence. (*2*) The complex binds to SecY, and the polypeptide substrate inserts as a loop into the channel. (*3*) The SecA polypeptide–binding groove opens and moves away from the channel, leaving a polypeptide segment in the channel. (*4*) The binding groove grabs the next section of the polypeptide chain and then closes. (*5*) The polypeptide-binding domains move toward the channel, pushing the polypeptide segment into the channel. Steps (*3*)–(*5*) are repeated until the polypeptide chain is fully translocated (not shown).

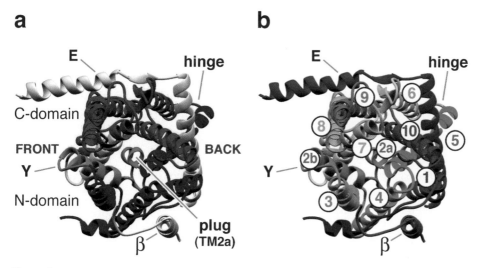

a

E
hinge
C-domain
FRONT   BACK
Y
N-domain
β   plug
(TM2a)

b

E
hinge
9 6
8 10
7 2a 5
2b 1
Y
3 4
β

Figure 4

(*a*) The structure of the *M. jannaschii* SecY complex viewed from the cytoplasm. The N-terminal domain of SecY (TM1–5) is shown in dark blue, with the exception of TM2b (*bright blue*). The C-terminal domain (TM6–10) is shown in red, with the exception of TM7 (*yellow*). The signal sequence intercalates at the front, between TM2b and TM7. The plug (TM2a), which blocks the pore of the closed channel, is shown in green. The proposed hinge region between TM segments 5 and 6 is labeled. (*b*) A cytoplasmic view of the *M. janaschii* SecY complex with individual helices colored and labeled.

located at the periphery of the complex. The SecE subunit contacts SecY extensively, occupying two sides of the square. The SecY subunit contains ten TM segments, organized into N- and C-terminal domains, comprising TM1–5 and TM6–10, respectively (**Figure 4**). The two domains are connected at the back of the complex by the loop between TM5 and TM6. SecY displays pseudosymmetry such that its C-terminal domain is essentially an upside-down version of its N-terminal domain. The domain organization of SecY and the locations of the two small subunits at the periphery leave the front of the complex as the only site that could open laterally toward the lipid phase. Such a lateral gate is necessary for the function of the SecY complex. The complex can therefore be likened to a clamshell that can open at the front toward the lipid, with the hinge located at the back of the complex between TM5 and TM6. The SecE subunit may serve as a brace that prevents the two domains from separating completely.

The X-ray structure suggests that the channel pore is located at the center of a single copy of the SecY complex (van den Berg et al. 2004) rather than at the interface of three or four complexes (Beckmann et al. 1997, Breyton et al. 2002, Manting et al. 2000, Morgan et al. 2002). Disulfide bridge formation between cysteines in a translocation substrate and cysteines in SecY support the notion that the polypeptide chain moves through the center of a single SecY molecule (Cannon et al. 2005). In addition, almost all of the conserved residues in the SecY complex are located not at the periphery but in the center of the complex (van den Berg et al. 2004). Mutations that allow proteins with defective or missing signal sequences to be transported (prl mutations; Bieker et al. 1990, Derman et al. 1993) are also located in the center of the SecY complex. Moreover, the interface of laterally associated complexes cannot form a hydrophilic pore; similar to all other membrane proteins, a single SecY complex has an entirely hydrophobic belt of ~25 Å width around it.

In a membrane, this belt would be exposed to the hydrophobic interior of the lipid bilayer. Together, these observations suggest that the pore is contained within a single SecY complex. Indeed, the structure shows a cytoplasmic funnel that may mark the channel entrance. The funnel tapers to a close in the middle of the membrane and is blocked on the extracytoplasmic side by the presence of a small helical segment (TM2a) dubbed the "plug" (**Figure 4a**) (van den Berg et al. 2004). The crystal structure of the archaebacterial SecY complex therefore corresponds to that of a closed channel; this is as expected, given that the complex was crystallized in the absence of translocation partners and substrate.

Opening of the channel appears to require movement of the plug (**Figure 5**). Cysteines introduced into the plug and into the TM segment of SecE of the *E. coli* SecY complex form a disulfide bridge in vivo (Harris & Silhavy 1999), suggesting that the plug moves toward the back of the complex, into a cavity at the extracellular side. Disulfide bridge formation cannot be explained by the structure of the closed channel, in which the cysteines would be too far apart (>20 Å). As expected, locking the channel into a permanently open state by inducing disulfide bridge formation is lethal to cells (Harris & Silhavy 1999).

The channel is probably in a dynamic equilibrium, with the plug moving between the closed and open positions. In the unoccupied channel, the equilibrium is on the side of the closed state, but it can be shifted toward the open state by the binding of a signal sequence or, in the case of many membrane proteins, a TM segment. Cross-linking experiments have shown that the hydrophobic core of a signal sequence forms a short helix containing about two turns. This helix intercalates between TM2b and TM7 of Sec61/SecY at the front of the molecule and contacts phospholipids (Plath et al. 1998). The translocation substrate is inserted as a loop; the signal sequence is intercalated into the channel wall, and the following polypeptide segment is located in the pore proper. Signal sequence intercalation requires a hinge motion at the back of Sec61/SecY to open the "mouth of

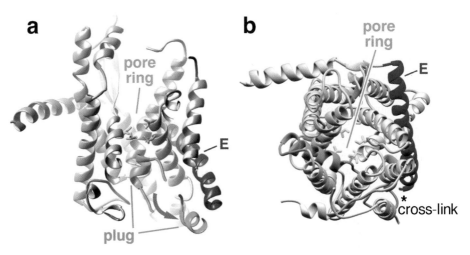

Figure 5

Plug movement leads to opening of the SecY channel. (*a*) View from the side of the channel with the front half of the model cut away. The modeled movement of the plug toward the SecE subunit is indicated by an arrow. The side chains of residues in the pore ring are colored in gold. (*b*) Cytosolic view, with the plug modeled in its open position. TM2b and TM7 located at the front of the complex are shown in blue and yellow, respectively. The asterisk indicates the region where introduced cysteines result in cross-links between the plug and the TM segment of SecE (Harris & Silhavy 1999).

the clamshell." The separation of the two halves of the molecule may destabilize interactions that keep the plug in the center of the molecule, thus promoting channel opening. In support of this model, many signal sequence suppressor mutations in SecY appear to destabilize the structure of the closed channel (van den Berg et al. 2004). Once the signal sequence is inserted into the channel walls, the polypeptide segment distal to the signal sequence may move through the pore and prevent the plug from returning to its closed-state position (**Figure 1**).

The binding of a channel partner (SecA or the ribosome) may also regulate channel opening. Support for the notion that ribosomes destabilize the closed state of the channel comes from electrophysiological experiments, in which increased ion conductance is observed when a nontranslating ribosome is bound to the channel (Simon & Blobel 1991). The ribosome binds exclusively to the cytosolic loops located in the C-terminal half of Sec61/SecY (Raden et al. 2000) and therefore does not prevent the separation of the two halves of the molecule.

The open channel may be shaped like an hourglass, with hydrophilic funnels on both sides of a constriction in the center of the membrane. This is consistent with the observation that a translocating polypeptide chain moves through the membrane in an aqueous environment (Crowley et al. 1993, Simon & Blobel 1991). During translocation, a substrate may primarily make contact with residues at the channel constriction, minimizing substrate-channel interactions. Restriction of contacts between the translocating chain and the channel to a narrow region is supported by recent experiments (Cannon et al. 2005).

The constriction point of the channel, or pore ring, consists of six hydrophobic amino acid residues, which in many species are isoleucines (**Figure 5**) (van den Berg et al. 2004). The pore ring may fit like a gasket around the translocating polypeptide chain, thereby providing a seal that restricts the passage of ions and other small molecules during protein translocation. In this model, the membrane barrier can be maintained in all modes of translocation. In an alternative model, the seal for small molecules is provided by the binding of a ribosome to the cytosolic side of the channel or by the binding of BiP to the ER lumenal side (Crowley et al. 1994, Hamman et al. 1998). This model is at odds with the available structural data (for further discussion see Rapoport et al. 2004). In addition, it does not explain how the membrane barrier is maintained in the absence of a ribosome (during posttranslational translocation) or in the absence of BiP (in prokaryotes).

In addition to plug movement, widening of the pore is likely required to allow polypeptide chain translocation. The diameter of the pore ring, as observed in the crystal structure, is too small to allow passage of even an unfolded, extended polypeptide chain. Widening of the channel may occur by movement of the helices to which the pore residues are attached. Flexible glycine-rich sequences in the cytosolic loops between TM4 and TM5 and between TM9 and TM10 may allow the channel to accommodate movement of these helices. Pore widening is required to explain the experimentally observed translocation of α-helices, a 13-residue disulfide-bonded polypeptide loop (Tani et al. 1990), or of amino acid side chains modified with bulky groups (De Keyzer et al. 2002, Kurzchalia et al. 1988). The flexibility of the pore region is supported by molecular dynamics simulations, which show that a ball of 10–12 Å or a helix with a diameter of 10 Å may move through the pore (P. Tian & I. Andricioaei, J. Gumbart & K. Schulten, personal communications). The intercalation of a signal sequence at the front of Sec61/SecY (opening of the clamshell) may cause additional widening of the pore, as is required for loop insertion of a polypeptide chain.

The estimated maximum dimensions of the pore based on the X-ray structure are ~15 × 20 Å. This is much smaller than the estimate of a pore diameter of at least 40 Å; this latter estimate derives from the observation that large reagents can pass through the membrane channel to quench fluorescent probes in a nascent polypeptide chain (Hamman et al. 1997). Such a large hydrophilic channel can be generated only if several Sec61/SecY molecules associate with their front surfaces and open to fuse their pores. However, at least in eubacteria, SecY molecules appear to associate back-to-back. This is the arrangement in the dimer seen by EM in two-dimensional crystals of the *E. coli* SecY complex (**Figure 6**) (Breyton et al. 2002), and the functional importance of this orientation in posttranslational translocation is supported by cross-linking experiments (Kaufmann et al. 1999). The fluorescence quenching data are also at odds with the X-ray structure of the large ribosomal subunit, because the same reagents quench probes inside the ribosomal tunnel, which has a diameter much narrower than 40 Å (Hamman et al. 1997). A relatively narrow pore is also consistent with the fact that even a small polypeptide domain can-

not fold inside the channel (Kowarik et al. 2002).

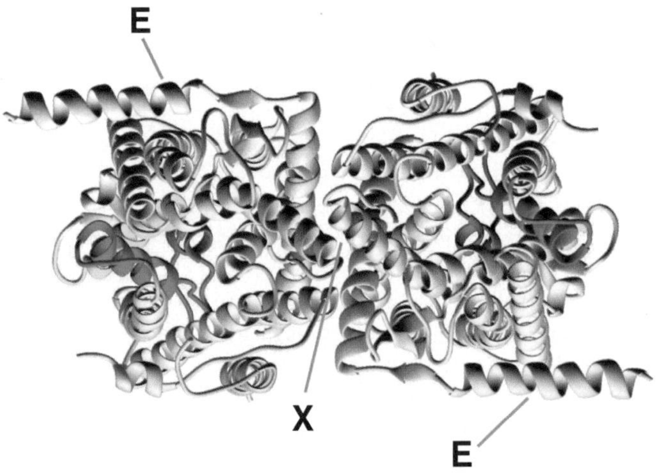

MECHANISM OF COTRANSLATIONAL TRANSLOCATION

Ribosome-Channel Interaction

The eukaryotic ribosome-channel complex has been visualized by single-particle EM (Beckmann et al. 1997, 2001, Menetret et al. 2000, Morgan et al. 2002). The ribosome likely is associated with four copies of the Sec61 complex. A low-density area in the center of the assembly was initially interpreted as a central pore, but in the most recent reconstructions at ~15–17 Å resolution, with an improved contour level, a pore is no longer visible (Beckmann et al. 2001, Morgan et al. 2002). Although the resolution of the EM data is insufficient to unambiguously dock the X-ray structure of the SecY complex, a plausible arrangement of the four Sec61 molecules consists of two side-by-side associated dimers, which in turn are formed by back-to-back assembled monomers (**Figure 7**) (Menetret et al. 2005). Such a side-by-side packing of dimers is seen in the two-dimensional crystals of the *E. coli* SecY complex (Breyton et al. 2002). This arrangement generates a low-density central region, but this region is entirely hydrophobic and may be filled with lipid or, after solubilization, with detergent. The β-subunits contribute significantly to the interface between the dimers (Bessonneau et al. 2002, Breyton et al. 2002), but are not essential, suggesting that ribosome binding may play a significant role in assembling the tetramer.

The linkage between the ribosome and the four copies of the Sec61 complex consists of approximately four to seven connections (Beckmann et al. 2001, Menetret et al. 2005, Morgan et al. 2002). Several ribosomal proteins and regions of ribosomal RNA, which may be involved in the interaction, have been identified (Beckmann et al. 2001, Morgan

Figure 6

Back-to-back arrangement of *E.coli* SecY complexes in the dimer as seen in the EM structure derived from two-dimensional crystals (Breyton et al. 2002). TM2b and TM7 at the front of the complexes are colored in blue and yellow, respectively. The plug is shown in dark green. Cysteines introduced at the positions indicated by the red spheres result in efficient disulfide formation (X) between two SecE subunits (Kaufmann et al. 1999).

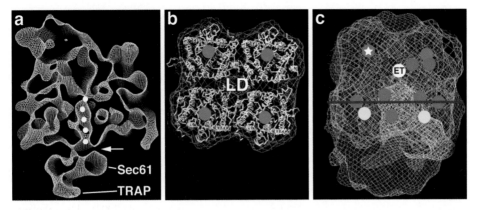

Figure 7

EM structure of ribosome-Sec61 channel complexes and possible oligomeric arrangement of Sec61 complexes (Menetret et al. 2005). (*a*) Ribosome-channel complexes derived from native ER membranes were analyzed by single-particle EM. A cross section through the ribosome-channel complex, viewed from the side, is shown. The ribosomal exit tunnel, from which a growing polypeptide chain would emerge, is marked by white dots. The gap between the ribosome and channel is labeled with an arrow. A lumenal protrusion is formed by the TRAP (translocon-associated protein) complex. (*b*) A tetrameric assembly of SecY complexes, as seen in the EM structure of two-dimensional crystals of the *E. coli* SecY complex (Breyton et al. 2002). The position of the pores within each SecY molecule is indicated by a blue dot. The expected position of a low-density central region is labeled LD. A mask generated to encompass the whole volume of the tetramer is shown as a blue mesh. (*c*) A lumenal view of the Sec61 channel bound to the ribosome. The blue mask enveloping the tetrameric SecY assembly shown in (*b*) is docked into the density. The ribosomal exit tunnel is marked by a white dot and labeled ET. Connections between the ribosome and the channel are marked by red dots. The prominent line of connections is indicated by a red line. The pores of the two copies of Sec61 that are separated from the ribosome exit site by the line of connections are labeled with yellow dots. One of the other two pores (in *blue*) may translocate a polypeptide chain. The Sec61 complex with only weak or no connections to the ribosome is additionally labeled with a white asterisk. The region of density not occupied by the blue meshwork is assigned to TRAP.

et al. 2002). Biochemical data suggest that RNA provides the major contacts with the channel (Prinz et al. 2000). As expected from the asymmetry of the ribosome, the four copies of Sec61 complex bind differently; one of them has no or only weak connections (**Figure 7c**), whereas the others have multiple linkages. Two of the Sec61 molecules are on one side of a line of connections, which separates them from the exit site where the nascent chain emerges from the ribosome, leaving one of the other two copies to form the active pore. The ribosome-channel junction is open and thus provides a path for polypeptides from the ribosomal exit site into the cytosol, as is required when the ribosome synthesizes cytosolic domains of membrane proteins. The gap of 12–15 Å width between the ri-

bosome and channel is consistent with the size of the cytosolic loops in the C-terminal half of Sec61/SecY (van den Berg et al. 2004), which provide the major ribosome-binding sites (Raden et al. 2000). The size of the gap may prevent many large cytosolic molecules from reaching the pore and passing through it, but the pore ring inside the channel is likely the main device that maintains the membrane barrier.

If the channel is formed from a single copy of the Sec61 complex, what is the role of oligomerization? The answer is not yet known, but one possibility is that oligomerization serves to create binding sites for the recruitment of other components. In eukaryotes, these other components may include signal peptidase, which cleaves signal sequences

TRAM:
translocating
chain–associating
membrane protein

TRAP: translocon-
associated
protein

from translocating polypeptides; oligosac-charyl transferase, which attaches carbohy-drate chains to them; and TRAM, a multi-spanning membrane protein that may serve as a membrane chaperone (see below). All of these proteins are close to the chan-nel, but have no strong affinity for either the isolated Sec61 complex or the ribosome (Gorlich & Rapoport 1993). Oligomerization may also be the trigger for the recruitment of the TRAP complex, a tetrameric membrane protein complex of unknown function. EM analysis of ribosome-channel complexes de-rived from native ER membranes shows that TRAP is bound to the two Sec61 complexes that are inaccessible to the nascent chain (**Figure 7c**) (Menetret et al. 2005). This sug-gests that the function of these Sec61 com-plexes is to recruit TRAP rather than to translocate a nascent chain. Oligomerization of the Sec61 complex may also regulate ribo-some binding. Tetramers may provide a larger number of linkages, resulting in strong ribo-some binding during translocation, whereas dissociation of the tetramers may weaken the interaction and facilitate ribosome release upon termination of translocation.

The ribosome-channel interaction in eu-bacteria and archaea has not been studied ex-tensively. It is unclear whether it is as tight as in eukaryotes or whether tetramers of the SecY complex are involved. Several bac-terial membrane proteins require SecA for translocation of their extracellular domains (Neumann-Haefelin et al. 2000). For steric reasons, the ribosome and SecA cannot bind simultaneously to the channel. This suggests that, in contrast with the situation in eukary-otes, the ribosome in eubacteria and archaea may dissociate during translocation.

Membrane Protein Integration

The integration of membrane proteins is more complicated than the translocation of soluble proteins, and many issues are still un-resolved. In the following section, we briefly summarize our current understanding (for

a more extensive discussion of controversial points, see Rapoport et al. 2004).

In contrast to a signal sequence, which al-ways has its N terminus in the cytosol, the first TM segment of a nascent membrane pro-tein can have its N terminus on either side of the membrane, depending on the amino acid sequence of the protein. In a multispanning protein, the first TM segment often deter-mines the orientation of the subsequent ones, which alternate correspondingly. A model for how the orientation of the first TM segment may be determined is depicted in **Figure 8**. A passive orientation of downstream TM seg-ments is suggested by the fact that many mem-brane proteins seem to have evolved by the fusion of two halves that have opposite ori-entations. In this respect, it is interesting to note that the transporter EmrE is proposed to be a dimer of identical subunits with op-posite topologies (Ma & Chang 2004), sim-ilar to the postulated evolutionary predeces-sors of current membrane proteins possessing pseudo twofold symmetry. There are, how-ever, exceptions in which internal TM seg-ments have a preferred orientation regardless of the behavior of preceding TM segments (Gafvelin & von Heijne 1994, Goder et al. 1999, Locker et al. 1992, McGovern et al. 1991, Nilsson et al. 2000, Sato et al. 1998).

During the synthesis of a membrane pro-tein, TM segments must move from the aque-ous interior of the channel through its lateral gate into the lipid phase. The lateral gate is formed by relatively short segments of TM8, TM7, TM2b, and TM3 (van den Berg et al. 2004). Because TM2b and TM3 are located in the N-terminal half of SecY and TM7 and TM8 are located in the C-terminal half, the gate may undergo "breathing," i.e., contin-uous opening and closure. This may be fa-cilitated in the open channel when the plug has moved toward the back of the channel and no longer contacts the gate's TM seg-ments. Breathing of the lateral gate would occasionally expose segments of a polypep-tide chain located in the aqueous channel to the hydrophobic interior of the lipid bilayer,

N-terminus of first TM segment is translocated

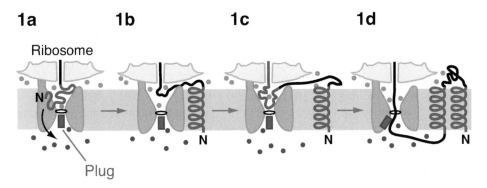

C-terminus of first TM segment is translocated

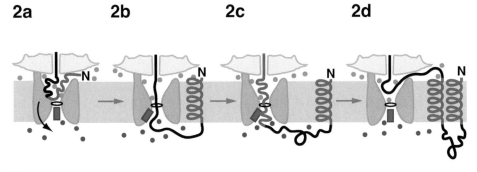

Figure 8

Model of membrane protein integration. (*1a, 1b*) When the first TM segment (*red*) of a membrane protein has fully emerged from the ribosome, the N-terminus can flip across the membrane (*arrow*) if the TM segment is long and hydrophobic and the preceding polypeptide segment is not positively charged or folded (N terminus translocated; *upper panel*) (Wahlberg & Spiess 1997). The N terminus may be translocated through the channel after a brief displacement of the plug, and the TM segment partitions into the lipid. (*1c, 1d*) The following hydrophilic polypeptide segment emerges into the cytosol through the gap between the ribosome and the channel. The next TM segment (*red*) inserts into the channel as a loop, destabilizing its closed state. The channel opens by movement of the plug, and the second TM segment of the polypeptide partitions into the lipid. As this occurs, the next hydrophilic segment enters the channel and will ultimately be translocated to the other side of the membrane. (*2a, 2b*) In the cases of other proteins with a short first TM segment or a preceding region that is either folded or positively charged, the N terminus may stay in the cytosol. The TM segment inserts into the channel as a loop, destabilizing its closed state. Upon chain elongation, the C-terminal end flips across the membrane (*arrow*), allowing the TM segment to partition into the lipid and leaving the channel occupied by the following hydrophilic region of the polypeptide chain. (*2c*) The N terminus of the second TM segment enters the open channel. (*2d*) When sufficient hydrophobic residues have emerged from the ribosome, they will exit laterally into the lipid, allowing the plug to return to its closed state position. The following hydrophilic segment will emerge into the cytosol through the gap between the ribosome and the channel. During translocation and membrane integration of a polypeptide, either the plug or the nascent chain hinders the passage of small molecules (*green* and *purple*) through the channel.

enabling them to equilibrate between the two phases. If sufficiently long and hydrophobic, a segment exits into the lipid phase (Duong & Wickner 1998, Heinrich et al. 2000). Because TM segments differ widely in sequence, they are unlikely to play an active role in opening of the lateral gate. A passive partitioning model is also supported by the observation that a hydrophobicity scale, derived from peptide interactions with an organic solvent, can be used to predict the tendency for a TM segment to integrate into the membrane (Hessa et al. 2005).

The open channel is most likely too small to allow "storage" of several TMs; during the synthesis of a multispanning membrane protein, the TMs leave the channel one by one, or perhaps in pairs. After moving through the lateral gate, some hydrophobic TMs are immediately surrounded by lipid, while other TMs that contain charges remain in proximity of the channel (Heinrich et al. 2000), sometimes until termination of translation (Do et al. 1996). Factors other than hydrophobicity of a TM, perhaps properties of the flanking region, may also influence how long a TM remains close to the channel (McCormick et al. 2003, Meacock et al. 2002). TMs that remain close to the channel for prolonged periods of time appear to be associated with TRAM, a protein located at the front of the Sec61 channel (Mothes et al. 1998). TRAM can be cross-linked to the signal sequences of secretory proteins and to charged TMs of nascent membrane proteins (Do et al. 1996, Görlich et al. 1992, Heinrich et al. 2000), and it is required for the translocation of secretory proteins with weakly hydrophobic signal sequences (Voigt et al. 1996). It may act as a membrane chaperone to stabilize TMs with hydrophilic residues and facilitate the association of these TMs until they can be released as a hydrophobic assembly into bulk lipid. The bacterial YidC protein, which has a similar topology as TRAM and is required for the integration and folding of some membrane proteins, may have an analogous function (Dalbey & Kuhn 2004).

During synthesis of a cytosolic domain of a membrane protein, the ribosome remains bound to the channel (Mothes et al. 1997). The nascent chain must therefore emerge between the ribosome and channel into the cytosol (**Figure 8**). Such a lateral path may be provided by the gap of 12–15 Å between the two partners, as seen in EM reconstructions. In contrast to models in which the ribosome-channel junction opens and closes (Johnson & van Waes 1999), the junction in the model proposed in **Figure 8** is always open, allowing a nascent chain to move sideways into the cytosol.

SecA-MEDIATED POSTTRANSLATIONAL TRANSLOCATION IN EUBACTERIA

The mechanism by which the cytoplasmic ATPase SecA moves polypeptide chains through the SecY channel is still poorly understood, but some new insights are provided by structural studies. SecA consists of five domains: two RecA-like folds, referred to as nucleotide-binding folds 1 and 2 (NBF1 and NBF2); the preprotein cross-linking domain (PPXD); the helical scaffold domain (HSD); and the helical wing domain (HWD) (**Figure 9a** and **b**) (Hunt et al. 2002, Osborne et al. 2004, Sharma et al. 2003). The ATPase site of SecA is similar to that in superfamily 1 and 2 helicases, with the nucleotide bound at the interface between NBF1 and NBF2. Nucleotide-dependent domain movements in SecA may therefore be similar to those seen in the helicase PcrA (Velankar et al. 1999). Contrary to earlier assumptions, there is only one nucleotide-binding site in SecA, with both NBF domains providing residues critical for ATP hydrolysis (Mitchell & Oliver 1993, Or et al. 2002, Papanikou et al. 2004, Schmidt et al. 2001, Sianidis et al. 2001). A conserved arginine residue in NBF2 likely senses the presence or absence of the γ-phosphate and triggers the appropriate domain movements (Or et al. 2002). These changes may be transmitted to

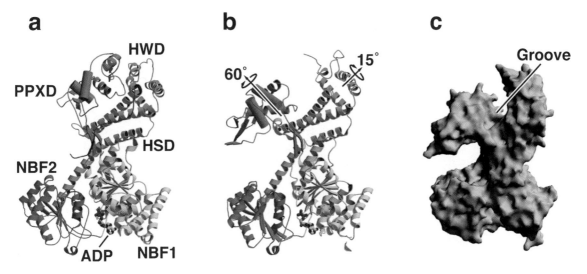

Figure 9

The structure of SecA. (*a*) Dimeric *Bacillus subtilis* SecA in a closed conformation (Hunt et al. 2002). A single subunit is shown. Nucleotide-binding fold 1 (NBF1) is shown in yellow, nucleotide-binding fold 2 (NBF2) in blue, the preprotein cross-linking domain (PPXD) in orange, the helical scaffold domain (HSD) in green, and the helical wing domain (HWD) in cyan. ADP is shown in a ball and stick representation. (*b*) Monomeric *B. subtilis* SecA in an open conformation (Osborne et al. 2004), colored as in (*a*). The arrows indicate movements required to convert the open conformation to the closed conformation. (*c*) A surface representation of SecA in the open conformation. The groove is proposed to close around the translocating polypeptide.

and amplified by the other domains that bind the polypeptide substrate and push it into the SecY channel.

SecA exists in equilibrium between monomeric and dimeric states (Benach et al. 2003, Ding et al. 2003, Or et al. 2002, Woodbury et al. 2002). When isolated, it is mostly a dimer. The X-ray structure by Hunt et al. (2002) may correspond to the physiological dimer, although other dimeric forms have been postulated (Sharma et al. 2003). Dissociation into monomers is stimulated upon interaction with ligands such as lipids (Benach et al. 2003, Bu et al. 2003, Or et al. 2002) or synthetic signal peptides (although the latter is controversial), thereby suggesting that the monomer is the active species in translocation. This is supported by the observation that cross-linked products corresponding to SecA dimers are lost upon interaction with the SecY complex (Or et al. 2002). A monomeric mutant of SecA retains some activity at least

under some conditions (Or et al. 2004), although the same mutant was found to be inactive in other studies (Jilaveanu et al. 2005, Randall et al. 2005). Upon solubilization, a single copy of SecA is found in a complex containing an arrested translocation substrate and SecY (Duong 2003). However, the exact nature of the complex during protein translocation is unclear, as it has been claimed that two SecAs may associate with two or four SecY complexes (Duong 2003, Manting et al. 2000, Tziatzios et al. 2004). SecA-induced tetramers of SecY complexes, observed by EM, may be arranged in a similar way as ribosome-associated Sec61/SecY complexes (Veenendaal et al. 2004).

Compared to the structure of the *Bacillus subtilis* SecA dimer, monomeric SecA is in an open conformation, in which the PPXD, HSD, and HWD have undergone dramatic movements, while the NBF domains have remained at the same position

(**Figure 9a** and **b**). In the open conformation, the HSD/HWD and the PPXD form a large groove (**Figure 9c**) (Osborne et al. 2004) that is likely the polypeptide-binding site, as indicated by cross-linking and mutagenesis studies (Kimura et al. 1991, Kourtz & Oliver 2000). The groove is similar in dimensions to those seen in other proteins—such as OppA, DnaK, and SecB—that interact with a wide range of peptide substrates (Sleigh et al. 1999, Xu et al. 2000, Zhu et al. 1996). In all of these proteins, a deep binding groove appears to wrap around the peptide, allowing binding to substrates that differ in sequence. Whether SecA binds signal sequences in a more specific way is unclear, but a potential binding site is a hydrophobic groove located at the interface of NBF1, the HSD, and the PPXD (Hunt et al. 2002).

It is likely that, as originally proposed, SecA pushes the polypeptide substrate through the SecY channel (Economou & Wickner 1994), but it is unclear exactly how this happens. A pushing mechanism implies that there are two polypeptide-binding sites that alternate in their affinities for the polypeptide substrate and that can move relative to each other. One possibility is that both sites are located in SecA, similar to helicases. However, as only one peptide-binding groove is apparent from the SecA structures, it seems more likely that SecY provides the second binding site.

It has been proposed that SecA inserts deeply into the SecY channel, reaching the other side of the membrane (Economou & Wickner 1994, Eichler & Wickner 1997, Kim et al. 1994, Ramamurthy & Oliver 1997, van der Does et al. 1996). This mechanism has been inferred from the fact that SecA is accessible to proteases and labeling reagents added from the outside of the cell. However, the structural data indicate that SecA is too big to insert into the channel. Thus, the previous data may be better explained if we assume that SecA adopts a protease-resistant conformation upon SecY binding (van der Does et al. 1998) and is accessible to labeling reagents

through the open SecY channel. The modification sites are indeed spread out over the entire SecA molecule (Hunt et al. 2002).

Taking into account the recent structural data, it seems likely that SecA pushes the polypeptide into the SecY channel without itself inserting deeply into the channel (**Figure 3**). In this model, SecA binds to a polypeptide segment, pushes it into the channel, and then releases it. Backsliding of the polypeptide substrate is reduced by its interactions with the SecY channel. Next, SecA releases the substrate and undergoes a conformational change, moving the peptide-binding site away from the channel to bind the next polypeptide segment. This cycle continues until the entire polypeptide is translocated. Although the current data suggest that SecA-mediated translocation is processive [i.e., a single SecA translocates each polypeptide substrate entirely (Joly & Wickner 1993, Schiebel et al. 1991)], the occasional disengagement of SecA, or even a nonprocessive mode of translocation, cannot be completely ruled out. In addition, although early experiments suggested that during each cycle SecA pushes 20–30 residues through the channel (Schiebel et al. 1991, Uchida et al. 1995, van der Wolk et al. 1997), this step size seems very large (it corresponds to ~100 Å of extended polypeptide). It is clear that further studies are required to resolve these issues.

PERSPECTIVES

Structural studies of ribosome/Sec61 complexes, of SecA, and particularly of the SecY channel have significantly advanced our understanding of the mechanism of protein translocation. Interpretation of these structures has been made possible by equally important genetic and biochemical data accumulated in many laboratories over the years. The recent data have led to new hypotheses that need to be tested experimentally. In addition, these data highlight a number of unresolved issues. For example, how exactly does SecA move polypeptides through the SecY channel? What is the role of the oligomerization of the Sec61/SecY channel? How do interacting partners of the Sec61/SecY channel regulate its function? How are membrane proteins integrated and folded? Progress will depend on a combination of different approaches, with the structure of an active channel being a major goal for the future.

SUMMARY POINTS

1. The protein-conducting channel, formed by the Sec61/SecY complex, is required for both the translocation of polypeptides across cellular membranes and for the integration of these polypeptides into lipid bilayers.

2. The X-ray structure of the SecY complex provides new insight into how the protein-conducting channel functions.

3. Polypeptide translocation may occur posttranslationally or cotranslationally.

4. Each different mode of translocation requires different channel partners.

ACKNOWLEDGMENTS

We thank K. Cannon and L. Pond for critically reading the manuscript and William Clemons for providing Figures 4, 5, and 6. We also thank P. Tian, I. Andricioaei, J. Gumbart, and K. Schulten for sharing unpublished results. T.A.R. is a Howard Hughes Medical Institute investigator and is also supported by an NIH grant.

LITERATURE CITED

Akimaru J, Matsuyama SI, Tokuda H, Mizushima S. 1991. Reconstitution of a protein translocation system containing purified SecY, SecE, and SecA from *Escherichia coli*. *Proc. Natl. Acad. Sci. USA* 88:6545–49

Beckmann R, Bubeck D, Grassucci R, Penczek P, Verschoor A, et al. 1997. Alignment of conduits for the nascent polypeptide chain in the ribosome-Sec61 complex. *Science* 19:2123–26

These authors, together with Morgan et al. 2002, have determined the structure of ribososome-translocation channel complexes by electron microscopy.

Beckmann R, Spahn CM, Eswar N, Helmers J, Penczek PA, et al. 2001. Architecture of the protein-conducting channel associated with the translating 80S ribosome. *Cell* 107:361–72

Benach J, Chou YT, Fak JJ, Itkin A, Nicolae DD, et al. 2003. Phospholipid-induced monomerization and signal-peptide-induced oligomerization of SecA. *J. Biol. Chem.* 278:3628–38

Bensing BA, Sullam PM. 2002. An accessory sec locus of *Streptococcus gordonii* is required for export of the surface protein GspB and for normal levels of binding to human platelets. *Mol. Microbiol.* 44:1081–94

Bessonneau P, Besson V, Collinson I, Duong F. 2002. The SecYEG preprotein translocation channel is a conformationally dynamic and dimeric structure. *EMBO J.* 21:995–1003

Bieker KL, Phillips GJ, Silhavy TJ. 1990. The sec and prl genes of *Escherichia coli*. *J. Bioenerg. Biomembr.* 22:291–310

Breyton C, Haase W, Rapoport TA, Kuhlbrandt W, Collinson I. 2002. Three-dimensional structure of the bacterial protein-translocation complex SecYEG. *Nature* 418:662–65

Brundage L, Hendrick JP, Schiebel E, Driessen AJM, Wickner W. 1990. The purified *E. coli* integral membrane protein SecY/E is sufficient for reconstitution of SecA-dependent precursor protein translocation. *Cell* 62:649–57

Bu Z, Wang L, Kendall DA. 2003. Nucleotide binding induces changes in the oligomeric state and conformation of Sec A in a lipid environment: a small-angle neutron-scattering study. *J. Mol. Biol.* 332:23–30

Cannon KS, Or E, Clemons WM Jr, Shibata Y, Rapoport TA. 2005. Disulfide bridge formation between SecY and a translocating polypeptide localizes the translocation pore to the center of SecY. *J. Cell Biol.* 169: 219–25

Crowley KS, Liao SR, Worrell VE, Reinhart GD, Johnson AE. 1994. Secretory proteins move through the endoplasmic reticulum membrane via an aqueous, gated pore. *Cell* 78:461–71

Crowley KS, Reinhart GD, Johnson AE. 1993. The signal sequence moves through a ribosomal tunnel into a noncytoplasmic aqueous environment at the ER membrane early in translocation. *Cell* 73:1101–15

Dalbey RE, Kuhn A. 2004. YidC family members are involved in the membrane insertion, lateral integration, folding, and assembly of membrane proteins. *J. Cell Biol.* 166:769–74

De Keyzer J, Van Der Does C, Driessen AJ. 2002. Kinetic analysis of the translocation of fluorescent precursor proteins into *Escherichia coli* membrane vesicles. *J. Biol. Chem.* 277:46059–65

Derman AI, Puziss JW, Bassford PJ, Beckwith J. 1993. A signal sequence is not required for protein export in prlA mutants of *Escherichia coli*. *EMBO J.* 12:879–88

Deshaies RJ, Sanders SL, Feldheim DA, Schekman R. 1991. Assembly of yeast Sec proteins involved in translocation into the endoplasmic reticulum into a membrane-bound multi-subunit complex. *Nature* 349:806–8

Ding H, Hunt JF, Mukerji I, Oliver D. 2003. *Bacillus subtilis* SecA ATPase exists as an antiparallel dimer in solution. *Biochemistry* 42:8729–38

Do H, Falcone D, Lin J, Andrews DW, Johnson AE. 1996. The cotranslational integration of membrane proteins into the phospholipid bilayer is a multistep process. *Cell* 85:369–78

Duong F. 2003. Binding, activation and dissociation of the dimeric SecA ATPase at the dimeric SecYEG translocase. *EMBO J.* 22:4375–84

Duong F, Wickner W. 1998. Sec-dependent membrane protein biogenesis: SecYEG, preprotein hydrophobicity and translocation kinetics control the stop-transfer function. *EMBO J.* 17:696–705

Economou A, Wickner W. 1994. SecA promotes preprotein translocation by undergoing ATP-driven cycles of membrane insertion and deinsertion. *Cell* 78:835–43

Eichler J, Wickner W. 1997. Both an N-terminal 65-kDa domain and a C-terminal 30-kDa domain of SecA cycle into the membrane at SecYEG during translocation. *Proc. Natl. Acad. Sci. USA* 94:5574–81

Finke K, Plath K, Panzner S, Prehn S, Rapoport TA, et al. 1996. A second trimeric complex containing homologs of the Sec61p complex functions in protein transport across the ER membrane of S. cerevisiae. *EMBO J.* 15:1482–94

Gafvelin G, von Heijne G. 1994. Topological "frustration" in multi-spanning *E. coli* inner membrane proteins. *Cell* 77:401–12

Goder V, Bieri C, Spiess M. 1999. Glycosylation can influence topogenesis of membrane proteins and reveals dynamic reorientation of nascent polypeptides within the translocon. *J. Cell Biol.* 147:257–66

Görlich D, Hartmann E, Prehn S, Rapoport TA. 1992. A protein of the endoplasmic reticulum involved early in polypeptide translocation. *Nature* 357:47–52

Gorlich D, Rapoport TA. 1993. Protein translocation into proteoliposomes reconstituted from purified components of the endoplasmic reticulum membrane. *Cell* 75:615–30

Halic M, Beckmann R. 2005. The signal recognition particle and its interactions during protein targeting. *Curr. Opin. Struct. Biol.* 15:116–25

Hamman BD, Chen JC, Johnson EE, Johnson AE. 1997. The aqueous pore through the translocon has a diameter of 40–60 Å during cotranslational protein translocation at the ER membrane. *Cell* 89:535–44

Hamman BD, Hendershot LM, Johnson AE. 1998. BiP maintains the permeability barrier of the ER membrane by sealing the lumenal end of the translocon pore before and early in translocation. *Cell* 92:747–58

Harris CR, Silhavy TJ. 1999. Mapping an interface of SecY (PrlA) and SecE (PrlG) by using synthetic phenotypes and in vivo cross-linking. *J. Bacteriol.* 181:3438–44

Hegde RS, Lingappa VR. 1997. Membrane protein biogenesis: regulated complexity at the endoplasmic reticulum. *Cell* 91:575–82

Heinrich SU, Mothes W, Brunner J, Rapoport TA. 2000. The Sec61p complex mediates the integration of a membrane protein by allowing lipid partitioning of the transmembrane domain. *Cell* 102:233–44

Hessa T, Kim H, Bihlmaier K, Lundin C, Boekel J, et al. 2005. Recognition of transmembrane helices by the endoplasmic reticulum translocon. *Nature* 433:377–81

Hunt JF, Weinkauf S, Henry L, Fak JJ, McNicholas P, et al. 2002. Nucleotide control of interdomain interactions in the conformational reaction cycle of SecA. *Science* 297:2018–26

Irihimovitch V, Eichler J. 2003. Post-translational secretion of fusion proteins in the halophilic archaea *Haloferax volcanii*. *J. Biol. Chem.* 278:12881–87

Jilaveanu LB, Zito CR, Oliver D. 2005. Dimeric SecA is essential for protein translocation. *Proc. Natl. Acad. Sci. USA* 102:7511–16

Johnson AE, van Waes MA. 1999. The translocon: a dynamic gateway at the ER membrane. *Annu. Rev. Cell Dev. Biol.* 15:799–842

Joly JC, Wickner W. 1993. The SecA and SecY subunits of translocase are the nearest neighbors of a translocating preprotein, shielding it from phospholipids. *EMBO J.* 12:255–63

Kaufmann A, Manting EH, Veenendaal AK, Driessen AJ, van der Does C. 1999. Cysteine-directed cross-linking demonstrates that helix 3 of SecE is close to helix 2 of SecY and helix 3 of a neighboring SecE. *Biochemistry* 38:9115–25

Kim YJ, Rajapandi T, Oliver D. 1994. SecA protein is exposed to the periplasmic surface of the *E. coli* inner membrane in its active state. *Cell* 78:845–53

Kimura E, Akita M, Matsuyama S, Mizushima S. 1991. Determination of a region in SecA that interacts with presecretory proteins in *Escherichia coli*. *J. Biol. Chem.* 266:6600–6

Kourtz L, Oliver D. 2000. Tyr-326 plays a critical role in controlling SecA-preprotein interaction. *Mol. Microbiol.* 37:1342–56

Kowarik M, Kung S, Martoglio B, Helenius A. 2002. Protein folding during cotranslational translocation in the endoplasmic reticulum. *Mol. Cell* 10:769–78

Kurzchalia TV, Wiedmann M, Breter H, Zimmermann W, Bauschke E, Rapoport TA. 1988. tRNA-mediated labelling of proteins with biotin A nonradioactive method for the detection of cell-free translation products. *Eur. J. Biochem.* 172:663–68

Liebermeister W, Rapoport TA, Heinrich R. 2001. Ratcheting in post-translational protein translocation: a mathematical model. *J. Mol. Biol.* 305:643–56

Locker JK, Rose JK, Horzinek MC, Rottier PJM. 1992. Membrane assembly of the triple-spanning coronavirus M-protein. Individual transmembrane domains show preferred orientation. *J. Biol. Chem.* 267:21911–18

By systematically substituting amino acids in a TM segment and determining the effect on integration, the authors establish rules for membrane protein integration that correspond remarkably well to a simple physico-chemical model.

This paper describes the first high-resolution structure of the cytoplasmic ATPase SecA, an essential component in posttranslational translocation in bacteria.

Luirink J, Sinning I. 2004. SRP-mediated protein targeting: structure and function revisited. *Biochim. Biophys. Acta* 1694:17–35

Ma C, Chang G. 2004. Structure of the multidrug resistance efflux transporter EmrE from *Escherichia coli. Proc. Natl. Acad. Sci. USA* 101:2852–57

Manting EH, van Der Does C, Remigy H, Engel A, Driessen AJ. 2000. SecYEG assembles into a tetramer to form the active protein translocation channel. *EMBO J.* 19:852–61

Matlack KE, Misselwitz B, Plath K, Rapoport TA. 1999. BiP acts as a molecular ratchet during posttranslational transport of prepro-α factor across the ER membrane. *Cell* 97:553–64

Matlack KES, Mothes W, Rapoport TA. 1998. Protein translocation: tunnel vision. *Cell* 92:381–90

McCormick PJ, Miao Y, Shao Y, Lin J, Johnson AE. 2003. Cotranslational protein integration into the ER membrane is mediated by the binding of nascent chains to translocon proteins. *Mol. Cells* 12:329–41

McGovern K, Ehrmann M, Beckwith J. 1991. Decoding signals for membrane protein assembly using alkaline phosphatase fusions. *EMBO J.* 10:2773–82

Meacock SL, Lecomte FJ, Crawshaw SG, High S. 2002. Different transmembrane domains associate with distinct endoplasmic reticulum components during membrane integration of a polytopic protein. *Mol. Biol. Cell* 13:4114–29

Menetret JF, Hegde RS, Heinrich SU, Chandramouli P, Ludtke SJ, et al. 2005. Architecture of the ribosome-channel complex derived from native membranes. *J. Mol. Biol.* 348:445–57

Menetret J, Neuhof A, Morgan DG, Plath K, Radermacher M, et al. 2000. The structure of ribosome-channel complexes engaged in protein translocation. *Mol. Cell* 6:1219–32

Meyer HA, Grau H, Kraft R, Kostka S, Prehn S, et al. 2000. Mammalian Sec61 is associated with Sec62 and Sec63. *J. Biol. Chem.* 275:14550–57

Misselwitz B, Staeck O, Rapoport TA. 1998. J proteins catalytically activate Hsp70 molecules to trap a wide range of peptide sequences. *Mol. Cell* 2:593–603

Mitchell C, Oliver D. 1993. Two distinct ATP-binding domains are needed to promote protein export by *Escherichia coli* SecA ATPase. *Mol. Microbiol.* 10:483–97

Morgan DG, Menetret JF, Neuhof A, Rapoport TA, Akey CW. 2002. Structure of the mammalian ribosome-channel complex at 17 Å resolution. *J. Mol. Biol.* 324:871–86

Mori H, Ito K. 2001. The Sec protein-translocation pathway. *Trends Microbiol.* 9:494–500

Mothes W, Heinrich SU, Graf R, Nilsson I, von Heijne G, et al. 1997. Molecular mechanism of membrane protein integration into the endoplasmic reticulum. *Cell* 89:523–33

Mothes W, Jungnickel B, Brunner J, Rapoport TA. 1998. Signal sequence recognition in cotranslational translocation by protein components of the ER membrane. *J. Cell Biol.* 142:355–64

Mothes W, Prehn S, Rapoport TA. 1994. Systematic probing of the environment of a translocating secretory protein during translocation through the ER membrane. *EMBO J.* 13:3937–82

Neumann-Haefelin C, Schafer U, Muller M, Koch HG. 2000. SRP-dependent co-translational targeting and SecA-dependent translocation analyzed as individual steps in the export of a bacterial protein. *EMBO J.* 19:6419–26

Ng DT, Brown JD, Walter P. 1996. Signal sequences specify the targeting route to the endoplasmic reticulum membrane. *J. Cell Biol.* 134:269–78

Nilsson I, Witt S, Kiefer H, Mingarro I, von Heijne G. 2000. Distant downstream sequence determinants can control N-tail translocation during protein insertion into the endoplasmic reticulum membrane. *J. Biol. Chem.* 275:6207–13

Or E, Boyd D, Gon S, Beckwith J, Rapoport T. 2005. The bacterial ATPase SecA functions as a monomer in protein translocation. *J. Biol. Chem.* 280:9097–105

Or E, Navon A, Rapoport T. 2002. Dissociation of the dimeric SecA ATPase during protein translocation across the bacterial membrane. *EMBO J.* 21:4470–79

Ortenberg R, Mevarech M. 2000. Evidence for post-translational membrane insertion of the integral membrane protein bacterioopsin expressed in the heterologous halophilic archaeon *Haloferax volcanii. J. Biol. Chem.* 275:22839–46

Osborne AR, Clemons WM Jr., Rapoport TA. 2004. A large conformational change of the translocation ATPase SecA. *Proc. Natl. Acad. Sci. USA* 101:10937–42

Panzner S, Dreier L, Hartmann E, Kostka S, Rapoport TA. 1995. Posttranslational protein transport in yeast reconstituted with a purified complex of Sec proteins and Kar2p. *Cell* 81:561–70

Papanikou E, Karamanou S, Baud C, Sianidis G, Frank M, Economou A. 2004. Helicase motif III in SecA is essential for coupling preprotein binding to translocation ATPase. *EMBO Rep.* 5:807–11

Plath K, Mothes W, Wilkinson BM, Stirling CJ, Rapoport TA. 1998. Signal sequence recognition in posttranslational protein transport across the yeast ER membrane. *Cell* 94:795–807

Plath K, Rapoport TA. 2000. Spontaneous release of cytosolic proteins from posttranslational substrates before their transport into the endoplasmic reticulum. *J. Cell Biol.* 151:167–78

Prinz A, Behrens C, Rapoport TA, Hartmann E, Kalies KU. 2000. Evolutionarily conserved binding of ribosomes to the translocation channel via the large ribosomal RNA. *EMBO J.* 19:1900–6

Raden D, Song W, Gilmore R. 2000. Role of the cytoplasmic segments of Sec61α in the ribosome-binding and translocation-promoting activities of the Sec61 complex. *J. Cell Biol.* 150:53–64

Randall LL, Crane JM, Lilly AA, Liu G, Mao C, et al. 2005. Asymmetric binding between SecA and SecB two symmetric proteins: implications for function in export. *J. Mol. Biol.* 348:479–89

Ramamurthy V, Oliver D. 1997. Topology of the integral membrane form of *Escherichia coli* SecA protein reveals multiple periplasmically exposed regions and modulation by ATP binding. *J. Biol. Chem.* 272:23239–46

Rapoport TA, Goder V, Heinrich SU, Matlack KE. 2004. Membrane-protein integration and the role of the translocation channel. *Trends Cell Biol.* 14:568–75

Rapoport TA, Jungnickel B, Kutay U. 1996. Protein transport across the eukaryotic endoplasmic reticulum and bacterial inner membranes. *Annu. Rev. Biochem.* 65:271–303

Sato M, Hresko R, Mueckler M. 1998. Testing the charge difference hypothesis for the assembly of a eucaryotic multispanning membrane protein. *J. Biol. Chem.* 273:25203–8

Schiebel E, Driessen AJM, Hartl F-U, Wickner W. 1991. $\Delta\mu H^+$ and ATP function at different steps of the catalytic cycle of preprotein translocase. *Cell* 64:927–39

Schmidt MO, Brosh RM Jr., Oliver DB. 2001. *Escherichia coli* SecA helicase activity is not required in vivo for efficient protein translocation or autogenous regulation. *J. Biol. Chem.* 276:37076–85

Sharma V, Arockiasamy A, Ronning DR, Savva CG, Holzenburg A, et al. 2003. Crystal structure of *Mycobacterium tuberculosis* SecA, a preprotein translocating ATPase. *Proc. Natl. Acad. Sci. USA* 100:2243–48

Sianidis G, Karamanou S, Vrontou E, Boulias K, Repanas K, et al. 2001. Cross-talk between catalytic and regulatory elements in a DEAD motor domain is essential for SecA function. *EMBO J.* 20:961–70

Simon SM, Blobel G. 1991. A protein-conducting channel in the endoplasmic reticulum. *Cell* 65:371–80

Sleigh SH, Seavers PR, Wilkinson AJ, Ladbury JE, Tame JR. 1999. Crystallographic and calorimetric analysis of peptide binding to OppA protein. *J. Mol. Biol.* 291:393–415

Tani K, Tokuda H, Mizushima S. 1990. Translocation of proOmpA possessing an intramolecular disulfide bridge into membrane vesicles of *Escherichia coli* Effect of membrane energization. *J. Biol. Chem.* 265:17341–47

Tyedmers J, Lerner M, Bies C, Dudek J, Skowronek MH, et al. 2000. Homologs of the yeast Sec complex subunits Sec62p and Sec63p are abundant proteins in dog pancreas microsomes. *Proc. Natl. Acad. Sci. USA* 97:7214–19

Tziatzios C, Schubert D, Lotz M, Gundogan D, Betz H, et al. 2004. The bacterial protein-translocation complex: SecYEG dimers associate with one or two SecA molecules. *J. Mol. Biol.* 340:513–24

Uchida K, Mori H, Mizushima S. 1995. Stepwise movement of preproteins in the process of translocation across the cytoplasmic membrane of *Escherichia coli*. *J. Biol. Chem.* 270:30862–68

van den Berg B, Clemons WM, Jr., Collinson I, Modis Y, Hartmann E, et al. 2004. X-ray structure of a protein-conducting channel. *Nature* 427:36–44

van der Does C, den Blaauwen T, de Wit JG, Manting EH, Groot NA, et al. 1996. SecA is an intrinsic subunit of the *Escherichia coli* preprotein translocase and exposes its carboxyl terminus to the periplasm. *Mol. Microbiol.* 22:619–29

van der Does C, Manting EH, Kaufmann A, Lutz M, Driessen AJ. 1998. Interaction between SecA and SecYEG in micellar solution and formation of the membrane-inserted state. *Biochemistry* 37:201–10

van der Wolk JP, de Wit JG, Driessen AJ. 1997. The catalytic cycle of the *Escherichia coli* SecA ATPase comprises two distinct preprotein translocation events. *EMBO J.* 16:7297–304

Veenendaal AK, van der Does C, Driessen AJ. 2004. The protein-conducting channel SecYEG. *Biochim. Biophys. Acta.* 1694:81–95

Velankar SS, Soultanas P, Dillingham MS, Subramanya HS, Wigley DB. 1999. Crystal structures of complexes of PcrA DNA helicase with a DNA substrate indicate an inchworm mechanism. *Cell* 97:75–84

Voigt S, Jungnickel B, Hartmann E, Rapoport TA. 1996. Signal sequence-dependent function of the TRAM protein during early phases of protein transport across the endoplasmic reticulum membrane. *J. Cell Biol.* 134:25–35

Wahlberg JM, Spiess M. 1997. Multiple determinants direct the orientation of signal-anchor proteins: the topogenic role of the hydrophobic signal domain. *J. Cell Biol.* 137:555–62

Woodbury RL, Hardy SJ, Randall LL. 2002. Complex behavior in solution of homodimeric SecA. *Protein Sci.* 11:875–82

Xu Z, Knafels JD, Yoshino K. 2000. Crystal structure of the bacterial protein export chaperone secB. *Nat. Struct. Biol.* 7:1172–77

Zhu X, Zhao X, Burkholder WF, Gragerov A, Ogata CM, et al. 1996. Structural analysis of substrate binding by the molecular chaperone DnaK. *Science* 272:1606–14

This paper describes the first high-resolution structure of the SecY protein translocation channel, which is a central and essential component of protein secretion and membrane protein integration in all cells.

Retinotectal Mapping: New Insights from Molecular Genetics

Greg Lemke[1] and Michaël Reber[2]

[1]Molecular Neurobiology Laboratory, The Salk Institute, La Jolla, California 92037;
email: lemke@salk.edu

[2]INSERM U.575-Centre de Neurochimie, 67084 Strasbourg, France;
email: michael.reber@inserm.u-strasbg.fr

Annu. Rev. Cell Dev. Biol.
2005. 21:551–80

First published online as a
Review in Advance on
June 29, 2005

The *Annual Review of
Cell and Developmental
Biology* is online at
http://cellbio.annualreviews.org

doi: 10.1146/
annurev.cellbio.20.022403.093702

Copyright © 2005 by
Annual Reviews. All rights
reserved

1081-0706/05/1110-
0551$20.00

Key Words

topographic mapping, Eph receptors, ephrins, mathematical modeling

Abstract

The sensory and motor components of nervous systems are connected topographically and contain neural maps of the external world. The paradigm for such maps is the precisely ordered wiring of the output cells of the eye to their synaptic targets in the tectum of the midbrain. The retinotectal map is organized in development through the graded activity of Eph receptor tyrosine kinases and their ephrin ligands. These signaling proteins are arrayed in complementary expression gradients along the orthogonal axes of the retina and tectum, and provide both input and recipient cells with Cartesian coordinates that specify their location. Molecular genetic studies in the mouse indicate that these coordinates are interpreted in the context of neuronal competition for termination sites in the tectum. They further suggest that order in the retinotectal map is determined by ratiometric rather than absolute difference comparisons in Eph signaling along the temporal-nasal and dorsal-ventral axes of the eye.

Contents

INTRODUCTION

Vertebrate nervous systems carry multiple, reiterated representations, or maps, of the external world (Kaas 1997). These representations translate stimulus features into a coherent neural code, allowing for the interpretation of what are often complex streams of sensory information. Many maps are organized "topographically" with regard to a stimulus parameter: Auditory neurons that respond to adjacent tonal frequencies, somatosensory neurons that respond to adjacent points on the body surface, and visual system neurons that respond to adjacent points in the visual field are all positioned at adjacent locations within their respective sensory maps. Nervous systems use topographic maps to reduce the

RGC (retinal
ganglion cell)
neurons: the
primary output
neurons of the eye.
RGC axons exit the
eye at a single central
point (the optic disk)
and are bundled
together to form the
optic nerve

consequences of noise in the transmission of quantized information, to interpolate between data points, and to facilitate the interpretation of coincident stimuli.

How topographic maps are specified and organized during development has been a central preoccupation of neurobiologists for decades. Despite a great deal of effort, it has nonetheless proven difficult to formulate quantitatively rigorous models of map development that can be tested experimentally. Recently, a series of conceptual and experimental advances, derived mainly from molecular genetic analyses, have made this possible for the paradigmatic topographic map: the connection of RGC neurons in the eye to their targets in the optic tectum/SC of the midbrain.

TOPOGRAPHIC MAPPING

Most topographic maps may be viewed as an ordered array of connections in which the spatial relationships of a set of input neurons are preserved in their synaptic connection to their targets. Frequently, this involves preservation of the Cartesian (x,y) positional coordinates within a two-dimensional sheet of sensory neurons—across the surface of the skin, for example—in their connection to a similar two-dimensional sheet of target neurons. Within visual, somatosensory, olfactory, and auditory sensory modalities, multiple such maps are, at successive levels of information processing, repeatedly employed to represent sensory space.

Vertebrate maps are formed over an extended time frame in late embryonic and early postnatal development. In nearly all instances, the process is a dynamic one in which a course initial map is progressively refined into a fine-grained mature map. Map development often begins before the onset of coordinated neuronal activity, shortly after axons initially invade their targets (Holt & Harris 1998, Friauf & Lohman 1999, Rubel & Cramer 2002). Nonetheless, it is now well established that the firing of neurons—spatially correlated within the input ensemble—is crucial for both the progressive refinement and mature precision of topographic maps (Holt & Harris 1993, Debski & Cline 2002). Although we do not discuss the role of correlated electrical activity in map refinement, recent molecular genetic studies have resulted in important advances in this area (e.g., Huh et al. 2000, McLaughlin et al. 2003).

Sensory and Motor Maps

There are many topographic maps. In the somatosensory system, whose modalities include discriminative touch, pain, temperature, and proprioception, orderly connections are established between peripheral nerve endings and the CNS. In the reciprocal motor system, similarly ordered connections are established between motor neurons in the CNS and neuromuscular junctions in the periphery. Somatotopic organization is thus a hallmark of both input sensory and output motor pathways. The detailed description of this coupled organization by Penfield and colleagues led to the promulgation of the two best-known topographic maps: the cortical maps of the body surface referred to as the motor and somatosensory homunculi, which provide point-to-point cortical representations of the body surface (Penfield & Rasmussen 1950, Kaas 1997).

Auditory Maps

A second well-known instance of reiterated topographic mapping is seen in the connections that underlie the auditory and vestibular pathways. Auditory connections, for example, are topographic along the low- to high-frequency (distal to proximal) axis of the basilar membrane of the cochlea. These connections encode frequency-specific sound processing and form a "tonotopic" map that is conserved within the anteroventral, posteroventral, and dorsal cochlear nuclei (Echteler & Nofsinger 2000, Rubel & Fritsch 2002). Tonotopy is evident in the projection

of these cochlear nuclei to the superior olivary complex as well as in the inferior colliculus in the midbrain, the medial geniculate nucleus in the auditory thalamus, and the primary auditory cortex in the forebrain. At the level of the inferior colliculus and primary auditory cortex, further topographic connections are made to the SC and the frontal eye field, respectively, where representations of auditory space are integrated (Knudsen 1982, King & Palmer 1983, Cohen & Knudsen 1999).

Visual Maps

Although this review focuses on a single visual projection—from the retina to the midbrain—topographic mapping is a prominent feature of nearly all the sensory pathways that originate in the eye. Axons from RGCs, the output cells of the retina, exit the eye and fasiculate to form the optic nerve, which enters the brain proper near the ventral hypothalamus. In creatures with forward-facing eyes, roughly half of these axons cross the midline at the optic chiasm and project to the contralateral hemisphere of the brain. (In animals whose eyes are positioned laterally, e.g., mice, nearly all RGC axons make this midline crossing.) As they leave the optic chiasm, retinal axons travel together in the optic tract, which terminates in both the LGN of the thalamus and in the SC of the midbrain (or, in nonmammalian vertebrates, the optic tectum of the midbrain) (Kanaseki & Sprague 1974, Rodieck 1979). Axons from the LGN project to the primary visual (striate) cortex, and from there to a panoply of extrastriate visual areas. Although there is only course retinotopic order apparent within the optic nerve, optic tract, and thalamocortical and corticocortical axon tracts in most species, the synaptic terminations of these pathways are precisely retinotopic: The spatial arrangement of RGCs in the retina is preserved in the order of axon terminations in the LGN, the SC, and the striate and extrastriate cortices.

Optic tectum: a region of the midbrain, one of two primary recipients of RGC input; the other is the lateral geniculate nucleus of the thalamus. In mammals, the optic tectum is referred to as the superior colliculus

SC: superior colliculus

CNS: central nervous system

LGN: lateral geniculate nucleus

THE RETINOTECTAL MAP AS AN EXPERIMENTAL PARADIGM

Classical Studies

tn: temporal-nasal

Chemoaffinity hypothesis: the idea that the establishment of connections between neurons is directed by "chemical codes under genetic control." Modern versions of the chemoaffinity hypothesis are predicated on expression gradients of these chemical codes

The retinotectal (or retinocollicular) map has been the most intensively analyzed of all topographic maps. Historically, frogs, newts, and fishes have been the favored models for investigating both the configuration of the map and the mechanisms that underlie the map's formation. The visual responses of these creatures can be assessed using both electrophysiological and behavioral assays, and their optic nerves regenerate following surgical injuries such as transection of the nerve or partial ablation of the retina or tectum. This has allowed investigators to use the regenerating projections of large animals as models for what first occurs in development (Meyer 1982).

In addition to electrophysiology, anatomical methods such as tritiated proline autoradiography, horseradish-peroxidase histochemistry, and focal tectal degeneration following focal ablation of RGCs—pioneered by Sperry and colleagues using the optic system of fishes and amphibians (Attardi & Sperry 1960, Arora & Sperry 1962, Attardi & Sperry 1963)—have been extensively used to study map development. Modern anatomical methods, especially the use of fluorescent axonal tracers such as DiI, have come to dominate studies over the past two decades. At the same time, analyses of the retinotectal projection of chicks and of the equivalent retinocollicular projection of mammals, especially mice, have assumed increasing prominence owing to the considerable interpretive power of gain-of-function studies in the chick and of genetic perturbations, i.e., knockouts and knock-ins, in the mouse. There are important differences in the details of retinotectal mapping in frogs and fishes as opposed to that in mammals (see below), but the basic rules governing map formation now appear to be universal.

Configuration of the map. In the retinotectal projection, the Cartesian coordinates of the eye are mapped onto those of the SC/tectum. Thus, axons that arise from RGCs in the extreme nasal retina (near the nose) project to targets at the far caudal (posterior) end of the SC/tectum; conversely, axons from RGCs in the extreme temporal retina project to the rostral (anterior) end (**Figure 1**). RGCs located at intermediate tn retinal positions project to correspondingly intermediate positions along the rostral-caudal axis of the SC/tectum. The dorsal-ventral axis of the retina, orthogonal to the tn axis, is mapped with equal precision onto the lateral-medial axis of the SC/tectum (**Figure 1**).

Chemoaffinity. The dominant model for understanding retinotectal map development has been the chemoaffinity hypothesis promulgated by Sperry (1963). This hypothesis was developed through a series of experiments performed in the regenerating retinotectal projections of amphibians and fish. In one of the best-known experiments, Sperry cut the optic nerves of newts and rotated their eyes 180°. After regeneration of the connection of retinal axons to the tectum, the operated animals behaved as if they saw the world upside down and back to front. Because this response was 180° out of phase with respect to visual experience, it seemed inconsistent with models of retinotectal mapping determined solely by neuronal activity. Sperry postulated that retinal cells carry stable positional chemical labels, or chemoaffinity tags, that determine their proper tectal termination (Sperry 1943, 1944, 1963). He proposed that the simplest configuration of tags, notably with respect to the number of tags required, would occur if the chemical labels took the form of gradients that would mark retinal cells with appropriate x,y (in his words, "latitude and longitude") coordinates (Sperry 1963).

Map perturbations. Many lines of experiment substantiate the basic concepts in the chemoaffinity model, including the existence of fixed chemoaffinity tags (Fraser & Hunt 1980, Holt & Harris 1993). As

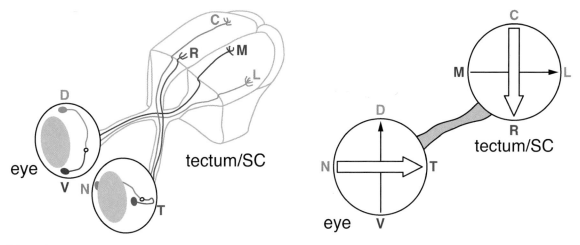

Figure 1

Configuration of the retinotectal map. (*Left*) Axonal connections from the eye to the SC, or tectum in nonmammalian vertebrates. As described in the text, the dorsal(D)-ventral(V) axis of the retina is mapped onto the lateral(L)-medial(M) axis of the tectum, and the temporal(T)-nasal(N) axis of the retina is mapped onto the rostral(R)-caudal(C) axis of the tectum. (*Right*) A schematic representation of the map.

noted above, if the optic tectum is rotated after the retinotectal map is mature, the map that regenerates will have nasal RGCs connected to the new rostral end of the rotated tectum. Although experimental phenomena of this sort argued strongly for fixed chemoaffinity labels, many other observations also argued for a considerable measure of plasticity with regard to the expression, stability, or interpretation of these labels. When the nasal half of a retina is ablated, for example, RGC axons from the remaining temporal half sprout, eventually filling the entire tectum. This filling is topographic with respect to position in the remaining retina: The nasal-most remaining RGCs send their axons to the caudal-most tectum (**Figure 2**). This generates an expanded map (Schmidt et al. 1978). Similarly, when the caudal half of the tectum is ablated, retinal axons redistribute to evenly and topographically fill the remaining tectum (**Figure 2**), leading to a compressed map (Sharma 1972, Yoon 1976, Cook 1979). These and related perturbation phenomena led to a great deal of interpretive controversy in the 1970s and 80s not only with regard to the potential plasticity

of positional gradients, but also with respect to the requirement for additional developmental mechanisms that may mediate topographic mapping. Favored among proponents of the latter was the idea that competition among RGC axons for limiting tectal innervation space may, together with chemoaffinity, drive mapping (Prestige & Willshaw 1975). A key problem with regard to deciding between alternative developmental mechanisms was that Sperry's hypothesized chemical labels eluded identification for decades.

The Cell Biology of Mapping

Membrane stripe assays and chemorepellents. In a series of illuminating studies, Bonhoeffer and his colleagues developed and exploited a set of in vitro membrane stripe assays to demonstrate that an important bioactivity guiding the ordered projection of RGC axons along the tn axis of the retina is a chemorepellent that is maximally expressed in the caudal tectum (Walter et al. 1987b). After depositing membrane fractions from rostral and caudal chick tectum in alternating stripes on an in vitro substrate, these workers

Chemorepellant: an activity (generally a protein) that, unlike a chemoattractant, inhibits axonal growth, extension, and branching, and promotes disintegration of the actin cytoskeleton and growth cone collapse

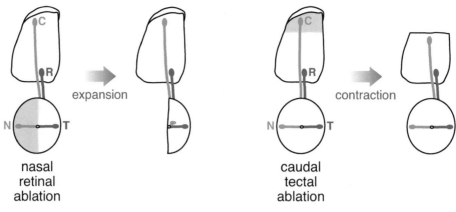

Figure 2

Plasticity of the retinotectal map. (*Left*) Map expansion. When the nasal half of the retina is ablated in amphibians, the termination zones of the ablated nasal RGCs are lost from the caudal tectum. The vacated tectal sites are eventually occupied by the axons of the RGCs that remain, and this occupation is retinotopic. (*Right*) Map contraction. When a caudal segment of the tectum is removed, RGC axons redistribute to evenly and topographically fill the tectum that remains.

GPI: glycosylphos-phatidylinositol

RAGS: repulsive axon guidance signal

Receptor protein-tyrosine kinases: transmembrane cell-surface receptors, such as the Eph proteins, that contain a protein-tyrosine kinase domain within their cytoplasmic regions

placed explants of nasal or temporal retina orthogonal to these stripes and then observed the frequency and orientation of RGC axonal outgrowth along the stripes. When given a choice, axons arising from RGCs in the temporal retina consistently displayed a strong preference for growth on rostral, as opposed to caudal, tectal membranes (Walter et al. 1987a). In contrast, RGCs in the extreme nasal retina exhibited no strong preference for rostral versus caudal membranes (Walter et al. 1987a, Cox et al. 1990).

Bonhoeffer and colleagues proceeded to show that the chemorepellent activity of the caudal tectal membranes was associated with a protein attached to the membrane surface through a GPI linkage (Stahl et al. 1990, Drescher et al. 1995). They used biochemical fractionation and purification methods to identify and clone cDNAs for one GPI-linked chemorepellent, which they termed RAGS (Drescher et al. 1995). As discussed below, this protein is now referred to as ephrin-A5. It is expressed in concert with the closely related ephrin-A2, which also inhibits axon outgrowth and which is also graded across both the mouse SC and the chick tectum (Cheng et al. 1995, Drescher et al. 1995, Zhang et al. 1996, Frisén et al. 1998). Given both their graded tectal expression and the principles of chemoaffinity, these caudal tectal chemorepellents were presumed to be sensed by a receptor, or receptors, that would be preferentially expressed in the temporal retina.

Eph receptors and ephrins. Studies of retinotectal mapping over the past decade have been greatly stimulated by the identification of the Eph receptors, proteins that were originally isolated from an erythropoietin-producing hepatocellular carcinoma line. The 14 Eph receptors—by far the largest family of receptor protein-tyrosine kinases—are divided in two subfamilies of 8 EphAs (A1–A8) and 6 EphBs (B1–B6). Their ligands, the ephrins, are also divided into two subfamilies, the GPI-anchored ephrin-As (A1–A6) and transmembrane ephrin-Bs (B1–B3).

Interactions of the Eph receptors and their ephrin ligands require cell-cell contact. Indeed, Eph receptors are activated i.e., their protein-tyrosine kinase activity is stimulated, only by multimerized membrane-bound ephrins (or when soluble forms of ephrins are artificially clustered). In vivo, receptor-ligand engagement invariably occurs at the surface of apposed cells or their processes. EphA receptors bind and are activated by only ephrin-As, and EphBs by only ephrin-Bs, with the exception of EphA4, which binds both ephrin-As and -Bs (Brambilla et al. 1995, Gale et al. 1996). There is considerable promiscuity in ligand-receptor pairing within the A and B subgroups, as measured by in vitro binding assays, although certain features of preferred ligand-receptor pairing within each subgroup have been discerned (Flanagan & Vanderhaegen 1998).

The Eph receptors display structurally distinctive extracellular domains, composed of a ligand-binding globular domain, a cysteine-rich region, and two fibronectin type III repeats. Their cytoplasmic regions are composed of a juxtamembrane domain containing two conserved tyrosine residues, a canonical protein-tyrosine kinase domain, and several protein-protein interaction domains (including a sterile α-motif and a PDZ-domain binding motif) (Flanagan & Vanderhaegen 1998, Kullander & Klein 2002, Murai & Pasquale 2003).

Depending on cellular context, activation of Eph receptors—as monitored by their autophosphorylation—may lead to either chemorepulsion or chemoattraction. Signal transduction downstream of Eph receptor engagement runs prominently through the small G proteins Rho, Rac, and cdc42 and, in turn, through regulators that directly control the polymerization state of actin. There is evidence for forward and reverse signaling within both the EphA and EphB systems, in that the ephrin-A and ephrin-B "ligands" appear to function as receptors in certain contexts and to signal cell-autonomously (Flanagan & Vanderhagen 1998, Hindges et al. 2002, Mann et al. 2002). These features of bidirectional signaling and regulation of the actin cytoskelton have been recently reviewed (Kullander & Klein 2002, Murai & Pasquale 2003). Although there are exceptions (see below), in the context of retinotec-

tal mapping, the preponderance of evidence suggests that forward signaling through Eph receptor–expressing RGCs predominates and that this forward signaling induces chemorepulsion (actin bundling, depolymerization, and growth cone collapse) as a consequence of EphA activation, and chemoattraction (actin polymerization and growth cone extension) as a consequence of EphB activation.

Gradients of Eph receptors and ephrins. In RGCs, an EphA3 gradient running from low nasally to high temporally was first reported in the chick; and two favored ligands for EphA3, ephrin-A2 and ephrin-A5, were at the same time found to be expressed in a low-rostral-to-high-caudal gradient in the chick tectum (Cheng et al. 1995, Drescher et al. 1995, Winslow et al. 1995). Both ephrin-A2 and ephrin-A5 are expressed in tectal gradients, but the two ligands display slightly different distributions. In the chick and mouse, ephrin-A2 expression is highest just rostral to the caudal end of the SC/tectum, and decreases smoothly toward the rostral end (Cheng et al. 1995, Zhang et al. 1996, Brennan et al. 1997, Marin et al. 2001). Ephrin-A5 is most highly expressed at the extreme caudal end and decreases more markedly toward the rostral end (Drescher et al. 1995, Monschau et al. 1997). In addition to ephrin-A5 expression in the SC, there is a sharp increase in its expression at the border between the SC and inferior colliculus (Donoghue et al. 1996, Zhang et al. 1996, Hansen et al. 2004).

Mice display a similar low-nasal-to-high-temporal EphA gradient in RGCs (**Figure 3**), but the graded receptors are EphA5 and EphA6, rather than EphA3 (Cheng et al. 1995, Connor et al. 1998, Brown et al. 2000, Marin et al. 2001, Reber et al. 2004). (EphA3 is not expressed by mouse RGCs.) In addition to these graded retinal EphA receptors, there are, in the mouse, chick, and all other species thus far examined, one or more EphA receptors whose expression is ungraded in RGCs along the tn axis (**Figure 3**). EphA4 is expressed but ungraded in the mouse, and both

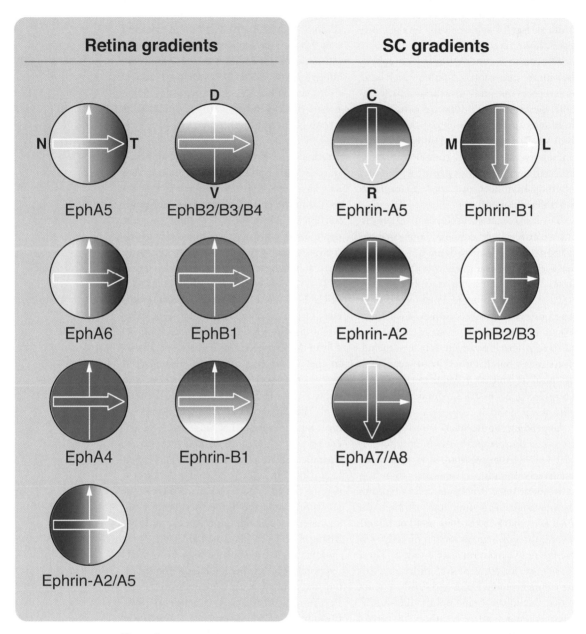

Figure 3

Gradients of Ephs and ephrins in the mouse. Retinal gradients (*left set of diagrams*) along the TN axis are depicted in blue, and those along the DV axis are depicted in red. Similarly, collicular gradients (*right set of diagrams*) along the RC axis are depicted in blue, and those along the LM axis are depicted in red. Increasing color indicates increasing expression. Orientation of the retinal and collicular axes are as depicted in **Figure 1** (*right*). Similar gradients exist in other species, although the specific Eph receptor or ephrin ligand that is graded varies between species.

EphA4 and EphA5 are similarly expressed but ungraded in the chick (Marcus et al. 1996, Feldheim et al. 1998, Brown et al. 2000, Reber et al. 2004).

These distributions of receptors and ligands mean that nasal RGCs that have the fewest EphA receptors project to caudal tectal (collicular) targets with the highest levels of ephrin-As and, conversely, that temporal RGCs with the most EphAs project to rostral targets that contain the lowest levels of ephrin-As. This pattern is in keeping with the finding that the GPI-linked ephrin-As are chemorepellents for EphA-expressing axons and with in vitro observations that temporal RGCs, which carry the highest levels of EphA, are the most sensitive to the GPI-linked chemorepellent activity of the caudal tectum.

The orthogonal axes of the retinotectal projection—the dorsal-ventral axis of the retina and the lateral-medial axis of the SC/tectum—have received less experimental attention but have not been neglected. Eph proteins and their ligands are also distributed in gradients along these axes, but it is the EphBs and the ephrin-Bs that form these gradients, rather than the EphAs and ephrin-As (**Figure 3**). EphB2, EphB3, and EphB4 are each expressed in a high-ventral-to-low-dorsal gradient in the retina (Holash & Pasquale 1995, Henkemeyer et al. 1996, Connor et al. 1998, Birgbauer et al. 2000, Hindges et al. 2002), and ephrin-B1 is present in a high-medial-to-low-lateral gradient in the mouse SC/tectum (Braisted et al. 1997, Hindges et al. 2002). These gradients mean that ventral RGCs with the most EphBs project to medial collicular targets where ephrin-B levels are highest, the exact opposite of the situation with the EphAs/ephrin-As on the orthogonal axes. This configuration of EphB/ephrin-B gradients led workers to speculate that forward signaling through EphBs in RGCs would result in chemoattraction, rather than chemorepulsion, during retinotectal mapping. This possibility has recently been tested experimentally (see below).

Just to make things more complicated, Eph proteins—both EphAs and EphBs—are also expressed in gradients in the SC/tectum, and ephrins—both ephrin-As and ephrin-Bs—are also expressed in gradients in the retina (Marcus et al. 1996, Brennan et al. 1997, Connor et al. 1998, Hornberger et al. 1999, Marin et al. 2001, Menzel et al. 2001, Hindges et al. 2002). In general, the gradient of a given ligand or receptor runs counter to that of its cognate receptor or ligand (**Figure 3**). Thus, in the chick retina, ephrin-A2, -A5, and -A6 are expressed in a high-nasal-to-low-temporal RGC gradient, counter to the gradient of EphA3 (Connor et al. 1998, Hornberger et al. 1999, Menzel et al. 2001). In the chick tectum, EphA3 and EphA5 are expressed in a high-rostral-to-low-caudal gradient, counter to the gradient of ephrin-A2 and -A5 (Connor et al. 1998, Marin et al. 2001). Similarly, ephrin-A2 and -A5 have been detected in a high-nasal-to-low-temporal gradient in the mouse retina (Gale et al. 1996, Marcus et al. 1996, Brennan et al. 1997, Monschau et al. 1997), and EphA7 and EphA8 have been found in high-rostral-to-low-caudal collicular gradients (Park et al. 1997, Rogers et al. 1999, Knoll et al. 2001). In the mouse retina, ephrin-B1 and -B2 are expressed in a high-dorsal-to-low-ventral RGC gradient (Marcus et al. 1996, Birgbauer et al. 2000, Hindges et al. 2002), and EphB2 and B3 are expressed in a high-lateral-to-low-medial gradient in the SC (Hindges et al. 2002).

These Eph/ephrin countergradients cannot be ignored as they are also seen in other regions in the developing CNS. However, their biological significance is at present unresolved and they remain the subject of active study. There are several alternative hypotheses as to the role of ephrins in the retina and Eph receptors in the SC/tectum (Connor et al. 1998, Hornberger et al. 1999, McLaughlin & O'Leary 1999, Menzel et al. 2001, Yates et al. 2001). These include the idea that retinal ephrin-As (a) serve to "shape" the retinal EphA gradient by binding to EphA4 locally in the nasal retina and inhibiting EphA4 action in the tectum, (b) are transported along RGC

TZs: termination zones

axons to the SC/tectum where they might function in map refinement, or (c) may mediate "reverse-signaling" events in both locations. Some of these possibilities are discussed in the context of the molecular genetic experiments summarized below.

Chemorepellent activity of ephrin-As. As noted above, most studies of retinotectal mapping have focused on the tn to rostral-caudal projection. In this projection, forward signaling mediated by chemorepellents in the tectum appears to dominate. These chemorepellents are ephrin-A2 and -A5. In in vitro membrane stripe assays, they mediate the selective outgrowth of temporal axons onto rostral membranes as well as the selective sprouting of collateral axon branches that arise from a primary axon shaft (Walter et al. 1987a, Godement & Bonhoeffer 1989, Roskies & O'Leary 1994). Axons from temporal RGCs are strongly repelled by both ephrin-A5 and ephrin-A2 in the stripe assays. Ephrin-A5 also shows weak chemorepellent activity against nasal axons (Drescher et al. 1995, Monschau et al. 1997). Collateral axon-branch formation, the primary mechanism of topographic map development in higher vertebrates (Yates et al. 2001, Sakurai et al. 2002), is also inhibited by caudal collicular membranes and ephrin-As in these assays (Roskies & O'Leary 1994).

MODERN IN VIVO STUDIES OF RETINOCOLLICULAR MAPPING

Development Versus Regeneration

As noted above, the pioneering work on retinotectal map development was performed in young adult fish or amphibians, following nerve injuries or tissue displacements. But in these creatures, the retinotectal system continues to develop late into adult life. As they grow, their retinae and tecta also grow: Retinal cells are added at the periphery of the eye, and tectal cells are added at the caudal-medial edge of the target. Therefore, most of the clas-

sical mapping studies are in fact compound experiments that include both regeneration and de novo development, with regrown fibers and newly innervating fibers intermixed in the tectum. Active fiber guidance, involving axon-axon interactions between these fiber populations, is an important mechanism for regenerating retinotectal axons in these settings (Gaze & Fawcett 1983, Taylor & Gaze 1985). Regeneration studies therefore have limitations with regard to addressing map formation as it first occurs in development.

Fish Versus Fowl

More recent studies of initial retinotectal development in frogs (Holt & Harris 1983, Holt 1984, Sakaguchi & Murphey 1985) and fish (Stuermer 1988) indicate that topographic order is in fact present at early stages of development, with the growth cones of retinal axons reaching their topographically correct regions in the tectum with relatively high initial fidelity. That is, navigation of the tectal surface as well as the decision as to where a particular RGC should terminate and arborize appear to be executed by the growth cones of RGCs in frogs and fish. Eventual TZs are established and refined through the extension of many small branches at the base of the growth cone in a process termed "backbranching."

This situation with respect to frogs and fish appears to be superficially different from that of chicks and rodents, where the axonal growth cone plays essentially no role in deciding where a particular RGC should terminate. However, this difference may be deceptive. Studies of the development of retinotectal projections in chick (Nakamura & O'Leary 1989, Yates et al. 2001) and of retinocollicular projections in rat (Simon & O'Leary 1992a–c) and mouse (Hindges et al. 2002) demonstrate that primary RGC axons at first significantly overshoot their correct TZs along the rostral-caudal axis of the SC/tectum and that axon distribution along the medial-lateral axis is also initially diffuse. Indeed, the growth cones of most RGC axons—irrespective of where they arise along the tn axis of the retina—initially run all the way to the caudal end of the tectum/SC. Later on, these axons extend interstitial collateral axons branches, which over time become increasingly concentrated at the correct TZ. Collateral branches outside the TZs are less and less frequently observed as development proceeds and the primary axon shaft is eventually retracted to a tightly circumscribed TZ (Nakamura & O'Leary 1989, Simon & O'Leary 1992, Yates et al. 2001, Hindges et al. 2002). This process of branching, refinement, retraction, and consolidation takes place over the first postnatal week in mouse and by P8 the map resembles its mature form. The differences in map refinement between frogs and fish on the one hand and chicks and rodents on the other—particularly with respect to the role of the growth cone as a tectal sensor—may be more apparent than real. The rostral-caudal extent of the tectum is vastly (~50-fold) larger in the chick than the frog at the onset of mapping. The fraction of the total tectal surface that may be sampled by a single growth is therefore corresponding very much larger in the frog (McLaughlin et al. 2003). And from a cell biological perspective, the extensive and dynamic backbranching seen at the base of amphibian growth cones is essentially the same process of biased actin polymerization seen in the collateral axon branching of higher vertebrates. Eph/ephrin signaling regulates both.

Mis/Overexpression in the Chick

Gain-of-function studies were among the first to substantiate a chemorepellent role for ephrins in retinotectal mapping in vivo (Nakamoto et al. 1996). In these experiments, patches of exogenous ephrin-A2 were produced in the developing chick tectum through infection of embryos with a retrovirus that contained an ephrin-A2 cDNA. Via this manipulation, ephrin-A2 patches were superimposed in a random pattern on top of the gradient of endogenous ephrin-A2 and -A5. Small numbers of both mid-temporal and nasal

RGC axons, labeled by focal injection of DiI into the contralateral retina, were then monitored for their mapping behavior around, in, or through these ectopic ephrin-A2 patches. Mid-temporal axons, which express relatively high levels of EphA receptors and are particularly sensitive to the membrane-associated chemorepellent of the caudal tectum, were observed to avoid the patches and to map to rostrally shifted tectal sites. In contrast, axons from RGCs in the extreme nasal retina, which normally have a limited ability to distinguish between rostral and caudal tectal membranes, did not (Nakamoto et al. 1996).

More recently, the role of ephrin-B1 was also assessed by gain-of-function studies in the chick (McLaughlin et al. 2003). Tectal electroporation of a retrovirus carrying an ephrin-B1 cDNA again generated patches of high ephrin-B1 expression superimposed onto the endogenous ephrin-B1 gradient. These studies demonstrated that collateral branches arising from primary axons located laterally to their correct termination site were attracted up the ephrin-B1 gradient to medially positioned ephrin-B1 patches. Conversely, branches arising from axons medial to their termination site were repelled down the ephrin-B1 gradient to laterally positioned patches. Together with the results of *EphB2/B3* knockout studies (see below), these findings have been interpreted as indicative of a bifunctional role of ephrin-B1 as both a branch repellent and attractant in the control of mapping along the lateral-medial axis of the tectum. However, the biochemical mechanisms that may underlie such bidirectionality remain obscure (McLaughlin et al. 2003).

Loss-of-Function Studies in the Mouse

Ephrin-A5 knockouts. Frisen and colleagues (1998) studied the role of ephrin-A5 in vivo by generating knockout mice that lack this ligand. In these mice, retinal axons establish TZs at topographically incorrect sites (**Figure 4**). Some axons initially overshoot the

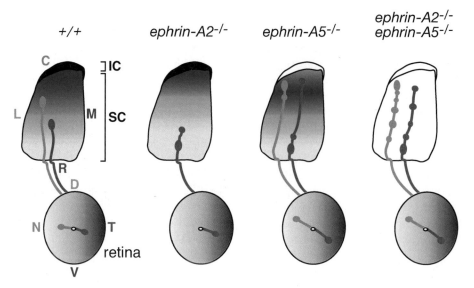

Figure 4

Mapping anomalies in mouse *ephrin-A* knockouts (*three right panels*) versus wild type (+/+, first panel). In *ephrin-A2* mutants (*second panel*), temporal RGCs terminate in their normal rostral position, but also sometimes in a second, more caudally positioned TZ. In *ephrin-A5* knockouts (*third panel*), temporal RGCs display similar mismapping phenomena, occasionally accompanied by a TZ in the far caudal SC. Nasal RGCs sometimes terminate in additional ectopic TZs that are positioned rostrally to their appropriate site. In mice lacking both ephrin-As (*fourth panel*), multiple TZs, arrayed all along the rc axis of the SC, are seen following DiI injections into either the nasal or temporal retina. Perturbations in ephrin-A gradients are depicted only for the SC, although as discussed in the text, these gradients are also lost from the retina in the knockouts. Gray shading in the retina depicts the aggregate EphA gradient in the mouse.

SC and make connections in the inferior colliculus. This suggests that the sharp increase in ephrin-A5 seen at the border between the SC and the inferior colliculus is important in the delimitation of this border. Other retinal axons transiently overshoot their correct TZ but remain in the caudal SC. Aberrant TZs are less prevalent in the middle of the SC's rostral-caudal axis, where ephrin-A2 is most highly expressed. Frisen and colleagues suggest that ectopic TZs in the caudal SC may be accounted for by the fact that the low level of ephrin-A2 present in this region is incapable (in the absence of ephrin-A5) of inhibiting the inappropriate branching of mid-temporal axons in the caudal SC.

Ephrin-A2/-A5 double knockouts. Analysis of double mutants that lack both ephrin-A2 and -A5 dramatically demonstrate

that their repellent activity is required for normal retinocollicular mapping (Feldheim et al. 2000). These double knockouts display marked abnormalities in the retinocollicular projection. Topographic mapping from the tn axis of the eye to the rostral-caudal axis of SC is almost entirely lost (**Figure 4**). RGCs from both nasal and temporal retinal locations are seen to terminate at multiple sites all along the rostral-caudal axis of the SC. The magnitude and extent of mapping anomalies in the double mutants is very much larger than in either the ephrin-A2 or ephrin-A5 single knockouts, consistent with complementary and overlapping roles for these ephrin-As in mapping (Feldheim et al. 2000).

The mapping abnormalities of the ephrin-A2/-A5 double mutants are most straightforwardly explained by the nearly total loss of chemorepellent activity in the SC, although an important complication of any standard knockout analysis is that the gene in question is inactivated everywhere in the organism. Thus, the normal high-nasal-to-low-temporal ephrin-A gradient in the retina is also eliminated in the double mutants. Feldheim and colleagues (2000) addressed this issue with "mix-and-match" in vitro membrane stripe assays in which RGCs from either mutant or wild-type retina were confronted with alternating stripes of rostral and caudal collicular membranes from either mutant or wild-type mice. Axons from wild-type temporal RGCs showed no preference for rostral versus caudal membranes when confronted with membranes prepared from the double mutants, consistent with the loss of topographic mapping being accounted for by a loss of chemorepellent activity in the target. However, a modest increase in the sensitivity of nasal double-mutant RGCs to chemorepulsion by caudal wild-type membranes was also noted in these in vitro assays. This finding in turn is consistent with the idea that the ephrin-As preferentially expressed in the nasal retina normally may dampen, through the retinal engagement of EphAs, the ability of nasal RGCs to use these EphAs to sense ephrin-As in the target (Connor et al. 1998, Dutting et al. 1999, Hornberger et al. 1999).

***EphA5* knockouts.** Loss-of-function analyses have also been reported for the mouse *EphA5* gene (Feldheim et al. 2004). Mapping abnormalities are only partially penetrant in these mutants. For example, approximately 50% of temporal RGCs project normally, with the remaining 50% displaying both a normal TZ and in addition one or more ectopic, caudally displaced TZs (**Figure 5**). This partial penetrance of the *EphA5* mutant phenotype is not unexpected, given that the retinal EphA gradient in the mouse is generated by three receptors: two that are graded (EphA5 and EphA6) and one that is not (EphA4). In general, the misprojections evident in the EphA5 mutant mice

are entirely consistent with the EphA5 receptor making a significant contribution to the ability of temporal RGCs to sense the ephrin-A chemorepellents that are maximally expressed in the caudal SC. Mix-and-match in vitro membrane stripe assays, similar to those performed for the *ephrin-A2/-A5* double knockouts, also support the conclusions that EphA5 acts primarily as a forward-signaling receptor in temporal RGCs and is necessary for these neurons to respond fully to ephrin-A chemorepellents in the SC (Feldheim et al. 2004).

***EphB2/B3* double knockouts.** Hindges and colleagues used double knockouts of the mouse *EphB2* and *EphB3* genes to address the role of EphB receptors in the mapping of the dorsal-ventral axis of the retina to the lateral-medial axis of the SC (Hindges et al. 2002). As noted above, the gradients of EphB2 and EphB3 (from low-dorsal-to-high-ventral in the retina) and of ephrin-B1 (from low-lateral-to-high-medial in the SC)—assuming that ephrin-Bs are chemoattractants—are consistent with such a role. Conventional mouse mutants lacking the EphB2 and EphB3 proteins entirely, as well as an EphB2 mutant with intact extracellular and transmembrane domains but without a tyrosine kinase domain, were employed. In the latter mutant protein, designated EphB2ki, the EphB2 kinase domain is replaced with β-galactosidase. This mutant protein therefore retains the capacity for reverse signaling but loses the capacity for forward signaling.

Analyses of the retinocollicular maps in these mutants demonstrated an important role for EphB2 and EphB3 forward signaling in topographic mapping, notably with regard to the medial-lateral collicular orientation of the interstitial branches that are extended from the primary shafts of RGC axons during map formation (Hindges et al. 2002). Ectopic, laterally shifted TZs were frequently observed along the medial-lateral axis of the SC in the *EphB2/B3* double mutants but

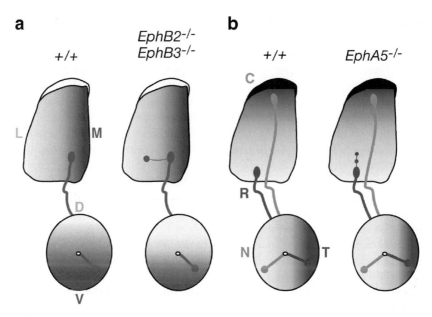

Figure 5

Mapping anomalies in mouse *EphB2/B3* double knockouts (*a, second panel*) and *EphA5* knockouts (*b, fourth panel*) versus wild type (*first and third panels*). In the *EphB2/B3* mutants, ventral-temporal RGCs project to a normal medial-rostral TZ in the SC and occasionally also to an abnormal, more lateral TZ (*second panel*). In the *EphA5* mutants, temporal RGCs project to their normal sites in the rostral SC and occasionally to more caudal TZs (*fourth panel*). Nasal *EphA5*⁻/⁻ RGCs occasionally display abnormal, rostrally positioned TZs, a phenotype that may be secondary to the development of ectopic caudally positioned TZs from temporal RGCs. Perturbations in EphB and EphA gradients are depicted only for the retina although, as discussed in the text, gradients are also lost from the SC in the knockouts. Gray shading in the SC depicts the aggregate gradients of ephrin-Bs (*a panels*) and ephrin-As (*b panels*) in the mouse.

rarely observed along the rostral-caudal axis (**Figure 5**). These laterally shifted TZs were always observed in addition to a correctly positioned TZ, and their appearance was, as for the *EphA5* mutants, not fully penetrant. This is again to be expected because, in addition to EphB2 and EphB3, EphB1 and EphB4 are also expressed in the mouse retina (Birgbauer et al. 2000, Hindges et al. 2002). Importantly, Hindges and colleagues demonstrated that the lateralization of mature TZs is accounted for by a perturbation of interstitial axon branching; the propensity for these branches to be oriented laterally (relative to the main axon shaft) increased significantly in the double mutants (Hindges et al. 2002; **Figure 5**). This disruption was even more pronounced in the *EphB2ki/B3* double mu-

tants than in the *EphB2/B3* double mutants. This suggests that the kinase inactive receptor may have a dominant-negative effect and, more importantly, demonstrates that EphB forward signaling (i.e., signaling in the receptor-expressing RGCs) is the dominant signaling pathway for regulation of the formation and orientation of interstitial branches along the medial-lateral axis. Given the distribution of EphBs in the retina and ephrin-Bs in the SC, these results suggest that EphB signaling promotes branch extension up the ephrin-B gradient. It is important to note that some medially oriented axon branches are still observed in the *EphB2/B3* double mutants and that evidence for reverse signaling through ephrin-Bs expressed by dorsal RGCs has been presented in *Xenopus* (Mann et al.

2002). Thus, additional mechanisms, perhaps involving a collicular chemorepellent activity distributed in a lateral-medial gradient similar to that of ephrin-B1, probably operate during topographic mapping.

QUANTITATIVE STUDIES

Gradient Perturbation

The hypothesis that gradients of Eph receptors and ephrins regulate topographic mapping is representative of a class of developmental models in which cells are ordered and distinguished from one another not by all-or-none differences in gene expression but rather by graded differences. As a means of testing such models, conventional loss-of-function analyses may provide a qualitative assessment of the importance of a particular signaling system. Such analyses, however, are limited with respect to providing a mechanistic understanding of the system. For example, the case for another candidate regulator of retinotectal mapping—RGM (Monnier et al. 2002)—was weakened when mouse knockouts of the RGM gene were found to display normal retinocollicular maps (Niederkofler et al. 2004). And conversely, the nearly complete loss of normal topography observed in the *ephrin-A2/-A5* double knockouts (Feldheim et al. 2000) greatly strengthens the case for EphA signaling playing a central role in retinocollicular mapping. However, such knockouts provide only a qualitative test of what is a quantitative model and do not yield direct insights into whether mapping actually requires the continuously graded expression of ligands and receptors. Such an assessment can only be made if the gradients in question are perturbed in a systematic and informative way. In the retinotectal system, this would require changing the slope, the shape, the orientation, or the periodicity of either the Eph receptor or ephrin ligand gradients in a quantifiable way such that predictions as to alterations in topographic mapping can be made and tested. This has been carried out in a set of

knock-in and compound mutant mice (Brown et al. 2000, Reber et al. 2004).

Isl2-EphA3 Knock-ins

RGM: repulsive guidance molecule

Perturbation of the EphA gradient in the mouse retina was achieved through the ectopic expression of EphA3 in a subset of RGCs (Brown et al. 2000). Although not expressed in mouse RGCs, EphA3, like EphA5 and EphA6, binds the collicular ephrin-A ligands with high affinity. A mouse EphA3 cDNA was inserted into the 3' untranslated region of the mouse *Islet-2* (*Isl2*) gene downstream of a viral internal ribosome entry site sequence. This construct brings EphA3 under the control of the regulatory elements of the *Isl2* gene but does not perturb endogenous *Isl2* expression. *Isl2* is expressed at a constant level in a subset (~40%) of RGCs that are randomly scattered across the retina in a salt-and-pepper fashion (except for the extreme ventral-temporal crescent of the retina, where the number of Isl2$^+$ cells is lower). Importantly, the *Isl2* gene is not expressed in the SC. Any perturbations of retinocollicular mapping are therefore specific to the retina (Brown et al. 2000). The *Isl2-EphA3* construct was "knocked-in" to the mouse *Isl2* locus through homologous recombination in embryonic stem cells (Brown et al. 2000).

Map duplication and collapse. A mixed population of RGCs was thereby generated in the mice produced from these cells. One of these RGC populations is wild type and carries the endogenous smooth gradient of EphA receptors; the other population has "spikes" of new EphA3 expression superimposed on top of this gradient (**Figure 6**). This generates a compound gradient that is quasi-oscillatory, with the oscillations—the EphA3 spikes—having twice the amplitude in *Isl2-EphA3* homozygotes as in heterozygotes. Brown et al. (2000) analyzed the retinocollicular maps in these knock-in mice relative to wild type by first making small focal injections of DiI into the retina and then plotting the

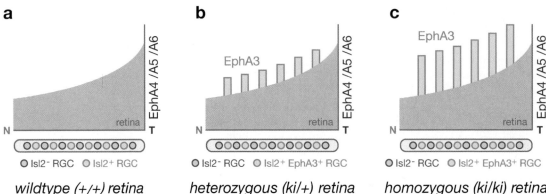

a	b	c
wildtype (+/+) retina	heterozygous (ki/+) retina	homozygous (ki/ki) retina

Figure 6

Systematic perturbation of the retinal EphA gradient in *Isl2-EphA3* knock-in mice. (*a*) The aggregate EphA gradient in wild-type mice increases smoothly from the nasal (N) to the temporal (T) end of the retina. This aggregate gradient is composed of two receptors (EphA5 and EphA6) whose expression varies and one (EphA4) whose expression does not. RGCs that express the transcription factor Islet-2 (Isl2$^+$, *green circles*) and RGCs that do not (*blue circles*) both contribute to the gradient. (*b*) The aggregate EphA gradient in mice heterozygous for the *Isl2-EphA3* knock-in. This gradient is now quasi-oscillatory and has constant but intermittent "spikes" of ectopic EphA3 expression—contributed by Isl2$^+$ RGCs—superimposed onto the normal aggregate EphA gradient. (*c*) The aggregate EphA gradient in mice homozygous for the *Isl2-EphA3* knock-in is the same as that for the heterozygous knock-ins except that the EphA3 spikes are twice as large. Modified from Brown et al. (2000), with permission.

position of these injections along the tn axis relative to the number and position of the TZs labeled (after 24 hours) in the contralateral SC (**Figure 7**).

These retinocollicular maps turned out to be especially informative. The wild-type mouse map is an apparently linear transfer of the retinal tn axis onto the collicular rostral-caudal axis (**Figure 7**). In marked contrast, *Isl2-EphA3* homozygotes display two maps rather than one. Thus, for each retinal location, single extracellular DiI injections into the retina—which always label a population of normal RGCs as well as an immediately adjacent population of EphA3$^+$ RGCs—yield two TZs in the SC rather than one (Brown et al. 2000). This means that the retinocollicular map is effectively duplicated: Every point in visual space along the tn axis of the eye is represented at two different locations in the SC. Brown et al. (2000) demonstrated that the more caudal of these TZs corresponds exclusively to wild-type RGCs and the more rostral to EphA3$^+$ RGCs. This is as ex-

pected because the EphA3$^+$ cells have higher summed EphA levels and therefore should be more sensitive to the chemorellent activity of the rostral-to-caudal collicular ephrin-A gradient.

Perhaps most importantly, neither of the retinocollicular maps of the homozygous *Isl2-EphA3* knock-ins is normal: The wild-type map does not take its wild-type form but rather is displaced caudally in the SC as a result of the preferential occupation of the rostral SC by EphA3$^+$ RGCs (**Figure 7**). This observation alone provides strong in vivo evidence for the hypothesis, debated for decades, that RGCs compete with one another as an ensemble for innervation space in the SC. It also demonstrates that RGCs do not terminate in the SC on the basis of a fixed or absolute level of EphA signaling but rather on the basis of an ensemble-wide comparison tied to relative levels of EphA signaling.

The retinocollicular map of the *Isl2-EphA3* heterozygotes displays two features that are distinct from the homozygote map. Like the

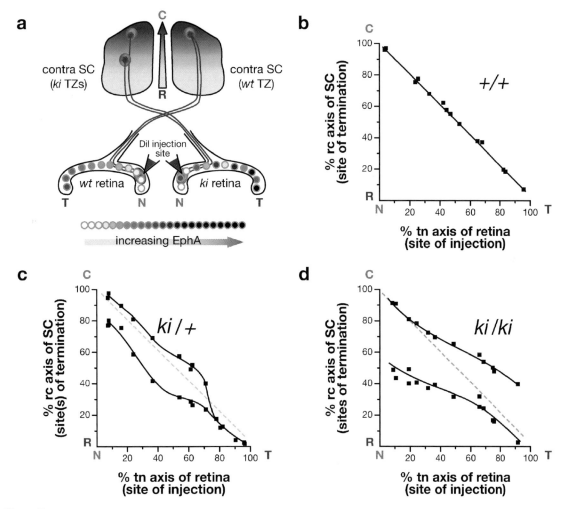

Figure 7

Retinocollicular maps in the *Isl2-EphA3* knock-ins. (*a*) Focal injection of DiI into the nasal retina of wild-type (*wt, left*) versus knock-in (*ki, right*) mice labels small clusters of cells. Adjacent wt RGCs with equivalent EphA project to one TZ in the caudal SC whereas adjacent ki RGCs with different levels of EphA project to two TZs, the more rostral of which corresponds to *EphA3+* RGCs. (*b*) The retinocollicular map of wild-type mice. The x axis is the tn axis of the retina (N = 0, T = 100) and the y axis is the rostral-caudal axis of the SC (R = 0, C = 100). Plotted is the rostral-caudal center of each DiI-labeled TZ as a function of the tn position of the retinal injection through which that TZ was labeled. (*c*) The retinocollicular map of *Isl2-EphA3* heterozygotes. DiI injections performed over the first 76% of the retinal tn axis always yield two TZs (as schematized in *a*), rather than one, whereas all injections performed over the remaining 24% of the axis always yield a single TZ. This results in partial map duplication and collapse. (*d*) The retinocollicular map of *Isl2-EphA3* homozygotes. DiI injections performed over the full retinal tn axis always yield two TZs rather than one. Dashed gray lines in *c* and *d* mark the wild-type map. Modified from Brown et al. (2000), with permission.

homozygote map, the retinocollicular map is duplicated but only for the nasal-most 76% of the retinal tn axis (where nasal = 0%, and temporal = 100%). All DiI injections performed in more temporal regions of the retina—from 76% to 100% of the tn axis—yield only a single TZ. Brown and colleagues refer to this as mapping "collapse," because the map changes suddenly from a duplicated to a single map. In addition, in the nasal-most 76% of the retina where the heterozygote map is duplicated, the extent of the duplication—that is, the separation between EphA3$^-$ and EphA3$^+$ TZs arising from each retinal location—is on average half that seen in homozygous *Isl2-EphA3* knock-ins.

Relativity in EphA signaling. These retinocollicular maps are not consistent with simple mass-action models of topographic mapping that are based on ligand-receptor matching or on topography being determined by a fixed value of EphA signaling (Nakamoto et al. 1996). Instead, order of termination within the SC appears to be determined by relative EphA levels, that is, the level of EphA signaling in a given RGC relative to that in all other RGCs. Thus, all temporal EphA3$^+$ RGCs still project to the SC in the knock-ins in spite of the fact that they have a much higher level of EphA receptors than does any wild-type RGC. Independent of their absolute level of EphA expression, axons with the highest EphA level always project to the most rostral site. Moreover, two RGCs located at the opposite end of the tn axis of the retina will project to the very same rostral-caudal position in the SC if they express the same level of EphA.

A Quantitative Relative Signaling Model

Exponential gradients. Reber and colleagues (2004) noted that these observations are most economically explained by a developmental program in which RGCs compete as an ensemble for termination sites along the rostral-caudal axis of the SC through comparison of relative, or ratio-based, differences in EphA signaling intensity—a program in which RGCs are ordered on the basis of a fold-difference in aggregate EphA (ΣEphA) signaling between themselves and all of their competitors. Reber et al. termed this ratiometric comparison "relative signaling." If such a comparison is in fact made, and if ΣEphA signaling intensity is proportional to ΣEphA expression, then the apparent linearity of the wild-type retinocollicular map (**Figure 6**) requires that, for the EphA receptors that vary (in the mouse EpA5 and EphA6), the rate of change in the concentration of these receptors per RGC (designated C_v) along the tn axis of the retina must be proportional to the receptor concentration itself (Reber et al. 2004). If the retinal axis is designated the x axis, then this is expressed as

$$dC_v = kC_v dx$$
$$\text{or } C_v(x) = C_{v0}e^{kx},$$

where k is a constant and C_{v0} is the value of C_v at x = 0, the nasal pole of the retina. (In the studies of Brown et al. (2000) and Reber et al. (2004), the temporal pole is assigned the value 100.) Because there is also an ungraded EphA receptor in the mouse (EphA4), the full aggregate EphA receptor distribution (C) should be described by an exponential plus a constant (C_4), i.e.,

$$C(x) = C_{v0}e^{kx} + C_4$$

(Reber et al. 2004). Accurate quantitation of EphA receptor protein levels as a function of position across the tn axis of the retina is rendered problematic, if not impossible, by two considerations. The more important of these is that the receptor proteins do not remain in RGC cell bodies. Instead, they are transported down RGC axons, which are both bundled together in the retina and, at the start of mapping, highly intermixed in the SC (where the receptor proteins act and where they should in principle be measured). The second consideration is that a set of antibodies that are both highly specific to the individual mouse EphAs, and that at the same time bind to these proteins with the same affinity and signal-to-noise properties, are not yet available. In contrast, as is the case for the mRNAs of most transmembrane proteins, a substantial fraction of the EphA mRNAs are retained within RGC cell bodies in close association with the ER. Reber et al. (2004) therefore performed quantitative in situ hybridization to determine the relative expression levels of the *EphA4*, *A5*, and *A6* mRNAs in RGCs across the tn axis of the mouse retina. In doing so, these researchers assumed that measured differences in mRNA expression translate into linearly related differences in protein expression. Their measurements demonstrated that the aggregate mouse EphA (ΣEphA) retinal gradient is indeed well fit by an exponential plus a constant, i.e.,

$$C(x) = 0.26e^{0.023X} + 1.05,$$

where C_{v0} is 0.26, k is 0.023 and C_4 is 1.05 (Reber et al. 2004). These investigators then used the same methods to measure the expression level of *EphA3* mRNA in the *EphA3+* (*Isl2+*) RGCs of both *Isl2-EphA3* heterozygous and homozygous knock-ins. These measurements demonstrated that the level of this mRNA is approximately constant across the tn axis of the knock-ins and that the heterozygote signal is ~50% of the homozygote signal (Reber et al. 2004). When the measured *EphA3* "spikes," corresponding to the subset of RGCs that are *Isl2+*, were added to the wild-type ΣEphA curve, the quasi-oscillatory EphA gradients of the *Isl2-EphA3* heterozygotes and homozygotes were obtained (**Figure 8**).

Molecular genetic tests of the relative signaling model. Reber and colleagues used these measured EphA gradients to examine mapping collapse in the heterozygous knock-ins. These investigators proposed that collapse occurs at the point at which the ratio of EphA signaling between an EphA3+ RGC and an immediately adjacent EphA3− RGC—designated the R_{lrs}—becomes too low for the mapping system to discriminate. The measured gradients in the knock-ins allowed these investigators to assign this "discrimination limit" ratio the value of 1.36: This is the value of R_{lrs} at 76% of the retinal axis in the heterozygous knock-ins, the point at which mapping collapse occurs (**Figure 8**). In this formulation, two RGCs whose signaling difference is below this value are topographically indistinguishable. The R_{lrs} function, which describes how R_{lrs} varies across the tn axis, never reaches this value in the homozygous knock-ins; this is consistent with the fact that the retinocollicular map in these mice does not collapse (**Figure 8**, Reber et al. 2004).

This R_{lrs} function and the discrimination limit turned out to have remarkable predictive power with respect to mapping behavior in the *Isl2-EphA3* knock-ins. Reber and colleagues generated compound mutant mice that were heterozygous or homozygous for the *Isl2-EphA3* knock-in and at the same time heterozygous or homozygous for the knockout of the *EphA4* gene. Each of these compound mutants has a distinct R_{lrs} function that predicts if, and if so, where, mapping collapse occurs (Reber et al. 2004). The authors anatomically measured the retinocollicular maps in each genotype and demonstrated that each of the predictions made by the R_{lrs} functions was confirmed in the measured map of each genotype. In addition to defining a minimum R_{lrs} value (1.36) below which two RGCs are topographically identical, Reber and colleagues also presented evidence from their mapping analyses in the compound *Isl2-EphA3/EphA4* mutant genotypes that there is a maximum value for R_{lrs} at which two RGCs are maximally different (maximally separated) in terms of their topographic mapping to the SC. Perhaps not surprisingly, this value turns out to be 2.75—the ratio in EphA expression observed between the temporal and nasal poles of the wild-type mouse retina.

R_{lrs}: local relative signaling ratio

The analysis of the compound mutants generated by Reber and colleagues—notably with respect to the rostral-caudal extent of TZ separation in the knock-ins—also addressed one possible role for ephrin-As in the retina. As mentioned above, these ephrin-As run in a countergradient, from high nasal to low temporal, to EphA5 and EphA6 (Marcus et al. 1996, Hornberger et al. 1999). Retinal ephrin-As have been proposed to act locally in retinal organization or to be delivered, via RGC axons, to the SC, where they can participate

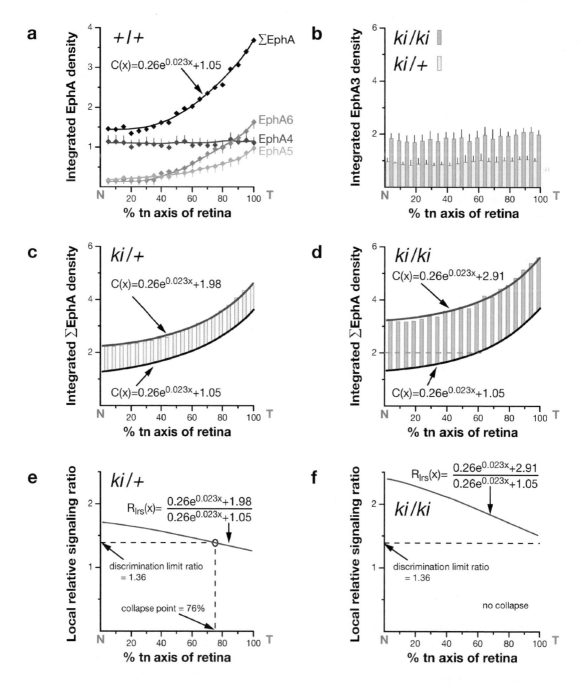

in map refinement (McLaughlin & O'Leary 1999, Yates et al. 2001). Another possibility is that these ligands sharpen the EphA functional gradient by inactivating nasally the EphA4 receptor (Connor et al. 1998, Hornberger et al. 1999, Menzel et al. 2001). Results from Reber and colleagues tend to argue against the latter possibility. If EphA4 function were fully graded—that is, if this receptor were completely inactivated in the nasal retina—then the removal of half or the entire EphA4 contribution to the ΣEphA gradient should have little or no effect on the projections of the nasal retinal axons. Alternatively, if EphA4 function were, like EphA4 expression, not graded in the retina, then removing some or all EphA4 should have very large effects on the projections of the nasal retinal axons. This latter phenomenon was in fact observed (Reber et al. 2004).

Relative signaling and the wild-type map. Reber and colleagues went on to propose a relative signaling explanation for the formation of the normal retinocollicular map. This explanation, which is again based on the hypoth- esis that topographic mapping order along the tn axis is dependent on a ratiometric comparison of EphA-signaling intensity, represents a generalized extension of the local relative signaling phenomena of the knock-ins to the entire RGC ensemble of the normal eye (Reber et al. 2004). In the case of the smoothly and gradually increasing ΣEphA gradient of the wild-type retina, the local RS ratio carries almost no information. This ratio is only trivially different from 1.0 as adjacent RGCs along the wild-type axis have essentially the same level of ΣEphA. So how are local RS comparisons made in the context of the ensemble-wide competition that leads to formation of the wild-type retinocollicular map?

As noted above, map formation in most vertebrates takes place during a week-long period of competition between RGC axons for innervation targets in the SC (Simon et al. 1992, Yates et al. 2001, Hindges et al. 2002). If ratiometric comparisons of ΣEphA are used to set the rules by which RGCs compete for a limiting feature of the SC during this time, the dominant cell in the ensemble will be the

---

Figure 8

Relative signaling in retinocollicular mapping. (*a*) Quantitation of the *EphA4*, *EphA5*, and *EphA6* mRNA gradients across the tn axis of the wild-type mouse retina, as described by Reber et al. (2004). Summing each of the measured values at each point yields ΣEphA, the aggregate level of *EphA* mRNA. As demanded by the relative signaling formulations, the ΣEphA gradient is described by the indicated equation: an exponential plus a constant. (*b*) Quantitation of the intermittent expression of *EphA3* mRNA in the Isl2[+] RGCs of *Isl2-EphA3* heterozygotes (ki/+, *yellow bars*) and homozygotes (ki/ki, *light blue bars*). (*c*) The ΣEphA gradient of *Isl2-EphA3* heterozygotes. The ki/+ *EphA3* "spikes" in *b* are superimposed onto the wild-type ΣEphA curve of *a* (*bottom equation*, *curve*). The red curve, which is described by the indicated equation, connects the tops of the spikes and corresponds to the ΣEphA gradient of the EphA3[+] RGCs. (*d*) The ΣEphA gradient of *Isl2-EphA3* homozygotes. The ki/ki *EphA3* "spikes" in *b* are superimposed onto the wild-type ΣEphA curve of *a* (*bottom equation*, *curve*). The red curve, which is described by the indicated equation, corresponds to the ΣEphA gradient of the EphA3[+] RGCs. (*e*) The local relative signaling ratio (R$_{lrs}$) function of *Isl2-EphA3* heterozygotes. This function describes how R$_{lrs}$—the ratiometric difference in ΣEphA expression between an EphA3[+] RGC and an immediately adjacent EphA3[−] RGC—varies with position along the tn axis of the retina. R$_{lrs}$ is described by dividing the right-hand sides of the upper and lower equations in *c*. The discrimination limit ratio (1.36) is the ratiometric difference limit below which two RGCs map to the same rostral-caudal collicular location and is defined by the collapse point of the *Isl2-EphA3* heterozygote map (**Figure 7c**). (*f*) The R$_{lrs}$ function of *Isl2-EphA3* homozygotes, generated as for the heterozygotes in *e*. (The homozygote R$_{lrs}$ function is described by dividing the right-hand sides of the upper and lower equations in *d*.) Modified from Reber et al. (2004), with permission from Nature (http://www.nature.com), copyright 2004 by Macmillan Publishers Ltd.

R_{grs}: general relative
signaling ratio

BDNF:
brain-derived
neurotrophic factor

neuron with the highest levels of ΣEphA signaling and the greatest sensitivity to the collicular ephrin-As, i.e., the RGC located at the temporal pole of the retina. (As also noted above, temporal RGCs are indeed most sensitive to chemorepulsion by ephrin-As.) In this milieu, all RGCs may compare themselves to all other RGCs by calculating a general signaling ratio relative to this same temporal reference.

Thirty years ago, Prestige & Wilshaw (1975) advanced, on purely theoretical grounds, a compelling case for a dominant polar reference cell of this form in so-called competition-by-exclusion models of retinotectal mapping. In the formulation of Reber et al. (2004), the highest R_{grs} will be the temporal/nasal ΣEphA ratio of RGCs at the nasal pole of the retina. Increasingly more temporal RGCs will have increasingly lower ratios, reaching 1.0 at the temporal pole. Most importantly, the model requires that the R_{grs} function—which describes the ratio between the temporal pole and RGCs all along the retinal axis—must yield, i.e., be identical to, the retinocollicular map of wild-type mice. Reber and colleagues demonstrated that this is indeed the case (**Figure 9**).

EphA ratios and axon competition in the SC. The ratiometric signaling parameters established by the ΣEphA gradient in the retina somehow must be transferred to competitive interactions between RGC axons that result in the topographically correct innervation of the axons' targets in the SC. Reber et al. (2004) speculated that competition of RGC axons for some limiting feature of the SC may itself be mediated by relative signaling and that a candidate for the limiting collicular feature is BDNF. This neurotrophin is expressed in the SC during the period of map formation and is a potent inducer of collateral branching by RGC axons. It is also a well-known stabilizer of nascent synaptic connections throughout the CNS (Alsina et al. 2001). RGC axons express TrkB, the receptor for BDNF (Suzuki et al. 1998).

Reber et al. (2004) suggest that relative signaling through the EphA receptor system might be translated into biased competition for BDNF through biased axon branching. As noted above, at the start of the axon competition events that mediate topographic mapping (around P0 in the mouse), nearly all primary axons occupy the full rostral-caudal extent of the SC. As competition ensues, RGC axons with low ΣEphA (i.e., nasal axons) are relatively insensitive to the inhibitory effects of collicular ephrin-As and therefore attempt to branch all along the collicular axis. In contrast, axons with high ΣEphA (i.e., temporal axons) extend many fewer branches and only in the most rostral regions of the axis. Reber and colleagues suggest that if the concentration of BDNF in the SC is limiting with respect to the number of RGC axons and their branches, then the difference in the extent of branch extension between nasal and temporal RGC axons is translated over time into a local difference in the intensity of BDNF/TrkB signaling at each location along the caudal-rostral axis of the SC. This occurs because the membrane density of TrkB receptors at any given point in a nasal axon is lower than in a temporal axon at the same position because the former axon has distributed its available TrkB protein over many more collateral branches that occupy more membrane area. According to this speculation, the local membrane density of TrkB receptors becomes progressively lower in progressively more caudal regions of the SC in exact proportion to the variation in the level of ΣEphA along the tn axis of the retina (Reber et al. 2004). An ΣEphA-directed biasing of local TrkB signaling of this form will occur even if all RGCs express the same level of TrkB per cell (Reber et al. 2004).

Relative signaling in relation to earlier models. The relative signaling model stands in contrast to earlier models of retinotectal mapping. Many of these were purely theoretical and, because the gradients of hypothesized signaling molecules were neither known nor measured, could not be tested

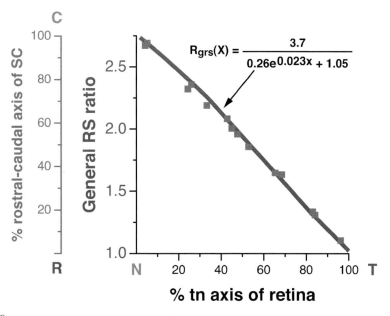

$$R_{grs}(X) = \frac{3.7}{0.26e^{0.023x} + 1.05}$$

Figure 9

Relative signaling yields the wild-type mouse retinocollicular map. The indicated equation is the R_{grs} function for wild-type mice and is plotted as the red curve. The equation describes the ratio of ΣEphA at the temporal pole of the retina (3.7) divided by the term that describes the ΣEphA gradient along the tn axis of the retina. R_{grs} varies from a maximum of 2.75 (temporal pole/nasal pole) to a minimum of 1 (temporal pole/temporal pole). The gray data points are those of **Figure 7b** (plotted on the *gray* y axis), and correspond to the wild-type mouse retinocollicular map. Modified from Reber et al. (2004) with permission from Nature (http://www.nature.com), copyright 2004 by Macmillan Publishers Ltd.

experimentally. They generally invoked the presence of multiple gradients (Gierer 1983, 1987). In the absence of RGC axon competition in the SC, even a rostral-caudal chemorepellent gradient in the SC coupled to a tn receptor gradient in the retina requires additional gradients for topographic specificity and tight TZs. These additional gradients might include a chemoattractant gradient in the SC that parallels the ephrin-A chemorepellent gradient (O'Leary et al. 1999). The evidence for additional gradients at this point remains circumstantial, although a role for ephrin-A-mediated reverse signaling, which is compatible with the relative signaling formulations of Reber et al. (2004), is certainly plausible.

As noted above, simple mass-action models of mapping, in which topographic mapping is determined by a threshold ligand (L)-receptor (R) interaction that yields a sig-

naling complex RL (e.g., Nakamoto et al. 1996), have not been supported by the results of molecular genetic manipulations of Eph receptors and their ligands (Feldheim et al. 2000, Brown et al. 2000). Instead, the phenotypes of *ephrin-A2/-A5* double knockouts and the *Isl2-EphA3* knock-ins are most straightforwardly explained by a scenario in which RGCs are forced to compete with one another for a limiting feature of the SC and in which the rules for this competition are set by ratiometric comparisons of EphA signaling intensity (Reber et al. 2004). The limiting collicular feature in theory might be space (innervation sites on postsynaptic cells) or a neurotrophic factor that stimulates branch formation and/or stabilizes synapses. Indeed, the behavior of wild-type axons in the *Isl2-EphA3* homozygous knock-ins—in which these axons are pushed to progressively more caudal locations in the SC as one moves to progressively more temporal retinal locations (Brown et al. 2000)—is difficult to account for by any mechanism other than simple competition.

The generality of relative signaling. Relative signaling may apply to many instances in development in which competition between cells operates. In the nervous system, survival of nascent neurons based on access to neurotrophins (Van Ooyen & Willshaw 1999, Van Ooyen 2001), stabilization of synapses in the neocortex (Vicario-Abejon et al. 2002), and innervation of muscle targets by motor neurons through the elimination of ectopic synapses (Van Ooyen 2001) are all thought to depend upon competition. Each of these competitive events is dynamic and takes place over an extended period of time. They result in the establishment of innervation order, the numerical matching of input and output cells, or both. The demonstration that developing ensembles of competing RGCs perform comparisons on the basis of ratios rather than absolute differences—that is, that competing RGCs divide rather than subtract to establish order—suggests that ratiometric comparisons of signaling intensity may operate in these other instances as well.

Computational Approaches

Few subjects in developmental neurobiology have been as extensively and repeatedly modeled as retinotectal mapping (Prestige & Willshaw 1975; Fraser & Hunt 1980; Gierer 1981; Whitelaw & Cowan 1981; Fraser & Perkel 1990; Goodhill 1998; Goodhill & Urbach 1999; Loschinger et al. 2000; Honda 2003, 2004; Reber et al. 2004; Yates et al. 2004). Beginning in the 1970s, many of these models were implemented in computer programs that dynamically simulated map development. Among the first of these was a simulation based on "systems matching": the idea that RGC axons compete for limiting tectal space. In such models, retinal and tectal gradients of hypothetical labels were configured to run in the same direction. All retinal ax-

ons competed for the tectal site displaying the highest label but only the RGCs carrying the highest cognate label synapsed (Prestige & Willshaw 1975). Several groups proposed and implemented more-abstract systems matching models that contained a variety of additional biological constraints. These were typically based on reaching an optimal balance between opposing forces of competition and systems matching; this balance was achieved through the action of stable linear or exponential gradients (Fraser & Hunt 1980, Gierer 1981, Whitelaw & Cowan 1981, Fraser & Perkel 1990). A major concern of the models was whether they could mimic the results of both retinal/tectal ablation and rotation perturbations as well as those of developmental studies (Holt & Harris 1983, Holt 1984, Sakaguchi & Murphey 1985, Stuemer 1988). Many researchers were, in fact, able to do this, but because they lacked any information as to the identity of the molecules that make up the gradients or their bioactivity, their expression level, or their graded distributions during development, these models remained theoretical exercises.

The identification of the Eph receptors and their ephrin ligands, the observation and quantitation of their graded distributions in the retina and tectum, and the discovery of their chemorepellent and chemoattractant activities have grounded the models in biological reality. Modifications of the mass-action model, including the mass-action/set-point rule model (Goodhill & Richards 1999), the imprinting-matching model (Loschinger et al. 2000), and the servomechanism-competition model (Honda 2003, 2004) have incorporated these known gradients and have included some biologically relevant constraints such as competition and colocalization of receptor and ligands. The latest of these efforts comes much closer to reproducing the mapping phenomena seen following molecular genetic manipulations, notably those evident in the *ephrin-A2/-A5* knockouts and the *Isl2-EphA3* knock-ins (Koulakov & Tsigankov 2004). The recent quantitation of each of the constituent receptors of the EphA gradient in the mouse retina, together with the molecular genetic demonstration that EphA signaling differences may be interpreted ratiometrically (Reber et al. 2004), should allow for more powerful refinement of these computational models of mapping.

CONCLUSIONS

The mouse mutant findings summarized above demonstrate that EphA signaling is essential to the mapping of the tn axis of the retina onto the rostral-caudal axis of the tectum and, similarly, that EphB signaling is required for the accurate mapping of the dorsal-ventral axis of the retina onto the lateral-medial axis of the tectum. They further show that these Eph signaling systems operate in the context of ensemble-wide neuronal competition. Analysis of the *Isl2-EphA3* knock-ins, both alone and in combination with *EphA4* knockouts, demonstrates that EphA signaling comparisons within the RGC ensemble are made—and retinotectal mapping is thereby ordered—on the basis of relative and not absolute differences in signaling

intensity. The relative signaling equations of Reber et al. (2004) demonstrate that ratiometric comparison of EphA signaling between the temporal pole RGC of the retina and all other RGCs along the tn axis yields the wild-type retinocollicular map.

Although each of these demonstrations represents significant advances, important questions remain: Do the orthogonally arrayed EphA and EphB signaling systems interact? How important is "reverse signaling" in either system? In what contexts might ephrin-As function as chemoattractants as well as chemorepellents? (See, for example, Hansen et al. 2004.) It is possible to address these questions using molecular genetics, but sophisticated approaches that go beyond "plain vanilla" knockouts will be required. Both Ephs and ephrins are expressed in both the retina and the tectum, can act as both chemoattractants and chemorepellents, and can serve as both receptors and ligands. These considerations argue for future conditional or inducible genetic manipulations that, as with the *Isl2-EphA3* knock-ins, are confined to a defined population of cells in either the retina or the tectum, but not both.

LITERATURE CITED

Alsina B, Vu T, Cohen-Cory S. 2001. Visualizing synapse formation in arborizing optic axons in vivo: dynamics and modulation by BDNF. *Nat. Neurosci.* 4:1093–101

Arora HL, Sperry RW. 1962. Optic nerve regeneration after surgical cross-union of medial and lateral optic tracts. *Am. Zool.* 2:389

Attardi DG, Sperry RW. 1963. Preferential selection of central pathways by regenerating optic fibers. *Exp. Neurol.* 7:46–64

Attardi DG, Sperry RW. 1960. Central routes taken by regenerating optic fibers. *Physiologist* 3:12

Birgbauer E, Cowan CA, Sretavan DW, Henkemeyer M. 2000. Kinase independent function of EphB receptors in retinal axon pathfinding to the optic disc from dorsal but not ventral retina. *Development* 127:1231–41

Braisted JE, McLaughlin T, Wang HU, Friedman GC, Anderson DJ, et al. 1997. Graded and lamina-specific distributions of ligands of EphB receptor tyrosine kinases in the developing retinotectal system. *Dev. Biol.* 191:14–28

Brambilla R, Schnapp A, Casagranda F, Labrador JP, Bergemann AD, et al. 1995. Membrane-bound LERK2 ligand can signal through three different Eph-related receptor tyrosine kinases. *EMBO J.* 14:3116–26

Brennan C, Monschau B, Lindberg R, Guthrie B, Drescher U, et al. 1997. Two Eph receptor tyrosine kinase ligands control axon growth and may be involved in the creation of the retinotectal map in the zebrafish. *Development* 124:655–64

Brown A, Yates PA, Burrola P, Ortuno D, Vaidya A, et al. 2000. Topographic mapping from the retina to the midbrain is controlled by relative but not absolute levels of EphA receptor signaling. *Cell* 102:77–88

Cheng HJ, Nakamoto M, Bergemann AD, Flanagan JG. 1995. Complementary gradients in expression and binding of ELF-1 and Mek4 in development of the topographic retinotectal projection map. *Cell* 82:371–81

Cohen YE, Knudsen EI. 1999. Maps versus clusters: different representations of auditory space in the midbrain and forebrain. *Trends Neurosci.* 22:128–35

Connor RJ, Menzel P, Pasquale EB. 1998. Expression and tyrosine phosphorylation of Eph receptors suggest multiple mechanisms in patterning of the visual system. *Dev. Biol.* 193:21–35

Cook JE. 1979. Interactions between optic fibres controlling the locations of their terminals in the goldfish optic tectum. *J. Embryol. Exp. Morphol.* 52:89–103

Cox EC, Muller B, Bonhoeffer F. 1990. Axonal guidance in the chick visual system: posterior tectal membranes induce collapse of growth cones from the temporal retina. *Neuron* 4:31–37

Debski EA, Cline HT. 2002. Activity-dependent mapping in the retinotectal projection. *Curr. Opin. Neurobiol.* 12:93–99

Donoghue MJ, Lewis RM, Merlie JP, Sanes JR. 1996. The Eph kinase ligand AL-1 is expressed by rostral muscles and inhibits outgrowth from caudal neurons. *Mol. Cell. Neurosci.* 8:185–98

Drescher U, Kremoser C, Handwerker C, Loschinger J, Noda M, et al. 1995. In vitro guidance of retinal ganglion cell axons by RAGS, a 25 kDa tectal protein related to ligands for Eph receptor tyrosine kinases. *Cell* 82:359–70

Dutting D, Handwerker C, Drescher U. 1999. Topographic targeting and pathfinding errors of retinal axons following overexpression of ephrinA ligands on retinal ganglion cell axons. *Dev. Biol.* 216:297–311

Echteler SM, Nofsinger YC. 2000. Development of ganglion cell topography in the postnatal cochlea. *J. Comp. Neurol.* 425:436–46

Feldheim DA, Kim YI, Bergemann AD, Frisen J, Barbacid M, et al. 2000. Genetic analysis of ephrin-A2 and ephrin-A5 shows their requirement in multiple aspects of retinocollicular mapping. *Neuron* 25:563–74

Feldheim DA, Nakamoto M, Osterfield M, Gale NW, DeChiara TM, et al. 2004. Loss-of-function analysis of EphA receptors in retinotectal mapping. *J. Neurosci.* 24:2542–50

Feldheim DA, Vanderhaeghen P, Hansen MJ, Frisen J, Lu Q, et al. 1998. Topographic guidance labels in a sensory projection to the forebrain. *Neuron* 21:1303–13

Flanagan JG, Vanderhaeghen P. 1998. The ephrins and Eph receptors in neural development. *Annu. Rev. Neurosci.* 21:309–45

Fraser SE, Hunt RK. 1980. Retinotectal specificity: models and experiments in search of a mapping function. *Annu. Rev. Neurosci.* 3:319–52

Fraser SE, Perkel DH. 1990. Competitive and positional cues in the patterning of nerve connections. *J. Neurobiol.* 21:51–72

Friauf E, Lohmann C. 1999. Development of auditory brainstem circuitry. Activity-dependent and activity-independent processes. *Cell Tissue Res.* 297:187–95

This paper provides strong evidence that topographic order is determined by relative rather than fixed EphA levels, and RGCs compete for a limiting feature of the SC during mapping.

This paper was among the first to demonstrate that EphA receptors and their ligands are graded in the retina and tectum in a manner consistent with the chemoaffinity hypothesis.

This paper describes the dramatic phenotypes apparent in mice doubly mutant for ephrin-A2 and ephrin-A5; this leads to a substantial degradation of topographic mapping along the retinal tn axis.

Frisen J, Yates PA, McLaughlin T, Friedman GC, O'Leary DD, et al. 1998. Ephrin-A5 (AL-1/RAGS) is essential for proper retinal axon guidance and topographic mapping in the mammalian visual system. *Neuron* 20:235–43

Gale NW, Holland SJ, Valenzuela DM, Flenniken A, Pan L, et al. 1996. Eph receptors and ligands comprise two major specificity subclasses and are reciprocally compartmentalized during embryogenesis. *Neuron* 17:9–19

Gaze RM, Fawcett JW. 1983. Pathways of Xenopus optic fibres regenerating from normal and compound eyes under various conditions. *J. Embryol. Exp. Morphol.* 73:17–38

Gierer A. 1981. Generation of biological patterns and form: some physical, mathematical, and logical aspects. *Prog. Biophys. Mol. Biol.* 37:1–47

Gierer A. 1983. Model for the retino-tectal projection. *Proc. R. Soc. London Ser. B* 218:77–93

Gierer A. 1987. Directional cues for growing axons forming the retinotectal projection. *Development* 101:479–89

Godement P, Bonhoeffer F. 1989. Cross-species recognition of tectal cues by retinal fibers in vitro. *Development* 106:313–20

Goodhill GJ. 1998. Mathematical guidance for axons. *Trends Neurosci.* 21:226–31

Goodhill GJ, Richards LJ. 1999. Retinotectal maps: molecules, models and misplaced data. *Trends Neurosci.* 22:529–34

Goodhill GJ, Urbach JS. 1999. Theoretical analysis of gradient detection by growth cones. *J. Neurobiol.* 41:230–41

Hansen MJ, Dallal GE, Flanagan JG. 2004. Retinal axon response to ephrin-As shows a graded, concentration-dependent transition from growth promotion to inhibition. *Neuron* 42:717–30

Henkemeyer M, Orioli D, Henderson JT, Saxton TM, Roder J, et al. 1996. Nuk controls pathfinding of commissural axons in the mammalian central nervous system. *Cell* 86:35–46

Hindges R, McLaughlin T, Genoud N, Henkemeyer M, O'Leary DD. 2002. EphB forward signaling controls directional branch extension and arborization required for dorsal-ventral retinotopic mapping. *Neuron* 35:475–87

Holash JA, Pasquale EB. 1995. Polarized expression of the receptor protein tyrosine kinase Cek5 in the developing avian visual system. *Dev. Biol.* 172:683–93

Holt CE, Harris WA. 1983. Order in the initial retinotectal map in Xenopus: a new technique for labelling growing nerve fibres. *Nature* 301:150–52

Holt CE. 1984. Does timing of axon outgrowth influence initial retinotectal topography in Xenopus? *J. Neurosci.* 4:1130–52

Holt CE, Harris WA. 1993. Position, guidance, and mapping in the developing visual system. *J. Neurobiol.* 24:1400–22

Holt CE, Harris WA. 1998. Target selection: invasion, mapping and cell choice. *Curr. Opin. Neurobiol.* 8:98–105

Honda H. 2003. Competition between retinal ganglion axons for targets under the servomechanism model explains abnormal retinocollicular projection of Eph receptor-overexpressing or ephrin-lacking mice. *J. Neurosci.* 23:10368–77

Honda H. 2004. Competitive interactions between retinal ganglion axons for tectal targets explain plasticity of retinotectal projection in the servomechanism model of retinotectal mapping. *Dev. Growth Differ.* 46:425–37

Hornberger MR, Dutting D, Ciossek T, Yamada T, Handwerker C, et al. 1999. Modulation of EphA receptor function by coexpressed ephrinA ligands on retinal ganglion cell axons. *Neuron* 22:731–42

This paper, analyzing mice doubly mutant for EphB2 and EphB3, indicates that the EphB signaling system is required for accurate mapping along the medial-lateral axis of the tectum, and that the system primarily acts to direct the medial growth up an ephrin-B gradient.

Huh GS, Boulanger LM, Du H, Riqueline PA, Brotz TM, et al. 2000. Functional requirement for class I MHC in CNS development and plasticity. *Science* 290:2155–59

Kaas JH. 1997. Topographic maps are fundamental to sensory processing. *Brain Res. Bull.* 44:107–12

Kanaseki T, Sprague JM. 1974. Anatomical organization of pretectal nuclei and tectal laminae in the cat. *J. Comp. Neurol.* 158:319–37

King AJ, Palmer AR. 1983. Cells responsive to free-field auditory stimuli in guinea-pig superior colliculus: distribution and response properties. *J. Physiol.* 342:361–81

Knoll B, Isenmann S, Kilic E, Walkenhorst J, Engel S, et al. 2001. Graded expression patterns of ephrin-As in the superior colliculus after lesion of the adult mouse optic nerve. *Mech. Dev.* 106:119–27

Knudsen EI. 1982. Auditory and visual maps of space in the optic tectum of the owl. *J. Neurosci.* 2:1177–94

Koulakov AA, Tsigankov DN. 2004. A stochastic model for retinocollilular map development. *BMC Neurosci.* 5:30–47

Kullander K, Klein R. 2002. Mechanisms and functions of Eph and ephrin signalling. *Nat. Rev. Mol. Cell Biol.* 3:475–86

Loschinger J, Weth F, Bonhoeffer F. 2000. Reading of concentration gradients by axonal growth cones. *Philos. Trans. R. Soc. London Ser. B* 355:971–82

Mann F, Ray S, Harris W, Holt C. 2002. Topographic mapping in dorsoventral axis of the Xenopus retinotectal system depends on signaling through ephrin-B ligands. *Neuron* 35:461–73

Marcus RC, Gale NW, Morrison ME, Mason CA, Yancopoulos GD. 1996. Eph family receptors and their ligands distribute in opposing gradients in the developing mouse retina. *Dev. Biol.* 180:786–89

Marin O, Blanco MJ, Nieto MA. 2001. Differential expression of Eph receptors and ephrins correlates with the formation of topographic projections in primary and secondary visual circuits of the embryonic chick forebrain. *Dev. Biol.* 234:289–303

McLaughlin T, Hindges R, Yates PA, O'Leary DD. 2003. Bifunctional action of ephrin-B1 as a repellent and attractant to control bidirectional branch extension in dorsal-ventral retinotopic mapping. *Development* 130:2407–18

McLaughlin T, O'Leary DD. 1999. Functional consequences of coincident expression of EphA receptors and ephrin-A ligands. *Neuron* 22:636–39

Menzel P, Valencia F, Godement P, Dodelet VC, Pasquale EB. 2001. Ephrin-A6, a new ligand for EphA receptors in the developing visual system. *Dev. Biol.* 230:74–88

Meyer RL. 1982. Ordering of retinotectal connections: a multivariate operational analysis. *Curr. Top. Dev. Biol.* 17:101–45

Monschau B, Kremoser C, Ohta K, Tanaka H, Kaneko T, et al. 1997. Shared and distinct functions of RAGS and ELF-1 in guiding retinal axons. *EMBO J.* 17:1258–67

Murai K, Pasquale EB. 2003. 'Eph'ective signaling: forward, reverse and crosstalk. *J. Cell Sci.* 116:2823–32

Nakamoto M, Cheng HJ, Friedman GC, McLaughlin T, Hansen MJ, et al. 1996. Topographically specific effects of ELF-1 on retinal axon guidance in vitro and retinal axon mapping in vivo. *Cell* 86:755–66

Nakamura H, O'Leary DD. 1989. Inaccuracies in initial growth and arborization of chick retinotectal axons followed by course corrections and axon remodeling to develop topographic order. *J. Neurosci.* 9:3776–95

O'Leary DD, Fawcett JW, Cowan WM. 1986. Topographic targeting errors in the retinocollicular projection and their elimination by selective ganglion cell death. *J. Neurosci.* 6:3692–705

O'Leary DD, Yates PA, McLaughlin T. 1999. Molecular development of sensory maps: representing sights and smells in the brain. *Cell* 96:255–69

Park S, Frisen J, Barbacid M. 1997. Aberrant axonal projections in mice lacking EphA8 (Eek) tyrosine protein kinase receptors. *EMBO J.* 16:3106–14

Penfield W, Rasmussen T. 1950. *The Cerebral Cortex of Man. A Clinical Study of Localization of Function.* New York: Macmillan

Prestige MC, Willshaw DJ. 1975. On a role for competition in the formation of patterned neural connexions. *Proc. R. Soc. London Ser. B* 190:77–98

Reber M, Burrola P, Lemke G. 2004. A relative signaling model for the formation of a topographic neural map. *Nature* 431:847–53

Rodieck RW. 1979. Visual pathways. *Annu. Rev. Neurosci.* 2:193–225

Rogers JH, Ciossek T, Ullrich A, West E, Hoare M, et al. 1999. Distribution of the receptor EphA7 and its ligands in development of the mouse nervous system. *Brain Res. Mol. Brain Res.* 74:225–30

Roskies AL, O'Leary DD. 1994. Control of topographic retinal axon branching by inhibitory membrane-bound molecules. *Science* 265:799–803

Rubel EW, Cramer KS. 2002. Choosing axonal real estate: location, location, location. *J. Comp. Neurol.* 448:1–5

Rubel EW, Fritzsch B. 2002. Auditory system development: primary auditory neurons and their targets. *Annu. Rev. Neurosci.* 25:51–101

Sachs GM, Jacobson M, Caviness VS Jr. 1986. Postnatal changes in arborization patterns of murine retinocollicular axons. *J. Comp. Neurol.* 246:395–408

Sakaguchi DS, Murphey RK. 1985. Map formation in the developing Xenopus retinotectal system: an examination of ganglion cell terminal arborizations. *J. Neurosci.* 5:3228–45

Sakurai T, Wong E, Drescher U, Tanaka H, Jay DG. 2002. Ephrin-A5 restricts topographically specific arborization in the chick retinotectal projection in vivo. *Proc. Natl. Acad. Sci. USA* 99:10795–800

Schmidt JT, Ciceron CM, Easter SS. 1978. Expansion of the half retina projection to the tectum in goldfish: an electrophysiological and anatomical study. *J. Comp. Neurol.* 177:257–78

Sharma SC. 1972. Reformation of retinotectal projections after various tectal ablations in adult goldfish. *Exp. Neurol.* 34:171–82

Simon DK, O'Leary DD. 1992a. Responses of retinal axons in vivo and in vitro to position-encoding molecules in the embryonic superior colliculus. *Neuron* 9:977–89

Simon DK, O'Leary DD. 1992b. Development of topographic order in the mammalian retinocollicular projection. *J. Neurosci.* 12:1212–32

Simon DK, O'Leary DD. 1992c. Influence of position along the medial-lateral axis of the superior colliculus on the topographic targeting and survival of retinal axons. *Brain Res. Dev. Brain Res.* 69:167–72

Sperry RW. 1943. Effect of 180 degree rotation of the retinal field on visuomotor coordination. *J. Exp. Zool.* 92:263–79

Sperry RW. 1944. Optic nerve regeneration with return of vision in anurans. *J. Neurophysiol.* 7:57–69

Sperry RW. 1963. Chemoaffinity in the orderly growth of nerve fiber patterns and connections. *Proc. Natl. Acad. Sci. USA* 50:703–10

Stahl B, Muller B, von Boxberg Y, Cox EC, Bonhoeffer F. 1990. Biochemical characterization of a putative axonal guidance molecule of the chick visual system. *Neuron* 5:735–43

Stuermer CA. 1988. Retinotopic organization of the developing retinotectal projection in the zebrafish embryo. *J. Neurosci.* 8:4513–30

This paper presents a quantitative analysis of the Isl2-EphA3 knock-ins, both alone and in combination with EphA4 knockouts. It provides the first comprehensive quantitative formulation of competitive neighbor interactions in retinotectal mapping.

This classic paper advances Roger Sperry's ideas with respect to chemoaffinity.

Taylor JS, Gaze RM. 1985. The effects of the fibre environment on the paths taken by regenerating optic nerve fibres in Xenopus. *J. Embryol. Exp. Morphol.* 89:383–401

van Ooyen A. 2001. Competition in the development of nerve connections: a review of models. *Network* 12:R1–47

van Ooyen A, Willshaw DJ. 1999. Competition for neurotrophic factor in the development of nerve connections. *Proc. R. Soc. London Ser. B* 266:883–92

Vicario-Abejon C, Owens D, McKay R, Segal M. 2002. Role of neurotrophins in central synapse formation and stabilization. *Nat. Rev. Neurosci.* 3:965–74

Walter J, Henke-Fahle S, Bonhoeffer F. 1987a. Avoidance of posterior tectal membranes by temporal retinal axons. *Development* 101:909–13

Walter J, Kern-Veits B, Huf J, Stolze B, Bonhoeffer F. 1987b. Recognition of position-specific properties of tectal cell membranes by retinal axons in vitro. *Development* 101:685–96

Whitelaw VA, Cowan JD. 1981. Specificity and plasticity of retinotectal connections: a computational model. *J. Neurosci.* 1:1369–87

Winslow JW, Moran P, Valverde J, Shih A, Yuan JQ, et al. 1995. Cloning of AL-1, a ligand for an Eph-related tyrosine kinase receptor involved in axon bundle formation. *Neuron* 14:973–81

Yates PA, Holub AD, McLaughlin T, Sejnowski TJ, O'Leary DD. 2004. Computational modeling of retinotopic map development to define contributions of EphA-ephrinA gradients, axon-axon interactions, and patterned activity. *J. Neurobiol.* 59:95–113

Yates PA, Roskies AL, McLaughlin T, O'Leary DD. 2001. Topographic-specific axon branching controlled by ephrin-As is the critical event in retinotectal map development. *J. Neurosci.* 21:8548–63

Yoon M. 1976. Progress of topographic regulation of the visual projection in the halved optic tectum of adult goldfish. *J. Physiol.* 257:621–43

Zhang JH, Cerretti DP, Yu T, Flanagan JG, Zhou R. 1996. Detection of ligands in regions anatomically connected to neurons expressing the Eph receptor Bsk: potential roles in neuron-target interaction. *J. Neurosci.* 16:7182–92

In Vivo Imaging of Lymphocyte Trafficking

Cornelia Halin, J. Rodrigo Mora, Cenk Sumen, and Ulrich H. von Andrian

The CBR Institute for Biomedical Research and the Department of Pathology, Harvard Medical School, Boston, Massachusetts, 02115; email: halin@cbr.med.harvard.edu; mora@cbr.med.harvard.edu; sumen@cbr.med.harvard.edu; uva@cbr.med.harvard.edu

Annu. Rev. Cell Dev. Biol. 2005. 21:581–603

First published online as a Review in Advance on June 29, 2005

The *Annual Review of Cell and Developmental Biology* is online at http://cellbio.annualreviews.org

doi: 10.1146/annurev.cellbio.21.122303.133159

Copyright © 2005 by Annual Reviews. All rights reserved

1081-0706/05/1110-0581$20.00

Key Words

intravital microscopy, two-photon microscopy, lymph node, immune response

Abstract

Over the past decades, intravital microscopy (IVM), the imaging of cells in living organisms, has become a valuable tool for studying the molecular determinants of lymphocyte trafficking. Recent advances in microscopy now make it possible to image cell migration and cell-cell interactions in vivo deep within intact tissues. Here, we summarize the principal techniques that are currently used in IVM, discuss options and tools for fluorescence-based visualization of lymphocytes in microvessels and tissues, and describe IVM models used to explore lymphoid and non-lymphoid organs. The latter will be introduced according to the physiologic itinerary of developing and differentiating T and B lymphocytes as they traffic through the body, beginning with their development in bone marrow and thymus and continuing with their migration to secondary lymphoid organs and peripheral tissues.

Contents

INTRODUCTION

Over 40 years of research on leukocyte trafficking (Gesner & Gowans 1962) have generated substantial insights into the pathologic and physiologic trafficking patterns of blood-borne leukocyte subsets and their role in health and disease (Butcher & Picker 1996, Springer 1994, von Andrian & Mackay 2000). This knowledge has been essential for the recent development of new anti-inflammatory therapies, such as monoclonal antibodies against LFA-1 or alpha-4 integrins, to treat psoriasis and multiple sclerosis, respectively (Miller et al. 2003a, Vugmeyster et al. 2004). Among the established techniques to study leukocyte trafficking, direct observation by intravital microscopy (IVM) is the oldest. Unlike other in vivo techniques that study leukocyte accumulation in tissues at the population level, IVM can analyze migratory behavior at the single-cell level and pinpoint the role of traffic molecules in multistep intravascular adhesion cascades.

However, intravascular adhesion represents only one aspect of leukocyte trafficking; many leukocytes spend only a short part of their life in the blood. The circulation half-life of naïve T cells, for example, is ~30 min, whereas they spend hours to days migrating within secondary lymphoid organs (SLO), querying dendritic cells (DCs) for cognate antigen (Ag). The rules governing interstitial cell migration, the adhesion molecules, and chemoattractants that determine cellular tissue localization and dwell-time are still scarcely understood compared with the rules governing intravascular trafficking.

Our current knowledge about cellular tissue distribution or cell-cell interactions during immune responses stems largely from investigations employing histology or in vitro tissue assays. Until recently, technical difficulties associated with imaging of events within solid organs have limited IVM observations to membranous tissues or superficial regions of non-translucent organs. These constraints are now overcome by combining IVM with two-photon microscopy (2P-IVM) (Denk et al. 1990). 2P-IVM allows immune cell imaging in solid organs at depths of up to 500 μm below the surface without interfering with blood or lymph flow, tissue oxygenation, or innervation (Sumen et al. 2004). Results from 2P-IVM have radically altered our perception of cellular dynamics within SLO, e.g., by revealing a much more dynamic behavior of naïve T cells than would be predicted from their sessile appearance ex vivo. Additionally, two-photon microscopy (2PM) has extended the application of IVM to immunological research that reaches beyond trafficking. It is now feasible to observe immune response at subcellular resolution in situ (Sumen et al. 2004).

In addition to advances in photonics, IVM has gained momentum recently by innovations in other areas; the availability of refined fluorescent probes, improved hardware and software for three-dimensional image analysis, and a diverse array of drugs, antibodies, recombinant proteins, and gene-targeted mice have contributed to the increasing acceptance and utilization of IVM as a powerful instrument for immunological research.

MODES OF VISUALIZATION

The first intravascular IVM studies of leukocytes employed brightfield microscopy (BFM) dating back to the 19th century (Cohnheim 1889, Wagner 1839). In BFM, white light illuminates the background behind a tissue, which must be sufficiently translucent, such as the frog mesentery and tongue (Cohnheim 1889), or rodent mesentery and cremaster muscle (Atherton & Born 1972, Baez 1973). BFM cannot distinguish different leukocyte subpopulations, and analysis of leukocyte be-

havior is largely limited to intravascular cells because leukocytes are difficult to track in the extravascular space.

Epifluorescence IVM (EF-IVM) is best-suited to study intravascular lymphocyte adhesion because it allows visualization of distinct leukocyte subpopulations in solid organs (**Table 1**). Typically, EF-IVM employs long, working-distance, water-immersion objectives for close access to surgically exposed tissues submerged in a physiologic buffer. To reduce phototoxicity and photobleaching, microscopes can be equipped with video-triggered xenon-arc stroboscopes that generate microsecond-pulsed excitation light and substantially reduce light exposure (30,000-fold versus continuous illumination). Most intravascular adhesion events occur within

Table 1 Murine IVM models employed for studying lymphocyte trafficking

Process	Organ	Imaging modality	Reference
Progenitor cell homing	BM	EFM	(Mazo et al. 1998)
Homing to SLO	Inguinal LN	EFM	(von Andrian 1996)
	Mesenteric LN	EFM	(Grayson et al. 2003)
	Peyer's Patch	EFM	(Bargatze et al. 1995)
	Spleen	EFM	(Grayson et al. 2003)
Migration within LN	Inguinal LN	2PM	(Miller et al. 2003b)
	Popliteal LN	2PM	(Mempel et al. 2004)
Inflammation and Immuno-surveillance	BM	2PM	(Mazo et al. 2005)
	CNS (spc)	EFM	(Vajkoczy et al. 2001)
	CNS (cwm)	EFF	(Yuan et al. 1994)
	CNS (brain)	EFM	(Piccio et al. 2002)
	Cremaster– Muscle	BFM	(Baez 1973)
		EFM	(Singbartl et al. 2001)
	Eye (iris)	EPF	(Becker et al. 2000)
	Intestine	EFM	(Massberg et al. 1998)
		EPF	(Fujimori et al. 2002)
	Liver	EFM	(Nakagawa et al. 1996)
		LSCM	(Hoffmeister et al. 2003)
	Pancreas	EFM	(Enghofer et al. 1995)
	Skin (ear)	EFM	(Reus et al. 1984)
	Skin (dsc)	EFM	(Lehr et al. 1994)
	Pancreas	EFM	(Enghofer et al. 1995)
	Tumor	EFM	(Leunig et al. 1992)

Abbreviations: EFM, epifluorescence microscopy; CFM, confocal microscopy; BFM, brightfield microscopy; 2PM, 2-photon microscopy; LSCM, laser-scanning fluorescent microscopy; spc, spinal cord; cwm, cranial window model; dsc; dorsal skinfold chamber.

seconds (or faster) and are analyzed off-line from two-dimensional recordings taken at video frame rates [30 fps in NTSC (National Television System Committee)]. The principles of intravascular IVM and associated measurement parameters are summarized in **Figure 1***a* and **Table 2**, respectively.

Although EF-IVM works well for superficial intravascular two-dimensional imaging, it is less suitable for deep tissue imaging. Nonconfocal EF-IVM below a sample's surface suffers from out-of-focus fluorescent signals, which reduce image quality and limit tissue penetration. Moreover, whereas blood flow dictates the directionality of cell movement in microvessels, extravascular cells can migrate freely in any direction, therefore requiring accurate three-dimensional tracking. The three-dimensional resolution needed to study interstitial cell behavior is not achievable with conventional fluorescence techniques. Confocal microscopy, which excludes out-of-focus fluorescence, was used successfully to generate three-dimensional images (see **Figure 1***b* for a schematic illustration). In intact lymph nodes (LN), confocal microscopy penetrates up to 80 μm deep, but cannot reach regions below the superficial cortex (Stoll et al. 2002). Recently, 2P-IVM has overcome many limitations of EF-IVM and confocal approaches (Denk et al. 1990). By generating interpretable images at $\leq$500 μm depth (**Figure 2**), 2P-IVM can visualize cell migration and cell-cell interactions in regions that were previously inaccessible. Technical aspects of 2P-IVM were reviewed recently (Cahalan et al. 2002, Sumen et al. 2004, Zipfel et al. 2003); measurement parameters to quantify cell migration from three-dimensional data sets are summarized in **Table 2**.

FLUORESCENT TAGS AND LABELS

IVM relies on tools, such as fluorescent compounds or proteins, for visualizing cell populations with high specificity, without altering their phenotype or function. Countless IVM experiments have used reagents such as rhodamine 6G or acridine red or orange, which, upon i.v. injection, accumulate in leukocytes (and to some extent in other cells), but not in red cells. A drawback to this approach is the potential excitation light-induced phototoxic effect that alters hemodynamics and cell adhesion (Saetzler et al. 1997). This issue can be partly alleviated by video-triggered xenon-arc stroboscopes (discussed above). One method for studying distinct leukocyte subsets consists of purifying cells from donor mice, labeling them fluorescently in vitro and re-injecting them into a recipient. Aside from the multitude of organic fluorophores, an interesting recent addition to the fluorescence tool box are the nanometer-scale semiconductor crystals called quantum dots, which emit in different colors (400–1350 nm), are photostable, and are bright enough for detection of single particles (Michalet et al. 2005).

Transgenic mice expressing green fluorescent protein (GFP) or its variants in a cell type-specific fashion are alternatives to ex vivo purified, fluorescently labeled cells. This approach subjects cells to less ex vivo handling (see "Technical Considerations," above), and genetically encoded fluorescent tags do not become diluted or fade in dividing cells. Additionally, GFP[+] transgenic leukocytes can be studied endogenously without ex vivo manipulation or adoptive transfer (Lindquist et al. 2004, Manjunath et al. 1999, Singbartl et al. 2001, Weninger et al. 2003). Alternatively, some cell populations can be specifically labeled in situ. For example, dermal DCs and Langerhans cells (LC) take up intracutaneously injected carboxy-fluorescein diacetate succinimidyl ester (CFSE) within tissues and subsequently migrate to draining LN where they can be imaged (Miller et al. 2004b).

Recently, GFP-like green, yellow, and red fluorescent proteins have been cloned (Matz et al. 1999), notably DsRed, a red fluorescent protein from Discosoma coral. Red light penetrates tissues better than green fluorescence,

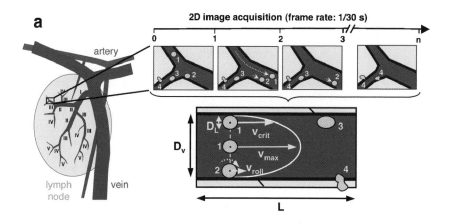

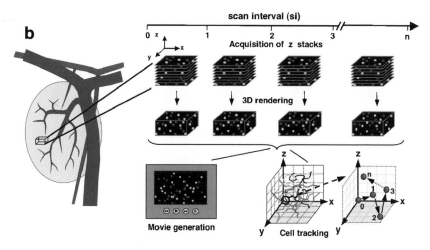

Figure 1

Principles of video-based IVM to analyze intravascular trafficking events and interstitial MP-IVM for two-dimensional recordings. Schematic representations of a murine LN with a typical venular tree (*a, b, left*). Venular order designations (in *a*) are assigned (Roman numerals) by counting successive generations of venular branches in upstream direction from the collecting venule draining into the larger vein. (*a, right*) Schematic representation of a vessel as seen in successive video frame segments (1, 2, 3...n). The different modes of intravascular cellular trafficking are illustrated by free flowing (1), rolling (2), sticking (3), and diapedesing (4) cells. The lower panel depicts the fluid dynamic parameters that determine leukocyte velocities in microvessels with laminar blood flow. For further definitions of each parameter refer to **Table 2**. (*b, right*) Schematic representation of the three-dimensional image acquisition process employed in confocal or MP-IVM. Stacks of optical sections are acquired by rapid incremental vertical repositioning of the objective along the z axis. In the example shown, one z stack is composed of eight sections at defined focal depths that are acquired during one scan interval (si). Image acquisition of z stacks is continuously repeated over a certain time period (si 1 → n). Each z stack is subsequently processed by image analysis software and rendered as a three-dimensional volume. Movies are typically generated as two-dimensional projections of successive three-dimensional image stacks. Using image analysis software, tracking identities are assigned to individual cells to follow their paths over time. The position of the tracked cell's centroid yields a series of xyz coordinates that correspond to different imaging time points. Detailed explanations of intravascular and interstitial measurement parameters are provided in **Table 2**.

Table 2 Measurement parameters used in intravascular and interstitial IVM

	Description	Unit
Intravascular 2D IVM		
Vascular dimensions: diameter (D_v), length (L)	Best determined after injection of a fluorescent plasma marker, e.g., 150 kD FITC dextran; L = distance of unbranched vascular segment between bifurcations	μm
Luminal surface area (A)	$A = (D_v/2)^2 * \pi * L$	μm^2
Maximal velocity (v_{max})	Velocity of the fastest flowing cell in a population of non-interacting cells, assumed to be flowing in the centerline	μm/s
Mean blood flow velocity (v_{blood})	Calculated from velocity analysis of non-interacting cells assuming a parabolic flow profile: $v_{blood} = v_{max}/(2 - (D_l/D_v)^2)$, where D_l = leukocyte diameter (7 μm)	μm/s
Blood flow (Q)	$Q = v_{blood} * \pi * D_v^2/4 * 10^{-6}$	nl/s
Critical velocity (v_{crit})	The lowest velocity that a non-interacting cell can assume in a microvessel. Can be estimated as $v_{crit} = v_{max} * (D_l/D_v) * (2 - (D_l/D_v))$, where D_l = leukocyte diameter (7 μm)	μm/s
Total cellular flux (TF)	Number of cells that pass a vessel during an observation period	min^{-1}
Rolling	Cellular movement (v_{roll}) along the vessel wall that is detectably slower than v_{crit}	
Rolling flux (RF)	Number of cells that roll in a vessel during an observation period	min^{-1}
Rolling fraction (RFx)	Percentage of cells that roll in a vessel relative to the total cellular flux: $RFx = RF/TF * 100$	%
Sticking	Cellular arrest for $\geq$30 s in a physiologically perfused vessel	
Sticking flux (SF)	Number of cells that stick during an observation period	min^{-1}
Sticking cell accumulation	SF/A	μm^{-2}
Sticking fraction (SFx)	Percentage of rolling cells that arrest in a vessel for $\geq$30 s: $SFx = SF/RF * 100$	%
Sticking efficiency (SFe)	Percentage of cells that arrest in a vessel for $\geq$30 s relative to the total cellular flux: $SFe = SF/TF * 100$	%
Interstitial 3D IVM		
Instantaneous velocity (v)	Velocity measured during a defined time interval (t), e.g., $v_{0\to1} = d_{0\to1}/t_{0\to1}$	μm/min
Mean velocity (V)	Mean velocity of a cell over the entire measurement period e.g., $V_{0\to n} = (v_{0\to1} + v_{1\to2} + v_{2\to3} + \ldots v_{n-1\to n})/n$	μm/min
Displacement (d)	Distance between first and last imaging point e.g., $d_{0\to n} = \text{sqrt}(\Delta x_{0\to n}^2 + \Delta y_{0\to n}^2 + \Delta z_{0\to n}^2)$	μm
Mean Displacement (MD)	Average distance migrated from arbitrary points of origin over a selected time (t) representing a multiple of the scan interval (si) e.g., $MD_{(t=1si)} = (d_{0\to1} + d_{1\to2} + d_{2\to3} + \ldots d_{n-1\to n})/n$; $MD_{(t=2si)} = (d_{0\to2} + d_{1\to3} + \ldots d_{n-2\to n})/(n - 1)$; $MD_{(t=3si)} = (d_{0\to3} + \ldots d_{n-3\to n})/(n - 2)$	μm
Chemotactic index (CI)	A measure for the directionality of cell migration. Represents the ratio of the displacement (d) over the total path length e.g., $CI = d_{0\to n}/(d_{0\to1} + d_{1\to2} + d_{2\to3} + \ldots d_{n-1\to n})$	no dimension
Turning angle	The angle at which a cell deviates from a straight line between subsequent measurement intervals	degree

(Continued)

Table 2 (*Continued*)

Interstitial 3D IVM	Description	Unit
Persistence time	Time during which a cell continues to migrate in the same direction before making a significant turn. For detailed explanation see Sumen et al. (2004)	min
Motility coefficient	A measure for a cell's propensity to move away from its point of origin, analogous to the diffusion coefficient of Brownian motion. For detailed explanation see Sumen et al. (2004)	$\mu m^2/min$

and its spectral separation from green emitters allows multicolor-imaging (Verkhusha & Lukyanov 2004). However, the propensity of DsRed to form aggregates plus the slow folding and maturation of the protein have hampered widespread use in IVM. Recently developed monomeric DsRed variants might solve these problems (Verkhusha & Lukyanov 2004).

Other interesting probes for IVM are fluorescent indicators of cell activation and/or signaling, particularly fluorescent proteins that redistribute intracellularly or change fluorescence intensity or color. Fluorescent reporters can also measure intracellular biochemical mediators or protease and kinase activities (Zhang et al. 2002). Similarly, fluorescent sensors that are based on FRET (fluorescence resonance energy transfer) hold promise for IVM. Furthermore, photoactivatable fluorescent proteins that either become fluorescent or undergo a color change upon UV illumination have been reported (Patterson & Lippincott-Schwartz 2004). The latter may become useful for photo-labeling cells during IVM for long-term tracking and/or subsequent isolation.

IVM MODELS: WATCHING THE LIFE-LONG JOURNEY OF T AND B LYMPHOCYTES

To protect the body from pathogens, the immune system must cope with the complex and challenging task of directing the right cells to the right place at the right time. Immunologists have learned to distinguish different lymphocyte subsets by distinct surface markers, which are telltale signs of a cell's prior exposure to Ag, the context in which the Ag was encountered (if at all), and the prevalent immunological function of the cell (see Supplemental Material **Table 1**. Follow the Supplemental Material link from the Annual Reviews home page at **http://www.annualreviews.org**). Most subset-defining markers are surface receptors involved in cell-cell recognition, such as adhesion molecules and chemoattractant receptors used by lymphocytes to gain access to different anatomic compartments (von Andrian & Mackay 2000). To understand the immune system, one must not only determine how cells detect and respond to outside stimuli, but also consider where each subset is exposed to such stimuli and what tissues(s) are accessible to the responding cells. Below, we discuss IVM models to study lymphocyte trafficking to and within lymphoid and non-lymphoid tissues (**Table 1**).

Bone Marrow: Cradle for Versatile Travelers

In the bone marrow (BM), hematopoietic stem cells (HSC) give rise to the various cell constituents of blood. Some BM-derived leukocytes then home quickly to other organs and tissues. However, the BM is not only a site of continuous unidirectional egress of blood cells and extravascular leukocytes; many circulating cells also (re-)enter the BM. The latter event has been documented for leukocytes such as naïve and memory T cells, memory B cells, plasma cells, and HSC (Sumen et al. 2004).

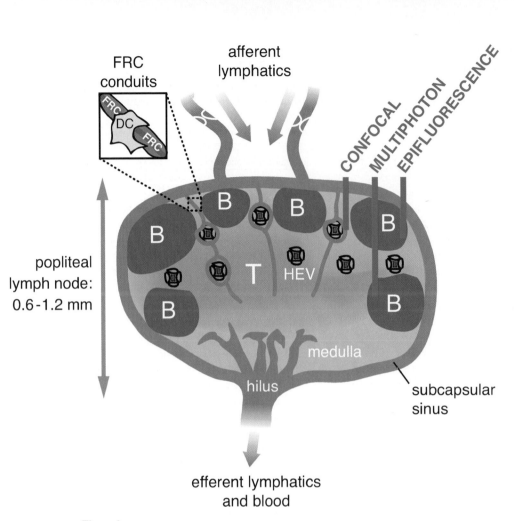

Figure 2

The lymph node. The diagram shows the distal B cell follicles and central T cell area of a schematized LN. Lymph flows into the node via the afferent lymphatics and courses around the node via the subcapsular sinus that surrounds the node. It can also flow into the medulla by traversing the cortex via the fibroblastic reticular cell (FRC) conduits that surround and interconnect a subset of high endothelial venules (HEVs), allowing sampling by resident DCs (*inset*). Lymph collects at the hilus and is removed from the node via the efferent lymphatic vessels. Approximate imaging depth for epifluorescence, confocal, and multiphoton imaging is also shown (murine popliteal lymph node).

The BM's inaccessible location has been a major obstacle for IVM: early studies in rabbit BM were performed over 30 years ago (reviewed in Mazo & von Andrian 1999), but technical challenges limited widespread application of this model. Recently, another technique for BM IVM was developed in mice (Mazo et al. 1998). This model exploits the fact that the bone covering BM cavities in the frontoparietal skull of mice is very thin and sufficiently transparent to visualize underlying BM. Thus only a small incision in the scalp is required, and the skull itself remains untouched. This model has been instrumental in analyzing adhesion events involved in homing of HSC (Katayama et al. 2003; Mazo et al. 1998, 2002) and central memory T cells (Mazo et al. 2005).

It is well established that the BM functions as a primary lymphoid organ, but recent reports suggest an additional role during primary and secondary T cell responses. The BM harbors naïve and memory T cells (Di Rosa & Santoni 2002), which are recruited from the blood (Koni et al. 2001, Mazo et al. 2005). Recent reports have established that the BM can support primary immune responses involving naïve T cells interacting with BM resident DCs (Feuerer et al. 2003, 2004). However, the BM can probably not support all types of adaptive immune responses, such as T cell–dependent antibody responses (Shinkura et al. 1996) or induction of allograft rejection (Lakkis et al. 2000). 2P-IVM may help reveal differences during T cell activation in BM versus SLO.

Distinct steps in lymphocyte differentiation are believed to take place within specialized supportive BM niches (Katsura 2002), but the microanatomic context of these events is unknown. HSC are concentrated in subendosteal regions within BM cavities (Calvi et al. 2003, Nilsson et al. 2001) and migrate to more central regions as they differentiate (Lord et al. 1975). This movement between niches is an active process that likely depends on attractants and stroma cells (Tokoyoda et al. 2004). 2P-IVM of skull BM offers a window to study the migration of lymphoid precursors. However, such studies require transgenic mice or transduced cells expressing fluorescent reporter genes under stage-specific promoters (Yu et al. 1999).

Thymus: Essential Detour for T Cell Differentiation

BM-derived lymphoid progenitors migrate via the blood to the thymus where they develop into mature T cells (**Figure 3**). Little is known so far about the traffic molecules governing progenitor cell homing to the thymus, but thymocyte migration within the thymus has been studied. Static snapshots of thymocyte development were obtained by im-

munohistochemistry (Lind et al. 2001), which revealed that thymocytes migrate to special regions fostering successive developmental stages (Petrie 2003). Cell-cell interactions are critical for thymocyte development: Newly arranged T cell receptor (TCR) specificities of thymocytes are tested for their affinity for self-peptide-MHC complexes on thymic stromal cells during positive and negative selection.

The inaccessible location of the thymus in the thorax has so far impeded IVM in situ. Thymocyte differentiation in the medulla occurs >500 μm below the capsule, which complicates imaging by 2P-IVM. Given the absence of in vivo models, thymocyte behavior and motility have presently been visualized only in vitro, such as in fetal thymic organ cultures (FTOC), three-dimensional reaggregated thymic organ cultures (RTOC) (Bousso et al. 2002, Richie et al. 2002), or in fresh thymic slices (Bhakta et al. 2005). In these settings, thymocytes were highly motile and migrated along seemingly random trajectories similar to T cells in LN, albeit more slowly (Bhakta et al. 2005, Mempel et al. 2004). Bousso et al. analyzed thymocyte-stromal cell interactions by reaggregating fluorescent TCR transgenic CD4$^+$CD8$^+$ thymocytes with differentially labeled MHC$^{+/+}$ or MHC$^{-/-}$ stromal cells (Bousso et al. 2002). MHC recognition during positive selection led to prolonged thymocyte-stromal cell interactions. Both short, highly dynamic interactions and long-lived stable contacts were observed. The meaning of this behavior is presently unclear, but is reminiscent of the first two phases of T cell-Ag presenting DC interactions in LN (Mempel et al. 2004).

Secondary Lymphoid Organs: Elementary School for Lymphocyte Activation and Differentiation

Once B and T cells have become fully differentiated in the BM and thymus, respectively, they embark on a relentless search for "their"

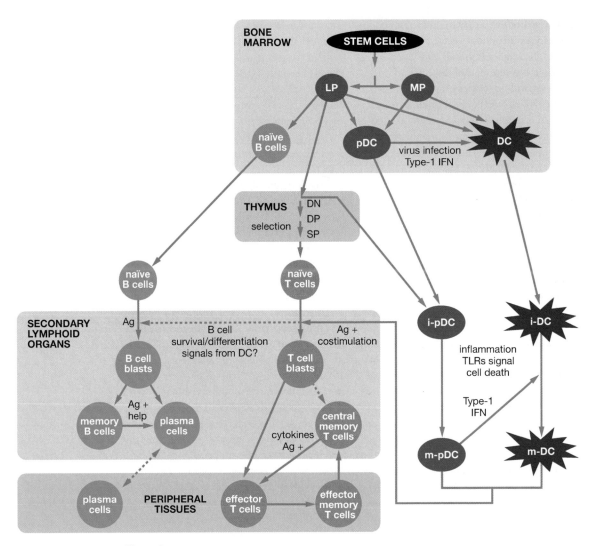

Figure 3

T cells, B cells, and dendritic cells: Lineage relationship, phenotype, and function. A common lymphoid progenitor (LP) gives rise to T cell precursors and B cells. T cell precursors undergo essential differentiation steps in the thymus, generating CD4 and CD8 T cells. Naïve T cells are activated by their cognate Ag presented by mature DCs, they proliferate (blasts) in the SLO (LN, spleen, Peyer's patches (PP)) and differentiate into effector (T_{EFF}) or central memory T cells (T_{CM}) (see Supplemental **Table 1**. Follow the Supplemental Material link from the Annual Reviews home page at **http://www. annualreviews.org**). T_{EFF} can either die or further differentiate to effector memory T cells (T_{EM}). T_{EM} can differentiate in vivo into T_{CM}. T_{CM} give rise to T_{EFF} upon re-encounter with their Ag. Conversely, naïve B cells are activated in a T cell-dependent or -independent fashion (depending on the Ag); they proliferate in SLO (blasts) and give rise to plasma and memory B cells. Memory B cells can differentiate into plasma cells when re-encountering their Ag. DCs differentiate from lymphoid or myeloid progenitors (MP). Immature plasmacytoid DCs (i-pDC; found in blood and SLO) or immature conventional DCs (i-DC; found in blood an peripheral tissues) differentiate into mature DCs (m-pDC or m-DC) when encountering appropriate environmental signals (usually from pathogens) and then migrate into secondary lymphoid organs where they activate naïve T cells. Ag: antigen(s); TLR: Toll-like receptors.

Ag. The anatomic compartments where naïve lymphocytes encounter Ag are the SLO (e.g., spleen, LN, and Peyer's patches). These structures have evolved specialized mechanisms to collect Ag and Ag-presenting cells from peripheral tissues or (in case of the spleen) from the blood. Therefore, SLO represent the staging ground for adaptive immune responses, which involve a choreographed series of interactions between lymphocytes and DCs. IVM has been performed in various SLO (**Table 1**), but most published studies have focused on peripheral LN (PLN) and Peyer's patch.

IVM models of peripheral lymph node and Peyer's patch for intravascular imaging. At steady state, LN contain mostly naïve T and B lymphocytes, which are concentrated within the deep paracortex and the superficial cortical B cell follicles, respectively (**Figure 2**). Lymphocyte recruitment to LN occurs in specialized microvessels, the high endothelial venules (HEV), and involves a classic multistep adhesion cascade (Butcher & Picker 1996, von Andrian & Mackay 2000). The first IVM studies of SLO in mice were performed in exteriorized small intestine Peyer's patch (Bargatze et al. 1995) and in the subiliac LN (von Andrian 1996), which is surgically exposed by carefully dissecting and stabilizing an abdominal skin flap.

IVM studies have been instrumental in characterizing the various adhesion molecules involved in lymphocytes homing to PLN (reviewed in von Andrian & Mempel 2003) and Peyer's patch (Bargatze et al. 1995, Okada et al. 2002). In peripheral LN, the initial adhesive interaction requires binding of L-selectin on lymphocyte microvilli to HEV-expressed sulfated Lewis x (sLe^x) glycoproteins, collectively termed peripheral node addressin (PNAd). The unique kinetics of these lectin-carbohydrate interactions allow leukocytes to tether and roll slowly within HEV, even at high shear flow. Subsequently, the rolling lymphocytes are exposed to chemokines that are non-covalently presented on the luminal surface of HEV and activate chemokine re-

ceptors [CCR7 on T cells, CCR7 or CXCR4 (or possibly CXCR5) on B cells]. The latter then trigger activation of the lymphocyte-expressed integrin LFA-1, which mediates firm arrest by engaging with HEV-expressed ICAM-1 or ICAM-2 (von Andrian & Mempel 2003). By contrast, Peyer's patch HEV do not express PNAd in their lumen, but mucosal addressin cell-adhesion molecule 1 (MAdCAM-1), which binds the lymphocyte integrin $\alpha_4\beta_7$. Rolling of lymphocytes in Peyer's patch HEV is mediated by both L-selectin and $\alpha_4\beta_7$ interacting with different domains of MAdCAM-1 (Bargatze et al. 1995, Berg et al. 1993). Firm arrest of rolling lymphocytes in Peyer's patch HEV requires CCR7 for T cells, whereas some B cells can alternatively use CXCR4 and CXCR5 (Okada et al. 2002).

Life within a lymph node: two-photon microscopy. Two different LN in mice have been employed for 2P-IVM. The inguinal LN model used in intravascular IVM (von Andrian 1996) has been adapted to 2P-IVM (Lindquist et al. 2004, Miller et al. 2003b): This preparation requires mechanical restraints applied to the surrounding soft tissue to inhibit sample movement. More recently, the popliteal LN behind the knee has been used (Mempel et al. 2004). This model employs skeletal pivots as fixation points for near-perfect immobilization without application of force to soft tissue, which may interfere with lymph flow. Moreover, the T cell area is more superficial in the smaller popliteal node and is more accessible to deep imaging. Both LN preparations have yielded similar measurements of DC migration (Lindquist et al. 2004, Mempel et al. 2004) and T cell motility (Mempel et al. 2004, Miller et al. 2003b). The in vivo measurements of cell migration and communication are also in good agreement with 2PM studies performed in vitro in excised, intact murine LN (Miller et al. 2002, 2004a; Stoll et al. 2002). However, such ex vivo imaging studies require an environment containing 95% O_2, and it is not yet clear whether this non-physiologic atmosphere and/or the absence of innervation, lymph, and blood flow affect cellular functions at a more subtle level.

Immunohistochemical analysis of histological sections have been instrumental in defining the different lymphocyte compartments in LN (**Figure 2**): B cells reside in subcapsular follicles, whereas T cells inhabit the deep cortex and the regions between and below B follicles called the cortical ridge, which also harbors a network of DCs (Lindquist et al. 2004). The medulla contains many lymphatic sinusoids and various macrophages, memory B cells, and plasma cells. Fibroblastic reticular cells (FRCs) form a meshwork throughout the cortex and are intimately associated with a web of collagen-rich extracellular matrix fibers that direct an ultra-filtrate of lymph from the subcapsular sinus to DCs in the T cell area (Gretz et al. 2000, Sixt et al. 2005). 2PM techniques, such as second harmonic generation of UV photons by collagen fibers, allow visualization of some of these fibers without specific staining (von Andrian & Mempel 2003). Putative gradients of chemokines and other factors that may affect cellular traffic and mobility in LN have not been visualized to date. In fact, it is currently unclear if such gradients really exist because thus far the migratory behavior of all observed intranodal leukocytes is best described as a random walk (Mempel et al. 2004, Miller et al. 2002).

Watching the dance: T cell-dendritic cell interactions. DCs are uniquely capable of both activating and inducing tolerance in naïve T cells (Banchereau et al. 2000, Steinman et al. 2003). Numerous DC subpopulations have been described in the mouse (Itano & Jenkins 2003). Most DCs arrive in LN via lymphatics that drain peripheral tissues. Several 2P-IVM studies have reported on DC motility and interactions with T cells in LN (Bousso & Robey 2003, Hugues et al. 2004, Lindquist et al. 2004, Mempel et al. 2004, Miller et al. 2004b, Stoll et al. 2002). Three primary means are known by which DCs acquire Ag for presentation in LN (Itano

et al. 2003): (*a*) Soluble Ag in afferent lymph drains into the subcapsular sinus, where DCs can take up, process, and then carry Ag into the T cell area. (*b*) Alternatively, lymph-borne molecules with a radius of ≤ 4 nm can enter the FRC conduits and are acquired by resident DCs in the deep cortex (Sixt et al. 2005). (*c*) Finally, peripheral DCs endocytose or phagocytose Ag and, upon maturation, migrate via lymphatics into draining LN. 2P-IVM indicates that the latter DCs are initially more motile than their LN-resident counterparts, which are organized in relatively stationary networks (Lindquist et al. 2004). However, all DCs constantly extend and retract rapidly moving dendrites or lamellipodia, which contact large numbers of lymphocytes surrounding them. Estimates of the average rate of contacts between a DC and T cells ranges from 500–5000/h (Bousso & Robey 2003, Miller et al. 2004b). This may explain how a few Ag-bearing DCs can rapidly be detected by the rare T cells expressing a suitable TCR among the vast, highly diverse T cell pool.

Several studies using 2P-IVM have analyzed the quality and duration of T cell–DC interactions in LN and established that priming of both CD8[+] (Hugues et al. 2004, Mempel et al. 2004) and CD4[+] T cells (Miller et al. 2004b) occurs in distinct sequential stages: During the first 8 h upon entering the T cell area, T cells engage in multiple short-lasting contacts with DCs (phase 1). This is followed by a period of ~12 h (phase 2) when T cells and DCs form long-lasting (>1 h) stable conjugates and begin to secrete cytokines. Finally, these conjugates dissociate, and T cells proliferate vigorously and resume their rapid migration while undergoing only brief interactions with DCs (phase 3).

DC-T cell interactions have mostly been analyzed during priming, but one recent report has also observed interactions during tolerance–induction (Hugues et al. 2004). Under these conditions, naïve T cells engaged in brief contacts with DCs but never entered into stable interactions. Thus the duration and nature of T cell-DC contacts during phase 2 is probably critical for the outcome of immune responses.

Homing preferences of antigen-experienced T cells. Once naïve T cells have become activated in SLO, they proliferate and differentiate. Depending on the strength of the signal they receive through their TCR, and on the effect of local cytokines, T cells differentiate into various specialized effector cells (T_{EFF}) and into the longer-lived effector memory cells (T_{EM}) and central memory cells (T_{CM}) (Sallusto et al. 2004). T_{EFF} were initially defined as CCR7$^-$L-selectin$^-$ memory marker–expressing cells that exert immediate effector/cytotoxic activity without the need for costimulation (see Supplemental **Table 1**. Follow the Supplemental Material link from the Annual Reviews home page at **http://www.annualreviews.org**) (Sallusto et al. 2004). In contrast, T_{CM} were CCR7$^+$L-selectin$^+$ cells without immediate effector/cytotoxic activity, which proliferated and secreted cytokines, such as IL-2, more readily in response to recall Ag compared with naïve T cells or T_{EFF} (Manjunath et al. 2001, Sallusto et al. 2004). However, virus-specific T_{CM} in virally infected mice were shown to possess immediate effector activity comparable to that of T_{EFF} or T_{EM} (Gourley et al. 2004). In the mouse, T_{EM} have been shown to convert slowly into a T_{CM} phenotype expressing L-selectin and CCR7. Given their high expression of these trafficking molecules, T_{CM} readily home into LN, whereas T_{EFF} and T_{EM} cannot access LN from the blood (Weninger et al. 2001). Some T_{CM} also express adhesion molecules for migration into non-lymphoid tissues, such as the gut and the skin (Campbell et al. 2001), and both T_{CM} and T_{EM} (but not naïve T cells) migrate to acutely inflamed tissues (Weninger et al. 2001).

The small number of endogenous Ag-specific T cells that typically can be isolated from mice has limited their study by IVM. To bypass this difficulty, protocols for the in vitro generation and expansion of the different Ag-experienced T cell subpopulations have recently been established. For example, our laboratory has described that Ag-stimulated lymphoblasts cultured in the presence of IL-15, IL-7, or low doses of IL-2 acquire the phenotype and function of T_{CM}, whereas culture in high concentrations of IL-2 (>20 ng/ml) drives these cells to acquire phenotypic and functional properties of T_{EFF} cells (**Figure 4**) (Manjunath et al. 2001). This simple approach has been exploited to produce the prerequisite numbers of effector/memory subsets for in vivo homing experiments and IVM (Mazo et al. 2005, Scimone et al. 2004, Weninger et al. 2001). These studies have demonstrated that in vitro–generated T_{CM} migrate preferentially to SLO and the BM, whereas T_{EFF} accumulate in certain extralymphoid tissues, such as liver and lung, and efficiently migrate to sites of inflammation (Weninger et al. 2001). However, neither ex vivo–generated T_{CM} nor T_{EFF} migrate to intestinal compartments. As is discussed below, T cells need additional tissue-specific imprinting signals to acquire gut-specific homing molecules.

Tissue-specific T cell migration and differentiation. DC from gut-associated lymphoid tissues imprint gut-homing potential on T cells upon activation (Iwata et al. 2004, Johansson-Lindbom et al. 2003, Mora et al. 2003). Conversely, non-gut associated DC from the spleen or cutaneous LN preferentially induce skin-homing T cells (Mora et al. 2005). Interestingly, gut- and skin-homing potentials of T cells represent dynamic functional states that can be rapidly modified when tissue-tropic memory T cells are restimulated in a different anatomic context (Mora et al. 2005). Recent evidence points to retinoic acid (RA), a vitamin A metabolite, as a central mediator in the imprinting of gut-homing T cells, and in fact gut-associated DCs, but not DCs from other SLO, express enzymes to convert food-derived vitamin A

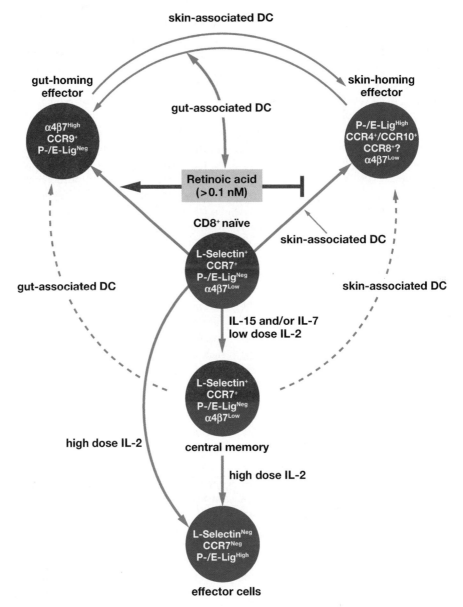

Figure 4

Ex vivo–generated gut- and skin-homing T cells and effector (T_{EFF}) and central memory (T_{CM}) CD8$^+$ T cells. To generate gut- or skin-homing T cells, naïve TCR–transgenic T cells are co-cultured during 4 to 5 days with Ag-pulsed PP- or PLN-DCs, respectively. Alternatively, T cells can be activated with plate-bound anti-CD3 + anti-CD28 in the presence or absence of *all-trans* retinoic acid (>0.1 nM), which generates gut- or skin-homing T cells, respectively. To generate T_{EFF} and T_{CM} CD8$^+$ T cells, total splenocytes from TCR transgenic mice are cultured for 2 h with the specific peptide, followed by 2 days without it. T cells are then cultured for 5 to 7 days in high-dose IL-2 (>20 ng/ml) to generate T_{EFF}, or in IL-15 and/or IL-7 (>5 ng/ml), or in low-dose IL-2, to obtain T_{CM}. Interestingly, gut- and skin-homing CD8$^+$ T cells can be significantly reversed in vitro by re-activating the effector T cells with the opposite DCs. Furthermore, ex vivo generated T_{CM} can be differentiated into gut- or skin-homing effector T cells by re-activating them with PP- or PLN-DC, respectively.

into RA (Iwata et al. 2004). Conversely, the induction of E- and P-selectin ligands for skin-homing appears to be a default pathway in T cells activated in the absence of RA or gut-DCs (**Figure 4**) (Iwata et al. 2004, Mora et al. 2005).

Consequently, T cells activated by intestinal DCs roll poorly in skin-associated post-capillary venules, which constitutively express P- and E-selectin (Weninger et al. 2000) but not the $\alpha_4\beta_7$ ligand MAdCAM-1. By contrast, effector cells activated by DCs from non-intestinal SLO roll in skin-associated venules at a higher frequency and lower velocity than gut-tropic T cells and accumulate to significantly large numbers in inflamed skin (Mora et al. 2005).

B cells and plasma cells. Some information has also been gathered on the trafficking of B cell subsets (Kunkel & Butcher 2003). B-1 B cells constitute a small subset with self-renewal capacity and reside primarily in the abdominal and thoracic cavities where they continuously secrete low affinity "natural antibodies" (Poletaev & Osipenko 2003). By contrast, conventional naïve B cells (B-2 cells) become activated by Ag in a T cell–dependent or –independent fashion in SLO. B-2 cells subsequently differentiate into antibody-secreting cells (plasma cells) and migrate via the blood to spleen, BM (mainly IgG-producing plasma cells), or mucosal tissues (IgA-producing cells). Similarly to T cells, B cell-derived plasma cells also demonstrate tissue–tropism, depending on the site of B cell activation (Kunkel & Butcher 2003). B cells activated in gut-associated lymphoid tissues upregulate $\alpha_4\beta_7$ and CCR9 and/or CCR10, which target them to mucosal sites (Kunkel & Butcher 2003). To date, no IVM studies have been performed with Ag-experienced B cells. Although gut-DCs and RA apparently induce IgA isotype switching in B cells upon activation (Spalding et al. 1984, Tokuyama & Tokuyama 1996), it is unknown if these factors also induce gut-tropism in B cells. If this were the case, this could greatly facilitate the

in vitro generation of gut-tropic plasma cells for use in IVM.

IVM models for imaging lymphocytes in peripheral tissues. Non-lymphoid tissues harbor the majority of early memory T lymphocytes that arise after systemic Ag challenge (Masopust et al. 2001, Reinhardt et al. 2001). These tissues must also recruit effector and memory lymphocytes generated in SLO in response to Ag challenge at a peripheral site. Indeed, effector cell recruitment is essential for lymphocytes to exert most of their protective and/or pathogenic activity. Some peripheral tissues (i.e., gut and skin) can recruit lymphocytes under non-inflammatory steady-state conditions because their microvessels constitutively express tissue-specific adhesion molecules and chemokines for homeostatic recruitment. In addition, all tissues can recruit lymphocytes upon induction by inflammatory signals, which upregulate a plethora of adhesion molecules recognized by most effector cells. 2P-IVM studies of interstitial lymphocyte migration in non-lymphoid tissues have been initiated only recently, but several epifluoresence-based IVM studies have already characterized the intravascular lymphocyte recruitment steps in such tissues under both non-inflammatory and inflammatory conditions. A representative (but invariably incomplete) selection of relevant examples in the mouse is briefly discussed below:

Skin. The skin is the body's most vulnerable organ and consequently requires specialized mechanisms of immune surveillance for protection against external hazards (Kupper & Fuhlbrigge 2004). IVM of the mouse ear microcirculation allows percutaneous observations of leukocyte adhesion in dermal microvessels without requiring surgical manipulation (Reus et al. 1984). The model has been valuable for studying physiologic cutaneous immune surveillance, in particular with regard to P- and E-selectin and the leukocyte-expressed glycoproteins and glycosyltransferases that contribute to selectin ligand activity (Maly et al. 1996, Weninger et al. 2000). Moreover, the skin is the target of acute and chronic inflammation. However, inflammatory events are difficult to visualize in the ear model because the intact epidermis impedes experimental application of many pro-inflammatory reagents, and inflammation-induced tissue swelling reduces visibility of dermal microvessels. Other IVM models that visualize dermal and subdermal microvessels, such as the dorsal skinfold chamber preparation, may be more suitable for studying inflamed skin (Lehr et al. 1994).

Cremaster muscle. The murine cremaster muscle consists of a thin sheet of striated muscle cells that surround the testis and is easily exposed. This tissue has been the prototypical peripheral site to perform IVM under acute inflammatory conditions (Baez 1973, Ley et al. 1995). The model features excellent optical properties, access to a dense, highly organized network of microvessels, and the possibility for easy experimental manipulation such as exposure to inflammatory mediators.

Small and large intestine. One IVM model to image the intestinal microvasculature has been described (Fujimori et al. 2002). It involves exteriorizing an intestinal loop and opening the intestine for visualization of microvessels from the luminal mucosal surface. This technique has so far not been employed to image highly dynamic events, such as lymphocyte tethering and rolling. A technical challenge for IVM of intestinal microvasculature is the intrinsic, peristaltic tissue movement, which can deteriorate the quality of IVM recordings, especially at high magnification. This latter problem could be alleviated by local drug treatment (e.g., with atropine).

Liver. The liver IVM model has been used to obtain unique insights into leukocyte and platelet interactions with highly specialized hepatic microvessels and intrahepatic macrophages, respectively (Hoffmeister et al. 2003, Nakagawa et al. 1996). A major technical difficulty of this model is reducing the breathing-associated movement of the liver to allow precise imaging. This problem is more acute with upright microscopes when imaging the liver from above. Using an inverted setup, the liver's own weight helps keep the tissue still and in focus, albeit at the expense of easy access to the tissue for rapid experimental manipulation.

Central nervous system. Three different models have been described for visualizing intravascular adhesion events in the central nervous system (CNS). A simple model exists for imaging brain microvessels in anaesthetized mice, by exposing the scalp in the parietal area where bones are sufficiently transparent for direct imaging of underlying superficial brain microvessels (Piccio et al. 2002). Furthermore, a cranial window model also exists (Yuan et al. 1994). These models have been used for studying T cell interactions with cortical venules under steady-state and inflammatory conditions. A technically more challenging approach is IVM of spinal cord microvessels, which requires sophisticated microsurgical techniques (Vajkoczy et al. 2001) but allows unique insight to elucidate the adhesion pathways involved in the recruitment of T cells to both the gray and white matter of the CNS.

TECHNICAL CONSIDERATIONS

Most questions in biology cannot be addressed experimentally without inflicting some degree of manipulation on the system under study. In IVM, these manipulations comprise the use of anesthetics and other drugs, invasive surgical procedures to expose tissues, purification and fluorescent labeling of cells, and tissue exposure to intense illumination causing potential phototoxicity. The manual or computer-assisted analysis of complex imaging data sets also represents a possible source of errors. To obtain IVM data that describe in vivo leukocyte behavior accurately, it is important to recognize possible caveats and to take appropriate precautions.

Surgical Preparation and Associated Issues

Most IVM models require tissue dissection to create optimal accessibility for imaging. This procedure likely induces local or systemic inflammation. Several measures can keep inflammation to a minimum: Surgical equipment and buffer solutions must be free of contamination with bacteria or bacterial byproducts (e.g., endotoxins); drying of exposed tissue must be prevented to avoid thrombosis, inflammation and cell necrosis; proper surgical techniques also require careful dosage of anesthetics for full analgesia without causing hypotension or respiratory depression; the temperature of the animal and the exposed tissue should be monitored because interstitial cell motility is compromised below 35°C (Miller et al. 2002).

IVM typically requires specialized stages for optimally positioning and immobilizing the animal. This is particularly important for three-dimensional imaging using 2P-IVM. Lymphocytes in tissue move two to three orders of magnitude more slowly than in microvessels (a few micrometers per minute compared with several millimeters per second for cells in arterioles). Thus, to accurately detect interstitial cell movement, long observation periods (up to hours) are needed to obtain three-dimensional information from stacks of multiple optical sections, which must be acquired over scan intervals lasting up to 1 min each (**Figure 1b**). Tissue movement by as little as a few micrometers can render three-dimensional image stacks uninterpretable (imagine holding a camera with a 1-min shutter speed). Conversely, application of mechanical force to immobilize the tissue must not interfere with physiologic functions such as lymph and blood flow.

Cell Purification

Intravascular IVM with ex vivo purified and fluorescently labeled cells requires that the cells be re-introduced into the circulation. To minimize cell loss in the circulation upon i.v. injection, cells are ideally injected directly into an artery upstream of the tissue under observation. Even so, intravascular IVM generally requires several million highly purified cells—a number difficult to obtain for rare primary cell types. Cell lines are often not a viable alternative because many long-term cultured cells are much larger than blood-borne leukocytes and can cause massive capillary plugging upon i.v. injection, sometimes with lethal consequences. When using fluorescently tagged primary cells, high purity is essential because contaminating cells will not be distinguishable. Additionally, cell purification and fluorescent labeling may damage or otherwise alter the cells. The latter concern especially holds true for DCs, which are easily induced to mature, but improper ex vivo handling can also affect other leukocytes, e.g., by inducing shedding of L-selectin or increasing apoptosis. The quality of purified and labeled cells should, therefore, be monitored in parallel assays, such as quantitative homing studies, chemotaxis experiments, or flow cytometry.

Data Acquisition, Analysis, and Interpretation

Compared with intravascular video-based epifluorescence IVM, three-dimensional imaging within living tissue is a very young discipline and, consequently, still faces many unresolved technical issues. Primarily, the transition from two-dimensional imaging of cells within vessels to three-dimensional imaging of cells moving freely in any direction adds considerable complexity to both imaging and analysis.

For example, T cell motility varies between shallow and deep T cell areas (relative to the capsule) of the same LN. Thus to compare results from different experiments, it is important to consider imaging depth and anatomic location (Mempel et al. 2004, Stoll et al. 2002). Furthermore, non-fluorescent structures such as unlabeled cells, connective tissue fibers, blood vessels, or lymphatics may influence cell motility. Lastly, highly motile cells may be over-represented in certain measurements as they may transiently leave a scanned volume of tissue and then re-enter it to be counted as a new cell.

Clearly, the unique challenges and limitations of three-dimensional imaging of the immune system in a living animal are only now beginning to be understood and will take some time to be adequately addressed. Similarly, at the level of data processing and presentation, substantial variability exists between the approaches undertaken by different research teams. Therefore, it will be important that measurement parameters are clearly defined and that investigators in the field agree on common procedures to generate a better understanding of the relevance of different measurement parameters.

LITERATURE CITED

Atherton A, Born GVR. 1972. Quantitative investigation of the adhesiveness of circulating polymorphonuclear leucocytes to blood vessel walls. *J. Physiol.* 222:447–74

Baez S. 1973. An open cremaster muscle preparation for the study of blood vessels by in vivo microscopy. *Microvasc. Res.* 5:384–94

Banchereau J, Briere F, Caux C, Davoust J, Lebecque S, et al. 2000. Immunobiology of dendritic cells. *Annu. Rev. Immunol.* 18:767–811

Bargatze RF, Jutila MA, Butcher EC. 1995. Distinct roles of L-selectin and integrins α 4β 7 and LFA-1 in lymphocyte homing to Peyer's patch-HEV in situ: the multistep model confirmed and refined. *Immunity* 3:99–108

Becker MD, Nobiling R, Planck SR, Rosenbaum JT. 2000. Digital video-imaging of leukocyte migration in the iris: intravital microscopy in a physiological model during the onset of endotoxin-induced uveitis. *J. Immunol. Methods* 240:23–37

Berg EL, McEvoy LM, Berlin C, Bargatze RF, Butcher EC. 1993. L-selectin-mediated lymphocyte rolling on MAdCAM-1. *Nature* 366:695–98

Bhakta NR, Oh DY, Lewis RS. 2005. Calcium oscillations regulate thymocyte motility during positive selection in the three-dimensional thymic environment. *Nat. Immunol.* 6:143–51

Bousso P, Bhakta NR, Lewis RS, Robey E. 2002. Dynamics of thymocyte-stromal cell interactions visualized by two-photon microscopy. *Science* 296:1876–80

Bousso P, Robey E. 2003. Dynamics of CD8[+] T cell priming by dendritic cells in intact lymph nodes. *Nat. Immunol.* 4:579–85

Butcher EC, Picker LJ. 1996. Lymphocyte homing and homeostasis. *Science* 272:60–66

Cahalan MD, Parker I, Wei SH, Miller MJ. 2002. Two-photon tissue imaging: seeing the immune system in a fresh light. *Nat. Rev. Immunol.* 2:872–80

Calvi LM, Adams GB, Weibrecht KW, Weber JM, Olson DP, et al. 2003. Osteoblastic cells regulate the haematopoietic stem cell niche. *Nature* 425:841–46

Campbell JJ, Murphy KE, Kunkel EJ, Brightling CE, Soler D, et al. 2001. CCR7 expression and memory T cell diversity in humans. *J. Immunol.* 166:877–84

Cohnheim J. 1889. *Lectures on General Pathology: A Handbook for Practitioners and Students.* London: New Sydenham Society

Denk W, Strickler JH, Webb WW. 1990. Two-photon laser scanning fluorescence microscopy. *Science* 248:73–76

Di Rosa F, Santoni A. 2002. Bone marrow CD8 T cells are in a different activation state than those in lymphoid periphery. *Eur. J. Immunol.* 32:1873–80

Enghofer M, Bojunga J, Usadel KH, Kusterer K. 1995. Intravital measurement of donor lymphocyte adhesion to islet endothelium of recipient animals in diabetes transfer experiments. *Exp. Clin. Endocrinol. Diabetes* 103(Suppl.) 2:99–102

Feuerer M, Beckhove P, Garbi N, Mahnke Y, Limmer A, et al. 2003. Bone marrow as a priming site for T-cell responses to blood-borne antigen. *Nat. Med.* 9:1151–57

Feuerer M, Beckhove P, Mahnke Y, Hommel M, Kyewski B, et al. 2004. Bone marrow microenvironment facilitating dendritic cell: CD4 T cell interactions and maintenance of CD4 memory. *Int. J. Oncol.* 25:867–76

Fujimori H, Miura S, Koseki S, Hokari R, Komoto S, et al. 2002. Intravital observation of adhesion of lamina propria lymphocytes to microvessels of small intestine in mice. *Gastroenterology* 122:734–44

Gesner BM, Gowans JL. 1962. The output of lymphocytes from the thoracic duct of unanaesthetized mice. *Br. J. Exp. Pathol.* 43:424

Gourley TS, Wherry EJ, Masopust D, Ahmed R. 2004. Generation and maintenance of immunological memory. *Semin. Immunol.* 16:323–33

Grayson MH, Hotchkiss RS, Karl IE, Holtzman MJ, Chaplin DD. 2003. Intravital microscopy comparing T lymphocyte trafficking to the spleen and the mesenteric lymph node. *Am. J. Physiol. Heart Circ. Physiol.* 284:H2213–26

Gretz JE, Norbury CC, Anderson AO, Proudfoot AE, Shaw S. 2000. Lymph-borne chemokines and other low molecular weight molecules reach high endothelial venules via specialized conduits while a functional barrier limits access to the lymphocyte microenvironments in lymph node cortex. *J. Exp. Med.* 192:1425–40

Haddad W, Cooper CJ, Zhang Z, Brown JB, Zhu Y, et al. 2003. P-selectin and P-selectin glycoprotein ligand 1 are major determinants for Th1 cell recruitment to nonlymphoid effector sites in the intestinal lamina propria. *J. Exp. Med.* 198:369–77

Hoffmeister KM, Felbinger TW, Falet H, Denis CV, Bergmeier W, et al. 2003. The clearance mechanism of chilled blood platelets. *Cell* 112:87–97

Hosoe N, Miura S, Watanabe C, Tsuzuki Y, Hokari R, et al. 2004. Demonstration of functional role of TECK/CCL25 in T lymphocyte-endothelium interaction in inflamed and uninflamed intestinal mucosa. *Am. J. Physiol. Gastrointest. Liver Physiol.* 286:G458–66

Hugues S, Fetler L, Bonifaz L, Helft J, Amblard F, Amigorena S. 2004. Distinct T cell dynamics in lymph nodes during the induction of tolerance and immunity. *Nat. Immunol.* 5:1235–42

Itano AA, Jenkins MK. 2003. Antigen presentation to naïve CD4 T cells in the lymph node. *Nat. Immunol.* 4:733–39

Itano AA, McSorley SJ, Reinhardt RL, Ehst BD, Ingulli E, et al. 2003. Distinct dendritic cell populations sequentially present a subcutaneous antigen to CD4 T cells and stimulate different aspects of cell-mediated immunity. *Immunity* 19:47–57

Iwata M, Hirakiyama A, Eshima Y, Kagechika H, Kato C, Young SY. 2004. Retinoic acid imprints gut-homing specificity on T cells. *Immunity* 21(4):527–38

Johansson-Lindbom B, Svensson M, Wurbel MA, Malissen B, Marquez G, Agace W. 2003. Selective generation of gut tropic T cells in gut-associated lymphoid tissue (GALT): requirement for GALT dendritic cells and adjuvant. *J. Exp. Med.* 198:963–69

Katayama Y, Hidalgo A, Furie BC, Vestweber D, Furie B, Frenette PS. 2003. PSGL-1 participates in E-selectin-mediated progenitor homing to bone marrow: evidence for cooperation between E-selectin ligands and alpha4 integrin. *Blood* 102:2060–67

Katsura Y. 2002. Redefinition of lymphoid progenitors. *Nat. Rev. Immunol.* 2:127–32

Koni PA, Joshi SK, Temann UA, Olson D, Burkly L, Flavell RA. 2001. Conditional vascular cell adhesion molecule 1 deletion in mice. Impaired lymphocyte migration to bone marrow. *J. Exp. Med.* 193:741–54

Kunkel EJ, Butcher EC. 2003. Plasma-cell homing. *Nat. Rev. Immunol.* 3:822–29

Kupper TS, Fuhlbrigge RC. 2004. Immune surveillance in the skin: mechanisms and clinical consequences. *Nat. Rev. Immunol.* 4:211–22

Lakkis FG, Arakelov A, Konieczny BT, Inoue Y. 2000. Immunologic 'ignorance' of vascularized organ transplants in the absence of secondary lymphoid tissue. *Nat. Med.* 6:686–88

Lehr HA, Olofsson AM, Carew TE, Vajkoczy P, von Andrian UH, et al. 1994. P-selectin mediates the interaction of circulating leukocytes with platelets and microvascular endothelium in response to oxidized lipoprotein in vivo. *Lab. Invest.* 71:380–86

Leunig M, Yuan F, Menger MD, Boucher Y, Goetz AE, et al. 1992. Angiogenesis, microvascular architecture, microhemodynamics, and interstitial fluid pressure during early growth of human adenocarcinoma LS174T in SCID mice. *Cancer Res.* 52:6553–60

Ley K, Bullard DC, Arbonés ML, Bosse R, Vestweber D, et al. 1995. Sequential contribution of L- and P-selectin to leukocyte rolling in vivo. *J. Exp. Med.* 181:669–75

Lind EF, Prockop SE, Porritt HE, Petrie HT. 2001. Mapping precursor movement through the postnatal thymus reveals specific microenvironments supporting defined stages of early lymphoid development. *J. Exp. Med.* 194:127–34

Lindquist RL, Shakhar G, Dudziak D, Wardemann H, Eisenreich T, et al. 2004. Visualizing dendritic cell networks in vivo. *Nat. Immunol.* 5:1243–50

Lord BI, Testa NG, Hendry JH. 1975. The relative spatial distributions of CFUs and CFUc in the normal mouse femur. *Blood* 46:65–72

Maly P, Thall AD, Petryniak B, Rogers CE, Smith PL, et al. 1996. The $\alpha_{1,3}$ fucosyltransferase Fuc-TVII controls leukocyte trafficking through an essential role in L-, E-, and P-selectin ligand biosynthesis. *Cell* 86:643–53

Manjunath N, Shankar P, Stockton B, Dubey PD, Lieberman J, von Andrian UH. 1999. A transgenic mouse model to analyze CD8$^+$ effector T cell differentiation in vivo. *Proc. Natl. Acad. Sci. USA* 96:13932–37

Manjunath N, Shankar P, Wan J, Weninger W, Crowley MA, et al. 2001. Effector differentiation is not prerequisite for generation of memory cytotoxic T lymphocytes. *J. Clin. Invest.* 108:871–78

Masopust D, Vezys V, Marzo AL, Lefrancois L. 2001. Preferential localization of effector memory cells in nonlymphoid tissue. *Science* 291:2413–17

Massberg S, Enders G, Leiderer R, Eisenmenger S, Vestweber D, et al. 1998. Platelet-endothelial cell interactions during ischemia/reperfusion: the role of P-selectin. *Blood* 92:507–15

Matz MV, Fradkov AF, Labas YA, Savitsky AP, Zaraisky AG, et al. 1999. Fluorescent proteins from nonbioluminescent Anthozoa species. *Nat. Biotechnol.* 17:969–73

Mazo IB, Gutierrez-Ramos J-C, Frenette PS, Hynes RO, Wagner DD, von Andrian UH. 1998. Hematopoietic progenitor cell rolling in bone marrow microvessels: parallel contributions by endothelial selectins and VCAM-1. *J. Exp. Med.* 188:465–74

Mazo IB, Honczarenko M, Leung H, Cavanagh LL, Bonasio R, et al. 2005. Bone marrow is a major reservoir and site of recruitment for central memory CD8$^+$ T Cells. *Immunity* 22:259–70

Mazo IB, Quackenbush EJ, Lowe JB, von Andrian UH. 2002. Total body irradiation causes profound changes in endothelial traffic molecules for hematopoietic progenitor cell recruitment to bone marrow. *Blood* 99:4182–91

Mazo IB, von Andrian UH. 1999. Adhesion and homing of blood-borne cells in bone marrow microvessels. *J. Leukoc. Biol.* 66:25–32

Mempel TR, Henrickson SE, von Andrian UH. 2004. T cell priming by dendritic cells in lymph nodes occurs in three distinct phases. *Nature* 427:154–59

Michalet X, Pinaud FF, Bentolila LA, Tsay JM, Doose S, et al. 2005. Quantum dots for live cells, in vivo imaging, and diagnostics. *Science* 307:538–44

Miller DH, Khan OA, Sheremata WA, Blumhardt LD, Rice GP, et al. 2003a. A controlled trial of natalizumab for relapsing multiple sclerosis. *N. Engl. J. Med.* 348:15–23

Miller MJ, Hejazi AS, Wei SH, Cahalan MD, Parker I. 2004a. T cell repertoire scanning is promoted by dynamic dendritic cell behavior and random T cell motility in the lymph node. *Proc. Natl. Acad. Sci. USA* 101:998–1003

Miller MJ, Safrina O, Parker I, Cahalan MD. 2004b. Imaging the single cell dynamics of CD4[+] T cell activation by dendritic cells in lymph nodes. *J. Exp. Med.* 200:847–56

Miller MJ, Wei SH, Cahalan MD, Parker I. 2003b. Autonomous T cell trafficking examined in vivo with intravital two-photon microscopy. *Proc. Natl. Acad. Sci. USA* 100:2604–9

Miller MJ, Wei SH, Parker I, Cahalan MD. 2002. Two-photon imaging of lymphocyte motility and antigen response in intact lymph node. *Science* 296:1869–73

Mora JR, Bono MR, Manjunath N, Weninger W, Cavanagh LL, et al. 2003. Selective imprinting of gut-homing T cells by Peyer's patch dendritic cells. *Nature* 424:88–93

Mora JR, Cheng G, Picarella D, Briskin M, Buchanan N, von Andrian UH. 2005. Reciprocal and dynamic control of CD8 T cell homing by dendritic cells from skin- and gut-associated lymphoid tissues. *J. Exp. Med.* 201:303–16

Nakagawa K, Miller FN, Sims DE, Lentsch AB, Miyazaki M, Edwards MJ. 1996. Mechanisms of interleukin-2-induced hepatic toxicity. *Cancer Res.* 56:507–10

Nilsson SK, Johnston HM, Coverdale JA. 2001. Spatial localization of transplanted hemopoietic stem cells: inferences for the localization of stem cell niches. *Blood* 97:2293–99

Okada T, Ngo VN, Ekland EH, Forster R, Lipp M, et al. 2002. Chemokine requirements for B cell entry to lymph nodes and Peyer's patches. *J. Exp. Med.* 196:65–75

Patterson GH, Lippincott-Schwartz J. 2004. Selective photolabeling of proteins using photoactivatable GFP. *Methods* 32:445–50

Petrie HT. 2003. Cell migration and the control of post-natal T-cell lymphopoiesis in the thymus. *Nat. Rev. Immunol.* 3:859–66

Piccio L, Rossi B, Scarpini E, Laudanna C, Giagulli C, et al. 2002. Molecular mechanisms involved in lymphocyte recruitment in inflamed brain microvessels: critical roles for P-selectin glycoprotein ligand-1 and heterotrimeric G_i-linked receptors. *J. Immunol.* 168:1940–49

Poletaev A, Osipenko L. 2003. General network of natural autoantibodies as immunological homunculus (Immunculus). *Autoimmun. Rev.* 2:264–71

Reinhardt RL, Khoruts A, Merica R, Zell T, Jenkins MK. 2001. Visualizing the generation of memory CD4 T cells in the whole body. *Nature* 410:101–5

Reus WF, Robson MC, Zachary L, Heggers JP. 1984. Acute effects of tobacco smoking on blood flow in the cutaneous micro-circulation. *Br. J. Plast. Surg.* 37:213–15

Richie LI, Ebert PJ, Wu LC, Krummel MF, Owen JJ, Davis MM. 2002. Imaging synapse formation during thymocyte selection: inability of CD3zeta to form a stable central accumulation during negative selection. *Immunity* 16:595–606

Saetzler RK, Jallo J, Lehr HA, Philips CM, Vasthare U, et al. 1997. Intravital fluorescence microscopy: impact of light-induced phototoxicity on adhesion of fluorescently labeled leukocytes. *J. Histochem. Cytochem.* 45:505–13

Sallusto F, Geginat J, Lanzavecchia A. 2004. Central memory and effector memory T cell subsets: function, generation, and maintenance. *Annu. Rev. Immunol.* 22:745–63

Scimone ML, Felbinger TW, Mazo IB, Stein JV, von Andrian UH, Weninger W. 2004. CXCL12 mediates CCR7-independent homing of central memory cells, but not naive T cells, in peripheral lymph nodes. *J. Exp. Med.* 199:1113–20

Shinkura R, Matsuda F, Sakiyama T, Tsubata T, Hiai H, et al. 1996. Defects of somatic hypermutation and class switching in alymphoplasia (*aly*) mutant mice. *Int. Immunol.* 8:1067–75

Singbartl K, Thatte J, Smith ML, Wethmar K, Day K, Ley K. 2001. A CD2-green fluorescence protein-transgenic mouse reveals very late antigen-4-dependent CD8$^+$ lymphocyte rolling in inflamed venules. *J. Immunol.* 166:7520–26

Sixt M, Kanazawa N, Selg M, Samson T, Roos G, et al. 2005. The conduit system transports soluble antigens from the afferent lymph to resident dendritic cells in the T cell area of the lymph node. *Immunity* 22:19–29

Spalding DM, Williamson SI, Koopman WJ, McGhee JR. 1984. Preferential induction of polyclonal IgA secretion by murine Peyer's patch dendritic cell-T cell mixtures. *J. Exp. Med.* 160:941–46

Springer TA. 1994. Traffic signals for lymphocyte recirculation and leukocyte emigration: the multi-step paradigm. *Cell* 76:301–14

Steinman RM, Hawiger D, Nussenzweig MC. 2003. Tolerogenic dendritic cells. *Annu. Rev. Immunol.* 21:685–711

Stoll S, Delon J, Brotz TM, Germain RN. 2002. Dynamic imaging of T cell-dendritic cell interactions in lymph nodes. *Science* 296:1873–76

Sumen C, Mempel TR, Mazo IB, Von Andrian UH. 2004. Intravital microscopy; visualizing immunity in context. *Immunity* 21:315–29

Tietz W, Allemand Y, Borges E, von Laer D, Hallmann R, et al. 1998. CD4$^+$ T cells migrate into inflamed skin only if they express ligands for E- and P-selectin. *J. Immunol.* 161:963–70

Tokoyoda K, Egawa T, Sugiyama T, Choi BI, Nagasawa T. 2004. Cellular niches controlling B lymphocyte behavior within bone marrow during development. *Immunity* 20:707–18

Tokuyama Y, Tokuyama H. 1996. Retinoids as Ig isotype-switch modulators. *Cell. Immunol.* 170:230–34

Vajkoczy P, Laschinger M, Engelhardt B. 2001. Alpha4-integrin-VCAM-1 binding mediates G protein-independent capture of encephalitogenic T cell blasts to CNS white matter microvessels. *J. Clin. Invest.* 108:557–65

Verkhusha VV, Lukyanov KA. 2004. The molecular properties and applications of Anthozoa fluorescent proteins and chromoproteins. *Nat. Biotechnol.* 22:289–96

von Andrian UH. 1996. Intravital microscopy of the peripheral lymph node microcirculation in mice. *Microcirculation* 3:287–300

von Andrian UH, Mackay CR. 2000. T-cell function and migration. Two sides of the same coin. *N. Engl. J. Med.* 343:1020–34

von Andrian UH, Mempel TR. 2003. Homing and cellular traffic in lymph nodes. *Nat. Rev. Immunol.* 3:867–78

Vugmeyster Y, Kikuchi T, Lowes MA, Chamian F, Kagen M, et al. 2004. Efalizumab (anti-CD11a)-induced increase in peripheral blood leukocytes in psoriasis patients is preferentially mediated by altered trafficking of memory CD8$^+$ T cells into lesional skin. *Clin. Immunol.* 113:38–46

Wagner R. 1839. *Erläuterungstafeln zur Physiologie und Entwicklungsgeschichte.* Leipzig: Voss

Weninger W, Carlsen HS, Goodarzi M, Moazed F, Crowley MA, et al. 2003. Naïve T cell recruitment to non-lymphoid tissues: a role for endothelium-expressed CCL21 in autoimmune disease and lymphoid neogenesis. *J. Immunol.* 170:4638–48

Weninger W, Crowley MA, Manjunath N, von Andrian UH. 2001. Migratory properties of naive, effector, and memory CD8$^+$ T cells. *J. Exp. Med.* 194:953–66

Weninger W, Ulfman LH, Cheng G, Souchkova N, Quackenbush EJ, et al. 2000. Specialized contributions by alpha$_{1,3}$-fucosyltransferase-IV and FucT-VII during leukocyte rolling in dermal microvessels. *Immunity* 12:665–76

Yu W, Nagaoka H, Jankovic M, Misulovin Z, Suh H, et al. 1999. Continued RAG expression in late stages of B cell development and no apparent re-induction after immunization. *Nature* 400:682–87

Yuan F, Salehi HA, Boucher Y, Vasthare US, Tuma RF, Jain RK. 1994. Vascular permeability and microcirculation of gliomas and mammary carcinomas transplanted in rat and mouse cranial windows. *Cancer Res.* 54:4564–68

Zhang J, Campbell RE, Ting AY, Tsien RY. 2002. Creating new fluorescent probes for cell biology. *Nat. Rev. Mol. Cell. Biol.* 3:906–18

Zipfel WR, Williams RM, Webb WW. 2003. Nonlinear magic: multiphoton microscopy in the biosciences. *Nat. Biotechnol.* 21:1369–77

Stem Cell Niche:
Structure and Function

Linheng Li and Ting Xie

Stowers Institute for Medical Research, Kansas City, Missouri 64110;
email: lil@stowers-institute.org, tgx@stowers-institute.org

Annu. Rev. Cell Dev. Biol.
2005. 21:605–31

First published online as a
Review in Advance on
July 1, 2005

The *Annual Review of
Cell and Developmental
Biology* is online at
http://cellbio.annualreviews.org

doi: 10.1146/
annurev.cellbio.21.012704.131525

Copyright © 2005 by
Annual Reviews. All rights
reserved

1081-0706/05/1110-
0605$20.00

Key Words

adult stem cell, self-renewal, differentiation, multipotentiality,
signaling

Abstract

Adult tissue-specific stem cells have the capacity to self-renew
and generate functional differentiated cells that replenish lost cells
throughout an organism's lifetime. Studies on stem cells from diverse systems have shown that stem cell function is controlled by
extracellular cues from the niche and by intrinsic genetic programs
within the stem cell. Here, we review the remarkable progress recently made in research regarding the stem cell niche. We compare
the differences and commonalities of different stem cell niches in
Drosophila ovary/testis and *Caenorhabditis elegans* distal tip, as well as
in mammalian bone marrow, skin/hair follicle, intestine, brain, and
testis. On the basis of this comparison, we summarize the common
features, structure, and functions of the stem cell niche and highlight
important niche signals that are conserved from *Drosophila* to mammals. We hope this comparative summary defines the basic elements
of the stem cell niche, providing guiding principles for identification
of the niche in other systems and pointing to areas for future studies.

Contents

INTRODUCTION

Stem Cell Behavior is Regulated by Both Extrinsic Signals and Intrinsic Programs

Stem cells are a subset of cells that have the unique ability to replenish themselves through self-renewal and the potential to differentiate into different types of mature cells. These characteristics therefore play essential roles in organogenesis during embryonic development and tissue regeneration. There are two main types of stem cells: embryonic and adult. The pluripotent embryonic stem cell is derived from the inner cell mass of blastocysts and has the ability to give rise to all three embryonic germ layers—ectoderm, endoderm, and mesoderm (Chambers & Smith 2004, Thomson et al. 1998). As development proceeds, the need for organogenesis arises,

and the embryo proper forms germ line stem cells (GSCs) for reproduction and somatic stem cells (SSCs) for organogenesis. Although diversified, GSCs and SSCs retain the feature of self-renewal. They either are progressively restricted in development, giving rise to multiple lineages (including tissue-specific cells), or are unipotent, giving rise to single lineage cells destined for certain tissues (Fuchs et al. 2004, Rossant 2004, Weissman 2000). After birth, adult stem cells, including both GSCs and SSCs, reside in a special microenvironment termed the "niche," which varies in nature and location depending on the tissue type. These adult stem cells are an essential component of tissue homeostasis; they support ongoing tissue regeneration, replacing cells lost due to natural cell death (apoptosis) or injury. To sustain this function throughout the organism's life span, a delicate balance between self-renewal and differentiation must be maintained. The underlying mechanisms that control this delicate balance are fundamental to understanding stem cell regulation, the nature of cancer/tumor formation, and the therapeutic use of stem cells in human disease.

There are various intrinsic programs that control stem cell self-renewal and potency (Morrison et al. 1997). For example, HoxB4 is sufficient to induce and expand hematopoietic stem cells (HSCs) when introduced into embryonic stem cells (Kyba et al. 2002, Sauvageau et al. 1995). Bmi, a member of the polycomb family, is required for self-renewal of stem cells in the hematopoietic and neural systems (Lessard & Sauvageau 2003, Molofsky et al. 2003, Park et al. 2003). To ensure appropriate control of stem cell behavior, these intrinsic genetic programs must be subject to environmental regulation. This is supported by many studies, some of which are discussed later. Therefore, both environmental regulatory signals and intrinsic programs are required to maintain stem cell properties and to direct stem cell proliferation and differentiation.

The Hypothesis of and Evidence for the Stem Cell Niche

In 1978, Schofield proposed the "niche" hypothesis to describe the physiologically limited microenvironment that supports stem cells (Schofield 1978). The niche hypothesis has been supported by a variety of coculture experiments in vitro and by bone marrow transplantation, in which the niche is first "emptied" through irradiation or drug treatments (Brinster & Zimmermann 1994, Dexter et al. 1977, Moore et al. 1997, Rios & Williams 1990, Roecklein & Torok-Storb 1995, Sitnicka et al. 1996). However, these studies did not resolve the issue of the exact stem cell location and niche structure in vivo (Simmons et al. 2001, Verfaillie et al. 1999).

Although locating and further identifying stem cell niches in mammals has been dif-

ficult owing to their extremely complicated anatomic structures, studies regarding stem cells and their location/niche in other genetic model systems, including those of *Drosophila* and *Caenorhabditis elegans*, have been fruitful. In *Drosophila*, GSCs were located in the anterior region of ovary germarium on the basis of lineage tracing and laser ablation (Lin & Spradling 1993, Wieschaus & Szabad 1979). In 2000, the germarial tip adjacent to GSCs was defined as the niche supporting GSCs in the *Drosophila* ovary (Xie & Spradling 2000), whereas the hub, located at the tip of *Drosophila* testis, served this function in testis (Kiger et al. 2001, Tulina & Matunis 2001). In *C. elegans*, a distal tip cell (DTC) located at the tip of the germ line organization region was found to function as the niche in supporting GSCs (Crittenden et al. 2002).

In mammals, the epithelial stem cell location was successfully identified in the bulge area of hair follicles, and the intestinal stem cell location was identified near the crypt base. These were based on the adult stem cell's ability to retain the BrdU or ^{3}H-thymidine labels (Cotsarelis et al. 1990, Potten et al. 2002). Recently, there has been significant progress regarding stem cells and their surrounding microenvironments in a variety of mammalian models. In 2003, two independent, simultaneous studies using genetic mutant mouse models led to the identification of osteoblastic cells, primarily those lining the trabecular bone surface, as the key component of the HSC niche (Calvi et al. 2003, Zhang et al. 2003). In the neural system, the stem cell niche was found in endothelial cells located at the base of the subventricular zone (SVZ) and subgranular zone (SGZ) (Doetsch et al. 1999, Palmer et al. 1997, Shen et al. 2004).

Historically, "niche" is generally used to describe the stem cell location. In our view, however, "niche" is composed of the cellular components of the microenvironment surrounding stem cells as well as the signals emanating from the support cells. In this review, we summarize the research defining the stem cell niche in *Drosophila* and mammals;

compare the differences and commonalities of stem cell niches in these different systems; and use this information to define the basic features, structures, and functions of the stem cell niche.

STEM CELL NICHES IN *DROSOPHILA* OVARY AND TESTIS

Germ Line Stem Cell and Somatic Stem Cell Niches in the *Drosophila* Ovary

Two or three GSCs are located at the tip of the ovariole in the structure referred to as the germarium. These GSCs are surrounded by three types of somatic cells: terminal filament, cap cells, and inner germarial sheath (IGS) cells (**Figure 1**). The stem cells are easily identified by their direct contact with cap cells and the presence of a spectrosome (Lin 2002, Xie & Spradling 2001). Normally, a GSC divides to generate two daughter cells: one daughter that stays in association with cap cells and another daughter that moves away from the cap cells to form a cystoblast, which eventually becomes, through incomplete cytokinesis, an interconnected 16-cell cyst. Genetic and cell biological studies demonstrate that cap cells are the niche for GSCs (Xie & Spradling 2000). The anchorage of GSCs to cap cells through E-cadherin-mediated cell adhesion is essential for maintaining GSCs (Song & Xie 2002). Also, the number of GSCs correlates with the number of cap cells (Xie & Spradling 2000). Finally, cap cells express genes, such as *dpp*, *gbb*, *hh*, *piwi*, and *Yb*, that are known to be important for maintaining GSCs (Cox et al. 2000, King et al. 2001, Song et al. 2004, Xie & Spradling 1998, 2000) (**Figure 1**).

BMP-, Hh-, and Piwi-mediated signaling pathways play an important role in the control of ovarian GSC self-renewal (**Figure 1**). Two BMP-like genes, *dpp* and *gbb*, are expressed in niche cells, and GSCs mutant for *dpp*, *gbb*, and their downstream components

are lost prematurely (Song et al. 2004, Xie & Spradling 1998, 2000). Dpp overexpression completely prevents GSC differentiation and thereby causes GSC-like tumor formation (Song et al. 2004, Xie & Spradling 1998). BMP signaling was recently shown to exert control of GSC self-renewal by repressing expression of *bam* (Chen & McKearin 2003, Song et al. 2004), which is necessary and sufficient for cystoblast differentiation (Ohlstein & McKearin 1997).

Piwi- and Yb-mediated signaling is also required for controlling ovarian GSC self-renewal (Cox et al. 2000, King et al. 2001, Lin & Spradling 1997). Interestingly, Yb regulates expression of *piwi* and *hh* in TF/cap cells; these genes in turn control GSC self-renewal (King et al. 2001). Yb-mediated signaling is also involved in repressing *bam* expression in GSCs (Chen & McKearin 2005, Szakmary et al. 2005). It would be interesting to know the relationship between BMP signaling and Piwi-mediated signaling in controlling GSC self-renewal. Zero population growth (a *Drosophila* homolog of mammalian innexin-4) is expressed in GSCs and is also required for GSC maintenance, although the underlying molecular mechanism for such maintenance is largely unknown (Gilboa et al. 2003, Tazuke et al. 2002).

Two or three SSCs located in the middle of the germarium are responsible for generating somatic follicle and stalk cells (**Figure 1**). The follicle cells encapsulate 16-cell cysts, whereas the stalk cells connect adjacent egg chambers. Although the ovarian SSCs lack a unique marker, they can be identified using lineage tracing (Margolis & Spradling 1995, Song & Xie 2002, 2003, Zhang & Kalderon 2001). SSCs have low levels of Fasciclin III (Fas 3) expression, whereas differentiated follicle cells have high levels of Fas 3 expression. Loss of adhesion between SSCs and IGS jeopardizes SSC self-renewal, suggesting that the proximal IGS cells are at least a part of the SSC niche, anchoring the SSCs (Song et al. 2002). Although cap cells are not physically associated with SSCs, they produce two

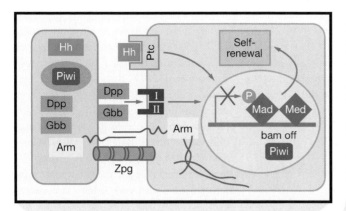

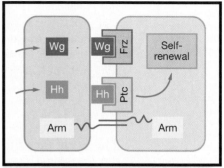

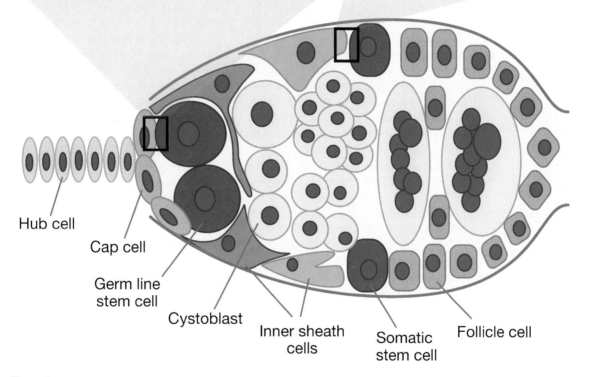

Hub cell

Cap cell

Germ line
stem cell

Cystoblast

Inner sheath
cells

Somatic
stem cell

Follicle cell

Figure 1

Drosophila germarium cross section showing the locations of germ line stem cells (GSCs), somatic stem
cells (SSCs), and their niches. Two or three GSCs (*red* cells, *left*) are situated in their niche, composed of
cap cells (*green* cells, *left*) and terminal filament cells (*light blue* cells, *left tip*), whereas their differentiated
progeny, including cystoblasts and differentiated cysts (*yellow* cells, *middle*), are surrounded by inner
sheath cells (*purple* cells and *green* cells, *bottom* and *top*). Two or three SSCs (*red* cells, *bottom* and *top*)
directly contact the posterior group of inner sheath cells (*green* cells, *bottom* and *top*) forming their niche,
whereas their differentiated progeny, also known as follicle progenitor cells (*orange* cells on *right*), further
proliferate and generate differentiated follicle cells. Two inserts depict major signaling pathways
controlling GSC (*top* and *left*) and SSC (*top* and *right*) self-renewal and proliferation; these inserts also
depict niche cells (*green*) and stem cells (*pink*).

diffusible growth factors, Hh and Wg, that are required for controlling SSC maintenance and proliferation (Forbes et al. 1996, King et al. 2001, Song & Xie 2003). This supports the hypothesis that these cap cells are also a part of the SSC niche (Forbes et al. 1996; King et al. 2001, Song & Xie 2003, Zhang & Kalderon 2001).

The Germ Line Stem Cell Niche in the *Drosophila* Testis

In the apical tip of the *Drosophila* testis, two types of stem cells, GSCs and SSCs (the latter are also known as cyst progenitor cells), are responsible for producing differentiated germ cells and somatic cyst cells, respectively (Fuller 1993, Kiger et al. 2001) (**Figure 2**). Seven to nine GSCs, each containing a spectrosome, are attached to the hub (Hardy et al. 1979, Lindsley & Tokuyasu 1980, Yamashita et al. 2003). A male GSC divides asymmetrically, giving rise to one stem cell that remains in contact with the hub and one gonialblast that moves away from the hub and differentiates (Hardy et al. 1979, Lindsley & Tokuyasu 1980, Yamashita et al. 2003). As a GSC divides to produce a gonialblast, the neighboring SSCs also divide to generate two cyst cells, which envelop the gonialblast. This process leads to production of 64 sperm (Gonczy & DiNardo 1996, Hardy et al. 1979). The hub generates signals, including Unpaired (Upd) and BMP, to control GSC self-renewal (Kawase et al. 2004, Kiger et al. 2001, Shivdasani & Ingham 2003, Tulina & Matunis 2001) (**Figure 2**).

Upd from the hub activates the JAK-STAT pathway in GSCs and promotes their self-renewal (Kiger et al. 2001, Tulina & Matunis 2001). Additionally, the activation of JAK-STAT signaling can reprogram mitotic germ cysts into GSCs (Brawley & Matunis 2004). As in the ovary, BMP signaling is required for controlling GSC self-renewal in the testis (Kawase et al. 2004, Schulz et al. 2004,

Shivdasani & Ingham 2003). Hub cells and somatic cyst cells express *gbb* at high levels and *dpp* at much lower levels; consequently, BMP downstream components are essential for controlling testicular GSC self-renewal (Kawase et al. 2004). Because *dpp* overexpression fails to suppress completely spermatogonial cell differentiation, BMP signaling likely plays a permissive role in controlling male GSC self-renewal. BMP and JAK-STAT signaling pathways are required for controlling male GSC self-renewal; thus, they must somehow interact with each other. The integration between these two pathways in male GSCs is an important area in need of future exploration.

Gonialblast differentiation is tightly controlled by unknown signals from SSCs and somatic cyst cells (Kiger et al. 2001, Tran et al. 2000). In somatic cells mutant for *Egfr* and *raf*, GSC- and gonialblast-like single germ cells are greatly increased in number and remain active longer than do wild type cells.

One mechanism ensuring that only one of the two stem cell daughters self-renews is control of the spindle orientation of the stem cell so as to place one self-renewing daughter in the niche and the other daughter destined to differentiate outside the niche (**Figure 2**). Cnn and APC1, centrosomal components in GSCs, control orientation of the spindle perpendicular to the hub. Mutation in these components leads to an increase in GSC number and subsequent crowding in the niche. APC2, which is concentrated at the junction between GSCs and hub cells, also controls correct GSC spindle orientation (Yamashita et al. 2003).

THE GERM LINE STEM CELL NICHE IN *C. ELEGANS*

In the *C. elegans* hermaphrodite gonad, only the 225 germ cells closest to the distal tip cell (DTC) are mitotic; those further proximal are arrested in meiotic pachytene (Crittenden et al. 1994) (**Figure 3**). Specific stem cells within the mitotic region have not been

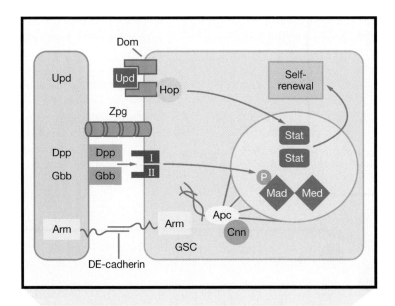

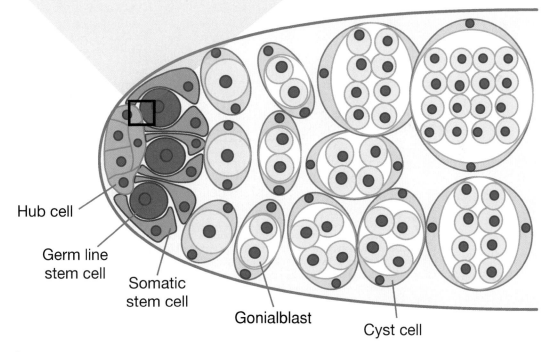

Hub cell

Germ line
stem cell

Somatic
stem cell

Gonialblast

Cyst cell

Figure 2

Cross section of the apical tip of the *Drosophila* testis, showing the locations of germ line stem cells
(GSCs), somatic stem cells (SSCs), and their niches. Hub cells (*green*) at the apical tip of the testis form
niches for both GSCs (*red*) and SSCs (*gray, left*), which generate, respectively, spermatogonial cells
(*yellow*) and somatic cyst cells (*light gray*) encapsulating differentiated spermatogonial cells. The insert on
top describes major signaling pathways involved in communication between GSCs and the niche cells for
controlling self-renewal and proliferation.

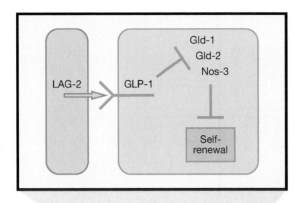

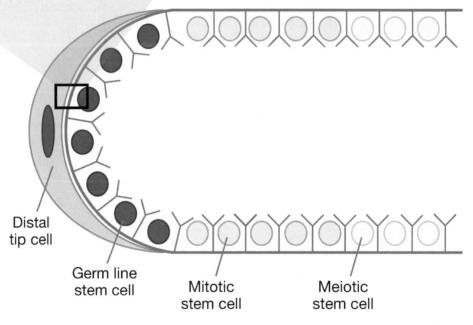

Figure 3

Cross section of the *C. elegans* hermaphrodite gonad. The putative germ line stem cells (GSCs) (*red*) are directly associated with their distal tip cell (DTC) niche cell (*green*), whereas their differentiated progeny (*light yellow*) move away from the DTC, progressing from the mitotic phase to the meiotic phase. The GLP-1 (Notch-like) signaling pathway is involved in communication between the DTC and GSCs and represses functions of differentiation-promoting gene products, such as Gld-1, Gld-2, and Nos-3, which regulate entry into meiosis (*insert*).

identified. The somatic DTC is required for maintaining these cells in mitosis (Kimble & White 1981). Although mitotic and meiotic germ cells in the tube share a central core of cytoplasm, only those mitotic germ cells located at the most distal tip (i.e., GSCs) adjacent to the DTC behave like stem cells, capable of self-renewing and generating differentiated gametes. The proximal mitotic neighbors behave more like transient amplifying cell populations, described in other systems. As germ cells move further away from the DTC, they terminate their mitotic activities and commit meiosis. Only those germ cells that physically interact with the DTC maintain their GSC identity; thus, the signal from the DTC either must be short ranging or mediated by a direct cell-cell interaction.

Signaling from the DTC to control GSC self-renewal is through a Notch-like cascade. The mitotic germ cells express the Notch-type receptor, GLP-1, which is activated by the Delta-like signal from DTC, LAG-2 (Crittenden et al. 1994, Henderson et al. 1994). Constitutive GLP-1 activity downregulates the meiosis-promoting genes *gld-1*, *gld-2*, and *nos-3* and thereby causes expansion of germ cell numbers (Berry et al. 1997). Because individual stem cells have not been identified in the mitotic region, it remains unclear whether they are maintained through a population mechanism or an asymmetric division mechanism.

KNOWN STEM CELL NICHES IN MAMMALIAN SYSTEMS

The stem cell and the niche hypothesis, first developed in the hematopoietic system in mammals, has provided the conceptual background for stem cell studies in *Drosophila* and *C. elegans* (Schofield 1978, Weissman 1994). Conversely, studies in *Drosophila* on the molecular pathways controlling the stem cell niche have provided important insight into identification of the stem cell niche in mammalian systems (Lin 2002, Spradling et al. 2001). In this section, we describe and compare the location and physical organization (if known) of adult stem cells in bone marrow, skin/hair follicle, intestine, neuron, and testis.

The Hematopoietic Stem Cell Niche

Bone marrow serves as the pioneer system for studying stem cells; the concept and basic features of stem cells were defined from studying hematopoietic stem cells (HSCs) (Orkin 2000, Till & McCulloch 1961, Weissman et al. 2001). However, the way in which HSCs interact with their local environment to pro-

mote stem cell maintenance has not been clear. Most studies of HSCs have examined their behavior in cell populations obtained from their natural niche in the bone marrow. Thus far, however, only limited culture systems exist that allow sustained maintenance and expansion of HSCs in vitro, attesting to the importance of as-yet poorly defined interactions in the bone marrow niche. Two independent studies recently 1) identified a subset of osteoblastic cells (N-cadherin$^+$CD45$^-$) to which HSCs physically attach in the bone marrow, 2) identified an N-cadherin/β-catenin adherens complex between HSCs and osteoblastic cells, 3) showed that Jagged1, generated from osteoblasts, influences HSCs by signaling through the Notch receptor, and 4) demonstrated that the number of N-cadherin$^+$ osteoblastic lining cells controls the number of HSCs (Calvi et al. 2003, Zhang et al. 2003). Homing studies to trace the location of GFP-labeled HSCs after transplantation also pointed to the endosteal surface as a possible stem cell niche (Nilsson et al. 2001). In vitro coculture of HSCs with osteoblasts can expand the HSC population (Taichman & Emerson 1998), and depletion of osteoblasts leads to loss of hematopoietic tissue (Visnjic et al. 2004). In addition, N-cadherin is a key target of Angiopoietin-1 (Ang-1)/Tie-2 signaling that maintains HSC quiescence (Arai et al. 2004) (**Figure 4**).

A primary function of the niche is to anchor stem cells. In addition to N-cadherin, other types of adhesion molecules, including integrin, play an important role in the microenvironment/stem cell interaction (Simmons et al. 1997). Stromal cell-derived factor-1 (SDF-1) and its receptor CXCR4 are involved in homing of HSCs (Lapidot & Kollet 2002) (**Figure 4**).

Although the analysis of the signals generated by the niche has just begun, gene expression profiling studies of HSCs have revealed which signals HSCs potentially receive from the niche. The components of evolutionally conserved and developmentally regulated pathways are prominent in stem cells and are indeed involved in the regulation of stem cell self-renewal or maintenance. These components include the Shh, Wnt, Notch, and TGF-β/BMP pathways (Akashi et al. 2003, Gomes et al. 2002, Ivanova et al. 2002, Park et al. 2002, Ramalho-Santos et al. 2002).

For example, the Wnt/β-catenin pathway is important for self-renewal of HSCs (Reya et al. 2003). The Notch pathway is required for maintaining HSCs in the undifferentiated state (Calvi et al. 2003, Duncan et al. 2005, Li et al. 1998, Varnum-Finney et al. 2000).

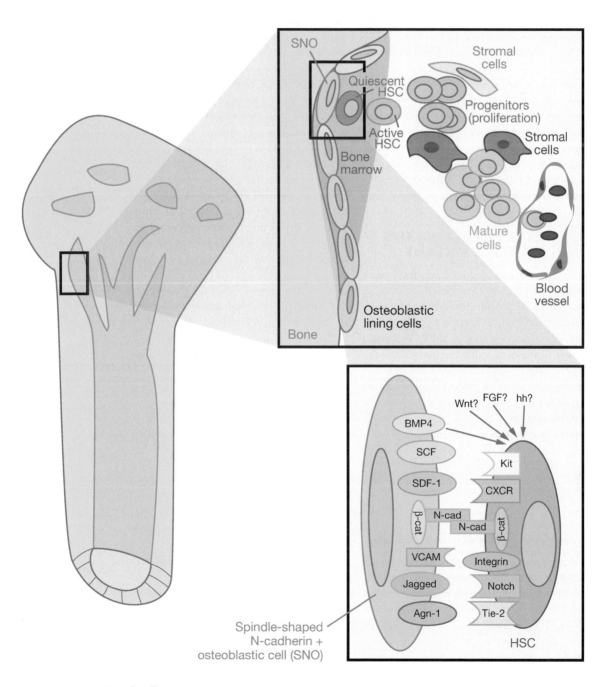

The BMP signal plays a role in control of HSC number (Zhang et al. 2003). The Shh signal mediated by the BMP pathway is able to maintain stem cells in vitro (Bhatia et al. 1999). (**Figure 4**).

The Epithelial Stem Cell Niche in Skin

Skin, with its appendix hair follicle structure, has well-organized architecture (**Figure 5**) and provides an excellent system for studying the molecular mechanisms that regulate stem cell self-renewal, proliferation, migration, and lineage commitment (Fuchs & Segre 2000). Each hair follicle is composed of a permanent portion, which includes sebaceous glands and the underlying bulge area, and a dynamic renewing portion, which undergoes cycles of anagen phase (a period of active growth), catagen phase (apoptosis-driven retraction), and telogen phase (a short period of rest) (Hardy 1992). The bulge area functions as a niche, where epithelial stem cells (Niemann & Watt 2002) are located and maintained (Cotsarelis et al. 1990, Sun et al. 1991). Epithelial stem cells are multipotent, giving rise to daughter cells that either migrate upward to serve as epidermal progenitors for generating epidermal cells during wound repair or migrate downward to convert to hair-matrix progenitors, which further give rise to the hair shaft (Niemann & Watt 2002, Oshima et al. 2001, Taylor et al. 2000).

During the early anagen phase, the dermal papilla region may provide the dynamic signals that activate stem cells; however, the cellular components of the niche in the bulge are yet to be defined other than as stem cells per se. The dermal sheath derived from mesenchymal cells adjacent to the epithelial stem cells in the bulge area most likely provides the niche function. The recent identification of markers for epithelial stem cells, such as CD34, will be helpful in further identifying the adjacent niche cells and the related niche structures, including adhesion molecules (i.e., $\alpha6$ integrin) (Blanpain et al. 2004).

Recent studies showed that label-retaining cells can regenerate the entire HF structure in transplantation experiments, thus demonstrating that these cells are bona fide epidermal stem cells (Blanpain et al. 2004, Braun et al. 2003). Molecular analysis of epithelial stem cells has revealed the following features: 1) the expression of adhesion molecules known to be involved in stem cell-niche interaction, 2) the presence of growth inhibition factors such as TGFβ/BMP molecules and cell cycle inhibitors, and 3) the components of Wnt signaling pathways, including receptors and inhibitors such as Dkk, sFRP, and WIF. Taken together, these molecular features indicate that the epithelial stem cell niche is a growth- and differentiation-restricted environment (Tumbar et al. 2004). This conclusion is, in general, consistent with the many previous studies that have used genetic

---

Figure 4

Illustration of the hematopoietic stem cell (HSC) niche. The HSC niche is located primarily on the surface of trabecular bone, where a small subset of spindle-shaped N-cadherin-positive osteoblastic cells (indicated as SNO cells) are the key component of the HSC niche. N-cadherin and β-catenin form an adherens complex at the interface between stem cells and niche cells, assisting stem cells in attaching to the niche. Multiple growth factors and cytokines are involved in stem-niche interaction. These include SCF/Kit, Jagged/Notch, SDF-1/CXCR4, and Ang1/Tie2. BMP4 is expressed in osteoblastic cells, but the type of receptor expressed in HSCs is unknown. The Wnt signal is important for stem cell self-renewal, but the Wnts present in the niche are unknown. The same is true for FGF and hedgehog. In vitro data suggest they affect HSC behavior; however, whether they are present as niche signals is unknown. Different types of stromal cells (illustrated as different colors and shapes) may regulate stem cell activation, proliferation, and differentiation by secreting different microenvironmental signals. Finally, maturated blood cells migrate and infiltrate into blood vessel.

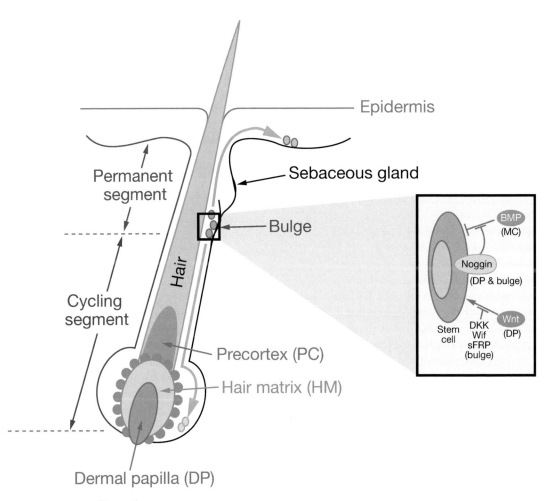

Figure 5

Illustration of the epidermal stem cells. Stem cells are located in the bulge region of the hair follicle beneath the sebaceous gland. Upon activation, stem cells undergo division; the daughter cells retained in the bulge remain as stem cells while other daughter cells migrate down to become hair-matrix progenitors responsible for hair regeneration. In neonatal mice or in damaged skin, stem cells can also migrate upward and convert to epidermal progenitors that replenish lost or damaged epidermis. The bulge area is an environment that restricts cell growth and differentiation by expressing Wnt inhibitors, including DKK, Wif, and sFRP as well as BMPs. During the early anagen phase, Wnts from dermal papilla (DP) and Noggin, which is derived from both DP and bulge (J. Zhang & L. Li, unpublished data), coordinate to overcome the restriction signals imposed by both BMPs and Wnt inhibitors; this leads to stem cell activation and subsequent hair regeneration. The FGF and Notch pathways are also involved in DP function for hair-matrix cell proliferation and lineage fate determination, but their influence on stem cells is not clear.

targeting and transgenic models to reveal that signaling molecules, including Wnts, Notch, and BMPs, have important roles in the regulation of HF development and regeneration (Fuchs et al. 2001, Jones et al. 1995, Lavker et al. 1993, Watt 2001).

Among these various signaling molecules, two family members are prominent, reflecting their important roles in controlling stem cell behavior. One is the Wnt signaling pathway which, through regulating β-catenin activity, controls stem cell activation, fate determination (by favoring HF over epidermal cell lineages), and differentiation (Gat et al. 1998, Huelsken et al. 2001, Merrill et al. 2001, Niemann et al. 2002). The second controlling pathway is the BMP signaling pathway (Hogan 1996). Although it is also required for HF differentiation at a later stage, BMP signaling, as opposed to Wnt signaling, restricts the activation of stem cells and favors epidermal cell fate (Botchkarev 2001, 2003, Kulessa et al. 2000). These observations also support the theory that coordination between Wnt and Noggin (through temporarily overriding the BMP restriction on stem cells) is required to initiate each hair growth cycle (Jamora et al. 2003 and J. Zhang & L. Li, unpublished data).

The Intestinal Stem Cell Niche

The intestinal architecture is composed of a sequential array of zones (or compartments) along the villus-crypt axis (**Figure 6**). Intestinal regeneration begins with intestinal stem cells (ISCs), which give rise to four different types of epithelial lineages: columnar enterocytes, mucin-producing goblet cells, Paneth cells, and enteroendocrine cells (Bjerknes & Cheng 1999, Hermiston & Gordon 1995, Winton 2000). ISCs are generally proposed to be located at the fourth or fifth position from the crypt bottom, above the Paneth cells (Booth & Potten 2000, He et al. 2004, Sancho et al. 2004), as evidenced either through a DNA-labeling retaining assay (Booth & Potten 2000, He et al. 2004, Potten et al. 1997, 2002) or through regeneration dynamics using chimeric mouse lines (Winton 2000, Bjerknes & Cheng 1999). A number of molecules—Telomerase, Tcf4, EphB3, P-PTEN, P-Akt, 14-3-3ζ, Noggin, and Musashi-1—are expressed in the proposed ISC position near the crypt base (Batlle et al. 2002, Booth & Potten 2000, He et al. 2004, Korinek et al. 1998, Nishimura et al. 2003). However, a combination of these markers and the cell position is required to locate ISCs more accurately.

During postnatal intestinal regeneration, mesenchymal cells subjacent to epithelial cells play a role in directing epithelial cell proliferation, differentiation, and apoptosis. BMP4, expressing in the ISC-adjacent mesenchymal cells, is one of the putative niche signals (He et al. 2004). However, the type of mesenchymal cells that expresses BMP4 adjacent to the ISCs is yet to be identified. Endothelial cells composed of vascular vessels have also been proposed to provide ISCs with survival signals such as FGF (Paris et al. 2001). Myofibroblasts that are distributed to the surrounding epithelial cells are proposed to be the candidate "niche" supporting ISCs and influencing other epithelial cells (Mills & Gordon 2001).

We have just begun to understand which niche signals regulate self-renewal and maintain the balance between self-renewal and differentiation of ISCs. An increasing number of molecules, including Wnt, BMP, FGF, Notch, and the underlying signal pathways, may play roles in this regard (Brittan & Wright 2002, Roberts et al. 1995, Sancho et al. 2004). Gene expression profiling revealed that Myc-related pathways and the PI3K/Akt pathway are predominantly present in these stem/progenitors (Mills et al. 2002, Stappenbeck et al. 2003). Inappropriate activation of the Wnt/β-catenin, which targets on Myc, results in the development of tumors as a consequence of an overproduction of stem cells (Clevers 2004). In addition, mutations in BMPRIA and its signaling mediator SMAD4 have been found in Juvenile polyposis syndrome (Howe et al. 1998a,b). Recent studies using gene targeting demonstrated that BMP signaling has a role in suppression of Wnt signaling and thereby maintains a balanced control of stem cell activation and self-renewal (Haramis et al. 2004, He et al. 2004). Mechanistically, inhibition of Wnt signaling by the BMP signal involves both the PTEN/PI-3k/Akt pathway and Smad-mediated transcriptional control (Haramis et al. 2004, He et al. 2004)

In summary, Wnt signaling plays a positive role in promoting ISC activation/self-renewal

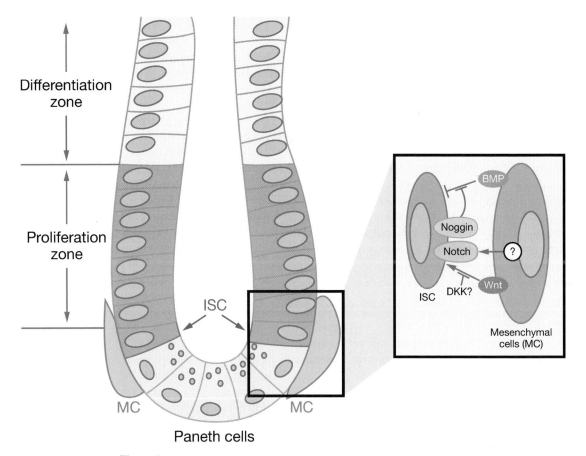

Figure 6

Illustration of the intestinal stem cell (ISC) niche. ISCs (*pink*) are located at the fourth or fifth position above the Paneth cells, as measured from the crypt base. Mesenchymal cells (*green*) adjacent to the ISCs function as the niche. BMP4 expressed from the niche influences the ISCs through its receptor Bmpr1a. Wnt signaling is present throughout the crypt, as revealed by phosphorylated coreceptor LRP6 (He et al. 2004). However, which Wnt receptor is expressed in stem cells is not yet well defined. Whether Wnt inhibitors, such as Dkk, are expressed in the ISC niche also is still unknown. The Notch pathway is known to affect stem/progenitor lineage fate. The expression patterns of the Notch receptor and ligand need to be determined. Noggin expression can be detected in stem cells, but its expression is transient. Noggin is proposed to be a molecular switch coordinating with Wnt signaling to fully activate stem cells by overriding BMP restriction signaling (He et al. 2004).

and crypt cell fate (van de Wetering et al. 2002); in contrast, BMP signaling restricts ISC activation/self-renewal and crypt cell fate (Haramis et al. 2004, He et al. 2004). Importantly, in intestine as well as in HF, overriding the restriction of BMP activity by Noggin as well as by active Wnt signaling is required to fully activate stem cells and support ongoing regeneration (He et al. 2004, Jamora et al. 2003).

The Neural Stem Cell Niche

In the 1990s, studies from a number of research groups led to the identification of neural stem cells (NSCs) (Alvarez-Buylla et al. 1990, Johe et al. 1996, Lois & Alvarez-Buylla 1993, Reynolds & Weiss 1992). NSCs can be isolated from various regions in the adult brain and peripheral nervous system. However, the subventricular zone (SVZ) and the subgranular zone (SGZ) of the hippocampus region are the primary and well-characterized germinal regions in which NSCs reside and support neurogenesis in the adult brain (Doetsch 1999, 2003, Lois & Alvarez-Buylla 1993, Palmer et al. 1997, Temple 2001).

There are four types of cells in the SVZ (**Figure 7**). A layer of ependymal cells lining the lateral ventricle (LV) region separates the SVZ from the LV. SVZ astrocytes are located adjacent to the ependymal cells, with a single cilium structure extending through the boundary of ependymal cells to contact the LV region and to form a glial tunnel that embraces a group of neuroblasts. Immature cells derived from SVZ astrocytes are precursors for neuroblasts. A specialized basal lamina extending from the blood vessels to the ependymal cells contacts all cell types in the SVZ. The SVZ astrocytes, which express astrocyte marker glial fibrillary acidic protein (GFAP), have stem cell features: They undergo self-renewal and give rise to transient amplifying precursor C cells, which further give rise to neuroblasts. Neuroblasts differentiate into neurons that migrate toward the olfactory bulb and other regions. In addition to producing neurons, SVZ astrocytes can also generate oligodendrocytes (Doetsch 2003, Mirescu & Gould 2003, Temple 2001). In the hippocampus, the SGZ is a germinal layer between the hilus and the dentate gyrus, and is responsible for generating dentate gyrus neurons (Palmer et al. 1997). In the SGZ, neurogenesis occurs locally in direct contact with blood vessels. SGZ astrocytes also express GFAP and function as stem cells, undergoing self-renewal and generating daughter cells that further produce granule neurons (**Figure 7**) (Doetsch 2003, Temple 2001).

In both the SVZ and SGZ structures, endothelial cells that form blood vessels and the specialized basal lamina are an essential component of the NSC niche: These endothelial cells provide attachment for SVZ and SGZ astrocytes and generate a variety of signals that control stem cell self-renewal and lineage commitment (Doetsch 2003, Shen et al.

2004). Signals generated from the niche include BMPs and their antagonist Noggin, FGFs, IGF, VEGF, TGFα, and BDNF. The BMP signal favors astrocyte lineage fate by inhibiting neuronal fate. In contrast, Noggin functions to inhibit BMP signaling and thereby favors neurogenesis (Temple 2001). An adherens junction composed of cadherins and β-catenin also plays a role in maintenance of stem cells. Interestingly, overexpression of β-catenin leads to expansion of the NSC population; this presumably reflects activation of Wnt signaling (Chenn & Walsh 2002). This phenotype is very similar to overexpression of IGF in transgenic mice, in which an increased brain size is also observed (Aberg et al. 2003). Both EGF and bFGF are able to expand NSCs in an in vitro culture system. In addition, signaling pathways, including Notch and PTEN/PI3K, are also involved in NSC regulation (Doetsch 2003, Temple 2001).

The Germ Line Stem Cell Niche in Mice

Stem cell transplantation capability, simple anatomy, and genetics make the mouse testis an attractive model for studying GSCs and their niche. The GSCs in mice are single cells that are located in the periphery of seminiferous tubules and that have the ability to self-renew and generate a large number of differentiated gametes (Brinster 2002) (**Figure 8**). GSCs in the mouse testis each divide asymmetrically to generate a GSC and a differentiated daughter, which forms an interconnected A_{pair} spermatogonial cell. The A_{pair} spermatogonial cell then divides synchronously to form a chain of interconnected spermatogonial cells. Stem cells, spermatogonia, spermatocytes, spermatids, and sperm cells can be distinguished by their spatial relation to differentiating sperm cells. GSCs are very rare and can be isolated using fluorescence-activated cell sorting (FACS) as a population of α_v-integrin$^{-/dim}$ α_6-integrin$^+$ Thy-1$^{lo/+}$ C-kit$^-$ cells (Kubota et al. 2003). Sertoli cells,

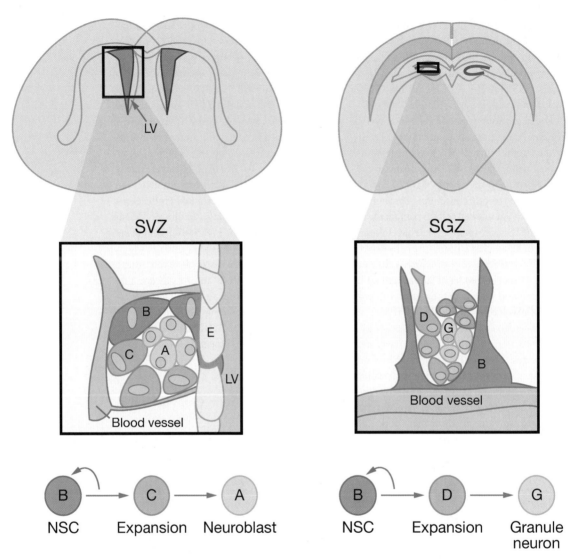

Figure 7

Illustration of the neural stem cell (NSC) niche. The subventricular zone (SVZ) and the subgranular zone (SGZ) are two well-characterized germinal regions in which NSCs (*pink*) are located. In the SVZ, astrocytes (*B*) lining the ependymal cells (*E*) function as NSCs; they give rise to transient amplifying cells (*C*) (*green*), which further produce neuroblast cells (*A*) (*blue*). Endothelial cells in the blood vessel/laminar maintain contact with astrocytes, which regulate NSC self-renewal and proliferation by generating different types of signals. In the SGZ, astrocytes (*B*) directly attach to the blood vessel and receive signals from the endothelial cells that direct NSCs to undergo self-renewal, proliferation (*D*), and differentiation (*G*). The figure is adapted and modified with permission from Doetsch 2003.

the somatic cells of the seminiferous tubules that physically interact with the stem cells, likely constitute functional niches for the stem cells by providing growth factors that promote stem cell self-renewal and/or proliferation.

Several studies support the idea that Sertoli cells regulate the maintenance of the stem cell pool (although little is known about the underlying molecular mechanisms). First, studies in which male GSCs and Sertoli cells

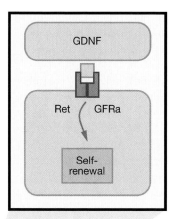

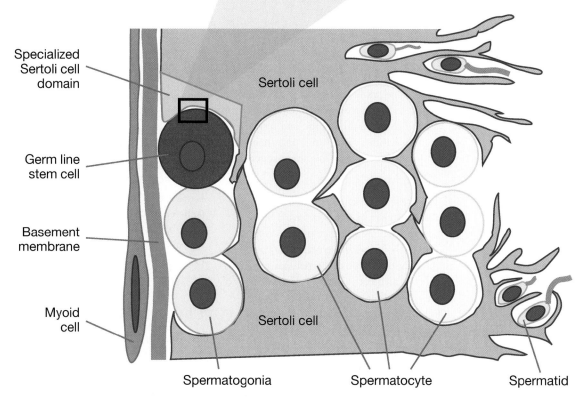

Figure 8

Cross section of a small section of mouse testis. A germ line stem cell (GSC) (*red*) directly contacts a Sertoli cell's (*purple*) basement membrane (*gray*) secreted by myoid cells (*pink*), and specialized region (*green*), which together may form a putative GSC niche. Myoid cells (*pink*) may also participate in niche function, as they are close to GSCs. The differentiated spermatogonial cells (*yellow*) are germ-line cysts that move through different domains formed by Sertoli cells toward the lumen, where mature sperm are released. The GDNF pathway, depicted in the insert (*top*), is a known major pathway for controlling GSC self-renewal in the mouse testis.

are transplanted into infertile mice show that Sertoli cells indeed can support GSC maintenance and spermatogenesis (Ogawa et al. 2000, Shinohara et al. 2000, 2003). Second, GDNF, a member of the TGF-β superfamily produced by Sertoli cells, can control GSC self-renewal and maintain GSCs in vitro (Kanatsu-Shinohara et al. 2004, Kubota et al. 2004). Therefore, Sertoli cells contribute to the function of the GSC niche. Future study is needed to define the physical structure of the GSC niche and its associated signals in the mouse testis.

CONCLUSION AND PROSPECTIVE

Common Features, Structures, and Functions of the Stem Cell Niche

After comparison of the stem cell niches in the ovary and testis of *Drosophila* and in *C. elegans*, as well as in mammalian bone marrow, hair follicle, intestine, brain, and testis, the common features, structures, and functions of the stem cell niche are summarized as follows:

1. The stem cell niche is composed of a group of cells in a special tissue location for the maintenance of stem cells. The niche's overall structure is variable, and different cell types can provide the niche environment. For example, N-cadherin-positive osteoblastic lining cells in the trabecular bone form the niche for HSCs, whereas endothelial cells form the NSC cell niche.

2. The niche functions as a physical anchor for stem cells. E-cadherin-mediated cell adhesion is required for anchoring GSCs and SSCs in *Drosophila*, and N-cadherin may be important for anchoring HSC in the bone marrow niche. Other adhesion molecules, such as integrins, may help anchor stem cells to extracellular matrixes.

3. The niche generates extrinsic factors that control stem cell fate and number. Many signal molecules have been shown to be involved in regulation of stem cell behavior, including hh, Wnts, BMPs, FGFs, Notch, SCF, Ang-1, and LIF or Upd through the JAK-Stat pathway. Among these, the BMP and Wnt signal pathways have emerged as common pathways for controlling stem cell self-renewal and lineage fate from *Drosophila* to mammals. Several pathways can be utilized to control self-renewal of one stem cell type, whereas one growth factor can regulate several different stem cell types. The presence of signaling components of multiple conserved developmental regulatory pathways in stem cells supports the ideas that stem cells retain the ability to respond to these embryonic regulatory signals and that orchestration of these signals is essential for proper regulation of stem cell self-renewal and lineage commitment.

4. In invertebrates and mammals, the stem cell niche exhibits an asymmetric structure. Upon division, one daughter cell is maintained in the niche as a stem cell (self-renewal); the other daughter cell leaves the niche to proliferate and differentiate, eventually becoming a functionally mature cell.

FUTURE DIRECTIONS

Recent studies regarding the stem cell niche in different organisms, including various mammalian organ systems, have resulted in significant progress; fundamental principles about the niche have been established. We hope that the knowledge gained from these studies discussed above will provide guidelines for defining the stem cell niche in other systems. Using a combination of genetic, molecular, and cell biological approaches, several important signaling pathways from the various niches have been identified for their ability to maintain and regulate self-renewal of stem cells. In general, multiple conserved developmental regulatory signals coexist; therefore, orchestration of these signals is essential for proper regulation of stem cell self-renewal and lineage commitment. Further studies of the cross-talk between these signal pathways and the relationship between these pathways and the intrinsic factors required for self-renewal and maintenance of stem cells will provide further insight into the molecular mechanisms governing stem cell self-renewal and differentiation.

Cellular and Molecular Components of the Stem Cell Niche

In uncovering other molecular components of the stem cell niche, genetic screening in *Drosophila* will continue to be an efficient method of identification of novel factors. In mammals, systematic analysis of gene expression in the niche cells (Hackney et al. 2002) will be as important and fruitful as analysis of gene expression in stem cells. For example, systematic analysis of the N-cadherin-positive osteoblastic lining cells, using gene array to compare to other types of marrow stromal cells, including N-cadherin-negative osteoblastic cells, is required to uncover any unique genes predominantly expressed in the HSC niche cells. Furthermore, comparisons of niche- and stem cell–specific gene profiles in different systems will provide important insight into the critical niche signals and intrinsic factors that potentially influence stem cell behavior. Thus, conserved signal molecules and intrinsic factors important for stem cell self-renewal and maintenance and specific factors unique to each stem cell niche can be identified.

Asymmetric Versus Symmetric Stem Cell Division

The stem cell niche exhibits structural asymmetry, and asymmetric division of stem cells is one of the proposed mechanisms controlling the balance between self-renewal and differentiation. This has been well illustrated in the *Drosophila* system. Whether this mechanism is preserved in the mammalian system needs to be determined. The centrosome-associated proteins APC1 and centrosomin are important in controlling spindle orientation during stem cell division in *Drosophila* (Yamashita et al. 2003). It is important to investigate whether control of spindle orientation is essential for asymmetric division of stem cells in other systems as well.

Stem Cell Maintenance and Reversion from Committed Daughter Cells

As described above, asymmetric stem cell division leads to the retention of one daughter cell in the niche (stem cell) and to the other daughter cell leaving the niche to become committed, an irreversible process in the normal physiological condition. Whether the committed daughter cell can revert to a stem cell if restored to the niche is an interesting and important question. Two recent studies in *Drosophila* provide solid evidence indicating that this may be possible (Brawley & Matunis 2004, Kai & Spradling 2004). It remains to be seen whether this is a general feature for different types of stem cells in invertebrates and mammals.

Normal Stem Cells and Cancer Stem Cells: Niche-Dependent or Niche-Independent

The concept of cancer stem cells has changed the perspective on cancer, in which stem cells and their underlying self-renewal is key. In adults, the niche prevents tumorigenesis by controlling stem cells in the arrested state and maintaining the balance between self-renewal and differentiation. In this context, any mutation that leads stem cells to escape from the niche control may result in tumorigenesis. It is therefore reasonable to hypothesize that one of the differences between normal stem cells and cancer stem cells is that cancer stem cells may no longer be dependent on niche signaling. This hypothesis needs to be tested.

CLOSING REMARKS

Stem cell behavior is regulated by coordination of environmental signals and intrinsic programs. Environmental signals are provided by the niche, which is composed of specialized cell populations located in unique topological relationships with the stem cells in different adult tissues. In this review, we compare the differences and commonalities of the niches in a variety of stem cell systems across different species and provide evidence demonstrating the impact of the niche on the homeostatic regulation of stem cells. Dissection of the niche's cellular and molecular components has revealed the basic features and functions of the stem cell niche and will provide important insights for identification of the stem cell niche in different systems. We believe that the ability to reconstitute the stem cell niche in vitro will open a new avenue for maintenance and expansion of adult stem cells. Uncovering the important signals generated by the niche will shed light on the mechanisms that regulate stem cell self-renewal and maintenance of stem cell multi-potentiality. Finally, understanding the interaction between stem cells and their natural partners will substantially benefit therapeutic approaches to human degenerative diseases.

ACKNOWLEDGMENTS

We thank L. Wiedemann for critical editing and D. di Natale for proofreading and manuscript organization. We are grateful for comments from P. Trainor. We apologize to those whose papers are not cited here due to limited space. Our work is supported by the Stowers Institute for Medical Research.

LITERATURE CITED

Aberg MA, Aberg ND, Palmer TD, Alborn AM, Carlsson-Skwirut C, et al. 2003. IGF-I has a direct proliferative effect in adult hippocampal progenitor cells. *Mol. Cell Neurosci.* 24:23–40

Akashi K, He X, Chen J, Iwasaki H, Niu C, et al. 2003. Transcriptional accessibility for genes of multiple tissues and hematopoietic lineages is hierarchically controlled during early hematopoiesis. *Blood* 101:383–89

Alvarez-Buylla A, Kirn JR, Nottebohm F. 1990. Birth of projection neurons in adult avian brain may be related to perceptual or motor learning. *Science* 249:1444–46

Arai F, Hirao A, Ohmura M, Sato H, Matsuoka S, et al. 2004. Tie2/angiopoietin-1 signaling regulates hematopoietic stem cell quiescence in the bone marrow niche. *Cell* 118:149–61

Batlle E, Henderson JT, Beghtel H, van den Born MMW, Sancho E, et al. 2002. Beta-catenin and TCF mediate cell positioning in the intestinal epithelium by controlling the expression of EphB/EphrinB. *Cell* 111:251–63

Berry LW, Westlund B, Schedl T. 1997. Germ-line tumor formation caused by activation of glp-1, a *Caenorhabditis elegans* member of the Notch family of receptors. *Development* 124:925–36

Bhatia M, Bonnet D, Wu D, Murdoch B, Wrana J, et al. 1999. Bone morphogenetic proteins regulate the developmental program of human hematopoietic stem cells. *J. Exp. Med.* 189:1139–48

Bjerknes M, Cheng H. 1999. Clonal analysis of mouse intestinal epithelial progenitors. *Gastroenterology* 116:7–14

Blanpain C, Lowry WE, Geoghegan A, Polak L, Fuchs E. 2004. Self-renewal, multipotency, and the existence of two cell populations within an epithelial stem cell niche. *Cell* 118:635–48

Booth C, Potten CS. 2000. Gut instincts: thoughts on intestinal epithelial stem cells. *J. Clin. Invest.* 105:1493–99

Botchkarev VA. 2003. Bone morphogenetic proteins and their antagonists in skin and hair follicle biology. *J. Invest. Dermatol.* 120:36–47

Botchkarev VA, Botchkareva NV, Nakamura M, Huber O, Funa K, et al. 2001. Noggin is required for induction of the hair follicle growth phase in postnatal skin. *FASEB J.* 15:2205–14

Braun KM, Niemann C, Jensen UB, Sundberg JP, Silva-Vargas V, Watt FM. 2003. Manipulation of stem cell proliferation and lineage commitment: visualization of label-retaining cells in whole mounts of mouse epidermis. *Development* 130:5241–55

Brawley C, Matunis E. 2004. Regeneration of male germline stem cells by spermatogonial dedifferentiation in vivo. *Science* 304:1331–34

Brinster RL. 2002. Germline stem cell transplantation and transgenesis. *Science* 296:2174–76

Brinster RL, Zimmermann JW. 1994. Spermatogenesis following male germ-cell transplantation. *Proc. Natl. Acad. Sci. USA* 91:11298–302

Brittan M, Wright NA. 2002. Gastrointestinal stem cells. *J. Pathol.* 197:492–509

Calvi LM, Adams GB, Weibrecht KW, Weber JM, Olson DP, et al. 2003. Osteoblastic cells regulate the haematopoietic stem cell niche. *Nature* 425:841–46

Chambers I, Smith A. 2004. Self-renewal of teratocarcinoma and embryonic stem cells. *Oncogene* 23:7150–60

Chen D, McKearin D. 2003. Dpp signaling silences bam transcription directly to establish asymmetric divisions of germline stem cells. *Curr. Biol.* 13:1786–91

Chen D, McKearin D. 2005. Gene circuitry controlling a stem cell niche. *Curr. Biol.* 15:179–84

Chenn A, Walsh CA. 2002. Regulation of cerebral cortical size by control of cell cycle exit in neural precursors. *Science* 297:365–69

Clevers H. 2004. At the crossroads of inflammation and cancer. *Cell* 118:671–74

Cotsarelis G, Sun TT, Lavker RM. 1990. Label-retaining cells reside in the bulge area of pilosebaceous unit: implications for follicular stem cells, hair cycle, and skin carcinogenesis. *Cell* 61:1329–37

Cox DN, Chao A, Baker J, Chang L, Qiao D, Lin HF. 1998. A novel class of evolutionarily conserved genes defined by piwi are essential for stem cell self-renewal. *Genes Dev.* 12:3715–27

Cox DN, Chao A, Lin H. 2000. piwi encodes a nucleoplasmic factor whose activity modulates the number and division rate of germline stem cells. *Development* 127:503–14

Crittenden SL, Bernstein DS, Bachorik JL, Thompson BE, Gallegos M, et al. 2002. A conserved RNA-binding protein controls germline stem cells in *Caenorhabditis elegans*. *Nature* 417:660–63

Crittenden SL, Troemel ER, Evans TC, Kimble J. 1994. GLP-1 is localized to the mitotic region of the *C. elegans* germ line. *Development* 120:2901–11

Dexter TM, Moore MA, Sheridan AP. 1977. Maintenance of hemopoietic stem cells and production of differentiated progeny in allogeneic and semiallogeneic bone marrow chimeras in vitro. *J. Exp. Med.* 145:1612–16

Doetsch F. 2003. A niche for adult neural stem cells. *Curr. Opin. Genet. Dev.* 13:543–50

Doetsch F, Caille I, Lim DA, Garcia-Verdugo JM, Alvarez-Buylla A. 1999. Subventricular zone astrocytes are neural stem cells in the adult mammalian brain. *Cell* 97:703–16

Duncan AW, Rattis FM, Dimascio LN, Congdon KL, Pazianos G, et al. 2005. Integration of Notch and Wnt signaling in hematopoietic stem cell maintenance. *Nat. Immunol.* 6:314–22

Forbes AJ, Lin H, Ingham PW, Spradlin AC. 1996. *hedgehog* is required for the proliferation and specification of ovarian somatic cells prior to egg chamber formation in *Drosophila*. *Development* 122:1125–35

Fuchs E, Merrill BJ, Jamora C, DasGupta R. 2001. At the roots of a never-ending cycle. *Dev. Cell* 1:13–25

Fuchs E, Segre JA. 2000. Stem cells: a new lease on life. *Cell* 100:143–55

Fuchs E, Tumbar T, Guasch G. 2004. Socializing with the neighbors: stem cells and their niche. *Cell* 116:769–78

Fuller MT. 1993. Spermatogenesis. In *The Development of Drosophila*, ed. M Bate, A Martinez-Arias, pp. 71–147. Cold Spring Harbor, NY: Cold Spring Harbor Lab. Press

Gat U, DasGupta R, Degenstein L, Fuchs E. 1998. De novo hair follicle morphogenesis and hair tumors in mice expressing a truncated beta-catenin in skin. *Cell* 95:605–14

Gilboa L, Forbes A, Tazuke SI, Fuller MT, Lehmann R. 2003. Germ line stem cell differentiation in Drosophila requires gap junctions and proceeds via an intermediate state. *Development* 130:6625–34

Gomes I, Sharma TT, Edassery S, Fulton N, Mar BG, Westbrook CA. 2002. Novel transcription factors in human CD34 antigen-positive hematopoietic cells. *Blood* 100:107–19

Gonczy P, DiNardo S. 1996. The germ line regulates somatic cyst cell proliferation and fate during Drosophila spermatogenesis. *Development* 122:2437–47

Hackney JA, Charbord P, Brunk BP, Stoeckert CJ, Lemischka IR, Moore KA. 2002. A molecular profile of a hematopoietic stem cell niche. *Proc. Natl. Acad. Sci. USA* 99:13061–66

Haramis AP, Begthel H, van den Born M, van Es J, Jonkheer S, et al. 2004. De novo crypt formation and juvenile polyposis on BMP inhibition in mouse intestine. *Science* 303:1684–86

Hardy MH. 1992. The secret life of the hair follicle. *Trends Genet.* 8:55–61

Hardy RW, Tokuyasu KT, Lindsley DL, Garavito M. 1979. The germinal proliferation center in the testis of *Drosophila melanogaster*. *J. Ultrastruct. Res.* 69:180–90

He XC, Zhang J, Tong WG, Tawfik O, Ross J, et al. 2004. BMP signaling inhibits intestinal stem cell self-renewal through suppression of Wnt-beta-catenin signaling. *Nat. Genet.* 36:1117–21

Helgason CD, Sauvageau G, Lawrence HJ, Largman C, Humphries RK. 1996. Overexpression of HOXB4 enhances the hematopoietic potential of embryonic stem cells differentiated in vitro. *Blood* 87:2740–49

Henderson ST, Gao D, Lambie EJ, Kimble J. 1994. lag-2 may encode a signaling ligand for the GLP-1 and LIN-12 receptors of *C. elegans*. *Development* 120:2913–24

Hermiston ML, Gordon JI. 1995. Organization of the crypt-villus axis and evolution of its stem cell hierarchy during intestinal development. *Am. J. Physiol. Gastrointest. Liver Physiol.* 268:G813–22

Hogan BL. 1996. Bone morphogenetic proteins in development. *Curr. Opin. Genet. Dev.* 6:432–38

Howe JR, Ringold JC, Summers RW, Mitros FA, Nishimura DY, Stone EM. 1998a. A gene for familial juvenile polyposis maps to chromosome 18q21.1. *Am. J. Hum. Genet.* 62:1129–36

Howe JR, Roth S, Ringold JC, Summers RW, Jarvinen HJ, et al. 1998b. Mutations in the SMAD4/DPC4 gene in juvenile polyposis. *Science* 280:1086–88

Huelsken J, Vogel R, Erdmann B, Cotsarelis G, Birchmeier W. 2001. Beta-catenin controls hair follicle morphogenesis and stem cell differentiation in the skin. *Cell* 105:533–45

Ivanova NB, Dimos JT, Schaniel C, Hackney JA, Moore KA, Lemischka IR. 2002. A stem cell molecular signature. *Science* 298:601–4

Jamora C, DasGupta R, Kocieniewski P, Fuchs E. 2003. Links between signal transduction, transcription and adhesion in epithelial bud development. *Nature* 422:317–22

Johe KK, Hazel TG, Muller T, Dugich-Djordjevic MM, McKay RD. 1996. Single factors direct the differentiation of stem cells from the fetal and adult central nervous system. *Genes Dev.* 10:3129–40

Jones PH, Harper S, Watt FM. 1995. Stem cell patterning and fate in human epidermis. *Cell* 80:83–93

Kai T, Spradling A. 2004. Differentiating germ cells can revert into functional stem cells in *Drosophila melanogaster* ovaries. *Nature* 428:564–69

Kanatsu-Shinohara M, Inoue K, Lee J, Yoshimoto M, Ogonuki N, et al. 2004. Generation of pluripotent stem cells from neonatal mouse testis. *Cell* 119:1001–12

Kawase E, Wong MD, Ding BC, Xie T. 2004. Gbb/Bmp signaling is essential for maintaining germline stem cells and for repressing bam transcription in the Drosophila testis. *Development* 131:1365–75

Kiger AA, Jones DL, Schulz C, Rogers MB, Fuller MT. 2001. Stem cell self-renewal specified by JAK-STAT activation in response to a support cell cue. *Science* 294:2542–45

Kimble JE, White JG. 1981. On the control of germ cell development in *Caenorhabditis elegans*. *Dev. Biol.* 81:208–19

King FJ, Lin H. 1999. Somatic signaling mediated by fs(1)Yb is essential for germline stem cell maintenance during Drosophila oogenesis. *Development* 126:1833–44

King FJ, Szakmary A, Cox DN, Lin H. 2001. Yb modulates the divisions of both germline and somatic stem cells through piwi- and hh-mediated mechanisms in the Drosophila ovary. *Mol. Cell* 7:497–508

Korinek V, Barker N, Moerer P, van Donselaar E, Huls G, et al. 1998. Depletion of epithelial stem-cell compartments in the small intestine of mice lacking Tcf-4. *Nat. Genet.* 19:379–83

Kubota H, Avarbock MR, Brinster RL. 2003. Spermatogonial stem cells share some, but not all, phenotypic and functional characteristics with other stem cells. *Proc. Natl. Acad. Sci. USA* 100:6487–92

Kubota H, Avarbock MR, Brinster RL. 2004. Growth factors essential for self-renewal and expansion of mouse spermatogonial stem cells. *Proc. Natl. Acad. Sci. USA* 101:16489–94

Kulessa H, Turk G, Hogan BL. 2000. Inhibition of Bmp signaling affects growth and differentiation in the anagen hair follicle. *EMBO J.* 19:6664–74

Kyba M, Perlingeiro RC, Daley GQ. 2002. HoxB4 confers definitive lymphoid-myeloid engraftment potential on embryonic stem cell and yolk sac hematopoietic progenitors. *Cell* 109:29–37

Lapidot T, Kollet O. 2002. The essential roles of the chemokine SDF-1 and its receptor CXCR4 in human stem cell homing and repopulation of transplanted immune-deficient NOD/SCID and NOD/SCID/B2m(null) mice. *Leukemia* 16:1992–2003

Lavker RM, Miller S, Wilson C, Cotsarelis G, Wei ZG, et al. 1993. Hair follicle stem cells: their location, role in hair cycle, and involvement in skin tumor formation. *J. Invest. Dermatol.* 101:16S–26S

Lessard J, Sauvageau G. 2003. Bmi-1 determines the proliferative capacity of normal and leukaemic stem cells. *Nature* 423:255–60

Li L, Huang GM, Banta AB, Deng Y, Smith T, et al. Cloning, characterization, and the complete 57-kilobase DNA sequence of the human Notch4 gene. *Genomics* 51:45–58

Lin H. 2002. The stem-cell niche theory: lessons from flies. *Nat. Rev. Genet.* 3:931–40

Lin H, Spradling AC. 1993. Germline stem cell division and egg chamber development in transplanted Drosophila germaria. *Dev. Biol.* 159:140–52

Lin H, Spradling AC. 1997. A novel group of pumilio mutations affects the asymmetric division of germline stem cells in the Drosophila ovary. *Development* 124:2463–76

Lin H, Yue L, Spradling AC. 1994. The Drosophila fusome, a germline-specific organelle, contains membrane skeletal proteins and functions in cyst formation. *Development* 120:947–56

Lindsley DT, Tokuyasu KT. 1980. Spermatogenesis. In *Genetics and Biology of Drosophila*, ed. M Ashburner, TRF Wright, pp. 225–94. New York: Acad. Press

Lois C, Alvarez-Buylla A. 1993. Proliferating subventricular zone cells in the adult mammalian forebrain can differentiate into neurons and glia. *Proc. Natl. Acad. Sci. USA* 90:2074–77

Margolis J, Spradling A. 1995. Identification and behavior of epithelial stem cells in the Drosophila ovary. *Development* 121:3797–807

Merrill BJ, Gat U, DasGupta R, Fuchs E. 2001. Tcf3 and Lef1 regulate lineage differentiation of multipotent stem cells in skin. *Genes Dev.* 15:1688–705

Mills JC, Andersson N, Hong CV, Stappenbeck TS, Gordon JI. 2002. Molecular characterization of mouse gastric epithelial progenitor cells. *Proc. Natl. Acad. Sci. USA* 99:14819–24

Mills JC, Gordon JI. 2001. The intestinal stem cell niche: there grows the neighborhood. *Proc. Natl. Acad. Sci. USA* 98:12334–36

Mirescu C, Gould E. 2003. Stem cells in the adult brain. In *Stem Cells: Adult and Fetal Stem Cells*, ed. R Lanza, pp. 219–24. Burlington, MA: Elsevier Acad.

Molofsky AV, Pardal R, Iwashita T, Park IK, Clarke MF, Morrison SJ. 2003. Bmi-1 dependence distinguishes neural stem cell self-renewal from progenitor proliferation. *Nature* 425:962–67

Moore KA, Ema H, Lemischka IR. 1997. In vitro maintenance of highly purified, transplantable hematopoietic stem cells. *Blood* 89:4337–47

Morrison SJ, Wright DE, Cheshier SH, Weissman IL. 1997. Hematopoietic stem cells: challenges to expectations. *Curr. Opin. Immunol.* 9:216–21

Niemann C, Owens DM, Hulsken J, Birchmeier W, Watt FM. 2002. Expression of DeltaNLef1 in mouse epidermis results in differentiation of hair follicles into squamous epidermal cysts and formation of skin tumours. *Development* 129:95–109

Niemann C, Watt FM. 2002. Designer skin: lineage commitment in postnatal epidermis. *Trends Cell Biol.* 12:185–92

Nilsson SK, Johnston HM, Coverdale JA. 2001. Spatial localization of transplanted hemopoietic stem cells: inferences for the localization of stem cell niches. *Blood* 97:2293–99

Nishimura S, Wakabayashi N, Toyoda K, Kashima K, Mitsufuji S. 2003. Expression of Musashi-1 in human normal colon crypt cells: a possible stem cell marker of human colon epithelium. *Dig. Dis. Sci.* 48:1523–29

Ogawa T, Dobrinski I, Avarbock MR, Brinster RL. 2000. Transplantation of male germ line stem cells restores fertility in infertile mice. *Nat. Med.* 6:29–34

Ohlstein B, McKearin D. 1997. Ectopic expression of the Drosophila Bam protein eliminates oogenic germline stem cells. *Development* 124:3651–62

Orkin SH. 2000. Diversification of haematopoietic stem cells to specific lineages. *Nat. Rev. Genet.* 1:57–64

Oshima H, Rochat A, Kedzia C, Kobayashi K, Barrandon Y. 2001. Morphogenesis and renewal of hair follicles from adult multipotent stem cells. *Cell* 104:233–45

Palmer TD, Takahashi J, Gage FH. 1997. The adult rat hippocampus contains primordial neural stem cells. *Mol. Cell Neurosci.* 8:389–404

Paris F, Fuks Z, Kang A, Capodieci P, Juan G, et al. 2001. Endothelial apoptosis as the primary lesion initiating intestinal radiation damage in mice. *Science* 293:293–97

Park IK, He Y, Lin F, Laerum OD, Tian Q, et al. 2002. Differential gene expression profiling of adult murine hematopoietic stem cells. *Blood* 99:488–98

Park IK, Qian D, Kiel M, Becker MW, Pihalja M, et al. 2003. Bmi-1 is required for maintenance of adult self-renewing haematopoietic stem cells. *Nature* 423:302–5

Potten CS, Booth C, Pritchard DM. 1997. The intestinal epithelial stem cell: the mucosal governor. *Int. J. Exp. Pathol.* 78:219–43

Potten CS, Owen G, Booth D. 2002. Intestinal stem cells protect their genome by selective segregation of template DNA strands. *J. Cell Sci.* 115:2381–88

Ramalho-Santos M, Yoon S, Matsuzaki Y, Mulligan RC, Melton DA. 2002. "Stemness": transcriptional profiling of embryonic and adult stem cells. *Science* 298:597–600

Reya T, Duncan AW, Ailles L, Domen J, Scherer DC, et al. 2003. A role for Wnt signalling in self-renewal of haematopoietic stem cells. *Nature* 423:409–14

Reynolds BA, Weiss S. 1992. Generation of neurons and astrocytes from isolated cells of the adult mammalian central nervous system. *Science* 255:1707–10

Rios M, Williams DA. 1990. Systematic analysis of the ability of stromal cell lines derived from different murine adult tissues to support maintenance of hematopoietic stem cells in vitro. *J. Cell Physiol.* 145:434–43

Roberts DJ, Johnson RL, Burke AC, Nelson CE, Morgan BA, Tabin C. 1995. Sonic hedgehog is an endodermal signal inducing Bmp-4 and Hox genes during induction and regionalization of the chick hindgut. *Development* 121:3163–74

Roecklein BA, Torok-Storb B. 1995. Functionally distinct human marrow stromal cell lines immortalized by transduction with the human papiloma virus E6/E7 genes. *Blood* 85:997–1005

Rossant J. 2004. Embryonic stem cells in prospective. In *Handbook of Stem Cells*, ed. R Lanza, J Gearhart, BL Hogan, D Melton, R Pedersen, et al. London: Elsevier Acad.

Sancho E, Batlle E, Clevers H. 2004. Signaling pathways in intestinal development and cancer. *Annu. Rev. Cell Dev. Biol.* 20:695–723

Sauvageau G, Thorsteinsdottir U, Eaves CJ, Lawrence HJ, Largman C, et al. 1995. Overexpression of HOXB4 in hematopoietic cells causes the selective expansion of more primitive populations in vitro and in vivo. *Genes Dev.* 9:1753–65

Schofield R. 1978. The relationship between the spleen colony-forming cell and the hamatopopietic stem cell. A hypothesis. *Blood Cells* 4:7–25

Schulz C, Kiger AA, Tazuke SI, Yamashita YM, Pantalena-Filho LC, et al. 2004. A misexpression screen reveals effects of bag-of-marbles and TGF beta class signaling on the Drosophila male germ-line stem cell lineage. *Genetics* 167:707–23

Shen Q, Goderie SK, Jin L, Karanth N, Sun Y, et al. 2004. Endothelial cells stimulate self-renewal and expand neurogenesis of neural stem cells. *Science* 304:1338–40

Shinohara T, Avarbock MR, Brinster RL. 2000. Functional analysis of spermatogonial stem cells in Steel and cryptorchid infertile mouse models. *Dev. Biol.* 220:401–11

Shinohara T, Orwig KE, Avarbock MR, Brinster RL. 2003. Restoration of spermatogenesis in infertile mice by Sertoli cell transplantation. *Biol. Reprod.* 68:1064–71

Shivdasani AA, Ingham PW. 2003. Regulation of stem cell maintenance and transit amplifying cell proliferation by tgf-beta signaling in Drosophila spermatogenesis. *Curr. Biol.* 13:2065–72

Simmons PJ, Gronthos S, Zannettino AC. 2001. The development of stromal cells. In *Hematopoiesis: A Developmental Approach*, ed. LI Zon, pp. 718–26. New York: Oxford Univ. Press

Simmons PJ, Levesque JP, Zannettino AC. 1997. Adhesion molecules in haemopoiesis. *Bailieres Clin. Haematol.* 10:485–505

Sitnicka E, Ruscetti FW, Priestley GV, Wolf NS, Bartelmez SH. 1996. Transforming growth factor beta 1 directly and reversibly inhibits the initial cell divisions of long-term repopulating hematopoietic stem cells. *Blood* 88:82–88

Song X, Wong MD, Kawase E, Xi R, Ding BC, et al. 2004. Bmp signals from niche cells directly repress transcription of a differentiation-promoting gene, bag of marbles, in germline stem cells in the Drosophila ovary. *Development* 131:1353–64

Song X, Xie T. 2002. DE-cadherin-mediated cell adhesion is essential for maintaining somatic stem cells in the Drosophila ovary. *Proc. Natl. Acad. Sci. USA* 99:14813–18

Song X, Xie T. 2003. Wingless signaling regulates the maintenance of ovarian somatic stem cells in Drosophila. *Development* 130:3259–68

Song X, Zhu CH, Doan C, Xie T. 2002. Germline stem cells anchored by adherens junctions in the Drosophila ovary niches. *Science* 296:1855–57

Spradling A, Drummond-Barbosa D, Kai T. 2001. Stem cells find their niche. *Nature* 414:98–104

Stappenbeck TS, Mills JC, Gordon JI. 2003. Molecular features of adult mouse small intestinal epithelial progenitors. *Proc. Natl. Acad. Sci. USA* 100:1004–9

Sun TT, Cotsarelis G, Lavker RM. 1991. Hair follicular stem cells: the bulge-activation hypothesis. *J. Invest. Dermatol.* 96:S77–78

Szakmary A, Cox DN, Wang Z, Lin H. 2005. Regulatory relationship among piwi, pumilio, and bag-of-marbles in Drosophila germline stem cell self-renewal and differentiation. *Curr. Biol.* 15:171–78

Taichman RS, Emerson SG. 1998. The role of osteoblasts in the hematopoietic microenvironment. *Stem Cells* 16:7–15

Taylor G, Lehrer MS, Jensen PJ, Sun TT, Lavker RM. 2000. Involvement of follicular stem cells in forming not only the follicle but also the epidermis. *Cell* 102:451–61

Tazuke SI, Schulz C, Gilboa L, Fogarty M, Mahowald AP, et al. 2002. A germline-specific gap junction protein required for survival of differentiating early germ cells. *Development* 129:2529–39

Temple S. 2001. The development of neural stem cells. *Nature* 414:112–17

Thomson JA, Itskovitz-Eldor J, Shapiro SS, Waknitz MA, Swiergiel JJ, et al. 1998. Embryonic stem cell lines derived from human blastocysts. *Science* 282:1145–47

Till JE, McCulloch EA. 1961. A direct measurement of the radiation sensitivity of normal mouse bone marrow cells. *Radiat. Res.* 14:213

Tran J, Brenner TJ, DiNardo S. 2000. Somatic control over the germline stem cell lineage during Drosophila spermatogenesis. *Nature* 407:754–57

Tulina N, Matunis E. 2001. Control of stem cell self-renewal in Drosophila spermatogenesis by JAK-STAT signaling. *Science* 294:2546–49

Tumbar T, Guasch G, Greco V, Blanpain C, Lowry WE, et al. 2004. Defining the epithelial stem cell niche in skin. *Science* 303:359–63

van de Wetering M, Sancho E, Verweij C, de Lau W, Oving I, et al. 2002. The beta-catenin/TCF-4 complex imposes a crypt progenitor phenotype on colorectal cancer cells. *Cell* 111:241–50

Varnum-Finney B, Xu L, Brashem-Stein C, Nourigat C, Flowers D, et al. 2000. Pluripotent, cytokine-dependent, hematopoietic stem cells are immortalized by constitutive Notch1 signaling. *Nat. Med.* 6:1278–81

Verfaillie CM, Gupta P, Prosper F, Hurley R, Lundell B, Bhatia R. 1999. The hematopoietic microenvironment: stromal extracellular matrix components as growth regulators for human hematopoietic progenitors. *Hematology* 4:321–33

Visnjic D, Kalajzic Z, Rowe DW, Katavic V, Lorenzo J, Aguila HL. 2004. Hematopoiesis is severely altered in mice with an induced osteoblast deficiency. *Blood* 103:3258–64

Watt FM. 2001. Stem cell fate and patterning in mammalian epidermis. *Curr. Opin. Genet. Dev.* 11:410–17

Weissman IL. 1994. Developmental switches in the immune system. *Cell* 76:207–18

Weissman IL. 2000. Translating stem and progenitor cell biology to the clinic: barriers and opportunities. *Science* 287:1442–46

Weissman IL, Anderson DJ, Gage F. 2001. Stem and progenitor cells: origins, phenotypes, lineage commitments, and transdifferentiations. *Annu. Rev. Cell Dev. Biol.* 17:387–403

Wieschaus E, Szabad J. 1979. The development and function of the female germ line in *Drosophila melanogaster*: a cell lineage study. *Dev. Biol.* 68:29–46

Winton D. 2000. Stem cells in the epithelium of the small intestine and colon. In *Stem Cell Biology*, ed. DR Marshak, RL Gardner, D Gottlieb, pp. 515–36. Cold Spring Harbor, NY: Cold Spring Harbor Lab. Press

Xie T, Spradling A. 2001. The Drosophila ovary: an in vivo stem cell system. In *Stem Cell Biology*, ed. DR Marshak, RL Gardner, D Gottlieb, pp. 129–48. Cold Spring Harbor, NY: Cold Spring Harbor Lab. Press

Xie T, Spradling AC. 1998. decapentaplegic is essential for the maintenance and division of germline stem cells in the Drosophila ovary. *Cell* 94:251–60

Xie T, Spradling AC. 2000. A niche maintaining germ line stem cells in the Drosophila ovary. *Science* 290:328–30

Yamashita YM, Jones DL, Fuller MT. 2003. Orientation of asymmetric stem cell division by the APC tumor suppressor and centrosome. *Science* 301:1547–50

Zhang J, Niu C, Ye L, Huang H, He X, et al. 2003. Identification of the haematopoietic stem cell niche and control of the niche size. *Nature* 425:836–41

Zhang Y, Kalderon D. 2001. Hedgehog acts as a somatic stem cell factor in the Drosophila ovary. *Nature* 410:599–604

Docosahexaenoic Acid, Fatty Acid–Interacting Proteins, and Neuronal Function: Breastmilk and Fish Are Good for You

Joseph R. Marszalek[1] and Harvey F. Lodish[1,2]

[1]Whitehead Institute for Biomedical Research, Cambridge, Massachusetts 02142;
email: joseph_marszalek@merck.com

[2]Department of Biology, Massachusetts Institute of Technology, Cambridge,
Massachusetts 02142; email: lodish@wi.mit.edu

Annu. Rev. Cell Dev. Biol.
2005. 21:633–57

First published online as a
Review in Advance on
July 1, 2005

The *Annual Review of
Cell and Developmental
Biology* is online at
http://cellbio.annualreviews.org

doi: 10.1146/
annurev.cellbio.21.122303.120624

Copyright © 2005 by
Annual Reviews. All rights
reserved

1081-0706/05/1110-
0633$20.00

Key Words

acyl CoA synthetase, fatty acid transport protein, fatty acid
internalization, neurodevelopment, polyunsaturated fatty acids

Abstract

In contrast to other tissues, the nervous system is enriched in the
polyunsaturated fatty acids (PUFAs): arachidonic acid (AA, 20:4 n-6)
and docosahexaenoic acid (DHA, 22:6 n-3). Despite their abundance
in the nervous system, AA and DHA cannot be synthesized de novo
by mammals; they, or their precursors, must be ingested from dietary
sources and transported to the brain. During late gestation and the
early postnatal period, neurodevelopment is exceptionally rapid, and
substantial amounts of PUFAs, especially DHA, are critical to ensure
neurite outgrowth as well as proper brain and retina development.
Here, we review the various functions of DHA in the nervous system, the proteins involved in its internalization and metabolism into
phospholipids, and its relationship to several neurological disorders,
including Alzheimer's disease and depression.

Contents

FATTY ACID BIOGENESIS AND METABOLISM

Before plunging into the details of DHA function in neuronal cells, we provide a background on saturated and unsaturated FAs, as this material is likely to be unfamiliar to most readers. Mammalian cells utilize three main types of FAs: saturated FAs, which do not contain any double bonds; monounsaturated FAs, which contain a single double bond; and polyunsaturated FAs, which contain multiple double bonds. The common FA nomenclature contains the number of carbons and the number of desaturations. For example, the most common saturated FA is stearic acid, which contains 18 carbons and no double bonds and is denoted as 18:0. An example of an unsaturated FA is DHA, which contains 22 carbons and 6 double bonds. It is therefore noted as 22:6. Unsaturated FAs are further divided into three groups according to the position of the first double bond at the third (n-3 or $\omega3$), sixth (n-6 or $\omega6$), or ninth (n-9 or $\omega9$) carbon from the FA's methyl end, which is designated as carbon 1.

Mammals are able to introduce, through the action of $\Delta9$-desaturase, a double bond after the ninth carbon from the methyl group. However, they are unable to introduce double bonds between carbons 1–9 because they lack the desaturase enzymes that introduce double bonds (or desaturations) at the n-3 or n-6 position (i.e., after either the third or sixth carbon from the methyl terminus) of the FA. Therefore, the n-3 and n-6 series FAs must be ingested in the physiologically active forms of DHA (22:6 n-3) and AA (20:4 n-6) or as these molecules' respective 18-carbon precursors LNA (18:3 n-3) and LA (18:2 n-6). Mammals can convert LNA to DHA and LA to AA by the action of several specific elongases and desaturases that add additional carbons and double bonds (**Figure 1**). n-3 and n-6 FAs compete at several steps for the same metabolic enzymes. $\Delta6$-desaturase is thought to be the rate-limiting enzyme involved in AA/DHA synthesis from their n-3

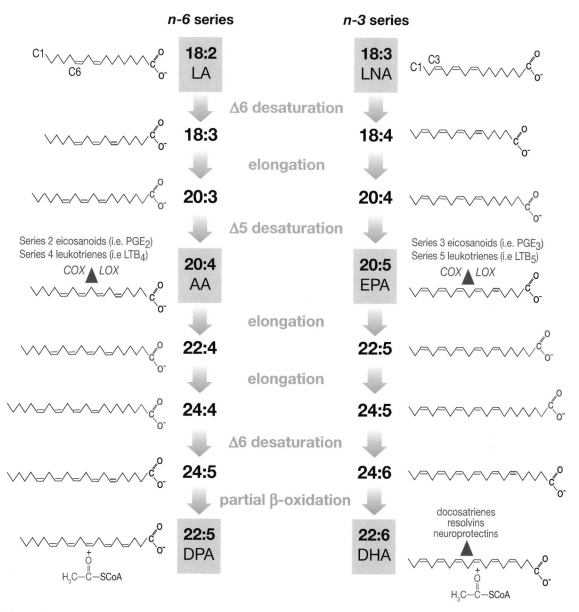

Figure 1

Biosynthesis of long-chain *n-3* and *n-6* series polyunsaturated FAs from their 18-carbon precursors. The terminal methyl group is carbon 1 and the *n-3* and *n-6* series of FAs are termed according to the position of the first double bond: after carbon 3 and carbon 6, respectively. Biologically important FAs are highlighted with a gray box. Newly added/removed carbons or double bonds introduced at each step are colored red. Signaling molecules derived from AA, EPA, and DHA are noted in blue. LA, linoleic acid; LNA, linolenic acid; AA, arachidonic acid; EPA, eicosapentanoic acid; DPA, docosapentanoic acid; DHA, docosahexaenoic acid; COX, cyclooxygenase; LOX, lipoxygenase; PG, prostaglandin; LT, leukotriene.

DHA:
docosahexaenoic acid

FA: fatty acid

AA: arachidonic acid

LNA: linolenic acid

LA: linoleic acid

PUFA:
polyunsaturated fatty
acid

PL: phospholipid

EPA:
eicosapentanoic acid

and *n-6* precursors; *n-3* FAs are the preferred substrate. The conversion of LNA to DHA (**Figure 1**) occurs primarily in the liver (Scott & Bazan 1989) and, to a lesser extent, in the astrocytes and vascular endothelial cells in the retina and brain (Moore 2001).

SOURCES OF POLYUNSATURATED FATTY ACIDS

Mother's Milk

During the last trimester of gestation, AA and DHA are transferred through the placenta from the mother to the developing neonate. During the first year of postnatal development, AA and DHA are provided via the mother's milk to the infant; these PUFAs are then avidly incorporated into membranes throughout the infant's nervous system (Martinez 1992b). Human milk contains significant amounts of AA and DHA. However, until 2002, infant formulas used in the United States contained considerable amounts of 18:2 *n-6* and 18:3 *n-3*, but no AA or DHA (Jensen & Heird 2002). Infants have limited capacity to convert LA to AA and LNA to DHA, as evidenced by the fact that the levels of AA and DHA in the PLs of erythrocytes are higher in breast-fed babies relative to infants fed a formula that was not supplemented with PUFAs (Jensen & Heird 2002). Therefore, it remains controversial whether infants, especially preterm infants, are capable of synthesizing enough DHA from LNA to meet the demands of neural development. A more detailed discussion of PUFAs in breast milk can be found in a review by Innis (2004), who provides an in-depth analysis of the effects the mother's diet has on the composition of her breast milk and how this composition affects the PUFA levels of breast-fed newborns.

Diet

The modern diet, especially in the United States, has shifted from one in which nearly equivalent amounts of *n-3* and *n-6* FAs are ingested to one in which the *n-6:n-3* ratio is approximately 15 to 1 (Simopoulos 2002). In general, metabolites of *n-6* and *n-3* FAs tend to have opposite biological effects: *n-6*-derived metabolites tend to be proinflammatory, whereas *n-3*-derived metabolites tend to be less or anti-inflammatory. This key point is discussed below in some detail. The enormous shift toward a diet enriched in *n-6* FAs is primarily the result of the extensive use of corn, safflower, and sunflower oils in the Western diet. These oils are each high in *n-6* FA content but low in *n-3* FA content. This shift to a higher ratio of *n-6* FAs has been further compounded by the low intake of fish, which is the primary source of the *n-3* FAs DHA and EPA, and the high intake of meats from cows, pigs, and chicken, which are typically fed diets that are based primarily on corn and thus have high *n-6:n-3* FA content. Oils that contain more *n-3* FAs include canola, soy, and flaxseed oils (**Table 1**).

POLYUNSATURATED FATTY ACIDS ARE ESSENTIAL FOR PROPER DEVELOPMENT

Research over the past 30 years has established that PUFAs are critical for proper infant growth and neurodevelopment. The nervous system, along with sperm, has the highest concentration of PUFAs, especially DHA. In the nervous system, the tissue most highly enriched in DHA is the photoreceptor outer segment, in which nearly 50% of the FAs in the PLs are DHA (Fliesler & Anderson 1983). In neurons, PUFAs internalized from the extracellular fluids are efficiently incorporated, primarily into PLs (Rapoport 2001), where they affect membrane fluidity and alter the function of many integral and membrane-associated proteins. Upon their liberation from the PLs by phospholipases, PUFAs can serve as signaling molecules or they can be converted into other powerful signaling molecules, such as prostaglandins, leukotrienes, and

Table 1 The *n-3* and *n-6* fatty acid (FA) content of vegetable oil, meat, and fish[a]

	n-3 Fatty Acids			*n-6* Fatty Acids		
	LNA (18:3)	EPA (20:5)	DHA (22:6)	LA (18:2)	AA (20:4)	*n-3/n-6* ratio
OILS						
Corn	0.8[b]	—	—	52	—	0.015
Safflower	0.1	—	—	77	—	0.0013
Olive	0.5	—	—	10	—	0.050
Soy	8	—	—	54	—	0.15
Canola	10	—	—	23	—	0.43
Linseed	48	—	—	17	—	2.82
FISH						
Salmon	1.1	13.5	18.9	1.6	0.7	14.57
Trout	1.7	7	20.4	4.8	0.8	5.20
Tuna	1.6	11.3	19.4	1.6	3.2	6.73
MEAT						
Chicken	0.9	0.3	0.6	12.2	0.5	0.14
Beef	0.3	Trace	Trace	2.1	0.4	0.12
Pork	0.5	Trace	0.4	8.1	0.5	0.10

[a]Values are g/100 g of total FA content. The ratio of total *n-3*/total *n-6* was calculated for each source. The FA levels are based on those reported by Innis (2004) and Larsson et al. (2004).
[b]Grams per 100 g of total FAs.

thromboxanes. These are collectively known as eicosanoids (derived from 20-carbon FAs) and the more recently identified docosanoids (derived from 22-carbon FAs); the latter group includes docosatrienes, neuroprotectins, and resolvins (**Figure 1**) (Serhan et al. 2004).

Deficiency of *n-6* FAs in humans results in several nonneuronal abnormalities, including reduced growth, reproductive failure, skin lesions, fatty liver, and polydipsia (excessive thirst). Deficiency of *n-3* FAs, on the other hand, results in several neuronal-specific defects that include reduced learning, abnormal electroretinogram, impaired vision, numbness, and leg pain (Connor et al. 1992, Holman et al. 1982). The different phenotypic defects that occur with either *n-3* or *n-6* deficiency clearly establish that these two classes of FA are not interchangeable. Fortunately, essential FA deficiency in humans is extremely rare and is usually reversible by supplying the missing FA to the diet.

Rodent models of essential FA deficiency are proving useful for assessing the effects of PUFA-deficient diets. Most of the work has been focused on the consequence of *n-3* deficiency because of the significant reduction in the intake of *n-3* FAs that has recently occurred in the Western diet and because of the numerous neurological deficits that occur with *n-3* FA deficiency in humans (see Diet section above). Because mother's milk is such an excellent source of DHA and DHA is avidly retained in the nervous system, the generation of *n-3*-deficient mice requires that multiple generations be fed an *n-3*-deficient diet to deplete *n-3* FAs from the nervous system. Several studies have shown that deprivation of DHA (22:6 *n-3*) leads to a compensatory increase in the amount of DPA (22:5 *n-6*) in brain PLs. Likewise, EPA (20:5 *n-3*) levels increase when there is an insufficient supply of AA (20:4 *n-6*).

Using various methods to deplete animals of *n-3* FAs, researchers have extensively

Eicosanoids: signaling molecules derived from 20-carbon AA or EPA

Docosanoids: signaling molecules derived from 22-carbon DHA

DPA: docosapentanoic acid

PS:
phosphatidylserine

Free FA:
unesterified FA that
contains a
hydrophobic acyl
chain and a
negatively charged
carboxyl group

OA: oleic acid

TAG:
triacylglyceride

PA: palmitic acid

sn position of PL:
positions of FAs in
the glycerol
backbone: *sn-1*,
carbon 1, typically
saturated FA; *sn-2*,
carbon 2, typically
PUFA; *sn-3*, carbon
3, phosphophate-
containing head
group

PI:
phosphatidylinositol

DAG:
diacylglyceride

PC:
phosphatidylcholine

PE: phos-
phatidylethanolamine

examined the physiological consequence of reduced DHA levels in the nervous system. The primary defects that occur with *n-3* deficiency include visual, sensory, and memory deficits (see Salem et al. 2001 for a detailed list). Recently, researchers have identified changes that occur at the cellular and molecular levels as a consequence of reduced DHA levels in the nervous system. Some interesting examples of physiological changes that occur in the nervous system of DHA-deficient mice include a decrease in neuronal cell-body size in several regions of the brain (Ahmad et al. 2002a), decreases in glucose uptake and cytochrome oxidase activity (Ximenes da Silva et al. 2002), and a significant decrease in G protein–coupled signaling efficiency in the photoreceptor outer segment (Niu et al. 2004). Additionally, the level of PS (whose importance in neuronal function is discussed below) is significantly reduced in *n-3*-deficient rodents (Hamilton et al. 2000).

FUNCTIONS OF DHA IN THE NERVOUS SYSTEM

Within neurons, DHA is highly enriched in the PLs of the synaptic plasma membrane and synaptic vesicles (Breckenridge et al. 1972). There, it may improve the efficiency of membrane fusion events. It may also function in synaptic signaling as either a metabolized or a free FA or as part of a PL. Additionally, during neurite outgrowth, high levels of DHA are found in the growth cone (Martin & Bazan 1992); DHA in the growth cone may be especially important for maximal neurite outgrowth during neurodevelopment. In the adult, DHA is found in the neuronal dendrites, where it may be involved in the extension and establishment of dendritic arborization, which occurs during memory formation. Additionally, DHA may be important for the efficient regeneration of axons and dendrites after neuronal injury.

Polyunsaturated Fatty Acids, Stimulation of Phospholipid Synthesis, and Neurite Outgrowth

A number of reports establish that supplementing cultured neuronal cell types with AA or DHA can increase neurite outgrowth (Calderon & Kim 2004, Dehaut et al. 1993, Furuya et al. 2002, Ikemoto et al. 1997, Marszalek et al. 2004, Okuda et al. 1994, Williams et al. 1994a,b). Low concentrations (1.5–60 μM) of AA and DHA supplementation significantly increase neurite outgrowth in several cell types (PC12, primary hippocampal) over different lengths of analysis (2–7 d). However, there is a limit to the amount of PUFA that can be used as a supplement because supplementation with high levels of AA (>60 μM) is cytotoxic.

PUFAs appear to enhance neurite outgrowth by several mechanisms that include increasing the synthesis and levels of PLs. In differentiated PC12 cells (a model neuronal cell line), a higher percentage of exogenous AA and DHA than of exogenous OA was incorporated into PLs relative to TAG (Marszalek et al. 2005). Similarly, in rats infused with palmitic acid (PA), AA, or DHA for 15 min, significantly more AA and DHA was incorporated into brain PLs relative to PA, even though each FA was equally incorporated into total brain lipids (Rapoport 2001). Together, these data suggest that in differentiating and mature neurons, PUFAs are preferentially incorporated into PLs rather than into TAGs, the other major class of cellular molecules that contain FAs.

During the synthesis of PLs, saturated FAs (i.e., 16:0 or 18:0) are typically acylated into the *sn-1* position of glycerol-3-phosphate, and PUFAs (i.e., AA or DHA) are primarily incorporated into the *sn-2* position to generate phosphatidic acid, which is the precursor for PI. However, most of the phosphatidic acid is subsequently dephosphorylated to generate DAG, which is further metabolized to PC, PE, or TAG. DAG molecules that contain PUFAs in the *sn-2* position are preferentially converted to PLs. To some extent, this appears to be due to the action of specific metabolism enzymes. For example, DAG species that contain DHA in the *sn-2* position are the preferred substrate of ethanolaminephosphotransferase, which converts DAG to PE (Holub 1978) by covalently linking etholamine to the *sn-3* position of DAG.

Every cell type typically maintains a constant ratio of PL species, and changes in one species often alters the levels of the other species (Araki & Wurtman 1998). For example, PC12 cells typically contain ~54% PC, ~11% PS, ~7% PI, and ~27% PE (Knapp & Wurtman 1999). Additionally, the FA compositions of PC, PS, PI, and PE species are significantly different from each other (Knapp & Wurtman 1999) (**Table 2**). The bulk of DHA is incorporated into PE and, to a lesser extent, in PS. Likewise, the majority of AA is incorporated into PI and PE, whereas the bulk of OA is incorporated into PC. There is a clear preference for 16:0 incorporation into PC, and 18:0 is preferentially incorporated into PS, PI, and PE. The ratio of 16:0 to 18:0 in PC is ~6 to 1, but the ratio is reversed in PS (1 to 7.5), PI (1 to 20), and PE (1 to 2.5), indicating that these species prefer 18:0 in the *sn-1* position (Knapp & Wurtman 1999). The differential incorporation of AA and DHA into PI, PS, and PE, compared with PC, may contribute to their functions in neurons (see below).

DHA and Membrane Function

The amount of desaturation in a FA is directly related to its flexibility. Saturated FAs such as PA (16:0) are straight and rigid. This rigidity allows saturated FAs to pack together tightly and form a solid at lower temperatures. Introduction of a double bond into the FA causes a "kink" to occur in the FA. Unsaturated FAs, such as DHA, adopt countless conformations because the FA can rotate around C–C bonds but not around the rigid C=C bonds (Feller et al. 2002). The flexible nature of DHA will not allow PLs containing DHA to pack tightly together, resulting in a significant increase in membrane fluidity relative to PLs comprised only of saturated FAs. Membranes high in DHA may also increase the efficiency of membrane fusion events (Teague et al. 2002). Additionally, increased fluidity of the membrane appears to be important for increasing the rate at which protein-protein interaction events occur within the PL bilayer. This is especially true in the outer segment of the photoreceptor, where activation of transducin by the rhodopsin-to-metarhodopsin conversion does not occur efficiently when the level of DHA in PLs is reduced (Niu et al. 2004). It is also interesting to note that the PLs of mitochondria are also enriched in DHA. High DHA content in mitochondrial membranes may increase the efficiency of electron transport by increasing the lateral movement of proteins within the bilayer, thus facilitating protein-protein interactions (Valentine & Valentine 2004). Additionally, there is a direct

Table 2   Distribution of fatty acids (FAs) within each phospholipid (PL) species of PC12 cells[a]

		PC	PS	PI	PE
PA	16:0	38.8[b]	6.2	2.8	9.8
SA	18:0	6.2	48.1	55.7	24.9
OA	18:1 (*n-9*)	54	41.6	12.3	37
AA	20:4 (*n-6*)	0.6	1.1	28.2	14.9
DHA	22:6 (*n-3*)	0.4	5.1	1	13.4

[a]Within each PL species, the distribution of palmitic acid (PA, 16:0), stearic acid (SA, 18:0), oleic acid (OA, 18:1, *n-9*), arachidonic acid (AA, 20:4 *n-6*), and docosahexaenoic acid (DHA, 22:6 *n-3*) was calculated. The percentages were based on the values reported by Knapp & Wurtman (1999).

[b]Relative percentage of FA for each PL class.

correlation between the DHA content of mitochondrial PLs and the permeability of the membrane to protons (Hulbert 2003). Thus, the DHA content of PLs influences mitochondrial function.

DHA and Modification of Enzyme Function

Receptor function and activation of signaling proteins can be influenced by DHA as a free FA and when it is incorporated into the PLs of the plasma membrane. DHA is concentrated in the PLs of neural tissues, including the hippocampus (Ahmad et al. 2002b), which is involved in learning as well as memory storage and retrieval. As part of a DAG molecule, DHA enhances diacylglycerol-dependent activation of protein kinase C (PKC) (Chen & Murakami 1994). In contrast, unesterified DHA has inhibitory effects on GABA responses in rat neurons (Hamano et al. 1996). It is interesting that PKC has an essential requirement for PS (Nishizuka 1995), which contains a high percentage of DHA. However, in vitro evidence suggests that unesterified EPA and DHA competitively inhibit PS-dependent PKC activation. Unesterified AA either stimulates or has no effect on PKC activity (Seung Kim et al. 2001). Another well-studied example of a protein whose function is modified by DHA is the Na+, K+ ATPase, which is an integral membrane protein found at high densities at the nodes of Ranvier in neurons. The primary neuronal function of this ATPase is to generate and maintain Na+ and K+ gradients necessary to maintain the resting membrane potential. The activity of Na+, K+ ATPase is increased in the sciatic nerve of rats that are supplemented with *n-3* FAs such as DHA (Gerbi et al. 1998).

Inhibition of Apoptosis in Neurons

The early signs of apoptosis include loss of intracellular water, increase in cytoplasmic calcium concentration, and translocation of PS to the outer leaflet of the plasma membrane (Sastry & Rao 2000). Activation of caspase-3 through self-cleavage (Thornberry & Lazebnik 1998) commits cells to death by apoptosis (Nagata 1997). Prevention of apoptosis by DHA incorporation into PLs has been reported for rat retinal photoreceptors (Rotstein et al. 1997), HL-60 cells (Kishida et al. 1998), and Neuro 2A cells (Kim et al. 2000). Additionally, increased dietary intake of DHA prevents apoptosis in mouse photoreceptors subjected to N-methyl-N-nitrosourea, a potent inducer of apoptosis (Moriguchi et al. 2003). DHA accumulation in PS appears to promote neuronal survival under adverse conditions (Kim et al. 2000).

As discussed above, in the nervous system, DHA is incorporated primarily into the anionic PLs, PS, and PE (Aid et al. 2003). PS is synthesized from PE or PC by serine replacement of ethanolamine or choline, respectively, in a base-exchange reaction. PS is involved in various cell-signaling events. Supplementation of cells with unesterified DHA promotes PS biosynthesis (Garcia et al. 1998, Hamilton et al. 2000, Kim & Hamilton 2000). The enrichment of DHA in PS and its effect on PS synthesis most likely are due to the fact that PL species containing DHA are the best substrates for PS synthesizing enzymes (Kim et al. 2004a). There is also a direct correlation between the level of PS and DHA content, which covaries by brain region, age (Wen & Kim 2004), type of cell, and subcellular region.

The antiapoptotic effects of DHA in neurons occur only after cultured cells or animals have been pretreated with DHA, suggesting that these effects are most likely due to DHA being metabolized into PLs. In a series of experiments, Kim and colleagues have begun to elucidate the molecular mechanism of how DHA prevents apoptosis. DHA enrichment in PS of Neuro 2A cells significantly reduces apoptosis induced by staurosporine (Akbar & Kim 2002) and serum deprivation (Kim et al. 2000). DHA incorporation into PS may prevent apoptosis by stimulating the PI_3-kinase/Akt pathway and enhancing the translocation of raf-1 kinase to the plasma membrane. High levels of PS, which are stimulated by DHA supplementation, enhance the translocation of raf-1 kinase to the plasma membrane; this translocation is necessary for raf-1 kinase activation. PI$_3$-kinase and Akt also localize to the plasma membrane, where their activity may be increased by higher levels of PS. Together, these events reduce caspase-3 activation.

It is interesting to note that in other non-neuronal cell types, DHA actually promotes apoptosis. For example, in CaCo-2 cells, a colon cancer cell line, DHA induces apoptosis by downregulating the expression of antiapoptotic genes and increasing the level of several proapoptotic genes (Narayanan et al. 2001). Therefore, the antiapoptotic effects of DHA-containing PS may be specific for neurons and critical for their long-term survival.

DHA and Metabolites as Signaling Molecules

PLs are also important sources of several potent second messengers that function in intracellular and intercellular signaling. There is evidence that different PL species are preferentially utilized to generate these signaling molecules. Altering the ratio of PL species or the FA composition of cells during differentiation or in adults in turn alters PUFA-dependent intracellular signaling.

AA and EPA/DHA are liberated, by various PLA$_2$ enzymes, from PLs and converted to potent autocrine and paracrine signaling molecules by the action of cyclooxygenases (COXs), lipoxygenases (LOXs), and cytochrome P450 monooxygenases. In general, n-6 and n-3 metabolites have opposite biological effects. For example, the n-6 metabolite PGE$_2$ is proinflammatory whereas the complementary n-3 metabolite PGE$_3$ is antiinflammatory (**Figure 1**). There is growing evidence that AA release from PLs is important for normal brain functions but is enhanced in inflammatory neuropathological conditions (Farooqui et al. 1997a). Because n-3 and n-6 FAs compete for incorporation into the PLs of neuronal membranes, increased levels of DHA/EPA in neurons result in a decrease in the level of AA in PLs. This reduces the synthesis of AA-derived eicosanoids, and thus inflammation. Additionally, DHA is converted to docosanoids, which protect neurons from oxidative stress (Bazan 2003, Mukherjee et al. 2004). DHA and docosanoids competitively inhibit AA conversion to eicosanoids by COX and LOX enzymes (Calder 1998, Hong et al. 2003). These eicosanoids and docosanoids are powerful signaling molecules because they can cross neural membranes from one cell to the next (Farooqui et al. 1997b).

In the nervous system, there are at least three forms of PLA$_2$ that may function in liberation of DHA from the sn-2 position of PLs—secretory PLA$_2$ (sPLA$_2$), cytosolic PLA$_2$ (cPLA$_2$), and Brain PlsEtn-PLA$_2$ (Farooqui & Horrocks 2004). The different isoforms of PLA$_2$ likely act on different pools of PLs located in different regions within a neuron or within different neurons located in different regions of the brain. Additionally, since these isoforms tend to be regulated by different mechanisms (Farooqui & Horrocks 2004), they may activate different second-messenger systems. This is important because different receptors are then activated. The activation of PLA$_2$ isoforms is rate limiting for the generation of AA and DHA derivatives that are used for signaling.

PLA$_2$ activity, which is detected primarily in mitochondrial and synaptosomal fractions (Bazan 1971, Goracci et al. 1978), is increased as PC12 cells differentiate (Akiyama et al. 2004, Li & Wurtman 1999). Inhibitors of PLA$_2$ have been shown to inhibit neurite outgrowth (Suburo & Cei de Job 1986, Tsukada et al. 1994).

sPLA$_2$ is associated with synaptosomes and synaptic vesicle fractions (Kim et al. 1995, Matsuzawa et al. 1996) and is found in differentiated PC12 cells (Matsuzawa et al. 1996). sPLA$_2$ may function in neurite outgrowth: When it is added to the medium of differentiating cells, neurite outgrowth is

PLA$_2$:
phospholipase A$_2$

AD: Alzheimer's disease

Acsl: acyl coenzyme A synthetase long-chain family member

CoA: coenzyme A

stimulated (Nakashima et al. 2003). Although sPLA$_2$ is not selective for any FA, it may contribute to eicosanoid/docosanoid signaling in the synapse because it is released from synaptosomes upon stimulation and inhibitors of sPLA$_2$ inhibit neurotransmitter release (Farooqui & Horrocks 2004).

cPLA$_2$ prefers AA in the *sn-2* position (Clark et al. 1995). In neural membranes, cPLA$_2$ activity is linked to several receptors through various coupling mechanisms (Farooqui et al. 2000). Interestingly, cPLA$_2$ colocalizes with COX-2 in the cerebellum (Pardue et al. 2003), suggesting that the release of AA from membrane lipids is integrally related to AA's conversion to an eicosanoid. Significantly, cPLA$_2$ localizes to the growth cone, where it may function to enhance neurite outgrowth or synaptic plasticity by stimulating PUFA release from the membrane lipids; these may then be converted to signaling molecules (Kishimoto et al. 1999, Negre-Aminou et al. 1996, Pardue et al. 2003, Stephenson et al. 1996, 1999). cPLA2 activity is increased in the brains of people with AD (Stephenson et al. 1996) and may contribute to the etiology of the disease.

Brain PlsEtn-PLA$_2$ hydrolyzes AA/EPA and DHA from the *sn-2* position of plasmalogens. Plasmalogens are a unique type of PL that contain a vinyl ether linkage at the *sn-1* position of the glycerol backbone and a high percentage of DHA in the *sn-2* position (Farooqui et al. 2003). PlsEtn-PLA$_2$ is enriched in both neurons and glia. Because PlsEtn-PLA$_2$ is expressed in neurons, it may be responsible for releasing significant amounts of EPA or DHA for the generation of *n-3* eicosanoids and docosanoids respectively (Hong et al. 2003, Serhan et al. 2004).

Gene Regulation

Numerous studies have shown that PUFAs modify gene expression by binding to specific transcription factors, including peroxisome proliferator activated receptors (PPARs), retinoid X receptor (RXRs), hepatic nuclear factor-4 (HNF-4), and sterol regulatory element binding proteins (SREBP). Activation of each of these proteins modulates the expression of genes involved in glucose, FA, TAG, and cholesterol metabolism. Of these, retinoid X receptor is present at significant levels in the brain; it is efficiently activated by DHA (de Urquiza et al. 2000, Lengqvist et al. 2004). In reaggregated rat brain cells, stimulation of peroxisome proliferator-activated protein β results in the upregulation of the mRNA encoding Acsl6 (Basu-Modak et al. 1999), a protein that preferentially converts DHA to the CoA derivative (Marszalek et al. 2005). Upon alteration of the expression of lipid metabolism genes, the optimal environment for neurite outgrowth can be achieved. For example, PUFAs have been shown to decrease the expression and activity of Δ9-desaturase, the enzyme that converts 18:0 to 18:1 (*n-9*) (Sessler et al. 1996). This effect may be important to ensure that saturated FAs, whether newly synthesized or taken up from the external environment, are available for insertion into the *sn-1* position of PLs as they are synthesized. Maintaining the proper PUFA/non-PUFA ratio in the PLs of the plasma membrane is critical for normal membrane fluidity and possibly for optimal neurite outgrowth. Additionally, several studies have shown that DHA increases neurite outgrowth and this may, in part, be a consequence of DHA stimulation of the expression of genes that promote PL synthesis. It is interesting that in microarray gene expression experiments using rat brains or retinas, fish oil or DHA supplementation alters the expression of many genes involved in signal transduction, synaptic plasticity, eicosanoid production, and energy metabolism (Barcelo-Coblijn et al. 2003a,b, Kitajka et al. 2002, Puskas et al. 2003, Rojas et al. 2003).

Fatty Acid Internalization and Utilization

As discussed earlier, the *n-3* and *n-6* series of PUFAs cannot be synthesized by mammals but rather must be internalized from the external environment. Despite this limitation, the nervous system is more highly enriched in PUFAs, especially DHA, than are almost all other cell types. This enrichment suggests that the nervous system possesses unique proteins that preferentially promote PUFA internalization and utilization by neurons. While diffusion of PUFAs through the PLs of the plasma membrane bilayer may contribute significantly to FA internalization by cells, diffusion, by itself, does not result in a net influx of FAs, nor can it provide the selectivity that may be required for the enrichment of AA or DHA that occurs in the nervous system. Rather, to promote efficient FA internalization, the cell employs several types of FA-interacting proteins.

Extracellular Fatty Acid–Binding Proteins

The primary extracellular FABP is albumin. As many as eight FAs can be complexed to a single albumin molecule, allowing the insoluble FAs to be transported within the aqueous environment of the bloodstream. However, at any given time, typically four or fewer FAs are bound to any albumin molecule. Additionally, other lipid-binding proteins, such as interphotoreceptor retinoid-binding protein, are present in the extracellular environment; such proteins transport other types of lipids, such as retinol. However, albumin is the primary extracellular protein, binding all types of free FAs.

Diffusion

Unlike aqueous molecules such as glucose and charged ions, which require transporters and channels, respectively, unesterified FAs can efficiently partition into the plasma membrane, owing to their extreme hydrophobicity. This was elucidated by Hamilton and coworkers in a series of experiments in which they determined that FAs are moved efficiently and rapidly from the outer leaflet of the plasma

membrane PL bilayer to the inner leaflet in a process termed "flip-flop" (Hamilton et al. 2002, Pownall & Hamilton 2003). This process is facilitated by the protonation of the FA to neutralize the negative charge of the carboxyl group, thus reducing the half-life of "flip-flop" by as much as three orders of magnitude (from >2 s to as little as a millisecond) (Kamp & Hamilton 1992). In order to produce net FA influx, the FA must desorb from the inner leaflet of the membrane and interact with an intracellular FA-interacting protein such as FATP, Acsl, or FABP (see next sections) to prevent it from repartitioning back into the membrane. If both of these events do not occur, the FA repartitions back into the membrane, where it "flip-flops" back to the outer leaflet and desorbs back to the plasma to interact once again with albumin. In general, the transfer rate (absorption/"flip-flop"/desorption) decreases as the FA chain length increases (Pownall & Hamilton 2003). However, this is counterbalanced by the fact that as the number of double bonds increases, so does the transfer rate. Therefore, the six double bonds in DHA significantly increase its ability to be internalized by diffusion.

Membrane Fatty Acid Transport Proteins

FATP1, along with Acsl1 (see next section), was identified, in an expression cloning screen, as a protein that, when overexpressed, increased the rate of FA internalization (Schaffer & Lodish 1994). Consistent with the role of FATP1 in FA internalization, a significant portion of FATP1 is localized at the plasma membrane (Schaffer & Lodish 1994). Subsequently, six other FATP family members with unique tissue expression patterns were also identified (Hirsch et al. 1998). FATPs were classified as FA transport proteins because, when overexpressed, they increase the rate of FA internalization, most notably at low concentrations when diffusion may not be sufficient.

FABP: fatty acid–binding protein

FATP: fatty acid transport protein

FATP proteins, along with Acsls, contain an ATP/AMP-binding motif that is found in all adenylate-forming enzymes. Additionally, FATPs contain a second region that is found in the family of very-long-chain acyl CoA synthetases, which preferentially activate very-long-chain FAs (FAs greater than 22 carbons in length). This region may contribute to FA specificity and the long-chain acyl CoA synthetase (ACS) activity (activation of FAs containing 16–22 carbons) that is observed in biochemical studies of FATPs (Black & DiRusso 2003, Coe et al. 1999, Watkins et al. 1998). The expression of FATP1 and FATP4 mRNAs in the brain (Fitscher et al. 1998, Hirsch et al. 1998, Schaffer & Lodish 1994, Utsunomiya et al. 1997) suggests that FATPs may be responsible for the activation and/or internalization of very-long-chain FAs that are important for brain function (Coe et al. 1999, Hall et al. 2003, Herrmann et al. 2001). However, FATP1 overexpression does not increase very-long-chain FA internalization, suggesting that the primary function of FATP1 does not concern very-long-chain internalization. Additionally, whereas FATP1 possesses ACS activity toward long-chain FAs (16–22 carbons), compared to that of Acsl1, the ACS activity of FATP1 is very low.

Mutation of serine 250 to alanine in FATP1 abrogates FA internalization, providing direct evidence that this region is essential for enzymatic activity (Stuhlsatz-Krouper et al. 1998). Depletion of ATP from the cell also reduces FA transport, suggesting FA transport requires ATP, perhaps because it is an ACS (Stuhlsatz-Krouper et al. 1998). Although its structure has not been established, a series of molecular and cell biology experiments suggest that FATP1 has only one membrane-spanning region and several membrane-associated regions (Lewis et al. 2001). Because this arrangement is not typical for a channel or transporter, FATP may increase FA internalization by increasing the rate of "flip-flop," trapping the FA in the inner leaflet of the plasma membrane, or activating the FA to its FA-CoA species.

Disruption of the FATP4 gene in mice established that it performs an essential function in normal mouse development. Three lines of mice, in which the FATP4 allele was either deleted or mutated, exhibited early embryonic lethality or died soon after they were born. Deletion of exons 1 and 2 resulted in early embryonic lethality at approximately embryonic day 9.5, but the cause of the lethality was not determined (Gimeno et al. 2003). In two studies in which the animals died shortly after birth, exon 3 was either deleted (Herrmann et al. 2003) or contained a spontaneous disruption mutation (Moulson et al. 2003). Regardless of the genotype, both disruptions lead to neonatal lethality owing to breathing difficulties secondary to tight skin that results from a disrupted skin barrier. This phenotype is similar to the human disease restrictive dermopathy. This suggests that FATP4 is required for proper development of the epidermal barrier (Herrmann et al. 2003, Moulson et al. 2003).

Surprisingly little has been done to determine the specificity of FATP enzymes for different FAs. Overexpression of FATP1 in 3T3-L1 cells results in 2.6-fold increases in the internalization of PA, OA, and AA, suggesting that FATP1 does have a preference for any of these FAs (Schaffer & Lodish 1994). However, in 293 cells, little, if any, AA is internalized, and overexpression of FATP4 in these cells has little effect on AA internalization (Stahl et al. 1999), suggesting FATP4 does not promote AA internalization.

Acyl CoA Synthetases

An Acsl specific for AA was first described in human platelets as an activity that was distinct from that of broadly activating ACS (Wilson et al. 1982). Subsequently, an AA-specific enzyme activity was separated from that of nonspecific ACS, establishing that there were at least two ACS enzymes (Laposata et al. 1985). The rat Acsl1 protein was purified biochemically and the amino acid and cDNA sequences were determined (Suzuki et al. 1990). Subsequently, four additional Acsl proteins (Acsl6,

Acsl3, Acsl4, and Acsl5), each possessing a unique tissue and developmental expression pattern, were identified (Fujino et al. 1996, Fujino & Yamamoto 1992, Kang et al. 1997, Oikawa et al. 1998). Of particular interest in the brain are Acsl6, Acsl3, and Acsl4. In rat brain, Acsl6 mRNA increases from postnatal day 15 through adulthood and Acsl3 mRNA expression is highest during early postnatal (5–20 d) development and lower in the adult, whereas Acsl1 mRNA level does not change during brain development. Additionally, in reaggregated rat brain cell cultures, Acsl6 and Acsl3 mRNAs are upregulated during differentiation, whereas the level of Acsl1 mRNA is unchanged (Basu-Modak et al. 1999). An important role for Acsl6 in PUFA internalization was established in PC12 cells: Overexpression of Acsl6 increased PUFA internalization, leading to increased PL synthesis and neurite outgrowth (Marszalek et al. 2004, 2005).

There is evidence that specific Acsl proteins, unlike FATPs, activate PUFAs with different efficiencies. Acsl1 and Acsl5 activate shorter-length FAs (16–18 carbons) with nearly equal efficiencies, Acsl3 and Acsl4 activate 20:4 (*n-6*) and 20:5 (*n-3*) with the greatest efficiency, and Acsl6 efficiently activates both DHA and AA. In PC12 cells, overexpression of Acsl6 preferentially increases the conversion of DHA to DHA-CoA, suggesting that Acsl6 may function primarily in DHA metabolism (Marszalek et al. 2005). Interestingly, a signature 25-amino-acid consensus sequence common to all fatty Acsls has been identified. Many mutations in this sequence in the bacterial homologue of Acsl either lower enzymatic activity and/or alter the substrate specificity (Black et al. 1997). However, it is unclear whether this sequence in mammalian Acsl proteins contributes to FA specificity.

Whereas recent evidence suggests that Acsl proteins are capable of promoting FA internalization by themselves (Schmelter et al. 2004), evidence is mounting that FATPs and Acsls may act in concert to facilitate FA internalization. Overexpression of FATP1 and Acsl1 in 3T3-L1 cells increases FA internal-

ization more than does either of these proteins alone; these proteins likely interact physically (Gargiulo et al. 1999). Additionally, FATP1 knockout mice are protected from deleterious rises in the levels of intramuscular acyl CoA that normally occur from a high-fat diet; this is consistent with FATPs and Acsls functioning together to internalize and convert FAs to their CoA derivatives (Kim et al. 2004b). Similarly, the yeast homologues of FATP (Fat1p) and Acsl (Faa1p and Faa4p) display similar propensities to interact physically and functionally to facilitate FA internalization (Zou et al. 2003). Therefore, FATP and Acsl proteins probably act in concert to efficiently import exogenous FAs.

An important function for Acsl4 in AA metabolism was established using Acsl4 knockout mice in which female fertility is impaired. In these mice, AA metabolism is altered in the uterus, resulting in elevated levels of several AA-derived PGs (Cho et al. 2001). Interestingly, Acsl4 mutations are found in several patients with Alport syndrome, elliptocytosis, and mental retardation (Meloni et al. 2002, Piccini et al. 1998). Additionally, an allele of Acsl4 is associated with increased prevalence of depression (Covault et al. 2004). These data suggest that Acsl4 may prevent AA, which is released from the membrane by PLA_2, from being converted to inflammatory prostaglandins because AA instead becomes activated to the CoA derivative, which is not a substrate for COX-2.

Intracellular Fatty Acid–Binding Proteins

FABPs are a multigene family of 15-kDa cytosolic proteins that are the counterpart to extracellular albumin. There are a total of nine FABPs, but only Heart-FABP (H-FABP), Epidermal-FABP (E-FABP), and Brain-FABP (B-FABP)—named after the tissue in which they were first identified—are expressed in the nervous system. (Myers-Payne et al. 1996, Owada 1996, Sellner et al. 1995). FABPs are believed to be involved in the uptake, transport, and targeting of long-chain FAs to specific intracellular organelles, where they can be metabolized. FABPs may promote FA internalization by removing FAs from the inner leaflet of the plasma membrane by collisional transfer with anionic PLs (Thumser et al. 2001) or by acting as a "sink" for unesterified FAs, thus decreasing the apparent intracellular FA concentration (**Figure 2**). Although there is some evidence that H-FABP prefers *n-6* FAs and B-FABP prefers *n-3* FAs (Hanhoff et al. 2002), overall there appears to be little FA specificity among FABP species (Veerkamp & Zimmerman 2001).

H-FABP expression is first detected postnatally, increases in gray matter during development (Owada et al. 1996), and is significantly decreased in aged mice (Pu et al. 1999). H-FABP concentration is highest in

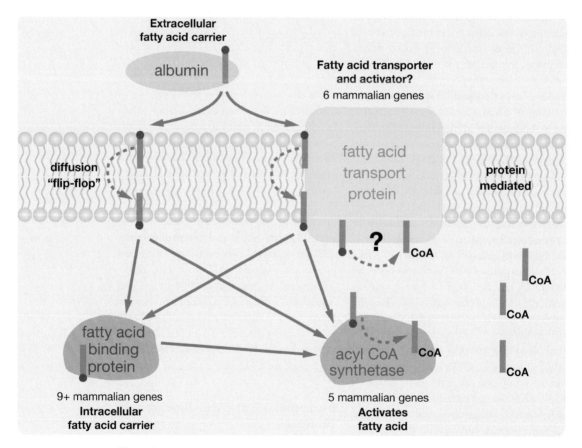

Figure 2

Proposed models for FA internalization in neurons. In the diffusion model of FA internalization, FAs are released from albumin and, owing to their hydrophobicity, partition into the outer leaflet of the plasma membrane. The FAs then "flip-flop" from the outer leaflet to the inner leaflet of the plasma membrane, where they can dissociate or be extracted by FABPs or Acsl proteins. FAs bound to FABPs are ultimately transferred to Acsl proteins, which catalyze the reaction that converts the free FAs to their CoA derivative. The FA-CoA cannot repartition into the membrane owing to the presence of the hydrophilic CoA. Additionally, FA internalization may be mediated by FATPs. FATPs may increase the rate or efficiency of FA "flip-flop" or increase FA stability in the inner leaflet, which increases the opportunity for FAs to interact with either FABPs or Acsl proteins. It remains controversial whether FATPs themselves convert the FAs to their FA-CoA derivative.

the synaptosomes of nerve endings, which suggests H-FABP has an important function at the synapse (Pu et al. 1999). Interestingly, H-FABP protein levels are decreased in patients with Down's syndrome and AD (Cheon et al. 2003). Although Binas et al. (1999) have generated a knockout of H-FABP, a thorough examination of the nervous system in these animals has not been performed.

E-FABP has a role during neurodevelopment: It is highly expressed during neurogenesis, neural migration, and differentiation of neurons and retinal ganglion cells (Liu et al. 1997, 2000, Allen 2001). Moreover, its expression is induced in dorsal root ganglion neurons after peripheral injury (De Leon et al. 1996). Although antisense knockdown of E-FABP in PC12 cells reduces neurite outgrowth (Allen et al. 2000), knockout of E-FABP does not result in any obvious neurological phenotype at the systemic level (Owada et al. 2002). However, as in the H-FABP knockout, a detailed analysis of neurodevelopment was not examined in these mice. Finally, whereas B-FABP is abundantly expressed in the human fetal brain (Shimizu et al. 1997), it is most highly expressed in glial cells (Owada et al. 1996).

Alterations in Polyunsaturated Fatty Acid Levels and Neuronal Function

As discussed above, DHA is especially abundant in the brain (Salem et al. 1986), the retina (Anderson et al. 1970), and sperm (Poulos et al. 1973). During periods of *n-3* deficiency, DHA is aggressively retained in the PLs of neurons through at least two mechanisms. First, DHA released from membrane lipids is rapidly reacylated into PLs. Second, there is a significant reduction in the rate of transfer of DHA out of the nervous system through the blood-brain barrier (Contreras et al. 2000). Many neurodegenerative conditions, such as the peroxisomal disorders (Martinez 1992a) and AD (Soderberg et al. 1991), are associated with reduced levels of *n-3* FAs.

Alzheimer's Disease

Data from numerous epidemiological studies suggest an inverse correlation between DHA intake and the likelihood of developing AD (Morris et al. 2003, Tully et al. 2003, and others). Other studies have shown that levels of PE, which is enriched in DHA (**Table 1**), and PI, which is enriched in AA, are significantly reduced in brain PLs of individuals affected by AD (Hashimoto et al. 2002). Specifically, there is a significant reduction in the amount of DHA in the hippocampus PLs of patients with AD (Soderberg et al. 1991). Pretreatment of rats with DHA protected the animals from the memory loss that typically occurs when they are infused with AD's Aβ peptide (1–40) (Hashimoto et al. 2002). In a transgenic mouse model of AD (Tg2576), which expresses high levels of a mutant amyloid precursor protein (APP), treatment for three months with a diet deficient in *n-3* FAs resulted in a decrease in the amount of DHA in the PLs of transgenic animals, but not in the normal controls (Calon et al. 2004). In cognitive tests, the mutant mice did not perform as well, and there were significant reductions in the levels of PI$_3$-kinase and the postsynaptic actin-regulating protein drebrin, similar to what occurs in AD brains. Additionally, cPLA$_2$ activity is increased in the brains of people with AD (Stephenson et al. 1996), suggesting that increased generation of AA-derived eicosanoids may contribute to the etiology of AD.

Depression

Depression is a multifactorial illness whose etiology is influenced by genetic, environmental, and nutritional factors. Support for a nutritional contribution derives from several studies reporting an inverse correlation between the level of DHA and/or EPA, as measured in either RBC-PLs or adipose tissue, and symptoms of depression. Over the past five years, numerous studies involving *n-3* FA supplementation show a reduction in many

ALS: amyotrophic lateral sclerosis

of the symptoms of many forms of depression, including bipolar disorder, agoraphobia, and anorexia nervosa (Logan 2004). Overall, EPA supplementation may be more effective than DHA supplementation in reducing the symptoms of depression. However, at this early stage of research, the mechanism(s) as to how *n-3* FAs may reduce the symptoms of depression are unclear. Increased levels of *n-3* FAs may alter the activity of integral membrane proteins (i.e., receptors), alter the fluidity of the plasma membrane, and counteract the proinflammatory functions of AA-derived eicosanoids.

Retinal Function

As stated earlier, DHA is essential for the proper development of the retina, in particular at the synapse and the outer segment of photoreceptors. Numerous studies have identified a correlation between retinitis pigmentosa and low DHA levels. There is some evidence that the ability to synthesize DHA is impaired in patients with X-linked retinitis pigmentosa (Hoffman et al. 2001). Additionally, DHA levels are low in the outer segments of rats carrying rhodopsin and peripherin mutations (Anderson et al. 2001, Anderson et al. 2002). Recently, double-blind controlled clinical studies designed to assess whether dietary supplement with large amounts of DHA reduced degeneration or increased retinal function were reported (Berson et al. 2004, Hoffman et al. 2004). Supplementation with 400 mg of DHA for 4 y clearly increased the mean concentration of DHA in red blood cells. In one positive study, the youngest patients showed a significant reduction in rod electroretinogram (which measures photoreceptor function) functional loss, suggesting that early intervention may be important in slowing the progression of the retinitis pigmentosa (Hoffman et al. 2004). In another positive study in which patients also received vitamin A in addition to DHA, the progression of the disease in these patients was slowed during the first two years of therapy (Berson et al. 2004).

Amyotrophic Lateral Sclerosis, Parkinson's Disease, and Inflammatory Eicosanoids

Recent evidence suggests that increased COX-2 activity is an important component to the etiology of the neurodegeneration that occurs during the progression of ALS and possibly Parkinson's disease and AD as well (Consilvio et al. 2004). This is most likely a consequence of the production of inflammatory eicosanoids primarily from PLA_2-liberated AA in neurons and supporting glia. In recent studies, the use of COX-2 inhibitors to decrease the synthesis of prostanoid products significantly reduced motor-neuron cell death in models of ALS (Drachman & Rothstein 2000). However, no studies have been reported that assess whether increased dietary intake of *n-3* FAs can counteract the proinflammatory effects of AA and ameliorate the symptoms of ALS.

CONCLUSION AND FUTURE DIRECTIONS

A significant body of data clearly establishes that DHA is important for proper neurodevelopment. Recent epidemiological evidence suggests that the decreasing DHA levels that normally occur during aging and that are exacerbated by reduced dietary intake may increase the prevalence of several neurological disorders. However, we are only beginning to understand the various functions that DHA performs as both a free FA and when it is incorporated into PLs. Research is needed to understand better the process of DHA internalization (i.e., the roles of FATPs and Acsls) as well as DHA metabolism into PLs. Additionally, the recent identification of several DHA metabolites likely involved in cell-cell signaling suggests that free DHA is utilized to perform many functions beyond a structural role in PLs. Investigation into the proteins that preferentially promote DHA internalization and metabolism is clearly needed to fully understand the importance of DHA in the developing and aging nervous system.

Establishing the function of DHA in the nervous system is critical to appreciate the possible health implications of the reduced dietary intake of DHA currently taking place in Western societies.

LITERATURE CITED

Ahmad A, Moriguchi T, Salem N. 2002a. Decrease in neuron size in docosahexaenoic acid-deficient brain. *Pediatr. Neurol.* 26:210–18

Ahmad A, Murthy M, Greiner RS, Moriguchi T, Salem N Jr. 2002b. A decrease in cell size accompanies a loss of docosahexaenoate in the rat hippocampus. *Nutr. Neurosci.* 5:103–13

Aid S, Vancassel S, Poumes-Ballihaut C, Chalon S, Guesnet P, Lavialle M. 2003. Effect of a diet-induced *n-3* PUFA depletion on cholinergic parameters in the rat hippocampus. *J. Lipid. Res.* 44:1545–51

Akbar M, Kim HY. 2002. Protective effects of docosahexaenoic acid in staurosporine-induced apoptosis: involvement of phosphatidylinositol-3 kinase pathway. *J. Neurochem.* 82:655–65

Akiyama N, Hatori Y, Takashiro Y, Hirabayashi T, Saito T, Murayama T. 2004. Nerve growth factor-induced up-regulation of cytosolic phospholipase $A_2\alpha$ level in rat PC12 cells. *Neurosci. Lett.* 365:218–22

Allen GW, Liu JW, De Leon M. 2000. Depletion of a fatty acid–binding protein impairs neurite outgrowth in PC12 cells. *Brain Res. Mol. Brain Res.* 76:315–24

Anderson RE, Maude MB, Bok D. 2001. Low docosahexaenoic acid levels in rod outer segment membranes of mice with rds/peripherin and P216L peripherin mutations. *Invest. Ophthalmol. Vis. Sci.* 42:1715–20

Anderson RE, Maude MB, Feldman GL. 1970. Lipids of ocular tissues. 3. The phospholipids of mature bovine iris. *Exp. Eye Res.* 9:281–84

Anderson RE, Maude MB, McClellan M, Matthes MT, Yasumura D, LaVail MM. 2002. Low docosahexaenoic acid levels in rod outer segments of rats with P23H and S334ter rhodopsin mutations. *Mol. Vis.* 8:351–58

Araki W, Wurtman RJ. 1998. How is membrane phospholipid biosynthesis controlled in neural tissues? *J. Neurosci. Res.* 51:667–74

Barcelo-Coblijn G, Hogyes E, Kitajka K, Puskas LG, Zvara A, et al. 2003a. Modification by docosahexaenoic acid of age-induced alterations in gene expression and molecular composition of rat brain phospholipids. *Proc. Natl. Acad. Sci. USA* 100:11321–26

Barcelo-Coblijn G, Kitajka K, Puskas LG, Hogyes E, Zvara A, et al. 2003b. Gene expression and molecular composition of phospholipids in rat brain in relation to dietary *n-6* to *n-3* fatty acid ratio. *Biochim. Biophys. Acta* 1632:72–79

Basu-Modak S, Braissant O, Escher P, Desvergne B, Honegger P, Wahli W. 1999. Peroxisome proliferator-activated receptor beta regulates acyl-CoA synthetase 2 in reaggregated rat brain cell cultures. *J. Biol. Chem.* 274:35881–88

Bazan NG. 2003. Synaptic lipid signaling: significance of polyunsaturated fatty acids and platelet-activating factor. *J. Lipid. Res.* 44:2221–33

Bazan NG Jr. 1971. Phospholipases A_1 and A_2 in brain subcellular fractions. *Acta Physiol. Lat. Am.* 21:101–6

Berson EL, Rosner B, Sandberg MA, Weigel-DiFranco C, Moser A, et al. 2004. Further evaluation of docosahexaenoic acid in patients with retinitis pigmentosa receiving vitamin A treatment: subgroup analyses. *Arch. Ophthalmol.* 122:1306–14

Binas B, Danneberg H, McWhir J, Mullins L, Clark AJ. 1999. Requirement for the heart-type fatty acid binding protein in cardiac fatty acid utilization. *FASEB J.* 13:805–12

Black PN, DiRusso CC. 2003. Transmembrane movement of exogenous long-chain fatty acids: proteins, enzymes, and vectorial esterification. *Microbiol. Mol. Biol. Rev.* 67:454–72

Black PN, Zhang Q, Weimar JD, DiRusso CC. 1997. Mutational analysis of a fatty acyl-coenzyme A synthetase signature motif identifies seven amino acid residues that modulate fatty acid substrate specificity. *J. Biol. Chem.* 272:4896–903

Breckenridge WC, Gombos G, Morgan IG. 1972. The lipid composition of adult rat brain synaptosomal plasma membranes. *Biochim. Biophys. Acta* 266:695–707

Calder PC. 1998. Dietary fatty acids and the immune system. *Nutr. Rev.* 56:S 70–83

Calderon F, Kim HY. 2004. Docosahexaenoic acid promotes neurite growth in hippocampal neurons. *J. Neurochem.* 90:979–88

Calon F, Lim GP, Yang F, Morihara T, Teter B, et al. 2004. Docosahexaenoic acid protects from dendritic pathology in an Alzheimer's disease mouse model. *Neuron* 43:633–45

Chen SG, Murakami K. 1994. Effects of cis-fatty acid on protein kinase C activation and protein phosphorylation in the hippocampus. *J. Pharm. Sci. Technol.* 48:71–75

Cheon MS, Kim SH, Fountoulakis M, Lubec G. 2003. Heart type fatty acid binding protein (H-FABP) is decreased in brains of patients with Down syndrome and Alzheimer's disease. *J. Neural. Transm. Suppl*: 225–34

Cho YY, Kang MJ, Sone H, Suzuki T, Abe M, et al. 2001. Abnormal uterus with polycysts, accumulation of uterine prostaglandins, and reduced fertility in mice heterozygous for acyl-CoA synthetase 4 deficiency. *Biochem. Biophys. Res. Commun.* 284:993–97

Clark JD, Schievella AR, Nalefski EA, Lin LL. 1995. Cytosolic phospholipase A_2. *J. Lipid Mediat. Cell Signal* 12:83–117

Coe NR, Smith AJ, Frohnert BI, Watkins PA, Bernlohr DA. 1999. The fatty acid transport protein (FATP1) is a very long chain acyl-CoA synthetase. *J. Biol. Chem.* 274:36300–4

Connor WE, Neuringer M, Reisbick S. 1992. Essential fatty acids: the importance of *n-3* fatty acids in the retina and brain. *Nutr. Rev.* 50:21–29

Consilvio C, Vincent AM, Feldman EL. 2004. Neuroinflammation, COX-2, and ALS–a dual role? *Exp. Neurol.* 187:1–10

Contreras MA, Greiner RS, Chang MC, Myers CS, Salem N Jr, Rapoport SI. 2000. Nutritional deprivation of alpha-linolenic acid decreases but does not abolish turnover and availability of unacylated docosahexaenoic acid and docosahexaenoyl-CoA in rat brain. *J. Neurochem.* 75:2392–400

Covault J, Pettinati H, Moak D, Mueller T, Kranzler HR. 2004. Association of a long-chain fatty acid-CoA ligase 4 gene polymorphism with depression and with enhanced niacin-induced dermal erythema. *Am. J. Med. Genet.* 127B: 42–47

De Leon M, Welcher AA, Nahin RH, Liu Y, Ruda MA, et al. 1996. Fatty acid binding protein is induced in neurons of the dorsal root ganglia after peripheral nerve injury. *J. Neurosci. Res.* 44:283–92

de Urquiza AM, Liu S, Sjoberg M, Zetterstrom RH, Griffiths W, et al. 2000. Docosahexaenoic acid, a ligand for the retinoid X receptor in mouse brain. *Science* 290:2140–44

Dehaut F, Bertrand I, Miltaud T, Pouplard-Barthelaix A, Maingault M. 1993. *n-6* polyunsaturated fatty acids increase the neurite length of PC12 cells and embryonic chick motoneurons. *Neurosci. Lett.* 161:133–36

Drachman DB, Rothstein JD. 2000. Inhibition of cyclooxygenase-2 protects motor neurons in an organotypic model of amyotrophic lateral sclerosis. *Ann. Neurol.* 48:792–95

Farooqui AA, Horrocks LA. 2004. Brain phospholipases A_2: a perspective on the history. *Prostaglandins Leukot. Essent. Fatty Acids* 71:161–69

Farooqui AA, Ong WY, Horrocks LA. 2003. Plasmalogens, docosahexaenoic acid and neurological disorders. *Adv. Exp. Med. Biol.* 544:335–54

Farooqui AA, Ong WY, Horrocks LA, Farooqui T. 2000. Brain cystosolic phospholipase A_2: localization, role, and involvement in neurological diseases. *Neuroscientist* 6:169–80

Farooqui AA, Rosenberger TA, Horrocks LA. 1997a. Arachidonic acid, neurotrauma, and neurodegenerative diseases. In *Handbook of Essential Fatty Acid Biology*, ed. S Yehuda, DI Mostofsky, pp. 277–95. Totowa, NJ: Humana Press

Farooqui AA, Yang HC, Rosenberger TA, Horrocks LA. 1997b. Phospholipase A_2 and its role in brain tissue. *J. Neurochem.* 69:889–901

Feller SE, Gawrisch K, MacKerell AD Jr 2002. Polyunsaturated fatty acids in lipid bilayers: intrinsic and environmental contributions to their unique physical properties. *J. Am. Chem. Soc.* 124:318–26

Fitscher BA, Riedel HD, Young KC, Stremmel W. 1998. Tissue distribution and cDNA cloning of a human fatty acid transport protein (hsFATP4). *Biochim. Biophys. Acta* 1443:381–85

Fliesler SJ, Anderson RE. 1983. Chemistry and metabolism of lipids in the vertebrate retina. *Prog. Lipid Res.* 22:79–131

Fujino T, Kang MJ, Suzuki H, Iijima H, Yamamoto T. 1996. Molecular characterization and expression of rat acyl-CoA synthetase 3. *J. Biol. Chem.* 271:16748–52

Fujino T, Yamamoto T. 1992. Cloning and functional expression of a novel long-chain acyl-CoA synthetase expressed in brain. *J. Biochem.* 111:197–203

Furuya H, Watanabe T, Sugioka Y, Inagaki Y, Okazaki I. 2002. Effect of ethanol and docosahexaenoic acid on nerve growth factor-induced neurite formation and neuron specific growth-associated protein gene expression in PC12 cells. *Nihon Arukoru Yakubutsu Igakkai Zasshi* 37:513–22

Garcia MC, Ward G, Ma YC, Salem N Jr, Kim HY. 1998. Effect of docosahexaenoic acid on the synthesis of phosphatidylserine in rat brain in microsomes and C6 glioma cells. *J. Neurochem.* 70:24–30

Gargiulo CE, Stuhlsatz-Krouper SM, Schaffer JE. 1999. Localization of adipocyte long-chain fatty acyl-CoA synthetase at the plasma membrane. *J. Lipid. Res.* 40:881–92

Gerbi A, Maixent JM, Barbey O, Jamme I, Pierlovisi M, et al. 1998. Alterations of Na,K-ATPase isoenzymes in the rat diabetic neuropathy: protective effect of dietary supplementation with *n-3* fatty acids. *J. Neurochem.* 71:732–40

Gimeno RE, Hirsch DJ, Punreddy S, Sun Y, Ortegon AM, et al. 2003. Targeted deletion of fatty acid transport protein-4 results in early embryonic lethality. *J. Biol. Chem.* 278:49512–16

Goracci G, Porcellati G, Woelk H. 1978. Subcellular localization and distribution of phospholipases A in liver and brain tissue. *Adv. Prostaglandin Thromboxane Res.* 3:55–67

Hall AM, Smith AJ, Bernlohr DA. 2003. Characterization of the acyl-CoA synthetase activity of purified murine fatty acid transport protein 1. *J. Biol. Chem.* 278:43008–13

Hamano H, Nabekura J, Nishikawa M, Ogawa T. 1996. Docosahexaenoic acid reduces GABA response in substantia nigra neuron of rat. *J. Neurophysiol.* 75:1264–70

Hamilton JA, Guo W, Kamp F. 2002. Mechanism of cellular uptake of long-chain fatty acids: do we need cellular proteins? *Mol. Cell Biochem.* 239:17–23

Hamilton L, Greiner R, Salem N Jr, Kim HY. 2000. *n-3* fatty acid deficiency decreases phosphatidylserine accumulation selectively in neuronal tissues. *Lipids* 35:863–69

Hanhoff T, Lucke C, Spener F. 2002. Insights into binding of fatty acids by fatty acid binding proteins. *Mol. Cell Biochem.* 239:45–54

Hashimoto M, Hossain S, Shimada T, Sugioka K, Yamasaki H, et al. 2002. Docosahexaenoic acid provides protection from impairment of learning ability in Alzheimer's disease model rats. *J. Neurochem.* 81:1084–91

Herrmann T, Buchkremer F, Gosch I, Hall AM, Bernlohr DA, Stremmel W. 2001. Mouse fatty acid transport protein 4 (FATP4): characterization of the gene and functional assessment as a very long chain acyl-CoA synthetase. *Gene* 270:31–40

Herrmann T, van der Hoeven F, Grone HJ, Stewart AF, Langbein L, et al. 2003. Mice with targeted disruption of the fatty acid transport protein 4 (Fatp 4, Slc27a4) gene show features of lethal restrictive dermopathy. *J. Cell Biol.* 161:1105–15

Hirsch D, Stahl A, Lodish HF. 1998. A family of fatty acid transporters conserved from mycobacterium to man. *Proc. Natl. Acad. Sci. USA* 95:8625–29

Hoffman DR, DeMar JC, Heird WC, Birch DG, Anderson RE. 2001. Impaired synthesis of DHA in patients with X-linked retinitis pigmentosa. *J. Lipid. Res.* 42:1395–401

Hoffman DR, Locke KG, Wheaton DH, Fish GE, Spencer R, Birch DG. 2004. A randomized, placebo-controlled clinical trial of docosahexaenoic acid supplementation for X-linked retinitis pigmentosa. *Am. J. Ophthalmol.* 137:704–18

Holman R, Johnson S, Hatch T. 1982. A case of human linolenic acid deficiency involving neurological abnormalities. *Am. J. Clin. Nutr.* 35:617–23

Holub BJ. 1978. Differential utilization of 1-palmitoyl and 1-stearoyl homologues of various unsaturated 1,2-diacyl-sn-glycerols for phosphatidylcholine and phosphatidylethanolamine synthesis in rat liver microsomes. *J. Biol. Chem.* 253:691–96

Hong S, Gronert K, Devchand PR, Moussignac RL, Serhan CN. 2003. Novel docosatrienes and 17S-resolvins generated from docosahexaenoic acid in murine brain, human blood, and glial cells. Autacoids in anti-inflammation. *J. Biol. Chem.* 278:14677–87

Hulbert AJ. 2003. Life, death and membrane bilayers. *J. Exp. Biol.* 206:2303–11

Ikemoto A, Kobayashi T, Watanabe S, Okuyama H. 1997. Membrane fatty acid modifications of PC12 cells by arachidonate or docosahexaenoate affect neurite outgrowth but not norepinephrine release. *Neurochem. Res.* 22:671–78

Innis SM. 2004. Polyunsaturated fatty acids in human milk: an essential role in infant development. *Adv. Exp. Med. Biol.* 554:27–43

Jensen CL, Heird WC. 2002. Lipids with an emphasis on long-chain polyunsaturated fatty acids. *Clin. Perinatol.* 29:261–81, vi

Kamp F, Hamilton JA. 1992. pH gradients across phospholipid membranes caused by fast flip-flop of un-ionized fatty acids. *Proc. Natl. Acad. Sci. USA* 89:11367–70

Kang MJ, Fujino T, Sasano H, Minekura H, Yabuki N, et al. 1997. A novel arachidonate-preferring acyl-CoA synthetase is present in steroidogenic cells of the rat adrenal, ovary, and testis. *Proc. Natl. Acad. Sci. USA* 94:2880–84

Kim DK, Rordorf G, Nemenoff RA, Koroshetz WJ, Bonventre JV. 1995. Glutamate stably enhances the activity of two cytosolic forms of phospholipase A$_2$ in brain cortical cultures. *Biochem. J.* 310 (Pt. 1):83–90

Kim HY, Akbar M, Lau A, Edsall L. 2000. Inhibition of neuronal apoptosis by docosahexaenoic acid (22:6 *n*-3). Role of phosphatidylserine in antiapoptotic effect. *J. Biol. Chem.* 275:35215–23

Kim HY, Bigelow J, Kevala JH. 2004a. Substrate preference in phosphatidylserine biosynthesis for docosahexaenoic acid containing species. *Biochemistry* 43:1030–36

Kim HY, Hamilton J. 2000. Accumulation of docosahexaenoic acid in phosphatidylserine is selectively inhibited by chronic ethanol exposure in C-6 glioma cells. *Lipids* 35:187–95

Kim JK, Gimeno RE, Higashimori T, Kim HJ, Choi H, et al. 2004b. Inactivation of fatty acid transport protein 1 prevents fat-induced insulin resistance in skeletal muscle. *J. Clin. Invest.* 113:756–63

Kishida E, Yano M, Kasahara M, Masuzawa Y. 1998. Distinctive inhibitory activity of docosahexaenoic acid against sphingosine-induced apoptosis. *Biochim. Biophys. Acta* 1391:401–8

Kishimoto K, Matsumura K, Kataoka Y, Morii H, Watanabe Y. 1999. Localization of cytosolic phospholipase A₂ messenger RNA mainly in neurons in the rat brain. *Neuroscience* 92:1061–77

Kitajka K, Puskas LG, Zvara A, Hackler L Jr, Barcelo-Coblijn G, et al. 2002. The role of *n-3* polyunsaturated fatty acids in brain: modulation of rat brain gene expression by dietary *n*-3 fatty acids. *Proc. Natl. Acad. Sci. USA* 99:2619–24

Knapp S, Wurtman RJ. 1999. Enhancement of free fatty acid incorporation into phospholipids by choline plus cytidine. *Brain Res.* 822:52–59

Laposata M, Reich EL, Majerus PW. 1985. Arachidonoyl-CoA synthetase. Separation from nonspecific acyl-CoA synthetase and distribution in various cells and tissues. *J. Biol. Chem.* 260:11016–20

Larsson SC, Kumlin M, Ingelman-Sundberg M, Wolk A. 2004. Dietary long-chain *n-3* fatty acids for the prevention of cancer: a review of potential mechanisms. *Am. J. Clin. Nutr.* 79:935–45

Lengqvist J, Mata De Urquiza A, Bergman AC, Willson TM, Sjovall J, et al. 2004. Polyunsaturated fatty acids including docosahexaenoic and arachidonic acid bind to the retinoid X receptor alpha ligand–binding domain. *Mol. Cell Proteomics* 3:692–703

Lewis SE, Listenberger LL, Ory DS, Schaffer JE. 2001. Membrane topology of the murine fatty acid transport protein 1. *J. Biol. Chem.* 276:37042–50

Li J, Wurtman RJ. 1999. Mechanisms whereby nerve growth factor increases diacylglycerol levels in differentiating PC12 cells. *Brain Res.* 818:252–59

Liu Y, Longo LD, De Leon M. 2000. In situ and immunocytochemical localization of E-FABP mRNA and protein during neuronal migration and differentiation in the rat brain. *Brain Res.* 852:16–27

Liu Y, Molina CA, Welcher AA, Longo LD, De Leon M. 1997. Expression of DA11, a neuronal-injury-induced fatty acid binding protein, coincides with axon growth and neuronal differentiation during central nervous system development. *J. Neurosci. Res.* 48:551–62

Logan AC. 2004. Omega-3 fatty acids and major depression: A primer for the mental health professional. *Lipids Health Dis.* 3:25

Marszalek JR, Kitidis C, Dararutana A, Lodish HF. 2004. Acyl-CoA synthetase 2 overexpression enhances fatty acid internalization and neurite outgrowth. *J. Biol. Chem.* 279:23882–91

Marszalek JR, Kitidis C, Dirusso CC, Lodish HF. 2005. Long-chain acyl CoA synthetase 6 preferentially promotes DHA metabolism. *J. Biol. Chem.* 280:10817–26

Martin RE, Bazan NG. 1992. Changing fatty acid content of growth cone lipids prior to synaptogenesis. *J. Neurochem.* 59:318–25

Martinez M. 1992a. Abnormal profiles of polyunsaturated fatty acids in the brain, liver, kidney and retina of patients with peroxisomal disorders. *Brain Res.* 583:171–82

Martinez M. 1992b. Tissue levels of polyunsaturated fatty acids during early human development. *J. Pediatr.* 120:S129–38

Matsuzawa A, Murakami M, Atsumi G, Imai K, Prados P, et al. 1996. Release of secretory phospholipase A₂ from rat neuronal cells and its possible function in the regulation of catecholamine secretion. *Biochem. J.* 318 (Pt. 2):701–9

Meloni I, Muscettola M, Raynaud M, Longo I, Bruttini M, et al. 2002. FACL4, encoding fatty acid-CoA ligase 4, is mutated in nonspecific X-linked mental retardation. *Nat. Genet.* 30:436–40

Moore SA. 2001. Polyunsaturated fatty acid synthesis and release by brain-derived cells in vitro. *J. Mol. Neurosci.* 16:195–200

Moriguchi K, Yuri T, Yoshizawa K, Kiuchi K, Takada H, et al. 2003. Dietary docosahexaenoic acid protects against N-methyl-N-nitrosourea-induced retinal degeneration in rats. *Exp. Eye Res.* 77:167–73

Morris MC, Evans DA, Bienias JL, Tangney CC, Bennett DA, et al. 2003. Consumption of fish and *n-3* fatty acids and risk of incident Alzheimer disease. *Arch. Neurol.* 60:940–46

Moulson CL, Martin DR, Lugus JJ, Schaffer JE, Lind AC, Miner JH. 2003. Cloning of wrinkle-free, a previously uncharacterized mouse mutation, reveals crucial roles for fatty acid transport protein 4 in skin and hair development. *Proc. Natl. Acad. Sci. USA* 100:5274–79

Mukherjee PK, Marcheselli VL, Serhan CN, Bazan NG. 2004. Neuroprotectin D1: a docosa-hexaenoic acid-derived docosatriene protects human retinal pigment epithelial cells from oxidative stress. *Proc. Natl. Acad. Sci. USA* 101:8491–96

Myers-Payne SC, Hubbell T, Pu L, Schnutgen F, Borchers T, et al. 1996. Isolation and charac-terization of two fatty acid binding proteins from mouse brain. *J. Neurochem.* 66:1648–56

Nagata S. 1997. Apoptosis by death factor. *Cell* 88:355–65

Nakashima S, Ikeno Y, Yokoyama T, Kuwana M, Bolchi A, et al. 2003. Secretory phospholipases A_2 induce neurite outgrowth in PC12 cells. *Biochem. J.* 376:655–66

Narayanan BA, Narayanan NK, Reddy BS. 2001. Docosahexaenoic acid regulated genes and transcription factors inducing apoptosis in human colon cancer cells. *Int. J. Oncol.* 19:1255–62

Negre-Aminou P, Nemenoff RA, Wood MR, de la Houssaye BA, Pfenninger KH. 1996. Char-acterization of phospholipase A_2 activity enriched in the nerve growth cone. *J. Neurochem.* 67:2599–608

Nishizuka Y. 1995. Protein kinase C and lipid signaling for sustained cellular responses. *FASEB J.* 9:484–96

Niu S-L, Mitchell DC, Lim S-Y, Wen Z-M, Kim H-Y, et al. 2004. Reduced G protein-coupled signaling efficiency in retinal rod outer segments in response to *n-3* fatty acid deficiency. *J. Biol. Chem.* 279:31098–104

Oikawa E, Iijima H, Suzuki T, Sasano H, Sato H, et al. 1998. A novel acyl-CoA synthetase, ACS5, expressed in intestinal epithelial cells and proliferating preadipocytes. *J. Biochem.* 124:679–85

Okuda S, Saito H, Katsuki H. 1994. Arachidonic acid: toxic and trophic effects on cultured hippocampal neurons. *Neuroscience* 63:691–99

Owada Y, Suzuki I, Noda T, Kondo H. 2002. Analysis on the phenotype of E-FABP-gene knockout mice. *Mol. Cell Biochem.* 239:83–86

Owada Y, Yoshimoto T, Kondo H. 1996. Spatio-temporally differential expression of genes for three members of fatty acid binding proteins in developing and mature rat brains. *J. Chem. Neuroanat.* 12:113–22

Pardue S, Rapoport SI, Bosetti F. 2003. Co-localization of cytosolic phospholipase A_2 and cyclooxygenase-2 in Rhesus monkey cerebellum. *Brain Res. Mol. Brain Res.* 116:106–14

Piccini M, Vitelli F, Bruttini M, Pober BR, Jonsson JJ, et al. 1998. FACL4, a new gene en-coding long-chain acyl-CoA synthetase 4, is deleted in a family with Alport syndrome, elliptocytosis, and mental retardation. *Genomics* 47:350–58

Poulos A, Darin-Bennett A, White IG. 1973. The phospholipid-bound fatty acids and aldehy-des of mammalian spermatozoa. *Comp. Biochem. Physiol. B.* 46:541–49

Pownall HJ, Hamilton JA. 2003. Energy translocation across cell membranes and membrane models. *Acta Physiol. Scand.* 178:357–65

Pu L, Igbavboa U, Wood WG, Roths JB, Kier AB, et al. 1999. Expression of fatty acid binding proteins is altered in aged mouse brain. *Mol. Cell Biochem.* 198:69–78

Puskas LG, Kitajka K, Nyakas C, Barcelo-Coblijn G, Farkas T. 2003. Short-term administration of omega 3 fatty acids from fish oil results in increased transthyretin transcription in old rat hippocampus. *Proc. Natl. Acad. Sci. USA* 100:1580–85

Rapoport SI. 2001. In vivo fatty acid incorporation into brain phosholipids in relation to plasma availability, signal transduction and membrane remodeling. *J. Mol. Neurosci.* 16:243–61

Rojas CV, Martinez JI, Flores I, Hoffman DR, Uauy R. 2003. Gene expression analysis in human fetal retinal explants treated with docosahexaenoic acid. *Invest. Ophthalmol. Vis. Sci.* 44:3170–77

Rotstein NP, Aveldano MI, Barrantes FJ, Roccamo AM, Politi LE. 1997. Apoptosis of retinal photoreceptors during development in vitro: protective effect of docosahexaenoic acid. *J. Neurochem.* 69:504–13

Salem N Jr, Litman B, Kim HY, Gawrisch K. 2001. Mechanisms of action of docosahexaenoic acid in the nervous system. *Lipids* 36:945–59

Salem NJ, Kim HY, Yergey JA. 1986. Docosahexaenoic acid: membrane function and metabolism. In *Health Effects of Polyunsaturated Fatty Acids in Seafood*, ed. AP Simopoulos, RE Martin, RR Kifer, pp. 263–317. New York: Academic

Sastry PS, Rao KS. 2000. Apoptosis and the nervous system. *J. Neurochem.* 74:1–20

Schaffer JE, Lodish HF. 1994. Expression cloning and characterization of a novel adipocyte long chain fatty acid transport protein. *Cell* 79:427–36

Schmelter T, Trigatti BL, Gerber GE, Mangroo D. 2004. Biochemical demonstration of the involvement of fatty acyl-CoA synthetase in fatty acid translocation across the plasma membrane. *J. Biol. Chem.* 279:24163–70

Scott BL, Bazan NG. 1989. Membrane docosahexaenoate is supplied to the developing brain and retina by the liver. *Proc. Natl. Acad. Sci. USA* 86:2903–7

Sellner PA, Chu W, Glatz JF, Berman NE. 1995. Developmental role of fatty acid–binding proteins in mouse brain. *Brain Res. Dev. Brain Res.* 89:33–46

Serhan CN, Gotlinger K, Hong S, Arita M. 2004. Resolvins, docosatrienes, and neuro-protectins, novel omega-3-derived mediators, and their aspirin-triggered endogenous epimers: an overview of their protective roles in catabasis. *Prostaglandins Other Lipid Mediat.* 73:155–72

Sessler AM, Kaur N, Palta JP, Ntambi JM. 1996. Regulation of stearoyl-CoA desaturase 1 mRNA stability by polyunsaturated fatty acids in 3T3-L1 adipocytes. *J. Biol. Chem.* 271:29854–58

Seung Kim HF, Weeber EJ, Sweatt JD, Stoll AL, Marangell LB. 2001. Inhibitory effects of omega-3 fatty acids on protein kinase C activity in vitro. *Mol. Psychiatry* 6:246–48

Shimizu F, Watanabe TK, Shinomiya H, Nakamura Y, Fujiwara T. 1997. Isolation and expression of a cDNA for human brain fatty acid–binding protein (B-FABP). *Biochim. Biophys. Acta* 1354:24–28

Simopoulos AP. 2002. The importance of the ratio of omega-6/omega-3 essential fatty acids. *Biomed. Pharmacother.* 56:365–79

Soderberg M, Edlund C, Kristensson K, Dallner G. 1991. Fatty acid composition of brain phospholipids in aging and in Alzheimer's disease. *Lipids* 26:421–25

Stahl A, Hirsch DJ, Gimeno RE, Punreddy S, Ge P, et al. 1999. Identification of the major intestinal fatty acid transport protein. *Mol. Cell* 4:299–308

Stephenson D, Rash K, Smalstig B, Roberts E, Johnstone E, et al. 1999. Cytosolic phospholipase A$_2$ is induced in reactive glia following different forms of neurodegeneration. *Glia* 27:110–28

Stephenson DT, Lemere CA, Selkoe DJ, Clemens JA. 1996. Cytosolic phospholipase A$_2$ (cPLA2) immunoreactivity is elevated in Alzheimer's disease brain. *Neurobiol. Dis.* 3:51–63

Stuhlsatz-Krouper SM, Bennett NE, Schaffer JE. 1998. Substitution of alanine for serine 250 in the murine fatty acid transport protein inhibits long chain fatty acid transport. *J. Biol. Chem.* 273:28642–50

Suburo A, Cei de Job C. 1986. The biphasic effect of phospholipase A$_2$ inhibitors on axon elongation. *Int. J. Dev. Neurosci.* 4:363–67

Suzuki H, Kawarabayasi Y, Kondo J, Abe T, Nishikawa K, et al. 1990. Structure and regulation of rat long-chain acyl-CoA synthetase. *J. Biol. Chem.* 265:8681–85

Teague WE, Fuller NL, Rand RP, Gawrisch K. 2002. Polyunsaturated lipids in membrane fusion events. *Cell Mol. Biol. Lett.* 7:262–64

Thornberry NA, Lazebnik Y. 1998. Caspases: enemies within. *Science* 281:1312–16

Thumser AE, Tsai J, Storch J. 2001. Collision-mediated transfer of long-chain fatty acids by neural tissue fatty acid–binding proteins (FABP): studies with fluorescent analogs. *J. Mol. Neurosci.* 16:143–50

Tsukada Y, Chiba K, Yamazaki M, Mohri T. 1994. Inhibition of the nerve growth factor-induced neurite outgrowth by specific tyrosine kinase and phospholipase inhibitors. *Biol. Pharm. Bull.* 17:370–75

Tully AM, Roche HM, Doyle R, Fallon C, Bruce I, et al. 2003. Low serum cholesteryl ester-docosahexaenoic acid levels in Alzheimer's disease: a case-control study. *Br. J. Nutr.* 89:483–89

Utsunomiya A, Owada Y, Yoshimoto T, Kondo H. 1997. Localization of mRNA for fatty acid transport protein in developing and mature brain of rats. *Brain Res. Mol. Brain Res.* 46:217–22

Valentine RC, Valentine DL. 2004. Omega-3 fatty acids in cellular membranes: a unified concept. *Prog. Lipid Res.* 43:383–402

Veerkamp JH, Zimmerman AW. 2001. Fatty acid–binding proteins of nervous tissue. *J. Mol. Neurosci.* 16:133–42

Watkins PA, Lu JF, Steinberg SJ, Gould SJ, Smith KD, Braiterman LT. 1998. Disruption of the *Saccharomyces cerevisiae* FAT1 gene decreases very long-chain fatty acyl-CoA synthetase activity and elevates intracellular very long-chain fatty acid concentrations. *J. Biol. Chem.* 273:18210–19

Wen Z, Kim HY. 2004. Alterations in hippocampal phospholipid profile by prenatal exposure to ethanol. *J. Neurochem.* 89:1368–77

Williams EJ, Furness J, Walsh FS, Doherty P. 1994a. Characterisation of the second messenger pathway underlying neurite outgrowth stimulated by FGF. *Development* 120:1685–93

Williams EJ, Walsh FS, Doherty P. 1994b. The production of arachidonic acid can account for calcium channel activation in the second messenger pathway underlying neurite outgrowth stimulated by NCAM, N-cadherin, and L1. *J. Neurochem.* 62:1231–34

Wilson DB, Prescott SM, Majerus PW. 1982. Discovery of an arachidonoyl coenzyme A synthetase in human platelets. *J. Biol. Chem.* 257:3510–15

Ximenes da Silva A, Lavialle F, Gendrot G, Guesnet P, Alessandri JM, Lavialle M. 2002. Glucose transport and utilization are altered in the brain of rats deficient in *n-3* polyunsaturated fatty acids. *J. Neurochem.* 81:1328–37

Zou Z, Tong F, Faergeman NJ, Borsting C, Black PN, DiRusso CC. 2003. Vectorial acylation in *Saccharomyces cerevisiae*. Fat1p and fatty acyl-CoA synthetase are interacting components of a fatty acid import complex. *J. Biol. Chem.* 278:16414–22

Specificity and Versatility in TGF-β Signaling Through Smads

Xin-Hua Feng[1] and Rik Derynck[2]

[1]Department of Molecular and Cellular Biology, Michael E. DeBakey Department of Surgery, Biology of Inflammation Center, Cancer Center, and Interdepartmental Program in Cell and Molecular Biology, Baylor College of Medicine, Houston, Texas 77030; email: xfeng@bcm.tmc.edu

[2]Departments of Cell and Tissue Biology and Anatomy, Programs in Cell Biology and Developmental Biology, University of California, San Francisco, California 94143; email: derynck@itsa.ucsf.edu

Annu. Rev. Cell Dev. Biol.
2005. 21:659–93

First published online as a Review in Advance on July 1, 2005

The *Annual Review of Cell and Developmental Biology* is online at http://cellbio.annualreviews.org

doi: 10.1146/annurev.cellbio.21.022404.142018

Copyright © 2005 by Annual Reviews. All rights reserved

1081-0706/05/1110-0659$20.00

Key Words

TGF-β receptor, bone morphogenetic protein, signal transduction, complexity, transcription

Abstract

The TGF-β family comprises many structurally related differentiation factors that act through a heteromeric receptor complex at the cell surface and an intracellular signal transducing Smad complex. The receptor complex consists of two type II and two type I transmembrane serine/threonine kinases. Upon phosphorylation by the receptors, Smad complexes translocate into the nucleus, where they cooperate with sequence-specific transcription factors to regulate gene expression. The vertebrate genome encodes many ligands, fewer type II and type I receptors, and only a few Smads. In contrast to the perceived simplicity of the signal transduction mechanism with few Smads, the cellular responses to TGF-β ligands are complex and context dependent. This raises the question of how the specificity of the ligand-induced signaling is achieved. We review the molecular basis for the specificity and versatility of signaling by the many ligands through this conceptually simple signal transduction mechanism.

Contents

INTRODUCTION

Members of the TGF-β family are secreted polypeptides that activate cellular responses during growth and differentiation. More than 60 TGF-β family members have been identified in multicellular organisms, with at least 29 and probably up to 42 proteins encoded by the human genome. Among these 60, there are three TGF-βs, five activins, and at least eight BMPs encoded by different genes.

TGF-β-related factors are made as precursors with a large propeptide and a C-terminal mature polypeptide that is proteolytically cleaved from the precursor (Annes et al. 2003). Mature TGF-β is a homodimer of two 12.5-kd polypeptides joined by a disulfide bond. Two copies of the propeptide remain associated with the TGF-β and maintain it in an inactive complex known as LTBP. An LTBP is often linked to the prosegment and plays a role in targeting the complex to the extracellular matrix, where TGF-β is activated and released by proteolytic cleavage of the prosegment (Annes et al. 2003). Other TGF-β family members are also expressed as disulfide-linked homodimers or heterodimers, are likely secreted as complexes, and undergo proteolytic activation. The activities of TGF-β family members are often regulated by secreted and matrix-associated proteins that bind the ligands in solution, thus sequestering the ligands from access to their receptors or helping to ensure ready availability (Annes et al. 2003). In addition, several cell surface proteins function as coreceptors and help in the presentation of the ligand to the receptor.

TGF-β family members are expressed in most cell types and play key roles in differentiation and tissue morphogenesis. TGF-β itself inhibits proliferation of many cell types, including epithelial and hematopoietic cells, and its signaling controls tumorigenesis. The cell's responses to TGF-β are complex as a result of differential transcriptional regulation and nontranscriptional effects that depend on the cell context and physiological environment.

The cell surface receptors for TGF-β-related proteins and Smads as intracellular effectors of TGF-β responses have been identified, and the general scheme for signaling from the cell surface to the nucleus has been established (Derynck & Zhang 2003, Shi & Massagué 2003). In addition, receptor activation induces non-Smad signaling

TGF-β:
transforming growth
factor-β

pathways that can regulate Smad signaling or lead to Smad-independent responses (Derynck & Zhang 2003). This review focuses on the question of how the Smad signaling mechanism, comprised of a limited number of receptors and Smads and a multiplicity of TGF-β ligands, exerts specificity and at the same time displays a considerable versatility in cellular responses.

THE GENERAL MODEL OF TGF-β-INDUCED SIGNALING THROUGH SMADS

TGF-β proteins signal through cell surface complexes of "type I" and "type II" receptors. These two types are structurally similar transmembrane serine/threonine kinases, but type I receptors have a conserved Gly/Ser-rich "GS sequence" immediately upstream from the kinase domain. Ligand binding allows the formation of a stable receptor complex consisting of two receptors of each type, allowing phosphorylation of the GS sequences by the type II receptor kinases. This phosphorylation activates the type I receptor kinases, resulting in autophosphorylation of the type I receptor and phosphorylation of Smad proteins (**Figure 1**) (Derynck & Zhang 2003, Shi & Massagué 2003).

The Smads are the only established intracellular effectors of TGF-β signaling. Smads exist as three subgroups: R-Smads, a common Smad (e.g., Smad4 in vertebrates), and inhibitory Smads. R-Smads and Smad4 contain two conserved polypeptide segments, the MH1 (N) and MH2 (C) domains linked by a less conserved linker region. The R-Smads have a C-terminal SXS motif in which both serines are targeted for direct phosphorylation by the type I receptors. Thus, upon ligand binding, the type I receptors recruit and phosphorylate R-Smads, i.e., Smad2 and Smad3, by the TβRI/ALK-5 type I receptor in response to TGF-β, and Smad1, Smad5, and Smad8 by BMP type I receptors. C-terminal SXS phosphorylation of the R-Smads leads to their conformational changes,

their dissociation from the type I receptors, and the formation of a trimeric complex consisting of two R-Smads and one Smad4. This trimeric complex translocates into the nucleus, where the Smads act as transcription factors (Derynck & Zhang 2003, Shi & Massagué 2003).

Smads act as ligand-induced transcription regulators of TGF-β responses. At the regulatory DNA sequences of genes, Smads activate transcription through assembly of a large nucleoprotein complex consisting of Smad-binding DNA elements, DNA-binding transcription factors, and the transcriptional coactivators. R-Smads and Smad4 have weak intrinsic DNA-binding ability and exhibit less stringent sequence requirements than the Smad-interacting transcription factors, which have high-affinity binding to a specific DNA sequence. Thus, a Smad-binding sequence in proximity to the cognate sequence for the interacting transcription factor allows for Smad-mediated transcriptional regulation. This mechanism explains why TGF-β activates only a select set of promoters with binding sites for the interacting transcription factor. The Smad interactions with coactivators CBP/p300 allow the Smad complex to enhance the inherent transcription activity of the interacting transcription factor(s). Therefore, the Smad complex may be considered as a coactivator complex for select transcription factors (Derynck & Zhang 2003, Shi & Massagué 2003).

COMBINATORIAL RECEPTOR ASSEMBLY AND LIGAND BINDING SPECIFICITY

In humans, *Drosophila melanogaster*, and *Caenorhabditis elegans*, the number of TGF-β ligands greatly exceeds the number of type II and type I receptors. For example, the human genome encodes at least 29 and probably up to 42 TGF-β ligands that form homodimers and possibly heterodimers, whereas only five type II and seven type I receptors have been identified. Combinatorial interactions of type

BMP: bone morphogenetic protein

LTBP: latent TGF-β-binding protein

Smad: composite name from Sma (*Caenorhabditis elegans*) and Mad (*Drosophila melanogaster*)

R-Smad: receptor-activated Smad

MH: mad homology

ALK: activin receptor-like kinase

CBP: CREB (cAMP-responsive element-binding protein) binding protein

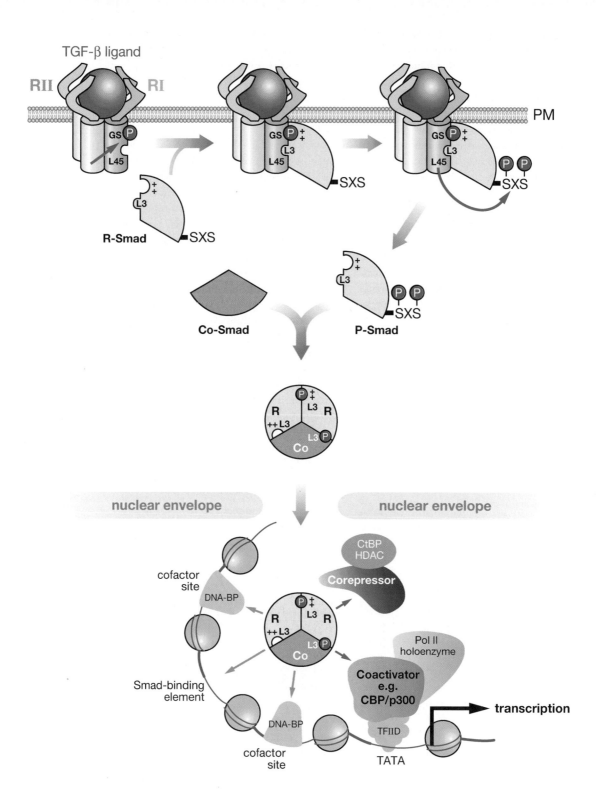

I and type II receptors in functional receptor complexes allow for diversity and selectivity in ligand binding as well as in intracellular signaling.

The high number of ligands is best explained by the need for finely tuned developmental patterns of receptor activation, which is achieved in part by differential regulation of ligand expression and activation from latent complexes. Thus, even though multiple ligands may activate the same receptor complexes and signaling pathways, their distinct expression patterns set the stage for multiple and highly restricted roles of TGF-β family ligands using a small number of receptor combinations. The restricted patterns of receptor activation during development are further specified by the limited diffusion of TGF-β ligands and their association with divergent propolypeptides and LTBPs, which may specify selective activation mechanisms.

Type I and type II receptors exist as homodimers at the cell surface in the absence of ligands, yet have an inherent heteromeric affinity for each other. While one may theorize that all type II receptors could combine with all type I receptors, only select combinations act as ligand-binding signaling complexes (**Figure 2**). The molecular basis of the selectivity of the type II-type I receptor interactions is largely unknown, but the structural complement at the interface of the ligand-receptor interactions may help define the selectivity of the receptor combinations (Greenwald et al. 2004). Most ligands bind with high affinity to the type II or type I receptor, while others bind efficiently only

to heteromeric receptor combinations. TGF-β1, TGF-β3 and activins bind efficiently to their respective type II receptors, TβRII and ActRII/ActRIIB, without the need for a type I receptor, yet the ligand contacts both receptor ectodomains to stabilize the type II-type I receptor complex (Boesen et al. 2002, Greenwald et al. 2004, Hart et al. 2002). In contrast, BMP-2 and -4 do not bind well to the type II receptor BMPRII, but bind efficiently to the type I receptors BMPRIA/ALK3 and BMPRIB/ALK6, and require the heteromeric complex for high affinity binding (Keller et al. 2004, Kirsch et al. 2000). Binding of TGF-β2 or BMP-7 requires both type II and type I receptor ectodomains (del Re et al. 2004, Greenwald et al. 2003). These and other observations provide evidence for the existence of unoccupied heteromeric receptor complexes at the cell surface.

In addition to binding of related ligands to the same receptor complex, a single ligand often activates several type II-type I receptor combinations. Dimeric TGF-β ligands have symmetric butterfly-like structures, whereby a monomer can be imagined as an open hand in which the central β-helix represents the wrist, the two aligned two-stranded β-sheets resemble four fingers, and the N-terminal sequence extends as a thumb (Shi & Massagué 2003). The BMP-2 homodimer complexed with two BMPRIA ectodomains shows two receptor-binding epitopes in the ligand that are conserved among BMPs (Keller et al. 2004, Kirsch et al. 2000). Superimposition of these data with the structure of BMP-7 in complex with ActRII ectodomains

---

Figure 1

The TGF-β signaling pathway. Ligands of the TGF-β superfamily first bind to the type II (RII) or type I (RI) homodimers or the RII-RI heterotetramer. Ligand binding stabilizes the receptor complex, in which RII phosphorylates the GS motif in the downstream type I receptor (RI) kinase. Following receptor activation, R-Smads are recruited to the receptor complex, primarily through an interaction between the L45 (on the RI) and L3 (on Smads) loops, and subsequently are phosphorylated in the SXS motif. Phosphorylated Smads (P-Smads) then form a trimeric complex with the common Smad4 in mammals. The Smad complex is then transported into the nucleus, where it interacts with DNA and transcription factors, including a large variety of DNA-binding transcription factors (DNA-BP) and coactivators or corepressors in a target gene–dependent manner.

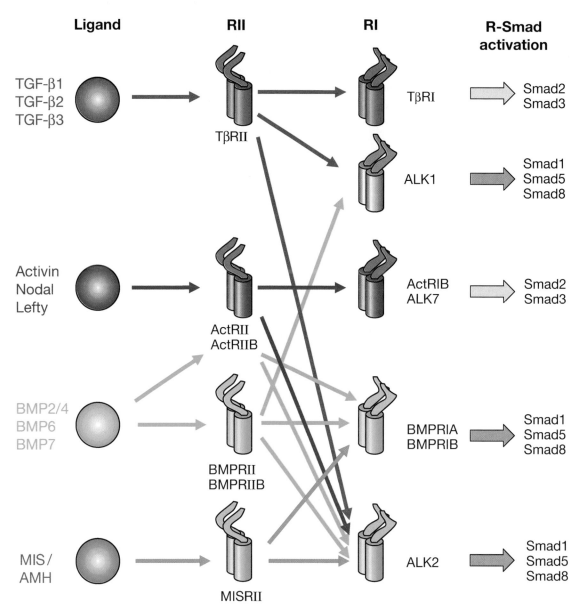

Figure 2

Heteromeric combination of TGF-β superfamily receptors. TGF-β ligands bind to specific combinations of RII-RI heterotetramers at the cell surface. Ligands, RII and RI are colorcoded for each pathway. Subsequent activation of R-Smads is shown at right. While ActRII and ActRIIB are encoded by different genes, BMPRII (long) and BMPRIIB (short) are two isoforms encoded by the same gene.

reveals that the type I and type II receptor extracellular domains in the tetrameric receptor complex do not interact with each other, yet allow cooperative ligand binding (Greenwald et al. 2004). This cooperativity in receptor binding may be modulated by the remarkable flexibility of the ligand (Sebald & Mueller 2003). The epitopes of TGF-β3 that bind TβRII are distinct from the receptor-binding epitopes in BMPs (Hart et al. 2002).

Furthermore, in the complex of TGF-β3 with two TβRII and two TβRI ectodomains, the TβRII and TβRI domains not only contact the ligand but also interact with each other (Hart et al. 2002). The differences in these complexes, together with the flexibility in ligand binding to the receptor, provide a structural basis for the versatility of ligand binding to receptor complexes.

Since the signaling responses are defined by the composition of the receptor complex, in particular that of the type I receptor, a ligand can induce different responses, depending on the nature of the activated receptor complexes. For example, in addition to the well-characterized TβRII-TβRI complex, TβRII forms functional complexes with ActRI/ALK2 or ALK1; these complexes signal differently from those involving TβRI (Goumans et al. 2002). The opposing activities of TGF-β signaling through TβRI and ALK1, in complex with TβRII, define the balance in endothelial cell migration and proliferation (Goumans et al. 2003). Similarly, ActRIB/ALK4, in combination with ActRII or ActRIIB, activates activin-induced gene responses, and the ActRI-ActRII complex transduces BMP7 signals (Macias-Silva et al. 1998). In addition, BMP-RIA and BMP-RIB combine not only with the "classical" BMP-RII but also with ActRII (Macias-Silva et al. 1998, Nishitoh et al. 1996). Consistent with the picture of differential signaling responses, BMP-RIA is able to promote adipogenic differentiation, whereas BMP-RIB may be more potent in osteoblast differentiation (Chen et al. 1998a). Combinatorial use of receptors also occurs in *Drosophila*, in which the type I receptors Tkv and Sax interact with the type II receptor Punt to bind one of three ligands: Dpp, Gbb, and Screw. In addition to differential Smad activation, differences in ligand-induced internalization and routing of the receptor complex, depending on the ligand and receptor composition, are likely to define the signaling responses as well. Finally, heterodimeric ligands, such as inhibins or BMP heterodimers, may activate asymmetric receptor combinations with two different type II and/or type I receptors.

Accessory cell surface proteins further define the binding efficiency and specificity of the ligand to the receptor complex. Betaglycan and endoglin bind TGF-β with high affinity, yet have no known role as signaling effectors. Coexpression of betaglycan or endoglin enhances TGF-β responsiveness and TGF-β binding to the TβRII-TβRI complex. While betaglycan strikingly enhances the minimal binding of TGF-β2 to TβRII (López-Casillas et al. 1993), endoglin is required for efficient TGF-β signaling through TβRII-ALK1 in endothelial cells (Lebrin et al. 2004). Regulation of betaglycan expression at the cell surface by the PDZ protein GIPC may further define the response to TGF-β (Blobe et al. 2001). Furthermore, the cytoplasmic domain of betaglycan interacts with TβRII, which phosphorylates this domain. This further triggers the interaction of betaglycan with β-arrestin, thereby modulating the internalization of TGF-β receptor complexes (Chen et al. 2003).

Other coreceptors also act as determinants of ligand binding and signaling. For example, nodal acts through ActRIIB and ActRIB, but efficient binding and signaling by nodal requires interaction of Cripto or the related EGF-CFC proteins Cryptic or FRL-1 with ActRIB (Yeo & Whitman 2001). Cripto binds nodal via its EGF domain and ActRIB through its CFC domain (Yeo & Whitman 2001). Like TGF-β and nodal, BMPs also have a coreceptor called DRAGON, which directly interacts with ligands and receptors to facilitate BMP signaling (Samad et al. 2005).

SPECIFICITY AND COMPLEXITY IN SMAD ACTIVATION

Upon ligand binding, the activated type I receptors specify the gene expression responses. In the case of tyrosine kinase receptors, the activated signaling events are largely dictated by cytoplasmic sequences outside the kinase domains. In contrast, the type I receptors specify their signaling largely through the L45 loop sequence located within the kinase domain. The L45 loop serves as the key determinant in the recruitment of Smads and the specificity of signaling (Chen & Massagué 1999, Chen et al. 1998b, Feng & Derynck 1997). Once phosphorylated by type II receptors in response to ligand binding, the GS motif immediately preceding the kinase domain also contributes to the strength of Smad binding. The short juxtamembrane segment that precedes the GS motif does not seem to contribute to the specificity of Smad activation, even though this segment shows considerable sequence divergence. Smad binding to and phosphorylation by the type I receptor are further modulated by interacting proteins such as SARA and the inhibitory Smads.

Smad-Receptor Interactions

The specificity in gene expression responses is defined by differential Smad recruitment and activation by the type I receptors. As discussed above, the Smad-receptor interaction involves the L45 loop and the phosphorylated GS motif in the type I receptor. This interaction allows the type I receptor to phosphorylate the C-terminal SXS motif of the R-Smad, resulting in a conformational change in and dissociation of the activated Smad, with subsequent heteromerization with Smad4. Smad2 and 3 are phosphorylated by TβRI and ActRIB, whereas Smad1, Smad5, and Smad8 are substrates for BMP-RIA, BMP-RIB, ALK-1, and ALK-2 (**Figure 2**). The efficiency of Smad activation at the endogenous receptor and Smad levels, the quantitative dependence of Smad phosphorylation on ligand stimulation, and the relative affinities of the Smads for the receptors are likely to differ significantly among related receptors and Smads, but these quantitative assessments await characterization.

The nine-amino acid L45 loop, which connects β strands 4 and 5 in the kinase do-

mains of the type I receptor, is accessible for protein interactions, as apparent from the structure of TβRI (Huse et al. 2001). Receptors with different signaling specificities have distinct L45 sequences (Chen et al. 1998b, Feng & Derynck 1997). The L45 sequences of TβRI and ActRIB, i.e., receptors that activate Smad2 and Smad3, are identical, but differ in four amino acids from the L45 sequences in BMP-RIA and -RIB, and in seven amino acids from the L45 loops of ALK-1 and ALK-2 (Chen et al. 1998b). Accordingly, replacement of the L45 sequence in ALK-2 with that of TβRI leads to TGF-β signaling (Feng & Derynck 1997). Similarly, TβRI with an L45 loop of BMP-RIB switches specificity to induce a BMP-like response, and a BMP-RIB with a TβRI L45 loop can activate TGF-β- and activin-like transcription, and does not activate BMP-inducible gene expression (Chen et al. 1998b).

The L45 loop interacts directly with the L3 loop in the MH2 domain of an R-Smad (Chen et al. 1998b, Lo et al. 1998). The L3 loop is located between two β sheets and is exposed in the trimeric Smad complexes (Shi & Massagué 2003). As with the L45 loop, only a few amino acids in the L3 loop define receptor-binding specificity. The L3 sequences are invariant between Smad2 and Smad3 as well as among BMP-activated Smad1, Smad5, and Smad8, but differ in two residues between both groups. Consequently, Smad1 with an L3 loop of Smad2 interacts with and is phosphorylated by TβRI, while Smad2 with the L3 loop of Smad1 no longer interacts with TGF-β receptors and is not phosphorylated by TβRI (Lo et al. 1998).

Adjacent sequences stabilize the interaction of the L3 and L45 loops and contribute to signaling specificity (Chen & Massagué 1999, Lagna & Hemmati-Brivanlou 1999, Lo et al. 1998). Smad1 with an L3 loop from Smad2 is still phosphorylated in response to BMP, albeit to a lesser extent than wild-type Smad1, and requires further replacement of the sequence downstream from the L3 loop with that of Smad2 to abolish BMP-induced phosphorylation. Such larger replacement fully switches receptor binding specificity and confers efficient Smad phosphorylation by TβRI; this result, which could not be achieved by replacing the L3 loop alone (Chen & Massagué 1999), is due to the interaction of the phosphorylated GS motif with the Smad sequence downstream from the L3 loop. This notion is consistent with the current structural model for activation of the signaling response (Huse et al. 2001). The interaction of the L45 loop of a receptor with the L3 loop of an R-Smad may play an initial role in the receptor-Smad selection, but the interaction of the phosphorylated GS motif with the Smad sequence downstream from the L3 loop stabilizes the receptor-Smad interaction (Wu et al. 2001b).

Accessory Proteins in Smad Activation

Efficient R-Smad recruitment and activation in response to TGF-β or activin require SARA, an FYVE domain-containing protein that interacts with the type I receptor and Smad2/3 (Tsukazaki et al. 1998, Wu et al. 2000). SARA is localized at the plasma membrane and concentrated in EEA1-positive early endosomes through the interaction of the FYVE domain with the membrane lipid PtdIns(3)P. Complex formation of the receptors with SARA and Smad2/3 in early endosomes may thus be essential to efficiently initiate TGF-β signaling (Di Guglielmo et al. 2003, Hayes et al. 2002, Panopoulou et al. 2002). The structural interface of SARA and the interacting sequence in the MH2 domain of Smad2 reveal critical determinants of SARA-Smad interaction, explaining the inability of BMP-activated Smads to interact with SARA (Wu et al. 2000). This model also explains the dissociation of Smad2 from SARA, following Smad2's C-terminal phosphorylation by the type I receptor. Hgs, another FYVE domain protein involved in endosomal trafficking, may play a role similar to that of SARA, since it also interacts with Smad2 and Smad3 and enhances ligand-induced Smad phosphorylation and gene expression (Miura et al. 2000). However, the FYVE finger of SARA has a higher affinity for PtdIns(3)P than does the FYVE finger of Hgs, suggesting a predominant role for SARA in TGF-β and activin signaling (Panopoulou et al. 2002). Truncated versions of SARA or Hgs impair TGF-β/activin signaling, underscoring the roles of SARA and Hgs in activation of signaling (Miura et al. 2000, Panopoulou et al. 2002, Tsukazaki et al. 1998). Additional FYVE proteins yet to be identified may play similar roles in the binding selectivity and receptor interaction of the BMP-activated Smads1, 5, and 8.

Disabled-2 (Dab2), a protein that plays a role in multiple signaling pathways, also interacts with the TβRII-TβRI complex and Smad2/3. TGF-β enhances Dab2's interaction with these Smads and may stabilize the receptor-Smad interaction (Hocevar et al. 2001). The ability of Dab2 to interact with clathrin and the clathrin adaptor AP-2, and Dab2's localization to clathrin-coated pits, may link Dab2 to clathrin-mediated endocytosis of the activated TGF-β receptor complexes. Absence of Dab2 renders cells insensitive to TGF-β-induced Smad activation (Hocevar et al. 2001). Finally, Dok-1, a rasGAP-binding protein that acts downstream from receptor tyrosine kinases, is required for activin-induced Smad signaling. Dok-1 interacts with type II and type I activin receptors and activin triggers association of Dok-1 with Smad3 (Yamakawa et al. 2002).

Inhibitory Smad Interactions

Smad6 and Smad7, and presumably also Dad in *Drosophila*, inhibit TGF-β family signaling primarily by interfering with the receptor-mediated activation of R-Smads. These inhibitory Smads associate with type I receptors, thus competitively interfering with R-Smad recruitment and phosphorylation (Hayashi et al. 1997, Imamura et al. 1997, Nakao et al. 1997). Smads1 and 5 induce Smad6 expression, whereas Smad3 induces Smad7 expression. Consequently, BMP signaling induces an inhibitory feedback loop through Smad6 expression, while TGF-β induces an inhibitory feedback loop through Smad7 expression, although BMPs and TGF-β can also induce Smad7 and Smad6 expression, respectively. Smad6 inhibits BMP and TGF-β signaling with similar potency, while Smad7 inhibits TGF-β signaling more efficiently than Smad6 (Miyazono 2000). Through their MH2 domains, Smad6 and Smad7 interact with the type I receptors, and the isolated MH2 domains interact with similar affinity with TβRI (Hanyu et al. 2001, Souchelnytskyi et al. 1998). Presence of the MH1 domain of Smad7, but not that of Smad6, increases the interaction of the Smad7 MH2 domain with TβRI (Hanyu et al. 2001). Structural predictions combined with mutation analysis have identified a basic surface in the MH2 domain as critical for the interaction of Smad7 with TβRI. Two Lys residues are essential for the binding of Smad7 to TGF-β receptor complexes and inhibition of TGF-β signaling by Smad7, whereas two other basic residues in the L3 loop are essential for inhibiting both TGF-β and BMP signaling (Mochizuki et al. 2004). The WD-repeat protein STRAP-1, which interacts with the TGF-β receptors, assists in the interaction of Smad7 with these receptors and thus cooperates with Smad7 to inhibit TGF-β signaling (Datta & Moses 2000).

In addition to the competitive interference of Smad6 and Smad7 with R-Smad binding to type I receptors, Smad6 also inhibits complex formation of BMP-activated Smad1 with Smad4 (Hata et al. 1998). Smad6 and Smad7 also inhibit TGF-β family signaling by interacting directly with Smurf E3 ubiquitin ligases and mobilizing these ligases to the type I receptors, leading to proteasomal degradation of the receptors (Ebisawa et al. 2001, Kavsak et al. 2000, Murakami et al. 2003). Conversely,

Smurf1 appears important in targeting Smad7 to the receptor complex (Suzuki et al. 2002).

Several signaling pathways lead to a rapid induction of Smad6 and Smad7 expression, which constitutes a critical point for negative regulation of TGF-β signaling. Most notably, TGF-β or BMP signaling induces Smad6 or Smad7 expression that can result in attenuation of ligand-induced Smad activation and gene expression. Smad7 is also induced by Jak/STAT signaling in response to interferon-γ (Ulloa et al. 1999) and by NF-κB signaling in response to inflammatory cytokines (e.g., TNF-α and IL-1) and lipopolysaccharide (Bitzer et al. 2000). In addition, fluid shear stress induces Smad6 and Smad7 expression in endothelial cells (Topper et al. 1997). The induced expression of inhibitory Smads consequently decreases receptor-mediated Smad activation and the cell's responsiveness to TGF-β ligands.

Heteromeric Smad Complex Formation

Upon release from the receptors, the phosphorylated R-Smads form complexes with Smad4 that act as effectors of ligand-induced signaling. Structural analyses have shown that the MH2 domains of Smad4 (Shi et al. 1997), pseudophosphorylated Smad3 (Chacko et al. 2001), and phosphorylated Smad2 form homotrimers (Wu et al. 2001b). In addition, the phosphorylated or pseudophosphorylated MH2 domains of Smad1 or Smad2/3 (Chacko et al. 2004, Qin et al. 2001, 2002) form heteromeric trimers with Smad4 consisting of two R-Smads and one Smad4. Smad homotrimerization and heterotrimerization in solution are also observed from biochemical analyses (Chacko et al. 2001, 2004, Jayaraman & Massagué 2000). High-affinity trimer formation is primarily mediated by the L3 loop in the MH2 domain and SXS phosphorylation (Chacko et al. 2001, 2004, Jayaraman & Massagué 2000). Thus, ligand-induced SXS phosphorylation of R-Smads may be a prerequisite for natural Smad trimerization.

This heterotrimeric Smad model is consistent with the requirement of Smad4 in most TGF-β-induced transcriptional responses. This structure also allows for combinatorial interactions and versatility and may explain the requirement of Smad2, Smad3, and Smad4 in the induction of transcription of the cdk inhibitors p15[Ink4B] (Feng et al. 2000) and p21[Cip1] (Pardali et al. 2000b). It is thus easily conceivable that two different BMP-activated Smads may combine with Smad4 or even that a BMP-activated Smad may combine with Smad2/3, and Smad4 as third partner, to activate or repress selective transcription responses. Competition of Smad3 with Smad2 for interaction with Smad4 may explain the ability of Smad3 to inhibit activin-induced goosecoid expression through Smad2/4 (Labbé et al. 1998).

The incorporation of Smad4, which has no SXS motif, into the activated R-Smad complex lends a possibility that Smad6 or Smad7 may form a complex with two R-Smads. This would be consistent with the observations that Smad6 interacts with Smad1 (Hata et al. 1998) and that Smad6 binding to the Id1 promoter requires the presence of Smad1 (Lin et al. 2003). Complex formation of Smad6 with R-Smads may provide a mechanism for functional repression of the effector functions of Smads. Replacement of Smad4 with Smad6 or Smad7 not only would eliminate Smad4 as coactivator, thus preventing gene activation, but would also recruit histone deacetylases to confer active gene repression. The coactivator role of Smad4 and possible roles of Smad6 and Smad7 in transcription repression will be discussed further below.

The evidence for Smad trimerization contrasts with some reports that Smads form dimers (Inman & Hill 2002, Jayaraman & Massagué 2000, Wu et al. 2001a). At promoter DNA, Smads may exist as trimers or dimers, depending on the interacting transcription factor (Inman & Hill 2002). Crystallographic analyses should provide insight into the characteristics of Smad complexes at DNA and their interactions with sequence-specific transcription factors.

Control of Smad Activation by Diverse Kinase Pathways

In addition to C-terminal SXS phosphorylation by type I receptors, R-Smad activation is regulated by cytoplasmic kinases. The linker regions of the R-Smads are targets for proline-directed kinases such as MAPKs and cyclin-dependent kinases. Erk MAPK, which is activated in response to mitogenic growth factors or oncogenic Ras mutants, can phosphorylate the linker regions of Smad1 and Smad2/3, thereby inhibiting ligand-induced nuclear translocation of Smads and consequently the TGF-β antiproliferative response (Kretzschmar et al. 1997, 1999, Pera et al. 2003). However, other studies did not observe impaired nuclear translocation of Smads in cells with activated Ras/MAPK signaling (de Caestecker et al. 1998, Engel et al. 1999). In addition, impaired Smad signaling in Ras transformed cells is not easily reconciled with the cooperation between Ras/MAPK and TGF-β signaling in tumor cell differentiation and behavior (Janda et al. 2002). Regulation of Smad activation by Erk MAPKs may also control developmental processes. In Xenopus embryos, FGF8, in combination with IGF2, induces a MAPK-dependent inhibitory phosphorylation of the Smad1 linker region, which contributes to the induction of a neural cell fate (Pera et al. 2003). It should be noted that TGF-β receptors can activate MAPK signaling (Derynck & Zhang 2003, Massagué 2003)

JNK, which is activated in response to mitogenic and stress signals, phosphorylates Smad3 outside its SXS motif and enhances activation and nuclear translocation of Smad3 (Engel et al. 1999). Furthermore, activation of MAPK kinase 1 (MEKK-1), an activator of JNK and Erk MAPK, leads to phosphorylation and activation of Smad2 (Brown et al. 1999). This mechanism may explain the ability of fluid shear stress or some growth factors to activate Smad2 (Brown et al. 1999,

MAPK: mitogen-activated protein kinase

JNK: Jun N-terminal kinase

PKC: protein kinase C

CaMKII: calmodulin-dependent kinase II

SBE: Smad-binding element

de Caestecker et al. 1998). This regulation can be further complemented by alterations in the stability of Smad4, e.g., through induction of proteolytic degradation in response to activated MAPK signaling (Liang et al. 2004, Saha et al. 2001).

The cyclin-dependent kinases CDK2 and CDK4 also phosphorylate the linker regions of Smads2 and 3, but at sites that differ from those targeted by Erk MAPK, and consequently inhibit Smad-dependent gene transcription and cell cycle arrest (Matsuura et al. 2004). Since tumor cells often activate these CDKs, inhibition of Smad activity by CDK-dependent phosphorylation may provide an escape from antiproliferative control by autocrine TGF-β signaling.

PKC and CaMKII also regulate Smad activation. PKC-dependent phosphorylation of the MH1 domain abolishes the DNA binding of Smad3 (Yakymovych et al. 2001). In mesangial cells, CaMKII phosphorylates Smad2, and to a lesser extent Smad3. One of the phosphoacceptor sites, Ser-240, which is in the linker region, is phosphorylated in response to EGF, PDGF, or TGF-β. CaMKII induces a Smad2-Smad4 complex independently of TGF-β receptor activation, but this complex may be inactive (Abdel-Wahab et al. 2002). Casein kinases I, which have been implicated in various processes, also control Smad activity. Casein kinase Iε associates with and can phosphorylate R-Smads, yet also interacts with TGF-β receptors. Consequently, casein kinase Iε may regulate TGF-β/Smad signaling (Waddell et al. 2004). Finally, Akt (protein kinase B), which can be activated in response to insulin, can associate directly with Smad2 and 3, and thus control their activation and the response to TGF-β. In response to insulin, Akt interacts with Smad3 that has not been phosphorylated in response to TGF-β. Consequently, Akt inhibits Smad3 activation by TGF-β and Smad3/4 complex formation and nuclear translocation, whereas TGF-β signaling decreases the partnering of Akt with Smad3. This balance results in the ability of Akt to decrease TGF-β/Smad3-mediated transcription and TGF-β-induced apoptosis (Conery et al. 2004, Remy et al. 2004).

Taken together, phosphorylation by MAPKs and kinases involved in other pathways exert differential effects by targeting distinct phosphorylation sites in the Smads, independently from C-terminal SXS phosphorylation by the type I receptor, but the outcome depends on the cell signaling context; depending on cell type and physiology, such phosphorylation even could exert opposite effects. The combination of these phosphorylation events greatly contribute to the final gene responses to Smad signaling (Massagué 2003).

SMADS IN THE NUCLEUS: SPECIFICITY AND VERSATILITY IN TRANSCRIPTIONAL CONTROL

Transcriptional activation by Smads is based on cooperation of the Smad complex with other DNA sequence-specific transcription factors at the promoter DNA. This interaction involves association of the Smad complex with the DNA-binding transcription factor, Smad binding to an adjacent DNA sequence and interaction of R-Smads with the CBP or p300 transcription coactivators. Smad4 then acts as Smad coactivator by stabilizing the interaction of activated R-Smads with CBP/p300. This mechanism allows for an extensive versatility, yet also confers specificity.

Smads as DNA-Binding Factors

Smads contact DNA selectively, with 5′-GTCTAGAC-3′ as the optimal sequence for Smad3 or Smad4 binding (Zawel et al. 1998). The Smad3 MH1 domain interacts through a β hairpin with the major groove of the DNA sequence 5′-GTCT-3′ and its reverse complement, 5′-CAGA-3′ (the SBE) (Shi et al. 1998). This interaction involves hydrogen bonds with the two G residues in the SBE. Since DNA binding of a Smad is marked by minimal sequence requirements and low affinity, multiple Smad binding sites are required for Smad3-mediated transcriptional activation in the absence of an interacting, sequence-specific transcription factor. At natural promoters, however, a Smad binding sequence that is adjacent to the sequence binding the Smad-interacting transcription factor with high affinity allows binding of Smad transcription complexes. The juxtaposition of both sequences may result in an affinity exceeding that of the interacting transcription factor for its cognate DNA sequence, which may explain why Sp1 and c-Jun, when interacting with Smad3, bind their cognate DNA sequences with higher affinity than in the absence of Smad3 (Feng et al. 2000, Qing et al. 2000).

Smad3 also binds a GGCGGG sequence in the c-myc promoter; binding to this sequence, which was shown to bind E2F, is required for the transcriptional repression of c-myc by TGF-β signaling (Frederick et al. 2004). Although able to bind to the SBE sequence, Smad1 and its *Drosophila* homolog Mad bind to a GCCG sequence with higher affinity, which consequently confers BMP responsiveness (Kim et al. 1997, Korchynskyi & ten Dijke 2002, Kusanagi et al. 2000). Smad4 and its *Drosophila* homolog Medea also bind to GC-rich sequences (Ishida et al. 2000).

In contrast to other R-Smads such as Smad3, Smad2 is unable to bind DNA owing to a sequence insert in the β hairpin (Shi et al. 1998). However, a splicing variant of Smad2 with a deletion of this insert has similar DNA-binding properties as Smad3 (Yagi et al. 1999). It is thought that Smad2/4 complexes bind DNA through Smad4.

Combinatorial Interactions of Smads with DNA-Binding Transcription Factors

Smads cooperate through physical interactions with a remarkable diversity of DNA sequence–binding transcription factors (**Table 1**). These interactions occur through either the Smads' MH1 or MH2 domains, depending on the transcription factor. The regulation of the activities of the interacting transcription factors by other signaling pathways further defines this cooperation. This versatility explains the complexity and cell context dependence of the transcription programs exerted by TGF-β ligands, as well as why no consensus TGF-β ligand response sequences can be defined.

FAST/FoxH1, a forkhead (Fox) transcription factor, was the first transcription factor reported to interact and cooperate with Smads in mediating TGF-β signals. In response to activin, Smad2/4 complexes interact with DNA-bound FoxH1 at an activin-response element and provide ligand-induced transcription (Chen et al. 1997). In this complex, Smad4 contacts the DNA while the MH2 domain of Smad2 interacts with FoxH1 (Labbé et al. 1998, Zhou et al. 1998). FoxH1 interacts with Smad2 using an SIM also present in Mix transcription factors and an FM uniquely present in FoxH1 (Randall et al. 2004). The SIM motif is also present in the Smad-binding domain of SARA, and is thus involved in the mutually exclusive Smad2-SARA and Smad2-FoxH1 interactions (Randall et al. 2002). Smad-FoxH1 cooperation mediates nodal signaling in endoderm and dorsal mesoderm formation in zebrafish (Sirotkin et al. 2000) and mice (Hoodless et al. 2001, Yamamoto et al. 2001). In mammals, FoxH1 cooperates with Nkx2.5 in Smad-dependent MEF2C expression, essential for heart looping morphogenesis (von Both et al. 2004).

Forkhead proteins also participate in the antiproliferative responses to TGF-β. In epithelial cells, TGF-β induces the expression of the CDK inhibitors p21[Cip1] and p15[Ink4B]. At the p21[Cip1] promoter, the Smad3/4 complex interacts with FoxO, a target of the PI3 kinase/Akt pathway, to induce transcription of the p21[Cip1] gene (Seoane et al. 2004). FoxO binds to a distal sequence of the p21[Cip1] promoter, but Smads also interact with Sp1 at a proximal sequence to regulate p21[Cip1] expression (Pardali et al. 2000b), suggesting the

SIM: Smad interaction motif

FM: FoxH1 motif

Table 1 Smad-interacting DNA-binding transcription factors in mammalian cells

Smad-binding partners	Interacting Smad and domains	Features/mechanisms of action	References
bHLH family			
E2F4/5	Smad3 (MH2)	Recruitment of p107 to Smad3 to repress the c-myc gene	Chen et al. 2002
Max	Smad3 (MH1)	Max inhibits transcription activation by Smad3	Grinberg & Kerppola 2003
MyoD	Smad3 (MH1-linker)	Interference of MyoD/E protein/DNA complex formation	Liu et al. 2001
TFE3	Smad3/4	Synergistic cooperation on TGF-β target genes such as PAI-1, Smad7	Hua et al. 1999, Huse et al. 2001, Kawata et al. 2002
bZIP family			
ATF2	Smad3/4 (MH1)	Stimulation of ATF2 transactivation	Sano et al. 1999
ATF3	Smad3 (MH2)	Repression of the Id1 promoter	Kang et al. 2003
c-Fos	Smad3 (MH2)	Cooperation on AP-1-dependent TGF-β target genes	Zhang et al. 1998
c-Jun, JunB, JunD	Smad3 (MH1), Smad4	Positively and negatively regulate Smad activity	Liberati et al. 1999, Zhang et al. 1998
CEBPα, β, δ	Smad3 (MH1)	Smad3 inhibits CEBP's transactivation	Choy & Derynck 2003, Coyle-Rink et al. 2002
Forkhead family			
FoxH1/FAST	Smad2/3	Formation of activin-responsive factors on the activin-responsive promoters	Chen et al. 1997, Labbé et al. 1998, Randall et al. 2002
FoxO	Smad2/3	Regulation of p21^{Cip1}	Seoane et al. 2002
Homeodomain protein family			
Dlx1	Smad4	Inhibits Smad4 signaling	Chiba et al. 2003
Hoxc-8	Smad1 (MH1-linker), Smad6 (MH2)	Relief of Hoxc-8-dependent repression Inhibition of Smad1-Hoxc-8 interaction	Shi et al. 1999
Milk/Mixer	Smad2 (MH2)	Recruitment of Smad2/Smad4 activators to the activin-responsive complex	Germain et al. 2000, Randall et al. 2002
Nuclear receptor family			
Androgen receptor (AR)	Smad3 (MH2)	Reciprocal inhibition of Smad3 DNA-binding activity and of AR activity	Chipuk et al. 2002, Hayes et al. 2001, Kang et al. 2002
Estrogen receptor	Smad1/3/4(MH2)	Repression of Smad target genes	Matsuda et al. 2001, Wu et al. 2000, Zhang et al. 2000
Glucocorticoid receptor	Smad3 (MH2)	Inhibition of Smad3 transactivation activity	Song et al. 1999
HNF4	Smad3/4	Cooperative activation	Chou et al. 2003
RXR	Smad3 (MH2)		Pendaries et al. 2003
Vitamin D3 receptor	Smad3 (MH1)	Coactivation of ligand-induced transactivation of vitamin D receptor	Yanagisawa et al. 1999
Runx family			
CBFA1/Runx2/AML	Smad1/2/3/5 (MH2)	Cooperative activation of BMP responses; regulation of immune responses	Hanai et al. 1999, Pardali et al. 2000a, Zhang & Derynck 2000, Zhang et al. 2000

(Continued)

Table 1   (*Continued*)

Smad-binding partners	Interacting Smad and domains	Features/mechanisms of action	References
Zinc finger protein family			
GATA3	Smad3	Recruits Smad3 to GATA sites to cooperatively activate transcription	Blokzijl et al. 2002
GATA4,5,6	Smad1	Cooperate in the regulation of Smad7 and Nkx2.5	Benchabane & Wrana 2003, Brown et al. 2004
GliΔC-ter	Smad1/2/3/4	Unknown	Liu et al. 1998
OAZ	Smad1/4 (MH2)	Formation of BMP-responsive activator complex	Hata et al. 2000
Sp1	Smad2 (MH1) Smad4 (MH2)	Cooperative activation of TGF-β target genes, e.g., p15^{Ink4B}, p21^{Cip1}, Smad7, PAI-1, and collagen	Feng et al. 2000, Pardali et al. 2000b
YY1	Smad1, Smad4 (MH1)	Complex with Smads and GATA	Kurisaki et al. 2003, Lee et al. 2004
ZNF198	Smad3 (MH2)	Unknown	Warner et al. 2003
Others			
β-catenin	Smad1/4	Wnt-dependent activation of LEF1 target genes	Hu et al. 2005, Hussein et al. 2003, Lei et al. 2004
HIF-1α	Smad3 (MH1, MH2)	Cooperation of TGF-β with hypoxia pathway and angiogenesis	Sanchez-Elsner et al. 2001
IRF-7 (IRFs)	Smad3 (MH2)	Smad3 activation of IRF-7 transactivation function	Qing et al. 2004
Lef1/TCF	Smad1/2/3/4 (MH1, MH2)	Smad coactivation of LEF1 signaling	Hu et al. 2005, Labbé et al. 2000, Nishita et al. 2000
MEF2 (MADS box)	Smad3	Smad3 represses the transcription activity of MEF2	Liu et al. 2004
Menin	Smad2/3 (MH2)	Facilitate Smad DNA binding	Kaji et al. 2001
NFκB p52	Smad3	Coactivation of κB site	Lopez-Rovira et al. 2000
NICD	Smad1/3 (MH2)	Coactivation of NICD-RBP-Jk complex to regulate the Notch targets	Blokzijl et al. 2003, Dahlqvist et al. 2003, Itoh et al. 2004, Zavadil et al. 2004
p53	Smad2/4	Synergism and antagonism	Chordenonsi et al. 2003, Takebayashi-Suzuki et al. 2003, Wilkinson et al. 2005
Pax8	Smad3	Smad3 reduces Pax8 DNA binding	Costamagna et al. 2003
SRF	Smad3	Mediate TGF-β-induced SM22α transcription	Qiu et al. 2003

formation of multiple Smad complexes in a single promoter.

The interaction of Smads with Sp1 illustrates their cooperation with Zn finger transcription factors. Sp1, which uses the Mediator complex as a coactivator, drives transcription of the p15^{Ink4B} and p21^{Cip1}

genes. At either promoter, TGF-β induces transcriptional cooperation of Smad2/3/4 complexes with Sp1 through association with a glutamine-rich domain in Sp1 (Feng et al. 2000, Pardali et al. 2000b). Smad-Sp1 interactions may also activate TGF-β-induced transcription of the α2(I) collagen, integrin β5,

Smad7, and PAI-1 genes. No interactions of Smads with the related Sp2 and Sp3 transcription factors have been reported.

Additional Zn finger proteins participate in BMP or TGF-β signaling. Smad1 can associate with OAZ in activation of the Xvent2 gene (Hata et al. 2000). GATA transcription factors, which regulate cell differentiation, also interact with Smad proteins and modulate responses to BMP. Smad1 interactions with GATA4, 5, or 6 regulate transcription of the Smad7 (Benchabane & Wrana 2003) and Nkx2.5 (Brown et al. 2004) genes. At the Nkx2.5 promoter, this cooperation also involves another Zn finger protein named YY1. YY1 associates with Smad1/4 at adjacent YY1- and Smad-binding sites, thereby constituting a minimal BMP-responsive enhancer; thus, a multicomponent complex consisting of Smads, YY1, and GATAs regulates the BMP-responsiveness of the Nkx2.5 gene (Lee et al. 2004). YY1 and GATA proteins also mediate TGF-β responses. At the interleukin 5 promoter, TGF-β induces Smad3 recruitment to GATA3 at GATA-binding sequences independently of Smad3 binding to DNA, and functional cooperation of Smad3 with GATA3 to activate transcription (Blokzijl et al. 2002). Also, YY1 interaction with the MH1 domain of Smad4 or other Smads inhibits TGF-β-activated transcription (Kurisaki et al. 2003). In *Drosophila*, the Zn finger protein Schnurri is targeted by Dpp-activated Mad. Their interaction allows for transcriptional activation (Dai et al. 2000), yet suppresses transcription of Brinker, a repressor of Mad-mediated transcription (Marty et al. 2000). Finally, the Zn finger proteins Evi-1 (Kurokawa et al. 1998, Alliston et al. 2005) and SIP1 (Postigo et al. 2003, Verschueren et al. 1999) interact with Smads to repress Smad-mediated transcription, as will be discussed later.

Smads also interact with select bZIP family transcription factors, which contain basic and leucine zipper domains involved in DNA binding and dimerization. Among these, Smad3 can interact with c-Jun, JunB, ATF-2, ATF3, and, with lower efficiency, c-Fos in response to TGF-β (Kang et al. 2003, Liberati et al. 1999, Sano et al. 1999, Zhang et al. 1998). c-Jun, JunB, and ATF-2 interact through their bZIP domains with the MH1 domain of Smad3, while c-Fos and ATF3 interact with the MH2 domain of Smad3. While the stoichiometry and configuration of these interactions at the DNA are unclear, the enhanced transcription presumably results from cooperative recruitment of CBP/p300. Smad3 and CREB similarly can cooperate at adjacent DNA sequences, even though no physical interaction is detected (Zhang & Derynck 2000). Since AP-1 complexes of c-Jun and c-Fos, or related dimers, mediate responses to mitogenic factors and stress, the cooperation of Smads with bZIP transcription complexes at TGF-β-responsive promoters provides a mechanism for convergence of both signaling pathways. Smad3 and Smad4 also associate with the C/EBP transcription factors. Interaction of Smad 3/4 with C/EBPβ mediates the TGF-β-dependent inhibition of adipocyte differentiation (Choy & Derynck 2003) and HIV Tat–mediated transcription (Coyle-Rink et al. 2002).

Several homeodomain transcription factors, which play crucial roles in patterning and tissue differentiation, are also targeted by Smad signaling. In *Drosophila*, Medea cooperates with Tinman, the homolog of Nkx2.5, to induce tinman transcription in response to Dpp. In Xenopus, Smad2/4 interacts with Mixer and Milk to activate activin-responsive transcription of the goosecoid gene (Germain et al. 2000). As in the case of FoxH1, these Mix proteins interact through their SIM sequences with the MH2 domain of Smad2 (Randall et al. 2002). Homeoproteins may also oppose TGF-β signaling. For example, Dlx1 interacts with Smad4 and blocks signals from TGF-β proteins in hematopoiesis and perhaps neurogenesis (Chiba et al. 2003). At the osteopontin promoter, Smad1/4 interacts with Hoxc-8 and blocks Hoxc-8 binding to the homeodomain-binding sequence, thereby preventing Hoxc-8-mediated transcriptional repression and allowing transcription in response to BMP (Shi et al. 1999).

In response to TGF-β, Smad3 can interact and cooperate with some bHLH transcription factors, which are characterized by a basic helix-loop-helix domain involved in DNA binding and dimerization. Smad3 cooperates with TFE3 in the transcription of the plasminogen activator inhibitor-1 (Hua et al. 1999), Smad7 (Hua et al. 2000), and laminin γ-chain (Kawata et al. 2002) genes. Smad3 also interacts with the myogenic bHLH transcription factors MyoD and myogenin (Liu et al. 2001), E2F4 (Chen et al. 2002), c-Myc (Feng et al. 2002), and Max (Grinberg & Kerppola 2003), but these interactions result in transcription repression, as will be discussed.

Smad3 also cooperates with Runt transcription factors. Runt proteins have a domain with homology to *Drosophila* Runt that interacts with DNA and promotes dimerization with a β subunit. Runx1/AML1 and Runx3/AML2 bind the germ line IgCα gene promoter at sequences adjacent to SBEs and cooperate with Smad3/4 to induce transcription in response to TGF-β, leading to IgA class switching (Hanai et al. 1999, Pardali et al. 2000a, Zhang & Derynck 2000). Smad3/4 also cooperates with Runx2/CBFA1 to induce transcription (Zhang et al. 2000), but this cooperation leads to repression of the Runx2 activity at the runx2 and osteocalcin promoters in mesenchymal cells (Alliston et al. 2001).

Smad3 also associates and cooperates with IRF-7, a member of the IRF transcription factors, which are involved in responses to viral and bacterial infection and inflammation. Smad3 cooperates with IRF-7 in the expression of interferon-β in response to polyI:C (Qing et al. 2004) through interaction of the MH2 domain of Smad3 with the transactivation domain of IRF-7. The transactivation domain of IRF-3, which resembles that of IRF-7, has a structure remarkably similar to the MH2 transactivation domain of Smads

(Qin et al. 2003), raising the possibility that a heteromeric Smad-IRF complex may reciprocally regulate the transcription of Smad and IRF target genes.

Several intracellular receptors are targeted by TGF-β-activated Smad3 for functional cooperativity. The interaction of Smad3 with the vitamin D3 receptor (Yanagisawa et al. 1999) or HNF-4 (Chou et al. 2003) can result in transcriptional activation, while the glucocorticoid (Song et al. 1999), estrogen (Matsuda et al. 2001), and retinoic acid receptors (Pendaries et al. 2003) can repress the transactivation function of Smad3. Smad3 also interacts with the androgen receptor (Chipuk et al. 2002, Hayes et al. 2001, Kang et al. 2002). Other Smads can crosstalk with nuclear receptors as well. Smad4 binds to estrogen receptor α and represses estrogen gene responses (Wu et al. 2003). Estrogen induces an interaction between the estrogen receptor and Smad1 to inhibit Smad activity (Yamamoto et al. 2002). The functional consequences of many of these interactions require further characterization.

The cooperation of activin and Wnt signaling in tissue differentiation can result from interactions of Smad signaling with Wnt signaling effectors. Wnt signaling is mediated by the HMG box domain transcription factors LEF1 or TCF and their coactivator β-catenin. Smad3 and Smad4 can associate and cooperate with LEF1/TCF at the Xenopus twin promoter (Labbé et al. 2000, Nishita et al. 2000). At the myc promoter, which contains Smad- and TCF-binding sites, BMP can induce interaction of Smad1 with β-catenin and TCF4 to stimulate myc transcription (Hu & Rosenblum 2005). Similar crosstalk of both pathways is likely to regulate other developmentally regulated genes (Hussein et al. 2003, Lei et al. 2004). Smad3 also interacts with axin, a negative regulator of Wnt signaling with which several Wnt signaling mediators interact. TGF-β induces dissociation of a Smad3/axin complex and axin enhances TGF-β/Smad3 signaling, suggesting a role for axin in TGF-β signaling (Furuhashi et al. 2001).

IRF: interferon regulatory factor

Like TGF-β and Wnt signaling, the Notch pathway controls cell differentiation. Activation of transmembrane Notch induces cytosolic release of its intracellular domain (NICD), which enters the nucleus where it interacts with the DNA-binding factor CSL/RBP-Jκ and activates Notch target genes repressed by CSL in the absence of Notch. TGF-β and BMP regulate Notch target gene expression through, respectively, the interaction of TGF-β-activated Smad3 and BMP-activated Smad1 with NICD (Blokzijl et al. 2003, Dahlqvist et al. 2003, Itoh et al. 2004, Zavadil et al. 2004). The Smad1-NICD interaction is further stabilized by associations with the p300/CBP and P/CAF coactivators (Itoh et al. 2004). This crosstalk leads to transcriptional cooperation or antagonism, depending on the gene and cell context. In myogenesis, upregulation of Hes and Hey1 expression by Notch signaling is required for TGF-β/BMP-mediated inhibition of differentiation (Blokzijl et al. 2003, Dahlqvist et al. 2003), whereas in endothelial cells, Herp2 expression in response to Notch inhibits cell migration by antagonizing BMP-induced Id1 function (Itoh et al. 2004).

TGF-β/Smad signaling also crosstalks with NF-κB signaling. NF-κB acts as a DNA-binding homodimer or heterodimer to induce transcription in response to inflammatory stimuli. TGF-β signaling can cooperate with NF-κB transcription through interaction of Smad3 with the p52 NF-κB subunit at adjacent NF-κB and Smad binding sites (Lopez-Rovira et al. 2000). Since NF-κB and R-Smads both interact with CBP/p300, their cooperation is likely a result of coordinately increased recruitment of CBP/p300, similar to the cooperation of Smad3 with many sequence-specific transcription factors.

Finally, TGF-β family signaling also synergizes with the p53 tumor suppressor, a regulator of cell proliferation, apoptosis, and differentiation. TGF-β/BMP signaling results in the formation of a p53-Smad complex that activates the transcription of target genes with distinct p53- and Smad-binding DNA sequences in their promoters. (Cordenonsi et al. 2003, Takebayashi-Suzuki et al. 2003). Furthermore, TGF-β treatment recruits p53, Smad2/4, and SnoN to adjacent SBE- and p53-binding sequences in the α-fetoprotein gene regulatory sequences, leading to transcription repression (Wilkinson et al. 2005).

In summary, the cooperation of Smads with DNA-binding transcription factors creates extensive versatility in the transcriptional regulation of target genes. Activated transcription often results from the interaction of the activated Smad complex with one DNA-binding transcription factor, but a higher level of complexity in which the Smad complex interacts with one or several DNA-binding transcription factors can occur, depending on the physiological context. This more complex scenario of transcriptional control with multiple Smad complexes or a larger complex may play out in the regulation of Smad7 transcription, through interactions of Smads with TFE3, AP-1, and Sp1 (Brodin et al. 2000, Hua et al. 2000); or the germ line IgCα promoter, through interactions of Smads with CREB and Runx proteins (Zhang & Derynck 2000). Such complex regulation may involve several Smad-binding sequences in addition to the DNA-binding sites for Smad-interacting transcription factors as in the promoter regions of the IgCα (Zhang & Derynck 2000), p15[Ink4B] (Feng et al. 2000, Seoane et al. 2001), and p21[Cip1] genes (Pardali et al. 2000b, Seoane et al. 2002).

Coactivators and Corepressors of Smads

In addition to interactions with DNA-binding transcription factors, Smads can recruit coactivators or corepressors into the transcription machinery that determine the amplitude of TGF-β/Smad-mediated transcriptional activation (**Table 2**).

Transcription coactivators, such as CBP/p300 and the Mediator complex, increase transcription by bringing the sequence-specific transcription factors into proximity

NICD: Notch intracellular domain

Table 2 Transcriptional coactivators and corepressors for Smads

Cofactors	Smad and Domains	Function	Reference
Coactivators			
ARC105	Smad2/3/4 (MH2)	Component of the ARC/Mediator	Kato et al. 2002
CBP/p300	Smad1/2/3 (MH2) Smad3 (linker) Smad4 (SAD)	Modulate chromatin structure and bridging TGF-β-independent transactivation function Smads with basic transcription machinery	Feng et al. 1998 Wang et al. 2005 de Caestecker et al. 2000
GCN5	Smad1/2/3/5	Modulation of chromatin structure	Kahata et al. 2004
MSG1	Smad4 (MH2)	Activation of CBP/p300-dependent transcription	Shioda et al. 1998
PCAF	Smad2/3 (MH2)	Modulation of chromatin structure and stimulation of CBP/p300-dependent transcription	Itoh et al. 2004
SKIP	Smad2/3 (Linker-MH2)	Derepression of Ski/SnoN?	Leong et al. 2001
SMIF	Smad4	Enhanced Smad4 coactivator function	Bai & Cao 2002
Swift	Smad2	Enhanced Smad2 transactivation function	Shimizu et al. 2001
ZEB1	Smad1/2/3/5 (MH2)	Promotes the formation of a p300-Smad transcriptional complex	Postigo 2003, Postigo et al. 2003
Corepressors			
c-Myc	Smad2/3 (MH2)	Inhibition of Smad-Sp1 activator complex	Feng et al. 2002
c-Ski, SnoN	Smad2/3/4 (MH2)	Recruits N-CoR, mSin3 and HADC	Luo et al. 1999, Wang et al. 2000, Wu et al. 2002
Evi-1 (ZF)	Smad1/2/3/4 (MH2)	Evi-1 is a zinc finger protein and recruits CtBP to repression complex	Izutsu et al. 2001, Kurokawa et al. 1998
SNIP1 (FHA)	Smad1/2/4	Inhibition of Smad4-p300 complex formation	Kim et al. 2000
TGIF (HD)	Smad2 (MH2)	Recruits CtBP and HDAC	Wotton & Massague 2001 and references therein
Tob	Smad1/5/8/4 Smad2/4	Targeting of BMP R-Smad to nuclear body Enhancement of Smad4 DNA-binding	Yoshida et al 2000 Tzachanis et al. 2001
YB-1	Smad3	Disrupt Smad3-DNA and Smad3-p300 interactions	Higashi et al. 2003
ZEB2/SIP1	Smad1/2/3/5 (MH2)	Recruitment of CtBP	Postigo 2003, Postigo et al. 2003, Verschueren et al. 1999

to the RNA polymerase II complex. Some coactivators, e.g., CBP and p300, possess histone acetyltransferase (HAT) activity to modify chromatin structure. Through their MH2 domains, R-Smads directly interact with CBP or p300; their efficient interaction requires C-terminal SXS phosphorylation. This interaction is required for the transactivation function of the MH2 domain. The ligand-independent interactions of CBP/p300 with the linker region of Smad3,

and possibly those of other Smads, contribute to full Smad3 activity (Wang et al. 2005).

The function of CBP as an R-Smad coactivator requires Smad4, which stabilizes the R-Smad interaction with CBP (Feng et al. 1998). The MH2 domain of Smad4 does not associate with CBP/p300 and has no transcription activity. However, inclusion of a proline-rich "SAD domain" upstream of the MH2 domain confers Smad4-dependent transcription (de Caestecker et al. 2000). This domain interacts with an N-terminal segment of p300 (de Caestecker et al. 2000) and also recruits SMIF, which has intrinsic transcription activity (Bai et al. 2002). Thus, a mutant Smad4 that does not interact with CBP/p300 yet retains SMIF binding is transcriptionally active (Bai et al. 2002). The interaction of SMIF with Smad4 suggests a function for SMIF in signaling by all TGF-β family members, irrespective of the nature of the activated R-Smad.

The coactivator functions of Smad4 and CBP/p300, and Smad-mediated transcription, can be further enhanced by MSG1. This coactivator interacts through a C-terminal domain with p300/CBP, and its N-terminal domain with the MH2 domain of Smad4 (Shioda et al. 1998). ZEB1, a Zn finger protein similar to ZEB1/SIP1, also enhances TGF-β signaling by promoting Smad3-p300/CBP interaction (Postigo et al. 2003).

The p300/CBP-associated PCAF and GCN5, two related coactivators, associate with Smad2 and Smad3 and potentiate TGF-β-induced transcription responses (Itoh et al. 2000, Kahata et al. 2004). GCN5, but not PCAF, also interacts with BMP-activated R-Smads and enhances BMP signaling (Kahata et al. 2004). Whether PCAF and GCN5 enhance Smad signaling through their ability to modify histones remains to be shown.

The ARC or Mediator complex acts as a coactivator in transcription through its interaction with RNA polymerase II, and may be a target of diverse regulatory circuits. ARC105, a component of this complex, is recruited to the Smad-responsive promoter in response to activin/nodal and binds Smad2/3 and Smad4, but not Smad1, in response to TGF-β (Kato et al. 2002). Thus, the Smad-ARC105 interaction mediates and relays TGF-β signaling to the Pol II machinery, which activates select genes. It is possible that BMP signals impinge on a distinct ARC component that interacts with a BMP-activated R-Smad and helps control BMP-responsive transcription.

Finally, the coactivator *Swift* interacts with Smad2 and has intrinsic transcription activity. Although it also interacts with Smad1, Swift enhances only activin/Smad2-mediated transcription and not BMP-induced responses in Xenopus embryos (Shimizu et al. 2001).

Corepressors that directly interact with Smads repress transcription induced by Smads. Several proto-oncogenes, including c-Ski/SnoN, c-Myc, and Evi-1, link repression of TGF-β/Smad signaling to malignant transformation. For example, c-Ski interacts with the MH2 domains of Smad2 and Smad3; increased expression of c-Ski or the related SnoN decreases activation of transcription by Smads (Luo 2004). In response to TGF-β, c-Ski inhibits both the induction of p15[Ink4B] and the downregulation of c-Myc expression, and consequently abolishes the growth inhibitory functions of TGF-β (Sun et al. 1999). c-Ski represses not only Smad2/3 responses but also BMP signaling through interaction with BMP-activated Smads and Smad4 (Wang et al. 2000). Additionally, c-Ski disrupts the functional complex of R-Smads with Smad4 (Wu et al. 2002), and recruits the nuclear N-CoR or mSin3 corepressors and interacting histone deacetylase(s) into the transcription complex (Luo et al. 1999), thus providing a dual mechanism of repression. The nuclear hormone receptor coactivator SKIP (Ski-interacting protein), which also interacts with the MH2 domain of Smad2 or Smad3, opposes the c-Ski-dependent repression of Smad transactivation and thus enhances Smad-mediated TGF-β responses (Leong et al. 2001).

c-Myc represses expression of p15[Ink4B] and p21[Cip1]. At the p15[Ink4B] promoter, c-Myc associates with Smad2 and Smad3 and does not interfere with the formation of the Smad-Sp1 activator complex (Feng et al. 2002). The interaction of c-Myc with Sp1 presumably helps stabilize the interaction of c-Myc with the Smad-Sp1 complex and represses the functional cooperation between the Smad complex and Sp1 (Feng et al. 2002). c-Myc also interacts with the Zn finger protein Miz-1 near the transcription initiation site of the p15[Ink4B] promoter and thereby represses the ability of Miz-1 to activate p15[Ink4B] expression.

Repression of c-myc expression in response to TGF-β results in decreased interaction of c-Myc with Miz-1 (Seoane et al. 2001), thus conferring derepression that allows for Smad/Sp1-mediated transcription activation.

Evi-1, a Zn finger transcription factor, also represses Smad-mediated signaling. The repression of growth inhibition by TGF-β is likely the basis of the oncogenic function of Evi-1. Evi-1 interacts with the MH2 domain of Smad3 and other R-Smads, and thereby represses their transactivation function (Kurokawa et al. 1998, Alliston et al. 2005). Consequently, Evi-1 represses gene expression that is activated by activin, TGF-β, and BMPs (Alliston et al. 2005). The repressor activity of Evi-1 requires direct association with the corepressor CtBP (Alliston et al. 2005, Izutsu et al. 2001).

The homeobox transcription factor TGIF can also interact with Smads to repress Smad-mediated transcription. TGIF recruits histone deacetylases through its interaction with mSin3 and CtBP and competes with CBP/p300 for the R-Smad interaction (Wotton & Massagué 2001). Thus, TGIF acts through histone deacetylase–dependent and –independent mechanisms to repress TGF-β/Smad-activated transcription. The corepressor activity of TGIF is not restricted to TGF-β/Smad signaling, since TGIF binds cognate DNA sequences via its homeodomain and thus represses transcription independently of its interactions with Smads (Wotton & Massagué 2001).

A similar mechanism may account for the corepressor function of ZEB2/SIP1, a Zn finger/homeodomain protein that binds E-box sequences. SIP1 interacts with the MH2 domains of Smads (Postigo 2003, Verschueren et al. 1999) and represses Smad-mediated transcription depending on a DNA sequence that allows SIP1 binding (Comijn et al. 2001). Interestingly, the related ZEB-1/δEF1 protein, which also binds E-box sequences and can interact with Smad MH2 domains, activates TGF-β/BMP signaling (Postigo 2003). SIP1 downregulates hTERT

(Lin & Elledge 2003) and E-cadherin expression (Comijn et al. 2001). Therefore, TGF-β-induced SIP1 expression can contribute to TGF-β-induced epithelial-to-mesenchymal transdifferentiation (Comijn et al. 2001) and inhibition of cellular transformation (Lin & Elledge 2003).

SNIP1 is yet another nuclear protein that can repress Smad-activated transcription. SNIP1 can interact with R-Smads and Smad4 as well as CBP/p300 (Kim et al. 2000). Thus, SNIP1 represses not only Smad-mediated transcription but also other responses that use CBP/p300 as coactivators.

Finally, Tob, a member of the Tob/BTG family of proteins with antiproliferative activities, participates in the regulation of both TGF-β and BMP signaling. Tob interacts with BMP-activated Smads and inhibits the stimulatory effect of BMPs on osteoblast function and bone deposition (Yoshida et al. 2000). In TGF-β signaling, interaction of Tob with Smad2 represses expression of interleukin-2 in T cells (Tzachanis et al. 2001). Tob proteins also bind inhibitory Smads and enhance their interactions with receptors, thereby inhibiting TGF-β signaling at the receptor level (Yoshida et al. 2003). Tob's mechanism of repression remains to be characterized.

Transcriptional Activation Versus Repression

Compared to Smad-mediated transcriptional activation, much less is known about the mechanisms of transcriptional repression by TGF-β family factors. Downregulation of c-myc expression has a key role in the growth inhibition response to TGF-β; preventing c-myc downregulation confers resistance to growth inhibition by TGF-β (Chen et al. 2001). The c-myc promoter contains a sequence that resembles the TIE in the promoter of the stromelysin 1 gene, which is also downregulated in response to TGF-β (Chen et al. 2002, Yagi et al. 2002). The c-myc TIE binds Smad3/4, E2F-4, and p107, and confers

TIE: TGF-β inhibitory element

a transcriptional repression response to TGF-β (Chen et al. 2002, Frederick et al. 2004). The sequence that binds Smad3 is distinct from an SBE and overlaps with a consensus E2F site that binds E2F4/5 and recruits p107 (Frederick et al. 2004). First, Smad3 forms a complex with E2F4/DP-1/p107 in the cytoplasm, and in response to TGF-β the complex associates with Smad4 and occupies the TIE. It is not clear whether or how Smad3 and E2F4 simultaneously bind to the TIE sequence to repress the c-myc promoter (Chen et al. 2002, Frederick et al. 2004).

The inhibition of osteoblast differentiation by TGF-β is mediated in part by the interaction of Smad3 with Runx2, leading to repression of Runx2 transcription activity. TGF-β/Smad3-mediated repression of Runx2 neither requires DNA binding of Smad3 to the promoter nor results from decreased Runx2 binding to its cognate DNA sequence (Alliston et al. 2001). On the basis of a comparison of the responses at the Runx2-binding sequences in the osteocalcin and IgCα promoter sequences, it is apparent that the DNA sequence and cell type are important determinants. Indeed, depending on the DNA sequence, Smad3 cooperates with Runx2 to enhance or repress transcription. In addition, at the Runx2 binding sequence of the osteocalcin promoter, TGF-β and Smad3 repress Runx2-mediated transcription in mesenchymal cells but enhance it in epithelial cells (Alliston et al. 2001). Thus, cell type–dependent factors are key determinants of Smad-dependent activation versus repression. In osteoblasts and other mesenchymal cells, this repression of Runx2 by Smad3 is mediated by the direct recruitment of class IIa histone deacetylases, specifically HDAC4 and HDAC5, by TGF-β-activated Smad3 to the Runx2-binding DNA sequence in the osteocalcin promoter, thus resulting in histone deacetylation (Kang et al. 2005).

Recruitment of histone deacetylases has also been invoked in TGF-β family–induced transcription repression. BMP signaling results in the formation of a complex of Nkx3.2, HDAC1, and Smad1, and represses the transcription activity of Nkx3.2. The interaction of Nkx3.2 with HDAC/Sin3A requires the interaction of Nkx3.2 with Smad1 and Smad4. Thus, as in the case of TGF-β, BMP-activated Smads support ligand-induced transcription repression (Kim & Lassar 2003).

A different mechanism of Smad-mediated repression operates in the inhibition of myogenic differentiation by TGF-β. In response to TGF-β, Smad3 represses the activity of MyoD and myogenin through its direct interaction with the HLH domains of MyoD or myogenin (Liu et al. 2001). As a consequence, Smad3 interferes with the heterodimerization of MyoD or myogenin with their obligatory partner E12/47, thereby decreasing the DNA binding of MyoD or myogenin. Smad3 also interacts with MEF2C, which is a direct DNA-binding transcription factor and also serves as a coactivator required for efficient transcription by myogenic bHLH transcription factors. This interaction of Smad3 prevents MEF2C from associating with the MyoD/E protein complex and GRIP1, a coactivator that is required for the transcription functions of MEF2C (Liu et al. 2004).

Finally, Smad3/4 repress C/EBPβ- and STAT-3-mediated transcription of the haptoglobin promoter (Zauberman et al. 2001) and Smad3 represses the transactivation functions of C/EBPs, leading to transcriptional repression of the PPAR-γ promoter (Choy & Derynck 2003). The mechanisms for these cases of repression have not been characterized.

Taken together, the mechanistic differences of Smad-mediated repression versus activation remain to be fully characterized, yet are determined by cell type– and DNA sequence–dependent factors. In some cases, histone deacetylase–independent mechanisms mediate Smad-dependent repression, as in the repression of myogenic bHLH transcription factors and MEF2 (Liu et al. 2001, 2004). In other cases, e.g., the repression of Runx2 and Nkx3.2, histone deacetylase recruitment is involved. The interaction of Smad3 with HDAC4 and HDAC5 (Kang et al. 2005) or with a different histone deacetylase activity (Liberati et al. 2001) illustrates the function of Smads as transcription repressors.

Inhibitory Smads as Transcription Regulators

Although inhibitory Smads interfere with receptor-mediated activation of R-Smads, several lines of evidence indicate that Smad6 and 7 also act as transcription regulators in the nucleus. Smad6 can physically interact with the corepressor CtBP; this interaction is mediated by Smad6's PLDLS motif, which is found in many repressors and confers intrinsic repressor activity to Smad6. Smad6-CtBP complexes are found at the BMP-responsive Id1 promoter and repress Id1 transcription (Lin et al. 2003). Smad6 can also interact with homeobox transcription factors at the DNA and thereby functions as corepressor (Bai et al. 2000). These interactions may also recruit class I histone deacetylases such as HDAC1 to repress BMP-induced gene transcription (Bai & Cao 2002). Although Smad7 does not interact with CtBP (Lin et al. 2003), it may also possess intrinsic transcription functions. Like Smad6, Smad7 is primarily localized in the nucleus. When fused to the DNA-binding domain of the Gal4 transcription factor,

Smad7 can transactivate a Gal4 reporter gene (Pulaski et al. 2001). Furthermore, Smad7 interacts with and can be acetylated by the coactivator p300, further implicating a possible function of Smad7 in the nucleus (Grönroos et al. 2002).

CONCLUSION

Although the signaling system through heteromeric TGF-β receptors and Smad complexes is conceptually simple, combinatorial interactions provide a high degree of signaling specificity and versatility. The signaling responses can be qualitatively and quantitatively regulated by differential type I-type II receptor interactions, Smad complex formation, receptor and Smad interactions with accessory proteins, and crosstalk of the Smads with other signaling pathways. The specificity and quantitative regulation of Smad signaling has additional levels of versatility dictated by the complex nature of the Smad activator complex. In this complex, functional and physical interactions of Smads with DNA-specific transcription factors, which themselves are regulated by other signaling pathways, and transcription coactivators or corepressors that link the Smad complex to the Pol II complex, confer both specificity and complexity in transcriptional responses to TGF-β family ligands.

SUMMARY POINTS

1. As central signal transducers in TGF-β signaling, Smads transduce the signals from ligand-receptor complexes at the cell surface to gene transcription in the nucleus. Specific Smad-protein interactions determine signaling specificity.

2. The L45 loop in the type I receptor and L3 loop in R-Smads are the key determinants in specifying signaling in response to specific ligands.

3. Smads are weak DNA-binding proteins and naturally function by cooperating with a large number of sequence-specific DNA-binding transcription factors, thus leading to signaling versatility in TGF-β gene responses.

4. Inhibitory Smads, coreceptors at the surface, and intracellular kinases can modify the signaling strength of Smads.

FUTURE ISSUES TO BE RESOLVED

1. A critical issue is how activated R-Smads are dephosphorylated, leading to recycling of Smads. What are the phosphatases?

2. It is important to solve the structures of full-length Smad proteins as well as those of Smad complexes with other transcriptional partners to understand how Smads function in transcriptional control.

3. Since Smads activate or repress transcription of genes in the context of chromatin, it is important to understand the effects of Smad signaling on chromatin remodeling.

4. Experimental approaches need to be improved to better understand the roles of endocytosis and intracellular routing in TGF-β signaling.

5. Are Smads the only signal transducers to receive signals directly from TGF-β receptors that lead to changes in transcription?

6. The mechanisms through which non-Smad signaling pathways are activated by the receptors and what these pathways contribute to the cellular response need to be better defined.

LITERATURE CITED

Abdel-Wahab N, Wicks SJ, Mason RM, Chantry A. 2002. Decorin suppresses transforming growth factor-β-induced expression of plasminogen activator inhibitor-1 in human mesangial cells through a mechanism that involves Ca2+-dependent phosphorylation of Smad2 at serine-240. *Biochem. J.* 362:643–49

Alliston T, Choy L, Ducy P, Karsenty G, Derynck R. 2001. TGF-β-induced repression of CBFA1 by Smad3 decreases cbfa1 and osteocalcin expression and inhibits osteoblast differentiation. *EMBO J.* 20:2254–72

Alliston T, Ko TC, Cao Y, Liang Y-Y, Feng X-H, Chang C, Derynck R. 2005. Repression of BMP- and activin-inducible transcription by Evi-1. *J. Biol. Chem.* 280:24227–37

Annes JP, Munger JS, Rifkin DB. 2003. Making sense of latent TGFβ activation. *J. Cell Sci.* 116:217–24

Bai RY, Koester C, Ouyang T, Hahn SA, Hammerschmidt M, et al. 2002. SMIF, a Smad4-interacting protein that functions as a co-activator in TGFβ signalling. *Nat. Cell Biol.* 4:181–90

Bai S, Cao X. 2002. A nuclear antagonistic mechanism of inhibitory Smads in transforming growth factor-β signaling. *J. Biol. Chem.* 277:4176–82

Bai S, Shi X, Yang X, Cao X. 2000. Smad6 as a transcriptional corepressor. *J. Biol. Chem.* 275:8267–70

Benchabane H, Wrana JL. 2003. GATA- and Smad1-dependent enhancers in the Smad7 gene differentially interpret bone morphogenetic protein concentrations. *Mol. Cell Biol.* 23:6646–61

Bitzer M, von Gersdorff G, Liang D, Dominguez-Rosales A, Beg AA, et al. 2000. A mechanism of suppression of TGF-β/SMAD signaling by NF-κB/RelA. *Genes Dev.* 14:187–97

Blobe GC, Liu X, Fang SJ, How T, Lodish HF. 2001. A novel mechanism for regulating transforming growth factor β (TGF-β) signaling. Functional modulation of type III

TGF-β receptor expression through interaction with the PDZ domain protein, GIPC. *J. Biol. Chem.* 276:39608–17

Blokzijl A, Dahlqvist C, Reissmann E, Falk A, Moliner A, et al. 2003. Cross-talk between the Notch and TGF-β signaling pathways mediated by interaction of the Notch intracellular domain with Smad3. *J. Cell Biol.* 163:723–28

Blokzijl A, ten Dijke P, Ibanez CF. 2002. Physical and functional interaction between GATA-3 and Smad3 allows TGF-β regulation of GATA target genes. *Curr. Biol.* 12:35–45

Boesen CC, Radaev S, Motyka SA, Patamawenu A, Sun PD. 2002. The 1.1 Å crystal structure of human TGF-β type II receptor ligand binding domain. *Structure* 10:913–19

Brodin G, Ahgren A, ten Dijke P, Heldin CH, Heuchel R. 2000. Efficient TGF-β induction of the Smad7 gene requires cooperation between AP-1, Sp1, and Smad proteins on the mouse Smad7 promoter. *J. Biol. Chem.* 275:29023–30

Brown CO 3rd, Chi X, Garcia-Gras E, Shirai M, Feng XH, et al. 2004. The cardiac determination factor, Nkx2-5, is activated by mutual cofactors GATA-4 and Smad1/4 via a novel upstream enhancer. *J. Biol. Chem.* 279:10659–69

Brown JD, DiChiara MR, Anderson KR, Gimbrone MA Jr, Topper JN. 1999. MEKK-1, a component of the stress (stress-activated protein kinase/c-Jun N-terminal kinase) pathway, can selectively activate Smad2-mediated transcriptional activation in endothelial cells. *J. Biol. Chem.* 274:8797–805

Chacko BM, Qin B, Correia JJ, Lam SS, de Caestecker MP, et al. 2001. The L3 loop and C-terminal phosphorylation jointly define Smad protein trimerization. *Nat. Struct. Biol.* 8:248–53

Chacko BM, Qin BY, Tiwari A, Shi G, Lam S, et al. 2004. Structural basis of heteromeric Smad protein assembly in TGF-β signaling. *Mol. Cell* 15:813–23

Chen CR, Kang Y, Massagué J. 2001. Defective repression of c-myc in breast cancer cells: A loss at the core of the transforming growth factor β growth arrest program. *Proc. Natl. Acad. Sci. USA* 98:992–99

Chen CR, Kang Y, Siegel PM, Massagué J. 2002. E2F4/5 and p107 as Smad cofactors linking the TGFβ receptor to c-myc repression. *Cell* 110:19–32

This study shows that Smads can recruit transcription repressors to the c-Myc promoter.

Chen D, Ji X, Harris MA, Feng JQ, Karsenty G, et al. 1998. Differential roles for bone morphogenetic protein (BMP) receptor type IB and IA in differentiation and specification of mesenchymal precursor cells to osteoblast and adipocyte lineages. *J. Cell Biol.* 142:295–305

Chen W, Kirkbride KC, How T, Nelson CD, Mo J, et al. 2003. β-arrestin 2 mediates endocytosis of type III TGF-β receptor and down-regulation of its signaling. *Science* 301:1394–97

Chen X, Weisberg E, Fridmacher V, Watanabe M, Naco G, et al. 1997. Smad4 and FAST-1 in the assembly of activin-responsive factor. *Nature* 389:85–89

This paper, together with Chen X et al. 1996, *Nature* 383:691–96, provides the first demonstration that Smad2/4 bind to a specific transcription factor at the target gene promoter.

Chen YG, Hata A, Lo RS, Wotton D, Shi Y, et al. 1998. Determinants of specificity in TGF-β signal transduction. *Genes Dev.* 12:2144–52

Chen YG, Massagué J. 1999. Smad1 recognition and activation by the ALK1 group of transforming growth factor-β family receptors. *J. Biol. Chem.* 274:3672–77

Chiba S, Takeshita K, Imai Y, Kumano K, Kurokawa M, et al. 2003. Homeoprotein DLX-1 interacts with Smad4 and blocks a signaling pathway from activin A in hematopoietic cells. *Proc. Natl. Acad. Sci. USA* 100:15577–82

Chipuk JE, Cornelius SC, Pultz NJ, Jörgensen JS, Bonham MJ, et al. 2002. The androgen receptor represses transforming growth factor-β signaling through interaction with Smad3. *J. Biol. Chem.* 277:1240–48

Chou WC, Prokova V, Shiraishi K, Valcourt U, Moustakas A, et al. 2003. Mechanism of a transcriptional cross talk between transforming growth factor-β-regulated Smad3 and Smad4 proteins and orphan nuclear receptor hepatocyte nuclear factor-4. *Mol. Biol. Cell* 14:1279–94

Choy L, Derynck R. 2003. Transforming growth factor-β inhibits adipocyte differentiation by Smad3 interacting with CCAAT/enhancer-binding protein (C/EBP) and repressing C/EBP transactivation function. *J. Biol. Chem.* 278:9609–19

Comijn J, Berx G, Vermassen P, Verschueren K, van Grunsven L, et al. 2001. The two-handed E box binding zinc finger protein SIP1 downregulates E-cadherin and induces invasion. *Mol. Cell* 7:1267–78

Conery AR, Cao Y, Thompson EA, Townsend CM Jr, Ko TC, et al. 2004. Akt interacts directly with Smad3 to regulate the sensitivity to TGF-β induced apoptosis. *Nat. Cell Biol.* 6:366–72

Cordenonsi M, Dupont S, Maretto S, Insinga A, Imbriano C, et al. 2003. Links between tumor suppressors: p53 is required for TGF-β gene responses by cooperating with Smads. *Cell* 113:301–14

Costamagna E, Garcia B, Santisteban P. 2003. The functional interaction between the paired domain transcription factor Pax8 and Smad3 is involved in the transforming growth factor-β repression of the sodium/iodide symporter gene. *J. Biol. Chem.* 279:3439–46

Coyle-Rink J, Sweet T, Abraham S, Sawaya B, Batuman O, et al. 2002. Interaction between TGFβ signaling proteins and C/EBP controls basal and Tat-mediated transcription of HIV-1 LTR in astrocytes. *Virology* 299:240–47

Dahlqvist C, Blokzijl A, Chapman G, Falk A, Dannaeus K, et al. 2003. Functional Notch signaling is required for BMP4-induced inhibition of myogenic differentiation. *Development* 130:6089–99

Dai H, Hogan C, Gopalakrishnan B, Torres-Vazquez J, Nguyen M, et al. 2000. The zinc finger protein schnurri acts as a Smad partner in mediating the transcriptional response to decapentaplegic. *Dev. Biol.* 227:373–87

Datta PK, Moses HL. 2000. STRAP and Smad7 synergize in the inhibition of transforming growth factor β signaling. *Mol. Cell Biol.* 20:3157–67

de Caestecker M, Parks W, Frank C, Castagnino P, Bottaro D, et al. 1998. Smad2 transduces common signals from receptor serine-threonine and tyrosine kinases. *Genes Dev.* 12:1587–92

de Caestecker MP, Yahata T, Wang D, Parks WT, Huang S, et al. 2000. The Smad4 activation domain (SAD) is a proline-rich, p300-dependent transcriptional activation domain. *J. Biol. Chem.* 275:2115–22

del Re E, Babitt JL, Pirani A, Schneyer AL, Lin HY. 2004. In the absence of type III receptor, the transforming growth factor (TGF)-β type II receptor requires the type I receptor to bind TGF-β2. *J. Biol. Chem.* 279:22765–72

Derynck R, Zhang YE. 2003. Smad-dependent and Smad-independent pathways in TGF-β family signalling. *Nature* 425:577–84

Di Guglielmo GM, Le Roy C, Goodfellow AF, Wrana JL. 2003. Distinct endocytic pathways regulate TGF-β receptor signalling and turnover. *Nat. Cell Biol.* 5:410–21

Ebisawa T, Fukuchi M, Murakami G, Chiba T, Tanaka K, et al. 2001. Smurf1 interacts with transforming growth factor-β type I receptor through Smad7 and induces receptor degradation. *J. Biol. Chem.* 276:12477–80

Engel ME, McDonnell MA, Law BK, Moses HL. 1999. Interdependent SMAD and JNK signaling in transforming growth factor-β-mediated transcription. *J. Biol. Chem.* 274:37413–20

Feng X-H, Derynck R. 1997. A kinase subdomain of transforming growth factor-β (TGF-β) type I receptor determines the TGF-β intracellular signaling specificity. *EMBO J.* 16:3912–23

Feng X-H, Liang Y-Y, Liang M, Zhai W, Lin X. 2002. Direct interaction of c-Myc with Smad2 and Smad3 to inhibit TGF-β-mediated induction of the CDK inhibitor p15^{Ink4B}. *Mol. Cell* 9:133–43

Feng X-H, Lin X, Derynck R. 2000. Smad2, Smad3 and Smad4 cooperate with Sp1 to induce p15^{Ink4B} transcription in response to TGF-β. *EMBO J.* 19:5178–93

Feng X-H, Zhang Y, Wu R-Y, Derynck R. 1998. The tumor suppressor Smad4/DPC4 and transcriptional adaptor CBP/p300 are coactivators for Smad3 in TGF-β-induced transcriptional activation. *Genes Dev.* 12:2153–63

Frederick JP, Liberati NT, Waddell DS, Shi Y, Wang XF. 2004. Transforming growth factor β-mediated transcriptional repression of c-myc is dependent on direct binding of Smad3 to a novel repressive Smad binding element. *Mol. Cell Biol.* 24:2546–59

Furuhashi M, Yagi K, Yamamoto H, Furukawa Y, Shimada S, et al. 2001. Axin facilitates Smad3 activation in the transforming growth factor β signaling pathway. *Mol. Cell Biol.* 21:5132–41

Germain S, Howell M, Esslemont GM, Hill CS. 2000. Homeodomain and winged-helix transcription factors recruit activated Smads to distinct promoter elements via a common Smad interaction motif. *Genes Dev.* 14:435–51

Goumans MJ, Valdimarsdottir G, Itoh S, Lebrin F, Larsson J, et al. 2003. Activin receptor-like kinase (ALK)1 is an antagonistic mediator of lateral TGFβ/ALK5 signaling. *Mol. Cell* 12:817–28

Goumans MJ, Valdimarsdottir G, Itoh S, Rosendahl A, Sideras P, et al. 2002. Balancing the activation state of the endothelium via two distinct TGF-β type I receptors. *EMBO J.* 21:1743–53

Greenwald J, Groppe J, Gray P, Wiater E, Kwiatkowski W, et al. 2003. The BMP7/ActRII extracellular domain complex provides new insights into the cooperative nature of receptor assembly. *Mol. Cell* 11:605–17

Greenwald J, Vega ME, Allendorph GP, Fischer WH, Vale W, et al. 2004. A flexible activin explains the membrane-dependent cooperative assembly of TGF-β family receptors. *Mol. Cell* 15:485–89

Grinberg AV, Kerppola T. 2003. Both Max and TFE3 cooperate with Smad proteins to bind the plasminogen activator inhibitor-1 promoter, but they have opposite effects on transcriptional activity. *J. Biol. Chem.* 278:11227–36

Grönroos E, Hellman U, Heldin CH, Ericsson J. 2002. Control of Smad7 stability by competition between acetylation and ubiquitination. *Mol. Cell* 10:483–93

Hanai J, Chen LF, Kanno T, Ohtani-Fujita N, Kim WY, et al. 1999. Interaction and functional cooperation of PEBP2/CBF with Smads. Synergistic induction of the immunoglobulin germline Cα promoter. *J. Biol. Chem.* 274:31577–82

Hanyu A, Ishidou Y, Ebisawa T, Shimanuki T, Imamura T, et al. 2001. The N domain of Smad7 is essential for specific inhibition of transforming growth factor-β signaling. *J. Cell Biol.* 155:1017–27

Hart PJ, Deep S, Taylor AB, Shu Z, Hinck CS, et al. 2002. Crystal structure of the human TβR2 ectodomain–TGF-β3 complex. *Nat. Struct. Biol.* 9:203–8

Hata A, Lagna G, Massagué J, Hemmati-Brivanlou A. 1998. Smad6 inhibits BMP/Smad1 signaling by specifically competing with the Smad4 tumor suppressor. *Genes Dev.* 12:186–97

This study identified the L45 loop in the kinase domain as a key determinant in specifying TGF-β receptor signaling.

This study proposes that CBP/p300 and the common Smad4 are essential coactivators in R-Smad signaling, and provides a molecular mechanism for these coactivation functions.

This paper, together with Goumans et al. (2002), shows that, in endothelial cells, TGF-β activates Smad2/3 and Smad1/5 pathways through the combination of the TβRI and ALK1 type I receptors.

Hata A, Seoane J, Lagna G, Montalvo E, Hemmati-Brivanlou A, et al. 2000. OAZ uses distinct DNA- and protein-binding zinc fingers in separate BMP-Smad and Olf signaling pathways. *Cell* 100:229–40

Hayashi H, Abdollah S, Qiu Y, Cai J, Xu Y-Y, et al. 1997. The MAD-related protein Smad7 associates with the TGFβ receptor and functions as an antagonist of the TGFβ signaling. *Cell* 89:1165–73

Hayes S, Chawla A, Corvera S. 2002. TGF β receptor internalization into EEA1-enriched early endosomes: role in signaling to Smad2. *J. Cell Biol.* 158:1239–49

Hayes SA, Zarnegar M, Sharma M, Yang F, Peehl DM, et al. 2001. SMAD3 represses androgen receptor-mediated transcription. *Cancer Res.* 61:2112–18

Higashi K, Inagaki Y, Fujimori K, Nakao A, Kaneko H, Nakatsuka I. 2003. Inteferon-γ interferes with transforming growth factor-β signaling through direct interaction of YB-1 with Smad3. *J. Biol. Chem.* 278:43470–79

Hocevar BA, Smine A, Xu XX, Howe PH. 2001. The adaptor molecule Disabled-2 links the transforming growth factor β receptors to the Smad pathway. *EMBO J.* 20:2789–801

Hoodless PA, Pye M, Chazaud C, Labbé E, Attisano L, et al. 2001. FoxH1 (Fast) functions to specify the anterior primitive streak in the mouse. *Genes Dev.* 15:1257–71

Hu MC, Rosenblum ND. 2005. Smad1, β-catenin and TCF4 associate in a molecular complex with the Myc promoter in dysplastic renal tissue and cooperate to control Myc transcription. *Development* 132:215–25

Hua X, Miller ZA, Benchabane H, Wrana JL, Lodish HF. 2000. Synergism between transcription factors TFE3 and Smad3 in TGF-β-induced transcription of the Smad7 gene. *J. Biol. Chem.* 275:33205–8

Hua X, Miller ZA, Wu G, Shi Y, Lodish HF. 1999. Specificity in transforming growth factor β-induced transcription of the plasminogen activator inhibitor-1 gene: interactions of promoter DNA, transcription factor μE3, and Smad proteins. *Proc. Natl. Acad. Sci. USA* 96:13130–35

Huse M, Muir TW, Xu L, Chen YG, Kuriyan J, et al. 2001. The TGF β receptor activation process: an inhibitor-to-substrate-binding switch. *Mol. Cell* 8:671–82

Hussein SM, Duff EK, Sirard C. 2003. Smad4 and β-catenin co-activators functionally interact with lymphoid-enhancing factor to regulate graded expression of Msx2. *J. Biol. Chem.* 278:48805–14

Imamura T, Takase M, Nishihara A, Oeda E, Hanai J, et al. 1997. Smad6 inhibits signalling by the TGF-β superfamily. *Nature* 389:622–26

Inman GJ, Hill CS. 2002. Stoichiometry of active Smad-transcription factor complexes on DNA. *J. Biol. Chem.* 277:51008–16

Ishida W, Hamamoto T, Kusanagi K, Yagi K, Kawabata M, et al. 2000. Smad6 is a Smad1/5-induced Smad inhibitor. Characterization of bone morphogenetic protein-responsive element in the mouse Smad6 promoter. *J. Biol. Chem.* 275:6075–79

Itoh F, Itoh S, Goumans MJ, Valdimarsdottir G, Iso T, et al. 2004. Synergy and antagonism between Notch and BMP receptor signaling pathways in endothelial cells. *EMBO J.* 23:541–51

Itoh S, Ericsson J, Nishikawa J, Heldin CH, ten Dijke P. 2000. The transcriptional co-activator P/CAF potentiates TGF-β/Smad signaling. *Nucleic Acids Res.* 28:4291–98

Izutsu K, Kurokawa M, Imai Y, Maki K, Mitani K, et al. 2001. The corepressor CtBP interacts with Evi-1 to repress transforming growth factor β signaling. *Blood* 97:2815–22

Janda E, Lehmann K, Killisch I, Jechlinger M, Herzig M, et al. 2002. Ras and TGF-β cooperatively regulate epithelial cell plasticity and metastasis: dissection of Ras signaling pathways. *J. Cell Biol.* 156:299–313

Jayaraman L, Massagué J. 2000. Distinct oligomeric states of SMAD proteins in the transforming growth factor-β pathway. *J. Biol. Chem.* 275:40710–17

Kahata K, Hayashi M, Asaka M, Hellman U, Kitagawa H, et al. 2004. Regulation of transforming growth factor-β and bone morphogenetic protein signalling by transcriptional coactivator GCN5. *Genes Cells* 9:143–51

Kaji H, Chanaff L, Lebrun JJ, Goltzman D, Hendy GN. 2001. Inactivation of menin, a Smad3-interacting protein, blocks transforming growth factor type β signaling. *Proc. Natl. Acad. Sci. USA* 98:3837–42

Kang HY, Huang KE, Chang SY, Ma WL, Lin WJ, et al. 2002. Differential modulation of androgen receptor-mediated transactivation by Smad3 and tumor suppressor Smad4. *J. Biol. Chem.* 277:43749–56

Kang JS, Alliston T, Delston R, Derynck R. 2005. Repression of Runx2 function by TGF-β through recruitment of class II histone deacetylases by Smad3. *EMBO J.* In press

Kang Y, Chen CR, Massagué J. 2003. A self-enabling TGFβ response coupled to stress signaling: Smad engages stress response factor ATF3 for Id1 repression in epithelial cells. *Mol. Cell* 11:915–26

Kato Y, Habas R, Katsuyama Y, Naar AM, He X. 2002. A component of the ARC/Mediator complex required for TGF β/Nodal signalling. *Nature* 418:641–46

Kavsak P, Rasmussen RK, Causing CG, Bonni S, Zhu H, et al. 2000. Smad7 binds to Smurf2 to form an E3 ubiquitin ligase that targets the TGF β receptor for degradation. *Mol. Cell* 6:1365–75

Kawata Y, Suzuki H, Higaki Y, Denisenko O, Schullery D, et al. 2002. bcn-1 element-dependent activation of the laminin γ1 chain gene by the cooperative action of transcription factor E3 (TFE3) and Smad proteins. *J. Biol. Chem.* 277:11375–84

Keller S, Nickel J, Zhang JL, Sebald W, Mueller TD. 2004. Molecular recognition of BMP-2 and BMP receptor IA. *Nat. Struct. Mol. Biol.* 11:481–88

Kim DW, Lassar AB. 2003. Smad-dependent recruitment of a histone deacetylase/Sin3A complex modulates the bone morphogenetic protein-dependent transcriptional repressor activity of Nkx3.2. *Mol. Cell Biol.* 23:8704–17

Kim J, Johnson K, Chen H, Carroll S, Laughon A. 1997. *Drosophila* Mad binds to DNA and directly mediates activation of *vestigial* by Decapentaplegic. *Nature* 388:304–8

Kim RH, Wang D, Tsang M, Martin J, Huff C, et al. 2000. A novel Smad nuclear interacting protein, SNIP1, suppresses p300-dependent TGF-β signal transduction. *Genes Dev.* 14:1605–16

Kirsch T, Sebald W, Dreyer MK. 2000. Crystal structure of the BMP-2-BRIA ectodomain complex. *Nat. Struct. Biol.* 7:492–96

Korchynskyi O, ten Dijke P. 2002. Identification and functional characterization of distinct critically important bone morphogenetic protein-specific response elements in the Id1 promoter. *J. Biol. Chem.* 277:4883–91

Kretzschmar M, Doody J, Massagué J. 1997. Opposing BMP and EGF signalling pathways converge on the TGF-β family mediator Smad1. *Nature* 389:618–22

Kretzschmar M, Doody J, Timokhina I, Massagué J. 1999. A mechanism of repression of TGFβ/Smad signaling by oncogenic Ras. *Genes Dev.* 13:804–16

Kurisaki K, Kurisaki A, Valcourt U, Terentiev AA, Pardali K, et al. 2003. Nuclear factor YY1 inhibits transforming growth factor β– and bone morphogenetic protein–induced cell differentiation. *Mol. Cell Biol.* 23:4494–510

Kurokawa M, Mitani K, Irie K, Matsuyama T, Takahashi T, et al. 1998. The oncoprotein Evi-1 represses TGF-β signalling by inhibiting Smad3. *Nature* 394:92–96

This report first demonstrated that MAP kinase pathways regulate the activation of TGF-β family–activated Smads through phosphorylation of Smads linker segments.

Kusanagi K, Inoue H, Ishidou Y, Mishima HK, Kawabata M, et al. 2000. Characterization of a bone morphogenetic protein-responsive Smad-binding element. *Mol. Biol. Cell* 11:555–65

Labbé E, Letamendia A, Attisano L. 2000. Association of Smads with lymphoid enhancer binding factor 1/T cell- specific factor mediates cooperative signaling by the transforming growth factor-β and Wnt pathways. *Proc. Natl. Acad. Sci. USA* 97:8358–63

Labbé E, Silvestri C, Hoodless PA, Wrana JL, Attisano L. 1998. Smad2 and Smad3 positively and negatively regulate TGF β-dependent transcription through the forkhead DNA-binding protein FAST2. *Mol. Cell* 2:109–20

Lagna G, Hemmati-Brivanlou A. 1999. A molecular basis for Smad specificity. *Dev. Dyn.* 214:269–77

Lebrin F, Goumans MJ, Jonker L, Carvalho RL, Valdimarsdottir G, et al. 2004. Endoglin promotes endothelial cell proliferation and TGF-β/ALK1 signal transduction. *EMBO J.* 23:4018–28

Lee KH, Evans S, Ruan TY, Lassar AB. 2004. SMAD-mediated modulation of YY1 activity regulates the BMP response and cardiac-specific expression of a GATA4/5/6-dependent chick Nkx2.5 enhancer. *Development* 131:4709–23

Lei S, Dubeykovskiy A, Chakladar A, Wojtukiewicz L, Wang TC. 2004. The murine gastrin promoter is synergistically activated by transforming growth factor-β/Smad and Wnt signaling pathways. *J. Biol. Chem.* 279:42492–502

Leong GM, Subramaniam N, Figueroa J, Flanagan JL, Hayman MJ, et al. 2001. Ski-interacting protein interacts with Smad proteins to augment transforming growth factor-β-dependent transcription. *J. Biol. Chem.* 276:18243–48

Liang M, Liang YY, Wrighton K, Ungermannova D, Wang XP, et al. 2004. Ubiquitination and proteolysis of cancer-derived Smad4 mutants by SCF^{Skp2}. *Mol. Cell Biol.* 24:7524–37

Liberati NT, Datto MB, Frederick JP, Shen X, Wong C, et al. 1999. Smads bind directly to the Jun family of AP-1 transcription factors. *Proc. Natl. Acad. Sci. USA* 96:4844–49

Liberati NT, Moniwa M, Borton AJ, Davie JR, Wang XF. 2001. An essential role for Mad homology domain 1 in the association of Smad3 with histone deacetylase activity. *J. Biol. Chem.* 276:22595–603

Lin SY, Elledge SJ. 2003. Multiple tumor suppressor pathways negatively regulate telomerase. *Cell* 113:881–89

Lin X, Liang Y-Y, Sun B, Liang M, Brunicardi FC, et al. 2003. Smad6 recruits transcription corepressor CtBP to repress bone morphogenetic protein–induced transcription. *Mol. Cell. Biol.* 23:9081–93

Liu D, Black BL, Derynck R. 2001. TGF-β inhibits muscle differentiation through functional repression of myogenic transcription factors by Smad3. *Genes Dev.* 15:2950–66

Liu D, Kang JS, Derynck R. 2004. TGF-β-activated Smad3 represses MEF2-dependent transcription in myogenic differentiation. *EMBO J.* 23:1557–66

Liu F, Massagué J, Ruiz i Altaba A. 1998. Carboxy-terminally truncated Gli proteins associate with Smads. *Nat. Genet.* 20:325–26

Lo R, Chen Y-G, Shi Y, Pavletich NP, Massagué J. 1998. The L3 loop: a structural motif determining specific interactions between SMAD proteins and TGF-β receptors. *EMBO J.* 17:996–1005

López-Casillas F, Wrana JL, Massagué J. 1993. Betaglycan presents ligand to the TGFβ signaling receptor. *Cell* 73:1435–44

Lopez-Rovira T, Chalaux E, Rosa JL, Bartrons R, Ventura F. 2000. Interaction and functional cooperation of NF-κB with Smads. Transcriptional regulation of the junB promoter. *J. Biol. Chem.* 275:28937–46

This paper, together with Feng & Derynck (1997), shows that the direct interaction of the L3 loop in the Smad MH2 domain with the type I receptor L45 loop confers specific Smad activation by a specific receptor.

Luo K. 2004. Ski and SnoN: negative regulators of TGF-β signaling. *Curr. Opin. Genet. Dev.* 14:65–70

Luo K, Stroschein SL, Wang W, Chen D, Martens E, et al. 1999. The Ski oncoprotein interacts with the Smad proteins to repress TGFβ signaling. *Genes Dev.* 13:2196–206

Macias-Silva M, Hoodless PA, Tang SJ, Buchwald M, Wrana JL. 1998. Specific activation of Smad1 signaling pathways by the BMP7 type I receptor, ALK2. *J. Biol. Chem.* 273:25628–36

Marty T, Muller B, Basler K, Affolter M. 2000. Schnurri mediates Dpp-dependent repression of brinker transcription. *Nat. Cell Biol.* 2:745–49

Massagué J. 2003. Integration of Smad and MAPK pathways: a link and a linker revisited. *Genes Dev.* 17:2993–97

Matsuda T, Yamamoto T, Muraguchi A, Saatcioglu F. 2001. Cross-talk between transforming growth factor-β and estrogen receptor signaling through Smad3. *J. Biol. Chem.* 276:42908–14

Matsuura I, Denissova NG, Wang G, He D, Long J, et al. 2004. Cyclin-dependent kinases regulate the antiproliferative function of Smads. *Nature* 430:226–31

Miura S, Takeshita T, Asao H, Kimura Y, Murata K, et al. 2000. Hgs (Hrs), a FYVE domain protein, is involved in Smad signaling through cooperation with SARA. *Mol. Cell Biol.* 20:9346–55

Miyazono K. 2000. Positive and negative regulation of TGF-β signaling. *J. Cell Sci.* 113:1101–9

Mochizuki T, Miyazaki H, Hara T, Furuya T, Imamura T, et al. 2004. Roles for the MH2 domain of Smad7 in the specific inhibition of transforming growth factor-β superfamily signaling. *J. Biol. Chem.* 279:31568–74

Murakami G, Watabe T, Takaoka K, Miyazono K, Imamura T. 2003. Cooperative inhibition of bone morphogenetic protein signaling by Smurf1 and inhibitory Smads. *Mol. Biol. Cell* 14:2809–17

Nakao A, Afrakhte M, Morén A, Nakayama T, Christian J, et al. 1997. Identification of Smad7, a TGFβ-inducible antagonist of TGF-β signalling. *Nature* 389:631–35

Nishita M, Hashimoto MK, Ogata S, Laurent MN, Ueno N, et al. 2000. Interaction between Wnt and TGF-β signalling pathways during formation of Spemann's organizer. *Nature* 403:781–85

Nishitoh H, Ichijo H, Kimura M, Matsumoto T, Makashima F, et al. 1996. Identification of type I and type II receptors for Growth/Differentiation Factor-5. *J. Biol. Chem.* 271:21345–52

Panopoulou E, Gillooly DJ, Wrana JL, Zerial M, Stenmark H, et al. 2002. Early endosomal regulation of Smad-dependent signaling in endothelial cells. *J. Biol. Chem.* 277:18046–52

Pardali E, Xie XQ, Tsapogas P, Itoh S, Arvanitidis K, et al. 2000. Smad and AML proteins synergistically confer transforming growth factor β1 responsiveness to human germ-line IgA genes. *J. Biol. Chem.* 275:3552–60

Pardali K, Kurisaki A, Moren A, ten Dijke P, Kardassis D, et al. 2000. Role of Smad proteins and transcription factor Sp1 in p21$^{Waf1/Cip1}$ regulation by transforming growth factor-β. *J. Biol. Chem.* 275:29244–56

Pendaries V, Verrecchia F, Michel S, Mauviel A. 2003. Retinoic acid receptors interfere with the TGF-β/Smad signaling pathway in a ligand-specific manner. *Oncogene* 22:8212–20

Pera EM, Ikeda A, Eivers E, De Robertis EM. 2003. Integration of IGF, FGF, and anti-BMP signals via Smad1 phosphorylation in neural induction. *Genes Dev.* 17:3023–28

Postigo AA. 2003. Opposing functions of ZEB proteins in the regulation of the TGFβ/BMP signaling pathway. *EMBO J.* 22:2443–52

Postigo AA, Depp JL, Taylor JJ, Kroll KL. 2003. Regulation of Smad signaling through a differential recruitment of coactivators and corepressors by ZEB proteins. *EMBO J.* 22:2453–62

Pulaski L, Landstrom M, Heldin CH, Souchelnytskyi S. 2001. Phosphorylation of Smad7 at Ser-249 does not interfere with its inhibitory role in transforming growth factor-β-dependent signaling but affects Smad7-dependent transcriptional activation. *J. Biol. Chem.* 276:14344–49

Qin BY, Chacko BM, Lam SS, de Caestecker MP, Correia JJ, et al. 2001. Structural basis of Smad1 activation by receptor kinase phosphorylation. *Mol. Cell* 8:1303–12

Qin BY, Lam SS, Correia JJ, Lin K. 2002. Smad3 allostery links TGF-β receptor kinase activation to transcriptional control. *Genes Dev.* 16:1950–63

Qin BY, Liu C, Lam SS, Srinath H, Delston R, et al. 2003. Crystal structure of IRF-3 reveals mechanism of autoinhibition and virus-induced phosphoactivation. *Nat. Struct. Biol.* 10:913–21

Qing J, Liu C, Choy L, Wu RY, Pagano JS, et al. 2004. Transforming growth factor β/Smad3 signaling regulates IRF-7 function and transcriptional activation of the β interferon promoter. *Mol. Cell Biol.* 24:1411–25

Qing J, Zhang Y, Derynck R. 2000. Structural and functional characterization of the TGF-β-induced Smad3/c-Jun transcriptional cooperativity. *J. Biol. Chem.* 275:38802–12

Qiu P, Feng X-H, Li L. 2003. Interaction of Smad3 and SRF-associated complex mediates TGF-β1 signals to regulate SM22 transcription during myofibroblast differentiation. *J. Mol. Cell. Cardiol.* 35:1407–20

Randall RA, Germain S, Inman GJ, Bates PA, Hill CS. 2002. Different Smad2 partners bind a common hydrophobic pocket in Smad2 via a defined proline-rich motif. *EMBO J.* 21:145–56

Randall RA, Howell M, Page CS, Daly A, Bates PA, et al. 2004. Recognition of phosphorylated-Smad2-containing complexes by a novel Smad interaction motif. *Mol. Cell Biol.* 24:1106–21

Remy I, Montmarquette A, Michnick SW. 2004. PKB/Akt modulates TGF-β signalling through a direct interaction with Smad3. *Nat. Cell Biol.* 6:358–65

Saha D, Datta PK, Beauchamp RD. 2001. Oncogenic ras represses transforming growth factor-β/Smad signaling by degrading tumor suppressor Smad4. *J. Biol. Chem.* 276:29531–37

Samad T, Rebbapragada A, Bell E, Zhang Y, Sidis Y, et al. 2005. DRAGON: a bone morphogenetic protein co-receptor. *J. Biol. Chem.* 280:14122–29

Sanchez-Elsner T, Botella LM, Velasco B, Corbi A, Attisano L, Bernabeu C. 2001. Synergistic cooperation between hypoxia and transforming growth factor-β pathways on human vascular endothelial growth factor gene expression. *J. Biol. Chem.* 276:38527–35

Sano Y, Harada J, Tashiro S, Gotoh-Mandeville R, Maekawa T, et al. 1999. ATF-2 is a common nuclear target of Smad and TAK1 pathways in transforming growth factor-β signaling. *J. Biol. Chem.* 274:8949–57

Sebald W, Mueller TD. 2003. The interaction of BMP-7 and ActRII implicates a new mode of receptor assembly. *Trends Biochem. Sci.* 28:518–21

Seoane J, Le HV, Massagué J. 2002. Myc suppression of the p21(Cip1) Cdk inhibitor influences the outcome of the p53 response to DNA damage. *Nature* 419:729–34

Seoane J, Le HV, Shen L, Anderson SA, Massagué J. 2004. Integration of Smad and forkhead pathways in the control of neuroepithelial and glioblastoma cell proliferation. *Cell* 117:211–23

Seoane J, Pouponnot C, Staller P, Schader M, Eilers M, et al. 2001. TGFβ influences Myc, Miz-1 and Smad to control the CDK inhibitor p15[INK4b]. *Nat. Cell Biol.* 3:400–8

Shi X, Yang X, Chen D, Chang Z, Cao X. 1999. Smad1 interacts with homeobox DNA–binding proteins in bone morphogenetic protein signaling. *J. Biol. Chem.* 274:13711–17

Shi Y, Hata A, Lo RS, Massagué J, Pavletich N. 1997. A structural basis for mutational inactivation of the tumour suppressor Smad4. *Nature* 388:87–93

Shi Y, Massagué J. 2003. Mechanisms of TGF-β signaling from cell membrane to the nucleus. *Cell* 113:685–700

Shi Y, Wang Y-F, Jayaraman L, Yang H, Massagué J, et al. 1998. Crystal structure of a Smad MH1 domain bound to DNA: insights on DNA binding in TGF-β signaling. *Cell* 94:585–94

Shimizu K, Bourillot PY, Nielsen SJ, Zorn AM, Gurdon JB. 2001. Swift is a novel BRCT domain coactivator of Smad2 in transforming growth factor β signaling. *Mol. Cell Biol.* 21:3901–12

Shioda T, Lechleider RJ, Dunwoodie SL, Li H, Yahata T, et al. 1998. Transcriptional activating activity of Smad4: roles of SMAD hetero-oligomerization and enhancement by an associating transactivator. *Proc. Natl. Acad. Sci. USA* 95:9785–90

Sirotkin HI, Gates MA, Kelly PD, Schier AF, Talbot WS. 2000. Fast1 is required for the development of dorsal axial structures in zebrafish. *Curr. Biol.* 10:1051–54

Song CZ, Tian X, Gelehrter TD. 1999. Glucocorticoid receptor inhibits transforming growth factor-β signaling by directly targeting the transcriptional activation function of Smad3. *Proc. Natl. Acad. Sci. USA* 96:11776–81

Souchelnytskyi S, Nakayama T, Nakao A, Morén A, Heldin CH, et al. 1998. Physical and functional interaction of murine and Xenopus Smad7 with bone morphogenetic protein receptors and transforming growth factor-β receptors. *J. Biol. Chem.* 273:25364–70

Sun Y, Liu X, Ng-Eaton E, Lodish HF, Weinberg RA. 1999. SnoN and Ski protooncoproteins are rapidly degraded in response to transforming growth factor β signaling. *Proc. Natl. Acad. Sci. USA* 96:12442–47

Suzuki C, Murakami G, Fukuchi M, Shimanuki T, Shikauchi Y, et al. 2002. Smurf1 regulates the inhibitory activity of Smad7 by targeting Smad7 to the plasma membrane. *J. Biol. Chem.* 277:39919–25

Takebayashi-Suzuki K, Funami J, Tokumori D, Saito A, Watabe T, et al. 2003. Interplay between the tumor suppressor p53 and TGF β signaling shapes embryonic body axes in Xenopus. *Development* 130:3929–39

Topper J, Cai J, Qiu Y, Anderson K, Xu Y-Y, et al. 1997. Vascular MADs: two novel MAD-related genes selectively inducible by flow in human vascular endothelium. *Proc. Natl. Acad. Sci. USA* 94:9314–19

Tsukazaki T, Chiang TA, Davison AF, Attisano L, Wrana JL. 1998. SARA, a FYVE domain protein that recruits Smad2 to the TGFβ receptor. *Cell* 95:779–91

Tzachanis D, Freeman GJ, Hirano N, van Puijenbroek AA, Delfs MW, et al. 2001. Tob is a negative regulator of activation that is expressed in anergic and quiescent T cells. *Nat. Immunol.* 2:1174–82

Ulloa L, Doody J, Massagué J. 1999. Inhibition of transforming growth factor-β/SMAD signalling by the interferon-γ/STAT pathway. *Nature* 397:710–13

Verschueren K, Remacle JE, Collart C, Kraft H, Baker BS, et al. 1999. SIP1, a novel Zinc finger/homeodomain repressor, interacts with Smad proteins and binds to 5′-CACCT sequences in candidate target genes. *J. Biol. Chem.* 274:20489–98

von Both I, Silvestri C, Erdemir T, Lickert H, Walls JR, et al. 2004. Foxh1 is essential for development of the anterior heart field. *Dev. Cell* 7:331–45

The first study that not only proposes a trimeric structure of Smads but also provides the basis for understanding MH2 domain functions in cancer.

Waddell DS, Liberati NT, Guo X, Frederick JP, Wang XF. 2004. Casein kinase Iε plays a functional role in the transforming growth factor-β signaling pathway. *J. Biol. Chem.* 279:29236–46

Wang G, Long J, Matsuura I, He D, Liu F. 2005. The Smad3 linker region contains a transcriptional activation domain. *Biochem. J.* 386:29–34

Wang W, Mariani FV, Harland RM, Luo K. 2000. Ski represses bone morphogenic protein signaling in Xenopus and mammalian cells. *Proc. Natl. Acad. Sci. USA* 97:14394–99

Warner DR, Roberts EA, Greene RM, Pisano MM. 2003. Identification of novel Smad binding proteins. *Biochem. Biophys. Res. Commun.* 312:1185–90

Wilkinson DS, Ogden SK, Stratton SA, Piechan JL, Nguyen TT, et al. 2005. A direct intersection between p53 and transforming growth factor β pathways targets chromatin modification and transcription repression of the α-fetoprotein gene. *Mol. Cell Biol.* 25:1200–12

Wotton D, Massagué J. 2001. Smad transcriptional corepressors in TGF β family signaling. *Curr. Top. Microbiol. Immunol.* 254:145–64

Wu G, Chen YG, Ozdamar B, Gyuricza CA, Chong PA, et al. 2000. Structural basis of Smad2 recognition by the Smad anchor for receptor activation. *Science* 287:92–97

Wu JW, Fairman R, Penry J, Shi Y. 2001. Formation of a stable heterodimer between Smad2 and Smad4. *J. Biol. Chem.* 276:20688–94

Wu JW, Hu M, Chai J, Seoane J, Huse M, et al. 2001. Crystal structure of a phosphorylated Smad2. Recognition of phosphoserine by the MH2 domain and insights on Smad function in TGF-β signaling. *Mol. Cell* 8:1277–89

Wu JW, Krawitz AR, Chai J, Li W, Zhang F, et al. 2002. Structural mechanism of Smad4 recognition by the nuclear oncoprotein Ski: insights on Ski-mediated repression of TGF-β signaling. *Cell* 111:357–67

Wu L, Wu Y, Gathings B, Wan M, Li X, et al. 2003. Smad4 as a transcription corepressor for estrogen receptor α. *J. Biol. Chem.* 278:15192–200

Yagi K, Furuhashi M, Aoki H, Goto D, Kuwano H, et al. 2002. c-myc is a downstream target of the Smad pathway. *J. Biol. Chem.* 277:854–61

Yagi K, Goto D, Hamamoto T, Takenoshita S, Kato M, et al. 1999. Alternatively spliced variant of Smad2 lacking exon 3. Comparison with wild-type Smad2 and Smad3. *J. Biol. Chem.* 274:703–9

Yakymovych I, ten Dijke P, Heldin CH, Souchelnytskyi S. 2001. Regulation of Smad signaling by protein kinase C. *FASEB J.* 15:553–55

Yamakawa N, Tsuchida K, Sugino H. 2002. The rasGAP-binding protein, Dok-1, mediates activin signaling via serine/threonine kinase receptors. *EMBO J.* 21:1684–94

Yamamoto M, Meno C, Sakai Y, Shiratori H, Mochida K, et al. 2001. The transcription factor FoxH1 (FAST) mediates Nodal signaling during anterior-posterior patterning and node formation in the mouse. *Genes Dev.* 15:1242–56

Yamamoto T, Saatcioglu F, Matsuda T. 2002. Cross-talk between bone morphogenic proteins and estrogen receptor signaling. *Endocrinology* 143:2635–42

Yanagisawa J, Yanagi Y, Masuhiro Y, Suzawa M, Watanabe M, et al. 1999. Convergence of transforming growth factor-β and vitamin D signaling pathways on SMAD transcriptional coactivators. *Science* 283:1317–21

Yeo C, Whitman M. 2001. Nodal signals to Smads through Cripto-dependent and Cripto-independent mechanisms. *Mol. Cell* 7:949–57

Yoshida Y, Tanaka S, Umemori H, Minowa O, Usui M, et al. 2000. Negative regulation of BMP/Smad signaling by Tob in osteoblasts. *Cell* 103:1085–97

Yoshida Y, von Bubnoff A, Ikematsu N, Blitz IL, Tsuzuku JK, et al. 2003. Tob proteins enhance inhibitory Smad-receptor interactions to repress BMP signaling. *Mech. Dev.* 120:629–37

Zauberman A, Lapter S, Zipori D. 2001. Smad proteins suppress CCAAT/enhancer-binding protein (C/EBP) β- and STAT3-mediated transcriptional activation of the haptoglobin promoter. *J. Biol. Chem.* 276:24719–25

Zavadil J, Cermak L, Soto-Nieves N, Böttinger EP. 2004. Integration of TGF-β/Smad and Jagged1/Notch signalling in epithelial-to-mesenchymal transition. *EMBO J.* 23:1155–65

Zawel L, Dai JL, Buckhaults P, Zhou S, Kinzler KW, et al. 1998. Human Smad3 and Smad4 are sequence-specific transcription activators. *Mol. Cell.* 1:611–17

Zhang Y, Derynck R. 2000. Transcriptional regulation of the transforming growth factor-β-inducible mouse germ line Igα constant region gene by functional cooperation of Smad, CREB, and AML family members. *J. Biol. Chem.* 275:16979–85

Zhang Y, Feng X-H, Derynck R. 1998. Smad3 and Smad4 cooperate with c-Jun/c-Fos to mediate TGF-β-induced transcription. *Nature* 394:909–13

Zhang YW, Yasui N, Ito K, Huang G, Fujii M, et al. 2000. A RUNX2/PEBP2αA/CBFA1 mutation displaying impaired transactivation and Smad interaction in cleidocranial dysplasia. *Proc. Natl. Acad. Sci. USA* 97:10549–54

Zhou S, Zawel L, Lengauer C, Kinzler KW, Vogelstein B. 1998. Characterization of human FAST-1, a TGF β and activin signal transducer. *Mol. Cell* 2:121–17

This paper, together with the data by Kim et al. (1997), established that Smads are DNA-binding transcription factors.

This study, along with Chen et al. (1997), provided the basis for the general mechanism of transcriptional activation by Smads.

The Great Escape: When Cancer Cells Hijack the Genes for Chemotaxis and Motility

John Condeelis,[1,2] Robert H. Singer,[1] and Jeffrey E. Segall[1]

[1]Anatomy and Structural Biology and [2]Analytical Imaging Facility, Albert Einstein College of Medicine, Bronx, New York 10461-1975; email: condeeli@aecom.yu.edu; rhsinger@aecom.yu.edu; segall@aecom.yu.edu

Annu. Rev. Cell Dev. Biol.
2005. 21:695–718

First published online as a Review in Advance on July 5, 2005

The *Annual Review of Cell and Developmental Biology* is online at http://cellbio.annualreviews.org

doi: 10.1146/annurev.cellbio.21.122303.120306

Copyright © 2005 by Annual Reviews. All rights reserved

1081-0706/05/1110-0695$20.00

Key Words

actin, cofilin, N-WASP, capping protein, Arp2/3 complex

Abstract

The combined use of the new technologies of multiphoton-based intravital imaging, the chemotaxis-mediated collection of invasive cells, and high sensitivity expression profiling has allowed the correlation of the behavior of invasive tumor cells in vivo with their gene expression patterns. New insights have resulted including a gene expression signature for invasive cells and the tumor microenvironment invasion model. This model proposes that tumor invasion and metastasis can be studied as a problem resembling normal morphogenesis. We discuss how these new insights may lead to a better understanding of the molecular basis of the invasive behavior of tumor cells in vivo, which may result in new strategies for the diagnosis and treatment of metastasis.

Contents

INTRODUCTION

The ability of tumor cells to spread from primary tumors (and metastatic tumors) is the major cause of death in cancer patients. Spreading of tumor cells relies upon cell motility, which results in the invasion of neighboring connective tissue and entry into lymphatics and blood vessels (intravasation) (Clark et al. 2000, Condeelis & Segall 2003, Woodhouse et al. 1997). We focus on tumor invasion and metastasis as a problem in cell motility. In this context, the motility behavior of tumor cells inside the tumor must be analyzed as carefully as the gene expression patterns displayed by invasive cancer cells. During motility, the microenvironment becomes a determinant in the success or failure of a cancer cell in its attempt to traverse the tumor and enter blood and lymphatic vessels (Liotta & Kohn 2001). Subtle changes in the cancer cell's interactions with extracellular matrix and gradients of growth factors and cytokines define whether a cell becomes invasive or remains stationary in the tumor mass. Only by understanding the basic biology of how the motility of cells inside the tumor is influenced by, and influences, the gene expression patterns of cancer cells and their microenvironment will it be possible to define strategies to impede the spread of cancer cells from tumors.

In this review three questions are considered: What motility behaviors contribute to invasion and intravasation? Is there an expression signature that correlates with these behaviors, that is, an invasion signature? How do the genes of the invasion signature contribute to invasion? Answering these questions to date has suggested a novel model for tumor invasion and metastasis, which is discussed at the end of the review.

WHAT MOTILITY BEHAVIORS CONTRIBUTE TO INVASION AND INTRAVASATION?

Intravital Imaging of Tumor Cell Behavior in Tumors In Vivo

An attempt to understand the behavior of tumor cells at single-cell resolution in vivo predates the introduction of green fluorescent protein (GFP) and its derivatives. Tumor cells were transiently labeled with vital dyes and observed with conventional transmitted and fluorescence microscopy (Chambers et al. 1995, Scherbarth & Orr 1997, Suzuki et al. 1996, Vajkoczy et al. 1999, Wood 1958, Yuan et al. 1995). This required the use of short-lived preparations in thin regions of tissue where light could pass efficiently and sometimes the use of viewing windows (Chambers et al. 1995, Wood 1958). These approaches usually limited the analysis of tumors to artificial locations and introduced the potential for artifact resulting from the viewing method. A major step was the introduction of stable GFP expression, which allowed genetic labeling of cells in tumors with tissue- and cell-type specificity without rejection of the GFP-tagged cells (Chishima et al. 1997, Farina et al. 1998). Thus more clinically relevant tumor models were developed that could be imaged in the location in the animal where the tumor naturally forms and progresses to different stages (Ahmed et al. 2002, Brown et al. 2001, Yang et al. 2000).

The introduction of the laser-scanning confocal microscope was an essential advance that made optical sectioning and single-cell resolution possible, essential capabilities for relating cell behavior to mechanisms of invasion (Chantrain et al. 2004, Farina et al. 1998). However, conventional one-photon laser-scanning confocal microscopy is limited by the relatively poor optical depth of penetration of short wavelength excitation light, photobleaching, and phototoxic damage to the whole tissue and not just at the focal point. The recent introduction of multiphoton microscopy, which uses 800–900 nm light from a pulsed laser, has largely solved the problems of photobleaching and toxicity and extended the depth of penetration by 20-fold (Condeelis & Segall 2003, Helmchen & Denk 2002, Jain et al. 2002, Wang et al. 2002, Williams et al. 2001, Zipfel et al. 2003b). In addition, the multiphoton excitation of tissue causes second harmonic scattering of photons from α-helix-containing proteins, thus allowing the imaging of extracellular matrix proteins such as collagen without the need for fluorescent labeling of the tissue matrix (Campagnola et al. 2001, Condeelis & Segall 2003, Zipfel et al. 2003a). This benefit of multiphoton excitation can be used to analyze cell-extracellular matrix interactions and matrix remodeling directly in live tissue (Condeelis & Segall 2003, Masters et al. 1997). The application of gradient index (GRIN) lens technology to multiphoton imaging holds the promise of extending intravital imaging to any depth within live mice, making systemic analysis of tumor invasion and metastasis possible (Levene et al. 2004). The tumor cell behaviors in live tumors discussed in the next section are derived from studies using laser-scanning confocal and multiphoton imaging in rats and mice.

Motility Behaviors Contributing to Invasion and Intravasation

An understanding of the motility of cancer cells and its contribution to metastasis has begun to emerge from intravital imaging of cells, at single-cell resolution, inside tumors within living animals. An important outcome of studying tumor cells within their normal tumor environment is that the behavior observed is an indication of what tumor cells actually do in vivo, not what they can do as inferred from in vitro and ectopic models in vivo. That is, artificial models such as ectopic growth of tumor cells in tissues in which tumors do not normally form, e.g., growing breast tumor cells in dermis instead of mammary gland, can lead to the observation of cell behaviors that do not occur in real breast tumors. Tumor models where tumor cells are grown in tissue that is the natural site for the tumor, e.g., breast in breast, are called orthotopic models. The behaviors discussed next are those seen in orthotopic models.

A number of behaviors have been observed in orthotopic models in vivo that relate to metastatic potential. These form a pattern common to a variety of tumor types and provide insight into mechanisms of invasion and intravasation (Condeelis & Segall 2003, Farina et al. 1998, Friedl & Wolf 2003, Sahai et al. 2005, Wang et al. 2002, Wyckoff et al. 2000a).

Tumor cells in primary mammary tumors move as solitary cells at up to 10 times their velocity in vitro. Tumor cell motility is characterized as solitary amoeboid movement, but it can also occur as cell streams and linear files suggesting the use of common paths on extracellular matrix (ECM) fibers (Farina et al. 1998, Friedl & Wolf 2003, Sahai et al. 2005, Wang et al. 2002, Wyckoff et al. 2000a). In fact, the highest velocities are observed for carcinoma cells in metastatic tumors that are moving along linear paths in association with ECM fibers, in particular, collagen fibers. These high-velocity linear excursions are unrestricted by networks of ECM in mammary tumors, except around blood vessels (Condeelis & Segall 2003).

Intravasation and invadopodia. Tumor cell motility is restricted at the basement membrane of blood vessels, where the cells must squeeze through small pores in the basement membrane/endothelium to gain access to the blood space. The degree to which the basement membrane of blood vessels represents a barrier has been documented by direct observations of cell behavior during intravasation. Carcinoma cells in nonmetastatic tumors are fragmented during intravasation as they squeeze across the basement membrane/endothelium indicating that the cell must be highly distended and under tension as it crosses. Remarkably, carcinoma cells in metastatic tumors cross this restriction as intact cells, possibly in large measure owing to the high levels of expression of cytokeratins in metastatic cells (Wang et al. 2002) and their ability to extend invadopodia (Condeelis & Segall 2003, Wang et al. 2002, Wyckoff et al. 2000a, Yamaguchi et al. 2005).

Chemotaxis to blood vessels. Carcinoma cells in metastatic tumors are attracted to blood vessels, where they form a layer of cells that are morphologically polarized toward the vessel. Chemotaxis to epidermal growth factor by carcinoma cells has been demonstrated in vivo (Wyckoff et al. 2004) and resembles that observed for tumor cells in vitro (Wyckoff et al. 2000b). Chemotaxis ability of tumor cells is highly correlated with their potential for invasion, intravasation, and metastasis and appears responsible for the attraction of carcinoma cells to blood vessels (Wyckoff et al. 2000a). Cell polarity toward blood vessels is correlated with increased intravasation and metastasis, indicating a local blood vessel associated source of chemoattractants (Pollard 2004, Wyckoff et al. 2004).

Blood vessel–associated macrophages are a source of EGF and other chemoattractants. An in vivo invasion assay (Wyckoff et al. 2000b) has been used to study the mechanism of chemotaxis in primary mammary tumors of rats and mice. These studies demonstrate that macrophages form a paracrine loop with invasive tumor cells (Wyckoff et al. 2004). Expression analysis of tumor cells and macrophages caught invading together indicates how these cells are attracted to each other and can invade jointly: Tumor cells express CSF-1, which stimulates macrophage chemotaxis, whereas macrophages express epidermal growth factor (EGF), which stimulates tumor cell chemotaxis (Wyckoff et al. 2004). Because metastatic mammary tumors contain large numbers of rapidly moving macrophages with many clustered near blood vessels (Condeelis & Segall 2003, Wyckoff et al. 2000a), they are a local source of chemotactic cytokines and chemotactic growth factors, such as EGF, within the tissue and near blood vessels (Lin et al. 2001, Pollard 2004, Wyckoff et al. 2004). The in vitro assay of invasive cell motility inside collagen matrices demonstrates that macrophages and tumor cells and the activity of their CSF-1 and EGF receptors, respectively, are necessary and sufficient for enhancement of invasion (Goswami et al. 2005).

IS THERE AN EXPRESSION SIGNATURE THAT CORRELATES WITH THESE BEHAVIORS?

The Concept of an Invasion Signature

Gene expression profiling has been used extensively in an attempt to sort tumors into subtypes that might be diagnosed and treated more effectively (Ramaswamy et al. 2003, van 't Veer et al. 2002). In addition, expression profiling has been used in an attempt to identify invasion- and metastasis-specific genes that might predict the metastatic potential of tumors and to gain insight into the mechanisms of invasion and metastasis. In general, studies involving (*a*) entire primary tumors, (*b*) laser capture microdissection of fixed primary tumors, and (*c*) cells isolated from metastases of bone marrow, lymphatics and distant solid organs have identified candidate genes that might be important for tumor cell invasion (reviewed in Wang et al. 2005). However, such approaches have had limitations:

- Expression analysis of whole primary tumors provides bulk tumor expression patterns, in which case the specific patterns of expression typical of invasive cells might be diluted.

- Laser capture microdissection studies must rely on morphology and histological location, an uncertain exercise, to select cells that might have been invading, thereby making the relevance to invasion of expression profiles from such cells questionable.

- Isolating cells from metastases are likely to produce expression profiles that are relevant to successful growth at the new site but not necessarily profiles indicative of invasion potential from the primary tumor.

In an alternative approach, an in vivo invasion assay, based on the chemotaxis of tumor cells to blood vessels seen in vivo (Wyckoff et al. 2000b), was used to collect invasive cells from live primary tumors in mice and rats. Because the in vivo invasion assay employs microneedles containing chemoattractants such as EGF and extracellular matrix that mimic conditions around blood vessels that are involved in chemotaxis (Wyckoff

et al. 2000b, 2004), the invasive cells collected are likely also to be the cells involved in intravasation. Invasive tumor cells collected by this method can then be interrogated directly relative to the tumor cells that remain behind in the primary tumor and, after subtraction of gene expression changes occurring in response to EGF and other collection conditions, this reveals the expression pattern unique to invasive tumor cells, an "invasion signature." The invasion signature shown in **Tables 1**, **2**, and **3** is derived from invasive cells collected in rat mammary tumors generated from carcinoma cell lines (Goswami et al. 2004; Wang et al. 2003, 2004). A similar invasion signature has been derived from mouse mammary tumors resulting from expression of the PyMT oncogene in situ (W. Wang, personal communication). This indicates that the invasion signature is common to mammary tumors in rats and mice regardless of the origins of the mammary tumor. The invasion signature shown in **Tables 1**, **2**, and **3** indicates that invasive cells are a population that is neither proliferating nor apoptotic but is highly motile (Goswami et al. 2004, Wang et al. 2004). The reduction in apoptosis is consistent with tumor cells having a survival advantage owing to suppression of apoptosis genes and up-regulation of prosurvival genes (**Table 2**). Furthermore, the pattern of expression of genes involved in proliferation suggests that invasive tumor cells are not proliferating (**Table 1**). This predicts that cancer treatments targeting cell growth may not be very effective at killing invasive tumor cells. This was tested by exposing invasive cells collected using the in vivo invasion assay to conventional chemotherapy that is directed at dividing cells. As predicted, the invasive cells survived better compared with non-invasive cells from the same tumor (Goswami et al. 2004).

Several of the genes of the invasion signature have been identified in clinical and conventional gene expression profiling studies. Clinical studies of bladder, breast,

Table 1 Genes of the cell proliferation part of the invasion signature[a]

Gene symbol	Gene description	Fold change[b]
Suppression of cell proliferation		
Psmc5	Protease (prosome, macropain) 26S subunit, ATPase 5	5.5
Rad9	Cell cycle checkpoint control protein (Rad9) mRNA	4.0
Hmg1	High mobility group protein 1	3.5
CKS2	Cyclin-dependent kinases regulatory subunit 2	3.4
Cks1	Cyclin-dependent kinase regulatory subunit 1	3.2
Fmo5	Flavin containing monooxygenase 5	3.0
GAS6	GAS 6 mRNA associated with growth arrest	2.8
Phb	Prohibitin	2.8
Mad2	Mitotic checkpoint component Mad2 mRNA	2.4
Madh3	MAD homolog 3	2.4
Hmg14	High mobility group protein 14	2.3
Enhancement of cell proliferation		
CGMC	Carcinoembryonic antigen CGM6 precursor	0.5
CPR2	Cell cycle progression 2 protein (CPR2)	0.4
Ask	Activator of S phase kinase	0.2

[a]To determine the significance of changes in gene expression in each of the functional categories of the genes represented in microarrays, the Student's t test, Chi-square, or SAM analysis were performed. The fold changes in gene expression of the invasion signature shown in **Tables 1–3** were found to be statistically significant in the invasive cells by Chi-square or SAM analysis. In addition, in all cases, $P < 0.05$. Random sets of equal numbers of genes did not generate the same pattern of up- and down-regulation, indicating that the pattern was not observed by chance ($P < 0.05$). Similarly, clustering the results from all genes of the general population in the same space of all genes on the microarray did not yield an outcome similar to the invasion signature. All results are from Goswami et al. 2004 and Wang et al. 2004.

[b]The fold change indicates the level of expression in the invasive tumor cells compared with the general population of tumor cells of the primary tumor.

and colorectal cancers have implicated Rho A, Rock (Kamai et al. 2003), Mena (Di Modugno et al. 2004), and the Arp2 and 3 subunits of the Arp2/3 complex (Otsubo et al. 2004), respectively, as up-regulated in these cancers. Studies have suggested that the elevated expression of Arp2/3 complex by both neoplastic and stromal cells contributes to the increased motility of both cell types and thus provides suitable conditions for invasion (Otsubo et al. 2004). In addition, LIM-kinase 1 is up-regulated in metastatic breast and prostate tumors (Davila et al. 2003, Yoshioka et al. 2003). Hence, the pathways involved in actin polymerization at the leading edge are implicated in invasion by different approaches (**Figure 1**).

HOW DO THE GENES OF THE INVASION SIGNATURE CONTRIBUTE TO INVASION?

Coordinate Regulation of Motility Pathways in Invasion

An important insight into the special motility properties of invasive cells, the high speeds of locomotion, chemotaxis, and invadopod formation, comes from the motility pathways portion of the invasion signature (**Table 3**). That is, the finding that the genes coding for the key effectors of the minimum motility machine (Loisel et al. 1999), i.e., the cofilin, capping protein, and Arp2/3 pathways, that regulate β-actin polymerization at the leading edge, are dramatically up-regulated

Table 2   Genes of the apoptosis and survival part of the invasion signature[a]

Gene	Description	Fold change[b]
Anti-apoptotic genes		
Ier3	Immediate early response 3	4.9
Ubl1a2	Ubiquitin-like 1 (sentrin) activating enzyme subunit 2	4.7
Txn	Thioredoxin	3.7
Hsp105	Heat shock protein, 105 kDa	3.5
Odc	Ornithine decarboxylase, structural	3.0
Dad1	Defender against cell death 1	2.7
Trp53	Transformation related protein 53	2.5
Hsp60	Heat shock protein, 60 kDa	2.4
Api4	Apoptosis inhibitor 4	2.3
Cldn3	Claudin 3	2.3
Api5	Apoptosis inhibitor 5	2.3
Hsp86-1	Heat shock protein, 86 kDa 1	2.1
Api1	Apoptosis inhibitor 1	2.0
Adam17	A disintegrin and metalloproteinase domain 17	2.0
Pro-apoptotic genes		
Pdcd4	Programmed cell death 4	0.1
Fem1b	Feminization 1 b homolog (*C. elegans*)	0.4
Apaf1	Apoptotic protease activating factor 1	0.6
Pdcd8	Programmed cell death 8 (apoptosis inducing factor)	0.8
	Cellular apoptosis susceptibility protein	1.1
	ESTs, highly similar to apoptosis specific protein	1.1
	Apoptosis-associated speck-like protein containing CARD	1.2
AIF	Apoptosis-inducing factor AIF	1.3

[a]For details on methodology and results, please see footnote a to **Table 1**.

[b]For more details, please see footnote b to **Table 1**.

(**Figure 1**) (Wang et al. 2004). Furthermore, the genes of the motility portion of the invasion signature (**Table 3**) can be organized into a series of converging pathways based on the known functions of the proteins for which they code (**Figure 2**) (Wang et al. 2004). The functions of each of these pathways and how they may contribute to the behavior of tumor cells during invasion and intravasation is considered next.

Cofilin pathway. The cofilin family of proteins in vertebrates consists of cofilin/ADF. The cofilin pathway for invasive carcinoma cells is summarized in **Figure 2a**, with the genes whose expression is altered in invasive tumor cells highlighted. The invasion signa-

ture indicates that the cofilin activity cycle has been impacted at several levels of regulation in invasive cells. Cofilin is the more abundant isoform of the family in carcinoma cells, and its expression is highly up-regulated in invasive cells (**Figure 2a**).

Cofilin's severing and depolymerization activities are inhibited by phosphorylation, G-actin binding, and binding to phosphatidylinositol (4,5)-bisphosphate (PIP_2) (Bamburg 1999, DesMarais et al. 2004a, Paavilainen et al. 2004). Changes in pH can also regulate the level of activity of cofilin, but over the physiological range of pH found in vertebrate cells (6.6–7.4) (Bernstein et al. 2000), the activities of cofilin are only graded, not inactivated, suggesting that pH may act more like a

Table 3 Genes of the motility part of the invasion signature[a]

Accession	Gene description	Fold change[b]
AA414612	Capping protein $\alpha 1^c$	4.00
AW556230	Cell division cycle 42[c]	3.96
AU015486	Capping protein α 2	3.89
C79581	Moesin[c]	3.67
C86972	Arp 2/3 complex subunit p16[c]	3.52
AW538432	Rho interactin protein 3[c]	3.33
AU015879	LIM-kinase 1[c]	3.24
AA285584	Palladin	3.12
AW555565	Zyxin	2.93
W10023	Catenin β	2.88
C76867	Tropomyosin α chain	2.86
AU023806	Rho-associated coiled-coil forming kinase 1[c]	2.71
AW536576	Testis expressed gene 9	2.67
AI324089	Phosphatidylinositol-4-phosphate 5-kinase type II α^c	2.60
AI427644	Epidermal growth factor receptor[c]	2.59
AW541453	Capping protein (actin filament), gelsolin-like	2.53
C86107	Actinin α 3[c]	2.52
AW543636	Annexin A5	2.47
AA052404	CRIPT protein	2.32
AA014771	Protein kinase C, ζ^c	2.30
AW546733	Arp 2/3 complex subunit p21[c]	2.22
AA538228	RAB25, member RAS oncogene family	2.19
AA275245	Vinculin	2.16
AA386680	Kinesin family member 5B	2.13
AW536843	Chaperonin subunit 4 (δ)	2.06
AW536183	Chaperonin subunit 3 (γ)	2.06
AI326287	Tubulin alpha-4 chain	2.05
AW553280	Integrin β 1 (fibronectin receptor β)	2.00
AW536098	Cofilin 1, nonmuscle[c]	2.00
AU017992	Kinectin 1	2.00
AW557123	Downstream of tyrosine kinase 1	2.00
AW549817	Burkitt lymphoma receptor 1	2.00
AA272097	Fibroblast growth factor receptor 1	0.54
AA073514	Zipcode-binding protein 1[c]	0.11

[a]For details on methodology and results, please see footnote a to **Table 1**.
[b]For more details, please see footnote b to **Table 1**.
[c]These results have been validated by quantitative real-time-PCR.

rheostat to regulate the amplitude of activity without acting like an on-off-switch.

The consequence of regulating cofilin by phosphorylation appears to differ by cell type. In some cell types, cofilin is almost 100% phosphorylated in resting cells and motility is stimulated by dephosphorylation (Kanamori et al. 1995, Okada et al. 1996). In carcinoma cells in serum, phospho-cofilin is less than half of the total cofilin at steady state (Zebda et al.

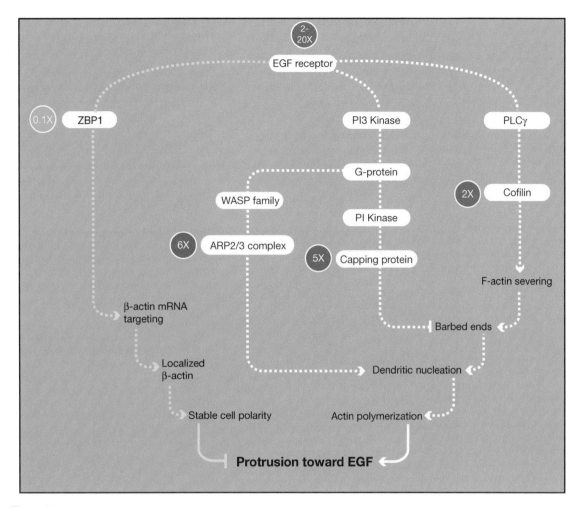

Figure 1

The four pathways leading to β-actin polymerization at the leading edge of carcinoma cells in response to EGF. The major effectors responsible for leading-edge polymerization are cofilin, capping protein, and the Arp2/3 complex. Zip-code-binding protein 1 (ZBP1) regulates chemotaxis to EGF through β-actin mRNA targeting. The fold changes in gene expression in this and **Figure 2** were determined by quantitative real-time PCR and are indicated as (nx). Cofilin and Arp2/3 complex are synergistic in the production of free barbed ends leading to dendritic nucleation and protrusive force. Capping protein funnels the available G-actin onto productive elongating barbed ends by capping nonproductive barbed ends. The four pathways therefore coordinately generate protrusions that act to steer the cells during chemotaxis and invasion.

2000), and in serum-starved cells, phospho-cofilin is as little as 10% of the total cofilin (X. Song & R. Eddy, personal communication). Even so, cofilin in both cases is mostly inactive (Chan et al. 2000), indicating that a mechanism other than phosphorylation must be at work to inhibit cofilin activity in carcinoma cells.

Another function of phosphorylation of cofilin in carcinoma cells is the recycling of cofilin from G-actin. Cofilin binds to G-actin with submicromolar affinity and the heterodimer is inactive in both severing and depolymerization (Bamburg 1999, Paavilainen et al. 2004). The release of cofilin from this heterodimer is crucial to the recycling of cofilin activity. Cyclase-associated protein (CAP) is capable of releasing cofilin from the heterodimer through a direct interaction with actin (Bertling et al. 2004, Paavilainen et al. 2004). In addition, because phospho-cofilin cannot bind to actin, LIM-kinase may also be involved in breaking the G-actin-cofilin heterodimer in vivo. Phosphorylation may also function to put limits on the amplitude, location, and duration of cofilin activity after its activation by EGF. Hence, while the phosphorylation/ dephosphorylation cycle of cofilin may not be directly involved in the activation of cofilin in carcinoma cells by EGF, LIM-kinase, along with CAP, may be crucial in regulating the localization and recycling of cofilin activity.

Four different kinases that appear to be downstream of the Rho-family GTPases have been shown to phosphorylate cofilin, LIM-kinase 1 and 2, and TES-kinase 1 and 2 (Arber et al. 1998, Dan et al. 2001, Rosok et al. 1999, Toshima et al. 2001, Yang et al. 1998). In invasive carcinoma cells, LIM-kinase 1 is

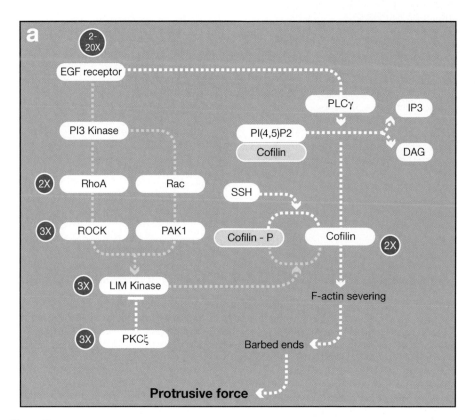

Figure 2

The pathways to barbed end generation and protrusive force. The fold changes in gene expression are indicated as (nx). (a) The cofilin pathway leading to barbed end production in response to EGF. Gene for both inhibitory (PAK, ROCK, LIM kinase) and stimulatory (PLC and PKCζ) inputs to cofilin are more highly expressed in invasive cells; these regulate the location, timing, and sharpness of cofilin-dependent actin polymerization transients that are required for chemotaxis. For a, the inhibitory parts are in yellow and the stimulatory parts in white. (b) The capping protein pathway leading to barbed end capping. Genes for both inhibition (Mena and PI5K) and stimulation (capping protein) of the capping activity of this pathway are more highly expressed in invasive cells. (c) The Arp2/3 complex pathway leading to dendritic nucleation in response to EGF. Genes coding for Arp2/3 complex subunits and upstream activators are more highly expressed in invasive cells.

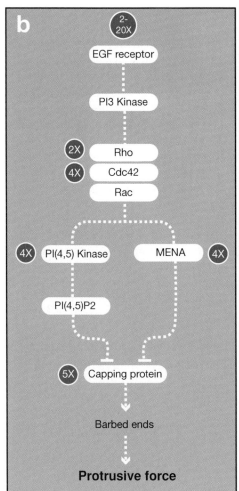

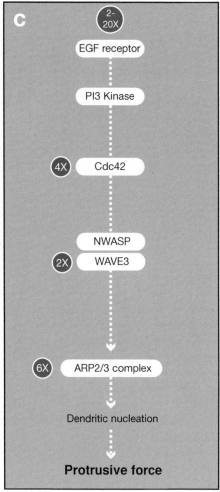

Figure 2
(*Continued*)

most prominently expressed, and its expression is up-regulated in invasive cells (Wang et al. 2004) (**Figure 2a**). Furthermore, the activation of LIM-kinase 1 occurs through the PI3K-induced activation of Rho-family G-proteins, which activate PAK and ROCK. Rho is highly expressed in invasive cells. Either PAK (Edwards et al. 1999) or ROCK (Ohashi et al. 2000) can phosphorylate LIM-kinase at threonine 508 thereby activating it to increase cofilin phosphorylation. Both kinases are also up-regulated in invasive cells (**Figure 2a**).

Inhibition of LIM-kinase activity is PKC dependent, and this involves one of the atypical PKC isoforms (Djafarzadeh & Niggli 1997, Kuroda et al. 1996). LIM-kinase and

PKCζ tightly associate via the interaction through the second LIM domain of LIM-kinase, which indicates direct phosphorylation of LIM-kinase (Kuroda et al. 1996). Additional studies have implicated the δ isoform of PKC as a negative regulator of LIM-kinase (Martiny-Baron et al. 1993). The expression of PKCζ is up-regulated in invasive cells (**Figure 2a**).

The general pattern of regulation in the cofilin pathway indicates that genes coding for proteins that both increase and decrease the activity of cofilin are coordinately up-regulated along with cofilin itself. This pattern may result from the toxicity of elevated cofilin expression (reviewed in Ghosh et al. 2004), where expression of inhibitory genes is essential to maintain higher levels of cofilin. Alternatively, the significance of this paradoxical pattern may be understood when one considers that the cofilin pathway is directly involved in sensing during chemotaxis of carcinoma cells to EGF (Mouneimne et al. 2004), and cofilin is sufficient to set the direction of cell movement (Ghosh et al. 2004). Directional sensing of EGF requires an early transient of free, actin filament barbed ends resulting from cofilin severing that causes localized actin polymerization (Chan et al. 2000, Mouneimne et al. 2004). If the free barbed ends of the early transient are either inhibited or sustained, then directional protrusion in response to EGF fails (Chan et al. 2000, Mouneimne et al. 2004, Zebda et al. 2000). That is, it is the generation of a transient of free barbed ends that is essential in directional sensing, not sustained polymerization. The up-regulation of genes that both increase and decrease cofilin severing activity, as seen in **Figure 2a**, is consistent with the enhanced ability of invasive cells to generate an early transient that is essential for chemotaxis. In addition, the localization and timing of the stimulatory and inhibitory branches of the cofilin pathway are believed to determine the precise location and duration of cofilin activity and its recycling to compartments where cofilin is inhibited in resting cells (DesMarais et al. 2004a).

Capping protein pathway. Capping protein binds to the growing barbed ends of actin filaments to prevent further elongation and regulate filament length. The patterns of regulation of genes of the capping protein pathway exhibit the same antagonistic relationships as seen in the cofilin pathway where expression of stimulatory and inhibitory branches are up-regulated together. Expression of both the α and β-subunits of capping protein is dramatically increased, suggesting higher capping protein activity in the pathway. However, the expression of genes that code for proteins that are inhibitory to capping protein activity, the type II-α isoform of PI4, 5 kinase (Cooper & Schafer 2000) and Mena (Bear et al. 2002), are also up-regulated (**Figure 2b**). Capping protein, like cofilin, is essential for viability, and large changes in its expression level may not be tolerated by cells over time (Cooper & Schafer 2000). Therefore, a more interesting interpretation of these results is that the amplitude and sharpness of capping protein activity as a transient is increased in invasive cells because of this antagonistic pattern of expression. The combination of heightened transient capping protein activity and changes in its timing and location could synergize with the barbed end generating activities of the cofilin and Arp2/3 pathways (**Figure 1**) to cause intense focal bursts of actin polymerization, as observed in in vitro experiments with purified proteins (Carlier 1998, Loisel et al. 1999).

Arp2/3 complex pathway. Both the cofilin and capping protein pathways converge on the Arp2/3 complex. Because the expression of key components of both pathways is up-regulated, it is interesting that the expression of several subunits of the Arp2/3 complex are also greatly up-regulated in invasive cells, as is the expression of upstream stimulators of the Arp2/3 complex, WAVE 3 and Cdc42 (**Figure 2c**).

Cofilin and Arp2/3 complex synergistically contribute to the nucleation of a dendritic array both in vitro (Ichetovkin et al. 2002) and in vivo (DesMarais et al. 2004b). This synergy results from the amplification of the Arp2/3 complex's nucleation activity by cofilin's severing activity, which creates barbed ends that elongate to form newly polymerized actin filaments (Ichetovkin et al. 2002). The newly polymerized filaments are the preferred filament type for Arp2/3 complex-mediated branching (DesMarais et al. 2004a,b; Ichetovkin et al. 2002). This synergistic amplification of the Arp2/3 complex activity has been proposed to explain the ability of cofilin to determine sites of protrusion and cell direction in uncaging

Stimulated protrusion
(G-actin abundant)

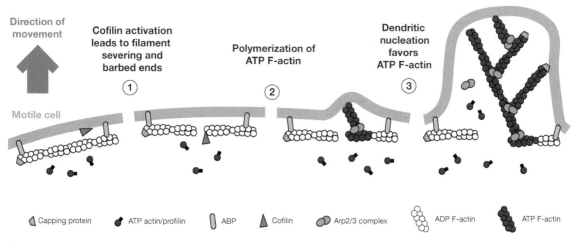

Figure 3

The stimulated protrusion model showing the role of cofilin severing in determining the site of dendritic nucleation, protrusion, and cell direction. Severing of actin filaments in the cortical actin cytoskeleton by cofilin creates free barbed ends that bias the location and the amount of dendritic nucleation by the Arp2/3 complex. Polymerization proceeds from a pool of pre-existing actin monomers, allowing the initiation of polymerization to occur without being tightly coupled to depolymerization. Redrawn from DesMarais et al. (2004a).

experiments (DesMarais et al. 2004a, Ghosh et al. 2004) (**Figure 3**).

Capping protein funnels actin monomers onto newly created free barbed ends by capping older filaments, thereby enhancing the formation of the short, branched filaments characteristic of Arp2/3 complex-nucleated dendritic arrays (Carlier 1998).

WAVE 3 is believed to activate the Arp2/3 complex, as do its relatives WAVEs 1 and 2 (Takenawa & Miki 2001). Both WAVEs 1 and 2 are regulated by Rac 1, which regulates their interaction with Arp2/3 complex to cause stimulation of the Arp2/3 complex's nucleation activity (Eden et al. 2002, Miki et al. 2000, Steffen et al. 2004). This in turn causes lateral (WAVE 2-dependent) and dorsal (WAVE 1-dependent) protrusions (Suetsugu et al. 2003). However, the molecular mechanism of regulation of WAVE 3 is unknown, as are the phenotypic consequences

on cell behavior of stimulating WAVE 3. More work will need to be done on the consequences of WAVE 3 expression and activation in carcinoma cells to understand its significance for tumor cell invasion.

An interesting finding is the coordinated up-regulation of expression of genes for several subunits of the Arp2/3 complex and Cdc42. Cdc42 regulates N-WASP, a ubiquitous member of the WASP family, which induces actin polymerization by activating Arp2/3 complex (Ho et al. 2004). Regulation of the activity of N-WASP involves an intramolecular interaction by which the VCA (verproline/cofilin/acidic) domain, the active site that binds to Arp2/3 complex, is masked by the N-terminal regulatory region of N-WASP (Kim et al. 2000, Rohatgi et al. 2000). The known regulators of N-WASP activity appear to operate by either stabilizing or destabilizing this autoinhibitory conformation (Ho et al. 2004). Elevated expression of Cdc42, as observed in invasive cells, in combination with the elevated expression of Arp2/3 complex (**Figure 2c**), would be expected to enhance the activity of the N-WASP-Arp2/3 complex pathway, thereby leading to increased invadopod production and cell invasion (Mizutani et al. 2002, Yamaguchi et al. 2005). Cofilin also has been found to amplify and stabilize N-WASP generated invadopods, suggesting that the synergistic interaction between the cofilin and Arp2/3 complex pathways described above is at work during invasion (Ghosh et al. 2004, Yamaguchi et al. 2005).

An additional consequence of increased Cdc42 expression might be its effect on the ability of tumor cells to acquire polarity to blood vessels, as observed during intravasation. Carcinoma cells in metastatic tumors are attracted to blood vessels, where they form a layer of cells that are morphologically polarized toward the vessel. This vessel-directed polarization is believed to be important for intravasation (Condeelis & Segall 2003, Wyckoff et al. 2000a). Chemotaxis undoubtedly contributes to the accumulation of cells around the vessels, but the acquisition of vessel-directed polarity might require additional steps in the reorganization of the cytoskeleton.

A consensus has been building that Cdc42 is involved in determining the direction of cell movement and cell polarity. Inhibition of Cdc42 prevents macrophage migration toward a chemotactic signal (Allen et al. 1998) and directional migration in astrocytes (Etienne-Manneville & Hall 2001). The development of stable cell polarity in astrocytes involves Cdc42, the orientation of the microtubule organizing center, and depends on microtubule dynamics (nocodazole-sensitive) but not on actin polymerization (Etienne-Manneville & Hall 2001, Gundersen et al. 2004). In microtubule-dependent cell polarization in astrocytes, Cdc42 activation involves recruitment of a GEF, FGD-1 and appears to operate through

Par6 to recruit PKCζ to inactivate GSK3b (Etienne-Manneville & Hall 2003). This pathway requires stimulation of integrins because arginine-glycine-aspartate (RGD) peptides inhibit the activation of Cdc42, PKC, and protrusion (Etienne-Manneville & Hall 2001).

However, chemotaxis by tumor cells requires actin polymerization and is unaffected by concentrations of nocodazole (100 nM) sufficient to block microtubule dynamics and inhibit cell polarity in astrocytes (Segall et al. 1996). In addition, N-WASP, a major effector of Cdc42, is required for invadopod formation (Yamaguchi et al. 2005) and chemotaxis in tumor cells (L. Soon, personal communication), indicating that cell polarization toward EGF in these cells is dependent on Cdc42 through N-WASP and actin polymerization. During N-WASP- and actin-dependent events at the leading edge of lamellipods, the Cdc42 GEF, intersectin 1, binds to and is activated by N-WASP (Hussain et al. 2001). Therefore, N-WASP may recruit, through intersectin 1, GDP-Cdc42 and activate it locally at the leading edge, making N-WASP function in the polarization to EGF of tumor cells analogous to that of Par6 in cell polarity during wounding. Furthermore, PKCζ may be involved in the regulation of LIM-kinase activity, which may regulate the amount of active cofilin at the leading edge and its synergy with N-WASP/Arp2/3 complex–mediated protrusion activity (**Figure 1**). These results illustrate parallels between how cells polarize (microtubule-dependent) and chemotax (actin-dependent) and suggest that N-WASP is involved in assembly of a compartment at the leading edge, analogous to the polarity complex in polarizing cells (Etienne-Manneville & Hall 2001), that is required for cell polarity during chemotaxis.

ZBP1 pathway. A gene whose expression is strongly down-regulated in invasive cells is ZBP1 (**Figure 1**). ZBP1 is a member of a family of RNA-binding proteins that contain four C-terminal hnRNP-K homology domains and two N-terminal RNA recognition motifs (Yaniv & Yisraeli 2002). ZBP1 is a 68-kDa RNA-binding protein that binds to the mRNA zip-code of β-actin mRNA and functions to localize β-actin mRNA to the leading edge of crawling cells. Because β-actin is the preferred isoform of actin for the polymerization of filaments at the leading edge of cells, it is acted on by the cofilin, capping protein, and Arp2/3 pathways (Shestakova et al. 2001). β-actin mRNA localization is required for the maintenance of stable cell polarity as observed in the absence of exogenous signals such as that seen in normal primary fibroblasts, epithelial cells, and tumor cells with differing metastatic potential in which actin polymerization is nucleated at only one pole of the cell in normal and nonmetastatic tumor cells (Shestakova et al. 1999). Disruption of ZBP1-mediated β-actin mRNA targeting in cultured cells leads to cells without cell polarity that are able to nucleate actin polymerization globally and exhibit amoeboid movement (Shestakova et al. 2001). Therefore, ZBP1 may determine the sites in cells where the Arp2/3 complex, capping protein, and cofilin pathways converge to determine the leading edge and cell polarity by controlling the sites of targeting of β-actin mRNA and the location of β-actin protein that is the common downstream effector of these pathways.

Tests of Function of Genes of the Invasion Signature in Chemotaxis, Invasion, and Metastasis

The genes of the motility part of the invasion signature can be organized into three converging pathways based on the known functions of the proteins for which they code (**Figure 2**). The functions of key gene products in these pathways and how they affect chemotaxis, invasion, and metastasis by carcinoma cells have been tested. The results of these tests are described next.

Cofilin and LIM-kinase. Direct tests of cofilin function are complicated by the fact

that cofilin is required for viability, which makes genetic approaches in carcinoma cells difficult to interpret. However, the acute inhibition of cofilin activity in carcinoma cells inhibits the generation of barbed ends and actin polymerization at the leading edge in response to EGF (Chan et al. 2000, DesMarais et al. 2004b). Inhibition of cofilin activity, through either the inhibition of PLCγ or direct inhibition using acute siRNA suppression of cofilin expression and cofilin function blocking antibodies, inhibits the early barbed end transient that is essential for the chemotaxis of carcinoma cells to EGF (Mouneimne et al. 2004). Furthermore, cofilin is required for the formation of the stable invadopods by carcinoma cells that are important in the invasion of dense extracellular matrix (Mullins et al. 1998, Yamaguchi et al. 2005), particularly that found around blood vessels (Condeelis & Segall 2003). Finally, the local activation of cofilin in carcinoma cells is sufficient to generate protrusive activity and determine cell direction (Ghosh et al. 2004). All these results indicate that cofilin is essential for the chemotaxis and invasion of mammary carcinoma cells to EGF through a mechanism involving the localized generation of barbed ends that causes the localized protrusion, which defines cell direction (**Figure 3**).

The effects of altering LIM-kinase expression have been studied in tumor cells by several groups, who have shown that overexpression of LIM-kinase 1 in tumor cell lines increases their motility and invasiveness in vitro (Davila et al. 2003, Yoshioka et al. 2003). Experimental reduction in the expression of LIM-kinase 1 in metastatic prostate cell lines decreased invasiveness in matrigel invasion assays. To study the effect of LIM-kinase 1 on metastasis in vivo, an experimental metastasis model was used where cells were injected directly into the left ventricle of mice (Yoshioka et al. 2003). In this case, the ability of cancer cells to survive in the blood, extravasate from blood vessels, and grow at metastatic sites all contribute to the metastasis score, so it is not clear how these results relate to invasion in the primary tumor. In general, these results are consistent with the observed overexpression of LIM-kinase 1 in invasive cells in mammary tumors and their invasion signature (**Table 3, Figure 2a**).

In a separate set of studies, the overexpression of either full-length-regulated LIM-kinase 1 or its constitutively active kinase domain has been reported to inhibit cofilin activity in vivo by phosphorylation. Overexpression also inhibits EGF-induced barbed end production, in particular the early barbed end transient, and lamellipod extension in culture (W. Wang, G. Mouneimne, J. Wyckoff, X. Chen, M. Sidani, and J. Condeelis, unpublished data; Zebda et al. 2000). Furthermore, the overexpression of full-length LIM-kinase 1 in carcinoma cells without altering cofilin expression is correlated with the inhibition of chemotaxis, invasion, intravasation, and metastasis of tumor cells in mammary tumors prepared from these carcinoma cells (W. Wang, G. Mouneimne, J. Wyckoff, X. Chen, M. Sidani, and J. Condeelis, unpublished data). Although these results appear contradictory to those described above, in fact they are consistent with the invasion signature associated with the cofilin pathway (**Figure 2a**). That is, highly invasive cells up-regulate LIM-kinase 1, cofilin, and their stimulatory and inhibitory effectors together (**Figure 2a**), consistent with the hypothesis that the up-regulation of both inhibitory and stimulatory branches of the cofilin pathway increases the amplitude and sharpness of cofilin-dependent actin polymerization transients that are essential for chemotaxis and invasion in carcinoma cells (Mouneimne et al. 2004; W. Wang, G. Mouneimne, J. Wyckoff, X. Chen, M. Sidani, and J. Condeelis, unpublished data). Therefore, to compare studies in which the expression of LIM-kinase, cofilin, or other members of this pathway are experimentally altered, it is essential to measure the output of the cofilin pathway as the timing and amplitude of cofilin-dependent barbed end production during chemotaxis. Manipulations that increase the cofilin-dependent barbed end production of the early transient during chemotaxis are predicted to increase invasiveness, and this predicts that studies in which cells are more invasive and metastatic after overexpression of LIM-kinase have associated compensatory increases in the expression of other members of the cofilin pathway so as to increase barbed end production in response to EGF. Additional work will be required to investigate this possibility.

N-WASP. N-WASP has been implicated in invasion of extracellular matrix in a number of studies. The invasion of Madin-Darby canine kidney cells during tubulogenesis in collagen gels is inhibited by expression of dominant-negative N-WASP (Yamaguchi et al. 2002). Furthermore, N-WASP, in cooperation with cofilin, is required for the formation of invadopods (Yamaguchi et al. 2005), and its activity is localized to nascent invadopods during the invasion of fibronectin gels (Lorenz et al. 2004). In particular, the depletion of N-WASP or the p34arc subunit of Arp2/3 complex by siRNA interference suppresses invadopod formation. In addition, siRNA interference and dominant-negative mutant expression analyses revealed that cofilin and the N-WASP regulators, Nck1, Cdc42, and WIP, but not Grb2 and WISH, are necessary for invadopod formation (Yamaguchi et al. 2005). EGF receptor kinase inhibitors block the formation of invadopods by carcinoma cells in the presence of serum, and EGF stimulation of serum-starved cells induces invadopod formation. These results indicate that EGF receptor-activated N-WASP and cofilin are required for the formation of invadopods and that Nck1 and Cdc42 mediate the signaling pathway.

A phenomenon that may be related to chemotaxis, invadopod formation, and pathfinding is the observation that the localized stimulation of the EGF receptor on carcinoma cells using EGF-bound beads results in localized actin polymerization and protrusion

(Kempiak et al. 2003). This highly focal actin polymerization requires the activation of the Arp2/3 complex by N-WASP and cofilin and is regulated by Grb2 and Nck2 (Kempiak et al. 2005). This phenomenon may be relevant to how EGF receptor ligands, which can bind to extracellular matrix, stimulate focal protrusions, invadopod formation, and adhesion in vivo (Kempiak et al. 2005). Additional work will be required to determine the effects of altering N-WASP activity on invasion, intravasation, and metastasis in vivo.

ZBP1. The targeting of β-actin mRNA to the leading lamella is essential for stable cell polarity during locomotion, and ZBP1 is required for mRNA targeting (Condeelis & Singer 2005). Highly metastatic cells lines have reduced levels of ZBP1, and this is consistent with the reduction in ZBP1 expression seen in invasive cells (Wang et al. 2004). Decreased β-actin mRNA targeting seen in cells with reduced ZBP1 is correlated with the loss of cell polarity and increased amoeboid movement in metastatic carcinoma cell lines in vitro and in vivo (Shestakova et al. 1999, Wang et al. 2002) and increased chemotaxis (Wang et al. 2004). Increasing the level of expression of ZBP1 in invasive carcinoma cells rescues the localization of β-actin mRNA to one pole of the cell and results in the inhibition of chemotaxis to EGF both in vitro and in vivo in tumors. In addition, tumors prepared from cells re-expressing ZBP1 are significantly less invasive and metastatic than their parental cell–generated counterparts (Wang et al. 2004). However, tumor growth is not significantly affected by increasing the expression of ZBP1. This suggests that the suppression of invasion and metastasis by ZBP1 is not related to growth of the tumor. These results are consistent with the observation that mouse mammary tumors that overexpress the ZBP1 homologue CRD-BP are not metastatic (Tessier et al. 2004).

The invasion and metastasis suppression activity of ZBP1 may result from its ability to suppress the chemotaxis of cancer cells by maintaining them in a polarized epithelial cell-like state. Cells that lack an intrinsic and stable polarity are more chemotactic to exogenous gradients, presumably because there is no intrinsic polarity to be overcome by the exogenous chemotactic signal and the cell can turn in any direction to respond to the gradient (Iijima et al. 2002, Parent & Devreotes 1999). This may account for the enhanced ability of invasive carcinoma cells to chemotax to blood vessels (Condeelis & Segall 2003, Wyckoff et al. 2000a). It also suggests that the generation of polarity in carcinoma cells that occurs around blood vessels is independent of ZBP1 activity, as discussed above.

NEW INSIGHTS INTO TUMOR INVASION AND METASTASIS

The identification of an invasion signature for mammary tumors that implicates the coordinate regulation of genes involved in functionally related activities presents a rich collection of targets for chemotherapy not previously detected in conventional expression profiling of whole tumors. The fact that the pathways are coordinately regulated in invasive cells suggests that combinations of therapeutics may be particularly effective.

An additional insight resulting from the study of invasive cells and their invasion signature comes from the comparison of expression profiles obtained from invasive cells with the conventional expression profiles of whole tumors. Gene expression profiles of whole tumors have shown promise in prognosis by identifying patterns of expression that are correlated with metastasis (Ramaswamy et al. 2003, van't Veer et al. 2002). However, unlike the invasion signature described for invasive cells, these patterns of expression appear as random sets of genes with unrelated functions and thus are difficult to interpret in terms of mechanisms of invasion and metastasis. This suggests that the invasion signature is either averaged out when interrogating the whole tumor because invasive cells are rare or that the changes in gene expression that represent the invasion signature are largely transient. Therefore, it is interesting that the expression profiles of whole tumors demonstrate that the invasive and metastatic potential of the primary tumor can be encoded early in the development of the tumor and throughout the bulk of the tumor including the stroma (Ramaswamy et al. 2003, van 't Veer et al. 2002). These results suggest that metastasis could occur early in tumor progression and that most cells in the tumor are potentially metastatic, thus favoring a "transient expression" model rather than an "averaged-out model" to explain the discordance between expression profile results. This conclusion is surprising because the traditional view of tumor progression is that tumors develop through a succession of stable genetic changes acquired through selection pressures, a process analogous to Darwinian evolution. According to the traditional view of tumor progression, the cells selected to be metastatic are very rare, and metastases arise from progressive genetic changes in these rare cells within a primary tumor delaying metastasis to late stages of tumor progression (Bernards & Weinberg 2002, Hanahan & Weinberg 2000).

The Tumor Microenvironment Invasion Model

A new model, the tumor microenvironment invasion model (TMIM), has been proposed to explain the relationship between the expression pattern of invasive cells and expression patterns of whole tumors and how these relate to the traditional view of tumor progression (Wang et al. 2005). In this model, the transient changes in gene expression leading to invasion (the invasion signature) result from microenvironments in the tumor that are defined by stable genetic changes in both stromal and tumor cells. That is, tumor progression, as described by traditional models (Hanahan & Weinberg 2000), leads to the development of microenvironments encoded within the tumor, which elicit the tran-

sient gene expression patterns that support invasion. In this context, invasion is similar to a morphogenetic program involving the transient expression of genes that lead to a change in the location of cells, a program that can occur repeatedly during tumor development and in any location in the tumor that has the microenvironment that elicits the morphogenetic program. The expression of genes that are synergistic for inducing microenvironments causing invasion could lead to the random appearance, in time and location, of these microenvironments during tumor progression leading to repeated episodes of invasion and metastasis throughout tumor progression.

TMIM is consistent with the finding that genes encoding the tumor microenvironment for invasion and metastasis appear to be expressed throughout the bulk of the tumor. It is also consistent with the ability to collect invasive cells by chemotaxis using needles that are placed in random locations in tumors if the growth factors inside the needles mimic microenvironments inducing invasion, as claimed (Wang et al. 2004, Wyckoff et al. 2004). Furthermore, the TMIM hypothesis is supported by intravital imaging of experimental tumors where only a small proportion of tumor cells are motile, and moving cells are not uniformly distributed but are observed in localized areas of the tumor (Condeelis & Segall 2003, Wang et al. 2002), and the observation that micrometastases are often genetically heterogeneous, suggesting that invasive behavior is not stably specified (Klein 2002). Finally, the TMIM hypothesis is consistent with our current understanding of how the tumor microenvironment contributes to invasion and metastasis (Bissell & Radisky 2001).

The exciting new technologies reviewed here have brought us to the point where tumor invasion and metastasis can be studied as a problem in morphogenesis. The future will reveal if the new insights that are emerging will lead to new strategies for the diagnosis and treatment of metastasis.

LITERATURE CITED

Ahmed F, Wyckoff J, Lin EY, Wang W, Wang Y, et al. 2002. GFP expression in the mammary gland for imaging of mammary tumor cells in transgenic mice. *Cancer Res.* 62:7166–69

Allen WE, Zicha D, Ridley AJ, Jones GE. 1998. A role for Cdc42 in macrophage chemotaxis. *J. Cell Biol.* 141:1147–57

Arber S, Barbayannis FA, Hanser H, Schneider C, Stanyon CA, et al. 1998. Regulation of actin dynamics through phosphorylation of cofilin by LIM-kinase. *Nature* 393:805–9

Bailly M, Yan L, Whitesides GM, Condeelis JS, Segall JE. 1998. Regulation of protrusion shape and adhesion to the substratum during chemotactic responses of mammalian carcinoma cells. *Exp. Cell Res.* 241:285–99

Bamburg JR. 1999. Proteins of the ADF/cofilin family: essential regulators of actin dynamics. *Annu. Rev. Cell Dev. Biol.* 15:185–230

Bear JE, Svitkina TM, Krause M, Schafer DA, Loureiro JJ, et al. 2002. Antagonism between Ena/VASP proteins and actin filament capping regulates fibroblast motility. *Cell* 109:509–21

Bernards R, Weinberg RA. 2002. A progression puzzle. *Nature* 418:823

Bernstein BW, Painter WB, Chen H, Minamide LS, Abe H, Bamburg JR. 2000. Intracellular pH modulation of ADF/cofilin proteins. *Cell Motil. Cytoskelet.* 47:319–36

Bertling E, Hotulainen P, Mattila PK, Matilainen T, Salminen M, Lappalainen P. 2004. Cyclase-associated protein 1 (CAP1) promotes cofilin-induced actin dynamics in mammalian nonmuscle cells. *Mol. Biol. Cell* 15:2324–34

Bissell MJ, Radisky D. 2001. Putting tumors in context. *Nat. Rev. Cancer* 1:46–54

Brown EB, Campbell RB, Tsuzuki Y, Xu L, Carmeliet P, et al. 2001. In vivo measurement of gene expression, angiogenesis and physiological function in tumors using multiphoton laser scanning microscopy. *Nat. Med.* 7:864–68

Campagnola PJ, Clark HA, Mohler WA, Lewis A, Loew LM. 2001. Second-harmonic imaging microscopy of living cells. *J. Biomed. Opt.* 6:277–86

Carlier MF. 1998. Control of actin dynamics. *Curr. Opin. Cell Biol.* 10:45–51

Chambers AF, MacDonald IC, Schmidt EE, Koop S, Morris VL, et al. 1995. Steps in tumor metastasis: new concepts from intravital videomicroscopy. *Cancer Metastasis Rev.* 14:279–301

Chan AY, Bailly M, Zebda N, Segall JE, Condeelis JS. 2000. Role of cofilin in epidermal growth factor-stimulated actin polymerization and lamellipod protrusion. *J. Cell Biol.* 148:531–42

Chantrain CF, Shimada H, Jodele S, Groshen S, Ye W, et al. 2004. Stromal matrix metalloproteinase-9 regulates the vascular architecture in neuroblastoma by promoting pericyte recruitment. *Cancer Res.* 64:1675–86

Chishima T, Miyagi Y, Wang X, Yamaoka H, Shimada H, et al. 1997. Cancer invasion and micrometastasis visualized in live tissue by green fluorescent protein expression. *Cancer Res.* 57:2042–47

Clark EA, Golub TR, Lander ES, Hynes RO. 2000. Genomic analysis of metastasis reveals an essential role for RhoC. *Nature* 406:532–35

Condeelis J, Segall JE. 2003. Intravital imaging of cell movement in tumours. *Nat. Rev. Cancer* 3:921–30

Condeelis J, Singer R. 2005. How and why does β-actin mRNA target? *Biol. Cell* 97:97–110

Condeelis J, Song X, Backer J, Wyckoff J, Segall J. 2003. *Chemotaxis of cancer cells during invasion and metastasis.* Presented at 5th Abercrombie Symp. Cell Behav., St. Catherine's College, Oxford, UK

Cooper JA, Schafer DA. 2000. Control of actin assembly and disassembly at filament ends. *Curr. Opin. Cell Biol.* 12:97–103

Dan C, Kelly A, Bernard O, Minden A. 2001. Cytoskeletal changes regulated by the PAK4 serine/threonine kinase are mediated by LIM kinase 1 and cofilin. *J. Biol. Chem.* 276:32115–21

Davila M, Frost AR, Grizzle WE, Chakrabarti R. 2003. LIM kinase 1 is essential for the invasive growth of prostate epithelial cells: Implications in prostate cancer. *J. Biol. Chem.* 278:36868–75

DesMarais V, Ghosh M, Eddy RE, Condeelis J. 2004a. Cofilin takes the lead. *J. Cell Sci.* 118:19–26

DesMarais V, Macaluso F, Condeelis J, Bailly M. 2004b. Synergistic interaction between the Arp2/3 complex and cofilin drives stimulated lamellipod extension. *J. Cell Sci.* 117:3499–510

Di Modugno F, Bronzi G, Scanlan MJ, Del Bello D, Cascioli S, et al. 2004. Human Mena protein, a serex-defined antigen overexpressed in breast cancer eliciting both humoral and CD8+ T-cell immune response. *Int. J. Cancer* 109:909–18

Djafarzadeh S, Niggli V. 1997. Signaling pathways involved in dephosphorylation and localization of the actin-binding protein cofilin in stimulated human neutrophils. *Exp. Cell Res.* 236:427–35

Eden S, Rohatgi R, Podtelejnikov AV, Mann M, Kirschner MW. 2002. Mechanism of regulation of WAVE 1. *Nature* 418:790–93

Edwards DC, Sanders LC, Bokoch GM, Gill GN. 1999. Activation of LIM-kinase by Pak1 couples Rac/Cdc42 GTPase signalling to actin cytoskeletal dynamics. *Nat. Cell Biol.* 1:253–59

Etienne-Manneville S, Hall A. 2001. Integrin-mediated activation of Cdc42 controls cell polarity in migrating astrocytes through PKCzeta. *Cell* 106:489–98

Etienne-Manneville S, Hall A. 2003. Cdc42 regulates GSK-3beta and adenomatous polyposis coli to control cell polarity. *Nature* 421:753–56

Farina KL, Wyckoff JB, Rivera J, Lee H, Segall JE, et al. 1998. Cell motility of tumor cells visualized in living intact primary tumors using green fluorescent protein. *Cancer Res.* 58:2528–32

Friedl P, Wolf K. 2003. Tumour-cell invasion and migration: diversity and escape mechanisms. *Nat. Rev. Cancer* 3:362–74

Ghosh M, Song X, Mouneimne G, Sidani M, Lawrence DS, Condeelis JS. 2004. Cofilin promotes actin polymerization and defines the direction of cell motility. *Science* 304:743–46

Goswami S, Sahai E, Wyckoff J, Cammer M, Cox D, Pixley F, et al. 2005. Macrophages promote the invasion of breast carcinoma cells via a paracrine loop. *Cancer Res.* 65:5278–83

Goswami S, Wang W, Wyckoff JB, Condeelis JS. 2004. Breast cancer cells isolated by chemotaxis from primary tumors show increased survival and resistance to chemotherapy. *Cancer Res.* 64:7664–67

Gundersen GG, Gomes ER, Wen Y. 2004. Cortical control of microtubule stability and polarization. *Curr. Opin. Cell Biol.* 16:106–12

Hanahan D, Weinberg RA. 2000. The hallmarks of cancer. *Cell* 100:57–70

Helmchen F, Denk W. 2002. New developments in multiphoton microscopy. *Curr. Opin. Neurobiol.* 12:593–601

Ho HY, Rohatgi R, Lebensohn AM, Le M, Li J, et al. 2004. Toca-1 mediates Cdc42-dependent actin nucleation by activating the N-WASP-WIP complex. *Cell* 118:203–16

Hussain NK, Jenna S, Glogauer M, Quinn CC, Wasiak S, et al. 2001. Endocytic protein intersectin-l regulates actin assembly via Cdc42 and N-WASP. *Nat. Cell Biol.* 3:927–32

Ichetovkin I, Grant W, Condeelis J. 2002. Cofilin produces newly polymerized actin filaments that are preferred for dendritic nucleation by the Arp2/3 complex. *Curr. Biol.* 12:79–84

Iijima M, Huang YE, Devreotes P. 2002. Temporal and spatial regulation of chemotaxis. *Dev. Cell* 3:469–78

Jain RK, Munn LL, Fukumura D. 2002. Dissecting tumour pathophysiology using intravital microscopy. *Nat. Rev. Cancer* 2:266–76

Kamai T, Tsujii T, Arai K, Takagi K, Asami H, et al. 2003. Significant association of Rho/ROCK pathway with invasion and metastasis of bladder cancer. *Clin. Cancer Res.* 9:2632–41

Kanamori T, Hayakawa T, Suzuki M, Titani K. 1995. Identification of two 17-kDa rat parotid gland phosphoproteins, subjects for dephosphorylation upon beta-adrenergic stimulation, as destrin- and cofilin-like proteins. *J. Biol. Chem.* 270:8061–67

Kempiak SJ, Yamaguchi H, Sarmiento C, Sidani M, Ghosh M, et al. 2005. An N-Wasp mediated pathway for localized activation of actin polymerization which is regulated by cortactin. *J. Biol. Chem.* 280(7):5836–42

Kempiak SJ, Yip SC, Backer JM, Segall JE. 2003. Local signaling by the EGF receptor. *J. Cell Biol.* 162:781–87

Kim AS, Kakalis LT, Abdul-Manan N, Liu GA, Rosen MK. 2000. Autoinhibition and activation mechanisms of the Wiskott-Aldrich syndrome protein. *Nature* 404:151–58

Klein CA, Blankenstein TJF, Schmidt-Kittler O, Petronio M, Polzer B, et al. 2002. Genetic heterogeneity of single disseminated tumour cells in minimal residual cancer. *Lancet* 360:683–89

Kuroda S, Tokunaga C, Kiyohara Y, Higuchi O, Konishi H, et al. 1996. Protein-protein interaction of zinc finger LIM domains with protein kinase C. *J. Biol. Chem.* 271:31029–32

Levene MJ, Dombeck DA, Kasischke KA, Molloy RP, Webb WW. 2004. In vivo multiphoton microscopy of deep brain tissue. *J. Neurophysiol.* 91:1908–12

Lin EY, Nguyen AV, Russell RG, Pollard JW. 2001. Colony-stimulating factor 1 promotes progression of mammary tumors to malignancy. *J. Exp. Med.* 193:727–40

Liotta LA, Kohn EC. 2001. The microenvironment of the tumour-host interface. *Nature* 411:375–79

Loisel TP, Boujemaa R, Pantaloni D, Carlier MF. 1999. Reconstitution of actin-based motility of Listeria and Shigella using pure proteins. *Nature* 401:613–16

Lorenz M, Yamaguchi H, Wang Y, Singer RH, Condeelis J. 2004. Imaging sites of N-WASP activity in lamellipodia and invadopodia of carcinoma cells. *Curr. Biol.* 14:697–703

Martiny-Baron G, Kazanietz MG, Mischak H, Blumberg PM, Kochs G, et al. 1993. Selective inhibition of protein kinase C isozymes by the indolocarbazole Go 6976. *J. Biol. Chem.* 268:9194–97

Masters BR, So PT, Gratton E. 1997. Multiphoton excitation fluorescence microscopy and spectroscopy of in vivo human skin. *Biophys. J.* 72:2405–12

Miki H, Yamaguchi H, Suetsugu S, Takenawa T. 2000. IRSp53 is an essential intermediate between Rac and WAVE in the regulation of membrane ruffling. *Nature* 408:732–35

Mizutani K, Miki H, He H, Maruta H, Takenawa T. 2002. Essential role of neural Wiskott-Aldrich syndrome protein in podosome formation and degradation of extracellular matrix in src-transformed fibroblasts. *Cancer Res.* 62:669–74

Mouneimne G, Soon L, DesMarais V, Sidani M, Song X, et al. 2004. Phospholipase C and cofilin are required for carcinoma cell directionality in response to EGF stimulation. *J. Cell Biol.* 166:697–708

Mullins RD, Heuser JA, Pollard TD. 1998. The interaction of Arp2/3 complex with actin: nucleation, high affinity pointed end capping, and formation of branching networks of filaments. *Proc. Natl. Acad. Sci. USA* 95:6181–86

Ohashi K, Nagata K, Maekawa M, Ishizaki T, Narumiya S, Mizuno K. 2000. Rho-associated kinase ROCK activates LIM-kinase 1 by phosphorylation at threonine 508 within the activation loop. *J. Biol. Chem.* 275:3577–82

Okada K, Takano-Ohmuro H, Obinata T, Abe H. 1996. Dephosphorylation of cofilin in polymorphonuclear leukocytes derived from peripheral blood. *Exp. Cell Res.* 227:116–22

Otsubo T, Iwaya K, Mukai Y, Mizokami Y, Serizawa H, et al. 2004. Involvement of Arp2/3 complex in the process of colorectal carcinogenesis. *Mod. Pathol.* 17:461–67

Paavilainen V, Bertling E, Falck S, Lappalainen P. 2004. Regulation of cytoskeletal dynamics by actin-monomer-binding proteins. *Trends Cell Biol.* 14:386–94

Parent CA, Devreotes PN. 1999. A cell's sense of direction. *Science* 284:765–70

Pollard JW. 2004. Tumour-educated macrophages promote tumour progression and metastasis. *Nat. Rev. Cancer* 4:71–78

Ramaswamy S, Ross KN, Lander ES, Golub TR. 2003. A molecular signature of metastasis in primary solid tumors. *Nat. Genet.* 33:49–54

Rohatgi R, Ho HY, Kirschner MW. 2000. Mechanism of N-WASP activation by CDC42 and phosphatidylinositol 4,5-bisphosphate. *J. Cell Biol.* 150:1299–310

Rosok O, Pedeutour F, Ree AH, Aasheim HC. 1999. Identification and characterization of TESK2, a novel member of the LIMK/TESK family of protein kinases, predominantly expressed in testis. *Genomics* 61:44–54

Sahai E, Wyckoff J, Philippar U, Gertler F, Segall J, Condeelis J. 2005. Simultaneous imaging of GFP, CFP and collagen in tumors in vivo. *BMC Biotechnol.* 5:14

Scherbarth S, Orr FW. 1997. Intravital videomicroscopic evidence for regulation of metastasis by the hepatic microvasculature: effects of interleukin-1alpha on metastasis and the location of B16F1 melanoma cell arrest. *Cancer Res.* 57:4105–10

Segall JE, Tyerech S, Boselli L, Masseling S, Helft J, et al. 1996. EGF stimulates lamellipod extension in metastatic mammary adenocarcinoma cells by an actin-dependent mechanism. *Clin. Exp. Metastasis* 14:61–72

Shestakova EA, Singer RH, Condeelis J. 2001. The physiological significance of beta -actin mRNA localization in determining cell polarity and directional motility. *Proc. Natl. Acad. Sci. USA* 98:7045–50

Shestakova EA, Wyckoff J, Jones J, Singer RH, Condeelis J. 1999. Correlation of beta-actin messenger RNA localization with metastatic potential in rat adenocarcinoma cell lines. *Cancer Res.* 59:1202–5

Steffen A, Rottner K, Ehinger J, Innocenti M, Scita G, et al. 2004. Sra-1 and Nap-1 link Rac to actin assembly driving lamellipodia formation. *EMBO J.* 23:749–59

Suetsugu S, Yamazaki D, Kurisu S, Takenawa T. 2003. Differential roles of WAVE1 and WAVE2 in dorsal and peripheral ruffle formation for fibroblast cell migration. *Dev. Cell* 5:595–609

Suzuki T, Yanagi K, Ookawa K, Hatakeyama K, Ohshima N. 1996. Flow visualization of microcirculation in solid tumor tissues: intravital microscopic observation of blood circulation by use of a confocal laser scanning microscope. *Front Med. Biol. Eng.* 7:253–63

Takenawa T, Miki H. 2001. WASP and WAVE family proteins: key molecules for rapid rearrangement of cortical actin filaments and cell movement. *J. Cell. Sci.* 114:1801–9

Tessier CR, Doyle GA, Clark BA, Pitot HC, Ross J. 2004. Mammary tumor induction in transgenic mice expressing an RNA-binding protein. *Cancer Res.* 64:209–14

Toshima J, Toshima JY, Amano T, Yang N, Narumiya S, Mizuno K. 2001. Cofilin phosphorylation by protein kinase testicular protein kinase 1 and its role in integrin-mediated actin reorganization and focal adhesion formation. *Mol. Biol. Cell* 12:1131–45

Vajkoczy P, Goldbrunner R, Farhadi M, Vince G, Schilling L, et al. 1999. Glioma cell migration is associated with glioma-induced angiogenesis in vivo. *Int. J. Dev. Neurosci.* 17:557–63

van 't Veer LJ, Dai H, van de Vijver MJ, He YD, Hart AA, et al. 2002. Gene expression profiling predicts clinical outcome of breast cancer. *Nature* 415:530–36

Wang W, Goswami S, Lapidus K, Wells AL, Wyckoff JB, et al. 2004. Identification and testing of a gene expression signature of invasive carcinoma cells within primary mammary tumors. *Cancer Res.* 64:8585–94

Wang W, Goswami S, Sahai E, Wyckoff J, Segall J, Condeelis J. 2005. Tumor cells caught in the act of invading: How they revealed their strategy for enhanced cell motility. *Trends Cell Biol.* 15(3):138–45

Wang W, Wyckoff JB, Frohlich VC, Oleynikov Y, Huttelmaier S, et al. 2002. Single cell behavior in metastatic primary mammary tumors correlated with gene expression patterns revealed by molecular profiling. *Cancer Res.* 62:6278–88

Wang W, Wyckoff JB, Wang Y, Bottinger EP, Segall JE, Condeelis JS. 2003. Gene expression analysis on small numbers of invasive cells collected by chemotaxis from primary mammary tumors of the mouse. *BMC Biotechnol.* 3:13–25

Williams RM, Zipfel WR, Webb WW. 2001. Multiphoton microscopy in biological research. *Curr. Opin. Chem. Biol.* 5:603–8

Wood S. 1958. Pathogenesis of metastasis formation observed in vivo in the rabbit ear chamber. *Arch. Pathol.* 66:550–68

Woodhouse EC, Chuaqui RF, Liotta LA. 1997. General mechanisms of metastasis. *Cancer* 80:1529–37

Wyckoff J, Wang W, Lin EY, Wang Y, Pixley F, et al. 2004. A paracrine loop between tumor cells and macrophages is required for tumor cell migration in mammary tumors. *Cancer Res.* 64:7022–29

Wyckoff JB, Jones JG, Condeelis JS, Segall JE. 2000a. A critical step in metastasis: in vivo analysis of intravasation at the primary tumor. *Cancer Res.* 60:2504–11

Wyckoff JB, Segall JE, Condeelis JS. 2000b. The collection of the motile population of cells from a living tumor. *Cancer Res.* 60:5401–4

Yamaguchi H, Lorenz M, Kempiak SJ, Sarmiento C, Coniglio S, et al. 2005. Molecular mechanism of invadopodium formation: The role of the N-WASP/Arp2/3 complex pathway and cofilin. *J. Cell Biol.* 168:441–52

Yamaguchi H, Miki H, Takenawa T. 2002. Neural Wiskott-Aldrich syndrome protein is involved in hepatocyte growth factor-induced migration, invasion, and tubulogenesis of epithelial cells. *Cancer Res.* 62:2503–9

Yang N, Higuchi O, Ohashi K, Nagata K, Wada A, et al. 1998. Cofilin phosphorylation by LIM-kinase 1 and its role in Rac-mediated actin reorganization. *Nature* 393:809–12

Yang M, Baranov E, Jiang P, Sun FX, Li XM, et al. 2000. Whole body optical imaging of green fluorescent protein expressing tumors and metastasis. *PNAS* 97:1206–11

Yaniv K, Yisraeli J. 2002. The involvement of a conserved family of RNA binding proteins in embryonic development and carcinogenesis. *Gene* 287:49–54

Yoshioka K, Foletta V, Bernard O, Itoh K. 2003. A role for LIM kinase in cancer invasion. *Proc. Natl. Acad. Sci. USA* 100:7247–52

Yuan F, Dellian M, Fukumura D, Leunig M, Berk DA, et al. 1995. Vascular permeability in a human tumor xenograft: molecular size dependence and cutoff size. *Cancer Res.* 55:3752–56

Zebda N, Bernard O, Bailly M, Welti S, Lawrence DS, Condeelis JS. 2000. Phosphorylation of ADF/cofilin abolishes EGF-induced actin nucleation at the leading edge and subsequent lamellipod extension. *J. Cell Biol.* 151:1119–28

Zipfel WR, Williams RM, Christie R, Nikitin AY, Hyman BT, Webb WW. 2003a. Live tissue intrinsic emission microscopy using multiphoton-excited native fluorescence and second harmonic generation. *Proc. Natl. Acad. Sci. USA* 100:7075–80

Zipfel WR, Williams RM, Webb WW. 2003b. Nonlinear magic: multiphoton microscopy in the biosciences. *Nat. Biotechnol.* 21:1369–77

Subject Index

TGF-β signaling through Smad complexes
and, 661, 663–66
Acyl coenzyme A synthetases
docosahexaenoic acid and, 633, 642, 644, 647
Adaptive immunity
phagocytosis and immunity, 511–23
Adenosine triphosphate (ATP)
nuclear envelope and, 350, 362
Adenovirus
Cajal bodies and, 116
Adhesion molecules
integrins and, 381, 383–85, 390, 403
stem cell niche and, 608, 615, 622
ADMIDAS ion
integrins and, 388–90, 392, 399, 401–2
Adrenal medulla
subcellular complexity and, 4
Aging
Cajal bodies and, 114–15
endoplasmic reticulum-associated degradation
and, 450
AGO genes
RNA silencing systems and plant development,
302–3
Agpt2 gene
lymphatic vasculature and, 463, 469
agr genes
quorum sensing and, 323–24
Agrobacterium tumefaciens
quorum sensing and, 323, 331, 336
RNA silencing systems and plant development,
299
aiiA gene
quorum sensing and, 335
ALADIN nucleoporin
nuclear envelope and, 349
Allostery
integrins and, 381, 386–87, 402
αA domain
integrins and, 383–86, 391
Alzheimer's disease
docosahexaenoic acid and, 633, 642, 647
endoplasmic reticulum-associated degradation
and, 450
Amoebae
phagocytosis and immunity, 522–23
Amyotrophic lateral sclerosis (ALS)
docosahexaenoic acid and, 648
Anagallis arvensis
anisotropic expansion of plant cell wall and,
210
Anagen phase

stem cell niche and, 615–16
Anaphase
centrosomes in cellular regeneration and, 414
Angioblasts
lymphatic vasculature and, 460
Angiogenesis
lymphatic vasculature and, 459–60
Angiopoietins
lymphatic vasculature and, 463, 469
stem cell niche and, 613–15
Angiosperms
root apical meristem development and,
485–502
angustifolia gene
spatial control of cell expansion by plant
cytoskeleton and, 279–82
Anionic phospholipids
lysosomal membrane digestion and, 94–97
Anisotropy
expansion of plant cell wall and
controls, 212–16
definitions, 204–5
degree, 205, 214–15
direction, 205, 212–14
force, 211
future research, 216
Green's degree of alignment hypothesis,
214–15
historical foundation, 205–6
introduction, 204
measuring, 206–7
microfibril synthesis, 212
microtubules, 215–16
multicellular cylindrical organs, 208–9
multicellular laminar organs, 209
multicellular organs, 210
occurrence, 206–10
perspectives, 216
quantification, 206–10
rates, 210–16
resistance, 211–12
roots, 208–9
shoot apical meristem, 210
stem epidermal cells, 212–14
stems, 208–9
strain rate, 205
summary, 216–17
tip-growing cells, 208
two-dimensional expansion of unit area of
cell wall, 204–5
Annexins
phosphinositide phosphates and, 66

Annulate lamellae
    nuclear envelope and, 365
Antigen cross-presentation
    phagocytosis and immunity, 511, 517–22
Antigen-experienced T cells
    in vivo imaging of lymphocyte trafficking and,
        593–94
Antigen presentation
    lysosomal membrane digestion and, 92–94
    phagocytosis and immunity, 517–18, 523
AP2 gene
    RNA silencing systems and plant development,
        301, 308
APC genes
    stem cell niche and, 611, 623
Apical-basal axis
    planar cell polarization and, 155–72
Apoptosis
    apoptosis mechanisms and structural biology,
        36–37, 49
    Cajal bodies and, 108, 113–14
    centrosomes in cellular regeneration and, 425
    chemotaxis and motility in
        invasion/intravasation of cancer cells, 701
    docosahexaenoic acid and, 640
    in vivo imaging of lymphocyte trafficking and,
        590, 597
    mechanisms and structural biology
        activation, 44–45, 49–51
        Bcl-2 protein family, 40–42
        BH3 domain, 42
        BIR domains, 48
        Caenorhabditis elegans, 50
        caspases, 43–47, 49–51
        conserved motifs, 48
        conserved pathway, 50
        death receptors, 38–40
        DFF40/CAD, 49–51
        Drice, 51–52
        Dronc, 50–51
        Drosophila spp., 50–51
        effectors, 43–47, 49
        extrinsic pathway, 38–40
        IAP-binding motif, 47–48
        inhibition, 46–47, 49
        initiators, 43
        intrinsic pathway, 40–50
        introduction, 36–38
        ligand binding, 38–40
        prospects, 52
        regulation, 50–52
        signaling motifs, 39–40

Smac/DIABLO, 47–48
Smac-mediated removal of caspase
    inhibition, 49
tetrapeptide IAP-binding motif, 48
viral proteins, 47
nuclear envelope and, 350, 367
phagocytosis and immunity, 517
stem cell niche and, 606, 615, 617
Arabidopsis thaliana
    anisotropic expansion of plant cell wall and,
        214
    Cajal bodies and, 118
    RNA silencing systems and plant development,
        300–10
    root apical meristem development and,
        486–87, 490–98
    spatial control of cell expansion by plant
        cytoskeleton and, 273, 282–85, 287
    steroid hormone signaling in plants and,
        177–96
    variant histone assembly into chromatin and,
        135
Arachidonic acid (AA, 20:4 n-6)
    docosahexaenoic acid and, 633, 636–37
Archaea
    protein translocation by Sec61/SecY channel
        and, 529–46
ARF6 protein
    phagocytosis and immunity, 514–15
Arp2/3 complex
    chemotaxis and motility in
        invasion/intravasation of cancer cells, 695,
        700, 703–8
    phosphinositide phosphates and, 62
    Rho GTPases and, 248–50
    spatial control of cell expansion by plant
        cytoskeleton and, 282–85, 289
ARR3/IAA17 repressor
    root apical meristem development and, 493
Arrest
    G_1
        centrosomes in cellular regeneration and,
            414–17
Arrestins
    planar cell polarization and, 164
Ascorbate
    root apical meristem development and,
        499–501
Asparagine endopeptidase
    phagocytosis and immunity, 518
Aspergillus nidulans
    centrosomes in cellular regeneration and, 414

nuclear envelope and, 356–57, 360

Ataxin-1
    Cajal bodies and, 119

ATPases
    endoplasmic reticulum-associated degradation and, 435, 445–47
    phagocytosis and immunity, 516
    protein translocation by Sec61/SecY channel and, 529, 531–34, 537, 542–44
    variant histone assembly into chromatin and, 139–40, 146

Auditory maps
    retinorectal mapping and, 553

Autoimmune disease
    apoptosis mechanisms and structural biology, 36

Autoinducers
    quorum sensing and, 319–38

Auxins
    root apical meristem development and, 485–502

Avidity
    integrins and, 402–3

Axons
    retinorectal mapping and, 551–75

axr6 gene
    root apical meristem development and, 496

Azurophil
    phagocytosis and immunity, 514

B

Bacillus subtilis
    protein translocation by Sec61/SecY channel and, 543
    quorum sensing and, 329–30, 334

Backbranching
    retinorectal mapping and, 560–61

Bacteria
    cell-to-cell communication in, 319–38
    lysosomal membrane digestion and, 88
    phagocytosis and immunity, 517–18, 520, 522
    protein translocation by Sec61/SecY channel and, 529–30, 534–36, 538, 542–44
    quorum sensing and, 319–38
    subcellular complexity and, 6–7, 11–12

BAK1 gene
    steroid hormone signaling in plants and, 181–85, 188, 196

bam genes
    stem cell niche and, 608

Barrier-to-autointegration factor (BAF)

nuclear envelope and, 364

Basic fibroblast growth factor (bFGF)
    stem cell niche and, 619

Basket structures
    nuclear envelope and, 349, 361

BC1 protein
    RNA transport and local control of translation, 226–27, 233–35

B cells
    in vivo imaging of lymphocyte trafficking and, 581, 587–96

Bcl-2 protein family
    apoptosis mechanisms and structural biology, 35, 40–42

bdl genes
    root apical meristem development and, 493

BDNF
    stem cell niche and, 619

BES1 gene
    steroid hormone signaling in plants and, 177, 186–90, 196

BH3 domain
    apoptosis mechanisms and structural biology, 42

Bid protein
    apoptosis mechanisms and structural biology, 37, 41

Bidirectional signaling
    integrins and, 381–403

BIN2 gene
    steroid hormone signaling in plants and, 177, 184–86

Biofilms
    quorum sensing and, 326–27

Biogenesis
    Cajal bodies and, 109–11
    docosahexaenoic acid and, 634–36
    phagocytosis and immunity, 511, 513–17

Bioinformatics
    quorum sensing and, 337

BiP protein
    endoplasmic reticulum-associated degradation and, 437, 439–40, 442
    nuclear envelope and, 362
    protein translocation by Sec61/SecY channel and, 529, 531–34, 537

BIR domains
    apoptosis mechanisms and structural biology, 45–48, 52

Bis(monoacylglycero)phosphatidic acid (BMP)
    lysosomal membrane digestion and, 81, 83–84, 92, 96–97

ubiquitylation, 447–48

Endosomes
lysosomal membrane digestion and, 94
phagocytosis and immunity, 514–16, 523

Endothelial cells
lymphatic vasculature and, 457, 459, 461–62,
464–68, 471–72

Entamoeba histolytica
lysosomal membrane digestion and, 89

Envelope viruses
subcellular complexity and, 1, 22–26

Ephrins
lymphatic vasculature and, 463, 470–71
retinorectal mapping and, 551, 556–75
stem cell niche and, 617

Epidermal cells
anisotropic expansion of plant cell wall and,
203, 212–14, 217
spatial control of cell expansion by plant
cytoskeleton and, 271, 273, 285–88
stem cell niche and, 617

Epidermal growth factor (EGF)
chemotaxis and motility in
invasion/intravasation of cancer cells,
698–99, 703–6
integrins and, 401
stem cell niche and, 619

Epifluorescence IVM (EF-IVM)
in vivo imaging of lymphocyte trafficking and,
583–84, 595

Epigenetics
RNA silencing systems and plant development,
297–98, 305, 309
variant histone assembly into chromatin and,
133, 139, 146

Epithelial cells
stem cell niche and, 615–17
subcellular complexity and, 1, 22–26

ERMAD domain
phosphinositide phosphates and, 63

ERM proteins
phosphinositide phosphates and, 63–64

ERS24 protein
phagocytosis and immunity, 515, 520

ERV genes
endoplasmic reticulum-associated degradation
and, 439

Erwinia carotovora
quorum sensing and, 334–35

Escherichia coli
lysosomal membrane digestion and, 88
phagocytosis and immunity, 517

protein translocation by Sec61/SecY channel
and, 530, 534, 536, 538
quorum sensing and, 334
subcellular complexity and, 6–7, 11–12

Eukaryotes
protein translocation by Sec61/SecY channel
and, 529–46
quorum sensing and, 319–38
RNA transport and local control of translation,
223–37

Euprymna scolopes
quorum sensing and, 320

Evolution
apoptosis mechanisms and structural biology,
36, 48, 50
lymphatic vasculature and, 458
phagocytosis and immunity, 523
quorum sensing and, 335–37
RNA silencing systems and plant development,
297
root apical meristem development and, 486–87
stem cell niche and, 613, 623
variant histone assembly into chromatin and,
135, 140

Exocytosis
phagocytosis and immunity, 514–15
phosphinositide phosphates and, 58, 71–72
spatial control of cell expansion by plant
cytoskeleton and, 280
subcellular complexity and, 4

Exohydrolases
lysosomal membrane digestion and, 81

Expansion
spatial control of cell expansion by plant
cytoskeleton and, 271–89
anisotropic expansion of plant cell wall and,
206–16

Exponential gradients
retinorectal mapping and, 568–69

Export
endoplasmic reticulum-associated degradation
and, 435, 443–47

Exportins
nuclear envelope and, 353

Expression signature
chemotaxis and motility in
invasion/intravasation of cancer cells,
698–700

Extracellular matrix
integrins and, 381–82
lymphatic vasculature and, 459–60

Extrinsic pathways

apoptosis mechanisms and structural biology, 38–40

Extrinsic signals
stem cell niche and, 605–7

Eye
in vivo imaging of lymphocyte trafficking and, 583
retinorectal mapping and, 551–75

Ezrin
phosphinositide phosphates and, 58–59, 63–64

F

FACT-facilitated transfer
variant histone assembly into chromatin and, 139

Fas-associated death domain (FADD)
apoptosis mechanisms and structural biology, 36

Fasciclin III
stem cell niche and, 608

fat genes
planar cell polarization and, 161

Fatty acid binding proteins (FABPs)
docosahexaenoic acid and, 643, 645–47

Fatty acid transport proteins (FATPs)
docosahexaenoic acid and, 633, 643–44

F box receptor
endoplasmic reticulum-associated degradation and, 448

FERM domains
integrins and, 397
phosphinositide phosphates and, 58–59, 63–64, 70

Festuca arundinacea
anisotropic expansion of plant cell wall and, 210

Fibrillarin
Cajal bodies and, 107, 109, 111, 116–17

Fibroblast growth factor (FGF) pathway
stem cell niche and, 616–17, 619

Fibronectin
integrins and, 399–403
retinorectal mapping and, 557

Field-emission scanning electron microscopy
anisotropic expansion of plant cell wall and, 212

Filaments
nuclear envelope and, 349–50
Rho GTPases and, 251
spatial control of cell expansion by plant cytoskeleton and, 271–89

Fish
docosahexaenoic acid and, 633, 636–37

Fission
phagocytosis and immunity, 516, 523

fj genes
planar cell polarization and, 161, 171

Fluorescence-based visualization
in vivo imaging of lymphocyte trafficking and, 581–98

Fluorescence recovery after photobleaching (FRAP)
nuclear envelope and, 354

Fluorescence resonance energy transfer (FRET)
integrins and, 396–97
in vivo imaging of lymphocyte trafficking and, 587
steroid hormone signaling in plants and, 182–83

Fluorescent in situ hybridization (FISH)
Cajal bodies and, 108, 114, 118

Fluorescent tags/labels
in vivo imaging of lymphocyte trafficking and, 584

fmi/stan genes
planar cell polarization and, 158, 161, 163, 167–69, 171

FMRP protein
RNA transport and local control of translation, 233–36

Focal exocytosis
phagocytosis and immunity, 514–15

Follicle cells
stem cell niche and, 608–9, 615–17

Force
anisotropic expansion of plant cell wall and, 211

Formins
Rho GTPases and, 250–51

Foxc2 gene
lymphatic vasculature and, 463, 470

FoxH1 motif
TGF-β signaling through Smad complexes and, 671

fra1 gene
spatial control of cell expansion by plant cytoskeleton and, 273

frizzled genes
planar cell polarization and, 158, 160–64, 166–71

Ft genes
planar cell polarization and, 158

Fungi

subcellular complexity and, 4

Gonialblasts

    stem cell niche and, 611

G proteins

    integrins and, 387

    nuclear envelope and, 350–53, 357, 360, 364–66

    phosphinositide phosphates and, 57, 59, 69

    planar cell polarization and, 164–65

    Rho GTPases and, 247–61

    spatial control of cell expansion by plant cytoskeleton and, 289

Gram-negative bacteria

    quorum sensing and, 320–23

Gram-positive bacteria

    quorum sensing and, 323–25, 330

Granule membranes

    subcellular complexity and, 4

Green fluorescent protein (GFP)

    centrosomes in cellular regeneration and, 414

    endoplasmic reticulum-associated degradation and, 437, 443

    integrins and, 396

    in vivo imaging of lymphocyte trafficking and, 584

    nuclear envelope and, 354, 360

    phagocytosis and immunity, 518

    spatial control of cell expansion by plant cytoskeleton and, 279

    stem cell niche and, 613

    steroid hormone signaling in plants and, 181

Green's degree of alignment hypothesis

    anisotropic expansion of plant cell wall and, 214–15

Growth

    anisotropic expansion of plant cell wall and, 206–16

    integrins and, 381–403

    steroid hormone signaling in plants and, 196

Growth cones

    retinorectal mapping and, 551–75

GRP78/Bip protein

    phagocytosis and immunity, 519

GTPases

    nuclear envelope and, 350–53, 357, 360, 364–66

    phagocytosis and immunity, 514–15

    phosphinositide phosphates and, 57, 59, 69

    Rho GTPases and, 247–61

    spatial control of cell expansion by plant cytoskeleton and, 289

Guide RNA

    Cajal bodies and, 123

H

Hair follicle

    stem cell niche and, 605, 615–17

Hematopoietic stem cells

    in vivo imaging of lymphocyte trafficking and, 587, 589–90

    stem cell niche and, 607, 613–15

Hemostasis

    integrins and, 381–403

HEN1 gene

    RNA silencing systems and plant development, 304, 311

Heterochromatin

    nuclear envelope and, 364

Heterodimers

    integrins and, 381–403

Heteromeric receptor complex

    TGF-β signaling through Smad complexes and, 659, 668–69

Heterotrimeric G proteins

    planar cell polarization and, 164–65

Heterozygosity

    retinorectal mapping and, 566–71

hh genes

    stem cell niche and, 608–9

HIRA complex

    variant histone assembly into chromatin and, 138–39

Histones

    centrosomes in cellular regeneration and, 425

    variant histone assembly into chromatin and, 133–47

HMG-CoA

    endoplasmic reticulum-associated degradation and, 440, 442

HOBBIT gene

    root apical meristem development and, 493, 496

Homeostasis

    apoptosis mechanisms and structural biology, 35–36

Homing

    in vivo imaging of lymphocyte trafficking and, 583, 593–94

Homologs

    centrosomes in cellular regeneration and, 423–26

Homozygosity

retinorectal mapping and, 566, 568–71

Host defense
    integrins and, 381–403

HP1 protein
    nuclear envelope and, 364
    variant histone assembly into chromatin and, 140

HSD domain
    protein translocation by Sec61/SecY channel and, 542–44

HSPC300 protein
    Rho GTPases and, 249
    spatial control of cell expansion by plant cytoskeleton and, 284

hTERT
    Cajal bodies and, 114–16

HTM1 gene
    endoplasmic reticulum-associated degradation and, 441

Hub cells
    stem cell niche and, 609, 611

Human immunodeficiency virus 1 (HIV-1)
    RNA transport and local control of translation, 227

HWD domain
    protein translocation by Sec61/SecY channel and, 542–44

Hydrolases
    phagocytosis and immunity, 517, 521
    spatial control of cell expansion by plant cytoskeleton and, 280

Hydrophobic membrane segments
    protein translocation by Sec61/SecY channel and, 529–46

Hydrophobicity
    apoptosis mechanisms and structural biology, 41–42, 46
    endoplasmic reticulum-associated degradation and, 446
    integrins and, 383–84, 386, 389, 392, 394
    lysosomal membrane digestion and, 89
    nuclear envelope and, 352
    phosphinositide phosphates and, 60
    protein translocation by Sec61/SecY channel and, 535–38, 540–42, 544

HYL1 gene
    RNA silencing systems and plant development, 304–5

Hypophysis
    root apical meristem development and, 493–94

I

IAP proteins
    apoptosis mechanisms and structural biology, 35–36, 45–52

ICAM-2 protein
    phosphinositide phosphates and, 64

Immune response
    in vivo imaging of lymphocyte trafficking and, 581–98
    lymphatic vasculature and, 472–73

Importins
    nuclear envelope and, 353, 360, 365–66

Inactivation
    centrosomes in cellular regeneration and, 411, 424, 426

Indole-3-acetic acid (IAA)
    root apical meristem development and, 490, 492–93, 501

Induction
    quorum sensing and, 331

Inflammation
    docosahexaenoic acid and, 648
    integrins and, 381–82
    in vivo imaging of lymphocyte trafficking and, 583, 590
    lymphatic vasculature and, 472–73
    phagocytosis and immunity, 517

Inhibition
    apoptosis mechanisms and structural biology, 46–47, 49
    endoplasmic reticulum-associated degradation and, 449
    nuclear envelope and, 353
    TGF-β signaling through Smad complexes and, 667–68, 681

Initiation
    apoptosis mechanisms and structural biology, 43
    endoplasmic reticulum-associated degradation and, 437
    planar cell polarization and, 160–61
    RNA transport and local control of translation, 223, 231–36
    spatial control of cell expansion by plant cytoskeleton and, 279–80, 282–83, 289

Innate immunity
    phagocytosis and immunity, 511–23

Inner nuclear membrane
    nuclear envelope and, 348–49, 354–56, 358, 362, 366, 368

Inside-out activation

Intrinsic pathways
   apoptosis mechanisms and structural biology,
      40–50
inturned genes
   planar cell polarization and, 162
Invadopodia
   chemotaxis and motility in
      invasion/intravasation of cancer cells, 698
Invariant chain (Ii)
   phagocytosis and immunity, 518
Invasion/intravasation of cancer cells
   chemotaxis and motility in
      Arp2/3 complex, 706–8
      blood vessels, 698
      capping protein, 706
      chemoattractants, 698
      chemotaxis, 709–11
      cofilin, 701–10
      coordinate regulation of motility pathways
         in invasion, 700–1
      epidermal growth factor, 698
      expression signature, 698–700
      genetics, 700–11
      intravital imaging of tumor cells in vivo,
         696–97
      introduction, 696
      invadopodia, 698
      invasion, 709–11
      invasion signature, 698–711
      LIM-kinase, 709–10
      macrophages, 698
      metastasis, 709–12
      motility behaviors, 696–700
      new insights, 711–12
      N-WASP, 710–11
      primary mammary tumors, 697–98
      solitary cells, 697–98
      test of function in genes, 709–11
      tumor microenvironment invasion model,
         712
      velocity in vitro, 697–98
      ZBP1, 708–11
Invasion signature
   chemotaxis and motility in
      invasion/intravasation of cancer cells,
      698–711
Isl2-EphA knock-ins
   retinorectal mapping and, 565–68
Isotropic expansion
   spatial control of cell expansion by plant
      cytoskeleton and, 274

J

Jagged protein
   stem cell niche and, 613–15
JAK-STAT pathway
   stem cell niche and, 610–11, 622
J-domain
   protein translocation by Sec61/SecY channel
      and, 532–33
Jun N-terminal kinase (JNK)
   planar cell polarization and, 163–64
   TGF-β signaling through Smad complexes
      and, 669
Juxtamembrane region
   steroid hormone signaling in plants and,
      179–80

K

K716-D740
   integrins and, 392
kar genes
   endoplasmic reticulum-associated degradation
      and, 437, 440
   nuclear envelope and, 362
Karyogamy
   nuclear envelope and, 362
Kc gene
   variant histone assembly into chromatin and,
      138
KHN protein
   endoplasmic reticulum-associated degradation
      and, 439
KIM127
   integrins and, 391
Kinases
   Cajal bodies and, 123
   centrosomes in cellular regeneration and, 413,
      418
   chemotaxis and motility in
      invasion/intravasation of cancer cells,
      704–5, 709–10
   integrins and, 398
   nuclear envelope and, 368
   phosphinoside phosphates and, 57–58, 61–63,
      70–71
   planar cell polarization and, 163–64
   retinorectal mapping and, 551, 556
   RNA transport and local control of translation,
      226–27, 231–34
   steroid hormone signaling in plants and,
      177–84, 196

TGF-β signaling through Smad complexes
and, 669–70

Mitosis
  Cajal bodies and, 109
  nuclear envelope and, 354–68
  Rho GTPases and, 255–56
  stem cell niche and, 611–13

Mitotic exit network (MEN)
  centrosomes in cellular regeneration and, 411,
    417, 419–23, 426

Mn^{2+}
  integrins and, 384, 386–87, 389–90, 397–400

mob1 gene
  centrosomes in cellular regeneration and, 421

Mobile silencing signals
  RNA silencing systems and plant development,
    306, 312

Moesin
  phosphinositide phosphates and, 58, 63–64

Molecular motors
  RNA transport and local control of translation,
    223, 229–31

Molecular switch
  Rho GTPases and, 247
  stem cell niche and, 618

monopterous gene
  root apical meristem development and, 493,
    496

mor genes
  anisotropic expansion of plant cell wall and,
    216
  spatial control of cell expansion by plant
    cytoskeleton and, 273–74

Morphogenesis
  anisotropic expansion of plant cell wall and,
    206–16
  Rho GTPases and, 247, 257–59
  spatial control of cell expansion by plant
    cytoskeleton and, 271, 278–88

Mother's milk
  docosahexaenoic acid and, 633, 636

Motility
  chemotaxis in invasion/intravasation of cancer
    cells and, 695–712
  Rho GTPases and, 247, 260

Motor maps
  retinorectal mapping and, 553

Motor proteins
  spatial control of cell expansion by plant
    cytoskeleton and, 271–89

mp genes
  root apical meristem development and, 493

MP-IVM
  in vivo imaging of lymphocyte trafficking and,
    585

msn genes
  planar cell polarization and, 158

mTOR serine/threonine kinase
  RNA transport and local control of translation,
    231–32, 234

Multiphoton-based intravital imaging
  chemotaxis and motility in
    invasion/intravasation of cancer cells,
    695–97

Multipotentiality
  stem cell niche and, 605–7, 615, 624

Musashi-1 gene
  stem cell niche and, 617

Mycobacterium spp.
  phagocytosis and immunity, 518, 520

Myelin basic protein (MBP)
  RNA transport and local control of translation,
    227, 236

Myosin
  phosphinositide phosphates and, 67–68
  RNA transport and local control of translation,
    229
  spatial control of cell expansion by plant
    cytoskeleton and, 281

Myxococcus xanthus
  quorum sensing and, 336

N

NADPH oxidase
  phagocytosis and immunity, 516

Nap proteins
  Rho GTPases and, 249
  spatial control of cell expansion by plant
    cytoskeleton and, 284

NBF1/NFB2 domains
  protein translocation by Sec61/SecY channel
    and, 542–44

Negative regulation
  steroid hormone signaling in plants and, 184

Network architecture
  quorum sensing and, 325–31

Network model
  RNA silencing systems and plant development,
    306–8

Neural cells
  stem cell niche and, 618–20

Neural maps
  retinorectal mapping and, 551–75

steroid hormone signaling in plants and, 185, 196

TGF-β signaling through Smad complexes and, 659, 661, 666–67, 670

Photobleaching
nuclear envelope and, 354

phrC gene
quorum sensing and, 330

Phycomyces blakesleeanus
anisotropic expansion of plant cell wall and, 208

PID genes
root apical meristem development and, 494

PIN genes
root apical meristem development and, 493–94, 497

PINOID genes
root apical meristem development and, 494, 496–97

pipetail gene
planar cell polarization and, 160

PIR121 protein
Rho GTPases and, 249

PirK/Akt pathway
stem cell niche and, 617

pirogi gene
spatial control of cell expansion by plant cytoskeleton and, 284

Pisum sativum
root apical meristem development and, 494

piwi genes
stem cell niche and, 608–9

pk genes
planar cell polarization and, 158, 161, 163, 167, 170

Planar cell polarization
conclusions, 171–72
conserved signaling cassette, 157, 159
Drosophila, 159–60
effectors, 164–65
emerging model, 165–71
evidence-based model, 166–71
future research, 171–72
genes, 158
heterotrimeric G proteins, 164–65
history of study, 156–57
introduction, 156
ligand, 160–61
molecular mechanisms, 160–65
signaling, 160–61, 164–65
vertebrates, 160

Plants

Cajal bodies and, 118

RNA silencing systems and plant development, 297–312

spatial control of cell expansion by, 271–89

steroid hormone signaling in plants and, 177–96

Plasma cells
in vivo imaging of lymphocyte trafficking and, 587, 595

Plasma membrane
integrins and, 382, 394
phagocytosis and immunity, 511, 513, 523
phosphinoside phosphates and, 53–73
spatial control of cell expansion by plant cytoskeleton and, 275
steroid hormone signaling in plants and, 177–96
subcellular complexity and, 4

Plasticity
retinorectal mapping and, 555–56

Plexin
integrins and, 386, 401

PLT genes
root apical meristem development and, 491, 494, 496–97

Plug
protein translocation by Sec61/SecY channel and, 535–38, 540–41

Plus end capture
microtubule
Rho GTPases and, 252

pmr1 gene
endoplasmic reticulum-associated degradation and, 441

PNG1 gene
endoplasmic reticulum-associated degradation and, 449

Podoplanin
lymphatic vasculature and, 463, 469–70

Polar auxin transport
root apical meristem development and, 490, 494

Polarization
planar cell, 155–72
Rho GTPases and, 247, 258
spatial control of cell expansion by plant cytoskeleton and, 289
subcellular complexity and, 1, 22–26

Polarized-light microscopy
anisotropic expansion of plant cell wall and, 212, 214

Pollen tubes

spatial control of cell expansion by plant
cytoskeleton and, 271, 277–78, 289

Polyubiquitylation
endoplasmic reticulum-associated degradation
and, 435, 439, 450

Polyunsaturated fatty acids (PUFAs)
docosahexaenoic acid and, 633–49

Pores
lysosomal membrane digestion and, 89
nuclear envelope and, 347–54
protein translocation by Sec61/SecY channel
and, 532, 536–39

Positive regulation
steroid hormone signaling in plants and,
186–87

Postmitotic cells
centrosomes in cellular regeneration and, 416

Postnatal period
docosahexaenoic acid and, 633

Postsynaptic microdomains
RNA transport and local control of translation,
223

Posttranslational modification
Cajal bodies and, 109
variant histone assembly into chromatin and,
133, 138

Posttranslational translocation
protein translocation by Sec61/SecY channel
and, 531, 537, 542–45

P-PTEN gene
stem cell niche and, 617

PPXD domain
protein translocation by Sec61/SecY channel
and, 542–44

PQS
quorum sensing and, 325

Pregnancy
docosahexaenoic acid and, 633

Pre-pore
nuclear envelope and, 359

Presynaptic domain
RNA transport and local control of translation,
234

prm3 gene
nuclear envelope and, 362

Profilin
phosphinositide phosphates and, 59–62
Rho GTPases and, 250

Progenitor cell homing
in vivo imaging of lymphocyte trafficking and,
583

Prokaryotes

quorum sensing and, 319–38

Proliferation
Cajal bodies and, 111–13
chemotaxis and motility in
invasion/intravasation of cancer cells,
699–700
in vivo imaging of lymphocyte trafficking and,
592
stem cell niche and, 609, 611, 615, 617–18,
620, 622

Prolifin
Cajal bodies and, 116–17

Promoters
root apical meristem development and, 492
variant histone assembly into chromatin and,
139–40, 146

Promyelocytic leukemia protein (PML) bodies
Cajal bodies and, 107, 109, 112, 122

Prosaposin
lysosomal membrane digestion and, 92

Proteases
apoptosis mechanisms and structural biology,
36
endoplasmic reticulum-associated degradation
and, 446
phagocytosis and immunity, 518, 520
protein translocation by Sec61/SecY channel
and, 544
subcellular complexity and, 15–16

Proteasomes
Cajal bodies and, 124
endoplasmic reticulum-associated degradation
and, 435, 437, 439, 447–50
phagocytosis and immunity, 519

Protein disulfide isomerase (PDI)
endoplasmic reticulum-associated degradation
and, 439–40

Protein kinase C (PKC)
chemotaxis and motility in
invasion/intravasation of cancer cells,
704–5
TGF-β signaling through Smad complexes
and, 670

Protein misfolding
endoplasmic reticulum-associated degradation
and, 435, 439, 450

Protein N-glycanase
endoplasmic reticulum-associated degradation
and, 449

Protein tyrosine kinases
phosphinositide phosphates and, 70–71
retinorectal mapping and, 551, 556

Strain rate
anisotropic expansion of plant cell wall and, 205

Streptococcus pneumoniae
quorum sensing and, 330–31

Stress response
anisotropic expansion of plant cell wall and, 208, 214
Cajal bodies and, 105, 113–15
centrosomes in cellular regeneration and, 411–12

Stromal cell-derived factor-1 (SDF-1)
stem cell niche and, 613–15

Structural biology
apoptosis mechanisms and, 35–52

Structure/function relationships
nuclear envelope and, 347–68
variant histone assembly into chromatin and, 141–43

Substrate recruitment
endoplasmic reticulum-associated degradation and, 450

Superior colliculus
retinorectal mapping and, 551–75

Surgical preparation
in vivo imaging of lymphocyte trafficking and, 597

SWI/SNF family
variant histone assembly into chromatin and, 139–40, 146

Switchblade model
integrins and, 398–401

Switches
Rho GTPases and, 247

swr1 gene
variant histone assembly into chromatin and, 140

Syk gene
lymphatic vasculature and, 463

Synaptic targets
retinorectal mapping and, 551, 574

Synaptotagmins
phagocytosis and immunity, 514–15
phosphinositide phosphates and, 71–72

Systems matching
retinorectal mapping and, 574

T

T1a gene
lymphatic vasculature and, 463, 469–70

Tails

cytoplasmic
integrins and, 391–94

Talin
integrins and, 392–94, 397
phosphinositide phosphates and, 65–66

Tapaisin
phagocytosis and immunity, 519

Targeting
endoplasmic reticulum-associated degradation and, 448–50
phosphinositide phosphates and, 53–73
retinorectal mapping and, 551, 574
RNA silencing systems and plant development, 297
steroid hormone signaling in plants and, 183, 194–95

T cells
endoplasmic reticulum-associated degradation and, 448–49
in vivo imaging of lymphocyte trafficking and, 581, 587–96, 598
phagocytosis and immunity, 517–20, 522
stem cell niche and, 615

Telogen phase
stem cell niche and, 615

Telomerase
Cajal bodies and, 114

Temporal-nasal axis
retinorectal mapping and, 551, 554–55, 558–59, 561, 568–70, 573, 575

Termination zones
retinorectal mapping and, 556, 560–64, 566–70, 573

Testis
stem cell niche and, 605, 607–11, 621–22

Tetrapeptide IAP-binding motif
apoptosis mechanisms and structural biology, 48

Tetraploidy
centrosomes in cellular regeneration and, 415

THMF
quorum sensing and, 333

Thymus
in vivo imaging of lymphocyte trafficking and, 581, 589–90

TIE element
stem cell niche and, 613–15
TGF-β signaling through Smad complexes and, 679

Tip growth
anisotropic expansion of plant cell wall and, 208

anisotropic expansion of plant cell wall and, 208

Vegetative growth factor (VGF)
stem cell niche and, 619

Vegfc gene
lymphatic vasculature and, 463

Vegfr3 gene
lymphatic vasculature and, 463

Velocity
chemotaxis and motility in
invasion/intravasation of cancer cells,
697–98

Vertebrates
nuclear envelope and, 349–51
phagocytosis and immunity, 517–18, 520,
522–23
planar cell polarization and, 160

Vesicles
nuclear envelope and, 356, 362–64
spatial control of cell expansion by plant
cytoskeleton and, 276, 280
subcellular complexity and, 4

VgRBP proteins
RNA transport and local control of translation,
226, 228

Vibrio spp.
quorum sensing and, 320–22, 325–27,
331–33

Video-based IVM
in vivo imaging of lymphocyte trafficking and,
585

Vigna radiata
steroid hormone signaling in plants and,
194

Vinculin
phosphinositide phosphates and, 64–65

Virulence
quorum sensing and, 326, 337

Viruses
apoptosis mechanisms and structural biology,
47
Cajal bodies and, 116
RNA transport and local control of translation,
227
subcellular complexity and, 1, 22–26

Virus-induced gene silencing (VIGS)
RNA silencing systems and plant development,
299–301, 312

Visual maps
retinorectal mapping and, 553

vWFA domain
integrins and, 382–86, 391

W

WASP proteins
phosphinoside phosphates and, 62–63
Rho GTPases and, 249–50

WAVE proteins
phosphinoside phosphates and, 62–63
Rho GTPases and, 249–50
spatial control of cell expansion by plant
cytoskeleton and, 283–85

WD-40 repeat proteins
nuclear envelope and, 350

WEX genes
RNA silencing systems and plant development,
302, 311

WIF protein
stem cell niche and, 615

Wild-type map
retinorectal mapping and, 571–72

WIP protein
Rho GTPases and, 249

Wnt genes
planar cell polarization and, 160–62, 165
stem cell niche and, 614–18

Wortmannin
phosphinoside phosphates and, 68

WOX genes
root apical meristem development and, 486,
491

WUS genes
root apical meristem development and, 492,
498

X

Xanthium pensylvanicum
anisotropic expansion of plant cell wall and,
209

Xanthomonas campestris
quorum sensing and, 334

Xenopus laevis
Cajal bodies and, 117–18
centrosomes in cellular regeneration and,
413
nuclear envelope and, 359, 361–65, 367
planar cell polarization and, 157, 160, 164–65,
169
retinorectal mapping and, 564
RNA transport and local control of translation,
224–26, 230, 233

X-ray crystallography
apoptosis mechanisms and structural biology,
38

integrins and, 381, 383–86, 390–96, 401–2
lysosomal membrane digestion and, 88–89
phosphinositide phosphates and, 64
protein translocation by Sec61/SecY channel
and, 534–38, 543, 545
XVA143
integrins and, 391
Xyloglucan endotransglucosylase/hydrolase
(XTH)
spatial control of cell expansion by plant
cytoskeleton and, 280, 282

Y

Yb genes
stem cell niche and, 608
Yeasts
Cajal bodies and, 118
centrosomes in cellular regeneration and, 412,
414, 417–23, 426
endoplasmic reticulum-associated degradation
and, 437, 449
nuclear envelope and, 350–51, 353, 357,
359–62, 367
phagocytosis and immunity, 515
phosphinositide phosphates and, 61

protein translocation by Sec61/SecY channel
and, 530–31
Rho GTPases and, 248
RNA silencing systems and plant development,
302
RNA transport and local control of translation,
225
variant histone assembly into chromatin and,
135, 138–41

Z

ZBP1/ZBP2 proteins
chemotaxis and motility in
invasion/intravasation of cancer cells, 703,
708–9, 711
RNA transport and local control of translation,
226, 228, 230, 236
Zea mays
root apical meristem development and,
488–89
zwi gene
spatial control of cell expansion by plant
cytoskeleton and, 279–82
ZYG-12 protein
nuclear envelope and, 354

Cumulative Indexes

Contributing Authors, Volumes 17–21

Goldman RD, 19:445–67
Goodrich J, 18:707–46
Gottesman S, 19:565–87
Graves BJ, 18:421–62
Gregorio CC, 18:637–706
Grishchuk EL, 18:193–219
Gurdon JB, 18:515–39
Gustafsson MK, 18:747–83

H

Hackstadt T, 18:221–45
Hake S, 20:125–51
Halin C, 21:581–603
Hall A, 21:247–69
Hampton RY, 18:345–78
Harding HP, 18:575–99
Harmer SL, 17:215–53
Hawley S, 20:525–58
Helfand BT, 19:445–67
Hemler ME, 19:397–422
Henikoff S, 21:133–53
Hepler PK, 17:159–87
Hetzer M, 21:347–80
Hicke L, 19:141–72
Higginbotham H,
  20:593–618
Holt C, 20:505–23
Holtan H, 20:125–51
Hong F, 21:177–201
Horvitz HR, 20:193–221
Howard L, 18:541–73
Hughson FM, 19:493–517
Hyman AA, 20:427–53

I

Iglesias PA, 20:223–53
Irish VF, 19:119–40
Irvine KD, 17:189–214
Izpisúa Belmonte JC,
  17:87–132

J

Jaffe AB, 21:247–69
Jaspersen SL, 20:1–28
Jiang K, 21:485–509
Johnson KR, 19:207–35
Jürgens G, 20:481–504
Jutras I, 21:511–27

K

Kappe SHI, 20:29–59
Katso R, 17:615–75
Kay SA, 17:215–53
Kendrick-Jones J, 20:649–76
Kindler S, 21:223–45
Kintner C, 19:367–95
Klein TJ, 21:155–76
Knuehl C, 17:517–68
Koehler CM, 20:309–35
Koizumi H, 20:593–618
Kolter T, 21:81–103
Koltunow A, 17:677–99
Krasnow MA, 19:623–47
Krause M, 19:541–64
Kuroda H, 20:285–308

L

Lamond AI, 21:105–31
Lang RA, 17:255–96
Lazarow PB, 17:701–52
Lee JY, 17:753–77
Lee MCS, 20:87–123
Lee S-J, 20:61–86
Legesse–Miller A, 20:559–91
Lemke G, 21:551–80
Levin M, 17:779–805
Li L, 21:605–31
Lo CW, 20:811–38
Lodish HF, 21:633–57
Logan CY, 20:781–810
Long RM, 17:297–310
Long Y, 20:223–53
Lord EM, 18:81–105
Loureiro JJ, 19:541–64
Luo L, 18:601–35
Luo M, 17:677–99
Luschnig S, 19:623–47

M

Magnani E, 20:125–51
Mahalingam B, 21:381–410
Manahan CL, 20:223–53
Mann MRW, 19:237–59
Marshall WF, 20:677–93
Marszalek JR, 21:633–57
Mattaj IW, 21:347–80
Matulef K, 19:23–44
Maxfield FR, 20:839–66

Mayer A, 18:289–314
McAinsh AD, 19:519–39
McCollum D, 21:411–34
McElhinny AS, 18:637–706
McIntosh JR, 18:193–219
McNeil PL, 19:697–731
Meins F Jr, 21:297–318
Mele G, 20:125–51
Melton DA, 19:71–89
Mercola M, 17:779–805
Metzstein MM, 19:623–47
Miller EA, 20:87–123
Miner JH, 20:255–84
Mlodzik M, 21:155–76
Mora JR, 21:581–603
Morita E, 20:395–425
Morrison D, 19:91–118
Mukherjee S, 20:839–66
Mullins RD, 18:247–88
Murtaugh LC, 19:71–89
Muth TR, 19:333–66

N

Nascimento AA, 19:469–91
Näthke IS, 20:337–66
Nemhauser JL, 21:177–201
Niggli V, 21:57–79
Novoa I, 18:575–99
Nusse R, 20:781–810
Nussenzweig V, 20:29–59

O

Okkenhaug K, 17:615–75
Oliver G, 21:457–83
Oppenheimer DG, 21:271–95
Orci L, 20:87–123
Orr-Weaver TL, 17:753–77
Osborne AR, 21:529–50
Owen DJ, 20:153–91

P

Page SL, 20:525–58
Palacios IM, 17:569–614
Panda S, 17:215–53
Parodi AJ, 19:649–76
Pasquinelli AE, 18:495–513
Payne T, 17:677–99
Peacock WJ, 17:677–99

Pearson BJ, 20:619–47
Perrimon N, 18:463–93
Piper M, 20:505–23
Pourquié O, 17:311–50
Pownall ME, 18:747–83
Pruyne D, 20:559–91
Pufall MA, 18:421–62
Purdue PE, 17:701–52

R

Raff M, 19:1–22
Ramamurthi KS, 18:107–33
Ramirez J, 20:125–51
Rapoport TA, 21:529–50
Rauskolb C, 17:189–214
Reber M, 21:551–80
Reddien PW, 20:193–221, 725–57
Richter D, 21:223–45
Riese MJ, 18:315–44
Roland JT, 19:469–91
Römisch K, 21:435–56
Ron D, 18:575–99
Rossant J, 18:541–73
Russell SD, 18:81–105
Ruvkun G, 18:495–513

S

Sabatini DD, 21:1–33
Sanchez Alvarado A, 20:725–57
Sancho E, 20:695–723
Sandhoff K, 21:81–103
Sandvig K, 18:1–24
Saxon E, 17:1–23
Schekman R, 20:87–123
Schier AF, 19:589–621
Schneewind O, 18:107–33
Schöck F, 18:463–93
Scholey JM, 19:423–43
Schroer TA, 20:759–79
Segall JE, 21:695–718
Shi Y, 21:35–56

Shogren–Knaak MA, 17:405–33
Shokat KM, 17:405–433
Shorter J, 18:379–420
Si-Ammour A, 21:297–318
Singer RH, 21:695–718
Singer RH, 17:297–310
Singh SB, 20:367–94
Smith AG, 17:435–62
Smith GA, 18:135–61
Smith HMS, 20:125–51
Smith LG, 21:271–95
Sorger PK, 19:519–39
Spudich G, 20:649–76
Steinhardt RA, 19:697–731
Sternlicht MD, 17:463–516
St. Johnston D, 17:569–614
Sumen C, 21:581–603
Sundquist WI, 20:395–425

T

Tanaka T, 20:593–618
Tarone G, 19:173–206
Theurkauf W, 21:411–34
Tiedge H, 21:223–45
Timms J, 17:615–75
Tontonoz P, 20:455–80
Towler MC, 17:517–68
Trombetta ES, 19:649–76
True JR, 18:53–80
Tucker MR, 17:677–99
Tweedie S, 18:707–46
Tytell JD, 19:519–39

U

Ungar D, 19:493–517
Urano F, 18:575–99

V

van den Berg B, 21:529–50
van Deurs B, 18:1–24

Vergne I, 20:367–94
Verona RI, 19:237–59
Vert G, 21:177–201
Vidali L, 17:159–87
von Andrian UH, 21:581–603

W

Wakeham DE, 17:517–68
Walther TC, 21:347–80
Wang H, 21:223–45
Warren G, 18:379–420
Waterfield MD, 17:615–75
Waters CM, 21:319–46
Wei C-J, 20:811–38
Wei N, 19:261–86
Weinmaster G, 19:367–95
Weissman IL, 17:387–403
Welch MD, 18:247–88
Werb Z, 17:463–516
West RR, 18:193–219
Wheelock MJ, 19:207–35
White S, 17:615–75
Winey M, 20:1–28

X

Xie T, 21:605–31
Xiong J-P, 21:381–410
Xu X, 20:811–38

Y

Yan N, 21:35–56
Young JAT, 19:45–70
Yurchenco PD, 20:255–84

Z

Zagotta WN, 19:23–44
Zheng Y, 20:867–94
Zhu C, 17:133–57
Zik M, 19:119–40

Chapter Titles, Volumes 17–21

Cell-Cell Interactions

Cell Cycle and its Regulation

Cytoskeleton

Protein and Membrane Traffic Control

ANNUAL REVIEWS
Intelligent Synthesis of the Scientific Literature

Annual Reviews – Your Starting Point for Research Online
http://arjournals.annualreviews.org

- Over 900 Annual Reviews volumes—more than 25,000 critical, authoritative review articles in 32 disciplines spanning the Biomedical, Physical, and Social sciences— available online, including all Annual Reviews back volumes, dating to 1932

- Current individual subscriptions include seamless online access to full-text articles, PDFs, Reviews in Advance (as much as 6 months ahead of print publication), bibliographies, and other supplementary material in the current volume and the prior 4 years' volumes

- All articles are fully supplemented, searchable, and downloadable — see http://cellbio.annualreviews.org

- Access links to the reviewed references (when available online)

- Site features include customized alerting services, citation tracking, and saved searches

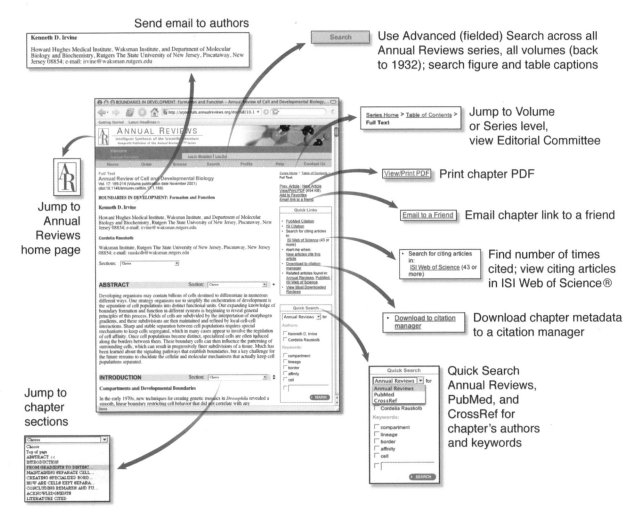

Send email to authors

Use Advanced (fielded) Search across all Annual Reviews series, all volumes (back to 1932); search figure and table captions

Jump to Volume or Series level, view Editorial Committee

Print chapter PDF

Email chapter link to a friend

Find number of times cited; view citing articles in ISI Web of Science®

Download chapter metadata to a citation manager

Quick Search Annual Reviews, PubMed, and CrossRef for chapter's authors and keywords

Jump to Annual Reviews home page

Jump to chapter sections

Copyright ® 2005 Annual Reviews, Nonprofit Publisher of the *Annual Review of* Series